CONVERSIONS BETWEEN U.S. CUSTOMARY UNITS AND SI UNITS (Continued)

U.S. Customary unit		Times conversion factor		Equals SI unit	
		Accurate	Practical		
Moment of inertia (area)					
inch to fourth power	in.^{4}	416,231	416,000	millimeter to fourth power	mm^4
inch to fourth power	in.^{4}	0.416231×10^{-6}	0.416×10^{-6}	meter to fourth power	m^4
Moment of inertia (mass)					
slug foot squared	slug-ft^2	1.35582	1.36	kilogram meter squared	kg·m^2
Power					
foot-pound per second	ft-lb/s	1.35582	1.36	watt (J/s or N·m/s)	W
foot-pound per minute	ft-lb/min	0.0225970	0.0226	watt	W
horsepower (550 ft-lb/s)	hp	745.701	746	watt	W
Pressure; stress					
pound per square foot	psf	47.8803	47.9	pascal (N/m^2)	Pa
pound per square inch	psi	6894.76	6890	pascal	Pa
kip per square foot	ksf	47.8803	47.9	kilopascal	kPa
kip per square inch	ksi	6.89476	6.89	megapascal	MPa
Section modulus					
inch to third power	in.^{3}	16,387.1	16,400	millimeter to third power	mm^3
inch to third power	in.^{3}	16.3871×10^{-6}	16.4×10^{-6}	meter to third power	m^3
Velocity (linear)					
foot per second	ft/s	0.3048*	0.305	meter per second	m/s
inch per second	in./s	0.0254*	0.0254	meter per second	m/s
mile per hour	mph	0.44704*	0.447	meter per second	m/s
mile per hour	mph	1.609344*	1.61	kilometer per hour	km/h
Volume					
cubic foot	ft^3	0.0283168	0.0283	cubic meter	m^3
cubic inch	in.^{3}	16.3871×10^{-6}	16.4×10^{-6}	cubic meter	m^3
cubic inch	in.^{3}	16.3871	16.4	cubic centimeter (cc)	cm^3
gallon (231 in.^{3})	gal.	3.78541	3.79	liter	L
gallon (231 in.^{3})	gal.	0.00378541	0.00379	cubic meter	m^3

*An asterisk denotes an *exact* conversion factor

Note: **To convert from SI units to USCS units, *divide* by the conversion factor**

Temperature Conversion Formulas

$$T(°C) = \frac{5}{9}[T(°F) - 32] = T(K) - 273.15$$

$$T(K) = \frac{5}{9}[T(°F) - 32] + 273.15 = T(°C) + 273.15$$

$$T(°F) = \frac{9}{5}T(°C) + 32 = \frac{9}{5}T(K) - 459.67$$

POWER SYSTEM ANALYSIS & DESIGN

SIXTH EDITION, SI

J. DUNCAN GLOVER
Failure Electrical, LLC

THOMAS J. OVERBYE
University of Illinois

MULUKUTLA S. SARMA
Northeastern University

Australia • Brazil • Japan • Korea • Mexico • Singapore • Spain • United Kingdom • United States

Power System Analysis & Design,
Sixth Edition, SI
J. Duncan Glover, Thomas J. Overbye, and Mulukutla S. Sarma

SI edition: Keith McIver

Product Director, Global Engineering: Timothy L. Anderson

Senior Content Developer: Mona Zeftel

Associate Media Content Developer: Ashley Kaupert

Product Assistant: Alexander Sham

Marketing Manager: Kristin Stine

Director, Content and Media Production: Sharon L. Smith

Content Project Manager: D. Jean Buttrom

Production Service: RPK Editorial Services, Inc.

Copyeditor: Warren Hapke

Proofreaders: Martha McMaster and Shelly Gerger-Knechtl

Indexer: RPK Editorial Services

Compositor: MPS Limited

Senior Art Director: Michelle Kunkler

Internal Designer: Carmela Pereira

Cover Designer: Itzhack Shelomi

Cover Image: High voltage power lines and pylon, kstudija/ShutterStock.com

Intellectual Property
Analyst: Christine Myaskovsky
Project Manager: Sarah Shainwald

Text and Image Permissions Researcher: Kristiina Paul

Manufacturing Planner: Doug Wilke

Printed in the United States of America
Print Number: 01 Print Year: 2016

For product information and technology assistance, contact us at **Cengage Learning Customer & Sales Support, 1-800-354-9706**.
For permission to use material from this text or product, submit all requests online at **www.cengage.com/permissions**.
Further permissions questions can be emailed to **permissionrequest@cengage.com**.

Library of Congress Control Number: 2016940942

ISBN: 978-1-305-63618-7

Cengage Learning
20 Channel Center Street
Boston, MA 02210
USA

Cengage Learning is a leading provider of customized learning solutions with employees residing in nearly 40 different countries and sales in more than 125 countries around the world. Find your local representative at **www.cengage.com.**

Cengage Learning products are represented in Canada by Nelson Education Ltd.

To learn more about Cengage Learning Solutions, visit **www.cengage.com/engineering**.

Purchase any of our products at your local college store or at our preferred online store **www.cengagebrain.com**.

In loving memory of my mentors Professor Fred C. Schweppe [1933–1988] and Dr. Alexander Kusko [1921–2013]. You taught me, you guided me, you set the bar for which I continue to strive. You shall not be forgotten.

My Guardian Poet[s]

A guardian poet you have been to me
Much like an angel, there protecting me
When I was silent, lost in dark of night
You read my words and brought me back to light

You told me that my words were ever true
That in my writes were thoughts profound and new
You would not let me simply drift away
A word of hope you'd send to greet each day

Your name is there below each thing I write
To tear dimmed eyes you brought a vision bright
"The Queen of Passion," how I love the name
You gave to me and life is not the same

To you, my Guardian Poet, thanks I bring
You fool me not; I see your angel wing

Eileen Manassian Ghali

To Jo, Tim, Hannah, and Amanda

Contents

Preface to the SI Edition x
Preface xi
List of Symbols, Units, and Notation xvii

CHAPTER 1 **Introduction 1**

Case Study: How the Free Market Rocked the Grid 2
1.1 History of Electric Power Systems 10
1.2 Present and Future Trends 17
1.3 Electric Utility Industry Structure 20
1.4 Computers in Power System Engineering 21
1.5 PowerWorld Simulator 22

CHAPTER 2 **Fundamentals 31**

Case Study: Key Connections 32
2.1 Phasors 40
2.2 Instantaneous Power in Single-Phase AC Circuits 42
2.3 Complex Power 47
2.4 Network Equations 52
2.5 Balanced Three-Phase Circuits 55
2.6 Power in Balanced Three-Phase Circuits 63
2.7 Advantages of Balanced Three-Phase versus Single-Phase Systems 68

CHAPTER 3 **Power Transformers 87**

Case Study: Power Transformers—Life Management and Extension 88
3.1 The Ideal Transformer 95
3.2 Equivalent Circuits for Practical Transformers 101
3.3 The Per-Unit System 107
3.4 Three-Phase Transformer Connections and Phase Shift 115
3.5 Per-Unit Equivalent Circuits of Balanced Three-Phase Two-Winding Transformers 120
3.6 Three-Winding Transformers 125
3.7 Autotransformers 129
3.8 Transformers with Off-Nominal Turns Ratios 131

CHAPTER 4 **Transmission Line Parameters 161**

Case Study: Integrating North America's Power Grid 162
Case Study: Grid Congestion - Unclogging the Arteries of North America's Power Grid 167

4.1 Transmission Line Design Considerations 173
4.2 Resistance 178
4.3 Conductance 181
4.4 Inductance: Solid Cylindrical Conductor 181
4.5 Inductance: Single-Phase Two-Wire Line and Three-Phase Three-Wire Line with Equal Phase Spacing 186
4.6 Inductance: Composite Conductors, Unequal Phase Spacing, Bundled Conductors 188
4.7 Series Impedances: Three-Phase Line with Neutral Conductors and Earth Return 196
4.8 Electric Field and Voltage: Solid Cylindrical Conductor 201
4.9 Capacitance: Single-Phase Two-Wire Line and Three-Phase Three-Wire Line with Equal Phase Spacing 204
4.10 Capacitance: Stranded Conductors, Unequal Phase Spacing, Bundled Conductors 206
4.11 Shunt Admittances: Lines with Neutral Conductors and Earth Return 210
4.12 Electric Field Strength at Conductor Surfaces and at Ground Level 215
4.13 Parallel Circuit Three-Phase Lines 218

CHAPTER 5 **Transmission Lines: Steady-State Operation 237**

Case Study: The ABCs of HVDC Transmission Technologies: An Overview of High Voltage Direct Current Systems and Applications 238
5.1 Medium and Short Line Approximations 258
5.2 Transmission-Line Differential Equations 265
5.3 Equivalent π Circuit 271
5.4 Lossless Lines 274
5.5 Maximum Power Flow 282
5.6 Line Loadability 284
5.7 Reactive Compensation Techniques 289

CHAPTER 6 **Power Flows 309**

Case Study: Finding Flexibility: Cycling the Conventional Fleet 310
6.1 Direct Solutions to Linear Algebraic Equations: Gauss Elimination 330
6.2 Iterative Solutions to Linear Algebraic Equations: Jacobi and Gauss-Seidel 334
6.3 Iterative Solutions to Nonlinear Algebraic Equations: Newton-Raphson 340
6.4 The Power Flow Problem 345
6.5 Power Flow Solution by Gauss-Seidel 351
6.6 Power Flow Solution by Newton-Raphson 353
6.7 Control of Power Flow 363
6.8 Sparsity Techniques 369
6.9 Fast Decoupled Power Flow 372
6.10 The "DC" Power Flow 372
6.11 Power Flow Modeling of Wind Generation 374
6.12 Economic Dispatch 376
6.13 Optimal Power Flow 389
Design Projects 1–3 404–412

CHAPTER 7 **Symmetrical Faults 415**

Case Study: Short-Circuit Modeling of a Wind Power Plant 416
7.1 Series R–L Circuit Transients 435
7.2 Three-Phase Short Circuit—Unloaded Synchronous Machine 438
7.3 Power System Three-Phase Short Circuits 442
7.4 Bus Impedance Matrix 445
7.5 Circuit Breaker and Fuse Selection 455
Design Project 3 (*continued*) 472

CHAPTER 8 **Symmetrical Components 475**

Case Study: Technological Progress in High-Voltage Gas-Insulated Substations 476
8.1 Definition of Symmetrical Components 493
8.2 Sequence Networks of Impedance Loads 499
8.3 Sequence Networks of Series Impedances 506
8.4 Sequence Networks of Three-Phase Lines 508
8.5 Sequence Networks of Rotating Machines 510
8.6 Per-Unit Sequence Models of Three-Phase Two-Winding Transformers 516
8.7 Per-Unit Sequence Models of Three-Phase Three-Winding Transformers 522
8.8 Power in Sequence Networks 524

CHAPTER 9 **Unsymmetrical Faults 539**

Case Study: Innovative Medium Voltage Switchgear for Today's Applications 540
9.1 System Representation 547
9.2 Single Line-to-Ground Fault 553
9.3 Line-to-Line Fault 557
9.4 Double Line-to-Ground Fault 560
9.5 Sequence Bus Impedance Matrices 567
Design Project 3 (*continued*) 588
Design Project 4 589

CHAPTER 10 **System Protection 593**

Case Study: Upgrading Relay Protection Be Prepared for the Next Replacement or Upgrade Project 594
10.1 System Protection Components 612
10.2 Instrument Transformers 614
10.3 Overcurrent Relays 620
10.4 Radial System Protection 625
10.5 Reclosers and Fuses 629
10.6 Directional Relays 633
10.7 Protection of a Two-Source System with Directional Relays 634
10.8 Zones of Protection 635
10.9 Line Protection with Impedance (Distance) Relays 639
10.10 Differential Relays 645
10.11 Bus Protection with Differential Relays 647
10.12 Transformer Protection with Differential Relays 648

10.13 Pilot Relaying 653
10.14 Numeric Relaying 654

CHAPTER 11 **Transient Stability 669**

Case Study: Down, but Not Out 671
11.1 The Swing Equation 689
11.2 Simplified Synchronous Machine Model and System Equivalents 695
11.3 The Equal-Area Criterion 697
11.4 Numerical Integration of the Swing Equation 707
11.5 Multimachine Stability 711
11.6 A Two-Axis Synchronous Machine Model 719
11.7 Wind Turbine Machine Models 724
11.8 Design Methods for Improving Transient Stability 730

CHAPTER 12 **Power System Controls 739**

Case Study: No Light in August: Power System Restoration Following the 2003 North American Blackout 742
12.1 Generator-Voltage Control 757
12.2 Turbine-Governor Control 761
12.3 Load-Frequency Control 767

CHAPTER 13 **Transmission Lines: Transient Operation 779**

Case Study: Surge Arresters 780
Case Study: Emergency Response 794
13.1 Traveling Waves on Single-Phase Lossless Lines 809
13.2 Boundary Conditions for Single-Phase Lossless Lines 813
13.3 Bewley Lattice Diagram 822
13.4 Discrete-Time Models of Single-Phase Lossless Lines and Lumped RLC Elements 828
13.5 Lossy Lines 834
13.6 Multiconductor Lines 838
13.7 Power System Overvoltages 841
13.8 Insulation Coordination 847

CHAPTER 14 **Power Distribution 859**

Case Study: It's All in the Plans 860
14.1 Introduction to Distribution 875
14.2 Primary Distribution 878
14.3 Secondary Distribution 885
14.4 Transformers in Distribution Systems 890
14.5 Shunt Capacitors in Distribution Systems 900
14.6 Distribution Software 905
14.7 Distribution Reliability 906
14.8 Distribution Automation 910
14.9 Smart Grids 913

Appendix 921
Index 925

Preface to the SI Edition

This edition of *POWER SYSTEM ANALYSIS & DESIGN* has been adapted to incorporate the International System of Units (*Le Système International d'Unités* or SI) throughout the book.

Le Système International d'Unités

The United States Customary System (USCS) of units uses FPS (foot–pound–second) units (also called English or Imperial units). SI units are primarily the units of the MKS (meter–kilogram–second) system. However, CGS (centimeter–gram–second) units are often accepted as SI units, especially in textbooks.

USING SI UNITS IN THIS BOOK

In this book, we have used both MKS and CGS units. USCS (U.S. Customary Units) or FPS (foot-pound-second) units used in the US Edition of the book have been converted to SI units throughout the text and problems. However, in case of data sourced from handbooks, government standards, and product manuals, it is not only extremely difficult to convert all values to SI, it also encroaches upon the intellectual property of the source. Some data in figures, tables, and references, therefore, remains in FPS units. For readers unfamiliar with the relationship between the USCS and the SI systems, a conversion table has been provided inside the front cover.

To solve problems that require the use of sourced data, the sourced values can be converted from FPS units to SI units just before they are to be used in a calculation. To obtain standardized quantities and manufacturers' data in SI units, readers may contact the appropriate government agencies or authorities in their regions.

INSTRUCTOR RESOURCES

The Instructors' Solution Manual in SI units is available through your Sales Representative or online through the book website at http://login.cengage.com. A digital version of the ISM, Lecture Note PowerPoint slides for the SI text, as well as other resources are available for instructors registering on the book website.

Feedback from users of this SI Edition will be greatly appreciated and will help us improve subsequent editions.

Cengage Learning

Preface

The objective of this book is to present methods of power system analysis and design, particularly with the aid of a personal computer, in sufficient depth to give the student the basic theory at the undergraduate level. The approach is designed to develop students' thinking processes, enabling them to reach a sound understanding of a broad range of topics related to power system engineering, while motivating their interest in the electrical power industry. Because we believe that fundamental physical concepts underlie creative engineering and form the most valuable and permanent part of an engineering education, we highlight physical concepts while giving due attention to mathematical techniques. Both theory and modeling are developed from simple beginnings so that they can be readily extended to new and complex situations.

NEW TO THIS EDITION

New chapter-opening case studies bring principles to life for students by providing practical, real-world engineering applications for the material discussed in each chapter.

Comprehensively revised problem sets ensure students have the practice they need to master critical skills.

Updated Instructor Resources

These resources include

- Instructor's Solutions Manual with solutions to all problems
- Sample Tests offering additional problems
- Annotated Lecture Note PowerPoint Slides
- Lesson Plans that detail how to most effectively use this edition
- Updated PowerWorld Simulator Software
- Student PowerPoint Notes

New design projects in this edition meet Accreditation Board for Engineering and Technology (ABET) requirements to provide valuable hands-on experience and to help ensure students are receiving an education that meets globally recognized accreditation standards.

The latest version of the valuable PowerWorld Simulator (version 19) is included and integrated throughout the text.

KEY FEATURES

The text presents present-day, practical applications and new technologies along with ample coverage of the ongoing restructuring of the electric utility industry. It is supported by an ample number of worked examples, including illustrations, covering most of the theoretical points raised. It also includes PowerWorld Simulator version 19 to extend fully worked examples into computer implementations of the solutions. Version 19 includes power flow, optimal power flow, contingency analysis, short circuit, and transient stability.

The text includes a chapter on Power Distribution with content on Smart Grids.

It also includes discussions on modeling of wind turbines in power flow and transient stability.

Four design projects are included, all of which meet ABET requirements.

POWERWORLD SIMULATOR

One of the most challenging aspects of engineering education is giving students an intuitive feel for the systems they are studying. Engineering systems are, for the most part, complex. While paper-and-pencil exercises can be quite useful for highlighting the fundamentals, they often fall short in imparting the desired intuitive insight. To help provide this insight, the book uses PowerWorld Simulator version 19 to integrate computer-based examples, problems, and design projects throughout the text.

PowerWorld Simulator was originally developed at the University of Illinois at Urbana-Champaign to teach the basics of power systems to nontechnical people involved in the electricity industry, with version 1.0 introduced in June 1994. The program's interactive and graphical design made it an immediate hit as an educational tool, but a funny thing happened—its interactive and graphical design also appealed to engineers doing analysis of real power systems. To meet the needs of a growing group of users, PowerWorld Simulator was commercialized in 1996 by the formation of PowerWorld Corporation. Thus while retaining its appeal for education, over the years PowerWorld Simulator has evolved into a top-notch analysis package, able to handle power systems of any size. PowerWorld Simulator is now used throughout the power industry, with a range of users encompassing universities, utilities of all sizes, government regulators, power marketers, and consulting firms.

In integrating PowerWorld Simulator with the text, our design philosophy has been to use the software to extend, rather than replace, the fully worked examples provided in previous editions. Therefore, except when the problem size makes it impractical, each PowerWorld Simulator example includes a fully worked hand solution of the problem along with a PowerWorld Simulator case. This format allows students to simultaneously see the details of how a problem is solved and a computer implementation of the solution. The added benefit from PowerWorld Simulator is its ability to easily extend the example. Through its interactive design, students can quickly vary example parameters and immediately see the impact such changes have on the solution. By reworking the examples with the new parameters, students get immediate feedback on whether they understand the solution process.

The interactive and visual design of PowerWorld Simulator also makes it an excellent tool for instructors to use for in-class demonstrations. With numerous examples utilizing PowerWorld Simulator instructors can easily demonstrate many of the text topics. Additional PowerWorld Simulator functionality is introduced in the text problems and design projects.

PREREQUISITES

As background for this course, it is assumed that students have had courses in electric network theory (including transient analysis) and ordinary differential equations and have been exposed to linear systems, matrix algebra, and computer programming. In addition, it would be helpful, but not necessary, to have had an electric machines course.

ORGANIZATION

The text is intended to be fully covered in a two-semester or three-quarter course offered to seniors and first-year graduate students. The organization of chapters and individual sections is flexible enough to give the instructor sufficient latitude in choosing topics to cover, especially in a one-semester course. The text is supported by an ample number of worked examples covering most of the theoretical points raised. The many problems to be worked with a calculator as well as problems to be worked using a personal computer have been revised in this edition.

After an introduction to the history of electric power systems along with present and future trends, *Chapter 2* orients the students to the terminology and serves as a brief review of fundamentals. The chapter reviews phasor concepts, power, and single-phase as well as three-phase circuits.

Chapters 3 through 5 examine power transformers including the per-unit system, transmission-line parameters, and steady-state operation of transmission lines. *Chapter 6* examines power flows including the Newton-Raphson method, power-flow modeling of wind generation, economic dispatch, and optimal power flow. These chapters provide a basic understanding of power systems under balanced three-phase, steady-state, normal operating conditions.

Chapters 7 through 10, which cover symmetrical faults, symmetrical components, unsymmetrical faults, and system protection, come under the general heading of power system short-circuit protection. *Chapter 11* examines transient stability, which includes the swing equation, the equal-area criterion, and multi-machine stability with modeling of wind-energy systems. *Chapter 12* covers power system controls, including generator-voltage control, turbine-governor control, and load-frequency control. *Chapter 13* examines transient operation of transmission lines including power system overvoltages and surge protection.

Chapter 14 introduces the basic features of primary and secondary distribution systems as well as basic distribution components including distribution substation transformers, distribution transformers, and shunt capacitors. We list some of the major distribution software vendors followed by an introduction to distribution reliability, distribution automation, and smart grids.

ADDITIONAL RESOURCES

Companion websites for this book are available for both students and instructors. These websites provide useful links and other support material.

Student Website

The **Student Companion Site** includes a link to download the free student version of PowerWorld Simulator software and Student PowerPoint Notes.

Instructor Resource Center

The **Instructor Companion Site** includes

- Instructor's Solutions Manual
- Annotated PowerPoint Slides
- Lecture Notes
- Sample Tests
- Link to PowerWorld Simulator software with examples and cases

To access the support material described here along with all additional course materials, please visit https://sso.cengage.com.

MINDTAP ONLINE COURSE AND READER

This textbook is also available online through Cengage Learning's MindTap, a personalized learning program. Students who purchase the MindTap have access to the book's multimedia-rich electronic Reader and are able to complete homework and assessment material online, on their desktops, laptops, or iPads. Instructors who use a Learning Management System (such as Blackboard, Canvas, or Moodle) for tracking course content, assignments, and grading, can seamlessly access the MindTap suite of content and assessments for this course.

With MindTap, instructors can

- Personalize the Learning Path to match the course syllabus by rearranging content or appending original material to the online content
- Connect a Learning Management System portal to the online course and Reader
- Customize online assessments and assignments
- Track student engagement, progress and comprehension
- Promote student success through interactivity, multimedia, and exercises

Additionally, students can listen to the text through ReadSpeaker, take notes in the digital Reader, study from and create their own Flashcards, highlight content for easy reference, and check their understanding of the material through practice quizzes and automatically-graded homework.

ACKNOWLEDGMENTS

The material in this text was gradually developed to meet the needs of classes taught at universities in the United States and abroad over the past 35 years. The original 13 chapters were written by the first author, J. Duncan Glover, *Failure Electrical LLC*, who is indebted to many people who helped during the planning and writing of this book. The profound influence of earlier texts written on power systems, particularly by W. D. Stevenson, Jr., and the developments made by various outstanding engineers are gratefully acknowledged. Details of sources can only be made through references at the end of each chapter, as they are otherwise too numerous to mention.

Chapter 14 *(Power Distribution)* was a collaborative effort between Dr. Glover (Sections 14.1-14.7) and Co-author Thomas J. Overbye (Sections 14.8 & 14.9). Professor Overbye, *University of Illinois at Urbana-Champaign* updated Chapter 6 *(Power Flows) and* Chapter 11 *(Transient Stability)*. He also provided the examples and problems using PowerWorld Simulator as well as three design projects. Co-author Mulukutla Sarma, *Northeastern University,* contributed to end-of-chapter multiple-choice questions and problems.

We commend the following Global Engineering team members at Cengage Learning: Timothy Anderson, Product Director; Mona Zeftel, Senior Content Developer; and Kristiina Paul, Freelance Permissions Researcher; as well as Rose Kernan of RPK Editorial Services, Inc. for their broad knowledge, skills, and ingenuity in publishing this edition. We also thank Jean Buttrom, Content Project Manager; Kristin Stine, Marketing Manager; Elizabeth Murphy, Engagement Specialist; Ashley Kaupert, Associate Media Content Developer; Teresa Versaggi and Alexander Sham, Product Assistants.

The reviewers for the sixth edition are as follows: Ross Baldick, *University of Texas at Austin*; François Bouffard, *McGill University*; Venkata Dinavahi, *University of Alberta*; Seyed Pouyan Jazayeri, *University of Calgary*; Bruno Osorno, *California State University at Northridge*, Zeb Tate, *University of Toronto*; and Mahyar Zarghami, *California State University at Sacramento.*

Substantial contributions to prior editions of this text were made by a number of invaluable reviewers, as follows:

Fifth Edition: Thomas L. Baldwin, *Florida State University;* Ali Emadi, *Illinois Institute of Technology;* Reza Iravani, *University of Toronto;* Surya Santoso, *University of Texas at Austin;* Ali Shaban, *California Polytechnic State University, San Luis Obispo;* and Dennis O. Wiitanen, *Michigan Technological University,* and Hamid Jaffari, *Danvers Electric.*

Fourth Edition: Robert C. Degeneff, *Rensselaer Polytechnic Institute;* Venkata Dina-vahi, *University of Alberta;* Richard G. Farmer, *Arizona State University;* Steven M. Hietpas, *South Dakota State University;* M. Hashem Nehrir, *Montana State University;* Anil Pahwa, *Kansas State University;* and Ghadir Radman, *Tennessee Technical University.*

Third Edition: Sohrab Asgarpoor, *University of Nebraska-Lincoln;* Mariesa L. Crow, *University of Missouri-Rolla;* Ilya Y. Grinberg, *State University of New York, College at Buffalo;* Iqbal Husain, *The University of Akron;* W. H. Kersting, *New Mexico State University;* John A. Palmer, *Colorado School of Mines;* Satish J. Ranada, *New Mexico State University;* and Shyama C. Tandon, *California Polytechnic State University.*

Second Edition: Max D. Anderson, *University of Missouri-Rolla;* Sohrab Asgarpoor, *University of Nebraska-Lincoln;* Kaveh Ashenayi, *University of Tulsa;* Richard D. Christie, Jr., *University of Washington;* Mariesa L. Crow, *University of Missouri-Rolla;* Richard G. Farmer, *Arizona State University;* Saul Goldberg, *California Polytechnic*

University; Clifford H. Grigg, *Rose-Hulman Institute of Technology;* Howard B. Hamilton, *University of Pittsburgh;* Leo Holzenthal, Jr., *University of New Orleans;* Walid Hubbi, *New Jersey Institute of Technology;* Charles W. Isherwood, *University of Massachusetts-Dartmouth;* W. H. Kersting, *New Mexico State University;* Wayne E. Knabach, *South Dakota State University;* Pierre-Jean Lagace, *IREQ Institut de Reserche a"Hydro-Quebec;* James T. Lancaster, *Alfred University;* Kwang Y. Lee, *Pennsylvania State University;* Mohsen Lotfalian, *University of Evansville;* Rene B. Marxheimer, *San Francisco State University,* Lamine Mili, *Virginia Polytechnic Institute and State University;* Osama A. Mohammed, *Florida International University;* Clifford C. Mosher, *Washington State University,* Anil Pahwa, *Kansas State University;* M. A. Pai, *University of Illinois at Urbana-Champaign;* R. Ramakumar, *Oklahoma State University;* Teodoro C. Robles, *Milwaukee School of Engineering,* Ronald G. Schultz, *Cleveland State University;* Stephen A. Sebo, *Ohio State University;* Raymond Shoults, *University of Texas at Arlington,* Richard D. Shultz, *University of Wisconsin at Platteville;* Charles Slivinsky, *University of Missouri-Columbia;* John P. Stahl, *Ohio Northern University;* E. K. Stanek, *University of Missouri-Rolla;* Robert D. Strattan, *University of Tulsa;* Tian-Shen Tang, *Texas A&M University-Kingsville;* S. S. Venkata, *University of Washington;* Francis M. Wells, *Vanderbilt University;* Bill Wieserman, *University of Pennsylvania-Johnstown;* Stephen Williams, *U.S. Naval Postgraduate School;* and Salah M. Yousif, *California State University-Sacramento.*

First Edition: Frederick C. Brockhurst, *Rose-Hulman Institute of Technology;* Bell A. Cogbill. *Northeastern University;* Saul Goldberg, *California Polytechnic State University;* Mack Grady, *University of Texas at Austin;* Leonard F. Grigsby, *Auburn University;* Howard Hamilton, *University of Pittsburgh;* William F. Horton, *California Polytechnic State University;* W. H. Kersting, *New Mexico State University;* John Pavlat, *Iowa State University;* R. Ramakumar, *Oklahoma State University;* B. Don Russell, *Texas A&M;* Sheppard Salon, *Rensselaer Polytechnic Institute;* Stephen A. Sebo, *Ohio State University;* and Dennis O. Wiitanen, *Michigan Technological University.*

In conclusion, the objective in writing this text and the accompanying software package will have been fulfilled if the book is considered to be student-oriented, comprehensive, and up to date, with consistent notation and necessary detailed explanation at the level for which it is intended.

J. Duncan Glover

Thomas J. Overbye

Mulukutla S. Sarma

List of Symbols, Units, and Notation

Symbol	Description
a	operator 1/120°
a_t	transformer turns ratio
A	area
A	transmission line parameter
$\boldsymbol{A}$	symmetrical components transformation matrix
B	loss coefficient
B	frequency bias constant
B	phasor magnetic flux density
B	transmission line parameter
C	capacitance
C	transmission line parameter
D	damping
D	distance
D	transmission line parameter
E	phasor source voltage
E	phasor electric field strength
f	frequency
G	conductance
G	conductance matrix
H	normalized inertia constant
H	phasor magnetic field intensity
$i(t)$	instantaneous current
I	current magnitude (rms unless otherwise indicated)
I	phasor current
$\boldsymbol{I}$	vector of phasor currents
j	operator 1/90°
J	moment of inertia
l	length
P	real power
q	Charge
Q	reactive power
r	radius
R	resistance
R	turbine-governor regulation constant
R	resistance matrix
s	Laplace operator
S	apparent power
S	complex power
t	time
T	period
T	temperature
T	torque
$v(t)$	instantaneous voltage
V	voltage magnitude (rms unless otherwise indicated)
V	phasor voltage
$\boldsymbol{V}$	vector of phasor voltages
X	reactance
X	reactance matrix
Y	phasor admittance
$\boldsymbol{Y}$	admittance matrix
Z	phasor impedance
$\boldsymbol{Z}$	impedance matrix
α	angular acceleration
α	transformer phase shift angle
β	current angle

Symbol	Description
l	length
L	inductance
L	inductance matrix
N	number (of buses, lines, turns, etc.)
p.f.	power factor
$p(t)$	instantaneous power
λ	magnetic flux linkage
λ	Penalty factor
Φ	magnetic flux
ρ	resistivity
τ	time in cycles
τ	transmission line transit time
β	area frequency response characteristic
δ	voltage angle
δ	torque angle
ε	permittivity
Γ	reflection or refraction coefficient
θ	impedance angle
θ	angular position
μ	permeability
υ	velocity of propagation
ω	radian frequency

SI Units

A	ampere
C	coulomb
F	farad
H	henry
Hz	hertz
J	joule
kg	kilogram
m	meter
N	newton
rad	radian
s	second
S	siemen
VA	voltampere
var	voltampere reactive
W	watt
Wb	weber
Ω	ohm

English Units

BTU	British thermal unit
Cmil	circular mil
ft	foot
hp	horsepower
in	inch
mi	mile

Notation

Lowercase letters such as v(t) and i(t) indicate instantaneous values.
Uppercase letters such as V and I indicate rms values.
Uppercase letters in italic such as *V* and *I* indicate rms phasors.
Matrices and vectors with real components such as **R** and **I** are indicated by boldface type.
Matrices and vectors with complex components such as ***Z*** and ***I*** are indicated by boldface italic type.
Superscript T denotes vector or matrix transpose.
Asterisk (*) denotes complex conjugate.
PW highlights problems that utilize PowerWorld Simulator.

1 Introduction

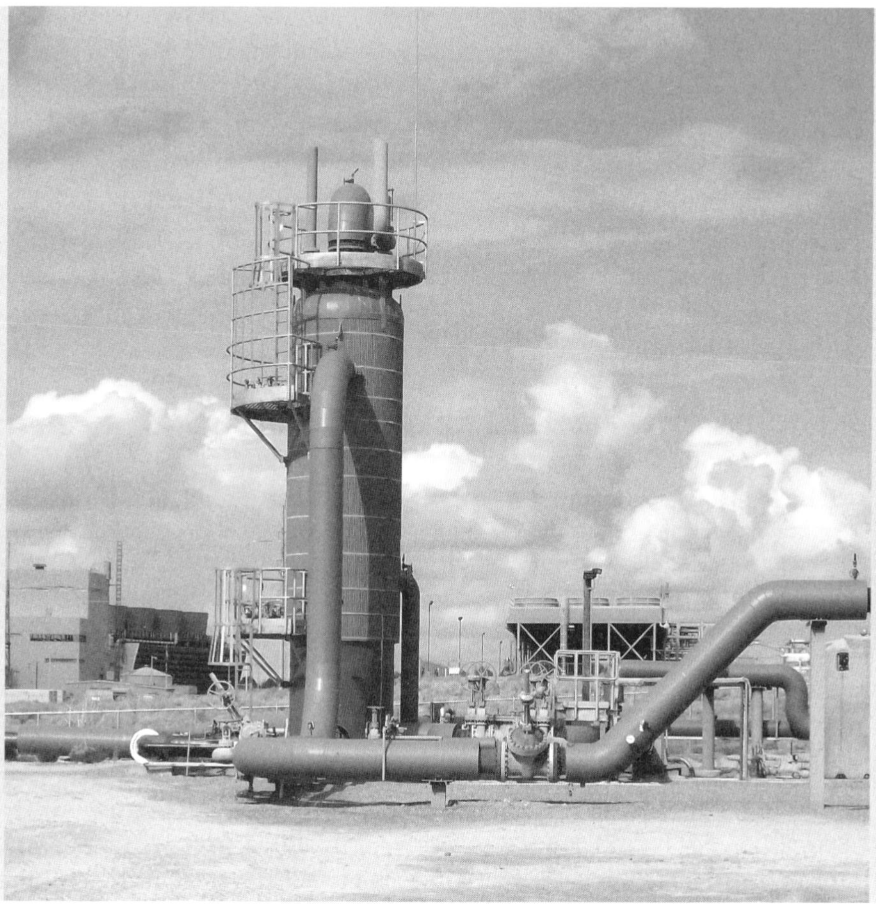

Blundell geothermal power plant near Milford, UT, USA. This 38-MW plant consists of two generating units powered by geothermal steam. Steam is created from water heated by magma at depths up to 6100 meters below Earth's surface. (Courtesy of PacifiCorp.)

Electrical engineers are concerned with every step in the process of generation, transmission, distribution, and utilization of electrical energy. The electric utility industry is probably the largest and most complex industry in the world. The electrical engineer who works in that industry encounters challenging problems in designing future power systems to deliver increasing amounts of electrical energy in a safe, clean, and economical manner.

The objectives of this chapter are to review briefly the history of the electric utility industry, to discuss present and future trends in electric power systems, to describe the restructuring of the electric utility industry, and to introduce PowerWorld Simulator—a power system analysis and simulation software package.

CASE STUDY

The following article describes the deregulation of the electric utility industry that has been taking place in the United States, including the benefits and problems that have been encountered with deregulation. During the last two decades, deregulation has had both good and bad effects. It has changed the mix of fuels in the U.S. generation fleet, shifting it away from coal and nuclear power toward natural gas and has opened the door to greener forms of electricity generation. It has also made many companies that provide electricity more efficient by increasing the reliability of power plants and reducing labor costs. However, wholesale prices of electricity have increased dramatically in some areas of the United States, market-based investments in transmission have been problematic, and rolling blackouts have been encountered [8].

How the Free Market Rocked the Grid*

Seth Blumsack Pennsylvania State University

It led to higher rates and rolling blackouts, but it also opened the door to greener forms of electricity generation.

Most of us take for granted that the lights will work when we flip them on, without worrying too much about the staggeringly complex things needed to make that happen. Thank the engineers who designed and built the power grids for that—but don't thank them too much. Their main goal was reliability; keeping the cost of electricity down was less of a concern. That's in part why so many people in the United States complain about high electricity prices. Some armchair economists (and a quite a few real ones) have long argued that the solution is deregulation. After all, many other U.S. industries have been deregulated—take, for instance, oil, natural gas, or trucking—and greater competition in those sectors swiftly brought prices down. Why not electricity?

Such arguments were compelling enough to convince two dozen or so U.S. states to deregulate their electric industries. Most began in the mid-1990s, and problems emerged soon after, most famously in the rolling blackouts that Californians suffered through in the summer of 2000 and the months that followed. At the root of these troubles is the fact that free markets can be messy and volatile, something few took into account when deregulation began. But the consequences have since proved so chaotic that a quarter of these states have now suspended plans to revamp the way they manage their electric utilities, and few (if any) additional states are rushing to jump on the deregulation bandwagon.

The United States is far from being the only nation that has struggled with electricity deregulation. But the U.S. experience is worth exploring because it highlights many of the challenges that can arise when complex industries such as electric power generation and distribution are subject to competition.

Unlike many other nations grappling with electricity deregulation, the United States has never had one governmentowned electric utility running the whole country. Instead, a patchwork of for-profit utilities, publicly owned municipal utilities, and electric cooperatives keeps the nation's lights on. The story of how that mixture has evolved over the last 128 years helps to explain why deregulation hasn't made electric power as cheap and plentiful as many had hoped.

The 1882 opening of Thomas Edison's Pearl Street generation station in New York City marks the birth of the American electric utility industry. That station produced low-voltage direct current, which had to be consumed close to the point of production, because sending it farther would have squandered most of the power as heat in the connecting wires.

Edison's approach prevailed for a while, with different companies scrambling to build neighborhood power stations. They were regulated only to the extent that their owners had to obtain licenses from local officials. Municipalities handed these licenses out freely, showing the prevailing laissez-faire attitude toward competition. Also, politicians wanted to see the cost of electricity drop. (A kilowatt-hour in the late 1800s cost about U.S. $5.00 in today's dollars; now it averages just 12 cents.)

It didn't take long, though, before Samuel Insull, a former Edison employee who became a utility entrepreneur in the Midwest, realized that the technology George Westinghouse was advocating—large steam or hydroelectric turbines linked to long-distance ac transmission lines—could provide electricity at lower cost. Using such equipment, his company soon drove its competitors out of business. Other big utilities followed Insull's lead and came to monopolize the electricity markets in New York, New Jersey, and the Southeast. But the rise of these companies was ultimately a bane to consumers, who had to pay exorbitant prices after the competition had been quashed.

Angered by the steep rates, consumers formed electricity cooperatives and municipal utilities. That in turn led Insull and his counterparts to plead with state officials for protection from this "ruinous" competition. Politicians complied, passing laws that granted the large electric power companies exclusive franchises in their areas in exchange for regulation of their prices and profits. The municipal utilities and electricity cooperatives continued to operate but in most cases never grew as large as the regulated for-profit (investor-owned) utilities.

This basic structure remained in place until the oil shocks of the 1970s. Real electricity prices rose by almost 50 percent during that troubled decade, despite having fallen virtually every year since the opening of Edison's Pearl Street station. One culprit was the widespread use of imported oil. The United States then generated almost 20 percent of its electricity using fuel oil; today that figure is less than 1 percent. And many utilities had made some poor investments—primarily in nuclear power—which their customers had to pay for.

The 1970s also exposed problems in how the electric power industry was regulated. Power grids were growing in complexity as different utilities began interconnecting, and many regulators—particularly those whose appointments were political favors—didn't understand the technical implications of their decisions. The combination of rising prices and obvious mismanagement led many large industrial consumers of electricity to push for deregulation.

The Public Utility Regulatory Policies Act of 1978 was the first shot fired in the ensuing battle. The new federal law allowed nonutility companies to generate electricity from "alternative" fuel sources (mostly natural gas), and it required utilities to sign long-term supply contracts with these new generating companies. The Energy Policy Act of 1992 expanded the pool of players in the wholesale electricity market by allowing financial institutions—Morgan Stanley being the first—to buy and sell bulk electric power. Yet neither act was effective in curbing electricity prices.

Two states, California and Pennsylvania, then decided to take more drastic measures. They established centralized spot markets for electricity and allowed individual customers to choose their electricity suppliers. While Pennsylvania's experiment has largely run smoothly, California's experience was quite different. After two years of reasonably stable operation, wholesale prices exploded in 2000, from a few cents per kilowatt-hour to more than a dollar per kilowatt-hour. One reason for those astronomical prices was that power-trading companies like Enron Corp. had figured out how to game the system. With retail prices capped by law at 6.7 cents per kilowatt-hour, two of the state's three investor-owned utilities, Pacific Gas & Electric and Southern California Edison, ran out of money to pay for electricity. That triggered a second power crisis the following year, which forced the state to buy electricity from producers. The long-term contracts signed during that period of panic buying saddled California taxpayers with a debt of some $40 billion.

■■■■

For Californians, at least, deregulation had lost its gloss. This turned out to be temporary: The state recently reintroduced centralized wholesale markets modeled after Pennsylvania's. But has deregulation on the whole made things better or worse? Dozens of studies have attempted to answer that question. But you can't simply compare states that have aggressively deregulated with ones that haven't. That would ignore the fact that some states have built-in advantages that keep prices low: proximity to natural resources, a large base of generation capacity, and so forth. It also ignores what utilities and regulators would have done if deregulation had never happened.

To answer the question properly, you'd need to figure out what things would have been like in the absence of deregulation. And that's well-nigh impossible. Of the various studies that have attempted to assess the impacts of deregulation, most have come from groups with a stake in the outcome of the regulatory reform process. So they tend to be either strongly for deregulation or strongly against it. In reality, deregulation has had both good and bad effects.

Consider a simple variable like the price of electricity. That competition will lead to lower prices is about as close to a universal truth as economics gets. But electricity seems to be an exception.

Here's why: Under regulation, each generating plant is paid for its electricity based on its average cost plus some prescribed rate of return. In a competitive market, supply and demand set the price. That means that the last plant coming online to handle the load determines the wholesale price of electricity. All generators in the system are then paid that same amount for each kilowatt-hour they inject into the grid.

That might seem only fair, but you have to remember that not all electricity generators are created equal. In most places, coal and nuclear plants, which can't be ramped up and down easily, produce the roughly constant baseload power feeding the grid. If more is needed, natural gas turbines then kick in. So in deregulated markets, the price of gas, which has historically been higher than that of coal or nuclear fuel, ends up controlling the wholesale price of electricity—allowing the owners of nuclear plants and efficient coal plants to earn much higher profits than they did under regulation. That's why electricity prices in many places rose so sharply when natural gas prices skyrocketed at the turn of the millennium.

Other strange dynamics also come into play. For example, state political leaders realize that escalating or erratic electricity prices are bad for economic development (and their own chances of reelection). So they've fought hard to keep them low and stable by imposing rate caps and freezes. But many of these same states also compelled their electric utilities to divest themselves of generating capacity in an attempt to spur competition. And when electricity demand is high and the utilities don't have enough of their own generating capacity, they're forced to buy more on the spot market, where prices are volatile. The results have not been pretty. In 2000, one of California's two largest utilities went bankrupt, and the other nearly did. And when regulators in Maryland finally allowed retail electricity rates in Baltimore to float with wholesale electricity prices, the local utility immediately announced a rate increase of 72 percent, leading to consumer outrage and eventually to the summary firing of the entire public utility commission.

DEREGULATION IS ALL OVER THE MAP
Countries have deregulated their electric power industries to different degrees, as these five examples show.

Argentina

Privatization of electricity generation in Argentina began in 1992, followed the next year by privatization of that nation's six transmission companies. Argentine law did not allow any of the resultant for-profit power companies to control more than 10 percent of the country's generation capacity, ensuring considerable competition among them.

United Kingdom

Electricity restructuring in the UK began under Margaret Thatcher, with the Electricity Act of 1983, which gave independent power producers access to the national grid. Government-owned generators were then fully privatized in the 1990s.

France

France began a very modest program of reform in 2001, but for the most part electricity supply remains completely dominated by the state electricity company, Électricité de France.

Germany

In response to a 1996 European directive, Germany abolished its law exempting electricity from competition in 1998. But most of that country's electricity still comes from just a few vertically integrated power companies, with comparatively little electricity trading on open exchanges.

Australia

The Australian state of Victoria privatized its electricity sector in 1994. Some other Australian states soon followed suit. And Australia established a national wholesale electricity market in December 1998.

Clearly, deregulation hasn't been at all successful in bringing prices down. But has it made the companies that provide electricity more efficient? Very probably. Their labor costs have fallen, mostly through reductions in staff, while the reliability of their power plants has improved. The champions in this regard are the nuclear power stations, whose uptimes have risen from around 65 percent in the 1980s to over 90 percent today. This shouldn't be a surprise. Because the construction costs of most of these plants have been paid off and because nuclear generators have very low operating expenses, the plants have become extraordinarily profitable. So their owners strive to have them online as much as possible, investing as needed to keep them well maintained.

Maintaining some other parts of the grid infrastructure has, however, proved to be more of a struggle. In the old days, investments in transmission lines and generating stations were determined by consensus between each utility and its regulator. Deregulation's architects envisioned a different scenario—that entrepreneurial firms would automatically make the needed investments in hopes of profiting from them. That didn't exactly happen. One thing deregulation definitely did do, though, was to change the mix of fuels in the U.S. generation fleet, shifting it away from coal and nuclear power toward natural gas. That's because gas units are quick to build, and many are flexible enough to operate only when prices are high enough to warrant throwing their switches on. It helps, too, that natural gas is a cleaner fuel than coal and less controversial than nuclear power, which helps with public approval. Also, because companies generating electricity in a free market need to demonstrate a return on investment within 5 to 10 years, building big nuclear and coal plants, which usually take over a decade to complete, just isn't an option. So more and more of the grid's power comes from gas turbines, despite the high fuel costs.

The changing investment environment has also inflated the cost of building new infrastructure. The reason is obvious once you think about it. Regulated utilities can spread the burden of investment among all their customers, and the government guarantees that these companies can charge enough to recover their initial outlay and make a decent profit on it. So there's little financial risk in building a new plant or transmission line, allowing the companies to attract low-priced capital. Not so with unregulated utilities, whose fortunes depend on an uncertain market. The greater risk they face means they must offer higher returns to attract investors, and these increased financing costs make capital projects more expensive.

Depending on market-based investment in transmission lines has proved especially problematic. Deregulation's proponents believed that for-profit companies would recover the money they invested in transmission lines through "congestion pricing"—charging more when demand for these lines is high. Instead, lucrative congestion revenues have only given the owners of existing transmission lines an incentive *not* to build more capacity. And the general aversion people have to high-tension cables nearby—the "not in my backyard" effect—has made it almost impossible to construct new lines.

No great wonder, then, that investment in transmission lines and equipment has mostly been falling since the 1970s. Many people paid little notice to that fact, but the Northeast blackout of 2003 was a wake-up call. It began on a hot August afternoon with several seemingly trivial outages of transmission lines in Ohio, but by nighttime a series of cascading failures grew to plunge more than 50 million people in the Midwest, the Northeast, and Ontario into darkness. This episode convinced even skeptics that investment in the nation's electricity grid was lagging.

■■■■

Given deregulation's checkered record, you have to wonder how well competitive electricity markets will handle upcoming challenges. In particular, how will they reconcile the need for reliable, low-cost power with the environmental costs of producing it?

One much-discussed way to use markets to benefit the environment is to put a price on emissions of carbon dioxide and other greenhouse gases. Many countries have already done this. But unless the price is set a lot higher than in Europe, U.S. utilities and generating companies aren't going to be abandoning their carbon-spewing coal plants anytime soon—they're just too profitable. Putting a dollar value on greenhouse gases might encourage some generators to invest in less carbon-intensive power sources where they can, but only if proper laws and regulations are in place to lower the risk. And that won't happen overnight.

In the meantime, 32 of the 50 U.S. states are trying to boost the use of renewables by mandating "renewable portfolio standards." These standards force utilities to buy considerable quantities of wind and solar power but also give them the freedom to shop for the least expensive sources. Also, the U.S. Department of Energy wants 20 percent of the nation's electricity to come from wind power by 2030. Government bodies are taking these actions because consumer demand alone hasn't sparked much renewable generation. That's not surprising. The wind and sun are notoriously fickle, which forces system operators to maintain plants that can fill in when necessary. Those backup generators are expensive, as are the transmission lines needed to link most renewable resources, which are located in sparsely populated areas, to the people using electricity. So the cost of generating "green" electricity is generally higher than the price it can command.

Renewable portfolio standards create a not-so-free market (but a market nevertheless) for wind and solar power while also pressuring these producers to keep their prices down.

Policymakers in both regulated and deregulated states are also hoping to harness other market-based approaches to reducing electricity consumption. Using less electricity not only helps the environment, it can be just as effective as increasing supply in maintaining the reliability of the grid. And it's less expensive to boot.

The most straightforward way to discourage electricity use is, of course, to charge a lot for it. But U.S. consumers, and the lawmakers who represent them, are never too keen on that. Another strategy now being explored—one that's less of a political hot potato—is to have utility operators offer their customers compensation for reducing their demand for electricity during times of peak use. A reduction in demand allows utilities to avoid having to buy so much electricity when wholesale costs are at their highest. This approach provides an enticement to consumers to react to market signals, even if they are not yet ready to face them squarely in the form of higher prices.

Another advance that probably wouldn't have come about without deregulation is the emergence of small-scale, distributed generation, particularly from renewable sources such as rooftop solar panels. What's happening in many places is that customers are producing some electricity on their own while still attached to the grid. So they can offset some of the electricity they would otherwise consume, perhaps even spinning their meters backward at times. Although this practice competes with the electricity that the utility sells, more and more utilities are nevertheless allowing it to a greater or lesser degree.

■■■■

In hindsight, the electricity crisis in California and the myriad problems with deregulation in other parts of the country could have been anticipated. Given the complex market rules, concentrated supply, and largely inelastic demand, it's really no wonder that Enron, other energy-trading companies, and the electricity suppliers themselves found clever ways to manipulate markets.

Would U.S. consumers have been better off if the industry had remained strictly regulated? It all depends. If your goal is low electricity rates, maybe the answer is yes—but don't forget that bad regulatory decisions helped drive up electricity prices in the first place. If, however, you want the ability to feed power from your rooftop solar panels into the grid, the answer is probably no.

The real question facing the United States now is whether it can maintain reliable electricity grids without building lots of new transmission lines and big power plants. The only realistic alternative to such massive construction projects is for the generation of electricity to become more widely distributed, coupled with substantial efforts in energy efficiency. Electricity markets will surely have to become more expansive and open to accommodate that inevitable evolution. And they will also require new technical standards and, yes, some new forms of regulation. ■

1.1 HISTORY OF ELECTRIC POWER SYSTEMS

In 1878, Thomas A. Edison began work on the electric light and formulated the concept of a centrally located power station with distributed lighting serving a surrounding area. He perfected his light by October 1879, and the opening of his historic Pearl Street Station in New York City on September 4, 1882, marked the beginning of the electric utility industry (see Figure 1.1). At Pearl Street, dc generators, then called dynamos, were driven by steam engines to supply an initial load of 30 kW for 110-V incandescent lighting to 59 customers in a 2.6-square-kilometer area. From this beginning in 1882 through 1972, the electric utility industry grew at a remarkable pace—a growth based on continuous reductions in the price of electricity due primarily to technological accomplishment and creative engineering.

The introduction of the practical dc motor by Sprague Electric, as well as the growth of incandescent lighting, promoted the expansion of Edison's dc systems. The development of three-wire, 220-V dc systems allowed load to increase somewhat, but as transmission distances and loads continued to increase, voltage problems were

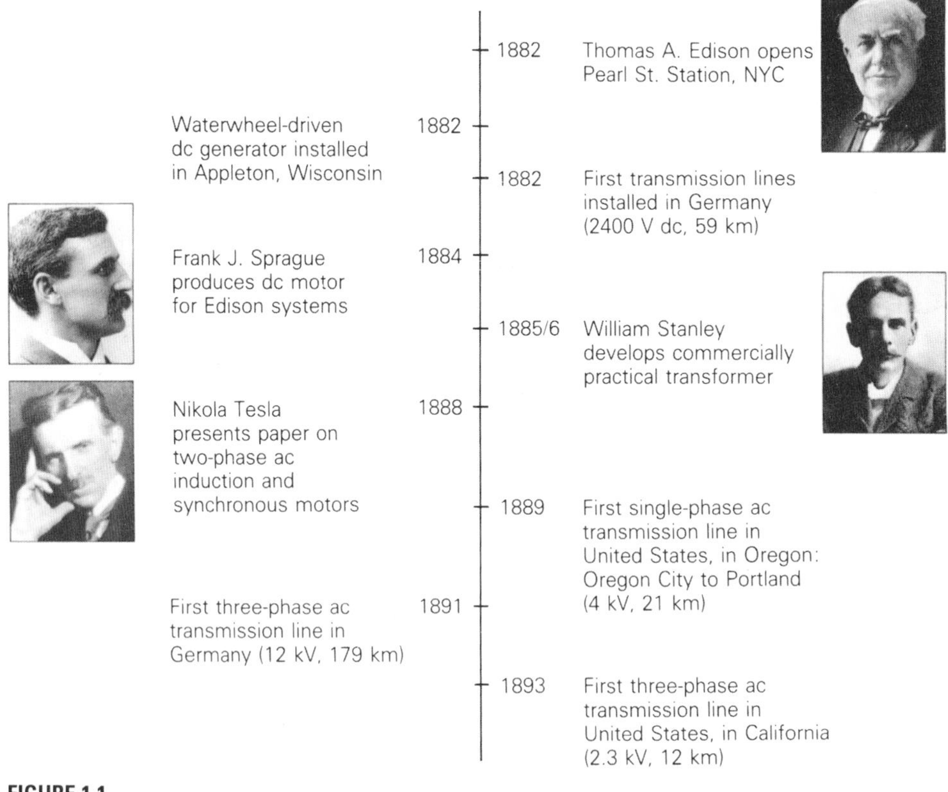

FIGURE 1.1

Milestones of the early electric utility industry [1]. (H.M. Rustebakke et al., *Electric Utility Systems Practice*, 4th Ed. (New York: Wiley, 1983). Reprinted with permission of John Wiley & Sons, Inc. Photos courtesy of Westinghouse Historical Collection.) (Photos courtesy of Westinghouse Historical Collection.)
(Based on H.M. Rustebakke et al., Electric Utility Systems Practice, 4th. Ed. (New York: Wiley, 1983).)

encountered. These limitations of maximum distance and load were overcome in 1885 by William Stanley's development of a commercially practical transformer. Stanley installed an ac distribution system in Great Barrington, Massachusetts, to supply 150 lamps. With the transformer, the ability to transmit power at high voltage with corresponding lower current and lower line-voltage drops made ac more attractive than dc. The first single-phase ac line in the United States operated in 1889 in Oregon, between Oregon City and Portland—21 km at 4 kV.

The growth of ac systems, further encouraged in 1888 when Nikola Tesla presented a paper at a meeting of the American Institute of Electrical Engineers describing two-phase induction and synchronous motors, made evident the advantages of polyphase versus single-phase systems. The first three-phase line in Germany became operational in 1891, transmitting power 179 km at 12 kV. The first three-phase line in the United States (in California) became operational in 1893, transmitting power 12 km at 2.3 kV. The three-phase induction motor conceived by Tesla went on to become the workhorse of the industry.

In the same year that Edison's steam-driven generators were inaugurated, a waterwheel-driven generator was installed in Appleton, Wisconsin. Since then, most electric energy has been generated in steam-powered and in water-powered (called hydro) turbine plants. Today, steam turbines account for more than 85% of U.S. electric energy generation, whereas hydro turbines account for about 7%. Gas turbines are used in some cases to meet peak loads. Also, the addition of wind turbines and solar cells into the bulk power system is expected to grow considerably in the near future.

Steam plants are fueled primarily by coal, gas, oil, and uranium. Of these, coal is the most widely used fuel in the United States due to its abundance in the country. Although many of these coal-fueled power plants were converted to oil during the early 1970s, that trend reversed back to coal after the 1973–74 oil embargo, which caused an oil shortage and created a national desire to reduce dependency on foreign oil. More recently, electric companies have been retiring older, coal-fueled power plants. In 2012, approximately 37% of electricity in the United States was generated from coal [2].

In 1957, nuclear units with 90-MW steam-turbine capacity, fueled by uranium, were installed, and today nuclear units with 1312-MW steam-turbine capacity are in service. In 2012, approximately 19% of electricity in the United States was generated from uranium from 104 nuclear power plants. However, the growth of nuclear capacity in the United States has been halted by rising construction costs, licensing delays, and public opinion. Although there are no emissions associated with nuclear power generation, there are safety issues and environmental issues, such as the disposal of used nuclear fuel and the impact of heated cooling-tower water on aquatic habitats. Future technologies for nuclear power are concentrated on safety and environmental issues [2, 3, 7].

Starting in the 1990s, the choice of fuel for new power plants in the United States has been natural gas due to its availability and low cost as well as the higher efficiency, lower emissions, shorter construction-lead times, safety, and lack of controversy associated with power plants that use natural gas. Natural gas is used to generate electricity by the following processes:

1. Gas combustion turbines use natural gas directly to fire the turbine.
2. Steam turbines burn natural gas to create steam in a boiler, which is then run through the steam turbine.

3. Combined cycle units use a gas combustion turbine by burning natural gas, and the hot exhaust gases from the combustion turbine are used to boil water that operates a steam turbine.
4. Fuel cells powered by natural gas generate electricity using electrochemical reactions by passing streams of natural gas and oxidants over electrodes that are separated by an electrolyte. In 2012, approximately 30% of electricity in the United States was generated from natural gas [2, 3, 7].

Renewable energy sources have the advantage of providing a more sustainable use of finite energy sources with little or no contribution to polluting emissions. Renewable sources include conventional hydroelectric (water power), geothermal, wood, wood waste, all municipal waste, landfill gas, other biomass, solar, and wind power. In 2012, in the United States, approximately 12% of electricity was generated by renewable sources [2, 3]. Germany, which has an official governmental goal to generate 100% of electricity from renewable sources by 2050, achieved 31% in 2014 [14].

Substantial research efforts have shown nuclear fusion energy to be a promising technology for producing safe, pollution-free, and economical electric energy later in the 21st century and beyond. The fuel consumed in a nuclear fusion reaction is deuterium, of which a virtually inexhaustible supply is present in seawater.

The early ac systems operated at various frequencies including 25, 50, 60, and 133 Hz. In 1891, it was proposed that 60 Hz be the standard frequency in the United States. In 1893, 25-Hz systems were introduced with the synchronous converter. However, these systems were used primarily for railroad electrification (and many are now retired) because they had the disadvantage of causing incandescent lights to flicker. In California, the Los Angeles Department of Power and Water operated at 50 Hz but converted to 60 Hz when power from the Hoover Dam became operational in 1937. In 1949, Southern California Edison also converted from 50 to 60 Hz. Today, the two standard frequencies for generation, transmission, and distribution of electric power in the world are 60 Hz (in the United States, Canada, Mexico, and Brazil) and 50 Hz (in Europe, the former Soviet republics, China, South America, except Brazil, and India). In Japan, the western part of the country, including Kyoto, uses 60 Hz, while the eastern part, including Tokyo, uses 50 Hz. The advantage of 60-Hz systems is that generators, motors, and transformers in these systems are generally smaller than 50-Hz equipment with the same ratings. The advantage of 50-Hz systems is that transmission lines and transformers have smaller reactances at 50 Hz than at 60 Hz.

As shown in Figure 1.2, the rate of growth of electric energy in the United States was approximately 7% per year from 1902 to 1972. This corresponds to a doubling of electric energy consumption every 10 years over the 70-year period. In other words, every 10 years, the industry installed a new electric system equal in energy-producing capacity to the total of what it had built since the industry began. The annual growth rate slowed after the oil embargo of 1973 to 1974. Kilowatt-hour consumption in the United States increased by 3.4% per year from 1972 to 1980, and by only 0.7% per year from 2000 to 2011. In 2012, total U.S. electricity generation was 4.05×10^{12} kilowatt-hours (kWh), a 1.3% decrease from 2011.

Along with increases in load growth, there have been continuing increases in the size of generating units (Table 1.1). The principal incentive to build larger units has been economy of scale—that is, a reduction in installed cost per kilowatt of capacity

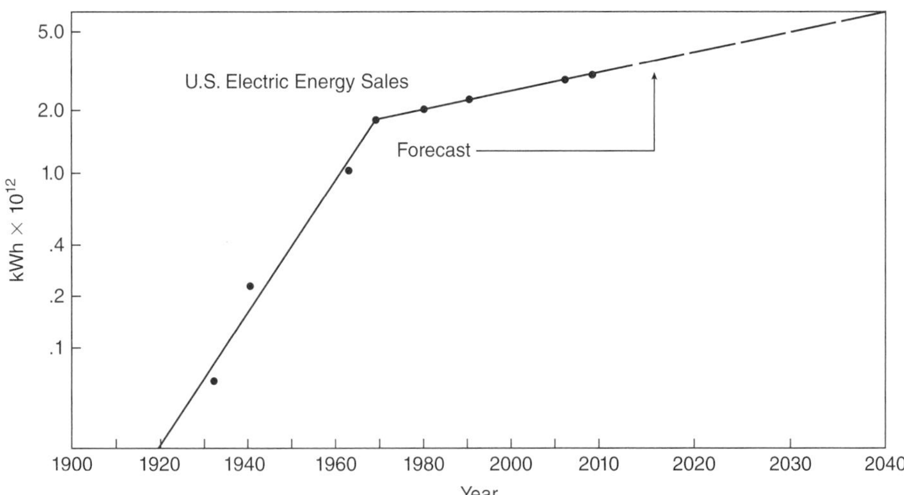

FIGURE 1.2

Growth of U.S. electric energy consumption [3]. (www.eia.gov, U.S. Energy Information Administration, Annual Energy Outlook 2010 Early Release Overview, M. P. Barhman and B.K. Johnson, "The ABC's of HVDC Transmission Technologies" IEEE Power & Energy Magazine, 5, 2, (March/April 2007) pp. 33–44.)

for larger units. However, there have also been steady improvements in generation efficiency. For example, in 1934 the average heat rate for steam generation in the U.S. electric industry was 18.94 MJ/kWh, which corresponds to 19% efficiency. By 1991, the average heat rate was 10.94 MJ/kWh, which corresponds to 33% efficiency. These improvements in thermal efficiency due to increases in unit size and in steam temperature and pressure, as well as to the use of steam reheat, have resulted in savings in fuel costs and overall operating costs.

There have been continuing increases, too, in transmission voltages (Table 1.2). From Edison's 220-V three-wire dc grid to 4-kV single-phase and 2.3-kV three-phase transmission, ac transmission voltages in the United States have risen progressively to 150, 230, 345, 500, and now 765 kV. And in 2009 in China, the first 1000 kV ultra-high voltage (UHV) ac transmission line, a 650-km line from Shaanxi Province

Hydroelectric Generators		Generators Driven by Single-Shaft, 3600 r/min Fossil-Fueled Steam Turbines	
Size (MVA)	Year of Installation	Size (MVA)	Year of Installation
4	1895	5	1914
108	1941	50	1937
158	1966	216	1953
232	1973	506	1963
615	1975	907	1969
718	1978	1120	1974

TABLE 1.1

Growth of generator sizes in the United States [1].

(*Source:* Based on H.M. Rustebakke et al., Electric Utility Systems Practice, 4th. Ed. (New York: Wiley, 1983).

Voltage (kV)	Year of Installation
2.3	1893
44	1897
150	1913
165	1922
230	1923
287	1935
345	1953
500	1965
765	1969

TABLE 1.2

History of increases in three-phase transmission voltages in the United States [1].

(*Source:* Based on H.M. Rustebakke et al., Electric Utility Systems Practice, 4th. Ed. (New York: Wiley, 1983).

to Hubei Province, began commercial operation [15]. The incentives for increasing transmission voltages have been: (1) increases in transmission distance and transmission capacity; (2) smaller line-voltage drops; (3) reduced line losses; (4) reduced right-of-way requirements per MW transfer; and (5) lower capital and operating costs of transmission. Today, one 765-kV three-phase line can transmit thousands of megawatts over hundreds of kilometers.

The technological developments that have occurred in conjunction with ac transmission, including developments in insulation, protection, and control, are in themselves important. The following examples are noteworthy:

1. The suspension insulator;
2. The high-speed relay system, currently capable of detecting short-circuit currents within one cycle (0.017 s);
3. High-speed, extra-high-voltage (EHV) circuit breakers, capable of interrupting up to 63-kA three-phase, short-circuit currents within two cycles (0.033 s);
4. High-speed reclosure of EHV lines, which enables automatic return to service within a fraction of a second after a fault has been cleared;
5. The EHV surge arrester, which provides protection against transient overvoltages due to lightning strikes and line-switching operations;
6. Power-line carrier, microwave, and fiber optics as communication mechanisms for protecting, controlling, and metering transmission lines;
7. The principle of insulation coordination applied to the design of an entire transmission system;
8. Energy control centers with supervisory control and data acquisition (SCADA) and with automatic generation control (AGC) for centralized computer monitoring and control of generation, transmission, and distribution;
9. Automated distribution features, including advanced metering infrastructure (AMI), reclosers, and remotely controlled sectionalizing switches with fault-indicating capability, along with automated mapping/facilities management (AM/FM) and geographic information systems (GIS) for quick isolation and identification of outages and for rapid restoration of customer services; and
10. Digital relays capable of circuit breaker control, data logging, fault locating, self-checking, fault analysis, remote query, and relay event monitoring/recording.

In 1954, the first modern high-voltage dc (HVDC) transmission line was put into commercial operation in Sweden between Vastervik and the island of Gotland in the Baltic Sea; it operated at 100 kV for a distance of 100 km. The first HVDC line in the United States was the ±400-kV (now ±500 kV), 1360-km Pacific Intertie line, installed between Oregon and California in 1970. As of 2014, seven other HVDC lines up to ±500 kV and 13 back-to-back ac-dc links had been installed in the

United States, and a total of 111 HVDC lines up to ±800 kV had been installed worldwide [4]. And in 2015 in China, the first ±1000 kV UHVDC transmission line, a 2600-km line from Xingjiang Province to Sichuan Province, is planned for commercial operation [16].

For an HVDC line embedded in an ac system, solid-state converters at both ends of the dc line operate as rectifiers and inverters. Since the cost of an HVDC transmission line is less than that of an ac line with the same capacity, the additional cost of converters for dc transmission is offset when the line is long enough. Studies have shown that overhead HVDC transmission is economical in the United States for transmission distances longer than about 600 km. However, HVDC also has the advantage that it may be the only feasible method to:

1. interconnect two asynchronous networks;
2. utilize long underground or underwater cable circuits;
3. bypass network congestion;
4. reduce fault currents;
5. share utility rights-of-way without degrading reliability; and mitigate environmental concerns [5].

In the United States, electric utilities grew first as isolated systems, with new ones continuously starting up throughout the country. Gradually, however, neighboring electric utilities began to interconnect, to operate in parallel. This improved both reliability and economy. Figure 1.3 shows major 115-kV and higher-voltage, interconnected transmission in the United States. An interconnected system has many advantages. An interconnected utility can draw upon another's rotating generator reserves during a time of need (such as a sudden generator outage or load increase), thereby maintaining continuity of service, increasing reliability, and reducing the total number of generators that need to be kept running under no-load conditions. Also, interconnected utilities can schedule power transfers during normal periods to take advantage of energy-cost differences in respective areas, load diversity, time zone differences, and seasonal conditions. For example, utilities whose generation is primarily hydro can supply low-cost power during high-water periods in spring/summer and can receive power from the interconnection during low-water periods in fall/winter. Interconnections also allow shared ownership of larger, more efficient generating units.

While sharing the benefits of interconnected operation, each utility is obligated to help neighbors who are in trouble, to maintain scheduled intertie transfers during normal periods, and to participate in system frequency regulation.

In addition to the benefits and obligations of interconnected operation, there are disadvantages. Interconnections, for example, have increased fault currents that occur during short circuits, thus requiring the use of circuit breakers with higher interrupting capability. Furthermore, although overall system reliability and economy have improved dramatically through interconnection, there is a remote possibility that an initial disturbance may lead to a regional blackout, such as the one that occurred in August 2003 in the northeastern United States and Canada.

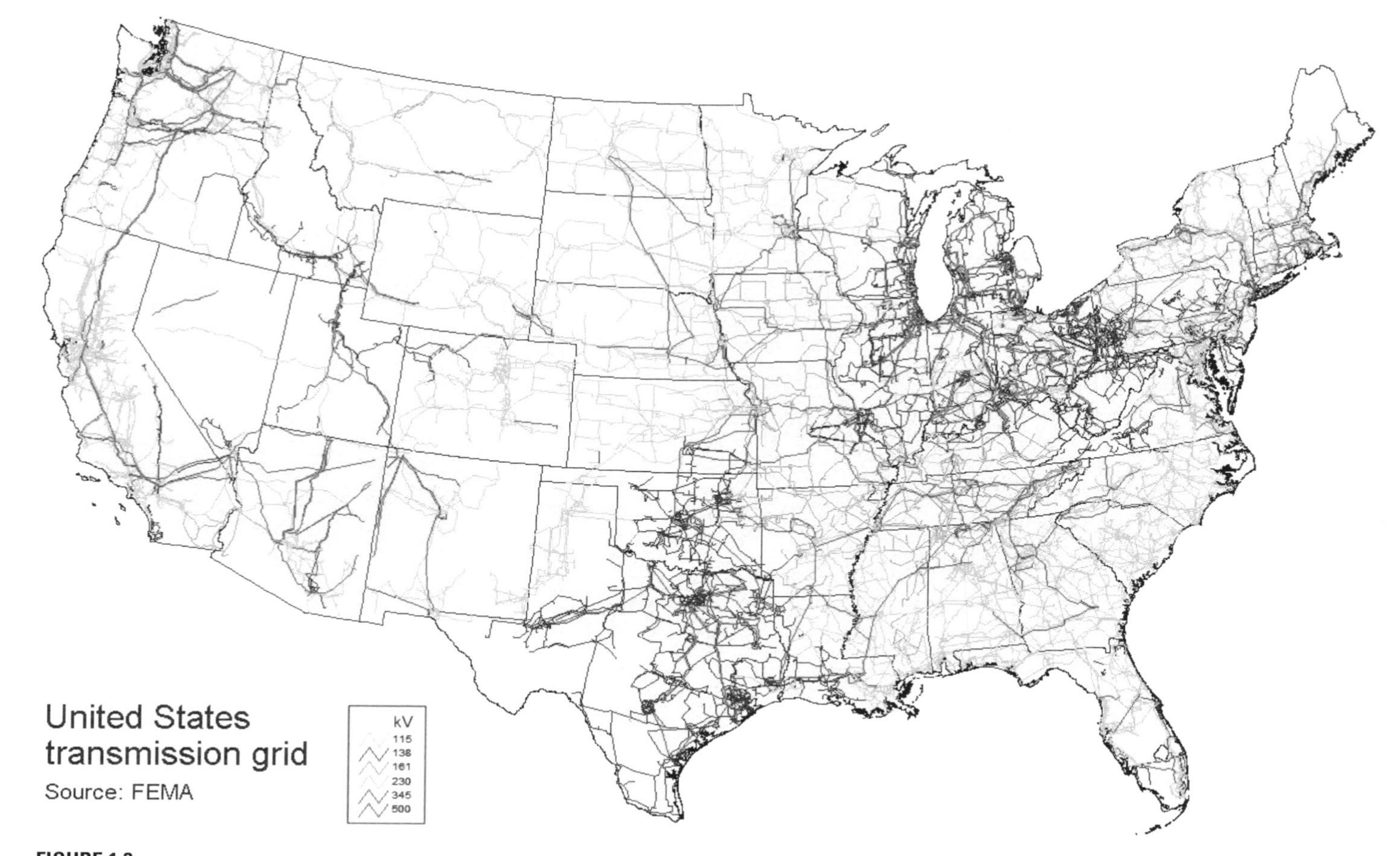

FIGURE 1.3

Major transmission in the United States (FEMA).

1.2 PRESENT AND FUTURE TRENDS

Present trends indicate that the United States is becoming more electrified as it shifts away from a dependence on the direct use of fossil fuels. The electric power industry advances economic growth, promotes business development and expansion, provides solid employment opportunities, enhances the quality of life for its users, and powers the world. Increasing electrification in the United States is evidenced in part by the ongoing digital revolution. Today the United States electric power industry is a robust, \$840-billion-plus industry that employs nearly 500,000 workers. In the United States economy, the industry represents approximately 3% of real gross domestic product (GDP) [6].

As shown in Figure 1.2, the growth rate in the use of electricity in the United States is projected to increase by about 1% per year from 2012 to 2025 [2]. Although electricity forecasts for the next thirteen years are based on economic and social factors that are subject to change, 1% annual growth rate is considered necessary to generate the GDP anticipated over that period. Variations in longer-term forecasts of 0.7 to 1.3% annual growth from 2012 to 2040 are based on low-to-high ranges in economic growth. Average end-use price of electricity is projected to increase from 9.8 cents to 11.1 cents per kilowatt-hour from 2012 to 2040 [2, 3].

Figure 1.4 shows the percentages of various fuels used to meet U.S. electric energy requirements for 2012 and those projected for 2025 and 2040. Several trends are apparent in the chart. One is the continuing use of coal. This trend is due primarily to the large amount of U.S. coal reserves, which, according to some estimates, is

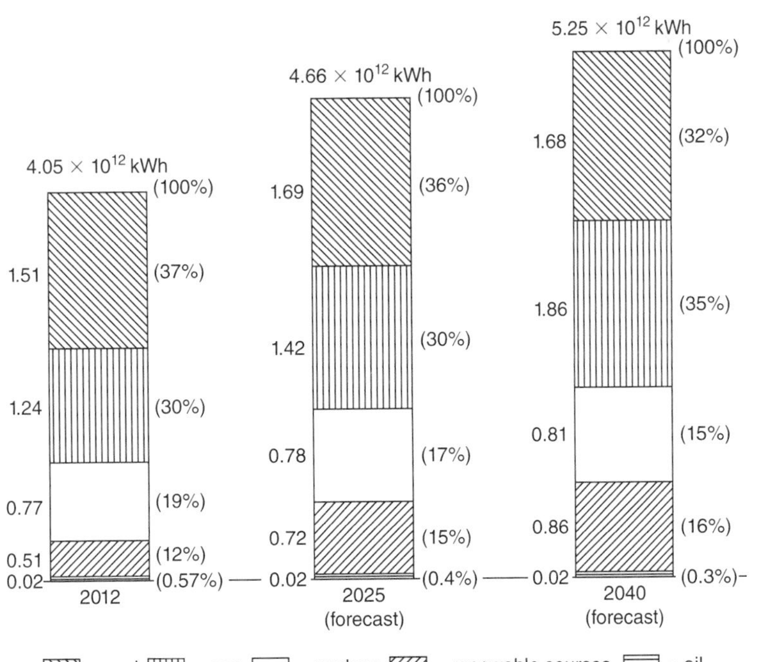

FIGURE 1.4 Electric energy generation in the United States, by principal fuel types [2, 3]. Renewable sources include conventional hydroelectric, geothermal, wood, wood waste, all municipal waste, landfill gas, other biomass, solar, and wind power. (U.S. Energy Information Administration, Electric power annual 2012, Dec 2013, U.S. Energy Information Administration, Annual Energy Outlook 2014, www.eia.gov.)

sufficient to meet U.S. energy needs for the next 500 years. Implementation of public policies that have been proposed to reduce carbon dioxide emissions and air pollution could reverse this trend. Another trend is the continuing consumption of natural gas in the long term with gas-fired turbines that are safe, clean, and more efficient than competing technologies. Regulatory policies to lower greenhouse gas emissions could accelerate a switchover from coal to gas, but that would require an increasing supply of deliverable natural gas. A slight percentage decrease in nuclear fuel consumption is also evident. No new nuclear plant has been ordered in the United States for more than 30 years. The projected growth from 0.77×10^{12} kWh in 2012 to 0.86×10^{12} kWh in 2040 in nuclear generation is based on uprates at existing plants and some new nuclear capacity that offsets early retirements. Safety concerns require passive or inherently safe reactor designs with standardized, modular construction of nuclear units. Also shown in Figure 1.4 is an accelerating increase in electricity generation from renewable resources in response to a variety of federal and state policies [2, 3].

Figure 1.5 shows the 2012 and projected 2025 U.S. generating capability by principal fuel type. As shown, total U.S. generating capacity is projected to reach 1110 GW (1 GW = 1000 MW) by the year 2025, which represents a 0.3% annual projected growth in generating capacity from 2013 to 2025. The projection assumes 131 GW of new capacity additions and 87 GW of cumulative capacity retirements from 2013 to 2025 [2, 3, 7].

As of 2012, there were 766,445 circuit kilometers of existing transmission (above 200 kV) in North America, which includes the continental United States,

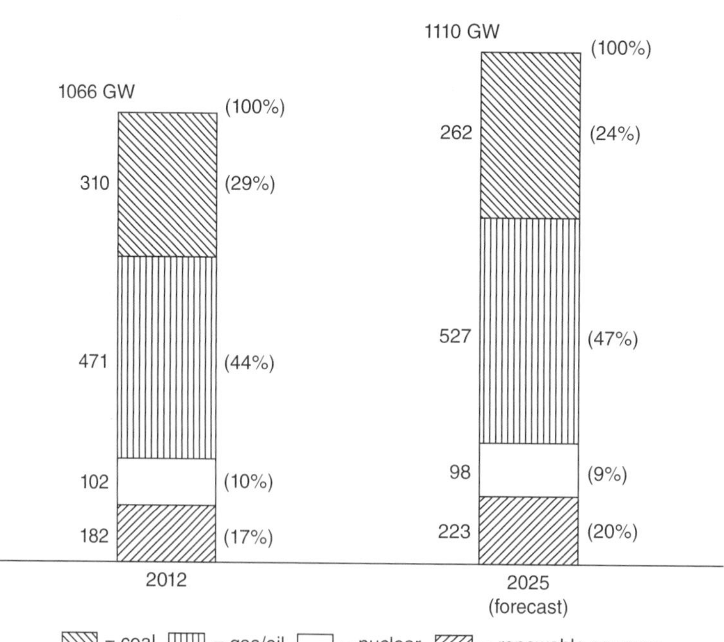

FIGURE 1.5 Installed generating capability in the United States by principal fuel types [2, 3]. Net summer capacities of renewable sources include conventional and pumped storage hydroelectric, geothermal, wood, wood waste, all municipal waste, landfill gas, other biomass, solar, and wind power. (U.S. Energy Information Administration, electric power annual 2012, Dec 2013, U.S. Energy Information Administration, Annual Energy Outlook 2014, www.eia.gov.)

Canada, and the northern portion of Baja California, Mexico. Currently, 8208 circuit kilometers are under construction and expected to be in service before 2015, and an additional 35,084 circuit kilometers are planned with in-service dates before 2023. According to the North American Electric Reliability Corporation (NERC), new transmission projects are being driven primarily to enhance reliability. Other reasons include congestion alleviation and integration of renewable energy sources [7].

According to NERC, building new transmission is an ongoing challenge for the electricity industry. Transmission right-of-way issues required in the siting of new transmission lines that span multiple states and provinces are highly visible and require increased coordination among multiple regulating agencies and authorities. A lack of coordination regarding cost allocation, coupled with public opposition due to land use and property valuation concerns, has often resulted in extended delays in transmission construction. Further, ongoing state and provincial policies require continued integration of renewable resources, including wind and solar. These variable resources are most commonly built in remote areas where wind power densities and solar development are favorable. In many cases, the existing transmission network needs to be expanded to integrate these renewable resources and meet other state-wide goals [7].

Growth in distribution construction roughly correlates with growth in electric energy consumption. During the last three decades, many U.S. utilities converted older 2.4-, 4.1-, and 5-kV primary distribution systems to 12 or 15 kV. The 15-kV voltage class is widely preferred by U.S. utilities for new installations; higher primary distribution voltages including 25 kV and 34.5 kV are also utilized. Secondary distribution reduces the voltage for utilization by commercial and residential customers. Common secondary distribution voltages in the United States are 240/120 V, single-phase, three-wire; 208 Y/120 V, three-phase, four-wire; and 480 Y/277 V, three-phase, four-wire.

Transmission and distribution grids in the United States as well as other industrialized countries are aging and being stressed by operational uncertainties and challenges never envisioned when they were developed many decades ago. There is a growing consensus in the power industry and among many governments that smart grid technology is the answer to the uncertainties and challenges. A smart grid is characterized by the following attributes:

1. Self-healing from power system disturbances;
2. Enables active participation by consumers in demand response;
3. Operates resiliency against both physical and cyber attacks;
4. Provides quality power that meets 21st century needs;
5. Accommodates all generation and energy storage technologies;
6. Enables new products, services, and markets; and
7. Optimizes asset utilization and operating efficiency.

The objective of a smart grid is to provide reliable, high-quality electric power to digital societies in an environmentally friendly and sustainable manner [11].

According to recent studies, almost 62% of electric utility employees in the United States have the potential to retire or leave over the ten-year period from 2014 to 2024.

According to the Power Engineering Society (PES) of the Institute of Electrical and Electronics Engineers (IEEE), of the 170 engineering faculty working full time in power engineering education and research in the United States, some 50 senior faculty members have retired during the 2009 to 2014 time frame. The continuing availability of qualified power system engineers is a critical resource to ensure that transmission and distribution systems are maintained and operated efficiently and reliably [9,10].

1.3 ELECTRIC UTILITY INDUSTRY STRUCTURE

The case study at the beginning of this chapter describes the restructuring of the electric utility industry that has been ongoing in the United States. The previous structure of large, vertically integrated monopolies that existed until the last decade of the twentieth century is being replaced by a horizontal structure with generating companies, transmission companies, and distribution companies as separate business facilities.

In 1992, the United States Congress passed the Energy Policy Act, which has shifted and continues to further shift regulatory power from the state level to the federal level. The 1992 Energy Policy Act mandates the Federal Energy Regulatory Commission (FERC) to ensure that adequate transmission and distribution access is available to exempt wholesale generators (EWGs) and nonutility generation (NUG). In 1996, FERC issued the "MegaRule," which regulates Transmission Open Access (TOA).

TOA was mandated in order to facilitate competition in wholesale generation. As a result, a broad range of independent power producers (IPPs) and cogenerators now submit bids and compete in energy markets to match electric energy supply and demand. Currently, the retail structure of power distribution is becoming similar to the existing structure of the telephone industry; that is, consumers choose which electric energy supplier to buy power from. Also, with demand-side metering, consumers know the retail price of electric energy at any given time and may choose when to purchase it.

Overall system reliability has become a major concern as the electric utility industry adapts to the new horizontal structure. The North American Electric Reliability Corporation (NERC) was created after the 1965 Northeast blackout and is responsible for maintaining system standards and reliability. NERC coordinates its efforts with FERC and governmental organizations in Canada [12].

As shown in Figure 1.3, the transmission system in North America is interconnected in a large power grid known as the North American Power Systems Interconnection. NERC divides this grid into eight regional entities including NPCC—the Northeast Power Coordinating Council, MRO—the Midwest Reliability Organization, and SPP—the Southwest Power Pool. Members of the regional entities come from all segments of the electric industry: investor-owned utilities; federal power agencies; rural electric cooperatives; state, municipal, and provincial utilities; independent power producers; power marketers; and end-use customers. These members jointly perform regional planning studies and operate jointly to schedule generation.

The basic premise of TOA is that transmission owners treat all transmission users on a nondiscriminatory and comparable basis. In December 1999, FERC issued

Order 2000, which calls for companies owning transmission systems to put transmission systems under the control of Regional Transmission Organizations (RTOs). Several of the NERC regions have either established independent system operators (ISOs) or planned for ISOs to operate the transmission system and facilitate transmission services. Maintenance of the transmission system remains the responsibility of the transmission owners.

At the time of the August 14, 2003, blackout in the northeastern United States and Canada, NERC reliability standards were voluntary. In August 2005, the U.S. Federal government passed the Energy Policy Act of 2005, which authorizes the creation of an electric reliability organization (ERO) with the statutory authority to enforce compliance with reliability standards among all market participants. As of June 18, 2007, FERC granted NERC the legal authority to enforce reliability standards with all users, owners, and operators of the bulk power system in the United States, and made compliance with those standards mandatory and enforceable. Currently, the legislative framework to make standards enforceable exists through action of the regulatory authority in the Canadian provinces of Alberta, British Columbia, Nova Scotia, and Quebec. In addition, standards become enforceable upon NERC board action in the Canadian provinces of Manitoba (Manitoba Hydro only), New Brunswick, Ontario, and Saskatchewan [12].

The objectives of electric utility restructuring are to increase competition, decrease regulation, and in the long run lower consumer prices. There is a concern that the benefits from breaking up the old, vertically integrated utilities are unrealized if the new unbundled generation and transmission companies are able to exert market power. Market power refers to the ability of one seller or group of sellers to maintain prices above competitive levels for a significant period of time, which could be done via collusion or by taking advantage of operational anomalies that create and exploit transmission congestion. Market power can be eliminated by independent supervision of generation and transmission companies, by ensuring that there are an ample number of generation companies, by eliminating transmission congestion, and by creating a truly competitive market, where the spot price at each node (bus) in the transmission system equals the marginal cost of providing energy at that node, where the energy provider is any generator bidding into the system [12].

1.4 COMPUTERS IN POWER SYSTEM ENGINEERING

As electric utilities have grown in size and the number of interconnections has increased, planning for future expansion has become increasingly complex. The increasing cost of additions and modifications has made it imperative that utilities consider a range of design options and perform detailed studies of the effects on the system of each option, based on a number of assumptions: normal and abnormal operating conditions, peak and off-peak loadings, and present and future years of operation. A large volume of network data must also be collected and accurately handled. To assist the engineer in this power system planning, digital computers and highly developed computer programs are used. Such programs include power-flow, stability, short-circuit, and transients programs.

Power-flow programs compute the voltage magnitudes, phase angles, and transmission-line power flows for a network under steady-state operating conditions. Other results, including transformer tap settings and generator reactive power outputs, are also computed. Today's computers have sufficient storage and speed to efficiently compute power-flow solutions for networks with 100,000 buses and 150,000 transmission lines. Results are usually viewed interactively on computer screens in the form of either tabular displays or single-line diagrams; the engineer uses these to modify the network with a mouse or from a keyboard and can readily visualize the results. The computer's large storage and high-speed capabilities allow the engineer to run the many different cases necessary to analyze and design transmission and generation-expansion options.

Stability programs are used to study power systems under disturbance conditions to determine whether synchronous generators and motors remain in synchronization. System disturbances can be caused by the sudden loss of a generator or transmission line, by sudden load increases or decreases, and by short circuits and switching operations. The stability program combines power-flow equations and machine-dynamic equations to compute the angular swings of machines during disturbances. The program also computes critical clearing times for network faults, and allows the engineer to investigate the effects of various machine parameters, network modifications, disturbance types, and control schemes.

Short-circuit programs are used to compute three-phase and line-to-ground faults in power system networks in order to select circuit breakers for fault interruption, select relays that detect faults and control circuit breakers, and determine relay settings. Short-circuit currents are computed for each relay and circuit-breaker location and for various system-operating conditions such as lines or generating units out of service in order to determine minimum and maximum fault currents.

Transients programs compute the magnitudes and shapes of transient overvoltages and currents that result from lightning strikes and line-switching operations. The planning engineer uses the results of a transients program to determine insulation requirements for lines, transformers, and other equipment, and to select surge arresters that protect equipment against transient overvoltages.

Other computer programs for power system planning include relay-coordination programs and distribution-circuits programs. Computer programs for generation-expansion planning include reliability analysis and loss-of-load probability (LOLP) programs, production cost programs, and investment cost programs.

1.5 POWERWORLD SIMULATOR

PowerWorld Simulator (PowerWorld) version 19 is a commercial-grade power system analysis and simulation package that accompanies this text. The purposes of integrating PowerWorld with the text are to provide computer solutions to examples in the text, to extend the examples, to demonstrate topics covered in the text, to provide a software tool for more realistic design projects, and to provide the readers with experience using a commercial grade power system analysis package. To use this software package, you must first install PowerWorld, along with all of the necessary case files onto your computer. The PowerWorld software and case files can be downloaded by going to the www.powerworld.com/gloversarmaoverbye Web page, and clicking on the **DownLoad PowerWorld Software and Cases for the 6th Edition** button. The remainder of this section provides the necessary details to get up and running with PowerWorld.

EXAMPLE 1.1

Introduction to PowerWorld Simulator

After installing PowerWorld, double-click on the PW icon to start the program. Power system analysis requires, of course, that the user provide the program with a model of the power system. With PowerWorld, you can either build a new case (model) from scratch or start from an existing case. Most of the examples in the book, as in this case, initially start from an existing case. PowerWorld uses the common ribbon user interface in which common commands, such as opening or saving a case, are available by clicking on the items on the top of the screen. So to open a case click **File** and then select **Open Case**. This displays the Open Dialog. Select the Example 1_1 case in the Chapter 1 directory, and then click Open. The display should look similar to Figure 1.6.

FIGURE 1.6 Example power system

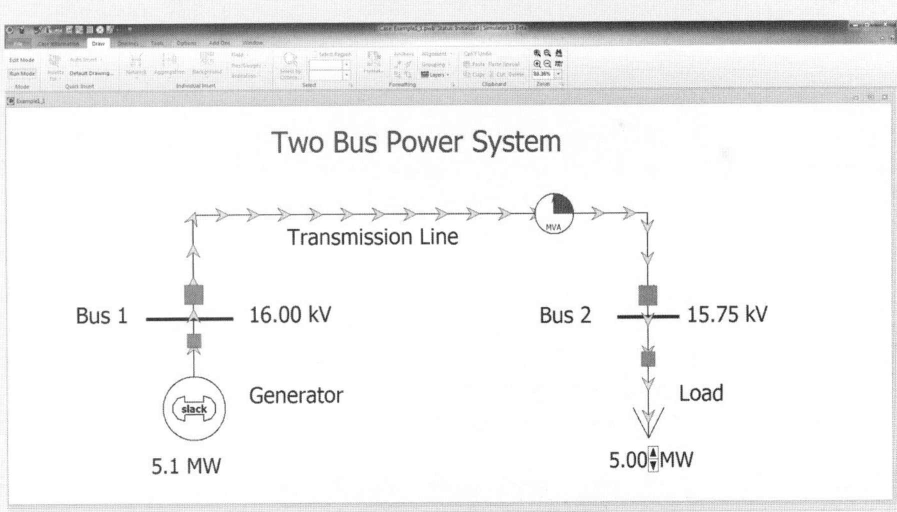

For users familiar with electric circuit schematics, it is readily apparent that Figure 1.6 does *not* look like a traditional schematic. This is because the system is drawn in what is called a oneline diagram form. A brief explanation is in order. Electric power systems range in size from small dc systems with peak power demands of perhaps a few milliwatts (mW) to large continent-spanning

(Continued)

interconnected ac systems with peak demands of hundreds of Giga-watts (GW) of demand (1 GW = 1×10^9 Watt). The subject of this book and also PowerWorld are the high-voltage, high-power, interconnected ac systems. Almost without exception these systems operate using three-phase ac power at either 50 or 60 Hz. As discussed in Chapter 2, a full analysis of an arbitrary three-phase system requires consideration of each of the three phases. Drawing such systems in full schematic form quickly gets excessively complicated. Thankfully, during normal operation, three-phase systems are usually balanced. This permits the system to be accurately modeled as an equivalent single-phase system (the details are discussed in Chapter 8, "Symmetrical Components"), Most power system analysis packages, including PowerWorld, use this approach. Then connections between devices are drawn with a single line joining the system devices, hence the term *oneline diagram*. However, do keep in mind that the actual systems are three-phase.

Figure 1.6 illustrates how the major power system components are represented in PowerWorld. Generators are shown as a circle with a "dog-bone" rotor, large arrows represent loads, and transmission lines are simply drawn as lines. In power system terminology, the nodes at which two or more devices join are called *buses*. In PowerWorld, thicker lines usually represent buses; the bus voltages are shown in kilovolts (kV) in the fields immediately to the right of the buses. In addition to voltages, power engineers are also concerned with how power flows through the system (the solution of the power flow problem is covered in Chapter 6, "Power Flows"). In PowerWorld, power flows can be visualized with arrows superimposed on the generators, loads, and transmission lines. The size and speed of the arrows indicate the direction of flow. One of the unique aspects of PowerWorld is its ability to animate power systems. To start the animation, select the **Tools** tab on the ribbon and then click on the green and black arrow button above **Solve** (i.e., the "Play" button). The oneline should come to life! While the oneline is being animated, you can interact with the system. Figure 1.6 represents a simple power system in which a generator is supplying power to a load through a 16-kV distribution system feeder. The solid red blocks on the line and load represent circuit breakers. To open a circuit breaker, simply click on it. Since the load is series-connected to the generator, clicking on any of the circuit breakers isolates the load from the generator resulting in a blackout. To restore the system, click again on the circuit breaker to close it and then again select the **Solve** button on the **Tools** ribbon. To vary the load, click on the up or down arrows between the load value and its "MW" field. Note that because of the impedance of the line, the load's voltage drops as its value is increased.

You can view additional information about most of the elements on the oneline by right-clicking on them. For example right-clicking on the generator symbol brings up a local menu of additional information about the generator, while right-clicking on the transmission line brings up local menu of information about the line. The meaning of many of these fields become clearer as you progress through the book. To modify the display itself, simply right-click on a blank area of the oneline. This displays the Onelines local menu. Select **Oneline Display Options** to display the Oneline Display Options dialog. From this dialog you can customize many of the display features. For example, to change the animated flow arrow color, select Animated Flows from the options shown on the left side of the dialog. Then click on the green colored box next to the Actual MW field (towards the bottom of the dialog) to change its color.

There are several techniques for panning and/or zooming on the oneline. One method to pan is to first click in an empty portion of the display and then press the keyboard arrow keys in the direction you would like to move. To zoom, just hold down the Ctrl key while pressing the up arrow to zoom in or the down arrow to zoom out. Alternatively you can drag the oneline by clicking and holding the left mouse button down and then moving the mouse—the oneline should follow. To go to a favorite view from the Onelines local menu, select the **Go To View** to view a list of saved views.

If you would like to retain your changes after you exit PowerWorld, you need to save the results. To do this, select **File** in the upper-left portion of the ribbon and then **Save Case As;** enter a different file name so as to not overwrite the initial case. One important note: PowerWorld actually saves the information associated with the power system model itself in a different file from the information associated with the oneline. The power system model is stored in *.pwb files (PowerWorld Binary file) while the Oneline Display information is stored in *.pwd files (PowerWorld Display file). For all the cases discussed in this book, the names of both files should be the same (except the different extensions). The reason for the dual file system is to provide flexibility. With large system models, it is quite common for a system to be displayed using multiple oneline diagrams. Furthermore, a single oneline diagram might be used at different times to display information about different cases.

EXAMPLE 1.2

PowerWorld Simulator—Edit Mode

PowerWorld has two major modes of operations. The Run Mode, which was just introduced, is used for running simulations and performing analysis. The Edit Mode, which is used for modifying existing cases and building new cases, is introduced in this example. To switch to the Edit Mode, click on the **Edit Mode** button, which is located in the upper-left portion of the display immediately below the PowerWorld icon. Use the Edit Mode to add an additional bus and load, as well as two new lines, to the Example 1_1 system.

When switching to the Edit Mode notice that the ribbon changes slightly, with several of the existing buttons and icons disabled and others enabled. Also, the

(*Continued*)

oneline now has a superimposed grid to help with alignment (the grid can be customized using the Grid/Highlight Unlinked options category on the Oneline Display Options dialog). In the Edit Mode, first add a new bus to the system. This can be done graphically by first selecting the **Draw** tab, then clicking on the **Network** button and selecting **Bus**. Once this is done, move the mouse to the desired oneline location and click (note the **Draw** tab is only available in the Edit Mode). The Bus Options dialog then appears. This dialog is used to set the bus parameters. For now leave all the bus fields at their default values, except set Bus Name to "bus 3" and set the nominal voltage to 16.0; note that the number for this new bus was automatically set to the one greater than the highest bus number in the case. The oneline should look similar to Figure 1.7. You may wish to save your case now to avoid losing your changes.

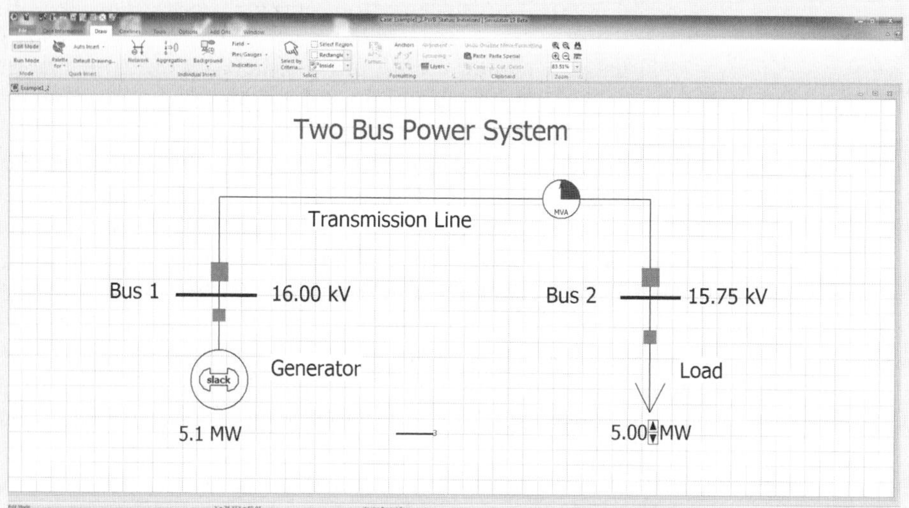

FIGURE 1.7 Example 1.2—Edit Mode view with new bus

By default, when a new bus is inserted, a bus field is also inserted. *Bus fields* are used to show information about buses on the onelines. In this case the new field shows the bus name, although initially in rather small fonts. To change the field's font size, click on the field to select it, and then select the **Format** button (on the Draw ribbon) to display the Format dialog. Click on the **Font** tab and change the font's size to a larger value to make it easier to see. You can also change the size of the bus itself using the Format dialog, Display/Size tab. To see the bus voltage magnitude, add an additional bus field. On the Draw ribbon select **Field, Bus Field**, and then click near bus 3. This displays the Bus Field Options dialog. Make sure the bus number is set to 3, and that the "Type of Field" is Bus Voltage. Again, resize with the **Format, Font** dialog.

Next, insert some load at bus 3. This can be done graphically by selecting **Network, Load** and then clicking on bus 3. The Load Options dialog appears, allowing you to set the load parameters. Note that the load was automatically assigned to bus 3. Leave all the fields at their default values, except set the orientation to "Down," and enter 10.0 in the Constant Power column MW Value field. As the name implies, a constant power load treats the load power as being independent of bus voltage; constant power load models are commonly used in power system analysis. By default, PowerWorld "anchors" each load symbol to its bus. This is a handy feature when changing a drawing, since when you drag the bus, the load and all associated fields move as well. Note that two fields showing the load's real (MW) and reactive (Mvar) power were also auto-inserted with the load. Select the reactive field right now, and then choose one. **Delete** (located towards the right side of the **Draw** ribbon) to remove it. You should also resize the MW field using the **Format, Font** command.

Now join the bus 3 load to the rest of the system. Do this by adding a line from bus 2 to bus 3. Select **Network, Transmission Line** and then click on bus 2. This begins the line drawing. During line drawing, PowerWorld adds a new line segment for each mouse click. After adding several segments place the cursor on bus 3 and double-click. The Transmission Line/Transformer Options dialog appears, allowing you to set the line's parameters. Note that PowerWorld should have automatically set the "from" and "to" bus numbers based upon the starting and ending buses (buses 2 and 3). If these values have not been set automatically, then you probably did not click exactly on bus 2 or bus 3; manually enter the values. Next, set the line's Series Resistance (R) field to 0.3, the Series Reactance (X) field to 0.6, and the MVA Limits Limit (A) field to 20 (the details of transformer and transmission line modeling is covered in Chapters 3 through 5). Select OK to close the dialog. Note that PowerWorld also auto-inserted two circuit breakers and a round "pie chart" symbol. The pie charts are used to show the percentage loading of the line. You can change the display size for these objects by right-clicking on them to display their option dialogs.

EXAMPLE 1.3

PowerWorld Simulator—Run Mode

Next, switch back to Run Mode to solve and animate the new system developed in Example 1.2. Click on the **Run Mode** button (immediately below the **Edit Mode** button), select the **Tools** on the ribbon, and then click the green-and-black button above **Solve** to start the simulation. You should see the arrows flow from bus 1 to bus 2 to bus 3. Note that the total generation is now about 15.4 MW, with 15 MW flowing to the two loads and 0.4 MW lost to the wire resistance. To add the load variation arrows to the bus 3 load, right click on the Load MW field (not the load

(Continued)

arrow itself) to display the field's local menu. Select **Load Field Information Dialog** to view the Load Field Options dialog. Set the Delta per Mouse Click field to 1.0, which changes the load by one MW per click on the up/down arrows. You may also like to set the Digits to Right of Decimal to 2 to see more digits in the load field. Be sure to save your case. The new system now has one generator and two loads. The system is still radial, meaning that a break anywhere on the wire joining bus 1 to bus 2 would result in a blackout of all the loads. Radial power systems are quite common in the lower voltage distribution systems. At higher voltage levels, networked systems are typically used. In a networked system, each load has at least two possible sources of power. Convert the system to a networked system simply by adding a new line from bus 1 to bus 3. To do this, switch back to Edit Mode, and then repeat the previous line insertion process except you should start at bus 1 and end at bus 3; use the same line parameters as for the bus 2 to 3 line. Also before returning to Run Mode, right click on the blue Two Bus Power System title and change it to Three Bus Power System. Return to Run Mode and again solve. Your final system should look similar to the system shown in Figure 1.8. Note that now you can open any single line and still supply both loads—a nice increase in reliability!

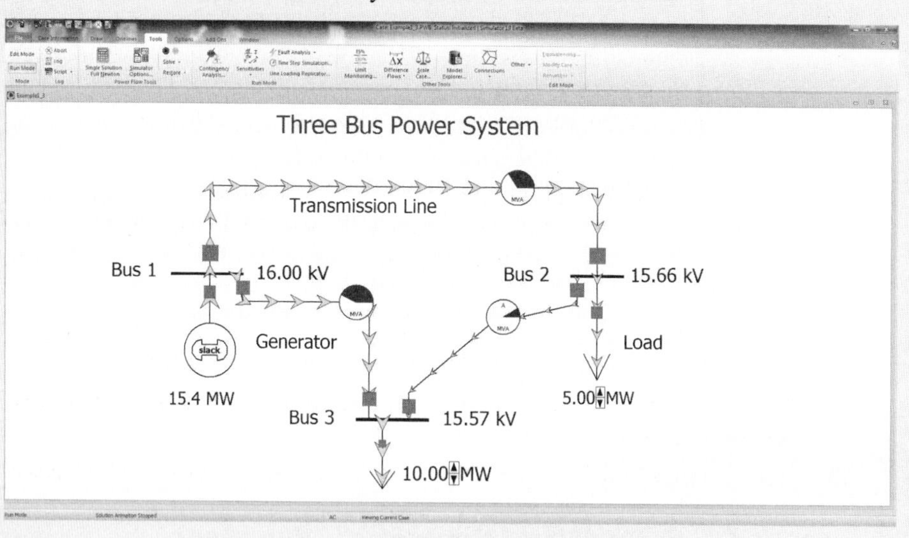

FIGURE 1.8 Example 1.3—New three-bus system

With this introduction, you now have the skills necessary to begin using PowerWorld to interactively learn about power systems. If you'd like to take a look at some of the larger systems you'll be studying, in the Chapter 6 directory open PowerWorld case Example 6_13. This case models a power system with 37 buses. Notice that when you open any line in the system, the flow of power immediately redistributes to continue to meet the total load demand. While the education version of PowerWorld is limited to 42 buses, the commercial version can handle cases with the tens of thousands of buses common in utility studies.

REFERENCES

1. H. M. Rustebakke et al., *Electric Utility Systems Practice,* 4th ed. (New York: Wiley. 1983). Photos courtesy of Westinghouse Historical Collection.
2. U.S. Energy Information Administration, *Electric Power Annual 2012—December 2013,* www.eia.gov.
3. U.S. Energy Information Administration, *Annual Energy Outlook 2014, April 2014,* www.eia.gov.
4. Wikipedia Encyclopedia, *List of HVDC Projects,* en.wikipedia.org.
5. M. P. Bahrman and B. K. Johnson, "The ABCs of HVDC Transmission Technologies," *IEEE Power & Energy Magazine,* 5,2 (March/April 2007), pp. 33–44.
6. Edison Electric Institute, *Key Facts about the Electric Power Industry,* May 2013, www.eei.org.
7. North American Electric Reliability Corporation (NERC), *2013 Long-Term Reliability Assessment* (Atlanta, GA: www.nerc.com, December 2013).
8. Seth Blumsack, "How the Free Market Rocked the Grid," *IEEE Spectrum,* 47, 12 (December, 2010), pp. 44–48, 57–59.
9. Utility Analytics Institute, *Aging Utility Workforce: Business Impacts and Strategies,* November 27, 2013, www.inin.com
10. IEEE Power & Energy Society, *Preparing the Foundation for Future Electric Energy Systems: A Strong Power and Energy Engineering Workforce,* U.S. Power & Energy Engineering Workforce Collaborative, April 2009, www.ieee-pes.org
11. E. Santacana, G. Rackliffe, L. Tang, and X. Feng, "Getting Smart," *IEEE Power & Energy Magazine,* 8,2 (March/April 2010), pp. 41–48.
12. North American Electric Reliability Corporation (NERC), *About NERC* (Princeton, NJ: www.nerc.com).
13. T.J. Overbye and J. Weber, "Visualizing the Electric Grid," *IEEE Spectrum,* 38,2 (February 2001), pp. 52–58.
14. *Solar Power in Germany,* en.wikipedia.com
15. "Focus on UHV AC: China shows the way by energising 1,000 kV line," *Global Transmission Report,* www.globaltransmission.info, March 2, 2009.
16. J. Cao and J.Y. Cai, "HVDC in China," *EPRI 2013 HVDC & Facts Conference,* Palo Alto, CA, www.cepri.com, August 28, and 29, 2013.

2 Fundamentals

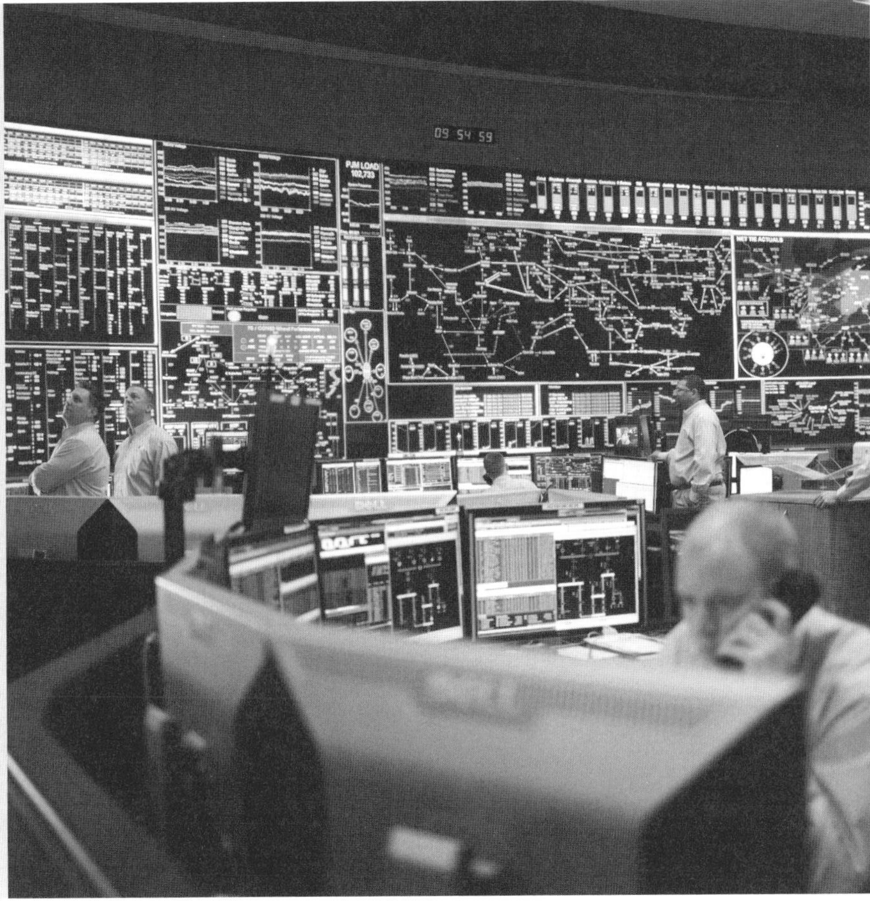

Operations Center at PJM. PJM is a regional transmission organization (RTO) that coordinates the movement of wholesale electricity and ensures the reliability of the high-voltage electric power system serving 61 million people in all or parts of Delaware, Illinois, Indiana, Kentucky, Maryland, Michigan, New Jersey, North Carolina, Ohio, Pennsylvania, Tennessee, Virginia, West Virginia and the District of Columbia, USA (Courtesy of PJM Interconnection, LLC.)

The objective of this chapter is to review basic concepts and establish terminology and notation. In particular, we review phasors, instantaneous power, complex power, network equations, and elementary aspects of balanced three-phase circuits. Students who have already had courses in electric network theory and basic electric machines should find this chapter to be primarily refresher material.

CASE STUDY

Throughout most of the 20th century, electric utility companies built increasingly larger generation plants, primarily hydro or thermal (using coal, gas, oil, or nuclear fuel). At the end of the 20th century, following the ongoing deregulation of the electric utility industry with increased competition in the United States and in other countries, smaller generation sources that connect directly to distribution systems have emerged. Distributed energy resources are sources of energy including generation and storage devices that are located near local loads. Distributed generation sources include renewable technologies (including geothermal, ocean tides, solar, and wind) and nonrenewable technologies (including internal combustion engines, combustion turbines, combined cycle, microturbines, and fuel cells). Microgrids are systems that have distributed energy resources and associated loads that can form intentional islands in distribution systems. The following article provides an overview of ongoing microgrid activities undertaken by the Office of Electricity Delivery and Energy Reliability of the U.S. Department of Energy [5].

KEY CONNECTIONS

By Merrill Smith and Dan Ton

U.S. Department of Energy

Microgrids have been identified as key components of the smart grid for improving power reliability and quality, increasing system energy efficiency, and providing the possibility of grid independence to individual end-user sites. The Microgrid Exchange Group, an ad hoc group of experts and implementers of microgrid technology, has defined a microgrid as "a group of interconnected loads and distributed energy resources within clearly defined electrical boundaries that acts as a single controllable entity with respect to the grid. A microgrid can connect and disconnect from the grid to enable it to operate in both grid-connected or island mode."

Microgrids have been described by other organizations using similar definitions, including the key concepts of a system made up of multiple loads along with generation and the ability to island from the grid. Microgrids provide multiple benefits that include:

- Enabling grid modernization and the integration of multiple smart grid technologies;

(Merrill Smith and Dan Ton, "Key Connections," U.S. Deptartment of Energy, Washington DC, USA)

- Enhancing and easing the integration of distributed and renewable energy sources that help reduce peak load and also reduce losses by locating generation near demand;
- Meeting end-user needs by ensuring energy supply for critical loads, controlling power quality and reliability at the local level, and promoting customer participation through demand-side management and community involvement in electricity supply; and
- Supporting the macrogrid by handling sensitive loads and the variability of renewables locally and supplying ancillary services to the bulk power system.

Within the Office of Electricity Delivery and Energy Reliability (OE) of the U.S. Department of Energy (DOE), the Smart Grid R&D Program was established to accelerate the deployment and integration of the advanced communication, control, and information technologies needed to modernize the nation's electric delivery network. This modernization includes preparing the U.S. electric infrastructure to meet the challenges of the 21st-century economy. The Smart Grid R&D Program has two goals: to dynamically optimize grid operations and resources for a robust, flexible, and secure electric grid; and to fully integrate demand response and consumer participation into grid resource planning and operations. The program's microgrid activities support the achievement of both of these goals.

According to the DOE's September 2011 update of the *Smart Grid Research and Development Multi-Year Program Plan: 2010–2014* (available at http://events.energetics.com/SmartGridPeerReview2012/pdfs/SG_MYPP_2011.pdf), the microgrid initiative has established the following 2020 performance targets for costs, reliability, system energy efficiencies, and emissions to advance these grid modernization goals: "to develop commercial-scale microgrid systems (capacity <10 MW) capable of reducing [the] outage time of required loads by >98% at a cost comparable to nonintegrated baseline solutions (uninterrupted power supply...plus diesel genset) while reducing emissions by >20% and improving system energy efficiencies by >20%."

This article provides an overview of ongoing microgrid activities being undertaken by OE and its Smart Grid R&D Program. In addition, it discusses the process OE has undertaken to engage microgrid stakeholders to jointly identify the remaining areas in which there are R&D gaps and develop a plan to address those gaps.

Ongoing Microgrid Activities

The bulk of the DOE's microgrid R&D efforts to date have focused on demonstration activities to meet niche application needs, such as those for meeting peak load reduction, renewable energy mandates and directives, and energy surety and reliability at certain critical facilities, including military installations. These ongoing microgrid demonstration projects consist of lab- and field-scale R&D test beds, renewable and distributed systems integration (RDSI) projects for peak load reduction, select Smart Grid Demonstration Program (SGDP) projects funded under the American Recovery and Reinvestment Act of 2009 (ARRA) as part of OE's implementation of grid modernization, and assessment and demonstration projects jointly supported by the U.S. Department of Defense (DOD) and the DOE.

OE is currently supporting nine RDSI projects with a total value exceeding U.S.$100 million (with approximately U.S.$55 million coming from the DOE). The two primary goals of these projects are to demonstrate at least 15% peak demand reduction at the distribution feeder or substation level through integrating distributed energy resources (DERs) and to demonstrate microgrids that can operate in both grid-parallel and islanded modes. These projects are proving out systems that can defer transmission and distribution investments and upgrades by utilizing local assets (generation and load reduction) in an integrated fashion. They are also increasing the reliability of the local distribution system by adding elements that make it more stable and reconfigurable. Other benefits being realized from these projects include addressing vulnerabilities in critical infrastructure, managing peak loads, lowering emissions, using fuel resources more efficiently, and helping customers manage energy costs. These RDSI projects are progressing toward achieving the goals of at least 15% in peak demand reductions and the ability to island. Some, such as the Illinois Institute of Technology, Chevron/Santa Rita Jail, and Ft. Collins FortZED projects, have already realized or exceeded the 15% reduction goal.

Under ARRA, SGDP has funded 16 smart grid regional demonstration projects to demonstrate emerging smart grid technologies and alternative architectures to validate business models and address regulatory and scalability issues. Among them, several projects are conducting demonstrations involving combinations of the following elements: renewable energy resources, distributed generation, energy storage, demand-side management, and charging schemes for plug-in electric vehicles. These projects include (but are not limited to) the Pacific Northwest Smart Grid Demonstration by Battelle Memorial Institute, which includes Portland General Electric's High Reliability Zone (a microgrid); the Energy Internet Demonstration by Pecan Street Project Inc. in Texas; and the Irvine Smart Grid Demonstration by Southern California Edison. Further information on the SGDP projects is available at http://www.smartgrid.gov/recovery_act/project_information?keys=&project%5B%5D=2.

There has also been a significant effort by national laboratories to produce microgrid designs, analyses, and demonstrations at test facilities and military bases. Lawrence Berkeley

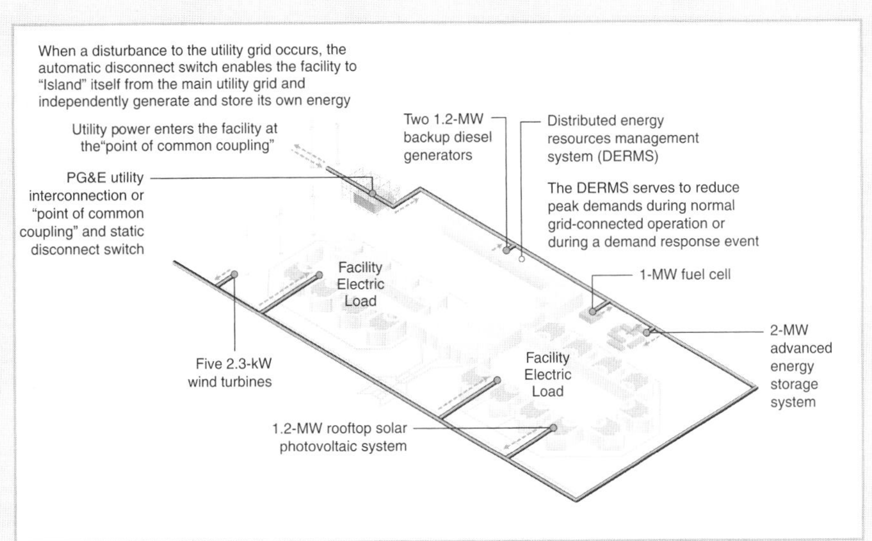
When a disturbance to the utility grid occurs, the automatic disconnect switch enables the facility to "Island" itself from the main utility grid and independently generate and store its own energy
Utility power enters the facility at the"point of common coupling"
PG&E utility interconnection or "point of common coupling" and static disconnect switch
Two 1.2-MW backup diesel generators
Distributed energy resources management system (DERMS)
The DERMS serves to reduce peak demands during normal grid-connected operation or during a demand response event
1-MW fuel cell
Facility Electric Load
Five 2.3-kW wind turbines
Facility Electric Load
2-MW advanced energy storage system
1.2-MW rooftop solar photovoltaic system

Figure 1 Chevron Energy Solutions' project at the Santa Rita Jail in Dublin, California, to demonstrate commercial application of a CERTS microgrid (used with permission) (Based on Chevron Energy Solutions.)

National Laboratory (LBNL) is teaming with American Electric Power (AEP), the University of Wisconsin, and Sandia National Laboratories (SNL) to apply Consortium for Electric Reliability Technology Solutions (CERTS) microgrid concepts in AEP's Dolan Technology Center at the Walnut Station Test Facility in Groveport, Ohio. The Sacramento Municipal Utility District, the Chevron Energy Solutions RDSI project (shown in Figure 1), and the DOD at Fort Sill and Maxwell Air Force Bases are also applying CERTS microgrid concepts in field demonstrations. LBNL has developed the distributed energy resources customer adoption model (DER-CAM), an economic model for predicting and optimizing the capacity and minimizing the cost of operating distributed generation in microgrids.

SNL has developed its energy surety microgrid (ESM) methodology, which uses cost and performance data from military bases to develop approaches for implementing high-reliability microgrids and to assist in planning for and analysis of the potential risks in future military and commercial projects. To date, 14 military bases have received assessments and/or conceptual microgrid designs using SNL's ESM methodology. From this work, SNL has developed a set of valuable lessons learned that, combined with its design methodology, provide a blueprint for future ESM microgrid implementation.

The ESM design work was the springboard for a larger joint microgrid deployment effort on the part of the DOE, the DOD, and the U.S. Department of Homeland Security (DHS). This joint effort, known as a joint capabilities technology demonstration (JCTD), is the first of its kind for the DOE and will ultimately deploy three smart power infrastructure demonstration for energy reliability and security (SPIDERS) microgrids. Five national laboratories—Idaho, National Renewable Energy Laboratory, Oak Ridge National Laboratory (ORNL), Pacific Northwest National Laboratory (PNNL), and SNL—supported by the DOE are playing a pivotal role in the SPIDERS deployment through their work on the microgrid designs, the cyber designs, and the development of a transition plan that can be used to design and deploy microgrids at additional military facilities and in the private sector. The three SPIDERS microgrids will be deployed in phases at Joint Base Pearl Harbor-Hickam (Hawaii), Fort Carson (Colorado), and Camp Smith (Hawaii), respectively. The first phase, at Pearl Harbor-Hickam, has completed successful technical and operational demonstrations. A key element of SPIDERS is standardization of the design approach, contracting, installation, security, and operation of these microgrids to support future applications.

Other work at the national laboratories is also supporting the microgrid effort. ORNL's Distributed Energy Communications & Controls laboratory is developing controls for inverter-based DERs to provide local voltage, power, and power quality support for the ORNL campus distribution system. On the simulation side, PNNL has been developing GridLAB-D as a distribution system simulation tool that integrates grid operations at several levels, including microgrids.

As OE moves forward with microgrid R&D, it is coordinating the national laboratories' efforts on microgrids so they will be more cohesive and unified. The labs will be working on three major tasks. The focus will be on:

- Developing use cases (based on existing microgrids) to define the performance requirements and technology specifications of microgrids;
- Performing cost and benefit analysis to identify high-impact R&D for additional investments; and
- Creating integrated tool sets for developing conceptual and preliminary designs and controls and operations architectures.

In addition to meeting individual niche applications, the various DOE demonstrations also field-prove capabilities of current technologies and unveil lessons learned, challenges, and needed but unmet capabilities. Current technologies will clearly not be enough to meet the 2020 performance targets established by the DOE for microgrids. As part of its continuing effort to engage stakeholders in joint planning and implementation of R&D activities, the Smart Grid R&D Program convened two microgrid workshops, in 2011 and 2012, to seek stakeholder input on key areas for R&D and performance

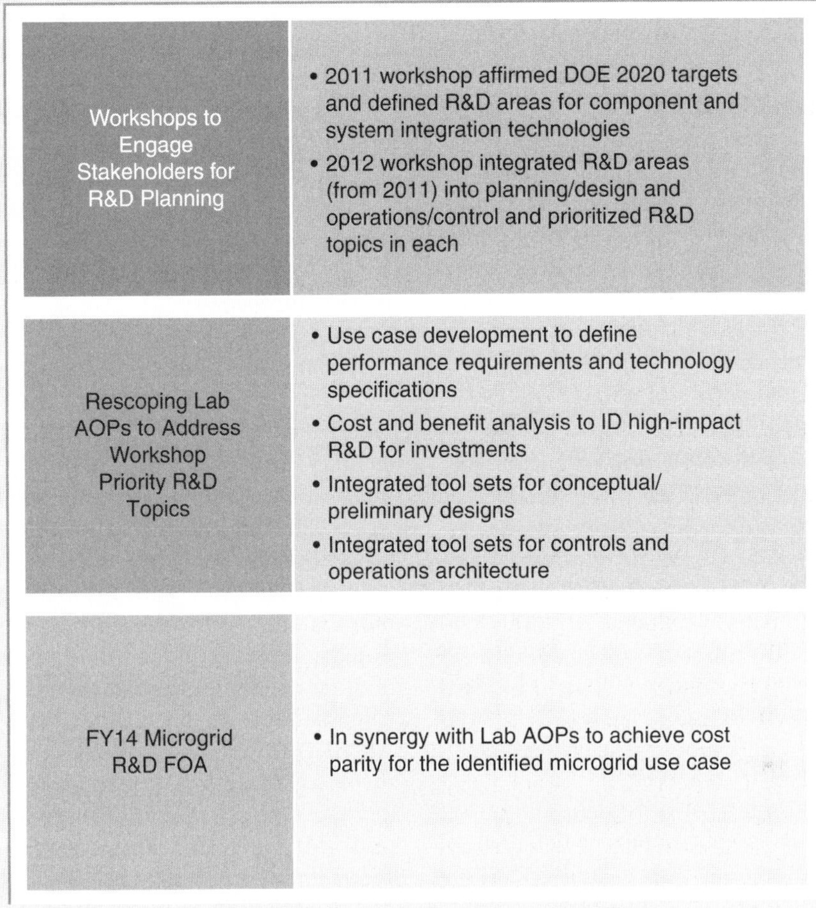

Workshops to Engage Stakeholders for R&D Planning
• 2011 workshop affirmed DOE 2020 targets and defined R&D areas for component and system integration technologies
• 2012 workshop integrated R&D areas (from 2011) into planning/design and operations/control and prioritized R&D topics in each
Rescoping Lab AOPs to Address Workshop Priority R&D Topics
• Use case development to define performance requirements and technology specifications
• Cost and benefit analysis to ID high-impact R&D for investments
• Integrated tool sets for conceptual/ preliminary designs
• Integrated tool sets for controls and operations architecture
FY14 Microgrid R&D FOA
• In synergy with Lab AOPs to achieve cost parity for the identified microgrid use case

Figure 2 R&D pathway toward the DOE's 2020 microgrid performance targets

baselines, targets, and actionable plans. This input was incorporated into the DOE's 2012 *Smart Grid Research and Development Multi-Year Program Plan* to guide current and future DOE R&D efforts in microgrids (see Figure 2).

Microgrid Workshops

The DOE held its first microgrid workshop on August 30–31, 2011 in San Diego, California; a follow-up workshop was held on July 30–31, 2012 in Chicago. The purpose of the first workshop was to convene experts and practitioners to assist the DOE in identifying and prioritizing R&D areas in the field of microgrids. The second workshop was held in response to path-forward discussions that called for sharing lessons learned and best practices for system integration from existing projects in the United States (including military microgrids) and elsewhere. In addition, the purpose of the second workshop was to delve more deeply into R&D topics gathered from the first workshop and subsequently determine system integration gap areas and functional requirements.

The 2011 Workshop

Two tracks were organized to address the potential cost reduction of major microgrid components and subcomponents. One track focused on microgrid components, with separate sessions on switch technologies, control and protection technologies, and inverters and converters. The second track focused on microgrid systems, with separate sessions on standards and protocols, system design and economic analysis tools, and system integration.

The 2012 Workshop

A working list of system integration issues, categorized in the two areas of planning and design and operations and control, was presented for input from the attendees, based on their experience. This brainstorming session resulted in 12 R&D topics for discussion. For each topic, participants discussed four aspects: the current status of the technology, needs and challenges, the scope of the needed R&D, and R&D metrics.

Conclusions from the breakout session discussions and the report-out presentations from the 2011 workshop were documented in the *DOE Microgrid Workshop Report*, available at http://energy.gov/sites/prod/files/Microgrid%20Workshop%20Report%20August%202011.pdf. The 2012 workshop program agenda with embedded presentation links is available at http://e2rg.com/events/agenda/. Table 1 lists the key R&D areas identified from the two workshops.

Conclusions and the Path Forward

The DOE's Smart Grid R&D Program considers microgrids to be key building blocks for smart grids and has established microgrid R&D as a key area of focus. A significant number of R&D needs and challenges have been identified for microgrids during the two workshops described above, with input from more than 170 experts and practitioners, representing a broad group of stakeholders in the United States, Canada, Europe, Japan, and Korea. At the two workshops, the scope of the R&D necessary to address the identified needs and challenges was outlined. The technical, economical, social, and environmental benefits that can result from successful development and deployment of microgrids became evident in the course of the workshop discussions and presentations.

Engaging stakeholders and knowledgeable practitioners to obtain input on R&D needs is a key part of the R&D topic development process. With the input collected, the Smart Grid R&D Program will further refine R&D requirements so as to plan and develop a competitive funding opportunity announcement, subject to available DOE funds. The DOE's

2011 Workshop		2012 Workshop	
R&D Areas			
Components	**Systems**	**Planning and Design**	**Operations and Controls**
Switch Technologies	**Standards and Protocols**	**System Architecture Development**	**Steady-State Control and Coordination**
Legacy grid-connection technologies to enable connection and disconnection from the grid	Universal microgrid communications and control standards	Definition of microgrid applications, interfaces, and services	Internal services within the microgrid
Requirements based on customer and utility needs	Microgrid protection, coordination, and safety	Open architectures that promote flexibility, scalability, and security	Interaction of the microgrid with utilities or other microgrids
Control and Protection Technologies	**System Design and Economic Analysis Tools**	**Modeling and Analysis**	**Transient State Control and Protection**
Best practices and specifications for protection and controls; information models	Microgrid multiobjective optimization framework	Performance optimization methods and uncertainty in the modeling and design process	Transient state control and protection
Reliable, low-cost protection	Designing an operations optimization methodology that takes uncertainty into account		
Switches to handle full fault current			
Inverters/Converters	**System Integration**	**Power System Design**	**Operational Optimization**
Topologies and control algorithms so that multiple inverters can operate in a microgrid	A common integration framework	DC power	Operational optimization of a single microgrid
Advanced power electronics technologies		Microgrid integration	Operational optimization of multiple microgrids

Table 1

Key R&D Areas Identified from the DOE's 2011 and 2012 Microgrid Workshops

microgrid R&D initiative hopes to advance microgrids, in partnership with industry and research experts, from conception through R&D execution.

For Further Reading

U.S. Department of Energy. Information/fact sheets on the smart grid projects (including the smart grid demonstration projects) funded through the American Recovery and Reinvestment Act of 2009. [Online]. Available: http://www.smartgrid.gov/recovery_act/project_information

U.S. Department of Energy. (2012, May 10). Smart grid R&D Program peer review June 7–8, 2012. [Online]. Available: http://energy.gov/oe/articles/smart-grid-rd-program-peer-review-june-7-8-2012

U.S. Department of Energy. Smart grid research & development multi-year program plan: 2010–2014. [Online]. Available: http://energy.gov/sites/prod/files/SG_MYPP_2012%20Update.pdf

U.S. Department of Energy. (2011). Microgrid workshop report. [Online]. Available: http://energy.gov/sites/prod/files/Microgrid%20Workshop%20Report%20August%202011.pdf

U.S. Department of Energy. (2012). Microgrid workshop agenda and presentations. [Online]. Available: http://e2rg.com/microgrid-2012/Workshop_Agenda-08092012.pdf

Sandia National Laboratories. SPIDERS: The smart power infrastructure demonstration for energy reliability and security. [Online]. Available: http://energy.sandia.gov/wp/wp-content/gallery/uploads/SPIDERS_Fact_Sheet_2012-1431P.pdf

Biographies

Merrill Smith is with the U.S. Department of Energy, Washington, D.C.

Dan Ton is with the U.S. Department of Energy, Washington, D.C. ■

2.1 PHASORS

A sinusoidal voltage or current at constant frequency is characterized by two parameters: a maximum value and a phase angle. A voltage

$$v(t) = V_{max} \cos(\omega t + \delta) \tag{2.1.1}$$

has a maximum value V_{max} and a phase angle δ when referenced to $\cos(\omega t)$. The root-mean-square (rms) value, also called *effective value*, of the sinusoidal voltage is

$$V = \frac{V_{max}}{\sqrt{2}} \tag{2.1.2}$$

Euler's identity, $e^{j\phi} = \cos\phi + j\sin\phi$, can be used to express a sinusoid in terms of a phasor. For the above voltage,

$$v(t) = \text{Re}\left[V_{\text{max}} e^{j(\omega t+\delta)}\right]$$
$$= \text{Re}\left[\sqrt{2}(Ve^{j\delta})e^{j\omega t}\right] \tag{2.1.3}$$

where $j = \sqrt{-1}$ and Re denotes "real part of." The rms phasor representation of the voltage is given in three forms—exponential, polar, and rectangular:

$$V = \underbrace{Ve^{j\delta}}_{\text{exponential}} = \underbrace{V\angle\delta}_{\text{polar}} = \underbrace{V\cos\delta + jV\sin\delta}_{\text{rectangular}} \tag{2.1.4}$$

A phasor can be easily converted from one form to another. Conversion from polar to rectangular is shown in the phasor diagram of Figure 2.1. Euler's identity can be used to convert from exponential to rectangular form. As an example, the voltage

$$v(t) = 169.7\cos(\omega t + 60°) \text{ volts} \tag{2.1.5}$$

has a maximum value $V_{\text{max}} = 169.7$ volts, a phase angle $\delta = 60°$ when referenced to $\cos(\omega t)$, and an rms phasor representation in polar form of

$$V = 120\angle 60° \quad \text{volts} \tag{2.1.6}$$

Also, the current

$$i(t) = 100\cos(\omega t + 45°) \text{ A} \tag{2.1.7}$$

has a maximum value $I_{\text{max}} = 100$ A, an rms value $I = 100/\sqrt{2} = 70.7$ A, a phase angle of 45°, and a phasor representation

$$I = 70.7\angle 45° = 70.7e^{j45} = 50 + j50 \quad \text{A} \tag{2.1.8}$$

The relationships between the voltage and current phasors for the three passive elements—resistor, inductor, and capacitor—are summarized in Figure 2.2, where sinusoidal-steady-state excitation and constant values of R, L, and C are assumed.

When voltages and currents are discussed in this text, lowercase letters such as $v(t)$ and $i(t)$ indicate instantaneous values, uppercase letters such as V and I indicate rms values, and uppercase letters in italics such as V and I indicate rms phasors. When voltage or current values are specified, they shall be rms values unless otherwise indicated.

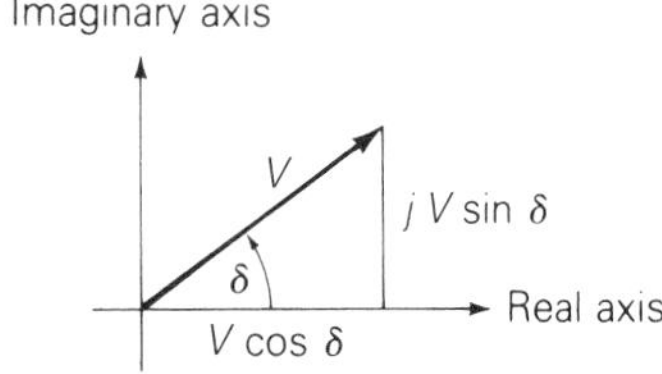

FIGURE 2.1

Phasor diagram for converting from polar to rectangular form

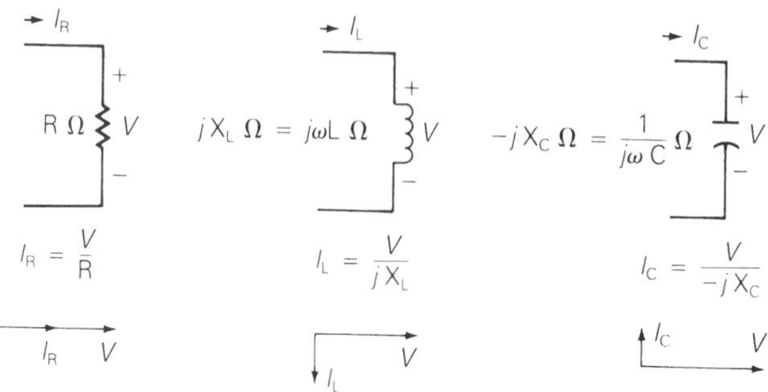

FIGURE 2.2 Summary of relationships between phasors V and I for constant R, L, and C elements with sinusoidal-steady-state excitation

2.2 INSTANTANEOUS POWER IN SINGLE-PHASE AC CIRCUITS

Power is the rate of change of energy with respect to time. The unit of power is the watt, which is a joule per second. Instead of saying that a load absorbs energy at a rate given by the power, it is common practice to say that a load absorbs power. The instantaneous power in watts absorbed by an electrical load is the product of the instantaneous voltage across the load in volts and the instantaneous current into the load in amperes. Assume that the load voltage is

$$v(t) = V_{max}\cos(\omega t + \delta) \text{ volts} \tag{2.2.1}$$

We now investigate the instantaneous power absorbed by purely resistive, purely inductive, purely capacitive, and general RLC loads. We also introduce the concepts of real power, power factor, and reactive power. The physical significance of real and reactive power is also discussed.

PURELY RESISTIVE LOAD

For a purely resistive load, the current into the load is in phase with the load voltage, $I = V/R$, and the current into the resistive load is

$$i_R(t) = I_{Rmax}\cos(\omega t + \delta) \quad \text{A} \tag{2.2.2}$$

where $I_{Rmax} = V_{max}/R$. The instantaneous power absorbed by the resistor is

$$\begin{aligned} p_R(t) &= v(t)i_R(t) = V_{max}I_{Rmax}\cos^2(\omega t + \delta) \\ &= \tfrac{1}{2}V_{max}I_{Rmax}\{1 + \cos[2(\omega t + \delta)]\} \\ &= VI_R\{1 + \cos[2(\omega t + \delta)]\} \quad \text{W} \end{aligned} \tag{2.2.3}$$

As indicated by (2.2.3), the instantaneous power absorbed by the resistor has an average value

$$P_R = VI_R = \frac{V^2}{R} = I_R^2 R \quad W \tag{2.2.4}$$

plus a double-frequency term $VI_R \cos[2(\omega t + \delta)]$.

PURELY INDUCTIVE LOAD

For a purely inductive load, the current lags the voltage by 90°, $I_L = V/(jX_L)$, and

$$i_L(t) = I_{Lmax} \cos(\omega t + \delta - 90°) \quad A \tag{2.2.5}$$

where $I_{Lmax} = V_{max}/X_L$, and $X_L = \omega L$ is the inductive reactance. The instantaneous power absorbed by the inductor is*

$$\begin{aligned} p_L(t) &= v(t)i_L(t) = V_{max} I_{Lmax} \cos(\omega t + \delta) \cos(\omega t + \delta - 90°) \\ &= \tfrac{1}{2} V_{max} I_{Lmax} \cos[2(\omega t + \delta) - 90°] \\ &= VI_L \sin[2(\omega t + \delta)] \quad W \end{aligned} \tag{2.2.6}$$

As indicated by (2.2.6), the instantaneous power absorbed by the inductor is a double-frequency sinusoid with *zero* average value.

PURELY CAPACITIVE LOAD

For a purely capacitive load, the current leads the voltage by 90°, $I_c = V/(-jX_C)$, and

$$i_C(t) = I_{Cmax} \cos(\omega t + \delta + 90°) \quad A \tag{2.2.7}$$

where $I_{Cmax} = V_{max}/X_C$, and $X_C = 1/(\omega C)$ is the capacitive reactance. The instantaneous power absorbed by the capacitor is

$$\begin{aligned} p_C(t) &= v(t)i_C(t) = V_{max} I_{Cmax} \cos(\omega t + \delta) \cos(\omega t + \delta + 90°) \\ &= \tfrac{1}{2} V_{max} I_{Cmax} \cos[2(\omega t + \delta) + 90°)] \\ &= -VI_C \sin[2(\omega t + \delta)] \quad W \end{aligned} \tag{2.2.8}$$

The instantaneous power absorbed by a capacitor is also a double-frequency sinusoid with *zero* average value.

GENERAL RLC LOAD

For a general load composed of RLC elements under sinusoidal-steady-state excitation, the load current is of the form

$$i(t) = I_{max} \cos(\omega t + \beta) \quad A \tag{2.2.9}$$

The instantaneous power absorbed by the load is then*

$$\begin{aligned} p(t) &= v(t)i(t) = V_{max}I_{max}\cos(\omega t + \delta)\cos(\omega t + \beta) \\ &= \tfrac{1}{2}V_{max}I_{max}\{\cos(\delta - \beta) + \cos[2(\omega t + \delta) - (\delta - \beta)]\} \\ &= VI\cos(\delta - \beta) + VI\cos(\delta - \beta)\cos[2(\omega t + \delta)] \\ &\quad + VI\sin(\delta - \beta)\sin[2(\omega t - \delta)] \\ p(t) &= VI\cos(\delta - \beta)\{1 + \cos[2(\omega t + \delta)]\} + VI\sin(\delta - \beta) \end{aligned}$$

Letting $I\cos(\delta - \beta) = I_R$ and $I\sin(\delta - \beta) = I_X$ gives

$$p(t) = \underbrace{VI_R\{1 + \cos[2(\omega t + \delta)]\}}_{p_R(t)} + \underbrace{VI_X\sin[2(\omega t + \delta)]}_{p_X(t)} \tag{2.2.10}$$

As indicated by (2.2.10), the instantaneous power absorbed by the load has two components: One can be associated with the power $p_R(t)$ absorbed by the resistive component of the load, and the other can be associated with the power $p_X(t)$ absorbed by the reactive (inductive or capacitive) component of the load. The first component $p_R(t)$ in (2.2.10) is identical to (2.2.3), where $I_R = I\cos(\delta - \beta)$ is the component of the load current in phase with the load voltage. The phase angle $(\delta - \beta)$ represents the angle between the voltage and current. The second component $p_X(t)$ in (2.2.10) is identical to (2.2.6) or (2.2.8), where $I_x = I\sin(\delta - \beta)$ is the component of load current 90° out of phase with the voltage.

REAL POWER

Equation (2.2.10) shows that the instantaneous power $p_R(t)$ absorbed by the resistive component of the load is a double-frequency sinusoid with average value P given by

$$P = VI_R = VI\cos(\delta - \beta)\ \text{W} \tag{2.2.11}$$

The *average power* P is also called *real power* or *active power*. All three terms indicate the same quantity P given by (2.2.11).

POWER FACTOR

The term $\cos(\delta - \beta)$ in (2.2.11) is called the *power factor*. The phase angle $(\delta - \beta)$, which is the angle between the voltage and current, is called the *power factor angle*. For dc circuits, the power absorbed by a load is the product of the dc load voltage and the dc load current; for ac circuits, the average power absorbed by a load is the product of the rms load voltage V, rms load current I, and the power factor $\cos(\delta - \beta)$, as shown by (2.2.11). For inductive loads, the current lags the voltage, which means β is less than δ, and the power factor is said to be *lagging*. For capacitive loads, the current leads the voltage, which means β is greater than δ, and the power factor is said to be *leading*. By convention, the power factor $\cos(\delta - \beta)$ is positive. If $|\delta - \beta|$

*Use the identity: $\cos A \cos B = \frac{1}{2}[\cos(A - B) + \cos(A + B)]$.

is greater than 90°, then the reference direction for current may be reversed, resulting in a positive value of $\cos(\delta - \beta)$.

REACTIVE POWER

The instantaneous power absorbed by the reactive part of the load, given by the component $p_X(t)$ in (2.2.10), is a double-frequency sinusoid with zero average value and with amplitude Q given by

$$Q = VI_X = VI \sin(\delta - \beta) \text{ var} \tag{2.2.12}$$

The term Q is given the name *reactive power.* Although it has the same units as real power, the usual practice is to define units of reactive power as volt-amperes reactive, or var.

EXAMPLE 2.1

Instantaneous, real, and reactive power and the power factor

The voltage $v(t) = 141.4\cos(\omega t)$ is applied to a load consisting of a 10-Ω resistor in parallel with an inductive reactance $X_L = \omega L = 3.77\ \Omega$. Calculate the instantaneous power absorbed by the resistor and by the inductor. Also calculate the real and reactive power absorbed by the load, and the power factor.

SOLUTION

The circuit and phasor diagram are shown in Figure 2.3(a). The load voltage is

$$V = \frac{141.4}{\sqrt{2}}\underline{/0^\circ} = 100\ \underline{/0^\circ} \text{ volts}$$

The resistor current is

$$I_R = \frac{V}{R} = \frac{100}{10}\underline{/0^\circ} = 10\underline{/0^\circ} \text{ A}$$

The inductor current is

$$I_L = \frac{V}{jX_L} = \frac{100}{(j3.77)}\underline{/0^\circ} = 26.53\underline{/-90^\circ} \text{ A}$$

The total load current is

$$I = I_R + I_L = 10 - j26.53 = 28.35\underline{/-69.34^\circ} \text{ A}$$

The instantaneous power absorbed by the resistor is, from (2.2.3),

$$\begin{aligned} p_R(t) &= (100)(10)[1 + \cos(2\omega t)] \\ &= 1000\,[1 + \cos(2\omega t)] \quad \text{W} \end{aligned}$$

(Continued)

FIGURE 2.3

Circuit and phasor diagram for Example 2.1

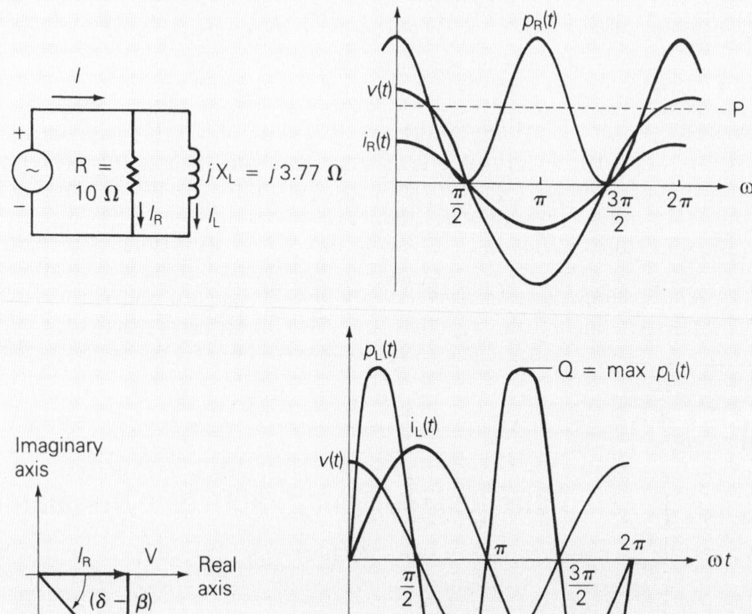

(a) Circuit and phasor diagram

(b) Waveforms

The instantaneous power absorbed by the inductor is, from (2.2.6),

$$
\begin{aligned}
p_L(t) &= (100)(26.53)\sin(2\omega t) \\
&= 2653 \sin(\omega t) \quad \text{W}
\end{aligned}
$$

The real power absorbed by the load is, from (2.2.11),

$$
\begin{aligned}
\text{P} &= \text{VI}\cos(\delta - \beta) = (100)(28.53)\cos(0^\circ + 69.34^\circ) \\
&= 1000 \text{ W}
\end{aligned}
$$

(*Note*: P is also equal to $\text{VI}_R = \text{V}^2/\text{R}$.)
The reactive power absorbed by the load is, from (2.2.12),

$$
\begin{aligned}
\text{Q} &= \text{VI}\sin(\delta - \beta) = (100)(28.53)\sin(0^\circ + 69.34^\circ) \\
&= 2653 \text{ var}
\end{aligned}
$$

(*Note*: Q is also equal to $\text{VI}_L = \text{V}^2/\text{X}_L$.)

The power factor is

$$\text{p.f.} = \cos(\delta - \beta) = \cos(69.34°) = 0.3528 \text{ lagging}$$

Voltage, current, and power waveforms are shown in Figure 2.3(b).

As shown for this parallel RL load, the resistor absorbs real power (1000 W) and the inductor absorbs reactive power (2653 vars). The resistor current $i_R(t)$ is in phase with the load voltage, and the inductor current $i_L(t)$ lags the load voltage by 90°. The power factor is lagging for an RL load.

Note $p_R(t)$ and $p_x(t)$, given by (2.2.10), are strictly valid only for a parallel R-X load. For a general RLC load, the voltages across the resistive and reactive components may not be in phase with the source voltage $v(t)$, resulting in additional phase shifts in $p_R(t)$ and $p_x(t)$ (see Problem 2.13). However, (2.2.11) and (2.2.12) for P and Q are valid for a general RLC load.

PHYSICAL SIGNIFICANCE OF REAL AND REACTIVE POWER

The physical significance of real power P is easily understood. The total energy absorbed by a load during a time interval T, consisting of one cycle of the sinusoidal voltage, is PT watt-seconds (Ws). During a time interval of n cycles, the energy absorbed is P(nT) watt-seconds, all of which is absorbed by the resistive component of the load. A kilowatt-hour meter is designed to measure the energy absorbed by a load during a time interval $(t_2 - t_1)$, consisting of an integral number of cycles, by integrating the real power P over the time interval $(t_2 - t_1)$.

The physical significance of reactive power Q is not as easily understood. Q refers to the maximum value of the instantaneous power absorbed by the reactive component of the load. The instantaneous reactive power, given by the second *term* $p_x(t)$ in (2.2.10), is alternately positive and negative, and it expresses the reversible flow of energy to and from the reactive component of the load. Q may be positive or negative, depending on the sign of $(\delta - \beta)$ in (2.2.12). Reactive power Q is a useful quantity when describing the operation of power systems (this will become evident in later chapters). As one example, shunt capacitors can be used in transmission systems to deliver reactive power and thereby increase voltage magnitudes during heavy load periods (see Chapter 5).

2.3 COMPLEX POWER

For circuits operating in sinusoidal-steady-state, real and reactive power are conveniently calculated from complex power, defined below. Let the voltage across a circuit element be $V = \text{V}\underline{/\delta}$, and the current into the element be $I = \text{I}\underline{/\beta}$. Then the complex power S is the product of the voltage and the conjugate of the current:

$$\begin{aligned} S &= VI^* = [\text{V}\underline{/\delta}][\text{I}\underline{/\beta}]^* = \text{VI}\underline{/\delta - \beta} \\ &= \text{VI}\cos(\delta - \beta) + j\text{VI}\sin(\delta - \beta) \end{aligned} \tag{2.3.1}$$

where $(\delta - \beta)$ is the angle between the voltage and current. Comparing (2.3.1) with (2.2.11) and (2.2.12), S is recognized as

$$S = \mathrm{P} + j\mathrm{Q} \tag{2.3.2}$$

The magnitude $\mathrm{S} = \mathrm{VI}$ of the complex power S is called the *apparent power.* Although it has the same units as P and Q, it is common practice to define the units of apparent power S as volt-amperes or VA. The real power P is obtained by multiplying the apparent power $\mathrm{S} = \mathrm{VI}$ by the power factor p.f. $= \cos(\delta - \beta)$.

The procedure for determining whether a circuit element absorbs or delivers power is summarized in Figure 2.4. Figure 2.4(a) shows the *load convention*, where the current *enters* the positive terminal of the circuit element, and the complex power *absorbed* by the circuit element is calculated from (2.3.1). This equation shows that, depending on the value of $(\delta - \beta)$, P may have either a positive or negative value. If P is positive, then the circuit element absorbs positive real power.

However, if P is negative, the circuit element absorbs negative real power, or alternatively, it delivers positive real power. Similarly, if Q is positive, the circuit element in Figure 2.4(a) absorbs positive reactive power. However, if Q is negative, the circuit element absorbs negative reactive power, or it delivers positive reactive power.

Figure 2.4(b) shows the *generator convention*, where the current *leaves* the positive terminal of the circuit element, and the complex power *delivered* is calculated from (2.3.1). When P is positive (negative) the circuit element *delivers* positive (negative) real power. Similarly, when Q is positive (negative), the circuit element *delivers* positive (negative) reactive power.

FIGURE 2.4

Load and generator conventions

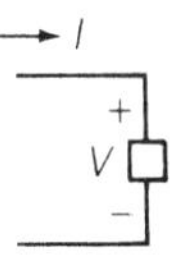

(a) *Load convention.* Current *enters* positive terminal of circuit element. If P is positive, then positive real power is *absorbed.* If Q is positive, then positive reactive power is *absorbed.* If P (Q) is negative, then positive real (reactive) power is *delivered.*

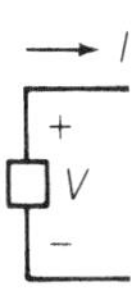

(b) *Generator convention.* Current *leaves* positive terminal of the circuit element. If P is positive, then positive real power is *delivered.* If Q is positive, then positive reactive power is *delivered.* If P (Q) is negative, then positive real (reactive) power is *absorbed.*

EXAMPLE 2.2

Real and reactive power, delivered or absorbed

A single-phase voltage source with $V = 100\angle 130^\circ$ volts delivers a current $I = 10\angle 10^\circ$ A, which leaves the positive terminal of the source. Calculate the source real and reactive power, and state whether the source delivers or absorbs each of these.

SOLUTION

Since I leaves the positive terminal of the source, the generator convention is assumed, and the complex power delivered is, from (2.3.1),

$$S = VI^* = [100\angle 130°][10\angle 10°]^*$$

$$S = 1000\angle 120° = -500 + j866$$

$$P = \text{Re}[S] = -500 \quad \text{W}$$

$$Q = \text{Im}[S] = +866 \quad \text{var}$$

where Im denotes "imaginary part of." The source absorbs 500 W and delivers 866 var. Readers familiar with electric machines will recognize that one example of this source is a synchronous motor. When a synchronous motor operates at a leading power factor, it absorbs real power and delivers reactive power.

The *load convention* is used for the RLC elements shown in Figure 2.2. Therefore, the complex power *absorbed* by any of these three elements can be calculated as follows. Assume a load voltage $V = V\angle\delta$. Then, from (2.3.1),

$$\text{resistor:} \quad S_R = VI_R^* = [V\angle\delta]\left[\frac{V}{R}\angle -\delta\right] = \frac{V^2}{R} \tag{2.3.3}$$

$$\text{indicator:} \quad S_L = VI_L^* = [V\angle\delta]\left[\frac{V}{-jX_L}\angle -\delta\right] = +j\frac{V^2}{X_L} \tag{2.3.4}$$

$$\text{capacitor:} \quad S_C = VI_C^* = [V\angle\delta]\left[\frac{V}{-jX_C}\angle -\delta\right] = -j\frac{V^2}{X_C} \tag{2.3.5}$$

From these complex power expressions, the following can be stated:

A (positive-valued) resistor absorbs (positive) real power, $P_R = V^2/R$ W, and zero reactive power, $Q_R = 0$ var.

An inductor absorbs zero real power, $P_L = 0$ W, and positive reactive power, $Q_L = V^2/X_L$ var.

A capacitor absorbs zero real power, $P_C = 0$ W, and *negative* reactive power, $Q_C = -V^2/X_C$ var. Alternatively, a capacitor *delivers positive* reactive power, $+V^2/X_C$.

For a general load composed of RLC elements, complex power S is also calculated from (2.3.1). The real power $P = \text{Re}(S)$ absorbed by a passive load is always positive. The reactive power $Q = \text{Im}(S)$ absorbed by a load may be either positive or negative. When the load is inductive, the current lags the voltage, which means β is less than δ in (2.3.1), and the reactive power absorbed is positive. When the load is

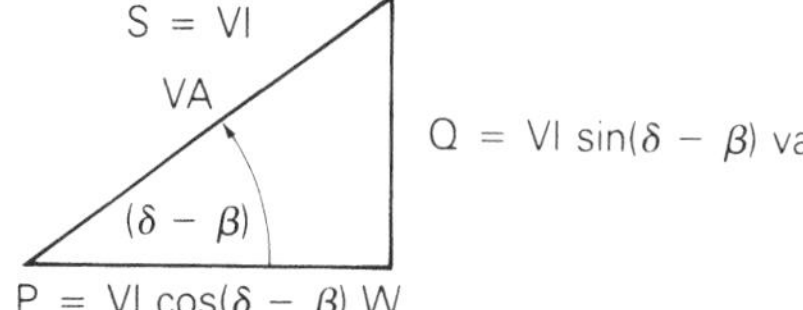

FIGURE 2.5 Power triangle

capacitive, the current leads the voltage, which means β is greater than δ, and the reactive power absorbed is negative; or, alternatively, the capacitive load delivers positive reactive power.

Complex power can be summarized graphically by use of the power triangle shown in Figure 2.5. As shown, the apparent power S, real power P, and reactive power Q form the three sides of the power triangle. The power factor angle $(\delta - \beta)$ is also shown, and the following expressions can be obtained:

$$S = \sqrt{P^2 + Q^2} \tag{2.3.6}$$

$$(\delta - \beta) = \tan^{-1}(Q/P) \tag{2.3.7}$$

$$Q = P\tan(\delta - \beta) \tag{2.3.8}$$

$$\text{p.f.} = \cos(\delta - \beta) = \frac{P}{S} = \frac{P}{\sqrt{P^2 + Q^2}} \tag{2.3.9}$$

EXAMPLE 2.3

Power triangle and power factor correction

A single-phase source delivers 100 kW to a load operating at a power factor of 0.8 lagging. Calculate the reactive power to be delivered by a capacitor connected in parallel with the load in order to raise the source power factor to 0.95 lagging. Also draw the power triangle for the source and load. Assume that the source voltage is constant, and neglect the line impedance between the source and load.

SOLUTION

The circuit and power triangle are shown in Figure 2.6. The real power $P = P_S = P_R$ delivered by the source and absorbed by the load is not changed when the capacitor is connected in parallel with the load, since the capacitor delivers only reactive power Q_C. For the load, the power factor angle, reactive power absorbed, and apparent power are

$$\theta_L = (\delta - \beta_L) = \cos^{-1}(0.8) = 36.87°$$

$$Q_L = P\tan\theta_L = 100\tan(36.87°) = 75 \quad \text{kvar}$$

$$S_L = \frac{P}{\cos\theta_L} = 125 \quad \text{kVA}$$

FIGURE 2.6

Circuit and power triangle for Example 2.3

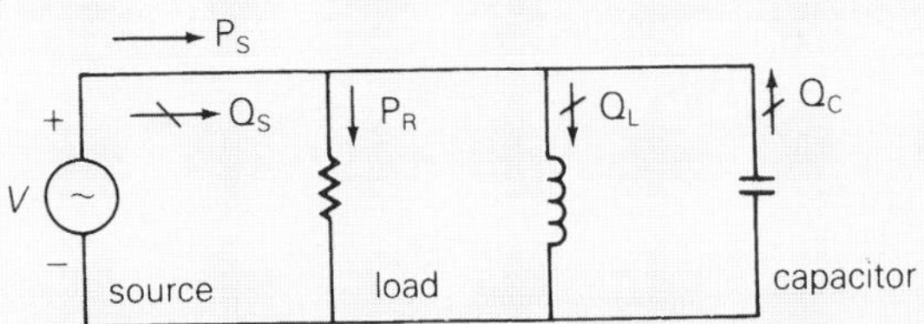

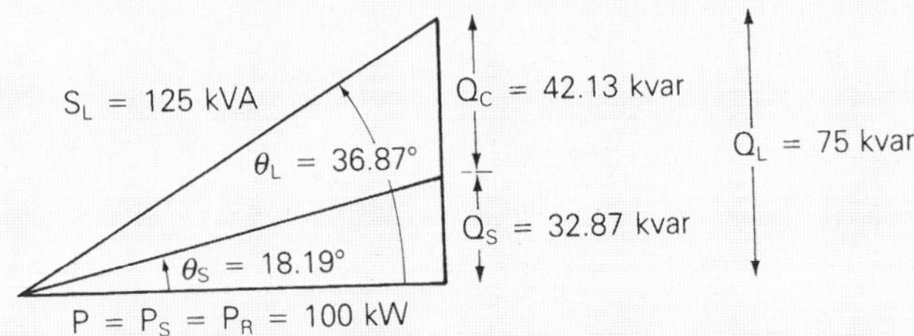

After the capacitor is connected, the power factor angle, reactive power delivered, and apparent power of the source are

$$\theta_S = (\delta - \beta_S) = \cos^{-1}(0.95) = 18.19°$$

$$Q_S = P \tan \theta_S = 100 \tan(18.19°) = 32.87 \quad \text{kvar}$$

$$S_S = \frac{P}{\cos \theta_S} = \frac{100}{0.95} = 105.6 \quad \text{kVA}$$

The capacitor delivers

$$Q_C = Q_L - Q_S = 75 - 32.87 = 42.13 \text{ kVA}$$

The method of connecting a capacitor in parallel with an inductive load is known as *power factor correction.* The effect of the capacitor is to increase the power factor of the source that delivers power to the load. Also, the source apparent power S_S decreases. As shown in Figure 2.6, the source apparent power for this example decreases from 125 kVA without the capacitor to 105.3 kVA with the capacitor. The source current $I_S = S_S/V$ also decreases. When line impedance between the source and load is included, the decrease in source current results in lower line losses and lower line-voltage drops. The end result of power factor correction is improved efficiency and improved voltage regulation.

To see an animated view of this example, open PowerWorld Simulator case Example 2_3 (see Figure 2.7). From the ribbon select the green-and-black Play button to begin the simulation. The speed and size of the green arrows are

(*Continued*)

FIGURE 2.7 Screen for Example 2.3

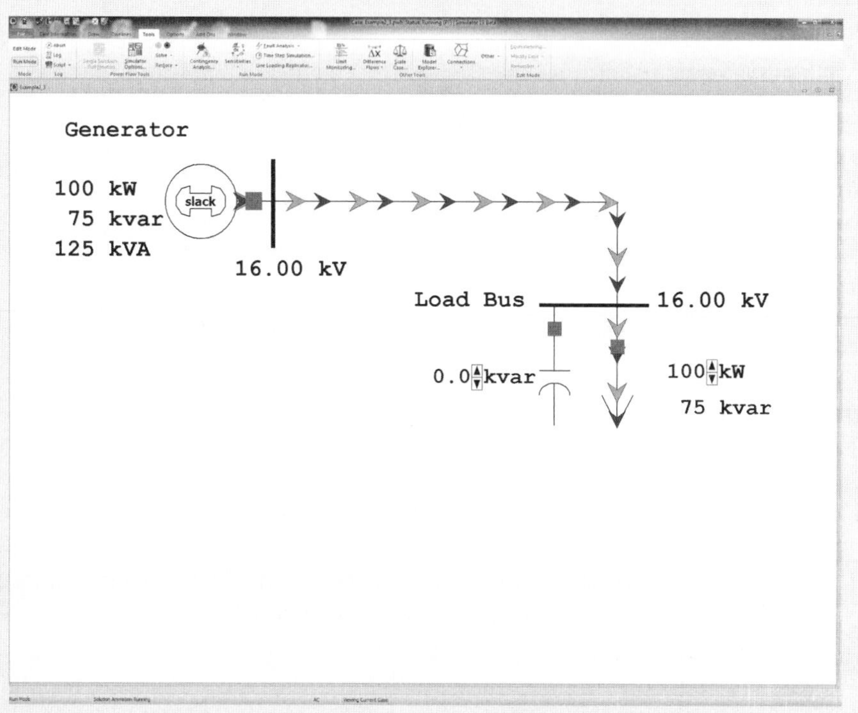

proportional to the real power supplied to the load bus, and the blue arrows are proportional to the reactive power. Here, reactive compensation can be supplied in discrete 20-kvar steps by clicking on the arrows in the capacitor's kvar field, and the load can be varied by clicking on the arrows in the load field. Notice that increasing the reactive compensation decreases both the reactive power flow on the supply line and the kVA power supplied by the generator; the real power flow is unchanged.

2.4 NETWORK EQUATIONS

For circuits operating in sinusoidal-steady-state, Kirchhoff's current law (KCL) and voltage law (KVL) apply to phasor currents and voltages. Thus the sum of all phasor currents entering any node is zero and the sum of the phasor voltage drops around any closed path is zero. Network analysis techniques based on Kirchhoff's laws, including nodal analysis, mesh or loop analysis, superposition, source transformations, and Thevenin's theorem or Norton's theorem, are useful for analyzing such circuits.

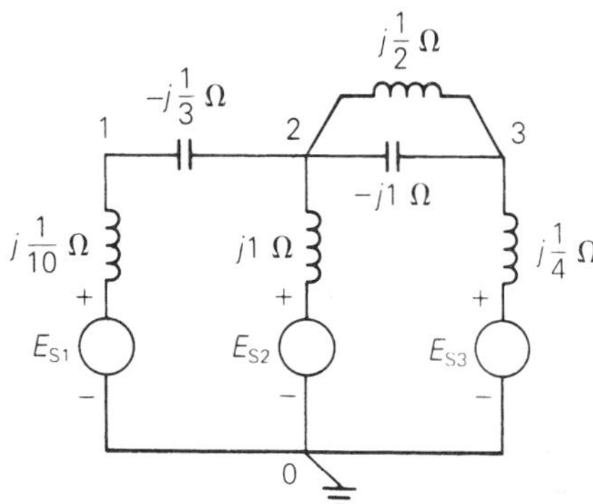

FIGURE 2.8

Circuit diagram for reviewing nodal analysis

Various computer solutions of power system problems are formulated from nodal equations, which can be systematically applied to circuits. The circuit shown in Figure 2.8, which is used here to review nodal analysis, is assumed to be operating in sinusoidal-steady-state; source voltages are represented by phasors E_{S1}, E_{S2}, and E_{S3}; circuit impedances are specified in ohms. Nodal equations are written in the following three steps:

STEP 1 For a circuit with $(N + 1)$ nodes (also called buses), select one bus as the reference bus and define the voltages at the remaining buses with respect to the reference bus.

The circuit in Figure 2.8 has four buses—that is, $N + 1 = 4$ or $N = 3$. Bus 0 is selected as the reference bus, and bus voltages V_{10}, V_{20}, and V_{30} are then defined with respect to bus 0.

STEP 2 Transform each voltage source in series with an impedance to an equivalent current source in parallel with that impedance. Also, show admittance values instead of impedance values on the circuit diagram. Each current source is equal to the voltage source divided by the source impedance.

In Figure 2.9, equivalent current sources I_1, I_2, and I_3 are shown, and all impedances are converted to corresponding admittances.

STEP 3 Write nodal equations in matrix format as follows:

$$\begin{bmatrix} Y_{11} & Y_{12} & Y_{13} & \cdots & Y_{1N} \\ Y_{21} & Y_{22} & Y_{23} & \cdots & Y_{2N} \\ Y_{31} & Y_{32} & Y_{33} & \cdots & Y_{3N} \\ \vdots & \vdots & \vdots & & \vdots \\ Y_{N1} & Y_{N2} & Y_{N3} & \cdots & Y_{NN} \end{bmatrix} \begin{bmatrix} V_{10} \\ V_{20} \\ V_{30} \\ \vdots \\ V_{N0} \end{bmatrix} = \begin{bmatrix} I_1 \\ I_2 \\ I_3 \\ \vdots \\ I_N \end{bmatrix} \tag{2.4.1}$$

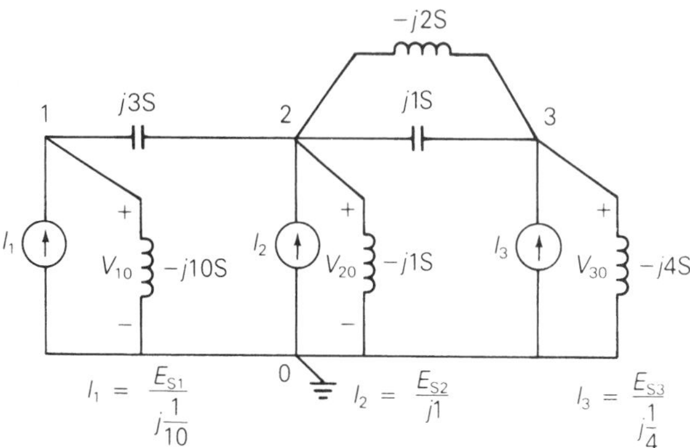

FIGURE 2.9 Circuit of Figure 2.8 with equivalent current sources replacing voltage sources. Admittance values are also shown.

Using matrix notation, (2.4.1) becomes

$$\boldsymbol{YV} = \boldsymbol{I} \tag{2.4.2}$$

where $\boldsymbol{Y}$ is the $N \times N$ bus admittance matrix, $\boldsymbol{V}$ is the column vector of N bus voltages, and $\boldsymbol{I}$ is the column vector of N current sources. The elements Y_{kn}, of the bus admittance matrix $\boldsymbol{Y}$ are formed as follows:

diagonal elements: Y_{kk} = sum of admittances connected to bus k $(k = 1, 2, \ldots, N)$ (2.4.3)

off-diagonal elements: Y_{kn} = −(sum of admittances connected between buses k and n) $(k \neq n)$ (2.4.4)

The diagonal element Y_{kk} is called the *self-admittance* or the *driving-point admittance* of bus k, and the off-diagonal element Y_{kn} for $k \neq n$ is called the *mutual admittance* or the *transfer admittance* between buses k and n. Since $Y_{kn} = Y_{kn}$, the matrix $\boldsymbol{Y}$ is symmetric.

For the circuit of Figure 2.9, (2.4.1) becomes

$$\begin{bmatrix} (j3 - j10) & -(j3) & 0 \\ -(j3) & (j3 - j1 + j1 - j2) & -(j1 - j2) \\ 0 & -(j1 - j2) & (j1 - j2 - j4) \end{bmatrix} \begin{bmatrix} V_{10} \\ V_{20} \\ V_{30} \end{bmatrix}$$

$$= \begin{bmatrix} I_1 \\ I_2 \\ I_3 \end{bmatrix}$$

$$j\begin{bmatrix} -7 & -3 & 0 \\ -3 & 1 & 1 \\ 0 & 1 & -5 \end{bmatrix} \begin{bmatrix} V_{10} \\ V_{20} \\ V_{30} \end{bmatrix} = \begin{bmatrix} I_1 \\ I_2 \\ I_3 \end{bmatrix} \tag{2.4.5}$$

The advantage of this method of writing nodal equations is that a computer can be used both to generate the admittance matrix $\boldsymbol{Y}$ and to solve (2.4.2) for the unknown bus voltage vector $\boldsymbol{V}$. Once a circuit is specified with the reference bus and other buses identified, the circuit admittances and their bus connections become computer input data for calculating the elements Y_{kn} via (2.4.3) and (2.4.4). After $\boldsymbol{Y}$ is calculated and the current source vector $\boldsymbol{I}$ is given as input, standard computer programs for solving simultaneous linear equations can then be used to determine the bus voltage vector $\boldsymbol{V}$.

When double subscripts are used to denote a voltage in this text, the voltage shall be that at the node identified by the first subscript with respect to the node identified by the second subscript. For example, the voltage V_{10} in Figure 2.9 is the voltage at node 1 with respect to node 0. Also, a current I_{ab} shall indicate the current from node a to node b. Voltage polarity marks (+/−) and current reference arrows (→) are not required when double subscript notation is employed. The polarity marks in Figure 2.9 for V_{10}, V_{20}, and V_{30}, although not required, are shown for clarity. The reference arrows for sources I_1, I_2, and I_3 in Figure 2.9 are required, however, since single subscripts are used for these currents. Matrices and vectors shall be indicated in this text by boldface italic type (for example, $\boldsymbol{Y}$ or $\boldsymbol{V}$).

2.5 BALANCED THREE-PHASE CIRCUITS

In this section, we introduce the following topics for balanced three-phase circuits: Y connections, line-to-neutral voltages, line-to-line voltages, line currents, Δ loads, Δ−Y conversions, and equivalent line-to-neutral diagrams.

BALANCED Y CONNECTIONS

Figure 2.10 shows a three-phase, Y-connected (or "wye-connected") voltage source feeding a balanced-Y-connected load. For a Y connection, the neutrals of each phase are connected. In Figure 2.10 the source neutral connection is labeled bus n and the load neutral connection is labeled bus N. The three-phase source is assumed to be ideal since source impedances are neglected. Also neglected are the

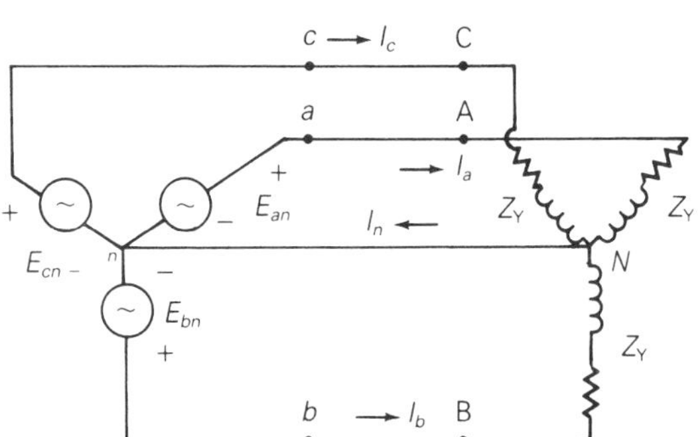

FIGURE 2.10

Circuit diagram of a three-phase Y-connected source feeding a balanced-Y load line impedances between the source and load terminals, and the neutral impedance between buses n and N. The three-phase load is *balanced*, which means the load impedances in all three phases are identical.

BALANCED LINE-TO-NEUTRAL VOLTAGES

In Figure 2.10, the terminal buses of the three-phase source are labeled a, b, and c, and the source line-to-neutral voltages are labeled E_{an}, E_{bn}, and E_{cn}. The source is *balanced* when these voltages have equal magnitudes and an equal 120°-phase difference between any two phases. An example of balanced three-phase line-to-neutral voltages is

$$\begin{aligned} E_{an} &= 10\underline{/0^\circ} \\ E_{bn} &= 10\underline{/-120^\circ} = 10\underline{/+240^\circ} \\ E_{cn} &= 10\underline{/+120^\circ} = 10\underline{/-240^\circ} \text{ volts} \end{aligned} \tag{2.5.1}$$

where the line-to-neutral voltage magnitude is 10 volts and E_{an} is the reference phasor. The phase sequence is called *positive sequence* or *abc* sequence when E_{an} leads E_{bn} by 120° and E_{bn} leads E_{cn} by 120°. The phase sequence is called *negative sequence* or *acb* sequence when E_{an} leads E_{cn} by 120° and E_{cn} leads E_{bn} by 120°. The voltages in (2.5.1) are positive-sequence voltages, since E_{an} leads E_{bn} by 120°. The corresponding phasor diagram is shown in Figure 2.11.

BALANCED LINE-TO-LINE VOLTAGES

The voltages E_{ab}, E_{bc}, and E_{ca} between phases are called line-to-line voltages. Writing a KVL equation for a closed path around buses a, b, and n in Figure 2.10.

$$E_{ab} = E_{an} - E_{bn} \tag{2.5.2}$$

For the line-to-neutral voltages of (2.5.1),

$$\begin{aligned} E_{ab} &= 10\underline{/0^\circ} - 10\underline{/-120^\circ} = 10 - 10\left[\frac{-1 - j\sqrt{3}}{2}\right] \\ E_{ab} &= \sqrt{3}(10)\left(\frac{\sqrt{3} + j1}{2}\right) = \sqrt{3}(10\underline{/30^\circ}) \quad \text{volts} \end{aligned} \tag{2.5.3}$$

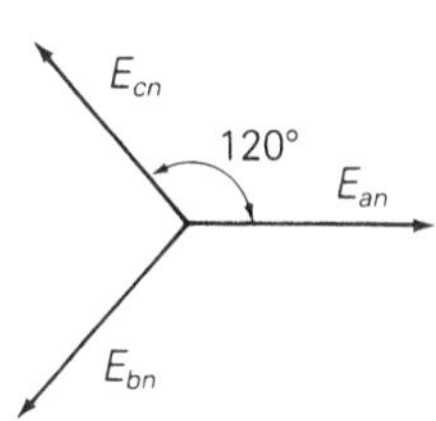

FIGURE 2.11 Phasor diagram of balanced positive-sequence line-to-neutral voltages with E_{an} as the reference

Similarly, the line-to-line voltages E_{bc} and E_{ca} are

$$\begin{aligned} E_{bc} &= E_{bn} - E_{cn} = 10\underline{/-120^\circ} - 10\underline{/+120^\circ} \\ &= \sqrt{3}(10\underline{/-90^\circ}) \quad \text{volts} \end{aligned} \tag{2.5.4}$$

$$\begin{aligned} E_{ca} &= E_{cn} - E_{an} = 10\underline{/+120^\circ} - 10\underline{/0^\circ} \\ &= \sqrt{3}(10\underline{/150^\circ}) \text{ volts} \end{aligned} \tag{2.5.5}$$

The line-to-line voltages of (2.5.3) through (2.5.5) are also balanced, since they have equal magnitudes of $\sqrt{3}$ (10) volts and 120° displacement between any two

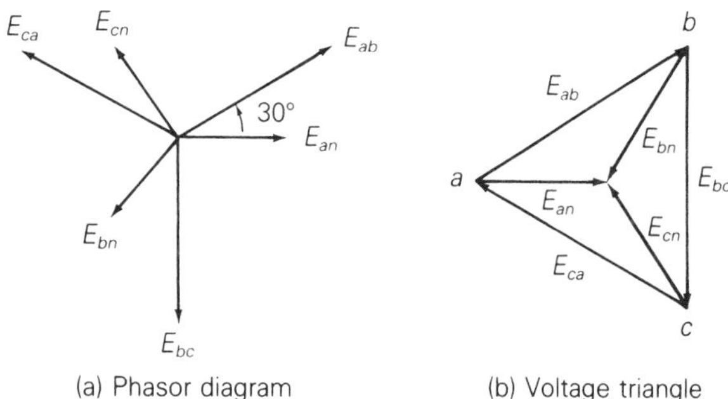

FIGURE 2.12 Positive-sequence line-to-neutral and line-to-line voltages in a balanced three-phase, Y-connected system

phases. Comparison of these line-to-line voltages with the line-to-neutral voltages of (2.5.1) leads to the following conclusion:

In a balanced three-phase Y-connected system with positive-sequence sources, the line-to-line voltages are $\sqrt{3}$ times the line-to-neutral voltages and lead by 30°. That is,

$$\begin{aligned} E_{ab} &= \sqrt{3}E_{an}\underline{/+30^\circ} \\ E_{bc} &= \sqrt{3}E_{bn}\underline{/+30^\circ} \\ E_{ca} &= \sqrt{3}E_{cn}\underline{/+30^\circ} \end{aligned} \tag{2.5.6}$$

This very important result is summarized in Figure 2.12. In Figure 2.12(a) each phasor begins at the origin of the phasor diagram. In Figure 2.12(b) the line-to-line voltages form an equilateral triangle with vertices labeled a, b, c corresponding to buses a, b, and c of the system; the line-to-neutral voltages begin at the vertices and end at the center of the triangle, which is labeled n for neutral bus n. Also, the clockwise sequence of the vertices abc in Figure 2.12(b) indicates positive-sequence voltages. In both diagrams, E_{an} is the reference. However, the diagrams could be rotated to align with any other reference.

Since the balanced line-to-line voltages form a closed triangle in Figure 2.12, their sum is zero. In fact, the sum of line-to-line voltages $(E_{ab} + E_{bc} + E_{ca})$ is *always* zero, even if the system is unbalanced, since these voltages form a closed path around buses a, b, and c. Also, in a balanced system, the sum of the line-to-neutral voltages $(E_{an} + E_{bn} + E_{cn})$ equals zero.

BALANCED LINE CURRENTS

Since the impedance between the source and load neutrals in Figure 2.10 is neglected, buses n and N are at the same potential, $E_{nN} = 0$. Accordingly, a separate KVL equation can be written for each phase, and the line currents can be written by inspection:

$$\begin{aligned} I_a &= E_{an}/Z_Y \\ I_b &= E_{bn}/Z_Y \\ I_c &= E_{cn}/Z_Y \end{aligned} \tag{2.5.7}$$

FIGURE 2.13

Phasor diagram of line currents in a balanced three-phase system

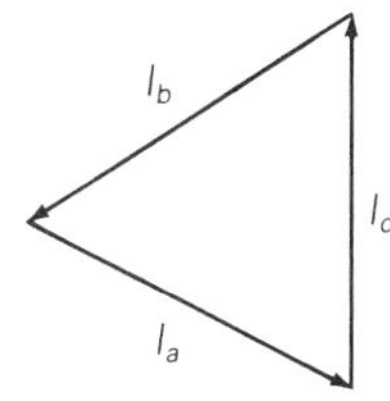

For example, if each phase of the Y-connected load has an impedance $Z_Y = 2\underline{/30^\circ}\ \Omega$, then

$$I_a = \frac{10\underline{/0^\circ}}{2\underline{/30^\circ}} = 5\underline{/-30^\circ}\ \text{A}$$
$$I_b = \frac{10\underline{/-120^\circ}}{2\underline{/30^\circ}} = 5\underline{/-150^\circ}\ \text{A} \qquad (2.5.8)$$
$$I_c = \frac{10\underline{/+120^\circ}}{2\underline{/30^\circ}} = 5\underline{/90^\circ}\ \text{A}$$

The line currents are also balanced, since they have equal magnitudes of 5 A and 120° displacement between any two phases. The neutral current I_n is determined by writing a KCL equation at bus N in Figure 2.10.

$$I_n = I_a + I_b + I_c \qquad (2.5.9)$$

Using the line currents of (2.5.8),

$$I_n = 5\underline{/-30^\circ} + 5\underline{/-150^\circ} + 5\underline{/90^\circ}$$
$$I_n = 5\left(\frac{\sqrt{3} - j1}{2}\right) + 5\left(\frac{-\sqrt{3} - j1}{2}\right) + j5 = 0 \qquad (2.5.10)$$

The phasor diagram of the line currents is shown in Figure 2.13. Since these line currents form a closed triangle, their sum, which is the neutral current I_n, is zero. In general, the sum of any balanced three-phase set of phasors is zero, since balanced phasors form a closed triangle. Thus, although the impedance between neutrals n and N in Figure 2.10 is assumed to be zero, the neutral current will be zero for *any* neutral impedance ranging from short circuit (0 Ω) to open circuit (∞ Ω), as long as the system is balanced. If the system is not balanced—which could occur if the source voltages, load impedances, or line impedances were unbalanced—then the line currents will not be balanced, and a neutral current I_n may flow between buses n and N.

BALANCED-Δ LOADS

Figure 2.14 shows a three-phase Y-connected source feeding a balanced-Δ-connected (or "delta-connected") load. For a balanced-Δ connection, equal load impedances Z_Δ are connected in a triangle whose vertices form the buses, labeled A, B, and C in Figure 2.14. The Δ connection does not have a neutral bus.

Since the line impedances are neglected in Figure 2.14, the source line-to-line voltages are equal to the load line-to-line voltages, and the Δ-load currents I_{AB}, I_{BC} and I_{CA} are

$$I_{AB} = E_{ab}/Z_\Delta$$
$$I_{BC} = E_{bc}/Z_\Delta \qquad (2.5.11)$$
$$I_{CA} = E_{ca}/Z_\Delta$$

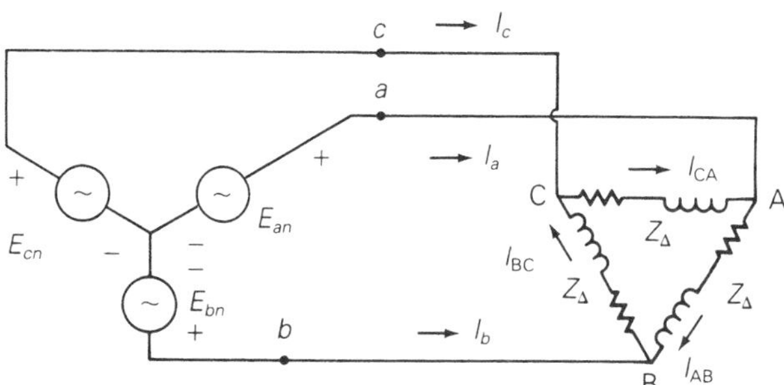

FIGURE 2.14

Circuit diagram of a Y-connected source feeding a balanced-Δ load

For example, if the line-to-line voltages are given by (2.5.3) through (2.5.5) and if $Z_\Delta = 5\underline{/30^\circ}\ \Omega$, then the Δ-load currents are

$$
\begin{aligned}
I_{AB} &= \sqrt{3}\left(\frac{10\underline{/-30^\circ}}{5\underline{/30^\circ}}\right) = 3.464\underline{/0^\circ}\ \text{A} \\
I_{BC} &= \sqrt{3}\left(\frac{10\underline{/-90^\circ}}{5\underline{/30^\circ}}\right) = 3.464\underline{/-120^\circ}\ \text{A} \\
I_{CA} &= \sqrt{3}\left(\frac{10\underline{/150^\circ}}{5\underline{/30^\circ}}\right) = 3.464\underline{/+120^\circ}\ \text{A}
\end{aligned}
\tag{2.5.12}
$$

Also, the line currents can be determined by writing a KCL equation at each bus of the Δ load, as follows:

$$
\begin{aligned}
I_a &= I_{AB} - I_{CA} = 3.464\underline{/0^\circ} - 3.464\underline{/120^\circ} = \sqrt{3}(3.464\underline{/-30^\circ}) \\
I_b &= I_{BC} - I_{AB} = 3.464\underline{/-120^\circ} - 3.464\underline{/0^\circ} = \sqrt{3}(3.464\underline{/-150^\circ}) \\
I_c &= I_{CA} - I_{BC} = 3.464\underline{/120^\circ} - 3.464\underline{/-120^\circ} = \sqrt{3}(3.464\underline{/+90^\circ})
\end{aligned}
\tag{2.5.13}
$$

Both the Δ-load currents given by (2.5.12) and the line currents given by (2.5.13) are balanced. Thus the sum of balanced Δ-load currents ($I_{AB} + I_{BC} + I_{CA}$) equals zero. The sum of line currents ($I_a + I_b + I_c$) is always zero for a Δ-connected load, even if the system is unbalanced, since there is no neutral wire. Comparison of (2.5.12) and (2.5.13) leads to the following conclusion:

For a balanced-Δ load supplied by a balanced positive-sequence source, the line currents into the load are $\sqrt{3}$ times the Δ-load currents and lag by 30°. That is,

$$
\begin{aligned}
I_a &= \sqrt{3}I_{AB}\underline{/-30^\circ} \\
I_b &= \sqrt{3}I_{BC}\underline{/-30^\circ} \\
I_c &= \sqrt{3}I_{CA}\underline{/-30^\circ}
\end{aligned}
\tag{2.5.14}
$$

This result is summarized in Figure 2.15.

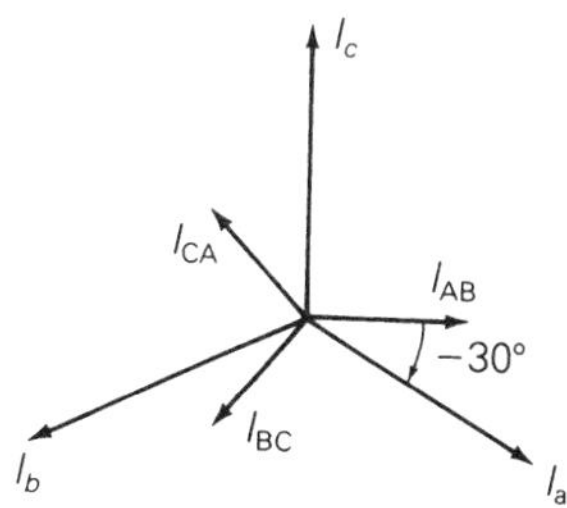

FIGURE 2.15 Phasor diagram of line currents and load currents for a balanced-Δ load

Δ–Y CONVERSION FOR BALANCED LOADS

Figure 2.16 shows the conversion of a balanced-Δ load to a balanced-Y load. If balanced voltages are applied, then these loads will be equivalent as viewed from their terminal buses A, B, and C when the line currents into the Δ load are the same as the line currents into the Y load. For the Δ load,

$$I_A = \sqrt{3} I_{AB} \underline{/-30°} = \frac{\sqrt{3} E_{AB} \underline{/-30°}}{Z_\Delta} \tag{2.5.15}$$

and for the Y load,

$$I_A = \frac{E_{AN}}{Z_Y} = \frac{E_{AB} \underline{/-30°}}{\sqrt{3} Z_Y} \tag{2.5.16}$$

Comparison of (2.5.15) and (2.5.16) indicates that I_A will be the same for both the Δ and Y loads when

$$Z_Y = \frac{Z_\Delta}{3} \tag{2.5.17}$$

Also, the other line currents I_B and I_C into the Y load will equal those into the Δ load when $Z_Y = Z_\Delta/3$, since these loads are balanced. Thus a balanced-Δ load can

FIGURE 2.16 Δ-Y conversion for balanced loads

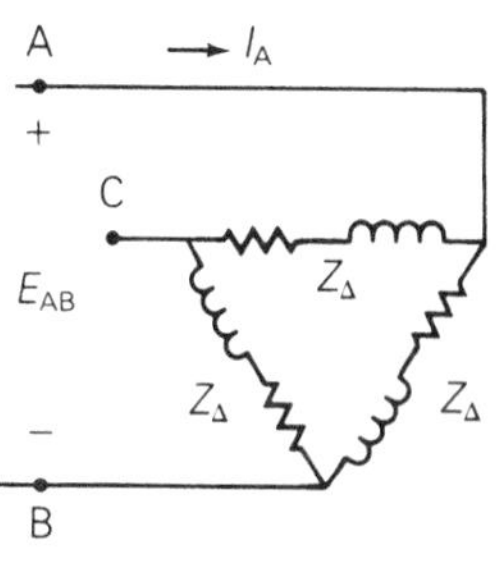

(a) Balanced-Δ load

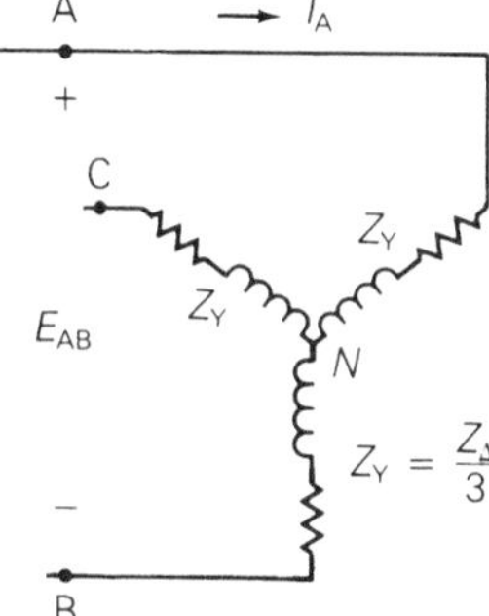

(b) Equivalent balanced-Y load

be converted to an equivalent balanced-Y load by dividing the Δ-load impedance by 3. The angles of these Δ- and equivalent Y-load impedances are the same. Similarly, a balanced-Y load can be converted to an equivalent balanced-Δ load using $Z_\Delta = 3Z_Y$.

EXAMPLE 2.4

Balanced-Δ and-Y loads

A balanced, positive-sequence, Y-connected voltage source with $E_{ab} = 480\underline{/0^\circ}$ volts is applied to a balanced-Δ load with $Z_\Delta = 30\underline{/40^\circ}\ \Omega$. The line impedance between the source and load is $Z_L = 1\underline{/85^\circ}\ \Omega$ for each phase. Calculate the line currents, the Δ-load currents, and the voltages at the load terminals.

SOLUTION

The solution is most easily obtained as follows. First, convert the Δ load to an equivalent Y. Then connect the source and Y-load neutrals with a zero-ohm neutral wire. The connection of the neutral wire has no effect on the circuit, since the neutral current $I_n = 0$ in a balanced system. The resulting circuit is shown in Figure 2.17. The line currents are

$$
\begin{aligned}
I_A &= \frac{E_{an}}{Z_L + Z_Y} = \frac{\dfrac{480}{\sqrt{3}}\underline{/-30^\circ}}{1\underline{/85^\circ} + \dfrac{30}{3}\underline{/40^\circ}} \\
&= \frac{277.1\underline{/-30^\circ}}{(0.0872 + j0.9962) + (7.660 + j6.428)} \\
&= \frac{277.1\underline{/-30^\circ}}{(7.748 + j7.424)} = \frac{277.1\underline{/-30^\circ}}{10.73\underline{/43.78^\circ}} = 25.83\underline{/-73.78^\circ}\ \text{A} \\
I_B &= 25.83\underline{/166.22^\circ}\ \text{A} \\
I_C &= 25.83\underline{/46.22^\circ}\ \text{A}
\end{aligned}
\tag{2.5.18}
$$

The Δ-load currents are, from (2.5.14),

$$
\begin{aligned}
I_{AB} &= \frac{I_a}{\sqrt{3}}\underline{/+30^\circ} = \frac{25.83}{\sqrt{3}}\underline{/-73.78^\circ + 30^\circ} = 14.91\underline{/-43.78^\circ}\ \text{A} \\
I_{BC} &= 14.91\underline{/-163.78^\circ}\ \text{A} \\
I_{CA} &= 14.91\underline{/+76.22^\circ}\ \text{A}
\end{aligned}
\tag{2.5.19}
$$

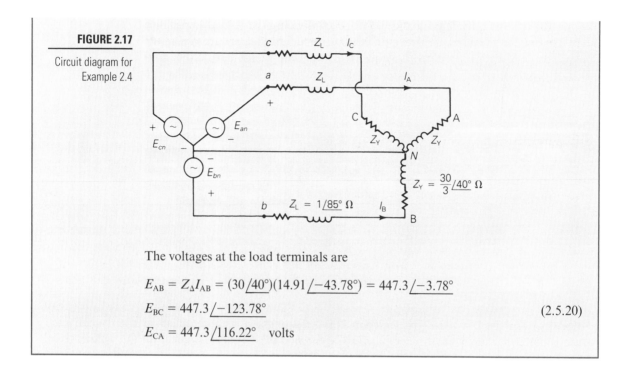

FIGURE 2.17

Circuit diagram for Example 2.4

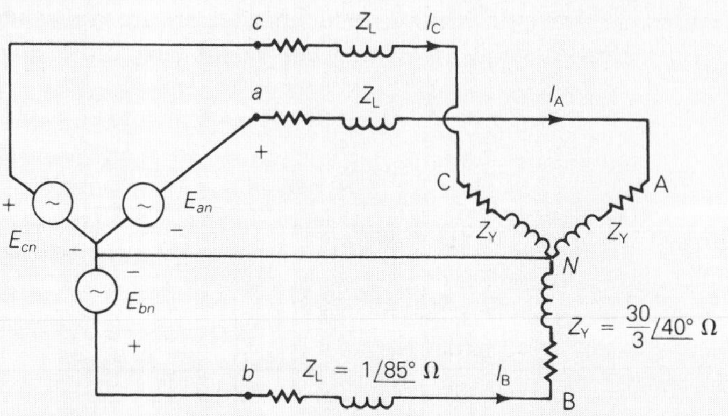

The voltages at the load terminals are

$$E_{AB} = Z_{\Delta}I_{AB} = (30\underline{/40^\circ})(14.91\underline{/-43.78^\circ}) = 447.3\underline{/-3.78^\circ}$$
$$E_{BC} = 447.3\underline{/-123.78^\circ} \qquad (2.5.20)$$
$$E_{CA} = 447.3\underline{/116.22^\circ} \quad \text{volts}$$

EQUIVALENT LINE-TO-NEUTRAL DIAGRAMS

When working with balanced three-phase circuits, only one phase need be analyzed. Δ loads can be converted to Y loads, and all source and load neutrals can be connected with a zero-ohm neutral wire without changing the solution. Then one phase of the circuit can be solved. The voltages and currents in the other two phases are equal in magnitude to and ±120° out of phase with those of the solved phase. Figure 2.18 shows an equivalent line-to-neutral diagram for one phase of the circuit in Example 2.4.

When discussing three-phase systems in this text, voltages shall be rms line-to-line voltages unless otherwise indicated. This is standard industry practice.

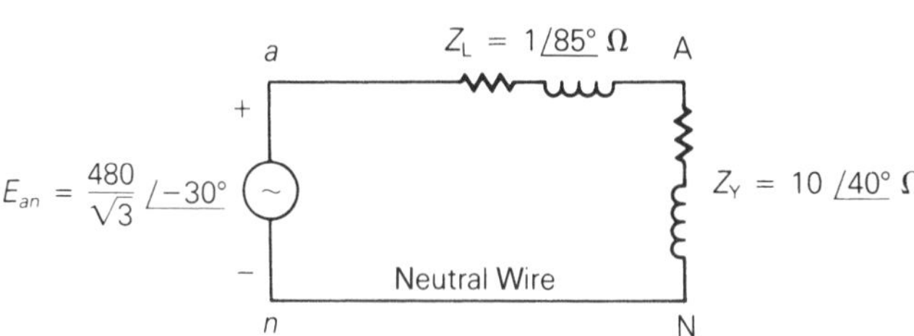

FIGURE 2.18

Equivalent line-to-neutral diagram for the circuit of Example 2.4

2.6 POWER IN BALANCED THREE-PHASE CIRCUITS

In this section, we discuss instantaneous power and complex power for balanced three-phase generators and motors and for balanced-Y and Δ-impedance loads.

INSTANTANEOUS POWER: BALANCED THREE-PHASE GENERATORS

Figure 2.19 shows a Y-connected generator represented by three voltage sources with their neutrals connected at bus n and by three identical generator impedances Z_g. Assume that the generator is operating under balanced steady-state conditions with the instantaneous generator terminal voltage given by

$$v_{an}(t) = \sqrt{2}V_{LN}\cos(\omega t + \delta) \quad \text{volts} \tag{2.6.1}$$

and with the instantaneous current leaving the positive terminal of phase a given by

$$i_a(t) = \sqrt{2}I_L \cos(\omega t + \beta) \text{ A} \tag{2.6.2}$$

where V_{LN} is the rms line-to-neutral voltage and I_L is the rms line current. The instantaneous power $p_a(t)$ delivered by phase a of the generator is

$$\begin{aligned} p_a(t) &= v_{an}(t)i_a(t) \\ &= 2V_{LN}I_L \cos(\omega t + \delta)\cos(\omega t + \beta) \\ &= V_{LN}I_L \cos(\delta - \beta) + V_{LN}I_L \cos(2\omega t + \delta + \beta) \quad \text{W} \end{aligned} \tag{2.6.3}$$

Assuming balanced operating conditions, the voltages and currents of phases b and c have the same magnitudes as those of phase a and are $\pm 120°$ out of phase with phase a. Therefore the instantaneous power delivered by phase b is

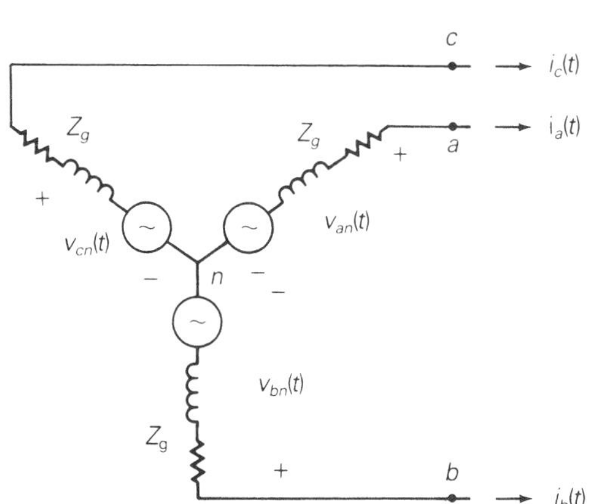

FIGURE 2.19

Y-connected generator

$$p_b(t) = 2\mathrm{V_{LN}I_L}\cos(\omega t + \delta - 120°)\cos(\omega t + \beta - 120°)$$
$$= \mathrm{V_{LN}I_L}\cos(\delta - \beta) + \mathrm{V_{LN}I_L}\cos(2\omega t + \delta + \beta - 240°) \quad \mathrm{W} \tag{2.6.4}$$

and by phase c,

$$p_c(t) = 2\mathrm{V_{LN}I_L}\cos(\omega t + \delta + 120°)\cos(\omega t + \beta + 120°)$$
$$= \mathrm{V_{LN}I_L}\cos(\delta - \beta) + \mathrm{V_{LN}I_L}\cos(2\omega t + \delta + \beta + 240°) \quad \mathrm{W} \tag{2.6.5}$$

The total instantaneous power $p_{3\phi}(t)$ delivered by the three-phase generator is the sum of the instantaneous powers delivered by each phase. Using (2.6.3) through (2.6.5):

$$p_{3\phi}(t) = p_a(t) + p_b(t) + p_c(t)$$
$$= 3\mathrm{V_{LN}I_L}\cos(\delta - \beta) + \mathrm{V_{LN}I_L}[\cos(2\omega t + \delta + \beta)$$
$$+ \cos(2\omega t + \delta + \beta - 240°)$$
$$+ \cos(2\omega t + \delta + \beta + 240°)] \ \mathrm{W} \tag{2.6.6}$$

The three cosine terms within the brackets of (2.6.6) can be represented by a *balanced* set of three phasors. Therefore, the sum of these three terms is zero for any value of δ, for any value of β, and for all values of t. Equation (2.6.6) then reduces to

$$p_{3\phi}(t) = \mathrm{P_{3\phi}} = 3\mathrm{V_{LN}I_L}\cos(\delta - \beta) \quad \mathrm{W} \tag{2.6.7}$$

Equation (2.6.7) can be written in terms of the line-to-line voltage $\mathrm{V_{LL}}$ instead of the line-to-neutral voltage $\mathrm{V_{LN}}$. Under balanced operating conditions.

$$\mathrm{V_{LN}} = \mathrm{V_{LL}}/\sqrt{3} \quad \text{and} \quad \mathrm{P_{3\phi}} = \sqrt{3}\mathrm{V_{LL}I_L}\cos(\delta - \beta) \quad \mathrm{W} \tag{2.6.8}$$

Inspection of (2.6.8) leads to the following conclusion:

> The total instantaneous power delivered by a three-phase generator under balanced operating conditions is not a function of time, but a *constant*, $p_{3\phi}(t) = \mathrm{P_{3\phi}}$.

INSTANTANEOUS POWER: BALANCED THREE-PHASE MOTORS AND IMPEDANCE LOADS

The total instantaneous power absorbed by a three-phase motor under balanced steady-state conditions is also a constant. Figure 2.19 can be used to represent a three-phase motor by reversing the line currents to enter rather than leave the positive terminals. Then (2.6.1) through (2.6.8), valid for power *delivered* by a generator, are also valid for power *absorbed* by a motor. These equations are also valid for the instantaneous power absorbed by a balanced three-phase impedance load.

COMPLEX POWER: BALANCED THREE-PHASE GENERATORS

The phasor representations of the voltage and current in (2.6.1) and (2.6.2) are

$$V_{an} = \mathrm{V_{LN}}\underline{/\delta} \quad \text{volts} \tag{2.6.9}$$
$$I_a = \mathrm{I_L}\underline{/\beta} \quad \mathrm{A} \tag{2.6.10}$$

where I_a leaves positive terminal "a" of the generator. The complex power S_a delivered by phase a of the generator is

$$S_a = V_{an}I_a^* = \text{V}_{\text{LN}}\text{I}_{\text{L}}\underline{/(\delta - \beta)}$$
$$= \text{V}_{\text{LN}}\text{I}_{\text{L}} \cos(\delta - \beta) + j\text{V}_{\text{LN}}\text{I}_{\text{L}} \sin(\delta - \beta) \tag{2.6.11}$$

Under balanced operating conditions, the complex powers delivered by phases b and c are identical to S_a, and the total complex power $S_{3\phi}$ delivered by the generator is

$$S_{3\phi} = S_a + S_b + S_c = 3S_a$$
$$= 3\text{V}_{\text{LN}}\text{I}_{\text{L}}\underline{/(\delta - \beta)}$$
$$= 3\text{V}_{\text{LN}}\text{I}_{\text{L}} \cos(\delta - \beta) + j3\text{V}_{\text{LN}}\text{I}_{\text{L}} \sin(\delta - \beta) \tag{2.6.12}$$

In terms of the total real and reactive powers,

$$S_{3\phi} = \text{P}_{3\phi} + j\text{Q}_{3\phi} \tag{2.6.13}$$

where

$$\text{P}_{3\phi} = \text{Re}(S_{3\phi}) = 3\text{V}_{\text{LN}}\text{I}_{\text{L}} \cos(\delta - \beta)$$
$$= \sqrt{3}\text{V}_{\text{LL}}\text{I}_{\text{L}} \cos(\delta - \beta) \quad \text{W} \tag{2.6.14}$$

and

$$\text{Q}_{3\phi} = \text{Im}(S_{3\phi}) = 3\text{V}_{\text{LN}}\text{I}_{\text{L}} \sin(\delta - \beta)$$
$$= \sqrt{3}\text{V}_{\text{LL}}\text{I}_{\text{L}} \sin(\delta - \beta) \quad \text{var} \tag{2.6.15}$$

Also, the total apparent power is

$$\text{S}_{3\phi} = |S_{3\phi}| = 3\text{V}_{\text{LN}}\text{I}_{\text{L}} = \sqrt{3}\text{V}_{\text{LL}}\text{I}_{\text{L}} \quad \text{VA} \tag{2.6.16}$$

COMPLEX POWER: BALANCED THREE-PHASE MOTORS

The preceding expressions for complex, real, reactive, and apparent power *delivered* by a three-phase generator are also valid for the complex, real, reactive, and apparent power *absorbed* by a three-phase motor.

COMPLEX POWER: BALANCED-Y AND BALANCED-Δ IMPEDANCE LOADS

Equations (2.6.13) through (2.6.16) are also valid for balanced-Y and -Δ impedance loads. For a balanced-Y load, the line-to-neutral voltage across the phase a load impedance and the current entering the positive terminal of that load impedance can be represented by (2.6.9) and (2.6.10). Then (2.6.11) through (2.6.16) are valid for the power absorbed by the balanced-Y load.

For a balanced-Δ load, the line-to-line voltage across the phase a-b load impedance and the current into the positive terminal of that load impedance can be represented by

$$V_{ab} = \mathrm{V_{LL}}\underline{/\delta} \quad \text{volts} \tag{2.6.17}$$

$$I_{ab} = \mathrm{I_\Delta}\underline{/\beta} \ \mathrm{A} \tag{2.6.18}$$

where $\mathrm{V_{LL}}$ is the rms line-to-line voltage and $\mathrm{I_\Delta}$ is the rms Δ-load current. The complex power S_{ab} absorbed by the phase $a-b$ load impedance is then

$$S_{ab} = V_{ab}I^*_{ab} = \mathrm{V_{LL}I_\Delta}\underline{/(\delta - \beta)} \tag{2.6.19}$$

The total complex power absorbed by the Δ load is

$$\begin{aligned} S_{3\phi} &= S_{ab} + S_{bc} + S_{ca} = 3S_{ab} \\ &= 3\mathrm{V_{LL}I_\Delta}\underline{/(\delta - \beta)} \\ &= 3\mathrm{V_{LL}I_\Delta}\cos(\delta - \beta) + j3\mathrm{V_{LL}I_\Delta}\sin(\delta - \beta) \end{aligned} \tag{2.6.20}$$

Rewriting (2.6.19) in terms of the total real and reactive power,

$$S_{3\phi} = \mathrm{P_{3\phi}} + j\mathrm{Q_{3\phi}} \tag{2.6.21}$$

$$\begin{aligned} \mathrm{P_{3\phi}} &= \mathrm{Re}(S_{3\phi}) = 3\mathrm{V_{LL}I_\Delta}\cos(\delta - \beta) \\ &= \sqrt{3}\mathrm{V_{LL}I_L}\cos(\delta - \beta) \ \mathrm{W} \end{aligned} \tag{2.6.22}$$

$$\begin{aligned} \mathrm{Q_{3\phi}} &= \mathrm{Im}(S_{3\phi}) = 3\mathrm{V_{LL}I_\Delta}\sin(\delta - \beta) \\ &= \sqrt{3}\mathrm{V_{LL}I_L}\sin(\delta - \beta) \quad \mathrm{var} \end{aligned} \tag{2.6.23}$$

where the Δ-load current $\mathrm{I_\Delta}$ is expressed in terms of the line current $\mathrm{I_L} = \sqrt{3}\mathrm{I_\Delta}$ in (2.6.22) and (2.6.23). Also, the total apparent power is

$$\mathrm{S_{3\phi}} = |S_{3\phi}| = 3\mathrm{V_{LL}I_\Delta} = \sqrt{3}\mathrm{V_{LL}I_L} \ \ \mathrm{VA} \tag{2.6.24}$$

Equations (2.6.21) through (2.6.24) developed for the balanced-Δ load are identical to (2.6.13) through (2.6.16).

EXAMPLE 2.5

Power in a balanced three-phase system

Two balanced three-phase motors in parallel, an induction motor drawing 400 kW at 0.8 power factor lagging and a synchronous motor drawing 150 kVA at 0.9 power factor leading, are supplied by a balanced, three-phase 4160-volt source. Cable impedances between the source and load are neglected. (a) Draw the power triangle for each motor and for the combined-motor load. (b) Determine the power factor of the combined-motor load. (c) Determine the magnitude of the line current delivered by the source. (d) A delta-connected capacitor bank is now installed in parallel with the combined-motor load. What value of capacitive reactance is required in each leg of the capacitor bank to make the source power factor unity? (e) Determine the magnitude of the line current delivered by the source with the capacitor bank installed.

SOLUTION

For the induction motor, P = 400 kW and

$S = P/\text{p.f.} = 400/0.8 = 500\ \text{kVA}$

$Q = \sqrt{S^2 - P^2} = \sqrt{(500)^2 - (400)^2} = 300\ \ \text{kvar absorbed}$

For the synchronous motor, S = 150 kVA and

$P = S(\text{p.f.}) = 150(0.9) = 135\ \text{kW}$

$Q = \sqrt{S^2 - P^2} = \sqrt{(150)^2 - (135)^2} = 65.4\ \ \text{kvar delivered}$

For the combined-motor load,

$P = 400 + 135 = 535\ \text{kW}$

$Q = 300 - 65.4 = 234.6\ \text{kvar absorbed}$

$S = \sqrt{P^2 + Q^2} = \sqrt{(535)^2 + (234.6)^2} = 584.2\ \ \text{kVA}$

a. The power triangles for each motor and the combined-motor load are shown in Figure 2.20.

b. The power factor of the combined-motor load is p.f. = P/S = 535/ 584.2 = 0.916 lagging.

c. The line current delivered by the source is $I = S/(\sqrt{3}\,V)$, where S is the three-phase apparent power of the combined-motor load and V is the magnitude of the line-to-line load voltage, which is the same as the source voltage for this example. $I = 584.2/(\sqrt{3}\,4160\ \text{V}) = 0.0811\ \text{kA} = 81.1$ per phase.

d. For unity power factor, the three-phase reactive power supplied by the capacitor bank should equal the three-phase reactive power absorbed by the combined-motor load. That is, $Q_c = 234.6$ kvar. For a Δ-connected capacitor bank, $Q_c = 3V^2/X_\Delta$ where V is the line-to-line voltage across the bank and X_Δ the capacitive reactance of each leg of the bank. The capacitive reactance of each leg is

$$X_\Delta = 3V^2/Q_c = 3(4160^2)/234.6 \times 10^3 = 221.3\ \Omega.$$

FIGURE 2.20

Power triangles for Example 2.5

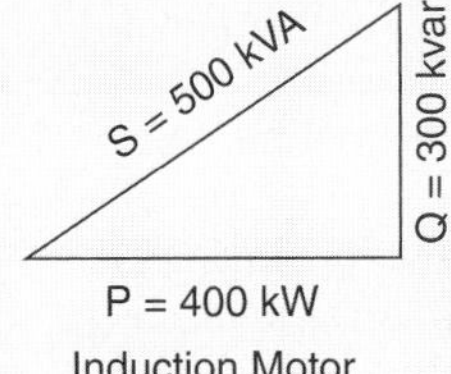

P = 135 kW

S = 150 kVA

Q = 65.4 kvar

Synchronous Motor

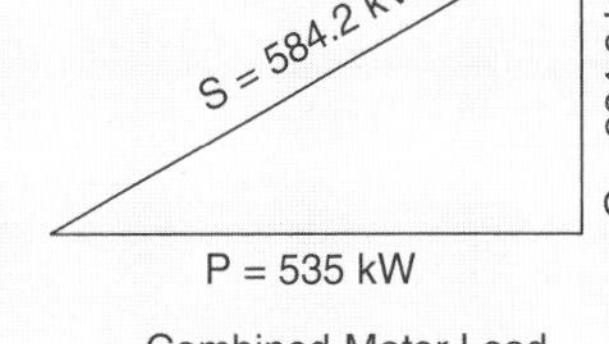

(Continued)

e. With the capacitor bank installed, the source power factor is unity, and the apparent power S delivered by the source is the same as the real power P delivered by the source. The line current magnitude is

$$I = S/(\sqrt{3}\,V) = P/(\sqrt{3}V) = 535/(\sqrt{3}\,4160) = 0.0743\text{kA} = 74.3\text{ A per phase.}$$

In this example, the source voltage of 4160 V is not specified as a line-to-line voltage or line-to-neutral voltage, RMS or peak. Therefore, it is assumed to be an RMS line-to-line voltage, which is the convention throughout this text and a standard practice in the electric power industry. The combined-motor load absorbs 535 kW of real power. The induction motor, which operates at lagging power factor, absorbs reactive power (300 kvar) and the synchronous motor, which operates at leading power factor, delivers reactive power (65.4 kvar). The capacitor bank also delivers reactive power (234.6 kvar). Note that the line current delivered by the source is reduced from 81.1 A without the capacitor bank to 74.3 A with the capacitor bank. Any I^2R losses due to cable resistances and voltage drops due to cable reactances between the source and loads (not included in this example) would also be reduced.

2.7 ADVANTAGES OF BALANCED THREE-PHASE VERSUS SINGLE-PHASE SYSTEMS

Figure 2.21 shows three separate single-phase systems. Each single-phase system consists of the following identical components: (1) a generator represented by a voltage source and a generator impedance Z_g; (2) a forward and return conductor represented by two series line impedances Z_L; (3) a load represented by an impedance Z_Y. The three single-phase systems, although completely separated, are drawn in a Y configuration in the figure to illustrate two advantages of three-phase systems.

Each separate single-phase system requires that *both* the forward and return conductors have a current capacity (or *ampacity*) equal to or greater than the load current. However, if the source and load neutrals in Figure 2.21 are connected to form a three-phase system, and if the source voltages are balanced with equal magnitudes and with 120° displacement between phases, then the neutral current will be zero [see (2.5.10)] and the three neutral conductors can be removed. Thus, the balanced three-phase system, while delivering the same power to the three load impedances Z_Y, requires only half the number of conductors needed for the three separate single-phase systems. Also, the total I^2R line losses in the three-phase system are only half those of the three separate single-phase systems, and the line-voltage drop between the source and load in the three-phase system is half that of each single-phase system. Therefore, one advantage of balanced three-phase systems over separate single-phase systems is reduced capital and operating costs of transmission and distribution, as well as better voltage regulation.

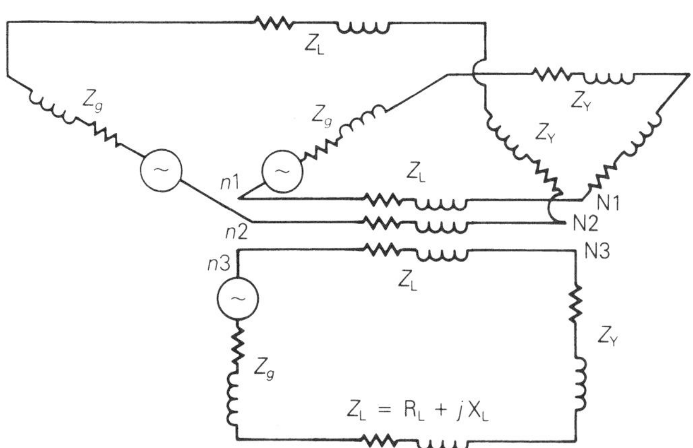

FIGURE 2.21

Three single-phase systems

Some three-phase systems such as Δ-connected systems and three-wire Y-connected systems do not have any neutral conductor. However, the majority of three-phase systems are four-wire Y-connected systems, where a grounded neutral conductor is used. Neutral conductors are used to reduce transient overvoltages, which can be caused by lightning strikes and by line-switching operations, and to carry unbalanced currents, which can occur during unsymmetrical short-circuit conditions. Neutral conductors for transmission lines are typically smaller in size and ampacity than the phase conductors because the neutral current is nearly zero under normal operating conditions. Thus, the cost of a neutral conductor is substantially less than that of a phase conductor. The capital and operating costs of three-phase transmission and distribution systems with or without neutral conductors are substantially less than those of separate single-phase systems.

A second advantage of three-phase systems is that the total instantaneous electric power delivered by a three-phase generator under balanced steady-state conditions is (nearly) constant, as shown in Section 2.6. A three-phase generator (constructed with its field winding on one shaft and with its three-phase windings equally displaced by 120° on the stator core) will also have a nearly constant mechanical input power under balanced steady-state conditions, since the mechanical input power equals the electrical output power plus the small generator losses. Furthermore, the mechanical shaft torque, which equals mechanical input power divided by mechanical radian frequency ($T_{mech} = P_{mech}/\omega_m$) is nearly constant.

On the other hand, the equation for the instantaneous electric power delivered by a single-phase generator under balanced steady-state conditions is the same as the instantaneous power delivered by one phase of a three-phase generator, given by $p_a(t)$ in (2.6.3). As shown in that equation, $p_a(t)$ has two components: a constant and a double-frequency sinusoid. Both the mechanical input power and the mechanical shaft torque of the single-phase generator will have corresponding double-frequency components that create shaft vibration and noise, which could cause shaft failure in large machines. Accordingly, most electric generators and motors rated 5 kVA and higher are constructed as three-phase machines in order to produce nearly constant torque and thereby minimize shaft vibration and noise.

MULTIPLE CHOICE QUESTIONS

SECTION 2.1

2.1 The rms value of $v(t) = V_{max} \cos(\omega t + \delta)$ is given by
(a) V_{max} (b) $V_{max}/\sqrt{2}$ (c) $2V_{max}$ (d) $\sqrt{2}V_{max}$

2.2 If the rms phasor of a voltage is given by $V = 120\angle 60°$ volts, then the corresponding $v(t)$ is given by
(a) $120\sqrt{2}\cos(\omega t + 60°)$
(b) $120\cos(\omega t + 60°)$
(c) $120\sqrt{2}\sin(\omega r + 60°)$

2.3 If a phasor representation of a current is given by $I = 70.7\angle 45°$ A, it is equivalent to
(a) $100\, e^{j45°}$ (b) $100 + j100$ (c) $50 + j50$

2.4 With sinusoidal-steady-state excitation, for a purely resistive circuit, the voltage and current phasors are
(a) In phase
(b) Perpendicular with each other with V leading I
(c) Perpendicular with each other with I leading V.

2.5 For a purely inductive circuit, with sinusoidal-steady-state excitation, the voltage and current phasors are
(a) In phase
(b) Perpendicular to each other with V leading I
(c) Perpendicular to each other with I leading V.

2.6 For a purely capacitive circuit, with sinusoidal-steady-state excitation, the voltage and current phasors are
(a) In phase
(b) Perpendicular to each other with V leading I
(c) Perpendicular to each other with I leading V.

SECTION 2.2

2.7 With sinusoidal-steady-state excitation, the average power in a single-phase ac circuit with a purely resistive load is given by
(a) $I^2_{rms}R$ (b) V^2_{max}/R (c) Zero

2.8 The average power in a single-phase ac circuit with a purely inductive load, for sinusoidal-steady-state excitation, is
(a) $I^2 X_L$ (b) V^2_{max}/X_L (c) Zero
[Note: $X_L = (\omega L)$ is the inductive reactance]

2.9 The average power in a single-phase ac circuit with a purely capacitive load, for sinusoidal-steady-state excitation, is
(a) Zero (b) V^2_{max}/X_C (c) $I^2_{rms}X_C$
[Note: $X_C = 1/(\omega C)$ is the capacitive reactance]

2.10 The average value of a double-frequency sinusoid, sin $2(\omega t + \delta)$, is given by
(a) 1 (b) δ (c) Zero

2.11 The power factor for an inductive circuit (R-L load), in which the current lags the voltage, is said to be
(a) Lagging (b) Leading (c) Zero

2.12 The power factor for a capacitive circuit (R-C load), in which the current leads the voltage, is said to be
(a) Lagging (b) Leading (c) One

SECTION 2.3

2.13 In a single-phase ac circuit, for a general load composed of RLC elements under sinusoidal-steady-state excitation, the average reactive power is given by
(a) $V_{rms} I_{rms} \cos \phi$ (b) $V_{rms} I_{rms} \sin \phi$ (c) Zero
[Note: ϕ is the power-factor angle]

2.14 The instantaneous power absorbed by the load in a single-phase ac circuit, for a general RLC load under sinusoidal-steady-state excitation, is
(a) Nonzero constant (b) Zero
(c) Containing double-frequency components

2.15 With load convention, where the current enters the positive terminal of the circuit element, if Q is positive then positive reactive power is absorbed.
(a) True (b) False

2.16 With generator convention, where the current leaves the positive terminal of the circuit element, if P is positive then positive real power is delivered.
(a) False (b) True

2.17 Consider the load convention that is used for the RLC elements shown in Figure 2.2 of the text.
A. If one says that an inductor absorbs zero real power and positive reactive power, is it
(a) True (b) False
B. If one says that a capacitor absorbs zero real power and negative reactive power (or delivers positive reactive power), is it
(a) False (b) True
C. If one says that a (positive-valued) resistor absorbs (positive) real power and zero reactive power, is it
(a) True (b) False

2.18 In an ac circuit, power factor improvement is achieved by
(a) Connecting a resistor in parallel with the inductive load.
(b) Connecting an inductor in parallel with the inductive load.
(c) Connecting a capacitor in parallel with the inductive load.

SECTION 2.4

2.19 The admittance of the impedance $-j\frac{1}{2}\,\Omega$ is given by

(a) $-j2S$ (b) $j2S^2$ (c) $-j4S$

2.20 Consider Figure 2.9 of the text. Let the nodal equations in matrix form be given by Eq. (2.4.1) of the text.

A. The element Y_{11} is given by

(a) 0 (b) $j13$ (c) $-j7$

B. The element Y_{31} is given by

(a) 0 (b) $-j5$ (c) $j1$

C. The admittance matrix is always symmetric square.

(a) False (b) True

SECTIONS 2.5 AND 2.6

2.21 The three-phase source line-to-neutral voltages are given by $E_{an} = 10\angle 0°$, $E_{bn} = 10\angle +240°$, and $E_{cn} = 10\angle -240°$ volts.

Is the source balanced?

(a) Yes (b) No

2.22 In a balanced three-phase Y-connected system with a positive-sequence source, the line-to-line voltages are $\sqrt{3}$ times the line-to-neutral voltages and lend by 30°.

(a) True (b) False

2.23 In a balanced system, the phasor sum of the line-to-line voltages and the phasor sum of the line-to-neutral voltages are always equal to zero.

(a) False (b) True

2.24 Consider a three-phase Y-connected source feeding a balanced-Δ load. The phasor sum of the line currents as well as the neutral current are always zero.

(a) True (b) False

2.25 For a balanced-Δ load supplied by a balanced positive-sequence source, the line currents into the load are $\sqrt{3}$ times the Δ-load currents and lag by 30°.

(a) True (b) False

2.26 A balanced Δ-load can be converted to an equivalent balanced-Y load by dividing the Δ-load impedance by

(a) $\sqrt{3}$ (b) 3 (c) 1/3

2.27 When working with balanced three-phase circuits, per-phase analysis is commonly done after converting Δ loads to Y loads, thereby solving only one phase of the circuit.

(a) True (b) False

2.28 The total instantaneous power delivered by a three-phase generator under balanced operating conditions is

(a) A function of time (b) A constant

2.29 The total instantaneous power absorbed by a three-phase motor (under balanced steady-state conditions) as well as a balanced three-phase impedance load is
(a) A constant (b) A function of time

2.30 Under balanced operating conditions, consider the three-phase complex power delivered by the three-phase source to the three-phase load. Match the following expressions, those on the left to those on the right.

(i) Real power, $P_{3\phi}$	(a) $(\sqrt{3}\ V_{LL}\ I_L)$VA
(ii) Reactive power, $Q_{3\phi}$	(b) $(\sqrt{3}\ V_{LL}\ I_L \sin\phi)$ var
(iii) Total apparent power, $S_{3\phi}$	(c) $(\sqrt{3}\ V_{LL}\ I_L \cos\phi)$ W
(iv) Complex power, $S_{3\phi}$	(d) $P_{3\phi} + jQ_{3\phi}$

Note that V_{LL} is the rms line-to-line voltage, I_L is the rms line current, and ϕ is the power-factor angle.

2.31 One advantage of balanced three-phase systems over separate single-phase systems is reduced capital and operating costs of transmission and distribution.
(a) True (b) False

2.32 While the instantaneous electric power delivered by a single-phase generator under balanced steady-state conditions is a function of time having two components of a constant and a double-frequency sinusoid, the total instantaneous electric power delivered by a three-phase generator under balanced steady-state conditions is a constant.
(a) True (b) False

PROBLEMS

SECTION 2.1

2.1 Given the complex numbers $A_1 = 6\angle 30$ and $A_2 = 4 + j5$, (a) convert A_1 to rectangular form; (b) convert A_2 to polar and exponential form; (c) calculate $A_3 = (A_1 + A_2)$, giving your answer in polar form; (d) calculate $A_4 = A_1A_2$, giving your answer in rectangular form; (e) calculate $A_5 = A_1/(A_2^*)$, giving your answer in exponential form.

2.2 Convert the following instantaneous currents to phasors, using $\cos(\omega t)$ as the reference. Give your answers in both rectangular and polar form.
(a) $i(t) = 500\sqrt{2}\cos(\omega t - 30)$
(b) $i(t) = 4\sin(\omega t + 30)$
(c) $i(t) = 5\cos(\omega t - 15) + 4\sqrt{2}\sin(\omega t + 30)$

2.3 The instantaneous voltage across a circuit element is $v(t) = 400\sin(\omega t + 30°)$ volts, and the instantaneous current entering the positive terminal of the circuit element is $i(t) = 100 \cos(\omega t + 10°)$ A. For both the current and voltage, determine (a) the maximum value, (b) the rms value, and (c) the phasor expression, using $\cos(\omega t)$ as the reference.

2.4 For the single-phase circuit shown in Figure 2.22, $I = 10\angle 0°$A. (a) Compute the phasors I_1, I_2, and V. (b) Draw a phasor diagram showing I, I_1, I_2, and V.

FIGURE 2.22

Circuit for Problem 2.4

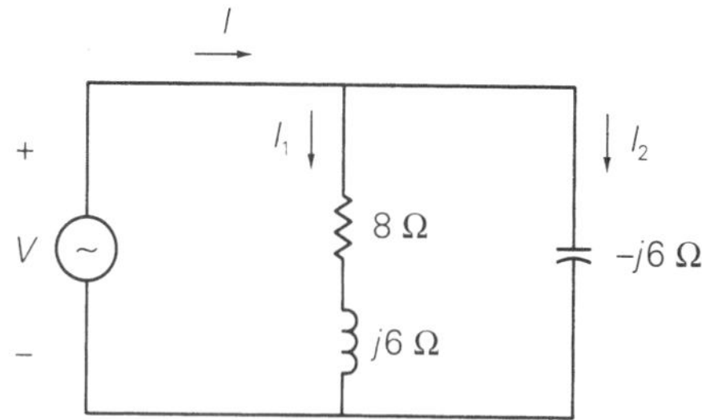

2.5 A 60-Hz, single-phase source with $V = 277\angle 30°$ volts is applied to a circuit element. (a) Determine the instantaneous source voltage. Also determine the phasor and instantaneous currents entering the positive terminal if the circuit element is (b) a 20-Ω resistor, (c) a 10-mH inductor, and (d) a capacitor with 25-Ω reactance.

2.6 (a) Transform $v(t) = 75 \cos(377t - 15°)$ to phasor form. Comment on whether $\omega = 377$ appears in your answer. (b) Transform $V = 50\angle 10°$ to instantaneous form. Assume that $\omega = 377$. (c) Add the two sinusoidal functions $a(t)$ and $b(t)$ of the same frequency given as follows: $a(t) = A\sqrt{2}\cos(\omega t + \alpha)$ and $b(t) = B\sqrt{2}\cos(\omega t + \beta)$. Use phasor methods and obtain the resultant $c(t)$. Does the resultant have the same frequency?

2.7 Let a 100-V sinusoidal source be connected to a series combination of a 3-Ω resistor, an 8-Ω inductor, and a 4-Ω capacitor. (a) Draw the circuit diagram. (b) Compute the series impedance. (c) Determine the current I delivered by the source. Is the current lagging or leading the source voltage? What is the power factor of this circuit?

2.8 Consider the circuit shown in Figure 2.23 in time domain. Convert the entire circuit into phasor domain.

FIGURE 2.23

Circuit for Problem 2.8

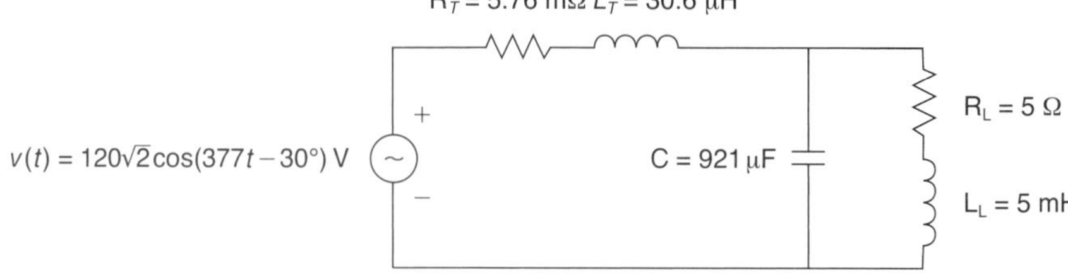

2.9 For the circuit shown in Figure 2.24, compute the voltage across the load terminals.

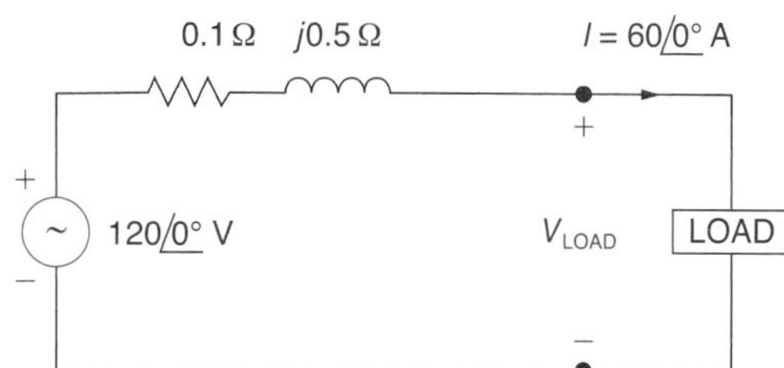

FIGURE 2.24

Circuit for Problem 2.9

SECTION 2.2

2.10 For the circuit element of Problem 2.3, calculate (a) the instantaneous power absorbed, (b) the real power (state whether it is delivered or absorbed), (c) the reactive power (state whether delivered or absorbed), (d) the power factor (state whether lagging or leading).
[*Note:* By convention the power factor $\cos(\delta - \beta)$ is positive. If $|\delta - \beta|$ is greater than 90°, then the reference direction for current may be reversed, resulting in a positive value of $\cos(\delta - \beta)$].

2.11 Referring to Problem 2.5, determine the instantaneous power, real power, and reactive power absorbed by (a) the 20-Ω resistor, (b) the 10-mH inductor, (c) the capacitor with 25-Ω reactance. Also determine the source power factor and state whether lagging or leading.

2.12 The voltage $v(t) = 359.3 \cos(\omega t)$ volts is applied to a load consisting of a 10- resistor in parallel with a capacitive reactance $X_C = 25\ \Omega$. Calculate (a) the instantaneous power absorbed by the resistor, (b) the instantaneous power absorbed by the capacitor, (c) the real power absorbed by the resistor, (d) the reactive power delivered by the capacitor, and (e) the load power factor.

2.13 Repeat Problem 2.12 if the resistor and capacitor are connected in series.

2.14 A single-phase source is applied to a two-terminal, passive circuit with equivalent impedance $Z = 3.0\underline{/-45°}\ \Omega$, measured from the terminals. The source current is $i(t) = 2\sqrt{2}\cos(\omega t)$ kA. Determine the (a) instantaneous power, (b) real power, (c) reactive power delivered by the source, and (d) source power factor.

2.15 Let a voltage source $v(t) = 4\cos(\omega t + 60°)$ be connected to an impedance $Z = 2\underline{/30°}\ \Omega$. (a) Given the operating frequency to be 60 Hz, determine the expressions for the current and instantaneous power delivered by the source as functions of time. (b) Plot these functions along with $v(t)$ on a single graph for comparison. (c) Find the frequency and average value of the instantaneous power.

2.16 A single-phase, 120-V (rms), 60-Hz source supplies power to a series R-L circuit consisting of R = 10 Ω and L = 40 mH. (a) Determine the power factor of the circuit and state whether it is lagging or leading. (b) Determine the real and reactive power absorbed by the load. (c) Calculate the peak magnetic energy W_{int} stored in the inductor by using the expression $W_{int} = L(I_{rms})^2$ and check whether the reactive power $Q = \omega W_{int}$ is satisfied. (*Note:* The instantaneous magnetic energy storage fluctuates between zero and the peak energy. This energy must be sent twice each cycle to the load from the source by means of reactive power flows.)

SECTION 2.3

2.17 Consider a load impedance of $Z = j\omega L$ connected to a voltage V and let the current drawn be I.
(a) Develop an expression for the reactive power Q in terms of ω, L, and I, from complex power considerations.
(b) Let the instantaneous current be $i(t) = \sqrt{2}I\cos(\omega t + \theta)$. Obtain an expression for the instantaneous power $p(t)$ into L, and then express it in terms of Q.
(c) Comment on the average real power P supplied to the inductor and the instantaneous power supplied.

2.18 Let a series RLC network be connected to a source voltage V, drawing a current I.
(a) In terms of the load impedance $Z = Z\,\underline{/Z}$, find expressions for P and Q, from complex power considerations.
(b) Express $p(t)$ in terms of P and Q, by choosing $i(t) = \sqrt{2}I\cos\omega t$.
(c) For the case of $Z = R + j\omega L + 1/j\omega C$, interpret the result of part (b) in terms of P, Q_L, and Q_C. In particular, if $\omega^2 LC = 1$, when the inductive and capacitive reactances cancel, comment on what happens.

2.19 Consider a single-phase load with an applied voltage $v(t) = 150\cos(\omega t + 10°)$ volts and load current $i(t) = 5\cos(\omega t - 50°)$ A. (a) Determine the power triangle. (b) Find the power factor and specify whether it is lagging or leading. (c) Calculate the reactive power supplied by capacitors in parallel with the load that correct the power factor to 0.9 lagging.

2.20 A circuit consists of two impedances, $Z_1 = 20\underline{/30°}\,\Omega$ and $Z_2 = 25\underline{/60°}\,\Omega$, in parallel, supplied by a source voltage $V = 100\underline{/60°}$ volts. Determine the power triangle for each of the impedances and for the source.

2.21 An industrial plant consisting primarily of induction motor loads absorbs 500 kW at 0.6 power factor lagging. (a) Compute the required kVA rating of a shunt capacitor to improve the power factor to 0.9 lagging. (b) Calculate the resulting power factor if a synchronous motor rated at 375 kW with 90% efficiency operating at rated load and at unity power factor is added to the plant instead of the capacitor. Assume constant voltage.

2.22 The real power delivered by a source to two impedances, $Z_1 = 4 + j5\ \Omega$ and $Z_2 = 10\ \Omega$, connected in parallel, is 1000 W. Determine (a) the real power absorbed by each of the impedances and (b) the source current.

2.23 A single-phase source has a terminal voltage $V = 120\underline{/0^\circ}$ volts and a current $I = 15\underline{/30^\circ}$ A, which leaves the positive terminal of the source. Determine the real and reactive power, and state whether the source is delivering or absorbing each.

2.24 A source supplies power to the following three loads connected in parallel: (1) a lighting load drawing 10 kW, (2) an induction motor drawing 10 kVA at 0.90 power factor lagging, and (3) a synchronous motor operating at 7.5 kW, 85% efficiency and 0.95 power factor leading. Determine the real, reactive, and apparent power delivered by the source. Also, draw the source power triangle.

2.25 Consider the series RLC circuit of Problem 2.7 and calculate the complex power absorbed by each of the R, L, and C elements, as well as the complex power absorbed by the total load. Draw the resultant power triangle. Check whether the complex power delivered by the source equals the total complex power absorbed by the load.

2.26 A small manufacturing plant is located 2 km down a transmission line, which has a series reactance of 0.5 Ω/km. The line resistance is negligible. The line voltage at the plant is $480\underline{/0^\circ}$ V (rms), and the plant consumes 120 kW at 0.85 power factor lagging. Determine the voltage and power factor at the sending end of the transmission line by using (a) a complex power approach and (b) a circuit analysis approach.

2.27 An industrial load consisting of a bank of induction motors consumes 50 kW at a power factor of 0.8 lagging from a 220-V, 60-Hz, single-phase source. By placing a bank of capacitors in parallel with the load, the resultant power factor is to be raised to 0.95 lagging. Find the net capacitance of the capacitor bank in μF that is required.

2.28 Three loads are connected in parallel across a single-phase source voltage of 240 V (RMS).
Load 1 absorbs 15 kW and 6.667 kvar;
Load 2 absorbs 3 kVA at 0.96PF leading;
Load 3 absorbs 15 kW at unity power factor.
Calculate the equivalent impedance, Z, for the three parallel loads, for two cases:
(i) Series combination of R and X, and (ii) parallel combination of R and X.

2.29 Modeling the transmission lines as inductors, with $S_{ij} = S_{ji}^*$, Compute S_{13}, S_{31}, S_{23}, S_{32}, and S_{G3} in Figure 2.25. (*Hint:* complex power balance holds good at each bus, satisfying KCL.)

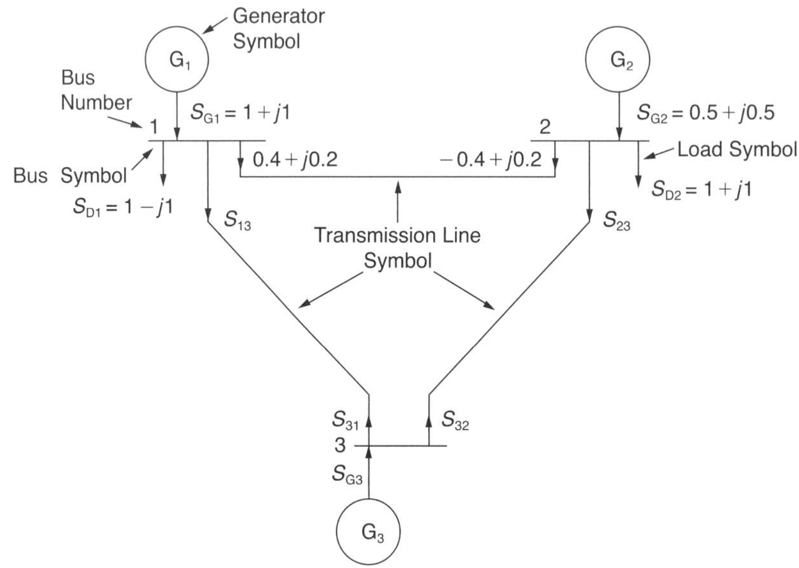

FIGURE 2.25 System diagram for Problem 2.29

2.30 Figure 2.26 shows three loads connected in parallel across a 1000-V (RMS), 60-Hz single-phase source.

Load 1: Inductive load, 125 kVA, 0.28PF lagging.

Load 2: Capacitive load, 10 kW, 40 kvar.

Load 3: Resistive load, 15 kW.

(a) Determine the total kW, kvar, kva, and supply power factor.

(b) In order to improve the power factor to 0.8 lagging, a capacitor of negligible resistance is connected in parallel with the above loads. Find the kvar rating of that capacitor and the capacitance in μF.

Comment on the magnitude of the supply current after adding the capacitor.

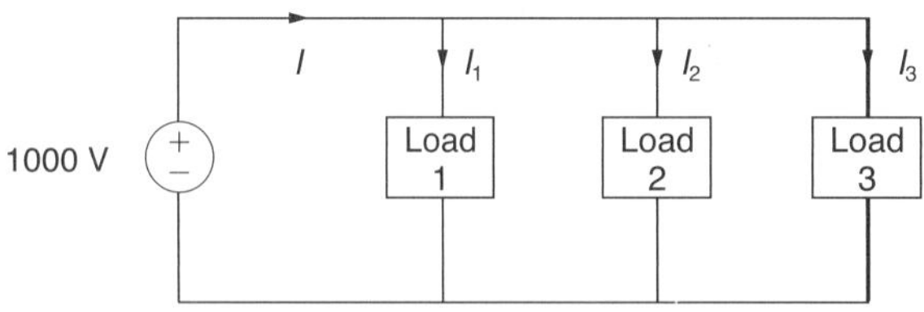

FIGURE 2.26 Circuit for Problem 2.30

2.31 Consider two interconnected voltage sources connected by a line of impedance $Z = j\text{X}\ \Omega$, as shown in Figure 2.27.

(a) Obtain expressions for P_{12} and Q_{12}.

(b) Determine the maximum power transfer and the condition for it to occur.

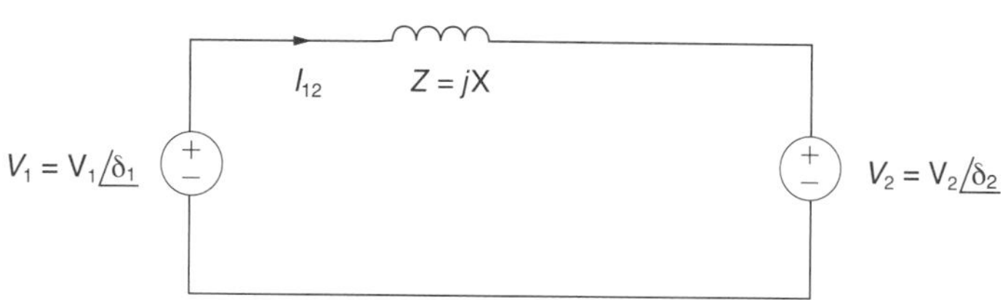

FIGURE 2.27 Circuit for Problem 2.31

2.32 In PowerWorld Simulator case Problem 2_32 (see Figure 2.28) a 8 MW and 4 Mvar load is supplied at 13.8 kV through a feeder with an impedance of $1 + j2\ \Omega$. The load is compensated with a capacitor whose output, Ω_{cap}, can be varied in 0.5 Mvar steps between 0 and 10.0 Mvars. What value of Ω_{cap} minimizes the real power line losses? What value of Ω_{cap} minimizes the MVA power flow into the feeder?

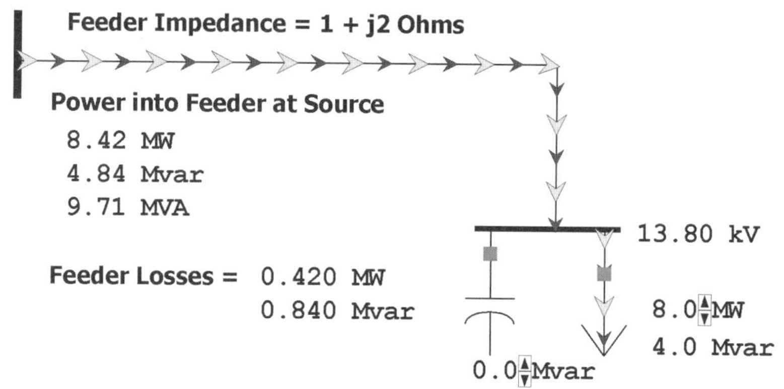

FIGURE 2.28 Screen for Problem 2.32

2.33 For the system from Problem 2.32, plot the real and reactive line losses as Ω_{cap} is varied between 0 and 10.0 Mvars.

2.34 For the system from Problem 2.32, assume that half the time the load is 10 MW and 5 Mvar, and for the other half it is 20 MW and 10 Mvar. What single value of Q_{cap} would minimize the average losses? Assume that Q_{cap} can only be varied in 0.5 Mvar steps.

SECTION 2.4

2.35 For the circuit shown in Figure 2.29, convert the voltage sources to equivalent current sources and write nodal equations in matrix format using bus 0 as the reference bus. Do not solve the equations.

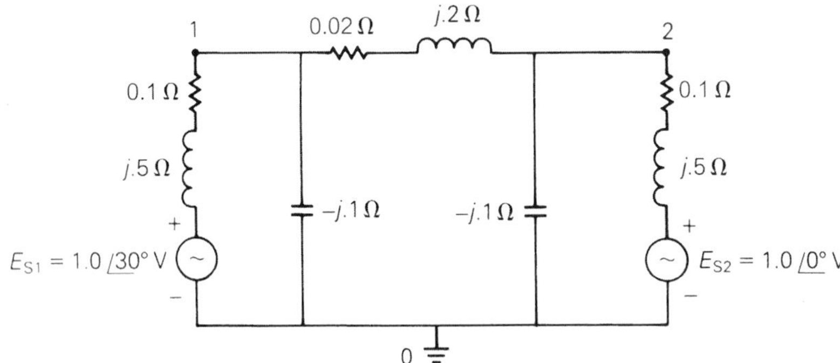

FIGURE 2.29 Circuit diagram for Problems 2.35 and 2.36

2.36 For the circuit shown in Figure 2.29, (a) determine the 2 × 2 bus admittance matrix $\boldsymbol{Y}_{\text{bus}}$, (b) convert the voltage sources to current sources and determine the vector of source currents into buses 1 and 2.

2.37 Determine the 4 × 4 bus admittance matrix $\boldsymbol{Y}_{\text{bus}}$ and write nodal equations in matrix format for the circuit shown in Figure 2.30. Do not solve the equations.

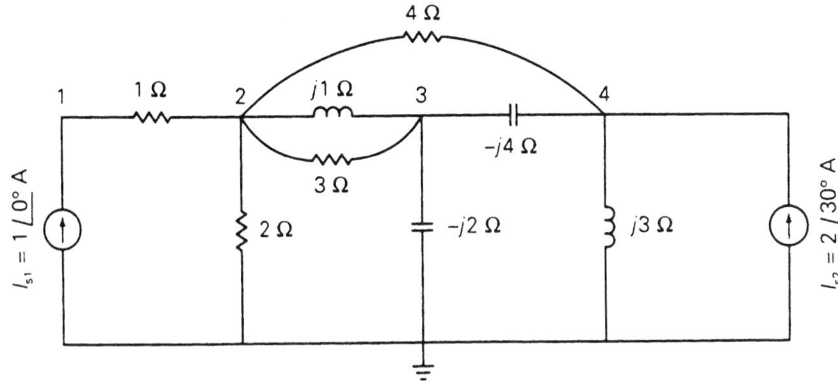

FIGURE 2.30 Circuit for Problem 2.37

2.38 Given the impedance diagram of a simple system as shown in Figure 2.31, draw the admittance diagram for the system and develop the 4 × 4 bus admittance matrix $\boldsymbol{Y}_{\text{bus}}$ by inspection.

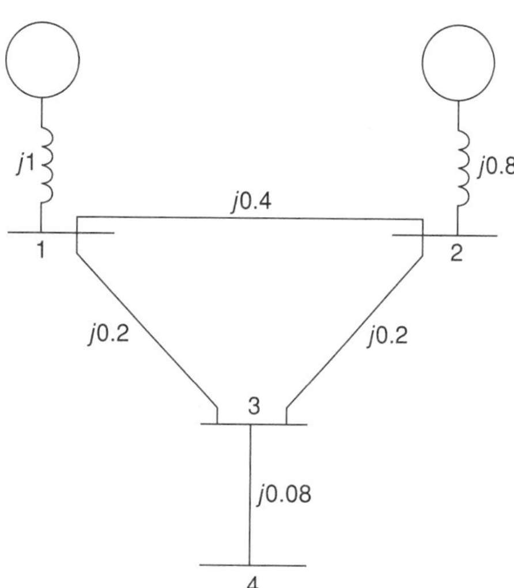

FIGURE 2.31

System diagram for Problem 2.38

2.39 (a) Given the circuit diagram in Figure 2.32 showing admittances and current sources at nodes 3 and 4, set up the nodal equations in matrix format. (b) If the parameters are given by: $Y_a = -j0.8$ S, $Y_b = -j4.0$ S, $Y_c = -j4.0$ S, $Y_d = -j8.0$ S, $Y_e = -j5.0$ S, $Y_f = -j2.5$ S, $Y_g = -j0.8$ S, $I_3 = 1.0\underline{/-90^\circ}$ A, and $I_4 = 0.62\underline{/-135^\circ}$ A, set up the nodal equations and suggest how you would go about solving for the voltages at the nodes.

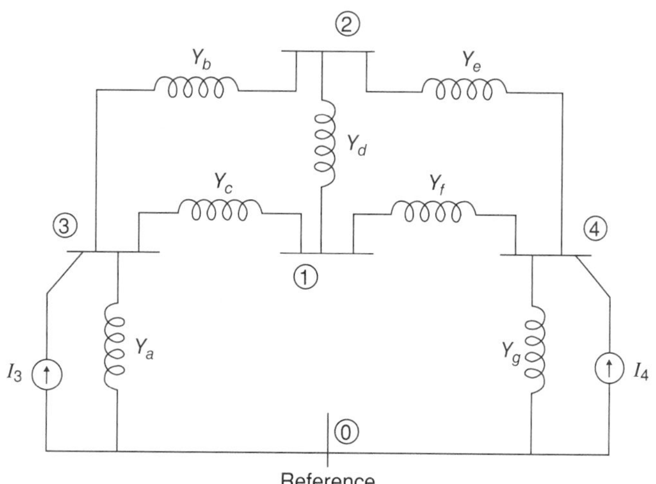

FIGURE 2.32

Circuit diagram for Problem 2.39

SECTIONS 2.5 AND 2.6

2.40 A balanced three-phase 240-V source supplies a balanced three-phase load. If the line current I_A is measured to be 15 A and is in phase with the line-to-line voltage, V_{BC}, find the per-phase load impedance if the load is (a) Y-connected, (b) Δ-connected.

2.41 A three-phase 25-kVA, 480-V, 60-Hz alternator, operating under balanced steady-state conditions, supplies a line current of 20 A per phase at a 0.8 lagging power factor and at rated voltage. Determine the power triangle for this operating condition.

2.42 A balanced Δ-connected impedance load with $(12 + j9)\ \Omega$ per phase is supplied by a balanced three-phase 60-Hz, 208-V source, (a) Calculate the line current, the total real and reactive power absorbed by the load, the load power factor, and the apparent load power, (b) Sketch a phasor diagram showing the line currents, the line-to-line source voltages, and the Δ-load currents. Use V_{ab} as the reference.

2.43 A three-phase line, which has an impedance of $(2 + j4)\ \Omega$ per phase, feeds two balanced three-phase loads that are connected in parallel. One of the loads is Y-connected with an impedance of $(30 + j40)\ \Omega$ per phase, and the other is Δ-connected with an impedance of $(60 - j45)\ \Omega$ per phase. The line is energized at the sending end from a 60-Hz, three-phase, balanced voltage source of $120\sqrt{3}$ V (rms, line-to-line). Determine (a) the current, real power, and reactive power delivered by the sending-end source; (b) the line-to-line voltage at the load; (c) the current per phase in each load; and (d) the total three-phase real and reactive powers absorbed by each load and by the line. Check that the total three-phase complex power delivered by the source equals the total three-phase power absorbed by the line and loads.

2.44 Two balanced three-phase loads that are connected in parallel are fed by a three-phase line having a series impedance of $(0.4 + j2.7)\ \Omega$ per phase. One of the loads absorbs 560 kVA at 0.707 power factor lagging, and the other 132 kW at unity power factor. The line-to-line voltage at the load end of the line is $2200\sqrt{3}$ V. Compute (a) the line-to-line voltage at the source end of the line, (b) the total real and reactive power losses in the three-phase line, and (c) the total three-phase real and reactive power supplied at the sending end of the line. Check that the total three-phase complex power delivered by the source equals the total three-phase complex power absorbed by the line and loads.

2.45 Two balanced Y-connected loads, one drawing 10 kW at 0.8 power factor lagging and the other 15 kW at 0.9 power factor leading, are connected in parallel and supplied by a balanced three-phase Y-connected, 480-V source. (a) Determine the source current. (b) If the load neutrals are connected to the source neutral by a zero-ohm neutral wire through an ammeter, what will the ammeter read?

2.46 Three identical impedances $Z_\Delta = 30\angle 30^\circ\Omega$ are connected in Δ to a balanced three-phase 208-V source by three identical line conductors with impedance $Z_L = (0.8 + j0.6)\ \Omega$ per line. (a) Calculate the line-to-line voltage at the load terminals. (b) Repeat part (a) when a Δ-connected capacitor bank with reactance $(-j60)\ \Omega$ per phase is connected in parallel with the load.

2.47 Two three-phase generators supply a three-phase load through separate three-phase lines. The load absorbs 30 kW at 0.8 power factor lagging. The line impedance is $(1.4 + j1.6)\ \Omega$ per phase between generator G_1 and the load, and $(0.8 + j1)\ \Omega$ per phase between generator G_2 and the load. If generator G_1 supplies 15 kW at 0.8 power factor lagging, with a terminal voltage of 460 V line-to-line, determine (a) the voltage at the load terminals, (b) the voltage at the terminals of generator G_2, and (c) the real and reactive power supplied by generator G_2. Assume balanced operation.

2.48 Two balanced Y-connected loads in parallel, one drawing 15 kW at 0.6 power factor lagging and the other drawing 10 kVA at 0.8 power factor leading, are supplied by a balanced, three-phase, 480-volt source. (a) Draw the power triangle for each load and for the combined load. (b) Determine the power factor of the combined load and state whether lagging or leading. (c) Determine the magnitude of the line current from the source. (d) Δ-connected capacitors are now installed in parallel with the combined load. What value of capacitive reactance is needed in each leg of the A to make the source power factor unity? Give your answer in Ω. (e) Compute the magnitude of the current in each capacitor and the line current from the source.

2.49 Figure 2.33 gives the general Δ-Y transformation. (a) Show that the general transformation reduces to that given in Figure 2.16 for a balanced three-phase load. (b) Determine the impedances of the equivalent Y for the following Δ impedances: $Z_{AB} = j10$, $Z_{BC} = j20$, and $Z_{CA} = -j25\ \Omega$.

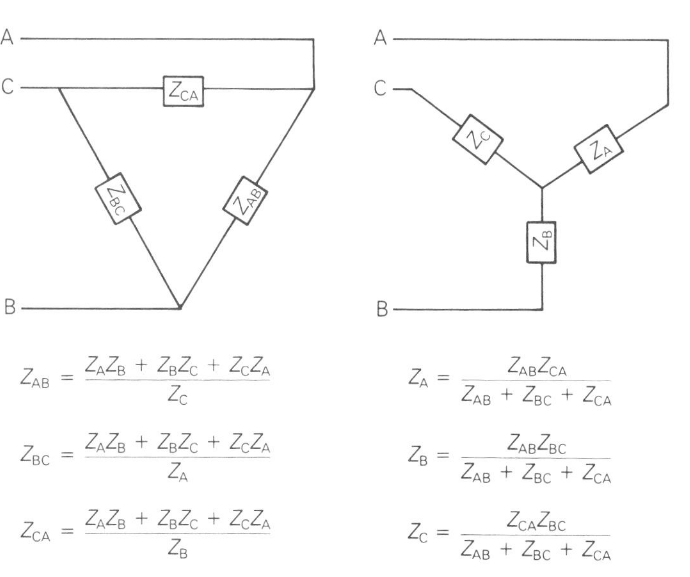

A
C
Z_{CA}
Z_{BC}
Z_{AB}
B
A
C
Z_C
Z_A
Z_B
B
$Z_{AB} = \frac{Z_A Z_B + Z_B Z_C + Z_C Z_A}{Z_C}$
$Z_{BC} = \frac{Z_A Z_B + Z_B Z_C + Z_C Z_A}{Z_A}$
$Z_{CA} = \frac{Z_A Z_B + Z_B Z_C + Z_C Z_A}{Z_B}$
$Z_A = \frac{Z_{AB} Z_{CA}}{Z_{AB} + Z_{BC} + Z_{CA}}$
$Z_B = \frac{Z_{AB} Z_{BC}}{Z_{AB} + Z_{BC} + Z_{CA}}$
$Z_C = \frac{Z_{CA} Z_{BC}}{Z_{AB} + Z_{BC} + Z_{CA}}$

FIGURE 2.33 General Δ-Y transformation

2.50 Consider the balanced three-phase system shown in Figure 2.34. Determine $v_1(t)$ and $i_2(t)$. Assume positive phase sequence.

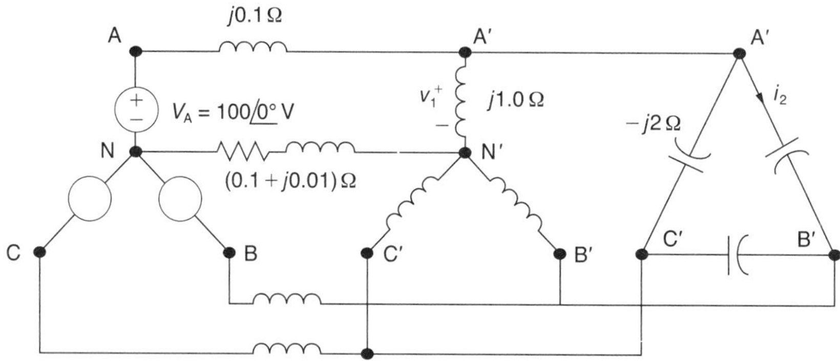

FIGURE 2.34 Circuit for Problem 2.50

2.51 A three-phase line with an impedance of $(0.2 + j1.0)\ \Omega$ /phase feeds three balanced three-phase loads connected in parallel.

Load 1: Absorbs a total of 150 kW and 120 kvar.

Load 2: Delta connected with an impedance of $(150 - j48)\ \Omega$ /phase.

Load 3: 120 kVA at 0.6 PF leading.

If the line-to-neutral voltage at the load end of the line is 2000 V (rms), determine the magnitude of the line-to-line voltage at the source end of the line.

2.52 A balanced three-phase load is connected to a 4.16-kV, three-phase, four-wire, grounded-Y dedicated distribution feeder. The load can be modeled by an impedance of $Z_L = (4.7 + j9)\ \Omega$ /phase, Y-connected. The impedance of the phase conductors is $(0.3 + j1)\ \Omega$. Determine the following by using the phase A to neutral voltage as a reference and assume positive phase sequence:

(a) Line currents for phases A, B, and C.

(b) Line-to-neutral voltages for all three phases at the load.

(c) Apparent, active, and reactive power dissipated per phase, and for all three phases in the load.

(d) Active power losses per phase and for all three phases in the phase conductors.

CASE STUDY QUESTIONS

a. What is a microgrid?

b. What are the benefits of microgrids?

c. What are the two primary goals of the U.S. Department of Energy's Smart Grid Research & Development Program?

d. Can smart grids defer transmission and distribution investments? If so, how?

REFERENCES

1. W. H. Hayt, Jr. and J. E. Kemmerly, *Engineering Circuit Analysis*, 7th ed. (New York: McGraw-Hill, 2006).
2. W. A. Blackwell and L. L. Grigsby, *Introductory Network Theory* (Boston: PWS, 1985).
3. A. E. Fitzgerald, D. E. Higginbotham, and A. Grabel, *Basic Electrical Engineering* (New York: McGraw-Hill, 1981).
4. W. D. Stevenson, Jr., *Elements of Power System Analysis*, 4th ed. (New York: McGraw-Hill, 1982).
5. M. Smith and D. Ton, "Key Connections," *IEEE Power and Energy Magazine*, 11,4 (July/August 2013), pp. 22–27.

3 Power Transformers

Core and coil assemblies of a three-phase 20.3 kVΔ/345kVY step-up transformer. This oil-immersed transformer is rated 325 MVA self-cooled (OA)/542 MVA forced oil, forced air-cooled (FOA)/607 MVA forced oil, forced air-cooled (FOA) (Courtesy of General Electric.)

The power transformer is a major power system component that permits economical power transmission with high efficiency and low series voltage drops. Since electric power is proportional to the product of voltage and current, low current levels (and therefore low I^2R losses and low IZ voltage drops) can be maintained for given power levels via high voltages. Power transformers transform ac voltage and current to optimum levels for generation, transmission, distribution, and utilization of electric power.

The development in 1885 by William Stanley of a commercially practical transformer was what made ac power systems more attractive than dc power systems.

The ac system with a transformer overcame voltage problems encountered in dc systems as load levels and transmission distances increased. Today's modern power transformers have nearly 100% efficiency, with ratings up to and beyond 1300 MVA.

In this chapter, basic transformer theory is reviewed and equivalent circuits for practical transformers operating under sinusoidal steady-state conditions are developed. Models of single-phase two-winding, three-phase two-winding, and three-phase three-winding transformers, as well as auto-transformers and regulating transformers are introduced. Also, the per-unit system, which simplifies power system analysis by eliminating the ideal transformer winding in transformer equivalent circuits, is introduced in this chapter and used throughout the remainder of the text.

CASE STUDY

The following article reviews methods of extending the life of a power transformer. Starting with specifying and purchasing; then manufacturing, installing, and commissioning; and finally operating and maintaining a transformer; options for extending life with the best cost-to-benefit ratio are explored [8].

POWER TRANSFORMERS

—*Life Management and Extension*

Carlos Gamez

TxMonitor

How Long is a Transformer Supposed to Last?

In this case study, the options to manage and, as far as possible, extend the life of these important assets in your system are explored. How long should a transformer last?

The life of a particular transformer depends on many factors, some of which are unpredictable in nature. In most circumstances there is not enough information to accurately predict the remaining life of a particular unit with any significant confidence. Current national and international standards and publications [1],[2],[3] and [4] favor the definition of life in "per unit" terms.

However, enough statistical data might be available in a particular system to be able to ascertain an estimated "average" life for a transformer in that system. An average life of 35 to 40 years is a reasonable number to be expected for transformers manufactured before the '90s and working under nominal conditions with some

Reprinted with permission from Carlos Gamez, "Power transformer life Part 3: Life management and extension," Transformers Magazine Vol. 1 Issue 3.

transformers, in very isolated cases, reaching into the 70 to 100 years of age mark [5].

So, if a transformer is to last 40 or more years of active service, what can you do to give it the best chances to do so?

The Life Cycle View of Life Extension

You might be tempted to think of the term life *extension* as something that is executed towards the end of the life of the asset in an attempt to extend its life. However, this would be a very shortsighted view of the topic. Life extension starts before the transformer is even manufactured. A holistic view of the complete life cycle of the unit will allow making the right decisions and putting the appropriate measures in place at every step of the way, from purchasing to disposal of the asset.

Just as with your own life and health, the most effective way of preventing a premature death is with prevention more than remediation. In the case of transformers this is not only logical, but considerably less costly. On a dollar per year of service of a particular asset, investing in prevention is almost always the wisest decision.

In this case study, the most important actions you can take to maximize the benefit that you get for your investment are laid out.

Specifying and Purchasing

A good start in life leads to a good-performing transformer during operation. By investing the time and resources in correctly specifying and sourcing the transformer most suitable for your application, you will save an incredible amount of money in the long run and a lot of headaches to your operations and maintenance team. A properly documented specification that adequately reflects your needs is essential to ensure that you are getting the right asset for the intended function. In many instances I have seen asset owners and buyers content with buying transformers that merely stick to the existing standards. While standards should be the starting point of any specification, they are certainly not sufficient in most cases. Standards are just the bare minimum that a transformer should comply with. By their very nature, standards cannot cater for the specific requirements that you might have in your specific operational context and circumstances.

Buying a transformer that simply meets the standards is like buying a car only specifying that you want it to have an engine, a body, and four wheels. The point is that you are likely to require certain characteristics in your transformers that are not listed or spelled out in the standards and that you will need to be explicit about when requesting a new unit. Things like specific types of bushing terminals, connection boxes, control wiring, and protective devices are but a few of the elements of your transformer design that you might want to stipulate to ensure compatibility with your environment. A well-crafted specification will also allow you to compare apples to apples when evaluating offers from various manufacturers. The best way to avoid unexpected outcomes is to be explicit about what you want.

Manufacturing

You have now ordered a new unit with a proper specification and from a manufacturer you feel comfortable with. Well, the next step is building the transformer. A good manufacture is made up of many small details that add up to a properly built and good quality finished transformer. From the quality of the raw materials, to the expertise and skill of the people manufacturing these machines, to the attention to detail on every step of the manufacturing process, it all counts.

The process of building a transformer is complex and hand-labor intensive where a lot of things can go wrong. In fact, almost always something will not go exactly as planned. In essence you will want to associate with a manufacturer that is not shy in acknowledging and correcting the issues that will inevitably arise during the manufacturing process.

In the realm of client-to-manufacturer relations, each company will have its own preference in how these are handled. Some customers prefer a pre-approval process to select the manufacturer or manufacturers with whom they plan to establish a long-term relationship. Others prefer to witness key milestones at the factory during the manufacturing process like tanking or final testing. It is in your best interest to ensure that you have a mechanism to guarantee that the manufacturing quality of your transformer is adequate and the final product fulfills your expectations.

Installation and Commissioning

Most power transformers of a certain size and above are like flat-packed furniture—some assembly is required. The level of assembly required varies with the size and manufacturer of the transformer. In some cases it is only necessary to fit a few of the components like radiators and conservator tanks and then "top up" the oil. In other cases a full assembly is required that finishes with the vacuum dry-out and hot oil-fill process.

In any case, executing the assembly procedure according to the manufacturer's recommendations will not only insure the unit is put together properly but also ensure your warranty is in full effect by the time the unit is placed in service. There are many factors that need to be taken care of during the final field assembly of the unit. You would want to ensure that a qualified team of technicians perform this assembly. Investing in the right service provider at this stage will increase your chances to avoid issues like oil leaks or any other type of assembly related malfunctions after energizing the transformer.

It is at this stage in the life of a transformer that the initial moisture level in the insulation system is established. It might be tempting to take shortcuts in this process. Sometimes the dry-out might take days to bring down the moisture to acceptable

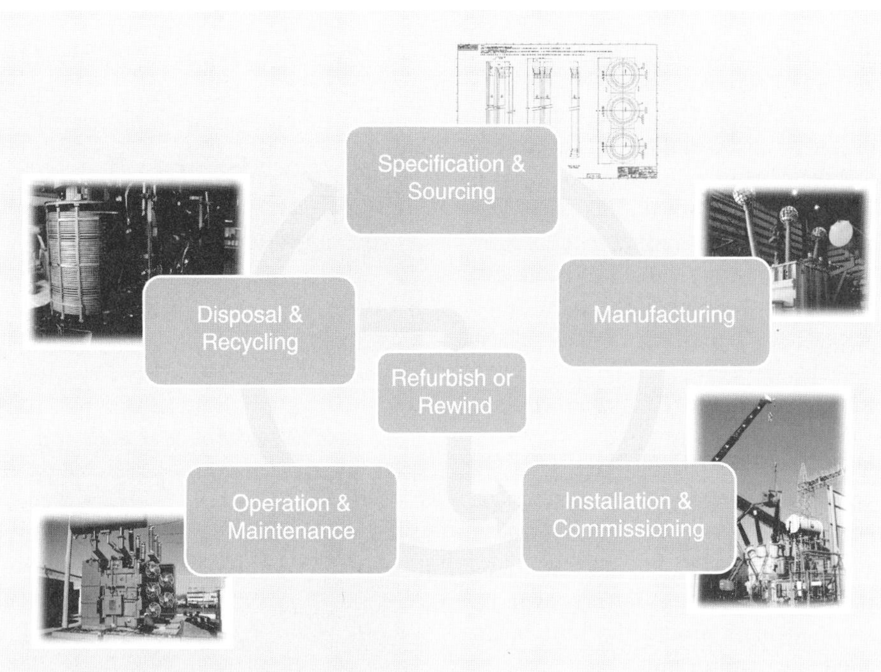

Figure 1 Typical transformer lifecycle

Figure 2 Good quality coils are essential to ensure the longevity of your transformer

levels (typically 0.5% by dry weight). But water is one of the main catalysts that accelerate the aging processes of the solid insulation of the transformer. Saving a few days or even a few hours of work at this stage has the potential to reduce the life by years in the long run. The economy is simply not there unless you want to operate your plant on a very short-term basis.

Figure 3 Different environments require different specifications

Operation and Maintenance

Since this stage in the life of a transformer is the longest (or at least it is supposed to be), much of your attention and day-to-day efforts will be spent in maintaining the unit in an adequate condition.

What you should focus your preventive maintenance efforts on during this period:

- Keep the transformer dry. Water is one of the main aging accelerating factors. In order to keep the transformer dry, you would want to ensure that there are no oil leaks (if oil can come out, moisture can come in). You will want to make sure that the oil preservation system is operating as the design intended. If silica gels are used, they should be dry and with enough capacity to dry the air that the transformer breathes. If diaphragms are used to separate the oil in the conservator tank from the ambient air, you want to make sure they are in good condition and free of ruptures. If automatic nitrogen equipment is used, it should be kept in good working order, with sufficient nitrogen in the supply and that the regulating system always ensures adequate positive pressure in the nitrogen chamber. In summary, ensure that water does not find its way into the coils of the unit.

- Keep the oxygen to a minimum. Similar to the point above, you would want to minimize the exposure of the oil to oxygen, which obviously accelerates the oxidation processes in the transformer. The same recommendations given above to keep the water out of the unit are applicable to keep the oxygen out, although in some cases, like a conservator tank without diaphragm, it is not always possible.
- Ensure nominal operating conditions. Temperature plays a major role in the aging processes. The unit is designed to operate within a certain temperature range, and the more you can do to keep it within that range, the better chances you are giving the unit a long useful life. Situations like overloading are sometimes unavoidable, and in these cases, there are clear guidelines available in the technical literature to allow you to estimate the impact of overloading in the life of a particular transformer. On a maintenance basis, you can help by ensuring that the cooling system is operating adequately and that the top oil or winding hot-spot temperature has not reached alarm levels. If you find that this has happened, you need to investigate the root cause so you can address it as soon as possible.

Now, what if you are not the person who has watched the transformer its entire life, and you just got handed a fleet of old transformers to maintain? (Hard to imagine right?) In this case you are in corrective maintenance territory, where the best course of action would be:

- Establish the condition of each unit in your fleet. You have many tools at your disposal for this purpose. Try to gather as much information as possible on each unit, and establish an effective information storage and retrieval mechanism. Include oil-analysis history and any other test and inspection performed on each transformer. This information will allow you to establish a preliminary condition ranking for each asset in the fleet.
- Once the above has been established, you will have a clearer idea of which units are priorities and which units can wait.
- Depending on the state of each unit, a number of actions can be taken in order to ameliorate the current condition or remove some of the agents that might be causing the accelerated aging.

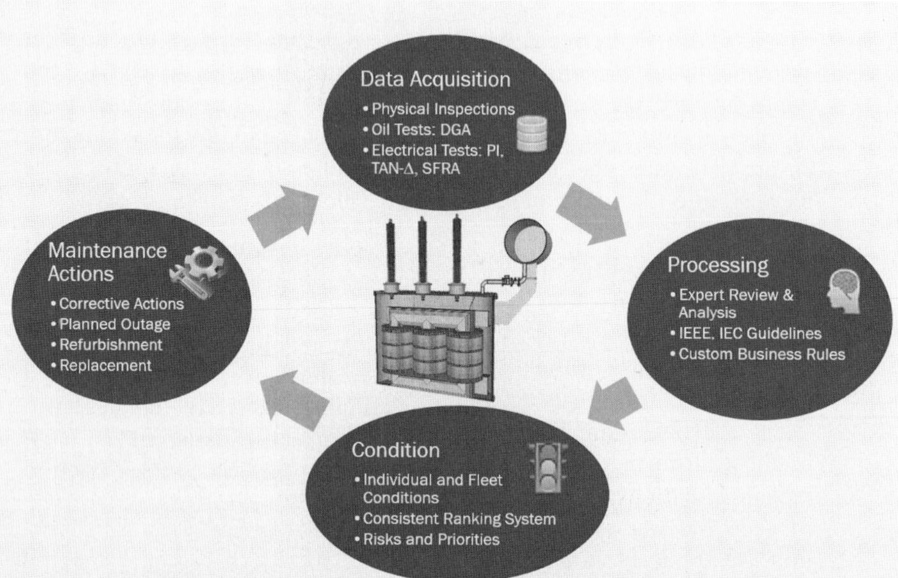

Figure 4 Example of a condition monitoring and maintenance program

- The oil can be processed to remove water and acids, which contribute to the aging processes.
- If any, more serious, failure modes are suspected, these need to be addressed on a case-by-case basis until you have satisfied the risk management policies of your company and you are aware of the condition of each unit.

Last but not least, an adequate condition monitoring program is essential to give you as much reaction time as possible if a failure mode starts to affect any particular transformer.

The condition monitoring method most commonly used is oil sampling and analysis. The analysis of the oil is a well-established technique that allows the early detection of incipient failures in the transformers. A suitable condition monitoring strategy will minimize the probability of unexpected failures and therefore minimize the overall operational risk of your transformer fleet.

References

[1] Institute of Electrical and Electronic Engineers (IEEE), C57.91-2011, *IEEE Guide for Loading Mineral-Oil-Immersed Transformers and Step-Voltage Regulators*. 2011. C57.91-2011.

[2] International Electrotechnical Commission (IEC), IEC 60076-7 ed1.0 - Power transformers—Part 7: *Loading guide for oil-immersed power transformers*. 2005. IEC 60076-7 ed1.0.

[3] Standards Australia, AS/NZS 60076.7: 2013 - *Power Transformers—Loading guide for oil-immersed power transformers*. s.l.: Standards Australia, Committee EL-008, 2013.

[4] CIGRE, *Cigré 227, guide for life management techniques for power transformer*. 2003. 227.
[5] ABB, Three workhorses retired after 100 years of active service in Australia! ABB, http://www.abbaustralia.com.au/cawp/seitp 202/95613170ca5ec79dc1257cca001471b7.aspx., current 08.09.2014.

Author

After graduating in Electrical and Mechanical Engineering in 1996, Carlos started working as a Transformer Design Engineer at PROLEC-GE, the biggest transformer factory for General Electric on the American continent.

Over the course of the following years, he gained expertise working in various roles such as product development, manufacturing improvements, technology and software development, field engineering, and customer service.

In early 2007, Carlos was seconded by General Electric to move to Perth, WA, to start up the Transformer Division to provide field and workshop maintenance and repair services to customers across Australia.

Having fulfilled this mission, in early 2011, Carlos accepted a principal consultant position with Assetivity, a consultancy firm leader in asset management.

In early 2013, he moved to TxMonitor, part of MM Group Holdings, where he currently works as a principal consultant and product manager in developing innovative solutions for the electrical asset management industry using both his technical and business acumen. ■

3.1 THE IDEAL TRANSFORMER

Figure 3.1 shows a basic single-phase two-winding transformer, where the two windings are wrapped around a magnetic core [1, 2, 3]. It is assumed here that the transformer is operating under sinusoidal steady-state excitation. Shown in the figure are the phasor voltages E_1 and E_2 across the windings, and the phasor

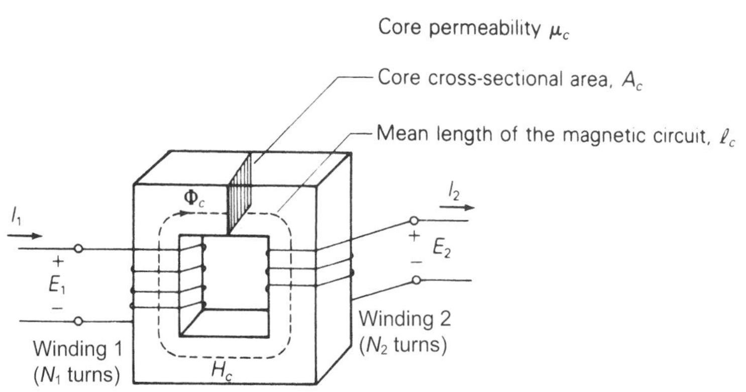

FIGURE 3.1 Basic single-phase two-winding transformer

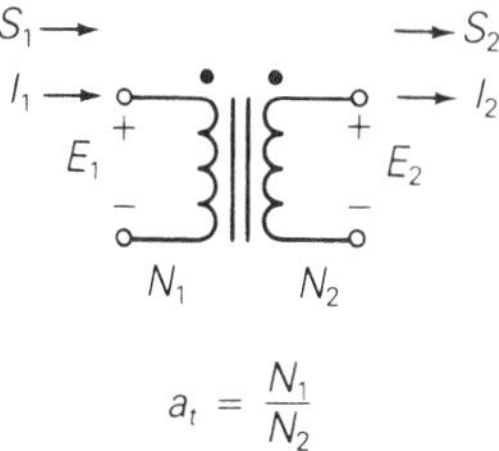

FIGURE 3.2 Schematic representation of a single-phase two-winding transformer

currents I_1 entering winding 1, which has N_1 turns, and I_2 leaving winding 2, which has N_2 turns. A phasor flux Φ_c set up in the core and a magnetic field intensity phasor H_c are also shown. The core has a cross-sectional area denoted A_c, a mean length of the magnetic circuit l_c, and a magnetic permeability μ_c, assumed constant.

For an ideal transformer, the following are assumed:

1. The windings have zero resistance; therefore, the I^2R losses in the windings are zero.
2. The core permeability μ_c is infinite, which corresponds to zero core reluctance.
3. There is no leakage flux; that is, the entire flux Φ_c is confined to the core and links both windings.
4. There are no core losses.

A schematic representation of a two-winding transformer is shown in Figure 3.2. Ampere's and Faraday's laws can be used along with the preceding assumptions to derive the ideal transformer relationships. *Ampere's law* states that the tangential component of the magnetic field intensity vector integrated along a closed path equals the net current enclosed by that path; that is,

$$\oint H_{\text{tan}}\, dl = I_{\text{enclosed}} \tag{3.1.1}$$

If the core center line shown in Figure 3.1 is selected as the closed path, and if H_c is constant along the path as well as tangent to the path, then (3.1.1) becomes

$$H_c l_c = N_1 I_1 - N_2 I_2 \tag{3.1.2}$$

Note that the current I_1 is enclosed N_1 times and I_2 is enclosed N_2 times, one time for each turn of the coils. Also, using the right-hand rule*, current I_1 contributes to clockwise flux, but current I_2 contributes to counterclockwise flux. Thus, in (3.1.2)

*The right-hand rule for a coil is as follows: Wrap the fingers of your right hand around the coil in the direction of the current. Your right thumb then points in the direction of the flux.

the net current enclosed is $N_1I_1 - N_2I_2$. For constant core permeability μ_c, the magnetic flux density B_c within the core, also constant, is

$$B_c = \mu_c H_c \quad \text{Wb/m}^2 \tag{3.1.3}$$

and the core flux Φ_c is

$$\Phi_c = B_C \mathrm{A}_C \quad \text{Wb} \tag{3.1.4}$$

Using (3.1.3) and (3.1.4) in (3.1.2) yields

$$N_1I_1 - N_2I_2 = l_cB_c/\mu_c = \left(\frac{l_c}{\mu_c \mathrm{A}_c}\right)\Phi_c \tag{3.1.5}$$

Core reluctance R_c is defined as

$$\mathrm{R}_c = \frac{l_c}{\mu_c \mathrm{A}_c} \tag{3.1.6}$$

Then (3.1.5) becomes

$$N_1I_1 - N_2I_2 = \mathrm{R}_c\Phi_c \tag{3.1.7}$$

Equation (3.1.7) can be called *Ohm's law* for the magnetic circuit, wherein the net magnetomotive force mmf $= N_1I_1 - N_2I_2$ equals the product of the core reluctance R_c and the core flux Φ_c. Reluctance R_c, which impedes the establishment of flux in a magnetic circuit, is analogous to resistance in an electric circuit. For an ideal transformer, μ_c is assumed infinite, which from (3.1.6) means that R_c is 0, and (3.1.7) becomes

$$N_1I_1 = N_2I_2 \tag{3.1.8}$$

In practice, power transformer windings and cores are contained within enclosures, and the winding directions are not visible. One way of conveying winding information is to place a dot at one end of each winding such that when current enters a winding at the dot, it produces an mmf acting in the *same* direction. This dot convention is shown in the schematic of Figure 3.2. The dots are conventionally called *polarity marks.*

Equation (3.1.8) is written for current I_1 *entering* its dotted terminal and current I_2 *leaving* its dotted terminal. As such, I_1 and I_2 are *in phase,* since $I_1 = (N_2/N_1)I_2$, If the direction chosen for I_2 were reversed, such that both currents entered their dotted terminals, then I_1 would be 180° *out of phase* with I_2.

Faraday's law states that the voltage $e(t)$ induced across an N-turn winding by a time-varying flux $\phi(t)$ linking the winding is

$$e(t) = N\frac{d\phi(t)}{dt} \tag{3.1.9}$$

Assuming a sinusoidal steady-state flux with constant frequency ω, and representing $e(t)$ and $\phi(t)$ by their phasors E and Φ, (3.1.9) becomes

$$E = N(j\omega)\Phi \tag{3.1.10}$$

For an ideal transformer, the entire flux is assumed to be confined to the core, linking both windings. From Faraday's law, the induced voltages across the windings of Figure 3.1 are

$$E_1 = N_1(j\omega)\Phi_c \tag{3.1.11}$$

$$E_2 = N_2(j\omega)\Phi_c \tag{3.1.12}$$

Dividing (3.1.11) by (3.1.12) yields

$$\frac{E_1}{E_2} = \frac{N_1}{N_2} \tag{3.1.13}$$

or

$$\frac{E_1}{N_1} = \frac{E_2}{N_2} \tag{3.1.14}$$

The dots shown in Figure 3.2 indicate that the voltages E_1 and E_2, both of which have their + polarities at the dotted terminals, are in phase. If the polarity chosen for one of the voltages in Figure 3.1 were reversed, then E_1 would be 180° out of phase with E_2.

The turns ratio a_t is defined as follows:

$$a_t = \frac{N_1}{N_2} \tag{3.1.15}$$

Using a_t in (3.1.8) and (3.1.14), the basic relations for an ideal single-phase two-winding transformer are

$$E_1 = \left(\frac{N_1}{N_2}\right)E_2 = a_t E_2 \tag{3.1.16}$$

$$I_1 = \left(\frac{N_2}{N_1}\right)I_2 = \frac{I_2}{a_t} \tag{3.1.17}$$

Two additional relations concerning complex power and impedance can be derived from (3.1.16) and (3.1.17) as follows. The complex power entering winding 1 in Figure 3.2 is

$$S_1 = E_1 I_1^* \tag{3.1.18}$$

Using (3.1.16) and (3.1.17),

$$S_1 = E_1 I_1^* = (a_t E_2)\left(\frac{I_2}{a_t}\right)^* = E_2 I_2^* = S_2 \tag{3.1.19}$$

As shown by (3.1.19), the complex power S_1 entering winding 1 equals the complex power S_2 leaving winding 2. That is, an ideal transformer has no real or reactive power loss.

If an impedance Z_2 is connected across winding 2 of the ideal transformer in Figure 3.2, then

$$Z_2 = \frac{E_2}{I_2} \tag{3.1.20}$$

This impedance, when measured from winding 1, is

$$Z_2' = \frac{E_1}{I_1} = \frac{a_t E_2}{I_2/a_t} = a_t^2 Z_2 = \left(\frac{N_1}{N_2}\right)^2 Z_2 \tag{3.1.21}$$

Thus, the impedance Z_2 connected to winding 2 is referred to winding 1 by multiplying Z_2 by a_t^2, which is the square of the turns ratio.

EXAMPLE 3.1

Ideal, single-phase two-winding transformer

A single-phase two-winding transformer is rated 20 kVA, 480/120 V, and 60 Hz. A source connected to the 480-V winding supplies an impedance load connected to the 120-V winding. The load absorbs 15 kVA at 0.8 p.f. lagging when the load voltage is 118 V. Assume that the transformer is ideal and calculate the following:

a. The voltage across the 480-V winding.
b. The load impedance.
c. The load impedance referred to the 480-V winding.
d. The real and reactive power supplied to the 480-V winding.

SOLUTION

a. The circuit is shown in Figure 3.3, where winding 1 denotes the 480-V winding and winding 2 denotes the 120-V winding. Selecting the load voltage E_2 as the reference,

$$E_2 = 118\underline{/0°} \text{ V}$$

FIGURE 3.3 Circuit for Example 3.1

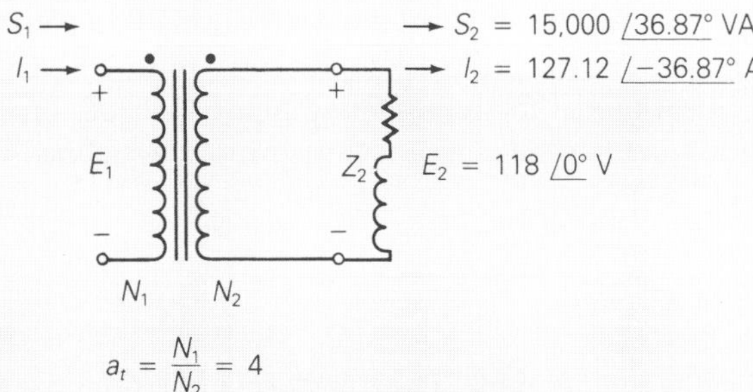

(Continued)

The turns ratio is, from (3.1.13),

$$a_t = \frac{N_1}{N_2} = \frac{E_{1\text{rated}}}{E_{2\text{rated}}} = \frac{480}{120} = 4$$

and the voltage across winding 1 is

$$E_1 = a_t E_2 = 4(118\underline{/0°}) = 472\underline{/0°}\ \text{V}$$

b. The complex power S_2 absorbed by the load is

$$S_2 = E_2 I_2^* = 118 I_2^* = 15{,}000\underline{/\cos^{-1}(0.8)} = 15{,}000\underline{/36.87°}\ \text{VA}$$

Solving, the load current I_2 is

$$I_2 = 127.12\underline{/-36.87°}\ \text{A}$$

The load impedance Z_2 is

$$Z_2 = \frac{E_2}{I_2} = \frac{118\underline{/0°}}{127.12\underline{/-36.87°}} = 0.9283\underline{/36.87°}\ \Omega$$

c. From (3.1.21), the load impedance referred to the 480-V winding is

$$Z_2' = a_t^2 Z_2 = (4)^2(0.9283\underline{/36.87°}) = 14.85\underline{/36.87°}\ \Omega$$

d. From (3.1.19)

$$S_1 = S_2 = 15{,}000\underline{/36.87°} = 12{,}000 + j9000$$

Thus, the real and reactive powers supplied to the 480-V winding are

$$\text{P}_1 = \text{Re}\, S_1 = 12{,}000\ \text{W} = 12\ \text{kW}$$
$$\text{Q}_1 = \text{Im}\, S_1 = 9000\ \text{var} = 9\ \text{kvar}$$

Figure 3.4 shows a schematic of a conceptual single-phase, phase-shifting transformer. This transformer is not an idealization of an actual transformer since it is physically impossible to obtain a complex turns ratio. It is used later in this chapter as a mathematical model for representing phase shift of three-phase transformers. As shown in Figure 3.4, the complex turns ratio a_t is defined for the phase-shifting transformer as

$$a_t = \frac{e^{j\phi}}{1} = e^{j\phi} \tag{3.1.22}$$

where ϕ is the phase-shift angle. The transformer relations are then

$$E_1 = a_t E_2 = e^{j\phi} E_2 \tag{3.1.23}$$

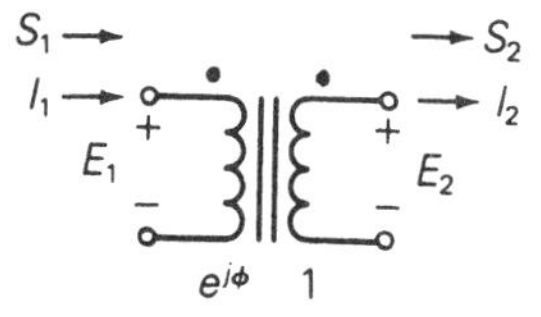

FIGURE 3.4

Schematic representation of a conceptual single-phase, phase-shifting transformer

$$I_1 = \frac{I_2}{a_t^*} = e^{j\phi} I_2 \tag{3.1.24}$$

Note that the phase angle of E_1 leads the phase angle of E_2 by ϕ. Similarly, I_1 leads I_2 by the angle ϕ. However, the magnitudes are unchanged; that is, $|E_1| = |E_2|$ and $|I_1| = |I_2|$.

From these two relations, the following two additional relations are derived:

$$S_1 = E_1 I_1^* = (a_t E_2)\left(\frac{I_2}{a_t^*}\right)^* = E_2 I_2^* = S_2 \tag{3.1.25}$$

$$Z_2' = \frac{E_1}{I_1} = \frac{a_t E_2}{\frac{1}{a_t^*} I_2} = |a_t|^2 Z_2 = Z_2 \tag{3.1.26}$$

Thus, impedance is unchanged when it is referred from one side of an ideal phase-shifting transformer to the other. Also, the ideal phase-shifting transformer has no real or reactive power losses since $S_1 = S_2$.

Note that (3.1.23) and (3.1.24) for the phase-shifting transformer are the same as (3.1.16) and (3.1.17) for the ideal physical transformer except for the complex conjugate (*) in (3.1.24). The complex conjugate for the phase-shifting transformer is required to make $S_1 = S_2$ (complex power into winding 1 equals complex power out of winding 2), as shown in (3.1.25).

3.2 EQUIVALENT CIRCUITS FOR PRACTICAL TRANSFORMERS

Figure 3.5 shows an equivalent circuit for a practical single-phase two-winding transformer, which differs from the ideal transformer as follows:

1. The windings have resistance.
2. The core permeability μ_c is finite.

FIGURE 3.5

Equivalent circuit of a practical single-phase two-winding transformer

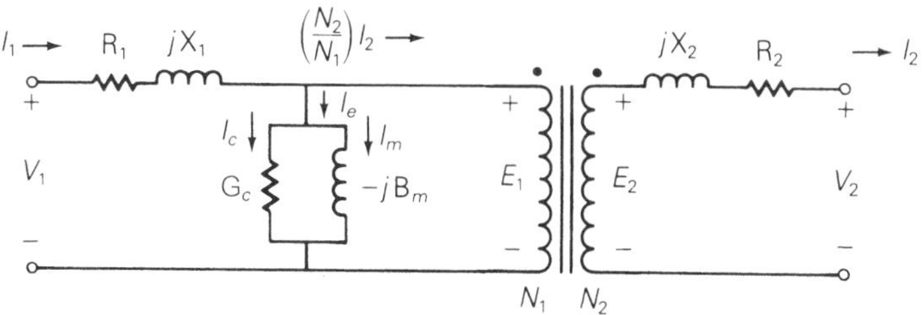

3. The magnetic flux is not entirely confined to the core.
4. There are real and reactive power losses in the core.

The resistance R_1 is included in series with winding 1 of the figure to account for I^2R losses in this winding. A reactance X_1, called the *leakage reactance* of winding 1, is also included in series with winding 1 to account for the leakage flux of winding 1. This leakage flux is the component of the flux that links winding 1 but does not link winding 2; it causes a voltage drop $I_1(jX_1)$, which is proportional to I_1 and leads I_1 by 90°. There is also a reactive power loss $I_1^2X_1$ associated with this leakage reactance. Similarly, there is a resistance R_2 and a leakage reactance X_2 in series with winding 2.

Equation (3.1.7) shows that for finite core permeability μ_c, the total mmf is not zero. Dividing (3.1.7) by N_1 and using (3.1.11), the result is

$$I_1 - \left(\frac{N_2}{N_1}\right)I_2 = \frac{R_c}{N_1}\Phi_c = \frac{R_c}{N_1}\left(\frac{E_1}{j\omega N_1}\right) = -j\left(\frac{R_c}{\omega N_1^2}\right)E_1 \tag{3.2.1}$$

Defining the term on the right-hand side of (3.2.1) to be I_m, called *magnetizing* current, it is evident that I_m lags E_1 by 90°, and can be represented by a shunt inductor with susceptance $B_m = \left(\frac{R_c}{\omega N_1^2}\right)$ S.* However, in reality, there is an additional shunt branch, represented by a resistor with conductance G_c siemens, which carries a current I_c, called the *core loss* current. I_c is in phase with E_1. When the core loss current I_c is included, (3.2.1) becomes

$$I_1 - \left(\frac{N_2}{N_1}\right)I_2 = I_c + I_m = (G_c - jB_m)E_1 \tag{3.2.2}$$

The equivalent circuit of Figure 3.5, which includes the shunt branch with admittance $(G_c - jB_m)$ siemens, satisfies the KCL equation (3.2.2). Note that when winding 2 is open ($I_2 = 0$) and when a sinusoidal voltage V_1 is applied to winding 1, then (3.2.2) indicates that the current I_1 will have two components: the core loss current I_c and the magnetizing current I_m. Associated with I_c is a real power loss $I_c^2/G_c = E_1^2G_c$ W. This real power loss accounts for both hysteresis and eddy current losses within the core. Hysteresis loss occurs because a cyclic variation of flux within the core requires

*The units of admittance, conductance, and susceptance, which in the SI system are siemens (with symbol S), are also called mhos (with symbol ℧) or ohms^{-1} (with symbol Ω^{-1}).

energy dissipated as heat. As such, hysteresis loss can be reduced by the use of special high grades of alloy steel as core material. Eddy current loss occurs because induced currents called eddy currents flow within the magnetic core perpendicular to the flux. As such, eddy current loss can be reduced by constructing the core with laminated sheets of alloy steel. Associated with I_m is a reactive power loss $I_m^2/B_m = E_1^2 B_m$ var. This reactive power is required to magnetize the core. The phasor sum $(I_c + I_m)$ is called the *exciting* current I_e.

Figure 3.6 shows three alternative equivalent circuits for a practical single-phase two-winding transformer. In Figure 3.6(a), the resistance R_2 and leakage reactance X_2 of winding 2 are referred to winding 1 via (3.1.21).

In Figure 3.6(b), the shunt branch is omitted, which corresponds to neglecting the exciting current. Since the exciting current is usually less than 5% of rated current, neglecting it in power system studies is often valid unless transformer efficiency or exciting current phenomena are of particular concern. For large power transformers rated more than 500 kVA, the winding resistances, which are small compared to the leakage reactances, often can be neglected, as shown in Figure 3.6(c).

Thus, a practical transformer operating in sinusoidal steady state is equivalent to an ideal transformer with external impedance and admittance branches, as shown in Figure 3.6. The external branches can be evaluated from short-circuit and open-circuit tests, as illustrated by the following example.

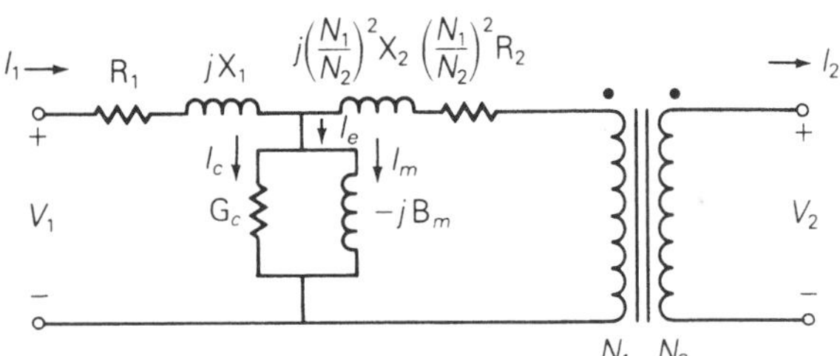

(a) R_2 and X_2 are referred to winding 1

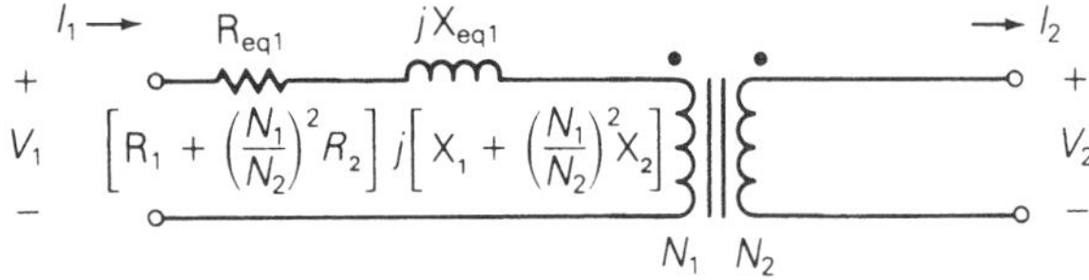

(b) Neglecting exciting current

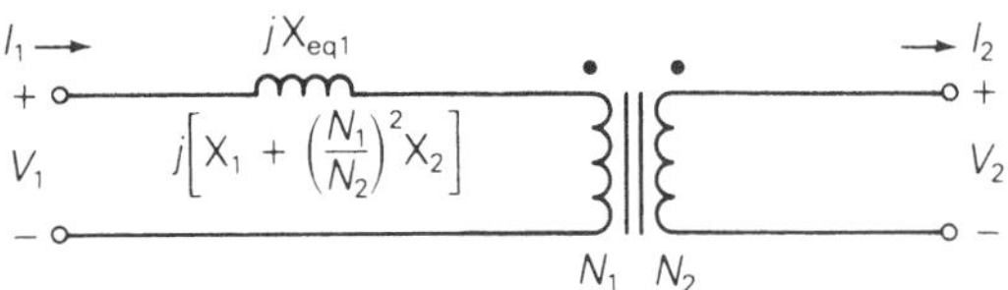

(c) Neglecting exciting current and I^2R winding loss

FIGURE 3.6

Equivalent circuits for a practical single-phase two-winding transformer

EXAMPLE 3.2

Transformer short-circuit and open-circuit tests

A single-phase two-winding transformer is rated 20 kVA, 480/120 volts, and 60 Hz. During a short-circuit test, where rated current at rated frequency is applied to the 480-volt winding (denoted winding 1), with the 120-volt winding (winding 2) shorted, the following readings are obtained: $V_1 = 35$ volts, $P_1 = 300$ W. During an open-circuit test, where rated voltage is applied to winding 2, with winding 1 open, the following readings are obtained: $I_2 = 12$ A, $P_2 = 200$ W.

a. From the short-circuit test, determine the equivalent series impedance $Z_{eq1} = R_{eq1} + jX_{eq1}$ referred to winding 1. Neglect the shunt admittance.

b. From the open-circuit test, determine the shunt admittance $Y_m = G_c - jB_m$ referred to winding 1. Neglect the series impedance.

SOLUTION

a. The equivalent circuit for the short-circuit test is shown in Figure 3.7(a), where the shunt admittance branch is neglected. Rated current for winding 1 is

$$I_{1rated} = \frac{S_{rated}}{V_{1rated}} = \frac{20 \times 10^3}{480} = 41.667 \quad \text{A}$$

R_{eq1}, Z_{eq1}, and X_{eq1} are then determined as follows:

$$R_{eq1} = \frac{P_1}{I^2_{1rated}} = \frac{300}{(41.667)^2} = 0.1728 \quad \Omega$$

$$|Z_{eq1}| = \frac{V_1}{I_{1rated}} = \frac{35}{41.667} = 0.8400 \quad \Omega$$

$$X_{eq1} = \sqrt{Z^2_{eq1} - R^2_{eq1}} = 0.8220 \quad \Omega$$

$$Z_{eq1} = R_{eq1} + jX_{eq1} + 0.1728 + j0.8220 = 0.8400\underline{/78.13^\circ} \quad \Omega$$

b. The equivalent circuit for the open-circuit test is shown in Figure 3.7(b), where the series impedance is neglected. From (3.1.16),

$$V_1 = E_1 = a_t E_2 = \frac{N_1}{N_2} V_{2rated} = \frac{480}{120}(120) = 480 \quad \text{volts}$$

G_c, Y_m, and B_m are then determined as follows:

$$G_c = \frac{P_2}{V_1^2} = \frac{200}{(480)^2} = 0.000868 \quad \text{S}$$

$$|Y_m| = \frac{\left(\frac{N_2}{N_1}\right) I_2}{V_1} = \frac{\left(\frac{120}{480}\right)(12)}{480} = 0.00625 \quad \text{S}$$

FIGURE 3.7

Circuits for Example 3.2

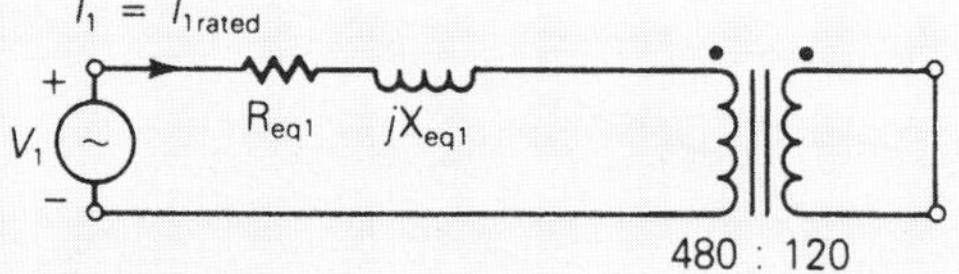

(a) Short-circuit test (neglecting shunt admittance)

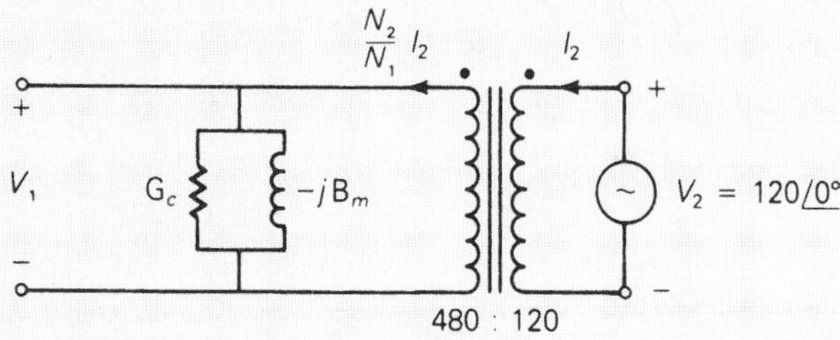

(b) Open-circuit test (neglecting series impedance)

$$\mathrm{B}_m = \sqrt{Y_m^2 - \mathrm{G}_c^2} = \sqrt{(0.00625)^2 - (0.000868)^2} = 0.00619 \quad \mathrm{S}$$

$$Y_m = \mathrm{G}_c - j\mathrm{B}_m = 0.000868 - j0.00619 = 0.00625\underline{/-82.02^\circ} \quad \mathrm{S}$$

Note that the equivalent series impedance is usually evaluated at rated current from a short-circuit test, and the shunt admittance is evaluated at rated voltage from an open-circuit test. For small variations in transformer operation near rated conditions, the impedance and admittance values are often assumed constant.

The following are not represented by the equivalent circuit of Figure 3.5:

1. Saturation
2. Inrush current
3. Nonsinusoidal exciting current
4. Surge phenomena

They are briefly discussed in the following sections.

SATURATION

In deriving the equivalent circuit of the ideal and practical transformers, constant core permeability μ_c and the linear relationship $B_c = \mu_c H_c$ of (3.1.3) were assumed. However, the relationship between B and H for ferromagnetic materials used for transformer cores is nonlinear and multivalued. Figure 3.8 shows a set of B–H curves for a grain-oriented electrical steel typically used in transformers. As shown, each curve is multivalued, which is caused by hysteresis. For many engineering applications, the

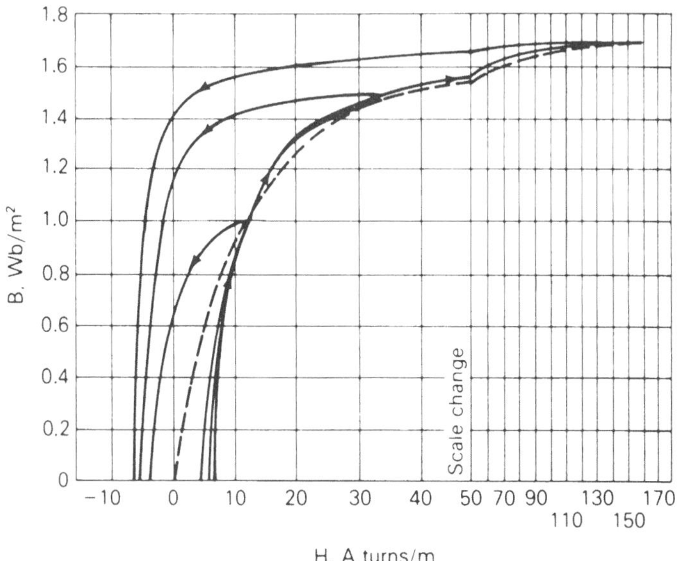

FIGURE 3.8 B–H curves for M-5 grain-oriented electrical steel 0.012 in. thick (Reprinted with permission of AK Steel Corporation.)

B–H curves can be adequately described by the dashed line drawn through the curves in Figure 3.8. Note that as H increases, the core becomes saturated; that is, the curves flatten out as B increases above 1 Wb/m^2. If the magnitude of the voltage applied to a transformer is too large, the core will saturate and a high magnetizing current will flow. In a well-designed transformer, the applied peak voltage causes the peak flux density in steady state to occur at the knee of the B–H curve, with a corresponding low value of magnetizing current.

INRUSH CURRENT

When a transformer is first energized, a transient current much larger than rated transformer current can flow for several cycles. This current, called *inrush current,* is nonsinusoidal and has a large dc component. To understand the cause of inrush, assume that before energization, the transformer core is magnetized with a residual flux density $B(0) = 1.5\ Wb/m^2$ (near the knee of the dotted curve in Figure 3.8). If the transformer is then energized when the source voltage is positive and increasing, Faraday's law, (3.1.9), will cause the flux density $B(t)$ to increase further, since

$$B(t) = \frac{\phi(t)}{A} = \frac{1}{NA}\int_0^t e(t)\,dt + B(0)$$

As $B(t)$ moves into the saturation region of the B–H curve, large values of $H(t)$ will occur, and, from Ampere's law, (3.1.1), corresponding large values of current $i(t)$ will flow for several cycles until it has dissipated. Since normal inrush currents can be as large as abnormal short-circuit currents in transformers, transformer protection schemes must be able to distinguish between these two types of currents.

NONSINUSOIDAL EXCITING CURRENT

When a sinusoidal voltage is applied to one winding of a transformer with the other winding open, the flux $\phi(t)$ and flux density $B(t)$ will, from Faraday's law, (3.1.9), be very nearly sinusoidal in steady state. However, the magnetic field intensity $H(t)$ and the resulting exciting current is not sinusoidal in steady state, due to the nonlinear B–H curve. If the exciting current is measured and analyzed by Fourier analysis techniques, one finds that it has a fundamental component and a set of odd harmonics. The principal harmonic is the third, whose rms value is typically about 40% of the total rms exciting current. However, the nonsinusoidal nature of exciting current is usually neglected unless harmonic effects are of direct concern, because the exciting current itself is usually less than 5% of rated current for power transformers.

SURGE PHENOMENA

When power transformers are subjected to transient overvoltages caused by lightning or switching surges, the capacitances of the transformer windings have important effects on transient response. Transformer winding capacitances and response to surges are discussed in Chapter 13.

3.3 THE PER-UNIT SYSTEM

Power-system quantities such as voltage, current, power, and impedance are often expressed in per-unit or percent of specified base values. For example, if a base voltage of 20 kV is specified, then the voltage 18 kV is (18/20) = 0.9 per unit or 90%. Calculations then can be made with per-unit quantities rather than with the actual quantities.

One advantage of the per-unit system is that by properly specifying base quantities, the transformer equivalent circuit can be simplified. The ideal transformer winding can be eliminated, such that voltages, currents, and external impedances and admittances expressed in per-unit do not change when they are referred from one side of a transformer to the other. This can be a significant advantage even in a power system of moderate size, where hundreds of transformers may be encountered. Using the per-unit system avoids the possibility of making serious calculation errors when referring quantities from one side of a transformer to the other. Another advantage of the per-unit system is that the per-unit impedances of electrical equipment of similar type usually lie within a narrow numerical range when the equipment ratings are used as base values. Because of this, per-unit impedance data can be checked rapidly for gross errors by someone familiar with per-unit quantities. In addition, manufacturers usually specify the impedances of machines and transformers in per-unit or percent of nameplate rating.

Per-unit quantities are calculated as follows:

$$\text{per-unit quantity} = \frac{\text{actual quantity}}{\text{base value of quantity}} \tag{3.3.1}$$

where *actual quantity* is the value of the quantity in the actual units. The base value has the same units as the actual quantity, thus making the per-unit quantity dimensionless. Also, the base value is always a real number. Therefore, the angle of the per-unit quantity is the same as the angle of the actual quantity.

Two independent base values can be arbitrarily selected at one point in a power system. Usually the base voltage V_{baseLN} and base complex power $S_{base1\phi}$ are selected for either a single-phase circuit or for one phase of a three-phase circuit. Then, in order for electrical laws to be valid in the per-unit system, the following relations must be used for other base values:

$$P_{base1\phi} = Q_{base1\phi} = S_{base1\phi} \tag{3.3.2}$$

$$I_{base} = \frac{S_{base1\phi}}{V_{baseLN}} \tag{3.3.3}$$

$$Z_{base} = R_{base} = X_{base} = \frac{V_{baseLN}}{I_{base}} = \frac{V^2_{baseLN}}{S_{base1\phi}} \tag{3.3.4}$$

$$Y_{base} = G_{base} = B_{base} = \frac{1}{Z_{base}} \tag{3.3.5}$$

In (3.3.2) through (3.3.5), the subscripts LN and 1ϕ denote "line-to-neutral" and "per-phase," respectively, for three-phase circuits. These equations are also valid for single-phase circuits, where subscripts can be omitted.

Convention requires adoption of the following two rules for base quantities:

1. The value of $S_{base1\phi}$ is the same for the entire power system of concern.
2. The ratio of the voltage bases on either side of a transformer is selected to be the same as the ratio of the transformer voltage ratings.

With these two rules, a per-unit impedance remains unchanged when referred from one side of a transformer to the other.

EXAMPLE 3.3

Per-unit impedance: single-phase transformer

A single-phase two-winding transformer is rated 20 kVA, 480/120 volts, and 60 Hz. The equivalent leakage impedance of the transformer referred to the 120-volt winding, denoted winding 2, is $Z_{eq2} = 0.0525\underline{/78.13^\circ}\ \Omega$. Using the transformer ratings as base values, determine the per-unit leakage impedance referred to winding 2 and referred to winding 1.

SOLUTION

The values of S_{base}, V_{base1}, and V_{base2} are, from the transformer ratings,

$$S_{base} = 20\ \text{kVA}, \quad V_{base1} = 480\ \text{volts}, \quad V_{base2} = 120\ \text{volts}$$

Using (3.3.4), the base impedance on the 120-volt side of the transformer is

$$Z_{\text{base2}} = \frac{V^2_{\text{base2}}}{S_{\text{base}}} = \frac{(120)^2}{20{,}000} = 0.72 \quad \Omega$$

Then, using (3.3.1), the per-unit leakage impedance referred to winding 2 is

$$Z_{\text{eq2p.u.}} = \frac{Z_{\text{eq2}}}{Z_{\text{base2}}} = \frac{0.0525\underline{/78.13^\circ}}{0.72} = 0.0729\underline{/78.13^\circ} \quad \text{per unit}$$

If Z_{eq2} is referred to winding 1,

$$Z_{\text{eq1}} = a_t^2 Z_{\text{eq2}} = \left(\frac{N_1}{N_2}\right)^2 Z_{\text{eq2}} = \left(\frac{480}{120}\right)^2 (0.0525\underline{/78.13^\circ})$$
$$= 0.84\underline{/78.13^\circ} \quad \Omega$$

The base impedance on the 480 volt side of the transformer is

$$Z_{\text{base1}} = \frac{V^2_{\text{base1}}}{S_{\text{base}}} = \frac{(480)^2}{20{,}000} = 11.52 \quad \Omega$$

and the per-unit leakage reactance referred to winding 1 is

$$Z_{\text{eq1p.u.}} = \frac{Z_{\text{eq1}}}{Z_{\text{base1}}} = \frac{0.84\underline{/78.13^\circ}}{11.52} = 0.0729 \ \underline{/78.13^\circ} \text{ per unit} = Z_{\text{eq2p.u.}}$$

Thus, the *per-unit* leakage impedance remains unchanged when referred from winding 2 to winding 1. This has been achieved by specifying

$$\frac{V_{\text{base1}}}{V_{\text{base2}}} = \frac{V_{\text{rated1}}}{V_{\text{rated2}}} = \left(\frac{480}{120}\right)$$

Figure 3.9 shows three per-unit circuits of a single-phase two-winding transformer. The ideal transformer, shown in Figure 3.9(a), satisfies the per-unit relations $E_{\text{1p.u.}} = E_{\text{2p.u.}}$ and $I_{\text{1p.u.}} = I_{\text{2p.u.}}$, which can be derived as follows. First divide (3.1.16) by V_{base1}:

$$E_{\text{1p.u.}} = \frac{E_1}{V_{\text{base1}}} = \frac{N_1}{N_2} \times \frac{E_2}{V_{\text{base1}}} \tag{3.3.6}$$

Then, using $V_{\text{base1}}/V_{\text{base2}} = V_{\text{rated1}}/V_{\text{rated2}} = N_1/N_2$,

$$E_{\text{1p.u.}} = \frac{N_1}{N_2} \frac{E_2}{\left(\frac{N_1}{N_2}\right) V_{\text{base2}}} = \frac{E_2}{V_{\text{base2}}} = E_{\text{2p.u.}} \tag{3.3.7}$$

FIGURE 3.9 Per-unit equivalent circuits of a single-phase two-winding transformer

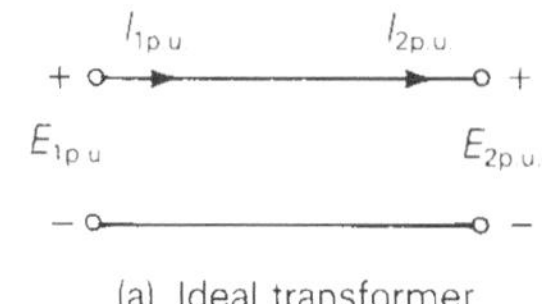

(a) Ideal transformer

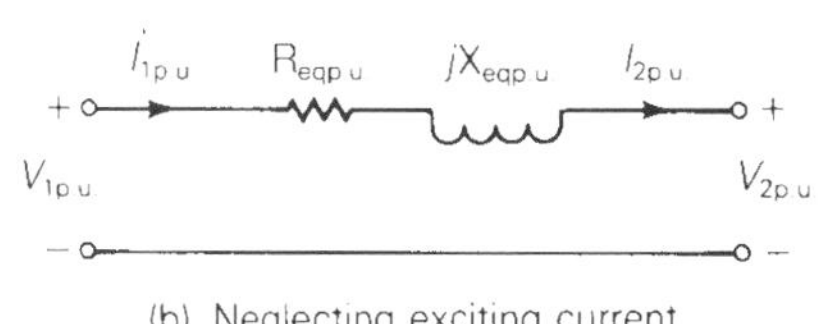

(b) Neglecting exciting current

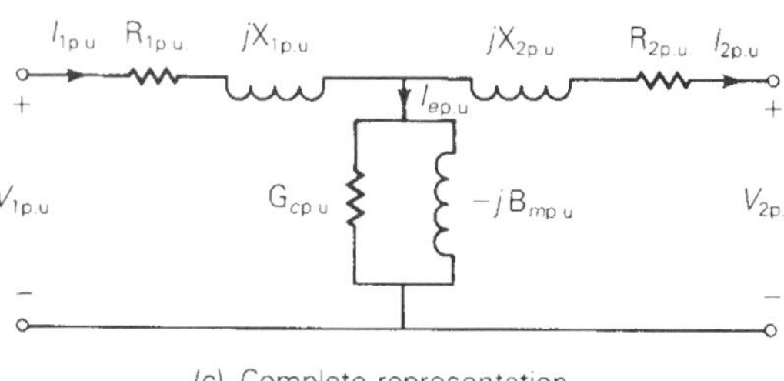

(c) Complete representation

Similarly, divide (3.1.17) by $\mathrm{I}_{\text{base1}}$:

$$I_{1\text{p.u.}} = \frac{I_1}{\mathrm{I}_{\text{base1}}} = \frac{N_2}{N_1}\frac{I_2}{\mathrm{I}_{\text{base1}}} \tag{3.3.8}$$

Then, using $\mathrm{I}_{\text{base1}} = \mathrm{S}_{\text{base}}/\mathrm{V}_{\text{base1}} = \mathrm{S}_{\text{base}}/[(N_1/N_2)\mathrm{V}_{\text{base2}}] = (N_2/N_1)\mathrm{I}_{\text{base2}}$

$$I_{1\text{p.u.}} = \frac{N_2}{N_1}\frac{I_2}{\left(\dfrac{N_2}{N_1}\right)\mathrm{I}_{\text{base2}}} = \frac{I_2}{\mathrm{I}_{\text{base2}}} = I_{2\text{p.u.}} \tag{3.3.9}$$

Thus, the ideal transformer winding in Figure 3.2 is eliminated from the per-unit circuit in Figure 3.9(a). The per-unit leakage impedance is included in Figure 3.9(b), and the per-unit shunt admittance branch is added in Figure 3.9(c) to obtain the complete representation.

When only one component, such as a transformer, is considered, the nameplate ratings of that component are usually selected as base values. When several components are involved, however, the system base values may be different from the nameplate ratings of any particular device. It is then necessary to convert the per-unit impedance of a device from its nameplate ratings to the system base values. To convert a per-unit impedance from "old" to "new" base values, use

$$Z_{\text{p.u.new}} = \frac{Z_{\text{actual}}}{Z_{\text{basenew}}} = \frac{Z_{\text{p.u.old}}Z_{\text{baseold}}}{Z_{\text{basenew}}} \tag{3.3.10}$$

or, from (3.3.4),

$$Z_{\text{p.u.new}} = Z_{\text{p.u.old}}\left(\frac{V_{\text{baseold}}}{V_{\text{basenew}}}\right)^2\left(\frac{S_{\text{basenew}}}{S_{\text{baseold}}}\right) \tag{3.3.11}$$

EXAMPLE 3.4

Per-unit circuit: three-zone single-phase network

Three zones of a single-phase circuit are identified in Figure 3.10(a). The zones are connected by transformers T_1 and T_2, whose ratings are also shown. Using base values of 30 kVA and 240 volts in zone 1, draw the per-unit circuit and determine the per-unit impedances and the per-unit source voltage. Then calculate the load current both in per-unit and in amperes. Transformer winding resistances and shunt admittance branches are neglected.

FIGURE 3.10 Circuit for Example 3.4

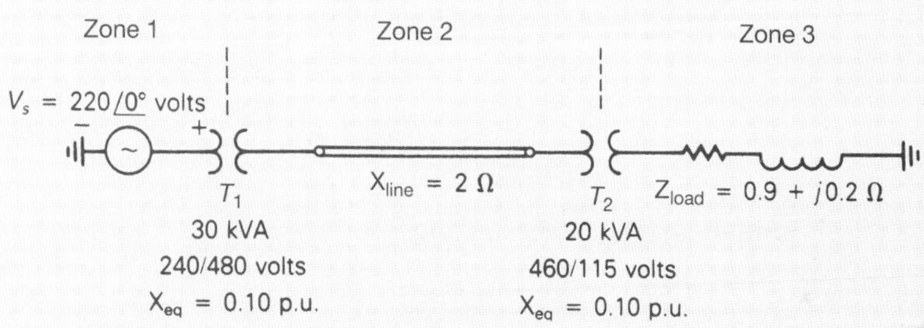

(a) Single-phase circuit

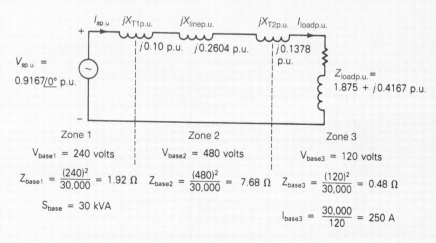

$I_{\text{sp.u.}}$ $jX_{\text{T1p.u.}}$ $jX_{\text{linep.u.}}$ $jX_{\text{T2p.u.}}$ $I_{\text{loadp.u.}}$

$j0.10$ p.u. $j0.2604$ p.u. $j0.1378$ p.u.

$V_{\text{sp.u.}} = 0.9167\underline{/0°}$ p.u.

$Z_{\text{loadp.u.}} = 1.875 + j0.4167$ p.u.

Zone 1

$V_{\text{base1}} = 240$ volts

$Z_{\text{base1}} = \frac{(240)^2}{30{,}000} = 1.92\ \Omega$

$S_{\text{base}} = 30$ kVA

Zone 2

$V_{\text{base2}} = 480$ volts

$Z_{\text{base2}} = \frac{(480)^2}{30{,}000} = 7.68\ \Omega$

Zone 3

$V_{\text{base3}} = 120$ volts

$Z_{\text{base3}} = \frac{(120)^2}{30{,}000} = 0.48\ \Omega$

$I_{\text{base3}} = \frac{30{,}000}{120} = 250$ A

(b) Per-unit circuit

(Continued)

SOLUTION

First the base values in each zone are determined. $S_{base} = 30$ kVA is the same for the entire network. Also, $V_{base1} = 240$ volts, as specified for zone 1. When moving across a transformer, the voltage base is changed in proportion to the transformer voltage ratings. Thus,

$$V_{base2} = \left(\frac{480}{240}\right)(240) = 480 \quad \text{volts}$$

and

$$V_{base3} = \left(\frac{115}{460}\right)(480) = 120 \quad \text{volts}$$

The base impedances in zones 2 and 3 are

$$Z_{base2} = \frac{V^2_{base2}}{S_{base}} = \frac{480^2}{30{,}000} = 7.86 \quad \Omega$$

and

$$Z_{base3} = \frac{V^2_{base3}}{S_{base}} = \frac{120^2}{30{,}000} = 0.48 \quad \Omega$$

and the base current in zone 3 is

$$I_{base3} = \frac{S_{base}}{V_{base3}} = \frac{30{,}000}{120} = 250 \quad A$$

Next, the per-unit circuit impedances are calculated using the system base values. Since $S_{base} = 30$ kVA is the same as the kVA rating of transformer T_1, and $V_{base1} = 240$ volts is the same as the voltage rating of the zone 1 side of transformer T_1, the per-unit leakage reactance of T_1 is the same as its nameplate value, $X_{T1p.u.} = 0.1$ per unit. However, the per-unit leakage reactance of transformer T_2 must be converted from its nameplate rating to the system base. Using (3.3.11) and $V_{base2} = 480$ volts,

$$X_{T2p.u.} = (0.10)\left(\frac{460}{480}\right)^2\left(\frac{30{,}000}{20{,}000}\right) = 0.1378 \quad \text{per unit}$$

Alternatively, using $V_{base3} = 120$ volts,

$$X_{T2p.u.} = (0.10)\left(\frac{115}{120}\right)^2\left(\frac{30{,}000}{20{,}000}\right) = 0.1378 \quad \text{per unit}$$

which gives the same result. The line, which is located in zone 2, has a per-unit reactance

$$X_{linep.u.} = \frac{X_{line}}{Z_{base2}} = \frac{2}{7.68} = 0.2604 \quad \text{per unit}$$

and the load, which is located in zone 3, has a per-unit impedance

$$Z_{\text{loadp.u.}} = \frac{Z_{\text{load}}}{Z_{\text{base3}}} = \frac{0.9 + j0.2}{0.48} = 1.875 + j0.4167 \quad \text{per unit}$$

The per-unit circuit is shown in Figure 3.10(b), where the base values for each zone, per-unit impedances, and the per-unit source voltage are shown. The per-unit load current is then easily calculated from Figure 3.10(b) as follows:

$$\begin{aligned} I_{\text{loadp.u.}} = I_{\text{sp.u.}} &= \frac{V_{\text{sp.u.}}}{j(X_{\text{T1p.u.}} + X_{\text{linep.u.}} + X_{\text{T2p.u.}}) + Z_{\text{loadp.u.}}} \\ &= \frac{0.9167\underline{/0^\circ}}{j(0.10 + 0.2604 + 0.1378) + (1.875 + j0.4167)} \\ &= \frac{0.9167\underline{/0^\circ}}{1.875 + j0.9149} = \frac{0.9167\underline{/0^\circ}}{2.086\underline{/26.01^\circ}} \\ &= 0.4395\underline{/-26.01^\circ} \quad \text{per unit} \end{aligned}$$

The actual load current is

$$I_{\text{load}} = (I_{\text{loadp.u.}})\text{I}_{\text{base3}} = (0.4395\underline{/-26.01^\circ})\,(250) = 109.9\underline{/-26.01^\circ} \quad \text{A}$$

Note that the per-unit equivalent circuit of Figure 3.10(b) is relatively easy to analyze, since ideal transformer windings have been eliminated by proper selection of base values.

Balanced three-phase circuits can be solved in per-unit on a per-phase basis after converting Δ-load impedances to equivalent Y impedances. Base values can be selected either on a per-phase basis or on a three-phase basis. Equations (3.3.1) through (3.3.5) remain valid for three-phase circuits on a per-phase basis. Usually $\text{S}_{\text{base}3\phi}$ and V_{baseLL} are selected, where the subscripts 3ϕ and LL denote "three-phase" and "line-to-line," respectively. Then the following relations must be used for other base values:

$$\text{S}_{\text{base}1\phi} = \frac{\text{S}_{\text{base}3\phi}}{3} \tag{3.3.12}$$

$$\text{V}_{\text{baseLN}} = \frac{\text{V}_{\text{baseLL}}}{\sqrt{3}} \tag{3.3.13}$$

$$\text{S}_{\text{base}3\phi} = \text{P}_{\text{base}3\phi} = \text{Q}_{\text{base}3\phi} \tag{3.3.14}$$

$$\text{I}_{\text{base}} = \frac{\text{S}_{\text{base}1\phi}}{\text{V}_{\text{baseLN}}} = \frac{\text{S}_{\text{base}3\phi}}{\sqrt{3}\text{V}_{\text{baseLL}}} \tag{3.3.15}$$

$$Z_{\text{base}} = \frac{\text{V}_{\text{baseLN}}}{\text{I}_{\text{base}}} = \frac{\text{V}^2_{\text{baseLN}}}{\text{S}_{\text{base}1\phi}} = \frac{\text{V}^2_{\text{baseLL}}}{\text{S}_{\text{base}3\phi}} \tag{3.3.16}$$

$$\text{R}_{\text{base}} = X_{\text{base}} = Z_{\text{base}} = \frac{1}{Y_{\text{base}}} \tag{3.3.17}$$

EXAMPLE 3.5

Per-unit and actual currents in balanced three-phase networks

As in Example 2.5, a balanced-Y-connected voltage source with $E_{ab} = 480\underline{/0°}$ volts is applied to a balanced-Δ load with $Z_\Delta = 30\underline{/40°}\ \Omega$. The line impedance between the source and load is $Z_L = 1\underline{/85°}\ \Omega$ for each phase. Calculate the per-unit and actual current in phase a of the line using $S_{base3\phi} = 10$ kVA and $V_{baseLL} = 480$ volts.

SOLUTION

First, convert Z_Δ to an equivalent Z_Y; the equivalent line-to-neutral diagram is shown in Figure 2.17. The base impedance is, from (3.3.16),

$$Z_{base} = \frac{V_{baseLL}^2}{S_{base3\phi}} = \frac{(480)^2}{10{,}000} = 23.04 \quad \Omega$$

The per-unit line and load impedances are

$$Z_{Lp.u.} = \frac{Z_L}{Z_{base}} = \frac{10\underline{/85°}}{23.04} = 0.04340\underline{/85°} \quad \text{per unit}$$

and

$$Z_{Yp.u.} = \frac{Z_Y}{Z_{base}} = \frac{10\underline{/40°}}{23.04} = 0.4340\underline{/40°} \quad \text{per unit}$$

Also,

$$V_{baseLN} = \frac{V_{baseLL}}{\sqrt{3}} = \frac{480}{\sqrt{3}} = 277 \quad \text{volts}$$

and

$$E_{anp.u.} = \frac{E_{an}}{V_{baseLN}} = \frac{277\underline{/-30°}}{277} = 1.0\underline{/-30°} \quad \text{per unit}$$

The per-unit equivalent circuit is shown in Figure 3.11. The per-unit line current in phase a is then

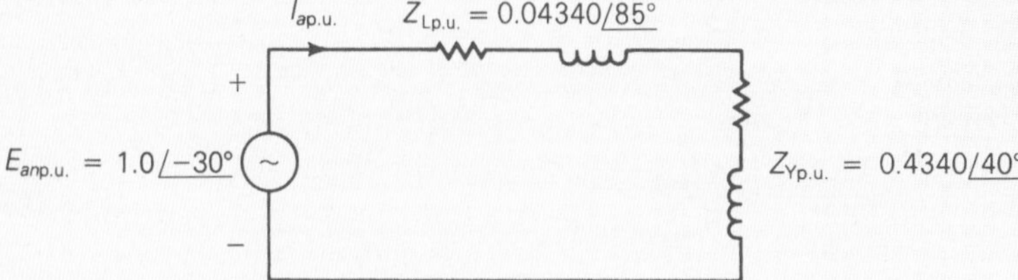

FIGURE 3.11 Circuit for Example 3.5

$$I_{ap.u.} = \frac{E_{anp.u.}}{Z_{Lp.u.} + Z_{Yp.u.}} = \frac{1.0\angle -30^\circ}{0.04340\angle 85^\circ + 0.4340\angle 40^\circ}$$

$$= \frac{1.0\angle -30^\circ}{(0.00378 + j0.04323) + (0.3325 + j0.2790)}$$

$$= \frac{1.0\angle -30^\circ}{0.3362 + j0.3222} = \frac{1.0\angle -30^\circ}{0.4657\angle 43.78^\circ}$$

$$= 2.147\angle -73.78^\circ \text{ per unit}$$

The base current is

$$I_{base} = \frac{S_{base3\phi}}{\sqrt{3}V_{baseLL}} = \frac{10{,}000}{\sqrt{3}(480)} = 12.03 \text{ A}$$

and the actual phase a line current is

$$I_a = (2.147\angle -73.78^\circ)(12.03) = 25.83\angle -73.78^\circ \text{ A}$$

3.4 THREE-PHASE TRANSFORMER CONNECTIONS AND PHASE SHIFT

Three identical single-phase two-winding transformers may be connected to form a three-phase bank. Four ways to connect the windings are Y–Y, Y–Δ, Δ–Y, and Δ–Δ. For example, Figure 3.12 shows a three-phase Y–Y bank. Figure 3.12(a) shows the core and coil arrangements. The American standard for marking three-phase transformers substitutes H1, H2, and H3 on the high-voltage terminals and X1, X2, and X3 on the low-voltage terminals in place of the polarity dots. Also, this text uses uppercase letters *ABC* to identify phases on the high-voltage side of the transformer and lowercase letters *abc* to identify phases on the low-voltage side of the transformer. In Figure 3.12(a), the transformer high-voltage terminals H1, H2, and H3 are connected to phases *A, B,* and *C*, and the low-voltage terminals X1, X2, and X3 are connected to phases *a, b,* and c, respectively.

Figure 3.12(b) shows a schematic representation of the three-phase Y–Y transformer. Windings on the same core are drawn in parallel, and the phasor relationship for balanced positive-sequence operation is shown. For example, high-voltage winding Hl–N is on the same magnetic core as low-voltage winding Xl–n in Figure 3.12(b). Also, V_{AN} is in phase with V_{an}. Figure 3.12(c) shows a single-line diagram of a Y–Y transformer. A single-line diagram shows one phase of a three-phase network with the neutral wire omitted and with components represented by symbols rather than equivalent circuits.

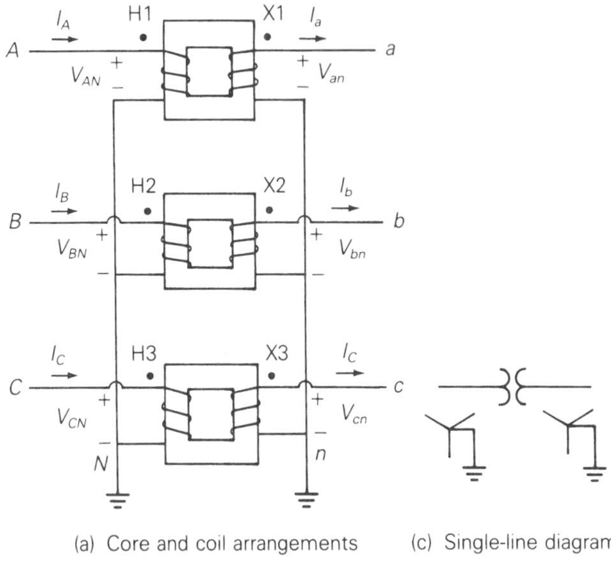

(a) Core and coil arrangements (c) Single-line diagram

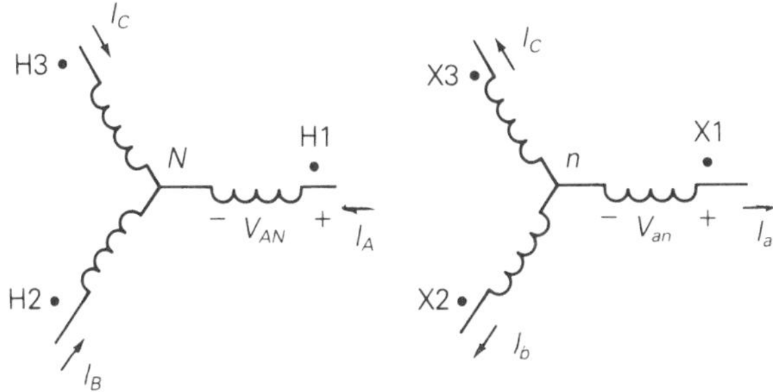

(b) Schematic representation showing phasor relationship for positive sequence operation

FIGURE 3.12 Three-phase two-winding Y–Y transformer bank

The phases of a Y–Y or a Δ–Δ transformer can be labeled so there is no phase shift between corresponding quantities on the low- and high-voltage windings. However, for Y–Δ and Δ–Y transformers, there is always a phase shift. Figure 3.13 shows a Y–Δ transformer. The labeling of the windings and the schematic representation are in accordance with the American standard, which is as follows:

> In either a Y–Δ or Δ–Y transformer, positive-sequence quantities on the high-voltage side shall lead their corresponding quantities on the low-voltage side by 30°.

As shown in Figure 3.13(b), V_{AN} leads V_{an} by 30°.

The positive-sequence phasor diagram shown in Figure 3.13(b) can be constructed via the following five steps, which are also indicated in Figure 3.13.

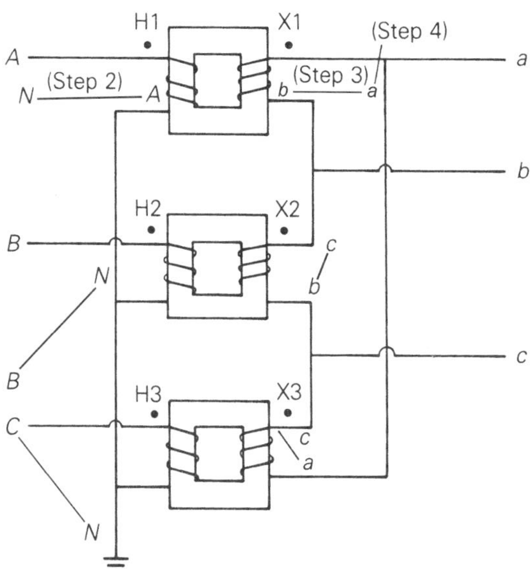

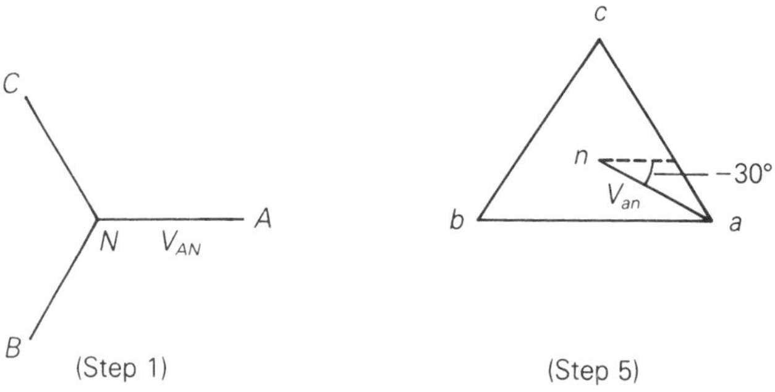

FIGURE 3.13 Three-phase two-winding Y–Δ transformer bank

STEP 1 Assume that balanced positive-sequence voltages are applied to the Y winding. Draw the positive-sequence phasor diagram for these voltages.

STEP 2 Move phasor *A–N* next to terminals *A–N* in Figure 3.13(a). Identify the ends of this line in the same manner as in the phasor diagram. Similarly, move phasors *B–N* and *C–N* next to terminals *B–N* and *C–N* in Figure 3.13(a).

STEP 3 For each single-phase transformer, the voltage across the low-voltage winding must be in phase with the voltage across the high-voltage winding, assuming an ideal transformer. Therefore, draw a line next to each low-voltage winding parallel to the corresponding line already drawn next to the high-voltage winding.

STEP 4 Label the ends of the lines drawn in Step 3 by inspecting the polarity marks. For example, phase *A* is connected to dotted terminal H1, and *A* appears on the *right* side of line *A–N*. Therefore, phase *a*, which is connected to dotted terminal X1, must be on the *right* side, and *b* on the left side of line *a–b*. Similarly, phase *B* is connected to dotted terminal H2, and *B* is *down* on line *B–N*. Therefore, phase *b*, connected to dotted terminal X2, must be *down* on line *b–c*. Similarly, *c* is *up* on line *c–a*.

STEP 5 Bring the three lines labeled in Step 4 together to complete the phasor diagram for the low-voltage *A* winding. Note that V_{AN} leads V_{an} by 30° in accordance with the American standard.

EXAMPLE 3.6

Phase shift in Δ–Y transformers

Assume that balanced negative-sequence voltages are applied to the high-voltage windings of the Y–Δ transformer shown in Figure 3.13. Determine the negative-sequence phase shift of this transformer.

SOLUTION

The negative-sequence diagram, shown in Figure 3.14, is constructed from the following five steps, as outlined previously:

STEP 1 Draw the phasor diagram of balanced negative-sequence voltages, which are applied to the Y winding.

STEP 2 Move the phasors *A–N*, *B–N*, and *C–N* next to the high-voltage Y windings.

STEP 3 For each single-phase transformer, draw a line next to the low-voltage winding that is parallel to the line drawn in Step 2 next to the high-voltage winding.

STEP 4 Label the lines drawn in Step 3. For example, phase *B*, which is connected to dotted terminal H2, is shown *up* on line *B–N*; therefore phase *b*, which is connected to dotted terminal X2, must be *up* on line *b–c*.

STEP 5 Bring the lines drawn in Step 4 together to form the negative-sequence phasor diagram for the low-voltage Δ winding.

As shown in Figure 3.14, the high-voltage phasors *lag* the low-voltage phasors by 30°. Thus the negative-sequence phase shift is the reverse of the positive-sequence phase shift.

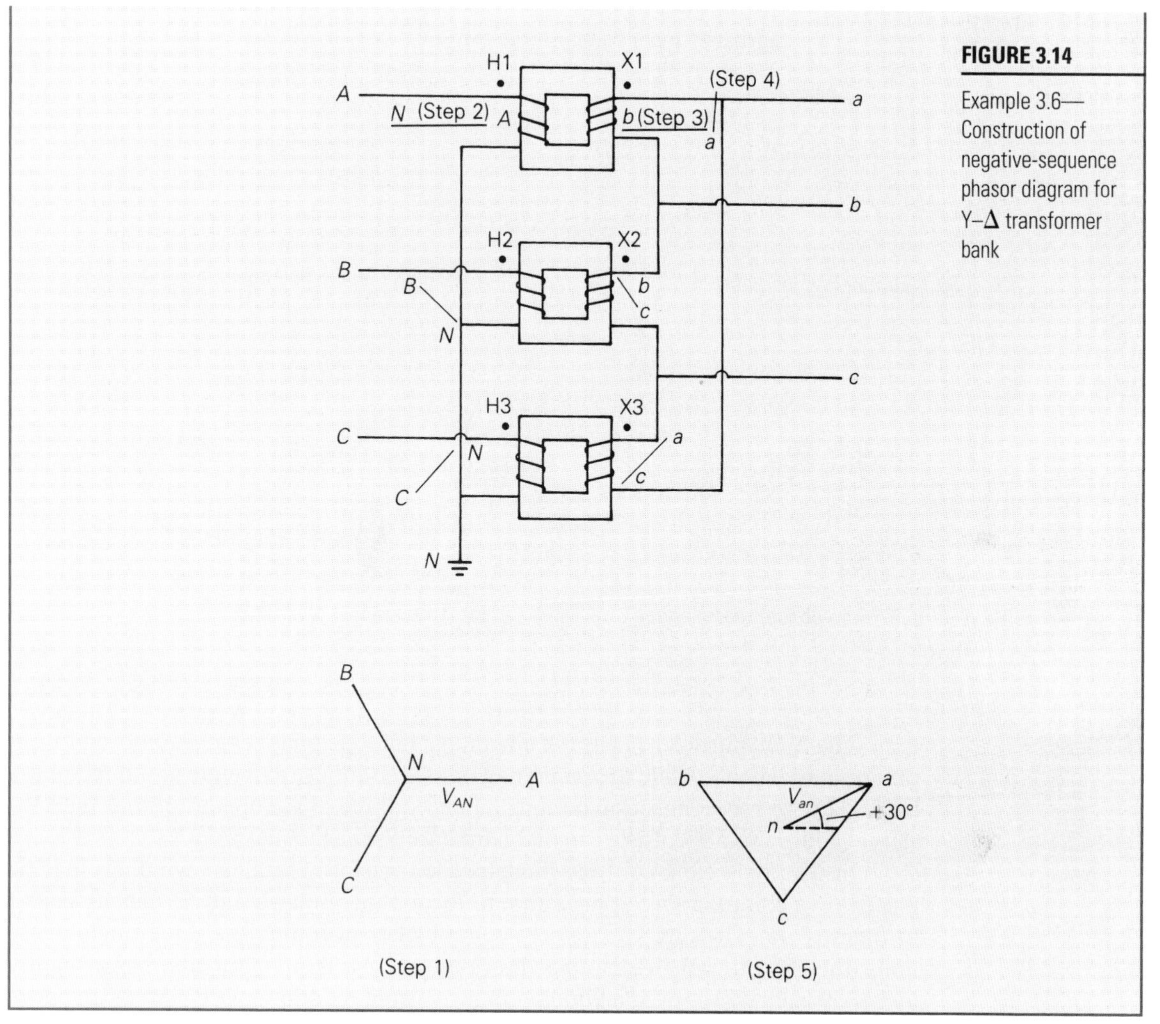

FIGURE 3.14

Example 3.6—Construction of negative-sequence phasor diagram for Y–Δ transformer bank

The Δ–Y transformer is commonly used as a generator step-up transformer, where the Δ winding is connected to the generator terminals and the Y winding is connected to a transmission line. One advantage of a high-voltage Y winding is that a neutral point N is provided for grounding on the high-voltage side. With a permanently grounded neutral, the insulation requirements for the high-voltage transformer windings are reduced. The high-voltage insulation can be graded or tapered from maximum insulation at terminals ABC to minimum insulation at grounded terminal N. One advantage of the Δ winding is that the undesirable third harmonic magnetizing current, caused by the nonlinear core B–H characteristic, remains trapped inside the Δ winding. Third harmonic currents are (triple-frequency) zero-sequence currents, which cannot enter or leave a Δ connection, but can flow within the Δ. The Y–Y transformer is seldom used because of difficulties with third harmonic exciting current.

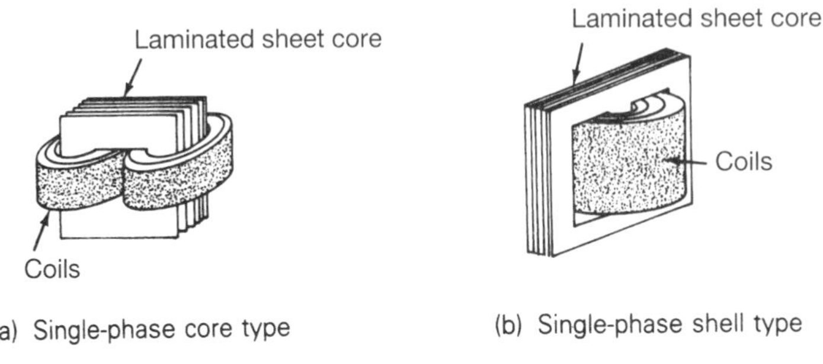

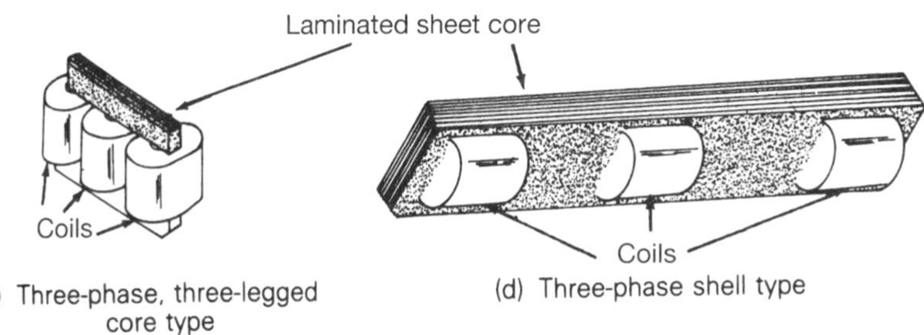

FIGURE 3.15 Transformer core configurations

The Δ–Δ transformer has the advantage that one phase can be removed for repair or maintenance while the remaining phases continue to operate as a three-phase bank. This *open*-Δ connection permits balanced three-phase operation with the kVA rating reduced to 58% of the original bank (see Problem 3.36).

Instead of a bank of three single-phase transformers, all six windings may be placed on a common three-phase core to form a three-phase transformer, as shown in Figure 3.15. The three-phase core contains less iron than the three single-phase units; therefore it costs less, weighs less, requires less floor space, and has a slightly higher efficiency. However, a winding failure would require replacement of an entire three-phase transformer, compared to replacement of only one phase of a three-phase bank.

3.5 PER-UNIT EQUIVALENT CIRCUITS OF BALANCED THREE-PHASE TWO-WINDING TRANSFORMERS

Figure 3.16(a) is a schematic representation of an ideal Y–Y transformer grounded through neutral impedances Z_N and Z_n. Figure 3.16(b) shows the per-unit equivalent circuit of this ideal transformer for balanced three-phase operation. Throughout the remainder of this text, per-unit quantities are used unless otherwise indicated. Also, the subscript "p.u.," used to indicate a per-unit quantity, is omitted in most cases.

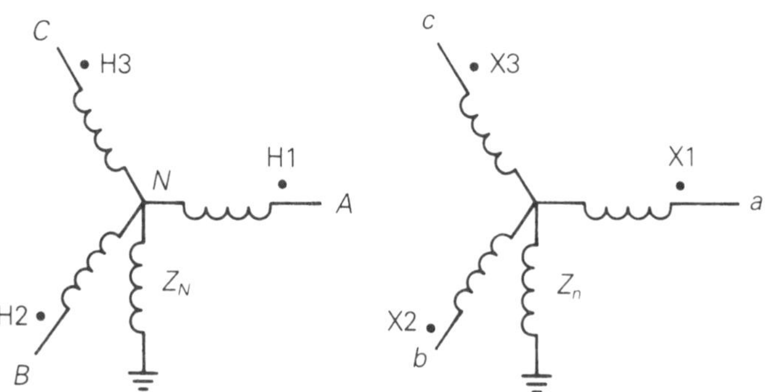

(a) Schematic representation

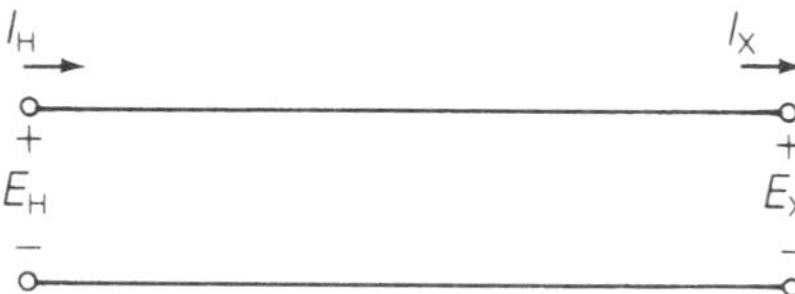

(b) Per-unit equivalent circuit for balanced three-phase operation

FIGURE 3.16

Ideal Y–Y transformer

The following are two conventional rules for selecting base quantities:

1. A common S_{base} is selected for both the H and X terminals.
2. The ratio of the voltage bases V_{baseH}/V_{baseX} is selected to be equal to the ratio of the rated line-to-line voltages $V_{ratedHLL}/V_{ratedXLL}$.

When balanced three-phase currents are applied to the transformer, the neutral currents are zero, and there are no voltage drops across the neutral impedances. Therefore, the per-unit equivalent circuit of the ideal Y–Y transformer, Figure 3.16(b), is the same as the per-unit single-phase ideal transformer, Figure 3.9(a).

The per-unit equivalent circuit of a practical Y–Y transformer is shown in Figure 3.17(a). This network is obtained by adding external impedances to the equivalent circuit of the ideal transformer, as in Figure 3.9(c).

The per-unit equivalent circuit of the Y–Δ transformer, shown in Figure 3.17(b), includes a phase shift. For the American standard, the positive-sequence voltages and currents on the high-voltage side of the Y–Δ transformer lead the corresponding quantities on the low-voltage side by 30°. The phase shift in the equivalent circuit of Figure 3.17(b) is represented by the phase-shifting transformer of Figure 3.4.

The per-unit equivalent circuit of the Δ–Δ transformer, shown in Figure 3.17(c), is the same as that of the Y–Y transformer. It is assumed that the windings are labeled so there is no phase shift. Also, the per-unit impedances do not depend on the winding connections, but the base voltages do.

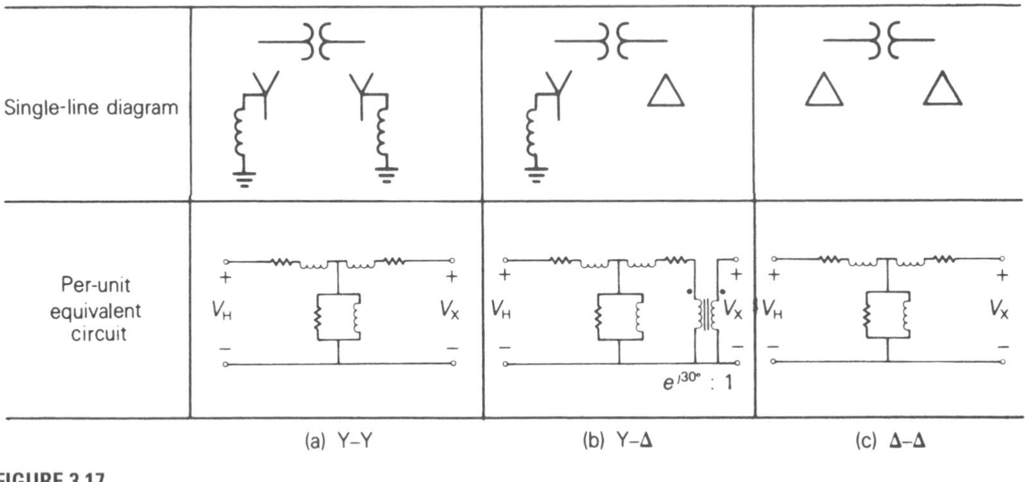

FIGURE 3.17

Per-unit equivalent circuits of practical Y–Y, Y–Δ, and Δ–Δ transformers for balanced three-phase operation

EXAMPLE 3.7

Voltage calculations: balanced Y–Y and Δ–Y transformers

Three single-phase two-winding transformers, each rated 400 MVA, 13.8/199.2 kV, with leakage reactance $X_{eq} = 0.10$ per unit, are connected to form a three-phase bank. Winding resistances and exciting current are neglected. The high-voltage windings are connected in Y. A three-phase load operating under balanced positive-sequence conditions on the high-voltage side absorbs 1000 MVA at 0.90 p.f. lagging, with $V_{AN} = 199.2\underline{/0°}$ kV. Determine the voltage V_{an} at the low-voltage bus if the low-voltage windings are connected (a) in Y and (b) in Δ.

SOLUTION

The per-unit network is shown in Figure 3.18. Using the transformer bank ratings as base quantities, $S_{base3\phi} = 1200$ MVA, $V_{baseHLL} = 345$ kV, and $I_{baseH} = 1200/(345\sqrt{3}) = 2.008$ kA. The per-unit load voltage and load current are then

$$V_{AN} = 1.0\underline{/0°} \quad \text{per unit}$$

$$I_A = \frac{1000/(345\sqrt{3})}{2.008}\underline{/-\cos^{-1}0.9} = 0.8333\underline{/-25.84°} \quad \text{per unit}$$

a. For the Y–Y transformer, Figure 3.18(a),

$$I_a = I_A = 0.8333\underline{/-25.84°} \quad \text{per unit}$$

FIGURE 3.18

Per-unit network for Example 3.7

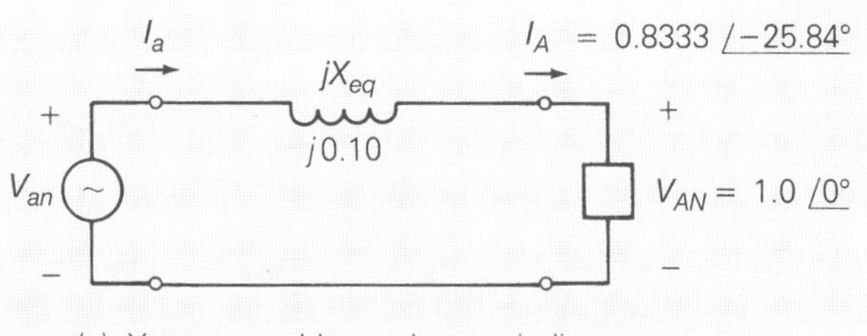

(a) Y-connected low-voltage windings

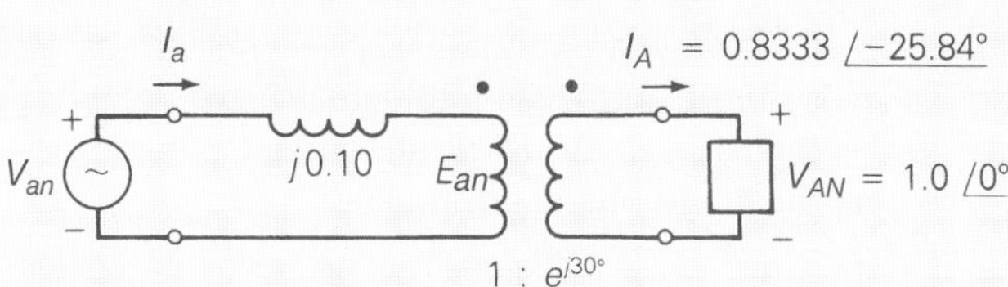

(b) Δ-connected low-voltage windings

$$\begin{aligned} V_{an} &= V_{AN} + (jX_{eq})I_A \\ &= 1.0\underline{/0°} + (j0.10)(0.8333\underline{/-25.84°}) \\ &= 1.0 + 0.08333\underline{/64.16°} = 1.0363 + j0.0750 = 1.039\underline{/4.139°} \\ &= 1.039\underline{/4.139°} \text{ per unit} \end{aligned}$$

Further, since $V_{\text{base}XLN} = 13.8$ kV for the low-voltage Y windings, $V_{an} = 1.039(13.8) = 14.34$ kV, and

$$V_{an} = 14.34\underline{/4.139°} \text{ kV}$$

b. For the Δ–Y transformer, Figure 3.18(b),

$$E_{an} = e^{-j30°}V_{AN} = 1.0\underline{/-30°} \text{ per unit}$$

$$I_a = e^{-j30°}I_A = 0.8333\underline{/-25.84° - 30°} = 0.8333\underline{/-55.84°} \text{ per unit}$$

$$V_{an} = E_{an} + (jX_{eq})I_a = 1.0\underline{/-30°} + (j0.10)(0.8333\underline{/-55.84°})$$

$$V_{an} = 1.039\underline{/-25.861°} \text{ per unit}$$

Further, since $V_{\text{baseXLN}} = 13.8/\sqrt{3} = 7.967$ kV for the low-voltage Δ windings, $V_{an} = (1.039)(7.967) = 8.278$ kV, and

$$V_{an} = 8.278\underline{/-25.861°} \text{ kV}$$

EXAMPLE 3.8

Per-unit voltage drop and per-unit fault current: balanced three-phase transformer

A 200-MVA, 345-kVΔ/34.5-kV Y substation transformer has an 8% leakage reactance. The transformer acts as a connecting link between 345-kV transmission and 34.5-kV distribution. Transformer winding resistances and exciting current are neglected. The high-voltage bus connected to the transformer is assumed to be an ideal 345-kV positive-sequence source with negligible source impedance. Using the transformer ratings as base values, determine:

a. The per-unit magnitudes of transformer voltage drop and voltage at the low-voltage terminals when rated transformer current at 0.8 p.f. lagging enters the high-voltage terminals.

b. The per-unit magnitude of the fault current when a three-phase-to-ground bolted short circuit occurs at the low-voltage terminals.

SOLUTION

In both parts (a) and (b), only balanced positive-sequence current will flow, since there are no imbalances. Because the only interest is in voltage and current magnitudes, the Δ–Y transformer phase shift can be omitted.

a. As shown in Figure 3.19(a),

$$V_{\text{drop}} = I_{\text{rated}}\text{X}_{\text{eq}} = (1.0)(0.08) = 0.08 \quad \text{per unit and}$$

$$\begin{aligned} V_{an} &= V_{AN} - (j\text{X}_{\text{eq}})I_{\text{rated}} \\ &= 1.0\underline{/0^\circ} - (j0.08)(1.0\underline{/-36.87^\circ}) \\ &= 1.0 - (j0.08)(0.8 - j0.6) = 0.952 - j0.064 \\ &= 0.954\underline{/-3.85^\circ} \quad \text{per unit} \end{aligned}$$

b. As shown in Figure 3.19(b),

$$I_{\text{SC}} = \frac{V_{AN}}{\text{X}_{\text{eq}}} = \frac{1.0}{0.08} = 12.5 \quad \text{per unit}$$

Under rated current conditions [part (a)], the 0.08 per-unit voltage drop across the transformer leakage reactance causes the voltage at the low-voltage terminals to be 0.954 per unit. Also, under three-phase short-circuit conditions [part (b)], the fault current is 12.5 times the rated transformer current. This example illustrates a compromise in the design or specification of transformer leakage reactance. A low value is desired to minimize voltage drops, but a high value is desired to limit fault currents. Typical transformer leakage reactances are given in Table A.2 in the Appendix.

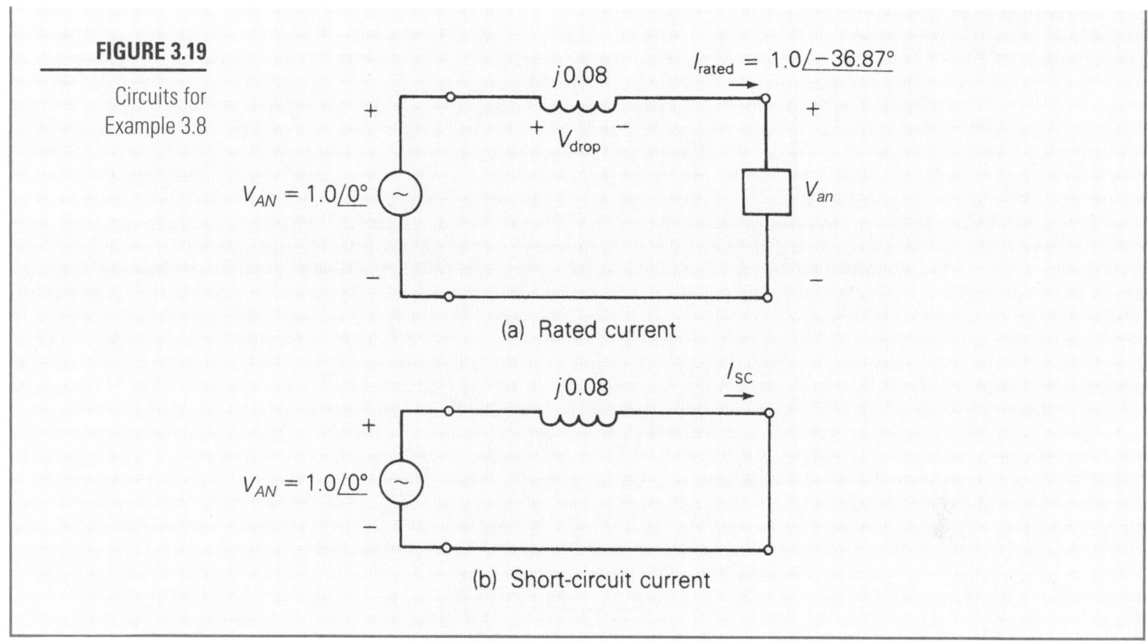

FIGURE 3.19 Circuits for Example 3.8

3.6 THREE-WINDING TRANSFORMERS

Figure 3.20(a) shows a basic single-phase three-winding transformer. The ideal transformer relations for a two-winding transformer, (3.1.8) and (3.1.14), easily can be extended to obtain corresponding relations for an ideal three-winding transformer. In actual units, these relations are

$$N_1 I_1 = N_2 I_2 + N_3 I_3 \tag{3.6.1}$$

$$\frac{E_1}{N_1} = \frac{E_2}{N_2} = \frac{E_3}{N_3} \tag{3.6.2}$$

where I_1 enters the dotted terminal, I_2 and I_3 leave dotted terminals, and E_1, E_2, and E_3 have their + polarities at dotted terminals. In per-unit, (3.6.1) and (3.6.2) are

$$I_{1\text{p.u.}} = I_{2\text{p.u.}} + I_{3\text{p.u.}} \tag{3.6.3}$$

$$E_{1\text{p.u.}} = E_{2\text{p.u.}} = E_{3\text{p.u.}} \tag{3.6.4}$$

where a common S_{base} is selected for all three windings, and voltage bases are selected in proportion to the rated voltages of the windings. These two per-unit relations are satisfied by the per-unit equivalent circuit shown in Figure 3.20(b). Also, external series impedance and shunt admittance branches are included in the practical three-winding transformer circuit shown in Figure 3.20(c). The shunt

FIGURE 3.20 Single-phase three-winding transformer

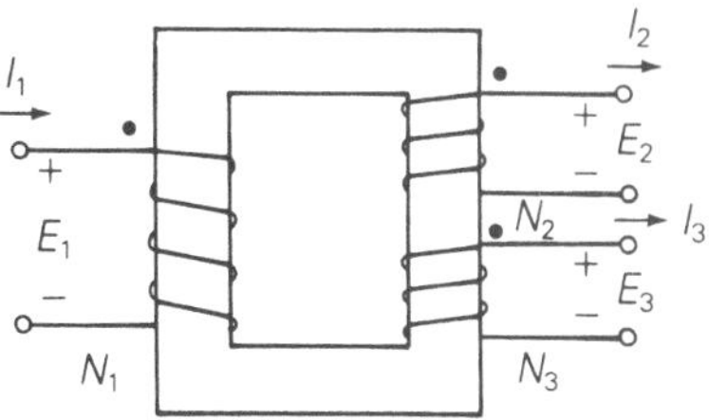

(a) Basic core and coil configuration

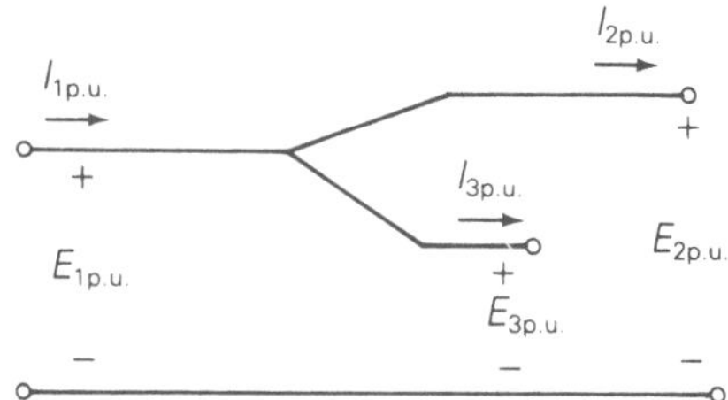

(b) Per-unit equivalent circuit—ideal transformer

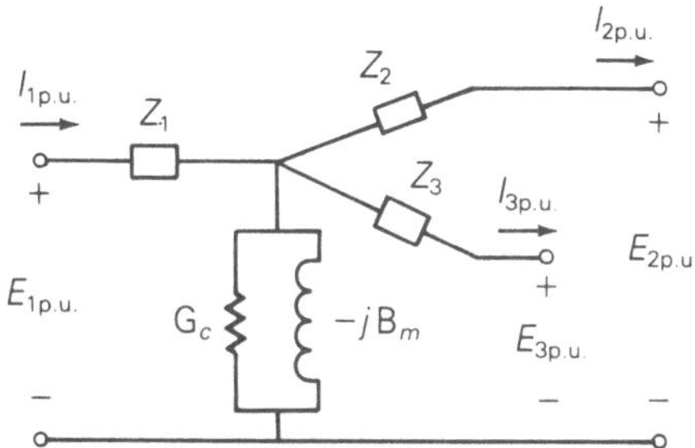

(c) Per-unit equivalent circuit—practical transformer

admittance branch, a core loss resistor in parallel with a magnetizing inductor, can be evaluated from an open-circuit test. Also, when one winding is left open, the three-winding transformer behaves as a two-winding transformer, and standard short-circuit tests can be used to evaluate per-unit leakage impedances, which are defined as follows:

Z_{12} = per-unit leakage impedance measured from winding 1, with winding 2 shorted and winding 3 open

Z_{13} = per-unit leakage impedance measured from winding 1, with winding 3 shorted and winding 2 open

Z_{23} = per-unit leakage impedance measured from winding 2, with winding 3 shorted and winding 1 open

From Figure 3.20(c), with winding 2 shorted and winding 3 open, the leakage impedance measured from winding 1 is, neglecting the shunt admittance branch,

$$Z_{12} = Z_1 + Z_2 \tag{3.6.5}$$

Similarly,

$$Z_{13} = Z_1 + Z_3 \tag{3.6.6}$$

and

$$Z_{23} = Z_2 + Z_3 \tag{3.6.7}$$

Solving (3.6.5) through (3.6.7),

$$Z_1 = \tfrac{1}{2}(Z_{12} + Z_{13} - Z_{23}) \qquad (3.6.8)$$

$$Z_2 = \tfrac{1}{2}(Z_{12} + Z_{23} - Z_{13}) \qquad (3.6.9)$$

$$Z_3 = \tfrac{1}{2}(Z_{13} + Z_{23} - Z_{12}) \qquad (3.6.10)$$

Equations (3.6.8) through (3.6.10) can be used to evaluate the per-unit series impedances Z_1, Z_2, and Z_3 of the three-winding transformer equivalent circuit from the per-unit leakage impedances Z_{12}, Z_{13}, and Z_{23}, which, in turn, are determined from short-circuit tests.

Note that each of the windings on a three-winding transformer may have a *different* kVA rating. If the leakage impedances from short-circuit tests are expressed in per-unit based on winding ratings, they must first be converted to per-unit on a common S_{base} *before* they are used in (3.6.8) through (3.6.10).

EXAMPLE 3.9

Three-winding single-phase transformer: per-unit impedances

The ratings of a single-phase three-winding transformer are

winding 1: 300 MVA, 13.8 kV
winding 2: 300 MVA, 199.2 kV
winding 3: 50 MVA, 19.92 kV

The leakage reactances, from short-circuit tests, are

$X_{12} = 0.10$ per unit on a 300-MVA, 13.8-kV base
$X_{13} = 0.16$ per unit on a 50-MVA, 13.8-kV base
$X_{23} = 0.14$ per unit on a 50-MVA, 199.2-kV base

Winding resistances and exciting current are neglected. Calculate the impedances of the per-unit equivalent circuit using a base of 300 MVA and 13.8 kV for terminal 1.

SOLUTION

$S_{base} = 300$ MVA is the same for all three terminals. Also, the specified voltage base for terminal 1 is $V_{base1} = 13.8$ kV. The base voltages for terminals 2 and 3 are then $V_{base2} = 199.2$ kV and $V_{base3} = 19.92$ kV, which are the rated voltages of these windings. From the data given, $X_{12} = 0.10$ per unit was measured from terminal 1 using the same base values as those specified for the circuit. However, $X_{13} = 0.16$ and $X_{23} = 0.14$ per unit on a 50-MVA base are first converted to the 300-MVA circuit base.

(Continued)

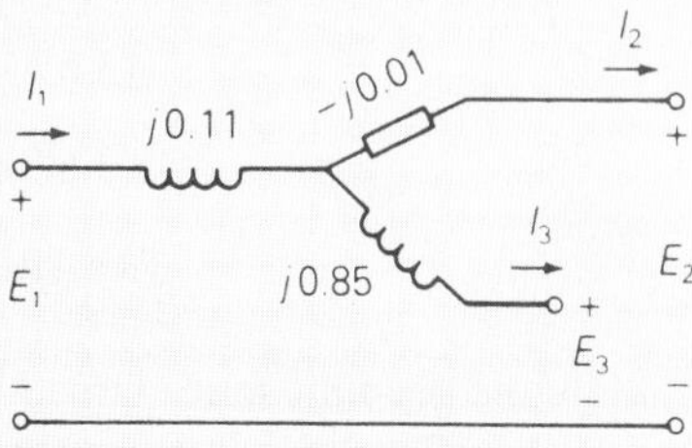

FIGURE 3.21 Circuit for Example 3.9

$$X_{13} = (0.16)\left(\frac{300}{50}\right) = 0.96 \quad \text{per unit}$$

$$X_{23} = (0.14)\left(\frac{300}{50}\right) = 0.84 \quad \text{per unit}$$

Then, from (3.6.8) through (3.6.10),

$$X_1 = \tfrac{1}{2}(0.10 + 0.96 - 0.84) = \quad 0.11 \quad \text{per unit}$$

$$X_2 = \tfrac{1}{2}(0.10 + 0.84 - 0.96) = -0.01 \quad \text{per unit}$$

$$X_3 = \tfrac{1}{2}(0.84 + 0.96 - 0.10) = \quad 0.85 \quad \text{per unit}$$

The per-unit equivalent circuit of this three-winding transformer is shown in Figure 3.21. Note that X_2 is negative. This illustrates the fact that X_1, X_2, and X_3 are *not* leakage reactances, but instead are equivalent reactances derived from the leakage reactances. Leakage reactances are always positive.

Note also that the node where the three equivalent circuit reactances are connected does not correspond to any physical location within the transformer. Rather, it is simply part of the equivalent circuit representation.

EXAMPLE 3.10

Three-winding three-phase transformer: balanced operation

Three transformers, each identical to that described in Example 3.9, are connected as a three-phase bank in order to feed power from a 900-MVA, 13.8-kV generator to a 345-kV transmission line and to a 34.5-kV distribution line. The transformer windings are connected as follows:

13.8-kV windings (X): Δ, to generator

199.2-kV windings (H): solidly grounded Y, to 345-kV line

19.92-kV windings (M): grounded Y through $Z_n = j0.10\ \Omega$, to 34.5-kV line

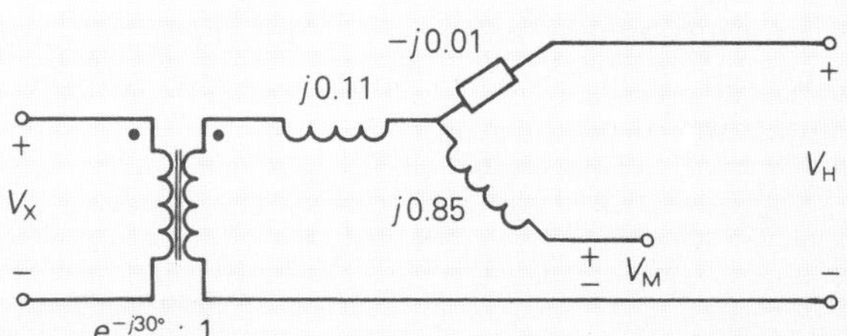

FIGURE 3.22 Per-unit network for Example 3.10

The positive-sequence voltages and currents of the high- and medium-voltage Y windings lead the corresponding quantities of the low-voltage Δ winding by 30°. Draw the per-unit network, using a three-phase base of 900 MVA and 13.8 kV for terminal X. Assume balanced positive-sequence operation.

SOLUTION

The per-unit network is shown in Figure 3.22. $V_{baseX} = 13.8$ kV, which is the rated line-to-line voltage of terminal X. Since the M and H windings are Y-connected, $V_{baseM} = \sqrt{3}(19.92) = 34.5$ kV, and $V_{baseH} = \sqrt{3}(199.2) = 345$ kV, which are the rated line-to-line voltages of the M and H windings. Also, a phase-shifting transformer is included in the network. The neutral impedance is not included in the network, since there is no neutral current under balanced operation.

3.7 AUTOTRANSFORMERS

A single-phase two-winding transformer is shown in Figure 3.23(a) with two separate windings, which is the usual two-winding transformer; the same transformer is shown in Figure 3.23(b) with the two windings connected in series, which is called an *autotransformer.* For the usual transformer [Figure 3.23(a)], the two windings are coupled magnetically via the mutual core flux. For the autotransformer [Figure 3.23(b)], the windings are both electrically and magnetically coupled. The autotransformer has smaller per-unit leakage impedances than the usual transformer; this results in both smaller series-voltage drops (an advantage) and higher short-circuit currents

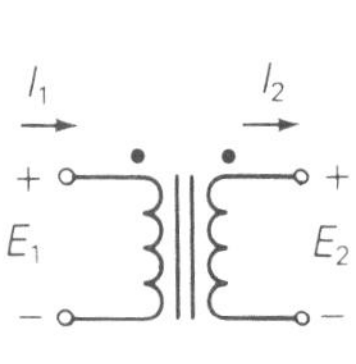

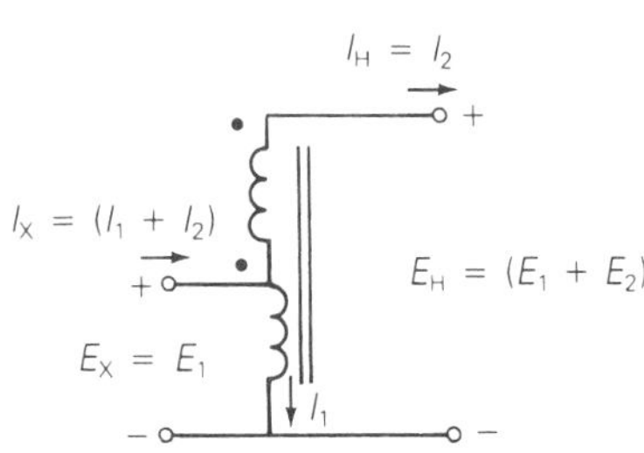

FIGURE 3.23 Ideal single-phase transformers

(a disadvantage). The autotransformer also has lower per-unit losses (higher efficiency), lower exciting current, and lower cost if the turns ratio is not too large. The electrical connection of the windings, however, allows transient overvoltages to pass through the autotransformer more easily.

EXAMPLE 3.11

Autotransformer: single-phase

The single-phase two-winding 20-kVA, 480/120-volt transformer of Example 3.3 is connected as an autotransformer, as in Figure 3.23(b), where winding 1 is the 120-volt winding. For this autotransformer, determine (a) the voltage ratings E_X and E_H of the low- and high-voltage terminals, (b) the kVA rating, and (c) the per-unit leakage impedance.

SOLUTION

a. Since the 120-volt winding is connected to the low-voltage terminal, $E_X =$ 120 volts. When $E_X = E_1 = 120$ volts is applied to the low-voltage terminal, $E_2 =$ 480 volts is induced across the 480-volt winding, neglecting the voltage drop across the leakage impedance. Therefore, $E_H = E_1 + E_2 = 120 + 480 = 600$ volts.

b. As a normal two-winding transformer rated 20 kVA, the rated current of the 480-volt winding is $I_2 = I_H = 20{,}000/480 = 41.667$ A. As an autotransformer, the 480-volt winding can carry the same current. Therefore, the kVA rating $S_H = E_H I_H = (600)\,(41.667) = 25$ kVA. Note also that when $I_H = I_2 = 41.667$ A, a current $I_1 = (480/120)(41.667) = 166.7$ A is induced in the 120-volt winding. Therefore, $I_x = I_1 + I_2 = 208.3$ A (neglecting exciting current) and $S_X = E_X I_X =$ $(120)(208.3) = 25$ kVA, which is the same rating as calculated for the high-voltage terminal.

c. From Example 3.3, the leakage impedance is $0.0729/\underline{78.13°}$ per unit as a normal, two-winding transformer. As an autotransformer, the leakage impedance *in ohms* is the same as for the normal transformer, since the core and windings are the same for both (only the external winding connections are different). However, the base impedances are different. For the high-voltage terminal, using (3.3.4),

$$Z_{\text{baseHold}} = \frac{(480)^2}{20{,}000} = 11.52\ \Omega \quad \text{as a normal transformer}$$

$$Z_{\text{baseHnew}} = \frac{(600)^2}{25{,}000} = 14.4\ \ \Omega \quad \text{as an autotransformer}$$

Therefore, using (3.3.10),

$$Z_{\text{p.u.new}} = (0.0729\underline{/78.13°})\left(\frac{11.52}{14.4}\right) = 0.05832\underline{/78.13°} \quad \text{per unit}$$

For this example, the rating is 25 kVA, 120/600 volts as an autotransformer versus 20 kVA, 120/480 volts as a normal transformer. The autotransformer has both a larger kVA rating and a larger voltage ratio for the same cost. Also, the per-unit leakage impedance of the autotransformer is smaller. However, the increased high-voltage rating as well as the electrical connection of the windings may require more insulation for both windings.

3.8 TRANSFORMERS WITH OFF-NOMINAL TURNS RATIOS

It has been shown that models of transformers that use per-unit quantities are simpler than those that use actual quantities. The ideal transformer winding is eliminated when the ratio of the selected voltage bases equals the ratio of the voltage ratings of the windings. In some cases, however, it is impossible to select voltage bases in this manner. For example, consider the two transformers connected in parallel in Figure 3.24. Transformer T_1 is rated 13.8/345 kV and T_2 is rated 13.2/345 kV. If $\text{V}_{\text{baseH}} = 345$ kV is selected, then transformer T_1 requires $\text{V}_{\text{baseX}} = 13.8$ kV and T_2 requires $\text{V}_{\text{baseX}} = 13.2$ kV. It is clearly impossible to select the appropriate voltage bases for both transformers.

To accommodate this situation, develop a per-unit model of a transformer whose voltage ratings are not in proportion to the selected base voltages. Such a transformer is said to have an off-nominal turns ratio. Figure 3.25(a) shows a transformer with rated voltages $V_{1\text{rated}}$ and $V_{2\text{rated}}$, which satisfy

$$V_{1\text{rated}} = a_t V_{2\text{rated}} \tag{3.8.1}$$

where a_t is assumed, in general, to be either real or complex. Suppose the selected voltage bases satisfy

$$\text{V}_{\text{base1}} = b\text{V}_{\text{base2}} \tag{3.8.2}$$

Defining $c = \dfrac{a_t}{b}$, (3.8.1) can be rewritten as

$$V_{1\text{rated}} = b\left(\frac{a_t}{b}\right) V_{2\text{rated}} = bc\ V_{2\text{rated}} \tag{3.8.3}$$

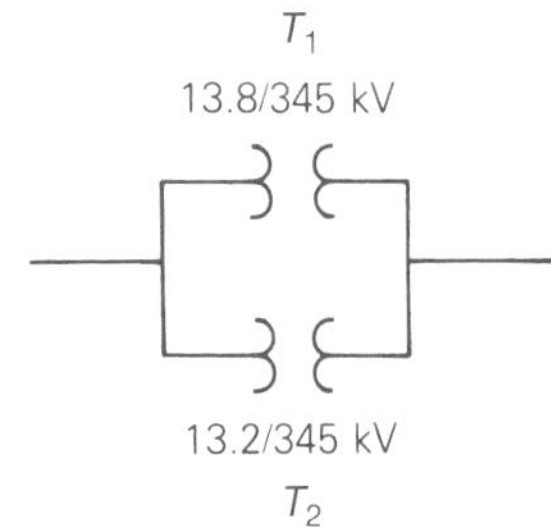

FIGURE 3.24

Two transformers connected in parallel

FIGURE 3.25 Transformer with off-nominal turns ratio

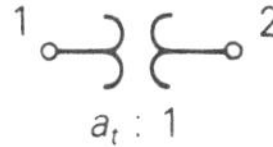

(a) Single-line diagram

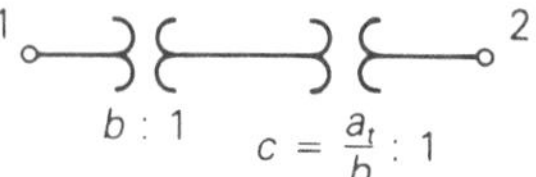

(b) Represented as two transformers in series

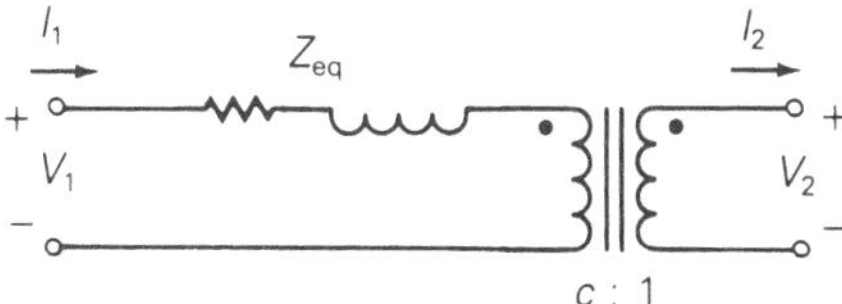

(c) Per-unit equivalent circuit (Per-unit impedance is shown)

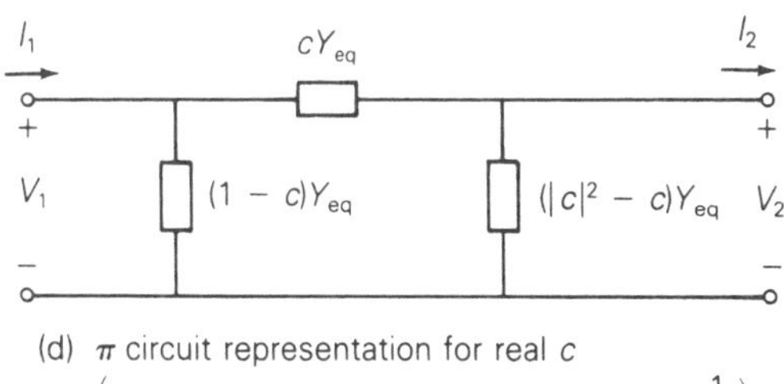

(d) π circuit representation for real c $\left(\text{Per-unit admittances are shown; } Y_{eq} = \frac{1}{Z_{eq}}\right)$

Equation (3.8.3) can be represented by two transformers in series, as shown in Figure 3.25(b). The first transformer has the same ratio of rated winding voltages as the ratio of the selected base voltages, b. Therefore, this transformer has a standard per-unit model, as shown in Figure 3.9 or 3.17. Assume that the second transformer is ideal, and all real and reactive losses are associated with the first transformer. The resulting per-unit model is shown in Figure 3.25(c), where, for simplicity, the shunt-exciting branch is neglected. Note that if $a_t = b$, then the ideal transformer winding shown in this figure can be eliminated, since its turns ratio $c = (a_t/b) = 1$.

The per-unit model shown in Figure 3.25(c) is perfectly valid, but it is not suitable for some of the computer programs presented in later chapters because these programs do not accommodate ideal transformer windings. An alternative representation can be developed, however, by writing nodal equations for this figure as follows:

$$\begin{bmatrix} I_1 \\ -I_2 \end{bmatrix} = \begin{bmatrix} Y_{11} & Y_{12} \\ Y_{21} & Y_{22} \end{bmatrix} \begin{bmatrix} V_1 \\ V_2 \end{bmatrix} \tag{3.8.4}$$

where both I_1 and $-I_2$ are referenced *into* their nodes in accordance with the nodal equation method (Section 2.4). Recalling two-port network theory, the admittance parameters of (3.8.4) are, from Figure 3.23(c)

$$Y_{11} = \left.\frac{I_1}{V_1}\right|_{V_2 = 0} = \frac{1}{Z_{eq}} = Y_{eq} \tag{3.8.5}$$

$$Y_{22} = \left.\frac{-I_2}{V_2}\right|_{V_1 = 0} = \frac{1}{Z_{eq}/|c|^2} = |c|^2 Y_{eq} \tag{3.8.6}$$

$$Y_{12} = \left.\frac{I_1}{V_2}\right|_{V_1 = 0} = \frac{-cV_2/Z_{eq}}{V_2} = -cY_{eq} \tag{3.8.7}$$

$$Y_{21} = \left.\frac{-I_2}{V_1}\right|_{V_2 = 0} = \frac{-c^*I_1}{V_1} = -c^*Y_{eq} \tag{3.8.8}$$

Equations (3.8.4) through (3.8.8) with real or complex c are convenient for representing transformers with off-nominal turns ratios in the computer programs presented later. Note that when c is complex, Y_{12} is not equal to Y_{21}, and the preceding admittance parameters cannot be synthesized with a passive RLC circuit. However, the π network shown in Figure 3.25(d), which has the same admittance parameters as (3.8.4) through (3.8.8), can be synthesized for real c. Note also that when $c = 1$, the shunt branches in this figure become open circuits (zero per unit siemens), and the series branch becomes Y_{eq} per unit siemens (or Z_{eq} per unit ohms).

EXAMPLE 3.12

Tap-changing three-phase transformer: per-unit positive-sequence network

A three-phase generator step-up transformer is rated 1000 MVA, 13.8 kV Δ/345 kV Y with $Z_{eq} = j0.10$ per unit. The transformer high-voltage winding has ±10% taps. The system base quantities are

$$S_{base3\phi} = 500 \quad \text{MVA}$$

$$V_{baseXLL} = 13.8 \quad \text{kV}$$

$$V_{baseHLL} = 345 \quad \text{kV}$$

Determine the per-unit equivalent circuit for the following tap settings:

a. Rated tap
b. –10% tap (providing a 10% voltage decrease for the high-voltage winding)

Assume balanced positive-sequence operation. Neglect transformer winding resistance, exciting current, and phase shift.

(Continued)

SOLUTION

a. Using (3.8.1) and (3.8.2) with the low-voltage winding denoted winding 1,

$$a_t = \frac{13.8}{345} = 0.04 \qquad b = \frac{V_{baseXLL}}{V_{baseHLL}} = \frac{13.8}{345} = a_t \qquad c = 1$$

From (3.3.11),

$$Z_{p.u.new} = (j0.10)\left(\frac{500}{1000}\right) = j0.05 \quad \text{per unit}$$

The per-unit equivalent circuit, not including winding resistance, exciting current, and phase shift is

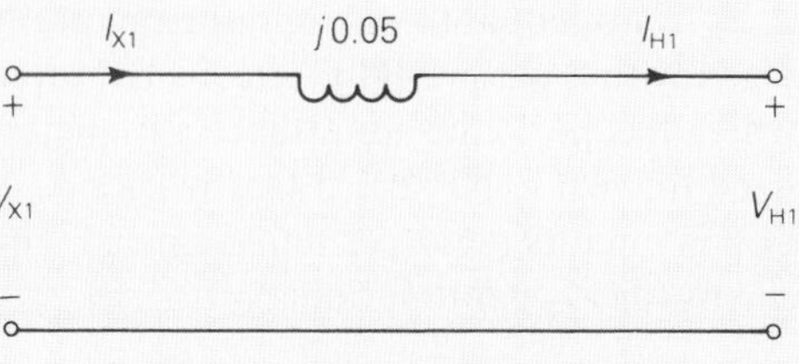

(Per-unit impedance is shown)

b. Using (3.8.1) and (3.8.2),

$$a_t = \frac{13.8}{345(0.9)} = 0.04444 \qquad b = \frac{13.8}{345} = 0.04$$

$$c = \frac{a_t}{b} = \frac{0.04444}{0.04} = 1.1111$$

From Figure 3.23(d),

$$cY_{eq} = 1.1111\left(\frac{1}{j0.05}\right) = -j22.22 \quad \text{per unit}$$

$$(1 - c)Y_{eq} = (-0.11111)(-j20) = +j2.222 \quad \text{per unit}$$

$$(|c|^2 - c)Y_{eq} = (1.2346 - 1.1)(-j20) = -j2.469 \quad \text{per unit}$$

The per-unit positive-sequence network is

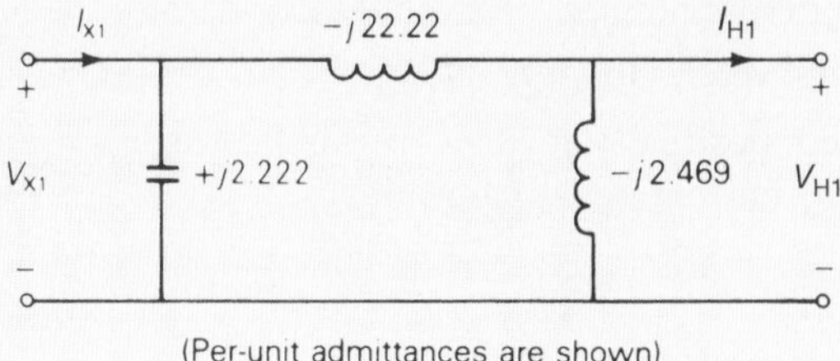

(Per-unit admittances are shown)

Open PowerWorld Simulator case Example 3_12 (see Figure 3.26) and select **Tools, Play** to see an animated view of this LTC transformer example. Initially the generator/step-up transformer feeds a 500 MW/100 Mvar load. As is typical in practice, the transformer's taps are adjusted in discrete steps, with each step changing the tap ratio by 0.625% (hence a 10% change requires 16 steps). Click on arrows next to the transformer's tap to manually adjust the tap by one step. Note that changing the tap directly changes the load voltage.

Because of the varying voltage drops caused by changing loads, LTCs are often operated to automatically regulate a bus voltage. This is particularly true when they are used as step-down transformers. To place the example transformer on automatic control, click on the LTC Control Status "No" field. This toggles the transformer control mode to automatic. Now the transformer manually changes its tap ratio to maintain the load voltage within a specified voltage range, between 0.995 and 1.005 per unit (343.3 to 346.7 kV) in this case. To see the LTC in automatic operation, use the load arrows to vary the load, particularly the Mvar field, noting that the LTC changes to keep the load's voltage within the specified deadband.

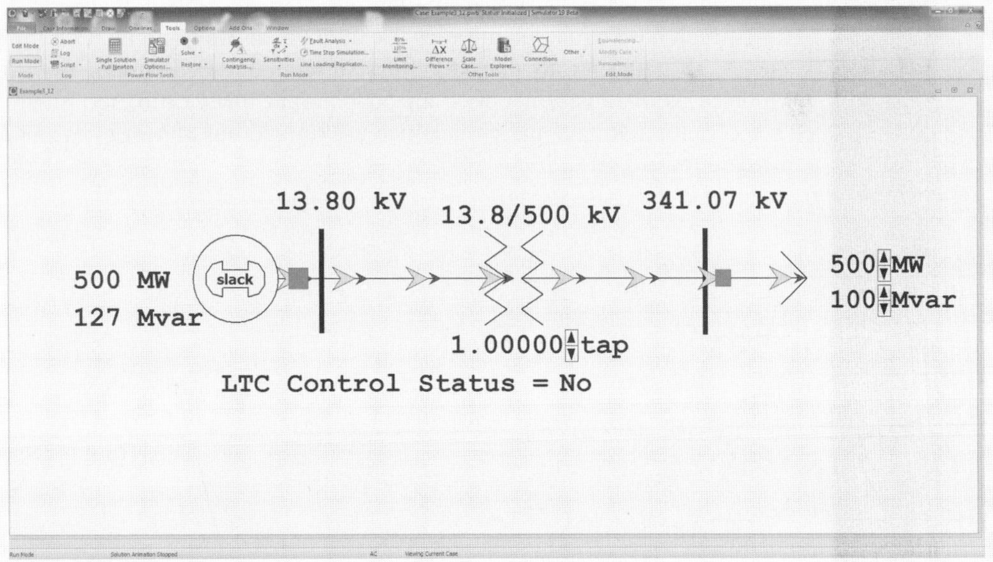

FIGURE 3.26

Screen for Example 3.12

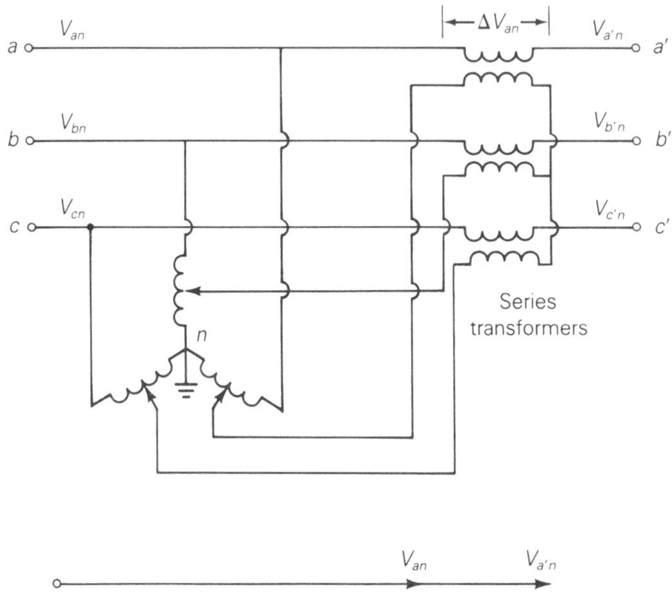

FIGURE 3.27 An example of a voltage-magnitude-regulating transformer

The three-phase regulating transformers shown in Figures 3.27 and 3.28 can be modeled as transformers with off-nominal turns ratios. For the voltage-magnitude-regulating transformer shown in Figure 3.27, adjustable voltages ΔV_{an}, ΔV_{bn}, and ΔV_{cn} which have equal magnitudes ΔV and which are in phase with the phase voltages V_{an}, V_{bn}, and V_{cn} are placed in the series link between buses a–a', b–b', and c–c'. Modeled as a transformer with an off-nominal turns ratio (see Figure 3.25), $c = (1 + \Delta V)$ for a voltage-magnitude increase toward bus abc, or $c = (1 + \Delta V)^{-1}$ for an increase toward bus $a'b'c'$.

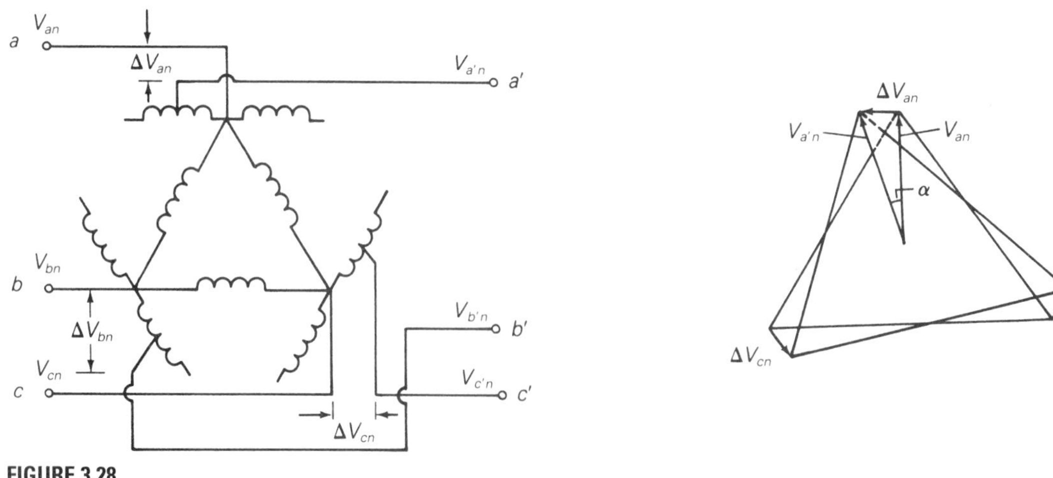

FIGURE 3.28

An example of a phase-angle-regulating transformer. Windings drawn in parallel are on the same core

For the phase-angle-regulating transformer in Figure 3.28, the series voltages ΔV_{an}, ΔV_{bn}, and ΔV_{cn} are ±90° out of phase with the phase voltages ΔV_{an}, V_{bn}, and V_{cn}. The phasor diagram in Figure 3.28 indicates that each of the bus voltages $V_{a'n}$, $V_{b'n}$, and $V_{c'n}$, has a phase shift that is approximately proportional to the magnitude of the added series voltage. Modeled as a transformer with an off-nominal turns ratio (see Figure 3.25), $c \approx 1\underline{/\alpha}$ for a phase increase toward bus *abc* or $c \approx 1\underline{/-\alpha}$ for a phase increase toward bus $a'b'c'$.

EXAMPLE 3.13

Voltage-regulating and phase-shifting three-phase transformers

Two buses *abc* and $a'b'c'$ are connected by two parallel lines L1 and L2 with positive-sequence series reactances $X_{L1} = 0.25$ and $X_{L2} = 0.20$ per unit. A regulating transformer is placed in series with line L1 at bus $a'b'c'$. Determine the 2×2 bus admittance matrix when the regulating transformer (a) provides a 0.05 per-unit increase in voltage magnitude toward bus $a'b'c'$ and (b) advances the phase 3° toward bus $a'b'c'$. Assume that the regulating transformer is ideal. Also, the series resistance and shunt admittance of the lines are neglected.

SOLUTION

The circuit is shown in Figure 3.29.

a. For the voltage-magnitude-regulating transformer, $c = (1 + \Delta V)^{-1} = (1.05)^{-1} = 0.9524$ per unit. From (3.8.5) through (3.8.8), the admittance parameters of the regulating transformer in series with line L1 are

$$Y_{11L1} = \frac{1}{j0.25} = -j4.0$$

$$Y_{22L1} = (0.9524)^2(-j4.0) = -j3.628$$

$$Y_{12L1} = Y_{21L1} = (-0.9524)(-j4.0) = j3.810$$

For line L2 alone,

$$Y_{11L2} = Y_{22L2} = \frac{1}{j0.20} = -j5.0$$

$$Y_{12L2} = Y_{21L2} = -(-j5.0) = j5.0$$

Combining the above admittances in parallel,

$$Y_{11} = Y_{11L1} + Y_{11L2} = -j4.0 - j5.0 = -j9.0$$

$$Y_{22} = Y_{22L1} + Y_{22L2} = -j3.628 - j5.0 = -j8.628$$

$$Y_{12} = Y_{21} = Y_{12L1} + Y_{12L2} = j3.810 + j5.0 = j8.810 \quad \text{per unit}$$

(*Continued*)

FIGURE 3.29 Positive-sequence circuit for Example 3.13

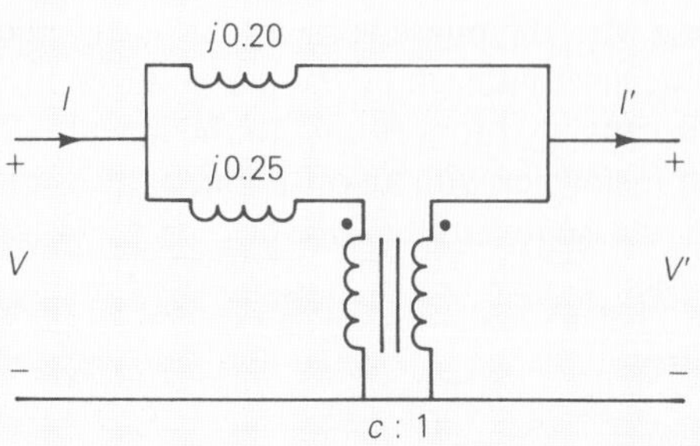

b. For the phase-angle-regulating transformer, $c = 1\underline{/-\alpha} = 1\underline{/-3^\circ}$. Then, for this regulating transformer in series with line L1,

$$Y_{11L1} = \frac{1}{j0.25} = -j4.0$$

$$Y_{22L1} = |1.0\underline{/-3^\circ}|^2(-j4.0) = -j4.0$$

$$Y_{12L1} = -(1.0\underline{/-3^\circ})(-j4.0) = 4.0\underline{/87^\circ} = 0.2093 + j3.9945$$

$$Y_{21L1} = -(1.0\underline{/-3^\circ})^*(-j4.0) = 4.0\underline{/93^\circ} = -0.2093 + j3.9945$$

The admittance parameters for line L2 alone are given in part (a). Combining the admittances in parallel,

$$Y_{11} = Y_{22} = -j4.0 - j5.0 = -j9.0$$

$$Y_{12} = 0.2093 + j3.9945 + j5.0 = 0.2093 + j8.9945$$

$$Y_{21} = -0.2093 + j3.9945 + j5.0 = -0.2093 + j8.9945 \quad \text{per unit}$$

To see this example in the PowerWorld Simulator, open case Example 3_13 (see Figure 3.30). In this case, the transformer and a parallel transmission line are assumed to be supplying power from a 345-kV generator to a 345-kV load. Initially, the off-nominal turns ratio is set to the value in part (a) of the example (PowerWorld has the off-nominal turns ratio on the load side [right-hand] so its tap value of $1.05 = c^{-1}$). To view the PowerWorld Simulator bus admittance matrix, select the **Case Information** ribbon, then **Solution Details**, **Ybus**. To see how the system flows vary with changes to the tap, select **Tools**, **Play**, and then click on the arrows next to the tap field to change the LTC tap in 0.625% steps. Next, to verify the results from part (b), change the tap

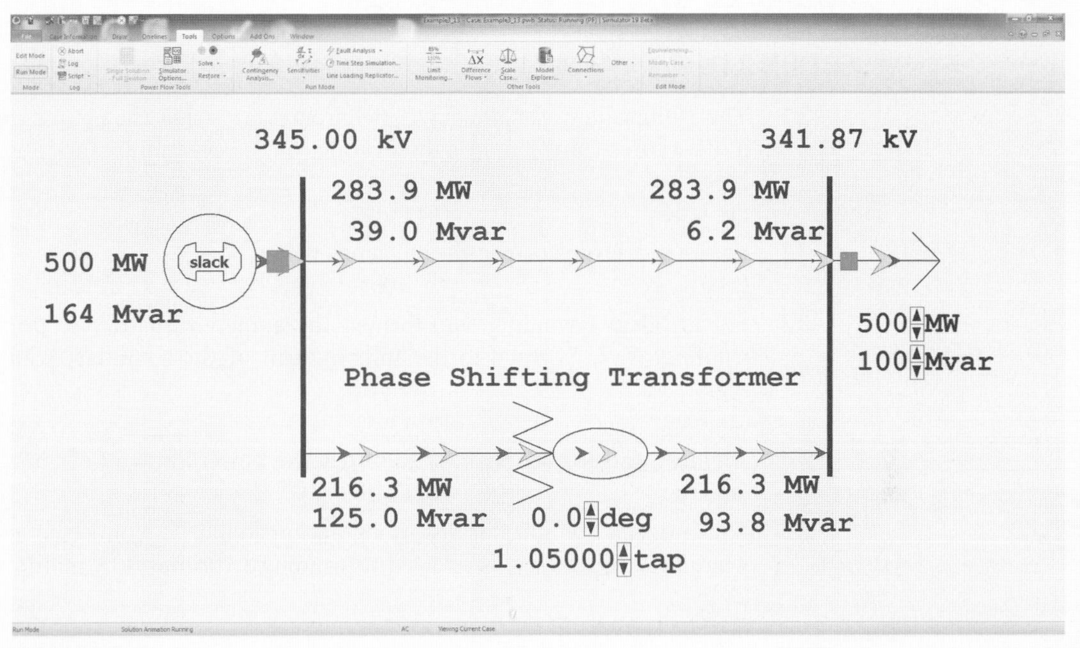

FIGURE 3.30

Screen for Example 3.13

field to 1.0 and the deg field to 3.0 degrees, and then again look at the bus admittance matrix. Click on the deg field arrow to vary the phase shift angle in one-degree steps. Notice that changing the phase angle primarily changes the real power flow, whereas changing the LTC tap changes the reactive power flow. In this example, the line flow fields show the absolute value of the real or reactive power flow; the direction of the flow is indicated with arrows. Traditional power flow programs usually indicate power flow direction using a convention that flow into a transmission line or transformer is assumed to be positive. You can display results in PowerWorld Simulator using this convention by first clicking on the **Onelines** ribbon and then selecting **Oneline Options**. Then on the **Display Options** tab uncheck the Use Absolute Values for MW/Mvar Line Flows fields.

Note that a voltage-magnitude-regulating transformer controls the *reactive* power flow in the series link in which it is installed, whereas a phase-angle-regulating transformer controls the *real* power flow (see Problem 3.59).

MULTIPLE CHOICE QUESTIONS

SECTION 3.1

3.1 The "Ohm's law" for the magnetic circuit states that the net magnetomotive force (mmf) equals the product of the core reluctance and the core flux.
(a) True (b) False

3.2 For an ideal transformer, the efficiency is
(a) 0% (b) 100% (c) 50%

3.3 For an ideal 2-winding transformer, the ampere-turns of the primary winding, $N_1 I_1$, is equal to the ampere-turns of the secondary winding, $N_2 I_2$
(a) True (b) False

3.4 An ideal transformer has no real or reactive power loss.
(a) True (b) False

3.5 For an ideal 2-winding transformer, an impedance Z_2 connected across winding 2 (secondary) is referred to winding 1 (primary) by multiplying Z_2 by
(a) The turns ratio (N_1/N_2)
(b) The square of the turns ratio $(N_1/N_2)^2$
(c) The cubed turns ratio $(N_1/N_2)^3$

3.6 Consider Figure 3.4. For an ideal phase-shifting transformer, the impedance is unchanged when it is referred from one side to the other.
(a) True (b) False

SECTION 3.2

3.7 Consider Figure 3.5. Match the following, those on the left to those on the right.
(i) I_m (a) Exciting current
(ii) I_c (b) Magnetizing current
(iii) I_e (c) Core loss current

3.8 The units of admittance, conductance, and susceptance are siemens.
(a) True (b) False

3.9 Match the following:
(i) Hysteresis loss (a) Can be reduced by constructing the core with laminated sheets of alloy steel
(ii) Eddy current loss (b) Can be reduced by the use of special high grades of alloy steel as core material.

3.10 For large power transformers rated more than 500 kVA, the winding resistances, which are small compared with the leakage reactances, can often be neglected.
(a) True (b) False

3.11 For a short-circuit test on a 2-winding transformer, with one winding shorted, can you apply the rated voltage on the other winding?
(a) Yes (b) No

SECTION 3.3

3.12 The per-unit quantity is always dimensionless.
(a) True (b) False

3.13 Consider the adopted per-unit system for the transformers. Specify true or false for each of the following statements:

(a) For the entire power system of concern, the value of S_{base} is not the same.
(b) The ratio of the voltage bases on either side of a transformer is selected to be the same as the ratio of the transformer voltage ratings.
(c) Per-unit impedance remains unchanged when referred from one side of a transformer to the other.

3.14 The ideal transformer windings are eliminated from the per-unit equivalent circuit of a transformer.
(a) True (b) False

3.15 To convert a per-unit impedance from "old" to "new" base values, the equation to be used is

(a) $Z_{\text{p.u.new}} = Z_{\text{p.u.old}}\left(\frac{V_{\text{baseold}}}{V_{\text{basenew}}}\right)^2\left(\frac{S_{\text{basenew}}}{S_{\text{baseold}}}\right)$

(b) $Z_{\text{p.u.new}} = Z_{\text{p.u.old}}\left(\frac{V_{\text{baseold}}}{V_{\text{basenew}}}\right)^2\left(\frac{S_{\text{basenew}}}{S_{\text{baseold}}}\right)$

(c) $Z_{\text{p.u.new}} = Z_{\text{p.u.old}}\left(\frac{V_{\text{baseold}}}{V_{\text{basenew}}}\right)^2\left(\frac{S_{\text{baseold}}}{S_{\text{basenew}}}\right)$

3.16 In developing per-unit circuits of systems such as the one shown in Figure 3.10, when moving across a transformer, the voltage base is changed in proportion to the transformer voltage ratings.
(a) True (b) False

3.17 Consider Figure 3.10 of the text. The per-unit leakage reactance of transformer T_1, given as 0.1 p.u., is based on the name plate ratings of transformer T_1.
(a) True (b) False

3.18 For balanced three-phase systems, Z_{base} is given by

$$Z_{\text{base}} = \frac{V^2_{\text{baseLL}}}{S_{\text{base3}}}$$

(a) True (b) False

SECTION 3.4

3.19 With the American Standard notation, in either a Y–Δ or Δ–Y transformer, positive- sequence quantities on the high-voltage side shall lead their corresponding quantities on the low-voltage side by 30°.
(a) True (b) False

3.20 In either a Y–Δ or Δ–Y transformer, as per the American Standard notation, the negative-sequence phase shift is the reverse of the positive-sequence phase shift.
(a) True (b) False

3.21 In order to avoid difficulties with third-harmonic exciting current, which three-phase transformer connection is seldom used for step-up transformers between a generator and a transmission line in power systems.
(a) Y–Δ (b) Δ–Y (c) Y–Y

3.22 Does an open-Δ connection permit balanced three-phase operation?
(a) Yes (b) No

3.23 Does an open-Δ operation, the kVA rating compared to that of the original three-phase bank is
(a) 2/3 (b) 58% (c) 1

SECTION 3.5

3.24 It is stated that
(i) balanced three-phase circuits can be solved in per unit on a per-phase basis after converting Δ-load impedances to equivalent Y impedances.
(ii) Base values can be selected either on a per-phase basis or on a three-phase basis.
(a) Both statements are true.
(b) Neither is true.
(c) Only one of the above is true.

3.25 In developing per-unit equivalent circuits for three-phase transformers, under balanced three-phase operation.
(i) A common S_{base} is selected for both the H and X terminals.
(ii) The ratio of the voltage bases V_{baseH}/V_{baseX} is selected to be equal to the ratio of the rated line-to-line voltages $V_{ratedHLL}/V_{ratedXLL}$.
(a) Only one of the above is true.
(b) Neither is true.
(c) Both statements are true.

3.26 In per-unit equivalent circuits of practical three-phase transformers, under balanced three-phase operation, in which of the following connections would a phase-shifting transformer come up?
(a) Y–Y (b) Y–Δ (c) Δ–Δ

3.27 A low value of transformer leakage reactance is desired to minimize the voltage drop, but a high value is desired to limit the fault current, thereby leading to a compromise in the design specification.
(a) True (b) False

SECTION 3.6

3.28 Consider a single-phase three-winding transformer with the primary excited winding of N_1 turns carrying a current I_1 and two secondary windings of N_2 and N_3 turns, delivering currents of I_2 and I_3 respectively. For an ideal case, how are the ampere-turns balanced?

(a) $N_1I_1 = N_2I_2 - N_3I_3$ (b) $N_1I_1 = N_2I_2 + N_3I_3$
(c) $N_1I_1 = -(N_2I_2 - N_3I_3)$

3.29 For developing per-unit equivalent circuits of single-phase three-winding transformer, a common S_{base} is selected for all three windings, and voltage bases are selected in proportion to the rated voltage of the windings.

(a) True (b) False

3.30 Consider the equivalent circuit of Figure 3.20(c) in the text. After neglecting the winding resistances and exciting current, could X_1, X_2, or X_3 become negative, even though the leakage reactance are always positive?

(a) Yes (b) No

SECTION 3.7

3.31 Consider an ideal single-phase 2-winding transformer of turns ratio $N_1/N_2 = a$. If it is converted to an autotransformer arrangement with a transformation ratio of $V_H/V_X = 1 + a$, (the autotransformer rating/two-winding transformer rating) would then be

(a) $1 + a$ (b) $1 + \frac{1}{a}$ (c) a

3.32 For the same output, the autotransformer (with not too large a turns ratio) is smaller in size than a two-winding transformer and has high efficiency as well as superior voltage regulation.

(a) True (b) False

3.33 The direct electrical connection of the windings allows transient over voltages to pass through the autotransformer more easily, and that is an important disadvantage of the autotransformer.

(a) True (b) False

SECTION 3.8

3.34 Consider Figure 3.25 of the text for a transformer with off-nominal turns ratio.

(i) The per-unit equivalent circuit shown in part (c) contains an ideal transformer which cannot be accommodated by some computer programs.

(a) True (b) False

(ii) In the π-circuit representation for real c in part (d), the admittance parameters Y_{11} and Y_{12} would be unequal.

(a) True (b) False

(iii) For complex c, can the admittance parameters be synthesized with a passive RLC circuit?

(a) Yes (b) No

PROBLEMS

SECTION 3.1

3.1 (a) An ideal single-phase two-winding transformer with turns ratio $a_t = N_1/N_2$ is connected with a series impedance Z_2 across winding 2. If one wants to replace Z_2, with a series impedance Z_1 across winding 1 and keep the terminal behavior of the two circuits to be identical, find Z_1 in terms of Z_2.

(b) Would the above result be true if instead of a series impedance there is a shunt impedance?

(c) Can one refer a ladder network on the secondary (2) side to the primary (1) side simply by multiplying every impendance by a_t^2?

3.2 An ideal transformer with $N_1 = 1000$ and $N_2 = 250$ is connected with an impedance Z_{22} across winding 2. If $V_1 = 1000\underline{/0°}$ V and $I_1 = 5\underline{/-30°}$ A, determine V_2, I_2, Z_2, and the impedance Z'_2, which is the value of Z_2 referred to the primary side of the transformer.

3.3 Consider an ideal transformer with $N_1 = 3000$ and $N_2 = 1000$ turns. Let winding 1 be connected to a source whose voltage is $e_1(t) = 100(1 - |t|)$ volts for $-1 \leq t \leq 1$ and $e_1(t) = 0$ for $|t| > 1$ second. A 2-farad capacitor is connected across winding 2. Sketch $e_1(t)$, $e_2(t)$, $i_1(t)$, and $i_2(t)$ versus time t.

3.4 A single-phase 100-kVA, 2400/240-volt, 60-Hz distribution transformer is used as a step-down transformer. The load, which is connected to the 240-volt secondary winding, absorbs 60 kVA at 0.8 power factor lagging and is at 230 volts. Assuming an ideal transformer, calculate the following: (a) primary voltage, (b) load impedance, (c) load impedance referred to the primary, and (d) the real and reactive power supplied to the primary winding.

3.5 Rework Problem 3.4 if the load connected to the 240-V secondary winding absorbs 110 kVA under short-term overload conditions at an 0.8 power factor leading and at 230 volts.

3.6 For a conceptual single-phase phase-shifting transformer, the primary voltage leads the secondary voltage by 30°. A load connected to the secondary winding absorbs 110 kVA at an 0.8 power factor leading and at a voltage $E_2 = 277\underline{/0°}$ volts. Determine (a) the primary voltage, (b) primary and secondary currents, (c) load impedance referred to the primary winding, and (d) complex power supplied to the primary winding.

3.7 Consider a source of voltage $v(t) = 10\sqrt{2}\sin(2t)$V, with an internal resistance of 1800 Ω. A transformer that can be considered as ideal is used to couple a 50-Ω resistive load to the source. (a) Determine the transformer primary-to-secondary turns ratio required to ensure maximum power transfer by matching the load and source resistances. (b) Find the average power delivered to the load, assuming maximum power transfer.

3.8 For the circuit shown in Figure 3.31, determine $v_{out}(t)$

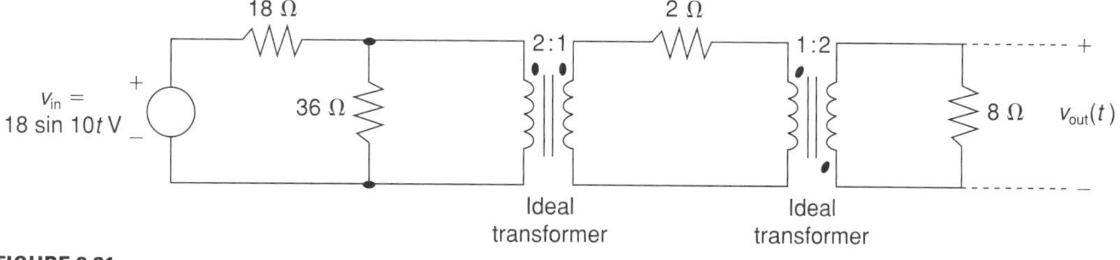

FIGURE 3.31

Problem 3.8

SECTION 3.2

3.9 A single-phase transformer has 2000 turns on the primary winding and 500 turns on the secondary. Winding resistances are $R_1 = 2\ \Omega$, and $R_2 = 0.125\ \Omega$; leakage reactances are $X_1 = 8\ \Omega$ and $X_2 = 0.5\ \Omega$. The resistance load on the secondary is 12 Ω.

(a) If the applied voltage at the terminals of the primary is 1000 V, determine V_2 at the load terminals of the transformer, neglecting magnetizing current.

(b) If the voltage regulation is defined as the difference between the voltage magnitude at the load terminals of the transformer at full load and at no load in percent of full-load voltage with input voltage held constant, compute the percent voltage regulation.

3.10 A single-phase step-down transformer is rated 13 MVA, 66 kV/11.5 kV. With the 11.5 kV winding short-circuited, rated current flows when the voltage applied to the primary is 5.5 kV. The power input is read as 100 kW. Determine R_{eq1} and X_{eq1} in ohms referred to the high-voltage winding.

3.11 For the transformer in Problem 3.10, the open-circuit test with 11.5 kV applied results in a power input of 65 kW and a current of 30 A. Compute the values for G_c and B_m in siemens referred to the high-voltage winding. Compute the efficiency of the transformer for a load of 10 MW at 0.8 p.f. lagging at rated voltage.

3.12 The following data are obtained when open-circuit and short-circuit tests are performed on a single-phase, 50-kVA, 2400/240-volt, 60-Hz distribution transformer.

	VOLTAGE (volts)	CURRENT (amperes)	POWER (watts)
Measurements on low-voltage side with high-voltage winding open	240	4.85	173
Measurements on high-voltage side with low-voltage winding shorted	52.0	20.8	650

(a) Neglecting the series impedance, determine the exciting admittance referred to the high-voltage side. (b) Neglecting the exciting admittance, determine the equivalent series impedance referred to the high-voltage side. (c) Assuming equal series impedances for the primary and referred secondary, obtain an equivalent T-circuit referred to the high-voltage side.

3.13 A single-phase 50-kVA, 2400/240-volt, 60-Hz distribution transformer has a 1-ohm equivalent leakage reactance and a 5000-ohm magnetizing reactance referred to the high-voltage side. If rated voltage is applied to the high-voltage winding, calculate the open-circuit secondary voltage. Neglect I^2R and G_c^2V losses. Assume equal series leakage reactances for the primary and the referred secondary.

3.14 A single-phase 50-kVA, 2400/240-volt, 60-Hz distribution transformer is used as a step-down transformer at the load end of a 2400-volt feeder whose series impedance is $(1.0 + j2.0)$ ohms. The equivalent series impedance of the transformer is $(1.0 + j2.5)$ ohms referred to the high-voltage (primary) side. The transformer is delivering rated load at a 0.8 power factor lagging and at a rated secondary voltage. Neglecting the transformer exciting current, determine (a) the voltage at the transformer primary terminals, (b) the voltage at the sending end of the feeder, and (c) the real and reactive power delivered to the sending end of the feeder.

3.15 Rework Problem 3.14 if the transformer is delivering rated load at rated secondary voltage and at (a) unity power factor, (b) 0.8 power factor leading. Compare the results with those of Problem 3.14.

3.16 A single-phase, 50-kVA, 2400/240-V, 60-Hz distribution transformer has the following parameters:

Resistance of the 2400-V winding: $R_1 = 0.75\ \Omega$

Resistance of the 240-V winding: $R_2 = 0.0075\ \Omega$

Leakage reactance of the 2400-V winding: $X_1 = 1.0\ \Omega$

Leakage reactance of the 240-V winding: $X_2 = 0.01\ \Omega$

Exciting admittance on the 240-V side $= 0.003 - j0.02$ S

(a) Draw the equivalent circuit referred to the high-voltage side of the transformer.

(b) Draw the equivalent circuit referred to the low-voltage side of the transformer. Show the numerical values of impedances on the equivalent circuits.

3.17 The transformer of Problem 3.16 is supplying a rated load of 50 kVA at a rated sec ondary voltage of 240 V and at 0.8 power factor lagging. Neglect the transformer exciting current. (a) Determine the input terminal voltage of the transformer on the high-voltage side. (b) Sketch the corresponding phasor diagram. (c) If the transformer is used as a step-down transformer at the load end of a feeder whose impedance is $0.5 + j2.0\ \Omega$, find the voltage V_S and the power factor at the sending end of the feeder.

SECTION 3.3

3.18 Using the transformer ratings as base quantities, work Problem 3.13 in per-unit.

3.19 Using the transformer ratings as base quantities, work Problem 3.14 in per-unit.

3.20 Using base values of 20 kVA and 115 volts in zone 3, rework Example 3.4.

3.21 Rework Example 3.5; using $S_{base3\phi} = 100$ kVA and $V_{baseLL} = 600$ volts.

3.22 A balanced Y-connected voltage source with $E_{ag} = 277\underline{/0^\circ}$ volts is applied to a balanced-Y load in parallel with a balanced-Δ load where $Z_Y = 20 + j10$ and $Z\Delta = 30 - j15$ ohms. The Y load is solidly grounded. Using base values of $S_{base1\phi} = 10$ kVA and $V_{baseLN} = 277$ volts, calculate the source current I_a in per-unit and in amperes.

3.23 Figure 3.32 shows the oneline diagram of a three-phase power system. By selecting a common base of 100 MVA and 22 kV on the generator side, draw an impedance diagram showing all impedances including the load impedance in per-unit. The data are given as follows:

G:	90 MVA	22 kV	x = 0.18 per unit
T_1:	50 MVA	22/220 kV	x = 0.10 per unit
T_2:	40 MVA	220/11 kV	x = 0.06 per unit
T_3:	40 MVA	22/110 kV	x = 0.064 per unit
T_4:	40 MVA	110/11 kV	x = 0.08 per unit
M:	66.5 MVA	10.45 kV	x = 0.185 per unit

Lines 1 and 2 have series reactances of 48.4 and 65.43 Ω, respectively. At bus 4, the three-phase load absorbs 57 MVA at 10.45 kV and 0.6 power factor lagging.

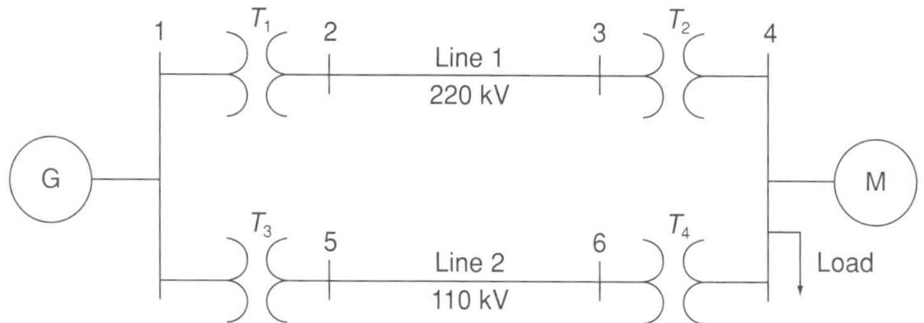

FIGURE 3.32 Problem 3.23

3.24 For Problem 3.18, the motor operates at full load, at 0.8 power factor leading, and at a terminal voltage of 10.45 kV. Determine (a) the voltage at bus 1, which is the generator bus, and (b) the generator and motor internal EMFs.

3.25 Consider a single-phase electric system shown in Figure 3.33. Transformers are rated as follows:

X–Y 15 MVA, 13.8/138 kV, leakage reactance 10%
Y–Z 15 MVA, 138/69 kV, leakage reactance 8%

With the base in circuit Y chosen as 15 MVA, 138 kV, determine the per-unit impedance of the 500 Ω resistive load in circuit Z, referred to circuits Z, Y, and X. Neglecting magnetizing currents, transformer resistances, and line impedances, draw the impedance diagram in per unit.

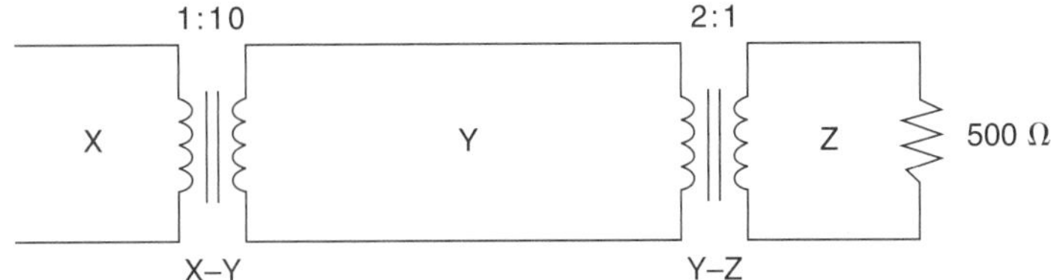

FIGURE 3.33 Single-phase electric system for Problem 3.25

3.26 A bank of three single-phase transformers, each rated 30 MVA, 38.1/3.81 kV, are connected in Y–Δ with a balanced load of three 1 Ω, Y-connected resistors. Choosing a base of 90 MVA, 66 kV for the high-voltage side of the three-phase transformer, specify the base for the low-voltage side. Compute the per-unit resistance of the load on the base for the low-voltage side. Also, determine the load resistance in ohms referred to the high-voltage side and the per-unit value on the chosen base.

3.27 A three-phase transformer is rated 1000 MVA, 220 Y/22 Δ kV. The Y-equivalent short-circuit impedance, considered equal to the leakage reactance, measured on the low-voltage side is 0.1 Ω. Compute the per-unit reactance of the transformer. In a system in which the base on the high-voltage side of the transformer is 100 MVA, 230 kV, what value of the per-unit reactance should be used to represent this transformer?

3.28 For the system shown in Figure 3.34, draw an impedance diagram in per unit by choosing 100 kVA to be the base kVA and 2400 V as the base voltage for the generators.

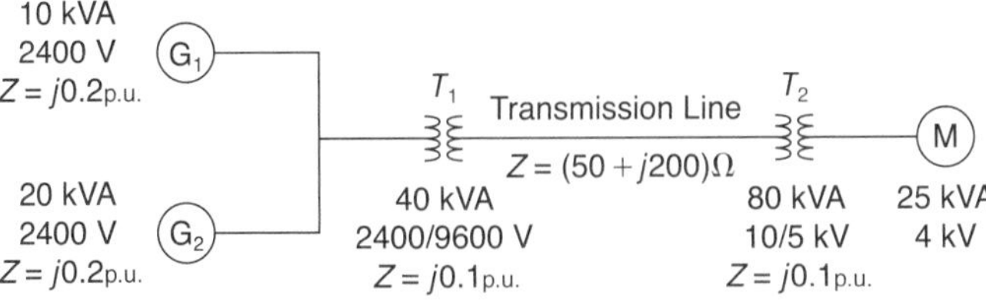

FIGURE 3.34 System for Problem 3.28

3.29 Consider three ideal single-phase transformers (with a voltage gain of η) put together as a Δ–Ω three-phase bank as shown in Figure 3.35. Assuming positive-sequence voltages for V_{an}, V_{bn}, and V_{cn}, find $V_{a'n'}$, $V_{b'n'}$, and $V_{C'n'}$ in terms of V_{an}, V_{bn}, and V_{cn}, respectively.

(a) Would such relationships hold for the line voltages as well?

(b) Looking into the current relationships, express I'_a, I'_b, and I'_c in terms of I_a, I_b, and I_c, respectively.

(c) Let S' and S be the per-phase complex power output and input, respectively. Find S′ in terms of S.

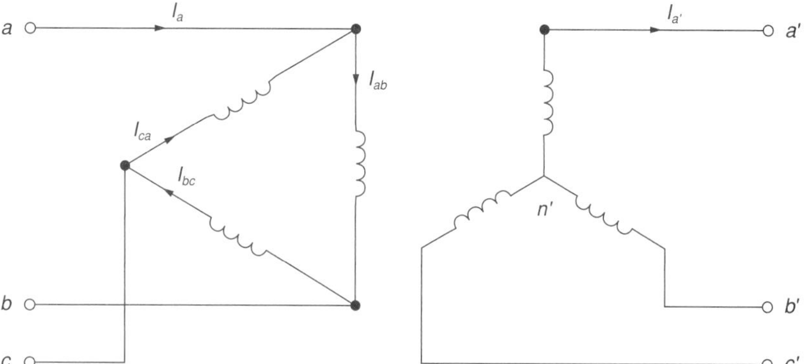

FIGURE 3.35

$\Delta-$Y connection for Problem 3.29

3.30 Reconsider Problem 3.29. If V_{an}, V_{bn}, and V_{cn} are a negative-sequence set, how would the voltage and current relationships change?

(a) If C_1 is the complex positive-sequence voltage gain in Problem 3.29 and (b) if C_2 is the negative sequence complex voltage gain, express the relationship between C_1 and C_2

3.31 If positive-sequence voltages are assumed and the Y-Δ connection is considered, again with ideal transformers as in Problem 3.29, find the complex voltage gain C_3.

(a) What would the gain be for a negative-sequence set?

(b) Comment on the complex power gain.

(c) When terminated in a symmetric Y-connected load, find the referred impedance Z'_L, the secondary impedance Z_L referred to primary (i.e., the per-phase driving-point impedance on the primary side), in terms of Z_L and the complex voltage gain C.

SECTION 3.4

3.32 Determine the positive- and negative-sequence phase shifts for the three-phase transformers shown in Figure 3.36.

3.33 Consider the three single-phase two-winding transformers shown in Figure 3.37. The high-voltage windings are connected in Y. (a) For the low-voltage side, connect the windings in Δ, place the polarity marks, and label the terminals *a*, *b*, and *c* in accordance with the American standard. (b) Relabel the terminals a', b', and c' such that V_{AN} is 90° out of phase with $V_{a'n}$ for positive sequence.

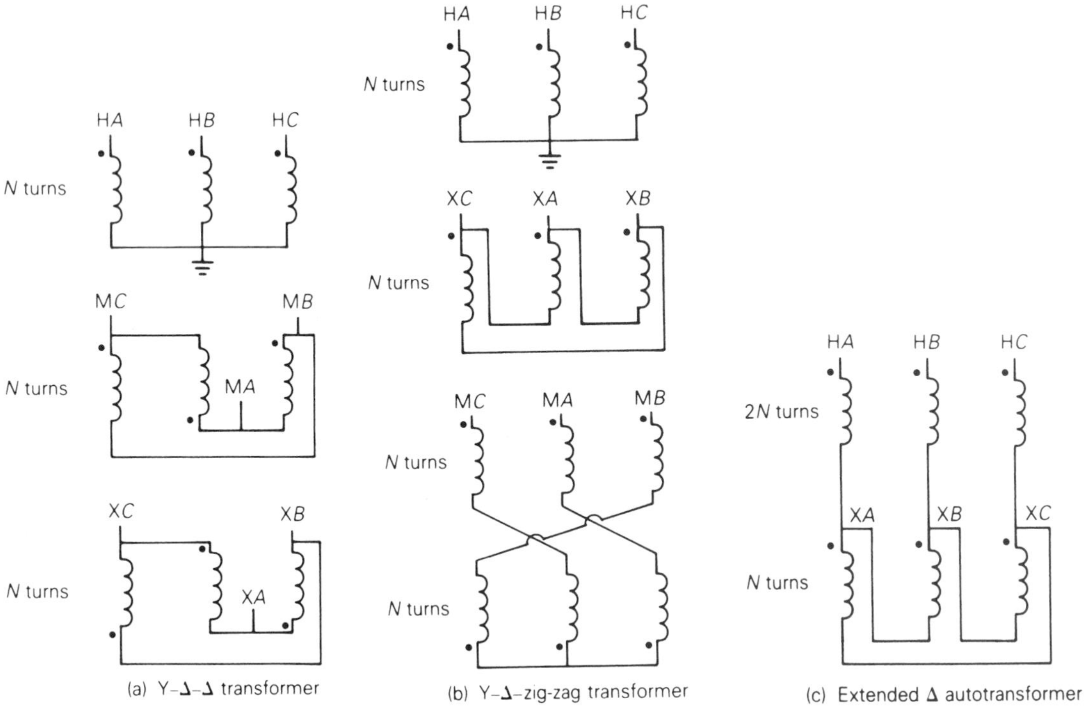

(a) Y–Δ–Δ transformer

(b) Y–Δ–zig-zag transformer

(c) Extended Δ autotransformer

FIGURE 3.36

Problems 3.32 and 3.52 (Coils drawn on the same vertical line are on the same core)

FIGURE 3.37

Problem 3.33

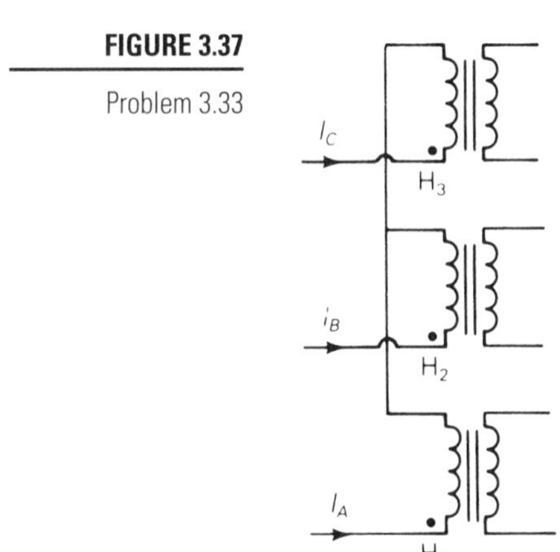

3.34 Three single-phase, two-winding transformers, each rated 450 MVA, 20 kV/288.7 kV, with leakage reactance $X_{eq} = 0.10$ per unit, are connected to form a three-phase bank. The high-voltage windings are connected in Y with a solidly grounded neutral. Draw the per-unit equivalent circuit if the low-voltage windings are connected (a) in Δ with American standard phase shift or (b) in Y with an open neutral. Use the transformer ratings as base quantities. Winding resistances and exciting current are neglected.

3.35 Consider a bank of three single-phase two-winding transformers whose high-voltage terminals are connected to a three-phase, 13.8-kV feeder. The low-voltage terminals are connected to a three-phase substation load rated 2.0 MVA and 2.5 kV. Determine the required voltage, current, and MVA ratings of both windings of each transformer, when the high-voltage/low-voltage windings are connected (a) Y–Δ, (b) Δ–Y, (c) Y–Y, and (d) Δ–Δ.

3.36 Three single-phase two-winding transformers, each rated 25 MVA, 34.5/13.8 kV, are connected to form a three-phase Δ–Δ bank. Balanced positive-sequence voltages are applied to the high-voltage terminals, and a balanced, resistive Y load connected to the low-voltage terminals absorbs 75 MW at 13.8 kV. If one of the single-phase transformers is removed (resulting in an open-Δ connection) and the balanced load is simultaneously reduced to 43.3 MW (57.7% of the original value), determine (a) the load voltages V_{an}, V_{bn}, and V_{cn}; (b) load currents I_a, I_b, and I_c; and (c) the MVA supplied by each of the remaining two transformers. Are balanced voltages still applied to the load? Is the open-Δ transformer overloaded?

3.37 Three single-phase two-winding transformers, each rated 25 MVA, 54.2/5.42 kV, are connected to form a three-phase Y–Δ bank with a balanced Y-connected resistive load of 0.6 Ω per phase on the low-voltage side. By choosing a base of 75 MVA (three phase) and 94 kV (line-to-line) for the high-voltage side of the transformer bank, specify the base quantities for the low-voltage side. Determine the per-unit resistance of the load on the base for the low-voltage side. Then determine the load resistance R_L in ohms referred to the high-voltage side and the per-unit value of this load resistance on the chosen base.

3.38 Consider a three-phase generator rated 300 MVA, 23 kV, supplying a system load of 240 MVA and 0.9 power factor lagging at 230 kV through a 330 MVA, 23 Δ/ 230 Y-kV step-up transformer with a leakage reactance of 0.11 per unit. (a) Neglecting the exciting current and choosing base values at the load of 100 MVA and 230 kV, find the phasor currents I_A, I_B, and I_C supplied to the load in per unit. (b) By choosing the load terminal voltage V_A as reference, specify the proper base for the generator circuit and determine the generator voltage V as well as the phasor currents I_A, I_B, and I_C, from the generator. (*Note:* Take into account the phase shift of the transformer.) (c) Find the generator terminal voltage in kV and the real power supplied by the generator in MW. (d) By omitting the transformer phase shift altogether, check to see whether you get the same magnitude of generator terminal voltage and real power delivered by the generator.

SECTION 3.5

3.39 The leakage reactance of a three-phase, 300-MVA, 230 Y/23 Δ-kV transformer is 0.06 per unit based on its own ratings. The Y winding has a solidly grounded neutral. Draw the per-unit equivalent circuit. Neglect the exciting admittance and assume the American Standard phase shift.

3.40 Choosing system bases to be 240/24 kV and 100 MVA, redraw the per-unit equivalent circuit for Problem 3.39.

3.41 Consider the single-line diagram of the power system shown in Figure 3.38. Equipment ratings are

Generator 1:	1000 MVA, 18 kV, X″ = 0.2 per unit
Generator 2:	1000 MVA, 18 kV, X″ = 0.2 p.u.
Synchronous motor 3:	1500 MVA, 20 kV, X″ = 0.2 p.u.
Three-phase Δ–Y transformers T_1, T_2, T_3, T_4:	1000 MVA, 500 kV Y/20 kV Δ, X = 0.1 p.u.
Three-phase Y–Y transformer T_5:	1500 MVA, 500 kV Y/20 kV Y, X = 0.1 p.u.

Neglecting resistance, transformer phase shift, and magnetizing reactance, draw the equivalent reactance diagram. Use a base of 100 MVA and 500 kV for the 50-ohm line. Determine the per-unit reactances.

FIGURE 3.38

Problems 3.41 and 3.42

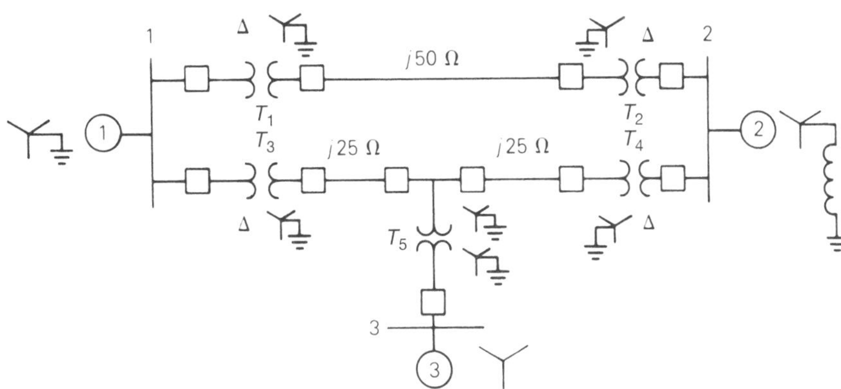

3.42 For the power system in Problem 3.41, the synchronous motor absorbs 1500 MW at 0.8 power factor leading with the bus 3 voltage at 18 kV. Determine the bus 1 and bus 2 voltages in kV. Assume that generators 1 and 2 deliver equal real powers and equal reactive powers. Also assume a balanced three-phase system with positive-sequence sources.

3.43 Three single-phase transformers, each rated 10 MVA, 66.4/12.5 kV, 60 Hz, with an equivalent series reactance of 0.1 per unit divided equally between primary and secondary, are connected in a three-phase bank. The high-voltage windings are Y-connected and their terminals are directly connected to a 115-kV three-phase bus. The secondary terminals are all shorted together. Find the currents entering the high-voltage terminals and leaving the low-voltage terminals if the low-voltage windings are (a) Y-connected and (b) Δ-connected.

3.44 A 130-MVA, 13.2-kV three-phase generator, which has a positive-sequence reactance of 1.5 per unit on the generator base, is connected to a 135-MVA, 13.2 Δ/115 Y-kV step-up transformer with a series impedance of $(0.005 + j0.1)$ per unit on its own base. (a) Calculate the per-unit generator reactance on the transformer base. (b) The load at the transformer terminals is 15 MW at unity power factor and at 115 kV. Choosing the transformer high-side voltage as the reference phasor, draw a phasor diagram for this condition. (c) For the condition of part (b), find the transformer low-side voltage and the generator internal voltage behind its reactance. Also compute the generator output power and power factor.

3.45 Figure 3.39 shows a oneline diagram of a system in which the three-phase generator is rated 300 MVA, 20 kV with a subtransient reactance of 0.2 per unit and with its neutral grounded through a 0.4-Ω reactor. The transmission line is 64 km long with a series reactance of 0.5 Ω/km. The three-phase transformer T_1 is rated 350 MVA, 230/ 20 kV with a leakage reactance of 0.1 per unit. Transformer T_2 is composed of three single-phase transformers, each rated 100 MVA, 127/13.2 kV with a leakage reactance of 0.1 per unit. Two 13.2-kV motors M_1 and M_2 with a subtransient reactance of 0.2 per unit for each motor represent the load. M_1 has a rated input of 200 MVA with its neutral grounded through a 0.4-Ω current-limiting reactor. M_2 has a rated input of 100 MVA with its neutral not connected to ground. Neglect phase shifts associated with the transformers. Choose the generator rating as base in the generator circuit and draw the positive-sequence reactance diagram showing all reactances in per unit.

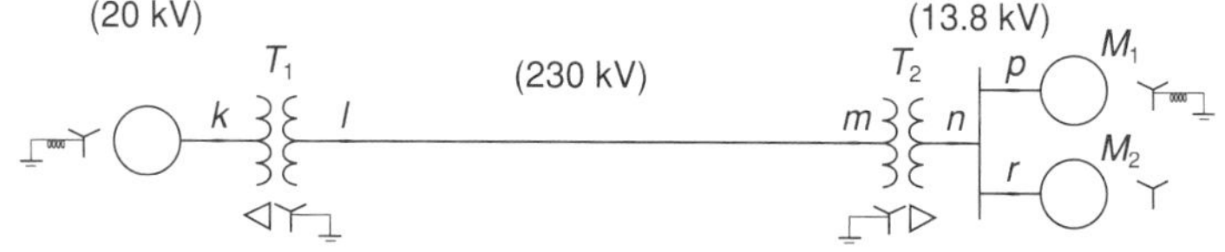

FIGURE 3.39

Problems 3.45 and 3.46

3.46 The motors M_1 and M_2 of Problem 3.45 have inputs of 120 and 60 MW, respectively, at 13.2 kV, and both operate at unity power factor. Determine the generator terminal voltage and voltage regulation of the line. Neglect transformer phase shifts.

3.47 Consider the oneline diagram shown in Figure 3.40. The three-phase transformer bank is made up of three identical single-phase transformers, each specified by $X_1 = 0.24\ \Omega$ (on the low-voltage side), negligible resistance and magnetizing current, and turns ratio $\eta = N_2/N_1 = 10$. The transformer bank is delivering 100 MW at 0.8 p.f. lagging to a substation bus whose voltage is 230 kV.

(a) Determine the primary current magnitude, primary voltage (line-to-line) magnitude, and the three-phase complex power supplied by the generator. Choose the line-to-neutral voltage at the bus, $V_{a'n'}$ as the reference. Account for the phase shift, and assume positive-sequence operation.

(b) Find the phase shift between the primary and secondary voltages.

FIGURE 3.40 Oneline diagram for Problem 3.47

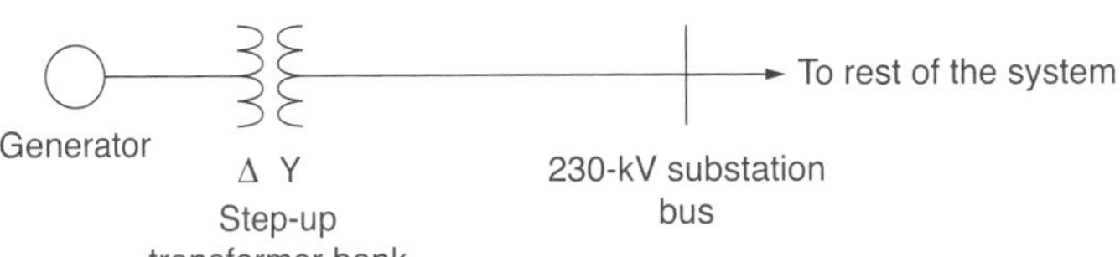

3.48 With the same transformer banks as in Problem 3.47, Figure 3.41 shows the oneline diagram of a generator, a step-up transformer bank, a transmission line, a step-down transformer bank, and an impedance load. The generator terminal voltage is 15 kV (line-to-line).

FIGURE 3.41 Oneline diagram for Problem 3.48

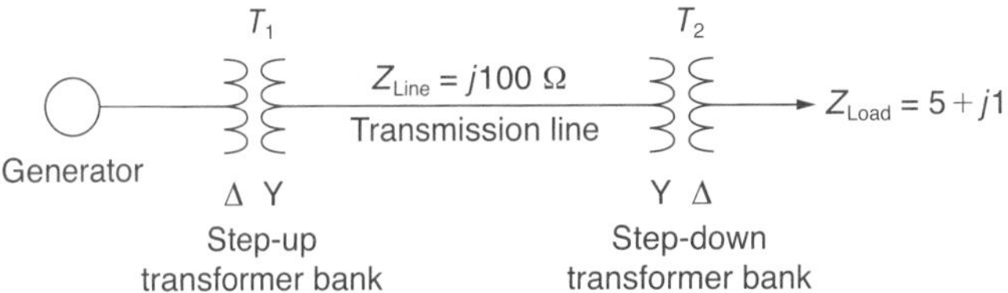

(a) Draw the per-phase equivalent circuit, accounting for phase shifts for positive-sequence operation.

(b) By choosing the line-to-neutral generator terminal voltage as the reference, determine the magnitudes of the generator current, transmission-line current, load current, and line-to-line load voltage. Also, find the three-phase complex power delivered to the load.

3.49 Consider the single-line diagram of a power system shown in Figure 3.42 with equipment ratings given:

Generator G_1:	50 MVA, 13.2 kV, $x = 0.15$ p.u.
Generator G_2:	20 MVA, 13.8 kV, $x = 0.15$ p.u.
Three-phase Δ–Y transformer T_1:	80MVA, 13.2Δ/165YkV, X=0.1p.u.
Three-phase Y–Δ transformer T_2:	40MVA, 165Y/13.8ΔkV, X=0.1p.u.
Load:	40 MVA, 0.8 PF lagging, operating at 150 kV

Choose a base of 100 MVA for the system and 132-kV base in the transmission-line circuit. Let the load be modeled as a parallel combination of resistance and inductance. Neglect transformer phase shifts. Draw a per-phase equivalent circuit of the system showing all impedances in per unit.

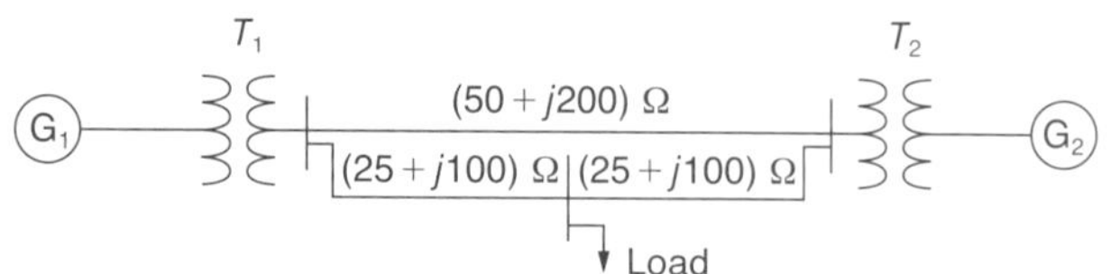

FIGURE 3.42 Oneline diagram for Problem 3.49

SECTION 3.6

3.50 A single-phase three-winding transformer has the following parameters: $Z_1 = Z_2 = Z_3 = 0 + j0.05$, $G_C = 0$, and $B_M = 0.2$ per unit. Three identical transformers, as described, are connected with their primaries in Y (solidly grounded neutral) and with their secondaries and tertiaries in Δ. Draw the per-unit sequence networks of this transformer bank.

3.51 The ratings of a three-phase three-winding transformer are

Primary (1): Y connected, 66 kV, 15 MVA

Secondary (2): Y connected, 13.2 kV, 10 MVA

Tertiary (3): Δ connected, 2.3 kV, 5 MVA

Neglecting winding resistances and exciting current, the per-unit leakage reactances are

$X_{12} = 0.08$ on a 15-MVA,66-kV base

$X_{13} = 0.10$ on a 15-MVA,66-kV base

$X_{23} = 0.09$ on a 10-MVA, 13.2-kV base

(a) Determine the per-unit reactances X_1, X_2, X_3 of the equivalent circuit on a 15-MVA, 66-kV base at the primary terminals. (b) Purely resistive loads of 7.5 MW at 13.2 kV and 5 MW at 2.3 kV are connected to the secondary and tertiary sides of the transformer, respectively. Draw the per-unit impedance diagram, showing the per-unit impedances on a 15-MVA, 66-kV base at the primary terminals.

3.52 Draw the per-unit equivalent circuit for the transformers shown in Figure 3.34. Include ideal phase-shifting transformers showing phase shifts determined in Problem 3.32. Assume that all windings have the same kVA rating and that the equivalent leakage reactance of any two windings with the third winding open is 0.10 per unit. Neglect the exciting admittance.

3.53 The ratings of a three-phase, three-winding transformer are

Primary: Y connected, 66 kV, 15 MVA

Secondary: Y connected, 13.2 kV, 10 MVA

Tertiary: Δ connected, 2.3 kV, 5 MVA

Neglecting resistances and exciting current, the leakage reactances are:

$X_{PS} = 0.09$ per unit on a 15-MVA, 66-kV base

$X_{PT} = 0.08$ per unit on a 15-MVA, 66-kV base

$X_{ST} = 0.05$ per unit on a 10-MVA, 13.2-kV base

Determine the per-unit reactances of the per-phase equivalent circuit using a base of 15 MVA and 66 kV for the primary.

3.54 An infinite bus, which is a constant voltage source, is connected to the primary of the three-winding transformer of Problem 3.53. A 7.5-MVA, 13.2-kV synchronous motor with a subtransient reactance of 0.2 per unit is connected to the transformer secondary. A 5-MW, 2.3-kV three-phase resistive load is connected to the tertiary. Choosing a base of 66 kV and 15 MVA in the primary, draw the impedance diagram of the system showing per-unit impedances. Neglect transformer exciting current, phase shifts, and all resistances except the resistive load.

SECTION 3.7

3.55 A single-phase 10-kVA, 2300/230-volt, 60-Hz two-winding distribution transformer is connected as an autotransformer to step up the voltage from 2300 to 2530 volts. (a) Draw a schematic diagram of this arrangement, showing all voltages and currents when delivering full load at rated voltage. (b) Find the permissible kVA rating of the autotransformer if the winding currents and voltages are not to exceed the rated values as a two-winding transformer. How much of this kVA rating is transformed by magnetic induction? (c) The following data are obtained from tests carried out on the transformer when it is connected as a two-winding transformer:

Open-circuit test with the low-voltage terminals excited:
Applied voltage = 230 V, input current = 0.45 A, input power = 70 W.

Short-circuit test with the high-voltage terminals excited:
Applied voltage = 120 V, input current = 4.5 A, input power = 240 W.

Based on the data, compute the efficiency of the autotransformer corresponding to full load, rated voltage, and 0.8 power factor lagging. Comment on why the efficiency is higher as an autotransformer than as a two-winding transformer.

3.56 Three single-phase two-winding transformers, each rated 3 kVA, 220/110 volts, 60 Hz, with a 0.10 per-unit leakage reactance, are connected as a three-phase extended Δ autotransformer bank, as shown in Figure 3.36 (c). The low-voltage Δ winding has a 110 volt rating. (a) Draw the positive-sequence phasor diagram and show that the high-voltage winding has a 479.5 volt rating. (b) A three-phase load connected to the low-voltage terminals absorbs 6 kW at 110 volts and at 0.8 power factor lagging. Draw the per-unit impedance diagram and calculate the voltage and current at the high-voltage terminals. Assume positive-sequence operation.

3.57 A two-winding single-phase transformer rated 60 kVA, 240/1200 V, 60 Hz, has an efficiency of 0.96 when operated at rated load, 0.8 power factor lagging. This transformer is to be utilized as a 1440/1200-V step-down autotransformer in a power distribution system. (a) Find the permissible kVA rating of the autotransformer if the winding currents and voltages are not to exceed the ratings as a two-winding transformer. Assume an ideal transformer. (b) Determine the efficiency of the autotransformer with the kVA loading of part (a) and 0.8 power factor leading.

3.58 A single-phase two-winding transformer rated 90 MVA, 80/120 kV is to be connected as an autotransformer rated 80/200 kV. Assume that the transformer is ideal. (a) Draw a schematic diagram of the ideal transformer connected as an autotransformer, showing the voltages, currents, and dot notation for polarity. (b) Determine the permissible kVA rating of the autotransformer if the winding currents and voltages are not to exceed the rated values as a two-winding transformer. How much of the kVA rating is transferred by magnetic induction?

SECTION 3.8

3.59 The two parallel lines in Example 3.13 supply a balanced load with a load current of $1.0\underline{/-30^\circ}$ per unit. Determine the real and reactive power supplied to the load bus from each parallel line with (a) no regulating transformer, (b) the voltage-magnitude-regulating transformer in Example 3.13(a), and (c) the phase-angle-regulating transformer in Example 3.13(b). Assume that the voltage at bus *abc* is adjusted so that the voltage at bus $a'b'c'$ remains constant at $1.0\underline{/0^\circ}$ per unit. Also assume positive sequence. Comment on the effects of the regulating transformers.

PW **3.60** PowerWorld Simulator case Problem 3_60 duplicates Example 3.13 except that a resistance term of 0.06 per unit has been added to the transformer and 0.05 per unit to the transmission line. Since the system is no longer lossless, a field showing the real power losses has also been added to the oneline. With the LTC tap fixed at 1.05, plot the real power losses as the phase shift angle is varied from -10 to $+10$ degrees. What value of phase shift minimizes the system losses?

PW **3.61** Repeat Problem 3.60, except keep the phase-shift angle fixed at 3.0 degrees while varying the LTC tap between 0.9 and 1.1. What tap value minimizes the real power losses?

3.62 Rework Example 3.12 for a +10% tap, providing a 10% increase for the high-voltage winding.

3.63 A 23/230-kV step-up transformer feeds a three-phase transmission line, which in turn supplies a 150-MVA, 0.8 lagging power factor load through a step-down 230/23-kV transformer. The impedance of the line and transformers at 230 kV is 18 + $j60\ \Omega$. Determine the tap setting for each transformer to maintain the voltage at the load at 23 kV.

3.64 The per-unit equivalent circuit of two transformers T_a and T_b connected in parallel, with the same nominal voltage ratio and the same reactance of 0.1 per unit on the same base, is shown in Figure 3.43. Transformer T_b has a voltage-magnitude step-up toward the load of 1.05 times that of T_a (that is, the tap on the secondary winding of T_b is set to 1.05). The load is represented by $0.8 + j0.6$ per unit at a voltage $V_2 = 1.0/0°$ per unit. Determine the complex power in per unit transmitted to the load through each transformer. Comment on how the transformers share the real and reactive powers.

FIGURE 3.43

Problem 3.64

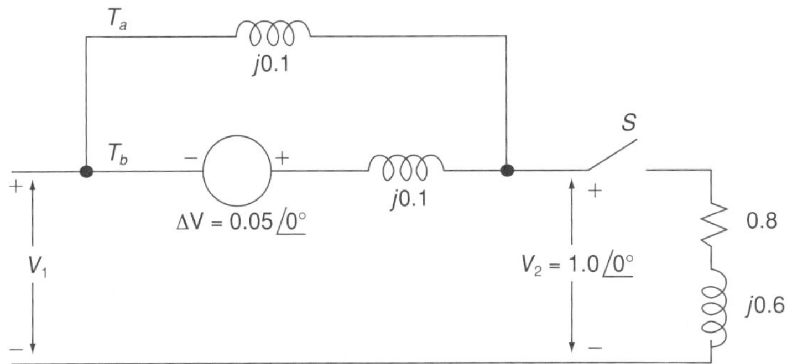

3.65 Reconsider Problem 3.64 with the change that now T_b includes both a transformer of the same turns ratio as T_a and a regulating transformer with a 4° phase shift. On the base of T_a, the impedance of the two components of T_b is $j0.1$ per unit. Determine the complex power in per unit transmitted to the load through each transformer. Comment on how the transformers share the real and reactive powers.

CASE STUDY QUESTIONS

a. What are the advantages of correctly specifying a transformer most suitable for its application?

b. Why is it important to reduce the moisture within a transformer to acceptable levels during transformer installation?

c. What should be the focus of transformer preventive maintenance efforts?

REFERENCES

1. R. Feinberg, *Modern Power Transformer Practice* (New York: Wiley, 1979).
2. A. C. Franklin and D. P. Franklin, *The J & P Transformer Book,* 11th ed. (London: Butterworths, 1983).
3. W. D. Stevenson, Jr., *Elements of Power System Analysis,* 4th ed. (New York: McGraw-Hill, 1982).
4. J. R. Neuenswander, *Modern Power Systems* (Scranton, PA: International Textbook Company, 1971).
5. M. S. Sarma, *Electric Machines* (Dubuque, IA: Brown, 1985).
6. A. E. Fitzgerald, C. Kingsley, and S. Umans, *Electric Machinery,* 4th ed. (New York: McGraw-Hill, 1983).
7. O. I. Elgerd, *Electric Energy Systems: An Introduction* (New York: McGraw-Hill, 1982).
8. C. Gamez, "Power Transformer–Part 3: Life Management and Extension," *Transformers Magazine*, 1, 3, October 2014, pp. 10–15 (www.transformers-magazine.com).

4 Transmission Line Parameters

765-kV transmission line with aluminum guyed-V towers (Courtesy of American Electric Power Company.)

This chapter discusses the four basic transmission-line parameters: series resistance, series inductance, shunt capacitance, and shunt conductance and investigates transmission-line electric and magnetic fields.

Series resistance accounts for ohmic (I^2R) line losses. Series impedance, including resistance and inductive reactance, gives rise to series-voltage drops along the line. Shunt capacitance gives rise to line-charging currents. Shunt conductance accounts for V^2G line losses due to leakage currents between conductors or between conductors and ground. Shunt conductance of overhead lines is usually neglected.

Although the ideas developed in this chapter can be applied to underground transmission and distribution, the primary focus here is on overhead lines. Underground transmission in the United States presently accounts for less than 1% of total transmission and is found mostly in large cities or under waterways. There is, however, a large application for underground cable in distribution systems.

CASE STUDY

Two transmission articles are presented here. The first article describes the integrated nature of the Canada-United States transmission grid and highlights its mutual benefits. The article identifies four strategic goals to further strengthen the bilateral electricity relationship [10]. The second article describes the ongoing congestion of the North American transmission grid, including its impacts on reliability, efficiency, and costs. The article emphasizes the need to upgrade and modernize the grid while leveraging new technologies [12].

Integrating North America's Power Grid

Patrick Brown

Canadian Electricity Association (CEA)

How the U.S. and Canada Maximize Their Working Relationship

North America is shifting towards a new energy paradigm. Where it was once viewed through the prism of scarcity, some are now beginning to speak of energy in the language of abundance.

The grounds for optimism are understandable. For instance, technological breakthroughs on the production side are unlocking previously inaccessible deposits of oil and natural gas, while advances in conservation and efficiency practices are affording customers greater control over their energy use.

"Integrating North America's Power Grid", by Patrick Brown, *Electricity Today T & D Magazine* (June 2013) Reprinted with Permission.

Still, the achievement of North American independence from volatile global energy markets is by no means a foregone conclusion. Economic recovery efforts remain fragile. The supply and delivery infrastructure for our resources is in urgent need of upgrade while an evolving landscape of risks poses significant threats to its security. And perhaps most importantly, transformative events in the market—such as the North American boom in natural gas production—have overtaken many of the policy structures in place across Canada and the United States.

Too much of the framework shaping governments' approach on energy issues is therefore not optimally suited to addressing the opportunities and challenges at hand. The backbone of any strategy to confront and leverage the new energy landscape must be an enhanced North American electricity system.

Electricity plays an integral role in the robust Canada-U.S. relationship on energy, which itself is a fundamental pillar of the broader flow of two-way trade that is without compare anywhere in the world. The linkages between the Canadian and U.S. grids—more than three dozen in total—offer several advantages to both countries, such as higher reliability and expanded access to low-emitting resources. These physical interconnections have facilitated steady growth in what has become a continent-wide marketplace for electricity, with supply fulfilling demand in an efficient, cost-effective manner across North America.

From the vantage point of the voice of the Canadian electricity sector, CEA views our vibrant, bilateral electricity relationship—which has long served as a central instrument for economic vitality in both countries—as being an ideal platform from which to launch a renewed push towards a more secure, prosperous, and autonomous continental energy posture.

This case study is intended to help guide such efforts. In order to set a proper context, the first section provides an overview of the interconnected and integrated nature of the Canada-U.S. relationship on electricity and highlights the mutual benefits thereof. The second section identifies four strategic goals to further strengthen the relationship.

The U.S.-Canada Electricity Relationship

The trading relationship enjoyed by Canada and the U.S. is a model of bilateral cooperation. According to the government of Canada, approximately 300,000 individuals and $1.6 billion (CAD) worth of goods and services move freely across the U.S.-Canada border daily, making each country the other's most important trading partner. The scope and scale of this two-way trading flow is without compare anywhere in the world. An integral part of this relationship is energy.

Energy serves as the main engine fueling economic activity and prosperity across North America, delivering a wealth of societal benefits and enabling a high quality of life on both sides of the border. The two countries have long acted as each other's best customers when it comes to the buying and selling of energy commodities. For example, research from the report, "*U.S. Natural Gas Exports by Country*," published by the United States Energy Information Administration, reveals that Canada receives the greatest share of U.S. natural gas exports (with volumes nearly tripling between 2006 and 2012). Also, the Embassy of Canada in the United States reports that Canada provides the United States with more crude oil and petroleum products, coal, uranium, and natural gas than any other foreign source.

Canada is the largest supplier to the U.S. of another vital commodity—electricity. Crucial in the day-to-day lives of North Americans, it is so reliable as to be taken for granted. In our homes, offices, factories, ports, and stock exchanges, it is instinctively assumed that the lights will turn on with a simple flick of the switch. Likewise, the origin of the electrons being used is rarely considered. The crowds cheering on the Vancouver Canucks might never contemplate that electricity generated in the United States could be illuminating the arena. And the car manufacturer in Michigan may be unaware that electricity from Canadian generators is powering its assembly line (see Figure 1).

The movement of electricity between Canada and the United States may go unnoticed by the public. However, this should not be viewed as a cause for concern. Instead, it reflects how routine and reliable a transaction the cross-border exchange of electricity has become in North America. The integration of electricity markets in Canada and the United States means supply can fulfill demand in an efficient, cost-effective manner across the continent.

North Americans benefit from a shared system which can generate and transmit electrons across vast distances to ensure a reliable, secure, and competitively priced supply of electricity, 24 hours a day, seven days a week. Figures 1 and 2 offer a clear visual indicator of the extent of the current integration of the electricity systems in Canada and the United States. By interconnecting into each other's networks at over 35 points, the two countries benefit from numerous advantages: a higher level of

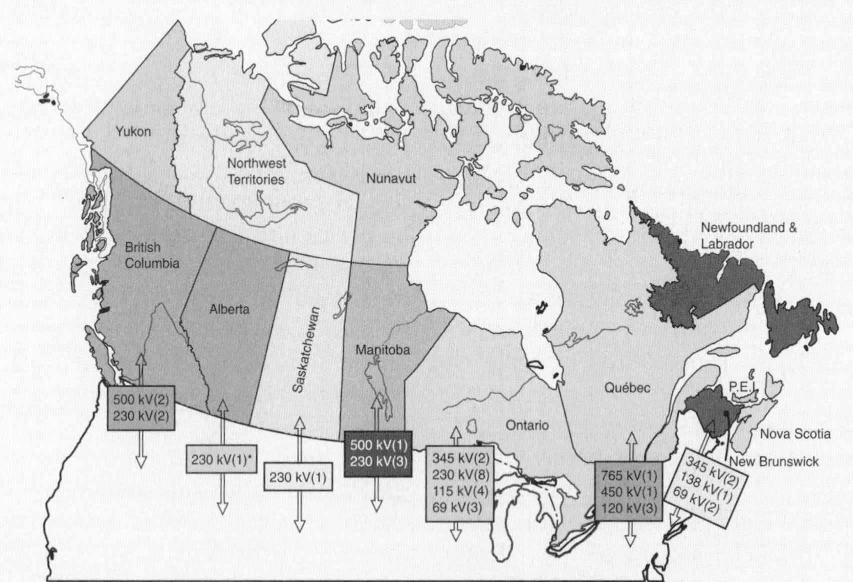

Figure 1 Major Transmission Interconnections
The physical linkages between the Canadian and U.S. grids have likewise enabled steady growth in what has become a robust, continent-wide marketplace for electricity. Electricity trade occurs at a range of points across and beyond the international border, reflecting the largely north–south nature of the networks in Canadian provinces, as they seamlessly tie into the denser web of transmission infrastructure in the United States.

reliable service for customers through enhanced system stability; efficiencies in system operation; efficiencies in fuel management; opportunities to use power from nearby markets to address local contingencies; and expanded access to low emitting and competitively priced resources.

Maximizing The Benefits of Grid Integration

Four Strategic Goals for Canada and The U.S. to Further Strengthen their Relationship

1. **Plugging the electric infrastructure gap**
   - Coordinate and cooperate on identifying new infrastructure needs and addressing aging infrastructure on the North American grid and effectively communicate such needs to the general public.
   - Remove regulatory impediments to much-needed electric infrastructure investments.
   - Update and enhance the efficiency of permitting procedures for international power line projects.

2. **Maximizing environmental sustainability**
   - Ensure that any market-based solution to combat climate change that incorporates the electricity sector recognizes the integrated nature of the North American electricity market, encourages the continued two-way flow in cross-border electricity trade, and contains credit provisions for early action taken to reduce emissions.
   - Support existing measures that are proven to cost-effectively deliver emission reductions and help enable intermittent renewable technologies.
   - Recognize hydropower as a clean, renewable resource across policy programs.
   - Maintain support for fostering cleaner ways to use North America's plentiful supplies of natural gas and coal for power generation, including the deployment of carbon capture and storage technology.

3. **Maintaining vigilance on reliability and security**
   - Support the existing framework in place at the North American Electric Reliability Corporation to develop reliability standards (including for cyber security protection) for the North American transmission grid.
   - Enhance public-private sector, as well as government-to-government coordination and sharing of timely and actionable threat information.
   - Address cyber and other security threats in a comprehensive manner across all industry sectors, with a focus on securing the most critical assets from the most urgent challenges.
   - Standardize and streamline border-crossing procedures for crews providing assistance to utilities in the other country.

Figure 2 Electricity Exports and Imports Between Canada and the U.S. (2012)

Source: National Energy Board, Electricity Exports and Imports, 2012. Retrieved February 21, 2013

4. **Ensuring free, fair, and fluid inter-jurisdictional trade**
   - Remain vigilant in avoiding erecting barriers that may inhibit inter-jurisdictional electricity trade.
   - Refrain from imposing unnecessarily burdensome over-the-counter derivatives of regulatory requirements on energy endusers which pose no systemic risk to the marketplace.
   - Update and enhance the efficiency of administrative procedures for authorizing exports of electricity across the international border.

Last Look

Against the backdrop of transformation and transition in the North American energy landscape, CEA believes that there are numerous opportunities for leveraging the legacy of cooperation and achievement in operating the world's largest integrated power grid, and expanding the suite of benefits which such integration offers to consumers in Canada and the United States.

Electricity Imports & Exports

Historically, electricity exports to the United States have represented anywhere from 5 to 10 percent of Canada's total production. The bulk of these exports involve the sale of surplus generation from major hydropower producing provinces such as Québec, British Columbia, and Manitoba. More recently, export volumes from Ontario have also risen, making the province the second largest exporter (on both a gross and net basis) for several consecutive years. While hydro enjoys a sizeable share of Ontario's supply mix, nuclear energy comprises approximately half of the province's portfolio.

Overall, the vast majority of electrons delivered across the border from Canadian generators to U.S. customers are derived from clean, nonemitting sources.

Patrick Brown is the director of U.S. Affairs for the Canadian Electricity Association. ■

Grid Congestion

Dave Bryant
CTC Global Corporation

Unclogging the Arteries of North America's Power Grid

Without access to affordable and reliable electricity, no society in today's world can possibly flourish. Canada and the United States rely heavily on electricity to pump water to grow crops, manufacture products competitively, and power North America's information-based infrastructure, wide-screen televisions, iPhones, and air conditioning. In many ways, the electric infrastructure could be considered society's most important artery.

Sadly, however, the American Society of Civil Engineers (ASCE) gave the U.S. energy infrastructure and electric artery a D+ rating in 2013. According to ASCE, the primary reason the United States received such a low grade is the fact that the country's electrical infrastructure is substantially aged and congested. The low rating is also a function of how difficult it will be for the United States to improve the power grid in light of its complexities, regulatory challenges, and implementation hurdles.

"Grid Congestion", by Dave Bryant, *Electricity Today T & D Magazine* (July/August 2013) Reprinted with Permission.

What is Grid Congestion?

Grid congestion is a situation wherein the existing transmission and/or distribution lines are unable to accommodate all required load during periods of high demand or during emergency load conditions, such as when an adjacent line is taken out of service or damaged by a storm, commonly referred to as an "N minus 1" or "N minus 2" condition—or worse. Grid congestion not only impacts reliability, it also reflects a decrease in efficiency. Under high load conditions, line losses escalate exponentially. If lines are congested and operating at or near their thermal limits, they would also be exhibiting significant line losses during high load conditions. Another significant problem with grid congestion is that during periods of high demand, electric retailers may not have access to the least expensive source of electricity which can "artificially" drive consumer prices to very high levels.

As North America has learned in the wake of several major blackouts, not unlike clogged arteries in the human body, a weak or congested transmission grid can lead to very unfavorable consequences.

Back Story

The Energy Policy Act of 1992 mandated that the Federal Energy Regulatory Commission (FERC) open up the national grid on a nondiscriminatory, non-preferential basis for the wholesale delivery of electric power. In April of 1996, FERC issued Orders 888 and 889 that encouraged wholesale competition. The primary objective of these orders was the elimination of the monopoly over the transmission of electricity. To achieve this objective, FERC required all public utilities that own, control, or operate facilities used for transmitting electric energy in interstate commerce to separate transmission from generation and marketing functions. FERC also required that these entities file open-access nondiscriminatory transmission tariffs.

In September, 1996, the State of California also put into law *The Electric Utility Industry Restructuring Act* (AB 1890) in an effort to make the generation of electricity more competitive. In 1999, FERC issued Order 2000 which asked all transmission-owning utilities, including non-public utilities, to place their transmission facilities under the control of an appropriate Regional Transmission Organization (RTO). This was done in an effort to better manage transmission congestion, oversee tariff management, and support the regional planning of system upgrades, among other objectives.

During the transition or "decoupling," as FERC later described in their staff report of March 26, 2003 (Docket No PA02-2-000), a supply-demand imbalance, flawed market design, and inconsistent rules made significant wholesale market manipulation possible. These, and several other factors, set the stage for the *Western energy crisis of 2000 to 2001*.

According to a report released by the Federal Energy Regulatory Commission, California's retail rate freeze (a component of new U.S. bill AB 1890 instituted by the California Legislature) and the inability to pass along price increases to customers not only bankrupted Pacific Gas & Electric (and nearly bankrupted Southern California Edison), the conditions also caused widespread rolling outages with an economic impact estimated at 40 to 45 billion dollars. In 1999, the retail rate freeze imposed on San Diego Gas & Electric (SDG&E) was temporarily lifted so retail customers in Southern California felt the brunt of cost increases immediately.

According to the final FERC report released in late March 2003, a number of factors contributed to the Western energy crisis of 2000 and 2001. While strong economic growth at that time led to increased demand for electricity, environmental concerns limited, prevented, or delayed new generation from being built locally.

Drought conditions in the Northwestern United States also resulted in reduced hydropower generation for the country while a ruptured natural gas line feeding California led to further generation constraints in a state that was highly dependent upon gas-fired generation. Unusually high temperatures, unplanned—or intentional—plant outages of older plants, and transmission capacity constraints also contributed to the crisis.

Congestion Costs

On the east coast, in 2002, the newly-established PJM Interconnection (a regional transmission organization) assessed the costs of congestion associated with individual transmission and distribution lines within their jurisdiction. PJM Interconnection reported that in several cases, over a period of several months, the combined costs of congestion of many transmission lines were measured to be over a billion dollars per year. This value doesn't include the costs of congestion from other regional transmission organizations.

The aforementioned costs were primarily a function of the difference in cost between a potentially available energy resource versus the cost paid for energy delivered from a more expensive but deliverable source (and/or as a function of market driven "supply and demand" pricing).

Impact of Congestion

On August 14, 2003, the Northeast United States and Ontario, Canada experienced the second most widespread blackout on record at that time (after Brazil in 1999), affecting 55 million people. Six weeks later, on September 28, 2003, a similar outage occurred in Europe that affected 56 million people.

According to the final report released by the North American Electric Reliability Corporation (NERC), the organization responsible for the adequacy and reliability of the bulk power transmission in North America, the U.S./Canada blackout of August 2003 was caused by a number of factors. The report stated that the causes of the blackout included inaccurate telemetry data used to operate the Midwest Independent Transmission System Operator (MISO) "State Estimator" (and a subsequent computer re-boot failure); a "race condition" computer bug in FirstEnergy's Energy Management System; and three 345 kilovolt (kV) transmission line trips (outages) due to excessive conductor sag, which led to a cascading of similar sag-trip outages on their 138-kV system.

These events and lack of effective communication between other utilities ultimately shut down 508 generating units at 265 power plants. According to the final NERC Report "*Technical Analysis of the August 14, 2003, Blackout: What Happened, Why, and What Did We Learn?*," the economic impact was estimated to approach 10 billion dollars. On July 30 and 31 of 2012, a similar series of cascading outages in India affected more than 670 million people. The Indian Government Ministry of Power also reported that substantial grid limitations and line tripping also played major roles in this event.

Steps Forward

Following the blackout of 2003, the Energy Policy Act of 2005 was established as a mechanism to help improve the efficiency, reliability, and capacity of the grid by creating incentives for utilities to invest in grid improvements and leverage new technologies including high-capacity low-sag conductors.

In 2009, the American Clean Energy and Security Act was approved by the House of Representatives but defeated in the Senate. While much of this bill was focused on climate change issues, many of its components also supported the modernization of the grid, primarily to enable the addition of new alternative "clean" sources of generation and develop and leverage new technology. Though the bill didn't pass, the message was clear and new cleaner sources of generation continue to be brought online.

The Challenges Continue

The addition of new sources of generation is further adding to the challenge of alleviating grid congestion. While new transmission lines are being proposed and/or built to link new generation, existing transmission and distribution lines may not be robust enough to handle additional load, especially during peak or emergency conditions, depending on where and how the new lines are integrated into the existing grid.

The Worst May Be Yet to Come

In the coming years, as outlined in the North American Reliability Corporation (NERC) "2012 Long-Term Reliability Assessment" report, the problem of congested transmission lines may become exacerbated not only by the increase in renewable resources that need to be integrated onto the grid, but with potentially greater impact, by the fact that the renewables may, under certain conditions, be unavailable at the same time.

The retirement of larger and/or strategically situated generation units will also impact the grid. Changes in generation type and location will require enhancements to the grid to provide reactive and voltage support, and address thermal (conductor sag) constraints, in order to insure system stability. This will be particularly challenging considering the timeframes required to make grid changes and the fact that, in many of the deregulated market areas, the system operator (performing the planning function) may have little insight into the retirement decisions of generation, which can be shut down with only a 90-day notice.

As the electric industry moves from conventional base load generation to various renewables that are dependent upon ambient conditions to perform (or not), and, if transmission enhancements are not implemented to accommodate the somewhat unpredictable changes, overall system reliability could be adversely affected.

Smart Grid Strategy Flaws

While utilities will continue to face significant challenges in securing permits to build new transmission lines, and incentives for utilities to invest in upgrading existing lines are not completely clear, another concern is the scope of the new Smart Grid policy enacted by the U.S. federal government.

The U.S. Department of Energy (DOE) defines Smart Grid "as an electrical grid that uses information and communications technology to gather and act on information, such as information about the behaviors of suppliers and consumers, in an automated fashion to improve the efficiency, reliability, economics, and sustainability of the production and distribution of electricity."

Surprisingly absent from the U.S. DOE definition is the word "capacity." The Energy Policy Act of 2005 clearly called for and provided a framework for incentives to leverage technology to "improve the efficiency, reliability and capacity of the grid." Capacity is not only a key element in mitigating congestion costs (which is a nice way of saying 'giving' the consumer access to the least expensive source of electricity—without artificial market manipulation), capacity is also extremely critical to grid reliability, not to mention North America's national security and financial wellbeing.

Lessons Learned

The key point in this feature is that utilities, consumers and regulatory bodies need to recognize the limitations of data and computers if they are serious about grid efficiency, capacity, and reliability—especially in this era of increasing cyber security risks. When telemetry devices and computers fail, thus leading to blackouts and outages, remember that the power grid is fundamentally a function of conductor capacity and thermal sag.

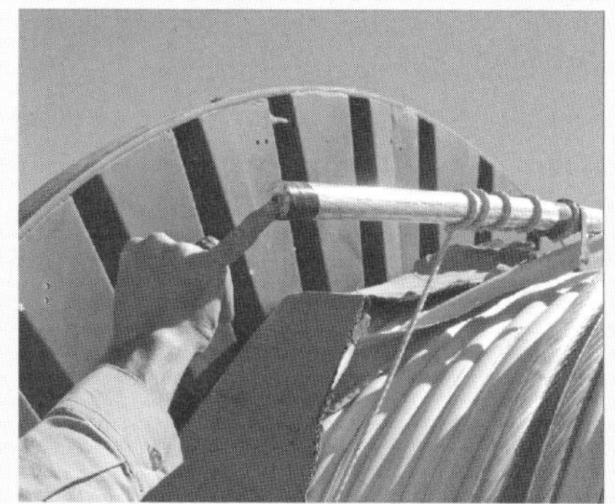

Figure 3 High-capacity low-sag conductor.

Modern Conductors

As increasingly higher levels of current are carried by overhead conductors, the electrical resistance of the wires causes them to heat up. Their coefficient of thermal expansion can then increase the conductor's sag, which can lead to short-circuit events and outages. While smart meters and other electronic devices may help utilities identify areas impacted by outages, strategies that outline prevention should become the priority.

The use of high-capacity, low-sag conductors (Figure 3) can alleviate grid congestion and serve to improve the efficiency, capacity, and reliability of the grid while also enabling utilities to more flexibly handle new sources of generation and storage technologies as they emerge and are integrated into the grid in varied locations.

Leveraging Technology

Today, a number of utilities such as American Electric Power (AEP), NV Energy, OG&E Energy Corporation, and others are discovering how to leverage new technologies including high-capacity, low-sag conductors.

These modern conductors use aerospace-derived technology to reduce conductor sag, increase line capacity, and reduce line losses.

The reduction of line losses has the same impact as building new generation. Thus, the economics of incorporating these technologies into the grid are very favorable—especially because existing structures rarely require modifications to accommodate the new wire. AEP is currently using a new high-capacity, low-sag conductor to increase the capacity of 240 circuit miles of a 345kV line near Corpus Christi, Texas, to accommodate new generation and growing demand (Figure 4). More than 100 other utilities in more than 25 countries have also deployed the technology for similar reasons at over 250 projects so far.

Role of Efficiency

For more than 100 years, utilities and their suppliers spent billions of dollars improving the efficiency of generators in an effort to squeak more out of less and improve economics. In the last 30 years, the push for improved efficiency has trickled downstream to the consumer where they too have discovered the benefits of using energy-efficient appliances to reduce their monthly bills. This effort has not only benefited consumers; it has also enabled utilities to postpone the development of new generation resources or retire old ones. As the saying goes, "it is cheaper to save a 'Negawatt' than it is to create a Megawatt."

Having spent billions of dollars improving the efficiency of generation and demand side appliances, perhaps it's time to focus on building more

Figure 4 Re-conductoring a 345-kV line.

efficiency into the grid itself where, in the United States, according to Jim Rogers, Chairman of Duke Energy, eight to 10 percent of all energy generated is lost during transmission and distribution.

Famous Last Words

What began as a function of old age—and continued growth in demand for affordable and reliable electricity—is now known as grid congestion. Congestion not only reflects capacity constraints, it also reflects substantial inefficiencies and decreased reliability. While electronics, smart meters and other technologies are certainly playing a role in decreasing stress on the grid and generation resources, to truly resolve the problem and prepare for a somewhat unknown future, the physical grid needs to be upgraded to combat grid congestion. The cost of upgrades will pale in comparison with the cost of ignoring this physical reality.

Later this year, the U.S. Department of Energy will be releasing its "2012 National Electric Transmission Congestion Study." It will be interesting to see how this report will impact the future of transmission congestion in North America. ■

4.1 TRANSMISSION LINE DESIGN CONSIDERATIONS

An overhead transmission line consists of conductors, insulators, support structures, and, in most cases, shield wires.

CONDUCTORS

Aluminum has replaced copper as the most common conductor metal for overhead transmission. Although a larger aluminum cross-sectional area is required to obtain the same loss as in a copper conductor, aluminum has a lower cost and lighter weight. Also, the supply of aluminum is abundant, whereas that of copper is limited.

One of the most common conductor types is aluminum conductor, steel-reinforced (ACSR), which consists of layers of aluminum strands surrounding a central core of steel strands (Figure 4.1). Stranded conductors are easier to manufacture, since larger conductor sizes can be obtained by simply adding successive layers of strands. Stranded conductors are also easier to handle and more flexible than solid conductors, especially in larger sizes. The use of steel strands gives ACSR conductors a high strength-to-weight ratio. For purposes of heat dissipation, overhead transmission-line conductors are bare (no insulating cover).

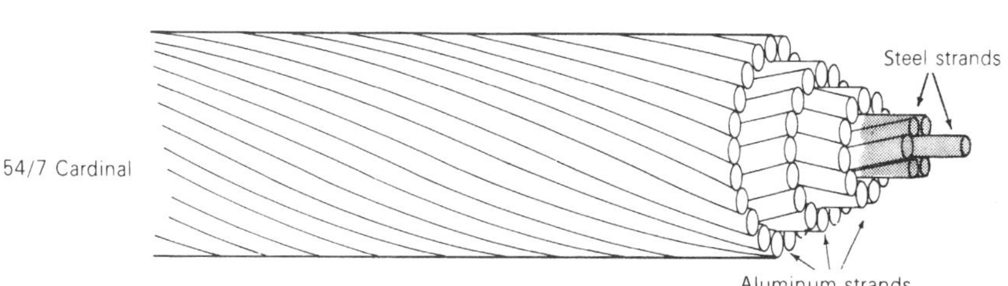

FIGURE 4.1 Typical ACSR conductor

FIGURE 4.2

A 765-kV transmission line with self-supporting lattice steel towers (Courtesy of the American Electric Power Company)

FIGURE 4.3

A 345-kV double-circuit transmission line with self-supporting lattice steel towers (Courtesy of NSTAR, formerly Boston Edison Company)

Otherconductortypesincludetheall-aluminumconductor(AAC),all-aluminum-alloy conductor (AAAC), aluminum conductor alloy-reinforced (ACAR), and aluminum-clad steel conductor (Alumoweld). Higher-temperature conductors capable of operation in excess of 150°C include the aluminum conductor steel supported (ACSS) that uses fully annealed aluminum around a steel core, and the gap-type ZT-aluminum conductor (GTZACSR) that uses heat-resistant aluminum over a steel core with a small annular gap between the steel and first layer of aluminum strands. Emerging technologies use composite materials, including the aluminum conductor carbon reinforced (ACFR), whose core is a resinmatrix composite containing carbon fiber, and the aluminum conductor composite reinforced (ACCR), whose core is an aluminum-matrix containing aluminum fibers [10].

EHV lines often have more than one conductor per phase; these conductors are called a *bundle.* The 765-kV line in Figure 4.2 has four conductors per phase, and the 345-kV double-circuit line in Figure 4.3 has two conductors per phase. Bundle conductors have a lower electric field strength at the conductor surfaces, thereby controlling corona. They also have a smaller series reactance.

INSULATORS

Insulators for transmission lines above 69 kV are typically suspension-type insulators, that consist of a string of discs constructed of porcelain, toughened glass, or polymer. The standard disc (Figure 4.4) has a 0.254-m diameter, 0.146-m spacing between centers of adjacent discs, and a mechanical strength of 7500 kg. The 765-kV line in Figure 4.2 has two strings per phase in a V-shaped arrangement, which helps to restrain conductor swings. The 345-kV line in Figure 4.5 has one vertical string per phase. The number of insulator discs in a string increases with line voltage

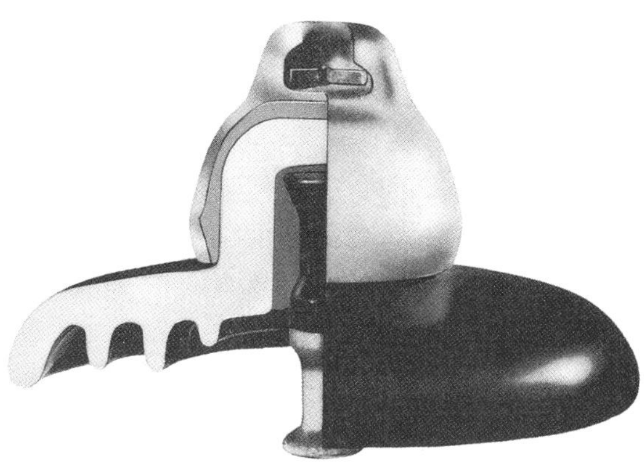

FIGURE 4.4

Cut-away view of a standard porcelain insulator disc for suspension insulator strings (Courtesy of Ohio Brass)

FIGURE 4.5

Wood frame structure for a 345-kV line (Courtesy of NSTAR, formerly Boston Edison Company)

(Table 4.1). Other types of discs include larger units with higher mechanical strength and fog insulators for use in contaminated areas.

SUPPORT STRUCTURES

Transmission lines employ a variety of support structures. Figure 4.2 shows a self-supporting, lattice steel tower typically used for 500- and 765-kV lines. Double-circuit 345-kV lines usually have self-supporting steel towers with the phases arranged either in a triangular configuration to reduce tower height or in a vertical configuration to reduce tower width (Figure 4.3). Wood frame configurations are commonly used for voltages of 345 kV and below (Figure 4.5).

SHIELD WIRES

Shield wires located above the phase conductors protect the phase conductors against lightning. They are usually high- or extra-high-strength steel, Alumoweld, or ACSR with much smaller cross section than the phase conductors. The number and location of the shield wires are selected so that almost all lightning strokes terminate on the shield wires rather than on the phase conductors. Figures 4.2, 4.3, and 4.5 have two shield wires. Shield wires are grounded to the tower. As such, when lightning strikes a shield wire, it flows harmlessly to ground, provided the tower impedance and tower footing resistance are small.

The decision to build new transmission is based on power-system planning studies to meet future system requirements of load growth and new generation. The points of interconnection of each new line to the system, as well as the power and voltage ratings of each, are selected based on these studies. Thereafter, transmission-line design is based on optimization of electrical, mechanical, environmental, and economic factors.

Nominal Voltage	Phase Conductors				
				Minimum Clearances	
(kV)	Number of Conductors per Bundle	Aluminum Cross-Section Area per Conductor (ACSR) (kcmil)*	Bundle Spacing (cm)	Phase-to-Phase (m)	Phase-to-Ground (m)
69	1	—	—	—	—
138	1	300–700	—	4 to 5	—
230	1	400–1000	—	6 to 9	—
345	1	2000–2500	—	6 to 9	7.6 to 11
345	2	800–2200	45.7	6 to 9	7.6 to 11
500	2	2000–2500	45.7	9 to 11	9 to 14
500	3	900–1500	45.7	9 to 11	9 to 14
765	4	900–1300	45.7	13.7	12.2

*1 kcmil = 0.5 mm^2

Nominal Voltage	Suspension Insulator String		Shield Wires		
(kV)	Number of Strings per Phase	Number of Standard Insulator Discs per Suspension String	Type	Number	Diameter (cm)
69	1	4 to 6	Steel	0, 1 or 2	—
138	1	8 to 11	Steel	0, 1 or 2	—
230	1	12 to 21	Steel or ACSR	1 or 2	1.1 to 1.5
345	1	18 to 21	Alumoweld	2	0.87 to 1.5
345	1 and 2	18 to 21	Alumoweld	2	0.87 to 1.5
500	2 and 4	24 to 27	Alumoweld	2	0.98 to 1.5
500	2 and 4	24 to 27	Alumoweld	2	0.98 to 1.5
765	2 and 4	30 to 35	Alumoweld	2	0.98

TABLE 4.1

Typical transmission-line characteristics [1, 2] (Electric Power Research Institute (EPRI), EPRI AC Transmission Line Reference Book—200 kV and Above (Palo Alto, CA: EPRI, www.epri.com, December 2005); Westinghouse Electric Corporation, Electrical Transmission and Distribution Reference Book, 4th ed. (East Pittsburgh, PA, 1964).)

ELECTRICAL FACTORS

Electrical design dictates the type, size, and number of bundle conductors per phase. Phase conductors are selected to have sufficient thermal capacity to meet continuous, emergency overload, and short-circuit current ratings. For EHV lines, the number of bundle conductors per phase is selected to control the voltage gradient at conductor surfaces, thereby reducing or eliminating corona.

Electrical design also dictates the number of insulator discs, vertical or V-shaped string arrangement, phase-to-phase clearance, and phase-to-tower clearance, all selected to provide adequate line insulation. Line insulation must withstand transient overvoltages due to lightning and switching surges, even when insulators are contaminated by fog, salt, or industrial pollution. Reduced clearances due to conductor swings during winds must also be accounted for.

The number, type, and location of shield wires are selected to intercept lightning strikes that would otherwise hit the phase conductors. Also, tower footing resistance can be reduced by using driven ground rods or a buried conductor (called *counterpoise*) running parallel to the line. Line height is selected to satisfy prescribed conductor-to-ground clearances and to control ground-level electric field and its potential shock hazard.

Conductor spacings, types, and sizes also determine the series impedance and shunt admittance. Series impedance affects line-voltage drops, I^2R losses, and stability limits (Chapters 5, 13). Shunt admittance, primarily capacitive, affects line-charging currents, which inject reactive power into the power system. Shunt reactors (inductors) are often installed on lightly loaded EHV lines to absorb part of this reactive power, thereby reducing overvoltages.

MECHANICAL FACTORS

Mechanical design focuses on the strength of the conductors, insulator strings, and support structures. Conductors must be strong enough to support a specified thickness of ice and a specified wind in addition to their own weight. Suspension insulator strings must be strong enough to support the phase conductors with ice and wind loadings from tower to tower (span length). Towers that satisfy minimum strength requirements, called suspension towers, are designed to support the phase conductors and shield wires with ice and wind loadings, and, in some cases, the unbalanced pull due to breakage of one or two conductors. Dead-end towers located every mile or so satisfy the maximum strength requirement of breakage of all conductors on one side of the tower. Angles in the line employ angle towers with intermediate strength. Conductor vibrations, which can cause conductor fatigue failure and damage to towers, are also of concern. Vibrations are controlled by adjustment of conductor tensions, use of vibration dampers, and—for bundle conductors—large bundle spacing and frequent use of bundle spacers.

ENVIRONMENTAL FACTORS

Environmental factors include land usage and visual impact. When a line route is selected, the effect on local communities and population centers, land values, access to property, wildlife, and use of public parks and facilities must all be considered. Reduction in visual impact is obtained by aesthetic tower design and by blending the line with the countryside. Also, the biological effects of prolonged exposure to electric and magnetic fields near transmission lines is of concern. Extensive research has been and continues to be done in this area.

ECONOMIC FACTORS

The optimum line design meets all the technical design criteria at lowest overall cost, which includes the total installed cost of the line as well as the cost of line losses over the operating life of the line. Many design factors affect cost. Utilities and consulting organizations use digital computer programs combined with specialized knowledge and physical experience to achieve optimum line design.

4.2 RESISTANCE

The dc resistance of a conductor at a specified temperature T is

$$R_{dc,T} = \frac{\rho_T l}{A} \quad \Omega \tag{4.2.1}$$

where ρ_T = conductor resistivity at temperature T
l = conductor length
A = conductor cross-sectional area

Two sets of units commonly used for calculating resistance, SI and English units, are summarized in Table 4.2. In this text we will use SI units throughout except where manufacturers' data is in English units. To interpret American manufacturers' data, it is useful to learn the use of English units in resistance calculations. In English units, conductor cross-sectional area is expressed in circular mils (cmil). One inch equals 1000 mils, and 1 cmil equals $\pi/4$ sq mil. A circle with diameter D in., or (D in.) (1000 mil/in.) = 1000 D mil = d mil, has an area

$$A = \left(\frac{\pi}{4} D^2 \text{ in.}^2\right)\left(1000 \frac{\text{mil}}{\text{in.}}\right)^2 = \frac{\pi}{4}(1000\ D)^2 = \frac{\pi}{4} d^2 \quad \text{sq mil}$$

or

$$A = \left(\frac{\pi}{4} d^2 \text{ sq mil}\right)\left(\frac{1 \text{ cmil}}{\pi/4 \text{ sq mil}}\right) = d^2 \quad \text{cmil} \tag{4.2.2}$$

1000 cmil or 1 kcmil is equal to 0.506 mm^2, often approximated to 0.5 mm^2.

Resistivity depends on the conductor metal. Annealed copper is the international standard for measuring resistivity ρ (or conductivity σ, where $\sigma = 1/\rho$). Resistivity of conductor metals is listed in Table 4.3. As shown, hard-drawn aluminum, which has 61% of the conductivity of the international standard, has a resistivity at 20 °C of 2.83×10^{-8} Ωm.

Quantity	Symbol	SI Units	English Units
Resistivity	ρ	Ωm	Ω-cmil/ft
Length	ℓ	m	ft
Cross-sectional area	A	m^2	cmil
dc resistance	$R_{dc} = \frac{\rho\ell}{A}$	Ω	Ω

TABLE 4.2

Comparison of SI and English units for calculating conductor resistance

Material	% Conductivity	$\rho_{20°C}$ Resistivity at 20°C $\Omega m \times 10^{-8}$	T Temperature Constant °C
Copper:			
Annealed	100%	1.72	234.5
Hard-drawn	97.3%	1.77	241.5
Aluminum			
Hard-drawn	61%	2.83	228.1
Brass	20–27%	6.4–8.4	480
Iron	17.2%	10	180
Silver	108%	1.59	243
Sodium	40%	4.3	207
Steel	2–14%	12–88	180–980

TABLE 4.3

% Conductivity, resistivity, and temperature constant of conductor metals

Conductor resistance depends on the following factors:

1. Spiraling
2. Temperature
3. Frequency ("skin effect")
4. Current magnitude—magnetic conductors

These are described in the following paragraphs.

For stranded conductors, alternate layers of strands are spiraled in opposite directions to hold the strands together. Spiraling makes the strands 1 or 2% longer than the actual conductor length. As a result, the dc resistance of a stranded conductor is 1 or 2% larger than that calculated from (4.2.1) for a specified conductor length.

Resistivity of conductor metals varies linearly over normal operating temperatures according to

$$\rho_{T2} = \rho_{T1}\left(\frac{T_2 + T}{T_1 + T}\right) \tag{4.2.3}$$

where ρ_{T2} and ρ_{T1} are resistivities at temperatures T_2 and T_1°C, respectively. T is a temperature constant that depends on the conductor material and is listed in Table 4.3.

The ac resistance or *effective* resistance of a conductor is

$$R_{ac} = \frac{P_{loss}}{|I|^2} \quad \Omega \tag{4.2.4}$$

where P_{loss} is the conductor real power loss in watts and I is the rms conductor current. For dc, the current distribution is uniform throughout the conductor cross section, and (4.2.1) is valid. However, for ac, the current distribution is nonuniform.

As frequency increases, the current in a solid cylindrical conductor tends to crowd toward the conductor surface with smaller current density at the conductor center. This phenomenon is called *skin effect.* A conductor with a large radius can even have an oscillatory current density versus the radial distance from the conductor center.

With increasing frequency, conductor loss increases, which, from (4.2.4), causes the ac resistance to increase. At power frequencies (60 Hz), the ac resistance is at most a few percent higher than the dc resistance. Conductor manufacturers normally provide dc, 50-Hz, and 60-Hz conductor resistance based on test data (see Appendix Tables A.3 and A.4).

For magnetic conductors, such as steel conductors used for shield wires, resistance depends on current magnitude. The internal flux linkages, and therefore the iron or magnetic losses, depend on the current magnitude. For ACSR conductors, the steel core has a relatively high resistivity compared to the aluminum strands, and therefore the effect of current magnitude on ACSR conductor resistance is small. Tables on magnetic conductors list resistance at two current levels (see Table A.4).

EXAMPLE 4.1

Stranded conductor: dc and ac resistance

Table A.3 lists a 4/0 copper conductor with 12 strands. Strand diameter is 0.1328 in. For this conductor:

a. Verify the total copper cross-sectional area of 107.2 mm^2.
b. Verify the dc resistance at 50°C of 0.188 Ω/km. Assume a 2% increase in resistance due to spiraling.
c. From Table A.3, determine the percent increase in resistance at 60 Hz versus dc.

SOLUTION

a. The strand diameter is $d = (0.1328 \text{ in.})(25.4 \text{ mm/in.}) = 3.373$ mm. Using four significant figures, the cross-sectional area of the 12-strand conductor is

$$A = 12\pi d^2/4 = 12\pi(3.373)^2/4 = 107.2 \text{ mm}^2$$

which agrees with the value given in Table A.3.

b. Using (4.2.3) and hard-drawn copper data from Table 4.3,

$$\rho_{50°C} = 1.77 \times 10^{-8}\left(\frac{50 + 241.5}{20 + 241.5}\right) = 1.973 \times 10^{-8} \quad \Omega\text{m}$$

From (4.2.1), the dc resistance at 50°C for a conductor length of 1 kilometer is

$$R_{dc,\,50°C} = \frac{(1.973 \times 10^{-8})(1000 \times 1.02)}{1.072 \times 10^{-4}} = 0.188 \quad \Omega/\text{km}$$

which agrees with the value listed in Table A.3.

c. From Table A.3,

$$\frac{R_{60\text{ Hz}, 50^\circ\text{C}}}{R_{\text{dc}, 50^\circ\text{C}}} = \frac{0.303}{0.302} = 1.003 \qquad \frac{R_{60\text{ Hz}, 25^\circ\text{C}}}{R_{\text{dc}, 25^\circ\text{C}}} = \frac{0.278}{0.276} = 1.007$$

Thus, the 60-Hz resistance of this conductor is about 0.3 to 0.7% higher than the dc resistance. The variation of these two ratios is due to the fact that resistance in Table A.3 is given to only three significant figures.

4.3 CONDUCTANCE

Conductance accounts for real power loss between conductors or between conductors and ground. For overhead lines, this power loss is due to leakage currents at insulators and to corona. Insulator leakage current depends on the amount of dirt, salt, and other contaminants that have accumulated on insulators, as well as on meteorological factors, particularly the presence of moisture. Corona occurs when a high value of electric field strength at a conductor surface causes the air to become electrically ionized and to conduct. The real power loss due to corona, called *corona loss*, depends on meteorological conditions, particularly rain, and on conductor surface irregularities. Losses due to insulator leakage and corona are usually small compared to conductor I^2R loss. Conductance is usually neglected in power system studies because it is a very small component of the shunt admittance.

4.4 INDUCTANCE: SOLID CYLINDRICAL CONDUCTOR

The inductance of a magnetic circuit that has a constant permeability μ can be obtained by determining the following:

1. Magnetic field intensity H, from Ampere's law
2. Magnetic flux density B $(B = \mu H)$
3. Flux linkages λ
4. Inductance from flux linkages per ampere $(L = \lambda/I)$

As a step toward computing the inductances of more general conductors and conductor configurations, first compute the internal, external, and total inductance of a solid cylindrical conductor. Also compute the flux linking one conductor in an array of current-carrying conductors.

Figure 4.6 shows a 1-meter section of a solid cylindrical conductor with radius r, carrying current I. For simplicity, assume that the conductor (1) is sufficiently long that end effects are neglected, (2) is nonmagnetic ($\mu = \mu_0 = 4\pi \times 10^{-7}$ H/m), and (3) has a uniform current density (skin effect is neglected). From (3.1.1), Ampere's law states that

FIGURE 4.6

Internal magnetic field of a solid cylindrical conductor

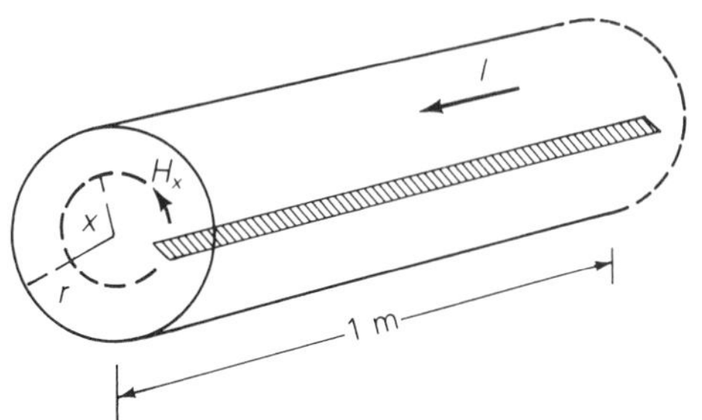

$$\oint H_{\tan}\, dl = I_{\text{enclosed}} \tag{4.4.1}$$

To determine the magnetic field inside the conductor, select the dashed circle of radius $x < r$ shown in Figure 4.6 as the closed contour for Ampere's law. Due to symmetry, H_x is constant along the contour. Also, there is no radial component of H_x, so H_x is tangent to the contour. That is, the conductor has a concentric magnetic field. From (4.4.1), the integral of H_x around the selected contour is

$$H_x(2\pi x) = I_x \qquad \text{for } x < r \tag{4.4.2}$$

where I_x is the portion of the total current enclosed by the contour. Solving (4.4.2),

$$H_x = \frac{I_x}{2\pi x} \quad \text{A/m} \tag{4.4.3}$$

Now assume a uniform current distribution within the conductor, that is

$$I_x = \left(\frac{x}{r}\right)^2 I \qquad \text{for } x < r \tag{4.4.4}$$

Using (4.4.4) in (4.4.3)

$$H_x = \frac{xI}{2\pi r^2} \quad \text{A/m} \tag{4.4.5}$$

For a nonmagnetic conductor, the magnetic flux density B_x is

$$B_x = \mu_0 H_x = \frac{\mu_0 x I}{2\pi r^2} \quad \text{Wb/m}^2 \tag{4.4.6}$$

The differential flux $d\Phi$ per-unit length of conductor in the cross-hatched rectangle of width dx shown in Figure 4.6 is

$$d\Phi = B_x\, dx \quad \text{Wb/m} \tag{4.4.7}$$

Computation of the differential flux linkage $d\lambda$ in the rectangle is tricky since only the fraction $(x/r)^2$ of the total current I is linked by the flux. That is,

$$d\lambda = \left(\frac{x}{r}\right)^2 d\Phi = \frac{\mu_0 I}{2\pi r^4}\, x^3\, dx \quad \text{Wb-t/m} \tag{4.4.8}$$

Integrating (4.4.8) from $x = 0$ to $x = r$ determines the total flux linkages λ_{int} inside the conductor

$$\lambda_{int} = \int_0^r d\lambda = \frac{\mu_0 I}{2\pi r^4}\int_0^r x^3\,dx = \frac{\mu_0 I}{8\pi} = \frac{1}{2}\times 10^{-7} I \quad \text{Wb-t/m} \tag{4.4.9}$$

The internal inductance L_{int} per-unit length of conductor due to this flux linkage is then

$$L_{int} = \frac{\lambda_{int}}{I} = \frac{\mu_0}{8\pi} = \frac{1}{2}\times 10^{-7} \quad \text{H/m} \tag{4.4.10}$$

Next, in order to determine the magnetic field outside the conductor, select the dashed circle of radius $x > r$ shown in Figure 4.7 as the closed contour for Ampere's law. Noting that this contour encloses the entire current I, integration of (4.4.1) yields

$$H_x(2\pi x) = I \tag{4.4.11}$$

which gives

$$H_x = \frac{I}{2\pi x} \quad \text{A/m} \quad x > r \tag{4.4.12}$$

Outside the conductor, $\mu = \mu_0$ and

$$B_x = \mu_0 H_x = (4\pi \times 10^{-7})\frac{I}{2\pi x} = 2\times 10^{-7}\frac{I}{x} \quad \text{Wb/m}^2 \tag{4.4.13}$$

$$d\Phi = B_x\,dx = 2\times 10^{-7}\frac{I}{x}\,dx \quad \text{Wb/m} \tag{4.4.14}$$

Since the entire current I is linked by the flux outside the conductor,

$$d\lambda = d\Phi = 2\times 10^{-7}\frac{I}{x}\,dx \quad \text{Wb-t/m} \tag{4.4.15}$$

Integrating (4.4.15) between two external points at distances D_1 and D_2 from the conductor center gives the external flux linkage λ_{12} between D_1 and D_2:

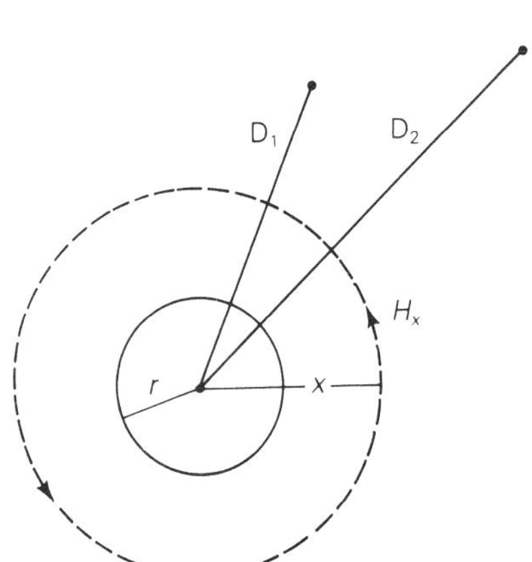

FIGURE 4.7 External magnetic field of a solid cylindrical conductor

$$\lambda_{12} = \int_{D_1}^{D_2} d\lambda = 2 \times 10^{-7} I \int_{D_1}^{D_2} \frac{dx}{x}$$

$$= 2 \times 10^{-7} I \ln\left(\frac{D_2}{D_1}\right) \quad \text{Wb-t/m} \tag{4.4.16}$$

The external inductance L_{12} per-unit length due to the flux linkages between D_1 and D_2 is then

$$L_{12} = \frac{\lambda_{12}}{I} = 2 \times 10^{-7} \ln\left(\frac{D_2}{D_1}\right) \quad \text{H/m} \tag{4.4.17}$$

The total flux λ_P linking the conductor out to external point P at distance D is the sum of the internal flux linkage, (4.4.9), and the external flux linkage, (4.4.16) from $D_1 = r$ to $D_2 = D$. That is

$$\lambda_P = \frac{1}{2} \times 10^{-7} I + 2 \times 10^{-7} I \ln \frac{D}{r} \tag{4.4.18}$$

Using the identity $\frac{1}{2} = 2 \ln e^{1/4}$ in (4.4.18), a more convenient expression for λ_P is obtained:

$$\lambda_p = 2 \times 10^{-7} I \left(\ln e^{1/4} + \ln \frac{D}{r}\right)$$

$$= 2 \times 10^{-7} I \ln \frac{D}{e^{-1/4} r}$$

$$= 2 \times 10^{-7} I \ln \frac{D}{r'} \quad \text{Wb-t/m} \tag{4.4.19}$$

where

$$r' = e^{-1/4} r = 0.7788 r \tag{4.4.20}$$

Also, the total inductance L_P due to both internal and external flux linkages out to distance D is

$$L_P = \frac{\lambda_P}{I} = 2 \times 10^{-7} \ln\left(\frac{D}{r'}\right) \quad \text{H/m} \tag{4.4.21}$$

Finally, consider the array of M solid cylindrical conductors shown in Figure 4.8. Assume that each conductor m carries current I_m referenced out of the page. Also assume that the sum of the conductor currents is zero—that is,

$$I_1 + I_2 + \cdots + I_M = \sum_{m=1}^{M} I_m = 0 \tag{4..4.22}$$

The flux linkage λ_{kPk}, which links conductor k out to point P due to current I_k, is, from (4.4.19),

$$\lambda_{kPk} = 2 \times 10^{-7} I_k \ln \frac{D_{Pk}}{r'_k} \tag{4.4.23}$$

Note that λ_{kPk} includes both internal and external flux linkages due to I_k. The flux linkage λ_{kPm}, which links conductor k out to P due to I_m, is, from (4.4.16),

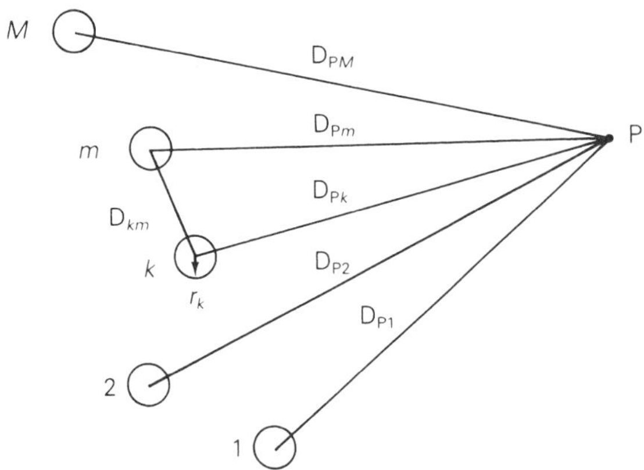

FIGURE 4.8 Array of *M* solid cylindrical conductors

$$\lambda_{kPm} = 2 \times 10^{-7} I_m \ln \frac{D_{Pm}}{D_{km}} \tag{4.4.24}$$

Equation (4.4.24) uses D_{km} instead of $(D_{km} - r_k)$ or $(D_{km} + r_k)$, which is a valid approximation when D_{km} is much greater than r_k. It can also be shown that this is a good approximation even when D_{km} is small. Using superposition, the total flux linkage λ_{kp}, which links conductor k out to P due to all the currents, is

$$\begin{aligned} \lambda_{kP} &= \lambda_{kP1} + \lambda_{kP2} + \cdots + \lambda_{kPM} \\ &= 2 \times 10^{-7} \sum_{m=1}^{M} I_m \ln \frac{D_{Pm}}{D_{km}} \end{aligned} \tag{4.4.25}$$

where $D_{kk} = r'_k = e^{-1/4} r_k$ is defined when $m = k$ in the above summation. Equation (4.4.25) is separated into two summations:

$$\lambda_{kP} = 2 \times 10^{-7} \sum_{m=1}^{M} I_m \ln \frac{1}{D_{km}} + 2 \times 10^{-7} \sum_{m=1}^{M} I_m \ln D_{Pm} \tag{4.4.26}$$

Removing the last term from the second summation results in

$$\lambda_{kP} = 2 \times 10^{-7} \left[\sum_{m=1}^{M} I_m \ln \frac{1}{D_{km}} + \sum_{m=1}^{M-1} I_m \ln D_{Pm} + I_M \ln D_{PM} \right] \tag{4.4.27}$$

From (4.4.22),

$$I_M = -(I_1 + I_2 + \cdots + I_{M-1}) = -\sum_{m=1}^{M-1} I_m \tag{4.4.28}$$

Using (4.4.28) in (4.4.27)

$$\begin{aligned} \lambda_{kP} &= 2 \times 10^{-7} \left[\sum_{m=1}^{M} I_m \ln \frac{1}{D_{km}} + \sum_{m=1}^{M-1} I_m \ln D_{Pm} - \sum_{m=1}^{M-1} I_m \ln D_{PM} \right] \\ &= 2 \times 10^{-7} \left[\sum_{m=1}^{M} I_m \ln \frac{1}{D_{km}} + \sum_{m=1}^{M-1} I_m \ln \frac{D_{Pm}}{D_{PM}} \right] \end{aligned} \tag{4.4.29}$$

Now, let λ_k equal the total flux linking conductor k out to infinity. That is, $\lambda_k = \lim_{P \to \infty} \lambda_{kP}$. As $P \to \infty$, all the distances D_{Pm} become equal, the ratios D_{Pm}/D_{PM} become unity, and $\ln(D_{Pm}/D_{PM}) \to 0$. Therefore, the second summation in (4.4.29) becomes zero as $P \to \infty$, and

$$\lambda_k = 2 \times 10^{-7} \sum_{m=1}^{M} I_m \ln \frac{1}{D_{km}} \quad \text{Wb-t/m} \tag{4.4.30}$$

Equation (4.4.30) gives the total flux linking conductor k in an array of M conductors carrying currents $I_1, I_2, \ldots, I_M$, whose sum is zero. This equation is valid for either dc or ac currents. λ_k is a dc flux linkage when the currents are dc, and λ_k is a phasor flux linkage when the currents are phasor representations of sinusoids.

4.5 INDUCTANCE: SINGLE-PHASE TWO-WIRE LINE AND THREE-PHASE THREE-WIRE LINE WITH EQUAL PHASE SPACING

The results of the previous section are used here to determine the inductances of two relatively simple transmission lines: a single-phase two-wire line and a three-phase three-wire line with equal phase spacing.

Figure 4.9(a) shows a single-phase two-wire line consisting of two solid cylindrical conductors x and y. Conductor x with radius r_x carries phasor current $I_x = I$ referenced out of the page. Conductor y with radius r_y carries return current $I_y = -I$. Since the sum of the two currents is zero, (4.4.30) is valid, from which the total flux linking conductor x is

$$\begin{aligned}\lambda_x &= 2 \times 10^{-7}\left(I_x \ln \frac{1}{D_{xx}} + I_y \ln \frac{1}{D_{xy}}\right) \\ &= 2 \times 10^{-7}\left(I \ln \frac{1}{r'_x} - I \ln \frac{1}{D}\right) \\ &= 2 \times 10^{-7} I \ln \frac{D}{r'_x} \quad \text{Wb-t/m}\end{aligned} \tag{4.5.1}$$

FIGURE 4.9 Single-phase two-wire line

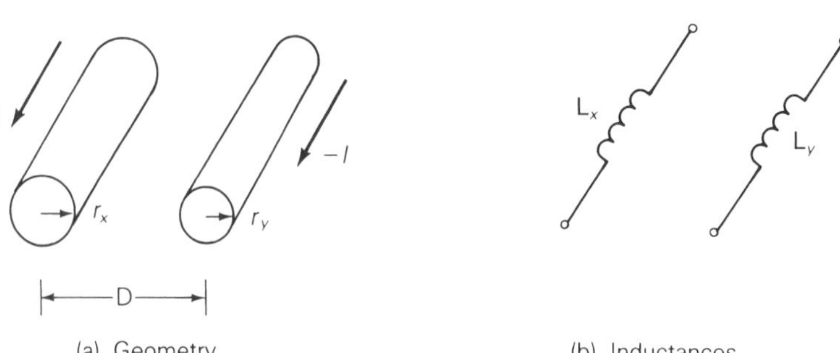

(a) Geometry (b) Inductances

Where $r_x' = e^{-1/4} r_x = 0.7788 r_x$.

The inductance of conductor x is then

$$L_x = \frac{\lambda_x}{I_x} = \frac{\lambda_x}{I} = 2 \times 10^{-7} \ln \frac{D}{r_x'} \quad \text{H/m per conductor} \tag{4.5.2}$$

Similarly, the total flux linking conductor y is

$$\begin{aligned}
\lambda_y &= 2 \times 10^{-7} \left(I_x \ln \frac{1}{D_{yx}} + I_y \ln \frac{1}{D_{yy}} \right) \\
&= 2 \times 10^{-7} \left(I \ln \frac{1}{D} - I \ln \frac{1}{r_y'} \right) \\
&= -2 \times 10^{-7} I \ln \frac{D}{r_y'}
\end{aligned} \tag{4.5.3}$$

and

$$L_y = \frac{\lambda_y}{I_y} = \frac{\lambda_y}{-I} = 2 \times 10^{-7} \ln \frac{D}{r_y'} \quad \text{H/m per conductor} \tag{4.5.4}$$

The total inductance of the single-phase circuit, also called *loop inductance,* is

$$\begin{aligned}
L = L_x + L_y &= 2 \times 10^{-7} \left(\ln \frac{D}{r_x'} + \ln \frac{D}{r_y'} \right) \\
&= 2 \times 10^{-7} \ln \frac{D^2}{r_x' r_y'} \\
&= 4 \times 10^{-7} \ln \frac{D}{\sqrt{r_x' r_y'}} \quad \text{H/m per circuit}
\end{aligned} \tag{4.5.5}$$

Also, if $r_x' = r_y' = r'$, the total circuit inductance is

$$L = 4 \times 10^{-7} \ln \frac{D}{r'} \quad \text{H/m per circuit} \tag{4.5.6}$$

The inductances of the single-phase two-wire line are shown in Figure 4.9(b).

Figure 4.10(a) shows a three-phase three-wire line consisting of three solid cylindrical conductors a, b, c, each with radius r, and with equal phase spacing D between any two conductors. To determine inductance, assume balanced positive-sequence currents I_a, I_b, I_c that satisfy $I_a + I_b + I_c = 0$. Then (4.4.30) is valid and the total flux linking the phase a conductor is

$$\begin{aligned}
\lambda_a &= 2 \times 10^{-7} \left(I_a \ln \frac{1}{r'} + I_b \ln \frac{1}{D} + I_c \ln \frac{1}{D} \right) \\
&= 2 \times 10^{-7} \left[I_a \ln \frac{1}{r'} + (I_b + I_c) \ln \frac{1}{D} \right]
\end{aligned} \tag{4.5.7}$$

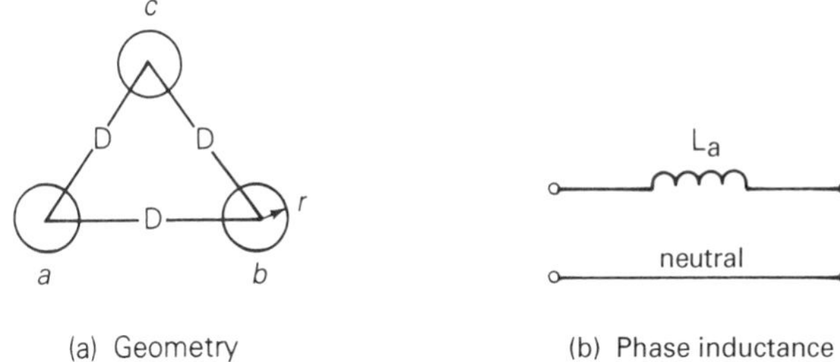

FIGURE 4.10 Three-phase three-wire line with equal phase spacing

Using $(I_b + I_c) = -I_a$,

$$\lambda_a = 2 \times 10^{-7}\left(I_a \ln \frac{1}{r'} - I_a \ln \frac{1}{D}\right)$$

$$= 2 \times 10^{-7}\, I_a \ln \frac{D}{r'} \quad \text{Wb-t/m} \tag{4.5.8}$$

The inductance of phase a is then

$$L_a = \frac{\lambda_a}{I_a} = 2 \times 10^{-7} \ln \frac{D}{r'} \quad \text{H/m per phase} \tag{4.5.9}$$

Due to symmetry, the same result is obtained for $L_b = \lambda_b/I_b$ and for $L_c = \lambda_c/I_c$. However, only one phase need be considered for balanced three-phase operation of this line, since the flux linkages of each phase have equal magnitudes and 120° displacement. The phase inductance is shown in Figure 4.10(b).

4.6 INDUCTANCE: COMPOSITE CONDUCTORS, UNEQUAL PHASE SPACING, BUNDLED CONDUCTORS

The results of Section 4.5 are extended here to include composite conductors, which consist of two or more solid cylindrical subconductors in parallel. A stranded conductor is one example of a composite conductor. For simplicity assume that for each conductor, the subconductors are identical and share the conductor current equally.

Figure 4.11 shows a single-phase two-conductor line consisting of two composite conductors x and y. Conductor x has N identical subconductors, each with radius r_x and with current (I/N) referenced out of the page. Similarly, conductor y consists of M identical subconductors, each with radius r_y and with return current $(-I/M)$. Since the sum of all the currents is zero, (4.4.30) is valid and the total flux Φ_k linking subconductor k of conductor x is

$$\Phi_k = 2 \times 10^{-7}\left[\frac{I}{N}\sum_{m=1}^{N} \ln \frac{1}{D_{km}} - \frac{1}{M}\sum_{m=1'}^{M} \ln \frac{1}{D_{km}}\right] \tag{4.6.1}$$

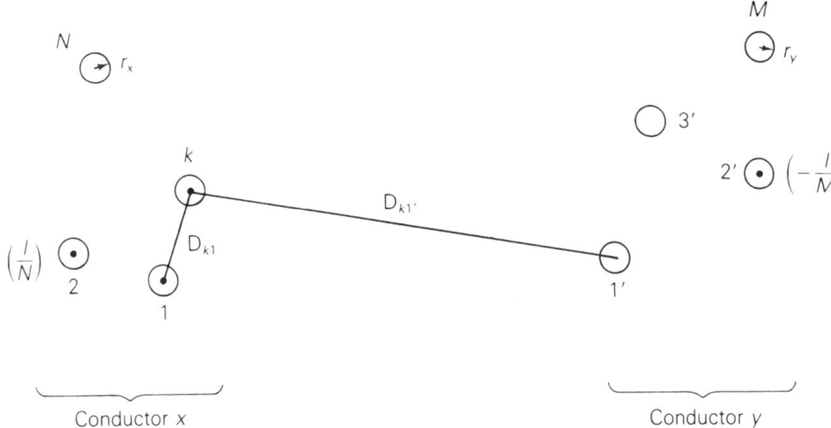

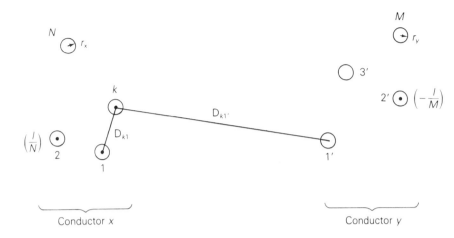

FIGURE 4.11 Single-phase two-conductor line with composite conductors

Since only the fraction $(1/N)$ of the total conductor current I is linked by this flux, the flux linkage λ_k of (the current in) subconductor k is

$$\lambda_k = \frac{\Phi_k}{N} = 2 \times 10^{-7} I \left[\frac{1}{N^2} \sum_{m=1}^{N} \ln \frac{1}{D_{km}} - \frac{1}{NM} \sum_{m=1'}^{M} \ln \frac{1}{D_{km}} \right] \tag{4.6.2}$$

The total flux linkage of conductor x is

$$\lambda_x = \sum_{k=1}^{N} \lambda_k = 2 \times 10^{-7} I \sum_{k=1}^{N} \left[\frac{1}{N^2} \sum_{m=1}^{N} \ln \frac{1}{D_{km}} - \frac{1}{NM} \sum_{m=1'}^{M} \ln \frac{1}{D_{km}} \right] \tag{4.6.3}$$

Using $\ln A^{\alpha} = \alpha \ln A$ and $\sum \ln A_k = \ln \prod A_k$ (sum of ln s = ln of products), (4.6.3) can be rewritten in the following form:

$$\lambda_x = 2 \times 10^{-7}\, I \ln \prod_{k=1}^{N} \frac{\left(\prod_{m=1'}^{M} D_{km} \right)^{1/NM}}{\left(\prod_{m=1}^{N} D_{km} \right)^{1/N^2}} \tag{4.6.4}$$

and the inductance of conductor x, $L_x = \dfrac{\lambda_x}{I}$, can be written as

$$L_x = 2 \times 10^{-7} \ln \frac{D_{xy}}{D_{xx}} \quad \text{H/m per conductor} \tag{4.6.5}$$

where

$$D_{xy} = \sqrt[MN]{\prod_{k=1}^{N} \prod_{m=1'}^{M} D_{km}} \tag{4.6.6}$$

$$D_{xx} = \sqrt[N^2]{\prod_{k=1}^{N} \prod_{m=1}^{N} D_{km}} \tag{4.6.7}$$

D_{xy}, given by (4.6.6), is the MNth root of the product of the MN distances from the subconductors of conductor x to the subconductors of conductor y. Associated with each subconductor k of conductor x are the M distances $D_{k1'}$, $D_{k2'}$, ..., D_{kM} to the subconductors of conductor y. For N subconductors in conductor x, there are therefore MN of these distances. D_{xy} is called the *geometric mean distance* or GMD between conductors x and y.

Also, D_{xx}, given by (4.6.7), is the N^2 root of the product of the N^2 distances between the subconductors of conductor x. Associated with each subconductor k are the N distances D_{k1}, D_{k2}, ..., $D_{kk} = r'$, ..., D_{kN}. For N subconductors in conductor x, there are therefore N^2 of these distances. D_{xx} is called the *geometric mean radius* or GMR of conductor x.

Similarly, for conductor y,

$$L_y = 2 \times 10^{-7} \ln \frac{D_{xy}}{D_{yy}} \quad \text{H/m per conductor} \tag{4.6.8}$$

where

$$D_{yy} = \sqrt[M^2]{\prod_{k=1'}^{M} \prod_{m=1'}^{M} D_{km}} \tag{4.6.9}$$

D_{yy}, the GMR of conductor y, is the M^2 root of the product of the M^2 distances between the subconductors of conductor y. The total inductance L of the single-phase circuit is

$$L = L_x + L_y \quad \text{H/m per circuit} \tag{4.6.10}$$

EXAMPLE 4.2

GMR, GMD, and inductance: single-phase two-conductor line

Expand (4.6.6), (4.6.7), and (4.6.9) for $N = 3$ and $M = 2'$. Then evaluate L_x, L_y, and L in H/m for the single-phase two-conductor line shown in Figure 4.12.

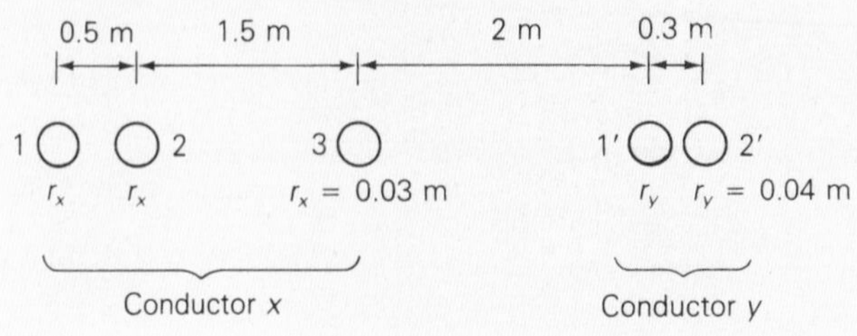

FIGURE 4.12 Single-phase two-conductor line for Example 4.2

SOLUTION

For $N = 3$ and $M = 2'$, (4.6.6) becomes

$$D_{xy} = \sqrt[6]{\prod_{k=1}^{3} \prod_{m=1'}^{2'} D_{km}}$$

$$= \sqrt[6]{\prod_{k=1}^{3} D_{k1'} D_{k2'}}$$

$$= \sqrt[6]{(D_{11'} D_{12'})(D_{21'} D_{22'})(D_{31'} D_{32'})}$$

Similarly, (4.6.7) becomes

$$D_{xx} = \sqrt[9]{\prod_{k=1}^{3} \prod_{m=1}^{3} D_{km}}$$

$$= \sqrt[9]{\prod_{k=1}^{3} D_{k1} D_{k2} D_{k3}}$$

$$= \sqrt[9]{(D_{11} D_{12} D_{13})(D_{21} D_{22} D_{23})(D_{31} D_{32} D_{33})}$$

and (4.6.9) becomes

$$D_{yy} = \sqrt[4]{\prod_{k=1'}^{2'} \prod_{m=1'}^{2'} D_{km}}$$

$$= \sqrt[4]{\prod_{k=1'}^{2'} D_{k1'} D_{k2'}}$$

$$= \sqrt[4]{(D_{1'1'} D_{1'2'})(D_{2'1'} D_{2'2'})}$$

Evaluating D_{xy}, D_{xx}, and D_{yy} for the single-phase two-conductor line shown in Figure 4.12,

$D_{11'} = 4$ m $\quad D_{12'} = 4.3$ m $\quad D_{21'} = 3.5$ m

$D_{22'} = 3.8$ m $\quad D_{31'} = 2$ m $\quad D_{32'} = 2.3$ m

$D_{xy} = \sqrt[6]{(4)(4.3)(3.5)(3.8)(2)(2.3)} = 3.189$ m

$D_{11} = D_{22} = D_{33} = r'_x = e^{-1/4} r_x = (0.7788)(0.03) = 0.02336$ m

$D_{21} = D_{12} = 0.5$ m

$D_{23} = D_{32} = 1.5$ m

$D_{31} = D_{13} = 2.0$ m

(Continued)

$$D_{xx} = \sqrt[9]{(0.02336)^3(0.5)^2(1.5)^2(2.0)^2} = 0.3128 \text{ m}$$

$$D_{1'1'} = D_{2'2'} = r'_y = e^{-1/4}r_y = (0.7788)(0.4) = 0.03115 \text{ m}$$

$$D_{1'2'} = D_{2'1'} = 0.3 \text{ m}$$

$$D_{yy} = \sqrt[4]{(0.03115)^2(0.3)^2} = 0.09667 \text{ m}$$

Then, from (4.6.5), (4.6.8), and (4.6.10):

$$L_x = 2 \times 10^{-7} \ln\left(\frac{3.189}{0.3128}\right) = 4.644 \times 10^{-7} \quad \text{H/m per conductor}$$

$$L_y = 2 \times 10^{-7} \ln\left(\frac{3.189}{0.09667}\right) = 6.992 \times 10^{-7} \quad \text{H/m per conductor}$$

$$L = L_x + L_y = 1.164 \times 10^{-6} \quad \text{H/m per circuit}$$

It is seldom necessary to calculate GMR or GMD for standard lines. The GMR of standard conductors is provided by conductor manufacturers and can be found in various handbooks (see Appendix Tables A.3 and A.4). Also, if the distances between conductors are large compared to the distances between subconductors of each conductor, then the GMD between conductors is approximately equal to the distance between conductor centers.

EXAMPLE 4.3

Inductance and inductive reactance: single-phase line

A single-phase line operating at 60 Hz consists of two 4/0 12-strand copper conductors with 1.5 m spacing between conductor centers. The line length is 32 km. Determine the total inductance in H and the total inductive reactance in Ω.

SOLUTION

The GMD between conductor centers is $D_{xy} = 1.5$ m. Also, from Table A.3, the GMR of a 4/0 12-strand copper conductor is $D_{xx} = D_{yy} = 0.005334$ m. From (4.6.5) and (4.6.8),

$$L_x = L_y = 2 \times 10^{-7} \ln\left(\frac{1.5}{0.005334}\right)\frac{\text{H}}{\text{m}} \times 32{,}000 \text{ m}$$

$$= 0.03609 \quad \text{H per conductor}$$

The total inductance is

$$L = L_x + L_y = 2 \times 0.03639 = 0.07279 \quad \text{H per circuit}$$

and the total inductive reactance is

$$X_L = 2\pi f L = (2\pi)(60)(0.07279) = 27.44 \quad \Omega \text{ per circuit}$$

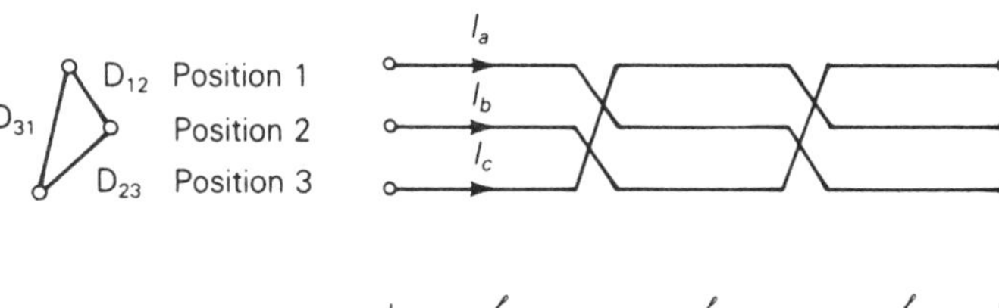

FIGURE 4.13 Completely transposed three-phase line

To calculate inductance for three-phase lines with stranded conductors and equal phase spacing, r' is replaced by the conductor GMR in (4.5.9). If the spacings between phases are unequal, then balanced positive-sequence flux linkages are not obtained from balanced positive-sequence currents. Instead, unbalanced flux linkages occur, and the phase inductances are unequal. However, balance can be restored by exchanging the conductor positions along the line, which is a technique called *transposition.*

Figure 4.13 shows a completely transposed three-phase line. The line is transposed at two locations such that each phase occupies each position for one-third of the line length. Conductor positions are denoted 1, 2, 3 with distances D_{12}, D_{23}, D_{31} between positions. The conductors are identical, each with GMR denoted D_S. To calculate inductance of this line, assume balanced positive-sequence currents I_a, I_b, I_c, for which $I_a + I_b + I_c = 0$. Again, (4.4.30) is valid, and the total flux linking the phase a conductor while it is in position 1 is

$$\lambda_{a1} = 2 \times 10^{-7}\left[I_a \ln\frac{1}{D_S} + I_b \ln\frac{1}{D_{12}} + I_c \ln\frac{1}{D_{31}}\right] \quad \text{Wb-t/m} \tag{4.6.11}$$

Similarly, the total flux linkage of this conductor while it is in positions 2 and 3 is

$$\lambda_{a2} = 2 \times 10^{-7}\left[I_a \ln\frac{1}{D_S} + I_b \ln\frac{1}{D_{23}} + I_c \ln\frac{1}{D_{12}}\right] \quad \text{Wb-t/m} \tag{4.6.12}$$

$$\lambda_{a3} = 2 \times 10^{-7}\left[I_a \ln\frac{1}{D_S} + I_b \ln\frac{1}{D_{31}} + I_c \ln\frac{1}{D_{23}}\right] \quad \text{Wb-t/m} \tag{4.6.13}$$

The average of the above flux linkages is

$$\lambda_a = \frac{\lambda_{a1}\left(\frac{l}{3}\right) + \lambda_{a2}\left(\frac{l}{3}\right) + \lambda_{a3}\left(\frac{l}{3}\right)}{l} = \frac{\lambda_{a1} + \lambda_{a2} + \lambda_{a3}}{3}$$

$$= \frac{2 \times 10^{-7}}{3}\left[3I_a \ln\frac{1}{D_S} + I_b \ln\frac{1}{D_{12}D_{23}D_{31}} + I_c \ln\frac{1}{D_{12}D_{23}D_{31}}\right] \tag{4.6.14}$$

Using $(I_b + I_c) = -I_a$ in (4.6.14),

$$\lambda_a = \frac{2 \times 10^{-7}}{3}\left[3I_a \ln\frac{1}{D_S} - I_a \ln\frac{1}{D_{12}D_{23}D_{31}}\right]$$

$$= 2 \times 10^{-7} I_a \ln\frac{\sqrt[3]{D_{12}D_{23}D_{31}}}{D_S} \quad \text{Wb-t/m} \tag{4.6.15}$$

and the average inductance of phase a is

$$L_a = \frac{\lambda_a}{I_a} = 2 \times 10^{-7} \ln \frac{\sqrt[3]{D_{12}D_{23}D_{31}}}{D_S} \quad \text{H/m per phase} \tag{4.6.16}$$

The same result is obtained for $L_b = \lambda_b/I_b$ and for $L_c = \lambda_c/I_c$. However, only one phase need be considered for balanced three-phase operation of a completely transposed three-phase line. Defining

$$D_{eq} = \sqrt[3]{D_{12}D_{23}D_{31}} \tag{4.6.17}$$

results in

$$L_a = 2 \times 10^{-7} \ln \frac{D_{eq}}{D_S} \quad \text{H/m} \tag{4.6.18}$$

D_{eq}, the cube root of the product of the three-phase spacings, is the geometric mean distance between phases. Also, D_S is the conductor GMR for stranded conductors, or r' for solid cylindrical conductors.

EXAMPLE 4.4

Inductance and inductive reactance: three-phase line

A completely transposed 60-Hz three-phase line has flat horizontal phase spacing with 10 m between adjacent conductors. The conductors are 1,590,000 cmil ACSR with 54/3 stranding. Line length is 200 km. Determine the inductance in H and the inductive reactance in Ω.

SOLUTION

From Table A.4, the GMR of a 1,590,000 cmil 54/3 ACSR conductor is

$$D_S = 0.0520 \text{ ft} \, \frac{1 \text{ m}}{3.28 \text{ ft}} = 0.0159 \text{ m}$$

Also, from (4.6.17) and (4.6.18),

$$D_{eq} = \sqrt[3]{(10)(10)(20)} = 12.6 \text{ m}$$

$$L_a = 2 \times 10^{-7} \ln \left(\frac{12.6}{0.0159}\right) \frac{\text{H}}{\text{m}} \times \frac{1000 \text{ m}}{\text{km}} \times 200 \text{ km}$$

$$= 0.267 \text{ H}$$

The inductive reactance of phase a is

$$X_a = 2\pi f L_a = 2\pi(60)(0.267) = 101 \quad \Omega$$

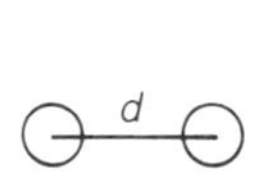

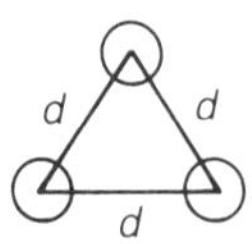

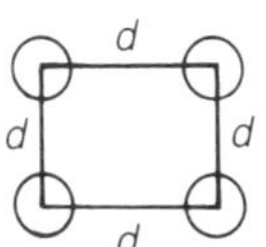

FIGURE 4.14

Bundle conductor configurations

It is common practice for EHV lines to use more than one conductor per phase, a practice called *bundling*. Bundling reduces the electric field strength at the conductor surfaces, which in turn reduces or eliminates corona and its results: undesirable power loss, communications interference, and audible noise. Bundling also reduces the series reactance of the line by increasing the GMR of the bundle.

Figure 4.14 shows common EHV bundles consisting of two, three, or four conductors. The three-conductor bundle has its conductors on the vertices of an equilateral triangle, and the four-conductor bundle has its conductors on the corners of a square. To calculate inductance, D_S in (4.6.18) is replaced by the GMR of the bundle. Since the bundle constitutes a composite conductor, calculation of bundle GMR is, in general, given by (4.6.7). If the conductors are stranded and the bundle spacing d is large compared to the conductor outside radius, each stranded conductor is first replaced by an equivalent solid cylindrical conductor with GMR $= D_S$. Then the bundle is replaced by one equivalent conductor with GMR $= D_{SL}$, given by (4.6.7) with $n = 2$, 3, or 4 as follows:

Two-conductor bundle:

$$D_{SL} = \sqrt[4]{(D_S \times d)^2} = \sqrt{D_S d} \tag{4.6.19}$$

Three-conductor bundle:

$$D_{SL} = \sqrt[9]{(D_S \times d \times d)^3} = \sqrt[3]{D_S d^2} \tag{4.6.20}$$

Four-conductor bundle:

$$D_{SL} = \sqrt[16]{(D_S \times d \times d \times d\sqrt{2})^4} = 1.091\sqrt[4]{D_S d^3} \tag{4.6.21}$$

The inductance is then

$$L_a = 2 \times 10^{-7} \ln \frac{D_{eq}}{D_{SL}} \quad \text{H/m} \tag{4.6.22}$$

If the phase spacings are large compared to the bundle spacing, then sufficient accuracy for D_{eq} is obtained by using the distances between bundle centers.

EXAMPLE 4.5

Inductive reactance: three-phase line with bundled conductors

Each of the 1,590,000 cmil conductors in Example 4.4 is replaced by two 795,000 cmil ACSR 26/2 conductors, as shown in Figure 4.15. Bundle spacing is

(*Continued*)

FIGURE 4.15 Three-phase bundled conductor line for Example 4.5

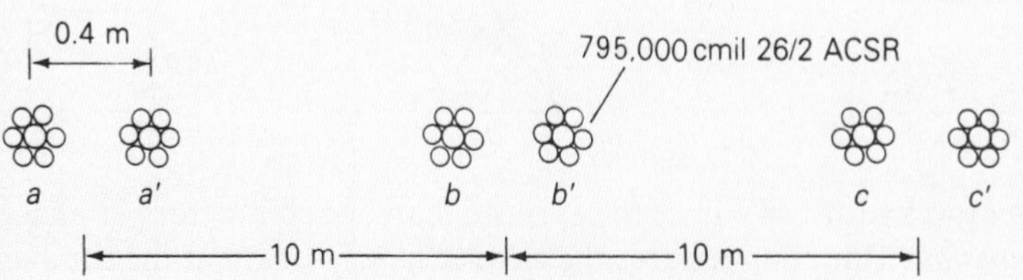

0.40 m. Flat horizontal spacing is retained, with 10 m between adjacent bundle centers. Calculate the inductive reactance of the line and compare it with that of Example 4.4.

SOLUTION

From Table A.4, the GMR of a 795,000 cmil 26/2 ACSR conductor is

$$D_S = 0.0375 \text{ ft} \times \frac{1 \text{ m}}{3.28 \text{ ft}} = 0.0114 \text{ m}$$

From (4.6.19), the two-conductor bundle GMR is

$$D_{SL} = \sqrt{(0.0114)(0.40)} = 0.0676 \text{ m}$$

Since $D_{eq} = 12.6$ m is the same as in Example 4.4,

$$L_a = 2 \times 10^{-7} \ln\left(\frac{12.6}{0.0676}\right)(1000)(200) = 0.209 \text{ H}$$

$$X_a = 2\pi f L_1 = (2\pi)(60)(0.209) = 78.8 \ \Omega$$

The reactance of the bundled line, 78.8 Ω, is 22% less than that of Example 4.4, even though the two-conductor bundle has the same amount of conductor material (that is, the same cmil per phase). One advantage of reduced series line reactance is smaller line-voltage drops. Also, the loadability of medium and long EHV lines is increased (see Chapter 5).

4.7 SERIES IMPEDANCES: THREE-PHASE LINE WITH NEUTRAL CONDUCTORS AND EARTH RETURN

This section develops equations suitable for computer calculation of the series impedances, including resistances and inductive reactances, for the three-phase overhead line shown in Figure 4.16. This line has three phase conductors *a*, *b*, and c, where bundled conductors, if any, have already been replaced by equivalent conductors, as described in Section 4.6. The line also has *N* neutral conductors denoted *n*1, *n*2, ..., *nN*.* All the neutral conductors are connected in parallel and are grounded

* Instead of *shield wire*, the term *neutral conductor* is used. It applies to distribution as well as transmission lines.

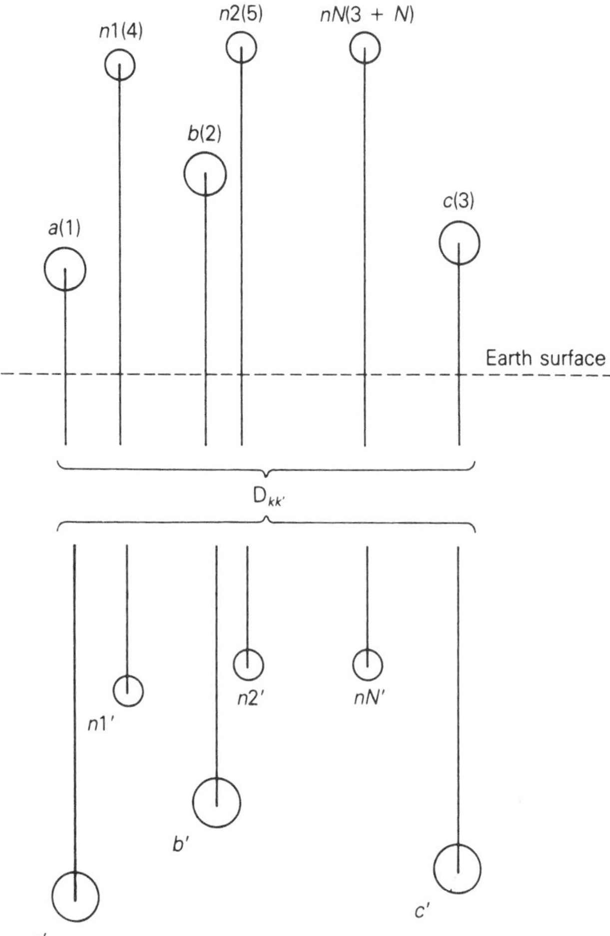

FIGURE 4.16 Three-phase transmission line with earth replaced by earth return conductors

to the earth at regular intervals along the line. Any isolated neutral conductors that carry no current are omitted. The phase conductors are insulated from each other and from the earth.

If the phase currents are not balanced, there may be a return current in the grounded neutral wires and in the earth. The earth return current will spread out under the line, seeking the lowest impedance return path. A classic paper by Carson [4], later modified by others [5, 6], shows that the earth can be replaced by a set of "earth return" conductors located directly under the overhead conductors, as shown in Figure 4.16. Each earth return conductor carries the negative of its overhead conductor current, has a GMR denoted $D_{k'k'}$, distance $D_{kk'}$ from its overhead conductor, and resistance $R_{k'}$ given by:

$$D_{k'k'} = D_{kk} \quad \text{m} \tag{4.7.1}$$

$$D_{kk'} = 658.5\sqrt{\rho/f} \quad \text{m} \tag{4.7.2}$$

$$R_{k'} = 9.869 \times 10^{-7} f \quad \Omega/\text{m} \tag{4.7.3}$$

TABLE 4.4 Earth resistivities and 60-Hz equivalent conductor distances

Type of Earth	Resistivity (Ωm)	$D_{kk'}$ (m)
Sea water	0.01–1.0	8.50–85.0
Swampy ground	10–100	269–850
Average damp earth	100	850
Dry earth	1000	2690
Pure slate	10^7	269,000
Sandstone	10^9	2,690,000

where ρ is the earth resistivity in ohm-meters and f is frequency in hertz. Table 4.4 lists earth resistivities and 60-Hz equivalent conductor distances for various types of earth. It is common practice to select $\rho = 100\ \Omega$m when actual data are unavailable.

Note that the GMR of each earth return conductor, $D_{k'k'}$, is the same as the GMR of its corresponding overhead conductor, D_{kk}. Also, all the earth return conductors have the same distance $D_{kk'}$ from their overhead conductors and the same resistance $R_{k'}$.

For simplicity, renumber the overhead conductors from 1 to $(3 + N)$, beginning with the phase conductors, then overhead neutral conductors, as shown in Figure 4.16. Operating as a transmission line, the sum of the currents in all the conductors is zero. That is,

$$\sum_{k=1}^{(6+2N)} I_k = 0 \tag{4.7.4}$$

Equation (4.4.30) is therefore valid, and the flux linking overhead conductor k is

$$\lambda_k = 2 \times 10^{-7} \sum_{m=1}^{(3+N)} I_m \ln \frac{D_{km'}}{D_{km'}} \quad \text{Wb-t/m} \tag{4.7.5}$$

In matrix format, (4.7.5) becomes

$$\boldsymbol{\lambda} = \mathbf{L}\boldsymbol{I} \tag{4.7.6}$$

where

$\boldsymbol{\lambda}$ is a $(3 + N)$ vector

$\boldsymbol{I}$ is a $(3 + N)$ vector

$\mathbf{L}$ is a $(3 + N) \times (3 + N)$ matrix whose elements are

$$L_{km} = 2 \times 10^{-7} \ln \frac{D_{km'}}{D_{km}} \tag{4.7.7}$$

When $k = m$, D_{kk} in (4.7.7) is the GMR of (bundled) conductor k. When $k \neq m$, D_{km} is the distance between conductors k and m.

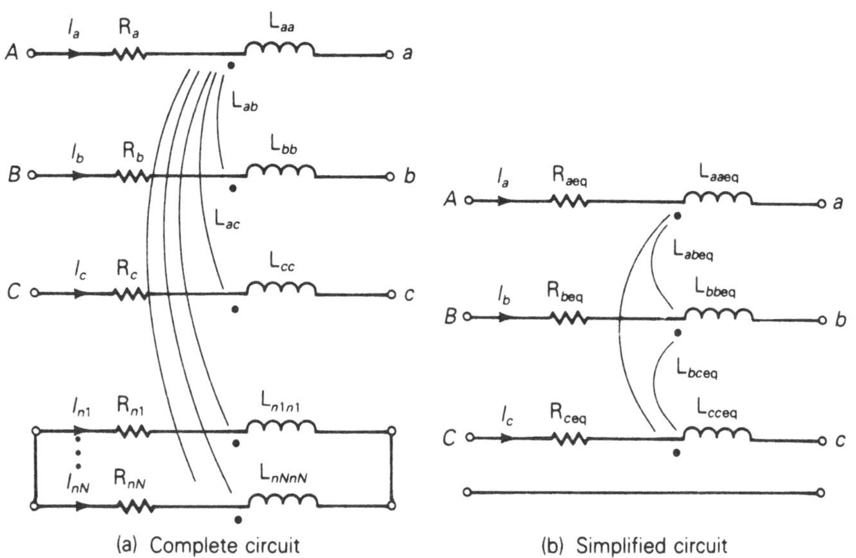

FIGURE 4.17 Circuit representation of series-phase impedances

A circuit representation of a 1-meter section of the line is shown in Figure 4.17(a). Using this circuit, the vector of voltage drops across the conductors is

$$\begin{bmatrix} E_{Aa} \\ E_{Bb} \\ E_{Cc} \\ 0 \\ 0 \\ \vdots \\ 0 \end{bmatrix} = (\mathbf{R} + j\omega\mathbf{L}) \begin{bmatrix} I_a \\ I_b \\ I_c \\ I_{n1} \\ \vdots \\ I_{nN} \end{bmatrix} \tag{4.7.8}$$

where **L** is given by (4.7.7) and **R** is a $(3 + N) \times (3 + N)$ matrix of conductor resistances.

$$\mathbf{R} = \begin{bmatrix} (R_a+R_{k'})R_{k'} \dots & & R_{k'} \\ R_{k'}(R_b+R_{k'})R_{k'} \cdots & & \vdots \\ \quad (R_c+R_{k'})R_{k'} \cdots & & \\ \qquad (R_{n1}+R_{k'})R_{k'} \cdots & & \\ \qquad \ddots & & \\ R_{k'} & & (R_{nN}+R_{K'}) \end{bmatrix} \Omega/\text{m} \tag{4.7.9}$$

The resistance matrix of (4.7.9) includes the resistance R_k of each overhead conductor and a mutual resistance $R_{k'}$ due to the image conductors. R_k of each overhead conductor is obtained from conductor tables such as Appendix Table A.3 or A.4, for a specified frequency, temperature, and current. $R_{k'}$ of all the image conductors is the same, as given by (4.7.3).

Our objective now is to reduce the (3 + N) equations in (4.7.8) to three equations, thereby obtaining the simplified circuit representations shown in Figure 4.17(b). We partition (4.7.8) as follows:

$$\begin{bmatrix} E_{Aa} \\ E_{Bb} \\ E_{Cc} \\ \hdashline 0 \\ \cdots \\ 0 \end{bmatrix} = \left[\begin{array}{ccc:ccc} \overbrace{Z_{11} \quad Z_{12} \quad Z_{13}}^{Z_A} & & & \overbrace{Z_{14} \quad \cdots \quad Z_{1(3+N)}}^{Z_B} & & \\ Z_{21} & Z_{22} & Z_{23} & Z_{24} & \cdots & Z_{2(3+N)} \\ Z_{31} & Z_{32} & Z_{33} & Z_{34} & \cdots & Z_{3(3+N)} \\ \hdashline Z_{41} & Z_{42} & Z_{43} & Z_{44} & \cdots & Z_{4(3+N)} \\ & & & & & \\ \underbrace{Z_{(3+N)1} \quad Z_{(3+N)2} \quad Z_{(3+N)3}}_{Z_C} & & & \underbrace{Z_{(3+N)4} \quad \cdots \quad Z_{(3+N)(3+N)}}_{Z_D} & & \end{array}\right] \begin{bmatrix} I_a \\ I_b \\ I_c \\ \hdashline I_{n1} \\ I_{nN} \\ \vdots \end{bmatrix} \tag{4.7.10}$$

The diagonal elements of this matrix are

$$Z_{kk} = \mathrm{R}_k + \mathrm{R}_{k'} + j\omega 2 \times 10^{-7} \ln \frac{\mathrm{D}_{kk'}}{\mathrm{D}_{kk}} \quad \Omega/\mathrm{m} \tag{4.7.11}$$

And the off-diagonal elements, for $k \neq m$, are

$$Z_{km} = \mathrm{R}_{k'} + j\omega 2 \times 10^{-7} \ln \frac{\mathrm{D}_{km'}}{\mathrm{D}_{km}} \quad \Omega/\mathrm{m} \tag{4.7.12}$$

Next, (4.7.10) is partitioned as shown above to obtain

$$\begin{bmatrix} \boldsymbol{E}_{\mathrm{P}} \\ \hline 0 \end{bmatrix} = \left[\begin{array}{c|c} \boldsymbol{Z}_A & \boldsymbol{Z}_B \\ \hline \boldsymbol{Z}_C & \boldsymbol{Z}_D \end{array}\right] \begin{bmatrix} \boldsymbol{I}_{\mathrm{P}} \\ \hline \boldsymbol{I}_n \end{bmatrix} \tag{4.7.13}$$

where

$$\boldsymbol{E}_{\mathrm{P}} = \begin{bmatrix} E_{Aa} \\ E_{Bb} \\ E_{Cc} \end{bmatrix}; \quad \boldsymbol{I}_{\mathrm{P}} = \begin{bmatrix} I_a \\ I_b \\ I_c \end{bmatrix}; \quad \boldsymbol{I}_n = \begin{bmatrix} I_{n1} \\ \vdots \\ I_{nN} \end{bmatrix}$$

$\boldsymbol{E}_{\mathrm{P}}$ is the three-dimensional vector of voltage drops across the phase conductors (including the neutral voltage drop), $\boldsymbol{I}_{\mathrm{P}}$ is the three-dimensional vector of phase currents and $\boldsymbol{I}_n$ is the N vector of neutral currents. Also, the $(3 + N) \times (3 + N)$ matrix in (4.7.10) is partitioned to obtain the following matrices:

$\boldsymbol{Z}_A$ with dimension 3×3

$\boldsymbol{Z}_B$ with dimension $3 \times N$

$\boldsymbol{Z}_C$ with dimension $N \times 3$

$\boldsymbol{Z}_D$ with dimension $N \times N$

Equation (4.7.13) is rewritten as two separate matrix equations:

$$\boldsymbol{E}_{\mathrm{P}} = \boldsymbol{Z}_A \boldsymbol{I}_{\mathrm{P}} + \boldsymbol{Z}_B \boldsymbol{I}_n \tag{4.7.14}$$

$$\boldsymbol{0} = \boldsymbol{Z}_C \boldsymbol{I}_{\mathrm{P}} + \boldsymbol{Z}_D \boldsymbol{I}_n \tag{4.7.15}$$

Solving (4.7.15) for $\boldsymbol{I}_n$,

$$\boldsymbol{I}_n = -\boldsymbol{Z}_D^{-1} \boldsymbol{Z}_C \boldsymbol{I}_{\mathrm{P}} \tag{4.7.16}$$

Using (4.7.16) in (4.7.14),

$$\boldsymbol{E}_{\mathrm{P}} = [\boldsymbol{Z}_A - \boldsymbol{Z}_B \boldsymbol{Z}_D^{-1} \boldsymbol{Z}_C] \boldsymbol{I}_{\mathrm{P}} \tag{4.7.17}$$

or

$$\boldsymbol{E}_{\mathrm{P}} = \boldsymbol{Z}_{\mathrm{P}} \boldsymbol{I}_{\mathrm{P}} \tag{4.7.18}$$

where

$$\boldsymbol{Z}_{\mathrm{P}} = [\boldsymbol{Z}_A - \boldsymbol{Z}_B \boldsymbol{Z}_D^{-1} \boldsymbol{Z}_C] \tag{4.7.19}$$

Equation (4.7.17), the desired result, relates the phase-conductor voltage drops (including neutral voltage drop) to the phase currents. $\boldsymbol{Z}_{\mathrm{P}}$ given by (4.7.19) is the 3×3 series-phase impedance matrix, whose elements are denoted

$$\boldsymbol{Z}_{\mathrm{p}} = \begin{bmatrix} Z_{aa\mathrm{eq}} & Z_{ab\mathrm{eq}} & Z_{ac\mathrm{eq}} \\ Z_{ab\mathrm{eq}} & Z_{bb\mathrm{eq}} & Z_{bc\mathrm{eq}} \\ Z_{ac\mathrm{eq}} & Z_{bc\mathrm{eq}} & Z_{cc\mathrm{eq}} \end{bmatrix} \quad \Omega/\mathrm{m} \tag{4.7.20}$$

If the line is completely transposed, the diagonal and off-diagonal elements are averaged to obtain

$$\hat{\boldsymbol{Z}}_{\mathrm{P}} = \begin{bmatrix} \hat{Z}_{aa\mathrm{eq}} & \hat{Z}_{ab\mathrm{eq}} & \hat{Z}_{ab\mathrm{eq}} \\ \hat{Z}_{ab\mathrm{eq}} & \hat{Z}_{aa\mathrm{eq}} & \hat{Z}_{ab\mathrm{eq}} \\ \hat{Z}_{ab\mathrm{eq}} & \hat{Z}_{ab\mathrm{eq}} & \hat{Z}_{aa\mathrm{eq}} \end{bmatrix} \quad \Omega/\mathrm{m} \tag{4.7.21}$$

where

$$\hat{Z}_{aa\mathrm{eq}} = \tfrac{1}{3}(Z_{aa\mathrm{eq}} + Z_{bb\mathrm{eq}} + Z_{cc\mathrm{eq}}) \tag{4.7.22}$$

$$\hat{Z}_{ab\mathrm{eq}} = \tfrac{1}{3}(Z_{ab\mathrm{eq}} + Z_{ac\mathrm{eq}} + Z_{bc\mathrm{eq}}) \tag{4.7.23}$$

4.8 ELECTRIC FIELD AND VOLTAGE: SOLID CYLINDRICAL CONDUCTOR

The capacitance between conductors in a medium with constant permittivity ε can be obtained by determining the following properties.

1. Electric field strength E, from Gauss's law
2. Voltage between conductors
3. Capacitance from charge per unit volt ($C = q/V$)

As a step toward computing capacitances of general conductor configurations, first compute the electric field of a uniformly charged, solid cylindrical conductor and the voltage between two points outside the conductor. Also compute the voltage between two conductors in an array of charged conductors.

Gauss's law states that the total electric flux leaving a closed surface equals the total charge within the volume enclosed by the surface. That is, the normal component of electric flux density integrated over a closed surface equals the charge enclosed:

$$\oiint D_{\perp}\, ds = \oiint \varepsilon E_{\perp}\, ds = Q_{\text{enclosed}} \tag{4.8.1}$$

where $D_{\perp}$ denotes the normal component of electric flux density, $E_{\perp}$ denotes the normal component of electric field strength, and ds denotes the differential surface area. From Gauss's law, electric charge is a source of electric fields. Electric field lines originate from positive charges and terminate at negative charges.

Figure 4.18 shows a solid cylindrical conductor with radius r and with charge q coulombs per meter (assumed positive in the figure), uniformly distributed on the conductor surface. For simplicity, assume that the conductor is (1) sufficiently long that end effects are negligible, and (2) a perfect conductor (that is, zero resistivity, $\rho = 0$).

Inside the perfect conductor, Ohm's law gives $E_{\text{int}} = \rho J = 0$. That is, the internal electric field E_{int} is zero. To determine the electric field outside the conductor, select the cylinder with radius $x > r$ and with 1-meter length, shown in Figure 4.18, as the closed surface for Gauss's law. Due to the uniform charge distribution, the

FIGURE 4.18

Perfectly conducting solid cylindrical conductor with uniform charge distribution

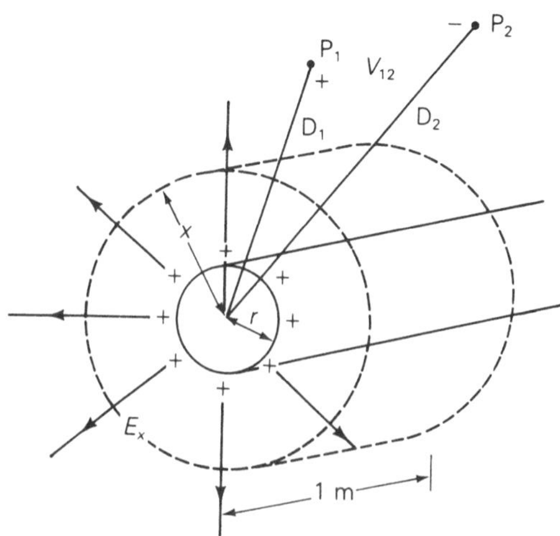

electric field strength E_x is constant on the cylinder. Also, there is no tangential component of E_x, so the electric field is radial to the conductor. Then, integration of (4.8.1) yields

$$\varepsilon E_x(2\pi x)(1) = q(1)$$

$$E_x = \frac{q}{2\pi\varepsilon x} \quad \text{V/m} \tag{4.8.2}$$

where, for a conductor in free space, $\varepsilon = \varepsilon_0 = 8.854 \times 10^{-12}$ F/m.

A plot of the electric field lines is also shown in Figure 4.18. The direction of the field lines, denoted by the arrows, is from the positive charges where the field originates, to the negative charges, which in this case are at infinity. If the charge on the conductor surface were negative, then the direction of the field lines would be reversed.

Concentric cylinders surrounding the conductor are constant potential surfaces. The potential difference between two concentric cylinders at distances D_1 and D_2 from the conductor center is

$$V_{12} = \int_{D_1}^{D_2} E_x dx \tag{4.8.3}$$

Using (4.8.2) in (4.8.1),

$$V_{12} = \int_{D_1}^{D_2} \frac{q}{2\pi\varepsilon x} dx = \frac{q}{2\pi\varepsilon} \ln \frac{D_2}{D_1} \quad \text{volts} \tag{4.8.4}$$

Equation (4.8.4) gives the voltage V_{12} between two points, P_1 and P_2, at distances D_1 and D_2 from the conductor center, as shown in Figure 4.18. Also, in accordance with the notation, V_{12} is the voltage at P_1 with respect to P_2. If q is positive and D_2 is greater than D_1, as shown in the figure, then V_{12} is positive; that is, P_1 is at a higher potential than P_2. Equation (4.8.4) is also valid for either dc or ac. For ac, V_{12} is a phasor voltage and q is a phasor representation of a sinusoidal charge.

Now apply (4.8.4) to the array of M solid cylindrical conductors shown in Figure 4.19. Assume that each conductor m has an ac charge q_m C/m uniformly

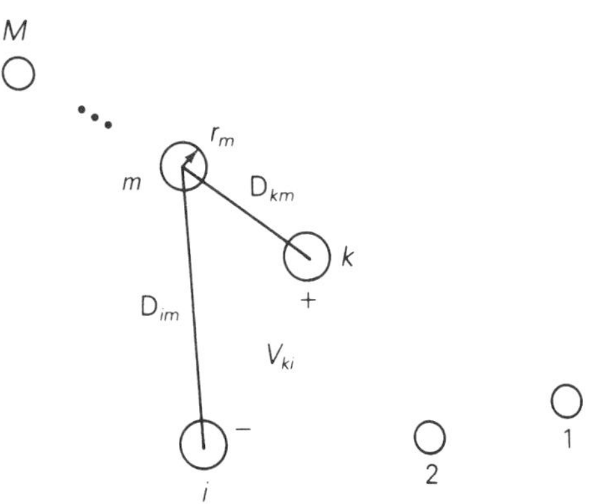

FIGURE 4.19

Array of M solid cylindrical conductors

distributed along the conductor. The voltage V_{ki} between conductors k and i due to the charge q_m acting alone is

$$V_{ki} = \frac{q_m}{2\pi\varepsilon} \ln \frac{D_{im}}{D_{km}} \quad \text{volts} \tag{4.8.5}$$

where $D_{mm} = r_m$ when $k = m$ or $i = m$. In (4.8.5), neglected is the distortion of the electric field in the vicinity of the other conductors, caused by the fact that the other conductors themselves are constant potential surfaces. V_{kim} can be thought of as the voltage between cylinders with radii D_{km} and D_{im} concentric to conductor m at points on the cylinders remote from conductors, where there is no distortion.

Using superposition, the voltage V_{ki} between conductors k and i due to all the changes is

$$V_{ki} = \frac{1}{2\pi\varepsilon} \sum_{m=1}^{M} q_m \ln \frac{D_{im}}{D_{km}} \quad \text{volts} \tag{4.8.6}$$

4.9 CAPACITANCE: SINGLE-PHASE TWO-WIRE LINE AND THREE-PHASE THREE-WIRE LINE WITH EQUAL PHASE SPACING

The results of the previous section are used here to determine the capacitances of the two relatively simple transmission lines considered in Section 4.5, a single-phase two-wire line and a three-phase three-wire line with equal phase spacing.

First consider the single-phase two-wire line shown in Figure 4.9. Assume that the conductors are energized by a voltage source such that conductor x has a uniform charge q C/m and, assuming conservation of charge, conductor y has an equal quantity of negative charge $-q$. Using (4.8.6) with $k = x$, $i = y$, and $m = x, y$,

$$V_{xy} = \frac{1}{2\pi\varepsilon}\left[q \ln\frac{D_{yx}}{D_{xx}} - q \ln \frac{D_{yy}}{D_{xy}}\right]$$

$$= \frac{q}{2\pi\varepsilon} \ln \frac{D_{yx}D_{xy}}{D_{xx}D_{yy}} \tag{4.9.1}$$

Using $D_{xy} = D_{yx} = D$, $D_{xx} = r_x$, and $D_{yy} = r_y$, (4.9.1) becomes

$$V_{xy} = \frac{q}{\pi\varepsilon} \ln \frac{D}{\sqrt{r_x r_y}} \quad \text{volts} \tag{4.9.2}$$

For a 1-meter line length, the capacitance between conductors is

$$C_{xy} = \frac{q}{V_{xy}} = \frac{\pi\varepsilon}{\ln\left(\frac{D}{\sqrt{r_x r_y}}\right)} \quad \text{F/m line-to-line} \tag{4.9.3}$$

and if $r_x = r_y = r$,

$$C_{xy} = \frac{\pi\varepsilon}{\ln(D/r)} \quad \text{F/m line-to-line} \tag{4.9.4}$$

If the two-wire line is supplied by a transformer with a grounded center tap, then the voltage between each conductor and ground is one-half that given by (4.9.2). That is,

$$V_{xn} = V_{yn} = \frac{V_{xy}}{2} \tag{4.9.5}$$

and the capacitance from either line to the grounded neutral is

$$C_n = C_{xn} = C_{yn} = \frac{q}{V_{xn}} = 2C_{xy}$$

$$= \frac{2\pi\varepsilon}{\ln(D/r)} \quad \text{F/m line-to-neutral} \tag{4.9.6}$$

Circuit representations of the line-to-line and line-to-neutral capacitances are shown in Figure 4.20. Note that if the neutral is open in Figure 4.20(b), the two line-to-neutral capacitances combine in series to give the line-to-line capacitance.

Next consider the three-phase line with equal phase spacing shown in Figure 4.10. Neglect the effect of earth and neutral conductors here. To determine the positive-sequence capacitance, assume positive-sequence charges q_a, q_b, q_c such that $q_a + q_b + q_c = 0$. Using (4.8.6) with $k = a$, $i = b$, and $m = a, b, c$, the voltage V_{ab} between conductors a and b is

$$V_{ab} = \frac{1}{2\pi\varepsilon}\left[q_a \ln\frac{D_{ba}}{D_{aa}} + q_b \ln\frac{D_{bb}}{D_{ab}} + q_c \ln\frac{D_{bc}}{D_{ac}}\right] \tag{4.9.7}$$

Using $D_{aa} = D_{bb} = r$, and $D_{ab} = D_{ba} = D_{ca} = D_{cb} = D$, (4.9.7) becomes

$$V_{ab} = \frac{1}{2\pi\varepsilon}\left[q_a \ln\frac{D}{r} + q_b \ln\frac{r}{D} + q_c \ln\frac{D}{D}\right]$$

$$= \frac{1}{2\pi\varepsilon}\left[q_a \ln\frac{D}{r} + q_b \ln\frac{r}{D}\right] \quad \text{volts} \tag{4.9.8}$$

Note that the third term in (4.9.8) is zero because conductors a and b are equidistant from conductor c. Thus, conductors a and b lie on a constant potential cylinder for the electric field due to q_c.

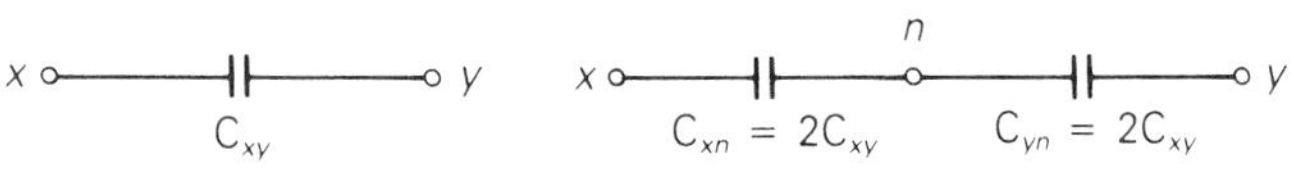

FIGURE 4.20 Circuit representation of capacitances for a single-phase two-wire line

Similarly, using (4.8.6) with $k = a$, $i = c$, and $m = a,b,c$, the voltage V_{ac} is

$$V_{ab} = \frac{1}{2\pi\varepsilon}\left[q_a \ln\frac{D_{ca}}{D_{aa}} + q_b \ln\frac{D_{cb}}{D_{ab}} + q_c \ln\frac{D_{cc}}{D_{ac}}\right]$$

$$= \frac{1}{2\pi\varepsilon}\left[q_a \ln\frac{D}{r} + q_b \ln\frac{D}{D} + q_c \ln\frac{r}{D}\right]$$

$$= \frac{1}{2\pi\varepsilon}\left[q_a \ln\frac{D}{r} + q_c \ln\frac{r}{D}\right] \quad \text{volts} \tag{4.9.9}$$

Recall that for balanced positive-sequence voltages,

$$V_{ab} = \sqrt{3}V_{an}\underline{/+30^\circ} = \sqrt{3}V_{an}\left[\frac{\sqrt{3}}{2} + j\frac{1}{2}\right] \tag{4.9.10}$$

$$V_{ac} = -V_{ca} = \sqrt{3}V_{an}\underline{/-30^\circ} = \sqrt{3}V_{an}\left[\frac{\sqrt{3}}{2} - j\frac{1}{2}\right] \tag{4.9.11}$$

Adding (4.9.10) and (4.9.11) yields

$$V_{ab} + V_{ac} = 3V_{an} \tag{4.9.12}$$

Using (4.9.8) and (4.9.9) in (4.9.12),

$$V_{an} = \frac{1}{3}\left(\frac{1}{2\pi\varepsilon}\right)\left[2q_a \ln\frac{D}{r} + (q_b + q_c)\ln\frac{r}{D}\right] \tag{4.9.13}$$

and with $q_b + q_c = -q_a$,

$$V_{an} = \frac{1}{2\pi\varepsilon} q_a \ln\frac{D}{r} \quad \text{volts} \tag{4.9.14}$$

The capacitance-to-neutral per line length is

$$C_{an} = \frac{q_a}{V_{an}} = \frac{2\pi\varepsilon}{\ln\left(\frac{D}{r}\right)} \quad \text{F/m line-to-neutral} \tag{4.9.15}$$

FIGURE 4.21 Circuit representation of the capacitance-to-neutral of a three-phase line with equal phase spacing

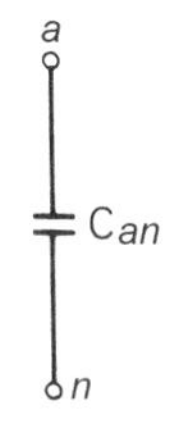

Due to symmetry, the same result is obtained for $C_{bn} = qb/V_{bn}$ and $C_{cn} = qc/V_{cn}$. For balanced three-phase operation, however, only one phase need be considered. A circuit representation of the capacitance-to-neutral is shown in Figure 4.21.

4.10 CAPACITANCE: STRANDED CONDUCTORS, UNEQUAL PHASE SPACING, BUNDLED CONDUCTORS

Equations (4.9.6) and (4.9.15) are based on the assumption that the conductors are solid cylindrical conductors with zero resistivity. The electric field inside these conductors is zero, and the external electric field is perpendicular to the conductor surfaces. Practical conductors with resistivities similar to those listed in Table 4.3 have a small internal electric field. As a result, the external electric field is slightly altered near the conductor surfaces. Also, the electric field near the surface of a stranded conductor is not the same as that of a solid cylindrical conductor. However, it is normal practice when calculating line capacitance to replace a stranded conductor by a perfectly conducting solid cylindrical conductor whose radius equals the outside radius of the stranded conductor. The resulting error in capacitance is small, since only the electric field near the conductor surfaces is affected.

Also, (4.8.2) is based on the assumption that there is uniform charge distribution. But conductor charge distribution is nonuniform in the presence of other charged conductors. Therefore (4.9.6) and (4.9.15), which are derived from (4.8.2), are not exact. However, the nonuniformity of conductor charge distribution can be shown to have a negligible effect on line capacitance.

For three-phase lines with unequal phase spacing, balanced positive-sequence voltages are not obtained with balanced positive-sequence charges. Instead, unbalanced line-to-neutral voltages occur, and the phase-to-neutral capacitances are unequal. Balance can be restored by transposing the line such that each phase occupies each position for one-third of the line length. If equations similar to (4.9.7) for V_{ab} as well as for V_{ac} are written for each position in the transposition cycle, and are then averaged and used in (4.9.12) through (4.9.14), the resulting capacitance becomes

$$C_{an} = \frac{2\pi\varepsilon}{\ln(D_{eq}/r)} \quad \text{F/m} \tag{4.10.1}$$

where

$$D_{eq} = \sqrt[3]{D_{ab}D_{bc}D_{ac}} \tag{4.10.2}$$

Figure 4.22 shows a bundled conductor line with two conductors per bundle. To determine the capacitance of this line, assume balanced positive-sequence charges q_a, q_b, q_c for each phase such that $q_a + q_b + q_c = 0$. Assume that the conductors in each bundle, which are in parallel, share the charges equally. Thus conductors a and a' each have the charge $q_a/2$. Also assume that the phase spacings are much larger than the bundle spacings so that D_{ab} may be used instead of $(D_{ab} - d)$ or $(D_{ab} + d)$. Then, using (4.8.6) with $k = a$, $i = b$, $m = a, a', b, b', c, c'$,

$$V_{ab} = \frac{1}{2\pi\varepsilon}\left[\frac{q_a}{2}\ln\frac{D_{ba}}{D_{aa}} + \frac{q_a}{2}\ln\frac{D_{ba'}}{D_{aa'}} + \frac{q_b}{2}\ln\frac{D_{bb}}{D_{ab}}\right.$$
$$\left. + \frac{q_b}{2}\ln\frac{D_{bb'}}{D_{ab'}} + \frac{q_c}{2}\ln\frac{D_{bc}}{D_{ac}} + \frac{q_c}{2}\ln\frac{D_{bc'}}{D_{ac'}}\right]$$
$$= \frac{1}{2\pi\varepsilon}\left[\frac{q_a}{2}\left(\ln\frac{D_{ab}}{r} + \ln\frac{D_{ab}}{d}\right) + \frac{q_b}{2}\left(\ln\frac{r}{D_{ab}} + \ln\frac{d}{D_{ab}}\right)\right.$$
$$\left. + \frac{q_c}{2}\left(\ln\frac{D_{bc}}{D_{ac}} + \ln\frac{D_{bc}}{D_{ac}}\right)\right]$$
$$= \frac{1}{2\pi\varepsilon}\left[q_a \ln\frac{D_{ab}}{\sqrt{rd}} + q_b \ln\frac{\sqrt{rd}}{D_{ab}} + q_c \ln\frac{D_{bc}}{D_{ac}}\right] \tag{4.10.3}$$

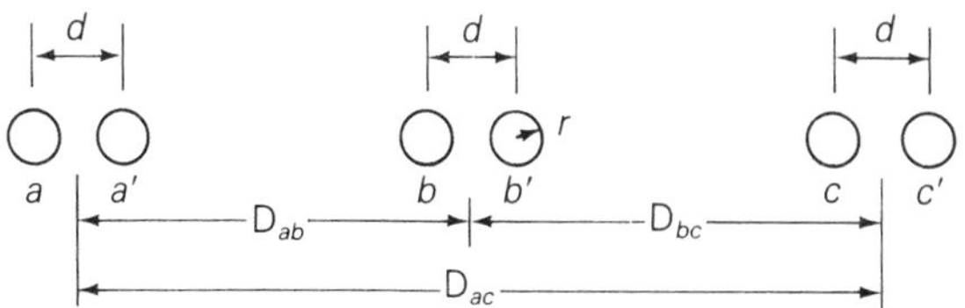

FIGURE 4.22 Three-phase line with two conductors per bundle

Equation (4.10.3) is the same as (4.9.7), except that D_{aa} and D_{bb}, in (4.9.7) are replaced by $\sqrt{rd}$ in this equation. Therefore, for a transposed line, derivation of the capacitance would yield

$$C_{an} = \frac{2\pi\varepsilon}{\ln(D_{eq}/D_{SC})} \quad \text{F/m} \tag{4.10.4}$$

where

$$D_{SC} = \sqrt{rd} \quad \text{for a two-conductor bundle} \tag{4.10.5}$$

Similarly,

$$D_{SC} = \sqrt[3]{rd^2} \quad \text{for a three-conductor bundle} \tag{4.10.6}$$

$$D_{SC} = 1.091\sqrt[4]{rd^3} \quad \text{for a four-conductor bundle} \tag{4.10.7}$$

Equation (4.10.4) for capacitance is analogous to (4.6.22) for inductance. In both cases D_{eq}, given by (4.6.17) or (4.10.2), is the geometric mean of the distances between phases. Also, (4.10.5)–(4.10.7) for D_{SC} are analogous to (4.6.19)−(4.6.21) for D_{SL}, except that the conductor outside radius r replaces the conductor GMR D_S.

The current supplied to the transmission-line capacitance is called *charging current*. For a single-phase circuit operating at line-to-line voltage $V_{xy} = V_{xy}\angle 0°$, the charging current is

$$I_{chg} = Y_{xy}V_{xy} = j\omega C_{xy}V_{xy} \quad \text{A} \tag{4.10.8}$$

As shown in Chapter 2, a capacitor delivers reactive power. From (2.3.5), the reactive power delivered by the line-to-line capacitance is

$$Q_C = \frac{V_{xy}^2}{X_c} = Y_{xy}V_{xy}^2 = \omega C_{xy}V_{xy}^2 \quad \text{var} \tag{4.10.9}$$

For a completely transposed three-phase line that has balanced positive-sequence voltages with $V_{an} = V_{LN}\angle 0°$, the phase a charging current is

$$I_{chg} = YV_{an} = j\omega C_{an}V_{LN} \quad \text{A} \tag{4.10.10}$$

and the reactive power delivered by phase a is

$$Q_{C1\phi} = YV_{an}^2 = \omega C_{an}V_{LN}^2 \quad \text{var} \tag{4.10.11}$$

The total reactive power supplied by the three-phase line is

$$Q_{C3\phi} = 3Q_{C1\phi} = 3\omega C_{an}V_{LN}^2 = \omega C_{an}V_{LL}^2 \quad \text{var} \tag{4.10.12}$$

EXAMPLE 4.6

Capacitance, admittance, and reactive power supplied: single-phase line

For the single-phase line in Example 4.3, determine the line-to-line capacitance in F and the line-to-line admittance in S. If the line voltage is 20 kV, determine the reactive power in kvar supplied by this capacitance.

SOLUTION

From Table A.3, the outside radius of a 4/0 12-strand copper conductor is

$$r = \frac{0.552}{2}\ \text{in.} \times \frac{0.0254\ \text{m}}{1\ \text{in.}} = 0.007010\ \text{m}$$

and from (4.9.4),

$$C_{xy} = \frac{\pi(8.854 \times 10^{-12})}{\ln\left(\frac{1.5}{0.007010}\right)} = 5.184 \times 10^{-12} \quad \text{F/m}$$

or

$$C_{xy} = 5.184 \times 10^{-12} \frac{\text{F}}{\text{m}} \times 32{,}000\ \text{m} = 1.66 \times 10^{-7} \quad \text{F}$$

and the shunt admittance is

$$Y_{xy} = j\omega C_{xy} = j(2\pi 60)(1.66 \times 10^{-7})$$

$$= j6.27 \times 10^{-5} \quad \text{S line-to-line}$$

From (4.10.9),

$$Q_C = (6.27 \times 10^{-5})(20 \times 10^3)^2 = 25.1 \quad \text{kvar}$$

EXAMPLE 4.7

Capacitance and shunt admittance; charging current and reactive power supplied: three-phase line

For the three-phase line in Example 4.5, determine the capacitance-to-neutral in F and the shunt admittance-to-neutral in S. If the line voltage is 345 kV, determine the charging current in kA per phase and the total reactive power in Mvar supplied by the line capacitance. Assume balanced positive-sequence voltages.

SOLUTION

From Table A.4, the outside radius of a 795,000 cmil 26/2 ACSR conductor is

$$r = \frac{1.108}{2}\ \text{in.} \times 0.0254 \frac{\text{m}}{\text{in.}} = 0.0141 \quad \text{m}$$

(Continued)

From (4.10.5), the equivalent radius of the two-conductor bundle is

$$D_{SC} = \sqrt{(0.0141)(0.40)} = 0.0750 \quad \text{m}$$

$D_{eq} = 12.6$ m is the same as in Example 4.5. Therefore, from (4.10.4),

$$C_{an} = \frac{(2\pi)(8.854 \times 10^{-12})}{\ln\left(\dfrac{12.6}{0.0750}\right)} \frac{\text{F}}{\text{m}} \times 1000 \frac{\text{m}}{\text{km}} \times 200 \text{ km}$$

$$= 2.17 \times 10^{-6} \quad \text{F}$$

The shunt admittance-to-neutral is

$$Y_{an} = j\omega C_{an} = j(2\pi 60)(2.17 \times 10^{-6})$$
$$= j8.19 \times 10^{-4} \quad \text{S}$$

From (4.10.10),

$$I_{chg} = |I_{chg}| = (8.19 \times 10^{-4})\left(\frac{345}{\sqrt{3}}\right) = 0.163 \text{ kA/phase}$$

and from (4.10.12),

$$Q_{C3\phi} = (8.19 \times 10^{-4})(345)^2 = 97.5 \quad \text{Mvar}$$

4.11 SHUNT ADMITTANCES: LINES WITH NEUTRAL CONDUCTORS AND EARTH RETURN

This section covers equations suitable for computer calculation of the shunt admittances for the three-phase overhead line shown in Figure 4.16. Approximate the earth's surface as a perfectly conducting horizontal plane, even though the earth under the line may have irregular terrain and resistivities as shown in Table 4.4.

The *method of images* accounts for the effect of the earth plane, described as follows. Consider a single conductor with uniform charge distribution and with height H above a perfectly conducting earth plane, as shown in Figure 4.23(a). When the conductor has a positive charge, an equal quantity of negative charge is induced on the earth. The electric field lines originate from the positive charges on the conductor and terminate at the negative charges on the earth. Also, the electric field lines are perpendicular to the surfaces of the conductor and earth.

Now replace the earth by the image conductor shown in Figure 4.23(b), which has the same radius as the original conductor, lies directly below the original conductor with conductor separation $H_{11} = 2H$, and has an equal quantity of negative charge. The electric field above the dashed line representing the location of the

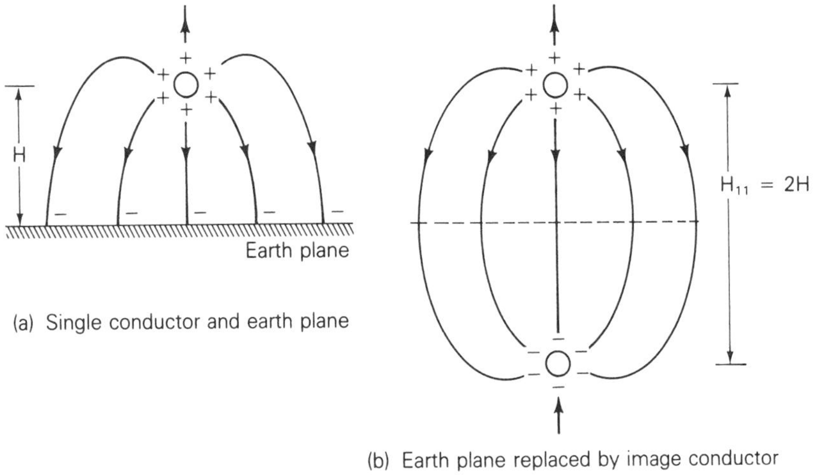

(a) Single conductor and earth plane

(b) Earth plane replaced by image conductor

FIGURE 4.23 Method of images

removed earth plane in Figure 4.23(b) is identical to the electric field above the earth plane in Figure 4.23(a). Therefore, the voltage between any two points above the earth is the same in both figures.

EXAMPLE 4.8

Effect of earth on capacitance: single-phase line

If the single-phase line in Example 4.6 has flat horizontal spacing with 5.5-m average line height, determine the effect of the earth on capacitance. Assume a perfectly conducting earth plane.

SOLUTION

The earth plane is replaced by a separate image conductor for each overhead conductor, and the conductors are charged as shown in Figure 4.24. From (4.8.6), the voltage between conductors x and y is

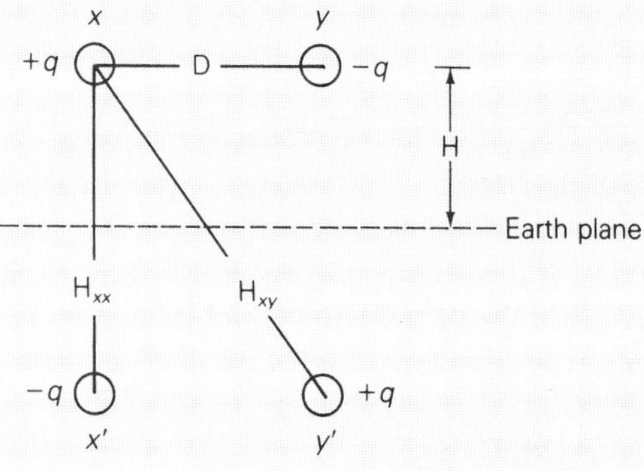

FIGURE 4.24 Single-phase line for Example 4.8

(Continued)

$$V_{xy} = \frac{q}{2\pi\varepsilon}\left[\ln\frac{D_{yx}}{D_{xx}} - \ln\frac{D_{yy}}{D_{xy}} - \ln\frac{H_{yx}}{H_{xx}} + \ln\frac{H_{yy}}{H_{xy}}\right]$$

$$= \frac{q}{2\pi\varepsilon}\left[\ln\frac{D_{yx}D_{xy}}{D_{xx}D_{yy}} - \ln\frac{H_{yx}H_{xy}}{H_{xx}H_{yy}}\right]$$

$$= \frac{q}{\pi\varepsilon}\left[\ln\frac{D}{r} - \ln\frac{H_{xy}}{H_{xx}}\right]$$

The line-to-line capacitance is

$$C_{xy} = \frac{q}{V_{xy}} = \frac{\pi\varepsilon}{\ln\frac{D}{r} - \ln\frac{H_{xy}}{H_{xx}}} \quad \text{F/m}$$

Using $D = 1.5$ m, $r = 0.007010$ m, $H_{xx} = 2H = 11.0$ m, and $H_{xy} = \sqrt{(11)^2 + (1.5)^2} = 11.102$ m,

$$C_{xy} = \frac{\pi(8.854 \times 10^{-12})}{\ln\frac{1.5}{0.007010\text{ m}} - \ln\frac{11.102}{11}} = 5.193 \times 10^{-12} \quad \text{F/m}$$

compared with 5.169×10^{-12} F/m in Example 4.6. The effect of the earth plane is to slightly increase the capacitance. Note that as the line height H increases, the ratio H_{xy}/H_{xx} approaches 1, $\ln(H_{xy}/H_{xx}) \to 0$, and the effect of the earth becomes negligible.

For the three-phase line with N neutral conductors shown in Figure 4.25, the perfectly conducting earth plane is replaced by a separate image conductor for each overhead conductor. The overhead conductors $a, b, c, n1, n2, \ldots, nN$ carry charges $q_a, q_b, q_c, q_{n1}, \ldots, q_{nN}$, and the image conductors $a', b', c', n1', \ldots, nN'$ carry charges $-q_a, -qb, -qc, -qn_1, \ldots, -q_{nN}$. Applying (4.8.6) to determine the voltage $V_{kk'}$ between any conductor k and its image conductor k',

$$V_{kk'} = \frac{1}{2\pi\varepsilon}\left[\sum_{m=a}^{nN} q_m \ln\frac{H_{km}}{D_{km}} - \sum_{m=a}^{nN} q_m \ln\frac{D_{km}}{H_{km}}\right]$$

$$= \frac{2}{2\pi\varepsilon}\sum_{m=a}^{nN} q_m \ln\frac{H_{km}}{D_{km}} \tag{4.11.1}$$

where $D_{kk} = r_k$ and D_{km} is the distance between overhead conductors k and m. H_{km} is the distance between overhead conductor k and image conductor m. By symmetry, the voltage V_{kn} between conductor k and the earth is one-half of $V_{kk'}$.

$$V_{kn} = \frac{1}{2}V_{kk'} = \frac{1}{2\pi\varepsilon}\sum_{m=a}^{nN} q_m \ln\frac{H_{km}}{D_{km}} \tag{4.11.2}$$

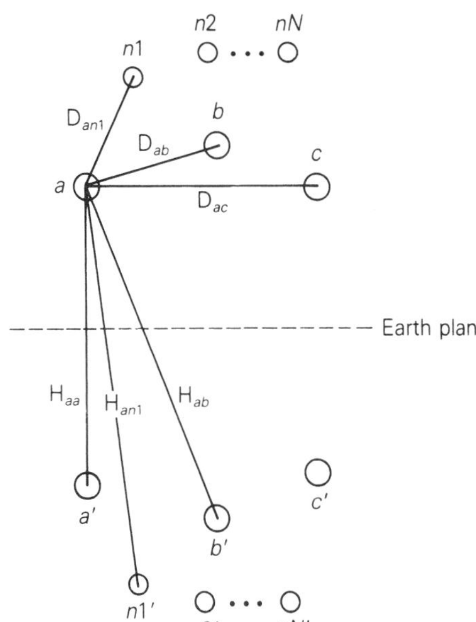

FIGURE 4.25

Three-phase line with neutral conductors and with earth plane replaced by image conductors

where

$$k = a, b, c, n1, n2, \ldots, nN$$
$$m = a, b, c, n1, n2, \ldots, nN$$

Since all the neutral conductors are grounded to the earth.

$$V_{kn} = 0 \qquad \text{for } k = n1, n2, \ldots, nN \tag{4.11.3}$$

In matrix format, (4.11.2) and (4.11.3) are

$$\begin{bmatrix} V_{an} \\ V_{bn} \\ V_{cn} \\ \hline 0 \\ \vdots \\ 0 \end{bmatrix} = \left[\begin{array}{ccc|ccc} \overbrace{P_{aa} \quad P_{ab} \quad P_{ac}}^{\mathbf{P}_A} & & & \overbrace{P_{an1} \quad \cdots \quad P_{anN}}^{\mathbf{P}_B} & & \\ P_{ba} & P_{bb} & P_{bc} & P_{bn1} & \cdots & P_{bnN} \\ P_{ca} & P_{cb} & P_{cc} & P_{cn1} & \cdots & P_{cnN} \\ \hline P_{n1a} & P_{n1b} & P_{n1c} & P_{n1n1} & \cdots & P_{n1nN} \\ \vdots & & & & & \\ \underbrace{P_{nNa} \quad P_{nNb} \quad P_{nNc}}_{\mathbf{P}_C} & & & \underbrace{P_{nNn1} \quad \cdots \quad P_{nNnN}}_{\mathbf{P}_D} & & \end{array}\right] \begin{bmatrix} q_a \\ q_b \\ q_c \\ \hline q_{n1} \\ \vdots \\ q_{nN} \end{bmatrix} \tag{4.11.4}$$

The elements of the $(3 + N) \times (3 + N)$ matrix $\mathbf{P}$ are

$$P_{km} = \frac{1}{2\pi\varepsilon} \ln \frac{H_{km}}{D_{km}} \quad \text{m/F} \tag{4.11.5}$$

where

$$k = a, b, c, n1, \ldots, nN$$
$$m = a, b, c, n1, \ldots, nN$$

Equation (4.11.4) is now partitioned as shown to obtain

$$\left[\begin{array}{c} V_P \\ \hline \mathbf{0} \end{array}\right] = \left[\begin{array}{c|c} \mathbf{P}_A & \mathbf{P}_B \\ \hline \mathbf{P}_C & \mathbf{P}_D \end{array}\right] \left[\begin{array}{c} \boldsymbol{q}_P \\ \hline \boldsymbol{q}_n \end{array}\right] \tag{4.11.6}$$

V_P is the three-dimensional vector of phase-to-neutral voltages. $\boldsymbol{q}_P$ is the three-dimensional vector of phase-conductor charges and $\boldsymbol{q}_n$ is the N vector of neutral conductor charges. The $(3 + N) \times (3 + N)\mathbf{P}$ matrix is partitioned as shown in (4.11.4) to obtain:

$\mathbf{P}_A$ with dimension 3×3

$\mathbf{P}_B$ with dimension $3 \times N$

$\mathbf{P}_C$ with dimension $N \times 3$

$\mathbf{P}_D$ with dimension $N \times N$

Equation (4.11.6) is rewritten as two separate equations:

$$V_P = \mathbf{P}_A \boldsymbol{q}_P + \mathbf{P}_B \boldsymbol{q}_n \tag{4.11.7}$$

$$\mathbf{0} = \mathbf{P}_C \boldsymbol{q}_P + \mathbf{P}_D \boldsymbol{q}_n \tag{4.11.8}$$

Then (4.11.8) is solved for $\boldsymbol{q}_n$, which is used in (4.11.7) to obtain

$$V_P = (\mathbf{P}_A - \mathbf{P}_B \mathbf{P}_D^{-1} \mathbf{P}_C)\boldsymbol{q}_P \tag{4.11.9}$$

or

$$\boldsymbol{q}_P = \mathbf{C}_P \mathbf{V}_P \tag{4.11.10}$$

where

$$\mathbf{C}_P = (\mathbf{P}_A - \mathbf{P}_B \mathbf{P}_D^{-1} \mathbf{P}_C)^{-1} \quad \text{F/m} \tag{4.11.11}$$

Equation (4.11.10), the desired result, relates the phase-conductor charges to the phase-to-neutral voltages. $\mathbf{C}_P$ is the 3×3 matrix of phase capacitances whose elements are denoted

$$\mathbf{C}_P = \begin{bmatrix} C_{aa} & C_{ab} & C_{ac} \\ C_{ab} & C_{bb} & C_{bc} \\ C_{ac} & C_{bc} & C_{cc} \end{bmatrix} \quad \text{F/m} \tag{4.11.12}$$

It can be shown that $\mathbf{C}_P$ is a symmetric matrix whose diagonal terms C_{aa}, C_{bb}, C_{cc} are positive, and whose off-diagonal terms C_{ab}, C_{bc}, C_{ac} are negative. This indicates that when a positive line-to-neutral voltage is applied to one phase, a positive charge is induced on that phase and negative charges are induced on the other phases, which is physically correct.

If the line is completely transposed, the diagonal and off-diagonal elements of $\mathbf{C}_P$ are averaged to obtain

$$\hat{\mathbf{C}}_P = \begin{bmatrix} \hat{C}_{aa} & \hat{C}_{ab} & \hat{C}_{ab} \\ \hat{C}_{ab} & \hat{C}_{aa} & \hat{C}_{ab} \\ \hat{C}_{ab} & \hat{C}_{ab} & \hat{C}_{aa} \end{bmatrix} \quad \text{F/m} \tag{4.11.13}$$

where

$$\hat{C}_{aa} = \tfrac{1}{3}(C_{aa} + C_{bb} + C_{cc}) \quad \text{F/m} \tag{4.11.14}$$

$$\hat{C}_{ab} = \tfrac{1}{3}(C_{ab} + C_{bc} + C_{ac}) \quad \text{F/m} \tag{4.11.15}$$

$\hat{\mathbf{C}}_P$ is a symmetrical capacitance matrix.

The shunt phase admittance matrix is given by

$$\boldsymbol{Y}_P = j\omega\mathbf{C}_P = j(2\pi f)\mathbf{C}_P \quad \text{S/m} \tag{4.11.16}$$

or, for a completely transposed line,

$$\hat{\boldsymbol{Y}}_P = j\omega\hat{\mathbf{C}}_P = j(2\pi f)\hat{\mathbf{C}}_P \quad \text{S/m} \tag{4.11.17}$$

4.12 ELECTRIC FIELD STRENGTH AT CONDUCTOR SURFACES AND AT GROUND LEVEL

When the electric field strength at a conductor surface exceeds the breakdown strength of air, current discharges occur. This phenomenon, called *corona*, causes additional line losses (corona loss), communications interference, and audible noise. Although breakdown strength depends on many factors, a rough value is 30 kV/cm in a uniform electric field for dry air at atmospheric pressure. The presence of water droplets or rain can lower this value significantly. To control corona, transmission lines are usually designed to maintain calculated values of conductor surface electric field strength below 20 kV_{rms}/cm.

When line capacitances are determined and conductor voltages are known, the conductor charges can be calculated from (4.9.3) for a single-phase line or from (4.11.10) for a three-phase line. Then the electric field strength at the surface of one phase conductor, neglecting the electric fields due to charges on other phase conductors and neutral wires, is, from (4.8.2),

$$E_r = \frac{q}{2\pi\varepsilon r} \quad \text{V/m} \tag{4.12.1}$$

where r is the conductor outside radius.

For bundled conductors with N_b conductors per bundle and with charge q C/m per phase, the charge per conductor is q/N_b and

$$E_{rave} = \frac{q/N_b}{2\pi\varepsilon r} \quad \text{V/m} \tag{4.12.2}$$

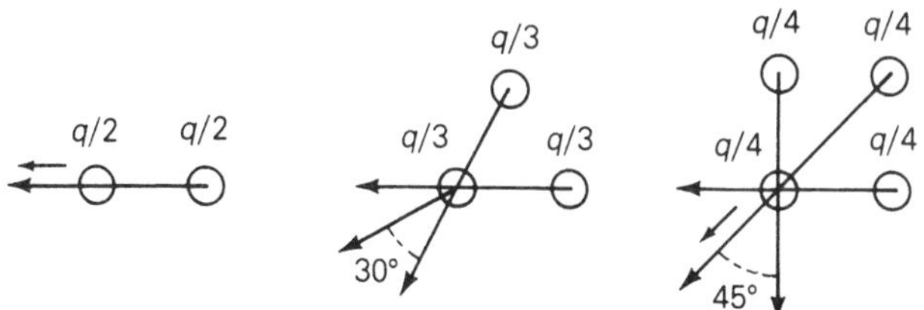

FIGURE 4.26 Vector addition of electric fields at the surface of one conductor in a bundle

Equation (4.12.2) represents an average value for an individual conductor in a bundle. The maximum electric field strength at the surface of one conductor due to all charges in a bundle, obtained by the vector addition of electric fields (as shown in Figure 4.26), is as follows:

Two-conductor bundle ($N_b = 2$):

$$E_{r\max} = \frac{q/2}{2\pi\varepsilon r} + \frac{q/2}{2\pi\varepsilon d} = \frac{q/2}{2\pi\varepsilon r}\left(1 + \frac{r}{d}\right)$$

$$= E_{r\text{ave}}\left(1 + \frac{r}{d}\right) \tag{4.12.3}$$

Three-conductor bundle ($N_b = 3$):

$$E_{r\max} = \frac{q/3}{2\pi\varepsilon}\left(\frac{1}{r} + \frac{2\cos 30^\circ}{d}\right) = E_{r\text{ave}}\left(1 + \frac{r\sqrt{3}}{d}\right) \tag{4.12.4}$$

Four-conductor bundle ($N_b = 4$):

$$E_{r\max} = \frac{q/4}{2\pi\epsilon}\left(\frac{1}{r} + \frac{1}{d\sqrt{2}} + \frac{2\cos 45^\circ}{d}\right) = E_{r\text{ave}}\left[1 + \frac{r}{d}(2.1213)\right] \tag{4.12.5}$$

Although the electric field strength at ground level is much less than at conductor surfaces where corona occurs, there are still capacitive coupling effects. Charges are induced on ungrounded equipment, such as vehicles with rubber tires located near a line. If a person contacts the vehicle and ground, a discharge current will flow to ground. Transmission-line heights are designed to maintain discharge currents below prescribed levels for any equipment that may be on the right-of-way. Table 4.5 shows examples of maximum ground-level electric field strength.

As shown in Figure 4.27, the ground-level electric field strength due to charged conductor k and its image conductor is perpendicular to the earth plane has a value of

$$E_k(w) = \left(\frac{q_k}{2\pi\varepsilon}\right)\frac{2\cos\theta}{\sqrt{y_k^2 + (w - x_k)^2}}$$

$$= \left(\frac{q_k}{2\pi\varepsilon}\right)\frac{2y_k}{y_k^2 + (w - x_k)^2} \quad \text{V/m} \tag{4.12.6}$$

where (x_k, y_k) are the horizontal and vertical coordinates of conductor k with respect to reference point R, w is the horizontal coordinate of the ground-level point where the electric field strength is to be determined, and q_k is the charge on conductor k.

Line Voltage (kV_{rms})	Maximum Ground-Level Electric Field Strength (kV_{rms}/m)
23 (1ϕ)	0.01–0.025
23 (3ϕ)	0.01–0.05
115	0.1–0.2
345	2.3–5.0
345 (double circuit)	5.6
500	8.0
765	10.0

TABLE 4.5

Examples of maximum ground-level electric field strength versus transmission-line voltage [1] (©Copyright 1987. Electric Power Research Institute (EPRI), Publication Number EL-2500. *Transmission Line Reference Book, 345-kV and Above, Second Edition, Revised.* Reprinted with permission.)

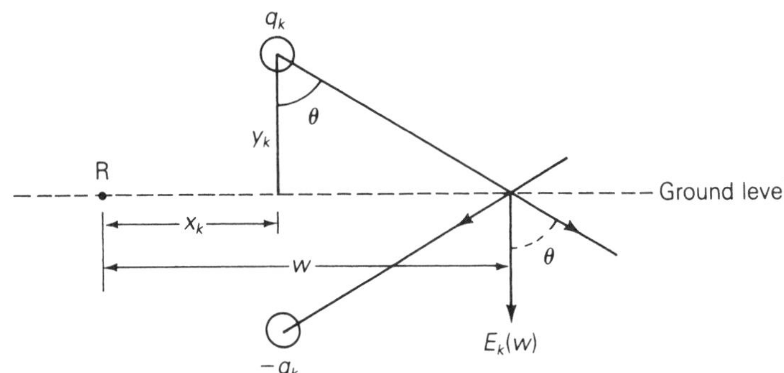

FIGURE 4.27

Ground-level electric field strength due to an overhead conductor and its image

The total ground-level electric field is the phasor sum of terms $E_k(w)$ for all overhead conductors. A lateral profile of ground-level electric field strength is obtained by varying w from the center of the line to the edge of the right-of-way.

EXAMPLE 4.9

Conductor surface and ground-level electric field strengths: single-phase line

For the single-phase line of Example 4.8, calculate the conductor surface electric field strength in kV_{rms}/cm. Also calculate the ground-level electric field in kV_{rms}/m directly under conductor x. The line voltage is 20 kV.

SOLUTION

From Example 4.8, $C_{xy} = 5.178 \times 10^{-12}$ F/m. Using (4.9.3) with $V_{xy} = 20\underline{/0^\circ}$ kV,

$$q_x = -q_y = (5.178 \times 10^{-12})(20 \times 10^3\underline{/0^\circ}) = 1.036 \times 10^{-7}\underline{/0^\circ} \quad \text{C/m}$$

(Continued)

From (4.12.1), the conductor surface electric field strength is, with $r = 0.00701$ m,

$$E_r = \frac{1.036 \times 10^{-7}}{(2\pi)(8.854 \times 10^{-12})(0.00701)} \frac{\text{V}}{\text{m}} \times \frac{\text{kV}}{1000\ \text{V}} \times \frac{\text{m}}{100\ \text{cm}}$$

$$= 2.66\ \text{kV}_{\text{rms}}/\text{cm}$$

Selecting the center of the line as the reference point R, the coordinates (x_x, y_x) for conductor x are $(-0.762$ m, 5.5 m$)$ and $(+0.762$ m, 5.5 m$)$ for conductor y. The ground-level electric field directly under conductor x, where $w = -0.762$ m, is, from (4.12.6),

$$E(-0.762) = E_x(-0.762) + E_y(-0.762)$$

$$= \frac{1.036 \times 10^{-7}}{(2\pi)(8.85 \times 10^{-12})}\left[\frac{(2)(5.5)}{(5.5)^2} - \frac{(2)(5.5)}{(5.5)^2 + (0.762 + 0.762)^2}\right]$$

$$= 1.862 \times 10^3(0.364 - 0.338) = 48.5\underline{/0°}\ \text{V/m} = 0.0485\ \text{kV/m}$$

For this 20-kV line, the electric field strengths at the conductor surface and at ground level are low enough to be of relatively small concern. For EHV lines, electric field strengths and the possibility of corona and shock hazard are of more concern.

4.13 PARALLEL CIRCUIT THREE-PHASE LINES

If two parallel three-phase circuits are close together, either on the same tower as in Figure 4.3, or on the same right-of-way, there are mutual inductive and capacitive couplings between the two circuits. When calculating the equivalent series impedance and shunt admittance matrices, these couplings should not be neglected unless the spacing between the circuits is large.

Consider the double-circuit line shown in Figure 4.28. For simplicity, assume that the lines are not transposed. Since both are connected in parallel. they have the same series-voltage drop for each phase. Following the same procedure as in Section 4.7, write $2(6 + N)$ equations similar to (4.7.6) through (4.7.9): six equations for the overhead phase conductors, N equations for the overhead neutral conductors,

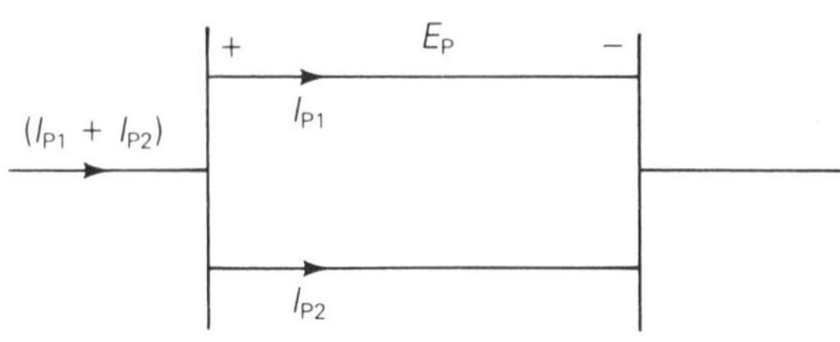

FIGURE 4.28 Single-line diagram of a double-circuit line

and $(6 + N)$ equations for the earth return conductors. After lumping the neutral voltage drop into the voltage drops across the phase conductors, and eliminating the neutral and earth return currents, resulting in

$$\begin{bmatrix} \mathbf{E}_P \\ \mathbf{E}_P \end{bmatrix} = \mathbf{Z}_P \begin{bmatrix} \mathbf{I}_{P1} \\ \mathbf{I}_{P2} \end{bmatrix} \tag{4.13.1}$$

where $\mathbf{E}_P$ is the vector of phase-conductor voltage drops (including the neutral voltage drop), and $\mathbf{I}_{P1}$ and $\mathbf{I}_{P2}$ are the vectors of phase currents for lines 1 and 2. $\mathbf{Z}_p$ is a 6×6 impedance matrix. Solving (4.13.1),

$$\begin{bmatrix} \mathbf{I}_{P1} \\ \hline \mathbf{I}_{P2} \end{bmatrix} = \mathbf{Z}_P^{-1} \begin{bmatrix} \mathbf{E}_P \\ \hline \mathbf{E}_P \end{bmatrix} = \left[\begin{array}{c|c} \mathbf{Y}_A & \mathbf{Y}_B \\ \hline \mathbf{Y}_C & \mathbf{Y}_D \end{array}\right] \begin{bmatrix} \mathbf{E}_P \\ \hline \mathbf{E}_P \end{bmatrix} = \begin{bmatrix} (\mathbf{Y}_A + \mathbf{Y}_B) \\ (\mathbf{Y}_C + \mathbf{Y}_D) \end{bmatrix} \mathbf{E}_P \tag{4.13.2}$$

where $\mathbf{Y}_A$, $\mathbf{Y}_B$, $\mathbf{Y}_C$, and $\mathbf{Y}_D$ are obtained by partitioning $\mathbf{Z}_P^{-1}$ into four 3×3 matrices. Adding $\mathbf{I}_{P1}$ and $\mathbf{I}_{P2}$,

$$(\mathbf{I}_{P1} + \mathbf{I}_{P2}) = (\mathbf{Y}_A + \mathbf{Y}_B + \mathbf{Y}_C + \mathbf{Y}_D)\mathbf{E}_P \tag{4.13.3}$$

and solving for $\mathbf{E}_P$,

$$\mathbf{E}_P = \mathbf{Z}_{Peq}(\mathbf{I}_{P1} + \mathbf{I}_{P2}) \tag{4.13.4}$$

where

$$\mathbf{Z}_{Peq} = (\mathbf{Y}_A + \mathbf{Y}_B + \mathbf{Y}_c + \mathbf{Y}_D)^{-1} \tag{4.13.5}$$

$\mathbf{Z}_{Peq}$ is the equivalent 3×3 series phase impedance matrix of the double-circuit line. Note that in (4.13.5) the matrices $\mathbf{Y}_B$ and $\mathbf{Y}_C$ account for the inductive coupling between the two circuits.

An analogous procedure can be used to obtain the shunt admittance matrix. Following the ideas of Section 4.11, write $(6 + N)$ equations similar to (4.11.4). After eliminating the neutral wire charges, the result is

$$\begin{bmatrix} \mathbf{q}_{P1} \\ \hline \mathbf{q}_{P2} \end{bmatrix} = \mathbf{C}_P \begin{bmatrix} \mathbf{V}_P \\ \hline \mathbf{V}_P \end{bmatrix} = \left[\begin{array}{c|c} \mathbf{C}_A & \mathbf{C}_B \\ \hline \mathbf{C}_C & \mathbf{C}_D \end{array}\right] \begin{bmatrix} \mathbf{V}_P \\ \hline \mathbf{V}_P \end{bmatrix} = \begin{bmatrix} (\mathbf{C}_A + \mathbf{C}_B) \\ (\mathbf{C}_C + \mathbf{C}_D) \end{bmatrix} \mathbf{V}_P \tag{4.13.6}$$

where $\mathbf{V}_p$ is the vector of phase-to-neutral voltages, and $\mathbf{q}_{P1}$ and $\mathbf{q}_{P2}$ are the vectors of phase-conductor charges for lines 1 and 2. $\mathbf{C}_p$ is a 6×6 capacitance matrix that is partitioned into four 3×3 matrices $\mathbf{C}_A$, $\mathbf{C}_B$, $\mathbf{C}_C$, and $\mathbf{C}_D$. Adding $\mathbf{q}_{P1}$ and $\mathbf{q}_{P2}$

$$(\mathbf{q}_{P1} + \mathbf{q}_{P2}) = \mathbf{C}_{Peq}\mathbf{V}_P \tag{4.13.7}$$

where

$$\mathbf{C}_{Peq} = (\mathbf{C}_A + \mathbf{C}_B + \mathbf{C}_C + \mathbf{C}_D) \tag{4.13.8}$$

Also,

$$\mathbf{Y}_{Peq} = j\omega\mathbf{C}_{Peq} \tag{4.13.9}$$

$\mathbf{Y}_{peq}$ is the equivalent 3×3 shunt admittance matrix of the double-circuit line. The matrices $\mathbf{C}_B$ and $\mathbf{C}_C$ in (4.13.8) account for the capacitive coupling between the two circuits.

These ideas can be extended in a straightforward fashion to more than two parallel circuits.

MULTIPLE CHOICE QUESTIONS

SECTION 4.1

4.1 ACSR stands for
(a) Aluminum-clad steel conductor
(b) Aluminum conductor steel supported
(c) Aluminum conductor steel reinforced

4.2 Overhead transmission-line conductors are bare with no insulating cover.
(a) True (b) False

4.3 Alumoweld is an aluminum-clad steel conductor.
(a) True (b) False

4.4 EHV lines often have more than one conductor per phase; these conductors are called a _______.

4.5 Shield wires located above the phase conductors protect the phase conductors against lightning.
(a) True (b) False

4.6 Conductor spacings, types, and sizes do have an impact on the series impedance and shunt admittance.
(a) True (b) False

SECTION 4.2

4.7 A circle with diameter D in. = 1000 D mil = d mil has an area of _______ c mil.

4.8 An ac resistance is higher than a dc resistance.
(a) True (b) False

4.9 Match the following for the current distribution throughout the conductor cross section:
(i) For dc (a) uniform
(ii) For ac (b) nonuniform

SECTION 4.3

4.10 Transmission line conductance is usually neglected in power system studies.
(a) True (b) False

SECTION 4.4

4.11 The internal inductance L_{int} per unit-length of a solid cylindrical conductor is a constant $\frac{1}{2} \times 10^{-7}$ H/m.
(a) True (b) False

4.12 The total inductance L_P of a solid cylindrical conductor (of radius r) due to both internal and external flux linkages out of distance D is given by (in H/m)

(a) 2×10^{-7} (b) $4 \times 10^{-7}\ln\left(\frac{D}{r}\right)$

(c) $2 \times 10^{-7}\ln\left(\frac{D}{r}\right)$

where $r' = e^{-\frac{1}{4}}r = 0.778r$.

SECTION 4.5

4.13 For a single-phase, two-wire line consisting of two solid cylindrical conductors of same radius, r, the total circuit inductance, also called loop inductance, is given by (in H/m)

(a) $2 \times 10^{-7}\ln\left(\frac{D}{r'}\right)$ (b) $4 \times 10^{-7}\ln\left(\frac{D}{r'}\right)$

where $r' = e^{-\frac{1}{4}}r = 0.778r$

4.14 For a three-phase three-wire line consisting of three solid cylindrical conductors, each with radius r and with equal phase spacing D between any two conductors, the inductance in H/m per phase is given by

(a) $2 \times 10^{-7}\ln\left(\frac{D}{r'}\right)$ (b) $4 \times 10^{-7}\ln\left(\frac{D}{r'}\right)$

(c) $6 \times 10^{-7}\ln\left(\frac{D}{r'}\right)$

where $r' = e^{-\frac{1}{4}}r = 0.778\,r$

4.15 For a balanced three-phase positive-sequence currents I_a, I_b, I_c, does the equation $I_a + I_b + I_c = 0$ hold good?

SECTION 4.6

4.16 A stranded conductor is an example of a composite conductor.

(a) True (b) False

4.17 $\Sigma \ln A_k = \ln \Pi A_k$

(a) True (b) False

4.18 Is Geometric Mean Distance (GMD) the same as Geometric Mean Radius (GMR)?

(a) Yes (b) No

4.19 Expand $6\sqrt{\Pi_{k=1}^{3} \Pi_{m=1'}^{2'} D_{km}}$.

4.20 If the distance between conductors are large compared to the distances between subconductors of each conductor, then the GMD between conductors is approximately equal to the distance between conductor centers.

(a) True (b) False

4.21 For a single-phase two-conductor line with composite conductors x and y, express the inductance of conductor x in terms of GMD and its GMR.

4.22 In a three-phase line, in order to avoid unequal phase inductances due to unbalanced flux linkages, what technique is used?

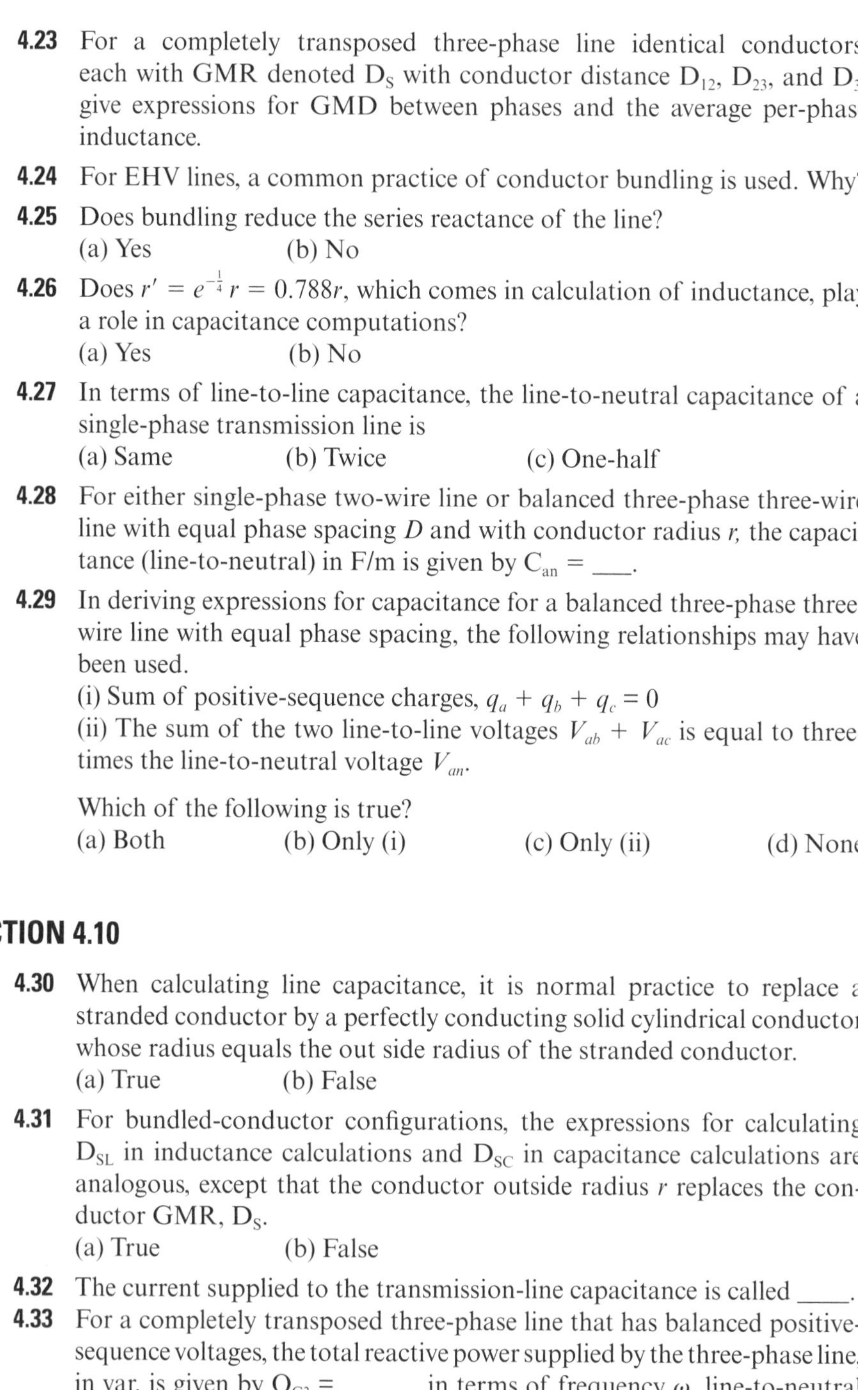

4.23 For a completely transposed three-phase line identical conductors, each with GMR denoted D_S with conductor distance D_{12}, D_{23}, and D_{31} give expressions for GMD between phases and the average per-phase inductance.

4.24 For EHV lines, a common practice of conductor bundling is used. Why?

4.25 Does bundling reduce the series reactance of the line?
(a) Yes (b) No

4.26 Does $r' = e^{-\frac{1}{4}} r = 0.788r$, which comes in calculation of inductance, play a role in capacitance computations?
(a) Yes (b) No

4.27 In terms of line-to-line capacitance, the line-to-neutral capacitance of a single-phase transmission line is
(a) Same (b) Twice (c) One-half

4.28 For either single-phase two-wire line or balanced three-phase three-wire line with equal phase spacing D and with conductor radius r, the capacitance (line-to-neutral) in F/m is given by C_{an} = ____.

4.29 In deriving expressions for capacitance for a balanced three-phase three-wire line with equal phase spacing, the following relationships may have been used.
(i) Sum of positive-sequence charges, $q_a + q_b + q_c = 0$
(ii) The sum of the two line-to-line voltages $V_{ab} + V_{ac}$ is equal to three-times the line-to-neutral voltage V_{an}.

Which of the following is true?
(a) Both (b) Only (i) (c) Only (ii) (d) None

SECTION 4.10

4.30 When calculating line capacitance, it is normal practice to replace a stranded conductor by a perfectly conducting solid cylindrical conductor whose radius equals the out side radius of the stranded conductor.
(a) True (b) False

4.31 For bundled-conductor configurations, the expressions for calculating D_{SL} in inductance calculations and D_{SC} in capacitance calculations are analogous, except that the conductor outside radius r replaces the conductor GMR, D_S.
(a) True (b) False

4.32 The current supplied to the transmission-line capacitance is called ____.

4.33 For a completely transposed three-phase line that has balanced positive-sequence voltages, the total reactive power supplied by the three-phase line, in var, is given by Q_{C3} = ______ in terms of frequency ω, line-to-neutral capacitance C_{an}, and line-to-line voltage V_{LL}.

SECTION 4.11

4.34 Considering lines with neutral conductors and earth return, the effect of earth plane is accounted for by the method of _______ with a perfectly conducting earth plane.

4.35 The affect of the earth plane is to slightly increase the capacitance, and as the line height increases, the effect of earth becomes negligible.
(a) True (b) False

SECTION 4.12

4.36 When the electric field strength at a conductor surface exceeds the breakdown strength of air, current discharges occur. This phenomenon is called ______.

4.37 To control corona, transmission lines are usually designed to maintain the calculated conductor surface electric field strength below _______ kV_{rms}/cm.

4.38 Along with limiting corona and its effects, particularly for EHV lines, the maximum ground-level electric field strength needs to be controlled to avoid the shock hazard.
(a) True (b) False

SECTION 4.13

4.39 Considering two parallel three-phase circuits that are close together, when calculating the equivalent series-impedance and shunt-admittance matrices, mutual inductive and capacitive couplings between the two circuits can be neglected.
(a) True (b) False

PROBLEMS

SECTION 4.2

4.1 The *Aluminum Electrical Conductor Handbook* lists a dc resistance of 0.01558 ohm per 1000 ft at 20°C and a 60-Hz resistance of 0.0956 ohm per mile at 50°C for the all-aluminum Marigold conductor, which has 61 strands and whose size is 1113 kcmil. Assuming an increase in resistance of 2% for spiraling, calculate and verify the dc resistance. Then calculate the dc resistance at 50°C, and determine the percentage increase due to skin effect.

4.2 The temperature dependence of resistance is also quantified by the relation $R_2 = R_1[1 + \alpha(T_2 - T_1)]$ where R_1 and R_2 are the resistances at temperatures T_1 and T_2, respectively, and α is known as the temperature coefficient of resistance. If a copper wire has a resistance of 55 Ω at 20°C, find the maximum permissible operating temperature of the wire if its resistance is to increase by at most 20%. Take the temperature coefficient at 20°C to be $\alpha = 0.00382$.

4.3 A transmission-line cable with a length of 2 km consists of 19 strands of identical copper conductors, each 1.5 mm in diameter. Because of the twist of the strands, the actual length of each conductor is increased by 5%. Determine the resistance of the cable if the resistivity of copper is 1.72 $\mu\Omega$-cm at 20°C.

4.4 One thousand circular mils or 1 kcmil is sometimes designated by the abbreviation MCM. Data for commercial bare-aluminum electrical conductors lists a 60 Hz resistance of 0.0080 ohm per kilometer at 75°C for a 793-MCM AAC conductor.

(a) Determine the cross-sectional conducting area of this conductor in square meters.

(b) Find the 60 Hz resistance of this conductor in ohms per kilometer at 50°C.

4.5 A 60-Hz, 765-kV, three-phase overhead transmission line has four ACSR 900 kcmil 54/3 conductors per phase. Determine the 60 Hz resistance of this line in ohms per kilometer per phase at 50°C.

4.6 A three-phase overhead transmission line is designed to deliver 190.5 MVA at 220 kV over a distance of 63 km, such that the total transmission line loss is not to exceed 2.5% of the rated line MVA. Given the resistivity of the conductor material to be 2.84×10^{-8} Ω-m, determine the required conductor diameter and the conductor size in circular mils. Neglect power losses due to insulator leakage currents and corona.

4.7 If the per-phase line loss in a 70-km-long transmission line is not to exceed 65 kW while it is delivering 100 A per phase, compute the required conductor diameter if the resistivity of the conductor material is 1.72×10^{-8} Ω-m.

SECTIONS 4.4 AND 4.5

4.8 A 60-Hz, single-phase two-wire overhead line has solid cylindrical copper conductors with a 1.5 cm diameter. The conductors are arranged in a horizontal configuration with 0.5 m spacing. Calculate in mH/km (a) the inductance of each conductor due to internal flux linkages only, (b) the inductance of each conductor due to both internal and external flux linkages, and (c) the total inductance of the line.

4.9 Rework Problem 4.8 if the diameter of each conductor is (a) increased by 20% to 1.8 cm or (b) decreased by 20% to 1.2 cm without changing the phase spacing. Compare the results with those of Problem 4.8.

4.10 A 60-Hz, three-phase three-wire overhead line has solid cylindrical conductors arranged in the form of an equilateral triangle with 1.22-m conductor spacing. The conductor diameter is 1.27 cm. Calculate the positive-sequence inductance in H/m and the positive-sequence inductive reactance in Ω/km.

4.11 Rework Problem 4.10 if the phase spacing is (a) increased by 20% to 1.464 m or (b) decreased by 20% to 0.976 m. Compare the results with those of Problem 4.10.

4.12 Find the inductive reactance per kilometer of a single-phase overhead transmission line operating at 60 Hz given the conductors to be *Partridge* and the spacing between centers to be 10 m.

4.13 A single-phase overhead transmission line consists of two solid aluminum conductors having a radius of 3 cm with a spacing 3.5 m between centers. (a) Determine the total line inductance in mH/m. (b) Given the operating frequency to be 60 Hz, find the total inductive reactance of the line in Ω/km. (c) If the spacing is doubled to 7 m, how does the reactance change?

4.14 (a) In practice, one deals with the inductive reactance of the line per phase per kilometer and use the logarithm to the base 10. Show that Eq. (4.5.9) of the text can be rewritten as

$$x = k \log \frac{D}{r'} \text{ ohms per kilometer per phase}$$
$$= x_d + x_a$$

where $x_d = k \log D$ is the inductive reactance spacing factor in ohms per kilometer

$x_a = k \log \frac{1}{r'}$ is the inductive reactance at 1-m spacing in ohms per kilometer

$k = 2.893 \times 10^{-3} f = 1.736$ at 60 Hz

(b) Determine the inductive reactance per kilometer per phase at 60 Hz for a single-phase line with phase separation of 3 m and conductor radius of 2 cm. If the spacing is doubled, how does the reactance change?

SECTION 4.6

4.15 Find the GMR of a stranded conductor consisting of six outer strands surrounding and touching one central strand, all strands having the same radius r.

4.16 A bundle configuration for UHV lines (above 1000 kV) has identical conductors equally spaced around a circle, as shown in Figure 4.29. N_b is the number of conductors in the bundle, A is the circle radius, and D_S is the conductor GMR. Using the distance D_{1n} between conductors 1 and n given by $D_{1n} = 2A \sin[(n-1)\pi/N_b]$ for $n = 1, 2, \ldots, N_b$, and the following trigonometric identity:

$$[2\sin(\pi/N_b)][2\sin(2\pi/N_b)][2\sin(3\pi/N_b)]\cdots[2\sin\{(N_b-1)\pi/N_b\}] = N_b$$

show that the bundle GMR, denoted D_{SL}, is

$$D_{SL} = [N_b D_S A^{(N_b-1)}]^{(1/N_b)}$$

Also show that the above formula agrees with (4.6.19) through (4.6.21) for EHV lines with $N_b = 2$, 3, and 4.

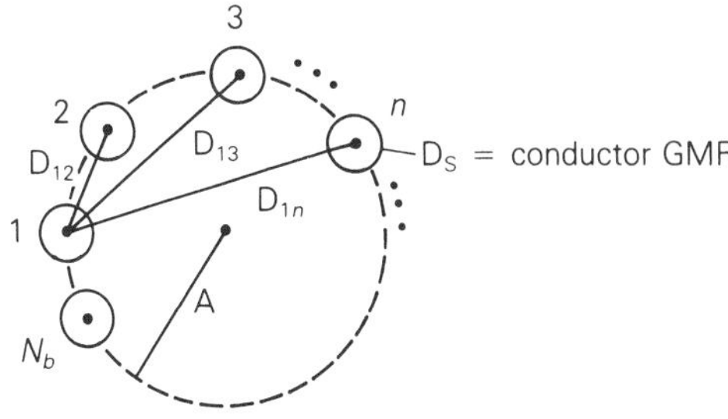

FIGURE 4.29 Bundle configuration for Problem 4.16

4.17 Determine the GMR of each of the unconventional stranded conductors shown in Figure 4.30. All strands have the same radius r.

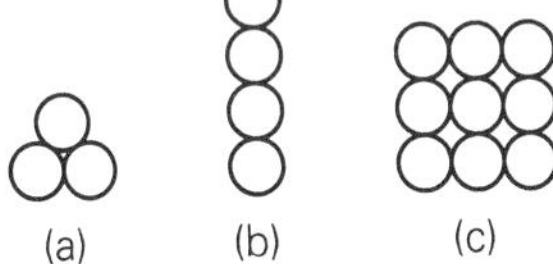

FIGURE 4.30 Unconventional stranded conductors for Problem 4.17

4.18 A 230-kV, 60-Hz, three-phase completely transposed overhead line has one ACSR 954 kcmil conductor per phase and flat horizontal phase spacing, with 7 m between adjacent conductors. Determine the inductance in H/m and the inductive reactance in Ω/km.

4.19 Rework Problem 4.18 if the phase spacing between adjacent conductors is (a) increased by 10% to 7.7 m or (b) decreased by 10% to 6.3 m. Compare the results with those of Problem 4.18.

4.20 Calculate the inductive reactance in Ω/km of a bundled 500-kV, 60-Hz, three-phase completely transposed overhead line having three ACSR 1113 kcmil conductors per bundle, with 0.5 m between conductors in the bundle. The horizontal phase spacings between bundle centers are 10, 10, and 20 m.

4.21 Rework Problem 4.20 if the bundled line has (a) three ACSR, 1351 kcmil conductors per phase or (b) three ACSR, 900 kcmil conductors per phase, without changing the bundle spacing or the phase spacings between bundle centers. Compare the results with those of Problem 4.20.

4.22 The conductor configuration of a bundled single-phase overhead transmission line is shown in Figure 4.31. Line X has its three conductors situated at the corners of an equilateral triangle with 10 cm spacing. Line Y has its three conductors arranged in a horizontal configuration with 10 cm spacing. All conductors are identical, solid-cylindrical conductors each with a radius of 2 cm. Find the equivalent representation in terms of the geometric mean radius of each bundle and a separation that is the geometric mean distance.

FIGURE 4.31

Problem 4.22

4.23 Figure 4.32 shows the conductor configuration of a completely transposed three-phase overhead transmission line with bundled phase conductors. All conductors have a radius of 0.74 cm with a 30-cm bundle spacing. (a) Determine the inductance per phase in mH/km. (b) Find the inductive line reactance per phase in Ω/km at 60 Hz.

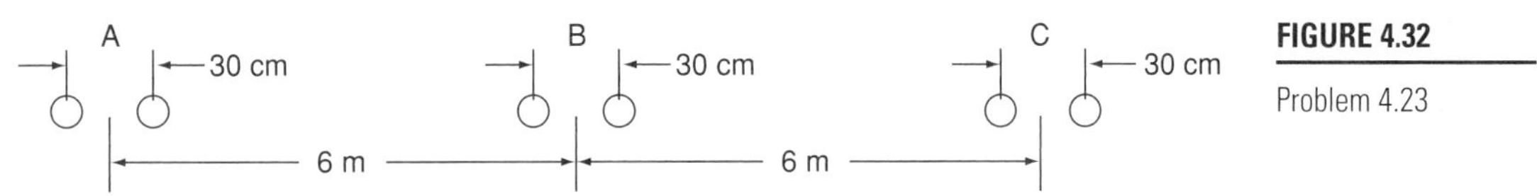

FIGURE 4.32

Problem 4.23

4.24 Consider a three-phase overhead line made up of three phase conductors: Linnet, 336.4 kcmil, and ACSR 26/7. The line configuration is such that the horizontal separation between center of C and that of A is 102 cm, and between that of A and B is also 102 cm in the same line; the vertical separation of A from the line of C–B is 41 cm. If the line is operated at 60 Hz at a conductor temperature of 75°C, determine the inductive reactance per phase in Ω/km,

(a) by using the formula given in Problem 4.14 (a), and

(b) by using (4.6.18) from the text.

4.25 For the overhead line of configuration shown in Figure 4.33 operating at 60 Hz and a conductor temperature of 70°C, determine the resistance per phase, inductive reactance in ohms/kilometer/phase, and the current-carrying capacity of the overhead line. Each conductor is ACSR Cardinal of Table A.4.

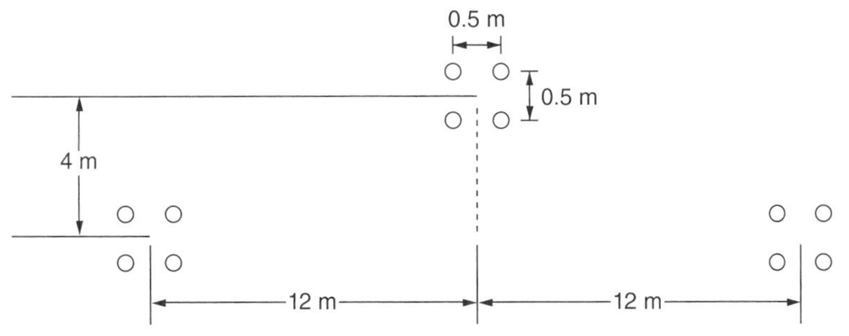

FIGURE 4.33

Line configuration for Problem 4.25

4.26 Consider a symmetrical bundle with N subconductors arranged in a circle of radius A. The inductance of a single-phase symmetrical bundle-conductor line is given by

$$L = 2 \times 10^{-7} \ln \frac{\text{GMD}}{\text{GMR}} \text{ H/m}$$

Where GMR is given by $[Nr'(A)^{N-1}]^{1/N}$ $r' = (e^{-1/4} r)$, r being the subconductor radius, and GMD is approximately the distance D between the bundle centers. Note that A is related to the subconductor spacing S in the bundle circle by $S = 2A \sin(\Pi/N)$

Now consider a 965-kV, single-phase, bundle-conductor line with eight subconductors per phase, phase spacing D = 20 m, and the subconductor spacing S = 45.72 cm. Each subconductor has a diameter of 4.572 cm. Determine the line inductance in H/m.

4.27 Figure 4.34 shows double-circuit conductors' relative positions in segment 1 of transposition of a completely transposed three-phase overhead transmission line. The inductance is given by

$$L = 2 \times 10^{-7} \ln \frac{\text{GMD}}{\text{GMR}} \text{ H/m/phase}$$

Where $\text{GMD} = (D_{AB_{eq}} D_{BC_{eq}} D_{AC_{eq}})^{1/3}$

With mean distances defined by equivalent spacings

FIGURE 4.34

For Problem 4.27 (Double-circuit conductor configuration)

A ● 1 3′ ● C′

B ● 2 2′ ● B′

C ● 3 1′ ● A′

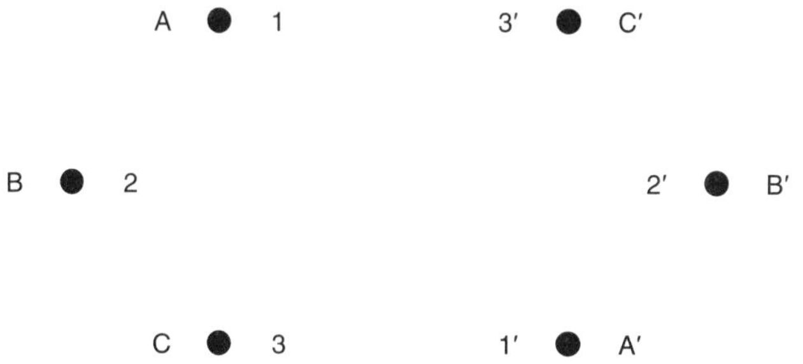

$$D_{AB_{eq}} = (D_{12} D_{1'2'} D_{12'} D_{1'2})^{1/4}$$

$$D_{BC_{eq}} = (D_{23} D_{2'3'} D_{2'3} D_{23'})^{1/4}$$

$$D_{AC_{eq}} = (D_{13} D_{1'3'} D_{13'} D_{1'3})^{1/4}$$

And $\text{GMR} = [(\text{GMR})_A (\text{GMR})_B (\text{GMR})_C]^{1/3}$ with phase GMRs defined by

$$(\text{GMR})_A = [r' D_{11'}]^{1/2}; \quad (\text{GMR})_B = [r' D_{22'}]^{1/2}; \quad (\text{GMR})_C = [r' D_{33'}]^{1/2}$$

and r' is the GMR of phase conductors.

Now consider a 345-kV, three-phase, double-circuit line with phase-conductor's GMR of 1.8 cm and the horizontal conductor configuration shown in Figure 4.35.

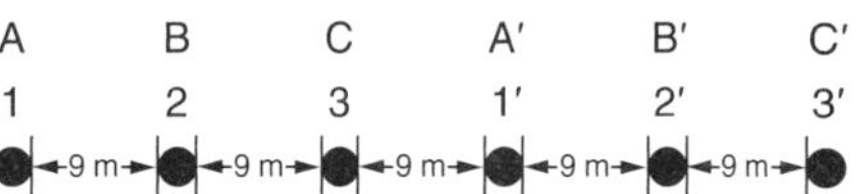

FIGURE 4.35 For Problem 4.27

(a) Determine the inductance per meter per phase in Henries (H).
(b) Calculate the inductance of just one circuit and then divide by 2 to obtain the inductance of the double circuit.

4.28 For the case of double-circuit, bundle-conductor lines, the same method indicated in Problem 4.27 applies with r' replaced by the bundle's GMR in the calculation of the overall GMR.

Now consider a double-circuit configuration shown in Figure 4.36 that belongs to a 500-kV, three-phase line with bundle conductors of three subconductors at 0.53 m spacing. The GMR of each subconductor is given to be 0.015 m.

Determine the inductive reactance of the line in ohms per kilometer per phase. You may use

$$X_L = 0.1786 \log \frac{\text{GMD}}{\text{GMR}} \ \Omega/\text{km/phase}$$

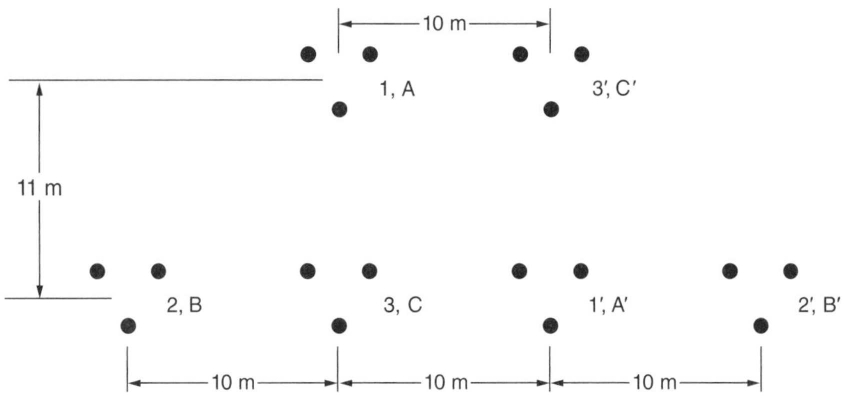

FIGURE 4.36 Configuration for Problem 4.28

4.29 Reconsider Problem 4.28 with an alternate phase placement given below:

	Physical Position					
	1	2	3	1′	2′	3′
Phase Placement	A	B	B′	C	C′	A′

Calculate the inductive reactance of the line in Ω/km/phase.

4.30 Reconsider Problem 4.28 with still another alternate phase placement shown below.

	Physical Position					
	1	2	3	1′	2′	3′
Phase Placement	C	A	B	B	A	C

Find the inductive reactance of the line in Ω/km/phase.

4.31 Figure 4.37 shows the conductor configuration of a three-phase transmission line and a telephone line supported on the same towers. The power line carries a balanced current of 250 A/phase at 60 Hz, while the telephone line is directly located below phase b. Assume balanced three-phase currents in the power line. Calculate the voltage per kilometer induced in the telephone line.

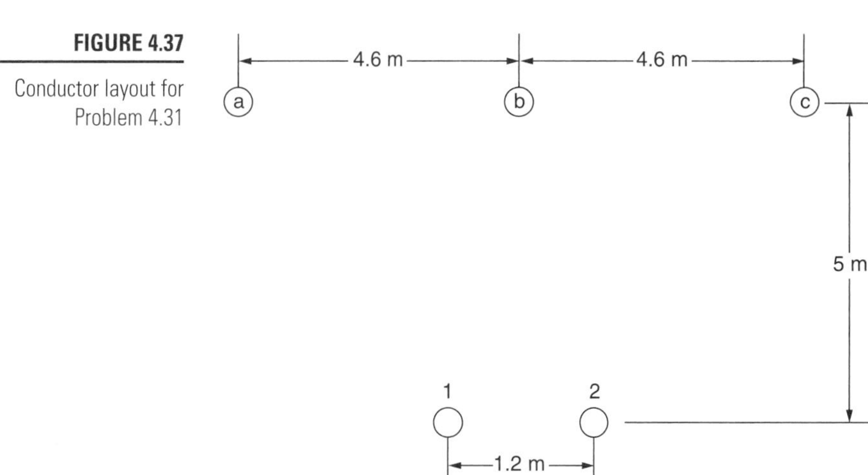

FIGURE 4.37 Conductor layout for Problem 4.31

SECTION 4.9

4.32 Calculate the capacitance-to-neutral in F/m and the admittance-to-neutral in S/km for the single-phase line in Problem 4.8. Neglect the effect of the earth plane.

4.33 Rework Problem 4.32 if the diameter of each conductor is (a) increased by 20% to 1.8 cm or (b) decreased by 20% to 1.2 cm. Compare the results with those of Problem 4.32.

4.34 Calculate the capacitance-to-neutral in F/m and the admittance-to-neutral in S/km for the three-phase line in Problem 4.10. Neglect the effect of the earth plane.

4.35 Rework Problem 4.34 if the phase spacing is (a) increased by 20% to 4.8 ft or (b) decreased by 20% to 3.2 ft. Compare the results with those of Problem 4.34.

4.36 The line of Problem 4.23 as shown in Figure 4.32 is operating at 60 Hz. Determine (a) the line-to-neutral capacitance in nF/km per phase; (b) the capacitive reactance in Ω-km per phase; and (c) the capacitive reactance in Ω per phase for a line length of 160 km.

4.37 (a) In practice, one deals with the capacitive reactance of the line in ohms · km to neutral. Show that Eq. (4.9.15) of the text can be rewritten as

$$\mathrm{X_C} = k' \log \frac{\mathrm{D}}{r} \quad \text{ohms} \cdot \text{km to neutral}$$

$$= x_d' + x_a'$$

where $x_d' = k' \log \mathrm{D}$ is the capacitive reactance spacing factor
$x_a' = k' \log \frac{1}{r}$ is the capacitive reactance at 1-m spacing
$k' = (21.65 \times 10^6)/f = 0.36 \times 10^6$ at $f = 60$ Hz

(b) Determine the capacitive reactance in Ω · km. for a single-phase line of Problem 4.14. If the spacing is doubled, how does the reactance change?

4.38 The capacitance per phase of a balanced three-phase overhead line is given by

$$\mathrm{C} = \frac{0.04217}{\log(\mathrm{GMD}/r)} \ \mu f/\text{km/phase}$$

For the line of Problem 4.24, determine the capacitive reactance per phase in Ω · km.

SECTION 4.10

4.39 Calculate the capacitance-to-neutral in F/m and the admittance-to-neutral in S/km for the three-phase line in Problem 4.18. Also calculate the line-charging current in kA/phase if the line is 110 km in length and is operated at 230 kV. Neglect the effect of the earth plane.

4.40 Rework Problem 4.39 if the phase spacing between adjacent conductors is (a) increased by 10% to 7.7 m or (b) decreased by 10% to 6.3 m. Compare the results with those of Problem 4.39.

4.41 Calculate the capacitance-to-neutral in F/m and the admittance-to-neutral in S/km for the line in Problem 4.20. Also calculate the total reactive power in Mvar/km supplied by the line capacitance when it is operated at 500 kV. Neglect the effect of the earth plane.

4.42 Rework Problem 4.41 if the bundled line has (a) three ACSR, 1351-kcmil conductors per phase or (b) three ACSR, 900 kcmil conductors per phase without changing the bundle spacing or the phase spacings between bundle centers.

4.43 Three ACSR *Drake* conductors are used for a three-phase overhead transmission line operating at 60 Hz. The conductor configuration is in the form of an isosceles triangle with sides of 6, 6, and 12 m. (a) Find the capacitance-to-neutral and capacitive reactance-to-neutral for each 1-kilometer length of line. (b) For a line length of 280 km and a normal operating voltage of 220 kV, determine the capacitive reactance-to-neutral for the entire line length as well as the charging current per mile and total three-phase reactive power supplied by the line capacitance.

4.44 Consider the line of Problem 4.25. Calculate the capacitive reactance per phase in $\Omega \cdot$ km.

SECTION 4.11

4.45 For an average line height of 10 m, determine the effect of the earth on capacitance for the single-phase line in Problem 4.32. Assume a perfectly conducting earth plane.

4.46 A three-phase 60-Hz, 125-km overhead transmission line has flat horizontal spacing with three identical conductors. The conductors have an outside diameter of 3.28 cm with 12 m between adjacent conductors. (a) Determine the capacitive reactance-to-neutral in Ω-m per phase and the capacitive reactance of the line in Ω per phase. Neglect the effect of the earth plane. (b) Assuming that the conductors are horizontally placed 20 m above ground, repeat part (a) while taking into account the effect of ground. Consider the earth plane to be a perfect conductor.

4.47 For the single-phase line of Problem 4.14 (b), if the height of the conductor above ground is 24 m, determine the line-to-line capacitance in F/m. Neglecting earth effect, evaluate the relative error involved. If the phase separation is doubled, repeat the calculations.

4.48 The capacitance of a single-circuit, three-phase transposed line with the configuration shown in Figure 4.38, including ground effect, and with conductors not equilaterally spaced is given by

$$C_{a\eta} \frac{2\pi\varepsilon_0}{\ln \dfrac{D_{eq}}{r} - \ln \dfrac{H_m}{H_s}} \text{ F/m line-to-neutral}$$

where $D_{eq} = \sqrt[3]{D_{12}D_{23}D_{13}} = \text{GMD}$

r = conductor's outside radius

$$H_m = (H_{12}H_{23}H_{13})^{1/3}$$

$$H_s = (H_1H_2H_3)^{1/3}$$

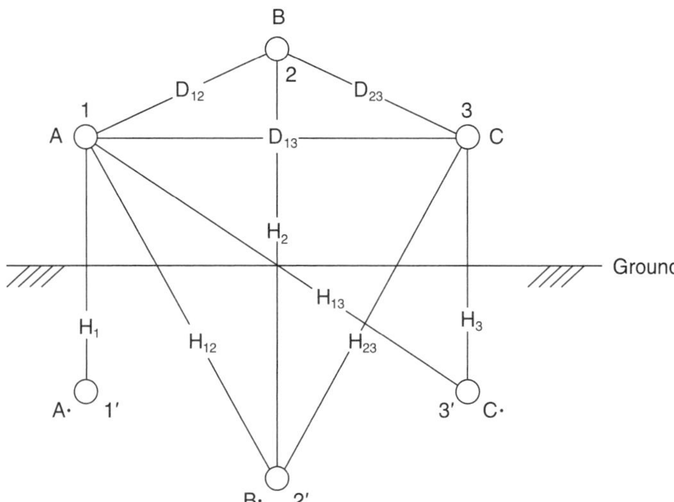

FIGURE 4.38 Three-phase single-circuit line configuration including ground effect for Problem 4.48

(a) Now consider Figure 4.39 in which the configuration of a three-phase, single circuit, 345-kV line with conductors having an outside diameter of 27.051 mm is shown. Determine the capacitance to neutral in F/m, including the ground effect.
(b) Next, neglecting the effect of ground, see how the value changes.

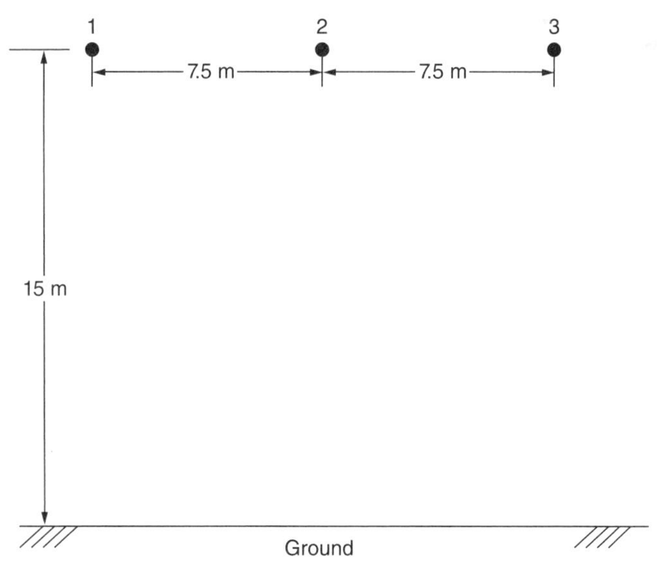

FIGURE 4.39 Configuration for Problem 4.48 (a)

4.49 The capacitance-to-neutral, neglecting the ground effect, for the three-phase, single-circuit, bundle-conductor line is given by

$$C_{an} = \frac{2\pi\varepsilon_0}{\ell n\left(\frac{\text{GMD}}{\text{GMR}}\right)} \text{ F/m line-to-neutral}$$

$$\text{Where GMD} = (D_{AB}D_{BC}D_{AC})^{1/3}$$
$$\text{GMR} = [rN(A)^{N-1}]^{1/N}$$

in which N is the number of subconductors of the bundle conductor on a circle of radius A, and each subconductor has an outside radius of r. The capacitive reactance in mega ohms for 1 km of line at 60 Hz can be shown to be

$$X_C = 0.11 \log\left(\frac{\text{GMD}}{\text{GMR}}\right) = X_a' + X_d'$$

where $X_a' = 0.11 \log\left(\frac{1}{\text{GMR}}\right)$ and $X_d' = 0.11 \log(\text{GMD})$.

Note that A is related to the bundle spacing S given by

$$A = \frac{S}{2\sin\left(\frac{\pi}{N}\right)} \quad \text{for } N > 1$$

Using the above information for the configuration shown in Figure 4.40, compute the capacitance-to-neutral in F/m and the capacitive reactance in $\Omega \cdot$ km to neutral for the three-phase, 765-kV, 60-Hz, single-circuit, bundle-conductor line ($N = 4$) with subconductor's outside diameter of 3 cm and subconductor spacing (S) of 46 cm.

FIGURE 4.40

Configuration for Problem 4.49

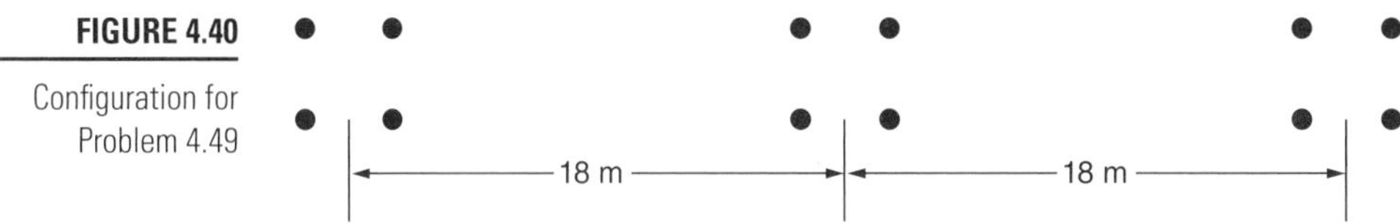

SECTION 4.12

4.50 Calculate the conductor surface electric field strength in kVrms/cm for the single-phase line in Problem 4.32 when the line is operating at 20 kV. Also calculate the ground-level electric field strength in kVrms/m directly under one conductor. Assume a line height of 10 m.

4.51 Rework Problem 4.50 if the diameter of each conductor is (a) increased by 25% to 1.875 cm or (b) decreased by 25% to 1.125 cm without changing the phase spacings. Compare the results with those of Problem 4.50.

CASE STUDY QUESTIONS

a. Approximately how many physical transmission interconnections are there between the United States and Canada? Across which states and provinces are the interconnections located?

b. How many kWhs of electrical energy was exported from Canada to the United States in 2012? Was the majority of that energy export derived from clean, non-emitting sources of electrical power?

c. What caused the August 2003 blackout in the United States and Canada? How many generating units were forced to shut down during that blackout?

d. What are the advantages of high-capacity, low-sag conductors? As of 2013, how many electric utilities in the world had deployed high-capacity, low-sag conductor technology?

REFERENCES

1. Electric Power Research Institute (EPRI), *EPRI AC Transmission Line Reference Book—200 kV and Above* (Palo Alto, CA: EPRI, www.epri.com, December 2005).
2. Westinghouse Electric Corporation, *Electrical Transmission and Distribution Reference Book*, 4th ed. (East Pittsburgh, PA, 1964).
3. General Electric Company, *Electric Utility Systems and Practices*, 4th ed. (New York: Wiley, 1983).
4. John R. Carson, "Wave Propagation in Overhead Wires with Ground Return," *Bell System Tech. J.* 5 (1926): 539–554.
5. C. F. Wagner and R. D. Evans, *Symmetrical Components* (New York: McGraw-Hill, 1933).
6. Paul M. Anderson, *Analysis of Faulted Power Systems* (Ames, IA: Iowa State Press, 1973).
7. M. H. Hesse, "Electromagnetic and Electrostatic Transmission Line Parameters by Digital Computer," *Trans. IEEE* PAS-82 (1963): 282–291.
8. W. D. Stevenson, Jr., *Elements of Power System Analysis,* 4th ed. (New York: McGraw-Hill, 1982).

9. C. A. Gross, *Power System Analysis* (New York: Wiley, 1979).

10. P. Brown, "Integrating North America's Power Grid," *Electricity Today T & D Magazine*, 26, 5, (June 2013), pp. 8–11 (www.electricity-today.com).

11. IEEE ANCI C2. *National Electrical Safety Code*, 2007 edition (New York: Institute of Electrical and Electronics Engineers).

12. D. Bryant, "Grid Congestion," *Electricity Today T & D Magazine*, 26, 6 (July/August 2013), pp. 58–60 (www.electricity-today.com).

5 Transmission Lines: Steady-State Operation

Series capacitor installation at Goshen Substation, Goshen, Idaho, USA rated at 395 kV, 965 Mvar (Courtesy of PacifiCorp.)

This chapter analyzes the performance of single-phase and balanced three-phase transmission lines under normal steady-state operating conditions. Expressions for voltage and current at any point along a line are developed where the distributed nature of the series impedance and shunt admittance is taken into account. A line is treated here as a two-port network for which the *ABCD* parameters and an equivalent π circuit are derived. Also, approximations are given for a medium-length line lumping the shunt admittance, for a short line neglecting the shunt admittance, and for a lossless line assuming zero series resistance and shunt conductance.

The concepts of *surge impedance loading* and transmission-line *wavelength* are also presented.

An important issue discussed in this chapter is *voltage regulation*. Transmission-line voltages are generally high during light load periods and low during heavy load periods. Voltage regulation, defined in Section 5.1, refers to the change in line voltage as line loading varies from no-load to full load.

Another important issue discussed here is line loadability. Three major line-loading limits are (1) the thermal limit, (2) the voltage-drop limit, and (3) the steady-state stability limit. Thermal and voltage-drop limits are discussed in Section 5.1. The theoretical steady-state stability limit, discussed in Section 5.4 for lossless lines and in Section 5.5 for lossy lines, refers to the ability of synchronous machines at the ends of a line to remain in synchronism. Practical line loadability is discussed in Section 5.6.

Section 5.7 covers line compensation techniques for improving voltage regulation and for raising line loadings closer to the thermal limit.

CASE STUDY

High voltage direct current (HVDC) applications embedded within ac power system grids have many benefits. A bipolar HVDC transmission line has only two insulated sets of conductors versus three for an ac transmission line. As such, HVDC transmission lines have smaller transmission towers, narrower rights-of-way, and lower line losses compared to ac lines with similar capacity. The resulting cost savings can offset the higher converter station costs of HVDC. Further, HVDC may be the only feasible method to (1) interconnect two asynchronous ac networks; (2) utilize long underground or underwater cable circuits; (3) bypass network congestion; (4) reduce fault currents; (5) share utility rights-of-way without degrading reliability; and (6) mitigate environmental concerns. The following article provides an overview of HVDC along with HVDC applications [6].

The ABCs of HVDC Transmission Technologies: An Overview of High Voltage Direct Current Systems and Applications

By Michael P. Bahrman and Brian K. Johnson

High voltage direct current (HVDC) technology has characteristics that make it especially attractive for certain transmission applications. HVDC transmission is widely recognized as being advantageous for long-distance bulk-power delivery, asynchronous interconnections, and long submarine cable crossings. The number of HVDC projects committed or under consideration globally has increased in recent years, reflecting a renewed interest in this mature technology. New converter designs have broadened the potential range of HVDC transmission to include applications for underground, offshore, economic replacement of reliability-must-run generation, and voltage stabilization. This broader range of applications has contributed to the recent growth of HVDC transmission. There are approximately ten new HVDC projects under construction or active consideration in North America along with many more projects underway globally. Figure 1 shows the Danish terminal for Skagerrak's pole 3, which is rated 440 MW. Figure 2 shows the ±500-kV HVDC transmission line for the 2000 MW Intermountain Power Project between Utah and California. This article discusses HVDC technologies, application areas where HVDC is favorable compared to ac transmission, system configuration, station design, and operating principles.

Core HVDC Technologies

Two basic converter technologies are used in modem HVDC transmission systems. These are conventional line-commutated current source converters (CSCs) and self-commutated voltage source converters (VSCs). Figure 3 shows a conventional HVDC converter station with CSCs while Figure 4 shows a HVDC converter station with VSCs.

Figure 1 HVDC converter station with ac filters in the foreground and valve hall in the background

Figure 2 A ±500-kV HVDC transmission line

Line-Commutated Current Source Converter

Conventional HVDC transmission employs line-commutated CSCs with thyristor valves. Such converters require a synchronous voltage source in order to operate. The basic building block used for HVDC conversion is the three-phase, full-wave bridge referred to as a six-pulse or Graetz bridge. The term six-pulse is due to six commutations or switching operations per period resulting in a characteristic harmonic ripple of six times the fundamental frequency in the dc output voltage. Each six-pulse bridge is comprised of six controlled switching elements or thyristor

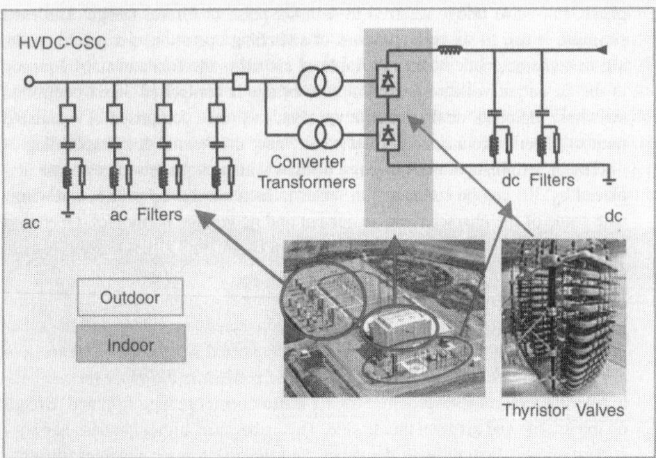

Figure 3 Conventional HVDC with current source converters

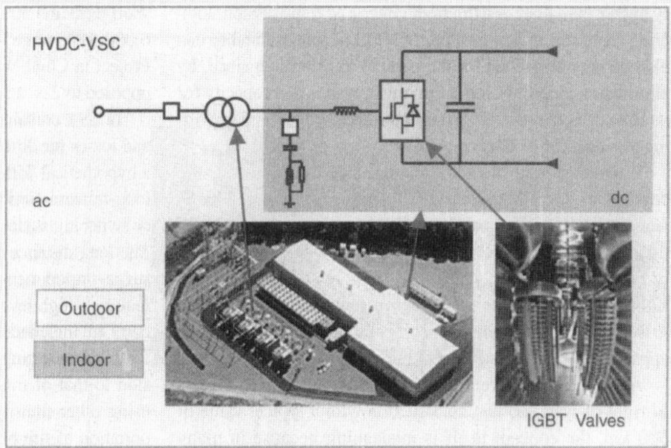

Figure 4 HVDC with voltage source converters

valves. Each valve is comprised of a suitable number of series-connected thyristors to achieve the desired dc voltage rating.

The dc terminals of two six-pulse bridges with ac voltage sources phase displaced by 30° can be connected in series to increase the dc voltage and eliminate some of the characteristic ac current and dc voltage harmonics. Operation in this manner is referred to as 12-pulse operation. In 12-pulse operation, the characteristic ac current and dc voltage harmonics have frequencies of $12n \pm 1$ and $12n$, respectively. The 30° phase displacement is achieved by feeding one bridge through a transformer with a wye-connected secondary and the other bridge through a transformer with a delta-connected secondary. Most modern HVDC transmission

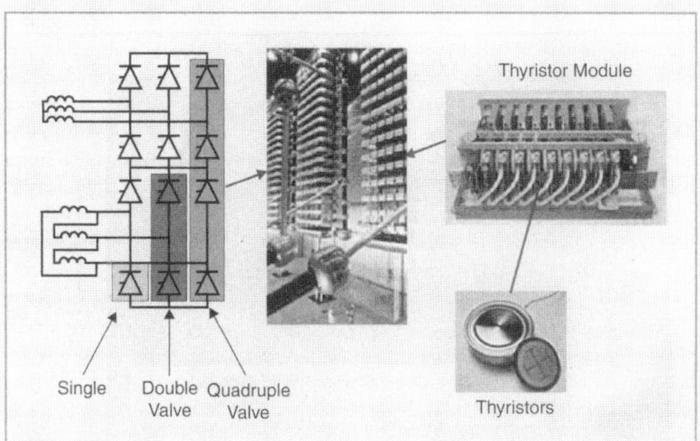

Figure 5 Thyristor valve arrangement for a 12-pulse converter with three quadruple valves, one for each phase

schemes utilize 12-pulse converters to reduce the harmonic filtering requirements required for six-pulse operation; for example, fifth and seventh on the ac side and sixth on the dc side. This is because, although these harmonic currents still flow through the valves and the transformer windings, they are 180° out of phase and cancel out on the primary side of the converter transformer. Figure 5 shows the thyristor valve arrangement for a 12-pulse converter with three quadruple valves, one for each phase. Each thyristor valve is built up with series-connected thyristor modules.

Line-commutated converters require a relatively strong synchronous voltage source in order to commutate. Commutation is the transfer of current from one phase to another in a synchronized firing sequence of the thyristor valves. The three-phase symmetrical short-circuit capacity available from the network at the converter connection point should be at least twice the converter rating for converter operation. Line-commutated CSCs can only operate with the ac current lagging the voltage, so the conversion process demands reactive power. Reactive power is supplied from the ac filters, which look capacitive at the fundamental frequency, shunt banks, or series capacitors that are an integral part of the converter station. Any surplus or deficit in reactive power from these local sources must be accommodated by the ac system. This difference in reactive power needs to be kept within a given band to keep the ac voltage within the desired tolerance. The weaker the ac system or the further the converter is away from generation, the tighter the reactive power exchange must be to stay within the desired voltage tolerance. Figure 6 illustrates the reactive power demand, reactive power compensation, and reactive power exchange with the ac network as a function of dc load current.

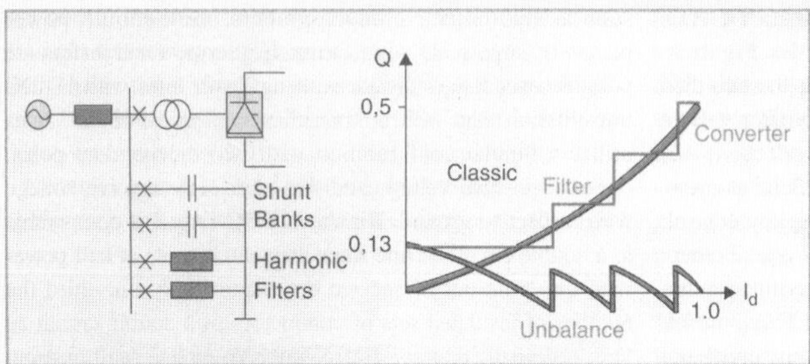

Figure 6 Reactive power compensation for conventional HVDC converter station

Converters with series capacitors connected between the valves and the transformers were introduced in the late 1990s for weak-system, back-to-back applications. These converters are referred to as capacitor-commutated converters (CCCs). The series capacitor provides some of the converter reactive power compensation requirements automatically with load current and provides part of the commutation voltage, improving voltage stability. The overvoltage protection of the series capacitors is simple since the capacitor is not exposed to line faults, and the fault current for internal converter faults is limited by the impedance of the converter transformers. The CCC configuration allows higher power ratings in areas were the ac network is close to its voltage stability limit. The asynchronous Garabi interconnection between Brazil and Argentina consists of four × 550 MW parallel CCC links. The Rapid City Tie between the Eastern and Western interconnected systems consists of 2 × 10 MW parallel CCC links (Figure 7). Both installations use a modular design with converter valves located within prefabricated electrical enclosures rather than a conventional valve hall.

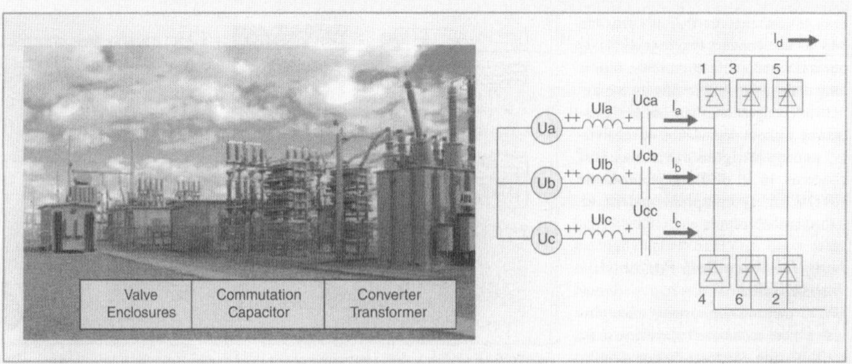

Figure 7 Asynchronous back-to-back tie with capacitor-commutated converter near Rapid City, South Dakota

Self-Commutated Voltage Source Converter

HVDC transmission using VSCs with pulse-width modulation (PWM), commercially known as HVDC Light, was introduced in the late 1990s. Since then, the progression to higher voltage and power ratings for these converters has roughly paralleled that for thyristor valve converters in the 1970s. These VSC-based systems are self-commutated with insulated-gate bipolar transistor (IGBT) valves and solid-dielectric extruded HVDC cables. Figure 8 illustrates solid-state converter development for the two different types of converter technologies using thyristor valves and IGBT valves.

HVDC transmission with VSCs can be beneficial to overall system performance. VSC technology can rapidly control both active and reactive power independently of one another. Reactive power can also be controlled at each terminal independent of the dc transmission voltage level. This control capability gives total flexibility to place converters anywhere in the ac network because there is no restriction on minimum network short-circuit capacity. Self-commutation with VSC even permits black start; that is, the converter can be used to synthesize a balanced set of three-phase voltages like a virtual synchronous generator. The dynamic support of the ac voltage at each converter terminal improves the voltage stability and can increase the transfer capability of the sending- and receiving-end ac systems, thereby leveraging the transfer capability of the dc link. Figure 9 shows the IGBT converter valve arrangement for a VSC station. Figure 10 shows the active and reactive power operating range for a converter station with a VSC. Unlike conventional HVDC transmission, the converters themselves have

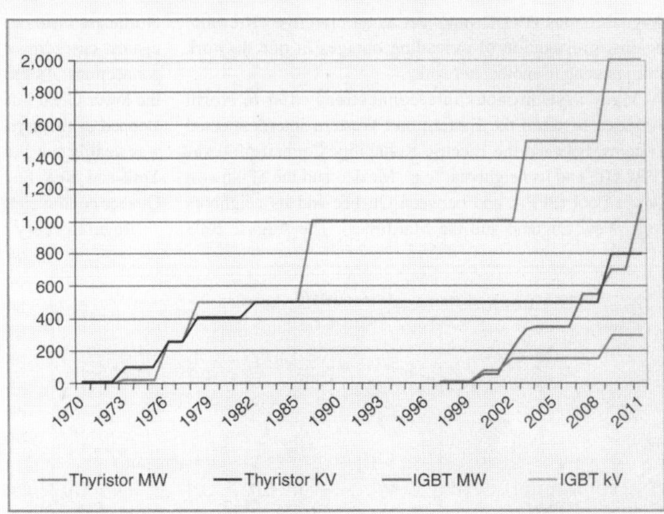

Figure 8 Solid-state converter development

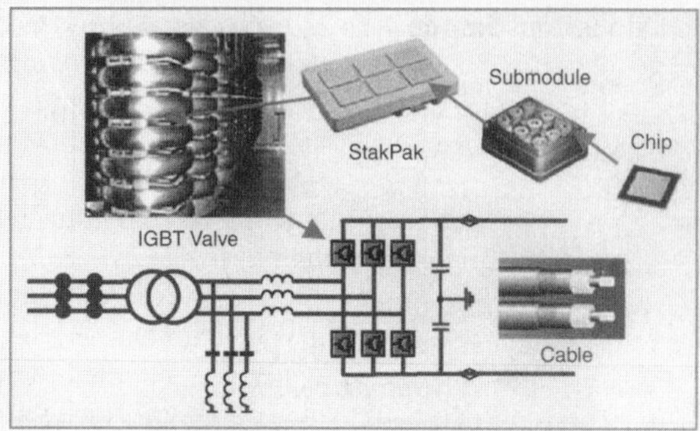

Figure 9 HVDC IGBT valve converter arrangement

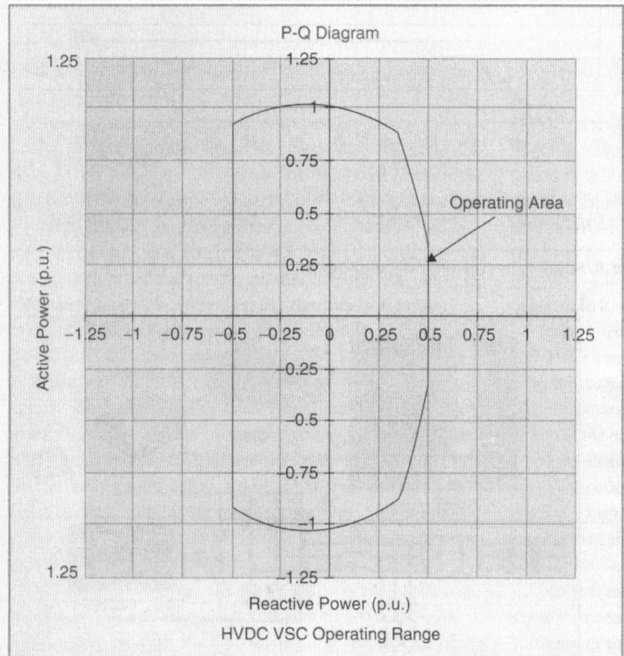

Figure 10 Operating range for voltage source converter HVDC transmission

no reactive power demand and can actually control their reactive power to regulate ac system voltage, just like a generator.

HVDC Applications

HVDC transmission applications can be broken down into different basic categories. Although the rationale for selection of HVDC is often economic, there may be other reasons for its selection. HVDC may be the only feasible way to interconnect two asynchronous networks, reduce fault currents, utilize long underground cable circuits, bypass network congestion, share utility rights-of-way without degradation of reliability, and to mitigate environmental concerns. In all of these applications, HVDC nicely complements the ac transmission system.

Long-Distance Bulk Power Transmission

HVDC transmission systems often provide a more economical alternative to ac transmission for long-distance bulk-power delivery from remote resources such as hydroelectric developments, mine-mouth power plants, or large-scale wind farms. Higher power transfers are possible over longer distances using fewer lines with HVDC transmission than with ac transmission. Typical HVDC lines utilize a bipolar configuration with two independent poles, one at a positive voltage and the other at a negative voltage with respect to ground. Bipolar HVDC lines are comparable to a double-circuit ac line since they can operate at half power with one pole out of service while requiring only one-third the number of insulated sets of conductors as a double circuit ac line. Automatic restarts from temporary dc line fault clearing sequences are routine, even for generator outlet transmission. No synchro-checking is required as for automatic reclosures following ac line faults since the dc restarts do not expose turbine generator units to high risk of transient torque amplification from closing into faults or across high phase angles. The controllability of HVDC links offer firm transmission capacity without limitation due to network congestion or loop flow on parallel paths. Controllability allows the HVDC to "leap-frog" multiple "choke-points" or bypass sequential path limits in the ac network. Therefore, the utilization of HVDC links is usually higher than that for extra high voltage ac transmission, lowering the transmission cost per MWh. This controllability can also be very beneficial for the parallel transmission since, by eliminating loop flow, it frees up this transmission capacity for its intended purpose of serving intermediate load and providing an outlet for local generation.

Whenever long-distance transmission is discussed, the concept of "break-even distance" frequently arises. This is where the savings in line costs offset the higher converter station costs. A bipolar HVDC line uses only two insulated sets of conductors rather than three. This results in narrower rights-of-way, smaller transmission towers, and lower line losses than with ac lines of comparable capacity. A rough approximation of the savings in line construction is 30%.

Although break-even distance is influenced by the costs of right-of-way and line construction with a typical value of 500 km, the concept itself is misleading because in many cases more ac lines are needed to deliver the same power over the same distance due to system stability limitations.

Furthermore, the long-distance ac lines usually require intermediate switching stations and reactive power compensation. This can increase the substation costs for ac transmission to the point where it is comparable to that for HVDC transmission.

For example, the generator outlet transmission alternative for the ±250-kV, 500-MW Square Butte project was two 345-kV series-compensated ac transmission lines. The 12,600-MW Itaipu project has half its power delivered on three 800-kV series-compensated ac lines (three circuits) and the other half delivered on two ±600-kV bipolar HVDC lines (four circuits). Similarly, the ±500-kV, 1600-MW Intermountain Power Project (IPP) ac alternative comprised two 500-kV ac lines. The IPP takes advantage of the double-circuit nature of the bipolar line and includes a 100% short-term and 50% continuous monopolar overload. The first 6000-MW stage of the transmission for the Three Gorges project in China would have required 5 × 500-kV ac lines as opposed to 2 × ±500-kV, 3000-MW bipolar HVDC lines.

Table 1 contains an economic comparison of capital costs and losses for different ac and dc transmission alternatives for a hypothetical 750-mile, 3000-MW transmission system. The long transmission distance requires intermediate substations or switching stations and shunt reactors for the ac alternatives. The long distance and heavy power transfer, nearly twice the surge-impedance loading on the 500-kV ac alternatives, require a high level of series compensation. These ac station costs are included in the cost estimates for the ac alternatives.

It is interesting to compare the economics for transmission to that of transporting an equivalent amount of energy using other transport methods, in this case using rail transportation of sub-bituminous western coal with a heat content of 19.8 MJ/kg to support a 3000-MW base load power plant with heat rate of 9.0 MJ/kWh operating at an 85% load factor. The rail route is assumed to be longer than the more direct transmission route—1400 km. Each unit train is comprised of 100 cars each carrying 90 tonnes of coal. The plant requires three unit trains per day. The annual coal transportation costs are about US$560 million per year at an assumed rate of US$55/tonne. This works out to be US$186 kW/year and US$25 per MWh. The annual diesel fuel consumed in the process is in excess of 76 million liters at 193 net tonne-km per liter. The rail transportation costs are subject to escalation and congestion whereas the transmission costs are fixed. Furthermore, transmission is the only way to deliver remote renewable resources.

Underground and Submarine Cable Transmission

Unlike the case for ac cables, there is no physical restriction limiting the distance or power level for HVDC underground or submarine cables. Underground cables can be used on shared rights-of-way with other utilities without impacting reliability concerns over use of common corridors.

	DC Alternatives				AC Alternatives			Hybrid AC/DC Alternative		
Alternative Capital Cost	+ 500 kV Bipole	2 × +500 kV 2 bipoles	+ 600 kV Bipole	+800 kV Bipole	500 kV 2 Single Ckt	500 kV Double Ckt	765 kV 2 Single Ckt	+500 kV Bipole	500 kV Single Ckt	Total AC + DC
Rated power (MW)	3000	4000	3000	3000	3000	3000	3000	3000	1500	4500
Station costs including reactive compensation (M$)	$420	$680	$465	$510	$542	$542	$630	$420	$302	$722
Transmission line cost (M$/mile)*	$1.60	$1.60	$1.80	$1.95	$2.00	$3.20	$2.80	$1.60	$2.00	
Distance in miles*	750	1500	750	750	1500	750	1500	750	750	1500
Transmission Line Cost (M$)	$1200	$2400	$1350	$1463	$3000	$2400	$4200	$1200	$1500	$2700
Total Cost (M$)	**$1620**	**$3080**	**$1815**	**$1973**	**$3542**	**$2942**	**$4830**	**$1620**	**$1802**	**$3422**
Annual Payment, 30 years @ 10%	$172	$327	$193	$209	$376	$312	$512	$172	$191	$363
Cost per kW-Yr	$57.28	$81.68	$64.18	$69.75	$125.24	$104.03	$170.77	$57.28	$127.40	$80.66
Cost per MWh @ 85% Utilization Factor	$7.69	$10.97	$8.62	$9.37	$16.82	$13.97	$22.93	$7.69	$17.11	$10.83
Losses @ full load	193	134	148	103	208	208	139	106	48	154
Losses at full load in %	6.44%	3.35%	4.93%	3.43%	6.93%	6.93%	4.62%	5.29%	4.79%	5.12%
Capitalized cost of losses @$1500kW(M$)	$246	$171	$188	$131	$265	$265	$177	$135	$61	$196
Parameters:										
Interest rate %	10%									
Capitalized cost of losses	$1500 $/kW									

Note:
AC current assumes 94% p.f.
Full load converter station losses = 0.75% per station
Total substation losses (transformers, reactors) assumed = 0.5% of rated power

TABLE 1

Comparative costs of HVDC and EHV AC transmission alternatives

*1 mile = 1.6 km

For underground or submarine cable systems, there is considerable savings in installed cable costs and cost of losses when using HVDC transmission. Depending on the power level to be transmitted, these savings can offset the higher converter station costs at distances of 40 km or more. Furthermore, there is a drop-off in cable capacity with ac transmission over distance due to its reactive component of charging current since cables have higher capacitances and lower inductances than ac overhead lines. Although this can be compensated by intermediate shunt compensation for underground cables at increased expense, it is not practical to do so for submarine cables.

For a given cable conductor area, the line losses with HVDC cables can be about half those of ac cables. This is due to ac cables requiring more conductors (three phases), carrying the reactive component of current, skin-effect, and induced currents in the cable sheath and armor.

With a cable system, the need to balance unequal loadings or the risk of postcontingency overloads often necessitates use of series-connected reactors or phase-shifting transformers. These potential problems do not exist with a controlled HVDC cable system.

Extruded HVDC cables with prefabricated joints used with VSC-based transmission are lighter, more flexible, and easier to splice than the mass-impregnated oil-paper cables (MINDs) used for conventional HVDC transmission, thus making them more conducive for land cable applications where transport limitations and extra splicing costs can drive up installation costs. The lower-cost cable installations made possible by the extruded HVDC cables and prefabricated joints makes long-distance underground transmission economically feasible for use in areas with rights-of-way constraints or subject to permitting difficulties or delays with overhead lines.

Asynchronous Ties

With HVDC transmission systems, interconnections can be made between asynchronous networks for more economic or reliable system operation. The asynchronous interconnection allows interconnections of mutual benefit while providing a buffer between the two systems. Often these interconnections use back-to-back converters with no transmission line. Asynchronous HVDC links act as an effective "firewall" against propagation of cascading outages in one network from passing to another network.

Many asynchronous interconnections exist in North America between the Eastern and Western interconnected systems, between the Electric Reliability Council of Texas (ERCOT) and its neighbors, [e.g., Mexico and the Southwest Power Pool (SPP)], and between Quebec and its neighbors (e.g., New England and the Maritimes). The August 2003 Northeast blackout provides an example of the "firewall" against cascading outages provided by asynchronous interconnections. As the outage expanded and propagated around the lower Great Lakes and through Ontario and New York, it stopped at the asynchronous interface with Quebec. Quebec was unaffected; the weak ac interconnections between New York and New England tripped, but the HVDC links from Quebec continued to deliver power to New England.

Regulators try to eliminate "seams" in electrical networks because of their potential restriction on power markets. Electrical seams, however, serve as natural points of separation by acting as "shear-pins," thereby reducing the impact of large-scale system disturbances. Asynchronous ties can eliminate market seams while retaining natural points of separation.

Interconnections between asynchronous networks are often at the periphery of the respective systems where the networks tend to be weak relative to the desired power transfer. Higher power transfers can be achieved with improved voltage stability in weak system applications using CCCs. The dynamic voltage support and improved voltage stability offered by VSC-based converters permits even higher power transfers without as much need for ac system reinforcement. VSCs do not suffer commutation failures, allowing fast recoveries from nearby ac faults. Economic power schedules that reverse power direction can be made without any restrictions since there is no minimum power or current restrictions.

Offshore Transmission

Self-commutation, dynamic voltage control, and black-start capability allow compact VSC HVDC transmission to serve isolated loads on islands or offshore production platforms over long-distance submarine cables. This capability can eliminate the need for running expensive local generation or provide an outlet for offshore generation such as that from wind. The VSCs can operate at variable frequency to more efficiently drive large compressor or pumping loads using high-voltage motors. Figure 11 shows the Troll A production platform in the North Sea where power to drive compressors is delivered from shore

Figure 11 VSC power supply to Troll A production platform

to reduce the higher carbon emissions and higher O&M costs associated with less efficient platform-based generation.

Large remote wind generation arrays require a collector system, reactive power support, and outlet transmission. Transmission for wind generation must often traverse scenic or environmentally sensitive areas or bodies of water. Many of the better wind sites with higher capacity factors are located offshore. VSC-based HVDC transmission allows efficient use of long-distance land or submarine cables and provides reactive support to the wind-generation complex. Figure 12 shows a design for an offshore converter station designed to transmit power from offshore wind generation.

Multiterminal Systems

Most HVDC systems are for point-to-point transmission with a converter station at each end. The use of intermediate taps is rare. Conventional HVDC transmission uses voltage polarity reversal to reverse the power direction. Polarity reversal requires no special switching arrangement for a two-terminal system where both terminals reverse polarity by control action with no switching to reverse power direction. Special dc-side switching arrangements are needed for polarity reversal in a multiterminal system, however, where it may be desired to reverse the power direction at a tap while maintaining the same power direction on the remaining terminals. For a bipolar system, this can be done by connecting the converter to the opposite pole. VSC HVDC transmission, however, reverses power through reversal of the current direction rather than voltage polarity. Thus, power can be reversed at an intermediate tap independently of the main power flow direction without switching to reverse voltage polarity.

Figure 12 VSC converter for offshore wind generation

Power Delivery to Large Urban Areas

Power supply for large cities depends on local generation and power import capability. Local generation is often older and less efficient than newer units located remotely. Often, however, the older, less-efficient units located near the city center must be dispatched out-of-merit because they must be run for voltage support or reliability due to inadequate transmission. Air-quality regulations may limit the availability of these units. New transmission into large cities is difficult to site due to right-of-way limitations and land-use constraints.

Compact VSC-based underground transmission circuits can be placed on existing dual-use rights-of-way to bring in power as well as to provide voltage support, allowing a more economical power supply without compromising reliability. The receiving terminal acts like a virtual generator, delivering power and supplying voltage regulation and dynamic reactive power reserve. Stations are compact and housed mainly indoors, making siting in urban areas somewhat easier. Furthermore, the dynamic voltage support offered by the VSC can often increase the capability of the adjacent ac transmission.

System Configurations and Operating Modes

Figure 13 shows the different common system configurations and operating modes used for HVDC transmission. Monopolar systems are the simplest and least expensive systems for moderate power transfers since only two converters and one high-voltage insulated cable or line conductor are required. Such systems have been used with low-voltage electrode lines and sea electrodes to carry the return current in submarine cable crossings.

In some areas, conditions are not conducive to monopolar earth or sea return. This could be the case in heavily congested areas, fresh-water cable crossings, or areas with high earth resistivity. In such cases a metallic neutral- or low-voltage cable is used for the return path, and the dc circuit uses a simple local ground connection for potential reference only. Back-to-back stations are used for interconnection of asynchronous networks and use ac lines to connect

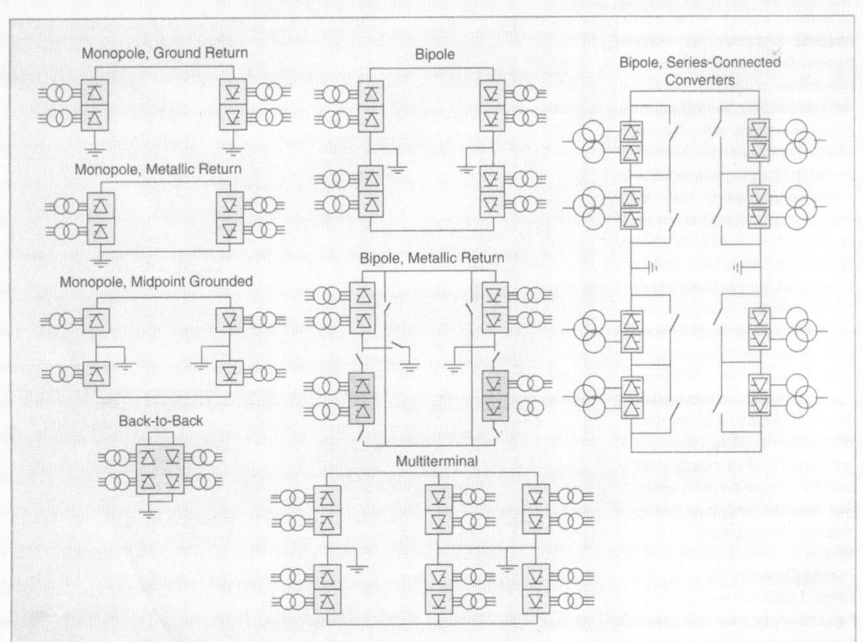

Figure 13 HVDC configurations and operating modes

on either side. In such systems, power transfer is limited by the relative capacities of the adjacent ac systems at the point of connection.

As an economic alternative to a monopolar system with metallic return, the midpoint of a 12-pulse converter can be connected to earth directly or through an impedance and two half-voltage cables or line conductors can be used. The converter is only operated in 12-pulse mode so there is never any stray earth current.

VSC-based HVDC transmission is usually arranged with a single converter connected pole-to-pole rather than pole-to-ground. The center point of the converter is connected to ground through a high impedance to provide a reference for the dc voltage. Thus, half the converter dc voltage appears across the insulation on each of the two dc cables, one positive and the other negative.

The most common configuration for modern overhead HVDC transmission lines is bipolar with a single 12-pulse converter for each pole at each terminal. This gives two independent dc circuits with each capable of half capacity. For normal balanced operation there is no earth current. Monopolar earth return operation, often with overload capacity, can be used during outages of the opposite pole.

Earth return operation can be minimized during monopolar outages by using the opposite pole line for metallic return via pole/converter bypass switches at each end. This requires a metallic-return transfer breaker in the ground electrode line at one of the dc terminals to commutate the current from the relatively low resistance of the earth into that of the dc line conductor. Metallic return operation capability is provided for most dc transmission systems. This not only is effective during converter outages but also during line insulation failures where the remaining insulation strength is adequate to withstand the low resistive voltage drop in the metallic return path.

For very-high-power HVDC transmission, especially at dc voltages above ± 500 kV (i.e., ± 600 kV or ± 800 kV), series-connected converters can be used to reduce the energy unavailability for individual converter outages or partial line insulation failure. By using two series-connected converters per pole in a bipolar system, only one quarter of the transmission capacity is lost for a converter outage or if the line insulation for the affected pole is degraded to where it can only support half the rated dc line voltage. Operating in this mode also avoids the need to transfer to monopolar metallic return to limit the duration of emergency earth return.

Station Design and Layout

Conventional HVDC

The converter station layout depends on a number of factors such as the dc system configuration (i.e., monopolar, bipolar, or back-to-back), ac filtering, and reactive power compensation requirements. The thyristor valves are air-insulated, water-cooled, and enclosed in a converter building often referred to as a valve hall. For back-to-back ties with their characteristically low dc voltage, thyristor valves can be housed in prefabricated electrical enclosures, in which case a valve hall is not required.

To obtain a more compact station design and reduce the number of insulated high-voltage wall bushings, converter transformers are often placed adjacent to the valve hall with valve winding bushings protruding through the building walls for connection to the valves. Double or quadruple valve structures, housing valve modules, are used within the valve hall. Valve arresters are located immediately adjacent to the valves. Indoor motor-operated grounding switches are used for personnel safety during maintenance. Closed-loop valve cooling systems are used to circulate the cooling medium of deionized water or water-glycol mix through the indoor thyristor valves with heat transfer to dry coolers located outdoors. Area requirements for conventional HVDC converter stations are influenced by the ac system voltage and reactive power compensation requirements where each individual bank rating may be limited by such system requirements as reactive power exchange and maximum voltage step on bank switching. The ac yard with filters and shunt compensation can take up as much as three quarters of the total area requirements of the converter station. Figure 14 shows a typical arrangement for an HVDC converter station.

VSC-Based HVDC

The transmission circuit consists of a bipolar two-wire HVDC system with converters connected pole-to-pole. DC capacitors are used to provide a

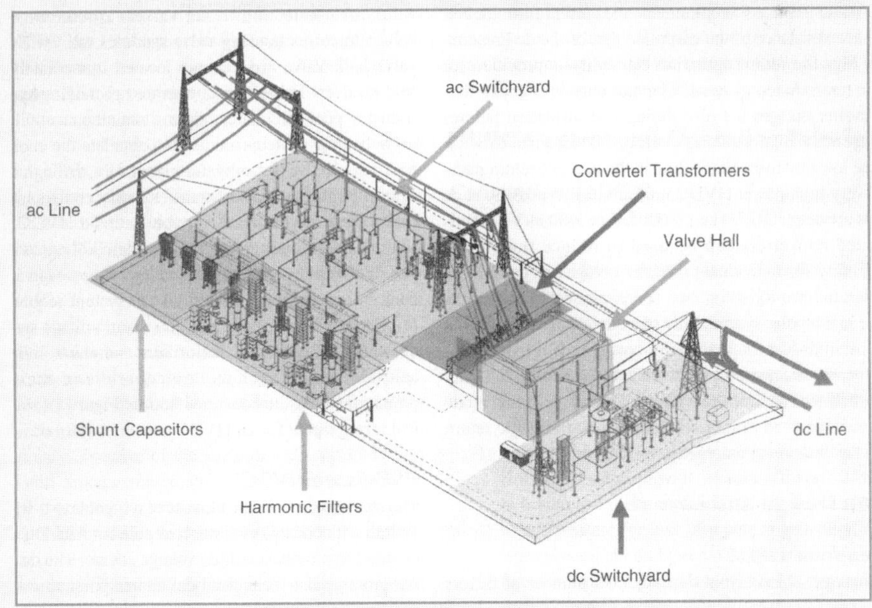

Figure 14 Monopolar HVDC converter station

stiff dc voltage source. The dc capacitors are grounded at their electrical center point to establish the earth reference potential for the transmission system. There is no earth return operation. The converters are coupled to the ac system through ac phase reactors and power transformers. Unlike most conventional HVDC systems, harmonic filters are located between the phase reactors and power transformers. Therefore, the transformers are exposed to no dc voltage stresses or harmonic loading, allowing use of ordinary power transformers. Figure 15 shows the station arrangement for a ±150-kV, 350 to 550-MW VSC converter station.

The IGBT valves used in VSC converters are comprised of series-connected IGBT positions. The IGBT is a hybrid device exhibiting the low forward drop of a bipolar transistor as a conducting device. Instead of the regular current-controlled base, the IGBT has a voltage-controlled capacitive gate, as in the MOSFET device.

A complete IGBT position consists of an IGBT, an anti-parallel diode, a gate unit, a voltage divider, and a water-cooled heat sink. Each gate unit includes gate-driving circuits, surveillance circuits, and optical interface. The gate-driving electronics control the gate voltage and current at turn-on and turn-off to achieve optimal turn-on and turn-off processes of the IGBTs.

To be able to switch voltages higher than the rated voltage of one IGBT, many positions are connected in series in each valve similar to thyristors in conventional HVDC valves. All IGBTs must turn on and off at the same moment to achieve an evenly distributed voltage across the valve. Higher currents are handled by paralleling IGBT components or press packs.

The primary objective of the valve dc-side capacitor is to provide a stiff voltage source and a low-inductance path for the turn-off switching currents and to provide

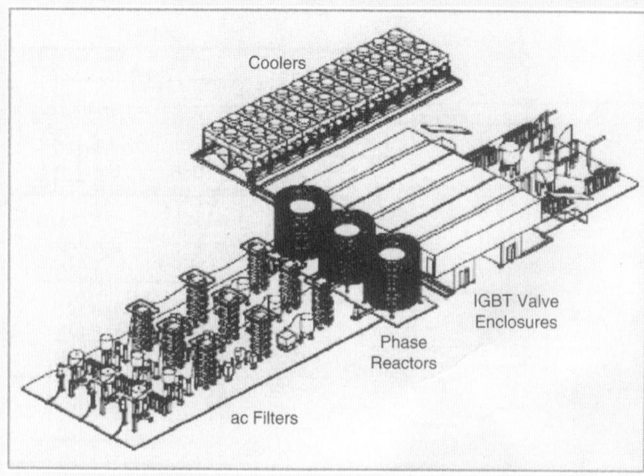

Figure 15 VSC HVDC converter station

energy storage. The capacitor also reduces the harmonic ripple on the dc voltage. Disturbances in the system (e.g., ac faults) cause dc voltage variations. The ability to limit these voltage variations depends on the size of the dc-side capacitor. Since the dc capacitors are used indoors, dry capacitors are used.

AC filters for VSC HVDC converters have smaller ratings than those for conventional converters and are not required for reactive power compensation. Therefore, these filters are always connected to the converter bus and not switched with transmission loading. All equipment for VSC-based HVDC converter stations, except the transformer, high-side breaker, and valve coolers, is located indoors.

HVDC Control and Operating Principles

Conventional HVDC

The fundamental objectives of an HVDC control system are as follows:

1. to control basic system quantities such as dc line current, dc voltage, and transmitted power accurately and with sufficient speed of response;
2. to maintain adequate commutation margin in inverter operation so that the valves can recover their forward blocking capability after conduction before their voltage polarity reverses;
3. to control higher-level quantities such as frequency in isolated mode or provide power oscillation damping to help stabilize the ac network;
4. to compensate for loss of a pole, a generator, or an ac transmission circuit by rapid readjustment of power;
5. to ensure stable operation with reliable commutation in the presence of system disturbances;
6. to minimize system losses and converter reactive power consumption; and
7. to ensure proper operation with fast and stable recoveries during ac system faults and disturbances.

For conventional HVDC transmission, one terminal sets the dc voltage level while the other terminal(s) regulates the (its) dc current by controlling its output voltage relative to that maintained by the voltage-setting terminal. Since the dc line resistance is low, large changes in current and hence power can be made with relatively small changes in firing angle (alpha). Two independent methods exist for controlling the converter dc output voltage. These are (1) by changing the ratio between the direct voltage and the ac voltage by varying the delay angle or (2) by changing the converter ac voltage via load tap changers (LTCs) on the converter transformer. Whereas the former method is rapid, the latter method is slow due to the limited speed of response of the LTC. Use of high-delay angles to achieve a larger dynamic range, however, increases the converter reactive power consumption.

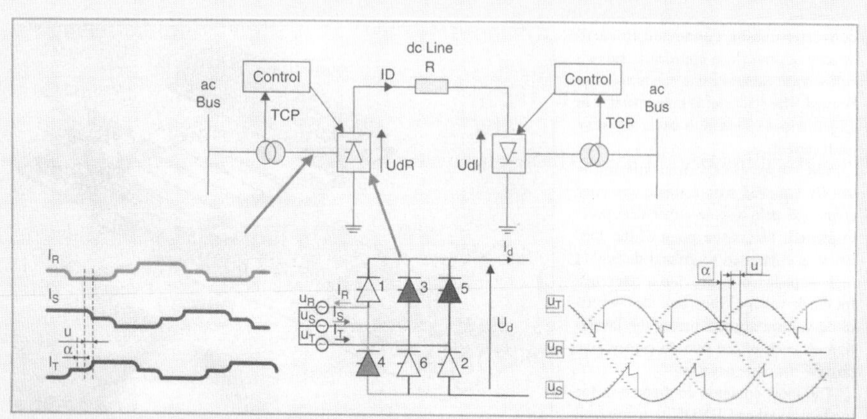

Figure 16 Conventional HVDC control

To minimize the reactive power demand while still providing adequate dynamic control range and commutation margin, the LTC is used at the rectifier terminal to keep the delay angle within its desired steady-state range (e.g., 13–18°) and at the inverter to keep the extinction angle within its desired range (e.g., 17–20°) if the angle is used for dc voltage control or to maintain rated dc voltage if operating in minimum commutation margin control mode. Figure 16 shows the characteristic transformer current and dc bridge voltage waveforms along with the controlled items Ud, Id, and tap changer position (TCP).

VSC-Based HVDC

Power can be controlled by changing the phase angle of the converter ac voltage with respect to the filter bus voltage, whereas the reactive power can be controlled by changing the magnitude of the fundamental component of the converter ac voltage with respect to the filter bus voltage. By controlling these two aspects of the converter voltage, operation in all four quadrants is possible. This means that the converter can be operated in the middle of its reactive power range near unity power factor to maintain dynamic reactive power reserve for contingency voltage support similar to a static var compensator. It also means that the real power transfer can be changed rapidly without altering the reactive power exchange with the ac network or waiting for switching of shunt compensation.

Being able to independently control ac voltage magnitude and phase relative to the system voltage allows use of separate active and reactive power control loops for HVDC system regulation. The active power control loop can be set to control either the active power or the dc-side voltage. In a dc link, one station is then selected to control the active power while the other must be set to control the dc-side voltage. The reactive power control loop can be set to control either the reactive power or the ac-side voltage. Either of these two modes can be selected independently at either end of the dc link.

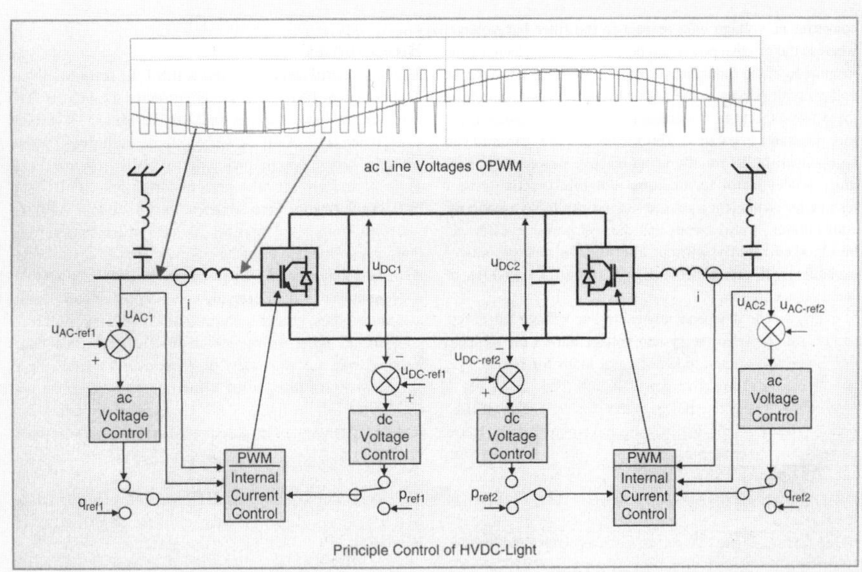

Figure 17 Control of VSC HVDC transmission

Figure 17 shows the characteristic ac voltage waveforms before and after the ac filters along with the controlled items Ud, Id, Q, and Uac.

Conclusions

The favorable economics of long-distance bulk-power transmission with HVDC together with its controllability make it an interesting alternative or complement to ac transmission. The higher voltage levels, mature technology, and new converter designs have significantly increased the interest in HVDC transmission and expanded the range of applications.

For Further Reading

B. Jacobson, Y. Jiang-Hafner, R. Rey, and G. Asplund, "HVDC with voltage source converters and extruded cables for up to ±300 kV and 1000 MW," in Proc. *CIGRE 2006*, Paris, France, pp. B4–105.

L. Ronstrom, B.D. Railing, J.J. Miller, R. Steckley, G. Moreau, R. Bard, and J. Lindberg, "Cross sound cable project second generation VSC technology for HVDC," *Proc. CIGRE 2006*, Paris, France, pp. B4–102.

M. Bahrman, D. Dickinson, P. Fisher, and M. Stoltz, "The Rapid City Tie—New Technology Tames the East-West Interconnection," in *Proc. Minnesota Power Systems Conf.*, St. Paul, MN, Nov. 2004.

D. McCallum, G. Moreau, J. Primeau, D. Soulier, M. Bahrman, and B. Ekehov, "Multiterminal Integration of the Nicolet Converter Station into the Quebec-New England Phase II Transmission System," in *Proc. CIGRE* 1994, Paris, France.

A. Ekstrom and G. Liss, "A refined HVDC control system," *IEEE Trans. Power Systems*, vol. PAS-89, pp. 723–732, May–June 1970.

Biographies

Michael P. Bahrman received a B.S.E.E. from Michigan Technological University. He is currently the U.S. HVDC marketing and sales manager for ABB Inc. He has 24 years of experience with ABB Power Systems, including system analysis, system design, multiterminal HVDC control development, and project management for various HVDC and FACTS projects in North America. Prior to joining ABB, he was with Minnesota Power for ten years where he held positions as transmission planning engineer, HVDC control engineer, and manager of system operations. He has been an active member of IEEE, serving on a number of subcommittees and working groups in the area of HVDC and FACTS.

Brian K. Johnson received the PhD in electrical engineering from the University of Wisconsin-Madison. He is currently a professor in the Department of Electrical and Computer Engineering at the University of Idaho. His interests include power system protection and the application of power electronics to utility systems, security, and survivability of ITS systems and power systems, distributed sensor and control networks, and real-time simulation of traffic systems. He is a member of the Board of Governors of the IEEE Intelligent Transportation Systems Society and the Administrative Committee of the IEEE Council on Superconductivity. ■

5.1 MEDIUM AND SHORT LINE APPROXIMATIONS

In this section, short and medium-length transmission-line approximations as a means of introducing *ABCD* parameters are presented. Some readers may prefer to start in Section 5.2, which presents the exact transmission-line equations.

It is convenient to represent a transmission line by the two-port network shown in Figure 5.1, where V_S and I_S are the sending-end voltage and current, and V_R and I_R are the receiving-end voltage and current.

The relation between the sending-end and receiving-end quantities can be written as

$$V_S = AV_R + BI_R \quad \text{volts} \tag{5.1.1}$$

$$I_S = CV_R + DI_R \quad \text{A} \tag{5.1.2}$$

FIGURE 5.1

Representation of two-port network

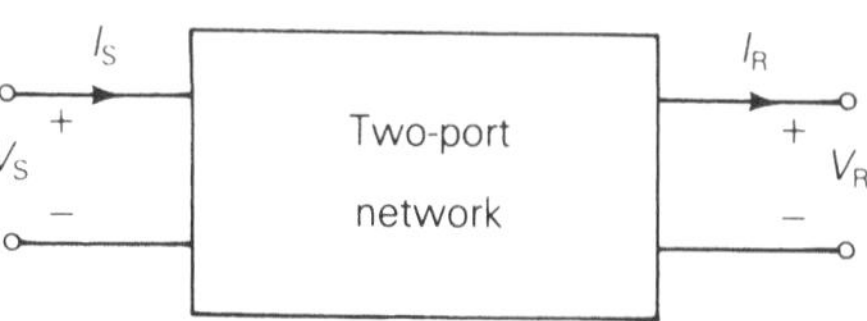

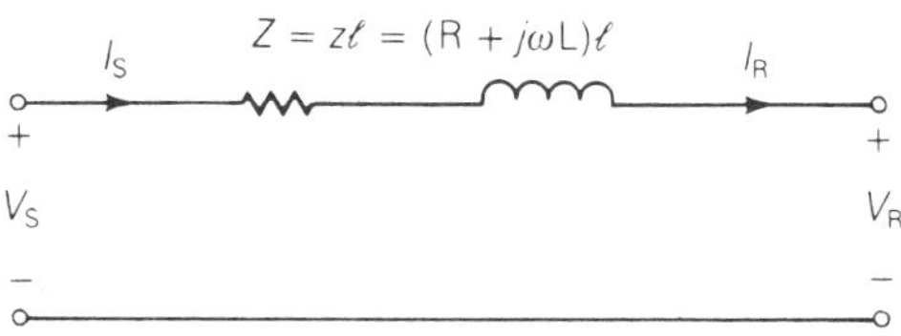

FIGURE 5.2

Short transmission line

or, in matrix format,

$$\begin{bmatrix} V_S \\ I_S \end{bmatrix} = \left[\begin{array}{c|c} A & B \\ \hline C & D \end{array}\right] \begin{bmatrix} V_R \\ I_R \end{bmatrix} \tag{5.1.3}$$

where A, B, C, and D are parameters that depend on the transmission-line constants R, L, C, and G. The $ABCD$ parameters are, in general, complex numbers. A and D are dimensionless. B has units of ohms, and C has units of siemens. Network theory texts [5] show that $ABCD$ parameters apply to linear, passive, bilateral two-port networks, with the following general relation:

$$AD - BC = 1 \tag{5.1.4}$$

The circuit in Figure 5.2 represents a short transmission line, usually applied to overhead 60-Hz lines less than 25 km in length. Only the series resistance and reactance are included. The shunt admittance is neglected. The circuit applies to either single-phase or completely transposed three-phase lines operating under balanced conditions. For a completely transposed three-phase line, Z is the series impedance, V_S and I_R are positive-sequence line-to-neutral voltages, and I_S and I_R are positive-sequence line currents.

To avoid confusion between total series impedance and series impedance per unit length, use the following notation:

$z = R + j\omega L \quad \Omega/m$, series impedance per unit length

$y = G + j\omega C \quad S/m$, shunt admittance per unit length

$Z = zl \quad \Omega$, total series impedance

$Y = yl \quad S$, total shunt admittance

$l =$ line length $\quad$ m

Recall that shunt conductance G is usually neglected for overhead transmission.

The $ABCD$ parameters for the short line in Figure 5.2 are easily obtained by writing a KVL and KCL equation as

$$V_S = V_R + ZI_R \tag{5.1.5}$$

$$I_S = I_R \tag{5.1.6}$$

or, in matrix format,

$$\begin{bmatrix} V_S \\ I_S \end{bmatrix} = \left[\begin{array}{c|c} 1 & Z \\ \hline 0 & 1 \end{array}\right] \begin{bmatrix} V_R \\ I_R \end{bmatrix} \tag{5.1.7}$$

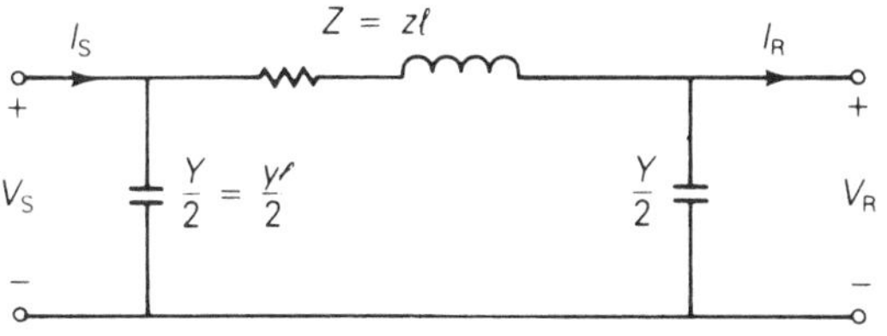

FIGURE 5.3 Medium-length transmission line—nominal π circuit

Comparing (5.1.7) and (5.1.3), the *ABCD* parameters for a short line are

$$A = D = 1 \text{ per unit} \tag{5.1.8}$$

$$B = Z\ \Omega \tag{5.1.9}$$

$$C = 0\ \text{S} \tag{5.1.10}$$

For medium-length lines, typically ranging from 25 to 250 km at 60 Hz, it is common to lump the total shunt capacitance and locate half at each end of the line. Such a circuit, called a *nominal π circuit*, is shown in Figure 5.3.

To obtain the *ABCD* parameters of the nominal π circuit, note first that the current in the series branch in Figure 5.3 equals $I_R + \frac{V_R Y}{2}$. Then, writing a KVL equation,

$$\begin{aligned} V_S &= V_R + Z\left(I_R + \frac{V_R Y}{2}\right) \\ &= \left(1 + \frac{YZ}{2}\right) V_R + Z I_R \end{aligned} \tag{5.1.11}$$

Also, writing a KCL equation at the sending end,

$$I_S = I_R + \frac{V_R Y}{2} + \frac{V_S Y}{2} \tag{5.1.12}$$

Using (5.1.11) in (5.1.12),

$$\begin{aligned} I_S &= I_R + \frac{V_R Y}{2} + \left[\left(1 + \frac{YZ}{2}\right) V_R + Z I_R\right] \frac{Y}{2} \\ &= Y\left(1 + \frac{YZ}{4}\right) V_R + \left(1 + \frac{YZ}{2}\right) I_R \end{aligned} \tag{5.1.13}$$

Writing (5.1.11) and (5.1.13) in matrix format,

$$\begin{bmatrix} V_S \\ \\ I_S \end{bmatrix} = \left[\begin{array}{c|c} \left(1 + \frac{YZ}{2}\right) & Z \\ \hline Y\left(1 + \frac{YZ}{4}\right) & \left(1 + \frac{YZ}{2}\right) \end{array}\right] \begin{bmatrix} V_R \\ \\ I_R \end{bmatrix} \tag{5.1.14}$$

Thus, comparing (5.1.14) and (5.1.3)

$$A = D = 1 + \frac{YZ}{2} \quad \text{per unit} \tag{5.1.15}$$

$$B = Z \quad \Omega \tag{5.1.16}$$

$$C = Y\left(1 + \frac{YZ}{4}\right) \quad \text{S} \tag{5.1.17}$$

Note that for both the short and medium-length lines, the relation $AD - BC = 1$ is verified. Note also that since the line is the same when viewed from either end, $A = D$.

Figure 5.4 gives the $ABCD$ parameters for some common networks, including a series impedance network that approximates a short line and a π circuit that approximates a medium-length line. A medium-length line could also be approximated by the T circuit shown in Figure 5.4, lumping half of the series impedance at each end

Circuit	*ABCD* Matrix
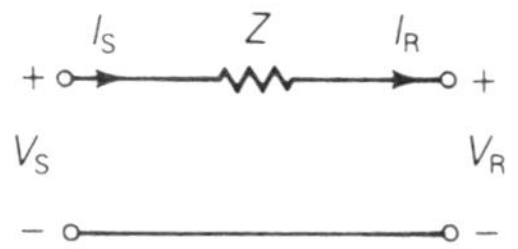 Series impedance (I_S, Z, I_R, V_S, V_R)	$\begin{bmatrix} 1 & Z \\ 0 & 1 \end{bmatrix}$
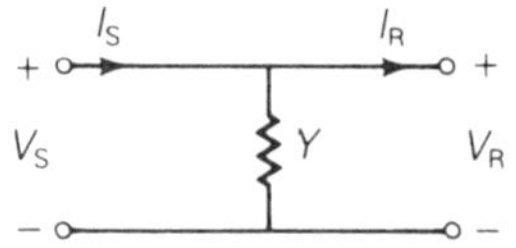 Shunt admittance (I_S, Y, I_R, V_S, V_R)	$\begin{bmatrix} 1 & 0 \\ Y & 1 \end{bmatrix}$
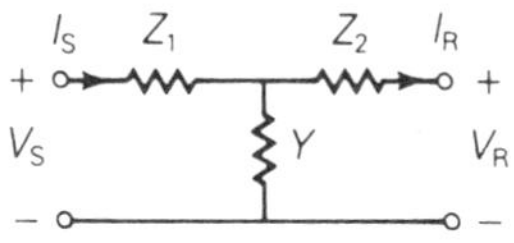 T circuit (I_S, Z_1, Y, Z_2, I_R, V_S, V_R)	$\begin{bmatrix} (1 + YZ_1) & (Z_1 + Z_2 + YZ_1Z_2) \\ Y & (1 + YZ_2) \end{bmatrix}$
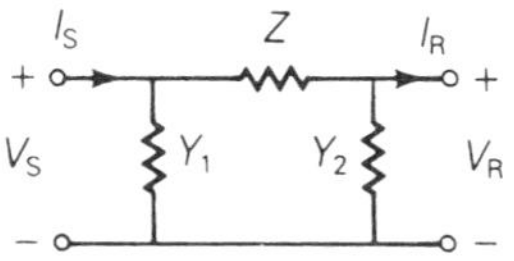 π circuit (I_S, Z, Y_1, Y_2, I_R, V_S, V_R)	$\begin{bmatrix} (1 + Y_2Z) & Z \\ (Y_1 + Y_2 + Y_1Y_2Z) & (1 + Y_1Z) \end{bmatrix}$
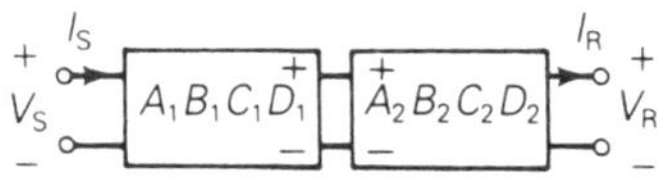 Series networks (I_S, $A_1B_1C_1D_1$, $A_2B_2C_2D_2$, I_R, V_S, V_R)	$\begin{bmatrix} A_1 & B_1 \\ C_1 & D_1 \end{bmatrix}\begin{bmatrix} A_2 & B_2 \\ C_2 & D_2 \end{bmatrix} = \begin{bmatrix} (A_1A_2 + B_1C_2) & (A_1B_2 + B_1D_2) \\ (C_1A_2 + D_1C_2) & (C_1B_2 + D_1D_2) \end{bmatrix}$

FIGURE 5.4 *ABCD* parameters of common networks

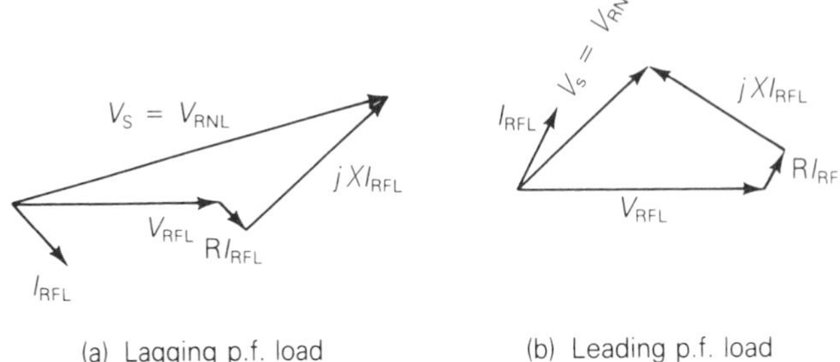

FIGURE 5.5

Phasor diagrams for a short transmission line

of the line. Also given are the *ABCD* parameters for networks in series, which are conveniently obtained by multiplying the *ABCD* matrices of the individual networks.

ABCD parameters can be used to describe the variation of line voltage with line loading. *Voltage regulation* is the change in voltage at the receiving end of the line when the load varies from no-load to a specified full load at a specified power factor, while the sending-end voltage is held constant. Expressed in percent of full-load voltage,

$$\text{percent VR} = \frac{|V_{\text{RNL}}| - |V_{\text{RFL}}|}{|V_{\text{RFL}}|} \times 100 \tag{5.1.18}$$

where percent VR is the percent voltage regulation, $|V_{\text{RNL}}|$ is the magnitude of the no-load receiving-end voltage, and $|V_{\text{RFL}}|$ is the magnitude of the full-load receiving-end voltage.

The effect of load power factor on voltage regulation is illustrated by the phasor diagrams in Figure 5.5 for short lines. The phasor diagrams are graphical representations of (5.1.5) for lagging and leading power factor loads. Note that, from (5.1.5) at no-load, $I_{\text{RNL}} = 0$ and $V_S = V_{\text{RNL}}$ for a short line. As shown, the higher (worse) voltage regulation occurs for the lagging p.f. load, where V_{RNL} exceeds V_{RFL} by the larger amount. A smaller or even negative voltage regulation occurs for the leading p.f. load. In general, the no-load voltage is, from (5.1.1) with $I_{\text{RNL}} = 0$,

$$V_{\text{RNL}} = \frac{V_S}{A} \tag{5.1.19}$$

which can be used in (5.1.18) to determine voltage regulation.

In practice, transmission-line voltages decrease when heavily loaded and increase when lightly loaded. When voltages on EHV lines are maintained within $\pm 5\%$ of rated voltage, corresponding to about 10% voltage regulation, unusual operating problems are not encountered. Ten percent voltage regulation for lower voltage lines including transformer-voltage drops is also considered good operating practice.

In addition to voltage regulation, line loadability is an important issue. Three major line-loading limits are (1) the thermal limit, (2) the voltage-drop limit, and (3) the steady-state stability limit.

The maximum temperature of a conductor determines its thermal limit. Conductor temperature affects the conductor sag between towers and the loss of conductor tensile strength due to annealing. If the temperature is too high, prescribed conductor-to-ground clearances may not be met, or the elastic limit of the conductor may be exceeded such that it cannot shrink to its original length when cooled. Conductor temperature depends on the current magnitude and its time duration, as well as on ambient temperature, wind velocity, and conductor surface conditions. Appendix Tables A.3 and A.4 give approximate current-carrying capacities of copper and ACSR conductors. The loadability of short transmission lines (less than 25 km in length for 60-Hz overhead lines) is usually determined by the conductor thermal limit or by ratings of line terminal equipment such as circuit breakers.

For longer line lengths (up to 300 km), line loadability is often determined by the voltage-drop limit. Although more severe voltage drops may be tolerated in some cases, a heavily loaded line with $V_R/V_S \geq 0.95$ is usually considered safe operating practice. For line lengths over 300 km, steady-state stability becomes a limiting factor. Stability, discussed in Section 5.4, refers to the ability of synchronous machines on either end of a line to remain in synchronism.

EXAMPLE 5.1

ABCD parameters and the nominal π circuit: medium-length line

A three-phase, 60-Hz, completely transposed 345-kV, 200-km line has two 795,000-cmil 26/2 ACSR conductors per bundle and the following positive-sequence line constants:

$$z = 0.032 + j0.35 \quad \Omega/\text{km}$$
$$y = j4.2 \times 10^{-6} \quad \text{S/km}$$

Full load at the receiving end of the line is 700 MW at 0.99 p.f. leading and at 95% of rated voltage. Assuming a medium-length line, determine the following:

a. *ABCD* parameters of the nominal π circuit
b. Sending-end voltage V_S, current I_S, and real power P_S
c. Percent voltage regulation
d. Thermal limit, based on the approximate current-carrying capacity listed in Table A.4
e. Transmission-line efficiency at full load

SOLUTION

a. The total series impedance and shunt admittance values are

$$Z = zl = (0.032 + j0.35)(200) = 6.4 + j70 = 70.29\underline{/84.78^\circ} \quad \Omega$$
$$Y = yl = (j4.2 \times 10^{-6})(200) = 8.4 \times 10^{-4}\underline{/90^\circ} \quad \text{S}$$

(*Continued*)

From (5.1.15) through (5.1.17),

$$A = D = 1 + (8.4 \times 10^{-4}\underline{/90°})(70.29\underline{/84.78°})\left(\tfrac{1}{2}\right)$$
$$= 1 + 0.02952\underline{/174.78°}$$
$$= 0.9706 + j0.00269 = 0.9706\underline{/0.159°} \quad \text{per unit}$$

$$B = Z = 70.29\underline{/84.78°} \quad \Omega$$

$$C = (8.4 \times 10^{-4}\underline{/90°})(1 + 0.01476\underline{/174.78°})$$
$$= (8.4 \times 10^{-4}\underline{/90°})(0.9853 + j0.00134)$$
$$= 8.277 \times 10^{-4}\underline{/90.08°} \quad \text{S}$$

b. The receiving-end voltage and current quantities are

$$\text{V}_\text{R} = (0.95)(345) = 327.8 \text{ kV}_\text{LL}$$

$$V_\text{R} = \frac{327.8}{\sqrt{3}}\underline{/0°} = 189.2\underline{/0°} \quad \text{kV}_\text{LN}$$

$$I_\text{R} = \frac{700\underline{/\cos^{-1}0.99}}{(\sqrt{3})(0.95 \times 345)(0.99)} = 1.246\underline{/8.11°} \quad \text{kA}$$

From (5.1.1) and (5.1.2), the sending-end quantities are

$$V_\text{S} = (0.9706\underline{/0.159°})(189.2\underline{/0°}) + (70.29\underline{/84.78})(1.246\underline{/8.11°})$$
$$= 183.6\underline{/0.159°} + 87.55\underline{/92.89°}$$
$$= 179.2 + j87.95 = 199.6\underline{/26.14°} \quad \text{kV}_\text{LN}$$

$$\text{V}_\text{S} = 199.6\sqrt{3} = 345.8 \text{ kV}_\text{LL} \approx 1.00 \quad \text{per unit}$$

$$I_\text{S} = (8.277 \times 10^{-4}\underline{/90.08°})(189.2\underline{/0°}) + (0.9706\underline{/0.159°})(1.246\underline{/8.11°})$$
$$= 0.1566\underline{/90.08°} + 1.209\underline{/8.27°}$$
$$= 1.196 + j0.331 = 1.241\underline{/15.5°} \quad \text{kA}$$

and the real power delivered to the sending end is

$$\text{P}_\text{S} = (\sqrt{3})(345.8)(1.241)\cos(26.14° - 15.5°)$$
$$= 730.5 \quad \text{MW}$$

c. From (5.1.19), the no-load receiving-end voltage is

$$\text{V}_\text{RNL} = \frac{\text{V}_\text{S}}{\text{A}} = \frac{345.8}{0.9706} = 356.3 \quad \text{kV}_\text{LL}$$

and, from (5.1.18),

$$\text{percent VR} = \frac{356.3 - 327.8}{327.8} \times 100 = 8.7\%$$

d. From Table A.4, the approximate current-carrying capacity of two 795,000-cmil 26/2 ACSR conductors is $2 \times 0.9 = 1.8$ kA.
e. The full-load line losses are $P_S - P_R = 730.5 - 700 = 30.5$ MW and the full-load transmission efficiency is

$$\text{percent EFF} = \frac{P_R}{P_S} \times 100 = \frac{700}{730.5} \times 100 = 95.8\%$$

Since $V_S = 1.00$ per unit, the full-load receiving-end voltage of 0.95 per unit corresponds to $V_R/V_S = 0.95$, considered in practice to be about the lowest operating voltage possible without encountering operating problems. Thus, for this 345-kV 200-km uncompensated line, voltage drop limits the full-load current to 1.246 kA at 0.99 p.f. leading, which is well below the thermal limit of 1.8 kA.

5.2 TRANSMISSION-LINE DIFFERENTIAL EQUATIONS

The line constants R, L, and C are derived in Chapter 4 as per-length values having units of Ω/m, H/m, and F/m. They are not lumped, but rather are uniformly distributed along the length of the line. In order to account for the distributed nature of transmission-line constants, consider the circuit shown in Figure 5.6, which represents a line section of length Δx. $V(x)$ and $I(x)$ denote the voltage and current at position x, which is measured in meters from the right, or receiving end of the line. Similarly, $V(x + \Delta x)$ and $I(x + \Delta x)$ denote the voltage and current at position $(x + \Delta x)$. The circuit constants are

$$z = R + j\omega L \quad \Omega/\text{m} \tag{5.2.1}$$

$$y = G + j\omega C \quad \text{S/m} \tag{5.2.2}$$

where G is usually neglected for overhead 60-Hz lines. Writing a KVL equation for the circuit

$$V(x + \Delta x) = V(x) + (z\Delta x)I(x) \quad \text{volts} \tag{5.2.3}$$

Rearranging (5.2.3),

$$\frac{V(x + \Delta x) - V(x)}{\Delta x} = zI(x) \tag{5.2.4}$$

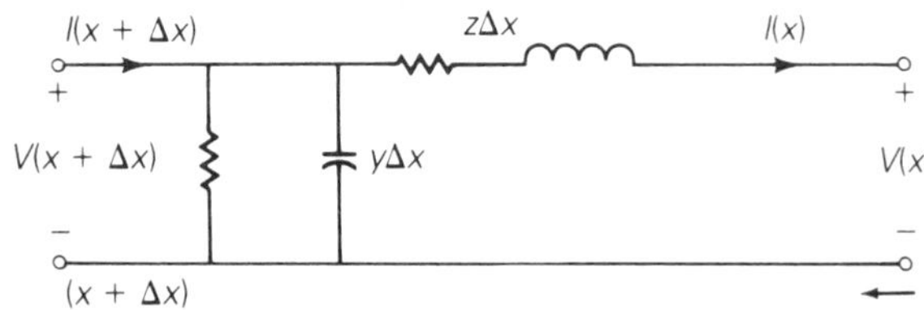

FIGURE 5.6 Transmission-line section of length Δx

and taking the limit as Δx approaches zero,

$$\frac{dV(x)}{dx} = zI(x) \tag{5.2.5}$$

Similarly, writing a KCL equation for the circuit,

$$I(x + \Delta x) = I(x) + (y\Delta x)V(x + \Delta x) \quad \text{A} \tag{5.2.6}$$

Rearranging,

$$\frac{I(x + \Delta x) - I(x)}{\Delta x} = yV(x) \tag{5.2.7}$$

and taking the limit as Δx approaches zero,

$$\frac{dI(x)}{dx} = yV(x) \tag{5.2.8}$$

Equations (5.2.5) and (5.2.8) are two linear, first-order, homogeneous differential equations with two unknowns, $V(x)$ and $I(x)$. Eliminate $I(x)$ by differentiating (5.2.5) and using (5.2.8) as follows:

$$\frac{d^2V(x)}{dx^2} = z\frac{dI(x)}{dx} = zyV(x) \tag{5.2.9}$$

or

$$\frac{d^2V(x)}{dx^2} - zyV(x) = 0 \tag{5.2.10}$$

Equation (5.2.10) is a linear, second-order, homogeneous differential equation with one unknown, $V(x)$. By inspection, its solution is

$$V(x) = A_1e^{\gamma x} + A_2e^{-\gamma x} \quad \text{volts} \tag{5.2.11}$$

where A_1 and A_2 are integration constants and

$$\gamma = \sqrt{zy} \quad \text{m}^{-1} \tag{5.2.12}$$

γ, whose units are m^{-1}, is called the *propagation constant*. By inserting (5.2.11) and (5.2.12) into (5.2.10), the solution to the differential equation can be verified.

Next, using (5.2.11) in (5.2.5),

$$\frac{dV(x)}{dx} = \gamma A_1e^{\gamma x} - \gamma A_2e^{-\gamma x} = zI(x) \tag{5.2.13}$$

Solving for $I(x)$,

$$I(x) = \frac{A_1e^{\gamma x} - A_2e^{-\gamma x}}{z/\gamma} \tag{5.2.14}$$

Using (5.2.12), $z/\gamma = z/\sqrt{zy} = \sqrt{z/y}$, (5.2.14) becomes

$$I(x) = \frac{A_1 e^{\gamma x} - A_2 e^{-\gamma x}}{Z_c} \tag{5.2.15}$$

where

$$Z_c = \sqrt{\frac{z}{y}} \quad \Omega \tag{5.2.16}$$

Z_c, whose units are ohms, is called the *characteristic impedance*.

Next, the integration constants A_1 and A_2 are evaluated from the boundary conditions. At $x = 0$, which is the receiving end of the line, the receiving-end voltage and current are

$$V_R = V(0) \tag{5.2.17}$$

$$I_R = I(0) \tag{5.2.18}$$

Also, at $x = 0$, (5.2.11) and (5.2.15) become

$$V_R = A_1 + A_2 \tag{5.2.19}$$

$$I_R = \frac{A_1 - A_2}{Z_c} \tag{5.2.20}$$

Solving for A_1 and A_2,

$$A_1 = \frac{V_R + Z_c I_R}{2} \tag{5.2.21}$$

$$A_2 = \frac{V_R - Z_c I_R}{2} \tag{5.2.22}$$

Substituting A_1 and A_2 into (5.2.11) and (5.2.15),

$$V(x) = \left(\frac{V_R + Z_c I_R}{2}\right) e^{\gamma x} + \left(\frac{V_R - Z_c I_R}{2}\right) e^{-\gamma x} \tag{5.2.23}$$

$$I(x) = \left(\frac{V_R + Z_c I_R}{2Z_c}\right) e^{\gamma x} - \left(\frac{V_R - Z_c I_R}{2Z_c}\right) e^{-\gamma x} \tag{5.2.24}$$

Rearranging (5.2.23) and (5.2.24),

$$V(x) = \left(\frac{e^{\gamma x} + e^{-\gamma x}}{2}\right) V_R + Z_c \left(\frac{e^{\gamma x} - e^{-\gamma x}}{2}\right) I_R \tag{5.2.25}$$

$$I(x) = \frac{1}{Z_c}\left(\frac{e^{\gamma x} - e^{-\gamma x}}{2}\right) V_R + \left(\frac{e^{\gamma x} + e^{-\gamma x}}{2}\right) I_R \tag{5.2.26}$$

Recognizing the hyperbolic functions cosh and sinh,

$$V(x) = \cosh(\gamma x)V_R + Z_c \sinh(\gamma x)I_R \tag{5.2.27}$$

$$I(x) = \frac{1}{Z_c}\sinh(\gamma x)\, V_R + \cosh(\gamma x)I_R \tag{5.2.28}$$

Equations (5.2.27) and (5.2.28) give the *ABCD* parameters of the distributed line. In matrix format,

$$\begin{bmatrix} V(x) \\ I(x) \end{bmatrix} = \left[\begin{array}{c|c} A(x) & B(x) \\ \hline C(x) & D(x) \end{array}\right] \begin{bmatrix} V_R \\ I_R \end{bmatrix} \tag{5.2.29}$$

where

$$A(x) = D(x) = \cosh(\gamma x) \text{ per unit} \tag{5.2.30}$$

$$B(x) = Z_c \sinh(\gamma x) \quad \Omega \tag{5.2.31}$$

$$C(x) = \frac{1}{Z_c}\sinh(\gamma x) \quad \text{S} \tag{5.2.32}$$

Equation (5.2.29) gives the current and voltage at any point x along the line in terms of the receiving-end voltage and current. At the sending end, where $x = l$, $V(l) = V_S$ and $I(l) = I_S$. That is,

$$\begin{bmatrix} V_S \\ I_S \end{bmatrix} = \left[\begin{array}{c|c} A & B \\ \hline C & D \end{array}\right] \begin{bmatrix} V_R \\ I_R \end{bmatrix} \tag{5.2.33}$$

where

$$A = D = \cosh(\gamma l) \text{ per unit} \tag{5.2.34}$$

$$B = Z_c \sinh(\gamma l) \quad \Omega \tag{5.2.35}$$

$$C = \frac{1}{Z_c}\sinh(\gamma l) \quad \text{S} \tag{5.2.36}$$

Equations (5.2.34) through (5.2.36) give the *ABCD* parameters of the distributed line. In these equations, the propagation constant γ is a complex quantity with real and imaginary parts denoted α and β. That is,

$$\gamma = \alpha + j\beta \quad \text{m}^{-1} \tag{5.2.37}$$

The quantity γl is dimensionless. Also

$$e^{\gamma l} = e^{(\alpha l + j\beta l)} = e^{\alpha l}e^{j\beta l} = e^{\alpha l}\underline{/\beta l} \tag{5.2.38}$$

Using (5.2.38), the hyperbolic functions cosh and sinh can be evaluated as follows:

$$\cosh(\gamma l) = \frac{e^{\gamma l} + e^{-\gamma l}}{2} = \frac{1}{2}\left(e^{\alpha l}\underline{/\beta l} + e^{-\alpha l}\underline{/-\beta l}\right) \tag{5.2.39}$$

Parameter	$A = D$	B	C
Units	per Unit	Ω	S
Short line (less than 25 km)	1	Z	0
Medium line—nominal π circuit (25 to 250 km)	$1 + \frac{YZ}{2}$	Z	$Y\left(1 + \frac{YZ}{4}\right)$
Long line—equivalent π circuit (more than 250 km)	$\cosh(\gamma\ell) = 1 + \frac{Y'Z'}{2}$	$Z_c \sinh(\gamma\ell) = Z'$	$(1/Z_c)\sinh(\gamma\ell) = Y'\left(1 + \frac{Y'Z'}{4}\right)$
Lossless line ($R = G = 0$)	$\cos(\beta\ell)$	$jZ_c \sin(\beta\ell)$	$\frac{j\sin(\beta\ell)}{Z_c}$

TABLE 5.1

Summary: Transmission-line *ABCD* parameters

and

$$\sinh(\gamma l) = \frac{e^{\gamma l} - e^{-\gamma l}}{2} = \frac{1}{2}(e^{\alpha l}\underline{/\beta l} - e^{-\alpha l}\underline{/-\beta l}) \tag{5.2.40}$$

Alternatively, the following identities can be used:

$$\cosh(\alpha l + j\beta l) = \cosh(\alpha l)\cos(\beta l) + j\sinh(\alpha l)\sin(\beta l) \tag{5.2.41}$$

$$\sinh(\alpha l + j\beta l) = \sinh(\alpha l)\cos(\beta l) + j\cosh(\alpha l)\sin(\beta l) \tag{5.2.42}$$

Note that in (5.2.39) through (5.2.42), the dimensionless quantity βl is in radians, not degrees.

The *ABCD* parameters given by (5.2.34) through (5.2.36) are exact parameters valid for any line length. For accurate calculations, these equations must be used for overhead 60-Hz lines longer than 250 km. The *ABCD* parameters derived in Section 5.1 are approximate parameters that are more conveniently used for hand calculations involving short- and medium-length lines. Table 5.1 summarizes the *ABCD* parameters for short, medium, long, and lossless (see Section 5.4) lines.

EXAMPLE 5.2

Exact *ABCD* parameters: long line

A three-phase 765-kV, 60-Hz, 300-km, completely transposed line has the following positive-sequence impedance and admittance:

$$z = 0.0165 + j0.3306 = 0.3310\underline{/87.14^\circ} \quad \Omega/\text{km}$$

$$y = j4.674 \times 10^{-6} \text{ S/km}$$

(*Continued*)

Assuming positive-sequence operation, calculate the exact $ABCD$ parameters of the line. Compare the exact B parameter with that of the nominal π circuit.

SOLUTION

From (5.2.12) and (5.2.16),

$$Z_c = \sqrt{\frac{0.3310\underline{/87.14^\circ}}{4.674 \times 10^{-6}\underline{/90^\circ}}} = \sqrt{7.082 \times 10^4\underline{/-2.86^\circ}}$$

$$= 266.1\underline{/-1.43^\circ} \quad \Omega$$

and

$$\gamma l = \sqrt{(0.3310\underline{/87.14^\circ})(4.674 \times 10^{-6}\underline{/90^\circ})} \times (300)$$

$$= \sqrt{1.547 \times 10^{-6}\underline{/177.14^\circ}} \times (300)$$

$$= 0.3731\underline{/88.57^\circ} = 0.00931 + j0.3730 \quad \text{per unit}$$

From (5.2.38),

$$e^{\gamma l} = e^{0.00931}e^{+j0.3730} = 1.0094\underline{/0.3730} \quad \text{radians}$$

$$= 0.9400 + j0.3678$$

and

$$e^{-\gamma l} = e^{-0.00931}e^{-j0.3730} = 0.9907\underline{/-0.3730} \quad \text{radians}$$

$$= 0.9226 - j0.3610$$

Then, from (5.2.39) and (5.2.40),

$$\cosh(\gamma l) = \frac{(0.9400 + j0.3678) + (0.9226 - j0.3610)}{2}$$

$$= 0.9313 + j0.0034 = 0.9313\underline{/0.209^\circ}$$

$$\sinh(\gamma l) = \frac{(0.9400 + j0.3678) - (0.9226 - j0.3610)}{2}$$

$$= 0.0087 + j0.3644 = 0.3645\underline{/88.63^\circ}$$

Finally, from (5.2.34) through (5.2.36),

$$A = D = \cosh(\gamma l) = 0.9313\underline{/0.209^\circ} \quad \text{per unit}$$

$$B = (266.1\underline{/-1.43^\circ})(0.3645\underline{/88.63^\circ}) = 97.0\underline{/87.2^\circ} \quad \Omega$$

$$C = \frac{0.3645\underline{/88.63^\circ}}{266.1\underline{/-1.43^\circ}} = 1.37 \times 10^{-3}\underline{/90.06^\circ} \quad \text{S}$$

Using (5.1.16), the B parameter for the nominal π circuit is

$$B_{\text{nominal } \pi} = Z = (0.3310\underline{/87.14°})(300) = 99.3\underline{/87.14°} \quad \Omega$$

which is 2% larger than the exact value.

5.3 EQUIVALENT π CIRCUIT

Many computer programs used in power system analysis and design assume circuit representations of components such as transmission lines and transformers. It is therefore convenient to represent the terminal characteristics of a transmission line by an equivalent circuit instead of its $ABCD$ parameters.

The circuit shown in Figure 5.7 is called an *equivalent π circuit*. It is identical in structure to the nominal π circuit of Figure 5.3, except that Z' and Y' are used instead of Z and Y. The objective is to determine Z' and Y' such that the equivalent π circuit has the same $ABCD$ parameters as those of the distributed line, (5.2.34) through (5.2.36). The $ABCD$ parameters of the equivalent π circuit, which has the same structure as the nominal π, are

$$A = D = 1 + \frac{Y'Z'}{2} \quad \text{per unit} \tag{5.3.1}$$

$$B = Z' \quad \Omega \tag{5.3.2}$$

$$C = Y'\left(1 + \frac{Y'Z'}{4}\right) \quad \text{S} \tag{5.3.3}$$

where Z and Y in (5.1.15) through (5.1.17) have been replaced with Z' and Y' in (5.3.1) through (5.3.3). Equating (5.3.2) to (5.2.35),

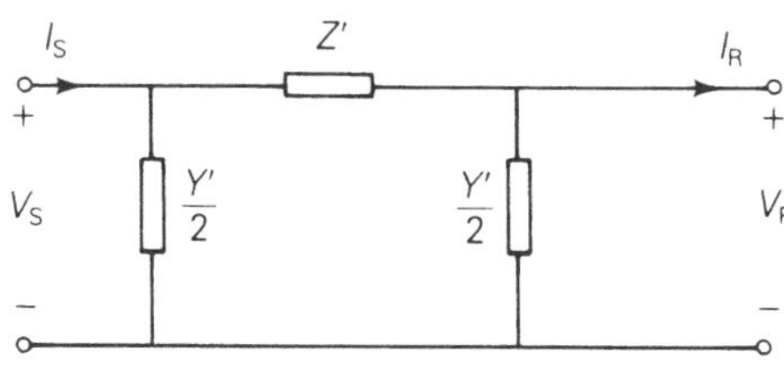

$$Z' = Z_c \sinh(\gamma\ell) = ZF_1 = Z\frac{\sinh(\gamma\ell)}{\gamma\ell}$$

$$\frac{Y'}{2} = \frac{\tanh(\gamma\ell/2)}{Z_c} = \frac{Y}{2}F_2 = \frac{Y}{2}\frac{\tanh(\gamma\ell/2)}{(\gamma\ell/2)}$$

FIGURE 5.7 Transmission-line equivalent π circuit

$$Z' = Z_c \sinh(\gamma l) = \sqrt{\frac{z}{y}} \sinh(\gamma l) \tag{5.3.4}$$

Rewriting (5.3.4) in terms of the nominal π circuit impedance $Z = zl$,

$$\begin{aligned} Z' &= zl\left[\sqrt{\frac{z}{y}}\frac{\sinh(\gamma l)}{zl}\right] = zl\left[\frac{\sinh(\gamma l)}{\sqrt{zy}\,l}\right] \\ &= ZF_1 \quad \Omega \end{aligned} \tag{5.3.5}$$

where

$$F_1 = \frac{\sinh(\gamma l)}{\gamma l} \quad \text{per unit} \tag{5.3.6}$$

Similarly, equating (5.3.1) to (5.2.34),

$$\begin{aligned} 1 + \frac{Y'Z'}{2} &= \cosh(\gamma l) \\ \frac{Y'}{2} &= \frac{\cosh(\gamma l) - 1}{Z'} \end{aligned} \tag{5.3.7}$$

Using (5.3.4) and the identity $\tanh\left(\frac{\gamma l}{2}\right) = \frac{\cosh(\gamma l) - 1}{\sinh(\gamma l)}$, (5.3.7) becomes

$$\frac{Y'}{2} = \frac{\cosh(\gamma l) - 1}{Z_c \sinh(\gamma l)} = \frac{\tanh(\gamma l/2)}{Z_c} = \frac{\tanh(\gamma l/2)}{\sqrt{\frac{z}{y}}} \tag{5.3.8}$$

Rewriting (5.3.8) in terms of the nominal π circuit admittance $Y = yl$,

$$\begin{aligned} \frac{Y'}{2} &= \frac{yl}{2}\left[\frac{\tanh(\gamma l/2)}{\sqrt{\frac{z}{y}}\frac{yl}{2}}\right] = \frac{yl}{2}\left[\frac{\tanh(\gamma l/2)}{\sqrt{zy}\,l/2}\right] \\ &= \frac{Y}{2}F_2 \quad \text{S} \end{aligned} \tag{5.3.9}$$

where

$$F_2 = \frac{\tanh(\gamma l/2)}{\gamma l/2} \quad \text{per unit} \tag{5.3.10}$$

Equations (5.3.6) and (5.3.10) give the correction factors F_1 and F_2 to convert Z and Y for the nominal π circuit to Z' and Y' for the equivalent π circuit.

EXAMPLE 5.3

Equivalent π circuit: long line

Compare the equivalent and nominal π circuits for the line in Example 5.2.

SOLUTION

For the nominal π circuit,

$$Z = zl = (0.3310\underline{/87.14^\circ})(300) = 99.3\underline{/87.14^\circ} \quad \Omega$$

$$\frac{Y}{2} = \frac{yl}{2} = \left(\frac{j4.674 \times 10^{-6}}{2}\right)(300) = 7.011 \times 10^{-4}\underline{/90^\circ} \quad \text{S}$$

From (5.3.6) and (5.3.10), the correction factors are

$$F_1 = \frac{0.3645\underline{/88.63^\circ}}{0.3731\underline{/88.57^\circ}} = 0.9769\underline{/0.06^\circ} \quad \text{per unit}$$

$$F_2 = \frac{\tanh(\gamma l/2)}{\gamma l/2} = \frac{\cosh(\gamma l) - 1}{(\gamma l/2)\sinh(\gamma l)}$$

$$= \frac{0.9313 + j0.0034 - 1}{\left(\frac{0.3731}{2}\underline{/88.57^\circ}\right)(0.3645\underline{/88.63^\circ})}$$

$$= \frac{-0.0687 + j0.0034}{0.06800\underline{/177.20^\circ}}$$

$$= \frac{0.06878\underline{/177.17^\circ}}{0.06800\underline{/177.20^\circ}} = 1.012\underline{/-0.03^\circ} \quad \text{per unit}$$

Then, from (5.3.5) and (5.3.9), for the equivalent π circuit,

$$Z' = (99.3\underline{/87.14^\circ})(0.9769\underline{/0.06^\circ}) = 97.0\underline{/87.2^\circ} \quad \Omega$$

$$\frac{Y'}{2} = (7.011 \times 10^{-4}\underline{/90^\circ})(1.012\underline{/-0.03^\circ}) = 7.095 \times 10^{-4}\underline{/89.97^\circ} \quad \text{S}$$

$$= 3.7 \times 10^{-7} + j7.095 \times 10^{-4} \quad \text{S}$$

Comparing these nominal and equivalent π circuit values, Z' is about 2% smaller than Z, and $Y'/2$ is about 1% larger than $Y/2$. Although the circuit values are approximately the same for this line, the equivalent π circuit should be used for accurate calculations involving long lines. Note the small shunt conductance, $G' = 3.7 \times 10^{-7}$ S, introduced in the equivalent π circuit. G' is often neglected.

5.4 LOSSLESS LINES

This section discusses the following concepts for lossless lines: surge impedance, *ABCD* parameters, equivalent π circuit, wavelength, surge impedance loading, voltage profiles, and steady-state stability limit.

When line losses are neglected, simpler expressions for the line parameters are obtained and the above concepts are more easily understood. Since transmission and distribution lines for power transfer generally are designed to have low losses, the equations and concepts developed here can be used for quick and reasonably accurate hand calculations leading to seat-of-the-pants analyses and to initial designs. More accurate calculations then can be made with computer programs for follow-up analysis and design.

SURGE IMPEDANCE

For a lossless line, R = G = 0, and

$$z = j\omega \mathrm{L} \quad \Omega/\mathrm{m} \tag{5.4.1}$$
$$y = j\omega \mathrm{C} \quad \mathrm{S/m} \tag{5.4.2}$$

From (5.2.12) and (5.2.16),

$$Z_c = \sqrt{\frac{z}{y}} = \sqrt{\frac{j\omega \mathrm{L}}{j\omega \mathrm{C}}} = \sqrt{\frac{\mathrm{L}}{\mathrm{C}}} \quad \Omega \tag{5.4.3}$$

and

$$\gamma = \sqrt{zy} = \sqrt{(j\omega \mathrm{L})(j\omega \mathrm{C})} = j\omega\sqrt{\mathrm{LC}} = j\beta \quad \mathrm{m}^{-1} \tag{5.4.4}$$

where

$$\beta = \omega\sqrt{\mathrm{LC}} \quad \mathrm{m}^{-1} \tag{5.4.5}$$

The characteristic impedance $Z_c = \sqrt{\mathrm{L}/\mathrm{C}}$, commonly called the *surge impedance* for a lossless line, is pure real—that is, resistive. The propagation constant $\gamma = j\beta$ is pure imaginary.

ABCD PARAMETERS

The *ABCD* parameters are, from (5.2.30) through (5.2.32),

$$\mathrm{A}(x) = \mathrm{D}(x) = \cosh(\gamma x) = \cosh(j\beta x)$$
$$= \frac{e^{j\beta x} + e^{-j\beta x}}{2} = \cos(\beta x) \quad \text{per unit} \tag{5.4.6}$$

$$\sinh(\gamma x) = \sinh(j\beta x) = \frac{e^{j\beta x} - e^{-j\beta x}}{2} = j\sin(\beta x) \quad \text{per unit} \tag{5.4.7}$$

$$B(x) = Z_c \sinh(\gamma x) = jZ_c \sin(\beta x) = j\sqrt{\frac{\mathrm{L}}{\mathrm{C}}}\sin(\beta x) \quad \Omega \tag{5.4.8}$$

$$C(x) = \frac{\sinh(\gamma x)}{Z_c} = \frac{j\sin(\beta x)}{\sqrt{\frac{\mathrm{L}}{\mathrm{C}}}} \quad \mathrm{S} \tag{5.4.9}$$

EQUIVALENT π CIRCUIT

For the equivalent π circuit, using (5.3.4).

$$Z' = jZ_c \sin(\beta l) = j\mathrm{X}' \quad \Omega \tag{5.4.10}$$

or, from (5.3.5) and (5.3.6),

$$Z' = (j\omega \mathrm{L}l)\left(\frac{\sin(\beta l)}{\beta l}\right) = j\mathrm{X}' \quad \Omega \tag{5.4.11}$$

Also, from (5.3.9) and (5.3.10),

$$\begin{aligned}\frac{Y'}{2} &= \frac{Y}{2}\frac{\tanh(j\beta l/2)}{j\beta l/2} = \frac{Y}{2}\frac{\sinh(j\beta l/2)}{(j\beta l/2)\cosh(j\beta l/2)}\\ &= \left(\frac{j\omega Cl}{2}\right)\frac{j\sin(\beta l/2)}{(j\beta l/2)\cos(\beta l/2)} = \left(\frac{j\omega Cl}{2}\right)\frac{\tan(\beta l/2)}{\beta l/2}\\ &= \left(\frac{j\omega C'l}{2}\right) \quad \mathrm{S}\end{aligned} \tag{5.4.12}$$

Z' and Y' are both pure imaginary. Also, for βl less than π radians, Z' is pure inductive and Y' is pure capacitive. Thus the equivalent π circuit for a lossless line, shown in Figure 5.8, is also lossless.

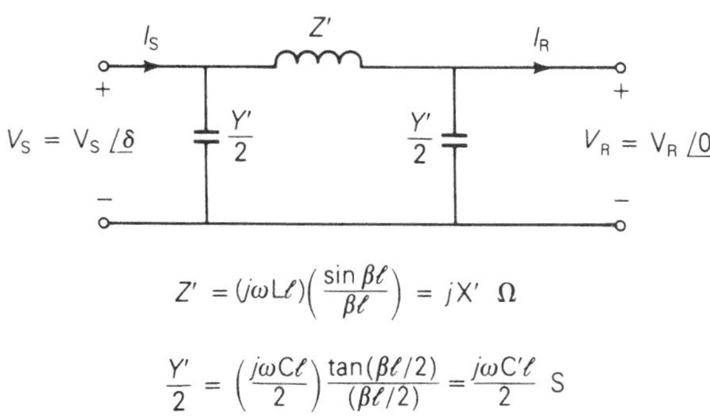

FIGURE 5.8 Equivalent π circuit for a lossless line ($\beta\ell$ less than π)

WAVELENGTH

A *wavelength* is the distance required to change the phase of the voltage or current by 2π radians or 360°. For a lossless line, using (5.2.29),

$$\begin{aligned} V(x) &= A(x)V_R + B(x)I_R \\ &= \cos(\beta x)\,V_R + jZ_c \sin(\beta x)I_R \end{aligned} \tag{5.4.13}$$

and

$$\begin{aligned} I(x) &= C(x)V_R + D(x)I_R \\ &= \frac{j\sin(\beta x)}{Z}V_R + \cos(\beta x)I_R \end{aligned} \tag{5.4.14}$$

From (5.4.13) and (5.4.14), $V(x)$ and $I(x)$ change phase by 2π radians when $x = 2\pi/\beta$. Denoting wavelength by λ, and using (5.4.5),

$$\lambda = \frac{2\pi}{\beta} = \frac{2\pi}{\omega\sqrt{LC}} = \frac{1}{f\sqrt{LC}} \quad \text{m} \tag{5.4.15}$$

or

$$f\lambda = \frac{1}{\sqrt{LC}} \tag{5.4.16}$$

The term $(1/\sqrt{LC})$ in (5.4.16) is the velocity of propagation of voltage and current waves along a lossless line, as shown in Chapter 13. For overhead lines, $(1/\sqrt{LC}) \approx 3 \times 10^8$ m/s, and for $f = 60$ Hz, (5.4.14) gives

$$\lambda \approx \frac{3 \times 10^8}{60} = 5 \times 10^6\,\text{m} = 5000\ \text{km}$$

Typical power-line lengths are only a small fraction of the above 60-Hz wavelength.

SURGE IMPEDANCE LOADING

Surge impedance loading (SIL) is the power delivered by a lossless line to a load resistance equal to the surge impedance $Z_c = \sqrt{L/C}$. Figure 5.9 shows a lossless line terminated by a resistance equal to its surge impedance. This line represents either

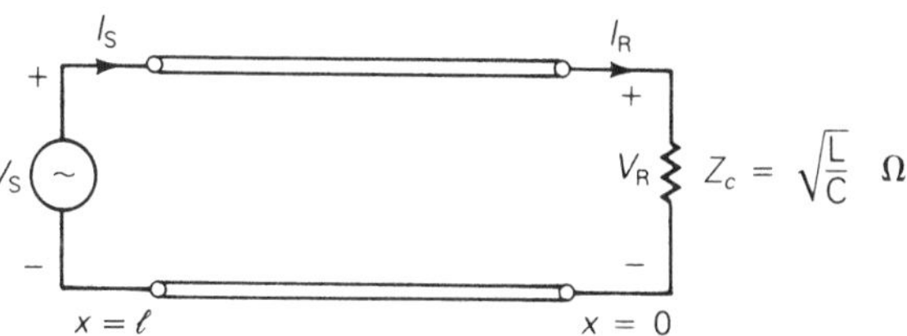

FIGURE 5.9 Lossless line terminated by its surge impedance

a single-phase line or one phase-to-neutral of a balanced three-phase line. At SIL, from (5.4.13),

$$\begin{aligned} V(x) &= \cos(\beta x)\, V_{\mathrm{R}} + jZ_c \sin(\beta x) I_{\mathrm{R}} \\ &= \cos(\beta x) V_{\mathrm{R}} + jZ_c \sin(\beta x)\left(\frac{V_{\mathrm{R}}}{Z_c}\right) \\ &= (\cos \beta x + j \sin \beta x)\, V_{\mathrm{R}} \\ &= e^{j\beta x} V_{\mathrm{R}} \quad \text{volts} \end{aligned} \tag{5.4.17}$$

$$|V(x)| = |V_{\mathrm{R}}| \quad \text{volts} \tag{5.4.18}$$

Thus, at SIL, the voltage profile is flat. That is, the voltage magnitude at any point x along a lossless line at SIL is constant. Also from (5.4.14) at SIL,

$$\begin{aligned} I(x) &= \frac{j \sin(\beta x)}{Z_c} V_{\mathrm{R}} + (\cos \beta x)\frac{V_{\mathrm{R}}}{Z_c} \\ &= (\cos \beta x + j \sin \beta x)\frac{V_{\mathrm{R}}}{Z_c} \\ &= (e^{j\beta x})\frac{V_{\mathrm{R}}}{Z_c} \quad \text{A} \end{aligned} \tag{5.4.19}$$

Using (5.4.17) and (5.4.19), the complex power flowing at any point x along the line is

$$\begin{aligned} S(x) &= \mathrm{P}(x) + j\mathrm{Q}(x) = V(x)I^*(x) \\ &= (e^{j\beta x} V_{\mathrm{R}})\left(\frac{e^{j\beta x} V_{\mathrm{R}}}{Z_c}\right)^* \\ &= \frac{|V_{\mathrm{R}}|^2}{Z_c} \end{aligned} \tag{5.4.20}$$

Thus the real power flow along a lossless line at SIL remains constant from the sending end to the receiving end. The reactive power flow is zero.

At rated line voltage, the real power delivered, or SIL, from (5.4.20), is

$$\mathrm{SIL} = \frac{\mathrm{V}_{\mathrm{rated}}^2}{Z_c} \tag{5.4.21}$$

where rated voltage is used for a single-phase line and rated line-to-line voltage is used for the total real power delivered by a three-phase line. Table 5.2 lists surge impedance and SIL values for typical overhead 60-Hz three-phase lines.

VOLTAGE PROFILES

In practice, power lines are not terminated by their surge impedance. Instead, loadings can vary from a small fraction of SIL during light load conditions up to multiples of SIL, depending on line length and line compensation, during heavy load conditions. If a line is not terminated by its surge impedance, then the voltage profile is not flat.

V_{rated} (kV)	$Z_C = \sqrt{L/C}$ (Ω)	$SIL = V_{rated}^2/Z_C$ (MW)
69	366–400	12–13
138	366–405	47–52
230	365–395	134–145
345	280–366	325–425
500	233–294	850–1075
765	254–266	2200–2300

TABLE 5.2

Surge impedance and SIL values for typical 60-Hz overhead lines [1, 2]
(Source: Electric Power Research Institute (EPRI), EPRI AC Transmission Line Reference Book—200 kV and Above (Palo Alto, CA: EPRI, *www.epri.com.* December 2005); Westinghouse Electric Corporation, *Electrical Transmission and Distribution Reference Book*, 4th ed. (East Pittsburgh, PA, 1964).)

Figure 5.10 shows voltage profiles of lines with a fixed sending-end voltage magnitude V_S for line lengths l up to a quarter wavelength. This figure shows four loading conditions: (1) no-load. (2) SIL, (3) short circuit, and (4) full load, which are described as follows:

1. At no-load, $I_{RNL} = 0$ and (5.4.13) yields

$$V_{NL}(x) = (\cos \beta x)V_{RNL} \tag{5.4.22}$$

   The no-load voltage increases from $V_S = (\cos \beta l)V_{RNL}$ at the sending end to V_{RNL} at the receiving end (where $x = 0$).

2. From (5.4.18), the voltage profile at SIL is flat.
3. For a short circuit at the load, $V_{RSC} = 0$ and (5.4.13) yields

$$V_{SC}(x) = (Z_c \sin \beta x)I_{RSC} \tag{5.4.23}$$

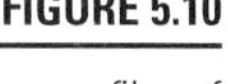

FIGURE 5.10

Voltage profiles of an uncompensated lossless line with fixed sending-end voltage for line lengths up to a quarter wavelength

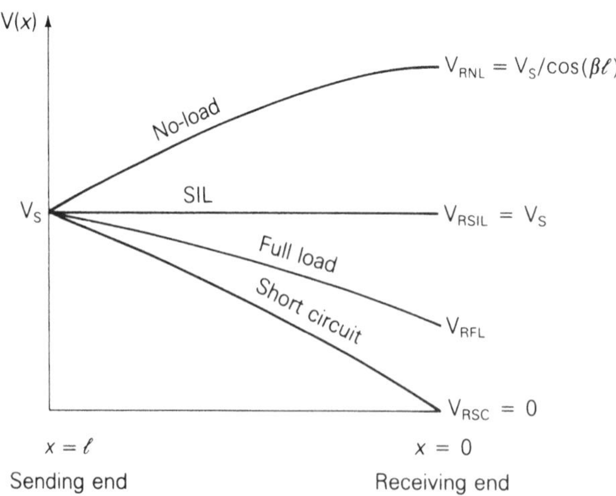

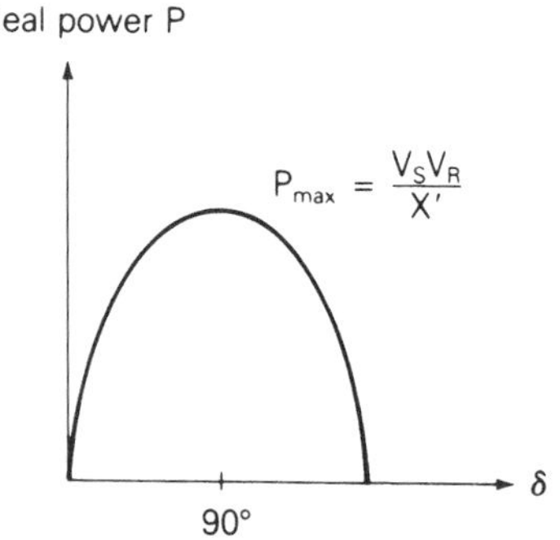

FIGURE 5.11

Real power delivered by a lossless line versus voltage angle across the line

The voltage decreases from $V_S = (\sin \beta l)(Z_c I_{RSC})$ at the sending end to $V_{RSC} = 0$ at the receiving end.

4. The full-load voltage profile, which depends on the specification of full-load current, lies above the short-circuit voltage profile.

Figure 5.10 summarizes these results, showing a high receiving-end voltage at no-load and a low receiving-end voltage at full load. This voltage regulation problem becomes more severe as the line length increases. In Section 5.6, shunt compensation methods to reduce voltage fluctuations are presented.

STEADY-STATE STABILITY LIMIT

The equivalent π circuit of Figure 5.8 can be used to obtain an equation for the real power delivered by a lossless line. Assume that the voltage magnitudes V_S and V_R at the ends of the line are held constant. Also, let δ denote the voltage-phase angle at the sending end with respect to the receiving end. From KVL, the receiving-end current I_R is

$$\begin{aligned} I_R &= \frac{V_S - V_R}{Z'} - \frac{Y'}{2} V_R \\ &= \frac{V_S e^{j\delta} - V_R}{jX'} - \frac{j\omega C' l}{2} V_R \end{aligned} \tag{5.4.24}$$

and the complex power S_R delivered to the receiving end is

$$\begin{aligned} S_R &= V_R I_R^* = V_R \left(\frac{V_S e^{j\delta} - V_R}{jX'} \right)^* + \frac{j\omega C' l}{2} V_R^2 \\ &= V_R \left(\frac{V_S e^{-j\delta} - V_R}{-jX'} \right) + \frac{j\omega C l}{2} V_R^2 \\ &= \frac{jV_R V_S \cos\delta + V_R V_S \sin\delta - jV_R^2}{X'} + \frac{j\omega C l}{2} V_R^2 \end{aligned} \tag{5.4.25}$$

FIGURE 5.12 Transmission-line loadability curve for 60-Hz overhead lines—no series or shunt compensation

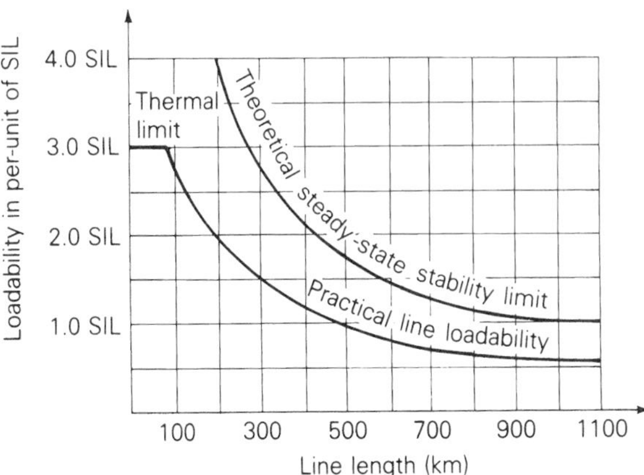

The real power delivered is

$$P = P_S = P_R = \text{Re}(S_R) = \frac{V_R V_S}{X'} \sin\delta \quad \text{W} \tag{5.4.26}$$

Note that since the line is lossless, $P_S = P_R$.

Equation (5.4.26) is plotted in Figure 5.11. For fixed voltage magnitudes V_S and V_R, the phase angle δ increases from 0 to 90° as the real power delivered increases. The maximum power that the line can deliver, which occurs when $\delta = 90°$, is given by

$$P_{max} = \frac{V_S V_R}{X'} \quad \text{W} \tag{5.4.27}$$

P_{max} represents the theoretical *steady-state stability* limit of a lossless line. If an attempt were made to exceed this steady-state stability limit, then synchronous machines at the sending end would lose synchronism with those at the receiving end. Stability is further discussed in Chapter 11.

It is convenient to express the steady-state stability limit in terms of SIL. Using (5.4.10) in (5.4.26),

$$P = \frac{V_S V_R \sin\delta}{Z_c \sin\beta l} = \left(\frac{V_S V_R}{Z_c}\right)\frac{\sin\delta}{\sin\left(\dfrac{2\pi l}{\lambda}\right)} \tag{5.4.28}$$

Expressing V_S and V_R in per-unit of rated line voltage,

$$P = \left(\frac{V_S}{V_{\text{rated}}}\right)\left(\frac{V_R}{V_{\text{rated}}}\right)\left(\frac{V_{\text{rated}}^2}{Z_c}\right)\frac{\sin\delta}{\sin\left(\dfrac{2\pi l}{\lambda}\right)}$$

$$= V_{S\text{p.u.}} V_{R\text{p.u.}} (\text{SIL}) \frac{\sin\delta}{\sin\left(\dfrac{2\pi l}{\lambda}\right)} \quad \text{W} \tag{5.4.29}$$

And for $\delta = 90°$, the theoretical steady-state stability limit is

$$P_{max} = \frac{V_{Sp.u.} V_{Rp.u.} (SIL)}{\sin\left(\frac{2\pi l}{\lambda}\right)} \quad W \tag{5.4.30}$$

Equations (5.4.27) through (5.4.30) reveal two important factors affecting the steady-state stability limit. First, from (5.4.27), it increases with the square of the line voltage. For example, a doubling of line voltage enables a fourfold increase in maximum power flow. Second, it decreases with line length. Equation (5.4.30) is plotted in Figure 5.12 for $V_{Sp.u.} = V_{Rp.u.} = 1$, $\lambda = 5000$ km, and line lengths up to 1100 km. As shown, the theoretical steady-state stability limit decreases from 4(SIL) for a 200 km line to about 2(SIL) for a 400 km line.

EXAMPLE 5.4

Theoretical steady-state stability limit: long line

Neglecting line losses, find the theoretical steady-state stability limit for the 300-km line in Example 5.2. Assume a 266.1-Ω surge impedance, a 5000 km wavelength, and $V_S = V_R = 765$ kV.

SOLUTION

From (5.4.21),

$$SIL = \frac{(765)^2}{266.1} = 2199 \quad MW$$

From (5.4.30) with $l = 300$ km and $\lambda = 5000$ km,

$$P_{max} = \frac{(1)(1)(2199)}{\sin\left(\frac{2\pi \times 300}{5000}\right)} = (2.716)(2199) = 5974 \quad MW$$

Alternatively, from Figure 5.12, for a 300 km line, the theoretical steady-state stability limit is (2.72)SIL = (2.72)(2199) = 5980 MW, which is about the same as the previous result (see Figure 5.13).

Open PowerWorld Simulator case Example 5_4 and select **Tools**, **Play** to see this example. When the load on a line is equal to the SIL, the voltage profile across the line is flat, and the line's net reactive power losses are zero. For loads above the SIL, the line consumes reactive power, and the load's voltage magnitude is below the sending-end value. Conversely, for loads below the SIL, the line actually generates reactive power, and the load's voltage magnitude is above the

(Continued)

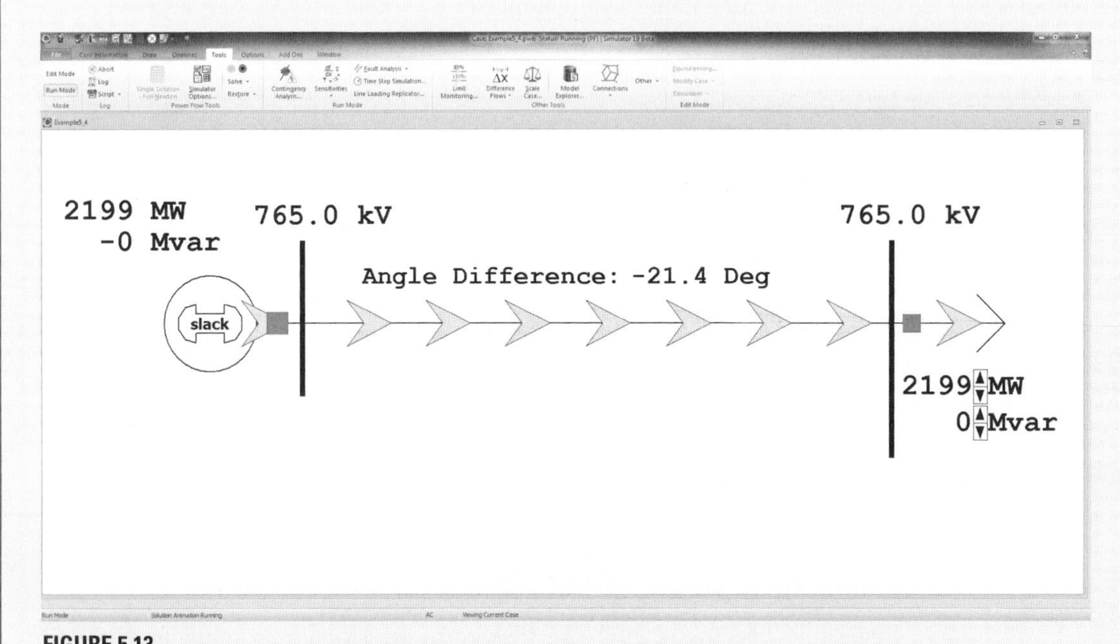

FIGURE 5.13

Screen for Example 5.4

sending-end value. Use the load arrow button to vary the load to see the changes in the receiving-end voltage and the line's reactive power consumption (indicated by the reactive power supplied by the generator).

5.5 MAXIMUM POWER FLOW

Maximum power flow, discussed in Section 5.4 for lossless lines, is derived here in terms of the *ABCD* parameters for lossy lines. The following notation is used:

$$A = \cosh(\gamma\ell) = \text{A}\underline{/\theta_\text{A}}$$
$$B = Z' = Z'\underline{/\theta_Z}$$
$$V_\text{S} = \text{V}_\text{S}\underline{/\delta} \quad V_\text{R} = \text{V}_\text{R}\underline{/0^\circ}$$

Solving (5.2.33) for the receiving-end current,

$$I_\text{R} = \frac{V_\text{S} - AV_\text{R}}{B} = \frac{\text{V}_\text{S}e^{j\delta} - \text{AV}_\text{R}e^{j\theta_\text{A}}}{\text{Z}'e^{j\theta_z}} \tag{5.5.1}$$

The complex power delivered to the receiving end is

$$S_R = P_R + jQ_R = V_R I_R^* = V_R \left[\frac{V_S e^{j(\delta-\theta_Z)} - AV_R e^{j(\theta_A-\theta_Z)}}{Z'} \right]^*$$

$$= \frac{V_R V_S}{Z'} e^{j(\theta_Z-\delta)} - \frac{AV_R^2}{Z'} e^{j(\theta_Z-\theta_A)} \tag{5.5.2}$$

The real and reactive power delivered to the receiving end are thus

$$P_R = \text{Re}(S_R) = \frac{V_R V_S}{Z'} \cos(\theta_Z - \delta) - \frac{AV_R^2}{Z'} \cos(\theta_Z - \theta_A) \tag{5.5.3}$$

$$Q_R = \text{Im}(S_R) = \frac{V_R V_S}{Z'} \sin(\theta_Z - \delta) - \frac{AV_R^2}{Z'} \sin(\theta_Z - \theta_A) \tag{5.5.4}$$

Note that for a lossless line, $\theta_A = 0°$, $B = Z' = jX'$, $Z' = X'$, $\theta_Z = 90°$, and (5.5.3) reduces to

$$P_R = \frac{V_R V_S}{X'} \cos(90 - \delta) - \frac{AV_R^2}{X'} \cos(90°)$$

$$= \frac{V_R V_S}{X'} \sin \delta \tag{5.5.5}$$

which is the same as (5.4.26).

The theoretical maximum real power delivered (or steady-state stability limit) occurs when $\delta = \theta z$ in (5.5.3):

$$P_{R\max} = \frac{V_R V_S}{Z'} - \frac{AV_R^2}{Z'} \cos(\theta_Z - \theta_A) \tag{5.5.6}$$

The second term in (5.5.6), and the fact that Z' is larger than X', reduce $P_{R\max}$ to a value somewhat less than that given by (5.4.27) for a lossless line.

EXAMPLE 5.5

Theoretical maximum power delivered: long line

Determine the theoretical maximum power, in MW and in per-unit of SIL, that the line in Example 5.2 can deliver. Assume $V_S = V_R = 765$ kV.

SOLUTION

From Example 5.2,

$A = 0.9313$ per unit; $\theta_A = 0.209°$

$B = Z' = 97.0\ \Omega$ $\theta_Z = 87.2°$

$Z_c = 266.1\ \Omega$

(Continued)

From (5.5.6) with $V_S = V_R = 765$ kV,

$$P_{Rmax} = \frac{(765)^2}{97} - \frac{(0.9313)(765)^2}{97}\cos(87.2° - 0.209°)$$

From (5.4.20),

$$SIL = \frac{(765)^2}{266.1} = 2199 \quad \text{MW}$$

Thus

$$P_{Rmax} = \frac{5738}{2199} = 2.61 \quad \text{per unit}$$

This value is about 4% less than that found in Example 5.4, where losses were neglected.

5.6 LINE LOADABILITY

In practice, power lines are not operated to deliver their theoretical maximum power, which is based on rated terminal voltages and an angular displacement $\delta = 90°$ across the line. Figure 5.12 shows a practical line loadability curve plotted below the theoretical steady-state stability limit. This curve is based on the voltage-drop limit $V_R/V_S \geq 0.95$ and on a maximum angular displacement of 30 to 35° across the line (or about 45° across the line and equivalent system reactances) in order to maintain stability during transient disturbances [1, 3]. The curve is valid for typical overhead 60-Hz lines with no compensation. Note that for short lines less than 25 km long, loadability is limited by the thermal rating of the conductors or by terminal equipment ratings, not by voltage drop or stability considerations. Section 5.7 investigates series and shunt compensation techniques to increase the loadability of longer lines toward their thermal limit.

EXAMPLE 5.6

Practical line loadability and percent voltage regulation: long line

The 300-km uncompensated line in Example 5.2 has four 1,272,000 cmil (644.5 mm^2) 54/3 ACSR conductors per bundle. The sending-end voltage is held constant at 1.0 per-unit of rated line voltage. Determine the following:

a. The practical line loadability. (Assume an approximate receiving-end voltage $V_R = 0.95$ per unit and $\delta = 35°$ maximum angle across the line.)

b. The full-load current at 0.986 p.f. leading based on the above practical line loadability
c. The exact receiving-end voltage for the full-load current found in part (b)
d. Percent voltage regulation for the above full-load current
e. Thermal limit of the line, based on the approximate current-carrying capacity given in Table A.4

SOLUTION

a. From (5.5.3), with $V_S = 765$, $V_R = 0.95 \times 765$ kV, and $\delta = 35°$, using the values of Z', θ_z, A, and θ_A from Example 5.5,

$$P_R = \frac{(765)(0.95 \times 765)}{97.0}\cos(87.2° - 35°) - \frac{(0.9313)(0.95 \times 765)^2}{97.0}\cos(87.2° - 0.209°)$$

$$= 3513 - 266 = 3247 \quad \text{MW}$$

$P_R = 3247$ MW is the practical line loadability, provided the thermal and voltage-drop limits are not exceeded. Alternatively, from Figure 5.12 for a 300 km line, the practical line loadability is (1.49)SIL= (1.49)(2199) = 3277 MW, which is about the same as the previous result.

b. For the loading at 0.986 p.f. leading and at 0.95 × 765 kV, the full-load receiving-end current is

$$I_{RFL} = \frac{P_R}{\sqrt{3}V_R(\text{p.f.})} = \frac{3247}{(\sqrt{3})(0.95 \times 765)(0.986)} = 2.616 \quad \text{kA}$$

c. From (5.1.1) with $I_{RFL} = 2.616\underline{/\cos^{-1} 0.986} = 2.616\underline{/9.599°}$ kA, using the A and B parameters from Example 5.2,

$$V_S = AV_{RFL} + BI_{RFL}$$

$$\frac{765}{\sqrt{3}}\underline{/\delta} = (0.9313\underline{/0.209°})\,(V_{RFL}\underline{/0°}) + (97.0\underline{/87.2°})\,(2.616\underline{/9.599°})$$

$$441.7\underline{/\delta} = (0.9313\,V_{RFL} - 30.04) + j(0.0034V_{RFL} + 251.97)$$

Taking the squared magnitude of the above equation,

$$(441.7)^2 = 0.8673V_{RFL}^2 - 54.24V_{RFL} + 64{,}391$$

Solving,

$$V_{RFL} = 420.7\ \text{kV}_{LN}$$

$$= 420.7\sqrt{3} = 728.7\ \text{kV}_{LL} = 0.953 \text{ per unit}$$

(Continued)

d. From (5.1.19), the receiving-end no-load voltage is

$$V_{RNL} = \frac{V_S}{A} = \frac{765}{0.9313} = 821.4 \quad kV_{LL}$$

And from (5.1.18),

$$\text{percent VR} = \frac{821.4 - 728.7}{728.7} \times 100 = 12.72\%$$

e. From Table A.4, the approximate current-carrying capacity of four 1,272,000-cmil 54/3 ACSR conductors is $4 \times 1.2 = 4.8$ kA.

Since the voltages $V_S = 1.0$ and $V_{RFL} = 0.953$ per unit satisfy the voltage-drop limit $V_R/V_S \geq 0.95$, the factor that limits line loadability is steady-state stability for this 300-km uncompensated line. The full-load current of 2.616 kA corresponding to loadability is also well below the thermal limit of 4.8 kA. The 12.7% voltage regulation is too high because the no-load voltage is too high. Compensation techniques to reduce no-load voltages are discussed in Section 5.7.

EXAMPLE 5.7

Selection of transmission line voltage and number of lines for power transfer

From a hydroelectric power plant, 9000 MW are to be transmitted to a load center located 500 km from the plant. Based on practical line loadability criteria, determine the number of three-phase, 60 Hz lines required to transmit this power, with one line out of service, for the following cases: (a) 345 kV lines with $Z_c = 297\ \Omega$; (b) 500-kV lines with $Z_c = 277\ \Omega$; and (c) 765-kV lines with $Z_c = 266\ \Omega$. Assume $V_S = 1.0$ per unit, $V_R = 0.95$ per unit, and $\delta = 35°$. Also assume that the lines are uncompensated and widely separated such that there is negligible mutual coupling between them.

SOLUTION

a. For 345 kV lines, (5.4.21) yields

$$SIL = \frac{(345)^2}{297} = 401 \quad MW$$

Neglecting losses, from (5.4.29), with $l = 500$ km and $\delta = 35°$,

$$P = \frac{(1.0)(0.95)(401)\sin(35°)}{\sin\left(\frac{2\pi \times 500}{5000}\right)} = (401)(0.927) = 372 \quad MW/line$$

Alternatively, the practical line loadability curve in Figure 5.12 can be used to obtain P = (0.93)SIL for typical 500-km overhead 60-Hz uncompensated lines.

In order to transmit 9000 MW with one line out of service,

$$\#345 \text{ kV lines} = \frac{9000 \text{ MW}}{372 \text{ MW/line}} + 1 = 24.2 + 1 \approx 26$$

b. For 500 kV lines,

$$\text{SIL} = \frac{(500)^2}{277} = 903 \quad \text{MW}$$

$$P = (903)(0.927) = 837 \quad \text{MW/line}$$

$$\#500 \text{ kV lines} = \frac{9000}{837} + 1 = 10.8 + 1 \approx 12$$

c. For 765 kV lines,

$$\text{SIL} = \frac{(765)^2}{266} = 2200 \quad \text{MW}$$

$$P = (2200)(0.927) = 2039 \quad \text{MW/line}$$

$$\#765 \text{ kV lines} = \frac{9000}{2039} + 1 = 4.4 + 1 \approx 6$$

Increasing the line voltage from 345 to 765 kV, which is a factor of 2.2, reduces the required number of lines from 26 to 6, which is a factor of 4.3.

EXAMPLE 5.8

Effect of intermediate substations on number of lines required for power transfer

Can five instead of six 765 kV lines transmit the required power in Example 5.7 if there are two intermediate substations that divide each line into three 167-km line sections, and if only one line section is out of service?

SOLUTION

The lines are shown in Figure 5.14. For simplicity, neglect line losses. The equivalent π circuit of one 500 km, 765 kV line has a series reactance from (5.4.10) and (5.4.15),

$$X' = (266) \sin\left(\frac{2\pi \times 500}{5000}\right) = 156.35 \quad \Omega$$

(Continued)

FIGURE 5.14

Transmission-line configuration for Example 5.8

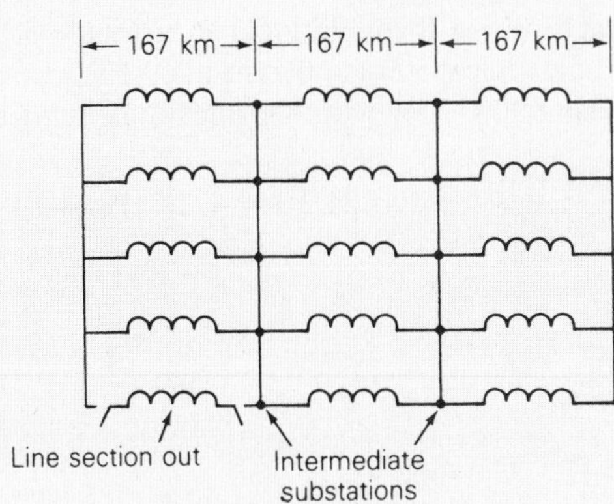

Combining series/parallel reactances in Figure 5.14, the equivalent reactance of five lines with one line section out of service is

$$X_{eq} = \frac{1}{5}\left(\frac{2}{3}X'\right) + \frac{1}{4}\left(\frac{X'}{3}\right) = 0.2167X' = 33.88 \quad \Omega$$

Then, from (5.4.26) with $\delta = 35°$,

$$P = \frac{(765)(765 \times 0.95)\sin(35°)}{33.88} = 9412 \quad \text{MW}$$

Inclusion of line losses would reduce the above value by 3 or 4% to about 9100 MW. Therefore, the answer is yes. Five 765-kV, 500-km uncompensated lines with two intermediate substations and with one line section out of service transmits 9000 MW. Intermediate substations are often economical if their costs do not outweigh the reduction in line costs.

This example is modeled in PowerWorld Simulator case Example 5_8 (see Figure 5.15). Each line segment is represented with the lossless line model from Example 5.4 with the π circuit parameters modified to exactly match those for a 167-km distributed line. The pie charts on each line segment show the percentage loading of the line, assuming a rating of 3500 MVA. The solid red squares on the lines represent closed circuit breakers, and the green squares correspond to open circuit breakers. Clicking on a circuit breaker toggles its status. The simulation results differ slightly from the simplified analysis done earlier in the example because the PowerWorld simulation includes the charging capacitance of the transmission lines. With all line segments in-service, use the load's arrow to verify that the SIL for this system is 11,000 MW, which is five times that of the single circuit line in Example 5.4.

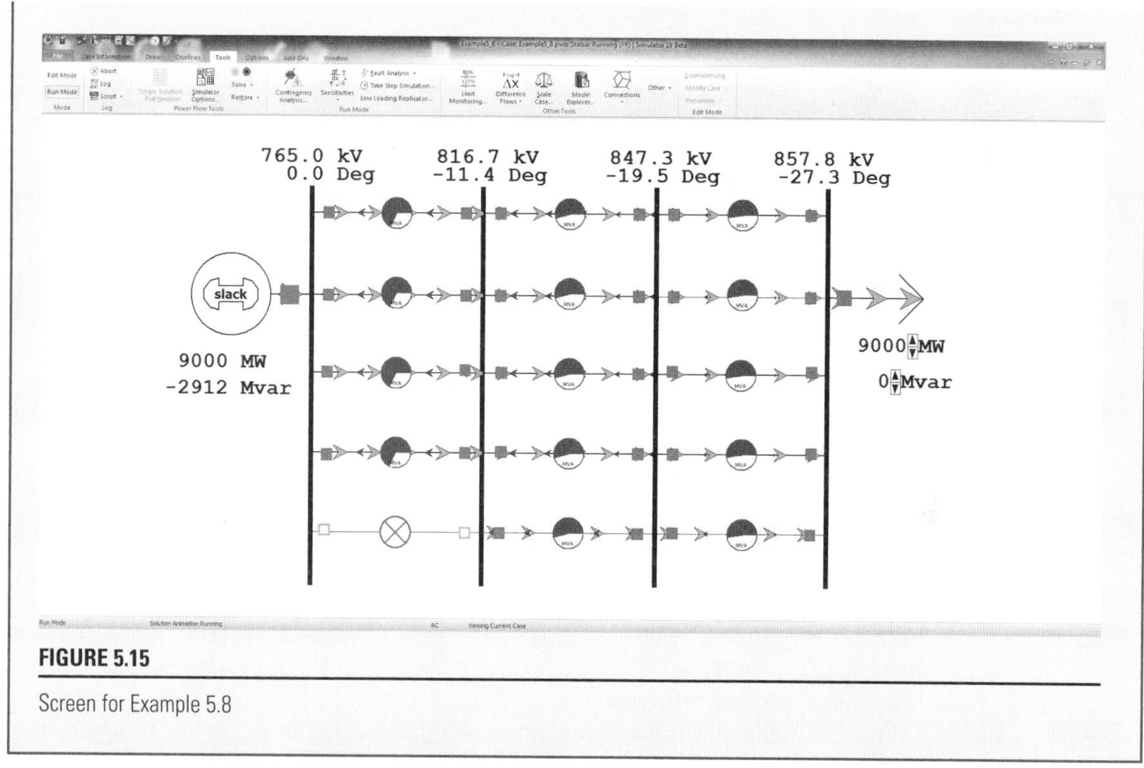

FIGURE 5.15

Screen for Example 5.8

5.7 REACTIVE COMPENSATION TECHNIQUES

Inductors and capacitors are used on medium-length and long transmission lines to increase line loadability and to maintain voltages near rated values.

Shunt reactors (inductors) are commonly installed at selected points along EHV lines from each phase to neutral. The inductors absorb reactive power and reduce overvoltages during light load conditions. They also reduce transient overvoltages due to switching and lightning surges. However, shunt reactors can reduce line loadability if they are not removed under full-load conditions.

In addition to shunt reactors, shunt capacitors are sometimes used to deliver reactive power and increase transmission voltages during heavy load conditions. Another type of shunt compensation includes thyristor-switched reactors in parallel with capacitors. These devices, called *static var compensators*, can absorb reactive power during light loads and deliver reactive power during heavy loads. Through automatic control of the thyristor switches, voltage fluctuations are minimized and line loadability is increased. Synchronous condensors (synchronous motors with no mechanical load) can also control their reactive power output, although more slowly than static var compensators.

Series capacitors are sometimes used on long lines to increase line loadability. Capacitor banks are installed in series with each phase conductor at selected points

FIGURE 5.16 Compensated transmission-line section

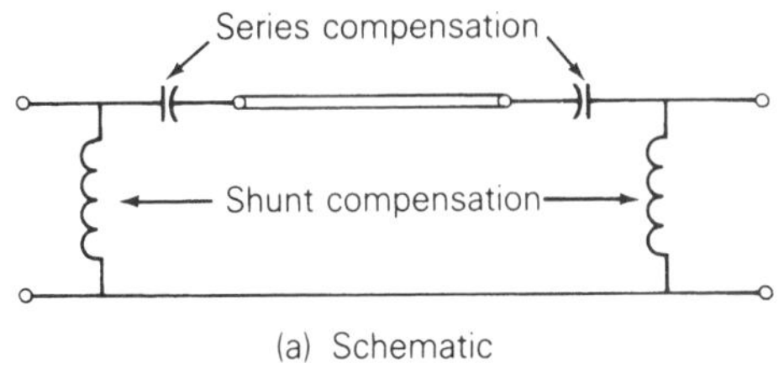

(a) Schematic

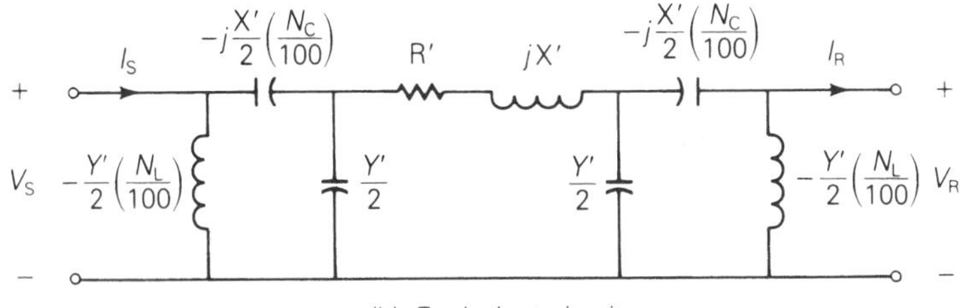

(b) Equivalent circuit

along a line. Their effect is to reduce the net series impedance of the line in series with the capacitor banks, thereby reducing line-voltage drops and increasing the steady-state stability limit. A disadvantage of series capacitor banks is that automatic protection devices must be installed to bypass high currents during faults and to reinsert the capacitor banks after fault clearing. Also, the addition of series capacitors can excite low-frequency oscillations, a phenomenon called *subsynchronous resonance*, which may damage turbine-generator shafts. Studies have shown, however, that series capacitive compensation can increase the loadability of long lines at only a fraction of the cost of new transmission [1].

Figure 5.16 shows a schematic and an equivalent circuit for a compensated line section, where N_C is the amount of series capacitive compensation expressed in percent of the positive-sequence line impedance and N_L is the amount of shunt reactive compensation in percent of the positive-sequence line admittance. It is assumed in Figure 5.16 that half of the compensation is installed at each end of the line section. The following two examples illustrate the effect of compensation.

EXAMPLE 5.9

Shunt reactive compensation to improve transmission-line voltage regulation

Identical shunt reactors (inductors) are connected from each phase conductor to neutral at both ends of the 300 km line in Example 5.2 during light load conditions, providing 75% compensation. The reactors are removed during heavy load conditions. Full load is 1.90 kA at unity p.f. and at 730 kV. Assuming that the sending-end voltage is constant, determine the following properties.

a. Percent voltage regulation of the uncompensated line

b. The equivalent shunt admittance and series impedance of the compensated line

c. Percent voltage regulation of the compensated line

SOLUTION

a. From (5.1.1) with $I_{RFL} = 1.9\underline{/0^\circ}$ kA, using the A and B parameters from Example 5.2,

$$
\begin{aligned}
V_S &= AV_{RFL} + BI_{RFL} \\
&= (0.9313\underline{/0.209^\circ})\left(\frac{730}{\sqrt{3}}\underline{/0^\circ}\right) + (97.0\underline{/87.2^\circ})(1.9\underline{/0^\circ}) \\
&= 392.5\underline{/0.209^\circ} + 184.3\underline{/87.2^\circ} \\
&= 401.5 + j185.5 \\
&= 442.3\underline{/24.8^\circ} \quad \text{kV}_{LN}
\end{aligned}
$$

$$V_S = 442.3\sqrt{3} = 766.0 \quad \text{kV}_{LL}$$

The no-load receiving-end voltage is, from (5.1.19),

$$V_{RNL} = \frac{766.0}{0.9313} = 822.6 \quad \text{kV}_{LL}$$

and the percent voltage regulation for the uncompensated line is, from (5.1.18),

$$\text{percent VR} = \frac{822.6 - 730}{730} \times 100 = 12.68\%$$

b. From Example 5.3, the shunt admittance of the equivalent π circuit without compensation is

$$
\begin{aligned}
Y' &= 2(3.7 \times 10^{-7} + j7.094 \times 10^{-4}) \\
&= 7.4 \times 10^{-7} + j14.188 \times 10^{-4} \quad \text{S}
\end{aligned}
$$

With 75% shunt compensation, the equivalent shunt admittance is

$$
\begin{aligned}
Y_{eq} &= 7.4 \times 10^{-7} + j14.188 \times 10^{-4}\left(1 - \tfrac{75}{100}\right) \\
&= 3.547 \times 10^{-4}\underline{/89.88^\circ} \quad \text{S}
\end{aligned}
$$

Since there is no series compensation, the equivalent series impedance is the same as without compensation:

$$Z_{eq} = Z' = 97.0\underline{/87.2^\circ} \quad \Omega$$

(*Continued*)

c. The equivalent A parameter for the compensated line is

$$A_{eq} = 1 + \frac{Y_{eq} Z_{eq}}{2}$$
$$= 1 + \frac{(3.547 \times 10^{-4}\underline{/89.88^\circ})(97.0\underline{/87.2^\circ})}{2}$$
$$= 1 + 0.0172\underline{/177.1^\circ}$$
$$= 0.9828\underline{/0.05^\circ} \quad \text{per unit}$$

Then, from (5.1.19),

$$V_{RNL} = \frac{766}{0.9828} = 779.4 \quad kV_{LL}$$

Since the shunt reactors are removed during heavy load conditions, $V_{RFL} = 730$ kV is the same as without compensation. Therefore

$$\text{percent VR} = \frac{779.4 - 730}{730} \times 100 = 6.77\%$$

The use of shunt reactors at light loads improves the voltage regulation from 12.68% to 6.77% for this line.

EXAMPLE 5.10

Series capacitive compensation to increase transmission-line loadability

Identical series capacitors are installed in each phase at both ends of the line in Example 5.2, providing 30% compensation. Determine the theoretical maximum power that this compensated line can deliver and compare with that of the uncompensated line. Assume $V_S = V_R = 765$ kV.

SOLUTION

From Example 5.3, the equivalent series reactance without compensation is

$$X' = 97.0 \sin 87.2^\circ = 96.88 \quad \Omega$$

Based on 30% series compensation, half at each end of the line, the impedance of each series capacitor is

$$Z_{cap} = -jX_{cap} = -j\left(\tfrac{1}{2}\right)(0.30)(96.88) = -j14.53 \quad \Omega$$

From Figure 5.4, the $ABCD$ matrix of this series impedance is

$$\left[\begin{array}{c|c} 1 & -j14.53 \\ \hline 0 & 1 \end{array}\right]$$

As also shown in Figure 5.4, the equivalent $ABCD$ matrix of networks in series is obtained by multiplying the $ABCD$ matrices of the individual networks. For this example there are three networks: the series capacitors at the sending end, the line, and the series capacitors at the receiving end. Therefore the equivalent $ABCD$ matrix of the compensated line is, using the $ABCD$ parameters, from Example 5.2,

$$\left[\begin{array}{c|c} 1 & -j14.53 \\ \hline 0 & 1 \end{array}\right] \left[\begin{array}{c|c} 0.9313\underline{/0.209^\circ} & 97.0\underline{/87.2^\circ} \\ \hline 1.37 \times 10^{-3}\underline{/90.06^\circ} & 0.9313\underline{/0.209^\circ} \end{array}\right] \left[\begin{array}{c|c} 1 & -j14.53 \\ \hline 0 & 1 \end{array}\right]$$

After performing these matrix multiplications, results in

$$\left[\begin{array}{c|c} A_{eq} & B_{eq} \\ \hline C_{eq} & D_{eq} \end{array}\right] = \left[\begin{array}{c|c} 0.9512\underline{/0.205^\circ} & 69.70\underline{/86.02^\circ} \\ \hline 1.37 \times 10^{-3}\underline{/90.06^\circ} & 0.9512\underline{/0.205^\circ} \end{array}\right]$$

Therefore

$$A_{eq} = 0.9512 \quad \text{per unit} \qquad \theta_{Aeq} = 0.205^\circ$$

$$B_{eq} = Z'_{eq} = 69.70 \quad \Omega \qquad \theta_{Zeq} = 86.02^\circ$$

From (5.5.6) with $V_S = V_R = 765$ kV,

$$P_{Rmax} = \frac{(765)^2}{69.70} - \frac{(0.9512)(765)^2}{69.70}\cos(86.02^\circ - 0.205^\circ)$$
$$= 8396 - 583 = 7813 \quad \text{MW}$$

which is 36.2% larger than the value of 5738 MW found in Example 5.5 without compensation. Note that the practical line loadability of this series compensated line is also about 35% larger than the value of 3247 MW found in Example 5.6 without compensation.

This example is modeled in PowerWorld Simulator case Example 5_10 (see Figure 5.17). When opened, both of the series capacitors are bypassed (i.e., they are modeled as short circuits), meaning this case is initially identical to the Example 5.4 case. Click on the blue Bypassed field to place each of the series capacitors into the circuit. This decreases the angle across the line, resulting in more net power transfer.

(*Continued*)

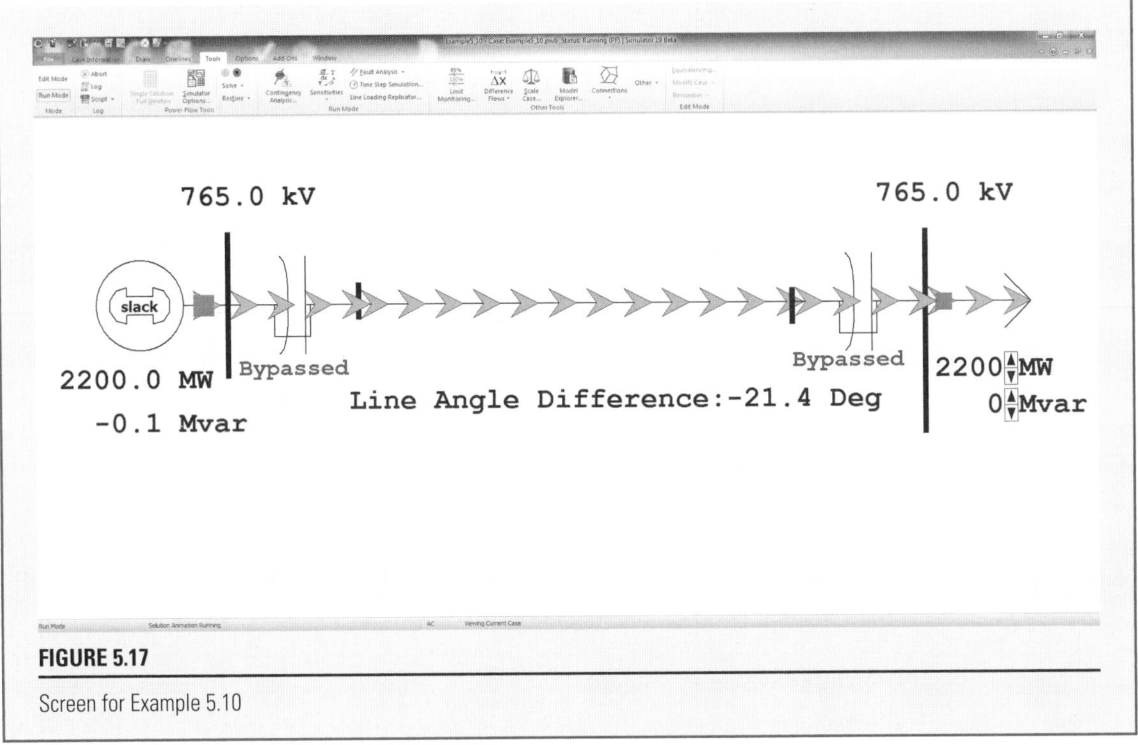

FIGURE 5.17

Screen for Example 5.10

MULTIPLE CHOICE QUESTIONS

SECTION 5.1

5.1 Representing a transmission line by the two-port network, in terms of *ABCD* parameters, (a) express V_S, which is the sending-end voltage, in terms of V_R, which is the receiving-end voltage, and I_R, the receiving-end current, and (b) express I_S, which is the sending-end current, in terms of V_R and I_R.
(a) $V_S =$ __________ (b) $I_S =$ __________

5.2 As applied to linear, passive, bilateral two-port networks, the *ABCD* parameters satisfy $AD - BC = 1$.
(a) True (b) False

5.3 Express the no-load receiving-end voltage V_{RNL} in terms of the sending-end voltage, V_S, and the *ABCD* parameters.

5.4 The *ABCD* parameters, which are in general complex numbers, have units of __________, __________, __________, and __________, respectively.

5.5 The loadability of short transmission lines (less than 25 km, represented by including only series resistance and reactance) is determined by __________; that of medium lines (less than 250 km, represented by nominal π circuit) is determined by __________; and that of long lines (more than 250 km, represented by equivalent π circuit) is determined by __________.

5.6 Can the voltage regulation, which is proportional to $(V_{RNL} - V_{RFL})$, be negative?
(a) Yes (b) No

SECTION 5.2

5.7 The propagation constant, which is a complex quantity in general, has units of ______, and the characteristic impedance has units of ______.

5.8 Express hyperbolic functions cosh $\sqrt{x}$ and sinh $\sqrt{x}$ in terms of exponential functions.

5.9 e^{γ}, where $\gamma = \alpha + j\beta$, can be expressed as $e^{\alpha l}\underline{/\beta l}$, in which αl is dimensionless and βl is in radians (also dimensionless).
(a) True (b) False

SECTION 5.3

5.10 The equivalent π circuit is identical in structure to the nominal π circuit.
(a) True (b) False

5.11 The correction factors $F_1 = \sinh(\gamma l)\ \gamma l$ and $F_2 = \tanh(\gamma l/2)/(\gamma l/2)$, which are complex numbers, have the units of ______.

SECTION 5.4

5.12 For a lossless line, the surge impedance is purely resistive and the propagation constant is pure imaginary.
(a) True (b) False

5.13 For equivalent π circuits of lossless lines, the A and D parameters are pure _______, whereas B and C parameters are pure _______.

5.14 In equivalent π circuits of lossless lines, Z' is pure __________, and Y' is pure __________.

5.15 Typical power-line lengths are only a small fraction of the 60-Hz wavelength.
(a) True (b) False

5.16 The velocity of propagation of voltage and current waves along a lossless overhead line is the same as speed of light.
(a) True (b) False

5.17 Surge Impedance Loading (SIL) is the power delivered by a lossless line to a load resistance equal to __________.

5.18 For a lossless line, at SIL, the voltage profile is __________, and the real power delivered, in terms of rated line voltage V and surge impedance Z_c, is given by __________.

5.19 The maximum power that a lossless line can deliver, in terms of the voltage magnitudes V_S and V_R (in volts) at the ends of the line held constant, and the series reactance X' of the corresponding equivalent π circuit, is given by ________, in watts.

SECTION 5.5

5.20 The maximum power flow for a lossy line is somewhat less than that for a lossless line.
(a) True (b) False

SECTION 5.6

5.21 For short lines less than 25 km long, loadability is limited by the thermal rating of the conductors or by terminal equipment ratings, not by voltage drop or stability considerations.
(a) True (b) False

5.22 Increasing the transmission line voltage reduces the required number of lines for the same power transfer.
(a) True (b) False

5.23 Intermediate substations are often economical from the viewpoint of the number of lines required for power transfer if their costs do not outweigh the reduction in line costs.
(a) True (b) False

SECTION 5.7

5.24 Shunt reactive compensation improves transmission-line ______, whereas series capacitive compensation increases transmission-line ______.

5.25 Static-var-compensators can absorb reactive power during light loads and deliver reactive power during heavy loads.
(a) True (b) False

PROBLEMS

SECTION 5.1

5.1 A 30-km, 34.5-kV, 60-Hz, three-phase line has a positive-sequence series impedance $z = 0.19 + j0.34\ \Omega$/km. The load at the receiving end absorbs 10 MVA at 33 kV. Assuming a short line, calculate: (a) the *ABCD* parameters, (b) the sending-end voltage for a load power factor of 0.9 lagging, and (c) the sending-end voltage for a load power factor of 0.9 leading.

5.2 A 200-km, 230-kV, 60-Hz, three-phase line has a positive-sequence series impedance $z = 0.08 + j0.48\ \Omega$/km and a positive-sequence shunt admittance $y = j3.33 \times 10^{-6}$ S/km. At full load, the line delivers 250 MW at

0.99 p.f. lagging and at 220 kV. Using the nominal π circuit, calculate: (a) the *ABCD* parameters, (b) the sending-end voltage and current, and (c) the percent voltage regulation.

5.3 Rework Problem 5.2 in per unit using 1000-MVA (three-phase) and 230-kV (line-to-line) base values. Calculate: (a) the per-unit *ABCD* parameters, (b) the per-unit sending-end voltage and current, and (c) the percent voltage regulation.

5.4 Derive the *ABCD* parameters for the two networks in series, as shown in Figure 5.4.

5.5 Derive the *ABCD* parameters for the T circuit shown in Figure 5.4.

5.6 (a) Consider a medium-length transmission line represented by a nominal π circuit shown in Figure 5.3 of the text. Draw a phasor diagram for lagging power-factor condition at the load (receiving end).

(b) Now consider a nominal T circuit of the medium-length transmission line shown in Figure 5.18.

First draw the corresponding phasor diagram for lagging power-factor load condition.

Then determine the *ABCD* parameters in terms of Y and Z for the nominal T circuit and for the nominal π circuit of part (a).

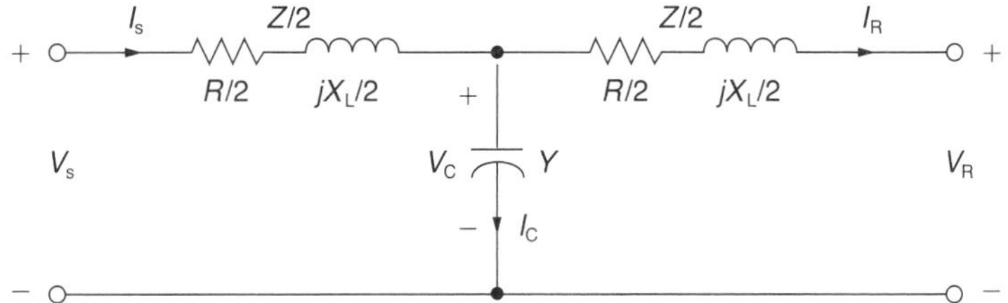

FIGURE 5.18

Nominal T-circuit for Problem 5.6

5.7 The per-phase impedance of a short three-phase transmission line is $0.5\angle 53.15°$ Ω. The three-phase load at the receiving end is 1200 kW at 0.8 p.f. lagging. If the line-to-line sending-end voltage is 3.3 kV, determine (a) the receiving-end line-to-line voltage in kV and (b) the line current. Draw the phasor diagram with the line current I, as reference.

5.8 Reconsider Problem 5.7 and find the following: (a) sending-end power factor, (b) sending-end three-phase power, and (c) the three-phase line loss.

5.9 The 100-km, 230-kV, 60-Hz, three-phase line in Problems 4.18 and 4.39 delivers 300 MVA at 218 kV to the receiving end at full load. Using the nominal π circuit, calculate the *ABCD* parameters, sending-end voltage, and percent voltage regulation when the receiving-end power factor is (a) 0.9 lagging, (b) unity, and (c) 0.9 leading. Assume a 50°C conductor temperature to determine the resistance of this line.

5.10 The 500-kV, 60-Hz, three-phase line in Problems 4.20 and 4.41 has a 180-km length and delivers 1600 MW at 475 kV and at 0.95 power factor leading to the receiving end at full load. Using the nominal π circuit, calculate the (a) $ABCD$ parameters, (b) sending-end voltage and current, (c) sending-end power and power factor, (d) full-load line losses and efficiency, and (e) percent voltage regulation. Assume a 50°C conductor temperature to determine the resistance of this line.

5.11 A 40-km, 220-kV, 60-Hz, three-phase overhead transmission line has a per-phase resistance of 0.15 Ω/km, a per-phase inductance of 1.3263 mH/km, and negligible shunt capacitance. Using the short line model, find the sending-end voltage, voltage regulation, sending-end power, and transmission line efficiency when the line is supplying a three-phase load of (a) 381 MVA at 0.8 power factor lagging and at 220 kV and (b) 381 MVA at 0.8 power factor leading and at 220 kV.

5.12 A 60-Hz, 100-km, three-phase overhead transmission line, constructed of ACSR conductors, has a series impedance of $(0.1826 + j0.784)$ Ω/km per phase and a shunt capacitive reactance-to-neutral of $185.5 \times 10^3 \underline{/-90^\circ}$ Ω-km per phase. Using the nominal π circuit for a medium-length transmission line, (a) determine the total series impedance and shunt admittance of the line; (b) compute the voltage, the current, and the real and reactive power at the sending end if the load at the receiving end draws 200 MVA at unity power factor and at a line-to-line voltage of 230 kV; and (c) find the percent voltage regulation of the line.

SECTION 5.2

5.13 Evaluate $\cosh(\gamma l)$ and $\tanh(\gamma l/2)$ for $\gamma l = 0.40\underline{/85^\circ}$ per unit.

5.14 A 500-km, 500-kV, 60-Hz, uncompensated three-phase line has a positive-sequence series impedance $z = 0.03 + j0.35$ Ω/km and a positive-sequence shunt admittance $y = j4.4 \times 10^{-6}$ S/km. Calculate: (a) Z_c, (b) (γl), and (c) the exact $ABCD$ parameters for this line.

5.15 At full load, the line in Problem 5.14 delivers 900 MW at unity power factor and at 475 kV. Calculate: (a) the sending-end voltage, (b) the sending-end current, (c) the sending-end power factor, (d) the full-load line losses, and (e) the percent voltage regulation.

5.16 The 500-kV, 60-Hz, three-phase line in Problems 4.20 and 4.41 has a 300-km length. Calculate: (a) Z_c, (b) (γl), and (c) the exact $ABCD$ parameters for this line. Assume a 50°C conductor temperature.

5.17 At full load, the line in Problem 5.16 delivers 1500 MVA at 480 kV to the receiving-end load. Calculate the sending-end voltage and percent voltage regulation when the receiving-end power factor is (a) 0.9 lagging, (b) unity, and (c) 0.9 leading.

5.18 A 60-Hz, 230-km, three-phase overhead transmission line has a series impedance $z = 0.8431\underline{/79.04^\circ}$ Ω/km and a shunt admittance $y = 5.105 \times 10^{-6}\underline{/90^\circ}$ S/km. The load at the receiving end is 125 MW at unity power factor and at 215 kV. Determine the voltage, current, and both real and reactive power at the sending end and the percent voltage regulation of the line. Also find the wavelength and velocity of propagation of the line.

5.19 Using per-unit calculations, rework Problem 5.18 to determine the sending-end voltage and current.

5.20 (a) The series expansions of the hyperbolic functions are given by

$$\cosh\theta = 1 + \frac{\theta^2}{2} + \frac{\theta^4}{24} + \frac{\theta^6}{720} + \cdots$$

$$\sinh\theta = 1 + \frac{\theta^2}{6} + \frac{\theta^4}{120} + \frac{\theta^6}{5040} + \cdots$$

For the *ABCD* parameters of a long transmission line represented by an equivalent π circuit, apply the above expansion considering only the first two terms, and express the result in terms of *Y* and *Z*.

(b) For the nominal π and equivalent π circuits shown in Figures 5.3 and 5.7 of the text, show that

$$\frac{A-1}{B} = \frac{Y}{2} \quad \text{and} \quad \frac{A-1}{B} = \frac{Y'}{2}$$

hold good, respectively.

5.21 Starting with (5.1.1) of the text, show that

$$A = \frac{V_S I_S + V_R I_R}{V_R I_S + V_S I_R} \quad \text{and} \quad B = \frac{V_S^2 - V_R^2}{V_R I_S + V_S I_R}$$

5.22 Consider the *A* parameter of the long line given by $\cosh\theta$, where $\theta = \sqrt{ZY}$. With $x = e^{-\theta} = x_1 + jx_2$ and $A = \mathrm{A}_1 + j\mathrm{A}_2$, show that x_1 and x_2 satisfy the following:

$$x_1^2 - x_2^2 - 2(\mathrm{A}_1 x_1 - \mathrm{A}_2 x_2) + 1 = 0$$

and $x_1 x_2 - (\mathrm{A}_2 x_1 + \mathrm{A}_1 x_2) = 0.$

SECTION 5.3

5.23 Determine the equivalent π circuit for the line in Problem 5.14 and compare it with the nominal π circuit.

5.24 Determine the equivalent π circuit for the line in Problem 5.16. Compare the equivalent π circuit with the nominal π circuit.

5.25 Let the transmission line of Problem 5.12 be extended to cover a distance of 200 km. Assume conditions at the load to be the same as in Problem 5.12. Determine the (a) sending-end voltage, (b) sending-end current, (c) sending-end real and reactive powers, and (d) percent voltage regulation.

SECTION 5.4

5.26 A 350-km, 500-kV, 60-Hz, three-phase uncompensated line has a positive-sequence series reactance $x = 0.34\ \Omega$/km and a positive-sequence shunt admittance $y = j4.5 \times 10^{-6}$ S/km. Neglecting losses, calculate: (a) Z_c, (b) γl, (c) the $ABCD$ parameters, (d) the wavelength λ of the line in kilometers, and (e) the surge impedance loading in MW.

5.27 Determine the equivalent π circuit for the line in Problem 5.26.

5.28 Rated line voltage is applied to the sending end of the line in Problem 5.26. Calculate the receiving-end voltage when the receiving end is terminated by (a) an open circuit, (b) the surge impedance of the line, and (c) one-half of the surge impedance. (d) Also calculate the theoretical maximum real power that the line can deliver when rated voltage is applied to both ends of the line.

5.29 Rework Problems 5.9 and 5.16, neglecting the conductor resistance. Compare the results with and without losses.

5.30 From (4.6.22) and (4.10.4), the series inductance and shunt capacitance of a three-phase overhead line are

$$L_a = 2 \times 10^{-7} \ln(D_{eq}/D_{SL}) = \frac{\mu_0}{2\pi} \ln(D_{eq}/D_{SL}) \quad \text{H/m}$$

$$C_{an} = \frac{2\pi\varepsilon_0}{\ln(D_{eq}/D_{SC})} \quad \text{F/m}$$

where $\mu_0 = 4\pi \times 10^{-7}$H/m and $\varepsilon_0 = \left(\frac{1}{36\pi}\right) \times 10^{-9}$ F/m.

Using these equations, determine formulas for surge impedance and velocity of propagation of an overhead lossless line. Then determine the surge impedance and velocity of propagation for the three-phase line given in Example 4.5. Assume positive-sequence operation. Neglect line losses as well as the effects of the overhead neutral wires and the earth plane.

5.31 A 500-kV, 300-km, 60-Hz, three-phase overhead transmission line, assumed to be lossless, has a series inductance of 0.97 mH/km per phase and a shunt capacitance of 0.0115 μF/km per phase. (a) Determine the phase constant β, the surge impedance Z_c, velocity of propagation v, and the wavelength λ of the line. (b) Determine the voltage, current, real and reactive power at the sending end, and the percent voltage regulation of the line if the receiving-end load is 800 MW at 0.8 power factor lagging and at 500 kV.

5.32 The following parameters are based on a preliminary line design: V_S = 1.0 per unit, V_R = 0.9 per unit, λ = 5000 km, Z_c = 320 Ω, δ = 36.8°.

A three-phase power of 700 MW is to be transmitted to a substation located 315 km from the source of power. (a) Determine a nominal voltage level for the three-phase transmission line, based on the practical line-loadability equation. (b) For the voltage level obtained in part (a), determine the theoretical maximum power that can be transferred by the line.

5.33 Consider a long radial line terminated in its characteristic impedance Z_c. Determine the following:

(a) V_1/I_1, known as the driving point impedance.

(b) $|V_2|/|V_1|$, known as the voltage gain, in terms of $\alpha\ell$.

(c) $|I_2|/|I_1|$, known as the current gain, in terms of $\alpha\ell$.

(d) The complex power gain, $-S_{21}/S_{12}$, in terms of $\alpha\ell$.

(e) The real power efficiency, $(-P_{21}/P_{12}) = \eta$, terms of $\alpha\ell$.

Note: 1 refers to sending end and 2 refers to receiving end. (S_{21}) is the complex power received at 2; S_{12} is sent from 1.

5.34 For the case of a lossless line, how would the results of Problem 5.33 change?

In terms of Z_c, which is a real quantity for this case, express P_{12} in terms $|I_1|$ and $|V_1|$.

5.35 For a lossless open-circuited line, express the sending-end voltage, V_1, in terms of the receiving-end voltage, V_2, for the three cases of short-line model, medium-length line model, and long-line model. Is it true that the voltage at the open receiving end of a long line is higher than that at the sending end, for small βl?

5.36 For a short transmission line of impedance $(R + jX)$ ohms per phase, show that the maximum power that can be transmitted over the line is

$$P_{max} = \frac{V_R^2}{Z^2}\left(\frac{ZV_S}{V_R} - R\right) \quad \text{where } Z = \sqrt{R^2 + X^2}$$

when the sending-end and receiving-end voltages are fixed, and for the condition

$$Q = \frac{-V_R^2 X}{R^2 + X^2} \quad \text{when } dP/dQ = 0$$

5.37 (a) Consider complex power transmission via the three-phase short line for which the per-phase circuit is shown in Figure 5.19. Express S_{12}, the complex power sent by bus 1 (or V_1), and $(-S_{21})$, the complex power received by bus 2 (or V_2), in terms of V_1, V_2, Z, $\angle Z$, and $\theta_{12} = \theta_1 - \theta_2$, which is the power angle.

(b) For a balanced three-phase transmission line in per-unit notation with $Z = 1\angle 85°$, $\theta_{12} = 10°$, determine S_{12} and $(-S_{21})$ for

(i) $V_1 = V_2 = 1.0$
(ii) $V_1 = 1.1$ and $V_2 = 0.9$
Comment on the changes of real and reactive powers from parts (i) to (ii).

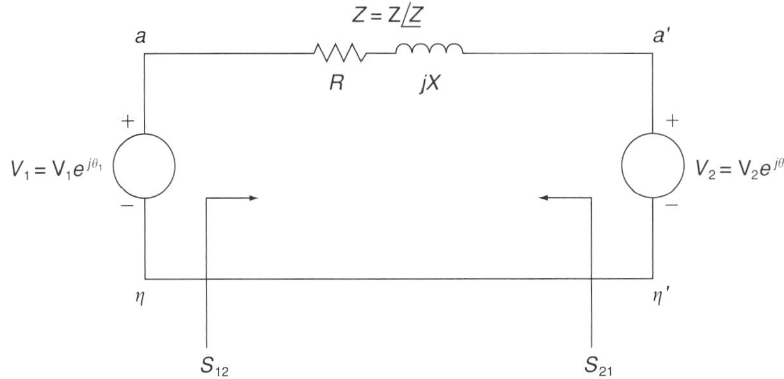

FIGURE 5.19 Per-phase circuit for Problem 5.37

SECTION 5.5

5.38 The line in Problem 5.14 has three ACSR 1113 kcmil (564 mm^2) conductors per phase. Calculate the theoretical maximum real power that this line can deliver and compare with the thermal limit of the line. Assume $V_S = V_R = 1.0$ per unit and unity power factor at the receiving end.

5.39 Repeat Problems 5.14 and 5.38 if the line length is (a) 200 km or (b) 600 km.

5.40 For the 500 kV line given in Problem 5.16, (a) calculate the theoretical maximum real power that the line can deliver to the receiving end when rated voltage is applied to both ends; (b) calculate the receiving-end reactive power and power factor at this theoretical loading.

5.41 A 230-kV, 100-km, 60-Hz, three-phase overhead transmission line with a rated current of 900 A/phase has a series impedance $z = 0.088 + j0.465$ Ω/km and a shunt admittance $y = j3.524$ μS/km. (a) Obtain the nominal π equivalent circuit in normal units and in per unit on a base of 100 MVA (three phase) and 230 kV (line-to-line). (b) Determine the three-phase rated MVA of the line. (c) Compute the *ABCD* parameters. (d) Calculate the SIL.

5.42 A three-phase power of 460 MW is transmitted to a substation located 500 km from the source of power. With $V_S = 1$ per unit, $V_R = 0.9$ per unit, $\lambda = 5000$ km, $Z_c = 500\ \Omega$, and $\delta = 36.87°$, determine a nominal voltage level for the lossless transmission line based on Eq. (5.4.29) of the text.

Using this result, find the theoretical three-phase maximum power that can be transferred by the lossless transmission line.

PW **5.43** Open PowerWorld Simulator case Example 5_4 and graph the load bus voltage as a function of load real power (assuming unity power factor at the load). What is the maximum amount of real power that can be transferred to the load at unity power factor if the load voltage always must be greater than 0.9 per unit?

PW **5.44** Repeat Problem 5.43, but now vary the load reactive power, assuming the load real power is fixed at 1499 MW.

SECTION 5.6

5.45 For the line in Problems 5.14 and 5.38, determine (a) the practical line loadability in MW, assuming $V_S = 1.0$ per unit, $V_R \approx 0.95$ per unit, and $\delta_{max} = 35°$; part (b) the full-load current at 0.99 p.f. leading, based on the above practical line loadability; (c) the exact receiving-end voltage for the full-load current in part (b); and (d) the percent voltage regulation. For this line, is loadability determined by the thermal limit, the voltage-drop limit, or steady-state stability?

5.46 Repeat Problem 5.45 for the 500 kV line given in Problem 5.10.

5.47 Determine the practical line loadability in MW and in per-unit of SIL for the line in Problem 5.14 if the line length is (a) 200 km or (b) 600 km. Assume $V_S = 1.0$ per unit, $V_R = 0.95$ per unit, $\delta_{max} = 35°$, and 0.99 leading power factor at the receiving end.

5.48 It is desired to transmit 2000 MW from a power plant to a load center located 300 km from the plant. Determine the number of 60 Hz, three-phase, uncompensated transmission lines required to transmit this power with one line out of service for the following cases: (a) 345 kV lines, $Z_c = 300\ \Omega$, (b) 500 kV lines, $Z_c = 275\ \Omega$, (c) 765 kV lines, $Z_c = 260\ \Omega$. Assume that $V_S = 1.0$ per unit, $V_R = 0.95$ per unit, and $\delta_{max} = 35°$.

5.49 Repeat Problem 5.48 if it is desired to transmit: (a) 3200 MW to a load center located 300 km from the plant or (b) 2000 MW to a load center located 400 km from the plant.

5.50 A three-phase power of 4000 MW is to be transmitted through four identical 60-Hz overhead transmission lines over a distance of 300 km. Based on a preliminary design, the phase constant and surge impedance of the line are $\beta = 9.46 \times 10^{-4}$ rad/km and $Z_c = 343\ \Omega$, respectively. Assuming $V_S = 1.0$ per unit, $V_R = 0.9$ per unit, and a power angle $\delta = 36.87°$, determine a suitable nominal voltage level in kV based on the practical line-loadability criteria.

5.51 The power flow at any point on a transmission line can be calculated in terms of the *ABCD* parameters. By letting $A = |A|\underline{/\alpha}$, $B = |B|\underline{/\beta}$, $V_R = |V_R|\underline{/0°}$, and $V_S = |V_S|\underline{/\delta}$, the complex power at the receiving end can be shown to be

$$P_R + jQ_R = \frac{|V_R||V_s|\underline{/\beta - \alpha}}{|B|} - \frac{|\delta||V_R^2|\underline{/\beta - \alpha}}{|B|}$$

(a) Draw a phasor diagram corresponding to the above equation. Let it be represented by a triangle O′OA with O′ as the origin and OA representing $P_R + jQ_R$.

(b) By shifting the origin from O′ to O, turn the result of part (a) into a power diagram, redrawing the phasor diagram. For a given fixed value of $|V_R|$ and a set of values for $|V_S|$, draw the loci of point A, thereby showing the so-called receiving-end circles.

(c) From the result of part (b) for a given load with a lagging power factor angle θ_R, determine the amount of reactive power that must be supplied to the receiving end to maintain a constant receiving-end voltage if the sending-end voltage magnitude decreases from $|V_{S1}|$ to $|V_{S2}|$

5.52 (a) Consider complex power transmission via the three-phase long line for which the per-phase circuit is shown in Figure 5.20. See Problem 5.37 in which the short-line case was considered. Show that

$$\text{sending-end power} = S_{12} = \frac{Y'^*}{2}V_1^2 + \frac{V_1^2}{Z'^*} - \frac{V_1V_2}{Z'^*}e^{j\theta_{12}}$$

$$\text{received power} = -S_{21} = -\frac{Y'^*}{2}V_2^2 - \frac{V_2^2}{Z'^*} + \frac{V_1V_2}{Z'^*}e^{-j\theta_{12}}$$

where $\theta_{12} = \theta_1 - \theta_2$.

(b) For a lossless line with equal voltage magnitudes at each end, show that

$$P_{12} = -P_{21} = \frac{V_1^2 \sin\theta_{12}}{Z_c \sin\beta\ell} = P_{SIL}\frac{\sin\theta_{12}}{\sin\beta\ell}$$

(c) For $\theta_{12} = 45°$ and $\beta = 0.002$ rad/km, find (P_{12}/P_{SIL}) as a function of line length in km, and sketch it.

(d) If a thermal limit of $(P_{12}/P_{SIL}) = 2$ is set, which limit governs for short lines and long lines?

FIGURE 5.20

Per-phase circuit for Problem 5.52

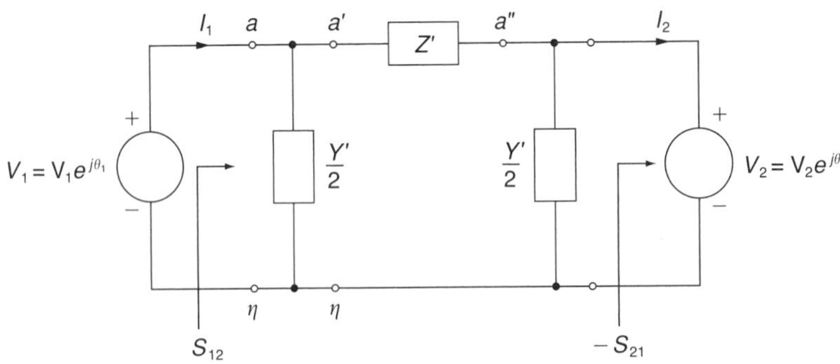

PW **5.53** Open PowerWorld Simulator case Example 5_8. If the load bus voltage is greater than or equal to 730 kV even with any line segment out of service, what is the maximum amount of real power that can be delivered to the load?

PW **5.54** Repeat Problem 5.53, but now assume any two line segments may be out of service.

SECTION 5.7

5.55 Recalculate the percent voltage regulation in Problem 5.15 when identical shunt reactors are installed at both ends of the line during light loads, providing 65% total shunt compensation. The reactors are removed at full load. Also calculate the impedance of each shunt reactor.

5.56 Rework Problem 5.17 when identical shunt reactors are installed at both ends of the line, providing 50% total shunt compensation. The reactors are removed at full load.

5.57 Identical series capacitors are installed at both ends of the line in Problem 5.14, providing 40% total series compensation. Determine the equivalent *ABCD* parameters of this compensated line. Also calculate the impedance of each series capacitor.

5.58 Identical series capacitors are installed at both ends of the line in Problem 5.16, providing 30% total series compensation. (a) Determine the equivalent *ABCD* parameters for this compensated line. (b) Determine the theoretical maximum real power that this series-compensated line can deliver when $V_S = V_R = 1.0$ per unit. Compare your result with that of Problem 5.40.

5.59 Determine the theoretical maximum real power that the series-compensated line in Problem 5.57 can deliver when $V_S = V_R = 1.0$ per unit. Compare your result with that of Problem 5.38.

5.60 What is the minimum amount of series capacitive compensation N_C in percent of the positive-sequence line reactance needed to reduce the number of 765-kV lines in Example 5.8 from five to four? Assume two intermediate substations with one line section out of service. Also, neglect line losses and assume that the series compensation is sufficiently distributed along the line so as to effectively reduce the series reactance of the equivalent π circuit to $X'(1-N_C/100)$.

5.61 Determine the equivalent *ABCD* parameters for the line in Problem 5.14 if it has 70% shunt reactive (inductors) compensation and 40% series capacitive compensation. Half of this compensation is installed at each end of the line, as in Figure 5.14.

5.62 Consider the transmission line of Problem 5.18. (a) Find the *ABCD* parameters of the line when uncompensated. (b) For a series capacitive compensation of 70% (35% at the sending end and 35% at the receiving end), determine the *ABCD* parameters. Comment on the relative change in the magnitude of the *B* parameter with respect to the relative changes in the magnitudes of the *A*, *C*, and *D* parameters. Also comment on the maximum power that can be transmitted when series compensated.

5.63 Given the uncompensated line of Problem 5.18, let a three-phase shunt reactor (inductor) that compensates for 70% of the total shunt admittance of the line be connected at the receiving end of the line during no-load conditions. Determine the effect of voltage regulation with the reactor connected at no load. Assume that the reactor is removed under full-load conditions.

5.64 Let the three-phase lossless transmission line of Problem 5.31 supply a load of 1000 MVA at 0.8 power factor lagging and at 500 kV. (a) Determine the capacitance/phase and total three-phase Mvars supplied by a three-phase, Δ-connected shunt-capacitor bank at the receiving end to maintain the receiving-end voltage at 500 kV when the sending end of the line is energized at 500 kV. (b) If series capacitive compensation of 40% is installed at the midpoint of the line, without the shunt capacitor bank at the receiving end, compute the sending-end voltage and percent voltage regulation.

PW **5.65** Open PowerWorld Simulator case Example 5_10 with the series capacitive compensation at both ends of the line in service. Graph the load bus voltage as a function of load real power (assuming unity power factor at the load). What is the maximum amount of real power that can be transferred to the load at unity power factor if the load voltage is always greater than 0.85 per unit?

PW **5.66** Open PowerWorld Simulator case Example 5_10 with the series capacitive compensation at both ends of the line in service. With the reactive power load fixed at 400 Mvar, graph the load bus voltage as the MW load is varied between 0 and 2600 MW in 200 MW increments. Then repeat with both of the series compensation elements out of service.

CASE STUDY QUESTIONS

a. For underground and underwater transmission, why are line losses for HVDC cables lower than those of ac cables with similar capacity?

b. Where are back-to-back HVDC converters (back-to-back HVDC links) currently located in North America? What are the characteristics of those locations that prompted the installation of back-to-back HVDC links?

c. Which HVDC technology can independently control both active (real) power flow and reactive power flow to and from the interconnected ac system?

REFERENCES

1. Electric Power Research Institute (EPRI), *EPRI AC Transmission Line Reference Book—200 kV and Above* (Palo Alto, CA: EPRI, *www.epri .com*, December 2005).

2. Westinghouse Electric Corporation, *Electrical Transmission and Distribution Reference Book*, 4th ed. (East Pittsburgh, PA, 1964).

3. R. D. Dunlop, R. Gutman, and P. P. Marchenko, "Analytical Development of Load-ability Characteristics for EHV and UHV Lines," *IEEE Trans. PAS*, Vol. PAS-98, No. 2 (March/April 1979): pp. 606–607.

4. W. D. Stevenson, Jr. *Elements of Power System Analysis*, 4th ed. (New York: McGraw-Hill, 1982).

5. W. H. Hayt, Jr. and J. E. Kemmerly, *Engineering Circuit Analysis*, 7th ed. (New York: McGraw-Hill, 2006).

6. M. P. Bahrman and B. K. Johnson, "The ABCs of HVDC Transmission Technologies," *IEEE Power & Energy Magazine*, 5, 2 (March/April 2007): pp. 32–44.

6 Power Flows

Tennessee Valley Authority (TVA) Regional Operations Center (Courtesy of TVA.)

Successful power system operation under normal balanced three-phase steady-state conditions requires the following:

1. Generation supplies the demand (load) plus losses.
2. Bus voltage magnitudes remain close to rated values.
3. Generators operate within specified real and reactive power limits.
4. Transmission lines and transformers are not overloaded.

The power flow (sometimes also called the *load flow*) is the basic tool for investigating these requirements. The power flow determines the voltage magnitude and angle at each bus in a power system under balanced three-phase steady-state conditions. It also computes real and reactive power flows for all equipment interconnecting the buses, as well as equipment losses.

Both existing power systems and proposed changes, including new generation and transmission, are of interest.

Conventional nodal or loop analysis is not suitable for power flow studies because the input data for loads are normally given in terms of power, not impedance. Also, generators are considered to be power sources, not voltage or current sources. The power flow problem is therefore formulated as a set of nonlinear algebraic equations suitable for computer solution.

Sections 6.1 through 6.3 review some basic methods, including direct and iterative techniques for solving algebraic equations. Then Sections 6.4 through 6.6 formulate the power flow problem, specify input data, and present two solution methods: Gauss-Seidel and Newton-Raphson. Means for controlling power flows are discussed in Section 6.7. Sections 6.8 and 6.9 introduce sparsity techniques and a fast decoupled power flow method, while Section 6.10 discusses the dc power flow, and Section 6.11 considers the power flow representation of wind turbine generators. Formulations for economic dispatch and optimal power flow are given in Sections 6.12 and 6.13

Since balanced three-phase steady-state conditions are assumed, this chapter uses only positive-sequence networks. Also, all power flow equations and input/output data are given in per unit (p.u.).

CASE STUDY

During this century, renewable energy sources, including solar and wind generation, are projected to substantially increase in the United States and worldwide. High penetrations of wind and solar generation can induce increased cycling of fossil-fueled power plants. The following case study examines the operational impacts of up to 35% wind and solar penetrations in the western United States. The primary impact of displacing fuel, as well as displacing fuel costs and emissions associated with fossil-fueled power plants, and the secondary impact of increased cycling, including costs and emissions associated with cycling of fossil-fueled power plants, are analyzed.

Finding Flexibility: Cycling the Conventional Fleet

By Debra Lew, Greg Brinkman, Nikhil Kumar, Steve Lefton, Gary Jordan, and Sundar Venkataraman

Adding new generation, load, or transmission to the grid changes the operation of the incumbent power system. Wind and solar generation plants are no different, but their impact on the rest of the grid is exacerbated by the facts that wind and solar energy is nondispatchable and such generators produce variable output. And because wind and solar effectively bid into the market at very low or negative cost, they are preferred resources in the dispatch stack. They are used by system operators whenever possible, unless there are generator operating limits or transmission constraints.

At low wind and solar penetrations or with low-variability resources (e.g., from high geographic diversity), the impact on the rest of the system may be small. But at high penetrations or with high-variability resources, wind and solar can induce increased cycling of the fossil-fueled fleet. This means that coal and gas generators may be asked to start up and shut down, ramp up and down, or operate at minimum generation levels more frequently. Cycling has impacts on emissions and on the wear-and-tear costs of the fossil-fueled fleet.

Coal and gas generators tend to have additional emissions at start-up and possibly also during ramps. Emissions are also affected by the output level of the generator. For example, units tend to be less efficient at partial load, thus increasing the carbon dioxide (CO_2) rates at minimum generation levels.

Starts and ramps also lead to increased wear and tear on the unit components and systems. Temperature and pressure changes lead to stresses that can result in premature component failure, an increased need for maintenance and overhauls, and more frequent repairs.

The primary impact of a MWh of wind or solar energy is to displace a MWh of other generation, typically fossil-fueled generation. Displacing a MWh of fossil-fueled generation displaces the costs and emissions associated with that fuel. But a secondary impact of this wind and solar energy can be to increase cycling of the fossil-fueled generators. As we have said, there are wear-and-tear costs and emissions impacts associated with cycling. This raises two questions: how big are those secondary impacts, and do they significantly negate the primary impacts that wind and solar energy bring to the table? Some recent analyses have claimed that wind actually increases overall emissions or that the avoided emissions from wind have been significantly overestimated.

Many integration studies have examined the impacts of high penetrations of wind power on particular systems and on the operation of the conventional fleet. Very few have considered cycling impacts in detail, partly because of the lack of data on cycling costs and emissions. This article describes the first study that combines detailed data on wear-and-tear costs and the emissions impacts of cycling with operational simulations of the entire western U.S. grid with high penetrations of wind and solar to determine these impacts.

The Western Wind and Solar Integration Study: Phase 2

In 2011, GE Energy, Intertek AIM, the National Renewable Energy Laboratory (NREL), and RePPAE began the Western Wind and Solar Integration Study Phase 2 (WWSIS-2). This study built on the Phase 1 effort, which had examined up to 35% wind and solar penetrations in subregions in the Western Interconnection and mitigation options for integrating those resources. The conclusion of Phase 1 was that several changes in operating practices would be needed to integrate the wind and solar energy. The two most important of these were increased balancing area coordination and intrahour scheduling. There are many efforts being considered now in the western United States to implement these changes, including the creation of an energy imbalance market.

The goal of WWSIS-2 was to investigate the cycling impacts on the fossil-fueled fleet in detail and to determine if the wear-and-tear costs and the impacts on emissions significantly reduced the benefits of wind and solar power. The point of this study was not to determine whether wind and solar plants should be built, but rather to understand what the impacts on the fossil-fueled fleet would be if wind and solar were built to a high penetration—especially the impacts on costs and emissions.

To do this, we simulated future power system operations under varying levels of wind and solar penetration. Any simulation of the grid requires simplifying assumptions in order to develop a model that can be run with a reasonable amount of computing power and time. Power system operation in a future scenario with high penetrations of wind and solar is likely to be different from today. Additionally, we did not have access to confidential information such as bilateral contracts for power or transmission flows. The key differences between our model and today's operation are:

- We assumed a natural gas price that varied subregionally but averaged US$4.36/GJ across the interconnection. We did not model a carbon tax or any renewable energy incentive, such as the production tax credit.
- We did not model bilateral contracts for power or transmission flows but assumed a least-cost unit commitment and economic dispatch.
- We modeled the nearly 40 standalone balancing authorities as 20 zones, each holding its own contingency and flexibility reserves, with no hurdle rates between them.

New Data Sets

NREL analyzed measured emissions from every power plant in the United States (using the Environmental Protection Agency's continuous emissions-monitoring data) to determine incremental emissions due to a start or ramp as well as emissions rates from part loading. This analysis was used to create a unit-specific emissions data set for cycling and noncyclic operation. These unit-specific emissions were used in the detailed operational simulations. For display purposes, however, we have averaged the data by unit type to show high-level results.

Heat rates as a function of load were examined for combined cycle (CC), combustion turbine (CT), coal, and gas steam units. Table 1 shows the resulting penalties for operating at part load. This penalty is defined as the percentage increase in emissions rate (kg/MWh) when the average unit is operating at 50% of maximum capacity. CC units are the most efficient at full load and part load; CC and CT units have the most significant emissions penalties for operating at 50% compared with 100% of maximum generation, however.

NO_x emissions as a function of load were analyzed similarly and are shown in Table 1. Steam units (coal and gas) emit approximately an order of magnitude more NO_x per MWh than gas CC units and CT units. Although part-load operation leads to an NO_x penalty for the CC and CT units, such operation was found to benefit the coal-fired steam units. For example, coal units operating at 50% were found to emit 14% less NO_x per MWh compared with full-loading operation; gas CC units were found to emit 22% more NO_x per MWh at 50% load compared with full load. Most of the NO_x from all units is created from nitrogen in the combustion air ("thermal" NO_x), as opposed to in the fuel, so flame temperature is likely a primary driver of NO_x emissions.

Because of the significant part-time usage of pollution control equipment for SO_2, it was impossible to create part-load emission curves.

Starting an off-line, fossil-fueled unit increases emissions for two reasons. First, it takes fuel to bring the unit online, and that fuel adds emissions without adding energy to the grid. Second, starts can increase emissions rates because most pollution-control equipment does not become fully effective until flame and flue gas temperatures are in the proper range. Table 2 shows the start penalties for different types of units and different emissions. For example, a coal unit emits 0.90 kg/MW of capacity of excess NO_x during start-up. This is equivalent to operating the unit at full load for approximately 30 minutes.

Ramping, or load-following, emissions were estimated in a similar way to the start-up emissions. A ramp was defined as an increase of output of 30% of maximum capacity (e.g., from 70% to 100% of maximum

Type of Unit	CO_2 (%)	NO_x (%)
Coal	5	−14
Gas CC	9	22
Gas CT	18	15
Gas steam	6	−14

TABLE 1

Emissions penalty for part-load operation

Type of Unit	Heat Input per Start (GJ/MW)	NO_x per Start (kg/MW)	SO_2 per Start (kg/MW)
Coal	17.4	0.90	1.8
Gas CC	2.1	0.24	n/a
Gas CT	3.7	0.36	n/a
Gas steam	14.5	0.38	n/a

TABLE 2

Start-up emissions per MW of capacity

capacity). Table 3 presents generation-weighted averages. Ramping emissions are much lower than start-up emissions. The most significant ramping emission impact is the NO_x emissions from coal units (equivalent to 10–15 minutes of full-load operation).

While emissions at various operating levels are reasonably well understood, many utilities in the western United States do not know the wear-and-tear costs of cycling their fossil-fueled units. First, they have not needed to know these costs because until very recently, many of these plants, which were designed as base load plants, were not cycled. Second, determining these costs is complex because cycling operation today may not have cost implications until several years in the future.

Intertek AIM studied wear-and-tear costs from cycling for hundreds of fossil-fueled units around the world. For each unit, Intertek AIM had determined a best fit and a lower-bound and upper-bound fit for cycling costs, where the bounds reflected the uncertainty range for that plant. While specific data from those studies were confidential, aggregated data from those studies could be used as generic wear-and-tear costs for similar units that have not been studied. In this way, we were able to define variable operations and maintenance (VOM) costs for a hot, warm, and cold start; a ramp (typical); and for noncyclic operation for different types of plants. Table 4 shows the lower-bound costs for the different plant types.

Upper-bound costs were also used in this study. The raw

Type of Unit	Heat Input per Ramp (GJ/MW)	NO_x per Ramp (kg/MW)	SO_2 per Ramp (kg/MW)
Coal	0.60	0.33	0.37
Gas CC	0.08	0.00	n/a
Gas CT	0.30	0.01	n/a
Gas steam	−0.09	0.02	n/a

TABLE 3

Ramping emissions per MW of capacity

	Small Subcritical Coal	Large Subcritical Coal	Super-Critical Coal	Gas, Combined Cycle	Gas, Large-Frame Combustion Turbine	Gas, Aero-derivative Combustion Turbine	Gas, Steam
Hot start (US$/MW)	94	59	54	35	32	19	36
Warm start (US$/MW)	157	65	64	55	126	24	58
Cold start (US$/MW)	147	105	104	79	103	32	75
Ramp (US$/MW)	3.34	2.45	1.96	0.64	1.59	0.63	1.92
Noncyclic operation (US$/MWh)	2.82	2.68	2.96	1.02	0.57	0.66	0.92

TABLE 4

Lower-bound median costs for cycling for various generation types

upper-bound data is confidential, however; only aggregated results can be presented here. All the cycling cost estimates used for this study are for typical units of various types; they are not unit-specific. The worst units for cycling are older base load power plants that should be retrofitted prior to significant cycling, using countermeasures such as procedure and chemistry changes and hardware retrofits. Without these measures, cycling could potentially lead to costly high-impact, low-probability events. Ongoing studies are examining the costs and benefits of retrofitting coal- and gas-fired power plants for increased flexibility.

Five Scenarios: Wind versus Solar

We used these new data sets in a commercial production simulation tool, PLEXOS, to model grid operations of the Western Interconnection on a 5-minute time step, because wind and solar output varies within the hour. We used the Western Electricity Coordinating Council (WECC) Transmission Expansion Policy Planning Committee (TEPPC) 2020 data set as a basis for our model, because it has been widely vetted among western U.S. utilities. A key change in assumptions was the natural gas price: the 2020 case average gas price was high, so we used the 2022 average gas price of US$4.36/GJ.

We created five scenarios to examine increasing penetration levels of variable generation and also to compare wind with solar. Solar included rooftop photovoltaics (PV), utilityscale PV, and concentrating solar power (CSP) with six hours of thermal storage. The scenarios were as follows:

- **No renewables:** 0 MW wind, 0 MW solar (0% wind, 0% solar)
- **TEPPC:** 27,900 MW wind; 7074 MW PV; 4352 MW CSP (9.4% wind, 3.6% solar)
- **High wind:** 63,840 MW wind; 20,064 MW PV; 6536 MW CSP (25% wind, 8% solar)
- **High mix:** 43,118 MW wind; 40,374 MW PV; 13,997 MW CSP (16.5% wind, 16.5% solar)
- **High solar:** 23,357 MW wind; 61,941 MW PV; 21,526 MW CSP (8% wind, 25% solar).

Penetration levels refer to energy, not capacity, and are penetration levels for the U.S. portion of the Western Interconnection only, because data for Canada and Mexico were lacking. The scenario without any renewables is an unrealistic one because existing wind and solar plants have been removed, but it was created to examine the impacts of all the wind and solar on the system.

Table 5 shows the penetration levels for each scenario. Wind and solar were nominally built to 33% energy penetration considering the historical weather patterns of 2004, 2005, and 2006. The analysis was for the load and weather pattern of 2006, which had a typical solar profile but better wind than the average of the years 2004–2006. After curtailment (curtailment includes CSP storage curtailment, some of which is built into the design of the generator), the 2006 penetration levels are 30–33% of U.S. load in the Western Interconnection and 24–26% of total load in the Western Interconnection. Because Canada and Mexico have relatively small connections with the United States in the Western Interconnection (compared with the total size of the U.S. portion), the impacts of renewables in the United States should be compared with the nominal 33% penetration numbers. Certain outputs, such as total change in emissions throughout the Western Interconnection, should be compared with the total Western Interconnection penetration levels (nominally 26%).

	2006 Penetration Level Across U.S. WECC After All Curtailment	2006 Penetration Level Across All WECC After All Curtailment
TEPPC	13.2	10.5
High wind	32.6	26.0
High mix	32.2	25.6
High solar	30.2	24.1

TABLE 5

Renewable energy penetration levels

To bring resources to load, we expanded the capabilities of existing transmission paths by iterating production cost runs until shadow prices across paths were within fixed, consistent cutoff values.

Simulating Western Grid Operations

We committed base load units one day ahead using synthesized day-ahead wind and solar forecasts. We then committed gas units four hours ahead using synthesized four-hour-ahead wind and solar forecasts. Finally, we ran a 5-minute real-time dispatch.

Balancing the system can be more challenging with higher penetrations of variable generation. We therefore increased operating reserves to accommodate the wind and solar, as shown in Table 6. We used dynamic reserve requirements, and Table 6 shows the average level of operating reserves held throughout the year. We did not increase contingency reserves, which were held zonally, because neither wind nor solar plants were the single largest contingency. We adjusted other operating reserves, however, based on the wind and solar output during each hour of the year. Regulating reserves were held equivalent to 1% of load plus 90% of the 10-minutes wind and solar variability. The high wind and solar penetrations required up to 10% more regulating reserves than the scenario without any renewables. We also held a new reserve category—"flexibility reserves"—to account for the 60-minutes variability of wind and solar. These amounted to 1–3% of the installed capacity of the wind and solar generating facilities.

All of this work was overseen by a technical review committee consisting of about 50 utility staff members, researchers, and power plant experts. They met approximately every other month to review inputs, assumptions, methodologies, and results.

System Operation with High Penetrations of Wind and Solar

We analyzed the 5-minute operational simulations, with a focus on cycling impacts of the coal and gas units. All results reflect the specific characteristics of the generators and transmission of the Western Interconnection.

During the summer, load is high, and the impacts of variable generation on cycling are modest. In the spring and fall, however, load is low, and both wind and solar output are high, resulting in challenging operational conditions. This makes for a low net load: net load is the load minus the variable generation. Five-minute dispatch from the most challenging week of the year, as defined by the minimum net load condition,

Scenario	Contingency (MW)	Regulation (MW)	Flexibility (MW)
No renewables	3361	1120	0
TEPPC	3361	1158	1193
High wind	3361	1236	2599
High mix	3361	1211	2035
High solar	3361	1207	1545

TABLE 6

Contingency, regulating, and flexibility reserves

is shown in Figure 1. At low penetrations, as shown in Figure 1(a), there is little impact on the rest of the system except on the minimum net load day of March 29. For the high-wind scenario, shown in Figure 1(b), however, most of the gas CC generation has been displaced; there is curtailment on March 29; and coal generation declines markedly over the week. In the high-solar scenario, shown in Figure 1(c), the high midday

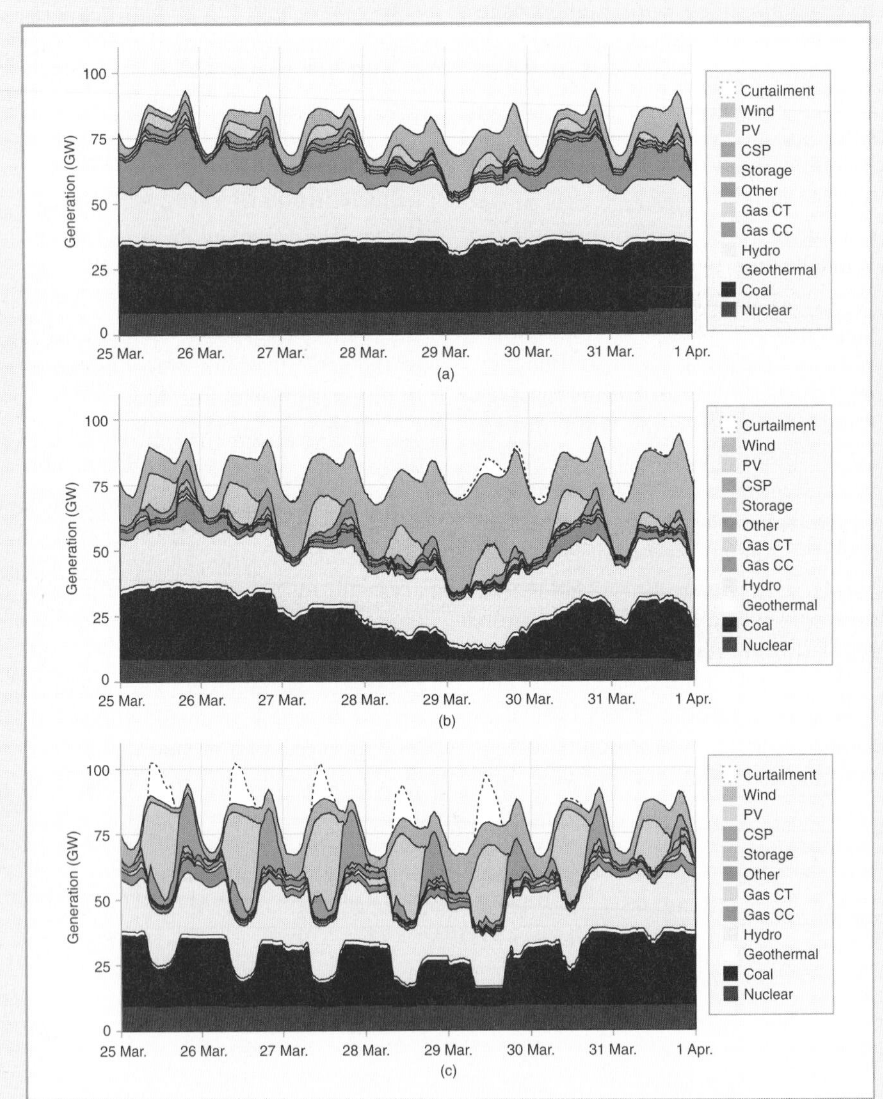

Figure 1 Five-minute dispatch stacks for the (a) TEPPC, (b) high-wind, and (c) high-solar scenarios for a week in March. This week represented the most challenging week, as defined by the minimum net load condition.

output leads to curtailment and the backing down of coal generation on a nearly daily basis in the middle of the day. Gas CT units turn on in the evening to help meet the evening load peak.

Despite these challenges, we find that the system can operate and balance load and generation. There were no regulating reserve violations and very few contingency reserve violations. Figure 2 shows that wind and solar mostly displace gas CC generation. Displacement of coal increased with increasing penetrations of wind because gas tends to be decommitted or backed down already at night, when there are high levels of wind. Curtailment of potential wind and solar generation on an annual basis was as much as 5% in the high-penetration scenarios. The high-mix scenario saw the least curtailment (1.6%).

The biggest impact from wind and solar on cycling of other generation is the increased ramping of the coal units, as shown in Figure 3(b). Starts for coal units, as shown in Figure 3(a), change little. Gas CCs start more with low wind and solar penetrations, but at high penetrations starts are similar to those in the scenario with no renewables. Gas CTs cycle more with high solar penetrations and less with high wind penetrations.

Figure 4 depicts coal starts and ramps for the challenging week in March. At low penetrations, there is little change in coal commitments, and the coal units are typically running at or near full output. In the high-penetration scenarios, coal capacity is shut down approximately each week, and the coal is ramped up and down each day, especially with high penetrations of solar.

Emissions Reductions Are Significant

We find that wind- and solar-induced cycling offsets a very small percentage of the wind- and solar-induced emissions reductions of CO_2, NO_x, and SO_2 across the Western Interconnection that are more than offset by the emissions reductions due

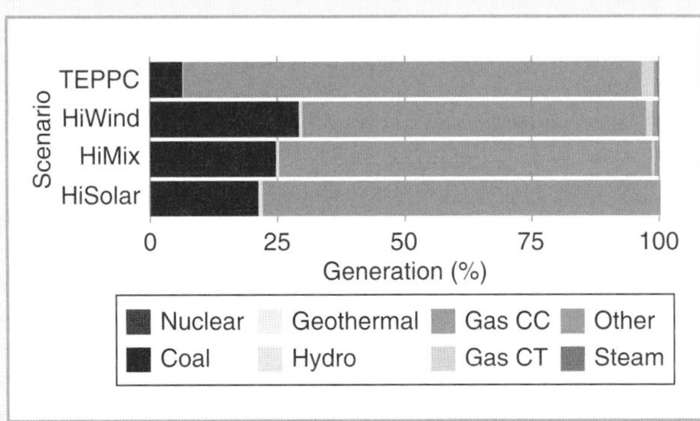

Figure 2 Generation displaced by wind and solar, as compared with the scenario without any renewables

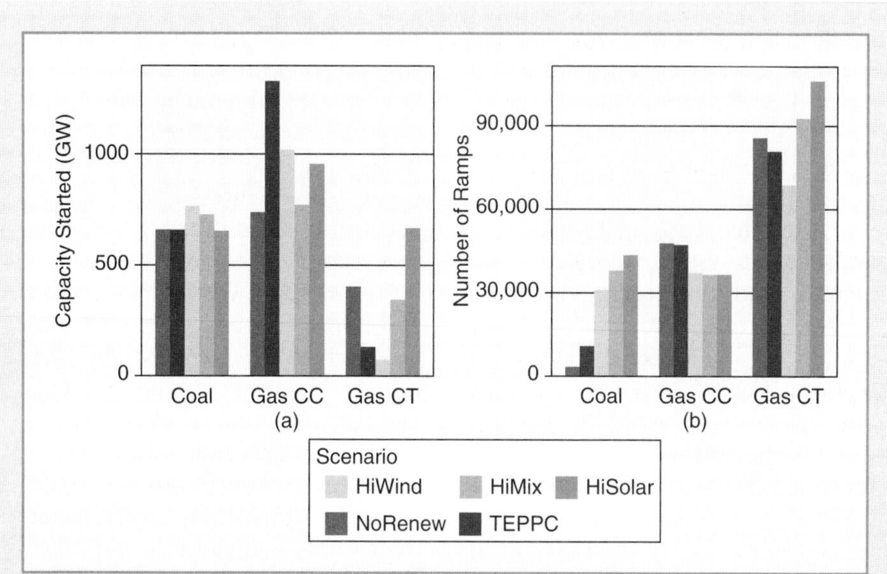

Figure 3 (a) Capacity started and (b) total number of ramps for each plant type by scenario for one year

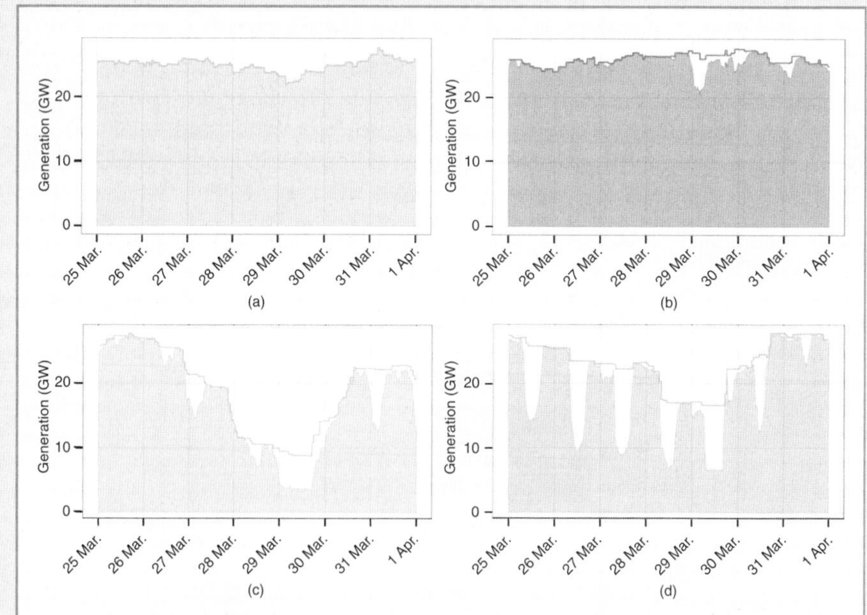

Figure 4 Capacity committed (solid line) and dispatched (shaded area) for coal units during a sample week in March. The white area between the solid line and the shaded area illustrates how far the capacity is backed down (a) no renewables, (b) TEPPC, (c) high wind, and (d) high solar.

to the displacement of fossil-fueled generation. Compared with the scenario without any renewables, the high-penetration scenarios (nominally 33% wind and solar in the U.S. portion of the Western Interconnection, resulting in 26% wind and solar across the interconnection) reduce CO_2 by 29–34%, NO_x by 16–22%, and SO_2 by 14–24%, including the cycling impacts. CO_2 emissions are reduced by a greater percentage than the wind and solar penetration level because wind and solar preferentially displace fossil-fueled generation, while typically the generation in the western United States is a combination of hydro, nuclear, fossil-fueled, and other renewable generation.

Figure 5(a) shows the total CO_2 emissions for each scenario. Ramping had no significant impact on CO_2 emissions, so those estimates are not shown. The start-up CO_2 emissions (shown by the thin, dark line at the top of each bar) were negligible in all cases. Figure 5(b) shows the CO_2 emissions saved by each MWh of wind/solar. Avoided CO_2—considering part-load, ramping, and start impacts—was 500–540 kg/MWh of wind and solar produced in the high-penetration scenarios (see Table 7). CO_2 emissions from starts were negligible here as well. We also calculated the part-load penalty— which was the incremental CO_2 emissions from part loading—as negligible. These values reflect aggregate emissions across the Western Interconnection; any specific plant might have lower or higher emissions than those shown here.

From the fossil-fueled plant perspective, average CO_2 emission rates of coal, CCs, and CTs change only slightly with wind and solar, as shown in Figure 6(a). Figure 6(b)

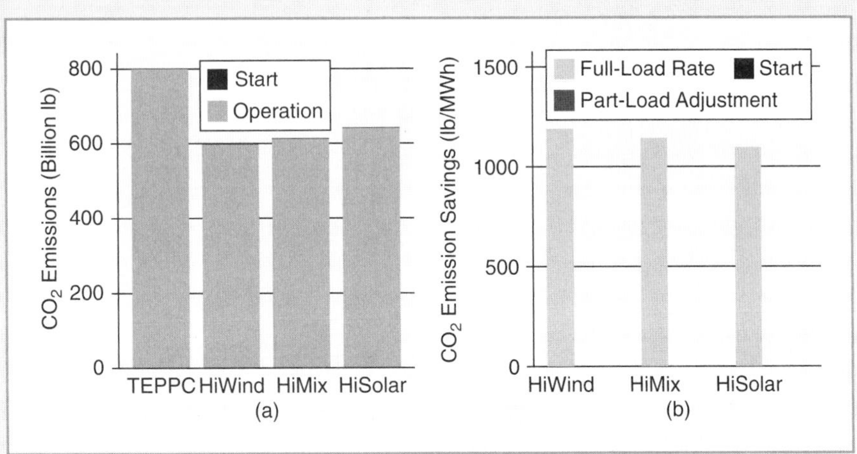

Figure 5 *CO_2 emissions by scenario: (a) absolute CO_2 emissions for operation and starts and (b) CO_2 emissions reductions compared with the scenario without renewables, separated into the constant emissions rate assumption and adjustments for part load and starts. Ramping emissions are excluded because they have no significant impact on CO_2 emissions. Note: 1 lb. = 0.45 kg*

Scenario	Avoided CO_2 (kg/MWh)	Avoided NO_x (kg/MWh)	Avoided SO_2 (kg/MWh)
High wind	540	0.42	0.25
High mix	522	0.36	0.20
High solar	500	0.33	0.16

TABLE 7

Emissions avoided per MWh of wind and solar, considering part-load, ramping, and start impacts. Part-load impacts were not included for SO_2 because of inadequate data

shows that adding wind and solar can positively or negatively affect emissions rates, depending on the plant type and scenario. For coal and CC units, wind and solar generally improves emissions rates by up to 1%. The largest negative impact of wind- and solar-induced cycling is found in the high-wind scenario and for the CT units, where the emissions rate increases by 2%. This is on average; individual units might be more or less affected.

Figure 7 shows the analysis for NO_x emissions. There was a negligible impact of starts on NO_x. Ramping

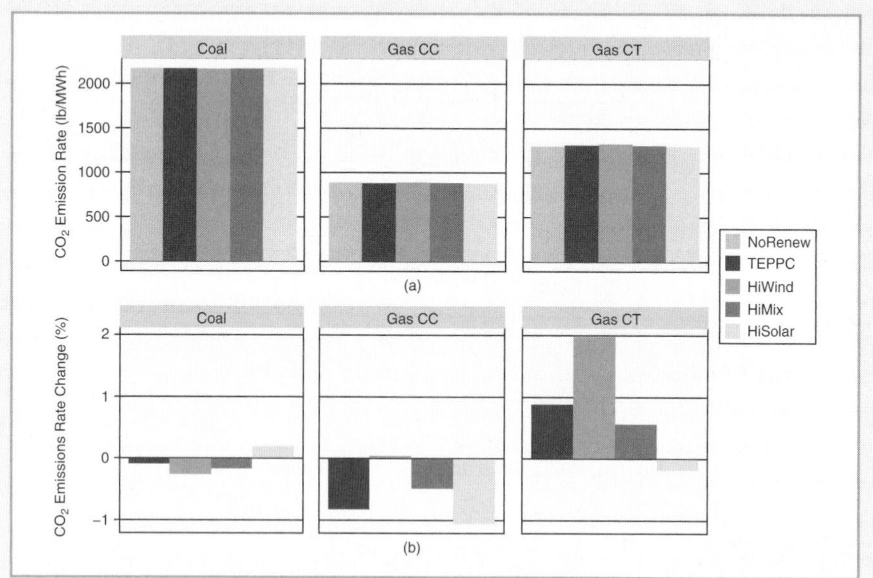

Figure 6 (a) Average CO_2 emission rates by plant type (defined as CO_2 emissions divided by MWh of coal, CC, or CT generation) for each scenario and (b) changes in emissions rates as compared with the scenario without renewables. Note: 1 lb. = 0.45 kg

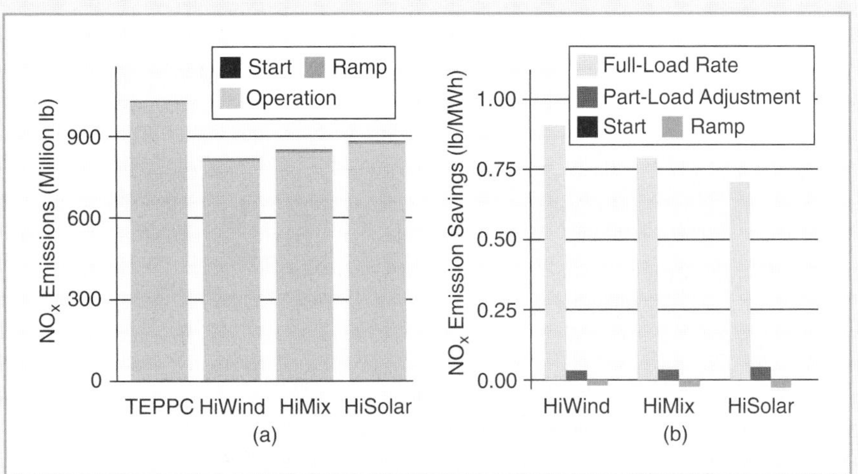

Figure 7 NO_X emissions by scenario: (a) absolute NO_X emissions for operation, ramps, and starts, and (b) NO_X emissions reductions compared with the scenario without renewables, separated into the constant emissions rate assumption and adjustments for part-load, ramps, and start impacts. Note: 1 lb. = 0.45 kg

reduced the avoided NO_x by 2–4%. This is shown in Figure 7(b) as a small negative contribution. Part-loading impacts, on the other hand, increased avoided NO_x by 4–6%. On average, coal units in the western United States emit less NO_x per MWh of generation at part load. The net impact of considering cycling improved avoided NO_x emissions from wind and solar by 1–2%.

Figure 8 shows the emissions analysis for SO_2. Because there were inadequate data to create SO_2 emission part-load curves, part-load impacts were not studied for SO_2. Ramping impacts on avoided SO_2 were modest for the high-penetration scenarios, reducing avoided SO_2 by 2–5%. Start-up emissions affected the avoided emissions rates by significantly less than 1%.

Wind- and Solar-Induced Cycling Operational Costs

The production simulation analysis undertaken in this study quantifies (1) the operational impacts of wind and solar displacing other generation and (2) the cycling costs induced by wind and solar. Operational costs (the industry term is production cost) include fuel, noncyclic VOM, and cycling costs. Cycling costs includes costs for starts, ramps and start-up fuel.

Under the scenarios studied, we find that the high wind and solar penetrations affect the grid by displacing US$7 billion/year in fuel costs and inducing an increase of US$35–157 million/year in cycling costs of fossil-fueled plants. The overall production cost for each scenario is shown in Figure 9, with the lower- and upper-bound uncertainty ranges shown.

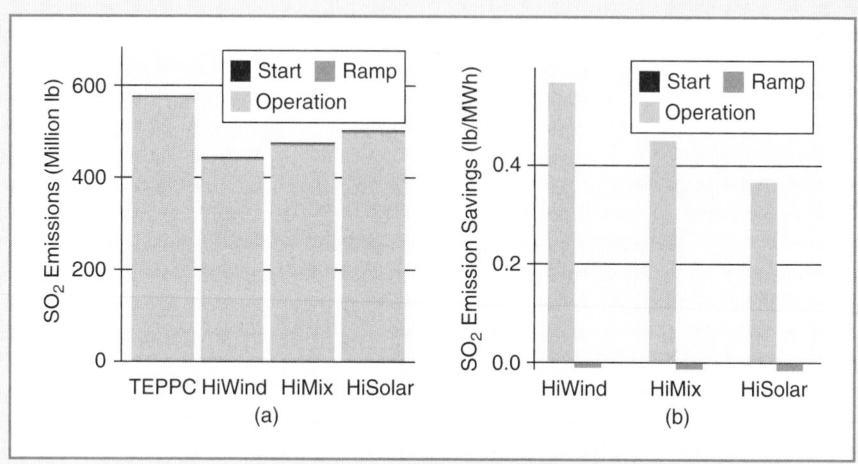

Figure 8 *SO_2 emissions by scenario: (a) absolute SO_2 emissions for operation, ramping, and starts and (b) SO_2 emission reductions compared with the scenario without renewables, separated into the constant emissions rate assumption and adjustments for ramps, and start impacts. Part-load impacts were not studied because of inadequate data. Note: 1 lb. = 0.45 kg*

At an average gas price of US$4.36/GJ, fuel dominates the production cost savings as wind and solar penetration increases. It is important to note that production cost does not include the capital costs of renewable or thermal generation or transmission.

Figure 10 details the cycling costs of each scenario. The cycling costs range from about US$270 million in the scenario without renewables, using lower-bound costs, to about US$800 million in the high-solar scenario, using upper-bound

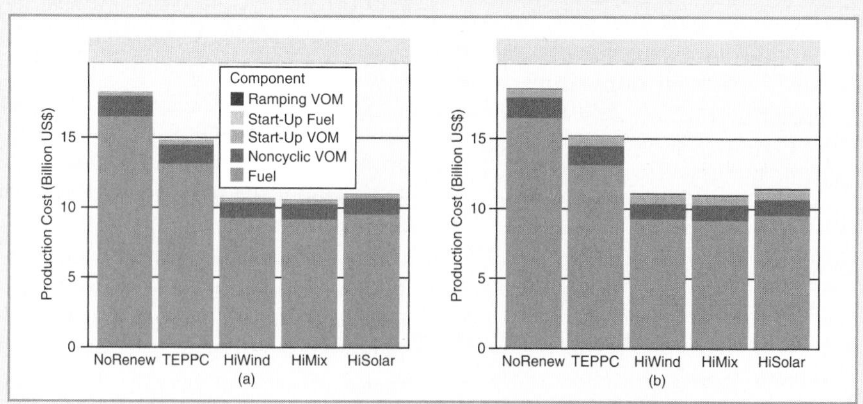

Figure 9 *Production cost for each scenario showing the (a) lower-bound and (b) upper-bound cycling costs. These costs do not include capital costs of renewable or thermal generation or transmission.*

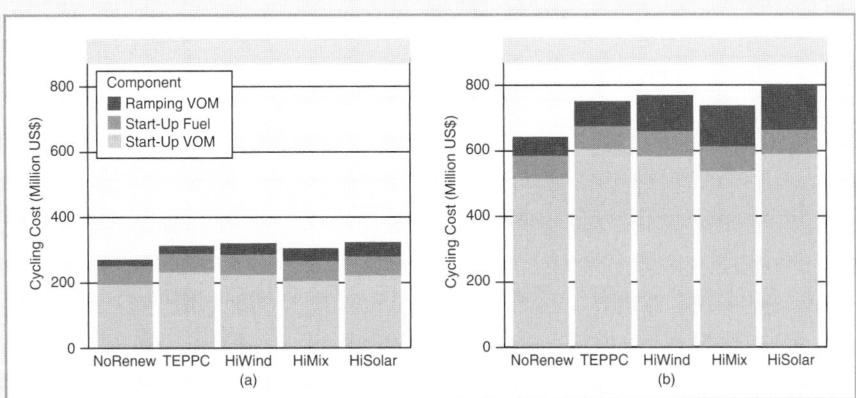

Figure 10 Production cost components resulting from cycling, showing the (a) lower-bound and (b) upper-bound wear-and-tear costs for each scenario. Cost components have been broken down into starts, start fuel, and ramping costs.

cycling costs. Interestingly, the high-mix scenario has a higher wind and solar penetration but lower cycling costs than the TEPPC scenario. In these scenarios, going from no wind and solar penetration to 13% nominal wind and solar penetration induces higher cycling costs than going from 13% to 33% penetration. On a per-MWh-of-fossil-fueled-generation basis, the increased cycling costs (compared with the scenario without any renewables) are US$0.18–0.44/MWh, US$0.52–1.24/MWh, US$0.47–1.14/MWh, and US$0.50–1.28/MWh for the TEPPC, high-wind, high-mix, and high-solar scenarios, respectively. The ranges represent the uncertainty range in the cycling cost inputs.

Figure 11 shows the lower-bound costs from the perspective of a coal, gas CC, or gas CT unit, in terms of cycling costs per MWh of that unit's generation. Must-run gas CT units were excluded from this plot, as they skew the results. Cycling costs for coal increase modestly with high penetrations of wind and solar. Cycling costs for gas CC units increase significantly with increased wind and solar penetration. But the largest cycling costs (per MWh) are borne by the gas CT units, which are operated as peakers and cycle the most. Interestingly, gas CT units see lower cycling costs in the TEPPC scenario than in the scenario without any renewables at all. And in the high-wind scenario, their cycling costs are similar to the scenario without renewables.

These cycling costs are comparable to the noncyclic VOM of approximately US$2/MWh. The impact of wind and solar on the economic viability of a fossil-fueled plant can be significant if that plant's energy production is displaced by wind and solar, its revenue is reduced, and it now bears an increased cycling cost. That fossil-fueled plant may still be needed to help balance the system or provide power on peak demand

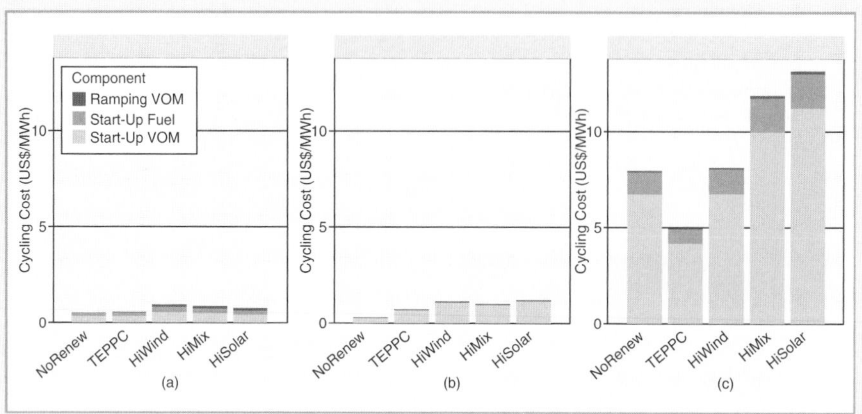

Figure 11 Lower-bound cycling cost for (a) coal, (b) gas CC units, and (c) gas CTs (excluding the must-run CTs). The total, system wide, lower-bound cycling costs were disaggregated by plant type and divided by MWh of generation for that plant type.

days. This raises a host of market and policy issues that require future analysis.

We performed a sensitivity analysis on the gas price to see if halving the price (to US$2.18/GJ) or doubling it (to US$8.72/GJ) significantly changed results. The US$8.72 gas price had little effect on operations. The US$2.18 gas price led to more gas usage and less coal usage, regardless of renewables on the system. The type of generation displaced by renewables was very similar for the three gas price estimates. Interestingly, the impact of wind and solar in the very low and very high gas price scenarios is to reduce overall cycling cost, because they displace starts for various unit types. The increase in cycling costs per MWh generated at fossil-fueled plants was similar in all the gas price scenarios. Units with increased cycling, however, will have a higher US$/MWh cost of generation and will probably see an increase in forced outage rates if adjustments to maintenance spending are not made.

An examination of the cycling impacts from a system perspective in Figure 12 shows the change in production cost (operational cost, not including capital or power purchase agreement costs) for each scenario as compared with the scenario without renewables. The primary operational impact of wind and solar is to displace a large amount of fuel cost (shown by the negative orange bars) and a small amount of noncyclic VOM (small negative blue bars). At this gas price (US$4.36/GJ), the secondary impact of wind and solar is to incur the startup VOM, start-up fuel, and ramping VOM shown by the small positive bars. While it is important to remember that these operational costs do not include the capital costs of any generator or transmission, one can see

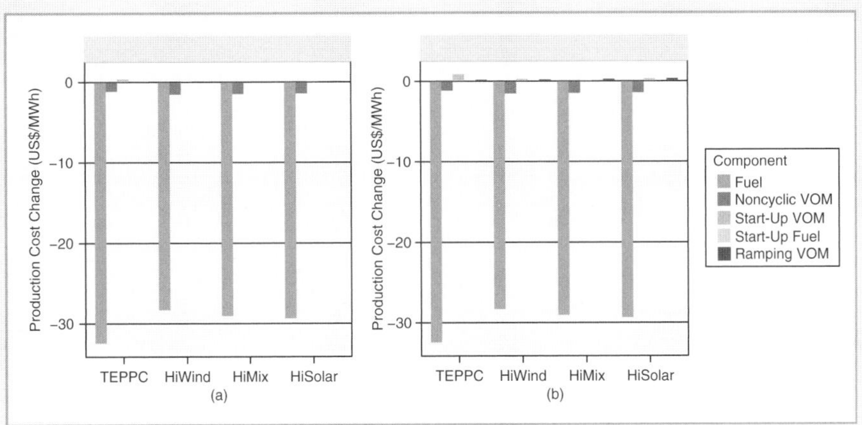

Figure 12 The change in production cost for each scenario relative to the scenario without renewables, per MWh of wind and solar generation, for the (a) lower-bound and (b) upper-bound wear-and-tear costs. Production costs do not include any fixed capital or power purchase agreement costs.

that the cycling costs are a small fraction of the costs of the fuel displaced. This cycling impact offsets the production cost reduction of wind and solar by US$0.14–0.67 per MWh of wind and solar generated. The net reduction in production cost compared with the scenario without renewables, including these cycling impacts, is approximately US$30 per MWh of wind and solar generated.

Wind Dominates Uncertainty, and Solar Dominates Variability

This study took advantage of recent advances in simulating large PV plants that allowed a comparison of the impacts of wind and solar on the grid. We conducted statistical analysis of variability and uncertainty (forecast error) to look at these impacts. We looked at extreme ramping events on an hourly and 5-minute basis and found that extreme ramping events were dominated by sunrise and sunset events. However, because we know when sunrise and sunset occur and, in fact, the path of the sun through the sky each day, we can plan for these events. When the solar diurnal variability is removed, PV variability due to weather is found to be similar to wind variability.

Statistical analysis of forecast errors showed that our day-ahead forecast errors were driven by wind uncertainty. Day-ahead solar forecasts were more accurate than day-ahead wind forecasts. Because forecast accuracy improves as one approaches the time in question, the four-hour-ahead forecasts were much more accurate, as shown in Figure 13. We used a four-hour-ahead unit commitment to commit gas units, similar to operations of the California Independent System Operator. This helped mitigate the uncertainty of the wind day-ahead forecasts.

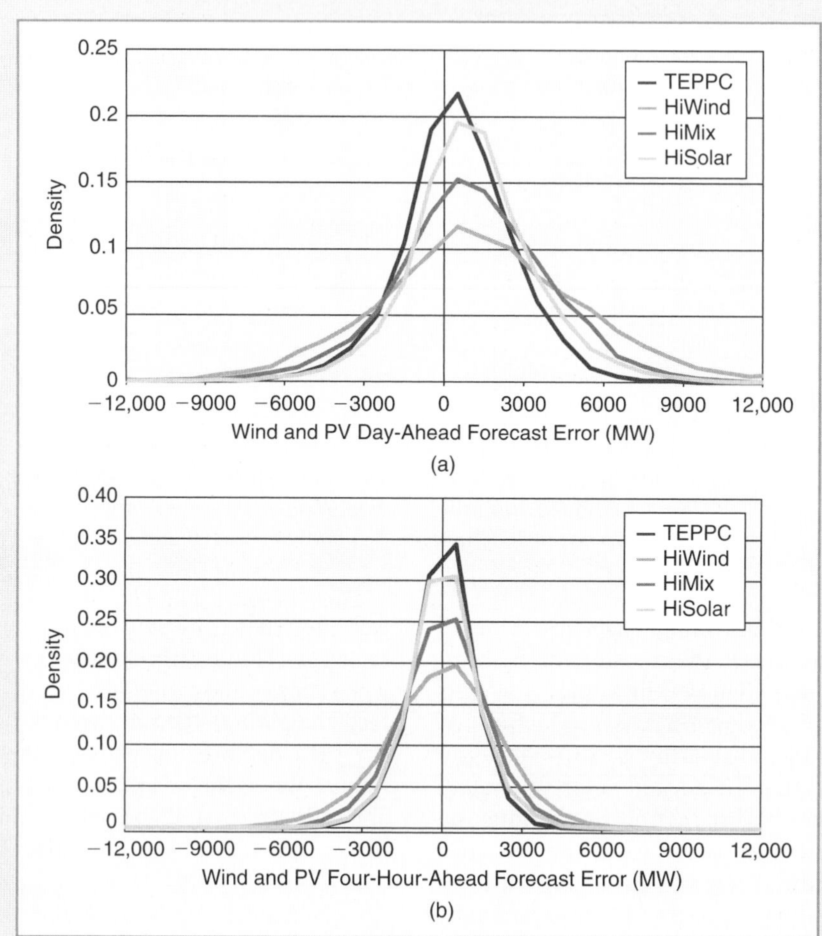

Figure 13 (a) Day-ahead and (b) four-hour-ahead wind and PV forecast errors for each scenario

Conclusions

For the first time, we have conducted an operational simulation of wind and solar impacts across the entire Western Interconnection using detailed data on cycling costs and cycling emissions. Our three high-penetration scenarios model a nominal 33% wind and solar penetration across the U.S. portion of the Western Interconnection, resulting in 26% nominal penetration across the entire Western Interconnection. We examine the primary operational impact of wind and solar in displacing fuel (and the costs and emissions associated with the fuel), along with the secondary impact of increased cycling (and the costs and emissions associated with cycling).

We found that wind- and solar-induced cycling has a small impact on avoided emissions of CO_2, NO_x, and SO_2. In our high-penetration scenarios, cycling reduces avoided CO_2 emissions by 0.2%, improves avoided NO_x by 1–2%, and lessens avoided SO_2 by up to 5%. The net result is that wind and solar in our high-penetration scenarios reduce CO_2 by 29–34%, NO_x by 16–22%, and SO_2 by 14–24%, inclusive of cycling impacts.

We also found that the secondary impact of increased cycling incurs costs that are a small fraction of the displaced fuel costs. In our high-penetration scenarios, wind and solar induce additional annual cycling costs of US$35–157 million. This same wind and solar also displaces about US$7 billion annually in fuel costs. In the high-penetration scenarios, the increase in cycling cost for the average fossil-fueled plant ranges from US$0.47–1.14 per MWh of fossil-fueled generation in the high-mix scenario to US$0.50–1.28 per MWh in the high-solar scenario. These additional costs, combined with reduced generation and revenue, beg the question of market and policy changes that may be required in a potential future with high levels of wind and solar energy.

For Further Reading

GE Energy. (2010). "Western wind and solar integration study." NREL. Golden, CO. NREL/SR-550-47434 [Online]. Available: http://www.nrel.gov/docs/fy10osti/47434.pdf.

N. Kumar, P. Besuner, S. Lefton, D. Agan, and D. Hilleman. (2012). "Power plant cycling costs." Intertek AIM. NREL. Sunnyvale, CA, Golden, CO. NREL/SR-5500-55433 [Online]. Available: http://www.nrel.gov/docs/fy12osti/55433.pdf.

D. Lew, G. Brinkman, E. Ibanez, A. Florita, M. Heaney, B.-M. Hodge, M. Hummon, G. Stark, J. King, S. Lefton, N. Kumar, D. Agan, G. Jordan, and S. Venkataraman. (2013). "The western wind and solar integration study phase 2." NREL. Golden, CO. NREL/TP-5500-55588. [Online]. Available: http://www.nrel.gov/docs/fy13osti/55588.pdf.

M. Milligan, K. Clark, J. King, B. Kirby, T. Guo, and G. Liu. (2013). "Examination of potential benefits of an energy imbalance market in the western interconnection." NREL. Golden, CO. NREL/TP-5500-57115. [Online]. Available: http://www.nrel.gov/docs/fy13osti/57115.pdf.

Biographies

Debra Lew is with the National Renewable Energy Laboratory, Golden, Colorado.

Greg Brinkman is with the National Renewable Energy Laboratory, Golden, Colorado.

Nikhil Kumar is with Intertek, Sunnyvale, California.

Steve Lefton is with Intertek and Global Utility Consultants LLC, Sunnyvale, California.

Gary Jordan is with GE Energy Consulting, Schenectady, New York.

Sundar Venkataraman is with GE Energy Consulting, Phoenix, Arizona. ■

6.1 DIRECT SOLUTIONS TO LINEAR ALGEBRAIC EQUATIONS: GAUSS ELIMINATION

Consider the following set of linear algebraic equations in matrix format:

$$\begin{bmatrix} A_{11} & A_{12} & \cdots & A_{1N} \\ A_{21} & A_{22} & \cdots & A_{2N} \\ \vdots & & \vdots & \\ A_{N1} & A_{N2} & \cdots & A_{NN} \end{bmatrix} \begin{bmatrix} x_1 \\ x_2 \\ \vdots \\ x_N \end{bmatrix} = \begin{bmatrix} y_1 \\ y_2 \\ \vdots \\ y_N \end{bmatrix} \tag{6.1.1}$$

or

$$\mathbf{Ax} = \mathbf{y} \tag{6.1.2}$$

where **x** and **y** are N vectors and **A** is an $N \times N$ square matrix. The components of **x**, **y**, and **A** may be real or complex. Given **A** and **y**, the objective is to solve for **x**. Assume the det(**A**) is nonzero, so a unique solution to (6.1.1) exists.

The solution **x** can be obtained easily when **A** is an upper triangular matrix with nonzero diagonal elements. Then (6.1.1) has the form

$$\begin{bmatrix} A_{11} & A_{12}\ldots & & A_{1N} \\ 0 & A_{22}\ldots & & A_{2N} \\ \vdots & & & \\ 0 & 0\ldots & A_{N-1,N-1} & A_{N-1,N} \\ 0 & 0\ldots 0 & & A_{NN} \end{bmatrix} \begin{bmatrix} x_1 \\ x_2 \\ \vdots \\ x_{N-1} \\ x_N \end{bmatrix} = \begin{bmatrix} y_1 \\ y_2 \\ \vdots \\ y_{N-1} \\ y_N \end{bmatrix} \tag{6.1.3}$$

Since the last equation in (6.1.3) involves only x_N,

$$x_N = \frac{y_N}{A_{NN}} \tag{6.1.4}$$

After x_N is computed, the next-to-last equation can be solved:

$$x_{N-1} = \frac{y_{N-1} - A_{N-1,N}x_N}{A_{N-1,N-1}} \tag{6.1.5}$$

In general, with $x_N, x_{N-1}, \ldots, x_{k+1}$ already computed, the kth equation can be solved as

$$x_k = \frac{y_k - \sum_{n=k+1}^{N} A_{kn}x_n}{A_{kk}} \qquad k = N, N-1, \ldots, 1 \tag{6.1.6}$$

This procedure for solving (6.1.3) is called *back substitution.*

If **A** is not an upper triangular, (6.1.1), it can be transformed to an equivalent equation with an upper triangular matrix. The transformation, called *Gauss* *elimination*, is described by the following $(N-1)$ steps. During Step 1, use the first equation in (6.1.1) to eliminate x_1 from the remaining equations. That is, Equation 1 is multiplied by A_{n1}/A_{11} and then subtracted from equation n, for $n = 2, 3, \ldots, N$. After completing Step 1, we have

$$\begin{bmatrix} A_{11} & A_{12} & \cdots & A_{1N} \\ 0 & \left(A_{22} - \frac{A_{21}}{A_{11}}A_{12}\right) & \cdots & \left(A_{2N} - \frac{A_{21}}{A_{11}}A_{1N}\right) \\ 0 & \left(A_{32} - \frac{A_{31}}{A_{11}}A_{12}\right) & \cdots & \left(A_{3N} - \frac{A_{31}}{A_{11}}A_{1N}\right) \\ \vdots & \vdots & & \vdots \\ 0 & \left(A_{N2} - \frac{A_{N1}}{A_{11}}A_{12}\right) & \cdots & \left(A_{NN} - \frac{A_{N1}}{A_{11}}A_{1N}\right) \end{bmatrix} \begin{bmatrix} x_1 \\ x_2 \\ x_3 \\ \vdots \\ x_N \end{bmatrix} = \begin{bmatrix} y_1 \\ y_2 - \frac{A_{21}}{A_{11}}y_1 \\ y_3 - \frac{A_{31}}{A_{11}}y_1 \\ \vdots \\ y_N - \frac{A_{N1}}{A_{11}}y_1 \end{bmatrix} \tag{6.1.7}$$

Equation (6.1.7) has the following form:

$$\begin{bmatrix} A_{11}^{(1)} & A_{12}^{(1)} & \cdots & A_{1N}^{(1)} \\ 0 & A_{22}^{(1)} & \cdots & A_{2N}^{(1)} \\ 0 & A_{32}^{(1)} & \cdots & A_{3N}^{(1)} \\ \vdots & \vdots & & \vdots \\ 0 & A_{N2}^{(1)} & \cdots & A_{NN}^{(1)} \end{bmatrix} \begin{bmatrix} x_1 \\ x_2 \\ x_3 \\ \vdots \\ x_N \end{bmatrix} = \begin{bmatrix} y_1^{(1)} \\ y_2^{(1)} \\ y_3^{(1)} \\ \vdots \\ y_N^{(1)} \end{bmatrix} \tag{6.1.8}$$

where the superscript (1) denotes Step 1 of the Gauss elimination.

During Step 2, use the second equation in (6.1.8) to eliminate x_2 from the remaining (third, fourth, fifth, and so on) equations. That is, Equation 2 is multiplied by $A_{n2}^{(1)}/A_{22}^{(1)}$ and subtracted from equation n, for $n = 3, 4, \ldots, N$.
After Step 2, there is

$$\begin{bmatrix} A_{11}^{(2)} & A_{12}^{(2)} & A_{13}^{(2)} & \cdots & A_{1N}^{(2)} \\ 0 & A_{22}^{(2)} & A_{23}^{(2)} & \cdots & A_{2N}^{(2)} \\ 0 & 0 & A_{33}^{(2)} & \cdots & A_{3N}^{(2)} \\ 0 & 0 & A_{43}^{(2)} & \cdots & A_{4N}^{(2)} \\ \vdots & \vdots & \vdots & & \vdots \\ 0 & 0 & A_{N3}^{(2)} & \cdots & A_{NN}^{(2)} \end{bmatrix} \begin{bmatrix} x_1 \\ x_2 \\ x_3 \\ x_4 \\ \vdots \\ x_N \end{bmatrix} = \begin{bmatrix} y_1^{(2)} \\ y_2^{(2)} \\ y_3^{(2)} \\ y_4^{(2)} \\ \vdots \\ y_N^{(2)} \end{bmatrix} \tag{6.1.9}$$

During step k, start with $\mathbf{A}^{(k-1)}\mathbf{x} = \mathbf{y}^{(k-1)}$. The first k of these equations, already triangularized, are left unchanged. Also, equation k is multiplied by $A_{nk}^{(k-1)}/A_{kk}^{(k-1)}$ and then subtracted from equation n, for $n = k+1, k+2, \ldots, N$.

After $(N-1)$ steps, the equivalent equation is $\mathbf{A}^{(N-1)}\mathbf{x} = \mathbf{y}^{(N-1)}$, where $\mathbf{A}^{(N-1)}$ is upper triangular.

EXAMPLE 6.1

Gauss elimination and back substitution: direct solution to linear algebraic equations

Solve

$$\left[\begin{array}{c|c} 10 & 5 \\ \hline 2 & 9 \end{array}\right]\begin{bmatrix} x_1 \\ x_2 \end{bmatrix} = \begin{bmatrix} 6 \\ 3 \end{bmatrix}$$

using Gauss elimination and back substitution.

SOLUTION

Since $N = 2$ for this example, there is $(N-1) = 1$ Gauss elimination step. Multiplying the first equation by $A_{21}/A_{11} = 2/10$ and then subtracting from the second,

$$\left[\begin{array}{c|c} 10 & 5 \\ \hline 0 & 9 - \frac{2}{10}(5) \end{array}\right]\begin{bmatrix} x_1 \\ x_2 \end{bmatrix} = \begin{bmatrix} 6 \\ 3 - \frac{2}{10}(6) \end{bmatrix}$$

or

$$\left[\begin{array}{c|c} 10 & 5 \\ \hline 0 & 8 \end{array}\right]\begin{bmatrix} x_1 \\ x_2 \end{bmatrix} = \begin{bmatrix} 6 \\ 1.8 \end{bmatrix}$$

which has the form $\mathbf{A}^{(1)}\mathbf{x} = \mathbf{y}^{(1)}$ where $\mathbf{A}^{(1)}$ is upper triangular. Now, using back substitution, (6.1.6) gives, for $k = 2$:

$$x_2 = \frac{y_2^{(1)}}{A_{22}^{(1)}} = \frac{1.8}{8} = 0.225$$

and, for $k = 1$,

$$x_1 = \frac{y_1^{(1)} - A_{12}^{(1)}x_2}{A_{11}^{(1)}} = \frac{6 - (5)(0.225)}{10} = 0.4875$$

EXAMPLE 6.2

Gauss elimination: triangularizing a matrix

Use Gauss elimination to triangularize

$$\begin{bmatrix} 2 & 3 & -1 \\ -4 & 6 & 8 \\ 10 & 12 & 14 \end{bmatrix} \begin{bmatrix} x_1 \\ x_2 \\ x_3 \end{bmatrix} = \begin{bmatrix} 5 \\ 7 \\ 9 \end{bmatrix}$$

SOLUTION

There are $(N-1) = 2$ Gauss elimination steps. During Step 1, subtract $A_{21}/A_{11} = -4/2 = -2$ times Equation 1 from Equation 2, and subtract $A_{31}/A_{11} = 10/2 = 5$ times Equation 1 from Equation 3, to give

$$\begin{bmatrix} 2 & 3 & -1 \\ 0 & 6-(-2)(3) & 8-(-2)(-1) \\ 0 & 12-(5)(3) & 14-(5)(-1) \end{bmatrix} \begin{bmatrix} x_1 \\ x_2 \\ x_3 \end{bmatrix} = \begin{bmatrix} 5 \\ 7-(-2)(5) \\ 9-(5)(5) \end{bmatrix}$$

or

$$\begin{bmatrix} 2 & 3 & -1 \\ 0 & 12 & 6 \\ 0 & -3 & 19 \end{bmatrix} \begin{bmatrix} x_1 \\ x_2 \\ x_3 \end{bmatrix} = \begin{bmatrix} 5 \\ 17 \\ -16 \end{bmatrix}$$

which is $\mathbf{A}^{(1)}\mathbf{x} = \mathbf{y}^{(1)}$. During Step 2, subtract $A_{32}^{(1)}/A_{22}^{(1)} = -3/12 = -0.25$ times Equation 2 from Equation 3, to give

$$\begin{bmatrix} 2 & 3 & -1 \\ 0 & 12 & 6 \\ 0 & 0 & 19-(-.25)(6) \end{bmatrix} \begin{bmatrix} x_1 \\ x_2 \\ x_3 \end{bmatrix} = \begin{bmatrix} 5 \\ 17 \\ -16-(-.25)(17) \end{bmatrix}$$

or

$$\begin{bmatrix} 2 & 3 & -1 \\ 0 & 12 & 6 \\ 0 & 0 & 20.5 \end{bmatrix} \begin{bmatrix} x_1 \\ x_2 \\ x_3 \end{bmatrix} = \begin{bmatrix} 5 \\ 17 \\ -11.75 \end{bmatrix}$$

which is triangularized. The solution $\mathbf{x}$ now can be easily obtained via back substitution.

Computer storage requirements for Gauss elimination and back substitution include N^2 memory locations for $\mathbf{A}$ and N locations for $\mathbf{y}$. If there is no further need to retain $\mathbf{A}$ and $\mathbf{y}$, then $A^{(k)}$ can be stored in the location of $\mathbf{A}$, and $\mathbf{y}^{(k)}$, as well as the solution $\mathbf{x}$, can be stored in the location of $\mathbf{y}$. Additional memory is also required for iterative loops, arithmetic statements, and working space.

Computer time requirements can be evaluated by determining the number of arithmetic operations required for Gauss elimination and back substitution. One can show that Gauss elimination requires $(N^3 - N)/3$ multiplications, $(N)(N-1)/2$ divisions, and $(N^3 - N)/3$ subtractions. Also, back substitution requires $(N)(N-1)/2$ multiplications, N divisions, and $(N)(N-1)/2$ subtractions. Therefore, for very large N, the approximate computer time for solving (6.1.1) by Gauss elimination and back substitution is the time required to perform $N^3/3$ multiplications and $N^3/3$ subtractions.

For example, consider a computer with a 2×10^{-9} s multiplication time and 1×10^{-9} s addition or subtraction time. Solving $N = 10{,}000$ equations would require

$$\frac{1}{3}N^3(2 \times 10^{-9}) + \frac{1}{3}N^3(1 \times 10^{-9}) = \frac{1}{3}(10{,}000)^3(3 \times 10^{-9}) = 1000 \quad \text{s}$$

approximately plus some additional bookkeeping time for indexing and managing loops.

Since the power flow problem often involves solving power systems with tens of thousands of equations, by itself Gauss elimination would not be a good solution. However, for matrixes that have relatively few nonzero elements, known as sparse matrices, special techniques can be employed to significantly reduce computer storage and time requirements. Since all large power systems can be modeled using sparse matrices, these techniques are briefly introduced in Section 6.8.

6.2 ITERATIVE SOLUTIONS TO LINEAR ALGEBRAIC EQUATIONS: JACOBI AND GAUSS-SEIDEL

A general iterative solution to (6.1.1) proceeds as follows. First select an initial guess $\mathbf{x}(0)$. Then use

$$\mathbf{x}(i+1) = \mathbf{g}[\mathbf{x}(i)] \qquad i = 0, 1, 2, \ldots \tag{6.2.1}$$

where $\mathbf{x}(i)$ is the ith guess and $\mathbf{g}$ is an N vector of functions that specify the iteration method. Continue this procedure until the following stopping condition is satisfied, as

$$\left| \frac{x_k(i+1) - x_k(i)}{x_k(i)} \right| < \varepsilon \qquad \text{for all } k = 1, 2, \ldots, N \tag{6.2.2}$$

where $x_k(i)$ is the kth component of $\mathbf{x}(i)$ and ε is a specified *tolerance level.*

The following questions are pertinent:

1. Will the iteration procedure converge to the unique solution?
2. What is the convergence rate (how many iterations are required)?
3. When using a computer, what are the computer storage and time requirements?

These questions are addressed for two specific iteration methods: *Jacobi* and *Gauss-Seidel.** The Jacobi method is obtained by considering the kth equation of (6.1.1), as follows:

$$y_k = A_{k1}x_1 + A_{k2}x_2 + \cdots + A_{kk}x_k + \cdots + A_{kN}x_N \tag{6.2.3}$$

Solving for x_k,

$$x_k = \frac{1}{A_{kk}}[y_k - (A_{k1}x_1 + \cdots + A_{k,k-1}x_{k-1} + A_{k,k+1}x_{k+1} + \cdots + A_{kN}x_N)]$$

$$= \frac{1}{A_{kk}}\left[y_k - \sum_{n=1}^{k-1} A_{kn}x_n - \sum_{n=k+1}^{N} A_{kn}x_n\right] \tag{6.2.4}$$

The Jacobi method uses the "old" values of $\mathbf{x}(i)$ at iteration i on the right side of (6.2.4) to generate the "new" value $x_k(i+1)$ on the left side of (6.2.4). That is,

$$x_k(i+1) = \frac{1}{A_{kk}}\left[y_k - \sum_{n=1}^{k-1} A_{kn}x_n(i) - \sum_{n=k+1}^{N} A_{kn}x_n(i)\right] k = 1, 2, \ldots, N \tag{6.2.5}$$

The Jacobi method given by (6.2.5) also can be written in the following matrix format:

$$\mathbf{x}(i+1) = \mathbf{M}\mathbf{x}(i) + \mathbf{D}^{-1}\mathbf{y} \tag{6.2.6}$$

where

$$\mathbf{M} = \mathbf{D}^{-1}(\mathbf{D} - \mathbf{A}) \tag{6.2.7}$$

and

$$\mathbf{D} = \begin{bmatrix} A_{11} & 0 & 0 & \cdots & 0 \\ 0 & A_{22} & 0 & \cdots & 0 \\ 0 & \vdots & \vdots & & \vdots \\ \vdots & & & & 0 \\ 0 & 0 & 0 & \cdots & A_{NN} \end{bmatrix} \tag{6.2.8}$$

For Jacobi, $\mathbf{D}$ consists of the diagonal elements of the $\mathbf{A}$ matrix.

The Gauss-Seidel method is given by

$$x_k(i+1) = \frac{1}{A_{kk}}\left[y_k - \sum_{n=1}^{k-1} A_{kn}x_n(i+1) - \sum_{n=k+1}^{N} A_{kn}x_n(i)\right] \tag{6.2.9}$$

*The Jacobi method is also called the Gauss method.

EXAMPLE 6.3

Jacobi method: iterative solution to linear algebraic equations

Solve Example 6.1 using the Jacobi method. Start with $x_1(0) = x_2(0) = 0$ and continue until (6.2.2) is satisfied for $\varepsilon = 10^{-4}$.

SOLUTION

From (6.2.5) with $N = 2$,

$$k = 1 \quad x_1(i+1) = \frac{1}{A_{11}}[y_1 - A_{12}x_2(i)] = \frac{1}{10}[6 - 5x_2(i)]$$

$$k = 2 \quad x_2(i+1) = \frac{1}{A_{22}}[y_2 - A_{21}x_1(i)] = \frac{1}{9}[3 - 2x_1(i)]$$

Alternatively, in matrix format using (6.2.6) through (6.2.8),

$$\mathbf{D}^{-1} = \left[\begin{array}{c|c} 10 & 0 \\ \hline 0 & 9 \end{array}\right]^{-1} = \left[\begin{array}{c|c} \frac{1}{10} & 0 \\ \hline 0 & \frac{1}{9} \end{array}\right]$$

$$\mathbf{M} = \left[\begin{array}{c|c} \frac{1}{10} & 0 \\ \hline 0 & \frac{1}{9} \end{array}\right] \left[\begin{array}{c|c} 0 & -5 \\ \hline -2 & 0 \end{array}\right] = \left[\begin{array}{c|c} 0 & -\frac{5}{10} \\ \hline -\frac{2}{9} & 0 \end{array}\right]$$

$$\begin{bmatrix} x_1(i+1) \\ x_2(i+1) \end{bmatrix} = \left[\begin{array}{c|c} 0 & -\frac{5}{10} \\ \hline -\frac{2}{9} & 0 \end{array}\right] \begin{bmatrix} x_1(i) \\ x_2(i) \end{bmatrix} + \left[\begin{array}{c|c} \frac{1}{10} & 0 \\ \hline 0 & \frac{1}{9} \end{array}\right] \begin{bmatrix} 6 \\ 3 \end{bmatrix}$$

The above two formulations are identical. Starting with $x_1(0) = x_2(0) = 0$, the iterative solution is given in the following table:

Jacobi

i	0	1	2	3	4	5	6	7	8	9	10
$x_1(i)$	0	0.60000	0.43334	0.50000	0.48148	0.48889	0.48683	0.48766	0.48743	0.48752	0.48749
$x_2(i)$	0	0.33333	0.20000	0.23704	0.22222	0.22634	0.22469	0.22515	0.22496	0.22502	0.22500

As shown, the Jacobi method converges to the unique solution obtained in Example 6.1.

The convergence criterion is satisfied at the 10th iteration, since

$$\left|\frac{x_1(10) - x_1(9)}{x_1(9)}\right| = \left|\frac{0.48749 - 0.48752}{0.48749}\right| = 6.2 \times 10^{-5} < \varepsilon$$

and

$$\left|\frac{x_2(10) - x_2(9)}{x_2(9)}\right| = \left|\frac{0.22500 - 0.22502}{0.22502}\right| = 8.9 \times 10^{-5} < \varepsilon$$

Comparing (6.2.9) with (6.2.5), note that Gauss-Seidel is similar to Jacobi except that during each iteration, the "new" values, $x_n(i + 1)$, for $n < k$ are used on the right side of (6.2.9) to generate the "new" value $x_k(i + 1)$ on the left side.

The Gauss-Seidel method of (6.2.9) also can be written in the matrix format of (6.2.6) and (6.2.7), where

$$\mathbf{D} = \begin{bmatrix} A_{11} & 0 & 0 & \cdots & 0 \\ A_{21} & A_{22} & 0 & \cdots & 0 \\ \vdots & \vdots & & & \vdots \\ A_{N1} & A_{N2} & \cdots & & A_{NN} \end{bmatrix} \tag{6.2.10}$$

For Gauss-Seidel, **D** in (6.2.10) is the lower triangular portion of **A**, whereas for Jacobi, **D** in (6.2.8) is the diagonal portion of **A**.

EXAMPLE 6.4

Gauss-Seidel method: iterative solution to linear algebraic equations

Rework Example 6.3 using the Gauss-Seidel method.

SOLUTION

From (6.2.9),

$$k = 1 \quad x_1(i + 1) = \frac{1}{A_{11}}[y_1 - A_{12}x_2(i)] = \frac{1}{10}[6 - 5x_2(i)]$$

$$k = 2 \quad x_2(i + 1) = \frac{1}{A_{22}}[y_2 - A_{21}x_1(i + 1)] = \frac{1}{9}[3 - 2x_1(i + 1)]$$

(Continued)

Using this equation for $x_1(i+1)$, $x_2(i+1)$ also can be written as

$$x_2(i+1) = \frac{1}{9}\left\{3 - \frac{2}{10}[6 - 5x_2(i)]\right\}$$

Alternatively, in matrix format, using (6.2.10), (6.2.6), and (6.2.7),

$$\mathbf{D}^{-1} = \left[\begin{array}{c|c} 10 & 0 \\ \hline 2 & 9 \end{array}\right]^{-1} = \left[\begin{array}{c|c} \frac{1}{10} & 0 \\ \hline -\frac{2}{90} & \frac{1}{9} \end{array}\right]$$

$$\mathbf{M} = \left[\begin{array}{c|c} \frac{1}{10} & 0 \\ \hline -\frac{2}{90} & \frac{1}{9} \end{array}\right]\left[\begin{array}{c|c} 0 & -5 \\ \hline 0 & 0 \end{array}\right] = \left[\begin{array}{c|c} 0 & -\frac{1}{2} \\ \hline 0 & \frac{1}{9} \end{array}\right]$$

$$\begin{bmatrix} x_1(i+1) \\ x_2(i+1) \end{bmatrix} = \left[\begin{array}{c|c} 0 & -\frac{1}{2} \\ \hline 0 & \frac{1}{9} \end{array}\right]\begin{bmatrix} x_1(i) \\ x_2(i) \end{bmatrix} + \left[\begin{array}{c|c} \frac{1}{10} & 0 \\ \hline -\frac{2}{90} & \frac{1}{9} \end{array}\right]\begin{bmatrix} 6 \\ 3 \end{bmatrix}$$

These two formulations are identical. Starting with $x_1(0) = x_2(0) = 0$, the solution is given in the following table:

Gauss-Seidel

I	0	1	2	3	4	5	6
$x_1(i)$	0	0.60000	0.50000	0.48889	0.48765	0.48752	0.48750
$x_2(i)$	0	0.20000	0.22222	0.22469	0.22497	0.22500	0.22500

For this example, Gauss-Seidel converges in 6 iterations, compared to 10 iterations with Jacobi.

The convergence rate is faster with Gauss-Seidel for some **A** matrices, but faster with Jacobi for other **A** matrices. In some cases, one method diverges while the other converges. In other cases both methods diverge, as illustrated by the next example.

EXAMPLE 6.5

Divergence of Gauss-Seidel method

Using the Gauss-Seidel method with $x_1(0) = x_2(0) = 0$, solve

$$\left[\begin{array}{c|c} 5 & 10 \\ \hline 9 & 2 \end{array}\right]\begin{bmatrix} x_1 \\ x_2 \end{bmatrix} = \begin{bmatrix} 6 \\ 3 \end{bmatrix}$$

SOLUTION

Note that these equations are the same as those in Example 6.1, except that x_1 and x_2 are interchanged. Using (6.2.9),

$$k = 1 \qquad x_1(i+1) = \frac{1}{A_{11}}[y_1 - A_{12}x_2(i)] = \frac{1}{5}[6 - 10x_2(i)]$$

$$k = 2 \qquad x_2(i+1) = \frac{1}{A_{22}}[y_2 - A_{21}x_1(i+1)] = \frac{1}{2}[3 - 9x_1(i+1)]$$

Successive calculations of x_1 and x_2 are shown in the following table:

Gauss-Seidel

i	0	1	2	3	4	5
$x_1(i)$	0	1.2	9	79.2	711	6397
$x_2(i)$	0	−3.9	−39	−354.9	−3198	−28,786

The unique solution by matrix inversion is

$$\begin{bmatrix} x_1 \\ x_2 \end{bmatrix} = \left[\begin{array}{c|c} 5 & 10 \\ \hline 9 & 2 \end{array}\right]^{-1}\begin{bmatrix} 6 \\ 3 \end{bmatrix} = \frac{-1}{80}\left[\begin{array}{c|c} 2 & -10 \\ \hline -9 & 5 \end{array}\right]\begin{bmatrix} 6 \\ 3 \end{bmatrix} = \begin{bmatrix} 0.225 \\ 0.4875 \end{bmatrix}$$

As shown, Gauss-Seidel does not converge to the unique solution; instead it diverges. Jacobi also diverges for this example.

If any diagonal element A_{kk} equals zero, then Jacobi and Gauss-Seidel are undefined, because the right-hand sides of (6.2.5) and (6.2.9) are divided by A_{kk}. Also, if any one diagonal element has too small a magnitude, these methods will diverge. In Examples 6.3 and 6.4, Jacobi and Gauss-Seidel converge, since the diagonals (10 and 9) are both large; in Example 6.5, however, the diagonals (5 and 2) are small compared to the off-diagonals, and the methods diverge.

In general, convergence of Jacobi or Gauss-Seidel can be evaluated by recognizing that (6.2.6) represents a digital filter with input y and output $x(i)$. The z-transform of (6.2.6) may be employed to determine the filter transfer function and its poles. The output $x(i)$ converges if and only if all the filter poles have magnitudes less than 1 (see Problems 6.16 and 6.17).

Rate of convergence is also established by the filter poles. Fast convergence is obtained when the magnitudes of all the poles are small. In addition, experience with specific **A** matrices has shown that more iterations are required for Jacobi and Gauss-Seidel as the dimension N increases.

Computer storage requirements for Jacobi include N^2 memory locations for the **A** matrix and $3N$ locations for the vectors $\mathbf{y}$, $\mathbf{x}(i)$, and $\mathbf{x}(i+1)$. Storage space is also required for loops, arithmetic statements, and working space to compute (6.2.5). Gauss-Seidel requires N fewer memory locations, since for (6.2.9) the new value $x_k(i+1)$ can be stored in the location of the old value $x_k(i)$.

Computer time per iteration is relatively small for Jacobi and Gauss-Seidel. Inspection of (6.2.5) or (6.2.9) shows that N^2 multiplications/divisions and $N(N-1)$ subtractions per iteration are required [one division, $(N-1)$ multiplications, and $(N-1)$ subtractions for each $k = 1, 2, \ldots, N$]. But as was the case with Gauss elimination, if the matrix is sparse (i.e., most of the elements are zero), special sparse matrix algorithms can be used to substantially decrease both the storage requirements and the computation time.

6.3 ITERATIVE SOLUTIONS TO NONLINEAR ALGEBRAIC EQUATIONS: NEWTON-RAPHSON

A set of nonlinear algebraic equations in matrix format is given by

$$\mathbf{f}(\mathbf{x}) = \begin{bmatrix} f_1(\mathbf{x}) \\ f_2(\mathbf{x}) \\ \vdots \\ f_N(\mathbf{x}) \end{bmatrix} = \mathbf{y} \tag{6.3.1}$$

where $\mathbf{y}$ and $\mathbf{x}$ are N vectors and $\mathbf{f}(\mathbf{x})$ is an N vector of functions. Given $\mathbf{y}$ and $\mathbf{f}(\mathbf{x})$, the objective is to solve for $\mathbf{x}$. The iterative methods described in Section 6.2 can be extended to nonlinear equations as follows. Rewriting (6.3.1).

$$0 = \mathbf{y} - \mathbf{f}(\mathbf{x}) \tag{6.3.2}$$

Adding $\mathbf{Dx}$ to both sides of (6.3.2), where $\mathbf{D}$ is a square $N \times N$ invertible matrix.

$$\mathbf{Dx} = \mathbf{Dx} + \mathbf{y} - \mathbf{f}(\mathbf{x}) \tag{6.3.3}$$

Premultiplying by $\mathbf{D}^{-1}$.

$$\mathbf{x} = \mathbf{x} + \mathbf{D}^{-1}\,[\mathbf{y} - \mathbf{f}(\mathbf{x})] \tag{6.3.4}$$

The old values $\mathbf{x}(i)$ are used on the right side of (6.3.4) to generate the new values $\mathbf{x}(i+1)$ on the left side. That is,

$$\mathbf{x}(i+1) = \mathbf{x}(i) + \mathbf{D}^{-1}\{\mathbf{y} - \mathbf{f}[\mathbf{x}(i)]\} \tag{6.3.5}$$

For linear equations, $\mathbf{f}(\mathbf{x}) = \mathbf{A}\mathbf{x}$ and (6.3.5) reduces to

$$\mathbf{x}(i+1) = \mathbf{x}(i) + \mathbf{D}^{-1}[\mathbf{y} - \mathbf{A}\mathbf{x}(z)] = \mathbf{D}^{-1}(\mathbf{D} - \mathbf{A})\mathbf{x}(i) + \mathbf{D}^{-1}\mathbf{y} \tag{6.3.6}$$

which is identical to the Jacobi and Gauss-Seidel methods of (6.2.6). For nonlinear equations, the matrix **D** in (6.3.5) must be specified.

One method for specifying **D**, called *Newton-Raphson*, is based on the following Taylor series expansion of $\mathbf{f}(\mathbf{x})$ about an operating point $\mathbf{x}_0$.

$$\mathbf{y} = \mathbf{f}(\mathbf{x}_0) + \left.\frac{d\mathbf{f}}{d\mathbf{x}}\right|_{\mathbf{x}=\mathbf{x}_0} (\mathbf{x} - \mathbf{x}_0)\cdots \tag{6.3.7}$$

Neglecting the higher order terms in (6.3.7) and solving for **x**,

$$\mathbf{x} = \mathbf{x}_0 + \left[\left.\frac{d\mathbf{f}}{d\mathbf{x}}\right|_{\mathbf{x}=\mathbf{x}_0}\right]^{-1} [\mathbf{y} - \mathbf{f}(\mathbf{x}_0)] \tag{6.3.8}$$

The Newton-Raphson method replaces $\mathbf{x}_0$ by the old value $\mathbf{x}(i)$ and **x** by the new value $\mathbf{x}(i+1)$ in (6.3.8). Thus,

$$\mathbf{x}(i+1) = \mathbf{x}(i) + \mathbf{J}^{-1}(i)\{\mathbf{y} - \mathbf{f}[\mathbf{x}(i)]\} \tag{6.3.9}$$

where

$$\mathbf{J}(i) = \left.\frac{d\mathbf{f}}{d\mathbf{x}}\right|_{\mathbf{x}=\mathbf{x}(i)} = \begin{bmatrix} \frac{\partial f_1}{\partial x_1} & \frac{\partial f_1}{\partial x_2} & \cdots & \frac{\partial f_1}{\partial x_N} \\ \frac{\partial f_2}{\partial x_1} & \frac{\partial f_2}{\partial x_2} & \cdots & \frac{\partial f_2}{\partial x_N} \\ \vdots & \vdots & & \vdots \\ \frac{\partial f_N}{\partial x_1} & \frac{\partial f_N}{\partial x_2} & \cdots & \frac{\partial f_N}{\partial x_N} \end{bmatrix}_{\mathbf{x}=\mathbf{x}(i)} \tag{6.3.10}$$

The $N \times N$ matrix $\mathbf{J}(i)$, whose elements are the partial derivatives shown in (6.3.10), is called the Jacobian matrix. The Newton-Raphson method is similar to extended Gauss-Seidel, except that **D** in (6.3.5) is replaced by $\mathbf{J}(i)$ in (6.3.9).

EXAMPLE 6.6

Newton-Raphson method: solution to polynomial equations

Solve the scalar equation $f(x) = y$, where $y = 9$ and $f(x) = x^2$. Starting with $x(0) = 1$, use (a) Newton-Raphson and (b) extended Gauss-Seidel with $\mathbf{D} = 3$ until (6.2.2) is satisfied for $\varepsilon = 10^{-4}$. Compare the two methods.

(Continued)

SOLUTION

a. Using (6.3.10) with $f(x) = x^2$,

$$\mathbf{J}(i) = \frac{d}{dx}(x^2)\bigg|_{x=x(i)} = 2x\bigg|_{x=x(i)} = 2x(i)$$

Using $\mathbf{J}(i)$ in (6.3.9),

$$x(i+1) = x(i) + \frac{1}{2x(i)}[9 - x^2(i)]$$

Starting with $x(0) = 1$, successive calculations of the Newton-Raphson equation are shown in the following table.

Newton–Raphson

i	0	1	2	3	4	5
$x(i)$	1	5.00000	3.40000	3.02353	3.00009	3.00000

b. Using (6.3.5) with $\mathbf{D} = 3$, the Gauss-Seidel method is

$$x(i+1) = x(i) + \frac{1}{3}[9 - x^2(i)]$$

The corresponding Gauss-Seidel calculations are as follows:

Gauss-Seidel (D = 3)

i	0	1	2	3	4	5	6
$x(i)$	1	3.66667	2.18519	3.59351	2.28908	3.54245	2.35945

As shown, Gauss-Seidel oscillates about the solution, slowly converging, whereas Newton-Raphson converges in five iterations to the solution $x = 3$. Note that if $x(0)$ is negative, Newton-Raphson converges to the negative solution $x = -3$. Also, it is assumed that the matrix inverse $\mathbf{J}^{-1}$ exists. Thus, the initial value $x(0) = 0$ should be avoided for this example.

EXAMPLE 6.7

Newton-Raphson method: solution to nonlinear algebraic equations

Solve

$$\begin{bmatrix} x_1 + x_2 \\ x_1 x_2 \end{bmatrix} = \begin{bmatrix} 15 \\ 50 \end{bmatrix} \qquad \mathbf{x}(0) = \begin{bmatrix} 4 \\ 9 \end{bmatrix}$$

Use the Newton-Raphson method starting with the above $\mathbf{x}(0)$ and continue until (6.2.2) is satisfied with $\varepsilon = 10^{-4}$.

SOLUTION

Using (6.3.10) with $f_1 = (x_1 + x_2)$ and $f_2 = x_1 x_2$,

$$\mathbf{J}(i)^{-1} = \left[\begin{array}{c|c} \dfrac{\partial f_1}{\partial x_1} & \dfrac{\partial f_1}{\partial x_2} \\ \hline \dfrac{\partial f_2}{\partial x_1} & \dfrac{\partial f_2}{\partial x_2} \end{array}\right]^{-1}_{\mathbf{x}=\mathbf{x}(i)} = \left[\begin{array}{c|c} 1 & 1 \\ \hline x_2(i) & x_1(i) \end{array}\right]^{-1} = \frac{\left[\begin{array}{c|c} x_1(i) & -1 \\ \hline -x_2(i) & 1 \end{array}\right]}{x_1(i) - x_2(i)}$$

Using $\mathbf{J}(i)^{-1}$ in (6.3.9),

$$\begin{bmatrix} x_1(i+1) \\ x_2(i+1) \end{bmatrix} = \begin{bmatrix} x_1(i) \\ x_2(i) \end{bmatrix} + \frac{\left[\begin{array}{c|c} x_1(i) & -1 \\ \hline -x_2(i) & 1 \end{array}\right]}{x_1(i) - x_2(i)} \begin{bmatrix} 15 - x_1(i) - x_2(i) \\ 50 - x_1(i)x_2(i) \end{bmatrix}$$

Writing the preceding as two separate equations,

$$x_1(i+1) = x_1(i) + \frac{x_1(i)[15 - x_1(i) - x_2(i)] - [50 - x_1(i)x_2(i)]}{x_1(i) - x_2(i)}$$

$$x_2(i+1) = x_2(i) + \frac{-x_2(i)[15 - x_1(i) - x_2(i)] + [50 - x_1(i)x_2(i)]}{x_1(i) - x_2(i)}$$

Successive calculations of these equations are shown in the following table.

Newton-Raphson

i	0	1	2	3	4
$x_1(i)$	4	5.20000	4.99130	4.99998	5.00000
$x_2(i)$	9	9.80000	10.00870	10.00002	10.00000

Newton-Raphson converges in four iterations for this example.

Equation (6.3.9) contains the matrix inverse $\mathbf{J}^{-1}$. Instead of computing $\mathbf{J}^{-1}$, (6.3.9) can be rewritten as follows:

$$\mathbf{J}(i)\Delta\mathbf{x}(i) = \Delta\mathbf{y}(i) \tag{6.3.11}$$

where

$$\Delta\mathbf{x}(i) = \mathbf{x}(i+1) - \mathbf{x}(i) \tag{6.3.12}$$

and

$$\Delta\mathbf{y}(i) = \mathbf{y} - \mathbf{f}[\mathbf{x}(i)] \tag{6.3.13}$$

Then, during each iteration, the following four steps are completed:

STEP 1 Compute $\Delta\mathbf{y}(i)$ from (6.3.13).

STEP 2 Compute $\mathbf{J}(i)$ from (6.3.10).

STEP 3 Using Gauss elimination and back substitution, solve (6.3.11) for $\Delta\mathbf{x}(i)$.

STEP 4 Compute $\mathbf{x}(i+1)$ from (6.3.12).

Experience from power flow studies has shown that Newton-Raphson converges in many cases where Jacobi and Gauss-Seidel diverge. Furthermore, the number of iterations required for convergence is independent of the dimension

EXAMPLE 6.8

Newton-Raphson method in four steps

Complete the above four steps for the first iteration of Example 6.7.

SOLUTION

STEP 1 $\Delta\mathbf{y}(0) = \mathbf{y} - \mathbf{f}[\mathbf{x}(0)] = \begin{bmatrix} 15 \\ 20 \end{bmatrix} - \begin{bmatrix} 4+9 \\ (4)(9) \end{bmatrix} = \begin{bmatrix} 2 \\ 14 \end{bmatrix}$

STEP 2 $\mathbf{J}(0) = \left[\begin{array}{c|c} 1 & 1 \\ \hline x_2(0) & x_1(0) \end{array}\right] = \left[\begin{array}{c|c} 1 & 1 \\ \hline 9 & 4 \end{array}\right]$

STEP 3 Using $\Delta\mathbf{y}(0)$ and $\mathbf{J}(0)$, (6.3.11) becomes

$$\left[\begin{array}{c|c} 1 & 1 \\ \hline 9 & 4 \end{array}\right] \begin{bmatrix} \Delta x_1(0) \\ \Delta x_2(0) \end{bmatrix} = \begin{bmatrix} 2 \\ 14 \end{bmatrix}$$

Using Gauss elimination, subtract $J_{21}/J_{11} = 9/1 = 9$ times the first equation from the second equation, giving

$$\left[\begin{array}{c|c} 1 & 1 \\ \hline 0 & -5 \end{array}\right] \begin{bmatrix} \Delta x_1(0) \\ \Delta x_2(0) \end{bmatrix} = \begin{bmatrix} 2 \\ -4 \end{bmatrix}$$

Solving by back substitution,

$$\Delta x_2(0) = \frac{-4}{-5} = 0.8$$

$$\Delta x_1(0) = 2 - 0.8 = 1.2$$

STEP 4 $\mathbf{x}(1) = \mathbf{x}(0) + \Delta\mathbf{x}(0) = \begin{bmatrix} 4 \\ 9 \end{bmatrix} + \begin{bmatrix} 1.2 \\ 0.8 \end{bmatrix} = \begin{bmatrix} 5.2 \\ 9.8 \end{bmatrix}$

This is the same as computed in Example 6.7.

N for Newton-Raphson, but increases with N for Jacobi and Gauss-Seidel. Most Newton-Raphson power flow problems converge in fewer than 10 iterations [1].

6.4 THE POWER FLOW PROBLEM

The power flow problem is the computation of voltage magnitude and phase angle at each bus in a power system under balanced three-phase steady-state conditions. As a by-product of this calculation, real and reactive power flows in equipment such as transmission lines and transformers, as well as equipment losses, can be computed.

The starting point for a power flow problem is a single-line (oneline) diagram of the power system, from which the input data can be obtained. Input data consist of bus data, transmission line data, and transformer data.

As shown in Figure 6.1, the following four variables are associated with each bus k: voltage magnitude V_k, phase angle δ_k, net real power P_k, and reactive power Q_k supplied to the bus. At each bus, two of these variables are specified as input data, and the other two are unknowns to be computed by the power flow program. For convenience, the power delivered to bus k in Figure 6.1 is separated into generator and load terms. That is,

$$
\begin{aligned}
P_k &= P_{Gk} - P_{Lk} \\
Q_k &= Q_{Gk} - Q_{Lk}
\end{aligned}
\tag{6.4.1}
$$

Each bus k is categorized into one of the following three bus types:

1. Swing bus (or slack bus)—There is only one swing bus, which for convenience is numbered bus 1 in this text. The swing bus is a reference bus for which $V_1\underline{/\delta_1}$, is input data with the angle typically zero degrees and the voltage magnitude close to 1.0 per unit. The power flow computes P_1 and Q_1.

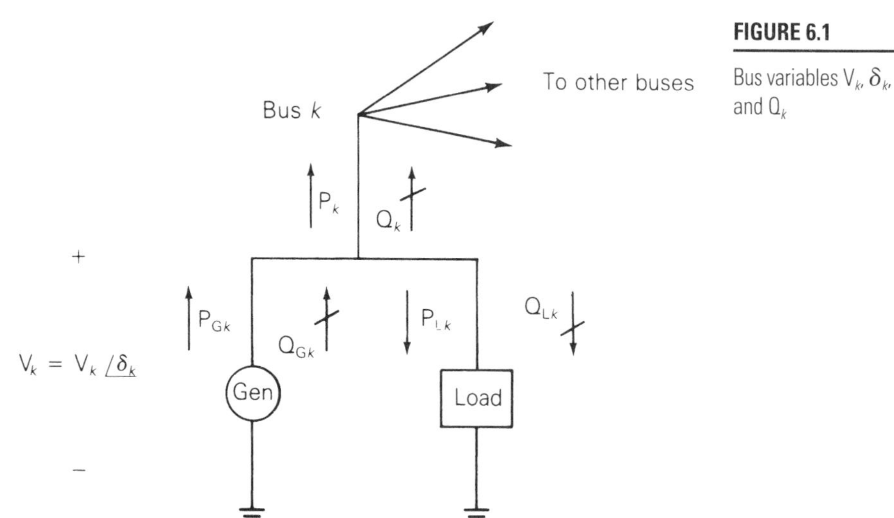

FIGURE 6.1

Bus variables V_k, δ_k, P_k, and Q_k

2. Load (PQ) bus—P_k and Q_k are input data. The power flow computes V_k and δ_k. Most buses in typical power flows are load buses.
3. Voltage controlled (PV) bus—P_k and V_k are input data. The power flow program computes Q_k and δ_k. Examples are buses to which generators, switched shunt capacitors, or static var systems are connected. Maximum and minimum reactive power (var) limits $Q_{Gk\max}$ and $Q_{Gk\min}$ that this equipment can supply are also input data. If an upper or lower reactive power limit is reached, then the reactive power output of the generator is held at the limit, and the bus is modeled as a PQ bus. Another example is a bus to which a tap-changing transformer is connected; the power flow then computes the tap setting.

Note that when bus k is a load bus with no generation, $P_k = -P_{Lk}$ is negative; that is, the real power supplied to bus k in Figure 6.1 is negative. If the load is inductive, $Q_k = -Q_{Lk}$ is negative.

Transmission lines are represented by the equivalent π circuit, shown in Figure 5.7. Transformers are also represented by equivalent circuits, as shown in Figure 3.9 for a two-winding transformer, Figure 3.20 for a three-winding transformer, or Figure 3.25 for a tap-changing transformer.

Input data for each transmission line include the per-unit equivalent π circuit series impedance Z' and shunt admittance Y', the two buses to which the line is connected, and maximum MVA rating. Similarly, input data for each transformer include per-unit winding impedances Z, the per-unit exciting branch admittance Y, the buses to which the windings are connected, and maximum MVA ratings. Input data for tap-changing transformers also include maximum tap settings.

The bus admittance matrix $\boldsymbol{Y}_{\text{bus}}$ can be constructed from the line and transformer input data. From (2.4.3) and (2.4.4), the elements of $\boldsymbol{Y}_{\text{bus}}$ are

$$\text{Diagonal elements: } Y_{kk} = \text{sum of admittances connected to bus } k$$

$$\text{Off-diagonal elements: } Y_{kn} = -(\text{sum of admittances connected between buses } k \text{ and } n)\ k \neq n \tag{6.4.2}$$

EXAMPLE 6.9

Power flow input data and $\boldsymbol{Y}_{\text{bus}}$

Figure 6.2 shows a single-line diagram of a five-bus power system. Input data are given in Tables 6.1, 6.2, and 6.3. As shown in Table 6.1, bus 1, to which a generator is connected, is the swing bus. Bus 3, to which a generator and a load are connected, is a voltage-controlled bus. Buses 2, 4, and 5 are load buses. Note that the loads at buses 2 and 3 are inductive since $Q_2 = -Q_{L2} = -2.8$ and $-Q_{L3} = -0.4$ are negative.

For each bus k, determine which of the variables V_k, δ_k, P_k, and Q_k are input data and which are unknowns. Also, compute the elements of the second row of $\boldsymbol{Y}_{\text{bus}}$.

FIGURE 6.2

Single-line diagram for Example 6.9

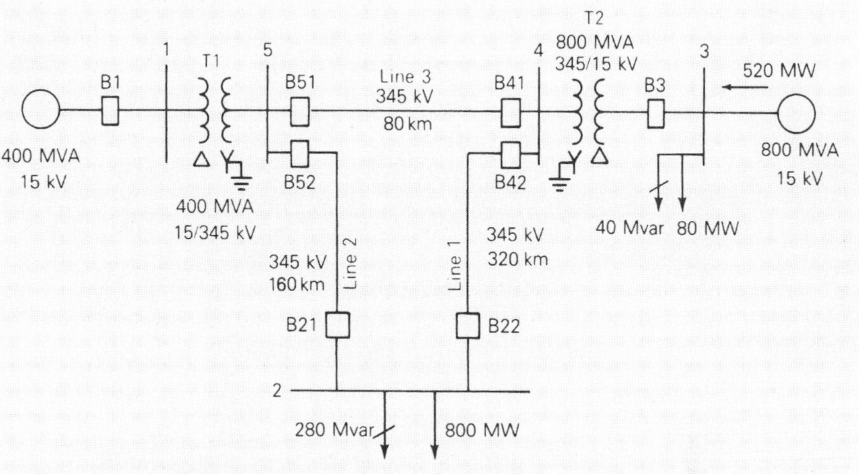

Bus	Type	V per unit	δ degrees	P_G per unit	Q_G per unit	P_L per unit	Q_L per unit	Q_{Gmax} per unit	Q_{Gmin} per unit
1	Swing	1.0	0	—	—	0	0	—	—
2	Load	—	—	0	0	8.0	2.8	—	—
3	Constant voltage	1.05	—	5.2	—	0.8	0.4	4.0	−2.8
4	Load	—	—	0	0	0	0	—	—
5	Load	—	—	0	0	0	0	—	—

TABLE 6.1

Bus input data for Example 6.9*

*S_{base} = 100 MVA, V_{base} = 15 kV at buses 1, 3, and 345 kV at buses 2, 4, 5

Bus-to-Bus	R′ per unit	X′ per unit	G′ per unit	B′ per unit	Maximum MVA per unit
2–4	0.0090	0.100	0	1.72	12.0
2–5	0.0045	0.050	0	0.88	12.0
4–5	0.00225	0.025	0	0.44	12.0

TABLE 6.2

Line input data for Example 6.9

(Continued)

Bus-to-Bus	R per unit	X per unit	G_c per unit	B_m per unit	Maximum MVA per unit	Maximum TAP Setting per unit
1–5	0.00150	0.02	0	0	6.0	—
3–4	0.00075	0.01	0	0	10.0	—

TABLE 6.3
Transformer input data for Example 6.9

Bus	Input Data	Unknowns
1	$V_1 = 1.0, \delta_1 = 0$	P_1, Q_1
2	$P_2 = P_{G2} - P_{L2} = -8$ $Q_2 = Q_{G2} - Q_{L2} = -2.8$	V_2, δ_2
3	$V_3 = 1.05$ $P_3 = P_{G3} - P_{L3} = 4.4$	Q_3, δ_4
4	$P_4 = 0, Q_4 = 0$	V_4, δ_4
5	$P_5 = 0, Q_5 = 0$	V_5, δ_5

TABLE 6.4
Input data and unknowns for Example 6.9

SOLUTION

The input data and unknowns are listed in Table 6.4. For bus 1, the swing bus, P_1 and Q_1 are unknowns. For bus 3, a voltage-controlled bus, Q_3 and δ_3 are unknowns. For buses 2, 4, and 5, load buses, V_2, V_4, V_5 and δ_2, δ_4, δ_5 are unknowns.

The elements of $\boldsymbol{Y}_{\text{bus}}$ are computed from (6.4.2). Since buses 1 and 3 are not directly connected to bus 2,

$$Y_{21} = Y_{23} = 0$$

Using (6.4.2),

$$Y_{24} = \frac{-1}{R'_{24} + jX'_{24}} = \frac{-1}{0.009 + j0.1} = -0.89276 + j9.91964 \quad \text{per unit}$$
$$= 9.95972\underline{/95.143^\circ} \quad \text{per unit}$$

$$Y_{25} = \frac{-1}{R'_{25} + jX'_{25}} = \frac{-1}{0.0045 + j0.05} = -1.78552 + j19.83932 \quad \text{per unit}$$

$$= 19.9195\underline{/95.143^\circ} \quad \text{per unit}$$

$$Y_{22} = \frac{1}{R'_{24} + jX'_{24}} + \frac{1}{R'_{25} + jX'_{25}} + j\frac{B'_{24}}{2} + j\frac{B'_{25}}{2}$$

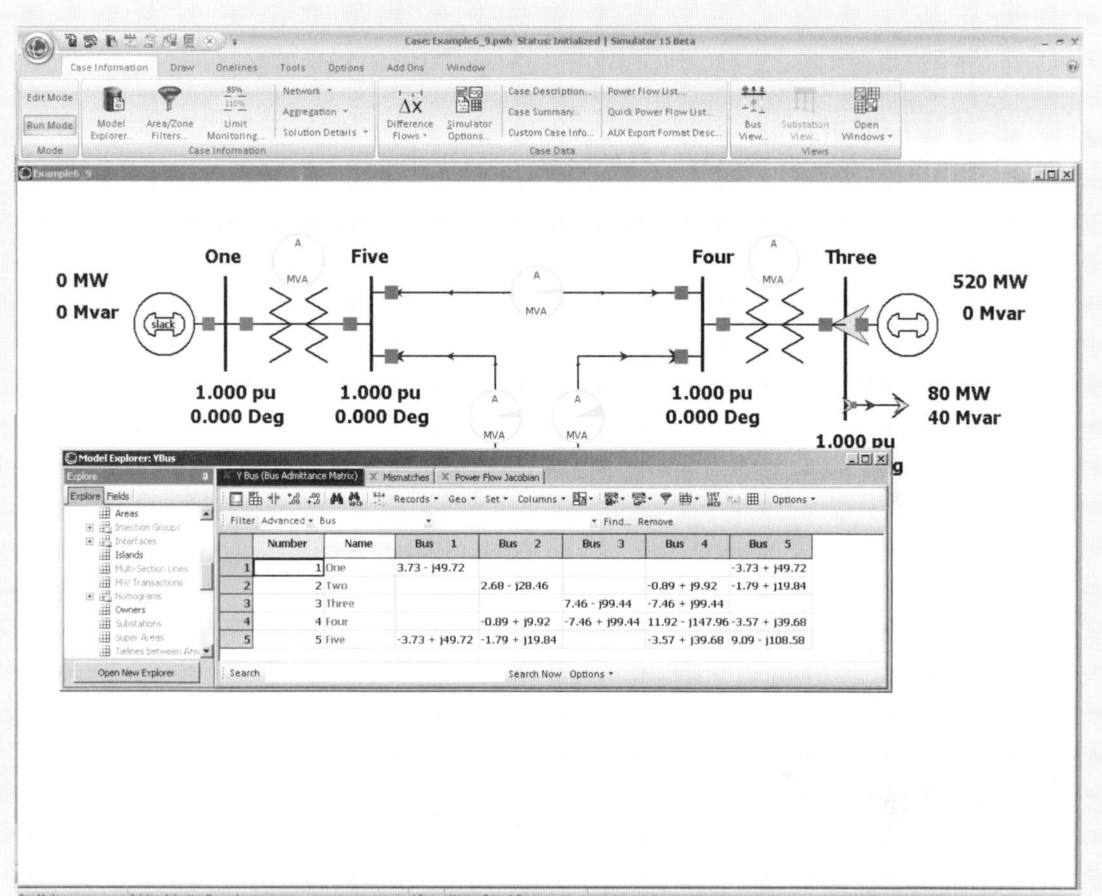

	Number	Name	Bus 1	Bus 2	Bus 3	Bus 4	Bus 5
1	1	One	3.73 - j49.72				-3.73 + j49.72
2	2	Two		2.68 - j28.46		-0.89 + j9.92	-1.79 + j19.84
3	3	Three			7.46 - j99.44	-7.46 + j99.44	
4	4	Four		-0.89 + j9.92	-7.46 + j99.44	11.92 - j147.96	-3.57 + j39.68
5	5	Five	-3.73 + j49.72	-1.79 + j19.84		-3.57 + j39.68	9.09 - j108.58

FIGURE 6.3

Screen for Example 6.9

$$= (0.89276 - j9.91964) + (1.78552 - j19.83932) + j\frac{1.72}{2} + j\frac{0.88}{2}$$

$$= 2.67828 - j28.4590 = 28.5847\underline{/-84.624^\circ} \quad \text{per unit}$$

where half of the shunt admittance of each line connected to bus 2 is included in Y_{22} (the other half is located at the other ends of these lines).

This five-bus power system is modeled in PowerWorld Simulator case Example 6_9 (see Figure 6.3). To view the input data, first click on the **Edit Mode** button (on the far left-hand side of the ribbon) to switch into the Edit mode (the Edit mode is used for modifying system parameters). Then by selecting the **Case Information** tab, you can view tabular displays showing the various parameters for

(Continued)

the system. For example, use **Network**, **Buses** to view the parameters for each bus, and **Network**, **Lines and Transformers** to view the parameters for the transmission lines and transformers. Fields shown in blue on the screen can be directly changed simply by typing over them, and those shown in green can be toggled by clicking on them. Note that the values shown on these displays match the per unit values from Tables 6.1 to 6.3, except the power values are shown in actual MW/Mvar units.

The elements of $\boldsymbol{Y}_{\text{bus}}$ also can be displayed by selecting **Solution Details**, $\boldsymbol{Y}_{\text{bus}}$. Since the $\boldsymbol{Y}_{\text{bus}}$ entries are derived from other system parameters, they cannot be changed directly. Notice that several of the entries are blank, indicating that there is no transmission line or transformer directly connecting these two buses (a blank entry is equivalent to zero). For larger networks, most of the elements of the $\boldsymbol{Y}_{\text{bus}}$ are zero since any single bus usually only has a few incident lines (such sparse matrices are considered in Section 6.8). The elements of the $\boldsymbol{Y}_{\text{bus}}$ can be saved in a Matlab compatible format by first right-clicking within the $\boldsymbol{Y}_{\text{bus}}$ matrix to display the local menu, and then selecting **Save $\boldsymbol{Y}_{\text{bus}}$ in Matlab Format** from the local menu.

Finally, notice that no flows are shown on the oneline because the nonlinear power flow equations have not yet been solved, and the solution of these equations are covered next.

Using $\boldsymbol{Y}_{\text{bus}}$, the nodal equations for a power system network are written as

$$\mathbf{I} = \boldsymbol{Y}_{\text{bus}}\mathbf{V} \tag{6.4.3}$$

where $\mathbf{I}$ is the N vector of source currents injected into each bus and $\mathbf{V}$ is the N vector of bus voltages. For bus k, the kth equation in (6.4.3) is

$$I_k = \sum_{n=1}^{N} Y_{kn} V_n \tag{6.4.4}$$

The complex power delivered to bus k is

$$S_k = \mathrm{P}_k + j\mathrm{Q}_k = V_k I_k^* \tag{6.4.5}$$

Power flow solutions by Gauss-Seidel are based on nodal equations, (6.4.4), where each current source I_k is calculated from (6.4.5). Using (6.4.4) in (6.4.5),

$$\mathrm{P}_k + j\mathrm{Q}_k = V_k \left[\sum_{n=1}^{N} Y_{kn} V_n \right]^* \qquad k = 1, 2, \ldots, N \tag{6.4.6}$$

With the following notation,

$$V_n = \mathrm{V}_n e^{j\delta_n} \tag{6.4.7}$$

$$Y_{kn} = \mathrm{Y}_{kn} e^{j\theta_{kn}} = G_{kn} + jB_{kn} \qquad k,n = 1, 2, \ldots, N \tag{6.4.8}$$

(6.4.6) becomes

$$\mathrm{P}_k + j\mathrm{Q}_k = \mathrm{V}_k \sum_{n=1}^{N} \mathrm{Y}_{kn} \mathrm{V}_n e^{j(\delta_k - \delta_n - \theta_{kn})} \tag{6.4.9}$$

Taking the real and imaginary parts of (6.4.9), the power balance equations are written as either

$$P_k = V_k \sum_{n=1}^{N} Y_{kn} V_n \cos(\delta_k - \delta_n - \theta_{kn}) \tag{6.4.10}$$

$$Q_k = V_k \sum_{n=1}^{N} Y_{kn} V_n \sin(\delta_k - \delta_n - \theta_{kn}) \qquad k = 1, 2, \ldots, N \tag{6.4.11}$$

or when the Y_{kn} is expressed in rectangular coordinates as

$$P_k = V_k \sum_{n=1}^{N} V_n [G_{kn} \cos(\delta_k - \delta_n) + B_{kn} \sin(\delta_k - \delta_n)] \tag{6.4.12}$$

$$Q_k = V_k \sum_{n=1}^{N} V_n [G_{kn} \sin(\delta_k - \delta_n) - B_{kn} \cos(\delta_k - \delta_n)] \qquad k = 1, 2, \ldots, N \tag{6.4.13}$$

Power flow solutions by Newton-Raphson are based on the nonlinear power flow equations given by (6.4.10) and (6.4.11) [or alternatively by (6.4.12) and (6.4.13)].

6.5 POWER FLOW SOLUTION BY GAUSS-SEIDEL

Nodal equations $\mathbf{I} = Y_{\text{bus}}\mathbf{V}$ are a set of linear equations analogous to $\mathbf{y} = \mathbf{A}\mathbf{x}$, solved in Section 6.2 using Gauss-Seidel. Since power flow bus data consists of P_k and Q_k for load buses or P_k and V_k for voltage-controlled buses, nodal equations do not directly fit the linear equation format; the current source vector **I** is unknown and the equations are actually nonlinear. For each load bus, I_k can be calculated from (6.4.5), giving

$$I_k = \frac{P_k - jQ_k}{V_k^*} \tag{6.5.1}$$

Applying the Gauss-Seidel method (6.2.9) to the nodal equations with I_k given above, obtain

$$V_k(i+1) = \frac{1}{Y_{kk}} \left[\frac{P_k - jQ_k}{V_k^*(i)} - \sum_{n=1}^{k-1} Y_{kn} V_n(i+1) - \sum_{n=k+1}^{N} Y_{kn} V_n(i) \right] \tag{6.5.2}$$

Equation (6.5.2) can be applied twice during each iteration for load buses, first using $V_k^*(i)$, then replacing $V_k^*(i)$, by $V_k^*(i+1)$ on the right side of (6.5.2).

For a voltage-controlled bus, Q_k is unknown but can be calculated from (6.4.11), giving

$$Q_k = V_k(i) \sum_{n=1}^{N} Y_{kn} V_n(i) \sin[\delta_k(i) - \delta_n(i) - \theta_{kn}] \tag{6.5.3}$$

Also,

$$Q_{Gk} = Q_k + Q_{Lk}$$

If the calculated value of Q_{Gk} does not exceed its limits, then Q_k is used in (6.5.2) to calculate $V_k(i+1) = V_k(i+1)\angle\delta_k(i+1)$. Then the magnitude $V_k(i+1)$ is changed to V_k, which is input data for the voltage-controlled bus. Thus, use (6.5.2) to compute only the angle $\delta_k(i+1)$ for voltage-controlled buses.

If the calculated value exceeds its limit $Q_{Gk\max}$ or $Q_{Gk\min}$ during any iteration, then the bus type is changed from a voltage-controlled bus to a load bus, with Q_{Gk} set to its limit value. Under this condition, the voltage-controlling device (e.g., generator, capacitor bank, static var compensator) is not capable of maintaining V_k as specified by the input data. The power flow then calculates a new value of V_k.

For the swing bus, denoted bus 1, V_1 and δ_1 *are* input data. As such, no iterations are required for the swing bus. After the iteration process has converged, one pass through (6.4.10) and (6.4.11) can be made to compute P_1 and Q_1.

EXAMPLE 6.10

Power flow solution by Gauss-Seidel

For the power system of Example 6.9, use Gauss-Seidel to calculate $V_2(1)$, the phasor voltage at bus 2 after the first iteration. Use zero initial phase angles and 1.0 per-unit initial voltage magnitudes (except at bus 3, where $V_3 = 1.05$) to start the iteration procedure.

SOLUTION

Bus 2 is a load bus. Using the input data and bus admittance values from Example 6.9 in (6.5.2),

$$V_2(1) = \frac{1}{Y_{22}}\left\{\frac{P_2 - jQ_2}{V_2^*(0)} - [Y_{21}V_1(1) + Y_{23}V_3(0) + Y_{24}V_4(0) + Y_{25}V_5(0)]\right\}$$

$$= \frac{1}{28.5847\angle -84.624^\circ}\left\{\frac{-8 - j(-2.8)}{1.0\angle 0^\circ}\right.$$

$$\left. - [(-1.78552 + j19.83932)(1.0) + (-0.89276 + j9.91964)(1.0)]\right\}$$

$$= \frac{(-8 + j2.8) - (-2.67828 + j29.7589)}{28.5847\angle -84.624^\circ}$$

$$= 0.96132\angle -16.543^\circ \quad \text{per unit}$$

Next, the above value is used in (6.5.2) to recalculate $V_2(1)$;

$$V_2(1) = \frac{1}{28.5847\angle -84.624^\circ}\left\{\frac{-8 + j2.8}{0.96132\angle 16.543^\circ} - [-2.67828 + j29.75829]\right\}$$

$$= \frac{-4.4698 - j24.5973}{28.5847\underline{/-84.624°}} = 0.87460\underline{/-15.675°} \quad \text{per unit}$$

Computations are next performed at buses 3, 4, and 5 to complete the first Gauss-Seidel iteration.

To see the complete convergence of this case, open PowerWorld Simulator case Example 6_10. By default, PowerWorld Simulator uses the Newton-Raphson method described in the next section since Gauss-Seidel, while being a useful technique for introducing the power flow to students, is now seldom used commercially. However, the free educational version of PowerWorld still allows cases to be solved with the Gauss-Seidel approach by selecting **Tools, Solve, Gauss-Seidel Power Flow**. To avoid getting stuck in an infinite loop if a case does not converge, PowerWorld Simulator places a limit on the maximum number of iterations. Usually for a Gauss-Seidel procedure this number is rather high, perhaps equal to 100 iterations. However, in this example to demonstrate the convergence characteristics of the Gauss-Seidel method, it has been set to a single iteration, allowing the voltages to be viewed after each iteration. To step through the solution one iteration at a time, just repeatedly select **Tools, Solve, Gauss-Seidel Power Flow**.

A common stopping criterion for the Gauss-Seidel is to use the scaled differences in the voltages from one iteration to the next (6.2.2). When the voltage differences for each bus are below a specified convergence tolerance ε, the problem is considered solved. An alternative approach, implemented in PowerWorld Simulator, is to examine the real and reactive mismatch equations, defined as the difference between the right- and left-hand sides of (6.4.10) and (6.4.11). PowerWorld Simulator continues iterating until all the bus mismatches are below an MVA (or kVA) tolerance. When single-stepping through the solution, the bus mismatches can be viewed after each iteration on the **Case Information, Mismatches** display. The solution mismatch tolerance can be changed on the Power Flow Solution page of the PowerWorld Simulator Options dialog (select **Tools, Simulator Options**, then select the **Power Flow Solution** category to view this dialog); the maximum number of iterations can also be changed from this page. A typical convergence tolerance is about 0.1 MVA.

6.6 POWER FLOW SOLUTION BY NEWTON-RAPHSON

Equations (6.4.10) and (6.4.11) are analogous to the nonlinear equation $\mathbf{y} = \mathbf{f}(\mathbf{x})$, solved in Section 6.3 by Newton-Raphson. The $\mathbf{x}$, $\mathbf{y}$, and $\mathbf{f}$ vectors for the power flow problem are defined as

$$\mathbf{x} = \begin{bmatrix} \delta \\ \mathbf{V} \end{bmatrix} = \begin{bmatrix} \delta_2 \\ \vdots \\ \delta_N \\ V_2 \\ \vdots \\ V_N \end{bmatrix}; \quad \mathbf{y} = \begin{bmatrix} \mathbf{P} \\ \mathbf{Q} \end{bmatrix} = \begin{bmatrix} P_2 \\ \vdots \\ P_N \\ Q_2 \\ \vdots \\ Q_N \end{bmatrix}$$

$$\mathbf{f}(\mathbf{x}) = \begin{bmatrix} \mathbf{P}(\mathbf{x}) \\ \mathbf{Q}(\mathbf{x}) \end{bmatrix} = \begin{bmatrix} P_2(\mathbf{x}) \\ \vdots \\ P_N(\mathbf{x}) \\ Q_2(\mathbf{x}) \\ \vdots \\ Q_N(\mathbf{x}) \end{bmatrix} \tag{6.6.1}$$

where all V, P, and Q terms are in per-unit and δ terms are in radians. The swing bus variables δ_1 and V_1 are omitted from (6.6.1), since they are already known. Equations (6.4.10) and (6.4.11) then have the following form:

$$y_k = P_k = P_k(x) = V_k \sum_{n=1}^{N} Y_{kn} V_n \cos(\delta_k - \delta_n - \theta_{kn}) \tag{6.6.2}$$

$$y_{k+N} = Q_k = Q_k(x) = V_k \sum_{n=1}^{N} Y_{kn} V_n \sin(\delta_k - \delta_n - \theta_{kn}) \qquad k = 2, 3, \ldots, N \tag{6.6.3}$$

The Jacobian matrix of (6.3.10) has the form

$$\mathbf{J} = \left[\begin{array}{ccc|ccc} \frac{\partial P_2}{\partial \delta_2} & \cdots & \frac{\partial P_2}{\partial \delta_N} & \frac{\partial P_2}{\partial V_2} & \cdots & \frac{\partial P_2}{\partial V_N} \\ \vdots & & & \vdots & & \\ \frac{\partial P_N}{\partial \delta_2} & \cdots & \frac{\partial P_N}{\partial \delta_N} & \frac{\partial P_N}{\partial V_2} & \cdots & \frac{\partial P_N}{\partial V_N} \\ \hline \frac{\partial Q_2}{\partial \delta_2} & \cdots & \frac{\partial Q_2}{\partial \delta_N} & \frac{\partial Q_2}{\partial V_2} & \cdots & \frac{\partial Q_2}{\partial V_N} \\ \vdots & & & \vdots & & \\ \frac{\partial Q_N}{\partial \delta_2} & \cdots & \frac{\partial Q_N}{\partial \delta_N} & \frac{\partial Q_N}{\partial V_2} & \cdots & \frac{\partial Q_N}{\partial V_N} \end{array}\right] \tag{6.6.4}$$

(Upper-left block: J1; upper-right block: J2; lower-left block: J3; lower-right block: J4.)

Equation (6.6.4) is partitioned into four blocks. The partial derivatives in each block, derived from (6.6.2) and (6.6.3), are given in Table 6.5.

$n \neq k$

$$J1_{kn} = \frac{\partial P_k}{\partial \delta_n} = V_k Y_{kn} V_n \sin(\delta_k - \delta_n - \theta_{kn})$$

$$J2_{kn} = \frac{\partial P_k}{\partial V_n} = V_k Y_{kn} \cos(\delta_k - \delta_n - \theta_{kn})$$

$$J3_{kn} = \frac{\partial Q_k}{\partial \delta_n} = -V_k Y_{kn} V_n \cos(\delta_k - \delta_n - \theta_{kn})$$

$$J4_{kn} = \frac{\partial Q_k}{\partial V_n} = V_k Y_{kn} \sin(\delta_k - \delta_n - \theta_{kn})$$

$n = k$

$$J1_{kk} = \frac{\partial P_k}{\partial \delta_k} = -V_k \sum_{\substack{n=1 \\ n \neq k}}^{N} Y_{kn} V_n \sin(\delta_k - \delta_n - \theta_{kn})$$

$$J2_{kk} = \frac{\partial P_k}{\partial V_k} = V_k Y_{kk} \cos \theta_{kk} + \sum_{n=1}^{N} Y_{kn} V_n \cos(\delta_k - \delta_n - \theta_{kn})$$

$$J3_{kk} = \frac{\partial Q_k}{\partial \delta_k} = V_k \sum_{\substack{n=1 \\ n \neq k}}^{N} Y_{kn} V_n \cos(\delta_k - \delta_n - \theta_{kn})$$

$$J4_{kk} = \frac{\partial Q_k}{\partial V_k} = -V_k Y_{kk} \sin \theta_{kk} + \sum_{n=1}^{N} Y_{kn} V_n \sin(\delta_k - \delta_n - \theta_{kn})$$

$$k, n = 2, 3, \ldots, N$$

TABLE 6.5

Elements of the Jacobian matrix

Now apply to the power flow problem the four Newton-Raphson steps outlined in Section 6.3, starting with $\mathbf{x}(i) = \begin{bmatrix} \boldsymbol{\delta}(i) \\ \mathbf{V}(i) \end{bmatrix}$ at the ith iteration.

STEP 1 Use (6.6.2) and (6.6.3) to compute

$$\Delta \mathbf{y}(i) = \begin{bmatrix} \Delta \mathbf{P}(i) \\ \Delta \mathbf{Q}(i) \end{bmatrix} = \begin{bmatrix} \mathbf{P} - \mathbf{P}[\mathbf{x}(i)] \\ \mathbf{Q} - \mathbf{Q}[\mathbf{x}(i)] \end{bmatrix} \tag{6.6.5}$$

STEP 2 Use the equations in Table 6.5 to calculate the Jacobian matrix.

STEP 3 Use Gauss elimination and back substitution to solve

$$\left[\begin{array}{c|c} \mathbf{J1}(i) & \mathbf{J2}(i) \\ \hline \mathbf{J3}(i) & \mathbf{J4}(i) \end{array}\right] \begin{bmatrix} \Delta \boldsymbol{\delta}(i) \\ \Delta \mathbf{V}(i) \end{bmatrix} = \begin{bmatrix} \Delta \mathbf{P}(i) \\ \Delta \mathbf{Q}(i) \end{bmatrix} \tag{6.6.6}$$

STEP 4 Compute

$$\mathbf{x}(i+1) = \begin{bmatrix} \boldsymbol{\delta}(i+1) \\ \mathbf{V}(i+1) \end{bmatrix} = \begin{bmatrix} \boldsymbol{\delta}(i) \\ \mathbf{V}(i) \end{bmatrix} + \begin{bmatrix} \Delta\boldsymbol{\delta}(i) \\ \Delta\mathbf{V}(i) \end{bmatrix} \tag{6.6.7}$$

Starting with initial value $\mathbf{x}(0)$, the procedure continues until convergence is obtained or until the number of iterations exceeds a specified maximum. Convergence criteria are often based on $\Delta\mathbf{y}(i)$ (called *power mismatches*) rather than on $\Delta\mathbf{x}(i)$ (phase angle and voltage magnitude mismatches).

For each voltage-controlled bus, the magnitude V_k is already known, and the function $Q_k(\mathbf{x})$ is not needed. Therefore, V_k from the $\mathbf{x}$ vector can be omitted, as well as Q_k from the $\mathbf{y}$ vector. The column corresponding to partial derivatives with respect to V_k and the row corresponding to partial derivatives of $Q_k(\mathbf{x})$ can also be omitted from the Jacobian matrix. Alternatively, rows and corresponding columns for voltage-controlled buses can be retained in the Jacobian matrix. Then during each iteration, the voltage magnitude $V_k(i+1)$ of each voltage-controlled bus is reset to V_k which is input data for that bus.

At the end of each iteration, $Q_k(\mathbf{x})$ is computed using (6.6.3) and $Q_{Gk} = Q_k(\mathbf{x}) + Q_{Lk}$ for each voltage-controlled bus. If the computed value of Q_{Gk} exceeds its limits, then the bus type is changed to a load bus with Q_{Gk} set to its limit value. The power flow program also computes a new value for V_k.

EXAMPLE 6.11

Jacobian matrix and power flow solution by Newton-Raphson

Determine the dimension of the Jacobian matrix for the power system in Example 6.9. Also calculate $\Delta P_2(0)$ in Step 1 and $J1_{24}(0)$ in Step 2 of the first Newton-Raphson iteration. Assume zero initial phase angles and 1.0 per-unit initial voltage magnitudes (except $V_3 = 1.05$).

SOLUTION

Since there are $N = 5$ buses for Example 6.9, (6.6.2) and (6.6.3) constitute $2(N-1) = 8$ equations, for which $\mathbf{J}(i)$ has dimension 8×8. However, there is one voltage-controlled bus, bus 3. Therefore, V_3 and the equation for $Q_3(\mathbf{x})$ could be eliminated, with $\mathbf{J}(i)$ reduced to a 7×7 matrix.

From Step 1 and (6.6.2),

$$\begin{aligned}\Delta P_2(0) = P_2 - P_2(x) = P_2 - V_2(0)\{&Y_{21}V_1\cos[\delta_2(0) - \delta_1(0) - \theta_{21}] \\ &+ Y_{22}V_2\cos[-\theta_{22}] + Y_{23}V_3\cos[\delta_2(0) - \delta_3(0) - \theta_{23}] \\ &+ Y_{24}V_4\cos[\delta_2(0) - \delta_4(0) - \theta_{24}] \\ &+ Y_{25}V_5\cos[\delta_2(0) - \delta_5(0) - \theta_{25}]\}\end{aligned}$$

$$\Delta P_2(0) = -8.0 - 1.0\{28.5847(1.0)\cos(84.624°)$$
$$+ 9.95972(1.0)\cos(-95.143°)$$
$$+ 19.9159(1.0)\cos(-95.143°)\}$$
$$= -8.0 - (-2.89 \times 10^{-4}) = -7.99972 \text{ per unit}$$

From Step 2 and J1 given in Table 6.5

$$J1_{24}(0) = V_2(0)Y_{24}V_4(0)\sin[\delta_2(0) - \delta_4(0) - \theta_{24}]$$
$$= (1.0)(9.95972)(1.0)\sin[-95.143°]$$
$$= -9.91964 \text{ per unit}$$

To see the complete convergence of this case, open PowerWorld Simulator case Example 6_11 (see Figure 6.4). Select **Case Information, Network, Mismatches** to see the initial mismatches, and **Case Information, Solution Details, Power Flow Jacobian** to view the initial Jacobian matrix. As is common in commercial power flows, PowerWorld Simulator actually includes rows in the Jacobian for voltage-controlled buses. When a generator is regulating its terminal voltage, this row corresponds to the equation setting the bus voltage magnitude equal to the generator voltage setpoint. However, if the generator hits a reactive power limit, the bus type is switched to a load bus.

To step through the Newton-Raphson solution, from the ribbon select **Solve, Single Solution–Full Newton**. Ordinarily this selection would perform a complete Newton-Raphson iteration, stopping only when all the mismatches are less than the desired tolerance. However, for this case, in order to allow you to see the solution process, the maximum number of iterations has been set to 1, allowing the voltages, mismatches and the Jacobian to be viewed after each iteration. To complete the solution, continue to select **Single Solution—Full Newton** until the solution convergence to the values shown in Tables 6.6, 6.7, and 6.8 (in about three iterations).

Bus #	Voltage Magnitude (per unit)	Phase Angle (degrees)	Generation		Load	
			PG (per unit)	QG (per unit)	PL (per unit)	QL (per unit)
1	1.000	0.000	3.948	1.144	0.000	0.000
2	0.834	−22.407	0.000	0.000	8.000	2.800
3	1.050	−0.597	5.200	3.376	0.800	0.400
4	1.019	−2.834	0.000	0.000	0.000	0.000
5	0.974	−4.548	0.000	0.000	0.000	0.000
		TOTAL	9.148	4.516	8.800	3.200

TABLE 6.6

Bus output data for the power system given in Example 6.9

(Continued)

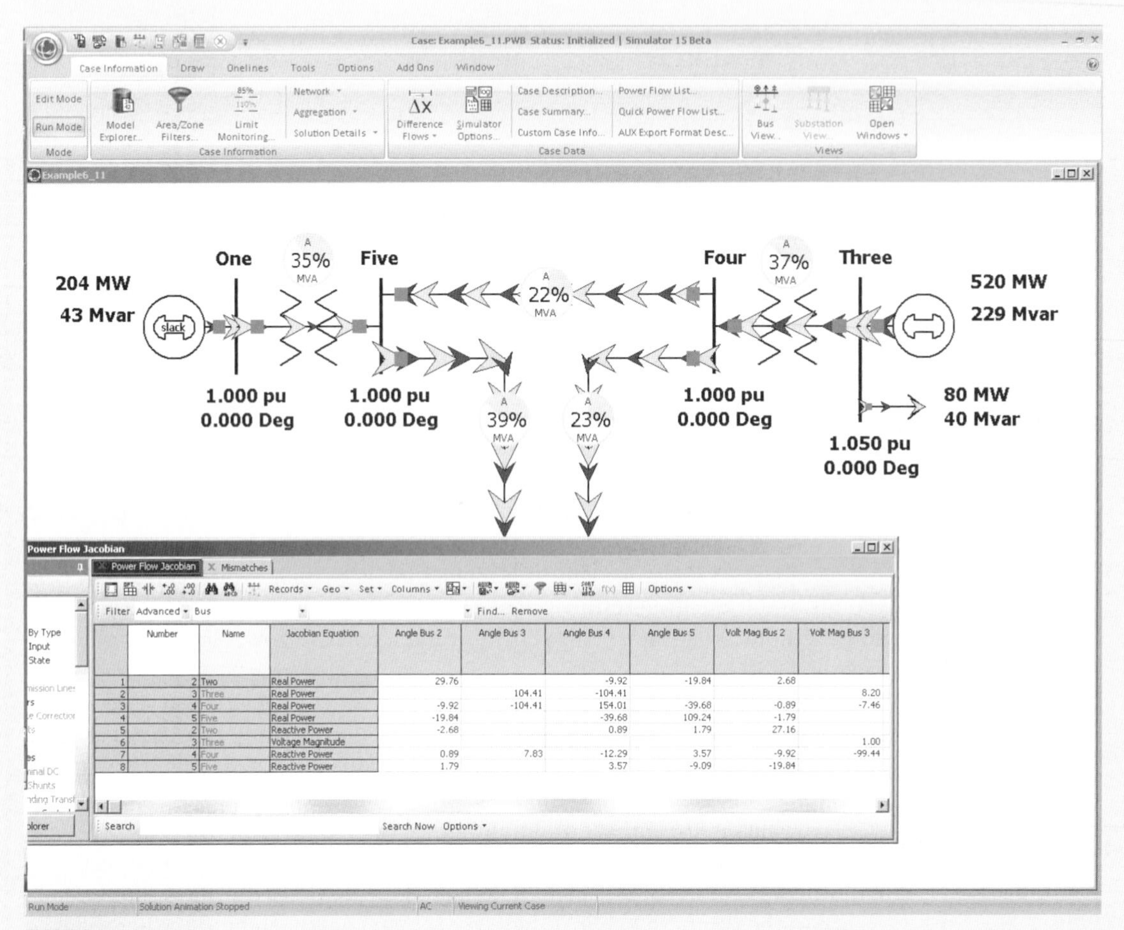

	Number	Name	Jacobian Equation	Angle Bus 2	Angle Bus 3	Angle Bus 4	Angle Bus 5	Volt Mag Bus 2	Volt Mag Bus 3
1	2	Two	Real Power	29.76		-9.92	-19.84	2.68	
2	3	Three	Real Power		104.41	-104.41			8.20
3	4	Four	Real Power	-9.92	-104.41	154.01	-39.68	-0.89	-7.46
4	5	Five	Real Power	-19.84		-39.68	109.24	-1.79	
5	2	Two	Reactive Power	-2.68		0.89	1.79	27.16	
6	3	Three	Voltage Magnitude						1.00
7	4	Four	Reactive Power	0.89	7.83	-12.29	3.57	-9.92	-99.44
8	5	Five	Reactive Power	1.79		3.57	-9.09	-19.84	

FIGURE 6.4

Screen for Example 6.11 showing Jacobian matrix at first iteration

Line #	Bus to Bus		P	Q	S
1	2	4	−2.920	−1.392	3.232
	4	2	3.036	1.216	3.272
2	2	5	−5.080	−1.408	5.272
	5	2	5.256	2.632	5.876
3	4	5	1.344	1.504	2.016
	5	4	−1.332	−1.824	2.260

TABLE 6.7

Line output data for the power system given in Example 6.9

Tran. #	Bus to Bus		P	Q	S
1	1	5	3.948	1.144	4.112
	5	1	−3.924	−0.804	4.004
2	3	4	4.400	2.976	5.312
	4	3	−4.380	−2.720	5.156

TABLE 6.8

Transformer output data for the power system given in Example 6.9

EXAMPLE 6.12

Power flow program: change in generation

Using the power-flow system given in Example 6.9, determine the acceptable generation range at bus 3, keeping each line and transformer loaded at or below 100% of its MVA limit.

SOLUTION

Load PowerWorld Simulator case Example 6_9. Select **Single Solution-Full Newton** to perform a single power flow solution using the Newton-Raphson approach. Then view the **Case Information** displays to verify that the PowerWorld Simulator solution matches the solution shown in Tables 6.6, 6.7, and 6.8. Additionally, the pie charts on the onelines show the percentage line and transformer loadings. Initially transformer Tl, between buses 1 and 5, is loaded at about 69% of its maximum MVA limit, while transformer T2, between buses 3 and 4, is loaded at about 53%.

Next, the bus 3 generation needs to be varied. This can be done a number of different ways in PowerWorld Simulator. The easiest (for this example) is to use the bus 3 generator MW oneline field to manually change the generation (see Figure 6.5). Right-click on the 520 MW field to the right of the bus 3 generator and select **Generator Field Information** dialog to view the **Generator Field Options** dialog. Set the Delta Per Mouse Click field to 10 and select OK. Small arrows are now visible next to this field on the oneline; clicking on the up arrow increases the generator's MW output by 10 MW, while clicking on the down arrow decreases the generation by 10 MW. Select **Tools, Play** to begin the simulation. Increase the generation until the pie chart for the transformer from bus 3 to 4 is loaded to 100%. This occurs at about 1000 MW. Notice that as the bus 3 generation is increased, the bus 1 slack generation decreases by a similar amount. Repeat the process, except now decreasing the generation. This unloads the transformer from bus 3 to 4, but increases the loading on the transformer from bus 1 to bus 5. The bus 1 to 5 transformer should reach 100% loading with the bus 3 generation equal to about 330 MW.

(*Continued*)

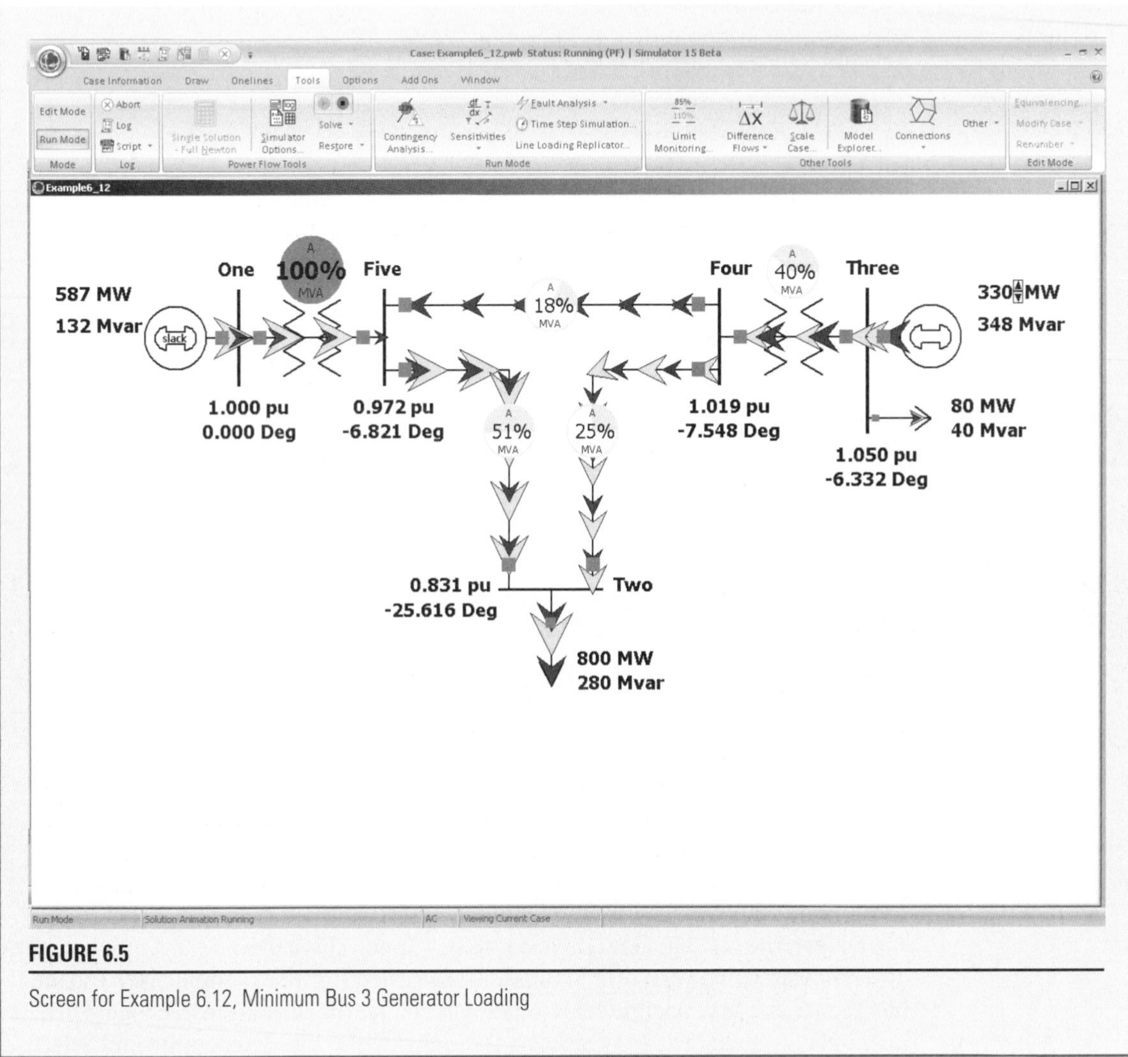

FIGURE 6.5

Screen for Example 6.12, Minimum Bus 3 Generator Loading

Voltage-controlled buses to which tap-changing or voltage-regulating transformers are connected can be handled by various methods. One method is to treat each of these buses as a load bus. The equivalent π circuit parameters (Figure 3.25) are first calculated with tap setting $c = 1.0$ for starting. During each iteration, the computed bus voltage magnitude is compared with the desired value specified by the input data. If the computed voltage is low (or high), c is increased (or decreased) to its next setting, and the parameters of the equivalent π circuit as well as Y_{bus} are recalculated. The procedure continues until the computed bus voltage magnitude equals the desired value within a specified tolerance or until the high or low tap-setting limit is reached. Phase-shifting transformers can be handled in a similar way by using a complex turns ratio $c = 1.0\underline{/\alpha}$ and by varying the phase-shift angle α.

A method with faster convergence makes c a variable and includes it in the $\mathbf{x}$ vector of (6.6.1). An equation is then derived to enter into the Jacobian matrix [4].

In comparing the Gauss-Seidel and Newton-Raphson algorithms, experience from power flow studies has shown that Newton-Raphson converges in many cases where Jacobi and Gauss-Seidel diverge. Furthermore, the number of iterations required for convergence is independent of the number of buses N for Newton-Raphson, but increases with N for Jacobi and Gauss-Seidel. The principal advantage of the Jacobi and Gauss-Seidel methods had been their more modest memory storage requirements and their lower computational requirements per iteration. However, with the vast increases in low-cost computer memory over the last several decades, coupled with the need to solve power flow problems with tens of thousands of buses, these advantages have been essentially eliminated. Therefore the Newton-Raphson, or one of the derivative methods discussed in Sections 6.9 and 6.10, are the preferred power flow solution approaches.

EXAMPLE 6.13

Power flow program: 37-bus system

To see a power flow example of a larger system, open PowerWorld Simulator case Example 6_13 (see Figure 6.6). This case models a 37-bus, 9-generator power system containing three different voltage levels (345 kV, 138 kV, and 69 kV) with 57 transmission lines or transformers. The oneline can be panned by pressing the arrow keys, and it can be zoomed by pressing the <ctrl> with the up arrow key to zoom in or with the down arrow key to zoom out. Use **Tools, Play** to animate the oneline and **Tools, Pause** to stop the animation.

Determine the lowest per-unit voltage and the maximum line/transformer loading both for the initial case and for the case with the line from bus OAK69 to WALNUT69 out of service.

SOLUTION

Use single solution to initially solve the power flow, and then **Case Information, Network, Buses...** to view a listing of all the buses in the case. To quickly determine the lowest per-unit voltage magnitude, left-click on the PU Volt column header to sort the column (clicking a second time reverses the sort). The lowest initial voltage magnitude is 0.990 at bus LOCUST69. Next, select **Case Information, Network, Lines and Transformers...** to view the Line and Transformer Records display. Left-click on % of Max Limit to sort the lines by percentage loading. Initially the highest percentage loading is 64.9% on the line between PEAR69 and PECAN69, circuit 1.

(Continued)

In the software, there are several ways to open the line between OAK69 and WALNUT69. One approach is to locate the line on the Line and Transformer Records display and then double-click on the Status field to change its value. An alternative approach is to find the line on the oneline (it is in the upper-lefthand portion) and then click on one of its circuit breakers. Once the line is removed, use single solution to resolve the power flow. The lowest per-unit voltage is now 0.911 at MAPLE69, and the highest percentage line loading is 135% on the line between PEACH69 to POPLAR69. Since there are now several bus and line violations, the power system is no longer at a reliable operating point. Control actions and/or design improvements are needed to correct these problems. The design projects at the end of the chapter discuss these options.

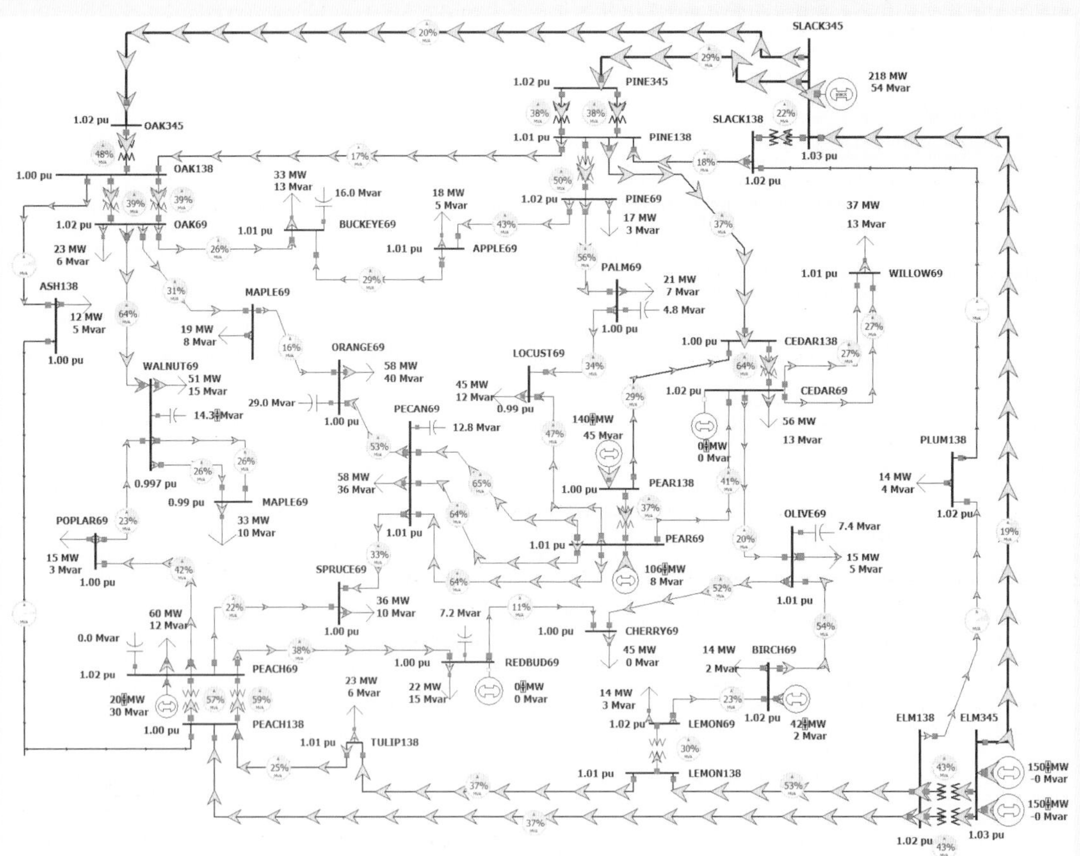

FIGURE 6.6

Screen for Example 6.13 showing the initial flows

6.7 CONTROL OF POWER FLOW

The following means are used to control system power flows:

1. Prime mover and excitation control of generators
2. Switching of shunt capacitor banks, shunt reactors, and static var systems
3. Control of tap-changing and regulating transformers

A simple model of a generator operating under balanced steady-state conditions is the Thévenin equivalent shown in Figure 6.7. V_t is the generator terminal voltage, E_g is the excitation voltage, δ is the power angle, and X_g is the positive-sequence synchronous reactance. From the figure, the generator current is

$$I = \frac{E_g e^{j\delta} - V_t}{jX_g} \tag{6.7.1}$$

and the complex power delivered by the generator is

$$S = P + jQ = V_t I^* = V_t \left(\frac{E_g e^{-j\delta} - V_t}{-jX_g} \right)$$

$$= \frac{V_t E_g (j \cos \delta + \sin \delta) - jV_t^2}{X_g} \tag{6.7.2}$$

The real and reactive powers delivered are then

$$P = \text{Re } S = \frac{V_t E_g}{X_g} \sin \delta \tag{6.7.3}$$

$$Q = \text{Im } S = \frac{V_t}{X_g} (E_g \cos \delta - V_t) \tag{6.7.4}$$

Equation (6.7.3) shows that the real power P increases when the power angle δ increases. From an operational standpoint, when the prime mover increases the power input to the generator while the excitation voltage is held constant, the rotor speed increases. As the rotor speed increases, the power angle δ also increases, causing an increase in generator real power output P. There is also a decrease in reactive power output Q, given by (6.7.4). However, when δ is less than 15°, the increase in P is much larger than the decrease in Q. From the power flow standpoint, an increase in prime-move power corresponds to an increase in P at the constant-voltage bus to

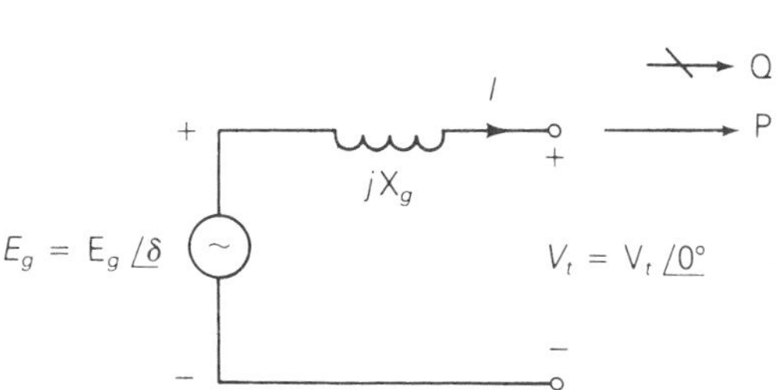

FIGURE 6.7 Generator Thévenin equivalent

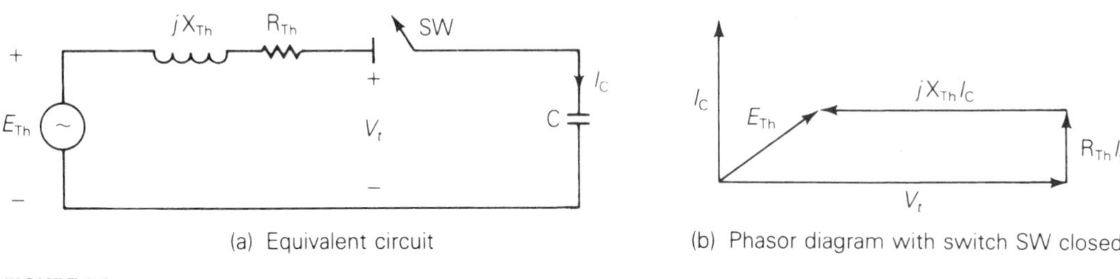

FIGURE 6.8

Effect of adding a shunt capacitor bank to a power system bus

which the generator is connected. The power flow program computes the increase in δ along with the small change in Q.

Equation (6.7.4) shows that reactive power output Q increases when the excitation voltage E_g increases. From the operational standpoint, when the generator exciter output increases while holding the prime-mover power constant, the rotor current increases. As the rotor current increases, the excitation voltage E_g also increases, causing an increase in generator reactive power output Q. There is also a small decrease in δ required to hold P constant in (6.7.3). From the power flow standpoint, an increase in generator excitation corresponds to an increase in voltage magnitude at the constant-voltage bus to which the generator is connected. The power flow program computes the increase in reactive power Q supplied by the generator along with the small change in δ.

Figure 6.8 shows the effect of adding a shunt capacitor bank to a power system bus. The system is modeled by its Thévenin equivalent. Before the capacitor bank is connected, the switch SW is open and the bus voltage equals E_{Th}. After the bank is connected, SW is closed, and the capacitor current I_C leads the bus voltage V_t by 90°. The phasor diagram shows that V_t is larger than E_{Th} when SW is closed. From the power flow standpoint, the addition of a shunt capacitor bank to a load bus corresponds to the addition of a negative reactive load, since a capacitor absorbs negative reactive power. The power flow program computes the increase in bus voltage magnitude along with the small change in δ. Similarly, the addition of a shunt reactor corresponds to the addition of a positive reactive load, wherein the power flow program computes the decrease in voltage magnitude.

Tap-changing and voltage-magnitude-regulating transformers are used to control bus voltages as well as reactive power flows on lines to which they are connected. Similarly, phase-angle regulating transformers are used to control bus angles as well as real power flows on lines to which they are connected. Both tap-changing and regulating transformers are modeled by a transformer with an off-nominal turns ratio c (Figure 3.25). From the power flow standpoint, a change in tap setting or voltage regulation corresponds to a change in c. The power flow program computes the changes in $\boldsymbol{Y}_{bus}$, bus voltage magnitudes and angles, and branch flows.

Besides the above controls, the power flow program can be used to investigate the effect of switching in or out lines, transformers, loads, and generators. Proposed system changes to meet future load growth, including new transmission, new transformers, and new generation can also be investigated. Power flow design studies are normally conducted by trial and error. Using engineering judgment, adjustments in generation levels and controls are made until the desired equipment loadings and voltage profile are obtained.

EXAMPLE 6.14

Power flow program: effect of shunt capacitor banks

Determine the effect of adding a 200-Mvar shunt capacitor bank at bus 2 on the power system in Example 6.9.

SOLUTION

Open PowerWorld Simulator case Example 6_14 (see Figure 6.9). This case is identical to Example 6.9 except that a 200-Mvar shunt capacitor bank has been added at bus 2. Initially this capacitor is open. Click on the capacitor's circuit to close the capacitor and then solve the case. The capacitor increases the bus 2 voltage from 0.834 per unit to a more acceptable 0.959 per unit. The insertion of the capacitor has also substantially decreased the losses, from 34.84 to 25.37 MW.

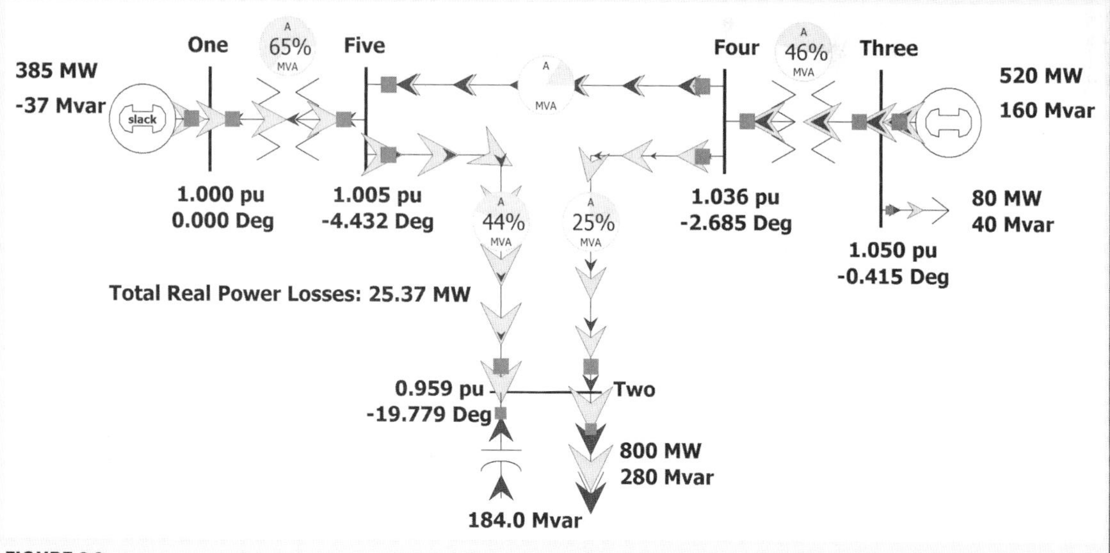

FIGURE 6.9

Screen for Example 6.14

(Continued)

Notice that the amount of reactive power actually supplied by the capacitor is only 184 Mvar. This discrepancy arises because a capacitor's reactive output varies with the square of the terminal voltage, $Q_{cap} = V_{cap}^2/X_c$ (see 2.3.5). A capacitor's Mvar rating is based on an assumed voltage of 1.0 per unit.

EXAMPLE 6.15

PowerWorld Simulator Case Example 6_15 (see Figure 6.10) modifies the Example 6.13 case by (1) opening one of the 138/69 kV transformers at the PEACH substation and (2) opening the 69-kV transmission line between CHERRY69 and OLIVE69. This causes a flow of 116 MVA on the remaining 138/69 kV

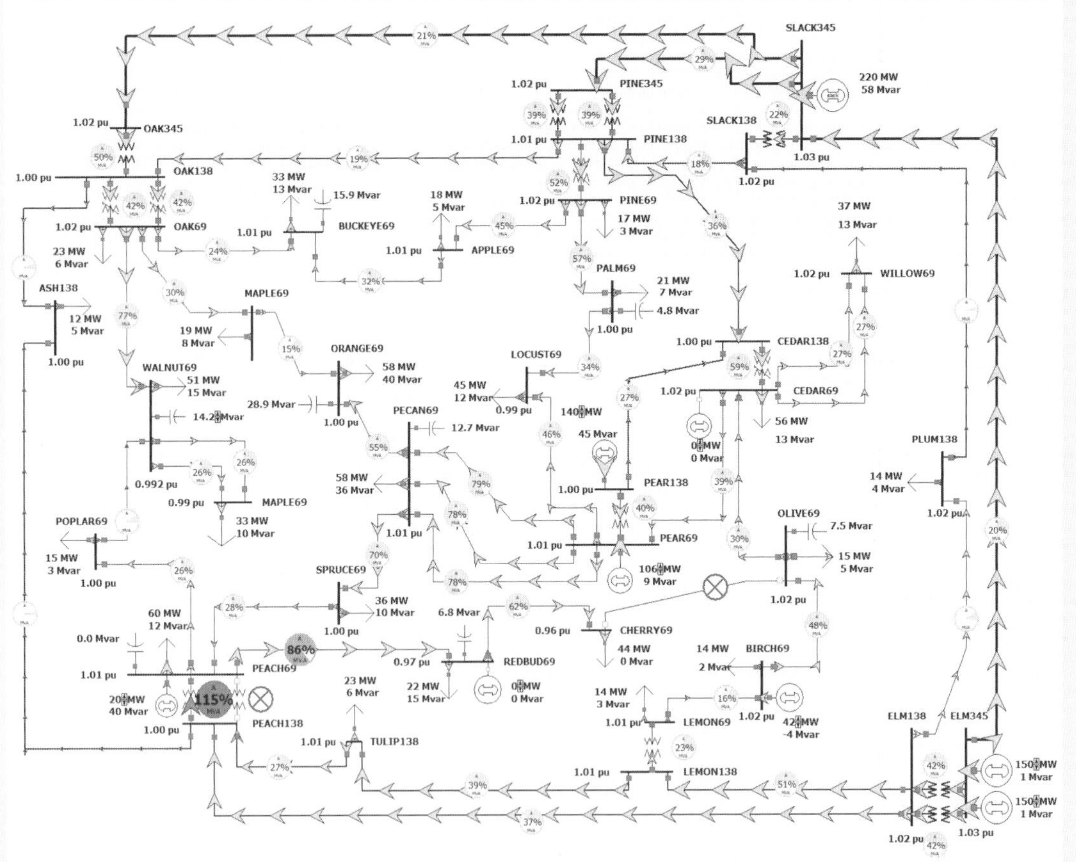

FIGURE 6.10

Screen for Example 6.15

transformer at PEACH. Since this transformer has a limit of 101 MVA, it results in an overload at 115%. Redispatch the generators in order to remove this overload.

SOLUTION

There are a number of solutions to this problem and several solution techniques. One solution technique would be to use engineering intuition, along with a trial-and-error approach (see Figure 6.11). Since the overload is from the 138 kV level to the 69 kV level and there is a generator directly connected to at the PEACH 69 kV bus, it stands to reason that increasing this generation would decrease the overload. Using this approach, the overload is removed by increasing the PEACH generation until the transformer flow is reduced to 100%. This occurs when the generation is increased from 20 MW to 51 MW. Notice that as the generation is

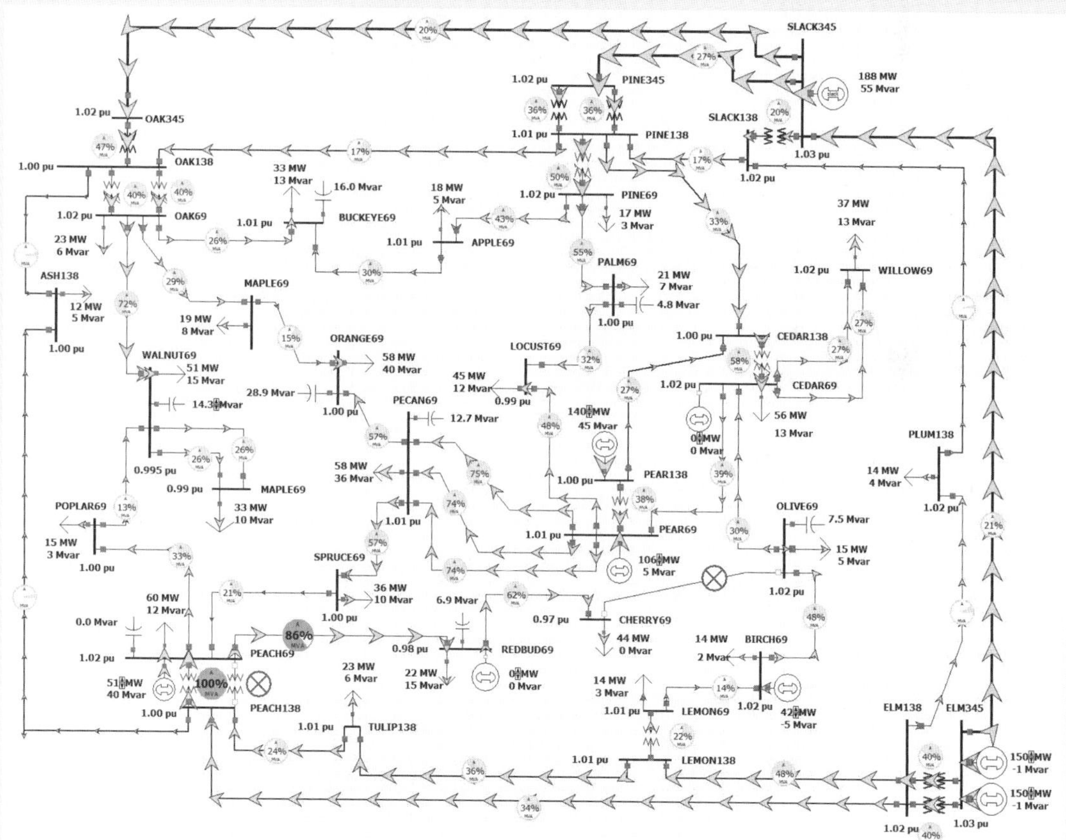

FIGURE 6.11

A solution to Example 6.15

(*Continued*)

increased, the slack bus (SLACK345) generation automatically decreases in order to satisfy the requirement that total system load plus losses must be equal to total generation.

An alternative possible solution is seen by noting that, since the overload is caused by power flowing from the 138 kV bus, decreasing the generation at ELM345 might also decrease this flow. This is indeed the case, but now the trial-and-error approach requires a substantial amount of work and ultimately doesn't solve the problem. Even when the total ELM345 generation is decreased from 300 MW to 0 MW, the overload is still present, albeit with its percentage decreased to 105%.

Another solution approach would be to first determine the generators with the most sensitivity to this violation and then adjust these (see Figure 6.12). This

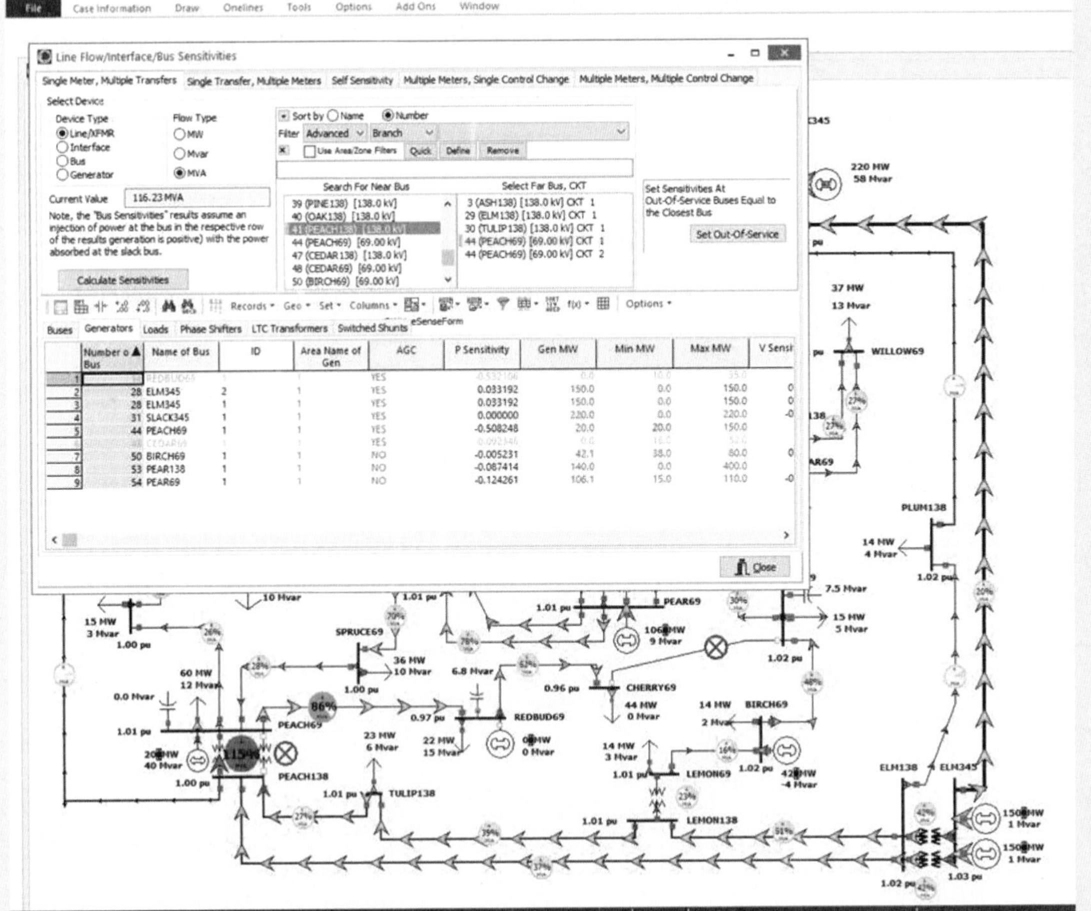

FIGURE 6.12

Example 6.15 Flow Sensitivities Dialog

can be done in PowerWorld Simulator by selecting **Tools, Sensitivities, Flows and Voltage Sensitivities** to calculate sensitivities. On the top of the page, select the PEACH138 to PEACH69 circuit 1 transformer; then click on the **Calculate Sensitivities** button, and select the Generator Sensitivities tab towards the bottom of the dialog. The "P Sensitivity" field tells how increasing the output of each generator by one MW would affect the MVA flow on this transformer. Note that the sensitivity for the PEACH69 generator is -0.508, indicating that if this generation is increased by 1 MW, the transformer MVA flow would decrease by 0.508 MVA. Hence, in order to decrease the flow by 15.2 MVA, increase the PEACH69 generator by 30 MW, which is exactly what was gotten by the trial-and-error approach. It is also clear that the ELM345 generators, with a sensitivity of just 0.033, would be relatively ineffective. In actual power system operation, these sensitivities, known as generator shift factors, are used extensively. These sensitivities are also used in the Optimal Power Flow (introduced in Section 6.13).

6.8 SPARSITY TECHNIQUES

A typical large power system has an average of fewer than three lines connected to each bus. As such, each row of Y_{bus} has an average of fewer than four nonzero elements: one off-diagonal for each line and the diagonal. Such a matrix, which has only a few nonzero elements, is said to be *sparse.*

Newton-Raphson power flow programs employ sparse matrix techniques to reduce computer storage and time requirements [2]. These techniques include compact storage of Y_{bus} and $\mathbf{J}(i)$ and reordering of buses to avoid fill-in of $\mathbf{J}(i)$ during Gauss elimination steps. Consider the following matrix:

$$\text{S} = \begin{bmatrix} 1.0 & -1.1 & -2.1 & -3.1 \\ -4.1 & 2.0 & 0 & -5.1 \\ -6.1 & 0 & 3.0 & 0 \\ -7.1 & 0 & 0 & 4.0 \end{bmatrix} \tag{6.8.1}$$

One method for compact storage of S consists of the following four vectors:

$$\textbf{DIAG} = [1.0\ 2.0\ 3.0\ 4.0] \tag{6.8.2}$$

$$\textbf{OFFDIAG} = [1.1\ -2.1\ -3.1\ -4.1\ -5.1\ -6.1\ -7.1] \tag{6.8.3}$$

$$\textbf{COL} = [2\ 3\ 4\ 1\ 4\ 1\ 1] \tag{6.8.4}$$

$$\textbf{ROW} = [3\ 2\ 1\ 1] \tag{6.8.5}$$

DIAG contains the ordered diagonal elements and **OFFDIAG** contains the nonzero off-diagonal elements of **S**. **COL** contains the column number of each off-diagonal element. For example, the *fourth* element in **COL** is 1, indicating that the *fourth* element of **OFFDIAG**, −4.1, is located in column 1. **ROW** indicates the number of off-diagonal elements in each row of **S**. For example, the *first* element of **ROW** is 3, indicating the *first* three elements of **OFFDIAG**, −1.1, −2.1, and −3.1, are located in the *first* row. The *second* element of **ROW** is 2, indicating the next two elements of **OFFDIAG**, −4.1 and −5.1, are located in the *second* row. The **S** matrix can be completely reconstructed from these four vectors. Note that the dimension of **DIAG** and **ROW** equals the number of diagonal elements of **S**, whereas the dimension of **OFFDIAG** and **COL** equals the number of nonzero off-diagonals.

Now assume that computer storage requirements are 4 bytes to store each magnitude and 4 bytes to store each phase entry in Y_{bus} in an N-bus power system. Also assume Y_{bus} has an average of $3N$ nonzero off-diagonals (three lines per bus) along with its N diagonals. Using the preceding compact storage technique, $(4 + 4)3N = 24N$ bytes are required for **OFFDIAG** and $(4 + 4)N = 8N$ bytes for **DIAG**. Also, assuming 2 bytes to store each integer, $6N$ bytes are required for **COL** and $2N$ bytes for **ROW**. Total computer memory required is then $(24 + 8 + 6 + 2)N = 40N$ bytes with compact storage of Y_{bus}, compared to $8N^2$ bytes without compact storage. For a 1000-bus power system, this means 40 instead of 8000 kilobytes to store Y_{bus}.

The Jacobian matrix is also sparse. From Table 6.5, whenever $Y_{kn} = 0$, $\text{J1}_{kn} = \text{J2}_{kn} = \text{J3}_{kn} = \text{J4}_{kn} = 0$. Compact storage of **J** for a 30,000-bus power system requires less than 10 megabytes with the above assumptions.

The other sparsity technique is to reorder buses. Suppose Gauss elimination is used to triangularize **S** in (6.8.1). After one Gauss elimination step, as described in Section 6.1, the result is

$$\mathbf{S}^{(1)} = \begin{bmatrix} 1.0 & -1.1 & -2.1 & -3.1 \\ 0 & -2.51 & -8.61 & -7.61 \\ 0 & -6.71 & -9.81 & -18.91 \\ 0 & -7.81 & -14.91 & -18.01 \end{bmatrix} \tag{6.8.6}$$

Note that the zeros in columns 2, 3, and 4 of **S** are filled in with nonzero elements in $\mathbf{S}^{(1)}$. The original degree of sparsity is lost.

One simple reordering method is to start with those buses having the fewest connected branches and to end with those having the most connected branches. For example, **S** in (6.8.1) has three branches connected to bus 1 (three off-diagonals in row 1), two branches connected to bus 2, and one branch connected to buses 3 and 4. Reordering the buses 4, 3, 2, 1 instead of 1, 2, 3, 4 results in

$$\mathbf{S}_{\text{reordered}} = \begin{bmatrix} 4.0 & 0 & 0 & -7.1 \\ 0 & 3.0 & 0 & -6.1 \\ -5.1 & 0 & 2.0 & -4.1 \\ -3.1 & -2.1 & -1.1 & 1.0 \end{bmatrix} \tag{6.8.7}$$

Now, after one Gauss elimination step,

$$\mathbf{S}^{(1)}_{\text{reordered}} = \begin{bmatrix} 4.0 & 0 & 0 & -7.1 \\ 0 & 3.0 & 0 & -6.1 \\ 0 & 0 & 2.0 & -13.15 \\ 0 & -2.1 & -1.1 & -4.5025 \end{bmatrix} \tag{6.8.8}$$

Note that the original degree of sparsity is not lost in (6.8.8).

Reordering buses according to the fewest connected branches can be performed once, before the Gauss elimination process begins. Alternatively, buses can be renumbered during each Gauss elimination step in order to account for changes during the elimination process.

Sparsity techniques similar to those described in this section are a standard feature of today's Newton-Raphson power flow programs. As a result of these techniques, typical 30,000-bus power flow solutions require less than 10 megabytes of storage, less than one second per iteration of computer time, and less than 10 iterations to converge.

EXAMPLE 6.16

Sparsity in a 37-bus system

To see a visualization of the sparsity of the power flow Y_{bus} and Jacobian matrices in a *37*-bus system, open PowerWorld Simulator case Example 6_13. Select **Case Information, Solution Details, Ybus** to view the bus admittance matrix. Then press <ctrl> Page Down to zoom the display out. Blank entries in the matrix correspond to zero entries. The 37 × 37 Y_{bus} has a total of 1369 entries, with only about 10% nonzero (see Figure 6.13). Select **Case Information, Solution Details, Power Flow Jacobian** to view the Jacobian matrix.

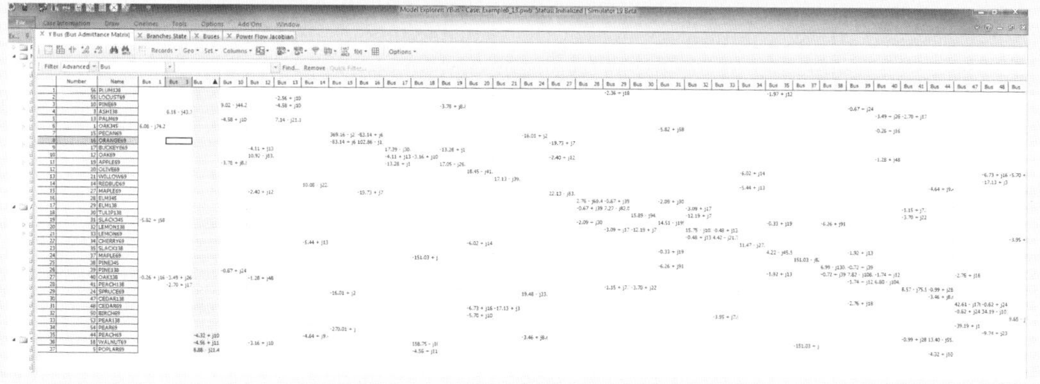

FIGURE 6.13

Screen for Example 6.16

6.9 FAST DECOUPLED POWER FLOW

Contingencies are a major concern in power system operations. For example, operating personnel need to know what power flow changes will occur due to a particular generator outage or transmission-line outage. Contingency information, when obtained in real time, can be used to anticipate problems caused by such outages and can be used to develop operating strategies to overcome the problems.

Fast power flow algorithms have been developed to give power flow solutions in seconds or less [8]. These algorithms are based on the following simplification of the Jacobian matrix. Neglecting $\mathbf{J}_2(i)$ and $\mathbf{J}_3(i)$, (6.6.6) reduces to two sets of decoupled equations:

$$\mathbf{J}_1(i)\,\mathbf{\Delta\delta}(i) = \Delta\mathrm{P}(i) \tag{6.9.1}$$

$$\mathbf{J}_4(i)\,\mathbf{\Delta V}(0) = \Delta\mathbf{Q}(i) \tag{6.9.2}$$

The computer time required to solve (6.9.1) and (6.9.2) is significantly less than that required to solve (6.6.6). Further reduction in computer time can be obtained from additional simplification of the Jacobian matrix. For example, assume $V_k \approx V_n \approx 1.0$ per unit and that the angle differences are small so the sin terms can be ignored. Then $\mathbf{J}_1$ and $\mathbf{J}_4$ are constant matrices whose elements in Table 6.5 are the negative of the imaginary components of Y_{bus}. As such, $\mathbf{J}_1$ and $\mathbf{J}_4$ do not have to be recalculated during successive iterations.

The above simplifications can result in rapid power flow solutions for most systems. While the fast decoupled power flow usually takes more iterations to converge, it is usually significantly faster than the Newton-Raphson algorithm since the Jacobian does not need to be recomputed each iteration. And since the mismatch equations themselves have not been modified, the solution obtained by the fast decoupled algorithm is the same as that found with the Newton-Raphson algorithm. However, in some situations in which only an approximate power flow solution is needed, the fast decoupled approach can be used with a fixed number of iterations (typically one) to give an extremely fast, albeit approximate solution.

6.10 THE "DC" POWER FLOW

The power flow problem can be further simplified by extending the fast decoupled power flow to completely neglect the Q-V equations by assuming that the voltage magnitudes are constant at 1.0 per unit. With these simplifications, the power flow on the line from bus j to bus k with reactance X_{jk} becomes

$$P_{jk} = \frac{\delta_j - \delta_k}{X_{jk}} \tag{6.10.1}$$

and the real power balance equations reduce to a completely linear problem

$$-\mathbf{B}\boldsymbol{\delta} = \mathbf{P} \tag{6.10.2}$$

where **B** is the imaginary component of the of Y_{bus} calculated neglecting line resistance and excepting the slack bus row and column and **P** is the vector of real power injections (with generation assumed positive).

Because (6.10.2) is a linear equation with a form similar to that found in solving dc resistive circuits, this technique is referred to as the dc power flow. However, in contrast to the previous power flow algorithms, the dc power flow only gives an approximate solution with the degree of approximation system dependent. Nevertheless, with the advent of power system restructuring, the dc power flow has become a commonly used analysis technique.

EXAMPLE 6.17

Determine the dc power flow solution for the five bus system from Example 6.9.

SOLUTION

With bus 1 as the system slack, the **B** matrix and **P** vector for this system are

$$\mathbf{B} = \begin{bmatrix} -30 & 0 & 10 & 20 \\ 0 & -100 & 100 & 0 \\ 10 & 100 & -150 & 40 \\ 20 & 0 & 40 & -110 \end{bmatrix} \quad \mathbf{P} = \begin{bmatrix} -8.0 \\ 4.4 \\ 0 \\ 0 \end{bmatrix}$$

$$\boldsymbol{\delta} = -\mathbf{B}^{-1}\mathbf{P} = \begin{bmatrix} -0.3263 \\ 0.0091 \\ -0.0349 \\ -0.0720 \end{bmatrix} \text{radians} = \begin{bmatrix} -18.70 \\ 0.5214 \\ -2.000 \\ -4.125 \end{bmatrix} \text{degrees}$$

To view this example in PowerWorld Simulator, open case Example 6_17, which has this example solved using the dc power flow (see Figure 6.14). To view the dc power flow options, select **Options, Simulator Options** to show the PowerWorld Simulator Options dialog. Then select the Power Flow Solution category and the DC Options page.

(Continued)

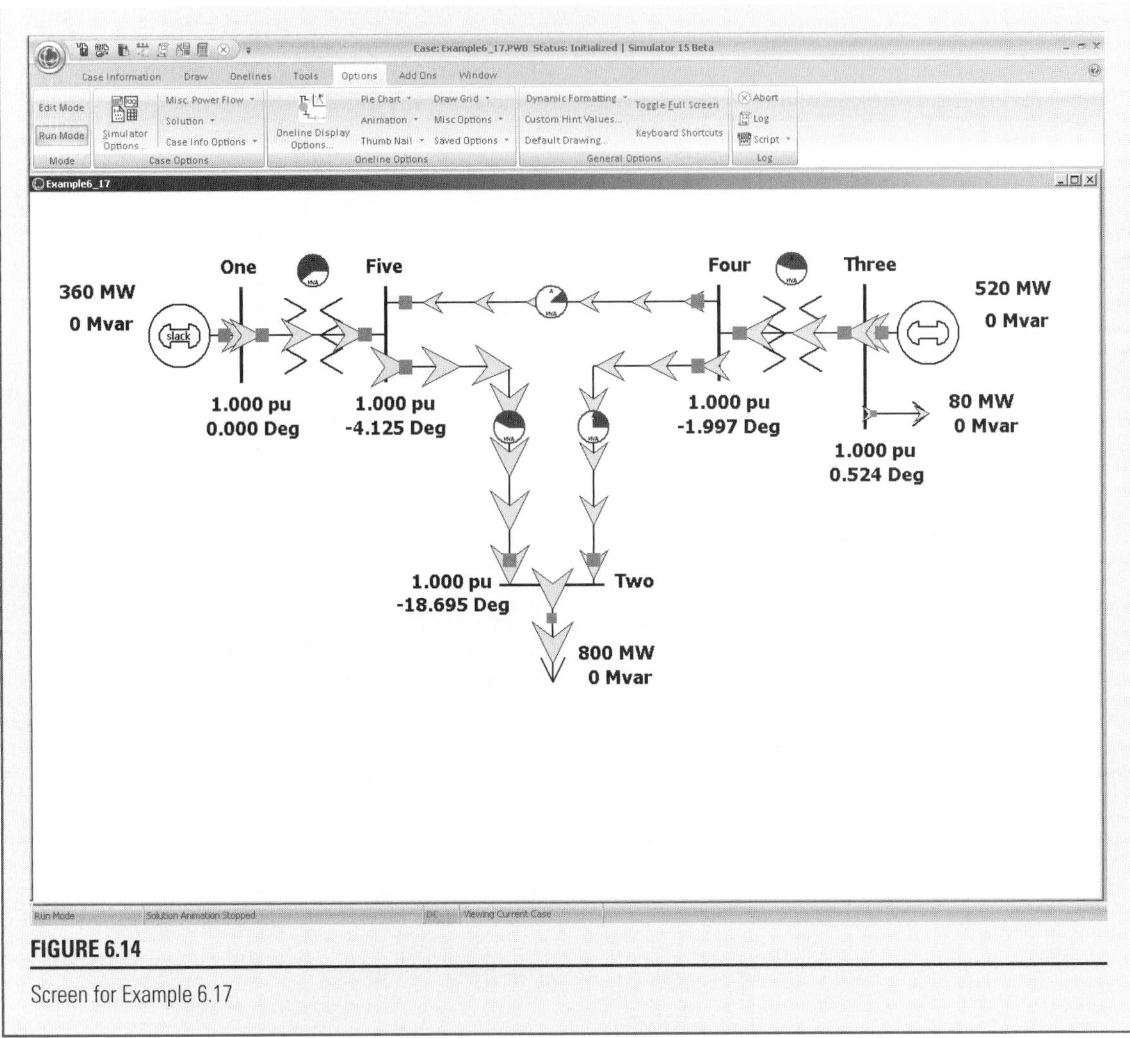

FIGURE 6.14

Screen for Example 6.17

6.11 POWER FLOW MODELING OF WIND GENERATION

As was mentioned in Chapter 1, the amount of renewable generation, particularly wind, being integrated into electric grids around the world is rapidly growing. For example, in 2013 Denmark obtained 33% of their total electric energy from wind while Spain got over 20%. In the United States the amount of wind capacity has been rapidly escalating from less than 2.5 GW in 2000 to more than 68 GW in 2015 (out of a total generation capacity of about 1065 GW), while worldwide, it was more than 360 GW in 2014.

Whereas most energy from traditional synchronous generators comes from large units with ratings of hundreds of MWs, comparatively speaking, individual wind turbine generator (WTG) power ratings are quite low, with common values for new land-based WTGs between one to three MWs, and offshore WTGs up to 6 MWs. This power is generated at low voltage (e.g., 600 V) and then usually stepped-up with a pad-mounted transformer at the base of the turbine to a distribution-level voltage (e.g., 34.5 kV). Usually dozens or even hundreds of individual WTGs are located in wind "farms" or "parks" that cover an area of many square miles, with most of the land still available for other uses such as farming. An underground and/or overhead collector system is used to transmit the power to a single interconnection point at which its voltage is stepped-up to a transmission level voltage ($>$ 100 kV). The layout of such a system is shown in Figure 6.15.

From a power system analysis perspective for large-scale studies, the entire wind farm can usually be represented as a single equivalent generator that is either directly connected at the interconnection point transmission system bus or connected to this bus through an equivalent impedance that represents the impedance of the collector system and the step-up transformers. The parameters associated with the equivalent generator are usually just scaled values of the parameters for the individual WTGs.

There are four main types of WTGs [13,15] with more details on each type provided in Chapter 11. Here the focus is on their power flow characteristics. As is the case with traditional synchronous generators, the real power outputs for all the WTG types are considered to be a constant value in power flow studies. Of course how much real power a wind farm can actually produce at any moment depends upon the wind speed, with a typical wind speed versus power curve shown in Figure 6.16.

Type 1 WTGs are squirrel-cage induction machines. Since induction machines consume reactive power and their reactive power output cannot be independently

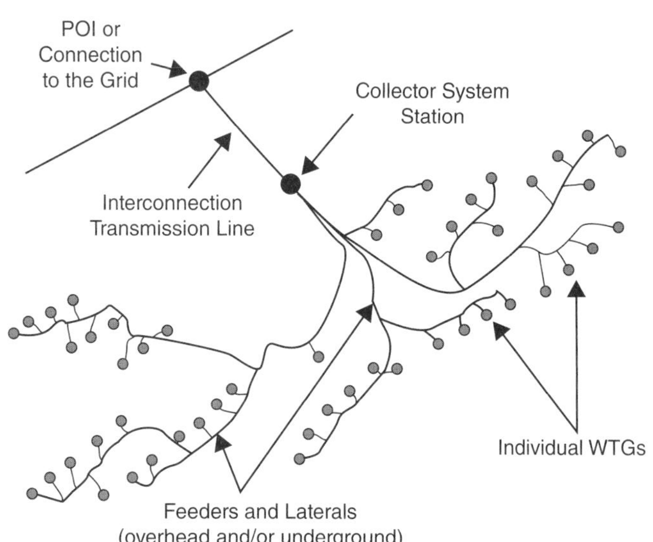

FIGURE 6.15 Wind power plant collector system topology [14] (Figure 1 from WECC Wind Generation Modeling Group, "WECC Wind Power Plant Power Flow Model Guide," WECC, May 2008, p. 2.)

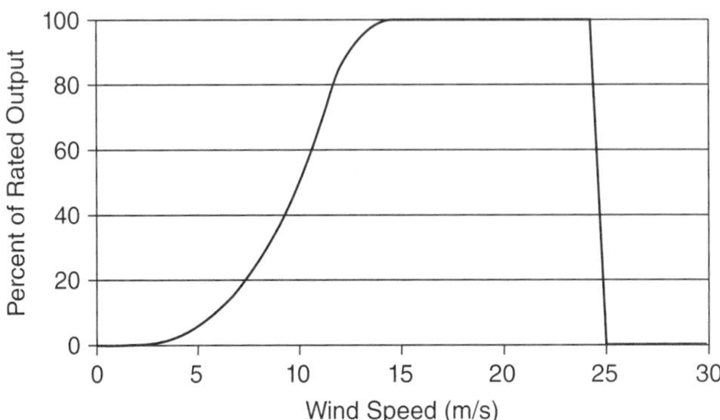

FIGURE 6.16 Typical wind speed versus power curve

controlled, typically these machines are modeled as a constant power factor PQ bus. By themselves these machines have under-excited (consuming reactive power) power factors of between 0.85 and 0.9, but banks of switched capacitors are often used to correct the wind farm power factor. Type 2 WTGs are wound rotor induction machines in which the rotor resistance can be controlled. The advantages of this approach are discussed in Chapter 11; from a power flow perspective, they perform like Type 1 WTGs.

Most of the installed wind capacity and almost all new WTGs are either Type 3 or Type 4. Type 3 wind turbines are used to represent doubly-fed asynchronous generators (DFAGs), also sometimes referred to as doubly-fed induction generators (DFIGs). This type models induction machines in which the rotor circuit is also connected to the ac network through an ac-dc-ac converter, allowing for much greater control of the WTG. Type 4 wind turbines are fully asynchronous machines in which the full power output of the machine is coupled to the ac network through an ac-dc-ac converter. From a power flow perspective, both types are capable of full voltage control like a traditional bus generator with reactive power control between a power factor of up to ± 0.9. However, like traditional synchronous generators, how their reactive power is actually controlled depends on commercial considerations, with many generator owners desiring to operate at unity power factor to maximize their real power outputs.

6.12 ECONOMIC DISPATCH

This section describes how the real power output of a controlled generating unit is selected to meet a given load and to minimize the total operating costs. This is the *economic dispatch* problem [16]. In interconnected power systems, economic dispatch is often solved for smaller portions of the system, known as *areas*, in which the total generation in each area is controlled to match the total area load; further details are provided in Chapter 12.

This section begins by considering only fossil-fuel generating units, with no constraints on maximum and minimum generator outputs, and with no transmission losses. The economic dispatch problem is first solved for this idealized case. Then it is expanded to include inequality constraints on generator outputs and to consider the impact of transmission losses. Finally, the dispatch of other types of units including solar and wind, nuclear, pumped-storage hydro, and hydro units is briefly discussed.

FOSSIL-FUEL UNITS, NO INEQUALITY CONSTRAINTS, NO TRANSMISSION LOSSES

Figure 6.17 shows the operating cost C_i, of a fossil-fuel generating unit versus its real power output P_i. Fuel cost is the major portion of the variable cost of operation, although other variable costs, such as maintenance, could have been included in the figure. Fixed costs, such as the capital cost of installing the unit, are not included. Only those costs that are a function of unit power output—that is, those costs that can be controlled by operating strategy—enter into the economic dispatch formulation.

In practice, C_i is constructed of piecewise continuous functions valid for ranges of output P_i based on empirical data. The discontinuities in Figure 6.17 may be due to the firing of equipment such as additional boilers or condensers as power output is increased. It is often convenient to express C_i in terms of J/hr, which is relatively constant over the lifetime of the unit, rather than \$/hr, which can change monthly or daily. C_i can be converted to \$/hr by multiplying the fuel input in J/hr by the cost of fuel in \$/J.

Figure 6.18 shows the unit incremental operating cost dC_i/dP_i versus unit output P_i, which is the slope or derivative of the C_i versus P_i curve in Figure 6.17. When C_i consists of only fuel costs, dC_i/dP_i is the ratio of the incremental fuel energy input in joules to incremental energy output in kWh, which is called the incremental *heat rate*. Note that the reciprocal of the heat rate, which is the ratio of output energy to input energy, gives a measure of fuel efficiency for the unit. For the unit shown in Figure 6.17, maximum efficiency occurs at $P_i = 600$ MW, where the heat rate is $C_i/P_i = 5.4 \times 10^{12}/600 \times 10^3 = 9$ MJ/kWh. The efficiency at this output is

$$\text{percentage efficency} = \left(\frac{1}{9{,}000{,}000}\,\frac{\text{kWh}}{\text{J}}\right)\left(3{,}600{,}000\,\frac{\text{J}}{\text{kWh}}\right) \times 100 = 40\%$$

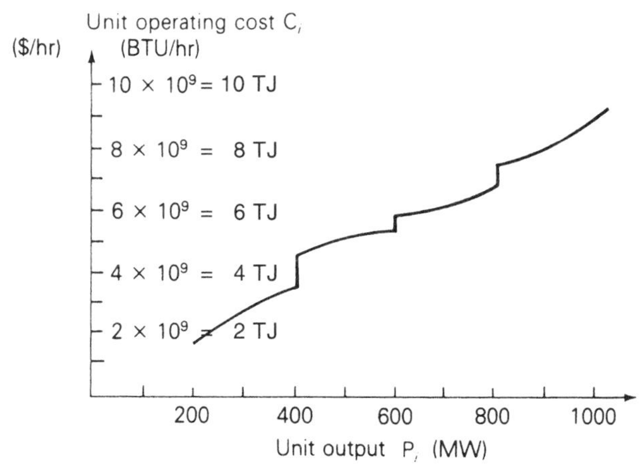

FIGURE 6.17

Unit operating cost versus real power output—fossil-fuel generating unit

FIGURE 6.18

Unit incremental operating cost versus real power output—fossil-fuel generating unit

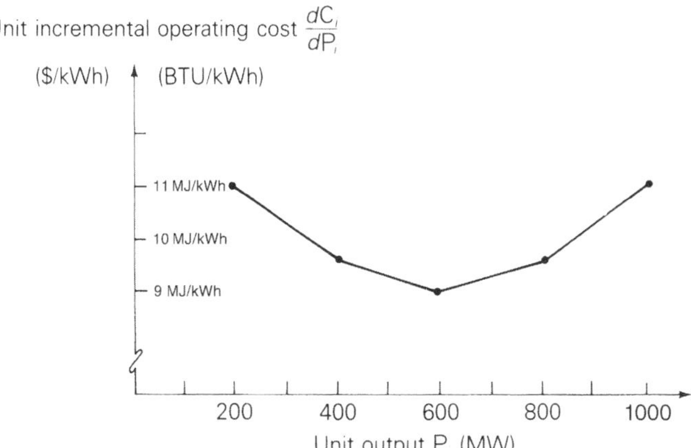

The dC_i/dP_i curve in Figure 6.18 is also represented by piecewise continuous functions valid for ranges of output P_i. For analytical work, the actual curves are often approximated by straight lines. The ratio dC_i/dP_i can also be converted to $/kWh by multiplying the incremental heat rate in J/kWh by the cost of fuel in $/J.

For the area of an interconnected power system consisting of N units operating on economic dispatch, the total variable cost C_T of operating these units is

$$C_T = \sum_{i=i}^{N} C_i$$
$$= C_1(P_1) + C_2(P_2) + \cdots + C_N(P_N) \quad \$/\text{hr} \tag{6.12.1}$$

where C_i, expressed in $/hr, includes fuel cost as well as any other variable costs of unit i. Let P_T equal the total load demand in the area. Neglecting transmission losses.

$$P_1 + P_2 + \cdots + P_N = P_T \tag{6.12.2}$$

Due to relatively slow changes in load demand, P_T may be considered constant for periods of 2 to 10 minutes. The economic dispatch problem can be stated as follows:

> Find the values of unit outputs $P_1, P_2, \ldots, P_N$ that minimize C_T given by (6.12.1), subject to the equality constraint given by (6.12.2).

A criterion for the solution to this problem is: All units on economic dispatch should operate at equal incremental operating cost. That is,

$$\frac{dC_1}{dP_1} = \frac{dC_2}{dP_2} = \cdots = \frac{dC_N}{dP_N} \tag{6.12.3}$$

An intuitive explanation of this criterion is the following. Suppose one unit is operating at a higher incremental operating cost than the other units. If the output power of that unit is reduced and transferred to units with lower incremental operating costs, then the total operating cost C_T decreases. That is, reducing the output of the unit with the *higher* incremental cost results in a *greater cost decrease* than the cost increase of adding that same output reduction to units with lower incremental costs. Therefore, all units must operate at the same incremental operating cost (the economic dispatch criterion).

A mathematical solution to the economic dispatch problem also can be given. The minimum value of C_T occurs when the total differential dC_T is zero. That is,

$$dC_T = \frac{\partial C_T}{\partial P_1} dP_1 + \frac{\partial C_T}{\partial P_2} dP_2 + \cdots + \frac{\partial C_T}{\partial P_N} dP_N = 0 \qquad (6.12.4)$$

Using (6.12.1), (6.12.4) becomes

$$dC_T = \frac{dC_1}{dP_1} dP_1 + \frac{dC_2}{dP_2} dP_2 + \cdots + \frac{dC_N}{dP_N} dP_N = 0 \qquad (6.12.5)$$

Also, assuming P_T is constant, the differential of (6.12.2) is

$$dP_1 + dP_2 + \cdots + dP_N = 0 \qquad (6.12.6)$$

Multiplying (6.12.6) by λ and subtracting the resulting equation from (6.12.5),

$$\left(\frac{dC_1}{dP_1} - \lambda\right) dP_1 + \left(\frac{dC_2}{dP_2} - \lambda\right) dP_2 + \cdots + \left(\frac{dC_N}{dP_N} - \lambda\right) dP_N = 0 \qquad (6.12.7)$$

Equation (6.12.7) is satisfied when each term in parentheses equals zero. That is,

$$\frac{dC_1}{dP_1} = \frac{dC_2}{dP_2} = \cdots = \frac{dC_N}{dP_N} = \lambda \qquad (6.12.8)$$

Therefore, all units have the same incremental operating cost, denoted here by λ, in order to minimize the total operating cost C_T.

EXAMPLE 6.18

Economic dispatch solution neglecting generator limits and line losses

An area of an interconnected power system has two fossil-fuel units operating on economic dispatch. The variable operating costs of these units are given by

$$C_1 = 10P_1 + 8 \times 10^{-3} P_1^2 \quad \$/\text{hr}$$

$$C_2 = 8P_2 + 9 \times 10^{-3} P_2^2 \quad \$/\text{hr}$$

where P_1 and P_2 are in megawatts. Determine the power output of each unit, the incremental operating cost, and the total operating cost C_T that minimizes C_T as the total load demand P_T varies from 500 to 1500 MW. Generating unit inequality constraints and transmission losses are neglected.

(Continued)

SOLUTION

The incremental operating costs of the units are

$$\frac{dC_1}{dP_1} = 10 + 16 \times 10^{-3}P_1 \quad \$/\text{MWh}$$

$$\frac{dC_2}{dP_2} = 8 + 18 \times 10^{-3}P_2 \quad \$/\text{MWh}$$

Using (6.12.8), the minimum total operating cost occurs when

$$\frac{dC_1}{dP_1} = 10 + 16 \times 10^{-3}P_1 = \frac{dC_2}{dP_2} = 8 + 18 \times 10^{-3}P_2$$

Using $P_2 = P_T - P_1$, the preceding equation becomes

$$10 + 16 \times 10^{-3}P_1 = 8 + 18 \times 10^{-3}(P_T - P_1)$$

Solving for P_1,

$$P_1 = \frac{18 \times 10^{-3}P_T - 2}{34 \times 10^{-3}} = 0.5294P_T - 58.82 \quad \text{MW}$$

Also, the incremental operating cost when C_T is minimized is

$$\frac{dC_2}{dP_2} = \frac{dC_1}{dP_1} = 10 + 16 \times 10^{-3}P_1 = 10 + 16 \times 10^{-3}(0.5294P_T - 58.82)$$

$$= 9.0589 + 8.4704 \times 10^{-3}P_T \quad \$/\text{MWh}$$

P_T MW	P_1 MW	P_2 MW	dC_i/dP_1 \$/MWh	C_T \$/hr
500	206	294	13.29	5529
600	259	341	14.14	6901
700	312	388	14.99	8358
800	365	435	15.84	9899
900	418	482	16.68	11,525
1000	471	529	17.53	13,235
1100	524	576	18.38	15,030
1200	576	624	19.22	16,910
1300	629	671	20.07	18,875
1400	682	718	20.92	20,924
1500	735	765	21.76	23,058

TABLE 6.9

Economic dispatch solution for Example 6.18

and the minimum total operating cost is

$$C_T = C_1 + C_2 = (10P_1 + 8 \times 10^{-3}P_1^2) + (8P_2 + 9 \times 10^{-3}P_2^2) \quad \$/\text{hr}$$

The economic dispatch solution is shown in Table 6.9 for values of P_T from 500 to 1500 MW.

EFFECT OF INEQUALITY CONSTRAINTS

Each generating unit must not operate above its rating or below some minimum value. That is.

$$P_{i\min} < P_i < P_{i\max} \qquad i = 1, 2, \ldots, N \tag{6.12.9}$$

Other inequality constraints also may be included in the economic dispatch problem. For example, some unit outputs may be restricted so that certain transmission lines or other equipment are not overloaded. Also, under adverse weather conditions, generation at some units may be limited to reduce emissions.

When inequality constraints are included, modify the economic dispatch solution as follows. If one or more units reach their limit values, then these units are held at their limits, and the remaining units operate at equal incremental operating cost λ. The incremental operating cost of the area equals the common λ for the units that are not at their limits.

EXAMPLE 6.19

Economic dispatch solution including generator limits

Rework Example 6.18 if the units are subject to the following inequality constraints:

$$100 \leqq P_1 \leqq 600 \quad \text{MW}$$

$$400 \leqq P_2 \leqq 1000 \quad \text{MW}$$

SOLUTION

At light loads, unit 2 operates at its lower limit of 400 MW, where its incremental operating cost is $dC_2/dP_2 = 15.2$ \$/MWh. Additional load comes from unit 1 until $dC_1/dP_1 = 15.2$ \$/MWh, or

$$\frac{dC_1}{dP_1} = 10 + 16 \times 10^{-3}P_1 = 15.2$$

$$P_1 = 325 \quad \text{MW}$$

(Continued)

Various methods of evaluating B coefficients from power flow studies are available [17]. In practice, more than one set of B coefficients may be used during the daily load cycle.

When the unit incremental cost curves are linear, an analytic solution to the economic dispatch problem is possible, as illustrated by Examples 6.18 through 6.20. However, in practice, the incremental cost curves are nonlinear and contain discontinuities. In this case, an iterative solution can be obtained. Given the load demand P_T, the unit incremental cost curves, generator limits, and B coefficients, such an iterative solution can be obtained by the following nine steps. Assume that the incremental cost curves are stored in tabular form, such that a unique value of P_i can be read for each dC_i/dP_i.

STEP 1 Set iteration index $m = 1$.

STEP 2 Estimate mth value of λ.

STEP 3 Skip this step for all $m > 1$. Determine initial unit outputs P_i, $(i = 1, 2, \ldots, N)$. Use $dC_i/dP_i = \lambda$ and read P_i from each incremental operating cost table. Transmission losses are neglected here.

STEP 4 Compute $\partial P_L/\partial P_i$ from (6.12.15) $(i = 1, 2, \ldots, N)$.

STEP 5 Compute dC_i/dP_i from (6.12.13) $(i = 1, 2, \ldots, N)$.

STEP 6 Determine updated values of unit output P_i $(i = 1, 2, \ldots, N)$. Read P_i from each incremental operating cost table. If P_i exceeds a limit value, set P_i to the limit value.

STEP 7 Compare P_i determined in Step 6 with the previous value $(i = 1, 2, \ldots, N)$. If the change in each unit output is less than a specified tolerance ε_1, go to Step 8. Otherwise, return to Step 4.

STEP 8 Compute P_L from (6.12.14).

STEP 9 If $\left|\left(\sum_{i=1}^{N} P_i\right) - P_L - P_T\right|$ is less than a specified tolerance ε_2, stop. Otherwise, set $m = m + 1$ and return to Step 2.

Instead of having their values stored in tabular form for this procedure, the incremental cost curves instead could be represented by nonlinear functions such as polynomials. Then, in Step 3 and Step 5, each unit output P_i would be computed from the nonlinear functions instead of being read from a table. Note that this procedure assumes that the total load demand P_T is constant. In practice, this economic dispatch program is executed every few minutes with updated values of P_T.

OTHER TYPES OF UNITS

The economic dispatch criterion has been derived for a power system area consisting of fossil-fuel generating units. In practice, however, an area has a mix of different types of units including fossil-fuel, nuclear, pumped-storage hydro, hydro, wind, and other types.

Wind and solar generation, which have no fuel costs, are represented with very low or negative cost. As such, they are preferred sources for economic dispatch and

and the minimum total operating cost is

$$C_T = C_1 + C_2 = (10P_1 + 8 \times 10^{-3}P_1^2) + (8P_2 + 9 \times 10^{-3}P_2^2) \quad \$/\text{hr}$$

The economic dispatch solution is shown in Table 6.9 for values of P_T from 500 to 1500 MW.

EFFECT OF INEQUALITY CONSTRAINTS

Each generating unit must not operate above its rating or below some minimum value. That is.

$$P_{i\min} < P_i < P_{i\max} \qquad i = 1, 2, \ldots, N \tag{6.12.9}$$

Other inequality constraints also may be included in the economic dispatch problem. For example, some unit outputs may be restricted so that certain transmission lines or other equipment are not overloaded. Also, under adverse weather conditions, generation at some units may be limited to reduce emissions.

When inequality constraints are included, modify the economic dispatch solution as follows. If one or more units reach their limit values, then these units are held at their limits, and the remaining units operate at equal incremental operating cost λ. The incremental operating cost of the area equals the common λ for the units that are not at their limits.

EXAMPLE 6.19

Economic dispatch solution including generator limits

Rework Example 6.18 if the units are subject to the following inequality constraints:

$$100 \leqq P_1 \leqq 600 \quad \text{MW}$$

$$400 \leqq P_2 \leqq 1000 \quad \text{MW}$$

SOLUTION

At light loads, unit 2 operates at its lower limit of 400 MW, where its incremental operating cost is $dC_2/dP_2 = 15.2$ \$/MWh. Additional load comes from unit 1 until $dC_1/dP_1 = 15.2$ \$/MWh, or

$$\frac{dC_1}{dP_1} = 10 + 16 \times 10^{-3}P_1 = 15.2$$

$$P_1 = 325 \quad \text{MW}$$

(Continued)

For P_T less than 725 MW, where P_1 is less than 325 MW, the incremental operating cost of the area is determined by unit 1 alone.

At heavy loads, unit 1 operates at its upper limit of 600 MW, where its incremental operating cost is $dC_1/dP_1 = 19.60$ \$/MWh. Additional load comes from unit 2 for all values of dC_2/dP_2 greater than 19.60 \$/MWh. At $dC_2/dP_2 =$ 19.60 \$/MWh,

$$\frac{dC_2}{dP_2} = 8 + 18 \times 10^{-3} P_2 = 19.60$$

$$P_2 = 644 \quad \text{MW}$$

For P_T greater than 1244 MW, where P_2 is greater than 644 MW, the incremental operating cost of the area is determined by unit 2 alone.

For $725 < P_T < 1244$ MW, neither unit has reached a limit value, and the economic dispatch solution is the same as that given in Table 6.9.

The solution to this example is summarized in Table 6.10 for values of P_T from 500 to 1500 MW.

P_T MW	P_1 MW	P_2 MW	dC/dP \$/MWh	C_T \$/hr
500	100	400	dC_1/dP_1: 11.60	5720
600	200	400	13.20	6960
700	300	400	14.80	8360
725	325	400	15.20	8735
800	365	435	15.84	9899
900	418	482	16.68	11,525
1000	471	529	17.53	13,235
1100	524	576	18.38	15,030
1200	576	624	19.22	16,910
1244	600	644	dC_2/dP_2: 19.60	17,765
1300	600	700	20.60	18,890
1400	600	800	22.40	21,040
1500	600	900	24.20	23,370

TABLE 6.10

Economic dispatch solution for Example 6.19

EXAMPLE 6.20

PowerWorld Simulator–economic dispatch, including generator limits

PowerWorld Simulator case Example 6_20 uses a five-bus, three-generator lossless case to show the interaction between economic dispatch and the transmission system (see Figure 6.19). The variable operating costs for each of the units are given by

$$C_1 = 10P_1 + 0.016P_1^2 \quad \$/\text{hr}$$

$$C_2 = 8P_2 + 0.018P_2^2 \quad \$/\text{hr}$$

$$C_4 = 12P_4 + 0.018P_4^2 \quad \$/\text{hr}$$

where P_1, P_2, and P_4 are the generator outputs in megawatts. Each generator has minimum/maximum limits of

$$100 \leq P_1 \leq 400 \quad \text{MW}$$

$$150 \leq P_2 \leq 500 \quad \text{MW}$$

$$50 \leq P_4 \leq 300 \quad \text{MW}$$

In addition to solving the power flow equations, PowerWorld Simulator can simultaneously solve the economic dispatch problem to optimally allocate the generation in an area. To turn on this option, select **Case Information, Aggregation, Areas...** to view a list of each of the control areas in a case (just one in this example). Then toggle the AGC Status field to ED. Now anytime the power flow equations are solved, the generator outputs are also changed using the economic dispatch.

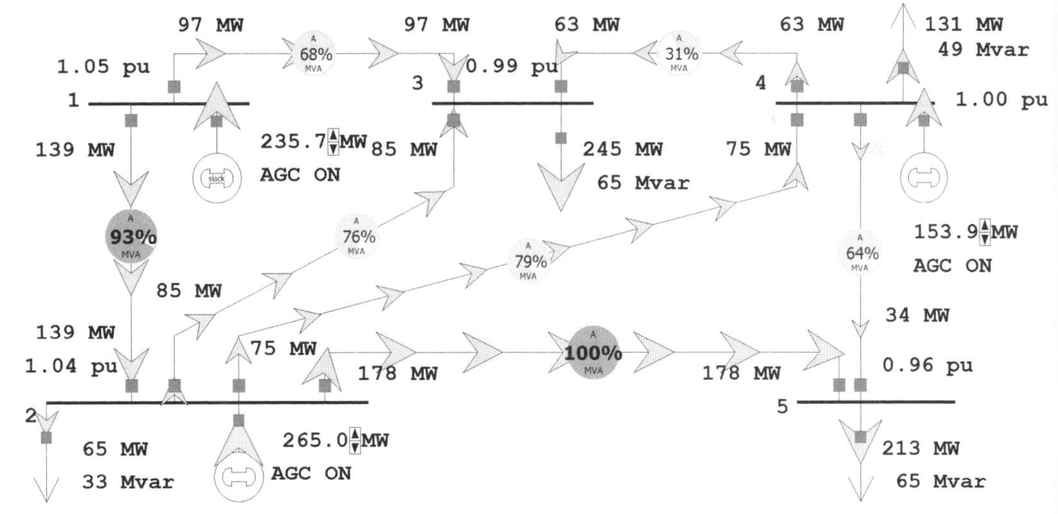

FIGURE 6.19

Example 6.20 with maximum economic loading

(Continued)

Initially, the case has a total load of 392 MW with an economic dispatch of $P_1 = 141$ MW, $P_2 = 181$, and $P_4 = 70$, and an incremental operating cost, λ, of 14.52 \$/MWh. To view a graph showing the incremental cost curves for all of the area generators, right-click on any generator to display the generator's local menu, and then select All Area Gen IC Curves (right-click on the graph's axes to change their scaling).

To see how changing the load impacts the economic dispatch and power flow solutions, first select **Tools, Play** to begin the simulation. Then, on the oneline, click on the up/down arrows next to the Load Scalar field. This field is used to scale the load at each bus in the system. Notice that the change in the Total Hourly Cost field is well approximated by the change in the load multiplied by the incremental operating cost.

Determine the maximum amount of load this system can supply without overloading any transmission line with the generators dispatched using economic dispatch.

SOLUTION

The maximum system economic loading is determined numerically to be 655 MW (which occurs with a Load Scalar of 1.67) with the line from bus 2 to bus 5 being the critical element.

EFFECT OF TRANSMISSION LOSSES

Although one unit may be very efficient with a low incremental operating cost, it also may be located far from the load center. The transmission losses associated with this unit may be so high that the economic dispatch solution requires the unit to decrease its output, while other units with higher incremental operating costs but lower transmission losses increase their outputs.

When transmission losses are included in the economic dispatch problem, (6.12.2) becomes

$$P_1 + P_2 + \cdots + P_N - P_L = P_T \tag{6.12.10}$$

where P_T is the total load demand and P_L is the total transmission loss in the area. In general, P_L is not constant but depends on the unit outputs $P_1, P_2, \ldots, P_N$. The total differential of (6.12.10) is

$$(dP_1 + dP_2 + \cdots + dP_N) - \left(\frac{\partial P_L}{\partial P_1} dP_1 + \frac{\partial P_L}{\partial P_2} dP_2 + \cdots + \frac{\partial P_L}{\partial P_N} dP_N\right) = 0 \tag{6.12.11}$$

Multiplying (6.12.11) by λ and subtracting the resulting equation from (6.12.5),

$$\left(\frac{dC_1}{dP_1} + \lambda\frac{\partial P_L}{\partial P_1} - \lambda\right) dP_1 + \left(\frac{dC_2}{dP_2} + \lambda\frac{\partial P_L}{\partial P_2} - \lambda\right) dP_2$$

$$+ \cdots + \left(\frac{dC_N}{dP_N} + \lambda\frac{\partial P_L}{\partial P_N} - \lambda\right)dP_N = 0 \tag{6.12.12}$$

Equation (6.12.12) is satisfied when each term in parentheses equals zero. That is.

$$\frac{dC_i}{dP_i} + \lambda\frac{\partial P_L}{\partial P_i} - \lambda = 0$$

or

$$\lambda = \frac{dC_i}{dP_i}(L_i) = \frac{dC_i}{dP_i}\left(\frac{1}{1 - \dfrac{\partial P_L}{\partial P_i}}\right) \qquad i = 1, 2, \ldots, N \tag{6.12.13}$$

Equation (6.12.13) gives the economic dispatch criteria, including transmission losses. Each unit that is not at a limit value operates such that its incremental operating cost dC_i/dP_i multiplied by the *penalty factor* L_i is the same. Note that when transmission losses are negligible $\partial P_L/\partial P_i = 0$, $L_i = 1$, and (6.12.13) reduces to (6.12.8).

EXAMPLE 6.21

Economic dispatch solution including generator limits and line losses

Total transmission losses for the power system area given in Example 6.18 are given by

$$P_L = 1.5 \times 10^{-4}P_1^2 + 2 \times 10^{-5}P_1P_2 + 3 \times 10^{-5}P_2^2 \text{ MW}$$

where P_1 and P_2 are given in megawatts. Determine the output of each unit, total transmission losses, total load demand, and total operating cost C_T when the area $\lambda = 16.00$ \$/MWh.

SOLUTION

Using the incremental operating costs from Example 6.18 in (6.12.13),

$$\frac{dC_1}{dP_1}\left(\frac{1}{1 - \dfrac{\partial P_L}{\partial P_1}}\right) = \frac{10 + 16 \times 10^{-3}P_1}{1 - (3 \times 10^{-4}P_1 + 2 \times 10^{-5}P_2)} = 16.00$$

$$\frac{dC_2}{dP_2}\left(\frac{1}{1 - \dfrac{\partial P_L}{\partial P_2}}\right) = \frac{8 + 18 \times 10^{-3}P_2}{1 - (6 \times 10^{-5}P_2 + 2 \times 10^{-5}P_1)} = 16.00$$

(Continued)

Rearranging the two equations,

$$20.8 \times 10^{-3}P_1 + 32 \times 10^{-5}P_2 = 6.00$$
$$32 \times 10^{-5}P_1 + 18.96 \times 10^{-3}P_2 = 8.00$$

Solving,

$$P_1 = 282 \quad \text{MW} \qquad P_2 = 417 \quad \text{MW}$$

Using the equation for total transmission losses,

$$P_L = 1.5 \times 10^{-4}(282)^2 + 2 \times 10^{-5}(282)(417) + 3 \times 10^{-5}(417)^2$$
$$= 19.5 \quad \text{MW}$$

From (6.12.10), the total load demand is

$$P_T = P_1 + P_2 P_L = 282 + 417 - 19.5 = 679.5 \quad \text{MW}$$

Also, using the cost formulas given in Example 6.18, the total operating cost is

$$C_T = C_1 + C_2 = 10(282) + 8 \times 10^{-3}(282)^2 + 8(417) + 9 \times 10^{-3}(417)^2$$
$$= 8357 \quad \$/\text{h}$$

Note that when transmission losses are included, λ given by (6.12.13) is no longer the incremental operating cost of the area. Instead, λ is the unit incremental operating cost dC_i/dP_i multiplied by the unit penalty factor L_i.

EXAMPLE 6.22

PowerWorld Simulator—economic dispatch, including generator limits and line losses

This example repeats the Example 6.19 power system, except that now losses are included with each transmission line modeled with an R/X ratio of 1/3 (see Figure 6.20). The current value of each generator's loss sensitivity, $\partial P_L/\partial P_i$, is shown immediately below the generator's MW output field. Calculate the penalty factors L_i, and verify that the economic dispatch shown in the figure is optimal. Assume a Load Scalar of 1.0.

SOLUTION

From (6.12.13), the condition for optimal dispatch is

$$\lambda = dC_i/dP_i(1/(1 - \partial P_L/\partial P_i) = dC_i/dP_i L_i \qquad i = 1, 2, \ldots, N$$

with

$$L_i = 1/(1 - \partial P_L/\partial P_i)$$

Therefore, $L_1 = 1.0$, $L_2 = 0.9733$, and $L_4 = 0.9238$.

$$\text{with } P_1 = 130.1 \text{ MW}, \quad dC_i/dP_1 \times L_1 = (10 + 0.032 \times 130.1) \times 1.0$$
$$= 14.16 \quad \$/\text{MWh}$$

$$\text{With } P_2 = 181.8 \text{ MW}, \quad dC_2/dP_2 * L_2 = (8 + 0.036 * 181.8) * 0.9733 = 14.16 \ \$/\text{MWh}$$

$$\text{With } P_4 = 92.4 \text{ MW}, \quad dC_4/dP_4 * L_4 = (12 + 0.036 * 92.4) * 0.9238 = 14.16 \ \$/\text{MWh}$$

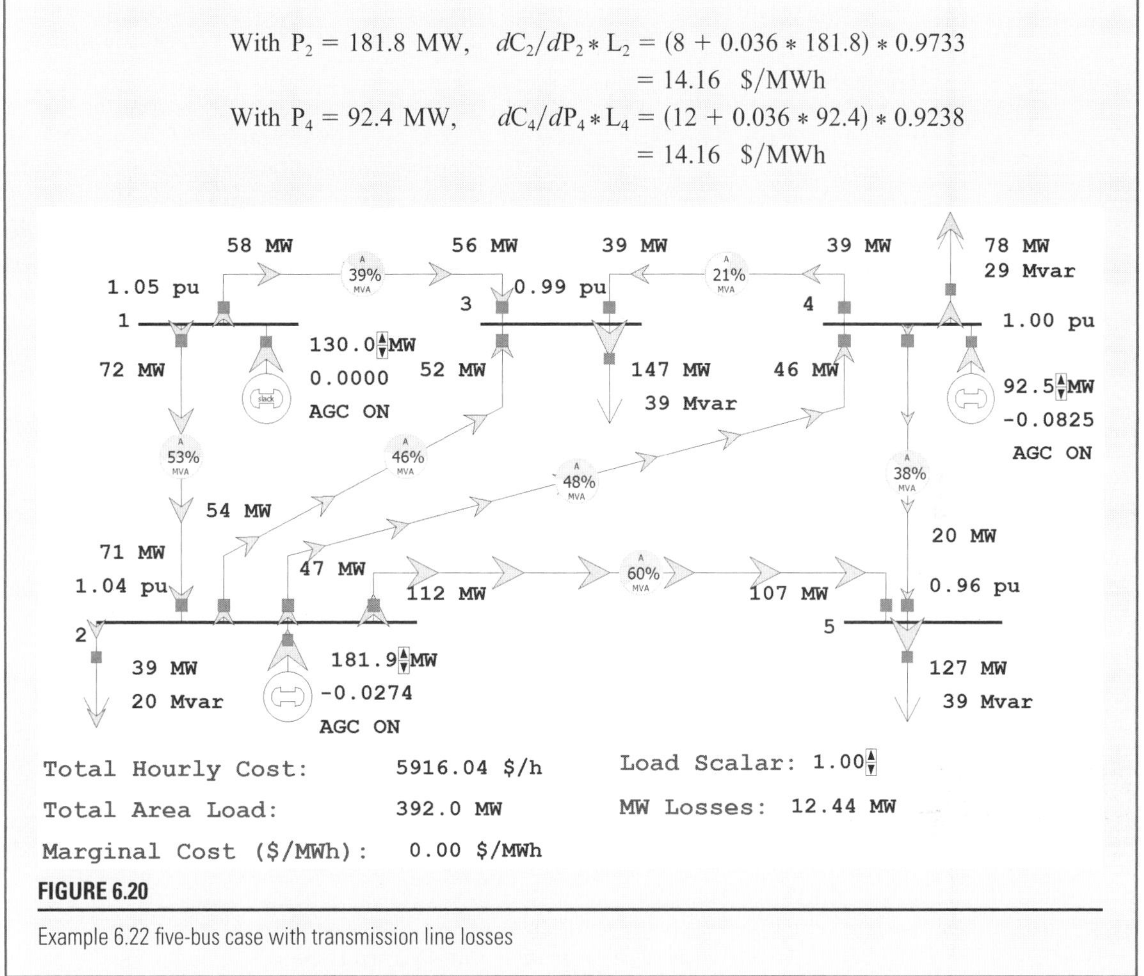

FIGURE 6.20

Example 6.22 five-bus case with transmission line losses

In Example 6.21, total transmission losses are expressed as a quadratic function of unit output powers. For an area with N units, this formula generalizes to

$$P_L = \sum_{i=1}^{N} \sum_{j=1}^{N} P_i B_{ij} P_j \tag{6.12.14}$$

where the B_{ij} terms are called *loss coefficients* or B *coefficients.* The B coefficients are not truly constant but vary with unit loadings. However, the B coefficients are often assumed constant in practice since the calculation of $\partial P_L/\partial P_i$ is thereby simplified. Using (6.12.14),

$$\frac{\partial P_L}{\partial P_i} = 2\sum_{j=1}^{N} B_{ij} P_j \tag{6.12.15}$$

This equation can be used to compute the penalty factor L_i, in (6.12.13).

Various methods of evaluating B coefficients from power flow studies are available [17]. In practice, more than one set of B coefficients may be used during the daily load cycle.

When the unit incremental cost curves are linear, an analytic solution to the economic dispatch problem is possible, as illustrated by Examples 6.18 through 6.20. However, in practice, the incremental cost curves are nonlinear and contain discontinuities. In this case, an iterative solution can be obtained. Given the load demand P_T, the unit incremental cost curves, generator limits, and B coefficients, such an iterative solution can be obtained by the following nine steps. Assume that the incremental cost curves are stored in tabular form, such that a unique value of P_i can be read for each dC_i/dP_i.

STEP 1 Set iteration index $m = 1$.

STEP 2 Estimate mth value of λ.

STEP 3 Skip this step for all $m > 1$. Determine initial unit outputs P_i, ($i = 1, 2, \ldots, N$). Use $dC_i/dP_i = \lambda$ and read P_i from each incremental operating cost table. Transmission losses are neglected here.

STEP 4 Compute $\partial P_L/\partial P_i$ from (6.12.15) ($i = 1, 2, \ldots, N$).

STEP 5 Compute dC_i/dP_i from (6.12.13) ($i = 1, 2, \ldots, N$).

STEP 6 Determine updated values of unit output P_i ($i = 1, 2, \ldots, N$). Read P_i from each incremental operating cost table. If P_i exceeds a limit value, set P_i to the limit value.

STEP 7 Compare P_i determined in Step 6 with the previous value ($i = 1, 2, \ldots, N$). If the change in each unit output is less than a specified tolerance ε_1, go to Step 8. Otherwise, return to Step 4.

STEP 8 Compute P_L from (6.12.14).

STEP 9 If $\left| \left(\sum_{i=1}^{N} P_i \right) - P_L - P_T \right|$ is less than a specified tolerance ε_2, stop. Otherwise, set $m = m + 1$ and return to Step 2.

Instead of having their values stored in tabular form for this procedure, the incremental cost curves instead could be represented by nonlinear functions such as polynomials. Then, in Step 3 and Step 5, each unit output P_i would be computed from the nonlinear functions instead of being read from a table. Note that this procedure assumes that the total load demand P_T is constant. In practice, this economic dispatch program is executed every few minutes with updated values of P_T.

OTHER TYPES OF UNITS

The economic dispatch criterion has been derived for a power system area consisting of fossil-fuel generating units. In practice, however, an area has a mix of different types of units including fossil-fuel, nuclear, pumped-storage hydro, hydro, wind, and other types.

Wind and solar generation, which have no fuel costs, are represented with very low or negative cost. As such, they are preferred sources for economic dispatch and are used by system operators whenever possible, unless there are generator operating limits or transmission constraints.

Although the fixed costs of a nuclear unit may be high, their operating costs are low due to inexpensive nuclear fuel. As such, nuclear units are normally base-loaded at their rated outputs. That is, the reference power settings of turbine-governors for nuclear units are held constant at rated output; therefore, these units do not participate in economic dispatch.

Pumped-storage hydro is a form of energy storage. During off-peak hours, these units are operated as synchronous motors to pump water to a higher elevation. Then during peak-load hours the water is released, and the units are operated as synchronous generators to supply power. As such, pumped-storage hydro units are used for light-load build-up and peak-load shaving. Economic operation of the area is improved by pumping during off-peak hours when the area λ is low, and by generating during peak-load hours when λ is high. Techniques are available for incorporating pumped-storage hydro units into economic dispatch of fossil-fuel units [18].

In an area consisting of hydro plants located along a river, the objective is to maximize the energy generated over the yearly water cycle rather than to minimize total operating costs. Reservoirs are used to store water during high-water or light-load periods, although some water may have to be released through spillways. Also, there are constraints on water levels due to river transportation, irrigation, or fishing requirements. Optimal strategies are available for coordinating outputs of plants along a river [19]. Economic dispatch strategies for mixed fossil-fuel/hydro systems are also available [20, 21, 22].

Techniques are also available for including reactive power flows in the economic dispatch formulation, whereby both active and reactive powers are selected to minimize total operating costs. In particular, reactive injections from generators, switched capacitor banks, and static var systems, along with transformer tap settings, can be selected to minimize transmission-line losses [22]. However, electric utility companies usually control reactive power locally. That is, the reactive power output of each generator is selected to control the generator terminal voltage, and the reactive power output of each capacitor bank or static var system located at a power system bus is selected to control the voltage magnitude at that bus. In this way, the reactive power flows on transmission lines are low, and the need for central dispatch of reactive power is eliminated.

6.13 OPTIMAL POWER FLOW

Economic dispatch has one significant shortcoming—it ignores the limits imposed by the devices in the transmission system. Each transmission line and transformer has a limit on the amount of power that can be transmitted through it, with the limits arising because of thermal, voltage, or stability considerations (Section 5.6). Traditionally, the transmission system was designed so that when the generation was dispatched economically there would be no limit violations. Hence, just solving economic dispatch was usually sufficient. However, with the worldwide trend toward deregulation of the electric utility industry, the transmission system is becoming increasingly constrained (with these constraints sometimes called congestion). For example, in the PJM power market in the eastern United States, the costs associated with active transmission line and transformer limit violations (congestion) increased from \$65 million in 1999 to almost \$2.1 billion in 2005 and have averaged about \$1 billion per year from 2008 to 2013 [23].

The solution to the problem of optimizing the generation while enforcing the transmission lines is to combine economic dispatch with either the full ac power flow, or a dc power flow. The result is known as the optimal power flow (OPF). There are several methods for solving the OPF with [24] providing a nice summary. One common approach is sequential linear programming (LP); this is the technique used with the PowerWorld Simulator. The LP OPF solution algorithm iterates between solving the power flow to determine the flow of power in the system devices and solving an LP to economically dispatch the generation (and possibility other controls) subject to the transmission system limits. In the absence of system elements loaded to their limits, the OPF generation dispatch is identical to the economic dispatch solution, and the marginal cost of energy at each bus is identical to the system λ. However, when one or more elements are loaded to their limits, the economic dispatch becomes constrained, and the bus marginal energy prices are no longer identical. In some electricity markets, these marginal prices are known as the Locational Marginal Prices (LMPs) and are used to determine the wholesale price of electricity at various locations in the system. For example, the real-time LMPs for the Midcontinent ISO (MISO) are available online at www.misoenergy.org/MarketsOperations.

EXAMPLE 6.23

PowerWorld Simulator—optimal power flow

PowerWorld Simulator case Example 6_23 duplicates the five-bus case from Example 6.20, except that the case is solved using PowerWorld Simulator's LP OPF algorithm (see Figure 6.21). To turn on the OPF option, first select **Case Information, Aggregation, Areas...,** and toggle the AGC Status field to OPF. Then, rather than solving the case with the "Single Solution" button, select **Add-ons, Primal LP** to solve using the LP OPF. Initially the OPF solution matches the ED solution from Example 6.20 since there are no overloaded lines. The green-colored fields on the screen immediately to the right of the buses show the marginal cost of supplying electricity to each bus in the system (i.e., the bus LMPs). With the system initially unconstrained, the bus marginal prices are all identical at \$14.5/MWh, with a Load Scalar of 1.0.

Now increase the Load Scalar field from 1.00 to the maximum economic loading value, determined to be 1.67 in Example 6.20, and again select **Add-ons, Primal LP**. The bus marginal prices are still all identical, now at a value of \$17.5/MWh, and with the line from bus 2 to 5 just reaching its maximum value. For load scalar values above 1.67, the line from bus 2 to bus 5 becomes constrained, with a result that the bus marginal prices on the constrained side of the line become higher than those on the unconstrained side.

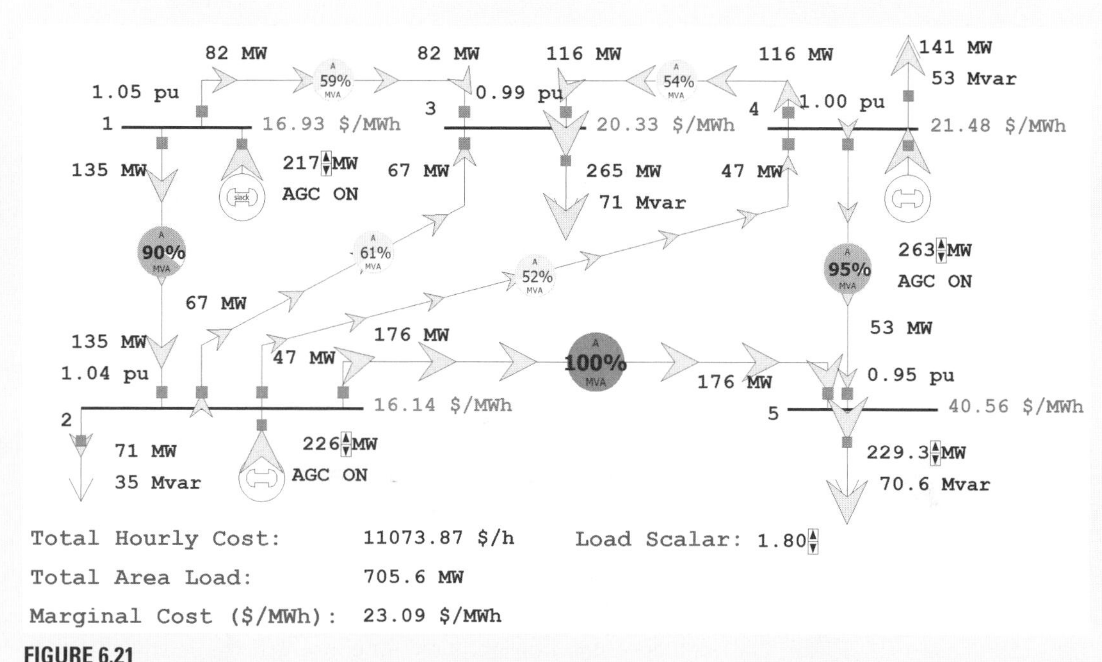

FIGURE 6.21

Example 6.23 optimal power flow solution with load multiplier = 1.80

With the load scalar equal to 1.80, numerically verify that the price of power at bus 5 is approximately \$40.60/MWh.

SOLUTION

The easiest way to numerically verify the bus 5 price is to increase the load at bus 5 by a small amount and compare the change in total system operating cost. With a load scalar of 1.80, the bus 5 MW load is 229.3 MW with a case hourly cost of \$11,073.90. Increasing the bus 5 load by 0.9 MW and resolving the LP OPF gives a new cost of \$11,110.40, which is a change of about \$40.60/MWh (note that this increase in load also increases the bus 5 price to over \$41/MWh). Because of the constraint, the price of power at bus 5 is actually more than double the incremental cost of the most expensive generator!

MULTIPLE CHOICE QUESTIONS

SECTION 6.1

6.1 For a set of linear algebraic equations in matrix format, $\mathbf{Ax} - \mathbf{y}$, for a unique solution to exist, det($\mathbf{A}$) should be ________.

6.2 For an $N \times N$ square matrix $\mathbf{A}$, in $(N - 1)$ steps, the technique of Gauss elimination can transform into an ________ matrix.

SECTION 6.2

6.3 For the iterative solution to linear algebraic equations **Ax** − **y**, the **D** matrix in the Jacobi method is the ________ portion of **A**, whereas **D** for Gauss-Siedel is the ________ portion of **A**.

6.4 Is convergence guaranteed always with Jacobi and Gauss-Siedel methods, as applied to iterative solutions of linear algebraic equations?
(a) Yes (b) No

SECTION 6.3

6.5 For the iterative solutions to nonlinear algebraic equations with the Newton-Raphson method, the Jacobian matrix **J** (i) consists of the partial derivatives. Write down the elements of first row of **J** (i).

6.6 For the Newton-Raphson method to work, one should make sure that $\mathbf{J}^{-1}$ exists.
(a) True (b) False

6.7 The Newton-Raphson method in four steps makes use of Gauss elimination and back substitution.
(a) True (b) False

6.8 The number of iterations required for convergence is dependent/independent of the dimension N for Newton-Raphson method. Choose one.

SECTION 6.4

6.9 The swing bus or slack bus is a reference bus for which $V_1\underline{/\delta_1}$, typically $1.0\underline{/0°}$ per unit, is input data. The power flow program computes ________. Fill in the blank.

6.10 Most buses in a typical power flow program are load buses, for which P_k and Q_k are input data. The power flow program computes ________.

6.11 For a voltage-controlled bus k, ________ are input data, while the power flow program computes ________.

6.12 When the bus k is a load bus with no generation and inductive load, in terms of generation and load, P_k = ________, and Q_k = ________.

6.13 Starting from a single-line diagram of a power system, the input data for a power flow problem consists of ________, ________, and ________.

SECTION 6.5

6.14 Nodal equations $I = Y_{\text{bus}}\, V$ are a set of linear equations analogous to $y = Ax$.
(a) True (b) False

6.15 Because of the nature of the power flow bus data, nodal equations do not directly fit the linear-equation format, and power flow equations are

actually nonlinear. However, the Gauss-Siedel method can be used for the power flow solution.
(a) True (b) False

SECTION 6.6

6.16 The Newton-Raphson method is most well suited for solving the nonlinear power flow equations.
(a) True (b) False

6.17 By default, PowerWorld Simulator uses ___________ method for the power flow solution.

SECTION 6.7

6.18 Prime-mover control of a generator is responsible for a significant change in ___________, whereas excitation control significantly changes ___________.

6.19 From the power flow standpoint, the addition of a shunt-capacitor bank to a load bus corresponds to the addition of a positive/negative reactive load. Choose the right word.

6.20 Tap-changing and voltage-magnitude-regulating transformers are used to control bus voltages and reactive power flows on lines to which they are connected.
(a) True (b) False

SECTION 6.8

6.21 A matrix, which has only a few nonzero elements, is said to be ________.

6.22 Sparse-matrix techniques are used in Newton-Raphson power flow programs in order to reduce computer ___________ and ___________ requirements.

6.23 Reordering buses can be an effective sparsity technique in power flow solutions.
(a) True (b) False

SECTION 6.9

6.24 While the fast decoupled power flow usually takes more iterations to converge, it is usually significantly faster than the Newton-Raphson method.
(a) True (b) False

SECTION 6.10

6.25 The "dc" power flow solution, giving approximate answers, is based on completely neglecting the Q-V equation and solving the linear real-power balance equations.
(a) True (b) False

PROBLEMS

SECTION 6.1

6.1 Using Gauss elimination, solve the following linear algebraic equations:

$$-25x_1 + 10x_2 + 10x_3 + 10x_4 = 0$$
$$5x_1 - 10x_2 + 10x_3 = 2$$
$$10x_1 + 5x_2 - 10x_3 + 10x_4 = 1$$
$$10x_1 - 20x_4 = -2$$

6.2 Using Gauss elimination and back substitution, solve

$$\begin{bmatrix} 8 & 2 & 1 \\ 4 & 6 & 2 \\ 3 & 4 & 14 \end{bmatrix} \begin{bmatrix} x_1 \\ x_2 \\ x_3 \end{bmatrix} = \begin{bmatrix} 3 \\ 4 \\ 2 \end{bmatrix}$$

6.3 Rework Problem 6.2 with the value of 8 changed to 4.

6.4 What is the difficulty in applying Gauss elimination to the following linear algebraic equations?

$$-5x_1 + 5x_2 = 5$$
$$10x_1 - 10x_2 = -5$$

6.5 Show that, after triangularizing $\mathbf{Ax} = \mathbf{y}$, the back substitution method of solving $\mathbf{A}^{(N-1)}\mathbf{x} = \mathbf{y}^{(N-1)}$ requires N divisions, $N(N-1)/2$ multiplications, and $N(N-1)/2$ subtractions. Assume that all the elements of $\mathbf{A}^{(N-1)}$ and $\mathbf{y}^{(N-1)}$are nonzero and real.

SECTION 6.2

6.6 Solve Problem 6.2 using the Jacobi iterative method. Start with $x_1(0) = x_2(0) = x_3(0) = 0$, and continue until (6.2.2) is satisfied with $\varepsilon = 0.01$.

6.7 Repeat Problem 6.6 using the Gauss-Seidel iterative method. Which method converges more rapidly?

6.8 Express the following set of equations in the form of (6.2.6), and then solve using the Jacobi iterative method with $\varepsilon = 0.05$ and with $x_1(0) = 1$, and $x_2(0) = 1$, $x_3(0) = 0$.

$$\begin{bmatrix} 10 & -2 & -4 \\ -2 & 6 & -2 \\ -4 & -2 & 10 \end{bmatrix} \begin{bmatrix} x_1 \\ x_2 \\ x_3 \end{bmatrix} = \begin{bmatrix} -2 \\ 3 \\ -1 \end{bmatrix}$$

6.9 Solve for x_1 and x_2 in the system of equations given by

$$x_2 - 3x_1 + 1.9 = 0$$

$$x_2 + x_1^2 - 3.0 = 0$$

using the Gauss method with an initial guess of $x_1 = 1$ and $x_2 = 1$.

6.10 Solve $x^2 - 4x + 1 = 0$ using the Jacobi iterative method with $x(0) = 1$. Continue until (6.2.2) is satisfied with $\varepsilon = 0.01$. Check using the quadratic formula.

6.11 Try to solve Problem 6.2 using the Jacobi and Gauss-Seidel iterative methods with the value of A_{33} changed from 14 to 0.14 and with $x_1(0) = x_2(0) = x_3(0) = 0$. Show that neither method converges to the unique solution.

6.12 Using the Jacobi method (also known as the Gauss method), solve for x_1 and x_2 in the following system of equations.

$$x_2 - 3x_1 + 1.9 = 0$$

$$x_2 + x_1^2 - 1.8 = 0$$

Use an initial guess of $x_1(0) = 1.0$ and $x_2 = (0) = 1.0$. Also, see what happens when you choose an uneducated initial guess of $x_1(0) = x_2(0) = 100$.

6.13 Use the Gauss-Seidel method to solve the following equations that contain terms that are often found in power flow equations.

$$x_1 = (1/(-20j)) * [(-1 + 0.5j)/(x_1)^* - (j10) * x_2 - (j10)]$$

$$x_2 = (1/(-20j)) * [(-3 + j)/(x_2)^* - (j10) * x_1 - (j10)]$$

Use an initial estimate of $x_1(0) = 1$ and $x_2(0) = 1$, and a stopping of $\varepsilon = 0.05$.

6.14 Find a root of the following equation by using the Gauss-Seidel method: (use an initial estimate of $x = 2$) $f(x) = x^3 - 6x^2 + 9x - 4 = 0$.

6.15 Use the Jacobi method to find a solution to $x^2 \cos x - x + 0.5 = 0$. Use $x(0) = 1$ and $\varepsilon = 0.01$. Experimentally determine the range of initial values that results in convergence.

6.16 Take the z-transform of (6.2.6) and show that $\mathbf{X}(z) = \mathbf{G}(z)\mathbf{Y}(z)$, where $\mathbf{G}(z) = (z\mathbf{U} - \mathbf{M})^{-1}\,\mathbf{D}^{-1}$ and $\mathbf{U}$ is the unit matrix.

Note: $\mathbf{G}(z)$ is the matrix transfer function of a digital filter that represents the Jacobi or Gauss-Seidel methods. The filter poles are obtained by solving $\det(z\mathbf{U} - \mathbf{M}) = 0$. The filter is stable if and only if all the poles have magnitudes less than 1.

6.17 Determine the poles of the Jacobi and Gauss-Seidel digital filters for the general two-dimensional problem ($N = 2$):

$$\left[\begin{array}{c|c} A_{11} & A_{12} \\ \hline A_{21} & A_{22} \end{array}\right]\begin{bmatrix} x_1 \\ x_2 \end{bmatrix} = \begin{bmatrix} y_1 \\ y_2 \end{bmatrix}$$

Then determine a necessary and sufficient condition for convergence of these filters when $N = 2$.

SECTION 6.3

6.18 Use Newton-Raphson to find a solution to the polynomial equation $f(x) = y$ where $y = 0$ and $f(x) = x^3 + 8x^2 + 2x - 40$. Start with $x(0) = 1$ and continue until (6.2.2) is satisfied with $\varepsilon = 0.001$.

6.19 Repeat 6.19 using $x(0) = -2$.

6.20 Use Newton-Raphson to find one solution to the polynomial equation $f(x) = y$, where $y = 7$ and $f(x) = x^4 + 3x^3 - 15x^2 - 19x + 30$. Start with $x(0) = 0$ and continue until (6.2.2) is satisfied with $\varepsilon = 0.001$.

6.21 Repeat Problem 6.20 with an initial guess of $x(0) = 4$.

6.22 For Problem 6.20, plot the function $f(x)$ between $x = 0$ and 4. Then provide a graphical interpretation why points close to $x = 2.2$ would be poorer initial guesses.

6.23 Use Newton-Raphson to find a solution to

$$\begin{bmatrix} e^{x_1 x_2} \\ \cos(x_1 + x_2) \end{bmatrix} = \begin{bmatrix} 1.2 \\ 0.5 \end{bmatrix}$$

where x_1 and x_2 are in radians. (a) Start with $x_1(0) = 1.0$ and $x_2(0) = 0.5$ and continue until (6.2.2) is satisfied with $\varepsilon = 0.005$. (b) Show that Newton-Raphson diverges for this example if $x_1(0) = 1.0$ and $x_2(0) = 2.0$.

6.24 Solve the following equations by the Newton-Raphson method:

$$2x_1 + x_2^2 - 8 = 0$$

$$x_1^2 - x_2^2 + x_1 x_2 - 3 = 0$$

Start with an initial guess of $x_1 = 1$ and $x_2 = 1$.

6.25 The following nonlinear equations contain terms that are often found in the power flow equations:

$$f_1(x) = 10x_1 \sin x_2 + 2 = 0$$

$$f_2(x) = 10\,(x_1)^2 - 10x_1 \cos x_2 + 1 = 0$$

Solve using the Newton-Raphson method starting with an initial guess of $x_1\,(0) = 1$ and $x_2(0) = 0$ radians and a stopping criteria of $\varepsilon = 10^{-4}$.

6.26 Repeat Problem 6.25 except using $x_1(0) = 0.25$ and $x_2(0) = 0$ radians as an initial guess.

6.27 For the Newton-Raphson method, the *region of attraction* (or *basin of attraction)* for a particular solution is the set of all initial guesses that converge to that solution. Usually initial guesses close to a particular solution will converge to that solution. However, for all but the simplest of multi-dimensional, nonlinear problems, the region of attraction boundary is often fractal. This makes it impossible to quantify the region of attraction and hence to guarantee convergence. Problem 6.25 has two solutions when x_2 is restricted to being between $-\pi$ and π. With the x_2 initial guess fixed at 0 radians, numerically determine the values of the x_1 initial guesses that converge to the Problem 6.25 solution. Restrict your search to values of x_1 between 0 and 1.

SECTION 6.4

6.28 Consider the simplified electric power system shown in Figure 6.22 for which the power flow solution can be obtained without resorting to iterative techniques. (a) Compute the elements of the bus admittance matrix Y_{bus}. (b) Calculate the phase angle δ_2 by using the real power equation at bus 2 (voltage-controlled bus). (c) Determine $|V_3|$ and δ_3 by using both the real and reactive power equations at bus 3 (load bus). (d) Find the real power generated at bus 1 (swing bus). (e) Evaluate the total real power losses in the system.

6.29 In Example 6.9, double the impedance on the line from bus 2 to bus 5. Determine the new values for the second row of Y_{bus}. Verify your result using PowerWorld Simulator case Example 6_9.

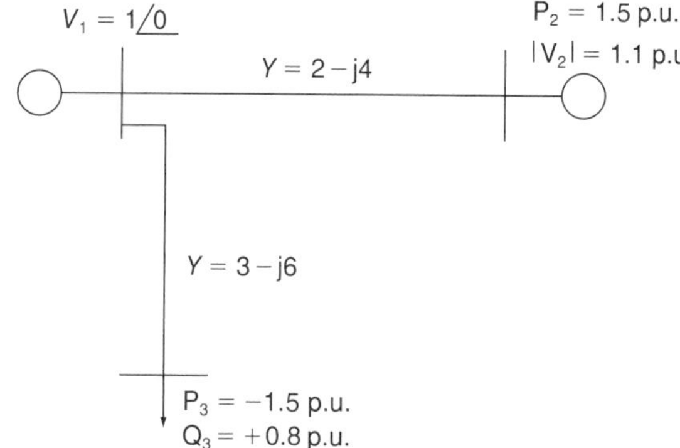

FIGURE 6.22 Problem 6.28

6.30 Determine the bus admittance matrix (Y_{bus}) for the three-phase power system shown in Figure 6.23 with input data given in Table 6.11 and partial results in Table 6.12. Assume a three-phase 100 MVA per unit base.

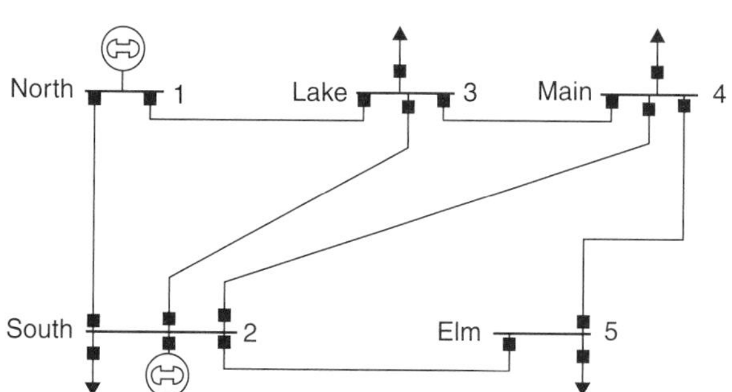

FIGURE 6.23 Sample System Diagram for Problem 6.30

Bus-to-Bus	R per unit	X per unit	B per unit
1-2	0.02	0.06	0.06
1-3	0.08	0.24	0.05
2-3	0.06	0.18	0.04
2-4	0.08	0.24	0.05
2-5	0.02	0.06	0.02
3-4	0.01	0.04	0.01
4-5	0.03	0.10	0.04

TABLE 6.11

Bus input data for Problem 6.30

$6.25 - j18.695$	$-5.00 + j15.00$	$-1.25 + j3.75$	0	0
$-5.00 + j15.00$				

TABLE 6.12

Partially Completed Bus Admittance Matrix ($\boldsymbol{Y}_{\text{bus}}$) for Problem 6.30

6.31 For the system from Problem 6.30, assume that a 75-Mvar shunt capacitance (three phase assuming one per unit bus voltage) is added at bus 4. Calculate the new value of Y_{44}.

SECTION 6.5

6.32 For a two-bus power system, a $0.7 + j0.4$ per unit load at bus 2 is supplied by a generator at bus 1 through a transmission line with series impedance of $0.05 + j0.1$ per unit. With bus 1 as the slack bus with a fixed per-unit voltage of $1.0\angle 0$, use the Gauss-Seidel method to calculate the voltage at bus 2 after three iterations.

6.33 Repeat Problem 6.32 with the slack bus voltage changed to $1.0\angle 30^\circ$ per unit.

6.34 For the three-bus system whose Y_{bus} is given, calculate the second iteration value of V_3 using the Gauss-Seidel method. Assume bus 1 as the slack (with $V_1 = 1.0\angle 0^\circ$), and buses 2 and 3 are load buses with a per-unit load of $S_2 = 1 + j0.5$ and $S_3 = 1.5 + j0.75$. Use voltage guesses of $1.0\angle 0^\circ$ at both buses 2 and 3. The bus admittance matrix for a three-bus system is

$$Y_{\text{bus}} = \begin{bmatrix} -j10 & j5 & j5 \\ j5 & -j10 & j5 \\ j5 & j5 & -j10 \end{bmatrix}$$

6.35 Repeat Problem 6.34 except assume the bus 1 (slack bus) voltage of $V_1 = 1.05\underline{/0^\circ}$.

6.36 The bus admittance matrix for the power system shown in Figure 6.24 is given by

$$Y_{\text{bus}} = \begin{bmatrix} 3 - j9 & -2 + j6 & -1 + j3 & 0 \\ -2 + j6 & 3.666 - j11 & -0.666 + j2 & -1 + j3 \\ -1 + j3 & -0.666 + j2 & 3.666 - j11 & -2 + j6 \\ 0 & -1 + j3 & -2 + j6 & 3 - j9 \end{bmatrix} \text{ per unit}$$

With the complex powers on load buses 2, 3, and 4 as shown in Figure 6.24, determine the value for V_2 that is produced by the first and second iterations of the Gauss-Seidel procedure. Choose the initial guess $V_2(0) = V_3(0) = V_4(0) = 1.0\underline{/0^\circ}$ per unit.

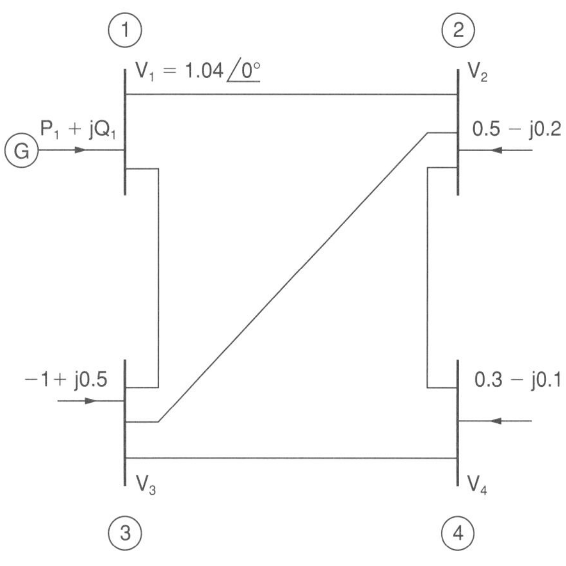

FIGURE 6.24
Problem 6.36

6.37 The bus admittance matrix of a three-bus power system is given by

$$Y_{\text{bus}} = -j\begin{bmatrix} 7 & -2 & -5 \\ -2 & 6 & -4 \\ -5 & -4 & 9 \end{bmatrix} \text{ per unit}$$

with $V_1 = 1.0\underline{/0^\circ}$ per unit; $V_2 = 1.0$ per unit; $P_2 = 60$ MW; $P_3 = -80$ MW; $Q_3 = -60$ Mvar (lagging) as a part of the power flow solution of the system. Find V_2 and V_3 within a tolerance of 0.01 per unit by using the Gauss-Seidel iteration method. Start with $\delta_2 = 0$, $V_3 = 1.0$ per unit, and $\delta_3 = 0$.

SECTION 6.6

6.38 A generator bus (with a 1.0 per unit voltage) supplies a 180 MW, 60 Mvar load through a lossless transmission line with per unit (100 MVA base) impedance of $j0.1$ and no line charging. Starting with an initial voltage guess of $1.0\angle 0°$, iterate until converged using the Newton-Raphson power flow method. For convergence criteria, use a maximum power flow mismatch of 0.1 MVA.

6.39 Repeat Problem 6.38 except use an initial voltage guess of $1.0\angle 30°$.

6.40 Repeat Problem 6.38 except use an initial voltage guess of $0.25\angle 0°$.

6.41 Determine the initial Jacobian matrix for the power system described in Problem 6.34.

6.42 Use the Newton-Raphson power flow to solve the power system described in Problem 6.34. For convergence criteria, use a maximum power flow mismatch of 0.1 MVA.

6.43 For a three-bus power system, assume bus 1 is the slack with a per unit voltage of $1.0\angle 0°$, bus 2 is a PQ bus with a per unit load of $2.0 + j0.5$, and bus 3 is a PV bus with 1.0 per unit generation and a 1.0 voltage setpoint. The per unit line impedances are $j0.1$ between buses 1 and 2, $j0.4$ between buses 1 and 3, and $j0.2$ between buses 2 and 3. Using a flat start, use the Newton-Raphson approach to determine the first iteration phasor voltages at buses 2 and 3.

6.44 Repeat Problem 6.43 except with the bus 2 real power load changed to 1.0 per unit.

PW **6.45** Load PowerWorld Simulator case Example 6_11; this case is set to perform a single iteration of the Newton-Raphson power flow each time **Single Solution** is selected. Verify that initially the Jacobian element J_{33} is 104.41. Then, give and verify the value of this element after each of the next three iterations (until the case converges).

PW **6.46** Load PowerWorld Simulator case Problem 6_46. Using a 100 MVA base, each of the three transmission lines have an impedance of $0.05 + j0.1$ p.u. There is a single 180 MW load at bus 3, while bus 2 is a PV bus with generation of 80 MW and a voltage setpoint of 1.0 p.u. Bus 1 is the system slack with a voltage setpoint of 1.0 p.u. Manually solve this case using the Newton-Raphson approach with a convergence criteria of 0.1 MVA. Show all your work. Then verify your solution by solving the case with PowerWorld Simulator.

PW **6.47** As was mentioned in Section 6.4, if a generator's reactive power output reaches its limit, then it is modeled as though it were a PQ bus. Repeat Problem 6.46, except assume the generator at bus 2 is operating with its reactive power limited to a maximum of 50 Mvar. Then verify your solution by solving the case with PowerWorld Simulator. To increase the reactive power output of the bus 2 generator, select **Tools, Play** to begin the power flow simulation, then click on the up arrow on the bus 2 magenta voltage setpoint field until the reactive power output reaches its maximum.

PW **6.48** Load PowerWorld Simulator case Problem 6_46. Plot the reactive power output of the generator at bus 2 as a function of its voltage setpoint value in 0.005 p.u. voltage steps over the range between its lower limit of −50 Mvar and its upper limit of 50 Mvar. To change the generator 2 voltage set point first select **Tools, Play** to begin the power flow simulation, and then click on the up/down arrows on the bus 2 magenta voltage setpoint field.

SECTION 6.7

PW **6.49** Open PowerWorld Simulator case Problem 6_49. This case is identical to Example 6.9, except that the transformer between buses 1 and 5 is now a tap-changing transformer with a tap range between 0.9 and 1.1 and a tap step size of 0.00625. The tap is on the high side of the transformer. As the tap is varied between 0.975 and 1.1, show the variation in the reactive power output of generator 1, V_5, V_2, and the total real power losses.

PW **6.50** Use PowerWorld Simulator to determine the Mvar rating of the shunt capacitor bank in the Example 6_14 case that increases V_2 to 1.0 per unit. Also determine the effect of this capacitor bank on line loadings and the total real power losses (shown immediately below bus 2 on the oneline). To vary the capacitor's nominal Mvar rating, right-click on the capacitor symbol to view the Switched Shunt Dialog, and then change Nominal Mvar field.

PW **6.51** Use PowerWorld Simulator to modify the Example 6_9 case by inserting a second line between bus 2 and bus 5. Give the new line a circuit identifier of "2" to distinguish it from the existing line. The line parameters of the added line should be identical to those of the existing lines 2 to 5. Determine the new line's effect on V_2, the line loadings, and on the total real power losses.

PW **6.52** Open PowerWorld Simulator case Problem 6_52. Open the 69 kV line between buses REDBUD69 and PEACH69 (shown towards the bottom of the oneline). With the line open, determine the amount of Mvar (to the nearest 1 Mvar) needed from the REDBUD69 capacitor bank to correct the REDBUD69 voltage to at least 1.0 p.u.

PW **6.53** Open PowerWorld Simulator case Problem 6_53. Plot the variation in the total system real power losses as the generation at bus PEAR138 is varied in 20 MW blocks between 0 MW and 400 MW. What value of PEAR138 generation minimizes the total system losses?

PW **6.54** Repeat Problem 6.53, except first remove the 69 kV line between LOCUST69 and PEAR69.

SECTION 6.8

6.55 Using the compact storage technique described in Section 6.8, determine the vectors **DIAG**, **OFFDIAG**, **COL**, and **ROW** for the following matrix:

$$\mathbf{S} = \begin{bmatrix} 17 & -9.1 & 0 & 0 & -2.1 & -7.1 \\ -9.1 & 25 & -8.1 & -1.1 & -6.1 & 0 \\ 0 & -8.1 & 9 & 0 & 0 & 0 \\ 0 & -1.1 & 0 & 2 & 0 & 0 \\ -2.1 & -6.1 & 0 & 0 & 14 & -5.1 \\ -7.1 & 0 & 0 & 0 & -5.1 & 15 \end{bmatrix}$$

6.56 For the triangular factorization of the corresponding Y_{bus}, number the nodes of the graph shown in Figure 6.9 in an optimal order.

SECTION 6.10

6.57 Compare the angles and line flows between the Example 6_17 case and results shown in Tables 6.6, 6.7, and 6.8.

6.58 Redo Example 6.17 with the assumption that the per-unit reactance on the line between buses 2 and 5 is changed from 0.05 to 0.03.

PW **6.59** Open PowerWorld Simulator case Problem 6_59, which models a seven-bus system using the dc power flow approximation. Bus 7 is the system slack. The real power generation/load at each bus is as shown, while the per-unit reactance of each of the lines (on a 100 MVA base) is as shown in yellow on the oneline. (a) Determine the six-by-six **B** matrix for this system and the **P** vector. (b) Use a matrix package such as Matlab to verify the angles as shown on the oneline.

PW **6.60** Using the PowerWorld Simulator case from Problem 6.59, if the rating on the line between buses 1 and 2 is 150 MW, the current flow is 101 MW (from bus 1 to bus 3), and the bus 1 generation is 160 MW, analytically determine the amount this generation can increase until this line reaches 100% flow. Assume any change in the bus 1 generation is absorbed at the system slack.

SECTION 6.11

PW **6.61** PowerWorld Simulator cases Problem 6_61_PQ and 6_61_PV model a seven-bus power system in which the generation at bus 4 is modeled as a Type 1 or 2 wind turbine in the first case and as a Type 3 or 4 wind turbine in the second. A shunt capacitor is used to make the net reactive power injection at the bus the same in both cases. Compare the bus 4 voltage between the two cases for a contingency in which the line between buses 2 and 4 is opened. What is an advantage of a Type 3 or 4 wind turbine with respect to voltage regulation following a contingency? What is the variation in the Mvar output of a shunt capacitor with respect to bus voltage magnitude?

SECTION 6.12

6.62 The fuel-cost curves for two generators are given as follows:

$$C_1(P_1) = 600 + 18 \cdot P_1 + 0.04 \cdot (P_1)^2$$

$$C_2(P_2) = 700 + 20 \cdot P_2 + 0.03 \cdot (P_2)^2$$

Assuming the system is lossless, calculate the optimal dispatch values of P_1 and P_2 for a total load of 1000 MW, the incremental operating cost, and the total operating cost.

6.63 Rework Problem 6.62, except assume that the limit outputs are subject to the following inequality constraints:

$$200 \le P_1 \le 800 \text{ MW}$$

$$100 \le P_2 \le 400 \text{ MW}$$

6.64 Rework Problem 6.62, except assume the 1000 MW value also includes losses, and the penalty factor for the first unit is 1.0 and for the second unit 0.95.

6.65 The fuel-cost curves for a two-generator power system are given as follows:

$$C_1(P_1) = 600 + 15 \cdot P_1 + 0.05 \cdot (P_1)^2$$

$$C_2(P_2) = 700 + 20 \cdot P_2 + 0.04 \cdot (P_2)^2$$

while the system losses can be approximated as

$$P_L = 2 \times 10^{-4}(P_1)^2 + 3 \times 10^{-4}(P_2)^2 - 4 \times 10^{-4}P_1P_2 \quad \text{MW}$$

If the system is operating with a marginal cost (λ) of \$60/hr, determine the output of each unit, the total transmission losses, the total load demand, and the total operating cost.

6.66 Expand the summations in (6.12.14) for $N = 2$, and verify the formula for $\partial P_L/\partial P_i$ given by (6.12.15). Assume $B_{ij} = B_{ji}$.

6.67 Given two generating units with their respective variable operating costs as

$$C_1 = 0.01P_{G1}^2 + 2P_{G1} + 100 \quad \$/\text{hr} \qquad \text{for } 25 \le P_{G1} \le 150 \quad \text{MW}$$
$$C_2 = 0.004P_{G2}^2 + 2.6P_{G2} + 80 \quad \$/\text{hr} \qquad \text{for } 30 \le P_{G2} \le 200 \quad \text{MW}$$

determine the economically optimum division of generation for $55 \le P_L \le 350$ MW. In particular, for $P_L = 282$ MW, compute P_{G1} and P_{G2}. Neglect transmission losses.

PW **6.68** Resolve Example 6.20, except with the generation at bus 2 set to a fixed value (i.e., modeled as off of Automatic Generation Control). Plot the variation in the total hourly cost as the generation at bus 2 is varied between 1000 and 200 MW in 5 MW steps, resolving the economic dispatch at each step. What is the relationship between bus 2 generation at the minimum point on this plot and the value from economic dispatch in Example 6.20? Assume a Load Scalar of 1.0.

PW **6.69** Using PowerWorld case Example 6_22 with the Load Scalar equal to 1.0, determine the generation dispatch that minimizes system losses *(Hint:* Manually vary the generation at buses 2 and 4 until their loss sensitivity values are zero). Compare the operating cost between this solution and the Example 6_22 economic dispatch result. Which is better?

PW **6.70** Repeat Problem 6.69, except with the Load Scalar equal to 1.4.

SECTION 6.13

PW **6.71** Using LP OPF with PowerWorld Simulator case Example 6_23, plot the variation in the bus 5 marginal price as the Load Scalar is increased from 1.0 in steps of 0.02. What is the maximum possible load scalar without overloading any transmission line? Why is it impossible to operate without violations above this value?

PW **6.72** Load PowerWorld Simulator case Problem 6_72. This case models a slightly modified, lossless version of the 37-bus case from Example 6.13 with generator cost information, but also with the transformer between buses PEAR138 and PEAR69 open. When the case is loaded, the "Total Cost" field shows the economic dispatch solution, which results in an overload on several lines. Before solving the case, select **Add-Ons, OPF Case Info, OPF Buses** to view the bus LMPs, noting that they are all identical. Then Select **Add-Ons, Primal LP** to solve the case using the OPF, and again view the bus LMPs. Verify the LMP at the PECAN69 bus by manually changing the load at the bus by one MW, and then noting the change in the Total Cost field. Repeat for the LOCUST69 bus. *Note*: Because of solution convergence tolerances, the manually calculated results may not exactly match the OFP calculated bus LMPs.

CASE STUDY QUESTIONS

a. What are the operational impacts on fossil-fueled power plants due to high penetrations of wind and solar generation into a power grid?

b. Do high penetrations of wind and solar generation increase the wear and tear costs of fossil-fueled generation? Why?

c. Which has more forecast uncertainty, wind generation or solar generation? Why?

DESIGN PROJECT 1: NEW LOAD

As a result of the low electric rates from the local utility, Metropolis Light and Power (MLP), several large server farms and a new factory are going to be built in the eastern portion of the MLP service territory (see Figure 6.25). With an anticipated peak load of about 75 MW and 20 Mvar, this new load also brings additional revenue to MLP. However, in order to supply this additional load, the new TULIP substation will need to be constructed. While they would like to receive electricity at the

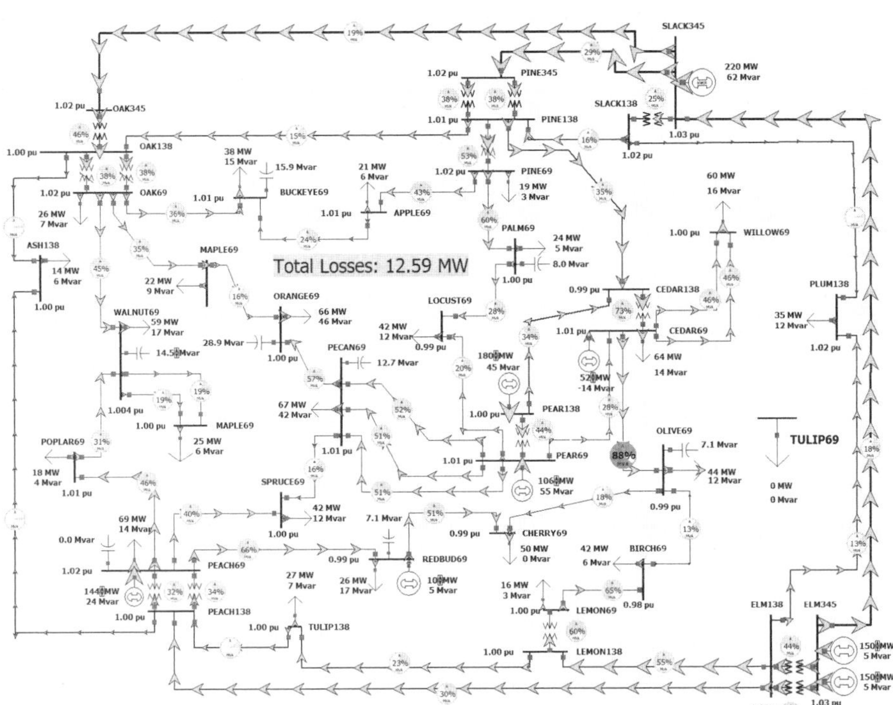

FIGURE 6.25

Design Case 1 System Oneline Diagram

69 kV level, the new substation location is large enough to accommodate a 138/69 kV transformer if needed. Additionally, for reliability purposes, the TULIP substation needs to have at least two separate lines into their substation.

As a planning engineer for MLP, your job is to make recommendations to ensure that, with new TULIP loads under peak loading conditions, the transmission system in the eastern region is adequate for any base case or first contingency loading situation. This is also a good opportunity not only to meet the new load, but also to fix some existing first contingency violations in the eastern portion of the MLP service territory. Table 6.13 shows the available right-of-way distances that

Right-of-Way/Substation	Right-of-Way Distance (km)
TULIP to ELM	15
TULIP to PLUM	12
TULIP to OLIVE	8
TULIP to CEDAR	10
TULIP to BIRCH	14
CEDAR to PLUM	13
WILLOW to PLUM	8
OLIVE TO CEDAR	10

TABLE 6.13

Available New Rights-of-Way

can be used for the construction of new 69 kV and/or new 138 kV lines. All existing 69 kV-only substations are large enough to accommodate 138 kV as well. The DesignCase1_2015 provides a power flow model of the initial conditions.

Design Procedure

1. Load DesignCase1_2015 into PowerWorld Simulator. Perform an initial power-flow solution to verify the base case system operation without the TULIP load. Note that the entire line flows and bus voltage magnitudes are within their limits. Assume all line MVA flows must be at or below 100% of the limit A values, and all voltages must be between 0.95 and 1.10 per unit.
2. Repeat the above analysis considering the impact of any single transmission line or transformer outage. This is known as contingency analysis. To simplify this analysis, PowerWorld Simulator has the ability to automatically perform a contingency analysis study. Select **Tools, Contingency Analysis** to show the Contingency Analysis display. Note that the 57 single line/transformer contingencies are already defined. Select **Start Run** to automatically see the impact of removing any single element. This system has line violations for several different contingencies.
3. Using the rights-of-way and the transmission line parameters/costs given in the cost section (see page 409), iteratively determine the least expensive system additions so that the base case and all the contingences result in reliable operation points (i.e., one with no violations) with the new TULIP load. The parameters of the new transmission lines(s) need to be derived using the tower configurations and conductor types. Tower configurations are provided by the instructor with default values given with the cost data. Several different conductor types are available in the cost section. The total cost of an addition is defined as the construction costs minus the savings associated with any decrease in system losses over the next 5 years.
4. Write a detailed report discussing the initial system problems, your approach to optimally solving the system problems, and the justification for your final recommendation.

Simplifying Assumptions

To simplify the analysis, several assumptions are made:

1. You need only consider the base case loading level given in DesignCase1_2015. In a real design, typically a number of different operating points/loading levels must be considered.
2. You should consider the generator outputs as fixed values; any changes in the losses and the new TULIP load are always picked up by the system slack generator.
3. You should not modify the status of the capacitors or the transformer taps.
4. You should assume that the system losses remain constant over the 5-year period and need only consider the impact the new design has on the base case losses. Electricity is priced at \$60/MWh.

DESIGN PROJECT 2: NEW WIND GENERATION AND GENERATION RETIREMENT

As a planning engineer for Island Electric Company (IEC), you have been tasked with determining the transmission system changes required to locate a new 600 MW wind farm in the western portion of your service territory (see Figure 6.26). IEC uses 345 and 161 kV transmission grids, so your changes are restricted to these existing voltages. The wind farm would like to connect at the 161 kV level and requires at least two transmission lines into the NewWind substation (which can be at either 161 and/or 345 kV). Since the location is usually quite windy, it is expected to have a capacity factor of at least 40%. However, the wind also can be quite variable, including during times of maximum system loading, so this generation cannot be counted on for firm capacity. Simultaneous with the addition of the new wind farm, IEC would like to retire the existing 300 MW generator at the Pheasant substation.

Hence, your job is to make recommendations on the least-cost design for the construction of new lines and transformers to ensure that the transmission system in the IEC system is adequate for any base case or first-contingency loading situation when then wind farm is installed and operating at either its maximum output of 600 or 0 MW and with the Pheasant generator removed from service. Note, this will also involve fixing some existing first-contingency violations. Since the wind farm will be built with Type 3 DFAG wind turbines, you can model the wind farm in the power flow as a single equivalent, traditional PV bus generator with a fixed output of either 0 or 600 MW, a voltage setpoint of 1.03 per unit, and with reactive power limits of ±250 Mvar.

Table 6.14 shows the right-of-way distances that are available for the construction of new 161 kV and/or new 345 kV lines. All existing 161-kV only substations are large enough to accommodate 345 kV as well, as is the NewWind substation.

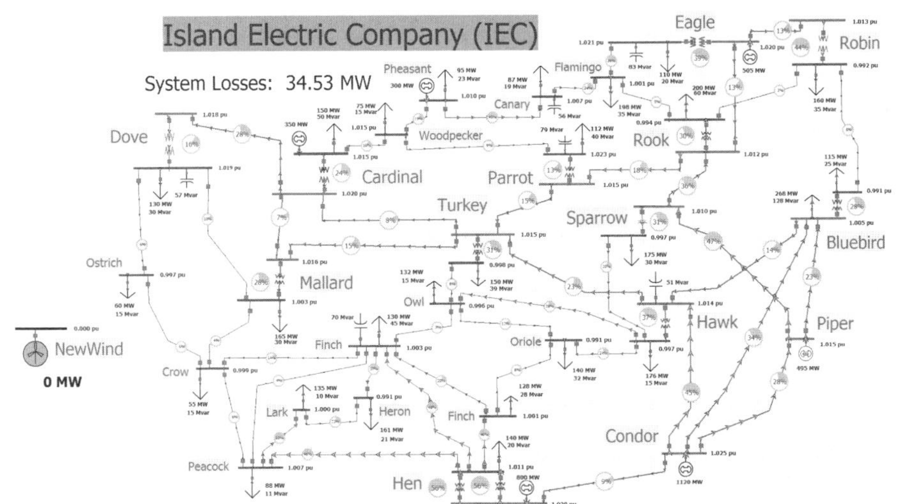

FIGURE 6.26

Design Case 2 System Oneline Diagram

Right-of-Way/Substation	Right-of-Way Distance (km)
NewWind to Ostrich	15
NewWind to Dove	55
NewWind to Crow	30
NewWind to Peacock	53
NewWind to Hen	70
Ostrich to Mallard	45
Peakcock to Hen	20
Dove to Cardinal	40

TABLE 6.14

Available New Rights-of-Way

Design Procedure

1. Load DesignCase2_2015 into PowerWorld Simulator, which contains the system power flow case and the disconnected NewWind generator and bus. Perform an initial power flow solution to verify the base case system operation. Note that all of the line flows and bus voltage magnitudes are within their limits. Assume all line MVA flows must be at or below 100% of their limit values, and all voltages must be between 0.92 and 1.10 per unit.
2. Repeat the above analysis, considering the impact of any single transmission line or transformer outage. This is known as contingency analysis. To simplify this analysis, PowerWorld Simulator has the ability to automatically perform a contingency analysis study. Select **Tools, Contingency Analysis** to show the Contingency Analysis display. Note that the 60 single line/transformer contingencies are already defined. Select **Start Run** (towards the bottom right corner of the display) to automatically see the impact of removing any single element. Note that there are several existing violations.
3. Open the existing 300 MW generator at the Pheasant substation, and repeat parts 1 and 2.
4. Using the rights-of-way given in Table 6.14 and the transmission line parameters/costs, iteratively determine the least-expensive system additions so that the base case and all the contingences result in reliable operation points with the NewWind generation connected with an output of either 0 or 600 MW. When the output is at 0 MW, the wind farm is still considered on-line and hence should be modeled as a PV bus regulating its voltage to 1.03 p.u. The parameters of the new transmission lines(s) need to be derived using the tower configurations and conductor types provided by the instructor. In addition, the transmission changes you propose will modify the total system losses, indicated by the large field

on the oneline diagram. In your design, you should consider the impact on total system losses in the studied condition for the next 5 years. Hence, you should minimize the total construction costs minus the savings associated with any decrease in system losses over the next 5 years.

5. Write a detailed report discussing the initial system problems, your approach to optimally solving the system problems, and the justification for your final recommendation.

Simplifying Assumptions

To simplify the analysis, several assumptions are made:

1. You need only consider the base case loading level given with the modification of opening the Pheasant generation. In a real design, typically a number of different operating points/loading levels must be considered.
2. You should consider all the generator real power outputs as fixed values with the exception that the NewWind generator should be studied at both 0 and 600 MW. The change in the total system generation and any changes in the system losses are always picked up by the system slack.
3. You should not modify the status of the capacitors or the transformer taps.
4. You should assume that the system losses remain constant over the 5-year period and need only consider the impact the new design has on the base case losses, assuming the NewWind generation is at 600 MW. The price for losses can be assumed to be \$50/MWh.
5. You do not need to consider contingencies involving the new transmission lines and possibly any transformers you may be adding.
6. While an appropriate control response to a contingency might be to decrease the wind farm output (by changing the pitch on the wind turbine blades), your supervisor has specifically asked you not to consider this possibility. Therefore the NewWind generator should always have either a 0 or 600 MW output.

DESIGN PROJECTS 1 AND 2: SAMPLE TRANSMISSION SYSTEM DESIGN COSTS

Transmission Lines (69 kV, 138 kV, 161 kV, 345 kV)

New transmission lines include a fixed cost and a variable cost. The fixed cost is for the design work, the purchase/installation of the three-phase circuit breakers, associated relays, and changes to the substation bus structure. The fixed costs are \$2,400,000 for a 345 kV line, \$1,100,000 for at 161 kV line, \$850,000 for a 138 kV line, and \$500,000 for a 69 kV line. The variable costs depend on the type of conductor and the length of the line. The assumed cost in \$/km are given in Table 6.15.

Conductor Type	Current Rating (Amps)	345-kV Lines	161-kV Lines	138-kV Lines	69-kV Lines
Rook	770				$200,000/km
Crow	830		$390,000/km	$330,000/km	$220,000/km
Condor	900		$410,000/km	$350,000/km	$240,000/km
Cardinal	1110	$600,000/km	$430,000/km	$370,000/km	
Pheasant	1200	$650,000/km	$450,000/km		
Falcon	1380	$700,000/km			

TABLE 6.15
Assumed costs

Lined impedance data and MVA ratings are determined based on the conductor type and tower configuration. The conductor characteristics are given in Table A.4 of the book. For these design problems, assume a symmetric tower configurations. Often the spacings between conductors are provided by the instructor and may be student specific. If no values are given, assume a GMD of 2 m for 69 kV, 4 m for 138 kV, 5 m for 161 kV, and 8 m for 345 kV.

Transformers

Transformer costs include associated circuit breakers, relaying, and installation.

138/69 kV, 101 MVA	$1,500,000
138/69 kV, 168 MVA	$1,800,000
345/161 kV, 560 MVA	$7,500,000

Assume any new 138/69 kV transformer has 0.0025 per unit resistance and 0.07 per unit reactance, and any new 345/161 kV transformer has 0.0004 per unit resistance and 0.025 per unit reactance (all on a 100 MVA base).

Bus Work

Upgrade 69-kV substation to also include 138 kV	$900,000
Upgrade 161-kV substation to also include 345 kV	$3,500,000

DESIGN PROJECT 3: POWER FLOW/SHORT CIRCUITS

Time given: 3 weeks
Approximate time required: 15 hours
Each student is assigned one of the single-line diagrams shown in Figures 6.27 and 6.28. Also, the length of line 2 in these figures is varied for each student.

Assignment I: Power Flow Preparation

For the single-line diagram that you have been assigned (Figure 6.27 or 6.28), convert all positive-sequence impedance, load, and voltage data to per unit using the given system base quantities. Then using the PowerWorld Simulator, create three input data files: bus input data, line input data, and transformer input data. Note that bus 1 is the swing bus. Your output for this assignment consists of three power-flow input data files.

The purpose of this assignment is to get started and to correct errors before going to the next assignment. It requires a knowledge of the per-unit system, which was covered in Chapter 3, but may need review.

Assignment 2: Power Flow

Case 1. Run the power flow program and obtain the bus, line, and transformer input/output data files that you prepared in Assignment 1.

Case 2. Suggest one method of increasing the voltage magnitude at bus 4 by 5%. Demonstrate the effectiveness of your method by making appropriate changes to the input data of case 1 and by running the power flow program.

Your output for this assignment consists of 12 data files, 3 input, and 3 output data files for each case, along with a one-paragraph explanation of your method for increasing the voltage at bus 4 by 5%.

During this assignment, course material contains voltage control methods, including use of generator excitation control, tap changing and regulating transformers, static capacitors, static var systems, and parallel transmission lines.

This project continues in Chapters 7 and 9.

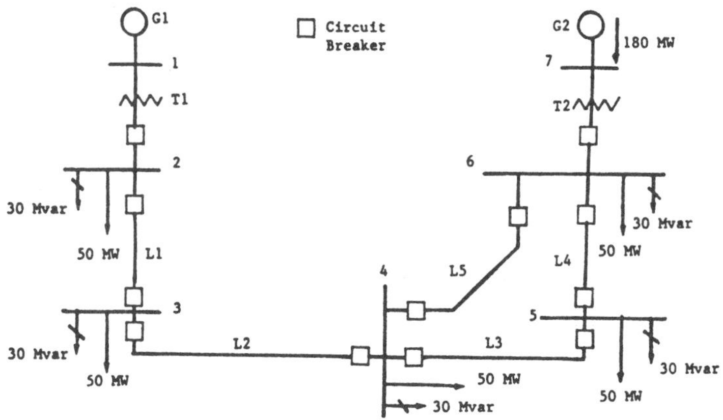

FIGURE 6.27 Single-line diagram for Design Project 3—transmission loop

<u>Generator Ratings</u>

G1: 100 MVA, 13.8 kV, $x'' = 0.12, x_2 = 0.14, x_0 = 0.05$ per unit

G2: 200 MVA, 15.0 kV, $x'' = 0.12, x_2 = 0.14, x_0 = 0.05$ per unit

The generator neutrals are solidly grounded

<u>Transformer Ratings</u>

T1: 100 MVA, 13.8 kVΔ/230 kVY, x = 0.1 per unit

T2: 200 MVA, 15 kVΔ/230 kVY, x = 0.1 per unit

The transformer neutrals are solidly grounded

<u>Transmission Line Ratings</u>

All Lines: 230 kV, $z_1 = 0.08 + j0.5$ Ω/km,

$z_0 = 0.2 + j1.5$ Ω/km, $y_1 = j3.3$ E-6 S/km,

Maximum MVA = 400

Line Lengths: $L_1 = 15$ km, L_2 assigned by the instructor (20 to 50 km), $L_3 = 40$ km, $L_4 = 15$ km, $L_5 = 50$ km.

<u>Power Flow Data</u>

Bus 1 : Swing bus, $V_1 = 13.8$ kV, $\delta_1 = 0°$

Bus 2,3,4,5,6 : Load buses

Bus 7 : Constant voltage magnitude bus, $V_7 = 15$ kV,

$P_{G7} = 180$ MW, -87 Mvar $< Q_{G7} < +87$ Mvar

<u>System Base Quantities</u>

$S_{base} = 100$ MVA (three-phase)

$V_{base} = 13.8$ kV (line-to-line) in the zone of G1

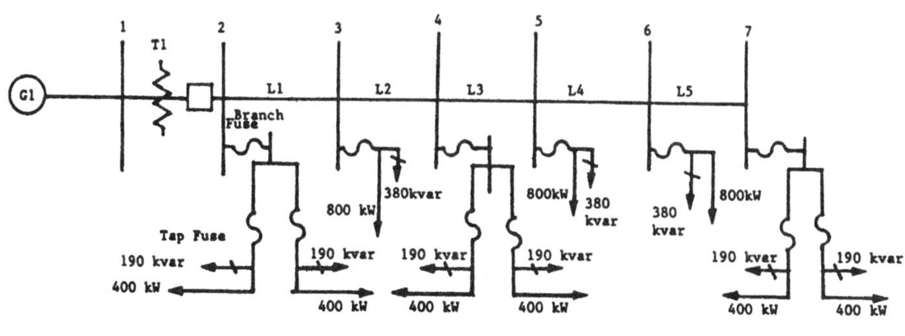

FIGURE 6.28 Single-line diagram for Design Project 3—radial distribution feeder

Circuit Breaker

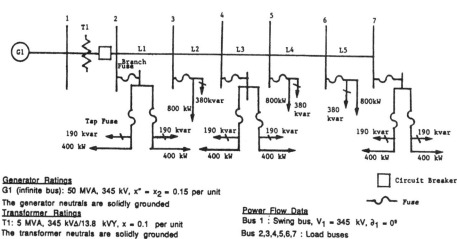

Fuse

Generator Ratings
G1 (infinite bus): 50 MVA, 345 kV, $x'' = x_2 = 0.15$ per unit
The generator neutrals are solidly grounded
Transformer Ratings
T1: 5 MVA, 345 kVΔ/13.8 kVY, $x = 0.1$ per unit
The transformer neutrals are solidly grounded
Transmission Line Ratings
All Lines: 13.8 kV, $z_1 = 0.19 + j0.38\ \Omega$/km,
$z_0 = 0.6 + j1.0\ \Omega$/km, $y_1 = j4.0$ E-6 S/km,
Maximum MVA = 5
Line Lengths: $L_1 = 2$ km, L_2 assigned by the instructor (1 to 5 km), $L_3 = L_4 = L_5 = 2$ km.

Power Flow Data
Bus 1 : Swing bus, $V_1 = 345$ kV, $\delta_1 = 0°$
Bus 2,3,4,5,6,7 : Load buses
System Base Quantities
S_{base} = 10 MVA (three-phase)
V_{base} = 13.8 kV (line-to-line) in the zone of the lines

REFERENCES

1. W. F. Tinney and C. E. Hart, "Power Flow Solutions by Newton's Method," *IEEE Trans. PAS*, *86* (November 1967), p. 1449.
2. W. F. Tinney and J. W. Walker, "Direct Solution of Sparse Network Equations by Optimally Ordered Triangular Factorization," *Proc. IEEE*, *55* (November 1967), pp. 1801–1809.
3. Glenn W. Stagg and Ahmed H. El-Abiad, *Computer Methods in Power System Analysis* (New York: McGraw-Hill, 1968).
4. N. M. Peterson and W. S. Meyer, "Automatic Adjustment of Transformer and Phase Shifter Taps in Newton Power Flow," *IEEE Trans. PAS*, *90* (January-February 1971), pp. 103–108.
5. W. D. Stevenson, Jr., *Elements of Power Systems Analysis*, 4th ed. (New York: McGraw-Hill, 1982).
6. A. Bramellar and R. N. Allan, *Sparsity* (London: Pitman, 1976).
7. C. A. Gross, *Power Systems Analysis* (New York: Wiley, 1979).
8. B. Stott, "Fast Decoupled Load Flow," *IEEE Trans. PAS*, Vol. PAS 91 (September–October 1972), pp. 1955–1959.
9. T. Overbye and J. Weber, "Visualizing the Electric Grid," *IEEE Spectrum*, 38, 2 (February 2001), pp. 52–58.
10. Westinghouse Electric Corporation, *Transmission and Distribution Reference Book*, 4th ed. (Pittsburgh: Westinghouse, 1964).

11. Aluminum Association, *The Aluminum Electrical Conductor Handbook* (Washington, D.C.: Aluminum Association).

12. A. J. Wood and B. F. Wollenberg, *Power Generation, Operation and Control*, 2nd ed. (New York: John Wiley & Sons, 1996).

13. A. Ellis, "Wind Power Plant Models for System Studies," Tutorial on Fundamentals of Wind Energy, Section V, *IEEE PES GM* (Calgary, AB: July 2009).

14. WECC Wind Generator Modeling Group, "WECC Wind Power Plant Power Flow Modeling Guide," *WECC*, May 2008.

15. E.H. Camm et al., "Characteristics of Wind Turbine Generators for Wind Power Plants," *Proc. IEEE* 2009 General Meeting (Calgary, AB: July 2009).

16. L. K. Kirchmayer, *Economic Operation of Power Systems* (New York: Wiley, 1958).

17. L. K. Kirchmayer and G. W. Stagg, "Evaluation of Methods of Coordinating Incremental Fuel Costs and Incremental Transmission Losses," *Transactions AIEE*, vol. 71, part III (1952), pp. 513–520.

18. G. H. McDaniel and A. F. Gabrielle, "Dispatching Pumped Storage Hydro," *IEEE Transmission PAS*, vol. PAS-85 (May 1966), pp. 465–471.

19. E. B. Dahlin and E. Kindingstad, "Adaptive Digital River Flow Predictor for Power Dispatch," *IEEE Transactions PAS*, vol. PAS-83 (April 1964), pp. 320–327.

20. L. K. Kirchmayer, *Economic Control of Interconnected Systems* (New York: Wiley, 1959).

21. J. H. Drake et al., "Optimum Operation of a Hydrothermal System," *Transactions AIEE (Power Apparatus and Systems)*, vol. 62 (August 1962), pp. 242–250.

22. A. J. Wood and B. F. Wollenberg, *Power Generation, Operation, and Control* (New York: Wiley, 1989).

23. 2005 and 2014 PJM State of the Market Report, available online at http://www.monitoringanalytics.com/reports/PJM_State_of_the_Market/2014.shtml.

24. A. Castillo and R. P. O'Neill, "Survey of Approaches to Solving the ACOPF," *U.S. FERC* (March 2013).

7 Symmetrical Faults

345-kV SF6 circuit breaker installation at Goshen Substation, Idaho Falls, Idaho, USA. This circuit breaker has a continuous current rating of 2000A and an interrupting current rating of 40 kA (Courtesy of PacifiCorp.)

Short circuits occur in power systems when equipment insulation fails due to system overvoltages caused by lightning or switching surges, to insulation contamination (salt spray or pollution), or to other mechanical causes. The resulting short circuit or "fault" current is determined by the internal voltages of the synchronous machines and by the system impedances between the machine voltages and the fault. Short-circuit currents may be several orders of magnitude larger than normal operating currents and, if allowed to persist, may cause thermal damage to equipment. Windings and busbars may also suffer mechanical damage due to high magnetic forces during faults. It is therefore necessary to remove faulted sections of a power system from service as soon as possible. Standard EHV protective equipment is designed to clear faults within 3 cycles (50 ms at 60 Hz). Lower voltage protective equipment operates more slowly (for example, 5 to 20 cycles).

Section 7.1 begins by reviewing series R–L circuit transients followed in Section 7.2 by a description of three-phase short-circuit currents at unloaded synchronous machines, analyzing both the ac component, including subtransient, transient, and steady-state currents, and the dc component of fault current. These results are extended in Sections 7.3 and 7.4 to power system three-phase short circuits by means of the superposition principle. We observe that the bus impedance matrix is the key to calculating fault currents. Section 7.5 discusses circuit breaker and fuse selection.

Balanced three-phase power systems are assumed throughout this chapter, working in per unit.

CASE STUDY

The following case study investigates the short-circuit behavior of wind power plants (WPPs) [11]. Conventional power plants including fossil-fueled, nuclear, and hydro plants consist of single or several synchronous generating units, wherein for each unit the rotational speed is fixed and the magnetic flux is controlled via exciter windings; the magnetic flux and the rotor rotate synchronously. A WPP consists of several wind turbine generators (WTGs), presently available in sizes between 1 and 5 MW, which are dispersed over a wide geographical area. There are four main types of WTGs: Type 1—fixed-speed turbine with a squirrel-cage induction generator; Type 2—variable-speed turbine with a wound-rotor induction generator that has a variable resistor in series with the rotor winding; Type 3—variable-speed turbine with a doubly fed induction generator; and Type 4—variable-speed turbine with a permanent-magnet synchronous generator and ac-dc-ac power-electronic converter. The short-circuit current characteristics for both symmetrical and unsymmetrical faults are examined for each of the four types of WTGs.

Short-Circuit Modeling of a Wind Power Plant

E. Muljadi
Fellow, IEEE

V. Gevorgian
Member, IEEE

National Renewable Energy Laboratory

Introduction

Energy and environmental issues have become one of the biggest challenges facing the world. In response to energy needs and environmental concerns, renewable energy technologies are considered the future technologies of choice [1, 2]. Renewable energy is harvested from nature, and it is clean and free. However, it is widely accepted that renewable energy is not a panacea that comes without challenges. With the federal government's aggressive goal of achieving 20% wind energy penetration by 2030, it is necessary to understand the challenges that must be overcome when using renewable energy.

In the years to come, there will be more and more wind power plants (WPPs) connected to the grid. With the goal of 20% wind penetration by 2030, the WPP's operation should be well planned. The power system switchgear and power system protection for WPPs should be carefully designed to be compatible with the operation of conventional synchronous generators connected to the same grid. This paper attempts to illustrate the behavior of short-circuit current (SCC) contributions for different types of WTGs.

Conventional Power Plants versus Wind Power Plants

A conventional power plant consists of a single or several large (e.g., 100 MW) generators. The prime mover of the generator can be steam, gas, or a combustion engine. The generator is controllable and is adjustable up to a maximum limit and down to minimum limit. The power output is dispatched according to the load forecast, influenced by human operation, and is based on optimum operation (i.e., scheduled operation). The power plant is usually located relatively close to the load center.

The typical conventional generator used is a synchronous generator. The rotational speed is fixed—no slip; and the flux is controlled via exciter windings. The magnetic flux and the rotor rotate synchronously.

A WPP consists of many (hundreds) of wind turbine generators (WTGs). Currently, available WTG sizes are between 1 MW and 5 MW. The prime mover of the WTG is wind, and it is free, natural, and pollution-free. The controllability of the WPP is typically curtailment (spilling the wind). The energy production of a WPP depends on the wind variability, and its dispatch capability is based on wind forecasting. It is influenced more by nature (wind) than human factors, with the goal set on maximizing energy production from renewable resources (i.e., unscheduled operation). Large-scale WPPs are located in high-wind resource regions, and these may be far from the load center.

Because a WPP covers a very large area, there are power output diversities found in a typical WPP. Each WTG in a WPP will be located at different electrical distances from the substation (diversity in line impedance). Each turbine may be driven by different instantaneous wind speeds.

Short-Circuit Modeling of a Wind Power Plant by E. Muljadi and V. Gevorgian, Conference Paper NREL/CP-5500-50632, *Power & Energy Society General Meeting*, July 24–29, 2011, Detroit MI, p i–ii, 1–10.

Thus, the operating condition of each turbine may be slightly different from others within the WPP.

Operation of Wind Turbine Generators

The generator at each turbine should be protected individually and independently because of the electrical diversity of the WPP. In practice, this is an advantage of a WPP compared to a conventional power plant. During a disturbance, the electrical characteristics at each terminal of the turbine is different from the other turbines, and only the most affected WTGs will be disconnected from the grid. For general faults (distance faults at the transmission point), only 5%–15% of the turbines are disconnected from the grid [3]. This is partially because WPPs are required to have zero voltage ride-through capability. Thus, the loss of generation is not as severe as in a power plant with large generators.

At the turbine level, the WTG generates at low voltage levels (480 V to 690 V). For the Type 1 and Type 2 WTGs, it is typically compensated by switched capacitor banks to generate at a unity power factor. Type 3 and Type 4 generators are operated to generate a constant voltage at a designated bus, or may be operated at constant power factor or constant reactive power. The generator is connected to a pad-mounted transformer to step up the voltage to 34.5 kV.

Collector System

The collector system consists of miles of line feeders connecting the high side of the pad-mounted transformer to the substation. Usually, wind turbines are divided into groups of turbines connected in a daisy-chain fashion using underground cables. These groupings are then connected to the substation by either underground cables or overhead lines at 34.5 kV. Since it is not practical to model hundreds of tur bines in a power flow calculation or in a dynamic simulation, it is common to find the equivalent of the turbines as either a single equivalent turbine representation or multiple turbine representation [4, 5].

Short Circuit Behavior under Symmetrical Faults

A utility-sized wind turbine is larger than non-grid wind turbine applications. In the early days, the turbines were sized from 10 kW to 100 kW. Nowadays, wind turbines are sized above 1000 kW (1 MW).

R–L Circuits

Short-circuit faults can occur in various locations of the power system in a number of different ways including line-to-ground and line-to-line faults. For simplicity purposes, we'll consider a symmetrical three-phase fault since it is the easiest to analyze. A simple equivalent diagram of a power system under such fault conditions is shown in Figure 1(a).

The fault in Figure 1(a) is represented by a shorting switch. Immediately after the fault, the SCC contribution from the generator can be found using the following equation:

$$u_g = L\frac{di}{dt} + iR \qquad (1)$$

Where u_g is the instantaneous voltage on the generator terminals, and

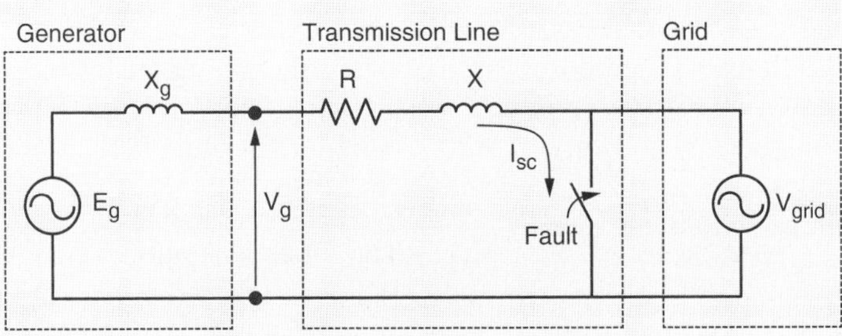

(a) Equivalent diagram of symmetrical fault

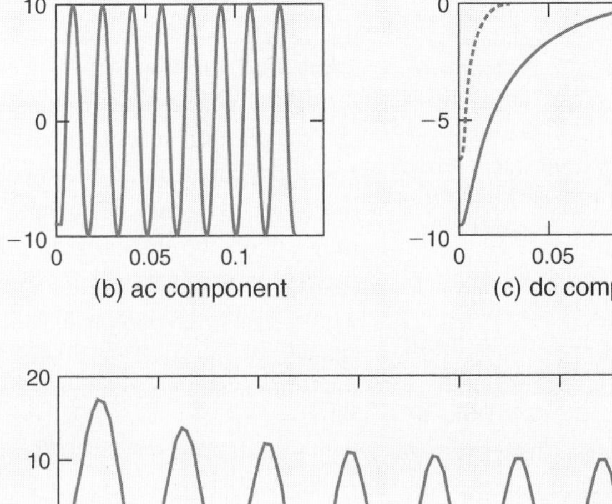

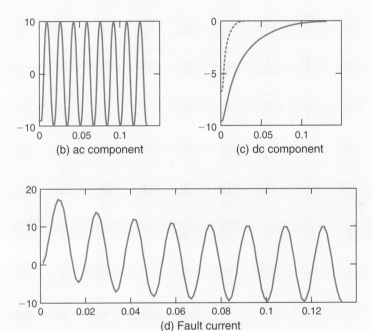

(b) ac component

(c) dc component

(d) Fault current

Figure 1 Symmetrical Three-Phase Fault

R and L are line resistance and inductance. Solving equation (1) for current

$$i = \frac{V_g}{Z}\sin\left(\omega t + \alpha - \operatorname{atan}\left(\frac{X}{R}\right)\right) - e^{-\frac{R}{L}t}\left[\frac{V_g}{Z}\sin\left(\alpha - \operatorname{atan}\left(\frac{X}{R}\right)\right)\right] \quad (2)$$

Where V_g is peak generator voltage, $Z = \sqrt{R^2 + X^2}$ is line impedance, and α is the voltage phase. The solution (2) has two components; the first component is stationary and varies sinusoidally with time as shown in Figure 1(b). It represents the steady SCC driven by the voltage source E_g.

The second component decays exponentially [as shown in Figure 1(c)] with a time constant equal to $\frac{R}{L}$. It represents the dc component of the current and the natural response of the circuit without the excitation provided by E_g.

The steady-state symmetrical fault rms value of the SCC I_{sc} from the generator can be calculated from the first component of equation (2) and is shown in

$$I_{sc} = \frac{V_g/\sqrt{2}}{\sqrt{R^2 + X^2}} \tag{3}$$

Obviously, the steady-state fault current depends on the impedance of the line. The closer the fault occurrence location to generator terminals, the larger the SCC contributed to the fault.

The peak magnitude of the transient component in equation (2) depends on line impedance as well, but it also depends on impedance angle $\varphi = \text{atan}\left(\frac{X}{R}\right)$ at the point of the fault. The dc term does not exist if $\phi = 0$, and will have its maximum initial value of $\frac{V_g}{Z}$, where $\alpha - \varphi = \pm\frac{\pi}{2}$.

The worst case scenario for the SCC peak value (including the dc component) for the circuit presented in Figure 1(a) is shown in equation (3).

So, depending on the time when the fault occurs, the circuit characteristics and the transient current waveform will be different. This means that in three-phase systems, the phase transient currents will have different peaks due to a 120° shift in voltages.

In large power systems with many generators and transmission lines, the actual fault current at any location in the grid will be the sum of collective contributions from all generators, making the above described analysis extremely complicated. So, some sort of simplification is needed for the fault current calculation in such a case.

SCC from a Type 1 WTG

The first generation of utility-sized WTGs is a fixed-speed turbine with a squirrel-cage induction generator (SCIG) and is called a Type 1 generator in wind-related applications. The SCIG generates electricity when it is driven above synchronous speed. The difference between the synchronous speed and the operating speed of the induction generator is measured by its slip (in per unit or in percent). A negative slip indicates that the wind turbine operates in generating mode. Normal operating slips for an induction generator are between 0% and −1%. The simplified single-phase equivalent circuit of a squirrel-cage induction machine is shown in Figure 2 [6].

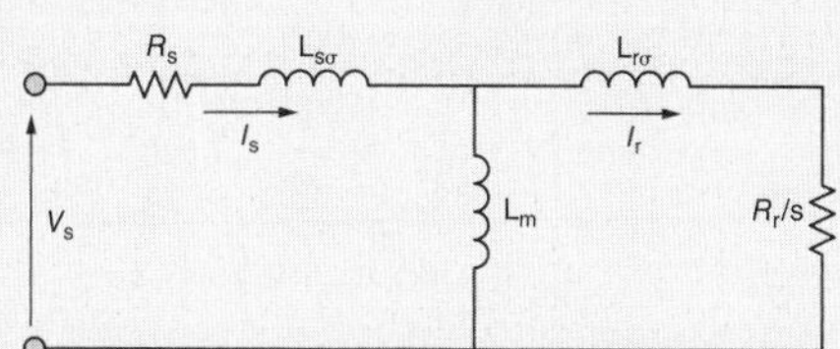

Figure 2 Equivalent circuit of a Type 1 generator

The circuit in Figure 2 is referred to the stator where R_S and R_r are stator and rotor resistances, $L_{s\sigma}$ and $L_{r\sigma}$ are stator and rotor leakage inductances, L_m is magnetizing reactance, and s is rotor slip. The example single-line connection diagram of a Type 1 generator is shown in Figure 3. In the case of a voltage fault, the inertia of the wind rotor drives the generator after the voltage drops at the generator

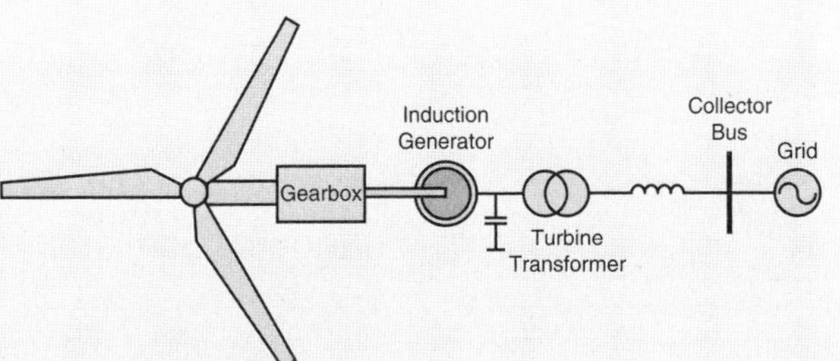

Figure 3 Type 1 WTG

terminals. The rotor flux may not change instantaneously right after the voltage drop due to a fault. Therefore, voltage is produced at the generator terminals causing fault current flow into the fault until the rotor flux decays to zero. This process takes a few electrical cycles. The fault current produced by an induction generator must be considered when selecting the rating for circuit breakers and fuses. The fault current is limited by generator impedance (and can be calculated from parameters in Figure 2) and impedance of the system from the short circuit to the generator terminals.

The initial value of fault current fed in by the induction generator is close to the locked rotor-inrush current. Assuming a three-phase symmetrical fault, an analytical solution can be found to estimate the current contribution of the generator. The SCC of an induction generator can be calculated as [7]:

$$i(t) = \frac{\sqrt{2}V_S}{Z'_S}\left[e^{-\frac{t}{T'_S}}\sin(\alpha) - (1-\sigma)e^{-\frac{t}{T'_r}}\sin(\omega t + \alpha)\right] \quad (4)$$

where α is the voltage phase angle for a given phase, σ is the leakage factor, $Z'_S = X'_S = \omega L'_S$ is stator transient reactance, and T'_S and T'_r are stator and rotor time constants representing the damping of the dc component in stator and rotor windings. The transient stator and rotor inductances L'_S and L'_r can be determined from the circuits shown in Figure 4.

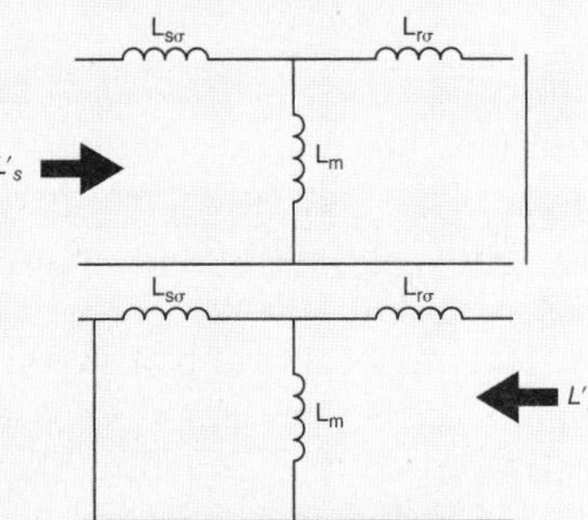

Figure 4 Stator and rotor transient inductances

$$L'_S = L_{S\sigma} + \frac{L_{r\sigma}L_m}{L_{r\sigma} + L_m}$$
$$L'_r = L_{r\sigma} + \frac{L_{S\sigma}L_m}{L_{S\sigma} + L_m} \quad (5)$$

$$T'_S = \frac{L'_S}{R_S} \qquad T'_r = \frac{L'_r}{R_r} \quad (6)$$

$$\sigma = 1 - \frac{L_m^2}{L_S L_r} \quad (7)$$

$$L_S = L_{S\sigma} + L_m$$
$$L_r = L_{r\sigma} + L_m \quad (8)$$

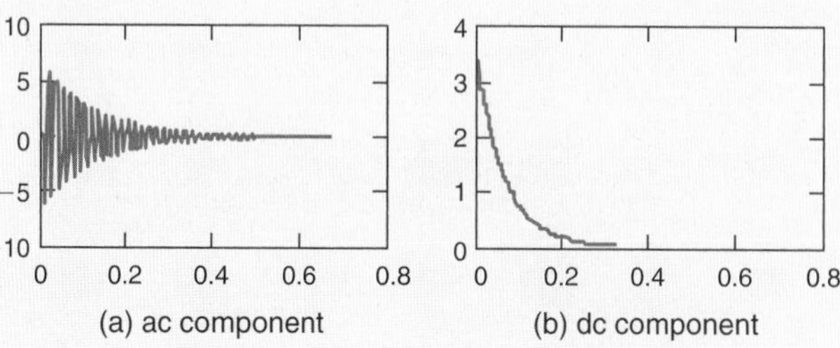

Figure 5 The two components of the SCC for a Type 1 WTG

Equation (4) is different from equation (2). The first distinction is that there is no voltage source driving the fault current as in equation (2). The fault current is driven by the decaying flux trapped in the rotor winding as represented by the right portion of the equation (4). The second distinction is the rotor time constant T'_r governs the dynamic of the decaying rotor flux [ac component of the SCC as shown in Figure 5(a)] and the decaying dc component of the fault current [refer to Figure 5(b)] is governed by the stator time constant T'_S. The larger the leakage inductances (σ), the smaller is the fault current amplitude. The third distinction is that the fault current dies out after the flux driving the fault current depleted to zero. Note, the dc and ac transient components of the SCC flowing out of the stator windings induce fault currents in the rotor winding and vice versa until the magnetic flux is depleted.

The current calculated from equation (4) is shown in Figure 6 using parameters for a typical 2-MW induction generator when the prefault voltage is 0.7 p.u. As can be seen from Figure 6, the current reaches the maximum value at π (first half a period). Therefore, it may be a good approximation to calculate the maximum (peak) current by substituting t = T/2 into equation (4). The resulting equation for peak current will be

$$i_{\max} = \frac{\sqrt{2}V_S}{Z'_S}\left[e^{-\frac{T}{2T'_S}} + (1-\sigma)e^{-\frac{T}{2T'_r}}\right] \quad (9)$$

It was demonstrated experimentally in [8] that equation (9) gives satisfactory accuracy for peak current assessment. The resulting current is shown in Figure 7. A detailed dynamic model of a Type 1 WTG is simulated in PSCAD™. A symmetrical three-phase fault is simulated and the resulting SCC is compared to the simulation result of the simplified representation as described in equation (4). It is shown in Figure 7 that the two traces are very closely matched.

From equation (4), it is shown that the operating slip does not influence the short-circuit transient behavior. To check the influence of the slip, we performed symmetrical three-phase faults on a Type 1 WTG for two different slips using the detailed model. As shown in Figure 8, the prefault current and the postfault current for the two different operating slips are very distinct. Similarly, the frequencies of the

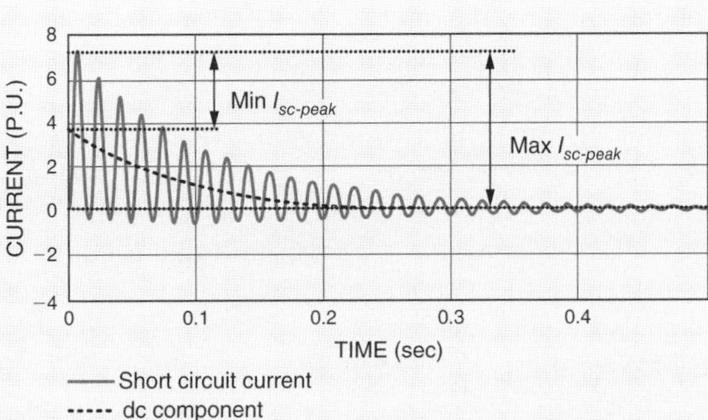

Figure 6 SCC from a Type 1 WTG

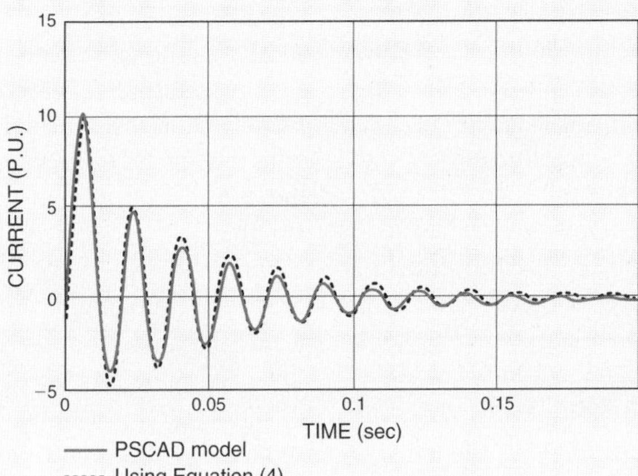

Figure 7 SCC comparison between the output of the detailed model and the output of the simplified model simulated for a Type 1 WTG

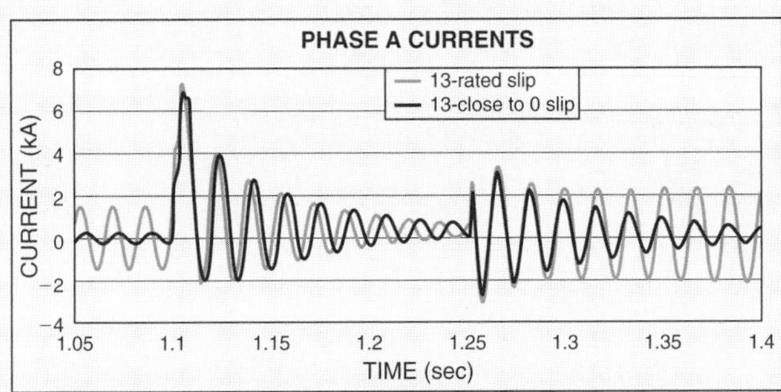

Figure 8 SCC comparison for two different slips

SCC during the fault are not the same for two different operating slips. However, the peak values of the SCC of the induction generator operating at two different slips are very closely matched.

SCC from a Type 2 WTG

The variable slip generator is essentially a wound-rotor induction generator with a variable resistor connected in series to the rotor winding (Type 2 generator). This external resistor is controlled by a high-frequency switch. Below rated power, the resistor control is inactive, so the system operates as a conventional induction generator. Above rated power, the resistor control allows the slip to vary, so variable speed operation is possible for a speed range of about 10% [9]. If the blade pitch angle is kept constant at zero degrees, the rotor speed, and thus the slip, will vary with wind speed. However, operation at higher slips generate a lot of loss because of the rotor resistance. Thus, the heat loss can be excessive. On the other hand, if the blade pitch angle is controlled to keep the rotor speeds within a small deviation from the rated slip, the losses in rotor resistance can be minimized. An equivalent electrical diagram of a variable-slip induction generator is shown in Figure 9, with a variable external resistor R_{ext}.

The connection diagram example for this type of generator is shown in Figure 10. In case of three-phase symmetrical fault, the same equations as for a Type 1 generator are applied. The only difference is for rotor time constant that needs to account for additional external resistance.

The modified rotor time constant can be calculated by adding the effect of the external resistor R_{ext} (refer to Table 2, p. 429), where R_{ext} is the value of external resistance that happens to be in the circuit at the time of the fault. The effect of such additional resistance on SCC is shown in Figure 11. So, adding the external resistors doubles the overall rotor resistance. The modified equation for SCC, maximum current, and the rotor transient time constant can then be derived using the values shown in Table 1 (see p. 428).

The maximum current occurs at ΔT, the time after a fault when current reaches its first peak. In this case, this additional resistance decreases the overall ac component in current, but does not much affect the first peak value of the current since the increase in resistance is relatively small. The same conclusion can be made by analyzing equations (4) and (9), where the additional external resistance has an effect on a second term that represents

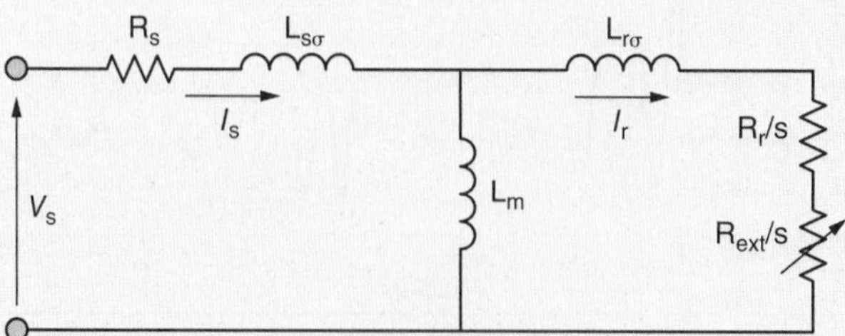

Figure 9 Equivalent circuit for a Type 2 generator

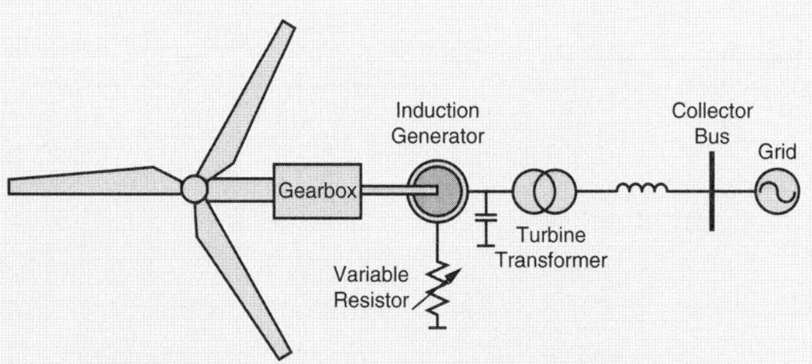

Figure 10 Connection diagram for a Type 2 WTG

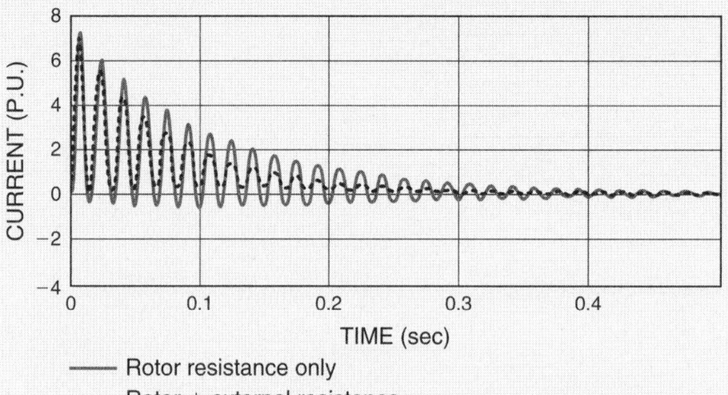

Figure 11 Effect of external resistance for a Type 2 WTG

the ac component of the current. As shown in Tables 1 and 2, the impact of the external rotor resistance on the SCC is two-fold: it reduces the SCC magnitude, and it shortens the rotor time constant (decay time of the SCC).

SCC from a Type 3 WTG

A Type 3 WTG is implemented by a doubly fed induction generator (DFIG). It is a variable-speed WTG where the rotor speed is allowed to vary within a slip range of ±30%. Thus, the power converter can be sized to about 30% of rated power. The equivalent electrical diagram of a DFIG is shown in Figure 12.

It is similar to one for a regular induction generator except for additional rotor voltage, representing voltage produced by a power converter. Under normal operation, this voltage is actually from a current-controlled power converter with the ability to control the real and reactive power output instantaneously and independently. The capability to control flux (flux-oriented controller—FOC) in induction machines has been used in the motor drive industry since the seventies.

The typical connection diagram for a DFIG (Type 3) WTG is shown in Figure 12. In an ideal situation, the power converter connected to the rotor winding should be able to withstand the currents induced by the dc and ac components flowing in the stator winding. However, the components of the power converter (IGBT, diode, capacitor, etc.) are designed to handle only normal currents and normal dc bus voltage. A crowbar system is

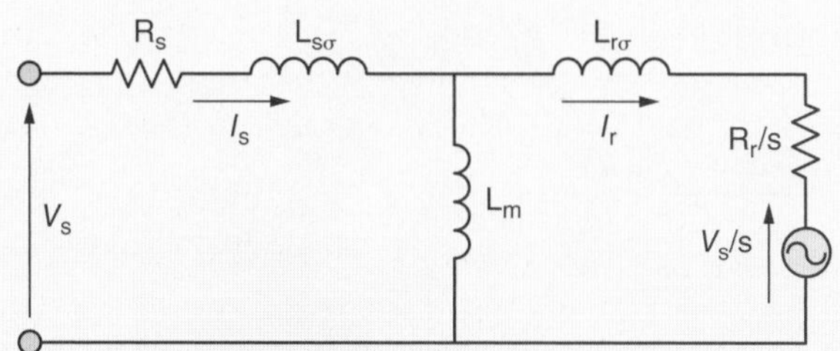

Figure 12 Simplified equivalent circuit of a DFIG

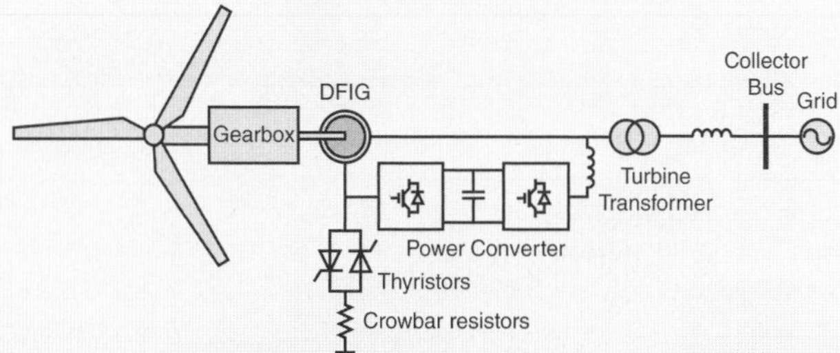

Figure 13 Connection diagram for a Type 3 WTG

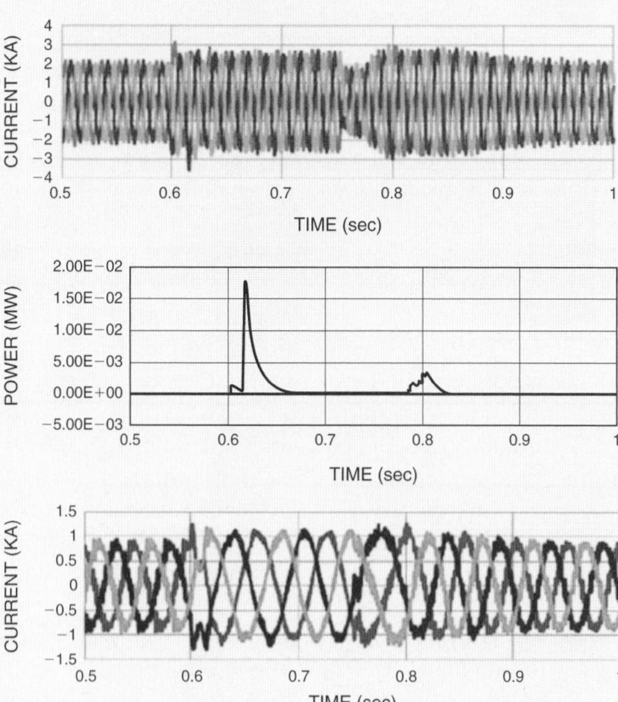

Figure 14 The fault currents, the rotor currents, and the power consumed by the crowbar circuits in a Type 3 WTG

usually used for protecting the power electronics converter from overvoltage and thermal breakdown during short-circuit faults. A crowbar is usually implemented to allow the insertion of additional resistance into the rotor winding to divert the SCC in the rotor winding from damaging the power converter. Additional dynamic braking on the dc bus is also used to limit the dc bus voltage.

During faults, the rotor windings are essentially short circuited by an equivalent adjustable crowbar resistance R_{CB}. The modified equation for SCC, maximum current, and the rotor transient time constant can then be derived using the values shown in Table 1 [7]. In a Type 3 WTG, however, the size of the crowbar is usually controlled, such that the actual fault current is more controllable than the simplified assumption. In Figure 14, three-phase fault currents are shown to be well regulated by proper control of the crowbar resistance. In this case, the crowbar circuit installed on the rotor winding is controlled to maintain the dc bus voltage constant.

A dynamic braking resistor is also installed on the dc bus to help regulate the dc bus. Figure 14 shows the size of the real power modulated in the crowbar during the faults. There are also dynamic braking resistors and a dc chopper installed on the dc bus to help regulate the dc bus voltage during transients. The corresponding rotor currents are also shown in Figure 14. Because of differences in crowbar implementation from one turbine manufacturer to the other, a protection engineer should evaluate the recommended value provided by the manufacturers. However, if none is available, the values of minimum and maximum SCCs presented in Table 1 can be used.

Type 4 WTGs

An example of a Type 4 direct-drive WTG with permanent-magnet synchronous generator (PMSG) is shown in Figure 15. This is a variable-speed WTG implemented with full power conversion. Recent advances and lower cost of power electronics make it feasible to build variable-speed wind turbines with power converters with the same rating as the turbines. The full power conversion allows separation between the WTG and the grid, thus, the mechanical dynamic can be buffered from entering the grid and the transient dynamic on the grid can be buffered from entering the wind turbine dynamic. Thus, while the grid is at 60 Hz, the stator winding of the generator may operate at variable frequencies. The temporary imbalance between aerodynamic power and generated power during a transient is handled by the pitch control, dynamic brake, and power converter control.

The SCC contribution for a three-phase fault is limited to its rated current or a little above its rated current. It is common to design a power converter for a Type 4 wind turbine with an overload capability of 10% above rated. Note that in any fault condition, the generator stays connected to the power converter and is buffered from the faulted lines on the grid. Thus, although there is a fault on the grid, the generator output current is controlled to stay within the current limit (e.g., 1.1 p.u.). However, keep in mind that with a fault on the grid, the

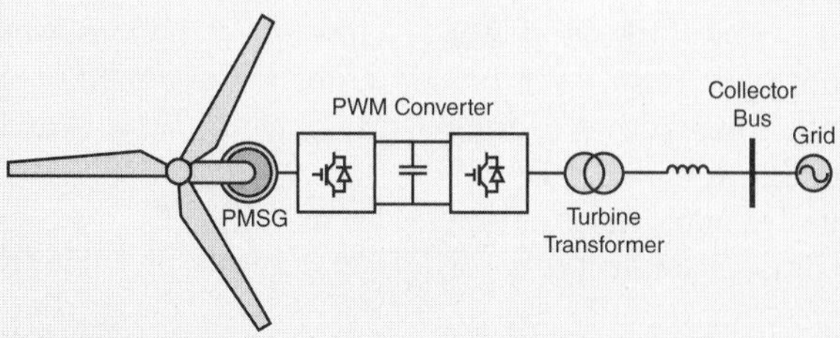

Figure 15 PMSG direct-drive WTG diagram

output power delivered to the grid is less than rated power. Although the currents can be made to balance, due to reduced voltage and/or unbalanced voltage, only a reduced output power can be delivered. In Type 4 WTGs, the SCC is a controlled parameter. So, such WTGs can be represented as a constant three-phase balanced current source in a short-circuit models. The priority of the real power versus reactive power during the fault depends on the prior setting of the controller. However, the current limit of the power converter must be followed to protect the power switches.

SCC Comparison for Symmetrical Faults

The SCCs for different types of wind turbines are not the same. For each turbine type, the peak value of the magnitude of the SCC is affected by the transient reactance, the prefault voltage, the effective rotor resistances, and the instant the fault occurs.

For turbine Types 1 through 3, the SCC declines as the fault progresses and eventually ceases as the rotor flux is depleted. For Type 4 WTGs, the SCC can be maintained constant.

The SCC transient behavior is affected by the stator time constant and the rotor time constant for Type 1 through Type 3 WTGs. The Type 4 generators can generate constant current during the fault.

In Table 1, the list parameters are shown. These parameters can be used to substitute the parameters from equation (4) and equation (6) for different types of WTGs. Table 2 lists the maximum and minimum possible values of the peak of SCC. It is shown that the Type 1 WTG can produce the largest SCC. The instant of the fault has affects on the magnitude of the SCC. The maximum value is based on the peak of the ac component and the highest value of the dc component, and the minimum value is based on the peak value of the ac component only.

For a Type 2 WTG, the maximum value is computed when $R_{ext} = 0\ \Omega$.

WTG	Type 1	Type 2	Type 3
$\mathbf{Z}'_S$	$X'_S = \omega L'_S$	$\sqrt{X'^2_S + R^2_{ext}}$	$\sqrt{X'^2_S + R^2_{CB}}$
$\mathbf{T}'_r$	$\dfrac{L'_r}{R_r}$	$\dfrac{L'_r}{R_r + R_{ext}}$	$\dfrac{L'_r}{R_r + R_{CB}}$

TABLE 1

Modified Values for SCC Calculation for Different Types of WTGs

WTG	Type 1	Type 2	Type 3	Type 4
Max I_{SC_PEAK}	$2\frac{\sqrt{2}V_s}{X_S'}$	$2\frac{\sqrt{2}V_s}{X_S'}$	$2\frac{\sqrt{2}V_s}{X_S'}$	1.1 I_{RATED}
Min I_{SC_PEAK}	$\frac{\sqrt{2}V_s}{X_S'}$	$\frac{\sqrt{2}V_s}{\sqrt{X_S'^2 + (9R_r')^2}}$	1.1 I_{RATED}	0

TABLE 2

Maximum and Minimum Possible Value of the SCC

The minimum value is computed when the slip reaches 10% above synchronous speed. And for a Type 3 WTG, the maximum value is computed when the crowbar shorts the rotor winding and the minimum value is computed when the power converter can follow the commanded current (i.e., in case the fault occurs far away from the point of interconnection, the remaining terminal voltage is relatively high enough to let the power converter operate normally and supply the commanded currents). Note, that for a symmetrical fault, the actual fault current for each phase is different from the other phases due to the fact that the time of the fault occurs at a different phase angle for different phases, thus affecting the dc offset. For a Type 4 WTG, the stator current can always be controlled because of the nature of power converter which is based on current controlled voltage source converter.

An example of SCC for Type 4 WTGs is given in the next section. In this section, the SCC is analyzed at the terminals of the generator. In an actual WPP, the faults will likely occur at the transmission side. Thus, the impact of the cable capacitance, plant level reactive compensation, and wind plant transformer connections are not included. References [10–14] provide good sources of information for the WPP environment.

Unsymmetrical Faults

The nature of the fault produces a different response for different wind turbine types. In this section, the observation of the short-circuit behavior for unsymmetrical faults on different types of WTGs is presented. Note that operating an induction generator under an unbalanced condition creates torque pulsation and unbalanced currents. If this condition persists for a long period of time, it may excite other parts of the wind turbine, and the unbalanced currents may create unequal heating in the three-phase windings and, thus, shorten the life of the winding insulation.

Unlike in a symmetrical three-phase fault, the positive-sequence voltage source continues to drive the fault current until the fault or the generator is removed from the circuit. The remaining un-faulted (normal) phases continue to maintain the air gap flux. The initial conditions of the fault currents are different for each phase. The three line currents usually show a different dc offset, which eventually settles out over time.

To explore the short-circuit behavior of unsymmetrical faults presented in this section, a detailed model of the system is developed in PSCAD™.

Single Line-to-Ground (SLG) Faults

The single line-to-ground fault is the most likely to occur in the power system. The magnetic flux in the air gap, although smaller than normal and unbalanced, is maintained by the remaining un-faulted lines. Thus, the short circuit in SLG faults will continue to flow until the circuit breaker removes the fault from the circuit.

Figure 16 shows the SCC of a Type 1 WTG for three lines-to-ground (3LG) and an SLG fault. In the symmetrical fault, the SCC dies out rather quickly, while in a SLG fault, the SCC is driven by the remaining two phases and it continues to flow until the short circuit is removed from the circuit. The peak current during a SLG fault is typically higher than for a 3LG fault (there is a quicker decay of current during symmetrical a 3LG fault due to magnetic field collapse). The difference in peak currents for both 3LG and SLG faults depends on generator parameters, fault location, etc. Note also that the presence of the positive sequence, the negative sequence, and zero sequence currents in the unsymmetrical faults influence the size of the SCC. No comparison has been made between dynamic simulation results and results obtained via symmetrical component calculations (this is planned for future work).

In Figure 17, the SCC for a Type 3 WTG is shown both for the three-phase currents and the corresponding sequence components. The changes in positive sequence and the sudden appearance of the negative sequence are also shown. The absence of the zero sequence current is a consequence of transformer winding connections.

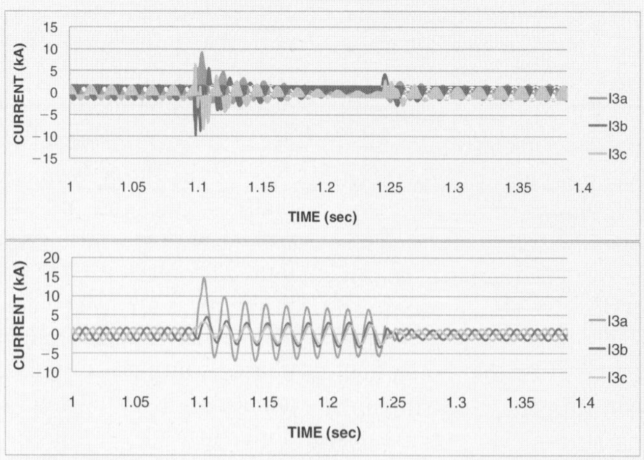

Figure 16 Voltage and SCC for 3LG and SLG for a Type 1 WTG

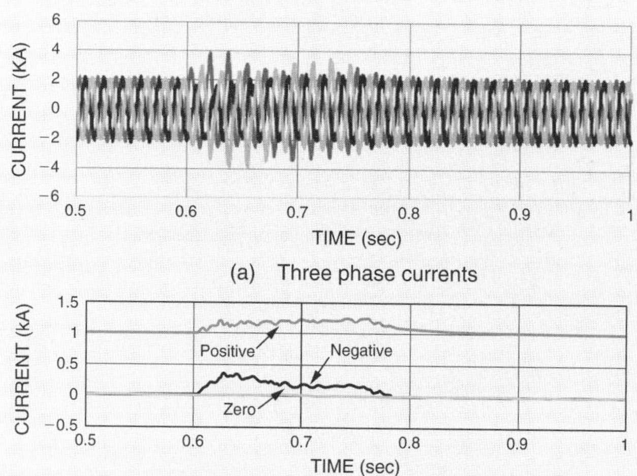

Figure 17 SCC for SLG for a Type 3 WTG

Line-to-Line (LL) and Line-to-Line-to-Ground (LLG) Faults

The line-to-line fault and the line-to-line-to-ground fault also maintain the air-gap flux during the fault. Output power of the generator will be limited and pulsating due to an unbalanced condition. The SCC will continue to flow until the circuit breaker removes the fault from the circuit.

As shown on Figures 18 and 19, the type of fault affects the existence of the zero sequence component in the SCC of the WPP. Thus, the line currents in the three phases are distributed differently based on its positive sequence, negative sequence, and the zero sequence magnitudes and phase angles.

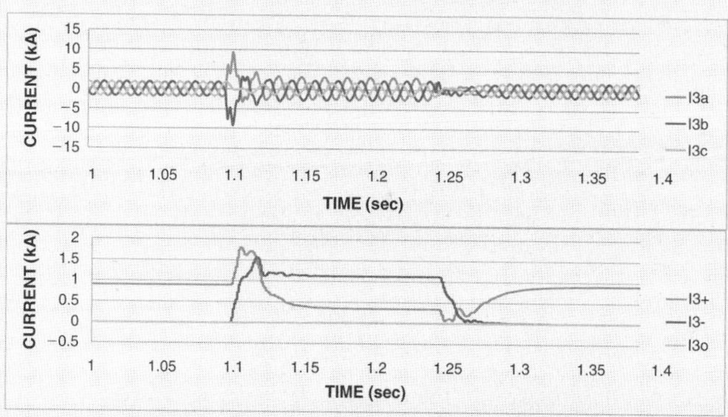

Figure 18 The SCC for a LL fault of a Type 2 WTG

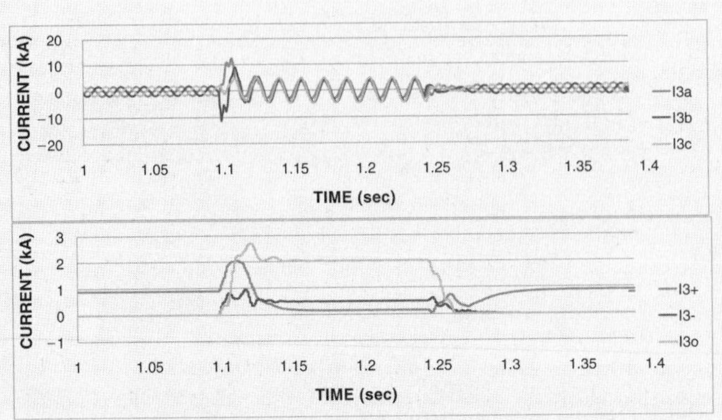

Figure 19 The SCC for LLG fault of a Type 2 WTG

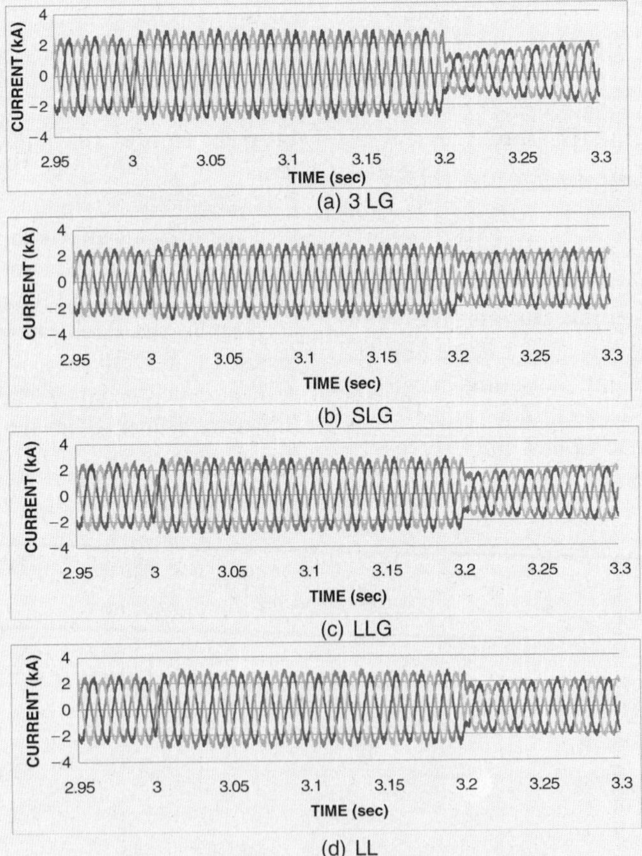

Figure 20 SCC for Type 4 WTG for different types of faults

SCC for the Type 4 WTG under Different Faults

In Figure 20, the fault currents for a Type 4 WTG are shown. Note, that the power converter buffers the generator from the grid. The SCC is basically controlled by the power converter. Hence, the line currents are symmetrical currents at different types of unsymmetrical faults. The postfault recovery may slightly differ for different faults.

Conclusions

In this case study, the SCC contributions of different WTGs for faults at the terminal of the generator were simulated using simplified model to determine SCC characteristics for symmetrical faults. The simplified model represents the size and the time constants governing the SCC behavior. A table summarizing different fault impedance and transient rotor time constants is provided. Another table summarizing the range of SCCs for different types of WTGs is presented in Table 2. For Type 1 and Type 2 WTGs, the maximum and the minimum values depend on timing of the fault and the parameters and the operating condition of the induction generator. For Type 3 WTGs, the control and the operation of the crowbar and dynamic braking affects the characteristics of the SCC. For Type 4 WTGs, the SCC is controllable by the power converter.

To compute unsymmetrical faults, detailed models were used to demonstrate the behavior of SCCs of different WTGs. As expected, the SCC continues to flow until the fault is cleared from the circuit or the generator is disconnected from the grid. The terminal voltage and currents are sustained longer because the line voltages, except from the faulted phase, are able to sustain air gap flux. The nature of SCC is not only affected by the type of WTG, but also by the nature of the faults, the winding connections of the generator, and the transformers between the fault and the generator. Auxiliary components (reactive compensations), cable length and capacitance, and the diversity of the WPP contribute to the size and nature of the SCC, one way or another.

Each WPP is unique. Therefore, recommended practice from local reliability organizations, the manufacturers, transmission planners, wind plant developers, and the local utilities should be followed very closely.

Acknowledgment

This work is supported by the U.S. Department of Energy and California Energy Commission.

References

1. U.S. Department of Energy—Energy Efficiency and Renewable Energy, "20% Wind Energy by 2030—Increasing Wind Energy's Contribution to U.S. Electricity Supply," May, 2008.
2. J. Charles Smith, Michael R. Milligan, Edgar A. DeMeo, and Brian Parsons, "Utility wind Integration and operating impact state of the art," *IEEE Trans. Power Systems,* vol. 22, pp. 900–908, Aug. 2007.

3. E. Muljadi, Z. Mills, A. Ellis, and R. Foster, "Fault Analysis at a Wind Power Plant for a One Year of Observation," presented at the *IEEE Power Engineering Society, General Meeting*, Pittsburgh, PA, July 20–24, 2008.
4. E. Muljadi, S. Pasupulati, A. Ellis, and D. Kosterev, "Method of Equivalencing for a Large Wind Power Plant with Multiple Turbine Representation," presented at the *IEEE Power Engineering Society, General Meeting*, Pittsburgh, PA, July 20–24, 2008.
5. E. Muljadi, C. P. Butterfield, A. Ellis, J. Mechenbier, J. Hochheimer, R. Young, N. Miller, R. Delmerico, R. Zavadil, and J. C. Smith, "Equivalencing the collector system of a large wind power plant," in *Proc. 2006 IEEE Power Engineering Society General Meeting.* June 18–22, 2006.
6. Nader Samaan, Robert Zavadil, J. Charles Smith, and Jose Conto, "Modeling of Wind Power Plants for Short Circuit Analysis in the Transmission Network," in *Proc. of IEEE/PES Transmission and Distribution Conference*, Chicago, USA, April 2008.
7. J. Moren, and S.W.H. de Haan, "Short-Circuit Current of Wind Turbines with Doubly Fed Induction Generator," *IEEE Transactions on Energy Conversion*, Vol. 22, No. 1, March 2007.
8. Sulawa, Zabara, et al. "Short circuit current of induction generators." *IEEE ISCAS 2007 proceedings.*
9. O. Anaya-Lara, N. Jenkins, et al. *Wind energy generation: modeling and control.* Wiley. ISBN 9780470714331.
10. IEEE PES Wind Plant Collector System Design Working Group,"Wind Power Plant Substation and Collector System Redundancy, Reliability and Economics," *Proceedings of the 2009 IEEE Power and Energy* Society General Meeting.
11. IEEE PES Wind Plant Collector System Design Working Group, "Power Transformer Application for Wind Plant Substation," *Proceedings of the 2009 IEEE Power and Energy Society General Meeting.*
12. IEEE PES Wind Plant Collector System Design Working Group, "Wind Power Plant Grounding, Overvoltage Protection, and Insulation Coordination," *Proceedings of the 2009 IEEE Power and Energy Society General Meeting.*
13. WECC Modeling and Validation Work Group, "WECC Wind Power Plant Power Flow Modeling Guide Prepared." May 2008.
14. "Short-circuit currents in three-phase a.c. systems–Part 0: Calculation of currents," *IEC Standard* 60909-0, July 2001.

Biographies

Eduard Muljadi (M'82-SM'94-F'10) received his PhD (in Electrical Engineering) from the University of Wisconsin, Madison. From 1988 to 1992, he taught at California State University, Fresno, California. In June 1992, he joined the National Renewable Energy Laboratory in Golden, Colorado. His current research interests are in the fields of electric machines, power electronics, and power systems in general with emphasis on renewable energy applications. He is a member of Eta Kappa Nu, Sigma Xi, and a Fellow of the IEEE. He is involved in the activities of the IEEE Industry Application Society (IAS), Power Electronics Society, and Power and Energy Society (PES).

He is currently a member of various committees of the IAS, and a member of the Working Group on Renewable Technologies and Dynamic Performance Wind Generation Task Force of the PES. He holds two patents in power conversion for renewable energy.

Vahan Gevorgian (M'97) graduated from the Yerevan Polytechnic Institute (Armenia) in 1986. During his studies, he concentrated on electrical machines. His thesis research dealt with doubly fed induction generators for stand-alone power systems. He obtained his PhD degree in electrical engineering from the State Engineering University of Armenia in 1993. His dissertation was devoted to a modeling of electrical transients in large wind turbine generators.

Dr. Gevorgian is currently working at the National Wind Technology Center (NWTC) of National Renewable Energy Laboratory (NREL) in Golden, Colorado, USA, as a research engineer. His current interests include modeling and testing of various applications of small wind turbine based power systems. ■

7.1 SERIES R–L CIRCUIT TRANSIENTS

Consider the series R–L circuit shown in Figure 7.1. The closing of switch SW at $t = 0$ represents to a first approximation a three-phase short circuit at the terminals of an unloaded synchronous machine. For simplicity, assume zero fault impedance; that is, the short circuit is a solid or "bolted" fault. The current is assumed to be zero before SW closes, and the source angle α determines the source voltage at $t = 0$. Writing a KVL equation for the circuit,

$$\frac{\mathrm{L}di(t)}{dt} + \mathrm{R}i(t) = \sqrt{2}V \sin(\omega t + \alpha) \quad t \geq 0 \tag{7.1.1}$$

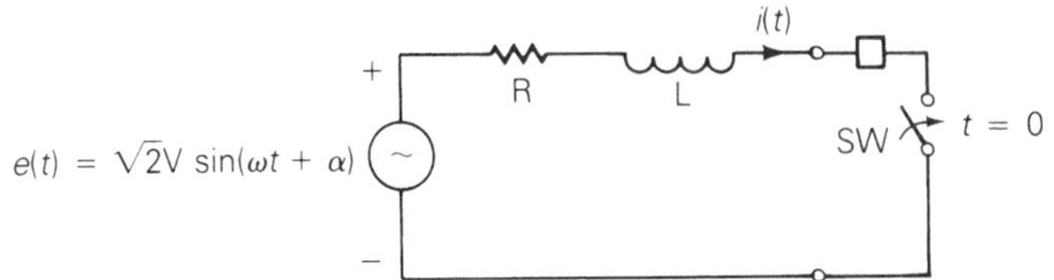

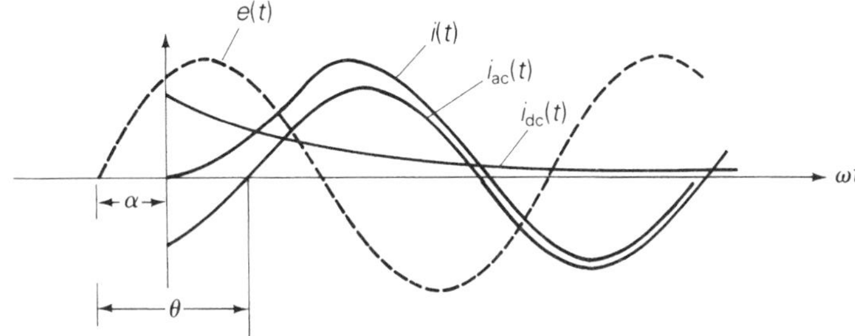

FIGURE 7.1 Current in a series R–L circuit with ac voltage source

The solution to (7.1.1) is

$$i(t) = i_{ac}(t) + i_{dc}(t)$$
$$= \frac{\sqrt{2}V}{Z}[\sin(\omega t + \alpha - \theta) - \sin(\alpha - \theta)e^{-t/T}] \quad \text{A} \tag{7.1.2}$$

where

$$i_{ac}(t) = \frac{\sqrt{2}V}{Z}\sin(\omega t + \alpha - \theta) \quad \text{A} \tag{7.1.3}$$

$$i_{dc}(t) = -\frac{\sqrt{2}V}{Z}\sin(\alpha - \theta)e^{-t/T} \quad \text{A} \tag{7.1.4}$$

$$Z = \sqrt{R^2 + (\omega L)^2} = \sqrt{R^2 + X^2} \quad \Omega \tag{7.1.5}$$

$$\theta = \tan^{-1}\frac{\omega L}{R} = \tan^{-1}\frac{X}{R} \tag{7.1.6}$$

$$T = \frac{L}{R} = \frac{X}{\omega R} = \frac{X}{2\pi f R} \quad \text{s} \tag{7.1.7}$$

The total fault current in (7.1.2), called the *asymmetrical fault current*, is plotted in Figure 7.1 along with its two components. The ac fault current (also called *symmetrical* or *steady-state fault current*), given by (7.1.3), is a sinusoid. The *dc offset current*, given by (7.1.4), decays exponentially with time constant $T = L/R$.

The rms ac fault current is $I_{ac} = V/Z$. The magnitude of the dc offset, which depends on α, varies from 0 when $\alpha = \theta$ to $\sqrt{2}I_{ac}$ when $\alpha = (\theta \pm \pi/2)$. Note that a short circuit may occur at any instant during a cycle of the ac source; that is, α can have any value. To find the largest fault current, choose $\alpha = (\theta - \pi/2)$. Then (7.1.2) becomes

$$i(t) = \sqrt{2}\mathrm{I}_{ac}[\sin(\omega t - \pi/2) + e^{-t/\mathrm{T}}] \quad \mathrm{A} \tag{7.1.8}$$

where

$$\mathrm{I}_{ac} = \frac{\mathrm{V}}{\mathrm{Z}} \quad \mathrm{A} \tag{7.1.9}$$

The rms value of $i(t)$ is of interest. Since $i(t)$ in (7.1.8) is not strictly periodic, its rms value is not strictly defined. However, to calculate the rms asymmetrical fault current with maximum dc offset, treat the exponential term as a constant, stretching the rms concept as follows:

$$\begin{aligned}\mathrm{I}_{rms}(t) &= \sqrt{[\mathrm{I}_{ac}]^2 + [\mathrm{I}_{dc}(t)]^2} \\ &= \sqrt{[\mathrm{I}_{ac}]^2 + [\sqrt{2}\mathrm{I}_{ac}e^{-t/\mathrm{T}}]^2} \\ &= \mathrm{I}_{ac}\sqrt{1 + 2e^{-2t/\mathrm{T}}} \quad \mathrm{A}\end{aligned} \tag{7.1.10}$$

It is convenient to use $\mathrm{T} = \mathrm{X}/(2\pi f\,\mathrm{R})$ and $t = \tau/f$, where τ is time in cycles, and write (7.1.10) as

$$\mathrm{I}_{rms}(\tau) = \mathrm{K}(\tau)\mathrm{I}_{ac} \quad \mathrm{A} \tag{7.1.11}$$

where

$$\mathrm{K}(\tau) = \sqrt{1 + 2e^{-4\pi\tau/(\mathrm{X/R})}} \quad \text{per unit} \tag{7.1.12}$$

From (7.1.11) and (7.1.12), the rms asymmetrical fault current equals the rms ac fault current times an "asymmetry factor," $\mathrm{K}(\tau)$. $\mathrm{I}_{rms}(\tau)$ decreases from $\sqrt{3}\mathrm{I}_{ac}$ when $\tau = 0$ to I_{ac} when τ is large. Also, higher X to R ratios (X/R) give higher values of $\mathrm{I}_{rms}(\tau)$. The above series R–L short-circuit currents are summarized in Table 7.1.

Component	Instantaneous Current (A)	rms Current (A)
Symmetrical (ac)	$i_{ac}(t) = \frac{\sqrt{2}V}{Z}\sin(\omega t + \alpha - \theta)$	$I_{ac} = \frac{V}{Z}$
dc offset	$i_{dc}(t) = \frac{-\sqrt{2}V}{Z}\sin(\alpha - \theta)e^{-t/T}$	
Asymmetrical (total)	$i(t) = i_{ac}(t) + i_{dc}(t)$	$I_{rms}(t) = \sqrt{I_{ac}^2 + i_{dc}(t)^2}$ with maximum dc offset: $I_{rms}(\tau) = K(\tau)I_{ac}$

TABLE 7.1

Short-circuit current—series R–L circuit*

*See Figure 7.1 and (7.1.1) through (7.1.12).

EXAMPLE 7.1

Fault currents: R–L circuit with ac source

A bolted short circuit occurs in the series R–L circuit of Figure 7.1 with V = 20 kV, X = 8 Ω, R = 0.8 Ω, and with maximum dc offset. The circuit breaker opens 3 cycles after fault inception. Determine (a) the rms ac fault current, (b) the rms "momentary" current at $\tau = 0.5$ cycle, which passes through the breaker before it opens, and (c) the rms asymmetrical fault current that the breaker interrupts.

SOLUTION

a. From (7.1.9),

$$I_{ac} = \frac{20 \times 10^3}{\sqrt{(8)^2 + (0.8)^2}} = \frac{20 \times 10^3}{8.040} = 2.488 \quad \text{kA}$$

b. From (7.1.11) and (7.1.12) with (X/R) = 8/(0.8) = 10 and $\tau = 0.5$ cycle,

$$K(0.5 \text{ cycle}) = \sqrt{1 + 2e^{-4\pi(0.5)/10}} = 1.438$$

$$I_{momentary} = K(0.5 \text{ cycle})I_{ac} = (1.438)(2.488) = 3.576 \quad \text{kA}$$

c. From (7.1.11) and (7.1.12) with (X/R) = 10 and $\tau = 3$ cycles,

$$K(3 \text{ cycles}) = \sqrt{1 + 2e^{-4\pi(3)/10}} = 1.023$$

$$I_{rms}(3 \text{ cycles}) = (1.023)(2.488) = 2.544 \quad \text{kA}$$

7.2 THREE-PHASE SHORT CIRCUIT—UNLOADED SYNCHRONOUS MACHINE

One way to investigate a three-phase short circuit at the terminals of a synchronous machine is to perform a test on an actual machine. Figure 7.2 shows an oscillogram of the ac fault current in one phase of an unloaded synchronous machine during such a test. The dc offset has been removed from the oscillogram. As shown, the amplitude of the sinusoidal waveform decreases from a high initial value to a lower steady-state value.

A physical explanation for this phenomenon is that the magnetic flux caused by the short-circuit armature currents (or by the resultant armature MMF) is initially forced to flow through high-reluctance paths that do not link the field winding or damper circuits of the machine. This is a result of the theorem of constant flux linkages that states that the flux linking a closed winding cannot change instantaneously. The armature inductance, which is inversely proportional to reluctance, is therefore

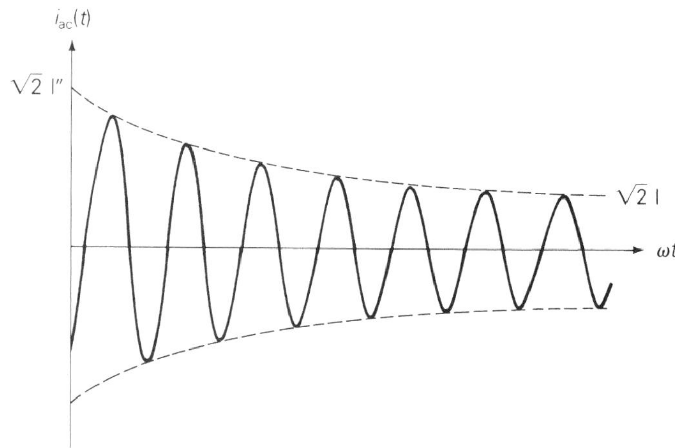

FIGURE 7.2

The ac fault current in one phase of an unloaded synchronous machine during a three-phase short circuit (the dc offset current is removed)

initially low. As the flux then moves toward the lower reluctance paths, the armature inductance increases.

The ac fault current in a synchronous machine can be modeled by the series R–L circuit of Figure 7.1 if a time-varying inductance $L(t)$ or reactance $X(t) = \omega L(t)$ is employed. In standard machine theory texts [3, 4], the following reactances are defined:

X''_d = direct axis subtransient reactance
X'_d = direct axis transient reactance
X_d = direct axis synchronous reactance

where $X''_d < X'_d < X_d$. The subscript d refers to the direct axis. There are similar quadrature axis reactances X''_q, X'_q, and X_q [3, 4]. However, if the armature resistance is small, the quadrature axis reactances do not significantly affect the short-circuit current. Using the above direct axis reactances, the instantaneous ac fault current can be written as

$$i_{ac}(t) = \sqrt{2}E_g\left[\left(\frac{1}{X''_d} - \frac{1}{X'_d}\right)e^{-t/T''_d} + \left(\frac{1}{X'_d} - \frac{1}{X_d}\right)e^{-t/T'_d} + \frac{1}{X_d}\right]\sin\left(\omega t + \alpha - \frac{\pi}{2}\right) \tag{7.2.1}$$

where E_g is the rms line-to-neutral prefault terminal voltage of the unloaded synchronous machine. Armature resistance is neglected in (7.2.1). Note that at $t = 0$, when the fault occurs, the rms value of $i_{ac}(t)$ in (7.2.1) is

$$I_{ac}(0) = \frac{E_g}{X''_d} = I'' \tag{7.2.2}$$

which is called the rms *subtransient fault current*, I″. The duration of I″ is determined by the time constant T''_d, called the *direct axis short-circuit sub-transient time constant*.

At a later time, when t is large compared to T''_d but small compared to the *direct axis short-circuit transient time constant* T'_d, the first exponential term in (7.2.1) has decayed almost to zero, but the second exponential has not decayed significantly. The rms ac fault current then equals the rms *transient fault current*, given by

$$I' = \frac{E_g}{X'_d} \tag{7.2.3}$$

When t is much larger than T'_d, the rms ac fault current approaches its steady-state value, given by

$$I_{ac}(\infty) = \frac{E_g}{X_d} = I \tag{7.2.4}$$

Since the three-phase no-load voltages are displaced 120° from each other, the three-phase ac fault currents are also displaced 120° from each other. In addition to the ac fault current, each phase has a different dc offset. The maximum dc offset in any one phase, which occurs when $\alpha = 0$ in (7.2.1), is

$$i_{dcmax}(t) = \frac{\sqrt{2}E_g}{X''_d} e^{-t/T_A} = \sqrt{2}I'' e^{-t/T_A} \tag{7.2.5}$$

where T_A is called the *armature time constant*. Note that the magnitude of the maximum dc offset depends only on the rms subtransient fault current I''. The above synchronous machine short-circuit currents are summarized in Table 7.2.

Component	Instantaneous Current (A)	rms Current (A)
Symmetrical (ac)	(7.2.1)	$I_{ac}(t) = E_g\left[\left(\frac{1}{X''_d} - \frac{1}{X'_d}\right)e^{-t/T''_d} + \left(\frac{1}{X'_d} - \frac{1}{X_d}\right)e^{-t/T'_d} + \frac{1}{X_d}\right]$
Subtransient		$I'' = E_g/X''_d$
Transient		$I' = E_g/X'_d$
Steady-state		$I = E_g/X_d$
Maximum dc offset	$i_{dc}(t) = \sqrt{2}I''e^{-t/T_A}$	
Asymmetrical (total)	$i(t) = i_{ac}(t) + i_{dc}(t)$	$I_{rms}(t) = \sqrt{I_{ac}(t)^2 + i_{dc}(t)^2}$ with maximum dc offset: $I_{rms}(t) = \sqrt{I_{ac}(t)^2 + [\sqrt{2}I''e^{-t/T_A}]^2}$

TABLE 7.2

Short-circuit current—unloaded synchronous machine*

*See Figure 7.2 and (7.2.1) through (7.2.5).

Machine reactances X''_d, X'_d, and X_d as well as time constants T''_d, T'_d, and T_A are usually provided by synchronous machine manufacturers. They also can be obtained from a three-phase short-circuit test by analyzing an oscillogram such as that in Figure 7.2 [2]. Typical values of synchronous machine reactances and time constants are given in Appendix Table A.1.

EXAMPLE 7.2

Three-phase short-circuit currents, unloaded synchronous generator

A 500-MVA, 20-kV, 60-Hz synchronous generator with reactances $X''_d = 0.15$, $X'_d = 0.24$, and $X_d = 1.1$ per unit and time constants $T''_d = 0.035$, $T'_d = 2.0$, $T_A = 0.20$ s is connected to a circuit breaker. The generator is operating at 5% above rated voltage and at no-load when a bolted three-phase short circuit occurs on the load side of the breaker. The breaker interrupts the fault three cycles after fault inception. Determine (a) the subtransient fault current in per-unit and kA rms; (b) maximum dc offset as a function of time; and (c) rms asymmetrical fault current, which the breaker interrupts, assuming maximum dc offset.

SOLUTION

a. The no-load voltage before the fault occurs is $E_g = 1.05$ per unit. From (7.2.2), the subtransient fault current that occurs in each of the three phases is

$$I'' = \frac{1.05}{0.15} = 7.0 \quad \text{per unit}$$

The generator base current is

$$I_{base} = \frac{S_{rated}}{\sqrt{3}V_{rated}} = \frac{500}{(\sqrt{3})(20)} = 14.43 \quad \text{kA}$$

The rms subtransient fault current in kA is the per-unit value multiplied by the base current:

$$I'' = (7.0)(14.43) = 101.0 \quad \text{kA}$$

b. From (7.2.5), the maximum dc offset that may occur in any one phase is

$$i_{dcmax}(t) = \sqrt{2}(101.0)e^{-t/0.20} = 142.9e^{-t/0.20} \quad \text{kA}$$

(Continued)

c. From (7.2.1), the rms ac fault current at $t = 3$ cycles $= 0.05$ s is

$$\begin{aligned} I_{ac}(0.05\text{ s}) &= 1.05\left[\left(\frac{1}{0.15} - \frac{1}{0.24}\right)e^{-0.05/0.035}\right. \\ &\quad \left. + \left(\frac{1}{0.24} - \frac{1}{1.1}\right)e^{-0.05/2.0} + \frac{1}{1.1}\right] \\ &= 4.920 \quad \text{per unit} \\ &= (4.920)(14.43) = 71.01 \quad \text{kA} \end{aligned}$$

Modifying (7.1.10) to account for the time-varying symmetrical component of fault current,

$$\begin{aligned} I_{rms}(0.05) &= \sqrt{[I_{ac}(0.05)]^2 + [\sqrt{2}I''e^{-t/T_a}]^2} \\ &= I_{ac}(0.05)\sqrt{1 + 2\left[\frac{I''}{I_{ac}(0.05)}\right]^2 e^{-2t/T_a}} \\ &= (71.01)\sqrt{1 + 2\left[\frac{101}{71.01}\right]^2 e^{-2(0.05)/0.20}} \\ &= (71.01)(1.8585) \\ &= 132 \quad \text{kA} \end{aligned}$$

7.3 POWER SYSTEM THREE-PHASE SHORT CIRCUITS

In order to calculate the subtransient fault current for a three-phase short circuit in a power system, make the following assumptions:

1. Transformers are represented by their leakage reactances. Winding resistances, shunt admittances, and Δ–Y phase shifts are neglected.
2. Transmission lines are represented by their equivalent series reactances. Neglect series resistances and shunt admittances.
3. Synchronous machines are represented by constant-voltage sources behind subtransient reactances. Neglect armature resistance, saliency, and saturation.
4. Neglect all nonrotating impedance loads.
5. Especially for small motors rated less than 40 kW either neglect induction motors or represent them in the same manner as synchronous machines.

These assumptions are made for simplicity in this text, and in practice they should not be made for all cases. For example, in distribution systems, resistances of primary and secondary distribution lines may in some cases significantly reduce fault current magnitudes.

Figure 7.3 shows a single-line diagram consisting of a synchronous generator feeding a synchronous motor through two transformers and a transmission line.

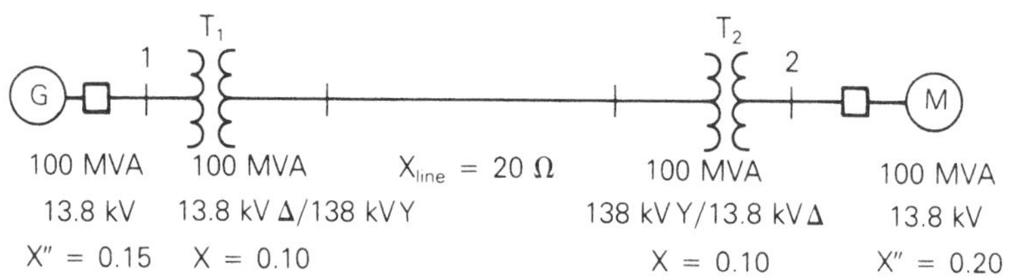

FIGURE 7.3 Single-line diagram of a synchronous generator feeding a synchronous motor

Consider a three-phase short circuit at bus 1. The positive-sequence equivalent circuit is shown in Figure 7.4(a), where the voltages E_g'' and E_m'' are the prefault internal voltages behind the subtransient reactances of the machines, and the closing of switch SW represents the fault. For purposes of calculating the subtransient fault current, E_g'' and E_m'' are assumed to be constant-voltage sources.

In Figure 7.4(b) the fault is represented by two opposing voltage sources with equal phasor values V_F. Using superposition, the fault current then can be calculated from the two circuits shown in Figure 7.4(c). However, if V_F equals the prefault

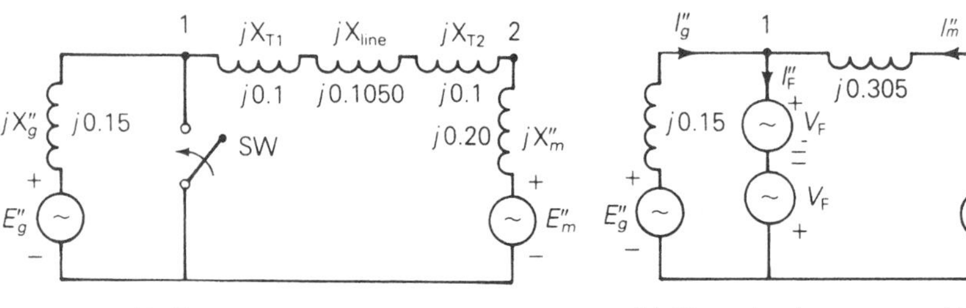

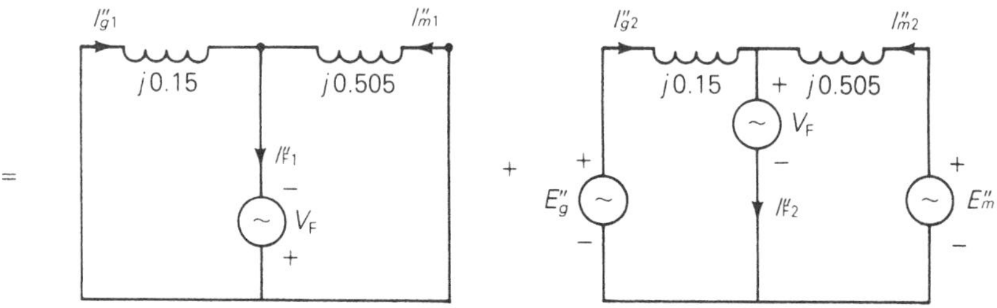

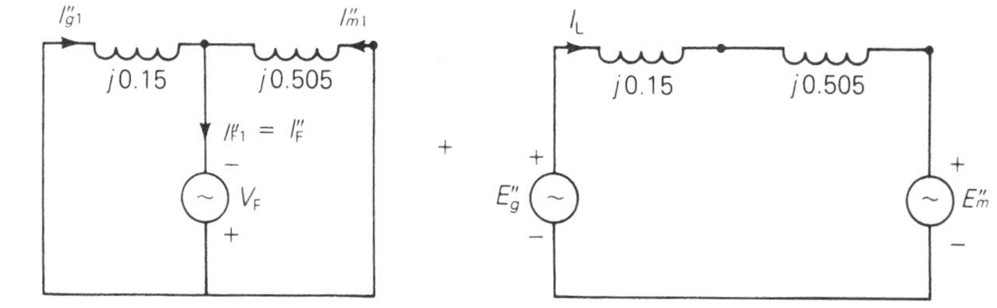

FIGURE 7.4 Application of superposition to a power system three-phase short circuit

voltage at the fault, then the second circuit in Figure 7.4(c) represents the system before the fault occurs. As such, $I''_{F2} = 0$ and V_F, which has no effect, can be removed from the second circuit, as shown in Figure 7.4(d). The subtransient fault current is then determined from the first circuit in Figure 7.4(d), $I''_F = I''_{F1}$. The contribution to the fault from the generator is $I''_g = I''_{g1} + I''_{g2} = I''_{g1} + I_L$, where I_L is the prefault generator current. Similarly, $I''_m = I''_{m1} - I_L$.

EXAMPLE 7.3

Three-phase short-circuit currents, power system

The synchronous generator in Figure 7.3 is operating at rated MVA, 0.95 p.f. lagging and at 5% above rated voltage when a bolted three-phase short circuit occurs at bus 1. Calculate the per-unit values of (a) subtransient fault current; (b) subtransient generator and motor currents neglecting prefault current; and (c) subtransient generator and motor currents including prefault current.

SOLUTION

a. Using a 100-MVA base, the base impedance in the zone of the transmission line is

$$Z_{\text{base, line}} = \frac{(138)^2}{100} = 190.44 \quad \Omega$$

and

$$X_{\text{line}} = \frac{20}{190.44} = 0.1050 \quad \text{per unit}$$

Figure 7.4 shows the per-unit reactances. From the first circuit in Figure 7.4(d), the Thévenin impedance as viewed from the fault is

$$Z_{\text{Th}} = jX_{\text{Th}} = j\frac{(0.15)(0.505)}{(0.15 + 0.505)} = j0.11565 \quad \text{per unit}$$

and the prefault voltage at the generator terminals is

$$V_F = 1.05\underline{/0^\circ} \quad \text{per unit}$$

The subtransient fault current is then

$$I''_F = \frac{V_F}{Z_{\text{Th}}} = \frac{1.05\underline{/0^\circ}}{j0.11565} = -j9.079 \quad \text{per unit}$$

b. Using current division in the first circuit of Figure 7.4(d),

$$I''_{g1} = \left(\frac{0.505}{0.505 + 0.15}\right) I''_F = (0.7710)(-j9.079) = -j7.000 \quad \text{per unit}$$

$$I''_{m1} = \left(\frac{0.15}{0.505 + 0.15}\right) I''_F = (0.2290)(-j9.079) = -j2.079 \quad \text{per unit}$$

c. The generator base current is

$$I_{\text{base,gen}} = \frac{100}{(\sqrt{3})(13.8)} = 4.1837 \quad \text{kA}$$

and the prefault generator current is

$$I_L = \frac{100}{(\sqrt{3})(1.05 \times 13.8)} \underline{/-\cos^{-1} 0.95} = 3.9845\underline{/-18.19°} \quad \text{kA}$$

$$= \frac{3.9845\underline{/-18.19°}}{4.1837} = 0.9524\underline{/-18.19°}$$

$$= 0.9048 - j0.2974 \quad \text{per unit}$$

The subtransient generator and motor currents, including prefault current, are then

$$I''_g = I''_{g1} + I_L = -j7.000 + 0.9048 - j0.2974$$

$$= 0.9048 - j7.297 = 7.353\underline{/-82.9°} \quad \text{per unit}$$

$$I''_m = I''_{m1} - I_L = -j2.079 - 0.9048 + j0.2974$$

$$= -0.9048 - j1.782 = 1.999\underline{/243.1°} \quad \text{per unit}$$

An alternate method of solving Example 7.3 is to first calculate the internal voltages E''_g and E''_m using the prefault load current I_L. Then, instead of using superposition, the fault currents can be resolved directly from the circuit in Figure 7.4(a) (see Problem 7.11). However, in a system with many synchronous machines, the superposition method has the advantage that all machine voltage sources are shorted, and the prefault voltage is the only source required to calculate the fault current. Also, when calculating the contributions to fault current from each branch, prefault currents are usually small, and hence can be neglected. Otherwise, prefault load currents could be obtained from a power-flow program.

7.4 BUS IMPEDANCE MATRIX

Now to extend the results of the previous section to calculate subtransient fault currents for three-phase faults in an N-bus power system, the system is modeled by its positive-sequence network, where lines and transformers are represented by series reactances and synchronous machines are represented by constant-voltage sources behind subtransient reactances. As before, all resistances, shunt admittances, and nonrotating impedance loads, and also for simplicity prefault load currents, are neglected.

Consider a three-phase short circuit at any bus n. To analyze two separate circuits, use the superposition method described in Section 7.3. [For example, see Figure 7.4(d).]

In the first circuit, all machine-voltage sources are short-circuited, and the only source is due to the prefault voltage at the fault. Writing nodal equations for the first circuit,

$$\boldsymbol{Y}_{\text{bus}}\boldsymbol{E}^{(1)} = \boldsymbol{I}^{(1)} \tag{7.4.1}$$

where $\boldsymbol{Y}_{\text{bus}}$ is the positive-sequence bus admittance matrix, $\boldsymbol{E}^{(1)}$ is the vector of bus voltages, and $\boldsymbol{I}^{(1)}$ is the vector of current sources. The superscript (1) denotes the first circuit. Solving (7.4.1),

$$\boldsymbol{Z}_{\text{bus}}\boldsymbol{I}^{(1)} = \boldsymbol{E}^{(1)} \tag{7.4.2}$$

where

$$\boldsymbol{Z}_{\text{bus}} = \boldsymbol{Y}_{\text{bus}}^{-1} \tag{7.4.3}$$

$\boldsymbol{Z}_{\text{bus}}$, the inverse of $\boldsymbol{Y}_{\text{bus}}$, is called the positive-sequence *bus impedance matrix*. Both $\boldsymbol{Z}_{\text{bus}}$ and $\boldsymbol{Y}_{\text{bus}}$ are symmetric matrices.

Since the first circuit contains only one source, located at faulted bus n, the current source vector contains only one nonzero component, $I_n^{(1)} = -I''_{\text{F}n}$. Also, the voltage at faulted bus n in the first circuit is $E_n^{(1)} = -V_{\text{F}}$. Rewriting (7.4.2),

$$\begin{bmatrix} Z_{11} & Z_{12} & \cdots & Z_{1n} & \cdots & Z_{1N} \\ Z_{21} & Z_{22} & \cdots & Z_{2n} & \cdots & Z_{2N} \\ \vdots & & & & & \\ Z_{n1} & Z_{n2} & \cdots & Z_{nn} & \cdots & Z_{nN} \\ \vdots & & & & & \\ Z_{N1} & Z_{N2} & \cdots & Z_{Nn} & \cdots & Z_{NN} \end{bmatrix} \begin{bmatrix} 0 \\ 0 \\ \vdots \\ -I''_{\text{F}n} \\ \vdots \\ 0 \end{bmatrix} = \begin{bmatrix} E_1^{(1)} \\ E_2^{(1)} \\ \vdots \\ -V_{\text{F}} \\ \vdots \\ E_N^{(1)} \end{bmatrix} \tag{7.4.4}$$

The minus sign associated with the current source in (7.4.4) indicates that the current injected into bus n is the negative of $I''_{\text{F}n}$, since $I''_{\text{F}n}$ flows away from bus n to the neutral. From (7.4.4), the subtransient fault current is

$$I''_{\text{F}n} = \frac{V_{\text{F}}}{Z_{nn}} \tag{7.4.5}$$

Also from (7.4.4) and (7.4.5), the voltage at any bus k in the first circuit is

$$E_k^{(1)} = Z_{kn}(-I''_{\text{F}n}) = \frac{-Z_{kn}}{Z_{nn}}V_{\text{F}} \tag{7.4.6}$$

The second circuit represents the prefault conditions. Neglecting prefault load current, all voltages throughout the second circuit are equal to the prefault voltage; that is, $E_k^{(2)} = V_{\text{F}}$ for each bus k. Applying superposition,

$$\begin{aligned} E_k = E_k^{(1)} + E_k^{(2)} &= \frac{-Z_{kn}}{Z_{nn}}V_{\text{F}} + V_{\text{F}} \\ &= \left(1 - \frac{Z_{kn}}{Z_{nn}}\right)V_{\text{F}} \qquad k = 1, 2, \ldots, N \end{aligned} \tag{7.4.7}$$

EXAMPLE 7.4

Using Z_{bus} to compute three-phase short-circuit currents in a power system

The faults at bus 1 and 2 in Figure 7.3 are of interest. The prefault voltage is 1.05 per unit, and prefault load current is neglected. (a) Determine the 2 × 2 positive-sequence bus impedance matrix. (b) For a bolted three-phase short circuit at bus 1, use $\boldsymbol{Z}_{bus}$ to calculate the subtransient fault current and the contribution to the fault current from the transmission line. (c) Repeat part (b) for a bolted three-phase short circuit at bus 2.

SOLUTION

a. The circuit of Figure 7.4(a) is redrawn in Figure 7.5 showing per-unit admittance rather than per-unit impedance values. Neglecting prefault load current, $E_g'' = E_m'' = V_F = 1.05\underline{/0°}$ per unit. From Figure 7.5, the positive-sequence bus admittance matrix is

$$\boldsymbol{Y}_{bus} = -j\begin{bmatrix} 9.9454 & -3.2787 \\ -3.2787 & 8.2787 \end{bmatrix} \quad \text{per unit}$$

Inverting $\boldsymbol{Y}_{bus}$,

$$\boldsymbol{Z}_{bus} = \boldsymbol{Y}_{bus}^{-1} = +j\begin{bmatrix} 0.11565 & 0.04580 \\ 0.04580 & 0.13893 \end{bmatrix} \quad \text{per unit}$$

b. Using (7.4.5), the subtransient fault current at bus 1 is

$$I_{F1}'' = \frac{V_F}{Z_{11}} = \frac{1.05\underline{/0°}}{j0.11565} = -j9.079 \quad \text{per unit}$$

which agrees with the result in Example 7.3, part (a). The voltages at buses 1 and 2 during the fault are, from (7.4.7),

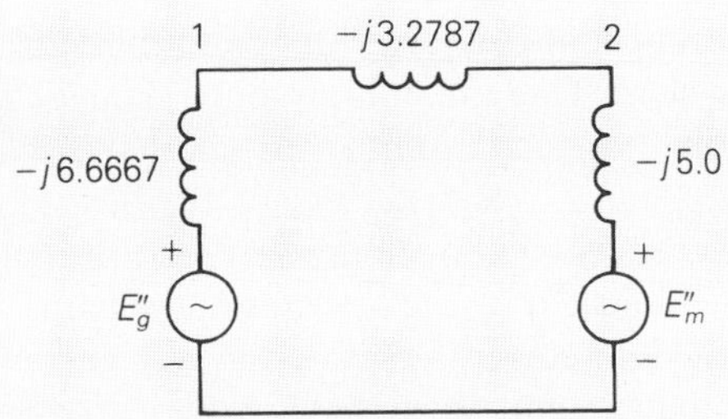

FIGURE 7.5 Circuit of Figure 7.4(a) showing per-unit admittance values

(Continued)

$$E_1 = \left(1 - \frac{Z_{11}}{Z_{11}}\right)V_F = 0$$

$$E_2 = \left(1 - \frac{Z_{21}}{Z_{11}}\right)V_F = \left(1 - \frac{j0.04580}{j0.11565}\right)1.05\underline{/0^\circ} = 0.6342\underline{/0^\circ}$$

The current to the fault from the transmission line is obtained from the voltage drop from bus 2 to 1 divided by the impedance of the line and transformers T_1 and T_2:

$$I_{21} = \frac{E_2 - E_1}{j(X_{line} + X_{T1} + X_{T2})} = \frac{0.6342 - 0}{j0.3050} = -j2.079 \quad \text{per unit}$$

which agrees with the motor current calculated in Example 7.3, part (b), where prefault load current is neglected.

c. Using (7.4.5), the subtransient fault current at bus 2 is

$$I''_{F2} = \frac{V_F}{Z_{22}} = \frac{1.05\underline{/0^\circ}}{j0.13893} = -j7.558 \quad \text{per unit}$$

and from (7.4.7),

$$E_1 = \left(1 - \frac{Z_{12}}{Z_{22}}\right)V_F = \left(1 - \frac{j0.04580}{j0.13893}\right)1.05\underline{/0^\circ} = 0.7039\underline{/0^\circ}$$

$$E_2 = \left(1 - \frac{Z_{22}}{Z_{22}}\right)V_F = 0$$

The current to the fault from the transmission line is

$$I_{12} = \frac{E_1 - E_2}{j(X_{line} + X_{T1} + X_{T2})} = \frac{0.7039 - 0}{j0.3050} = -j2.308 \quad \text{per unit}$$

Figure 7.6 shows a bus impedance equivalent circuit that illustrates the short-circuit currents in an N-bus system. This circuit is given the name *rake equivalent* in Neuenswander [5] due to its shape, which is similar to a garden rake.

The diagonal elements Z_{11}, Z_{22}, ..., Z_{NN} of the bus impedance matrix, which are the *self-impedances*, are shown in Figure 7.6. The off-diagonal elements, or the *mutual impedances*, are indicated by the brackets in the figure.

Neglecting prefault load currents, the internal voltage sources of all synchronous machines are equal both in magnitude and phase. As such, they can be connected, as shown in Figure 7.7, and replaced by one equivalent source V_F from neutral bus 0 to a references bus, denoted r. This equivalent source is also shown in the rake equivalent of Figure 7.6.

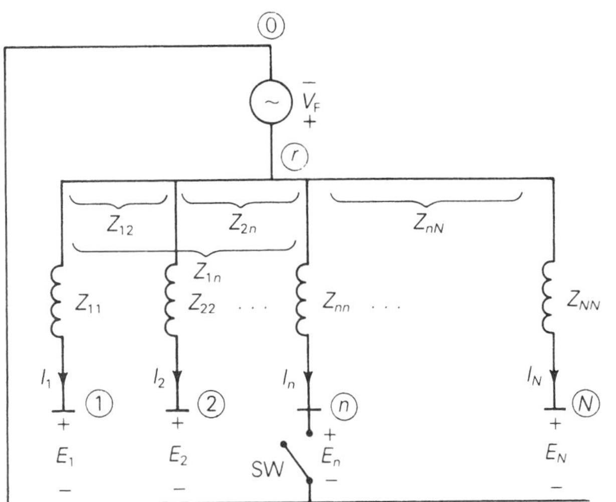

FIGURE 7.6

Bus impedance equivalent circuit (*rake equivalent*)

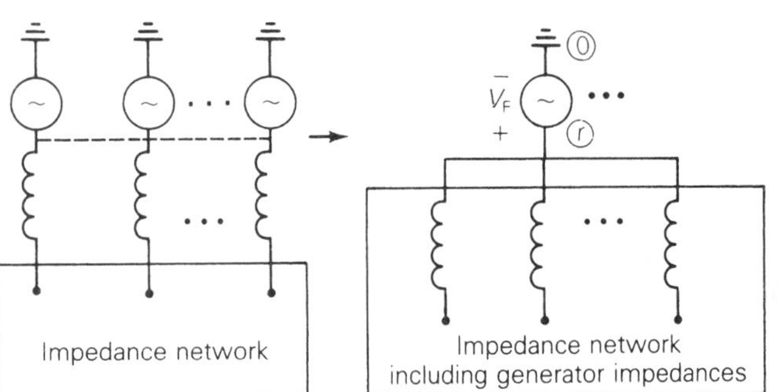

FIGURE 7.7

Parallel connection of unloaded synchronous machine internal-voltage sources

Using Z_{bus}, the fault currents in Figure 7.6 are given by

$$\begin{bmatrix} Z_{11} & Z_{12} & \cdots & Z_{1n} & \cdots & Z_{1N} \\ Z_{21} & Z_{22} & \cdots & Z_{2n} & \cdots & Z_{2N} \\ \vdots & & & & & \\ Z_{n1} & Z_{n2} & \cdots & Z_{nn} & \cdots & Z_{nN} \\ \vdots & & & & & \\ Z_{N1} & Z_{N2} & \cdots & Z_{Nn} & \cdots & Z_{NN} \end{bmatrix} \begin{bmatrix} I_1 \\ I_2 \\ \vdots \\ I_n \\ \vdots \\ I_N \end{bmatrix} = \begin{bmatrix} V_F - E_1 \\ V_F - E_2 \\ \vdots \\ V_F - E_n \\ \vdots \\ V_F - E_N \end{bmatrix} \tag{7.4.8}$$

where I_1, I_2, ... are the branch currents and $(V_F - E_1)$, $(V_F - E_2)$, ... are the voltages across the branches.

If switch SW in Figure 7.6 is open, all currents are zero and the voltage at each bus with respect to the neutral equals V_F. This corresponds to prefault conditions, neglecting prefault load currents. If switch SW is closed, corresponding to a short circuit at bus n, $E_n = 0$ and all currents except I_n remain zero. The fault current is

$I''_{Fn} = I_n = V_F/Z_{nn}$, which agrees with (7.4.5). This fault current also induces a voltage drop $Z_{kn}I_n = (Z_{kn}/Z_{nn})V_F$ across each branch k. The voltage at bus k with respect to the neutral then equals V_F minus this voltage drop, which agrees with (7.4.7).

As shown by Figure 7.6 as well as (7.4.5), subtransient fault currents throughout an N-bus system can be determined from the bus impedance matrix and the prefault voltage. $\boldsymbol{Z}_{bus}$ can be computed by first constructing $\boldsymbol{Y}_{bus}$, via nodal equations, and then inverting $\boldsymbol{Y}_{bus}$. Once $\boldsymbol{Z}_{bus}$ has been obtained, these fault currents are easily computed.

EXAMPLE 7.5

PowerWorld Simulator case Example 7_5 models the 5-bus power system whose oneline diagram is shown in Figure 6.2. Machine, line, and transformer data are given in Tables 7.3, 7.4, and 7.5. This system is initially unloaded. Prefault voltages at all the buses are 1.05 per unit. Use PowerWorld Simulator to determine the fault current for three-phase faults at each of the buses.

Bus	Machine Subtransient Reactance—X''_d
1	0.045
3	0.0225

TABLE 7.3

Synchronous machine data for the SYMMETRICAL SHORT CIRCUITS program*

*S_{base} = 100MVA
V_{base} = 15 kV at buses 1, 3
= 345 kV at buses 2, 4, 5

Bus-to-Bus	Equivalent Positive-Sequence Series Reactance (per unit)
2–4	0.1
2–5	0.05
4–5	0.025

TABLE 7.4

Line data for the SYMMETRICAL SHORT CIRCUITS program

Bus-to-Bus	Leakage Reactance—X (per unit)
1–5	0.02
3–4	0.01

TABLE 7.5

Transformer data for the SYMMETRICAL SHORT CIRCUITS program

SOLUTION

To fault a bus from the oneline, first right-click on the bus symbol to display the local menu, and then select **Fault** to display the Fault Analysis dialog (see Figure 7.8). In the list on the left side of the dialog, set the **Fault Definitions** to Single Fault; the Faulted Bus should be automatically selected to the selected bus location. Verify that the Fault Location field is Bus Fault and the Fault Type is 3-Phase Balanced (unbalanced faults are covered in Chapter 9). Then select **Calculate**, located in the upper left portion of the dialog, to determine the fault currents and voltages. The results are shown in the tables at the bottom of the dialog. Additionally, the fault currents can be shown on the oneline by selecting **Options** in the left list, and changing the Oneline Display Field value. Since with a three-phase fault the system remains balanced, the magnitudes of the a phase, b phase, and c phase values are identical. The 5×5 $\boldsymbol{Z}_{\text{bus}}$ matrix for this system is shown in Table 7.6, and the fault currents and bus voltages for faults at each of the buses are given in Table 7.7. Note that these fault currents are subtransient fault currents, since the machine reactance input data consist of direct axis subtransient reactances.

$$
j\begin{bmatrix}
0.0279725 & 0.0177025 & 0.0085125 & 0.0122975 & 0.020405 \\
0.0177025 & 0.0569525 & 0.0136475 & 0.019715 & 0.02557 \\
0.0085125 & 0.0136475 & 0.0182425 & 0.016353 & 0.012298 \\
0.0122975 & 0.019715 & 0.016353 & 0.0236 & 0.017763 \\
0.020405 & 0.02557 & 0.012298 & 0.017763 & 0.029475
\end{bmatrix}
$$

TABLE 7.6

Z_{bus} for Example 7.5

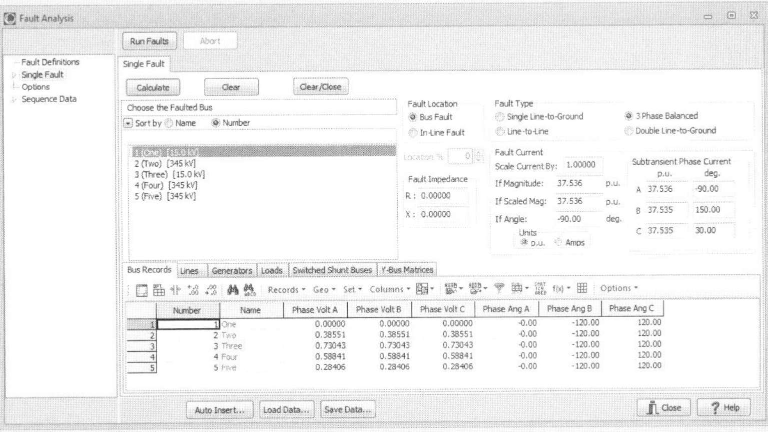

FIGURE 7.8

Fault Analysis Dialog for Example 7.5—fault at bus 1

(Continued)

Fault Bus	Fault Current (per unit)	Contributions to Fault Current		
		Gen Line or TRSF	Bus-to-Bus	Current (per unit)
1	37.536			
		G 1	GRND–1	23.332
		T 1	5–1	14.204
2	18.436			
		L 1	4–2	6.864
		L 2	5–2	11.572
3	57.556			
		G 2	GRND-3	46.668
		T 2	4–3	10.888
4	44.456			
		L 1	2–4	1.736
		L 3	5–4	10.412
		T 2	3–4	32.308
5	35.624			
		L 2	2–5	2.78
		L 3	4–5	16.688
		T 1	1–5	16.152

$V_F = 1.05$	Per-Unit Bus Voltage Magnitudes during the Fault				
Fault Bus	Bus 1	Bus 2	Bus 3	Bus 4	Bus 5
1	0.0000	0.7236	0.5600	0.5033	0.3231
2	0.3855	0.0000	0.2644	0.1736	0.1391
3	0.7304	0.7984	0.0000	0.3231	0.6119
4	0.5884	0.6865	0.1089	0.0000	0.4172
5	0.2840	0.5786	0.3422	0.2603	0.0000

TABLE 7.7

Fault currents and bus voltages for Example 7.5

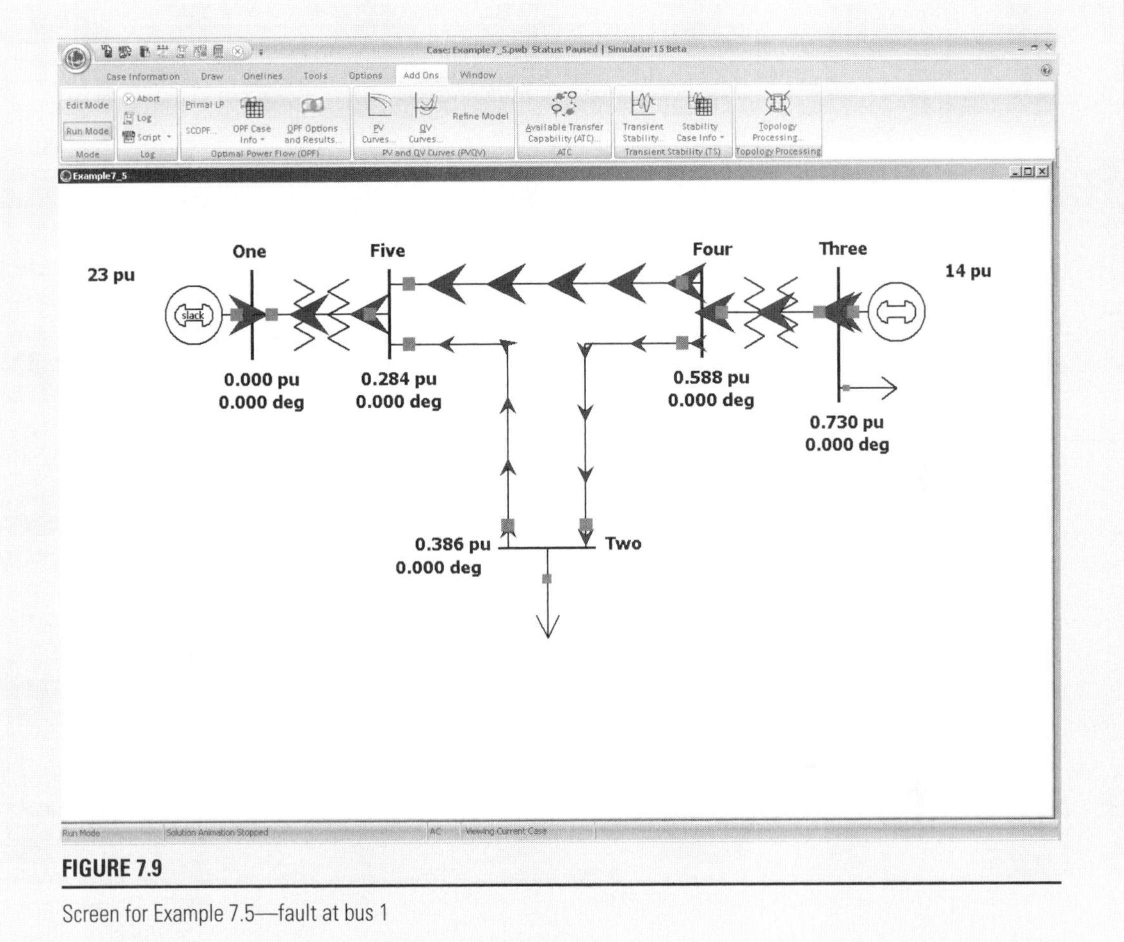

FIGURE 7.9

Screen for Example 7.5—fault at bus 1

EXAMPLE 7.6

Redo Example 7.5 with an additional line installed between buses 2 and 4. This line, whose reactance is 0.075 per unit, is not mutually coupled to any other line.

SOLUTION

The modified system is contained in PowerWorld Simulator case Example 7_6. The $\boldsymbol{Z}_{\text{bus}}$ along with the fault currents and bus voltages are shown in Tables 7.8 and 7.9. Notice the fault currents have increased, particularly at bus 2, since with the new line the total impedance between the faulted bus and the generators has decreased.

(Continued)

$$
j\begin{bmatrix}
0.027723 & 0.01597 & 0.00864 & 0.01248 & 0.02004 \\
0.01597 & 0.04501 & 0.01452 & 0.02097 & 0.02307 \\
0.00864 & 0.01452 & 0.01818 & 0.01626 & 0.01248 \\
0.01248 & 0.02097 & 0.01626 & 0.02349 & 0.01803 \\
0.02004 & 0.02307 & 0.01248 & 0.01803 & 0.02895
\end{bmatrix}
$$

TABLE 7.8

Z_{bus} for Example 7.6

Fault Bus	Fault Current (per unit)	Contributions to Fault Current: Gen Line or TRSF	Bus-to-Bus	Current (per unit)
1	37.872			
		G 1	GRND–1	23.332
		T 1	5–1	14.544
2	23.328			
		L 1	4–2	5.608
		L 2	5–2	10.24
		L 4	4–2	7.48
3	57.756			
		G 2	GRND–3	46.668
		T 2	4–3	11.088
4	44.704			
		L 1	2–4	1.128
		L 3	5–4	9.768
		L 4	2–4	1.504
		T 2	3–4	32.308
5	36.268			
		L 2	2–5	4.268
		L 3	4–5	15.848
		T 1	1–5	16.152

$V_F = 1.05$ Fault Bus	Per-Unit Bus Voltage Magnitudes during the Fault: Bus 1	Bus 2	Bus 3	Bus 4	Bus 5
1	0.0000	0.6775	0.5510	0.4921	0.3231
2	0.4451	0.0000	0.2117	0.1127	0.2133
3	0.7228	0.7114	0.0000	0.3231	0.5974
4	0.5773	0.5609	0.1109	0.0000	0.3962
5	0.2909	0.5119	0.3293	0.2442	0.0000

TABLE 7.9

Fault currents and bus voltages for Example 7.6

7.5 CIRCUIT BREAKER AND FUSE SELECTION

A SHORT CIRCUITS computer program may be utilized in power system design to select, set, and coordinate protective equipment such as circuit breakers, fuses, relays, and instrument transformers. In this section, basic principles of circuit breaker and fuse selection are discussed.

AC CIRCUIT BREAKERS

A *circuit breaker* is a mechanical switch capable of interrupting fault currents and of reclosing. When circuit-breaker contacts separate while carrying current, an arc forms. The breaker is designed to extinguish the arc by elongating and cooling it. The fact that ac arc current naturally passes through zero twice during its 60 Hz cycle aids the arc extinction process.

Circuit breakers are classified as *power* circuit breakers when they are intended for service in ac circuits above 1500 V and as *low-voltage* circuit breakers in ac circuits up to 1500 V. There are different types of circuit breakers depending on the medium—air, oil, SF_6 gas, or vacuum—in which the arc is elongated. Also, the arc can be elongated either by a magnetic force or by a blast of air.

Some circuit breakers are equipped with a high-speed automatic reclosing capability. Since most faults are temporary and self-clearing, reclosing is based on the idea that, if a circuit is deenergized for a short time, it is likely that whatever caused the fault has disintegrated and the ionized arc in the fault has dissipated.

When reclosing breakers are employed in EHV systems, standard practice is to reclose only once, approximately 15 to 50 cycles (depending on operating voltage) after the breaker interrupts the fault. If the fault persists and the EHV breaker recloses into it, the breaker reinterrupts the fault current and then "locks out," requiring operator resetting. Multiple-shot reclosing in EHV systems is not standard practice because transient stability (Chapter 11) may be compromised. However, for distribution systems (2.4 to 46 kV) where customer outages are of concern, standard reclosers are equipped for two or more reclosures.

For low-voltage applications, molded case circuit breakers with dual trip capability are available. There is a magnetic instantaneous trip for large fault currents above a specified threshold and a thermal trip with time delay for smaller fault currents.

Modern circuit-breaker standards are based on symmetrical interrupting current (Table 7.10). It is usually necessary to calculate only symmetrical fault current at a system location, and then select a breaker with a symmetrical interrupting capability equal to or above the calculated current. The breaker has the additional capability to interrupt the asymmetrical (or total) fault current if the dc offset is not too large.

Recall from Section 7.1 that the maximum asymmetry factor K ($\tau = 0$) is $\sqrt{3}$, which occurs at fault inception ($\tau = 0$). After fault inception, the dc fault current decays exponentially with time constant $T = (L/R) = (X/\omega R)$, and the asymmetry factor decreases. Power circuit breakers with a 2-cycle rated interruption time are designed for an asymmetrical interrupting capability up to 1.4 times their symmetrical interrupting capability, whereas slower circuit breakers have a lower asymmetrical interrupting capability.

Identification		Voltage		Rated Values			
				Insulation Level		Current	
				Rated Withstand Test Voltage		Rated Continuous Current at 60 Hz (Amperes, rms)	Rated Short-Circuit Current at Rated Max kV (kA, rms)
Nominal Voltage Class (kV, rms)	Nominal 3-Phase MVA Class	Rated Max Voltage (kV, rms)	Rated Voltage Range Factor (*K*)	Low Frequency (kV, rms)	Impulse (kV, Crest)		
Col 1	Col 2	Col 3	Col 4	Col 5	Col 6	Col 7	Col 8
14.4	250	15.5	2.67			600	8.9
14.4	500	15.5	1.29			1200	18
23	500	25.8	2.15			1200	11
34.5	1500	38	1.65			1200	22
46	1500	48.3	1.21			1200	17
69	2500	72.5	1.21			1200	19
115		121	1.0			1200	20
115		121	1.0			1600	40
115		121	1.0			2000	40
115		121	1.0			2000	63
115		121	1.0			3000	40
115		121	1.0			3000	63
138		145	1.0			1200	20
138	Not	145	1.0			1600	40
138		145	1.0			2000	40
138		145	1.0			2000	63
138		145	1.0			2000	80
138	Applicable	145	1.0			3000	40
138		145	1.0			3000	63
138		145	1.0			3000	80
161		169	1.0			1200	16
161		169	1.0			1600	31.5
161		169	1.0			2000	40
161		169	1.0			2000	50
230		242	1.0			1600	31.5
230		242	1.0			2000	31.5
230		242	1.0			3000	31.5
230		242	1.0			2000	40
230		242	1.0			3000	40
230		242	1.0			3000	63
345		362	1.0			2000	40
345		362	1.0			3000	40
500		550	1.0			2000	40
500		550	1.0			3000	40
700		765	1.0			2000	40
700		765	1.0			3000	40

TABLE 7.10

Preferred ratings for outdoor circuit breakers (symmetrical current basis of rating) [10]
Source: ANSI C 37.010-1999 (R 2005) Application Guide for AC High-Voltage Circuit Breakers Rated on a Symmetrical Current Basis. Reprinted with permission from IEEE.

Rated Values			Related Required Capabilities		
			Current Values		
			Max Symmetrical Interrupting Capability	3-Second Short-Time Current Carrying Capability	Closing and Latching Capability 1.6*K* Times Rated Short-Circuit Current
Rated Interrupting Time (Cycles)	Rated Permissible Tripping Delay (Seconds)	Rated Max Voltage Divided by *K* (*k* V, rms)	*K* Times Rated Short-Circuit Current (kA, rms)	(kA, rms)	(kA, rms)
Col 9	Col 10	Col 11	Col 12	Col 13	Col 14
5	2	5.8	24	24	38
5	2	12	23	23	37
5	2	12	24	24	38
5	2	23	36	36	58
5	2	40	21	21	33
5	2	60	23	23	37
3	1	121	20	20	32
3	1	121	40	40	64
3	1	121	40	40	64
3	1	121	63	63	101
3	1	121	40	40	64
3	1	121	63	63	101
3	1	145	20	20	32
3	1	145	40	40	64
3	1	145	40	40	64
3	1	145	63	63	101
3	1	145	80	80	128
3	1	145	40	40	64
3	1	145	63	63	101
3	1	145	80	80	128
3	1	169	16	16	26
3	1	169	31.5	31.5	50
3	1	169	40	40	64
3	1	169	50	50	80
3	1	242	31.5	31.5	50
3	1	242	31.5	31.5	50
3	1	242	31.5	31.5	50
3	1	242	40	40	64
3	1	242	40	40	64
3	1	242	63	63	101
3	1	362	40	40	64
3	1	362	40	40	64
2	1	550	40	40	64
2	1	550	40	40	64
2	1	765	40	40	64
2	1	765	40	40	64

A simplified method for breaker selection is called the "E/X simplified method" [1, 7]. The maximum symmetrical short-circuit current at the system location in question is calculated from the prefault voltage and system reactance characteristics using computer programs. Resistances, shunt admittances, nonrotating impedance loads, and prefault load currents are neglected. Then, if the X/R ratio at the system location is less than 15, a breaker with a symmetrical interrupting capability equal to or above the calculated current at the given operating voltage is satisfactory. However, if X/R is greater than 15, the dc offset may not have decayed to a sufficiently low value. In this case, a method for correcting the calculated fault current to account for dc and ac time constants as well as breaker speed can be used [10]. If X/R is unknown, the calculated fault current should not be greater than 80% of the breaker interrupting capability.

When selecting circuit breakers for generators, two cycle breakers are employed in practice, and the subtransient fault current is calculated; therefore subtransient machine reactances X_d'' are used in fault calculations. For synchronous motors, subtransient reactances X_d'' or transient reactances X_d' are used, depending on breaker speed. Also, induction motors can momentarily contribute to fault current. Large induction motors are usually modeled as sources in series with X_d'' or X_d', depending on breaker speed. Smaller induction motors (below 40 kW) are often neglected entirely.

Table 7.10 shows a schedule of preferred ratings for outdoor power circuit breakers. Some of the more important ratings shown are described next.

Voltage ratings

Rated maximum voltage: Designates the maximum rms line-to-line operating voltage. The breaker should be used in systems with an operating voltage less than or equal to this rating.

Rated low frequency withstand voltage: The maximum 60-Hz rms line-to-line voltage that the circuit breaker can withstand without insulation damage.

Rated impulse withstand voltage: The maximum crest voltage of a voltage pulse with standard rise and delay times that the breaker insulation can withstand.

Rated voltage range factor K: The range of voltage for which the symmetrical interrupting capability times the operating voltage is constant.

Current ratings

Rated continuous current: The maximum 60-Hz rms current that the breaker can carry continuously while it is in the closed position without overheating.

Rated short-circuit current: The maximum rms symmetrical current that the breaker can safely interrupt at rated maximum voltage.

Rated momentary current: The maximum rms asymmetrical current that the breaker can withstand while in the closed position without damage. Rated momentary current for standard breakers is 1.6 times the symmetrical interrupting capability.

Rated interrupting time: The time in cycles on a 60-Hz basis from the instant the trip coil is energized to the instant the fault current is cleared.

Rated interrupting MVA: For a three-phase circuit breaker, this is $\sqrt{3}$ times the rated maximum voltage in kV times the rated short-circuit current in

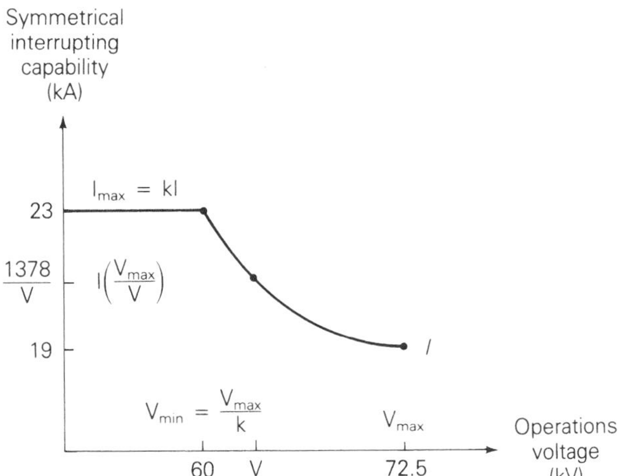

FIGURE 7.10

Symmetrical interrupting capability of a 69-kV class breaker

kA. It is more common to work with current and voltage ratings than with MVA rating.

As an example, the symmetrical interrupting capability of the 69-kV class breaker listed in Table 7.10 is plotted versus operating voltage in Figure 7.10. As shown, the symmetrical interrupting capability increases from its rated short-circuit current I = 19 kA at rated maximum voltage $V_{max} = 72.5$ kV up to $I_{max} = KI = (1.21)(19) = 23$ kA at an operating voltage $V_{max} = V_{max}/K = 72.5/1.21 = 60$ kV. At operating voltages V between V_{min} and V_{max}, the symmetrical interrupting capability is $I \times V_{max}/V = 1378/V$ kA. At operating voltages below V_{min}, the symmetrical interrupting capability remains at $I_{max} = 23$ kA.

Breakers of the 115 kV class and higher have a voltage range factor K = 1.0; that is, their symmetrical interrupting current capability remains constant.

EXAMPLE 7.7

Circuit breaker selection

The calculated symmetrical fault current is 17 kA at a three-phase bus where the operating voltage is 64 kV. The X/R ratio at the bus is unknown. Select a circuit breaker from Table 7.10 for this bus.

SOLUTION

The 69-kV-class breaker has a symmetrical interrupting capability $I(V_{max}/V) = 19(72.5/64) = 21.5$ kA at the operating voltage V = 64 kV. The calculated symmetrical fault current, 17 kA, is less than 80% of this capability (less than $0.80 \times 21.5 = 17.2$ kA), which is a requirement when X/R is unknown. Therefore, select the 69-kV-class breaker from Table 7.10.

FUSES

Figure 7.11(a) shows a cutaway view of a fuse, which is one of the simplest overcurrent devices. The fuse consists of a metal "fusible" link or links encapsulated in a tube, packed in filler material, and connected to contact terminals. Silver is a typical link metal, and sand is a typical filler material.

During normal operation, when the fuse is operating below its continuous current rating, the electrical resistance of the link is so low that it simply acts as a conductor. If an overload current from one to about six times its continuous current rating occurs and persists for more than a short interval of time, the temperature of the link eventually reaches a level that causes a restricted segment of the link to melt. As shown in Figure 7.11(b), a gap is then formed and an electric arc is established. As the arc causes the link metal to burn back, the gap width increases. The resistance of the arc eventually reaches such a high level that the arc cannot be sustained and it is extinguished, as in Figure 7.11(c). The current flow within the fuse is then completely cut off.

FIGURE 7.11 Typical fuse

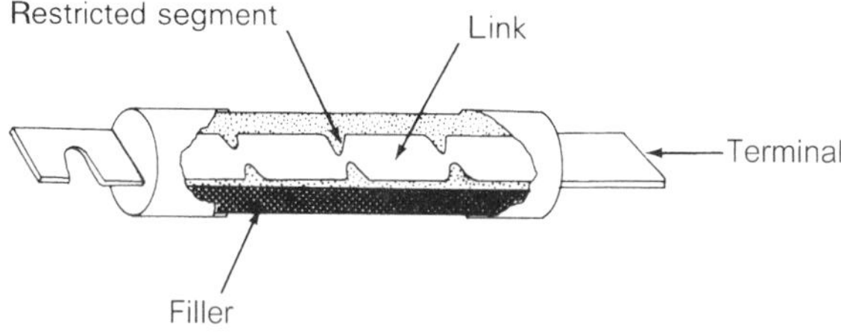

(a) Cutaway view

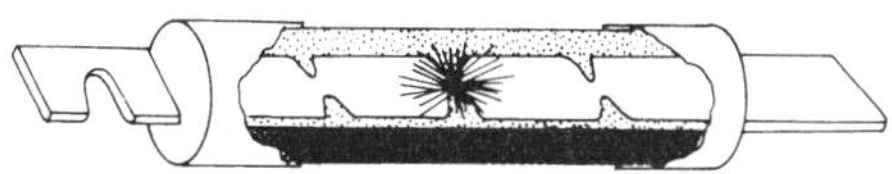

(b) The link melts and an arc is established under sustained overload current

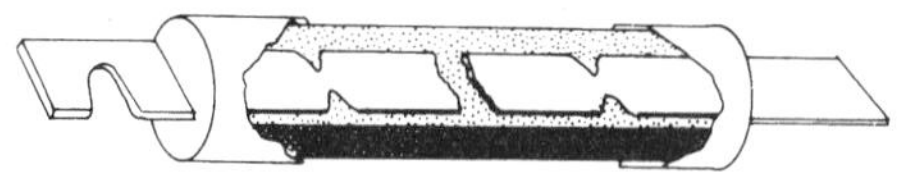

(c) The "open" link after clearing the overload current

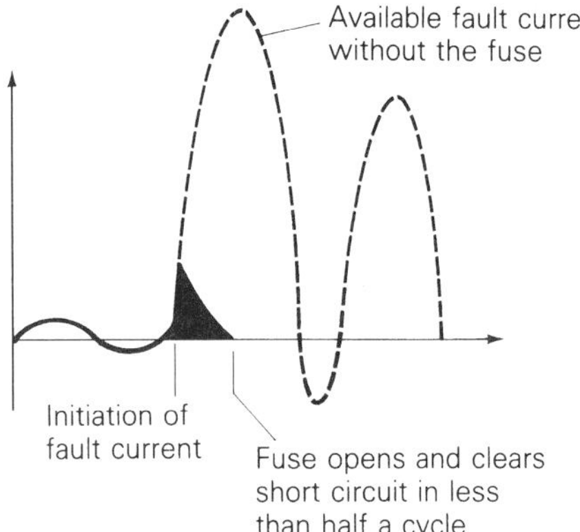

FIGURE 7.12

Operation of a current-limiting fuse

If the fuse is subjected to fault currents higher than about six times its continuous current rating, several restricted segments melt simultaneously, resulting in rapid arc suppression and fault clearing. Arc suppression is accelerated by the filler material in the fuse.

Many modern fuses are current limiting. As shown in Figure 7.12, a current-limiting fuse has such a high speed of response that it cuts off a high fault current in less than a half cycle—before it can build up to its full peak value. By limiting fault currents, these fuses permit the use of motors, transformers, conductors, and bus structures that could not otherwise withstand the destructive forces of high fault currents.

Fuse specification is normally based on the following four factors.

1. *Voltage rating*. This rms voltage determines the ability of a fuse to suppress the internal arc that occurs after the fuse link melts. A blown fuse should be able to withstand its voltage rating. Most low-voltage fuses have 250 or 600 V ratings. Ratings of medium-voltage fuses range from 2.4 to 34.5 kV.
2. *Continuous current rating*. The fuse should carry this rms current indefinitely, without melting and clearing.
3. *Interrupting current rating*. This is the largest rms asymmetrical current that the fuse can safely interrupt. Most modern, low-voltage current-limiting fuses have a 200-kA interrupting rating. Standard interrupting ratings for medium-voltage current-limiting fuses include 65, 80, and 100 kA.
4. *Time response*. The melting and clearing time of a fuse depends on the magnitude of the overcurrent or fault current and is usually specified by a "time-current" curve. Figure 7.13 shows the time-current curve of a 15.5-kV, 100-A (continuous) current-limiting fuse. As shown, the fuse link melts within 2 s and clears within 5 s for a 500-A current. For a 5 kA current, the fuse link melts in less than 0.01 s and clears within 0.015 s.

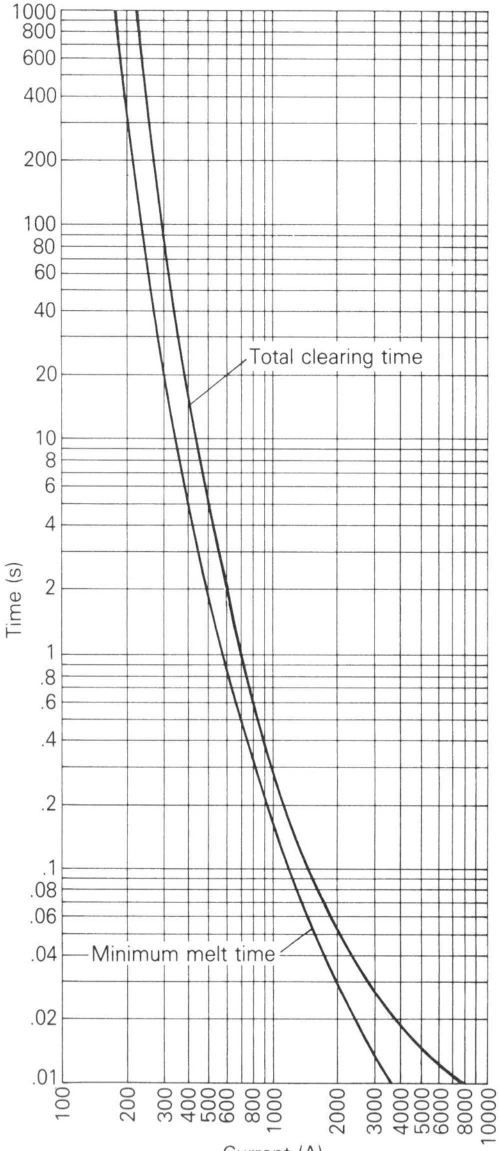

FIGURE 7.13 Time-current curves for a 15.5-kV, 100-A current-limiting fuse

It is usually a simple matter to coordinate fuses in a power circuit such that only the fuse closest to the fault opens the circuit. In a radial circuit, fuses with larger continuous-current ratings are located closer to the source, such that the fuse closest to the fault clears before other, upstream fuses melt.

Fuses are inexpensive, fast operating, easily coordinated, and reliable, and they do not require protective relays or instrument transformers. Their chief disadvantage is that the fuse or the fuse link must be manually replaced after it melts. They are basically one-shot devices that are, for example, incapable of high-speed reclosing.

MULTIPLE CHOICE QUESTIONS

SECTION 7.1

7.1 The asymmetrical short-circuit current in series R–L circuit for a simulated solid or "bolted fault" can be considered as a combination of symmetrical (ac) component that is a _____, and dc-offset current that decays _____ and depends on _____.

7.2 Even though the fault current is not symmetrical and not strictly periodic, the rms asymmetrical fault current is computed as the rms ac fault current times an "asymmetry factor," which is a function of _____.

SECTION 7.2

7.3 The amplitude of the sinusoidal symmetrical ac component of the three-phase short-circuit current of an unloaded synchronous machine decreases from a high initial value to a lower steady-state value, going through the stages of _____ and _____ periods.

7.4 The duration of subtransient fault current is dictated by _____ time constant and that of transient fault current is dictated by _____ time constant.

7.5 The reactance that plays a role under steady-state operation of a synchronous machine is called _____.

7.6 The dc-offset component of the three-phase short-circuit current of an unloaded synchronous machine is different in the three phases and its exponential decay is dictated by _____.

SECTION 7.3

7.7 Generally, in power-system short-circuit studies, for calculating subtransient fault currents, transformers are represented by their _____, transmission lines by their equivalent _____, and synchronous machines by _____ behind their subtransient reactances.

7.8 In power-system fault studies, all nonrotating impedance loads are usually neglected.

(a) True (b) False

7.9 Can superposition be applied in power-system short-circuit studies for calculating fault currents?

(a) Yes (b) No

7.10 Before proceeding with per-unit fault current calculations, based on the single-line diagram of the power system, a positive-sequence equivalent circuit is set up on a chosen base system.

(a) True (b) False

SECTION 7.4

7.11 The inverse of the bus-admittance matrix is called a _____ matrix.

7.12 For a power system, modeled by its positive-sequence network, both bus-admittance matrix and bus-impedance matrix are symmetric.
(a) True (b) False

7.13 The bus-impedance equivalent circuit can be represented in the form of a "rake" with the diagonal elements, which are _____, and the non-diagonal (off-diagonal) elements, which are _____.

SECTION 7.5

7.14 A circuit breaker is designed to extinguish the arc by _____.

7.15 Power-circuit breakers are intended for service in the ac circuit above _____ V.

7.16 In circuit breakers, besides air or vacuum, what gaseous medium, in which the arc is elongated, is used?

7.17 Oil can be used as a medium to extinguish the arc in circuit breakers.
(a) True (b) False

7.18 Besides a blast of air/gas, the arc in a circuit breaker can be elongated by _____.

7.19 For distribution systems, standard reclosers are equipped for two or more reclosures, whereas multiple-shot reclosing in EHV systems is not a standard practice.
(a) True (b) False

7.20 Breakers of the 115 kV class and higher have a voltage range factor K = _____, such that their symmetrical interrupting current capability remains constant.

7.21 A typical fusible link metal in fuses is _____, and a typical filler material is _____.

7.22 The melting and clearing time of a current-limiting fuse is usually specified by a _____ curve.

PROBLEMS

SECTION 7.1

7.1 In the circuit of Figure 7.1, $V = 277$ volts, $L = 2$ mH, $R = 0.4\ \Omega$, and $\omega = 2\pi 60$ rad/s. Determine (a) the rms symmetrical fault current; (b) the rms asymmetrical fault current at the instant the switch closes, assuming maximum dc offset; (c) the rms asymmetrical fault current five cycles after the switch closes, assuming maximum dc offset; and (d) the dc offset as a function of time if the switch closes when the instantaneous source voltage is 300 volts.

7.2 Repeat Example 7.1 with V = 4 kV, X = 2 Ω, and R = 1 Ω

7.3 In the circuit of Figure 7.1, let R = 0.125 Ω., L = 10 mH, and the source voltage is $e(t) = 151 \sin(377t + \alpha)$ V. Determine the current response after closing the switch for the following cases: (a) no dc offset or (b) maximum dc offset. Sketch the current waveform up to t = 0.10 s corresponding to parts (a) and (b).

7.4 Consider the expression for $i(t)$ given by

$$i(t) = \sqrt{2}I_{rms}[\sin(\omega t - \theta_z) + \sin \theta_z . e^{-(\omega R/X)t}]$$

where $\theta_z = \tan^{-1}(\omega L/R)$.

(a) For (X/R) equal to zero and infinity, plot $i(t)$ as a function of (ωt).

(b) Comment on the dc offset of the fault current waveforms.

(c) Find the asymmetrical current factor and the time of peak, t_p, in milliseconds, for (X/R) ratios of zero and infinity.

7.5 If the source impedance at a 13.2-kV distribution substation bus is (0.5 + j1.5) Ω per phase, compute the rms and maximum peak instantaneous value of the fault current for a balanced three-phase fault. For the system (X/R) ratio of 3.0, the asymmetrical factor is 1.9495 and the time of peak is 7.1 ms (see Problem 7.4). Comment on the withstanding peak current capability to which all substation electrical equipment need to be designed.

SECTION 7.2

7.6 A 1000-MVA, 20-kV, 60-Hz, three-phase generator is connected through a 1000-MVA, 20-kV, Δ/345-kV, Y transformer to a 345-kV circuit breaker and a 345-kV transmission line. The generator reactances are $X''_d = 0.17$, $X'_d = 0.30$, and $X_d = 1.5$ per unit, and its time constants are $T''_d = 0.05$, $T'_d = 1.0$, and $T_A = 0.10$ s. The transformer series reactance is 0.10 per unit; transformer losses and exciting current are neglected. A three-phase short-circuit occurs on the line side of the circuit breaker when the generator is operated at rated terminal voltage and at no-load. The breaker interrupts the fault three cycles after fault inception. Determine (a) the subtransient current through the breaker in per-unit and in kA rms and (b) the rms asymmetrical fault current the breaker interrupts, assuming maximum dc offset. Neglect the effect of the transformer on the time constants.

7.7 For Problem 7.6, determine (a) the instantaneous symmetrical fault current in kA in phase a of the generator as a function of time, assuming maximum dc offset occurs in this generator phase, and (b) the maximum dc offset current in kA as a function of time that can occur in any one generator phase.

7.8 A 300-MVA, 13.8-kV, three-phase, 60-Hz, Y-connected synchronous generator is adjusted to produce rated voltage on open circuit. A balanced three-phase fault is applied to the terminals at t = 0. After analyzing the raw data, the symmetrical transient current is obtained as

$$i_{ac}(t) = 10(1 + e^{-t/200} + 6e^{-t/15}) \cdot \sin\left(120\pi t - \frac{\pi}{4}\right) \text{ A}$$

(a) Sketch $i_{ac}(t)$ as a function of time for $0 \le t \le 500$ ms. (b) Determine X_d'' and X_d in per unit based on the machine ratings.

7.9 Two identical synchronous machines, each rated 60 MVA and 15 kV with a subtransient reactance of 0.1 p.u., are connected through a line of reactance 0.1 p.u. on the base of the machine rating. One machine is acting as a synchronous generator, while the other is working as a motor drawing 40 MW at 0.8 p.f. leading with a terminal voltage of 14.5 kV, when a symmetrical three-phase fault occurs at the motor terminals. Determine the subtransient currents in the generator, the motor, and the fault by using the internal voltages of the machines. Choose a base of 60 MVA and 15 kV in the generator circuit.

SECTION 7.3

7.10 Recalculate the subtransient current through the breaker in Problem 7.6 if the generator is initially delivering rated MVA at 0.80 p.f. lagging and at rated terminal voltage.

7.11 Solve Example 7.3 parts (a) and (c) without using the superposition principle. First calculate the internal machine voltages E_g'' and E_m'' using the prefault load current. Then determine the subtransient fault, generator, and motor currents directly from Figure 7.4(a). Compare your answers with those of Example 7.3.

7.12 Equipment ratings for the four-bus power system shown in Figure 7.14 are as follows:

Generator Gl:	500 MVA, 13.8 kV, X″ = 0.20 per unit
Generator G2:	750 MVA, 18 kV, X″ = 0.18 per unit
Generator G3:	1000 MVA, 20 kV, X″ = 0.17 per unit
Transformer Tl:	500 MVA, 13.8 Δ/500 Y kV, X = 0.12 per unit
Transformer T2:	750 MVA, 18 Δ/500 Y kV, X = 0.10 per unit
Transformer T3:	1000 MVA, 20 Δ/500 Y kV, X = 0.10 per unit
Each 500 kV line:	$X_1 = 50\ \Omega$.

A three-phase short circuit occurs at bus 1, where the prefault voltage is 525 kV. Prefault load current is neglected. Draw the positive-sequence reactance diagram in per unit on a 1000-MVA, 20-kV base in the zone of generator G3. Determine (a) the Thévenin reactance in per unit at the fault, (b) the subtransient fault current in per unit and in kA rms, and (c) contributions to the fault current from generator Gl and from line 1–2.

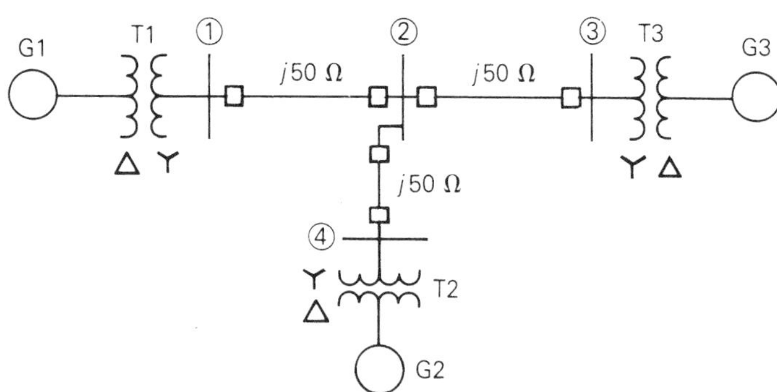

FIGURE 7.14

Problems 7.12, 7.13, 7.19, 7.24, 7.25, and 7.26

7.13 For the power system given in Problem 7.12, a three-phase short circuit occurs at bus 2, where the prefault voltage is 525 kV. Prefault load current is neglected. Determine the (a) Thévenin equivalent at the fault, (b) subtransient fault current in per unit and in kA rms, and (c) contributions to the fault from lines 1–2, 2–3, and 2–4.

7.14 Equipment ratings for the five-bus power system shown in Figure 7.15 are as follows:

Generator Gl:	50 MVA, 12 kV, X″ = 0.2 per unit
Generator G2:	100 MVA, 15 kV, X″ = 0.2 per unit
Transformer Tl:	50 MVA, 10 kV Y/138 kV Y, X = 0.10 per unit
Transformer T2:	100 MVA, 15 kV Δ/138 kV Y, X = 0.10 per unit
Each 138-kV line:	$X_1 = 40\ \Omega$.

A three-phase short circuit occurs at bus 5, where the prefault voltage is 15 kV. Prefault load current is neglected. (a) Draw the positive-sequence reactance diagram in per unit on a 100-MVA, 15-kV base in the zone of generator G2. Determine (b) the Thévenin equivalent at the fault, (c) the subtransient fault current in per unit and in kA rms, and (d) contributions to the fault from generator G2 and from transformer T2.

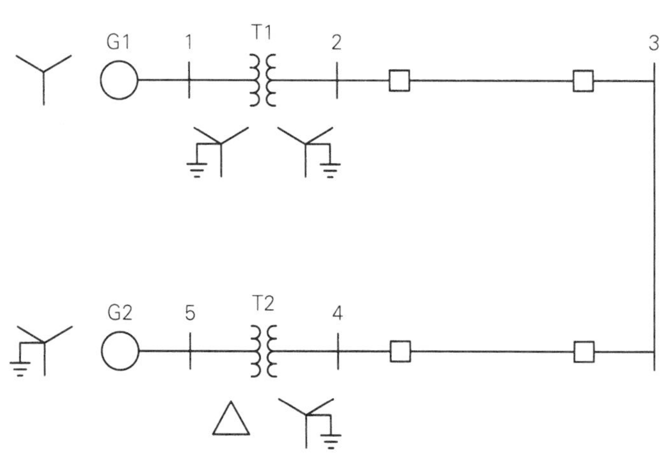

FIGURE 7.15

Problems 7.14, 7.15, 7.20

7.15 For the power system given in Problem 7.14, a three-phase short circuit occurs at bus 4, where the prefault voltage is 138 kV. Prefault load current is neglected. Determine (a) the Thévenin equivalent at the fault, (b) the subtransient fault current in per unit and in kA rms, and (c) contributions to the fault from transformer T2 and from line 3–4.

7.16 In the system shown in Figure 7.16, a three-phase short circuit occurs at point F. Assume that prefault currents are zero and that the generators are operating at rated voltage. Determine the fault current.

FIGURE 7.16

Problem 7.16

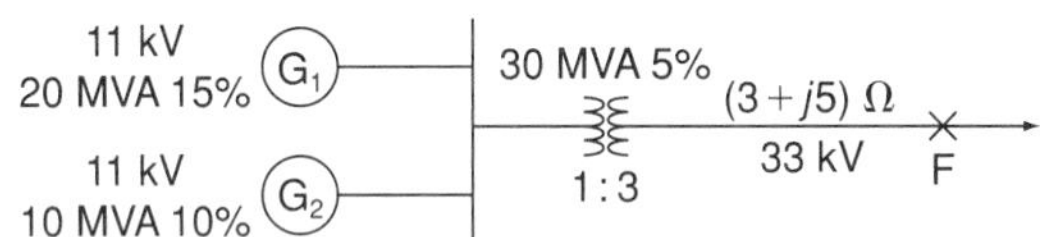

7.17 A three-phase short circuit occurs at the generator bus (bus 1) for the system shown in Figure 7.17. Neglecting prefault currents and assuming that the generator is operating at its rated voltage, determine the subtransient fault current using superposition.

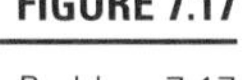

FIGURE 7.17

Problem 7.17

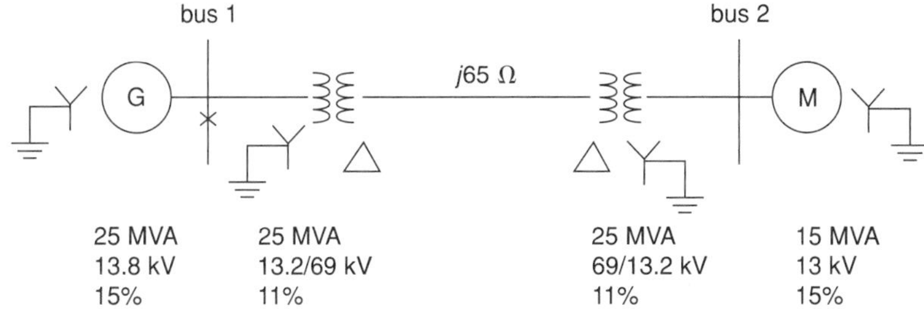

SECTION 7.4

7.18 The bus impedance matrix for a three-bus power system is

$$\boldsymbol{Z}_{\text{bus}} = j\begin{bmatrix} 0.12 & 0.08 & 0.04 \\ 0.08 & 0.12 & 0.06 \\ 0.04 & 0.06 & 0.08 \end{bmatrix} \quad \text{per unit}$$

where subtransient reactances were used to compute $\boldsymbol{Z}_{\text{bus}}$. Prefault voltage is 1.0 per unit and prefault current is neglected. (a) Draw the bus impedance matrix equivalent circuit (rake equivalent). Identify the per-unit self- and mutual impedances as well as the prefault voltage in the circuit. (b) A three-phase short circuit occurs at bus 2. Determine the subtransient fault current and the voltages at buses 1, 2, and 3 during the fault.

(c) Repeat for the case of

$$\boldsymbol{Z}_{\text{bus}} = j\begin{bmatrix} 0.4 & 0.1 & 0.3 \\ 0.1 & 0.8 & 0.5 \\ 0.3 & 0.5 & 1.2 \end{bmatrix} \quad \text{per unit}$$

7.19 Determine $\boldsymbol{Y}_{\text{bus}}$ in per unit for the circuit in Problem 7.12. Then invert $\boldsymbol{Y}_{\text{bus}}$ to obtain $\boldsymbol{Z}_{\text{bus}}$.

7.20 Determine $\boldsymbol{Y}_{\text{bus}}$ in per unit for the circuit in Problem 7.14. Then invert $\boldsymbol{Y}_{\text{bus}}$ to obtain $\boldsymbol{Z}_{\text{bus}}$.

7.21 Figure 7.18 shows a system reactance diagram. (a) Draw the admittance diagram for the system by using source transformations. (b) Find the bus admittance matrix $\boldsymbol{Y}_{\text{bus}}$. (c) Find the bus impedance $\boldsymbol{Z}_{\text{bus}}$ matrix by inverting $\boldsymbol{Y}_{\text{bus}}$.

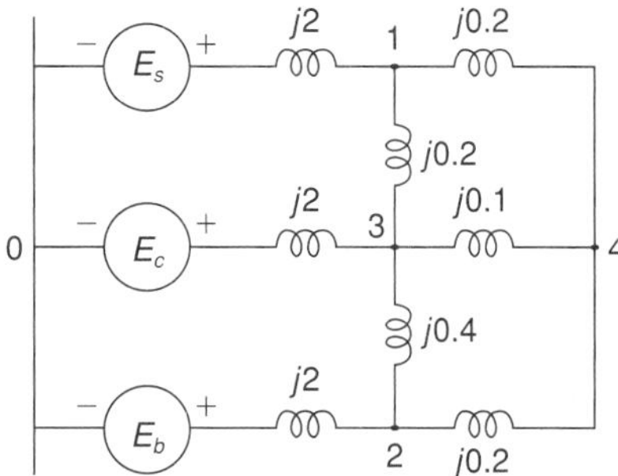

FIGURE 7.18

Problem 7.21

7.22 For the network shown in Figure 7.19, impedances labeled 1 through 6 are in per unit. (a) Determine $\boldsymbol{Y}_{\text{bus}}$, preserving all buses. (b) Using MATLAB or a similar computer program, invert $\boldsymbol{Y}_{\text{bus}}$ to obtain $\boldsymbol{Z}_{\text{bus}}$.

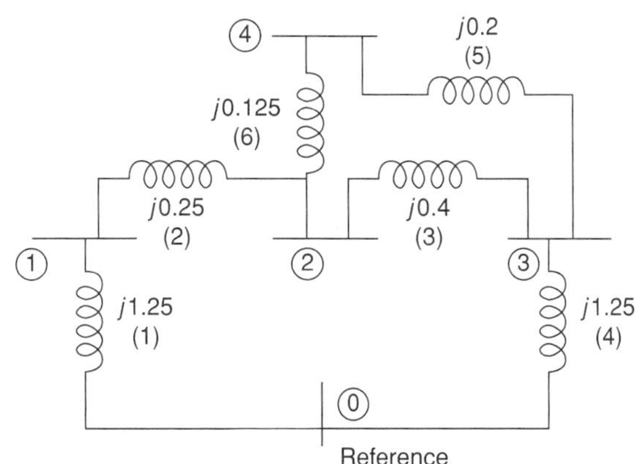

FIGURE 7.19

Problem 7.22

7.23 A single-line diagram of a four-bus system is shown in Figure 7.20, for which $\boldsymbol{Z}_{\text{bus}}$ is given below:

$$\boldsymbol{Z}_{\text{bus}} = j\begin{bmatrix} 0.25 & 0.2 & 0.16 & 0.14 \\ 0.2 & 0.23 & 0.15 & 0.151 \\ 0.16 & 0.15 & 0.196 & 0.1 \\ 0.14 & 0.151 & 0.1 & 0.195 \end{bmatrix} \text{ per unit}$$

Let a three-phase fault occur at bus 2 of the network.

(a) Calculate the initial symmetrical rms current in the fault.

(b) Determine the voltages during the fault at buses 1, 3, and 4.

(c) Compute the fault currents contributed to bus 2 by the adjacent unfaulted buses 1, 3, and 4.

(d) Find the current flow in the line from bus 3 to bus 1. Assume the prefault voltage V_f at bus 2 to be $1\underline{/0°}$ p.u., and neglect all prefault currents.

FIGURE 7.20

Single-line diagram for Problem 7.23

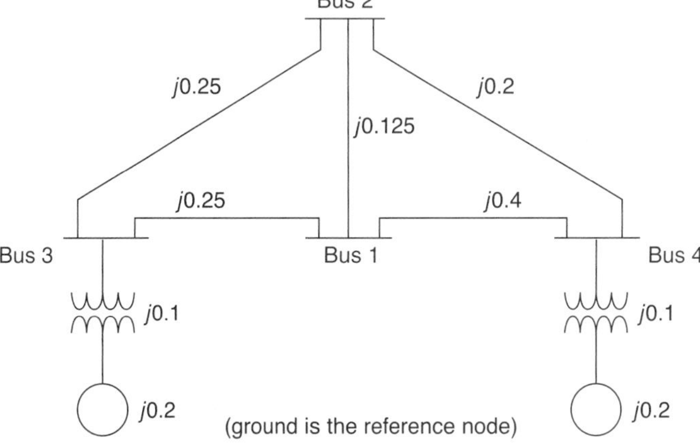

PW **7.24** PowerWorld Simulator case Problem 7_24 models the system shown in Figure 7.14 with all data on a 1000 MVA base. Using PowerWorld Simulator, determine the current supplied by each generator and the per-unit bus voltage magnitudes at each bus for a fault at bus 3.

PW **7.25** Repeat Problem 7.24, except place the fault at bus 4.

PW **7.26** Repeat Problem 7.24, except place the fault midway between buses 2 and 3. Determining the values for line faults requires that the line be split with a fictitious bus added at the point of the fault. The original line's impedance is then allocated to the two new lines based on the fault location, which is 50% each for this problem. Fault calculations are then the same as for a bus fault. This is done automatically in PowerWorld Simulator by first right-clicking on a line, and then selecting Fault. The Fault dialog appears as before, except now the fault type is changed to In-Line Fault.

Set the location percentage field to 50% to model a fault midway between buses 2 and 4.

PW **7.27** One technique for limiting fault current is to place reactance in series with the generators. Such reactance can be modeled in PowerWorld Simulator by increasing the value of the generator's positive sequence internal impedance. For the Problem 7.24 case, how much per-unit reactance must be added to G2 to limit its maximum fault current to 2.5 per unit for all three-phase bus faults? Where is the location of the most severe bus fault?

PW **7.28** Using PowerWorld Simulator case Example 6_13, determine the per-unit current and actual current in amps supplied by each of the generators for a fault at the POPLAR69 bus. During the fault, what percentage of the system buses have voltage magnitudes below 0.75 per unit?

PW **7.29** Repeat Problem 7.28, except place the fault at the REDBUD69 bus.

PW **7.30** Using PowerWorld Simulator case Example 7_5, open the line connecting buses 4 and 5. Then, determine the per unit current supplied by the generator at bus 3 due a fault at bus 2.

SECTION 7.5

7.31 A three-phase circuit breaker has a 15.5-kV rated maximum voltage, 9.0-kA rated short-circuit current, and a 2.50-rated voltage range factor. (a) Determine the symmetrical interrupting capability at 10-kV and 5-kV operating voltages. (b) Can this breaker be safely installed at a three-phase bus where the symmetrical fault current is 10 kA, the operating voltage is 13.8 kV, and the (X/R) ratio is 12?

7.32 A 345-kV, three-phase transmission line has a 2.2-kA continuous current rating and a 2.5-kA maximum short-time overload rating with a 356-kV maximum operating voltage. The maximum symmetrical fault current on the line is 30 kA. Select a circuit breaker for this line from Table 7.10.

7.33 A 69-kV circuit breaker has a voltage range factor K = 1.25, a continuous current rating of 1200 A, and a rated short-circuit current of 19,000 A at the maximum rated voltage of 72.5 kV. Determine the maximum symmetrical interrupting capability of the breaker. Also, explain its significance at lower operating voltages.

7.34 As shown in Figure 7.21, a 25-MVA, 13.8-kV, 60-Hz, synchronous generator with $X_d'' = 0.15$ per unit is connected through a transformer to a bus that supplies four identical motors. The rating of the three-phase transformer is 25 MVA and 13.8/6.9 kV with a leakage reactance of 0.1 per unit. Each motor has a subtransient reactance $X_d'' = 0.2$ per unit on a base of 5 MVA and 6.9 kV. A three-phase fault occurs at point P, when the bus voltage at the motors is 6.9 kV. Determine (a) the subtransient fault current, (b) the subtransient current through breaker A, and (c) the symmetrical short-circuit interrupting current (as defined for circuit breaker applications) in the fault and in breaker A.

FIGURE 7.21

Problem 7.34

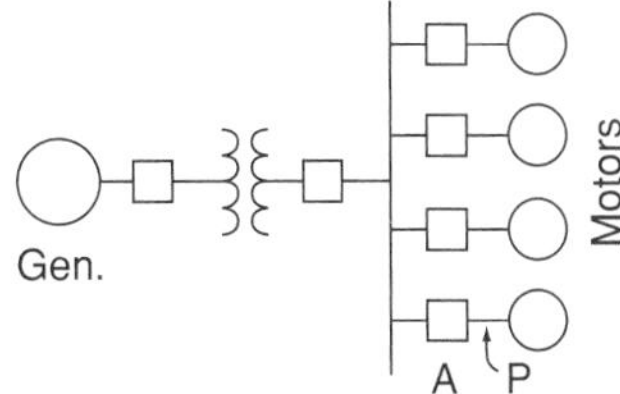

CASE STUDIES QUESTIONS

a. What are the four main types of wind-turbine generators (WTG)? How do WTGs differ from conventional generators?

b. Which type of WTG can produce the largest three-phase short-circuit current?

c. For which type of WTG can the short-circuit current be controlled? Why?

DESIGN PROJECT 3 (*CONTINUED*): POWER FLOW/ SHORT CIRCUITS

Additional time given: 3 weeks

Additional time required: 10 hours

This is a continuation of Design Project 3. Assignments 1 and 2 are given in Chapter 6.

Assignment 3: Symmetrical Short Circuits

For the single-line diagram that you have been assigned (Figure 6.13 or 6.14), convert the positive-sequence reactance data to per unit using the given base quantities. For synchronous machines, use subtransient reactance. Then using PowerWorld Simulator, create the machine, transmission line, and transformer input data files. Next, run the program to compute subtransient fault currents for a bolted three-phase-to-ground fault at bus 1, then at bus 2, then at bus 3, and so on. Also compute bus voltages during the faults and the positive-sequence bus impedance matrix. Assume 1.0 per-unit prefault voltage. Neglect prefault load currents and all losses.

Your output for this assignment consists of three input data files and three output data (fault currents, bus voltages, and the bus impedance matrix) files.

This project continues in Chapter 9.

REFERENCES

1. Westinghouse Electric Corporation, *Electrical Transmission and Distribution Reference Book*, 4th ed. (East Pittsburgh, PA: 1964).
2. E. W. Kimbark, *Power System Stability, Synchronous Machines*, vol. 3 (New York: Wiley, 1956).
3. A. E. Fitzgerald, C. Kingsley, and S. Umans, *Electric Machinery*, 5th ed. (New York: McGraw-Hill, 1990).
4. M. S. Sarma, *Electric Machines,* 2nd ed. (Boston: PWS Publishing, 1994).
5. J. R. Neuenswander, *Modern Power Systems* (New York: Intext Educational Publishers, 1971).
6. H. E. Brown, *Solution of Large Networks by Matrix Methods* (New York: Wiley, 1975).
7. G. N. Lester, "High Voltage Circuit Breaker Standards in the USA—Past, Present and Future," *IEEE Transactions PAS*, vol. PAS-93 (1974): pp. 590–600.
8. W. D. Stevenson, Jr., *Elements of Power System Analysis*, 4th ed. (New York: McGraw-Hill, 1982).
9. C. A. Gross, *Power System Analysis* (New York: Wiley, 1979).
10. *Application Guide for AC High-Voltage Circuit Breakers Rated on a Symmetrical Current Basis*, ANSI C 37.010 (New York: American National Standards Institute, 1999/R2005).
11. E. Muljadi and V. Grevorgian, "Short-Circuit Modeling of a Wind Power Plant," Conference Paper NREL/CP-5500-50632, *Power & Energy Society General Meeting*, July 24–29, 2011.

8 Symmetrical Components

Generator stator showing completed windings for a 757-MVA, 3600-RPM, 60-Hz synchronous generator (Courtesy of General Electric.)

The method of symmetrical components, first developed by C. L. Fortescue in 1918, is a powerful technique for analyzing unbalanced three-phase systems. Fortescue defined a linear transformation from phase components to a new set of components called *symmetrical components.* The advantage of this transformation is that for balanced three-phase networks the equivalent circuits obtained for the symmetrical components, called *sequence networks*, are separated into three uncoupled networks. Furthermore, for unbalanced three-phase systems, the three sequence networks are connected only at points of unbalance. As a result, sequence networks for many cases of unbalanced three-phase systems are relatively easy to analyze.

The symmetrical component method is basically a modeling technique that permits systematic analysis and design of three-phase systems. Decoupling a detailed three-phase network into three simpler sequence networks reveals complicated phenomena in more simplistic terms. Sequence network results then can be superposed to obtain three-phase network results. As an example, the application of symmetrical components to unsymmetrical short-circuit studies (see Chapter 9) is indispensable.

The objective of this chapter is to introduce the concept of symmetrical components in order to lay a foundation and provide a framework for later chapters covering both equipment models as well as power system analysis and design methods. Section 8.1 defines symmetrical components. In Sections 8.2 through 8.7, sequence networks of loads, series impedances, transmission lines, rotating machines, and transformers are presented. Complex power in sequence networks is presented in Section 8.8. Although Fortescue's original work is valid for polyphase systems with n phases, this chapter considers only three-phase systems.

CASE STUDY

The following case study traces the development of gas-insulated substations and switchgear (GIS) since GIS technology originated in 1936. GIS advantages include enhanced reliability, compact modular design, reduced maintenance and cost, prolonged life, and advanced monitoring capabilities. GIS circuit breakers are now available at voltages up to 1100 kV and interrupting currents up to 63 kA. Sulfur hexafluoride (SF_6) is the most commonly used gas for electrical insulation in GIS. This case study presents an overview of the environmental impacts of switchgear, where SF_6 as a greenhouse gas is an ongoing environmental concern [4].

Technological Progress in High-Voltage Gas-Insulated Substations

Ibrahim A. Metwally

In the last two decades, the evolutionary development of gas-insulated substations/switchgear (GIS) has resulted in higher integration of a number of new technologies to enhance performance and reliability by reducing defects, having more compact designs, and reducing maintenance intervals and costs. Incremental improvements are continuing in interrupter technology, such as self-extinguishing features at medium voltage (MV) and resistance interruption at extra- and ultra-high voltages (EHV and UHV). In addition, sulfur hexafluoride (SF_6) gas technology for circuit breakers, zinc oxide (ZnO) for arresters, radio communication for condition monitoring, and a choice of porcelain or polymer composite for the full range of equipment are also some of the technologies integrated or innovated by GIS manufacturers in recent years. Recently, ac GIS ratings have reached up to 1100-kV rated voltage and 50-kA_{rms} rated short-circuit breaking current. In addition, 1200-kV ac GIS are going to be visible very soon. Moreover, 500-kV dc GIS for dc transmission systems have become available.

GIS Construction

GIS is commonly used to designate gas-insulated, metal-clad electrical switchgear. GIS includes air entrance bushings, power cable connections, transformer connections, busbars, circuit breakers (CB), bus and cable isolators, earthing switches, current and voltage transformers "measuring devices" (CT and VT), and surge arresters. Figure 1 illustrates a single-line diagram and components of GIS. Many new SF_6-to-air bushings are now composed of composite construction, consisting of a fiberglass/epoxy inner cylinder that contains the SF_6 gas and provides structural strength. The external weather shed is made of silicone rubber.

SF_6 is the most common gas used for electrical insulation; its pressure values range from 0.29 MPa to 0.51 MPa (at 20 °C). Recently, SF_6 has been replaced with the 95% mixture of nitrogen (N_2) and 5% of SF_6 at 1.3 MPa pressure, or with a 90% mixture of N_2 and 10% of SF_6 at 0.94 MPa pressure, as well as with a 80% mixture of N_2 and 20% of SF_6 at 0.71 MPa pressure. The latter corresponds to the 0.4 MPa pressure when pure SF_6 is used.

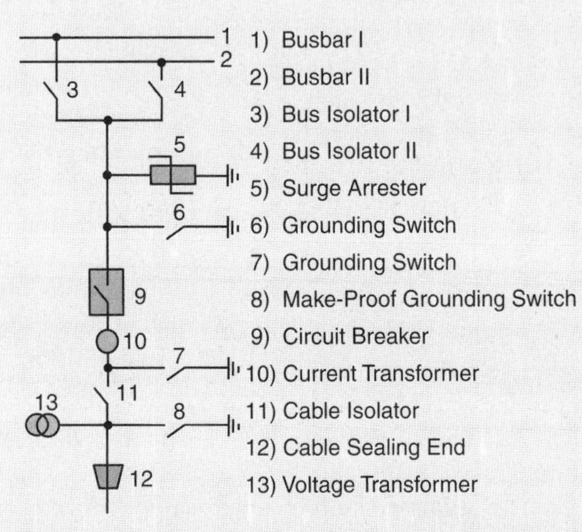

Figure 1 Single-line diagram and components of GIS

Today, the environmental consideration related to the "greenhouse" effect grows, especially for using SF_6 and its mixtures in the compressed-gas electric power apparatus. SF_6 gas is one of the strongest manmade "greenhouse" gases; its global warming potential is estimated to be approximately 25,000 times larger than that of carbon dioxide (CO_2) gas. At equal gas pressure, SF_6/N_2 mixtures are less sensitive to insulation defects than undiluted SF_6. The recent trend is to use ultra-diluted SF_6/N_2 gas mixtures with SF_6 content $\leq 1\%$.

Typically, solid insulators are required to provide support to stressed conductors in such systems. Solid insulating spacers represent the weakest points in these systems. Several troubles and system outages have been reported worldwide due to spacer failures. Normally, pure SF_6 or SF_6/N_2 mixtures at high pressures are used as an insulating medium. Complete failure or partial discharge may occur on the spacer surfaces, but they rarely occur at the highly pressurized gas unless there are solid conducting particles in the gas, as will be discussed later.

Spacers used in GIS are usually made of epoxy or cast resin. These spacers fundamentally are divided into two types according to their shapes, namely, discs and cones. The presence of spacers results in complex dielectric field distribution. It often intensifies the electric field, particularly on the spacer surface. The insulation ability of SF_6 is highly sensitive to the maximum electric field, and furthermore, the insulation strength along a spacer surface is usually lower than that in the gas space. Due to the previously mentioned spacer troubles, they should be precisely designed to realize a quasi-uniform electric field distribution along their surfaces. In addition, functionally graded materials (FGM) find extensive application in the insulation system such as GIS. FGM consist of materials of different permittivities (dielectric constants), i.e., with a spatial distribution of permittivity. This new spacer material can be optimized to control the electric field along the spacer surface, especially at the triple junction "gas/spacer/enclosure."

Advantages

Gas-insulated systems, such as gas-insulated lines and switchgear (GIL and GIS), are widely used in the electric power industry for transmitting and controlling bulk power, respectively. The concept of SF_6 HV GIS has proved itself in several thousands of installations worldwide. It offers the following outstanding advantages:

- Minimal space requirements, where the availability and price of land play an important part in selecting the type of switchgear to be used. It resolves the problems that arise in large towns, industrial conurbations, mountainous regions with narrow valleys, and underground power stations.
- Full protection against contact with live parts because the metal enclosure affords maximum safety for personnel under all operating and fault conditions.

- Protection against pollution due to the fact that the metal enclosure fully protects the switchgear internal components against environmental pollutants such as salt deposits in coastal regions, industrial vapors, and precipitates, as well as sandstorms. The compact switchgear can be installed in buildings of uncomplicated design in order to minimize the cost of cleaning and inspection and to make necessary repairs independent of weather conditions.
- Free choice of installation site that leads to potential savings for the expensive grading and foundation work, e.g., in permafrost zones, and short delivery and erection times for indoor switchgear installation regardless of the weather conditions.
- Protection of the environment because of its very flexible modular system that can meet all requirements of configuration given by network design and operating conditions.
- Longevity means an expected lifetime of at least 50 years and being maintenance-free for more than 20 years.

History and Technological Progress

Historically, air-insulated substations (AISs) were the only available technology until 1936. However, this resulted in a number of difficulties, namely, (1) pollution in desert areas or in close proximity to industrial or coastal areas, (2) insufficient space for AIS when constructing new substations/extensions (e.g., inside cities), and (3) restrictions by planning laws that only permit AIS substations where there is "no demonstrable alternative."

If GIS and AIS are compared in the need of space for the same function, the space reduction by using GIS is in the ratio of 1:5. That means less than 20% of the space of AISs is needed to install a GIS. In 2007, more than 20,000 bays in over 2000 substations are installed worldwide, in all kinds of environmental conditions and with the whole spectrum of voltage and current ratings.

Following the invention of SF_6 gas in 1900, the applications of this gas have significantly advanced since 1940. Alternating current (ac) GIS technology originated in 1936, when a Freon-GIS assembly, rated at 33 kV, was demonstrated in the United States. Later, in the mid-1950s, the excellent insulating and arc-extinguishing properties of SF_6 gas were recognized. By the mid-1960s, GIS was sufficiently well-developed to be commercially viable and appealing to a broader market. Over time, progressive innovative steps have allowed manufacturers to develop a range of GIS voltage ratings of 550 kV in 1976, 800 kV in 1979, and recently up to 1100 kV as shown in Figure 2 for the ac systems. The use of 1100-kV ac transmission lines, (i.e., doubling the voltage from

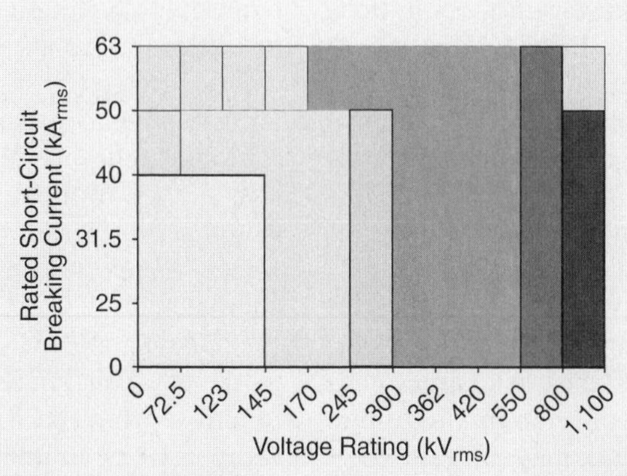

Figure 2 Rated short-circuit breaking current of ac GIS for a wide voltage range

the traditional level of 550 kV) reduces the transmission losses by a factor of four, which is a significant savings of energy. Moreover, the 1200-kV ac GIS are going to be seen very soon.

Recent trends in power transmission and distribution technologies involve high-voltage direct-current (HVDC) installations in the electric system. This improves the overall system reliability and achieves economical benefits. Direct current (dc) GIS are considered integral parts of these installations. Therefore, there has been a resurgence of interest in HVDC SF_6 equipment in recent years. Recently, 500-kV dc GIS for dc transmission systems have become available in Japan.

The phenomenon of accumulated charges on a solid insulator surface is one of the critical parameters at the insulation design stage even for both ac and dc all gas-insulated equipment (e.g., GIS and GIL). There are three kinds of electric charging mechanisms: volume conduction, surface conduction, and electric field emission. These mechanisms are characterized in terms of the time constant, applied voltage, and charge distribution. The behavior of metallic particles left on the enclosure is one of the largest differences. Also, degradation of metal-oxide arrester blocks under dc stress is of significant importance.

Figure 3 illustrates the percentage evolution and development of different types of HV substations in the market from 1960 to 2020. Highly integrated switchgear (HIS) is a compact switchgear solution for a rated voltage of up to 550 kV. HIS is mainly used for renewal or expansion of air-insulated outdoor and indoor substations, particularly if the operator wants to carry out modifications while the switchgear is in service. With the HIS solution, the circuit-breakers, disconnectors, earthing switches, and transformers are accommodated in compressed

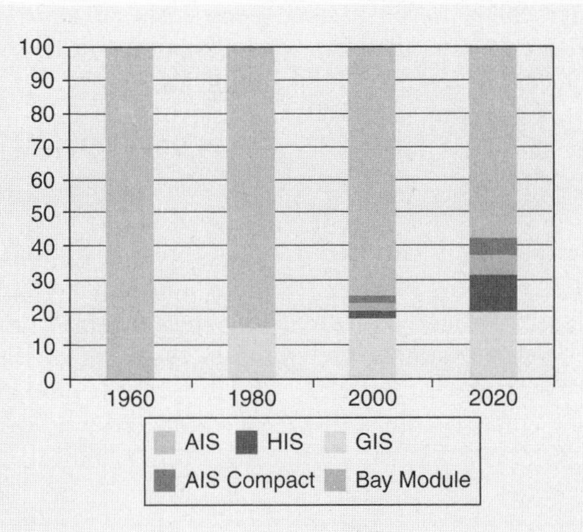

Figure 3 Percentage evolution and development of HV substations in the market

gas-tight enclosures and at a minimum number of independent gas compartments. This makes the switchgear extremely compact. The modularity and flexibility of today's switchgear allows designing highly optimized GIS using a reduced minimum number of junction elements (elbows and cross junctions, among others). This concept allows delivering subtransmission and distribution substations made of one or two shipping units only, fully assembled and tested in factory. This will result in quasielimination of the site assembly and tests, which is a real benefit of the customer's project management. These optimized GIS can shorten the time needed from ordering to commissioning to about 44% (i.e., by about 5–6 months). It is worth mentioning that the recent size of building, space requirement, and packing volume have been reduced for the 145 kV GIS, to be less than 20%, 15%, and 25% of those in 1968, respectively.

Gas-insulated modules have recently been seen at all voltage levels up to 550 kV, as an intermediary between GIS and AIS. They are suitable for the following categories of applications: standardization and optimization of new substation and/or extensions in large networks, reconstruction or refurbishment of AIS with operational constraints and/or space limitations, and the extension of AIS with space limitations. Therefore, the market of these applications will most likely increase as shown in Figure 3.

Recently, disconnecting circuit breaker technology, which integrates the disconnecting function into the circuit breaker and eliminates the need for two separate free-standing disconnectors, has led to shrinking footprint, equipment, and construction costs, and increases availability. The solution makes it possible—for the first time—to build compact EHV AIS substations and load hubs close to large cities and urban areas, where high land prices and space restrictions usually prohibit the construction of large AIS.

Transmission system growth may lead to higher requirements with regard to short-circuit current and/or nominal current ratings after the initial GIS installation. An upgrade of busbars, circuit breakers, and metering transformers may have to be investigated. Upgrades may also consist of an addition of isolated earthing switches for offline monitoring, voltage indicators, and partial discharge sensors. The planning for an extension or upgrade option should begin during the initial GIS design stage.

In some designs of bay module type used for GIS extensions and upgrades, the following elements are integrated into one bay: circuit breakers, bus disconnectors, earthing switches, voltage transformers, current transformers, outgoing feeder disconnectors with earthing switches, and surge arresters. In addition, the circuit breaker is integrated with current transformers, and disconnector contacts are directly mounted on circuit breaker. The key benefits of this compact design are: space-saving by optimized system design, modularity and flexibility for application in all possible substation designs, low overall height, optimized maintenance management, reduction in life cycle cost, and short on-site erection due to prefabricated and type-tested modules.

New Development Trends

Increased equipment loading and an inability to build new or expand conventional stations to accommodate additional loads has lowered the relative GIS costs compared with those of AIS. Manufacturing improvements have lowered the real GIS costs as well. Changes in the utility business environment and ownership have led to a greater acceptance of these changes. The above factors will increase the market share of GIS. This is expected even without further GIS technical improvements.

The evolutionary development of GIS has resulted in higher integration, reduced opportunity for defects, and more compact designs. Incremental improvements are continuing in interrupter technology, such as self-extinguishing features at MV and resistance interruption at EHV and UHV.

The latest GIS technological achievement is in the 1100-kV AC GIS. The 1100 kV system has the following features: multi-bundle conductors, larger line-charging capacity (approximately four times as large as that of the 550 kV system), larger power plants that are located relatively close to the UHV system, and large-capacity transformers (3000 MVA/bank) that interconnect the UHV system with the 550 kV system. To reduce corona noise, multibundle "eight" conductors with a large diameter are employed, which leads to a decrease in the line surge impedances and an increase in the time constant of dc component associated with fault current. In addition, the use of large power generators and large capacity power transformers lead to larger reactance-to-resistance ratio and contribute to an increase of the dc time constant in the fault currents. Consequently, the rate of decrease of current (di/dt) around current zero is taken as 28 A/μs, which is equivalent to 63 kA_{rms} with 10% component. As a result, the rated short-circuit breaking current, which is the highest short-circuit current that a circuit breaker is capable of breaking at the instant of contact separation in kA_{rms}, is taken as 50 kA_{rms} (see Figure 2).

Generally, there are two main problems for network expansion in the 550 kV to transfer power from remote plants:

- It is very difficult to secure multiple power transmission routes, which enable sufficient power transmission from remote power plants, thus, the construction of the 1100-kV transmission lines with a capability three to four times greater than that of 550-kV transmission lines.
- Short-circuit current in the 550-kV network will increase above the expected level (63 kA_{rms}).

Highly sophisticated system design technologies are necessary to deal with the following technical issues of the 1100-kV transmission system:

- Solutions for network problems and technology, such as secondary arc extinction. Secondary arc extinction is a serious concern in UHV systems and needs to be addressed at the design stage of such projects. Mitigation measures, such as neutral reactors, special switching schemes of shunt reactors, and high-speed grounding switches, have to be studied.
- Insulation coordination issues, such as overvoltages on transmission lines and in substations, can be effectively controlled by high-performance, heavy-duty surge arresters. Therefore, these arresters are selected with an individual energy rating of $\geqslant$ 55 MJ. These metal-oxide surge arresters are a key technology for 1100-kV insulation coordination, where their I-V characteristics become more flat (i.e., higher energy aborption capability). Lightning overvoltages dominate the nonself-restoring internal insulation design of substation equipment; it is important to suppress lightning overvoltages effectively by arranging metal-oxide surge arresters at adequate locations, such as at line entrances, busbars, transformers, and also within GIS. To control switching overvoltages below the ground fault overvoltage level, the closing/opening resistor can be employed as shown in Figure 4, where the total length and

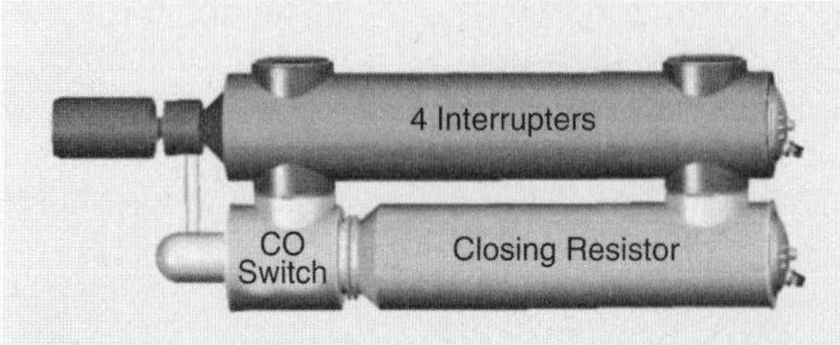

Figure 4 Four-interrupter and closing resistor and its close/open (CO) switch unit of the 1100-kV gas circuit breaker

the height of the 4-interrupter/closing resistor and its close/open (CO) switch unit are 10 m and 3.8 m, respectively.

For MV networks, on the other hand, the interrupting chamber of the self-extinguishing CB is divided into two main compartments at the same pressure of about 0.51 MPa during the closed condition. During opening, the gas pressure increases by heat generated in the arc, then the gas blasts from one arc compartment to the other one. By applying an external magnetic field, one arc rotates around the contact periphery and cools, and no pitting of contacts occurs.

Life Cycle Assessment

The switchgear only makes a minor contribution to the total global warming potential of a representative urban distribution grid. On the contrary, other grid components such as cables and transformers play the decisive role—regardless of whether AIS or GIS technology is used. Figure 5 shows an overview of the percentage environmental impact categories that were examined in a study at switchgear level. It is based on a representative mix of all switchgear types in the MV range. Clear advantages for GIS compared to AIS are also shown in regard to global warming potential, except the SF_6 emission as was explained earlier.

It is worth mentioning that the contributions to global warming potential of an urban power distribution grid is 92% from grid components, such as cables and lines, and 8% from switchgear, where the latter can be divided as 7% from switchgear in ring-main units and only 1% from switchgear of substation transformers.

GIS Failures

GIS have been in operation for more than 35 years, and they have shown a high level of reliability with extremely low failure rates. This is the result of quality assurance during the design and manufacturing processes as well as during the erection and on-site commissioning. However, the return of experiences has shown that some of the in-service failures are related to defects in the insulation system.

The main components involved in failures of 300–500 kV GIS are

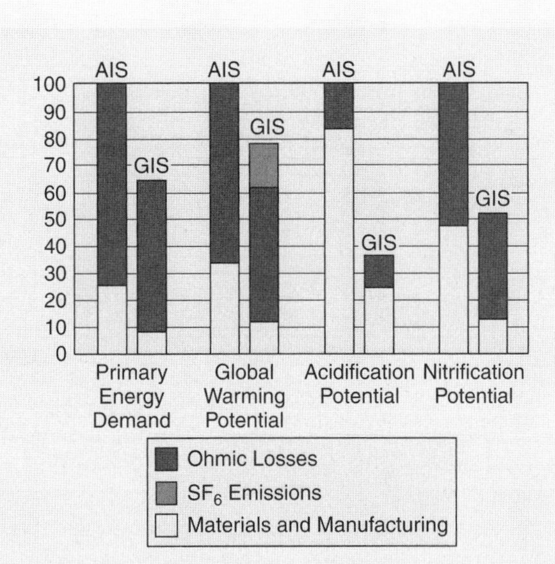

Figure 5 Environmental impact of AIS and GIS

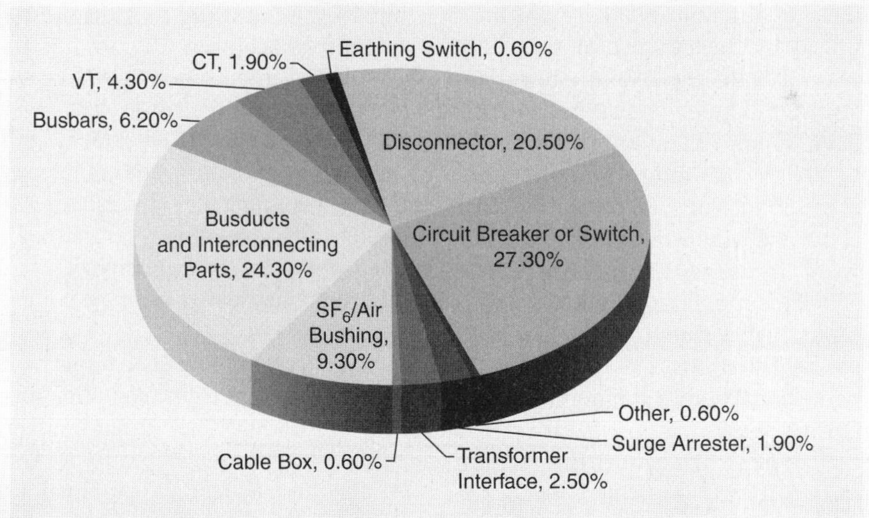

Figure 6 Main components involved in failures of 300–500 kV GIS

given in Figure 6, where the major components that cause failures by about 72% are circuit breakers (CB), disconnectors, and bus ducts and interconnecting parts. Major failure of GIS is a collapse of one of its major components or elements, which causes the lack of one or more of its

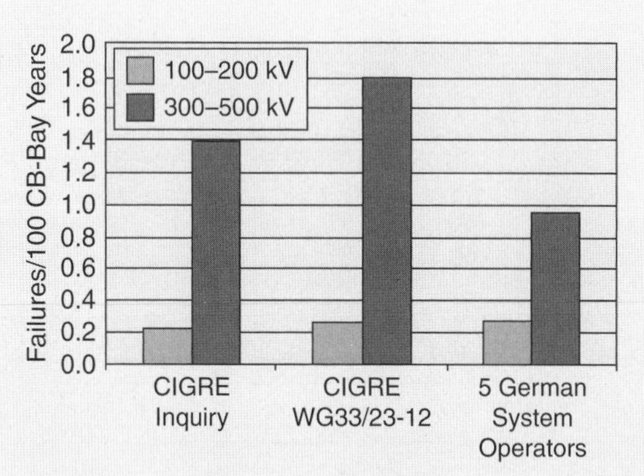

Figure 7 Rates of dielectric failures in GIS

fundamental functions. A major failure will result in an immediate change in the system operating conditions. The major failure rate is equal to the number of major failures divided by the sum of circuit-breaker bays, multiplied by the number of in-service years. The dielectric failure rates presented in Figure 7 are taken from different sources. In range 1 (from 100 kV to 200 kV), the failure rate is in good correspondence and amounts to be about 0.25 failures per 100 CB-bay years. In range 2 (from 300 kV to 500 kV), the failure rate is significantly higher due to the higher electric field strength. The deviations are between 0.95 and 1.8 failures per 100 CB-bay years.

When analyzing the failure causes, it must be stated that a lot of failures do not occur in GIS of modern design (e.g., insufficient insulation coordination of disconnectors and earthing switches or imperfections in solid material). Furthermore, a reduction of teething faults is likely due to the application of advanced testing methods. Therefore, a target failure rate of 0.1 failures per 100 CB-bay years should be achievable, where about 61% of the failures could have been detected and classified by monitoring and diagnostic systems, respectively.

Origin of PD in GIS

Partial discharges (PD) are electrical discharges that do not completely bridge between the electrodes. Although PD magnitudes are usually small, they cause progressive deterioration and may lead to ultimate failure. It is essential to detect their presence in a nondestructive controlled test. There are two types of the fine contaminants in both GIS and GIL systems: either insulating or metallic particles. The former has a relatively innocuous effect, while the latter drastically reduces the corona onset and breakdown voltages of the system.

GIS equipment is made compact; hence the working field strength within the equipment increases and becomes very sensitive to field perturbations due to defects. PD in compressed SF_6 GIS arise from protrusions, free conducting particles, floating components, and bulk insulation defects, such as voids and delaminations. PD resulting from the third and the fourth sources will, in turn, lead to failure of the GIS. In the case of a floating component (one not bonded to the conductor or sheath), the discharge magnitude is normally sufficient to decompose SF_6 in quantities. Treeing is a failure process in solid dielectrics, which, once initiated, will normally proceed to a failure through the bulk of dielectric. The most important defects are shown in Figure 8. Typical defects may result from errors in manufacturing, shipping, and assembly, including loose or electrically floating corona shields, undetected scratches, and poor electrical contacts.

Figure 9 illustrates the contribution of each of these defects and

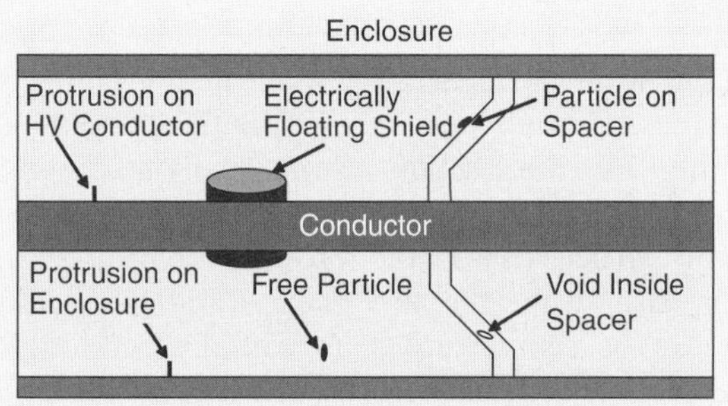

Figure 8 Possible defects in the insulation system of GIS

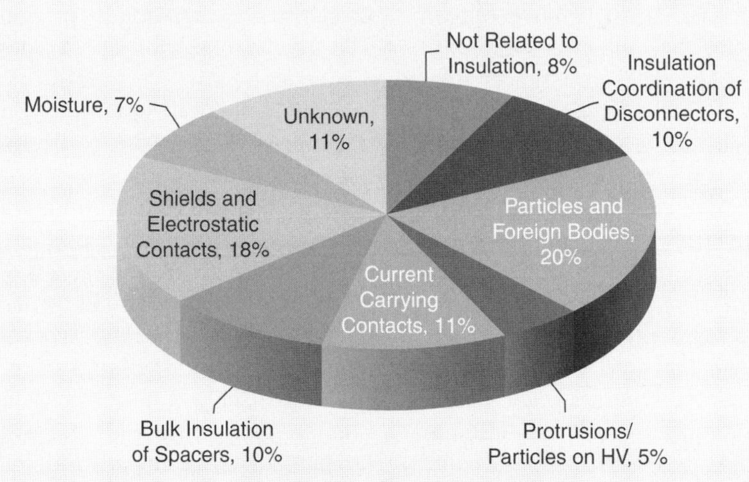

Figure 9 Main failure causes in GIS

other sources in the failure of GIS. It can be seen from Figure 9 that the defects shown in Figure 8 represent about 53% of the total main failure causes in GIS. A protrusion from live or grounded parts creates a local field enhancement. Such defects have little influence on the ac withstand level because the voltage varies slowly and the corona at the tip will have time to build up a space charge that shields the tip. Under switching and lightning surges, however, there is not enough time to build up such space charges. Consequently, the basic lightning impulse withstand level (BIL) will be drastically reduced. Usually, protrusions exceeding 1 mm are considered harmful.

Free moving particles have little impact on the BIL, while the ac withstand level can be significantly reduced. This reduction depends on their shape and position; the longer they are and the closer they get to the HV conductor, the more dangerous they become. If they move onto a spacer, they become even more dangerous. A particle on a spacer may, with time, lead to deterioration of the spacer surface. Under typical conditions for GIS, particles have inception electric fields of several kV/mm and true charges of several 10 pC to 100 pC. Wire particles approximately correspond to the type of particles encountered in practice.

Voids and defects inside spacers could create discharges once the initiation voltage is exceeded. Usually, such voids are found during quality control in the factory. A defect within a spacer will give rise to discharges, electrical trees, and eventually lead to breakdown. Since the sound absorption in epoxy is very high, the chance to detect them with acoustic measurement is small.

A floating component is a conducting element that is not bonded to, or in electrical contact with, the conductor or sheath. Generally, floating components should not be present in GIS. The most common types of components that may become floating are spacer inserts or field-grading shields at either the conductor or the sheath "enclosure." If a field-grading shield becomes mechanically loose, it may become electrically floating. A floating shield adjacent to an electrode could give rise to large discharges between the shield and the electrode, which can eventually lead to failure.

Floating components normally cause PD with magnitudes in the range of 10^4 to 10^6 pC/pulse with repetition rates of 120 to several thousand discharges per second, in multiples of 120 Hz (for 60 Hz power frequency). An electrically floating shield takes a potential, which is determined by the relationship between its capacitance to the conductor versus that to ground, which exceeds the insulation level to the conductor or to ground, and the capacitance will then discharge. Such discharges tend to be repetitive with a charge transfer in the range of nC to μC. The discharge pattern is usually regular and with PD magnitude larger than that for a void in an insulator.

PD Mitigation Methods

If the effects of defects/contaminants are mitigated or controlled, then improvement in the reliability of SF_6 GIS could be achieved. Moreover, this could lead to higher working stresses for future compressed gas apparatus, and consequently to a considerable reduction in SF_6 GIS size and cost. Some techniques that are used for the mitigation and control of particle contaminations in GIS are particle traps, dielectric coating of the electrodes, and the use of SF_6 gas mixtures.

GIS systems are typically conditioned before service by raising the voltage in discrete steps so as to move particles over a period of time into particle traps, but additional particles may be generated due to the switching operation of circuit breakers and conductor movement under load cycling. Particles adhering to the support insulators can result in significant reduction in the impulse flashover voltage of the system. In order to prevent the particles from interacting with solid support insulators, electrostatic particle drivers and traps are used. In ac GIS, the particle traps are usually fixed around, or near to, the insulators, which represent low-field areas, to prevent the attachment of the particles on the insulators and so reduce the chance of particle-initiated breakdowns associated with the insulators. In dc GIS, the bus consists of three regions, namely, a spacer region where an electric shield is installed at the triple junction, a particle scavenging region where both the particle driver and the particle trap exist, and a non-levitating region where the electrodes are coated. In such dc GIS, a shield ring "field-well ring" is placed at the end of the particle driver for two purposes: (1) to reduce the electric field and assist the particle trapping and (2) to prevent the firefly phenomenon, where the particle sometimes stays very close to the HV conductor under negative polarity.

Dielectric coatings of conductors in GIS systems improve the dielectric strength. This is due to coating over the conductor roughness and decreasing the high local field; coating resistance reduces the development of pre-discharges in the gas, and significance reduction in the particle charge during impacts and a consequent increase in the lifting field. However, coating the conductor can also create problems. If the coating is damaged, it could create particles. The coating lowers the particle mobility and thus, makes more difficult the use of particle traps. By increasing the lift-off voltage, it can be more dangerous because the particles can then lift off and induce direct breakdown without previous warning or possibility of PD detection "spring effect."

SF_6-N_2 is often considered to be the best substitute for SF_6 in both GIS and GIL because of the following reasons:

- N_2 is a cheap gas, and its dielectric strength in a uniform field is higher than that of gas mixtures of SF_6 with most common gases.

- It avoids SF_6 liquefaction at low ambient temperatures.
- It reduces the quantity of SF_6 and hence reduces environmental impact.
- At equal gas pressure, SF_6-N_2 mixtures are less sensitive to insulation defects than undiluted SF_6.

The effective ionization coefficient (α) for the SF_6-N_2 mixture decreases with the ratio of N_2/SF_6. In addition, the PD magnitude for fixed protrusions increases with gas pressure, where pure SF_6 gives the greatest increase. For free moving particles, the PD magnitude is independent of both the type and pressure of the gas.

Generally, the field strength near the triple junction can be reduced by devising the spacer shape such that an obtuse angle exists at each spacer-electrode interface, as well as by electrode-inserted spacers. In addition, FGM has recently been introduced, as was discussed earlier. Furthermore, the conducting particles can be effectively trapped and prevented from reaching the HV conductor by electrode-inserted rib spacers or trap-rib spacers with dielectric coating of the earthed enclosure.

Partial Discharge Monitoring Systems

Ultra-high frequency (UHF) technique can be applied for PD measurement in GIS, either using internal or external UHF couplers. The internally fitted UHF couplers have some limitations, such as the risk of breakdown (if they are not positioned in the hatchcover, where the field is very low), a large number of sensors is required to detect PD for a GIS (e.g., from six to nine per three-phase bay), and possible focusing of the UHF signal because the enclosure may act as a reflector. External UHF couplers also have some limitations, such as some loss of sensitivity as they may contain internal protection circuitry or preamplifiers that prevent pulse injection during the sensitivity verification test. However, external UHF couplers have many advantages such as: sensors are movable, the UHF technique is applicable to GIS having dielectric windows, it is a cost-effective method, and it offers a more effective and flexible PD location system.

The acoustic emission (AE) technique offers several advantages: movable sensors, good sensitivity, immunity to external noise, defects may be localized and recognized, and risk assessment based on source characterization. Its disadvantages are the high attenuation and, for some cast enclosures and defects in cast epoxy (e.g., voids) attenuation may be significant, requiring too many sensors. Therefore, the UHF technique has many characteristics that make it advantageous over that of the AE.

The world's first GIS online partial discharge monitoring (PDM) system was introduced in 1993, and they are now installed worldwide at voltages of 230–800 kV. Today, different PDM systems are available and utilize the same principle of UHF technology. Based on practical experiences with actual monitoring and diagnostic systems, the GIS system's reliability can be improved with exploitation of service life. The cost benefits of using PDM systems can be deduced from the dramatic reduction in failure rates when comparing the in-service failure statistics with and without using such systems.

A PDM system normally operates as a "black box" that captures UHF signals and submits warning and alarm signals to the substation control system only in the case of in-service relevant PD activity. Therefore, the most important PDM system features are the applied noise suppression techniques and the efficiency of the PD identification algorithms.

Today, the suppression of noise and other background signals like radar or mobile phone signals is realized by combined hardware and software filters. Actual PD identification algorithms are based on phase-resolved pulse sequence analysis. The applied redundant diagnosis systems (RDS) with hierarchical or hybrid structures consist of PD feature extraction and defect classification in combination with a proper reference data base to identify the type and nature of the insulation defect. The results from such RDS can have an accuracy of correct identification in the range of over 95%. Only a very small number of captured PD data sets are classified as unknown defect or identified in a wrong way.

Conclusion

GIS are widely used in the electric power industry as a key element for controlling bulk power from MV to UHV range. SF_6 and SF_6/N_2 mixture HV GIS have proved themselves in several thousands of installations worldwide because of their many outstanding advantages.

Epoxy or cast-resin solid insulators are used as spacers in GIS. They represent the weakest points in GIS systems as the electric field on their surfaces is higher than that in the gas space.

Historically, AIS were the only available technology until 1936, and it will continue to share more than 60% in the coming decade with the introduction of compact AIS and HIS types.

Highly optimized GIS, using a minimum number of junction elements (elbows and cross junctions), allows delivering much smaller subtransmission and distribution substations fully assembled and tested in factory (i.e., shortening the time to commissioning).

AC GIS rated 1100 kV and short-circuit breaking current of 50 kA_{rms} is the latest GIS technological achievement. The 1200-kV ac GIS are going to be visible very soon. DC GIS rated 500 kV for HVDC transmission systems have become available, too.

Circuit breakers, disconnectors, bus ducts, and interconnecting parts are the major components that cause failures by about 72% in GIS. Generally, the higher the operating voltage of GIS, the higher is the failure rate due to the higher electric field strength.

A target failure rate of 0.1 failures per 100 CB-bay years should be achievable, in particular by means of monitoring and diagnostic systems, as about 61% of the failures could have been detected.

PD in compressed SF_6 GIS arise from protrusions, free conducting particles, floating components, and bulk insulation defects (voids). These defects represent about 53% of the total main failure causes in GIS.

Some techniques used for the mitigation and control of particle contaminations in GIS are particle traps, dielectric coating of the electrodes, the use of SF_6 gas mixtures, and the use of FGM as solid spacers for optimizing its profile.

The ultra-high frequency and acoustic emission techniques can be used for GIS PD monitoring system, where the former has many advantageous characteristics over the latter. A dramatic reduction in failure rates can be achieved when using such systems.

Further Reading

I. A. Metwally, "Status review on partial discharge measurement techniques in gas-insulated switchgears/lines," *J. Electr. Power Syst. Res.*, vol. 69, no. 1, pp. 25–36, 2004.

IEEE Guide for Recommended Electrical Clearances and Insulation Levels in Air-Insulated Electrical Power Substations, IEEE Standard 2007.359971, May 4, 2007.

High-Voltage Switchgear and Control Gear—Part 205: Compact Switchgear Assemblies for Rated Voltages Above 52 kV, IEC Standard 62271-205, Edition 1.0, 2008.

J. D. McDonald, *Electric Power Substations Engineering*, 2nd ed. Boca Raton, FL: CRC, May 2007.

P. Ponchon, M. Bues, H. Bosia, and G. F. Montillet, "World applications of gas-insulated modules (GIMs) to HV and EHV substations," *IEEE Trans. Power Delivery*, vol. 21, no. 4, pp. 1935–1940, 2006.

S. Yanabu, E. Zaima, and T. Hasegawa, "Historical review of high voltage switchgear developments in the 20th century for power transmission and distribution system in Japan," *IEEE Trans. Power Delivery*, vol. 21, no. 2, pp. 659–664, 2006.

About the Author

Ibrahim A. Metwally (metwally@squ.edu.om) earned a B.Eng. in electrical engineering, and both an M.Eng. and a PhD in high-voltage engineering. He is a permanent professor with the Department of Electrical Engineering at Mansoura University, Egypt, and currently on leave as a professor with the Department of Electrical and Computer Engineering, Sultan Qaboos University, Oman. He is a senior member of the IEEE and a fellow of the Alexander von Humboldt Foundation in Bonn, Germany. ■

8.1 DEFINITION OF SYMMETRICAL COMPONENTS

Assume that a set of three-phase voltages designated V_a, V_b, and V_c is given. In accordance with Fortescue, these phase voltages are resolved into the following three sets of sequence components:

1. *Zero-sequence* components, consisting of three phasors with equal magnitudes and with zero phase displacement, as shown in Figure 8.1(a)
2. *Positive-sequence* components, consisting of three phasors with equal magnitudes, $+120°$ phase displacement, and positive sequence, as in Figure 8.1(b)
3. *Negative-sequence* components, consisting of three phasors with equal magnitudes, $+120°$ phase displacement, and negative sequence, as in Figure 8.1(c)

The zero-, positive-, and negative-sequence components of phase a, which are V_{a0}, V_{a1}, and V_{a2}, respectively, are presented in this section. For simplicity, drop the subscript a and denote these sequence components as V_0, V_1, and V_2. They are defined by the following transformation:

$$\begin{bmatrix} V_a \\ V_b \\ V_c \end{bmatrix} = \begin{bmatrix} 1 & 1 & 1 \\ 1 & a^2 & a \\ 1 & a & a^2 \end{bmatrix} \begin{bmatrix} V_0 \\ V_1 \\ V_2 \end{bmatrix} \tag{8.1.1}$$

where

$$a = 1\underline{/120°} = \frac{-1}{2} + j\frac{\sqrt{3}}{2} \tag{8.1.2}$$

V_{a0} V_{b0} V_{c0} $= V_0$

(a) Zero-sequence components

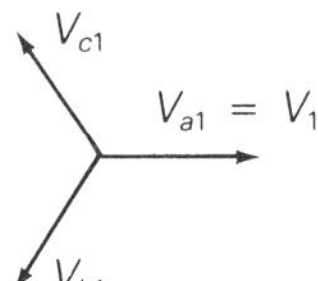

(b) Positive-sequence components

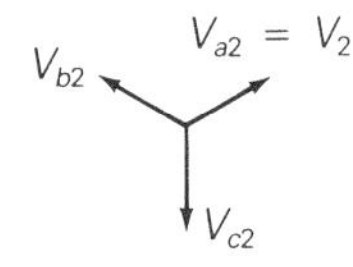

(c) Negative-sequence components

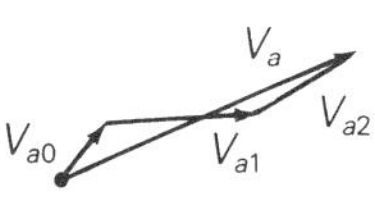

Phase a

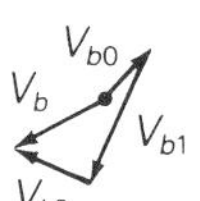

Phase b

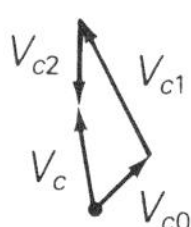

Phase c

FIGURE 8.1 Resolving phase voltages into three sets of sequence components

$$a^4 = a = 1\underline{/120^\circ}$$
$$a^2 = 1\underline{/240^\circ}$$
$$a^3 = 1\underline{/0^\circ}$$
$$1 + a + a^2 = 0$$
$$1 - a = \sqrt{3}\underline{/-30^\circ}$$
$$1 - a^2 = \sqrt{3}\underline{/+30^\circ}$$
$$a^2 - a = \sqrt{3}\underline{/270^\circ}$$
$$ja = 1\underline{/210^\circ}$$
$$1 + a = -a^2 = 1\underline{/60^\circ}$$
$$1 + a^2 = -a = 1\underline{/-60^\circ}$$
$$a + a^2 = -1 = 1\underline{/180^\circ}$$

TABLE 8.1

Common identities involving $a = 1\underline{/120^\circ}$

Writing (8.1.1) as three separate equations:

$$V_a = V_0 + V_1 + V_2 \tag{8.1.3}$$

$$V_b = V_0 + a^2V_1 + aV_2 \tag{8.1.4}$$

$$V_c = V_0 + aV_1 + a^2V \tag{8.1.5}$$

In (8.1.2), a is a complex number with unit magnitude and a 120° phase angle. When any phasor is multiplied by a, that phasor rotates by 120° (counterclockwise). Similarly, when any phasor is multiplied by $a^2 = (1\underline{/120^\circ})\,(1\underline{/120^\circ}) = 1\underline{/240^\circ}$, the phasor rotates by 240°. Table 8.1 lists some common identities involving a.

The complex number a is similar to the well-known complex number $j = \sqrt{-1} = 1\underline{/90^\circ}$, Thus, the only difference between j and a is that the angle of j is 90°, and that of a is 120°.

Equation (8.1.1) can be rewritten more compactly using matrix notation. Define the following vectors $\boldsymbol{V}_\text{P}$ and $\boldsymbol{V}_\text{S}$, and matrix $\boldsymbol{A}$:

$$\boldsymbol{V}_\text{P} = \begin{bmatrix} V_a \\ V_b \\ V_c \end{bmatrix} \tag{8.1.6}$$

$$\boldsymbol{V}_\text{S} = \begin{bmatrix} V_0 \\ V_1 \\ V_2 \end{bmatrix} \tag{8.1.7}$$

$$\boldsymbol{A} = \begin{bmatrix} 1 & 1 & 1 \\ 1 & a^2 & a \\ 1 & a & a^2 \end{bmatrix} \tag{8.1.8}$$

where $\boldsymbol{V}_P$ is the column vector of phase voltages, $\boldsymbol{V}_S$ is the column vector of sequence voltages, and $\boldsymbol{A}$ is a 3×3 transformation matrix. Using these definitions, (8.1.1) becomes

$$\boldsymbol{V}_P = \boldsymbol{A}\boldsymbol{V}_S \tag{8.1.9}$$

The inverse of the $\boldsymbol{A}$ matrix is

$$\boldsymbol{A}^{-1} = \frac{1}{3}\begin{bmatrix} 1 & 1 & 1 \\ 1 & a & a^2 \\ 1 & a^2 & a \end{bmatrix} \tag{8.1.10}$$

Equation (8.1.10) can be verified by showing that the product $\boldsymbol{A}\boldsymbol{A}^{-1}$ is the unit matrix. Also, premultiplying (8.1.9) by $\boldsymbol{A}^{-1}$ gives

$$\boldsymbol{V}_S = \boldsymbol{A}^{-1}\boldsymbol{V}_P \tag{8.1.11}$$

Using (8.1.6), (8.1.7), and (8.1.10), then (8.1.11) becomes

$$\begin{bmatrix} V_0 \\ V_1 \\ V_2 \end{bmatrix} = \frac{1}{3}\begin{bmatrix} 1 & 1 & 1 \\ 1 & a & a^2 \\ 1 & a^2 & a \end{bmatrix}\begin{bmatrix} V_a \\ V_b \\ V_c \end{bmatrix} \tag{8.1.12}$$

Writing (8.1.12) as three separate equations,

$$V_0 = \tfrac{1}{3}(V_a + V_b + V_c) \tag{8.1.13}$$

$$V_1 = \tfrac{1}{3}(V_a + aV_b + a^2V_c) \tag{8.1.14}$$

$$V_2 = \tfrac{1}{3}(V_a + a^2V_b + aV_c) \tag{8.1.15}$$

Equation (8.1.13) shows that there is no zero-sequence voltage in a *balanced* three-phase system because the sum of three balanced phasors is zero. In an unbalanced three-phase system, line-to-neutral voltages may have a zero-sequence component. However, line-to-line voltages never have a zero-sequence component, since by KVL, their sum is always zero.

The symmetrical component transformation also can be applied to currents, as follows. Let

$$\boldsymbol{I}_P = \boldsymbol{A}\boldsymbol{I}_S \tag{8.1.16}$$

where $\boldsymbol{I}_P$ is a vector of phase currents,

$$\boldsymbol{I}_P = \begin{bmatrix} I_a \\ I_b \\ I_c \end{bmatrix} \tag{8.1.17}$$

and $\boldsymbol{I}_S$ is a vector of sequence currents,

$$\boldsymbol{I}_S = \begin{bmatrix} I_0 \\ I_1 \\ I_2 \end{bmatrix} \tag{8.1.18}$$

Also,

$$\boldsymbol{I}_S = \boldsymbol{A}^{-1}\boldsymbol{I}_P \tag{8.1.19}$$

Equations (8.1.16) and (8.1.19) can be written as separate equations as follows. The phase currents are

$$I_a = I_0 + I_1 + I_2 \tag{8.1.20}$$

$$I_b = I_0 + a^2 I_1 + a I_2 \tag{8.1.21}$$

$$I_c = I_0 + a I_1 + a^2 I_2 \tag{8.1.22}$$

and the sequence currents are

$$I_0 = \tfrac{1}{3}(I_a + I_b + I_c) \tag{8.1.23}$$

$$I_1 = \tfrac{1}{3}(I_a + a I_b + a^2 I_c) \tag{8.1.24}$$

$$I_2 = \tfrac{1}{3}(I_a + a^2 I_b + a I_c) \tag{8.1.25}$$

In a three-phase Y-connected system, the neutral current I_n is the sum of the line currents:

$$I_n = I_a + I_b + I_c \tag{8.1.26}$$

Comparing (8.1.26) and (8.1.23),

$$I_n = 3I_0 \tag{8.1.27}$$

The neutral current equals three times the zero-sequence current. In a balanced Y-connected system, line currents have no zero-sequence component, since the neutral current is zero. Also, in any three-phase system with no neutral path, such as a Δ-connected system or a three-wire Y-connected system with an ungrounded neutral, line currents have no zero-sequence component.

The following three examples further illustrate symmetrical components.

EXAMPLE 8.1

Sequence components: balanced line-to-neutral voltages

Calculate the sequence components of the following balanced line-to-neutral voltages with *abc* sequence:

$$\boldsymbol{V}_P = \begin{bmatrix} V_{an} \\ V_{bn} \\ V_{cn} \end{bmatrix} = \begin{bmatrix} 277\angle 0^\circ \\ 277\angle -120^\circ \\ 277\angle +120^\circ \end{bmatrix} \text{ volts}$$

SOLUTION

Using (8.1.13) through (8.1.15),

$$V_0 = \tfrac{1}{3}[277\underline{/0^\circ} + 277\underline{/-120^\circ} + 277\underline{/+120^\circ}] = 0$$

$$V_1 = \tfrac{1}{3}[277\underline{/0^\circ} + 277\underline{/(-120^\circ + 120^\circ)} + 277\underline{/(120^\circ + 240^\circ)}]$$

$$= 277\underline{/0^\circ} \text{ volts} = V_{an}$$

$$V_2 = \tfrac{1}{3}[277\underline{/0^\circ} + 277\underline{/(-120^\circ + 240^\circ)} + 277\underline{/(120^\circ + 120^\circ)}]$$

$$= \tfrac{1}{3}[277\underline{/0^\circ} + 277\underline{/120^\circ} + 277\underline{/240^\circ}] = 0$$

This example illustrates the fact that balanced three-phase systems with *abc* sequence (or positive sequence) have no zero-sequence or negative-sequence components. For this example, the positive-sequence voltage V_1 equals V_{an}, and the zero-sequence and negative-sequence voltages are both zero.

EXAMPLE 8.2

Sequence components: balanced *acb* currents

A Y-connected load has balanced currents with *acb* sequence given by

$$I_P = \begin{bmatrix} I_a \\ I_b \\ I_c \end{bmatrix} = \begin{bmatrix} 10\underline{/0^\circ} \\ 10\underline{/+120^\circ} \\ 10\underline{/-120^\circ} \end{bmatrix} \text{ A}$$

Calculate the sequence currents.

SOLUTION

Using (8.1.23) through (8.1.25):

$$I_0 = \tfrac{1}{3}[10\underline{/0^\circ} + 10\underline{/120^\circ} + 10\underline{/-120^\circ}] = 0$$

$$I_1 = \tfrac{1}{3}[10\underline{/0^\circ} + 10\underline{/(120^\circ + 120^\circ)} + 10\underline{/(-120^\circ + 240^\circ)}]$$

$$= \tfrac{1}{3}[10\underline{/0^\circ} + 10\underline{/240^\circ} + 10\underline{/120^\circ}] = 0$$

$$I_2 = \tfrac{1}{3}[10\underline{/0^\circ} + 10\underline{/(120^\circ + 240^\circ)} + 10\underline{/(-120^\circ + 120^\circ)}]$$

$$= 10\underline{/0^\circ} \text{ A} = I_a$$

(*Continued*)

This example illustrates the fact that balanced three-phase systems with *acb* sequence (or negative sequence) have no zero-sequence or positive-sequence components. For this example, the negative-sequence current I_2 equals I_a, and the zero-sequence and positive-sequence currents are both zero.

EXAMPLE 8.3

Sequence components: unbalanced currents

A three-phase line feeding a balanced-Y load has one of its phases (phase *b*) open. The load neutral is grounded, and the unbalanced line currents are

$$\boldsymbol{I}_P = \begin{bmatrix} I_a \\ I_b \\ I_c \end{bmatrix} = \begin{bmatrix} 10\underline{/0^\circ} \\ 0 \\ 10\underline{/120^\circ} \end{bmatrix} \quad \text{A}$$

Calculate the sequence currents and the neutral current.

SOLUTION

The circuit is shown in Figure 8.2. Using (8.1.23) through (8.1.25):

$$I_0 = \tfrac{1}{3}[10\underline{/0^\circ} + 0 + 10\underline{/120^\circ}]$$
$$= 3.333\underline{/60^\circ} \quad \text{A}$$
$$I_1 = \tfrac{1}{3}[10\underline{/0^\circ} + 0 + 10\underline{/(120^\circ + 240^\circ)}] = 6.667\underline{/0^\circ} \quad \text{A}$$
$$I_2 = \tfrac{1}{3}[10\underline{/0^\circ} + 0 + 10\underline{/(120^\circ + 120^\circ)}]$$
$$= 3.333\underline{/-60^\circ} \quad \text{A}$$

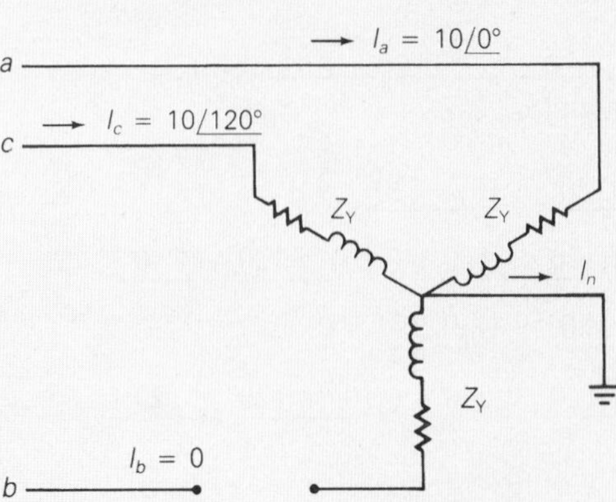

FIGURE 8.2 Circuit for Example 8.3

Using (8.1.26) the neutral current is

$$I_n = (10\underline{/0^\circ} + 0 + 10\underline{/120^\circ})$$
$$= 10\underline{/60^\circ}\ \text{A} = 3I_0$$

This example illustrates the fact that *unbalanced* three-phase systems may have nonzero values for all sequence components. Also, the neutral current equals three times the zero-sequence current, as given by (8.1.27).

8.2 SEQUENCE NETWORKS OF IMPEDANCE LOADS

Figure 8.3 shows a balanced-Y impedance load. The impedance of each phase is designated Z_Y, and a neutral impedance Z_n is connected between the load neutral and ground. Note from Figure 8.3 that the line-to-ground voltage V_{ag} is

$$\begin{aligned} V_{ag} &= Z_Y I_a + Z_n I_n \\ &= Z_Y I_a + Z_n(I_a + I_b + I_c) \\ &= (Z_Y + Z_n)I_a + Z_n I_b + Z_n I_c \end{aligned} \tag{8.2.1}$$

Similar equations can be written for V_{bg} and V_{cg}:

$$V_{bg} = Z_n I_a + (Z_Y + Z_n)I_b + Z_n I_c \tag{8.2.2}$$

$$V_{cg} = Z_n I_a + Z_n I_b + (Z_Y + Z_n)I_c \tag{8.2.3}$$

Equations (8.2.1) through (8.2.3) can be rewritten in matrix format:

$$\begin{bmatrix} V_{ag} \\ V_{bg} \\ V_{cg} \end{bmatrix} = \begin{bmatrix} (Z_Y + Z_n) & Z_n & Z_n \\ Z_n & (Z_Y + Z_n) & Z_n \\ Z_n & Z_n & (Z_Y + Z_n) \end{bmatrix} \begin{bmatrix} I_a \\ I_b \\ I_c \end{bmatrix} \tag{8.2.4}$$

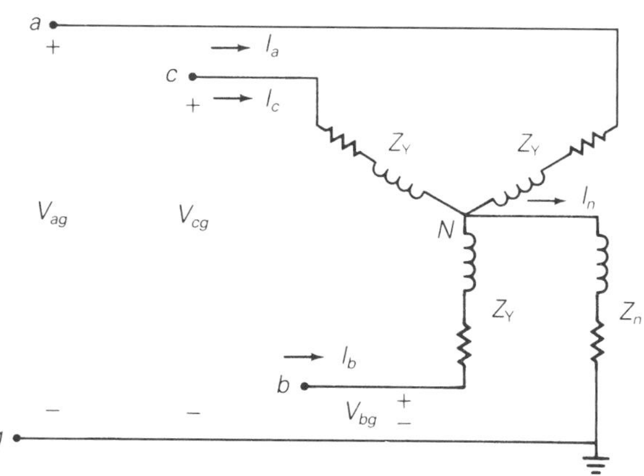

FIGURE 8.3
Balanced-Y impedance load

Equation (8.2.4) is written more compactly as

$$\boldsymbol{V}_P = \boldsymbol{Z}_P \boldsymbol{I}_P \tag{8.2.5}$$

where $\boldsymbol{V}_P$ is the vector of line-to-ground voltages (or phase voltages), $\boldsymbol{I}_P$ is the vector of line currents (or phase currents), and $\boldsymbol{Z}_P$ is the 3×3 phase impedance matrix shown in (8.2.4). Equations (8.1.9) and (8.1.16) can now be used in (8.2.5) to determine the relationship between the sequence voltages and currents, as follows:

$$\boldsymbol{A}\boldsymbol{V}_S = \boldsymbol{Z}_P \boldsymbol{A} \boldsymbol{I}_S \tag{8.2.6}$$

Premultiplying both sides of (8.2.6) of $\boldsymbol{A}^{-1}$ gives

$$\boldsymbol{V}_S = (\boldsymbol{A}^{-1}\boldsymbol{Z}_P \boldsymbol{A})\boldsymbol{I}_S \tag{8.2.7}$$

or

$$\boldsymbol{V}_S = \boldsymbol{Z}_S \boldsymbol{I}_S \tag{8.2.8}$$

where

$$\boldsymbol{Z}_S = \boldsymbol{A}^{-1}\boldsymbol{Z}_P \boldsymbol{A} \tag{8.2.9}$$

The impedance matrix $\boldsymbol{Z}_S$ defined by (8.2.9) is called the *sequence impedance matrix*. Using the definition of $\boldsymbol{A}$, its inverse $\boldsymbol{A}^{-1}$, and $\boldsymbol{Z}_P$ given by (8.1.8), (8.1.10), and (8.2.4), the sequence impedance matrix $\boldsymbol{Z}_S$ for the balanced-Y load is

$$\boldsymbol{Z}_S = \frac{1}{3}\begin{bmatrix} 1 & 1 & 1 \\ 1 & a & a^2 \\ 1 & a^2 & a \end{bmatrix}\begin{bmatrix} (Z_Y + Z_n) & Z_n & Z_n \\ Z_n & (Z_Y + Z_n) & Z_n \\ Z_n & Z_n & (Z_Y + Z_n) \end{bmatrix} \times \begin{bmatrix} 1 & 1 & 1 \\ 1 & a^2 & a \\ 1 & a & a^2 \end{bmatrix} \tag{8.2.10}$$

Performing the indicated matrix multiplications in (8.2.10), and using the identity $(1 + a + a^2) = 0$,

$$\boldsymbol{Z}_S = \frac{1}{3}\begin{bmatrix} 1 & 1 & 1 \\ 1 & a & a^2 \\ 1 & a^2 & a \end{bmatrix}\begin{bmatrix} (Z_Y + 3Z_n) & Z_Y & Z_Y \\ (Z_Y + 3Z_n) & a^2 Z_Y & a Z_Y \\ (Z_Y + 3Z_n) & a Z_Y & a^2 Z_Y \end{bmatrix}$$

$$= \begin{bmatrix} (Z_Y + 3Z_n) & 0 & 0 \\ 0 & Z_Y & 0 \\ 0 & 0 & Z_Y \end{bmatrix} \tag{8.2.11}$$

As shown in (8.2.11), the sequence impedance matrix $\boldsymbol{Z}_S$ for the balanced-Y load of Figure 8.3 is a diagonal matrix. Since $\boldsymbol{Z}_S$ is diagonal, (8.2.8) can be written as three *uncoupled* equations. Using (8.1.7), (8.1.18), and (8.2.11) in (8.2.8),

$$\begin{bmatrix} V_0 \\ V_1 \\ V_2 \end{bmatrix} = \begin{bmatrix} (Z_Y + 3Z_n) & 0 & 0 \\ 0 & Z_Y & 0 \\ 0 & 0 & Z_Y \end{bmatrix} \begin{bmatrix} I_0 \\ I_1 \\ I_2 \end{bmatrix} \tag{8.2.12}$$

Rewriting (8.2.12) as three separate equations,

$$V_0 = (Z_Y + 3Z_n)I_0 = Z_0 I_0 \tag{8.2.13}$$

$$V_1 = Z_Y I_1 = Z_1 I_1 \tag{8.2.14}$$

$$V_2 = Z_Y I_2 = Z_2 I_2 \tag{8.2.15}$$

As shown in (8.2.13), the zero-sequence voltage V_0 depends only on the zero-sequence current I_0 and the impedance $(Z_Y + 3Z_n)$. This impedance is called the *zero-sequence impedance* and is designated Z_0. Also, the positive-sequence voltage V_1 depends only on the positive-sequence current I_1 and an impedance $Z_1 = Z_Y$ called the *positive-sequence impedance.* Similarly, V_2 depends only on I_2 and the *negative-sequence impedance* $Z_2 = Z_Y$.

Equations (8.2.13) through (8.2.15) can be represented by the three networks shown in Figure 8.4. These networks are called the *zero-sequence*, *positive-sequence*, and *negative-sequence networks.* As shown, each sequence network is separate, uncoupled from the other two. The separation of these sequence networks is a consequence of the fact that $\boldsymbol{Z}_S$ is a diagonal matrix for a balanced-Y load. This separation underlies the advantage of symmetrical components.

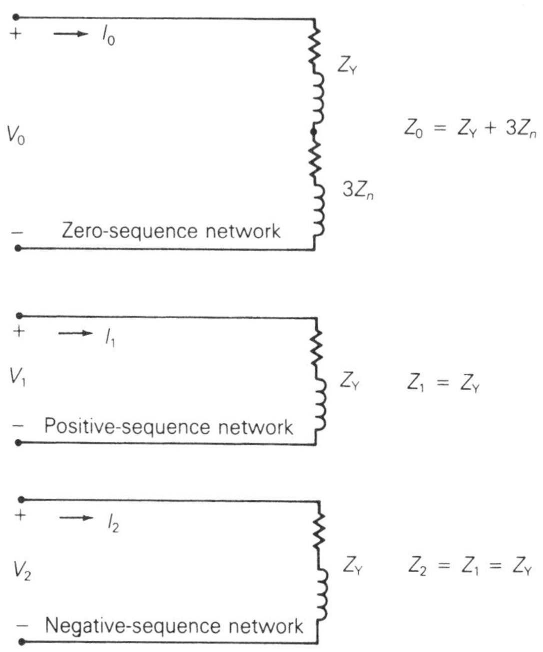

FIGURE 8.4

Sequence networks of a balanced-Y load

Note that the neutral impedance does not appear in the positive- and negative-sequence networks of Figure 8.4. This illustrates the fact that positive- and negative-sequence currents do not flow in neutral impedances. However, the neutral impedance is multiplied by 3 and placed in the zero-sequence network of the figure. The voltage $I_0(3Z_n)$ across the impedance $3Z_n$ is the voltage drop (I_nZ_n) across the neutral impedance Z_n in Figure 8.3, since $I_n = 3I_0$.

When the neutral of the Y load in Figure 8.3 has no return path, then the neutral impedance Z_n is infinite and the term $3Z_n$ in the zero-sequence network of Figure 8.4 becomes an open circuit. Under this condition of an open neutral, no zero-sequence current exists. However, when the neutral of the Y load is solidly grounded with a zero-ohm conductor, then the neutral impedance is zero and the term $3Z_n$ in the zero-sequence network becomes a short circuit. Under this condition of a solidly grounded neutral, zero-sequence current I_0 can exist when there is a zero-sequence voltage caused by unbalanced voltages applied to the load.

Figure 2.15 shows a balanced-Δ load and its equivalent balanced-Y load. Since the Δ load has no neutral connection, the equivalent Y load in Figure 2.15 has an open neutral. The sequence networks of the equivalent Y load corresponding to a balanced-Δ load are shown in Figure 8.5. As shown, the equivalent Y impedance $Z_Y = Z_\Delta/3$ appears in each of the sequence networks. Also, the zero-sequence network has an open circuit, since $Z_n = \infty$ corresponds to an open neutral. No zero-sequence current occurs in the equivalent Y load.

The sequence networks of Figure 8.5 represent the balanced-Δ load as viewed from its terminals, but they do not represent the internal load characteristics. The currents I_0, I_1, and I_2 in Figure 8.5 are the sequence components of the line currents feeding the Δ load, not the load currents within the Δ. The Δ load currents, which are related to the line currents by (2.5.14), are not shown in Figure 8.5.

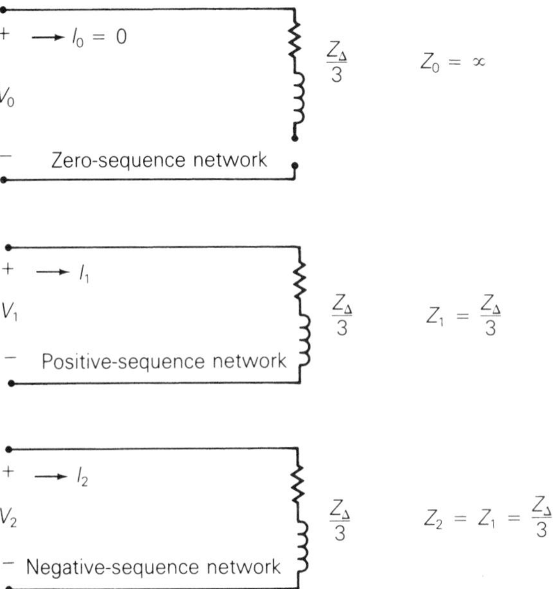

FIGURE 8.5 Sequence networks for an equivalent Y representation of a balanced-Δ load

EXAMPLE 8.4

Sequence networks: balanced-Y and balanced-Δ loads

A balanced-Y load is in parallel with a balanced-Δ-connected capacitor bank. The Y load has an impedance $Z_Y = (3 + j4)\ \Omega$ per phase, and its neutral is grounded through an inductive reactance $X_n = 2\ \Omega$. The capacitor bank has a reactance $X_c = 30\ \Omega$ per phase. Draw the sequence networks for this load and calculate the load-sequence impedances.

SOLUTION

The sequence networks are shown in Figure 8.6. As shown, the Y-load impedance in the zero-sequence network is in series with three times the neutral impedance. Also, the Δ-load branch in the zero-sequence network is open, since no zero-sequence current flows into the Δ load. In the positive- and negative-sequence circuits, the Δ-load impedance is divided by 3 and placed in parallel with the Δ-load impedance. The equivalent sequence impedances are

$$Z_0 = Z_Y + 3Z_n = 3 + j4 + 3(j2) = 3 + j10\ \ \Omega$$

$$Z_1 = Z_Y//(Z_\Delta/3) = \frac{(3 + j4)(-j30/3)}{3 + j4 - j(30/3)}$$

$$= \frac{(5\underline{/53.13^\circ})(10\underline{/-90^\circ})}{6.708\underline{/-63.43^\circ}} = 7.454\underline{/26.57^\circ}\ \ \Omega$$

$$Z_2 = Z_1 = 7.454\underline{/26.57^\circ}\ \ \Omega$$

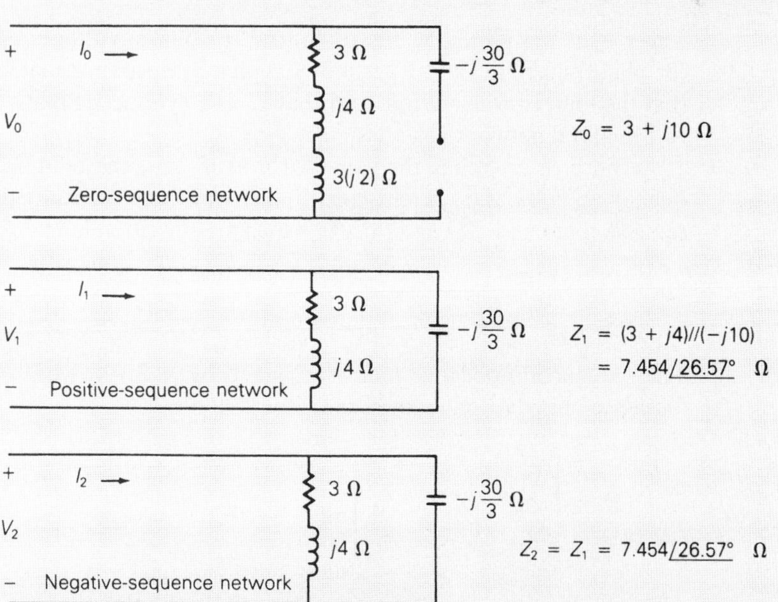

FIGURE 8.6 Sequence networks for Example 8.4

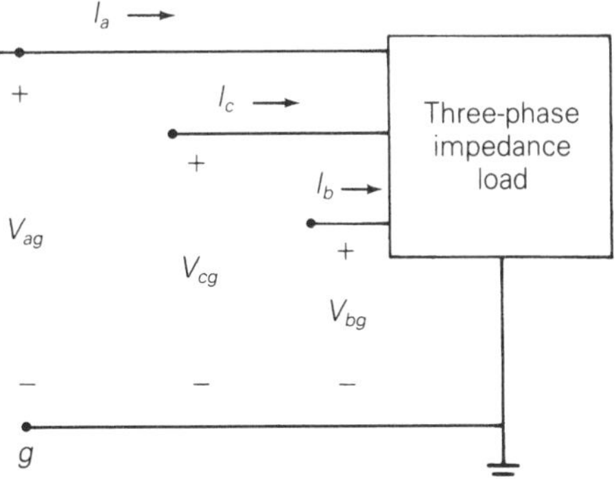

FIGURE 8.7 General three-phase impedance load (linear, bilateral network, nonrotating equipment)

Figure 8.7 shows a general three-phase linear impedance load. The load could represent a balanced load such as the balanced-Y or balanced-Δ load, or an unbalanced impedance load. The general relationship between the line-to-ground voltages and line currents for this load can be written as

$$\begin{bmatrix} V_{ag} \\ V_{bg} \\ V_{cg} \end{bmatrix} = \begin{bmatrix} Z_{aa} & Z_{ab} & Z_{ac} \\ Z_{ab} & Z_{bb} & Z_{bc} \\ Z_{ac} & Z_{bc} & Z_{cc} \end{bmatrix} \begin{bmatrix} I_a \\ I_b \\ I_c \end{bmatrix} \tag{8.2.16}$$

or

$$\boldsymbol{V}_{\mathrm{P}} = \boldsymbol{Z}_{\mathrm{P}}\boldsymbol{I}_{\mathrm{P}} \tag{8.2.17}$$

where $\boldsymbol{V}_{\mathrm{P}}$ is the vector of line-to-neutral (or phase) voltages, $\boldsymbol{I}_{\mathrm{P}}$ is the vector of line (or phase) currents, and $\boldsymbol{Z}_{\mathrm{P}}$ is a 3×3 phase impedance matrix. It is assumed here that the load is nonrotating, and that $\boldsymbol{Z}_{\mathrm{P}}$ is a symmetric matrix, which corresponds to a bilateral network.

Since (8.2.17) has the same form as (8.2.5), the relationship between the sequence voltages and currents for the general three-phase load of Figure 8.6 is the same as that of (8.2.8) and (8.2.9), which are rewritten here:

$$\boldsymbol{V}_{\mathrm{S}} = \boldsymbol{Z}_{\mathrm{S}}\boldsymbol{I}_{\mathrm{S}} \tag{8.2.18}$$

$$\boldsymbol{Z}_{\mathrm{S}} = \boldsymbol{A}^{-1}\boldsymbol{Z}_{\mathrm{P}}\boldsymbol{A} \tag{8.2.19}$$

The sequence impedance matrix $\boldsymbol{Z}_{\mathrm{S}}$ given by (8.2.19) is a 3×3 matrix with nine sequence impedances, defined as follows:

$$\boldsymbol{Z}_{\mathrm{S}} = \begin{bmatrix} Z_0 & Z_{01} & Z_{02} \\ Z_{10} & Z_1 & Z_{12} \\ Z_{20} & Z_{21} & Z_2 \end{bmatrix} \tag{8.2.20}$$

The diagonal impedances Z_0, Z_1, and Z_2 in this matrix are the self-impedances of the zero-, positive-, and negative-sequence networks. The off-diagonal impedances are the mutual impedances between sequence networks. Using the definitions of $\boldsymbol{A}$, $\boldsymbol{A}^{-1}$, $\boldsymbol{Z}_P$, and $\boldsymbol{Z}_S$, (8.2.19) is

$$\begin{bmatrix} Z_0 & Z_{01} & Z_{02} \\ Z_{10} & Z_1 & Z_{12} \\ Z_{20} & Z_{21} & Z_2 \end{bmatrix} = \frac{1}{3}\begin{bmatrix} 1 & 1 & 1 \\ 1 & a & a^2 \\ 1 & a^2 & a \end{bmatrix}\begin{bmatrix} Z_{aa} & Z_{ab} & Z_{ac} \\ Z_{ab} & Z_{bb} & Z_{bc} \\ Z_{ac} & Z_{cb} & Z_{cc} \end{bmatrix}\begin{bmatrix} 1 & 1 & 1 \\ 1 & a^2 & a \\ 1 & a & a^2 \end{bmatrix} \tag{8.2.21}$$

Performing the indicated multiplications in (8.2.21), and using the identity $(1 + a + a^2) = 0$, the following separate equations can be obtained (see Problem 8.18).

DIAGONAL SEQUENCE IMPEDANCES

$$Z_0 = \tfrac{1}{3}(Z_{aa} + Z_{bb} + Z_{cc} + 2Z_{ab} + 2Z_{ac} + 2Z_{bc}) \tag{8.2.22}$$

$$Z_1 = Z_2 = \tfrac{1}{3}(Z_{aa} + Z_{bb} + Z_{cc} - Z_{ab} - Z_{ac} - Z_{bc}) \tag{8.2.23}$$

OFF-DIAGONAL SEQUENCE IMPEDANCES

$$Z_{01} = Z_{20} = \tfrac{1}{3}(Z_{aa} + a^2Z_{bb} + aZ_{cc} - aZ_{ab} - a^2Z_{ac} - Z_{bc}) \tag{8.2.24}$$

$$Z_{02} = Z_{10} = \tfrac{1}{3}(Z_{aa} + aZ_{bb} + a^2Z_{cc} - a^2Z_{ab} - aZ_{ac} - Z_{bc}) \tag{8.2.25}$$

$$Z_{12} = \tfrac{1}{3}(Z_{aa} + a^2Z_{bb} + aZ_{cc} + 2aZ_{ab} + 2a^2Z_{ac} + 2Z_{bc}) \tag{8.2.26}$$

$$Z_{21} = \tfrac{1}{3}(Z_{aa} + aZ_{bb} + a^2Z_{cc} + 2a^2Z_{ab} + 2aZ_{ac} + 2Z_{bc}) \tag{8.2.27}$$

A *symmetrical load* is defined as a load whose sequence impedance matrix is diagonal; that is, all the mutual impedances in (8.2.24) through (8.2.27) are zero. Equating these mutual impedances to zero and solving, the following conditions for a symmetrical load are determined. When both

$$Z_{aa} = Z_{bb} = Z_{cc} \tag{8.2.28}$$

and

$$Z_{ab} = Z_{ac} = Z_{bc} \tag{8.2.29}$$

(conditions for a symmetrical load)

then

$$Z_{01} = Z_{10} = Z_{02} = Z_{20} = Z_{12} = Z_{21} = 0 \tag{8.2.30}$$

$$Z_0 = Z_{aa} + 2Z_{ab} \tag{8.2.31}$$

$$Z_1 = Z_2 = Z_{aa} - Z_{ab} \tag{8.2.32}$$

The conditions for a symmetrical load are that the diagonal phase impedances be equal and that the off-diagonal phase impedances be equal. These conditions can

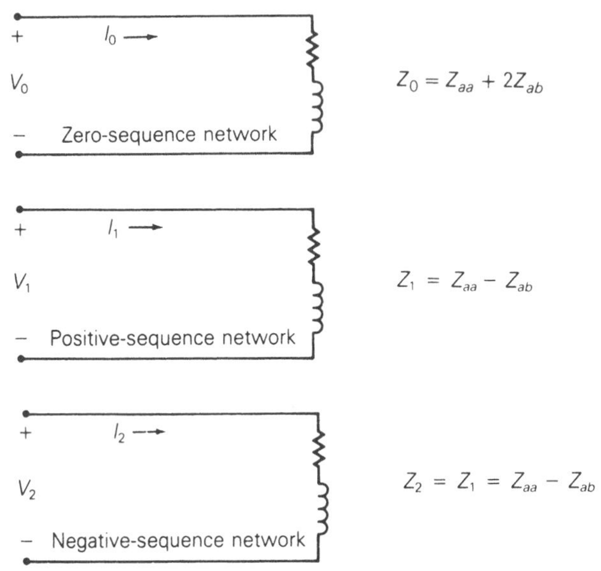

FIGURE 8.8 Sequence networks of a three-phase symmetrical impedance load (linear, bilateral network, nonrotating equipment)

be verified by using (8.2.28) and (8.2.29) with the identity $(1 + a + a^2) = 0$ in (8.2.24) through (8.2.27) to show that all the mutual sequence impedances are zero. Note that the positive- and negative-sequence impedances are equal for a symmetrical load, as shown by (8.2.32), and for a nonsymmetrical load, as shown by (8.2.23). This is always true for linear, symmetric impedances that represent nonrotating equipment such as transformers and transmission lines. However, the positive- and negative-sequence impedances of rotating equipment such as generators and motors are generally not equal. Note also that the zero-sequence impedance Z_0 is not equal to the positive- and negative-sequence impedances of a symmetrical load unless the mutual phase impedances $Z_{ab} = Z_{ac} = Z_{bc}$ are zero.

The sequence networks of a symmetrical impedance load are shown in Figure 8.8 Since the sequence impedance matrix $\boldsymbol{Z}_S$ is diagonal for a symmetrical load, the sequence networks are separate or uncoupled.

8.3 SEQUENCE NETWORKS OF SERIES IMPEDANCES

Figure 8.9 shows series impedances connected between two three-phase buses denoted abc and $a'b'c'$. Self-impedances of each phase are denoted Z_{aa}, Z_{bb}, and Z_{cc}. In general, the series network also may have mutual impedances between phases. The voltage drops across the series-phase impedances are given by

$$\begin{bmatrix} V_{an} - V_{a'n} \\ V_{bn} - V_{b'n} \\ V_{cn} - V_{c'n} \end{bmatrix} = \begin{bmatrix} V_{aa'} \\ V_{bb'} \\ V_{cc'} \end{bmatrix} = \begin{bmatrix} Z_{aa} & Z_{ab} & Z_{ac} \\ Z_{ab} & Z_{bb} & Z_{bc} \\ Z_{ac} & Z_{cb} & Z_{cc} \end{bmatrix} \begin{bmatrix} I_a \\ I_b \\ I_c \end{bmatrix} \tag{8.3.1}$$

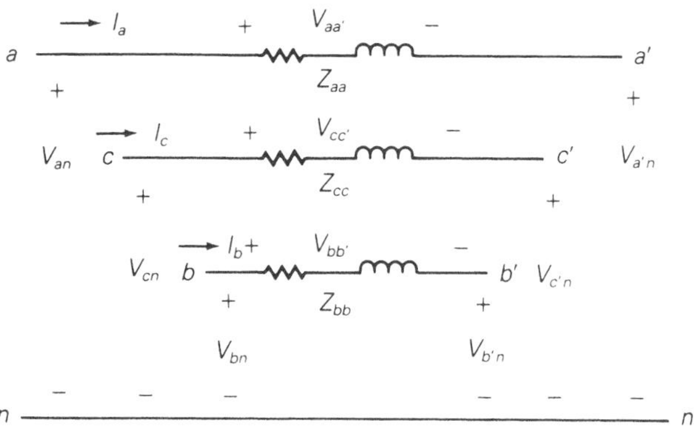

FIGURE 8.9 Three-phase series impedances (linear, bilateral network, nonrotating equipment)

Both self-impedances and mutual impedances are included in (8.3.1). It is assumed that the impedance matrix is symmetric, which corresponds to a bilateral network. It is also assumed that these impedances represent nonrotating equipment. Typical examples are series impedances of transmission lines and of transformers. Equation (8.3.1) has the following form:

$$\boldsymbol{V}_{\mathrm{P}} - \boldsymbol{V}_{\mathrm{P}'} = \boldsymbol{Z}_{\mathrm{P}}\boldsymbol{I}_{\mathrm{P}} \tag{8.3.2}$$

where $\boldsymbol{V}_{\mathrm{P}}$ is the vector of line-to-neutral voltages at bus abc, $\boldsymbol{V}_{\mathrm{P}'}$ the vector of line-to-neutral voltages at bus $a'b'c'$, $\boldsymbol{I}_{\mathrm{P}}$ is the vector of line currents, and $\boldsymbol{Z}_{\mathrm{P}}$ is the 3×3 phase impedance matrix for the series network. Equation (8.3.2) is now transformed to the sequence domain in the same manner that the load-phase impedances were transformed in Section 8.2. Thus,

$$\boldsymbol{V}_{\mathrm{S}} - \boldsymbol{V}_{\mathrm{S}'} = \boldsymbol{Z}_{\mathrm{S}}\boldsymbol{I}_{\mathrm{S}} \tag{8.3.3}$$

where

$$\boldsymbol{Z}_{\mathrm{S}} = \boldsymbol{A}^{-1}\,\boldsymbol{Z}_{\mathrm{P}}\boldsymbol{A} \tag{8.3.4}$$

From the results of Section 8.2, this sequence impedance $\boldsymbol{Z}_{\mathrm{S}}$ matrix is diagonal under the following conditions:

$$\left.\begin{array}{l} Z_{aa} = Z_{bb} = Z_{cc} \\ \text{and} \\ Z_{ab} = Z_{ac} = Z_{bc} \end{array}\right\} \begin{array}{l}\text{conditions for}\\ \text{symmetrical}\\ \text{series impedances}\end{array} \tag{8.3.5}$$

When the phase impedance matrix $\boldsymbol{Z}_{\mathrm{P}}$ of (8.3.1) has both equal self-impedances and equal mutual impedances, then (8.3.4) becomes

$$Z_{\mathrm{S}} = \begin{bmatrix} Z_0 & 0 & 0 \\ 0 & Z_1 & 0 \\ 0 & 0 & Z_2 \end{bmatrix} \tag{8.3.6}$$

where

$$Z_0 = Z_{aa} + 2Z_{ab} \tag{8.3.7}$$

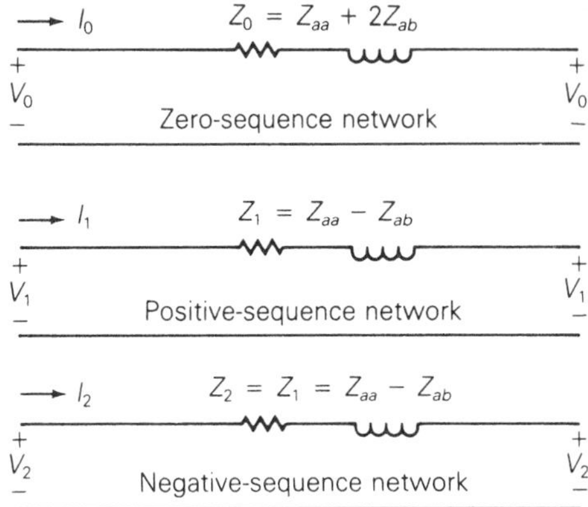

FIGURE 8.10 Sequence networks of three-phase symmetrical series impedances (linear, bilateral network, nonrotating equipment)

and

$$Z_1 = Z_2 = Z_{aa} - Z_{ab} \tag{8.3.8}$$

and (8.3.3) becomes three uncoupled equations, written as follows:

$$V_0 - V_{0'} = Z_0 I_0 \tag{8.3.9}$$

$$V_1 - V_{1'} = Z_1 I_1 \tag{8.3.10}$$

$$V_2 - V_{2'} = Z_2 I_2 \tag{8.3.11}$$

Equations (8.3.9) through (8.3.11) are represented by the three uncoupled sequence networks shown in Figure 8.10. From the figure it is apparent that, for symmetrical series impedances, positive-sequence currents produce only positive-sequence voltage drops. Similarly, negative-sequence currents produce only negative-sequence voltage drops, and zero-sequence currents produce only zero-sequence voltage drops. However, if the series impedances are not symmetrical, then $\boldsymbol{Z}_S$ is not diagonal, the sequence networks are coupled, and the voltage drop across any one sequence network depends on all three sequence currents.

8.4 SEQUENCE NETWORKS OF THREE-PHASE LINES

Section 4.7 develops equations suitable for computer calculation of the series phase impedances, including resistances and inductive reactances, of three-phase overhead transmission lines. The series phase impedance matrix $\boldsymbol{Z}_P$ for an untransposed line is given by Equation (4.7.19) and $\hat{\boldsymbol{Z}}_P$ for a completely transposed line is given by (4.7.21) through (4.7.23). Equation (4.7.19) can be transformed to the sequence domain to obtain

$$\boldsymbol{Z}_S = \boldsymbol{A}^{-1} \boldsymbol{Z}_P \boldsymbol{A} \tag{8.4.1}$$

$\boldsymbol{Z}_S$ is the 3×3 series sequence impedance matrix whose elements are

$$\boldsymbol{Z}_S = \begin{bmatrix} Z_0 & Z_{01} & Z_{02} \\ Z_{10} & Z_1 & Z_{12} \\ Z_{20} & Z_{21} & Z_2 \end{bmatrix} \quad \Omega/\text{m} \tag{8.4.2}$$

In general $\boldsymbol{Z}_S$ is not diagonal. However, if the line is completely transposed,

$$\hat{\boldsymbol{Z}}_S = \boldsymbol{A}^{-1}\hat{\boldsymbol{Z}}_P\boldsymbol{A} = \begin{bmatrix} \hat{Z}_0 & 0 & 0 \\ 0 & \hat{Z}_1 & 0 \\ 0 & 0 & \hat{Z}_2 \end{bmatrix} \tag{8.4.3}$$

where, from (8.3.7) and (8.3.8),

$$\hat{Z}_0 = \hat{Z}_{aaeq} + 2\hat{Z}_{abeq} \tag{8.4.4}$$

$$\hat{Z}_1 = \hat{Z}_2 = \hat{Z}_{aaeq} - \hat{Z}_{abeq} \tag{8.4.5}$$

A circuit representation of the series sequence impedances of a completely transposed three-phase line is shown in Figure 8.11.

Section 4.11 develops equations suitable for computer calculation of the shunt phase admittances of three-phase overhead transmission lines. The shunt admittance matrix $\boldsymbol{Y}_P$ for an untransposed line is given by Equation (4.11.16), and $\hat{\boldsymbol{Y}}_P$ for a completely transposed three-phase line is given by (4.11.17).

Equation (4.11.16) can be transformed to the sequence domain to obtain

$$\boldsymbol{Y}_S = \boldsymbol{A}^{-1}\boldsymbol{Y}_P\boldsymbol{A} \tag{8.4.6}$$

where

$$\boldsymbol{Y}_S = \mathbf{G}_S + j(2\pi f)\mathbf{C}_S \tag{8.4.7}$$

$$\mathbf{C}_S = \begin{bmatrix} C_0 & C_{01} & C_{02} \\ C_{10} & C_1 & C_{12} \\ C_{20} & C_{21} & C_2 \end{bmatrix} \ \text{F/m} \tag{8.4.8}$$

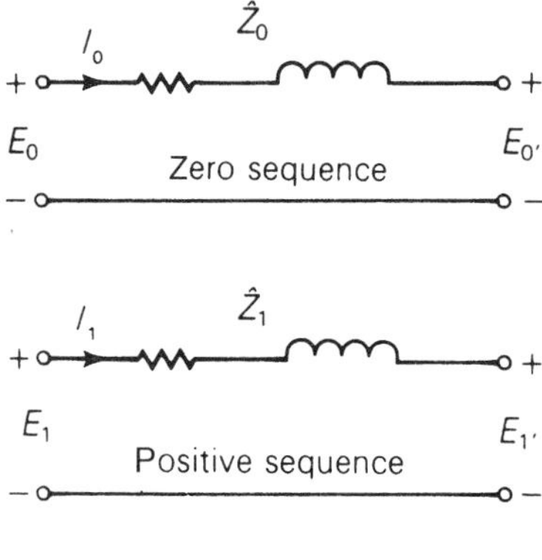

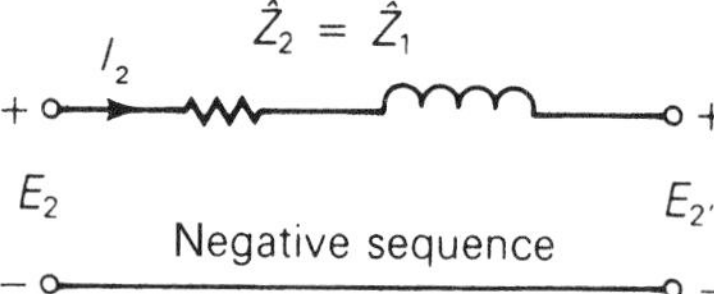

FIGURE 8.11

Circuit representation of the series sequence impedances of a completely transposed three-phase line

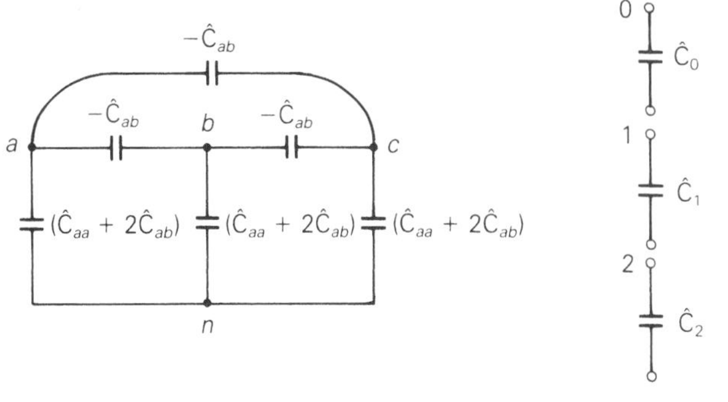

FIGURE 8.12 Circuit representations of the capacitances of a completely transposed three-phase line

In general, $\mathbf{C}_S$ is not diagonal. However, for the completely transposed line,

$$\hat{\mathbf{Y}}_S = A^{-1}\,\hat{\mathbf{Y}}_P A = \begin{bmatrix} \hat{y}_0 & 0 & 0 \\ 0 & \hat{y}_1 & 0 \\ 0 & 0 & \hat{y}_2 \end{bmatrix} = j(2\pi f)\begin{bmatrix} \hat{C}_0 & 0 & 0 \\ 0 & \hat{C}_1 & 0 \\ 0 & 0 & \hat{C}_2 \end{bmatrix} \tag{8.4.9}$$

where

$$\hat{C}_0 = \hat{C}_{aa} + 2\hat{C}_{ab} \quad \text{F/m} \tag{8.4.10}$$

$$\hat{C}_1 = \hat{C}_2 = \hat{C}_{aa} - \hat{C}_{ab} \quad \text{F/m} \tag{8.4.11}$$

Since $\hat{C}_{ab}$ is negative, the zero-sequence capacitance $\hat{C}_0$ is usually much less than the positive- or negative-sequence capacitance.

Circuit representations of the phase and sequence capacitances of a completely transposed three-phase line are shown in Figure 8.12.

8.5 SEQUENCE NETWORKS OF ROTATING MACHINES

A Y-connected synchronous generator grounded through a neutral impedance Z_n is shown in Figure 8.13. The internal generator voltages are designated E_a, E_b, and E_c, and the generator line currents are designated I_a, I_b, and I_c.

The sequence networks of the generator are shown in Figure 8.14. Since a three-phase synchronous generator is designed to produce balanced internal phase voltages E_a, E_b, and E_c with only a positive-sequence component, a source voltage E_{g1} is included only in the positive-sequence network. The sequence components of the line-to-ground voltages at the generator terminals are denoted V_0, V_1, and V_2 in Figure 8.14.

The voltage drop in the generator neutral impedance is $Z_n I_n$, which can be written as $(3Z_n)I_0$, since, from (8.1.27), the neutral current is three times the zero-sequence current. Since this voltage drop is due only to zero-sequence current, an impedance $(3Z_n)$ is placed in the zero-sequence network of Figure 8.14 in series with the generator zero-sequence impedance Z_{g0}.

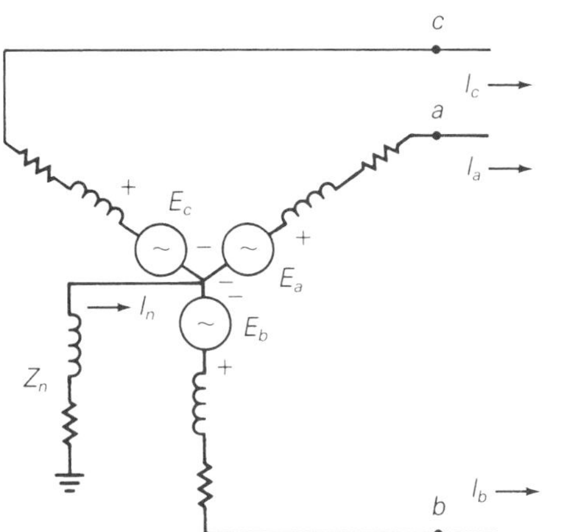

FIGURE 8.13

Y-connected synchronous generator

The sequence impedances of rotating machines are generally not equal. A detailed analysis of machine-sequence impedances is given in machine theory texts. Here is a brief explanation.

When a synchronous generator stator has balanced three-phase positive-sequence currents under steady-state conditions, the net mmf produced by these positive-sequence currents rotates at the synchronous rotor speed in the same direction as that of the rotor. Under this condition, a high value of magnetic flux penetrates the rotor, and the positive-sequence impedance Z_{g1} has a high value. Under steady-state conditions, the positive-sequence generator impedance is called the *synchronous impedance*.

When a synchronous generator stator has balanced three-phase negative-sequence currents, the net mmf produced by these currents rotates at synchronous speed in the direction opposite to that of the rotor. With respect to the rotor, the net mmf is not stationary but rotates at twice synchronous speed. Under this condition, currents are induced in the rotor windings that prevent the magnetic flux from penetrating the rotor. As such, the negative-sequence impedance Z_{g2} is less than the positive-sequence synchronous impedance.

When a synchronous generator has only zero-sequence currents, which are line (or phase) currents with equal magnitude and phase, then the net mmf produced by these currents is theoretically zero. The generator zero-sequence impedance Z_{g0} is the

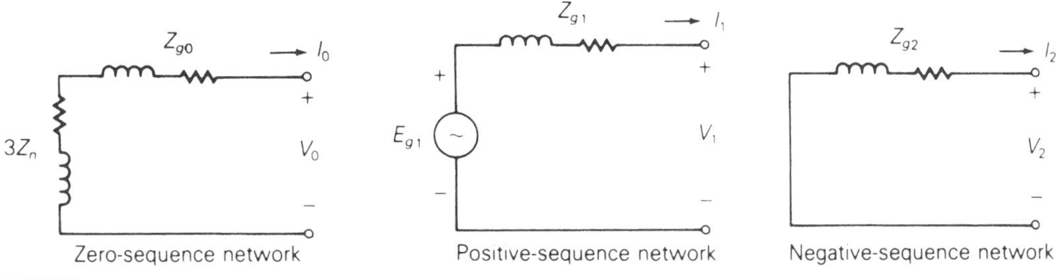

FIGURE 8.14

Sequence networks of a Y-connected synchronous generator

smallest sequence impedance and is due to leakage flux, end turns, and harmonic flux from windings that do not produce a perfectly sinusoidal mmf.

Typical values of machine-sequence impedances are listed in Table A.1 in the Appendix. The positive-sequence machine impedance is synchronous, transient, or subtransient. *Synchronous* impedances are used for steady-state conditions, such as in power-flow studies, which are described in Chapter 6. *Transient* impedances are used for stability studies, which are described in Chapter 13, and *subtransient* impedances are used for short-circuit studies, which are described in Chapters 7 and 9. Unlike the positive-sequence impedances, a machine has only one negative-sequence impedance and only one zero-sequence impedance.

The sequence networks for three-phase synchronous motors and for three-phase induction motors are shown in Figure 8.15. Synchronous motors have the same sequence networks as synchronous generators, except that the sequence currents for synchronous motors are referenced *into* rather than out of the sequence networks. Also, induction motors have the same sequence networks as synchronous motors, except that the positive-sequence voltage source E_{m1} is removed. Induction motors do not have a dc source of magnetic flux in their rotor circuits, and therefore E_{m1} is zero (or a short

FIGURE 8.15 Sequence networks of three-phase motors

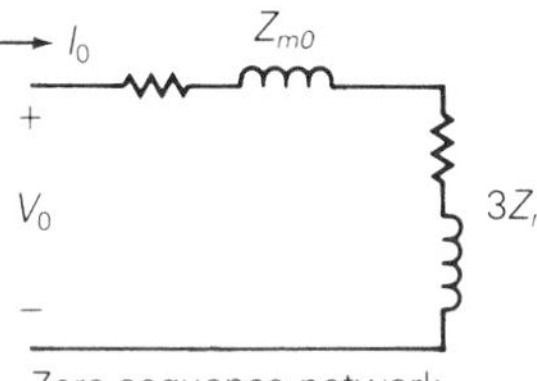

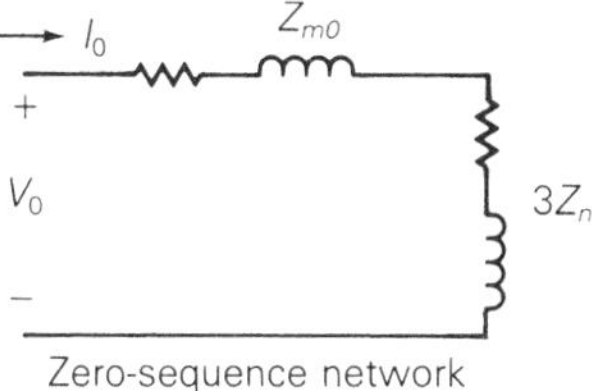

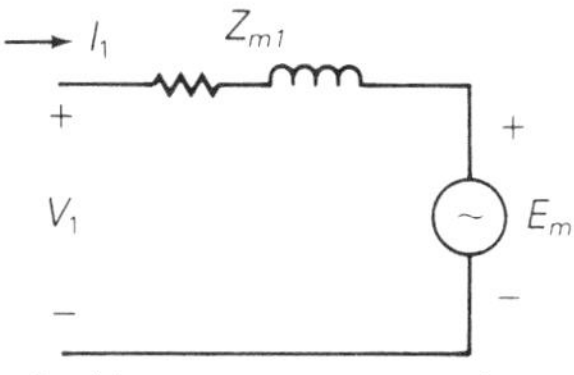

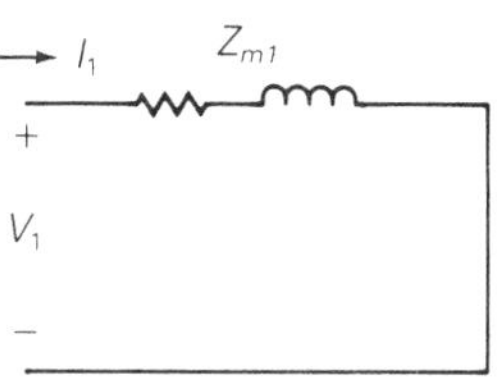

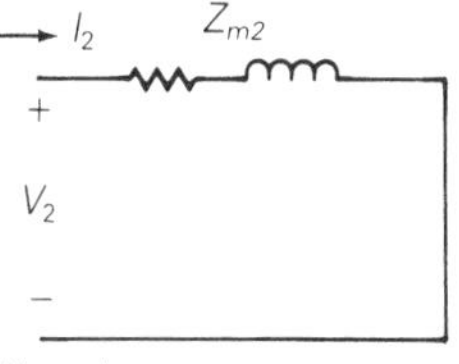

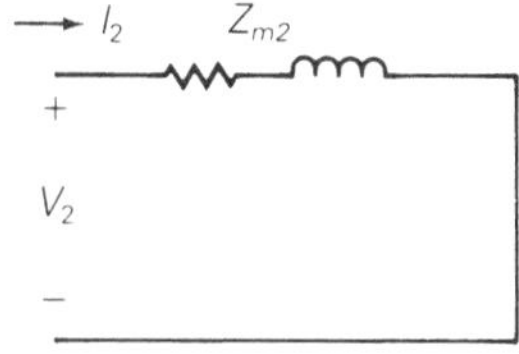

(a) Synchronous motor

(b) Induction motor

circuit). The sequence networks shown in Figures 8.14 and 8.15 are simplified networks for rotating machines. The networks do not take into account such phenomena as machine saliency, saturation effects, and more complicated transient effects. These simplified networks, however, are in many cases accurate enough for power system studies.

EXAMPLE 8.5

Currents in sequence networks

Draw the sequence networks for the circuit of Example 2.4 and calculate the sequence components of the line current. Assume that the generator neutral is grounded through an impedance $Z_n = j10\ \Omega$, and that the generator sequence impedances are $Z_{g0} = j1\ \Omega$, $Z_{g1} = j15\ \Omega$, and $Z_{g2} = j3\ \Omega$.

SOLUTION

The sequence networks are shown in Figure 8.16. They are obtained by interconnecting the sequence networks for a balanced-Δ load, for series-line

FIGURE 8.16 Sequence networks for Example 8.5

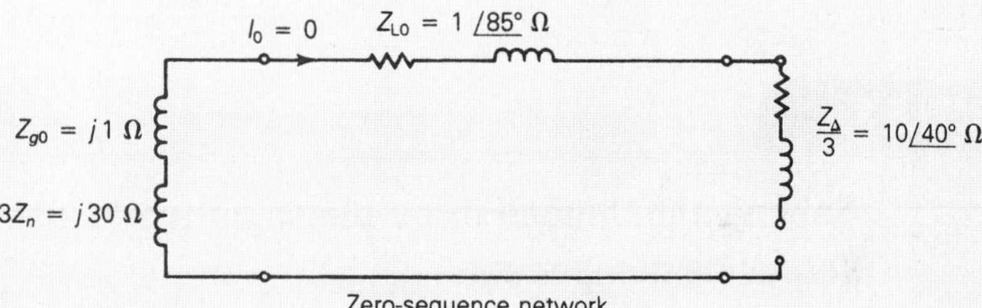

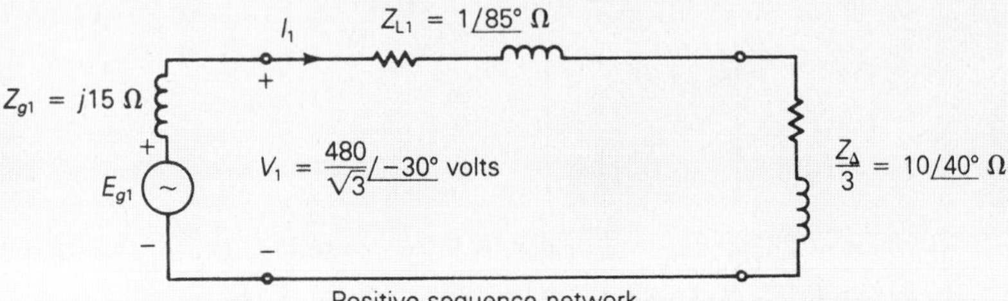

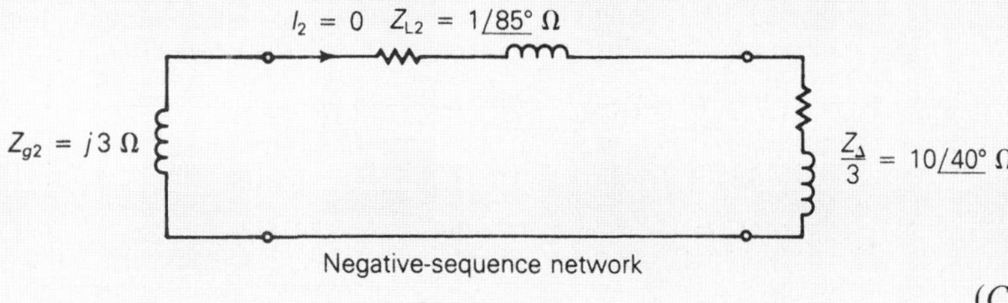

(Continued)

impedances, and for a synchronous generator, which are given in Figures 8.5, 8.10, and 8.14.

It is clear from Figure 8.16 that $I_0 = I_2 = 0$ since there are no sources in the zero- and negative-sequence networks. Also, the positive-sequence generator terminal voltage V_1 equals the generator line-to-neutral terminal voltage. Therefore, from the positive-sequence network shown in the figure and from the results of Example 2.4,

$$I_1 = \frac{V_1}{\left(Z_{L1} + \frac{1}{3}Z_\Delta\right)} = 25.83\underline{/-73.78^\circ}\text{ A} = I_a$$

Note that from (8.1.20), I_1 equals the line current I_a, since $I_0 = I_2 = 0$.

The following example illustrates the superiority of using symmetrical components for analyzing unbalanced systems.

EXAMPLE 8.6

Solving unbalanced three-phase networks using sequence components

A Y-connected voltage source with the following unbalanced voltage is applied to the balanced line and load of Example 2.4.

$$\begin{bmatrix} V_{ag} \\ V_{bg} \\ V_{cg} \end{bmatrix} = \begin{bmatrix} 277\underline{/0^\circ} \\ 260\underline{/-120^\circ} \\ 295\underline{/+115^\circ} \end{bmatrix} \text{ volts}$$

The source neutral is solidly grounded. Using the method of symmetrical components, calculate the source currents I_a, I_b, and I_c.

SOLUTION

Using (8.1.13) through (8.1.15), the sequence components of the source voltages are

$$V_0 = \tfrac{1}{3}(277\underline{/0^\circ} + 260\underline{/-120^\circ} + 295\underline{/115^\circ})$$

$$= 7.4425 + j14.065 = 15.912\underline{/62.11^\circ} \text{ volts}$$

$$V_1 = \tfrac{1}{3}(277\underline{/0^\circ} + 260\underline{/-120^\circ + 120^\circ} + 295\underline{/115^\circ + 240^\circ})$$
$$= \tfrac{1}{3}(277\underline{/0^\circ} + 260\underline{/0^\circ} + 295\underline{/-5^\circ})$$
$$= 276.96 - j8.5703 = 277.1\underline{/-1.772^\circ} \text{ volts}$$
$$V_2 = \tfrac{1}{3}(277\underline{/0^\circ} + 260\underline{/-120^\circ + 240^\circ} + 295\underline{/115^\circ + 120^\circ})$$
$$= \tfrac{1}{3}(277\underline{/0^\circ} + 260\underline{/120^\circ} + 295\underline{/235^\circ})$$
$$= -7.4017 - j5.4944 = 9.218\underline{/216.59^\circ} \text{ volts}$$

These sequence voltages are applied to the sequence networks of the line and load, as shown in Figure 8.17. The sequence networks of this figure are

FIGURE 8.17

Sequence networks for Example 8.6

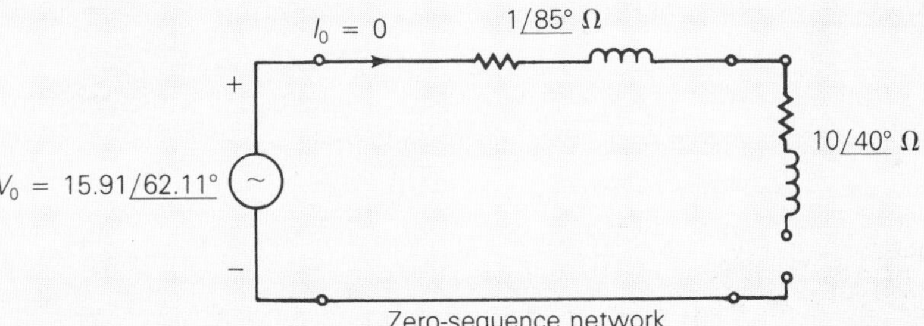

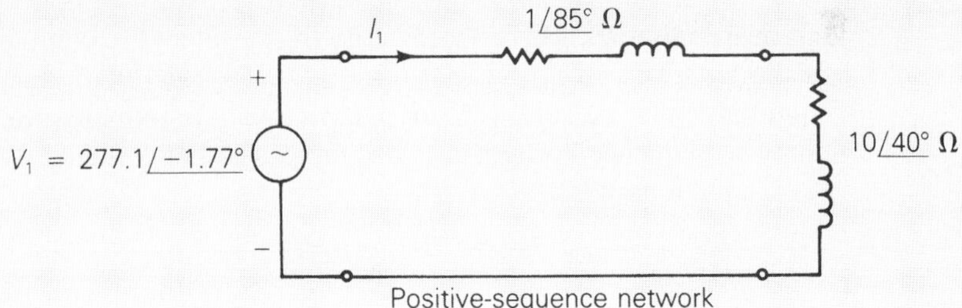

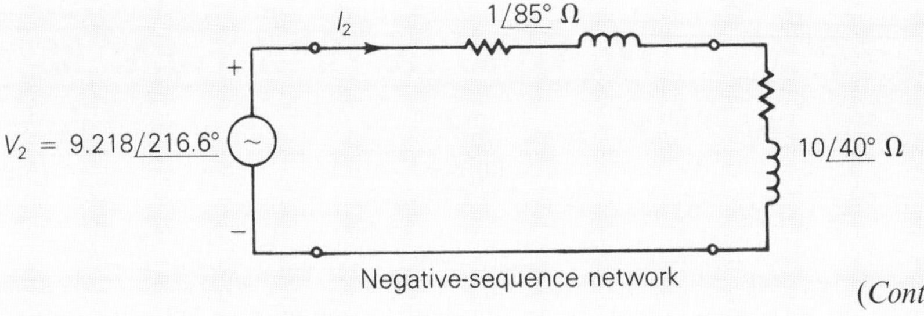

(Continued)

uncoupled, and the sequence components of the source currents are easily calculated as follows:

$$I_0 = 0$$

$$I_1 = \frac{V_1}{Z_{L1} + \frac{Z_\Delta}{3}} = \frac{277.1\angle{-1.772°}}{10.73\angle{43.78°}} = 25.82\angle{-45.55°} \quad \text{A}$$

$$I_2 = \frac{V_2}{Z_{L2} + \frac{Z_\Delta}{3}} = \frac{9.218\angle{216.59°}}{10.73\angle{43.78°}} = 0.8591\angle{172.81°} \quad \text{A}$$

Using (8.1.20) through (8.1.22), the source currents are

$$\begin{aligned} I_a &= (0 + 25.82\angle{-45.55°} + 0.8591\angle{172.81°}) \\ &= 17.23 - j18.32 = 25.15\angle{-46.76°} \quad \text{A} \\ I_b &= (0 + 25.82\angle{-45.55° + 240°} + 0.8591\angle{172.81° + 120°}) \\ &= (25.82\angle{194.45°} + 0.8591\angle{292.81°}) \\ &= -24.67 - j7.235 = 25.71\angle{196.34°} \quad \text{A} \\ I_c &= (0 + 25.82\angle{-45.55° + 120°} + 0.8591\angle{172.81° + 240°}) \\ &= (25.82\angle{74.45°} + 0.8591\angle{52.81°}) \\ &= 7.441 + j25.56 = 26.62\angle{73.77°} \quad \text{A} \end{aligned}$$

You should calculate the line currents for this example without using symmetrical components in order to verify this result and to compare the two solution methods (see Problem 8.33). Without symmetrical components, coupled KVL equations must be solved. With symmetrical components, the conversion from phase to sequence components decouples the networks as well as the resulting KVL equations, as shown above.

8.6 PER-UNIT SEQUENCE MODELS OF THREE-PHASE TWO-WINDING TRANSFORMERS

Figure 8.18(a) is a schematic representation of an ideal Y–Y transformer grounded through neutral impedances Z_N and Z_n. Figures 8.18(b through d) show the per-unit sequence networks of this ideal transformer.

When balanced positive-sequence currents or balanced negative-sequence currents are applied to the transformer, the neutral currents are zero, and there are no voltage drops across the neutral impedances. Therefore, the per-unit positive- and negative-sequence networks of the ideal Y–Y transformer, Figures 8.18(b) and (c), are the same as the per-unit single-phase ideal transformer, Figure 3.9(a).

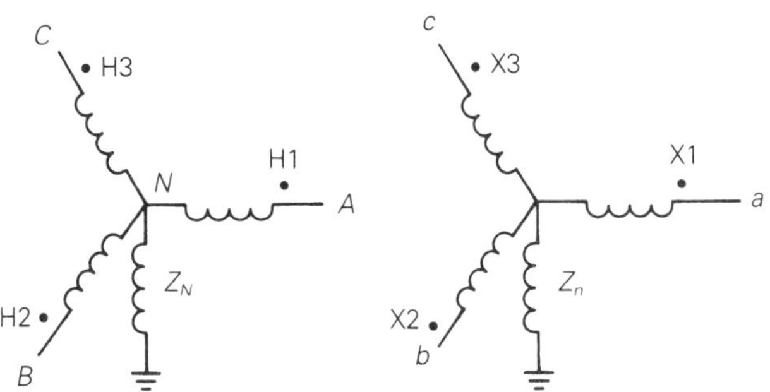

(a) Schematic representation

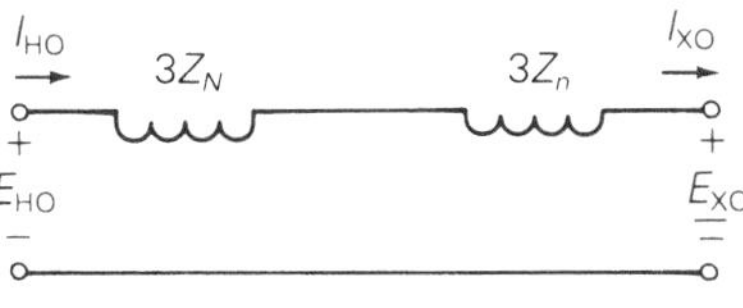

(b) Per-unit zero-sequence network

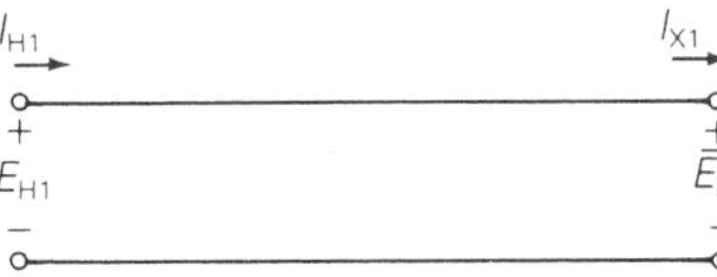

(c) Per-unit positive-sequence network

(d) Per-unit negative-sequence network

FIGURE 8.18

Ideal Y–Y transformer

Zero-sequence currents have equal magnitudes and equal phase angles. When per-unit sequence currents $I_{A0} = I_{B0} = I_{C0} = I_0$ are applied to the high-voltage windings of an ideal Y–Y transformer, the neutral current I_0 flows through the neutral impedance Z_N, with a voltage drop $(3Z_N)I_0$. Also, per-unit zero-sequence current I_0 flows in each low-voltage winding [from (3.3.9)], and therefore $3I_0$ flows through neutral impedance Z_n, with a voltage drop $(3I_0)Z_n$.

The per-unit zero-sequence network, which includes the impedances $(3Z_N)$ and $(3Z_n)$, is shown in Figure 8.18(b).

Note that if either one of the neutrals of an ideal transformer is ungrounded, then no zero sequence can flow in either the high- or low-voltage windings. For example, if the high-voltage winding has an open neutral, then $I_N = 3I_0 = 0$, which in turn forces $I_0 = 0$ on the low-voltage side. This can be shown in the zero-sequence network of Figure 8.18(b) by making $Z_N = \infty$, which corresponds to an open circuit.

The per-unit sequence networks of a practical Y–Y transformer are shown in Figure 8.19(a). These networks are obtained by adding external impedances to the sequence networks of the ideal transformer, as follows. The leakage impedances of the high-voltage windings are series impedances like the series impedances shown in Figure 8.9, with no coupling between phases ($Z_{ab} = 0$). If the phase a, b, and c

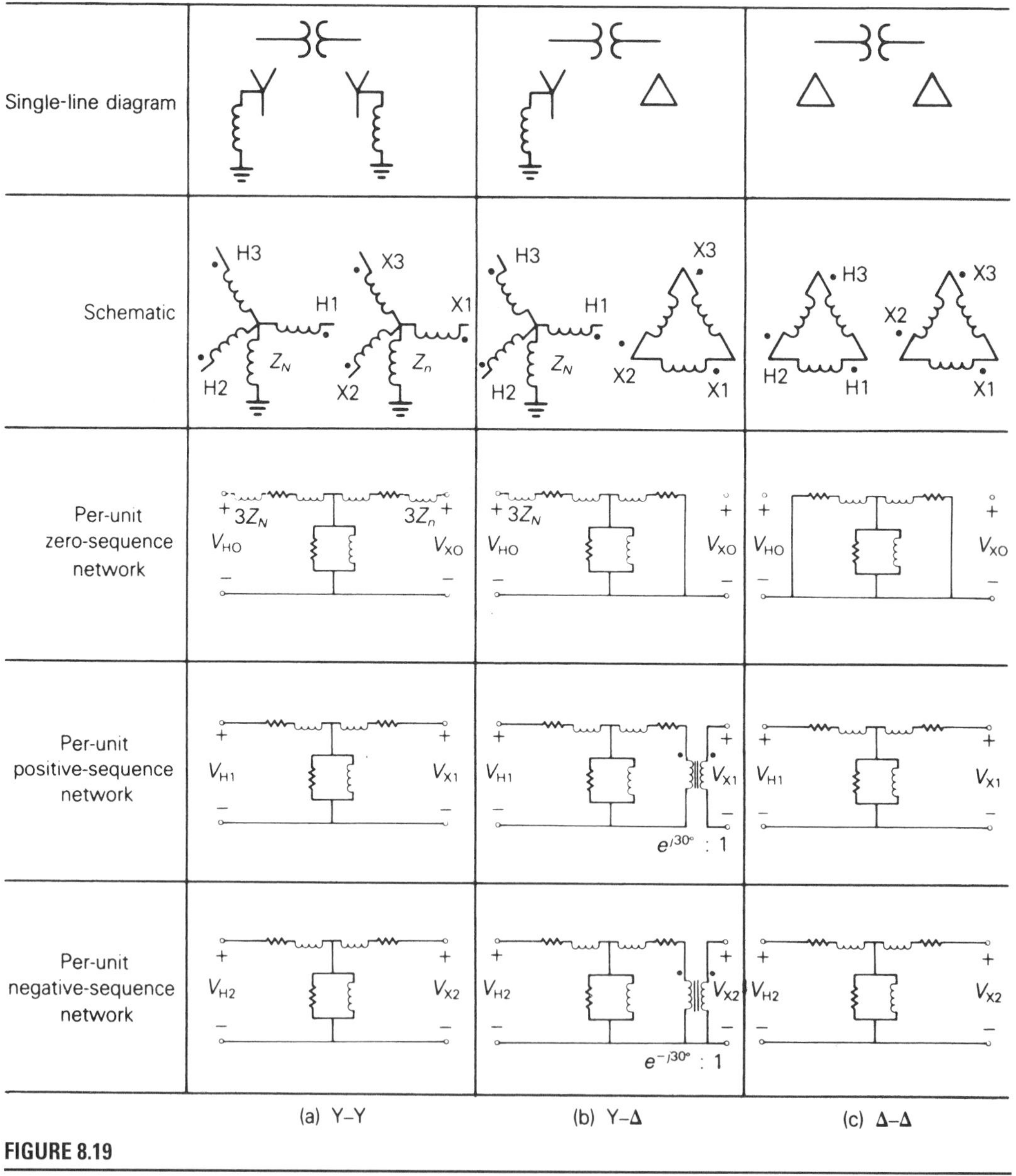

FIGURE 8.19

Per-unit sequence networks of practical Y–Y, Y–Δ, and Δ–Δ transformers

windings have equal leakage impedances $Z_H = R_H + jX_H$, then the series impedances are *symmetrical* with sequence networks, as shown in Figure 8.10, where $Z_{H0} = Z_{H1} = Z_{H2} = Z_H$. Similarly, the leakage impedances of the low-voltage windings are symmetrical series impedances with $Z_{X0} = Z_{X1} = Z_{X2} = Z_X$. These series leakage impedances are shown in per unit in the sequence networks of Figure 8.19(a).

The shunt branches of the practical Y–Y transformer, which represent exciting current, are equivalent to the Y load of Figure 8.3. Each phase in Figure 8.3 represents a core loss resistor in parallel with a magnetizing inductance. Assuming these are the same for each phase, then the Y load is *symmetrical*, and the sequence networks are shown in Figure 8.4. These shunt branches are also shown in Figure 8.19(a). Note that ($3Z_N$) and $3Z_n$) have already been included in the zero-sequence network.

The per-unit positive- and negative-sequence transformer impedances of the practical Y–Y transformer in Figure 8.19(a) are identical, which is always true for nonrotating equipment. The per-unit zero-sequence network, however, depends on the neutral impedances Z_N and Z_n.

The per-unit sequence networks of the Y–Δ transformer, shown in Figure 8.19(b), have the following features:

1. The per-unit impedances do not depend on the winding connections. That is, the per-unit impedances of a transformer that is connected Y–Y, Y–Δ, Δ–Y, or Δ–Δ are the same. However, the base voltages do depend on the winding connections.
2. A phase shift is included in the per-unit positive- and negative-sequence networks. For the American standard, the positive-sequence voltages and currents on the high-voltage side of the Y–Δ transformer lead the corresponding quantities on the low-voltage side by 30°. For negative sequence, the high-voltage quantities lag by 30°.
3. Zero-sequence currents can flow in the Y winding if there is a neutral connection, and corresponding zero-sequence currents flow within the Δ winding. However, no zero-sequence current enters or leaves the Δ winding.

The phase shifts in the positive- and negative-sequence networks of Figure 8.19(b) are represented by the phase-shifting transformer of Figure 3.4. Also, the zero-sequence network of Figure 8.19(b) provides a path on the Y side for zero-sequence current to flow, but no zero-sequence current can enter or leave the Δ side.

The per-unit sequence networks of the Δ–Δ transformer, shown in Figure 8.19(c), have the following features:

1. The positive- and negative-sequence networks, which are identical, are the same as those for the Y–Y transformer. It is assumed that the windings are labeled so there is no phase shift. Also, the per-unit impedances do not depend on the winding connections, but the base voltages do.
2. Zero-sequence currents *cannot* enter or leave either Δ winding, although they can circulate within the Δ windings.

EXAMPLE 8.7

Solving unbalanced three-phase networks with transformers using per-unit sequence components

A 75-kVA, 480-volt Δ/208-volt Y transformer with a solidly grounded neutral is connected between the source and line of Example 8.6. The transformer leakage reactance is $X_{eq} = 0.10$ per unit; winding resistances and exciting current are neglected. Using the transformer ratings as base quantities, draw the per-unit sequence networks and calculate the phase a source current I_a.

SOLUTION

The base quantities are $S_{base1\phi} = 75/3 = 25$ kVA, $V_{baseHLN} = 480/\sqrt{3} = 277.1$ volts, $V_{baseXLN} = 208/\sqrt{3} = 120.1$ volts, and $Z_{baseX} = (120.1)^2/25{,}000 = 0.5770\ \Omega$. The sequence components of the actual source voltages are given in Figure 8.17. In per unit, these voltages are

$$V_0 = \frac{15.91\underline{/62.11^\circ}}{277.1} = 0.05742\underline{/62.11^\circ} \text{ per unit}$$

$$V_1 = \frac{277.1\underline{/-1.772^\circ}}{277.1} = 1.0\underline{/-1.772^\circ} \text{ per unit}$$

$$V_2 = \frac{9.218\underline{/216.59^\circ}}{277.1} = 0.03327\underline{/216.59^\circ} \text{ per unit}$$

The per-unit line and load impedances, which are located on the low-voltage side of the transformer, are

$$Z_{L0} = Z_{L1} = Z_{L2} = \frac{1\underline{/85^\circ}}{0.577} = 1.733\underline{/85^\circ} \text{ per unit}$$

$$Z_{load1} = Z_{load2} = \frac{Z_\Delta}{3(0.577)} = \frac{10\underline{/40^\circ}}{0.577} = 17.33\underline{/40^\circ} \text{ per unit}$$

The per-unit sequence networks are shown in Figure 8.20. Note that the per-unit line and load impedances, when referred to the high-voltage side of the phase-shifting transformer, do not change [(see (3.1.26)]. Therefore, from Figure 8.20, the sequence components of the source currents are

$$I_0 = 0$$

$$I_1 = \frac{V_1}{jX_{eq} + Z_{L1} + Z_{load1}} = \frac{1.0\underline{/-1.772^\circ}}{j0.10 + 1.733\underline{/85^\circ} + 17.33\underline{/40^\circ}}$$

$$= \frac{1.0\underline{/-1.772^\circ}}{13.43 + j12.97} = \frac{1.0\underline{/-1.772^\circ}}{18.67\underline{/44.0^\circ}} = 0.05356\underline{/-45.77^\circ} \text{ per unit}$$

$$I_2 = \frac{V_2}{jX_{eq} + Z_{L2} + Z_{load2}} = \frac{0.03327\underline{/216.59^\circ}}{18.67\underline{/44.0^\circ}}$$

$$= 0.001782\underline{/172.59^\circ} \text{ per unit}$$

FIGURE 8.20

Per-unit sequence networks for Example 8.7

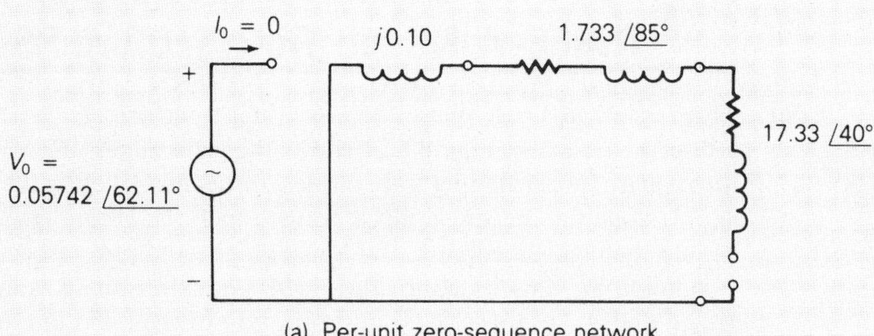

(a) Per-unit zero-sequence network

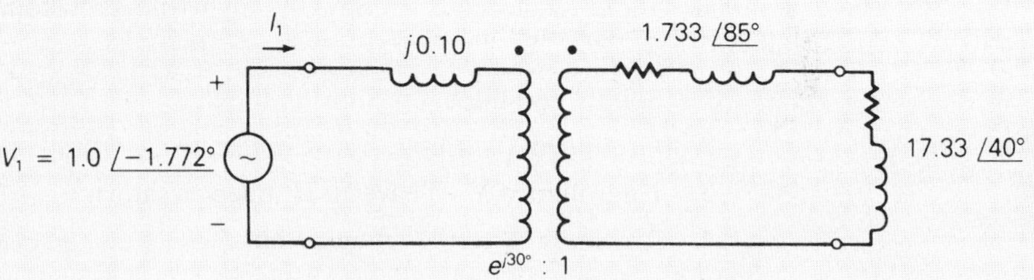

(b) Per-unit positive-sequence network

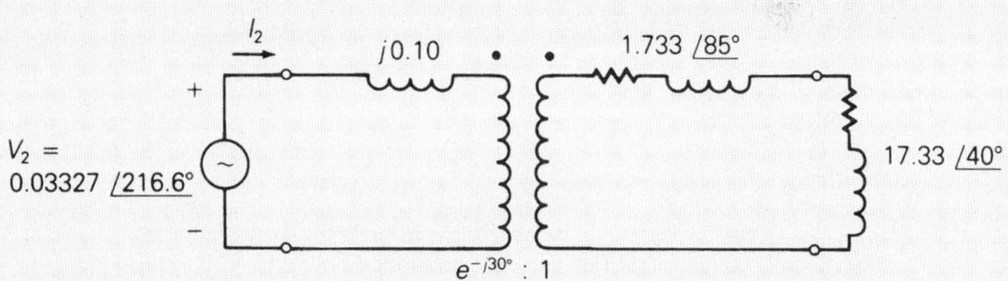

(c) Per-unit negative-sequence network

The phase a source current is then, using (8.1.20),

$$
\begin{aligned}
I_a &= I_0 + I_1 + I_2 \\
&= 0 + 0.05356\angle -45.77° + 0.001782\angle 172.59° \\
&= 0.03511 - j0.03764 = 0.05216\angle -46.19° \quad \text{per unit}
\end{aligned}
$$

Using $I_{\text{baseH}} = \dfrac{75{,}000}{480\sqrt{3}} = 90.21$ A,

$$I_a = (0.05216)(90.21)\angle -46.19° = 4.705\angle -46.19° \quad \text{A}$$

8.7 PER-UNIT SEQUENCE MODELS OF THREE-PHASE THREE-WINDING TRANSFORMERS

Three identical single-phase three-winding transformers can be connected to form a three-phase bank. Figure 8.21 shows the general per-unit sequence networks of a three-phase three-winding transformer. Instead of labeling the windings 1, 2, and 3, as was done for the single-phase transformer, the letters *H*, *M*, and *X* are used to denote the high-, medium-, and low-voltage windings, respectively. By convention, a common S_{base} is selected for the H, M, and X terminals, and voltage bases V_{baseH}, V_{baseM}, and V_{baseX} are selected in proportion to the rated line-to-line voltages of the transformer.

For the general zero-sequence network, Figure 8.21(a), the connection between terminals H and H′ depends on how the high-voltage windings are connected, as follows:

1. Solidly grounded Y—Short H to H′.
2. Grounded Y through Z_N—Connect ($3Z_N$) from H to H′.
3. Ungrounded Y—Leave H–H′ open as shown.
4. Δ—Short H′ to the reference bus.

FIGURE 8.21 Per-unit sequence networks of a three-phase three-winding transformer

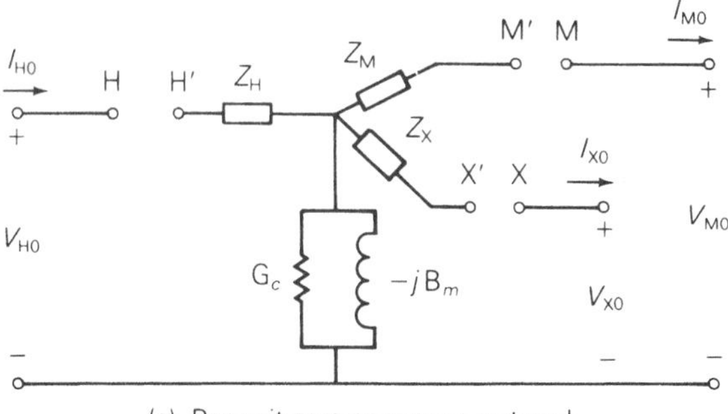

(a) Per-unit zero-sequence network

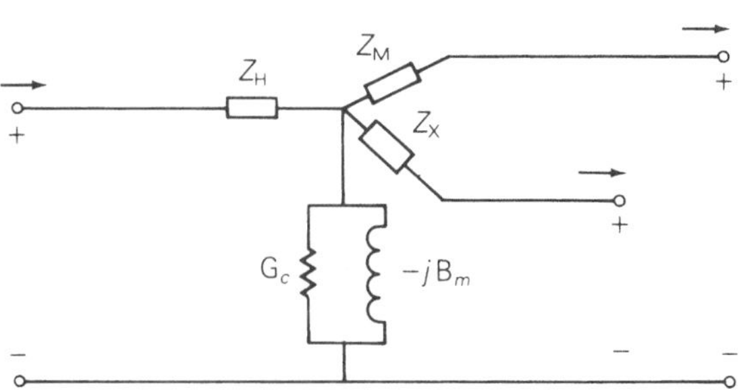

(b) Per-unit positive- or negative-sequence network (phase shift not shown)

Terminals X–X′ and M–M′ are connected in a similar manner.

The impedances of the per-unit negative-sequence network are the same as those of the per-unit positive-sequence network, which is always true for non-rotating equipment. Phase-shifting transformers, not shown in Figure 8.21(b), can be included to model phase shift between Δ and Y windings.

EXAMPLE 8.8

Three-winding three-phase transformer: per-unit sequence networks

Three transformers, each identical to that described in Example 3.9, are connected as a three-phase bank in order to feed power from a 900-MVA, 13.8-kV generator to a 345-kV transmission line and to a 34.5-kV distribution line. The transformer windings are connected as follows:

13.8-kV windings (X):	Δ, to generator
199.2-kV windings (H):	solidly grounded Y, to 345-kV line
19.92-kV windings (M):	grounded Y through $Z_n = j01.0\ \Omega$, to 34.5-kV line

The positive-sequence voltages and currents of the high- and medium-voltage Y windings lead the corresponding quantities of the low-voltage Δ winding by 30°. Draw the per-unit sequence networks, using a three-phase base of 900 MVA and 13.8 kV for terminal X.

SOLUTION

The per-unit sequence networks are shown in Figure 8.22. Since $V_{\text{baseX}} = 13.8$ kV is the rated line-to-line voltage of terminal X, $V_{\text{baseM}} = \sqrt{3}\,(19.92) = 34.5$ kV, which is the rated line-to-line voltage of terminal M. The base impedance of the medium-voltage terminal is then

$$Z_{\text{baseM}} = \frac{(34.5)^2}{900} = 1.3225 \quad \Omega$$

Therefore, the per-unit neutral impedance is

$$Z_n = \frac{j0.10}{1.3225} = j0.07561 \quad \text{per unit}$$

and $(3Z_n) = j0.2268$ is connected from terminal M to M′ in the per-unit zero-sequence network. Since the high-voltage windings have a solidly grounded neutral, H to H′ is shorted in the zero-sequence network. Also, phase-shifting transformers are included in the positive- and negative-sequence networks.

(*Continued*)

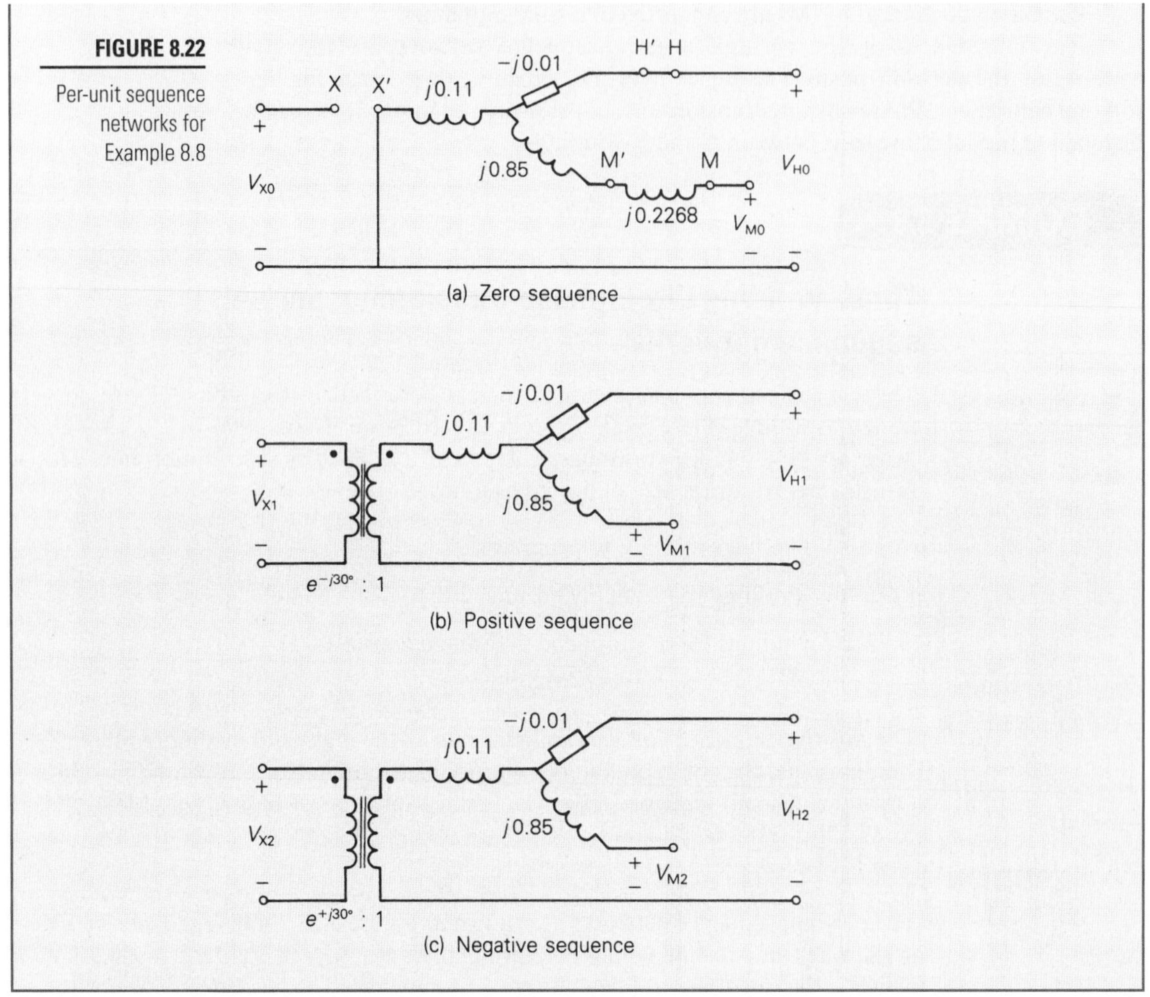

FIGURE 8.22 Per-unit sequence networks for Example 8.8

8.8 POWER IN SEQUENCE NETWORKS

The power delivered to a three-phase network can be determined from the power delivered to the sequence networks. Let S_P denote the total complex power delivered to the three-phase load of Figure 8.7, which can be calculated from

$$S_P = V_{ag}I_a^* + V_{bg}I_b^* + V_{cg}I_c^* \tag{8.8.1}$$

Equation (8.8.1) is also valid for the total complex power delivered by the three-phase generator of Figure 8.13, or for the complex power delivered to any three-phase bus. Rewriting (8.8.1) in matrix format,

$$S_P = [V_{ag} V_{bg} V_{cg}] \begin{bmatrix} I_a^* \\ I_b^* \\ I_c^* \end{bmatrix}$$
$$= V_P^T I_P^* \tag{8.8.2}$$

where T denotes transpose and * denotes complex conjugate. Now, using (8.1.9) and (8.1.16),

$$\begin{aligned} S_P &= (AV_s)^T (AI_s)^* \\ &= V_s^T [A^T A^*] I_s^* \end{aligned} \tag{8.8.3}$$

Using the definition of A, which is (8.1.8), to calculate the term within the brackets of (8.8.3), and noting that a and a^2 are conjugates,

$$A^T A^* = \begin{bmatrix} 1 & 1 & 1 \\ 1 & a^2 & a \\ 1 & a & a^2 \end{bmatrix}^T \begin{bmatrix} 1 & 1 & 1 \\ 1 & a^2 & a \\ 1 & a & a^2 \end{bmatrix}^*$$

$$= \begin{bmatrix} 1 & 1 & 1 \\ 1 & a^2 & a \\ 1 & a & a^2 \end{bmatrix} \begin{bmatrix} 1 & 1 & 1 \\ 1 & a & a^2 \\ 1 & a^2 & a \end{bmatrix}$$

$$= \begin{bmatrix} 3 & 0 & 0 \\ 0 & 3 & 0 \\ 0 & 0 & 3 \end{bmatrix} = 3\mathbf{U} \tag{8.8.4}$$

Equation (8.8.4) can now be used in (8.8.3) to obtain

$$S_P = 3V_S^T I_S^*$$
$$= 3[V_0 + V_1 + V_2] \begin{bmatrix} I_0^* \\ I_1^* \\ I_2^* \end{bmatrix} \tag{8.8.5}$$

$$\begin{aligned} S_P &= 3(V_0 I_0^* + V_1 I_1^* + V_2 I_2^*) \\ &= 3S_S \end{aligned} \tag{8.8.6}$$

Thus, the total complex power S_P delivered to a three-phase network equals *three* times the total complex power S_S delivered to the sequence networks.

The factor of 3 occurs in (8.8.6) because $A^T A^* = 3\mathbf{U}$ as shown by (8.8.4). It is possible to eliminate this factor of 3 by defining a new transformation matrix $A_1 = (1/\sqrt{3})\, A$ such that $A_1^T A_1^* = \mathbf{U}$, which means that A_1 is a *unitary* matrix. Using A_1 instead of A, the total complex power delivered to three-phase networks would equal the total complex power delivered to the sequence networks. However, standard industry practice for symmetrical components is to use A, defined by (8.1.8).

EXAMPLE 8.9

Power in sequence networks

Calculate S_P and S_S delivered by the three-phase source in Example 8.6. Verify that $S_P = 3S_S$.

SOLUTION

Using (8.5.1),

$$\begin{aligned} S_P &= (277\underline{/0^\circ})(25.15\underline{/+46.76^\circ}) + (260\underline{/-120^\circ})(25.71\underline{/-196.34^\circ}) \\ &\quad + (295\underline{/115^\circ})(26.62\underline{/-73.77^\circ}) \\ &= 6967\underline{/46.76^\circ} + 6685\underline{/43.66^\circ} + 7853\underline{/41.23^\circ} \\ &= 15{,}520 + j14{,}870 = 21{,}490\underline{/43.78^\circ} \quad \text{VA} \end{aligned}$$

In the sequence domain,

$$\begin{aligned} S_S &= V_0 I_0^* + V_1 I_1^* + V_2 I_2^* \\ &= 0 + (277.1\underline{/-1.77^\circ})(25.82\underline{/45.55^\circ}) \\ &\quad + (9.218\underline{/216.59^\circ})(0.8591\underline{/-172.81^\circ}) \\ &= 7155\underline{/43.78^\circ} + 7.919\underline{/43.78^\circ} \\ &= 5172 + j4958 = 7163\underline{/43.78^\circ} \quad \text{VA} \end{aligned}$$

Also,

$$3S_S = 3(7163\underline{/43.78^\circ}) = 21{,}490\underline{/43.78^\circ} = S_P$$

MULTIPLE CHOICE QUESTIONS

SECTION 8.1

8.1 Positive-sequence components consist of three phasors with _____ magnitudes and _____ phase displacement in positive sequence; negative-sequence components consist of three phasors with _____ magnitudes and _____ phase displacement in negative sequence; and zero-sequence components consist of three phasors with _____ magnitudes and _____ phase displacement.

8.2 In symmetrical-component theory, express the complex-number operator $a = 1\underline{/120^\circ}$ in exponential and rectangular forms.

8.3 In terms of sequence components of phase a given by $V_{a0} = V_0$, $V_{a1} = V_1$, and $V_{a2} = V_2$, give expressions for the phase voltages V_a, V_b, and V_c.
$V_a =$ _____________; $V_b =$ _____________; $V_c =$ _____________

8.4 The sequence components V_0, V_1, and V_2 can be expressed in terms of phase components V_a, V_b, and V_c.
$V_0 =$ _______________; $V_1 =$ _______________; $V_2 =$ _______________

8.5 In a balanced three-phase system, what is the zero-sequence voltage?
$V_0 =$ _______________

8.6 In an unbalanced three-phase system, line-to-neutral voltage ___________ have a zero-sequence component, whereas line-to-line voltages ___________ have a zero-sequence component.

8.7 Can the symmetrical component transformation be applied to currents, just as it is applied to voltages?
(a) Yes (b) No

8.8 In a three-phase Y-connected system with a neutral, express the neutral current in terms of phase currents and sequence-component terms.
$I_n =$ _______________ $=$ _______________

8.9 In a balanced Y-connected system, what is the zero-sequence component of the line currents?

8.10 In a Δ-connected three-phase system, line currents have no zero-sequence component.
(a) True (b) False

8.11 Balanced three-phase systems with positive sequence do not have zero-sequence and negative-sequence components.
(a) True (b) False

8.12 Unbalanced three-phase systems may have nonzero values for all sequence components.
(a) True (b) False

SECTION 8.2

8.13 For a balanced-Y impedance load with per-phase impedance of Z_Y and a neutral impedance Z_n connected between the load neutral and the close space ground, the 3×3 phase-impedance matrix consists of equal diagonal elements given by ___________ and equal nondiagonal elements given by ___________.

8.14 Express the sequence impedance matrix $\boldsymbol{Z}_S$ in terms of the phase-impedance matrix $\boldsymbol{Z}_P$, and the transformation matrix $\boldsymbol{A}$ which relates $\boldsymbol{V}_P = \boldsymbol{A}\boldsymbol{V}_S$ and $\boldsymbol{I}_P = \boldsymbol{A}\boldsymbol{I}_S$.
$\boldsymbol{Z}_S =$ ___________.

8.15 The sequence impedance matrix Z_S for a balanced-Y load is a diagonal matrix and the sequence networks are uncoupled.
(a) True (b) False

8.16 For a balanced-Y impedance load with per-phase impedance of Z_Y and a neutral impedance Z_n, the zero-sequence voltage $V_0 = Z_0 I_0$, where $Z_0 =$ ___________.

8.17 For a balanced-Δ load with per-phase impedance of Z_Δ, the equivalent Y-load has an open neutral; for the corresponding uncoupled sequence networks, $Z_0 =$ __________, $Z_1 =$ __________, and $Z_2 =$ __________.

8.18 For a three-phase symmetrical impedance load, the sequence impedance matrix is __________ and hence the sequence networks are (a) coupled or (b) uncoupled.

SECTION 8.3

8.19 Sequence networks for three-phase symmetrical series impedances are (a) coupled or (b) uncoupled; positive-sequence currents produce only __________ voltage drops.

SECTION 8.4

8.20 The series-sequence impedance matrix of a completely transposed three-phase line is __________ with its nondiagonal elements equal to __________.

SECTION 8.5

8.21 A Y-connected synchronous generator grounded through a neutral impedance Z_n with a zero-sequence impedance Z_{g0} has zero-sequence impedance $Z_0 =$ __________ in its zero-sequence network.

8.22 In sequence networks, a Y-connected synchronous generator is represented by its source per-unit voltage only in __________ network, while (a) synchronous, (b) transient or (c) sub-transient impedance is used in positive-sequence network for short-circuit studies.

8.23 In the positive-sequence network of a synchronous motor, a source voltage is represented, whereas in that of an induction motor, the source voltage (a) does or (b) does not come into picture.

8.24 With symmetrical components, the conversion from phase to sequence components decouples the networks and the resulting KVL equations.
(a) True (b) False

SECTION 8.6

8.25 Consider the per-unit sequence networks of Y–Y, Y–Δ, and Δ–Δ transformers with neutral impedances of Z_N on the high-voltage Y-side and Z_n on the low-voltage Y-side. Answer the following:

(i) Zero-sequence currents (a) can or (b) cannot flow in the Y winding with a neutral connection; corresponding zero-sequence currents (a) do or (b) do not flow within the delta winding; however zero-sequence current (a) does or (b) does not enter or leave the Δ winding. In a zero-sequence network, (a) 1, (b) 2, or (c) 3 times the neutral impedance comes into play in series.

(ii) In Y(HV)–Δ(LV) transformers, if a phase shift is included as per the American-standard notation, the ratio ___________ is used in positive-sequence network, and ___________ the ratio ___________ is used in the negative-sequence network.

(iii) The base voltages depend on the winding connections; the per-unit impedances (a) do or (b) do not depend on the winding connections.

SECTION 8.7

8.26 In per-unit sequence models of three-phase three-winding transformers, for the general zero-sequence network, the connection between terminals H and H′ depends on how the high-voltage windings are connected:

(i) For solidly grounded Y, ___________ H to H′.
(ii) For grounded Y through Z_n connect ___________ from H to H′.
(iii) For ungrounded Y, leave H–H′ ___________.
(iv) For Δ, ___________ H′ to the reference bus.

SECTION 8.8

8.27 The total complex power delivered to a three-phase network equals (a) 1, (b) 2, or (c) 3 times the total complex power delivered to the sequence networks.

8.28 Express the complex power S_S delivered to the sequence networks in terms of sequence voltages and sequence currents, where $S_S =$ ___________.

PROBLEMS

SECTION 8.1

8.1 Using the operator $a = 1\underline{/120°}$, evaluate the following in polar form: (a) $(a - 1)/(1 + a - a^2)$, (b) $(a^2 + a + j)/(ja + a^2)$, (c) $(1 + a)(1 + a^2)$, and (d) $(a - a^2)(a^2 - 1)$.

8.2 Using $a = 1\underline{/120°}$, evaluate the following in rectangular form:

a. a^{10}

b. $(ja)^{10}$

c. $(1 - a)^3$

d. e^a

Hint for (d): $e^{(x+jy)} = e^x e^{jy} = e^x\underline{/y}$, where y is in radians.

8.3 Determine the symmetrical components of the following line currents: (a) $I_a = 6\underline{/90°}$, $I_b = 6\underline{/320°}$, $I_c = 6\underline{/220°}$ A and (b) $I_a = j40$, $I_b = 40$, $I_c = 0$ A.

8.4 Find the phase voltages V_{an}, V_{bn}, and V_{cn} whose sequence components are $V_0 = 45\underline{/80°}$, $V_1 = 90\underline{/0°}$, $V_2 = 45\underline{/90°}$ V.

8.5 For the unbalanced three-phase system described by

$$I_a = 10\underline{/0°}\ \text{A},\ I_b = 8\underline{/-90°}\ \text{A},\ I_c = 6\underline{/150°}\ \text{A}$$

compute the symmetrical components I_0, I_1, and I_2.

8.6 (a) Given the symmetrical components to be

$$V_0 = 10\angle 0^\circ V,\ V_1 = 80\angle 30^\circ V,\ V_2 = 40\angle -30^\circ V$$

determine the unbalanced phase voltages V_a, V_b, and V_c. (b) Using the results of part (a), calculate the line-to-line voltages V_{ab}, V_{bc}, and V_{ca}. Then determine the symmetrical components of these line-to-line voltages, the symmetrical components of the corresponding phase voltages, and the phase voltages. Compare them with the result of part (a). Comment on why they are different, even though either set results in the same line-to-line voltages.

8.7 One line of a three-phase generator is open-circuited, while the other two are short-circuited to ground. The line currents are $I_a = 0$, $I_b = 1200\angle 150^\circ$, and $I_c = 1200\angle +30^\circ$ A. Find the symmetrical components of these currents. Also find the current into the ground.

8.8 Let an unbalanced, three-phase, Y-connected load (with phase impedances of Z_a, Z_b, and Z_c) be connected to a balanced three-phase supply, resulting in phase voltages of V_a, V_b, and V_c across the corresponding phase impedances.
Choosing V_{ab} as the reference, show that

$$V_{ab,0} = 0; \qquad V_{ab,1} = \sqrt{3}V_{a,\,1}e^{j30^\circ}; \qquad V_{ab,2} = \sqrt{3}V_{a,2}e^{-j30^\circ}$$

8.9 Reconsider Problem 8.8 and choosing V_{bc} as the reference, show that

$$V_{bc,0} = 0; \qquad V_{bc,1} = -j\sqrt{3}V_{a,\,1}; \qquad V_{bc,2} = j\sqrt{3}V_{a,2}$$

8.10 Given the line-to-ground voltages $V_{ag} = 280\angle 0^\circ$, $V_{bg} = 250\angle -110^\circ$, and $V_{cg} = 290\angle 130^\circ$ volts, calculate (a) the sequence components of the line-to-ground voltages, denoted V_{Lg0}, V_{Lg1} and V_{Lg2}; (b) line-to-line voltages V_{LL0}, V_{LL1}, and V_{LL2}; and (c) sequence components of the line-to-line voltages $V_{LL0} = 0$, V_{LL1}, and V_{LL2}. Also, verify the following general relation: $V_{LL0} = 0$, $V_{LL1} = \sqrt{3}V_{Lg1}\angle +30^\circ$, and $V_{LL2} = \sqrt{3}V_{Lg2}\angle -30^\circ$ volts.

8.11 A balanced Δ-connected load is fed by a three-phase supply for which phase C is open and phase A is carrying a current of $10\angle 0^\circ$ A. Find the symmetrical components of the line currents. (Note that zero-sequence currents are not present for any three-wire system.)

8.12 A Y-connected load bank with a three-phase rating of 500 kVA and 2300 V consists of three identical resistors of 10.58 Ω. The load bank has the following applied voltages: $V_{ab} = 1840\angle 82.8^\circ$, $V_{bc} = 2760\angle -41.4^\circ$, and $V_{ca} = 2300\angle 180^\circ$ V. Determine the symmetrical components of (a) the line-to-line voltages V_{ab0}, V_{ab1}, and V_{ab2}, (b) the line-to-neutral voltages V_{an0}, V_{an1}, and V_{an2}, (c) and the line currents I_{a0}, I_{a1}, and I_{a2}. (Note that the absence of a neutral connection means that zero-sequence currents are not present.)

SECTION 8.2

8.13 The currents in a Δ load are $I_{ab} = 10\angle 0^\circ$, $I_{bc} = 12\angle -90^\circ$, and $I_{ca} = 15\angle 90^\circ$ A. Calculate (a) the sequence components of the Δ-load currents, denoted $I_{\Delta 0}$, $I_{\Delta 1}$, and $I_{\Delta 2}$; (b) the line currents I_a, I_b, and I_c, which feed the Δ load;

and (c) sequence components of the line currents I_{L0}, I_{L1}, and I_{L2}. Also, verify the following general relation: $I_{L0} = 0$, $I_{L1} = \sqrt{3}I_{\Delta 1}\angle -30°$, and $I_{L2} = \sqrt{3}I_{\Delta 2}\angle +30°$.

8.14 The voltages given in Problem 8.10 are applied to a balanced-Y load consisting of $(12 + j16)$ ohms per phase. The load neutral is solidly grounded. Draw the sequence networks and calculate I_0, I_1, and I_2, the sequence components of the line currents. Then calculate the line currents I_a, I_b, and I_c.

8.15 Repeat Problem 8.14 with the load neutral open.

8.16 Repeat Problem 8.14 for a balanced-Δ load consisting of $(12 + j16)$ ohms per phase.

8.17 Repeat Problem 8.14 for the load shown in Example 8.4 (Figure 8.6).

8.18 Perform the indicated matrix multiplications in (8.2.21) and verify the sequence impedances given by (8.2.22) through (8.2.27).

8.19 The following unbalanced line-to-ground voltages are applied to the balanced-Y load shown in Figure 3.3: $V_{ag} = 100\angle 0°$, $V_{bg} = 75\angle 180°$, and $V_{cg} = 50\angle 90°$ volts. The Y load has $Z_Y = 3 + j4\ \Omega$ per phase with neutral impedance $Z_n = j1\ \Omega$. (a) Calculate the line currents I_a, I_b, and I_c without using symmetrical components, (b) Calculate the line currents I_a, I_b, and I_c using symmetrical components. Which method is easier?

8.20 (a) Consider three equal impedances of $(j27)\ \Omega$ connected in Δ. Obtain the sequence networks.
(b) Now, with a mutual impedance of $(j6)\ \Omega$ between each pair of adjacent branches in the Δ-connected load of part (a), how would the sequence networks change?

8.21 The three-phase impedance load shown in Figure 8.7 has the following phase impedance matrix:

$$Z_P = \begin{bmatrix} (5 + j10) & 0 & 0 \\ 0 & (5 + j10) & 0 \\ 0 & 0 & (5 + j10) \end{bmatrix} \Omega$$

Determine the sequence impedance matrix Z_S for this load. Is the load symmetrical?

8.22 The three-phase impedance load shown in Figure 8.7 has the following sequence impedance matrix:

$$Z_P = \begin{bmatrix} (6 + j10) & 0 & 0 \\ 0 & 5 & 0 \\ 0 & 0 & 5 \end{bmatrix} \Omega$$

Determine the phase impedance matrix Z_P for this load. Is the load symmetrical?

8.23 Consider a three-phase balanced Y-connected load with self and mutual impedances as shown in Figure 8.23. Let the load neutral be grounded through an impedance Z_n. Using Kirchhoff's laws, develop the equations for line-to-neutral voltages, and then determine the elements of the phase impedance matrix. Also find the elements of the corresponding sequence impedance matrix.

FIGURE 8.23

Problem 8.23

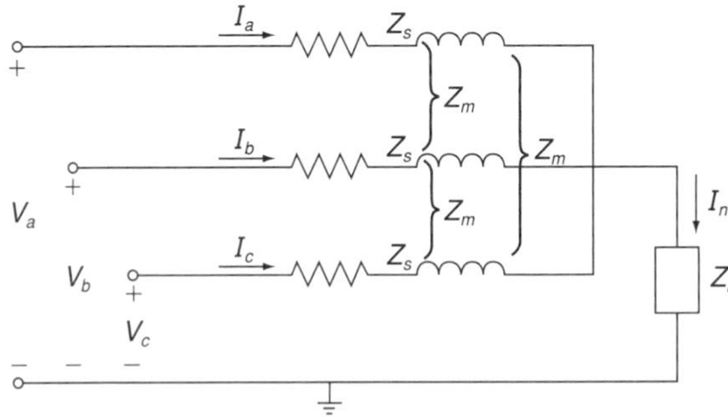

8.24 A three-phase balanced voltage source is applied to a balanced Y-connected load with ungrounded neutral. The Y-connected load consists of three mutually coupled reactances, where the reactance of each phase is $j12\ \Omega$, and the mutual coupling between any two phases is $j4\ \Omega$. The line-to-line source voltage is $100\sqrt{3}$ V. Determine the line currents (a) by mesh analysis without using symmetrical components and (b) using symmetrical components.

8.25 A three-phase balanced Y-connected load with series impedances of $(6+j24)\ \Omega$ per phase and mutual impedance between any two phases of $j3\ \Omega$ is supplied by a three-phase unbalanced source with line-to-neutral voltages of $V_{an} = 200\underline{/25°}$, $V_{bn} = 100\underline{/-155°}$, $V_{cn} = 80\underline{/100°}$ V. The load and source neutrals are both solidly grounded. Determine: (a) the load sequence impedance matrix, (b) the symmetrical components of the line-to-neutral voltages, (c) the symmetrical components of the load currents, and (d) the load currents.

SECTION 8.3

8.26 Repeat Problem 8.14 but include balanced three-phase line impedances of $(3 + j4)$ ohms per phase between the source and load.

8.27 Consider the flow of unbalanced currents in the symmetrical three-phase line section with neutral conductor as shown in Figure 8.24. (a) Express the voltage drops across the line conductors given by V_{aa}, V_{bb}, and V_{cc} in terms of line currents, self-impedances defined by $Z_S = Z_{aa} + Z_{nn} - 2Z_{an}$, and mutual impedances defined by $Z_m = Z_{ab} + Z_{nn} = 2Z_{an}$. (b) Show that the sequence components of the voltage drops between the ends of the line section can be written as $V_{aa'0} = Z_0 I_{a0}$, $V_{aa'1} = Z_1 I_{a1}$, and $V_{aa'2} = Z_2 I_{a2}$, where $Z_0 = Z_S + 2Z_m = Z_{aa} + 2Z_{ab} + 3Z_{nn} - 6Z_{an}$ and $Z_1 = Z_2 = Z_S = Z_m = Z_{aa} - Z_{ab}$.

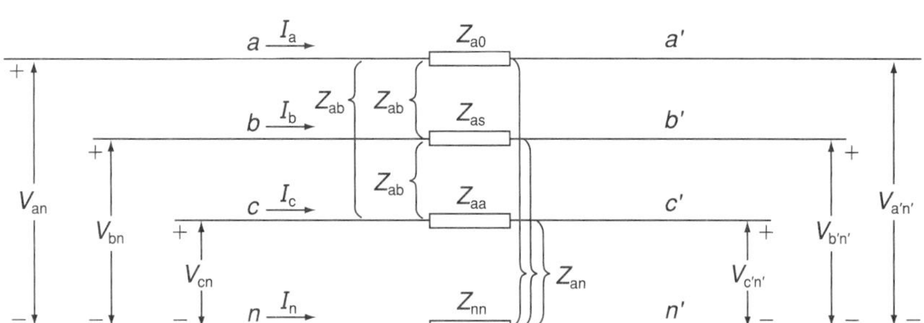

FIGURE 8.24

Problem 8.27

8.28 Let the terminal voltages at the two ends of the line section shown in Figure 8.24 be given by

$$V_{an} = (182 + j70)\ \text{kV} \qquad V_{an'} = (154 + j28)\ \text{kV}$$
$$V_{bn} = (72.24 - j32.62)\ \text{kV} \qquad V_{bn'} = (44.24 + j74.62)\ \text{kV}$$
$$V_{cn} = (-170.24 + j88.62)\ \text{kV} \quad V_{cn'} = (-198.24 + j46.62)\ \text{kV}$$

The line impedances are given by:

$Z_{aa} = j60\ \Omega \quad Z_{ab} = j20\ \Omega \qquad Z_{nn} = j80\ \Omega \qquad Z_{an} = 0$

(a) Compute the line currents using symmetrical components. (*Hint:* See Problem 8.27.) (b) Compute the line currents without using symmetrical components.

8.29 A completely transposed three-phase transmission line of 200 km in length has the following symmetrical sequence impedances and sequence admittances:

$$Z_1 = Z_2 = j0.5\ \Omega/\text{km}; \quad Z_0 = j2\ \Omega/\text{km}$$
$$Y_1 = Y_2 = j3 \times 10^{-9}\ \text{s/m}; \quad Y_0 = j1 \times 10^{-9}\ \text{s/m}$$

Set up the nominal π sequence circuits of this medium-length line.

SECTION 8.5

8.30 As shown in Figure 8.25, a balanced three-phase, positive-sequence source with $V_{AB} = 480\angle 0°$ volts is applied to an unbalanced Δ load. Note that one leg of the Δ is open. Determine (a) the load currents I_{AB} and I_{BC}; (b) the line currents I_A, I_B, and I_C, which feed the Δ load; and (c) the zero-, positive-, and negative-sequence components of the line currents.

8.31 A balanced Y-connected generator with terminal voltage $V_{bc} = 200\angle 0°$ volts is connected to a balanced-Δ load whose impedance is $10\angle 40°$ ohms per phase. The line impedance between the source and load is $0.5\angle 80°$ ohm for each phase. The generator neutral is grounded through an impedance of $j5$ ohms. The generator sequence impedances are given by $Z_{g0} = j7$, $Z_{g1} = j15$, and $Z_{g2} = j10$ ohms. Draw the sequence networks for this system and determine the sequence components of the line currents.

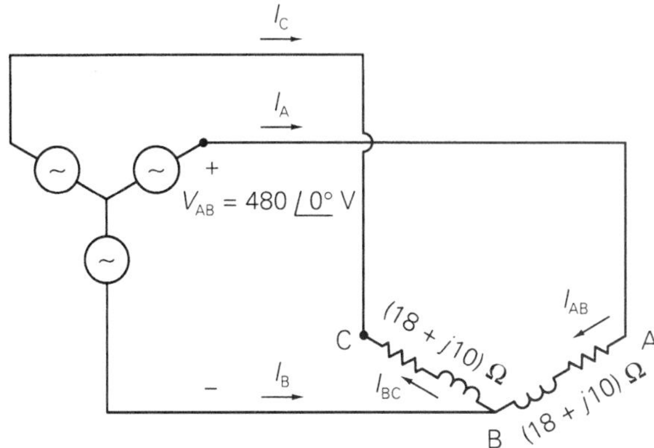

FIGURE 8.25

Problem 8.30

8.32 In a three-phase system, a synchronous generator supplies power to a 200-volt synchronous motor through a line having an impedance of $0.5\angle 80°$ ohm per phase. The motor draws 5 kW at 0.8 p.f. leading and at rated voltage. The neutrals of both the generator and motor are grounded through impedances of $j5$ ohms. The sequence impedances of both machines are $Z_0 = j5$, $Z_1 = j15$, and $Z_2 = j10$ ohms. Draw the sequence networks for this system and find the line-to-line voltage at the generator terminals. Assume balanced three-phase operation.

8.33 Calculate the source currents in Example 8.6 without using symmetrical components. Compare your solution method with that of Example 8.6. Which method is easier?

8.34 A Y-connected synchronous generator rated 20 MVA at 13.8 kV has a positive-sequence reactance of $j2.38$ Ω, negative-sequence reactance of $j3.33$ Ω, and zero-sequence reactance of $j0.95$ Ω. The generator neutral is solidly grounded. With the generator operating unloaded at rated voltage, a so-called single line-to-ground fault occurs at the machine terminals. During this fault, the line-to-ground voltages at the generator terminals are $V_{ag} = 0$, $V_{bg} = 8.071\angle -102.25°$, and $V_{cg} = 8.071\angle 102.25°$ kV. Determine the sequence components of the generator fault currents and the generator fault currents. Draw a phasor diagram of the prefault and postfault generator terminal voltages. (*Note:* For this fault, the sequence components of the generator fault currents are all equal to each other.)

8.35 Figure 8.26 shows a single-line diagram of a three-phase, interconnected generator-reactor system, in which the given per-unit reactances are based on the ratings of the individual pieces of equipment. If a three-phase short-circuit occurs at fault point F, obtain the fault MVA and fault current in kA if the prefault busbar line-to-line voltage is 13.2 kV. Choose 100 MVA as the base MVA for the system.

8.36 Consider Figures 8.13 and 8.14 of the text with reference to a Y-connected synchronous generator (grounded through a neutral impedance Z_n) operating at no load. For a line-to-ground fault occurring on phase a of the

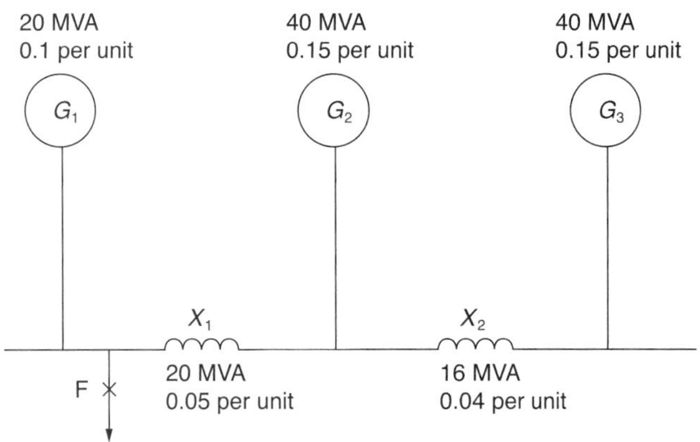

FIGURE 8.26

Oneline diagram for Problem 8.35

generator, list the constraints on the currents and voltages in the phase domain, transform those into the sequence domain, and then obtain a sequence-network representation. Also, find the expression for the fault current in phase *a*.

8.37 Reconsider the synchronous generator of Problem 8.36. Obtain sequence-network representations for the following fault conditions.
(a) A short-circuit between phases *b* and *c*.
(b) A double line-to-ground fault with phases *b* and *c* grounded.

SECTION 8.6

8.38 Three single-phase, two-winding transformers, each rated 450 MVA, 20 kV/288.7 kV, with leakage reactance $X_{eq} = 0.12$ per unit, are connected to form a three-phase bank. The high-voltage windings are connected in Y with a solidly grounded neutral. Draw the per-unit zero-, positive-, and negative-sequence networks if the low-voltage windings are connected (a) in Δ with American standard phase shift or (b) in Y with an open neutral. Use the transformer ratings as base quantities. Winding resistances and exciting current are neglected.

8.39 The leakage reactance of a three-phase, 500-MVA, 345 Y/23 Δ-kV transformer is 0.09 per unit based on its own ratings. The Y winding has a solidly grounded neutral. Draw the sequence networks. Neglect the exciting admittance and assume American standard phase shift.

8.40 Choosing system bases to be 360/24 kV and 100 MVA, redraw the sequence networks for Problem 8.39.

8.41 Draw the zero-sequence reactance diagram for the power system shown in Figure 3.38. The zero-sequence reactance of each generator and of the synchronous motor is 0.05 per unit based on equipment ratings. Generator 2 is grounded through a neutral reactor of 0.06 per unit on a 100-MVA, 18-kV base. The zero-sequence reactance of each transmission line is assumed to be three times its positive-sequence reactance. Use the same base as in Problem 3.41.

8.42 Three identical Y-connected resistors of $1.0\underline{/0°}$ per unit form a load bank that is supplied from the low-voltage Y-side of a Y–Δ transformer. The neutral of the load is not connected to the neutral of the system. The positive- and negative-sequence currents flowing toward the resistive load are given by

$$I_{a,1} = 1\underline{/4.5°} \text{ per unit}; \qquad I_{a,2} = 0.5\underline{/250°} \text{ per unit}$$

and the corresponding voltages on the low-voltage Y-side of the transformer are

$$V_{an,1} = 1\underline{/45°} \text{ per unit (Line-to-neutral voltage base)}$$

$$V_{an,2} = 0.5\underline{/250°} \text{ per unit (Line-to-neutral voltage base)}$$

Determine the line-to-line voltages and the line currents in per unit on the high-voltage side of the transformer. Account for the phase shift.

SECTION 8.7

8.43 Draw the positive-, negative-, and zero-sequence circuits for the transformers shown in Figure 3.34. Include ideal phase-shifting transformers showing phase shifts determined in Problem 3.32. Assume that all windings have the same kVA rating and that the equivalent leakage reactance of any two windings with the third winding open is 0.10 per unit. Neglect the exciting admittance.

8.44 A single-phase three-winding transformer has the following parameters: $Z_1 = Z_2 = Z_3 = 0 + j0.05$, $G_c = 0$, and $B_m = 0.2$ per unit. Three identical transformers, as described, are connected with their primaries in Y (solidly grounded neutral) and with their secondaries and tertiaries in Δ. Draw the per-unit sequence networks of this transformer bank.

SECTION 8.8

8.45 For Problem 8.14, calculate the real and reactive power delivered to the three-phase load.

8.46 A three-phase impedance load consists of a balanced-Δ load in parallel with a balanced-Y load. The impedance of each leg of the Δ load is $Z_\Delta = 6 + j6\ \Omega$, and the impedance of each leg of the Y load is $Z_Y = 2 + j2\ \Omega$. The Y load is grounded through a neutral impedance $Z_n = j1\ \Omega$. Unbalanced line-to-ground source voltages V_{ag}, V_{bg}, and V_{cg} with sequence components $V_0 = 10\underline{/60°}$, $V_1 = 100\underline{/0°}$, and $V_2 = 15\underline{/200°}$ volts are applied to the load. (a) Draw the zero-, positive-, and negative-sequence networks. (b) Determine the complex power delivered to each sequence network. (c) Determine the total complex power delivered to the three-phase load.

8.47 For Problem 8.12, compute the power absorbed by the load using symmetrical components. Then verify the answer by computing directly without using symmetrical components.

8.48 For Problem 8.25, determine the complex power delivered to the load in terms of symmetrical components. Verify the answer by adding up the complex power of each of the three phases.

8.49 Using the voltages of Problem 8.6(a) and the currents of Problem 8.5, compute the complex power dissipated based on (a) phase components and (b) symmetrical components.

CASE STUDIES QUESTIONS

a. What are the components of GIS?

b. What are the typical gas pressures in GIS equipment?

c. What is the environmental concern for SF_6 used in GIS? How is the environmental concern being addressed?

REFERENCES

1. Westinghouse Electric Corporation, *Applied Protective Relaying* (Newark, NJ: Westinghouse, 1976).
2. P. M. Anderson, *Analysis of Faulted Power Systems* (Ames, IA: Iowa State University Press, 1973).
3. W. D. Stevenson, Jr., *Elements of Power System Analysis*, 4th ed. (New York: McGraw-Hill, 1982).
4. I. A. Metwally, "Technological Progress in High-Voltage Gas-Insulated Substations, *IEEE Potentials Magazine*, 29, 6 (November/December 2010), pp. 25–32.

9 Unsymmetrical Faults

The converter switchyard at Bonneville Power Administration's Celilo Converter Station in 2009. This station converts ac power to HVDC for transmission up to 3150 MW at ±500kV over a 1360-km bipolar line in the USA between Dalles, OR and Los Angeles, CA. (Greg Wahl-Stephens/Stringer/Getty Images)

Short circuits occur in three-phase power systems as follows, in order of frequency of occurrence: single line-to-ground, line-to-line, double line-to-ground, and balanced three-phase faults. The path of the fault current may have either zero impedance, which is called a *bolted* short circuit, or nonzero impedance. Other types of faults include one-conductor-open and two-conductors-open, which can occur when conductors break or when one or two phases of a circuit breaker inadvertently open.

Although the three-phase short circuit occurs the least, it was covered first, in Chapter 7, because of its simplicity. When a balanced three-phase fault occurs in a balanced three-phase system, there is only positive-sequence fault current; the zero-, positive-, and negative-sequence networks are completely uncoupled.

When an unsymmetrical fault occurs in an otherwise balanced system, the sequence networks are interconnected only at the fault location. As such, the computation of fault currents is greatly simplified by the use of sequence networks.

As in the case of balanced three-phase faults, unsymmetrical faults have two components of fault current: an ac or symmetrical component—including subtransient, transient, and steady-state currents—and a dc component. The simplified E/X method for breaker selection described in Section 7.5 is also applicable to unsymmetrical faults. The dc offset current need not be considered unless it is too large—for example, when the X/R ratio is too large.

This chapter begins by using the per-unit zero-, positive-, and negative-sequence networks to represent a three-phase system. Also, certain assumptions are made to simplify fault-current calculations, and the balanced three-phase fault is briefly reviewed. Single line-to-ground, line-to-line, and double line-to-ground faults are presented in Sections 9.2, 9.3, and 9.4. The use of the positive-sequence bus impedance matrix for three-phase fault calculations in Section 7.4 is extended in Section 9.5 to unsymmetrical fault calculations by considering a bus impedance matrix for each sequence network. Examples using PowerWorld Simulator, which is based on the use of bus impedance matrices, are also included. The PowerWorld Simulator computes symmetrical fault currents for both three-phase and unsymmetrical faults. The Simulator may be used in power system design to select, set, and coordinate protective equipment.

CASE STUDY

The following article traces the development of medium voltage (MV) switchgear, starting with air-insulated masonry cubicles in the 1930s, to metal-enclosed drawout circuit breakers with air or oil insulation in the 1950s, to metal-enclosed drawout circuit breakers with compact SF_6 or vacuum interrupters in the late 1960s [8]. Since the 1990s, fixed circuit breakers have replaced drawout circuit breakers for many applications. Typical MV switchgear have rated voltages in the 5–40.5 kV range with rated interrupting currents up to 63 kA. Advantages of modern MV switchgear include safety, reliability, cost-effectiveness, long life, maintenance-free design, modularity, and flexibility. Current trends with smart grids require MV switchgear that have remote control and integrated intelligent devices.[8]

Innovative Medium Voltage Switchgear for Today's Applications

Canpeng Ma et al.

Introduction

Transmission system operators and distribution network operators need stability. The fact that the expected life duration of HV and MV switchgear is more than 30 years contributes to user's conservatism. Operational life seems easier to manage if there is no change in the habits of users. However, major evolutions in switchgear technology do occur every 20 or 30 years.

In recent years, the development of smart grids has brought in new requirements and ultimately new needs. Operators ask for maintenance-free circuit breakers able to be used in any environment with long life expectancy. With network evolution, operators require more flexibility and greater automation through remote control, integrated intelligent devices, and more metering points in the network. From an economical point of view, the acquisition cost is increasingly balanced by the notion of total cost of ownership, including construction, installation, maintenance, and end-of-life recovery costs. All these trends have to be considered when designing the new generation of MV switchgear for today's applications.

Historical Evolution of MV Switchgear Technology

The past decades have seen the trend of fixed devices replacing, slowly but surely, the drawout ones for many applications. Together with the technological evolution of MV switchgear, single line diagrams have regularly been challenged.

Breaking technologies have moved from air and oil technologies to SF_6 and vacuum technologies. Protection relays have moved from electromechanical to electronic and then to digital technologies. Current and voltage transformers have been adapted to the new relay designs and to switchgear architecture evolution.

Circuit Breaker Technological Evolution

The first technology used for breaking in circuit breakers was air. These circuit breakers were big as the principle of breaking was a large expansion of the arc in the air. Heavy maintenance was required, and for this reason they were abandoned. Trying to reduce the footprint came the oil circuit breakers. But frequent maintenance was still required; for example the oil needed to be changed after several operations due to carbonization

Figure 1 Drawout oil circuit breaker with arc control

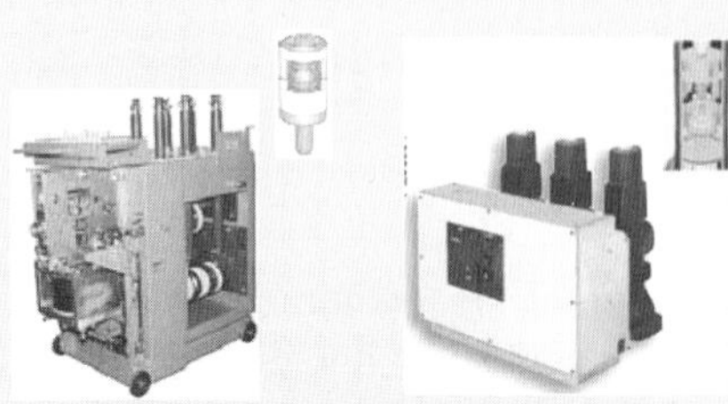

Figure 2 Drawout vacuum and SF_6 circuit breaker

by arcing. Additionally, oil circuit breaker failures can easily result in fatal accidents among operators and the public due to oil fires.

In the late sixties came SF_6 and vacuum circuit breakers. They are compact, thanks to SF_6 or vacuum insulation. They are much safer, drastically reducing fire risk. Reliability was increased with the improvement in electrical endurance in terms of fault breakings and load breakings. As a consequence, preventive and corrective maintenance is required less and less; thus state-of-the-art circuit breakers can now be called "maintenance free."

MV Switchgear and Single Line Diagram Evolution

From 1930 to 1950, most of the MV switchboards were air insulated masonry cubicles. Only simple wire fencing was used to prevent the access to the live parts.

Figure 3 Air-insulated masonry cubicles

Then with more safety awareness, they were gradually replaced by metal-enclosed Air Insulated Switchgear (AIS) with integrated removable circuit breakers from air or oil to SF_6 and vacuum technology.

The single line diagram with drawout circuit breakers is the oldest one and is still in use in some primary distribibution applications [1]. Disconnection is made by racking out the CB truck, providing visible disconnection. Earthing for safe access to internal components or cables is made by earthing trucks or dedicated earthing switches acting directly on cable ends.

Maintenance of the circuit breaker is very easy, and this was necessary for old circuit breakers.

However, some points have to be carefully considered. Remote control of the disconnector is not really practical because of the truck to be racked out. Earthing the busbar needs a dedicated earthing truck, which is heavy to handle. Testing the cables needs a direct access to cables, opening the cable compartment. And finally, the equipment must be installed in clean air rooms as it is sensitive to environmental conditions because of AIS technology.

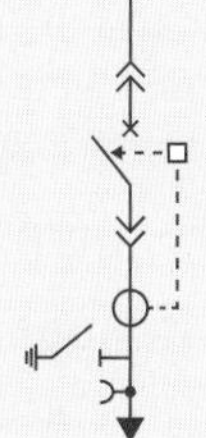

Figure 4 Typical AIS panel with drawout technology and single line diagram

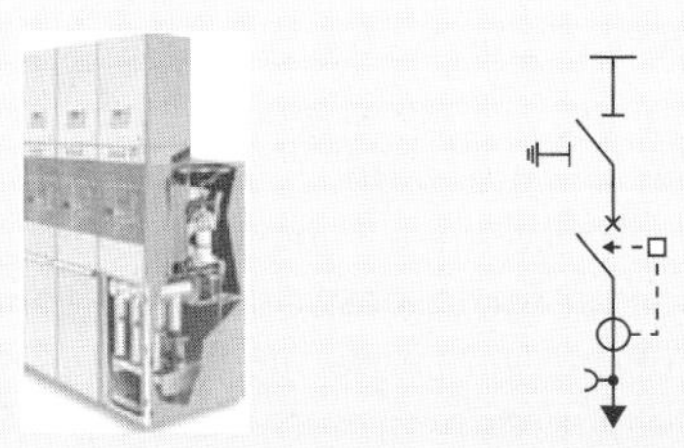

Figure 5 Typical primary GIS panel with fixed technology and single line diagram

More recently from the 1990s, fixed circuit breakers have been used more and more. There are some recent variants in metal enclosed cubicles with fixed circuit breakers where the insulation components are made of epoxy or other resin. These panels are generally called Solid Insulated Switchgear (SIS). However, both AIS and SIS panels could be sensitive to environmental conditions (dust, humidity, etc.) if not properly installed in protected rooms. That was the reason of the arrival of metal enclosed Gas Insulated Switchgear (GIS) in the nineties with sealed tanks filled with SF_6.

Among the points to be aware of is that the operation is not so intuitive because of five-position diagram mixing upstream three-position disconnector/earthing switch and downstream two-position circuit breaker. Particularly, cable earthing is made through circuit breaker closing that must remain closed to ensure operator safety when working.

For secondary applications, due to repetitive arrangement, the simplicity, insensitivity to severe installation conditions, and cost effectiveness became the criteria. The Ring Main Unit (RMU) configuration appeared in the fifties. The RMU evolved from oil type with switch fuse to SF_6 RMU with more and more circuit breakers for MV/LV transformer protection. Thanks to the all-in-one arrangement, breaking and disconnection are performed in a single operation, leading to the three-position scheme (line, open and disconnected, earthed). Local or remote operations become simple with an intuitive mimic diagram. Cable earthing is made directly, close to the cable connectors. Interlocking reliability is inherent between the different positions. It is very easy to implement a cable testing device, allowing access to cables without opening the cable box or interfering with the cable terminations.

The advantages of compact and repetitive Ring Main Units solution reach their limit when flexibility is needed, for larger switchboards over 4–5 functions or switchgear extension.

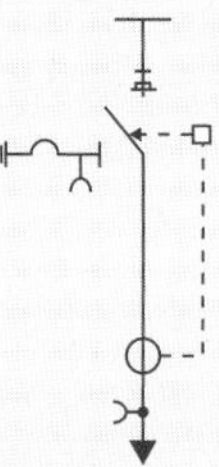

Figure 6 SF_6 Ring Main Unit with circuit breaker

Evolution of Protection Relays and Sensors

Protection relays have moved from electromechanical to electronic and then to digital technologies. For modern digital relays, the need of signal power from the sensors becomes very low. A new category of current sensors were developed, the low power current transformers (LPCT). They deliver a voltage signal representing the primary current, and offer great advantages in space and flexibility compared to traditional 5 A or 1 A current transformers.

Challenges for Future MV Switchgear

Strong Interests from Network Operators

A huge global market survey was undertaken to identify not only the current needs, but to anticipate the needs in the future, particularly in countries with fast growth and high development potential. Distribution network operators expressed strong interests in safety, reliability, cost-effectiveness, simplicity and ease of network evolution. Maintenance-free circuit breakers, able to be used in any environment with long life expectancy are more and more required. The network operators pay more and more attention to the total cost of ownership, including the acquisition, construction, installation, maintenance and end of life treatment costs.

The Challenge of Smart Grids

Smart grids have two main objectives. One is to optimize the relation between the demand and the offer of energy. The second one is to be able to integrate more and more distributed and renewable energies. The change in network management from one-way flow due to centralized energy production to two-way flow because of distributed generation is a big challenge. One question comes out: Are MV switchgear ready for this challenge?

Looking at the existing grids and some experimentations, it is possible to highlight some values of switchgear that will help to meet this challenge.

Smart grids will include more circuit breakers in the network to increase electrical energy availability [2].

Remote control will be mandatory for smart grids. One effect is that switchgear will include integrated intelligent devices providing better optimization. Low-power current transformers (LPCT) and low-power voltage transformers (LPVT) will be essential for the huge development of power management. Metering equipment will need to be cost-effective, compact, and reliable.

Modularity and flexibility are mandatory to meet the infinite number of different applications.

New Generation of MV Switchgear Through Innovations

To satisfy all these needs, it appears that an optimized solution for many applications relies on a breakthrough in switchgear design, getting benefits from a compact all-in-one three-position scheme and a shielded solid insulation system (2SIS).

New Three-Position Diagram

Nowadays, even if the technology is not changing much, the trend is to use vacuum breakers in secondary applications. Recent developments brought a new arrangement, keeping the same well-known benefits as the three-position diagram of GIS RMUs, but using vacuum breaking.

The new proposed arrangement includes an upstream vacuum disconnecting load-break switch or disconnecting circuit breaker and a downstream earthing switch providing double-gap isolation between cables and bus bars.

The three-position scheme, closed, open and disconnected, and earthed, with intuitive mimic diagram is the simplest and safest mode to operate the switchgear. The switching and disconnection of the circuit are done in a single operation within an innovative vacuum interrupter with a larger isolation gap compared to traditional design and with controlled surface for the arcing contacts. The downstream cable earthing is done in controlled air with a fast-acting earthing switch having making capacity and breaking capacity of capacitive cable currents.

Intuitiveness of mimic diagram makes the operation very easy. Safety interlocks are built in and positively driven. Operator safety is further enhanced with a dedicated cable testing device with front face access, complying with the new specifications on MV switchgear for cable test features defined in the recently revised switchgear standard IEC 62271-200 edition 2 [3].

Prior to the cable test, opening the disconnecting switch or disconnecting circuit breaker and closing the earthing switch provide a double gap of isolation between cable and busbar. Then a safe and fully interlocked earth link switch may be opened to give direct access to cable conductor. During testing, the cable box remains closed, the cable connections remain intact and the main contacts of the earthing switch remain in the same position. This recommended test procedure ensures the highest safety for test operators and also avoids any damage on the main circuit or cable connections.

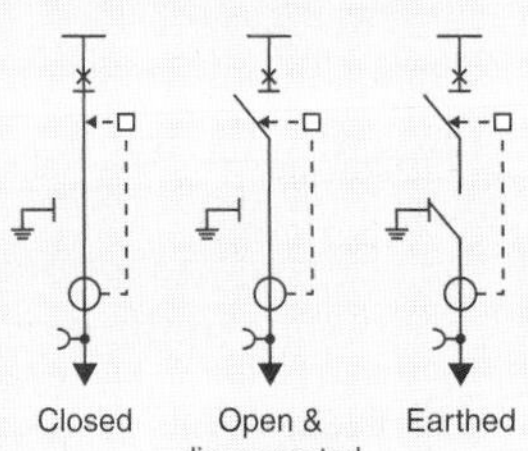

Figure 7 New three-position diagram and typical unit including vacuum disconnecting circuit breaker

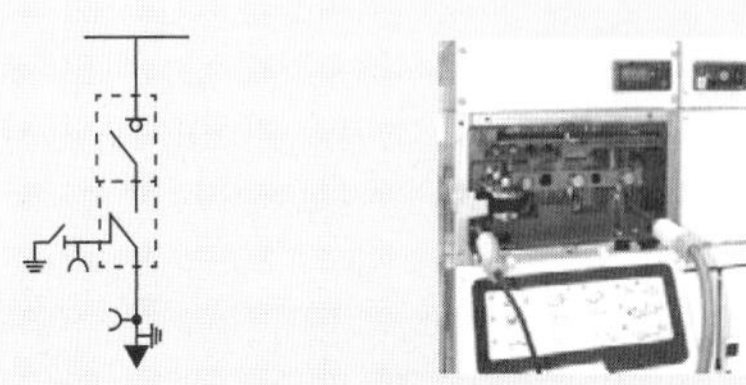

Figure 8 Dedicated cable test device

Figure 9 Shielded Solid Insulation for the whole MV circuit

Shielded Solid Insulation System (2SIS)

To meet the same advantages as GIS RMUs, the new arrangement must be insensitive to harsh environment and "maintenance free." This is ensured by an innovative Shielded Solid Insulation System (2SIS) for the whole MV circuit, using well-known technologies (epoxy, EPDM) in an original way. Busbars, vacuum interrupter encapsulation, and earthing switch enclosures are made of solid insulation that is covered by a conductive layer connected to the earth. While maintaining a constant electrical field within the solid insulation, all surfaces are at earth potential with protection grade PA according to IEC 62271-201 (can be accidentally touched by persons) [4]. The earth shield continuity is ensured by a patented flat interface, which combines installation simplicity, modularity, and flexibility of traditional connectors with the compact size and safety of the insulated and shielded connectors.

The 2SIS system improves safety by internal arc risk reduction, makes the switchgear totally insensitive to severe environments and extends service life with minimum maintenance and lower total cost of ownership. The 2SIS system is applicable for any switching function such as load break switches or circuit breakers, compact metering functions, or current and voltage transformers.

Smart Grid Ready

Compared to GIS RMUs, this new generation of switchgear with 2SIS technology and the new three-position scheme offers much better modularity as the general architecture is based on single units.

Modern control and monitoring devices and digital protection relays are integrated in the new switchgear with the possibility of using compact LPCTs and LPVTs, making the advanced management solutions much easier. The modularity of the switchgear is extended up to the architecture by linking together the various embedded intelligent electronic devices, in order to achieve the necessary functions of control and monitoring in the most flexible way.

Conclusion

The necessary evolution of the networks brings the opportunity to introduce new criteria for the choice of MV switchgear, such as flexibility, insensitivity to harsh environment, compactness, optimization of remote control, and so forth.

There is a great confidence that the 2SIS modular architecture using the three-position scheme and vacuum interrupters is well adapted for the coming deployment of smart grids. This architecture can address a large number of applications in MV secondary distribution and MV private networks but thanks to its modularity and flexibility, it also challenges some low-end applications where traditional drawout equipment is used. With this respect, this architecture blurs traditional MV primary and secondary boundaries.

Combining compactness and modularity through technological innovations, the new generation of switchgear creates a real breakthrough in MV switchgear habits and specifications.

References

[1] G. Moesch, J. M. Biasse, Y. Li, M. Adams, 2010: "Advent of fixed type Disconnecting Circuit Breakers in Medium Voltage secondary switchgear: a new arrangement of main circuit functions for simpler and safer operation," *Proceedings CICED 2010 paper FPOl15.*

[2] P. Deschamps, P. Ahbert, J. M. Biasse, 2011: "Reduce the number of outage by introducing circuit breaker in the distribution network: dream or reality?" *Proceedings CIRED 2011 paper. 0768.*

[3] IEC 62271-200 Ed2.0 2011-10, AC metal-enclosed switchgear and controlgear for rated voltages above 1 kV and up to and including 52 kV.

[4] IEC 62271-201 First Edition 2006-06, AC metal-enclosed switchgear and controlgear for rated voltages above 1 kV up to and including 52 kV. ■

9.1 SYSTEM REPRESENTATION

A three-phase power system is represented by its sequence networks in this chapter. The zero-, positive-, and negative-sequence networks of system components—generators, motors, transformers, and transmission lines—as developed in Chapter 8 can be used to construct system zero-, positive-, and negative-sequence networks. The following assumptions are made:

1. The power system operates under balanced steady-state conditions before the fault occurs. Thus the zero-, positive-, and negative-sequence networks are uncoupled before the fault occurs. During unsymmetrical faults, they are interconnected only at the fault location.
2. Prefault load current is neglected. Because of this, the positive-sequence internal voltages of all machines are equal to the prefault voltage V_F. Therefore, the prefault voltage at each bus in the positive-sequence network equals V_F.

3. Transformer winding resistances and shunt admittances are neglected.
4. Transmission-line series resistances and shunt admittances are neglected.
5. Synchronous machine armature resistance, saliency, and saturation are neglected.
6. All nonrotating impedance loads are neglected.
7. Induction motors are either neglected (especially for motors rated 40 kW or less) or represented in the same manner as synchronous machines.

Note that these assumptions are made for simplicity in this text, and in practice should not be made for all cases. For example, in primary and secondary distribution systems, prefault currents may be in some cases comparable to short-circuit currents, and in other cases line resistances may significantly reduce fault currents.

Although fault currents as well as contributions to fault currents on the fault side of Δ–Y transformers are not affected by Δ–Y phase shifts, contributions to the fault from the other side of such transformers are affected by Δ–Y phase shifts for unsymmetrical faults. Therefore, Δ–Y phase-shift effects are included in this chapter.

Consider faults at the general three-phase bus shown in Figure 9.1. Terminals *abc,* denoted the *fault terminals*, are brought out in order to make external connections that represent faults. Before a fault occurs, the currents I_a, I_b, and I_c are zero.

Figure 9.2 (a) shows general sequence networks as viewed from the fault terminals. Since the prefault system is balanced, these zero-, positive-, and negative-sequence networks are uncoupled. Also, the sequence components of the fault currents, I_0, I_1, and I_2, are zero before a fault occurs. The general sequence networks in Figure 9.2(a) are reduced to their Thévenin equivalents as viewed from the fault terminals in Figure 9.2(b). Each sequence network has a Thévenin equivalent impedance. Also, the positive-sequence network has a Thévenin equivalent voltage source, which equals the prefault voltage V_F.

FIGURE 9.1

General three-phase bus

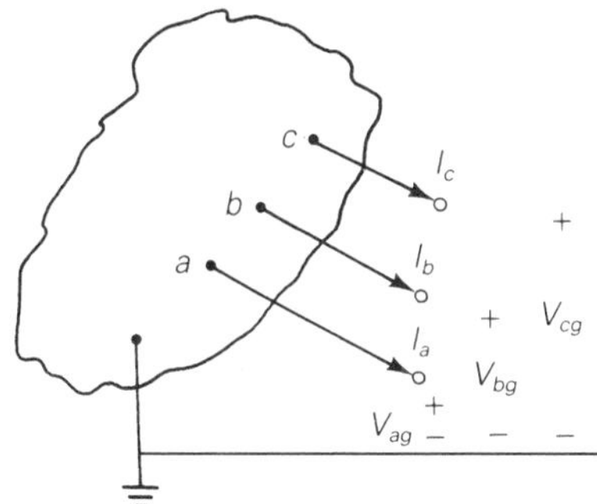

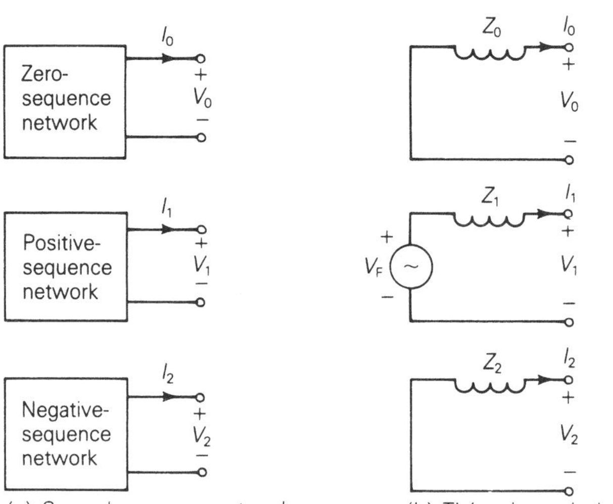

FIGURE 9.2

Sequence networks at a general three-phase bus in a balanced system

EXAMPLE 9.1

Power-system sequence networks and their Thévenin equivalents

A single-line diagram of the power system considered in Example 7.3 is shown in Figure 9.3, where negative- and zero-sequence reactances are also given. The neutrals of the generator and Δ–Y transformers are solidly grounded. The motor neutral is grounded through a reactance $X_n = 0.05$ per unit on the motor base. (a) Draw the per-unit zero-, positive-, and negative-sequence networks on a 100-MVA, 13.8-kV base in the zone of the generator. (b) Reduce the sequence networks to their Thévenin equivalents, as viewed from bus 2. Prefault voltage is $V_F = 1.05\underline{/0^\circ}$ per unit. The prefault load current and Δ–Y transformer phase shift are neglected.

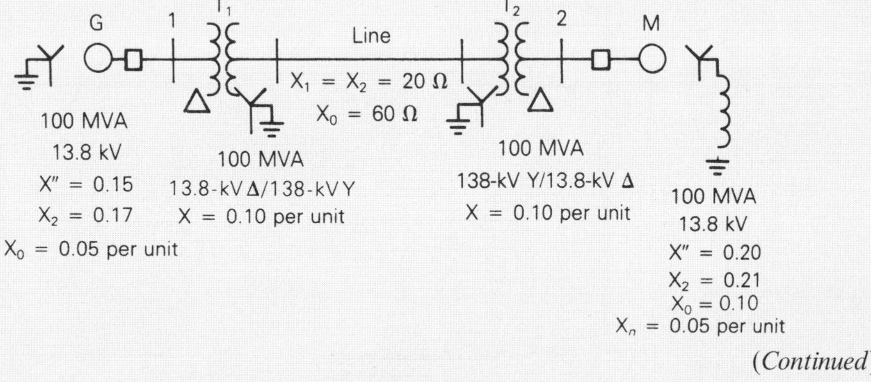

FIGURE 9.3

Single-line diagram for Example 9.1

(Continued)

SOLUTION

a. The sequence networks are shown in Figure 9.4. The positive-sequence network is the same as that shown in Figure 7.4(a). The negative-sequence network is similar to the positive-sequence network, except that it shows no sources but does show negative-sequence machine reactances. Δ–Y phase shifts are omitted from the positive- and negative-sequence networks for this example. In the zero-sequence network the zero-sequence generator, motor, and transmission-line reactances are shown. Since the motor neutral is grounded through a neutral reactance X_n, $3X_n$ is included in the zero-sequence motor circuit. Also, the zero-sequence Δ–Y transformer models are taken from Figure 8.19.

b. Figure 9.5 shows the sequence networks reduced to their Thévenin equivalents, as viewed from bus 2. For the positive-sequence equivalent, the Thévenin voltage source is the prefault voltage $V_F = 1.05\angle 0^\circ$ per unit.

FIGURE 9.4

Sequence networks for Example 9.1

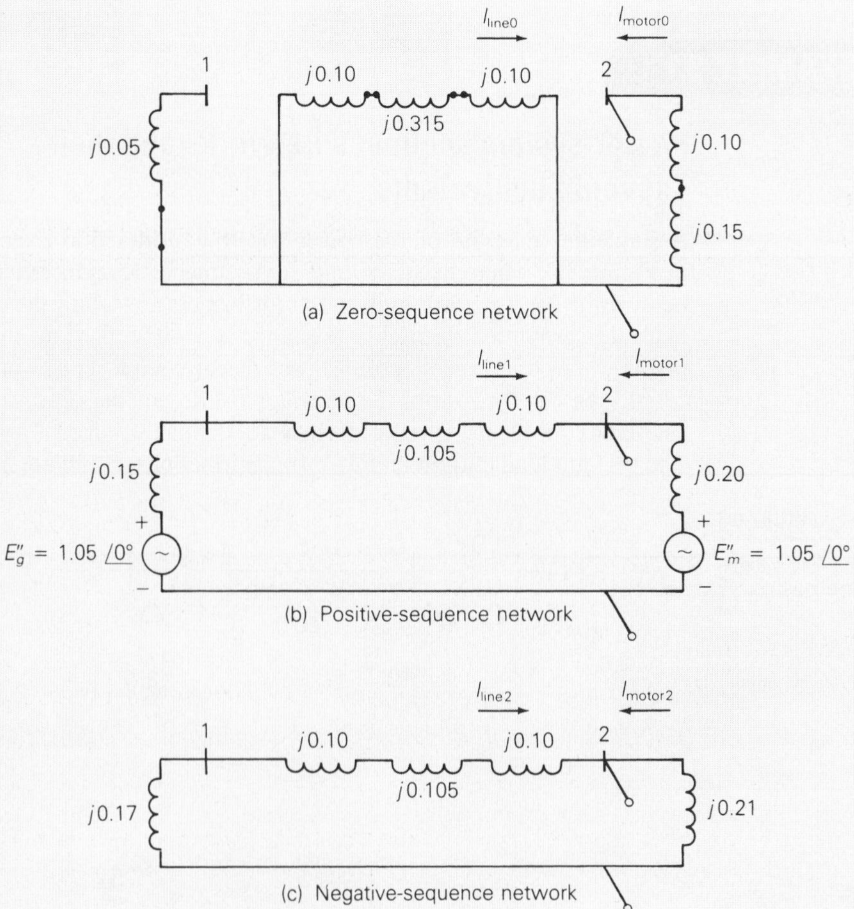

FIGURE 9.5 Thévenin equivalents of sequence networks for Example 9.1

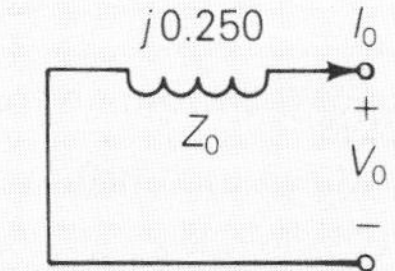

(a) Zero-sequence network

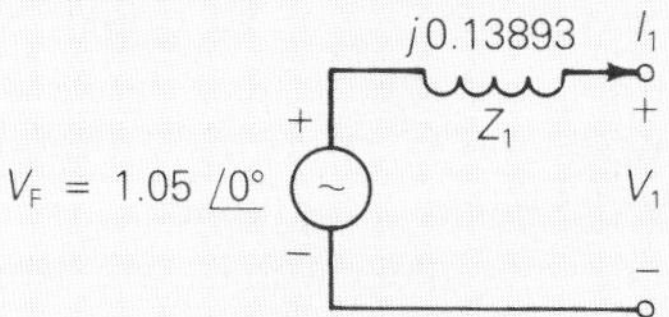

(b) Positive-sequence network

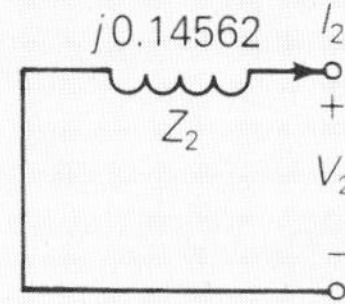

(c) Negative-sequence network

From Figure 9.4, the positive-sequence Thévenin impedance at bus 2 is the motor impedance $j0.20$, as seen to the right of bus 2, in parallel with $j(0.15 + 0.10 + 0.105 + 0.10) = j0.455$, as seen to the left; the parallel combination is $j0.20 // j0.455 = j0.13893$ per unit. Similarly, the negative-sequence Thévenin impedance is $j0.21 // j(0.17 + 0.10 + 0.105 + 0.10) = j0.21 // j0.475 = j0.14562$ per unit. In the zero-sequence network of Figure 9.4, the Thévenin impedance at bus 2 consists only of $j(0.10 + 0.15) = j0.25$ per unit, as seen to the right of bus 2; due to the Δ connection of transformer T_2, the zero-sequence network looking to the left of bus 2 is open.

Recall that for three-phase faults, as considered in Chapter 7, the fault currents are balanced and have only a positive-sequence component. Therefore, work only with the positive-sequence network when calculating three-phase fault currents.

EXAMPLE 9.2

Three-phase short-circuit calculations using sequence networks

Calculate the per-unit subtransient fault currents in phases *a, b,* and *c* for a bolted three-phase-to-ground short circuit at bus 2 in Example 9.1.

(*Continued*)

SOLUTION

The terminals of the positive-sequence network in Figure 9.5(b) are shorted, as shown in Figure 9.6. The positive-sequence fault current is

$$I_1 = \frac{V_F}{Z_1} = \frac{1.05\underline{/0°}}{j0.13893} = -j7.558 \quad \text{per unit}$$

which is the same result as obtained in part (c) of Example 7.4. Note that since subtransient machine reactances are used in Figures 9.4 through 9.6, the current calculated above is the positive-sequence subtransient fault current at bus 2. Also, the zero-sequence current I_0 and negative-sequence current I_2 are both zero. Therefore, the subtransient fault currents in each phase are, from (8.1.16),

$$\begin{bmatrix} I''_a \\ I''_b \\ I''_c \end{bmatrix} = \begin{bmatrix} 1 & 1 & 1 \\ 1 & a^2 & a \\ 1 & a & a^2 \end{bmatrix} \begin{bmatrix} 0 \\ -j7.558 \\ 0 \end{bmatrix} = \begin{bmatrix} 7.558\underline{/-90°} \\ 7.558\underline{/150°} \\ 7.558\underline{/30°} \end{bmatrix} \quad \text{per unit}$$

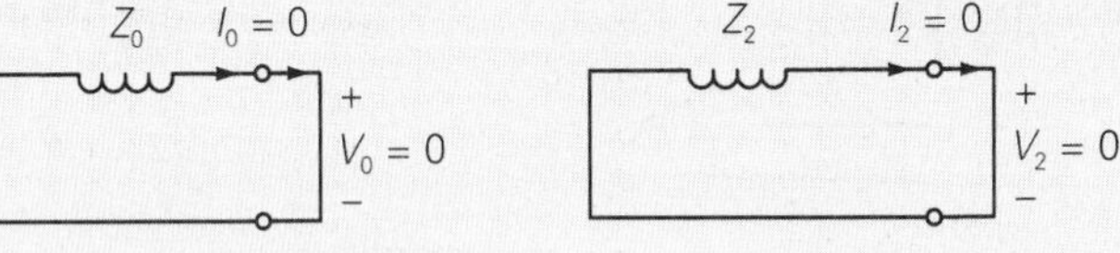

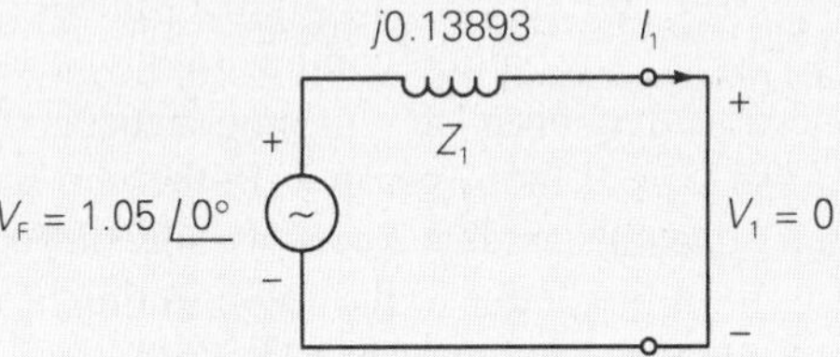

FIGURE 9.6 Example 9.2: Bolted three-phase-to-ground fault at bus 2

The sequence components of the line-to-ground voltages at the fault terminals are, from Figure 9.2(b),

$$\begin{bmatrix} V_0 \\ V_1 \\ V_2 \end{bmatrix} = \begin{bmatrix} 0 \\ V_F \\ 0 \end{bmatrix} - \begin{bmatrix} Z_0 & 0 & 0 \\ 0 & Z_1 & 0 \\ 0 & 0 & Z_2 \end{bmatrix} \begin{bmatrix} I_0 \\ I_1 \\ I_2 \end{bmatrix} \tag{9.1.1}$$

During a bolted three-phase fault, the sequence fault currents are $I_0 = I_2 = 0$ and $I_1 = V_F/Z_1$; therefore, from (9.1.1), the sequence fault voltages are $V_0 = V_1 = V_2 = 0$, which must be true since $V_{ag} = V_{bg} = V_{cg} = 0$. However, fault voltages need not be zero during unsymmetrical faults, which is considered next.

9.2 SINGLE LINE-TO-GROUND FAULT

Consider a single line-to-ground fault from phase a to ground at the general three-phase bus shown in Figure 9.7 (a). For generality, a fault impedance Z_F is included. In the case of a bolted fault, $Z_F = 0$, whereas for an arcing fault, Z_F is the arc impedance. In the case of a transmission-line insulator flashover, Z_F includes the total fault impedance between the line and ground, including the impedances of the arc and the transmission tower, as well as the tower footing if there are no neutral wires.

The relations to be derived here apply only to a single line-to-ground fault on phase a. However, since any of the three phases can be arbitrarily labeled phase a, single line-to-ground faults on other phases are not considered.

From Figure 9.7(a),

$$\left.\begin{array}{l}\text{Fault conditions in phase domain}\\ \text{Single line-to-ground fault}\end{array}\right\} \begin{array}{l} I_b = I_c = 0 \qquad (9.2.1)\\ V_{ag} = Z_F\, I_a \qquad (9.2.2)\end{array}$$

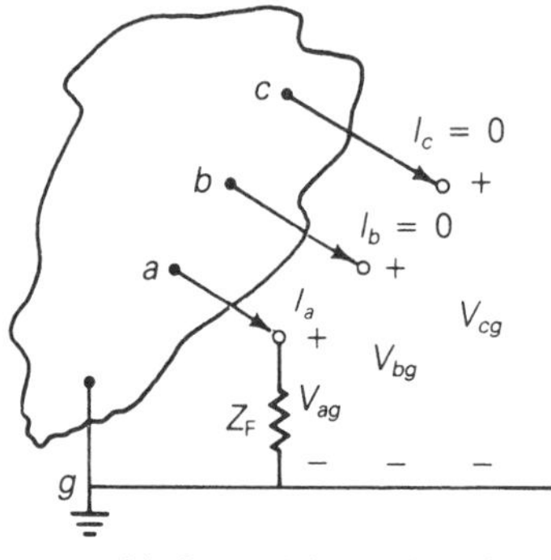

(a) General three-phase bus

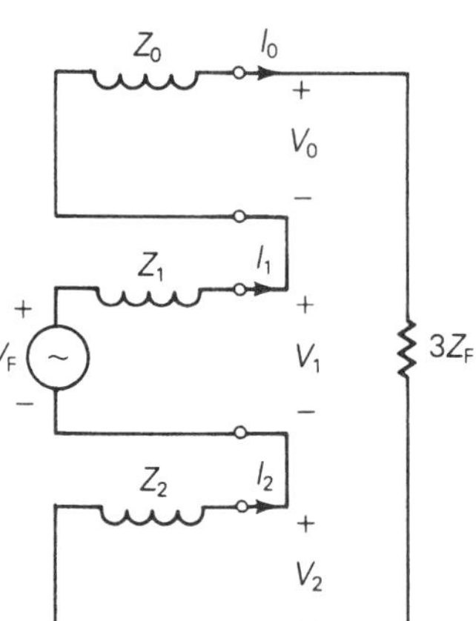

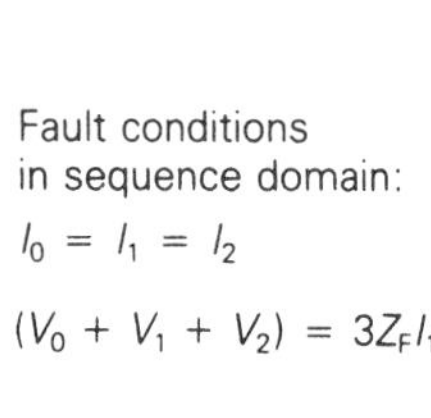

(b) Interconnected sequence networks

FIGURE 9.7 Single line-to-ground fault

Now transform (9.2.1) and (9.2.2) to the sequence domain. Using (9.2.1) in (8.1.19),

$$\begin{bmatrix} I_0 \\ I_1 \\ I_2 \end{bmatrix} = \frac{1}{3}\begin{bmatrix} 1 & 1 & 1 \\ 1 & a & a^2 \\ 1 & a^2 & a \end{bmatrix}\begin{bmatrix} I_a \\ 0 \\ 0 \end{bmatrix} = \frac{1}{3}\begin{bmatrix} I_a \\ I_a \\ I_a \end{bmatrix} \tag{9.2.3}$$

Also, using (8.1.3) and (8.1.20) in (9.2.2),

$$(V_0 + V_1 + V_2) = Z_F\,(I_0 + I_1 + I_2) \tag{9.2.4}$$

From (9.2.3) and (9.2.4),

$$\left.\begin{array}{l}\text{Fault conditions in sequence domain} \\ \text{Single line-to-ground fault}\end{array}\right\}\begin{array}{ll} I_0 = I_1 = I_2 & (9.2.5) \\ (V_0 + V_1 + V_2) = (3Z_F)I_1 & (9.2.6)\end{array}$$

Equations (9.2.5) and (9.2.6) can be satisfied by interconnecting the sequence networks in series at the fault terminals through the impedance $(3Z_F)$, as shown in Figure 9.7(b). From this figure, the sequence components of the fault currents are

$$I_0 = I_1 = I_2 = \frac{V_F}{Z_0 + Z_1 + Z_2 + (3Z_F)} \tag{9.2.7}$$

Transforming (9.2.7) to the phase domain via (8.1.20),

$$I_a = I_0 + I_1 + I_2 = 3I_1 = \frac{3V_F}{Z_0 + Z_1 + Z_2 + (3Z_F)} \tag{9.2.8}$$

Note also from (8.1.21) and (8.1.22),

$$I_b = (I_0 + a^2 I_1 + aI_2) = (1 + a^2 + a)I_1 = 0 \tag{9.2.9}$$

$$I_c = (I_0 + aI_1 + a^2 I_2) = (1 + a + a^2)I_1 = 0 \tag{9.2.10}$$

These are obvious, since the single line-to-ground fault is on phase a, not phase b or c.

The sequence components of the line-to-ground voltages at the fault are determined from (9.1.1). The line-to-ground voltages at the fault can then be obtained by transforming the sequence voltages to the phase domain.

EXAMPLE 9.3

Single line-to-ground short-circuit calculations using sequence networks

Calculate the subtransient fault current in per-unit and in kA for a bolted single line-to-ground short circuit from phase a to ground at bus 2 in Example 9.1. Also calculate the per-unit line-to-ground voltages at faulted bus 2.

SOLUTION

The zero-, positive-, and negative-sequence networks in Figure 9.5 are connected in series at the fault terminals, as shown in Figure 9.8.

Since the short circuit is bolted, $Z_F = 0$. From (9.2.7), the sequence currents are:

$$I_0 = I_1 = I_2 = \frac{1.05\underline{/0^\circ}}{j(0.25 + 0.13893 + 0.14562)}$$
$$= \frac{1.05}{j0.53455} = -j1.96427 \quad \text{per unit}$$

From (9.2.8), the subtransient fault current is

$$I''_a = 3(-j1.96427) = -j5.8928 \quad \text{per unit}$$

The base current at bus 2 is $100/(13.8\sqrt{3}) = 4.1837$ kA. Therefore,

$$I''_a = (-j5.8928)(4.1837) = 24.65\underline{/-90^\circ} \quad \text{kA}$$

From (9.1.1), the sequence components of the voltages at the fault are

$$\begin{bmatrix} V_0 \\ V_1 \\ V_2 \end{bmatrix} = \begin{bmatrix} 0 \\ 1.05\underline{/0^\circ} \\ 0 \end{bmatrix} - \begin{bmatrix} j0.25 & 0 & 0 \\ 0 & j0.13893 & 0 \\ 0 & 0 & j0.14562 \end{bmatrix} \begin{bmatrix} -j1.96427 \\ -j1.96427 \\ -j1.96427 \end{bmatrix}$$
$$= \begin{bmatrix} -0.49107 \\ 0.77710 \\ -0.28604 \end{bmatrix} \quad \text{per unit}$$

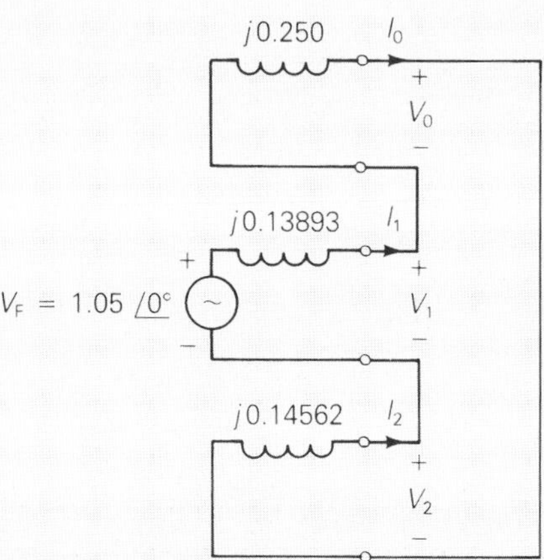

FIGURE 9.8 Example 9.3: Single line-to-ground fault at bus 2

(Continued)

Transforming to the phase domain, the line-to-ground voltages at faulted bus 2 are

$$\begin{bmatrix} V_{ag} \\ V_{bg} \\ V_{cg} \end{bmatrix} = \begin{bmatrix} 1 & 1 & 1 \\ 1 & a^2 & a \\ 1 & a & a^2 \end{bmatrix} \begin{bmatrix} -0.49107 \\ 0.77710 \\ -0.28604 \end{bmatrix} = \begin{bmatrix} 0 \\ 1.179\underline{/231.3^\circ} \\ 1.179\underline{/128.7^\circ} \end{bmatrix} \quad \text{per unit}$$

Note that $V_{ag} = 0$, as specified by the fault conditions. Also $I_b'' = I_c'' = 0$.

Open PowerWorld Simulator case Example 9_3 to see this example. This case includes the positive-, negative-, and zero-sequence values for the system devices. To see these values, right-click on an object on the oneline (e.g., a generator), open the object's Information dialog, and select the **Fault** tab. The process for

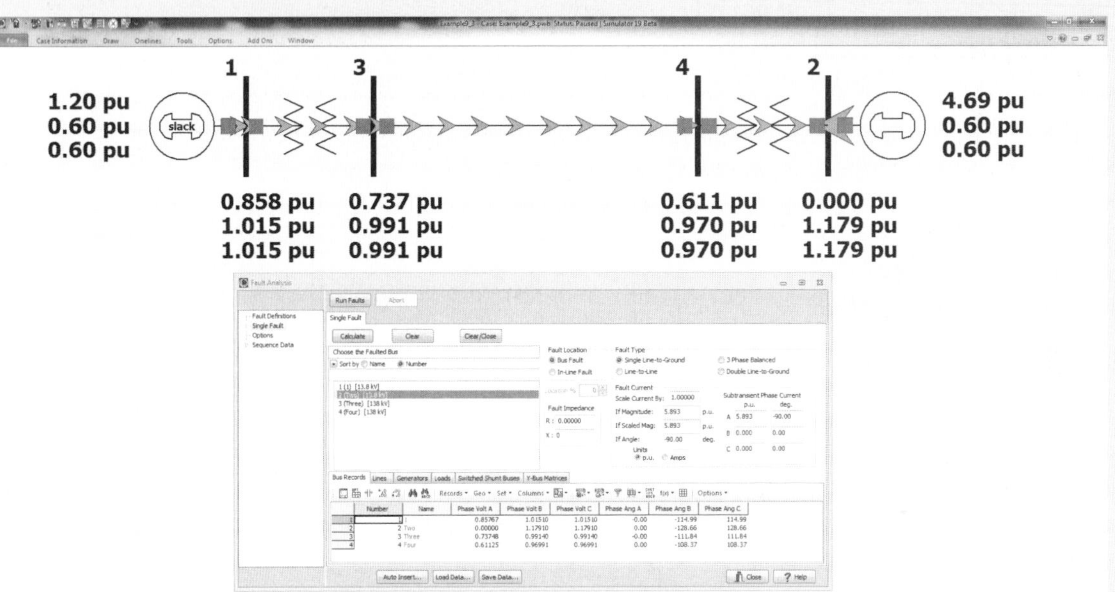

FIGURE 9.9

Screen for Example 9.3—fault at bus 2

simulating an unsymmetrical fault is almost identical to that for a balanced fault. That is, from the oneline, first right-click on the bus symbol corresponding to the fault location. This displays the local menu. Select **"Fault.."** to display the Fault Analysis dialog, and in the list on the left side of the dialog, set the **Fault Definitions** to Single Fault. Verify that the correct bus is selected, and then set the Fault Type field to Single Line-to-Ground. Finally, click on **Calculate** to determine the fault currents and voltages. The results are shown in the tables at the bottom of the dialog. Notice that with an unsymmetrical fault the phase magnitudes are no longer identical and the phase angles no longer differ by 120°. The fault values can be shown on the oneline by selecting **Options** in the left list, and changing the **Oneline Display** field value.

9.3 LINE-TO-LINE FAULT

Consider a line-to-line fault from phase b to c, shown in Figure 9.10 (a). Again, include a fault impedance Z_F for generality. From Figure 9.10(a),

Fault conditions in phase domain
Line-to-line fault

$$I_a = 0 \tag{9.3.1}$$

$$I_c = -I_b \tag{9.3.2}$$

$$V_{bg} - V_{cg} = Z_F I_b \tag{9.3.3}$$

Transform (9.3.1) through (9.3.3) to the sequence domain. Using (9.3.1) and (9.3.2) in (8.1.19),

$$\begin{bmatrix} I_0 \\ I_1 \\ I_2 \end{bmatrix} = \frac{1}{3}\begin{bmatrix} 1 & 1 & 1 \\ 1 & a & a^2 \\ 1 & a^2 & a \end{bmatrix}\begin{bmatrix} 0 \\ I_b \\ -I_b \end{bmatrix} = \begin{bmatrix} 0 \\ \frac{1}{3}(a - a^2)I_b \\ \frac{1}{3}(a^2 - a)I_b \end{bmatrix} \tag{9.3.4}$$

Using (8.1.4), (8.1.5), and (8.1.21) in (9.3.3),

$$(V_0 + a^2V_1 + aV_2) - (V_0 + aV_1 + a^2V_2) = Z_F(I_0 + a^2I_1 + aI_2) \tag{9.3.5}$$

Noting from (9.3.4) that $I_0 = 0$ and $I_2 = -I_1$, (9.3.5) simplifies to

$$(a^2 - a)V_1 - (a^2 - a)V_2 = Z_F(a^2 - a)I_1$$

or

$$V_1 - V_2 = Z_F I_1 \tag{9.3.6}$$

Therefore, from (9.3.4) and (9.3.6),

Fault conditions in sequence domain
Line-to-line fault

$$I_0 = 0 \tag{9.3.7}$$

$$I_2 = -I_1 \tag{9.3.8}$$

$$V_1 - V_2 = Z_F I_1 \tag{9.3.9}$$

FIGURE 9.10

Line-to-line fault

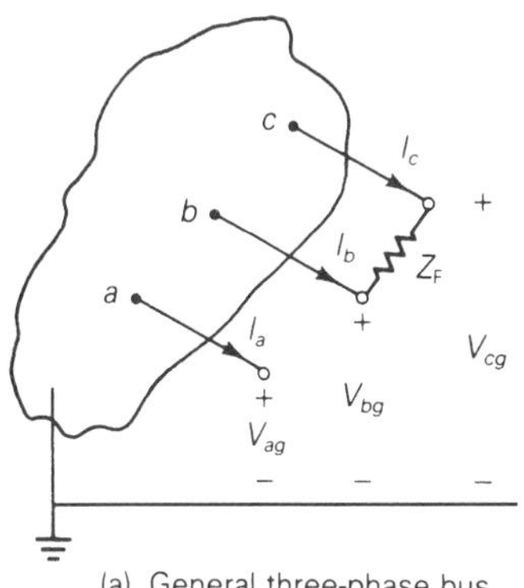

Fault conditions in phase domain:

$I_a = 0$

$I_c = -I_b$

$(V_{bg} - V_{cg}) = Z_F I_b$

(a) General three-phase bus

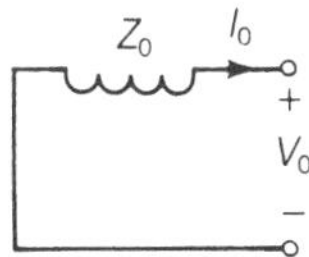

Fault conditions in sequence domain:

$I_0 = 0$

$I_2 = -I_1$

$(V_1 - V_2) = Z_F I_1$

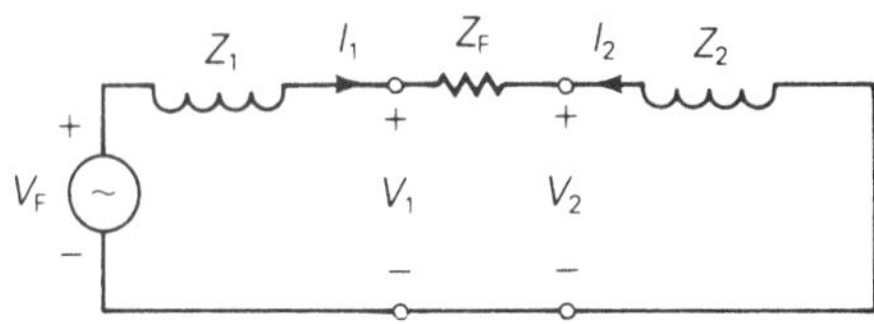

(b) Interconnected sequence networks

Equations (9.3.7) through (9.3.9) are satisfied by connecting the positive- and negative-sequence networks in parallel at the fault terminals through the fault impedance Z_F, as shown in Figure 9.10(b). From this figure, the fault currents are,

$$I_1 = -I_2 = \frac{V_F}{(Z_1 + Z_2 + Z_F)} \qquad I_0 = 0 \tag{9.3.10}$$

Transforming (9.3.10) to the phase domain and using the identity $(a^2 - a) = -j\sqrt{3}$, the fault current in phase b is

$$\begin{aligned} I_b &= I_0 + a^2 I_1 + a I_2 = (a^2 - a) I_1 \\ &= -j\sqrt{3} I_1 = \frac{-j\sqrt{3} V_F}{(Z_1 + Z_2 + Z_F)} \end{aligned} \tag{9.3.11}$$

Note also from (8.1.20) and (8.1.22) that

$$I_a = I_0 + I_1 + I_2 = 0 \tag{9.3.12}$$

and

$$I_c = I_0 + aI_1 + a^2I_2 = (a - a^2)I_1 = -I_b \tag{9.3.13}$$

which verify the fault conditions given by (9.3.1) and (9.3.2). The sequence components of the line-to-ground voltages at the fault are given by (9.1.1).

EXAMPLE 9.4

Line-to-line short-circuit calculations using sequence networks

Calculate the subtransient fault current in per-unit and in kA for a bolted line-to-line fault from phase b to c at bus 2 in Example 9.1.

SOLUTION

The positive- and negative-sequence networks in Figure 9.5 are connected in parallel at the fault terminals, as shown in Figure 9.11. From (9.3.10) with $Z_F = 0$, the sequence fault currents are

$$I_1 = -I_2 = \frac{1.05\underline{/0^\circ}}{j(0.13893 + 0.14562)} = 3.690\underline{/-90^\circ}$$

$$I_0 = 0$$

From (9.3.11), the subtransient fault current in phase b is

$$I_b'' = (-j\sqrt{3})(3.690\underline{/-90^\circ}) = -6.391 = 6.391\underline{/180^\circ} \quad \text{per unit}$$

Using 4.1837 kA as the base current at bus 2,

$$I_b'' = (6.391\underline{/180^\circ})(4.1837) = 26.74\underline{/180^\circ} \quad \text{kA}$$

Also, from (9.3.12) and (9.3.13),

$$I_a'' = 0 \qquad I_c'' = 26.74\underline{/0^\circ} \quad \text{kA}$$

The line-to-line fault results for this example can be shown in PowerWorld Simulator by repeating the Example 9.3 procedure, with the exception that the Fault Type field value should be Line-to-Line.

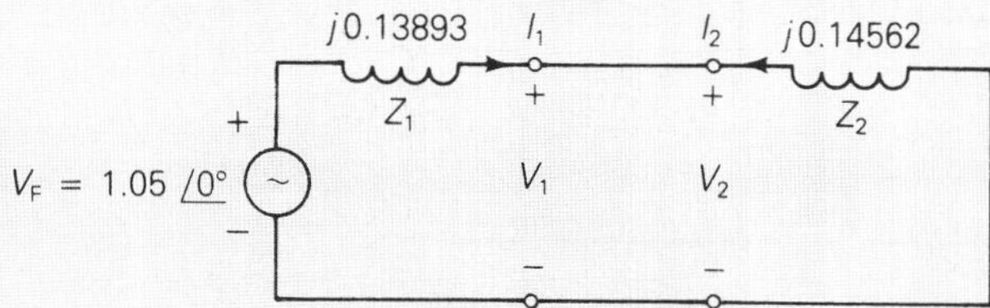

FIGURE 9.11 Example 9.4: Bolted three-phase-to-ground fault at bus 2

9.4 DOUBLE LINE-TO-GROUND FAULT

A double line-to-ground fault from phase b to phase c through fault impedance Z_F to ground is shown in Figure 9.12 (a). From this figure,

Fault conditions in the phase domain
Double line-to-ground fault

$$I_a = 0 \tag{9.4.1}$$
$$V_{cg} = V_{bg} \tag{9.4.2}$$
$$V_{bg} = Z_F(I_b + I_c) \tag{9.4.3}$$

Transforming (9.4.1) to the sequence domain via (8.1.20),

$$I_0 + I_1 + I_2 = 0 \tag{9.4.4}$$

Also, using (8.1.4) and (8.1.5) in (9.4.2),

$$(V_0 + aV_1 + a^2V_2) = (V_0 + a^2V_1 + aV_2)$$

Simplifying,

$$(a^2 - a)V_2 = (a^2 - a)V_1$$

FIGURE 9.12 Double line-to-ground fault

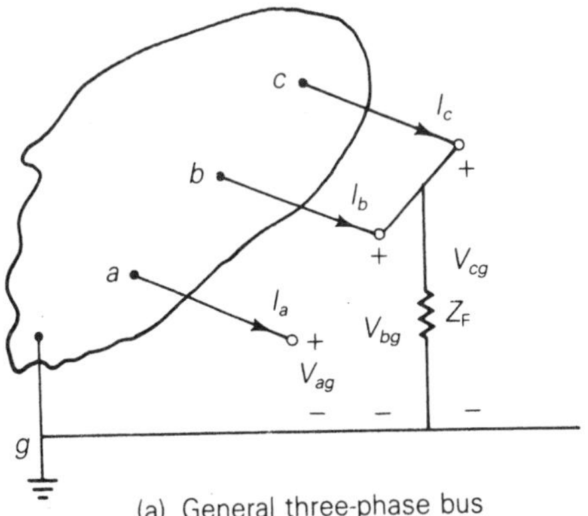

(a) General three-phase bus

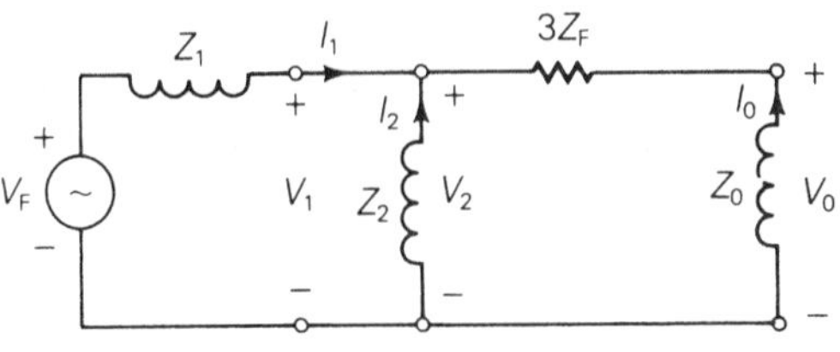

(b) Interconnected sequence networks

or

$$V_2 = V_1 \tag{9.4.5}$$

Now, using (8.1.4), (8.1.21), and (8.1.22) in (9.4.3),

$$(V_0 + a^2V_1 + aV_2) = Z_F(I_0 + a^2I_1 + aI_2 + I_0 + aI_1 + a^2I_2) \tag{9.4.6}$$

Using (9.4.5) and the identity $a^2 + a = -1$ in (9.4.6),

$$(V_0 - V_1) = Z_F(2I_0 - I_1 - I_2) \tag{9.4.7}$$

From (9.4.4), $I_0 = -(I_1 + I_2)$; therefore, (9.4.7) becomes

$$V_0 - V_1 = (3Z_F)I_0 \tag{9.4.8}$$

In summary, from (9.4.4), (9.4.5), and (9.4.8),

Fault conditions in the sequence domain
Double line-to-ground fault

$$I_0 + I_1 + I_2 = 0 \tag{9.4.9}$$

$$V_2 = V_1 \tag{9.4.10}$$

$$V_0 - V_1 = (3Z_F)I_0 \tag{9.4.11}$$

Equations (9.4.9) through (9.4.11) are satisfied by connecting the zero-, positive-, and negative-sequence networks in parallel at the fault terminal; additionally, $(3Z_F)$ is included in series with the zero-sequence network. This connection is shown in Figure 9.12(b). From this figure the positive-sequence fault current is

$$I_1 = \frac{V_F}{Z_1 + [Z_2//(Z_0 + 3Z_F)]} = \frac{V_F}{Z_1 + \left[\dfrac{Z_2(Z_0 + 3Z_F)}{Z_2 + Z_0 + 3Z_F}\right]} \tag{9.4.12}$$

Using current division in Figure 9.12(b), the negative- and zero-sequence fault currents are

$$I_2 = (-I_1)\left(\frac{Z_0 + 3Z_F}{Z_0 + 3Z_F + Z_2}\right) \tag{9.4.13}$$

$$I_0 = (-I_1)\left(\frac{Z_2}{Z_0 + 3Z_F + Z_2}\right) \tag{9.4.14}$$

These sequence fault currents can be transformed to the phase domain via (8.1.16). Also, the sequence components of the line-to-ground voltages at the fault are given by (9.1.1).

EXAMPLE 9.5

Double line-to-ground short-circuit calculations using sequence networks

Calculate (a) the subtransient fault current in each phase, (b) neutral fault current, and (c) contributions to the fault current from the motor and from the transmission line, for a bolted double line-to-ground fault from phase b to c to ground at bus 2 in Example 9.1. Neglect the Δ–Y transformer phase shifts.

SOLUTION

a. The zero-, positive-, and negative-sequence networks in Figure 9.5 are connected in parallel at the fault terminals in Figure 9.13. From (9.4.12) with $Z_F = 0$,

$$I_1 = \frac{1.05\underline{/0^\circ}}{j\left[0.13893 + \dfrac{(0.14562)(0.25)}{0.14562 + 0.25}\right]} = \frac{1.05\underline{/0^\circ}}{j0.23095}$$

$$= -j4.5464 \quad \text{per unit}$$

From (9.4.13) and (9.4.14),

$$I_2 = (+j4.5464)\left(\frac{0.25}{0.25 + 0.14562}\right) = j2.8730 \quad \text{per unit}$$

$$I_0 = (+j4.5464)\left(\frac{0.14562}{0.25 + 0.14562}\right) = j1.6734 \quad \text{per unit}$$

Transforming to the phase domain, the subtransient fault currents are:

$$\begin{bmatrix} I''_a \\ I''_b \\ I''_c \end{bmatrix} = \begin{bmatrix} 1 & 1 & 1 \\ 1 & a^2 & a \\ 1 & a & a^2 \end{bmatrix} \begin{bmatrix} +j1.6734 \\ -j4.5464 \\ +j2.8730 \end{bmatrix} = \begin{bmatrix} 0 \\ 6.8983\underline{/158.66^\circ} \\ 6.8983\underline{/21.34^\circ} \end{bmatrix} \quad \text{per unit}$$

Using the base current of 4.1837 kA at bus 2,

$$\begin{bmatrix} I''_a \\ I''_b \\ I''_c \end{bmatrix} = \begin{bmatrix} 0 \\ 6.8983\underline{/158.66^\circ} \\ 6.8983\underline{/21.34^\circ} \end{bmatrix}(4.1837) = \begin{bmatrix} 0 \\ 28.86\underline{/158.66^\circ} \\ 28.86\underline{/21.34^\circ} \end{bmatrix} \quad \text{kA}$$

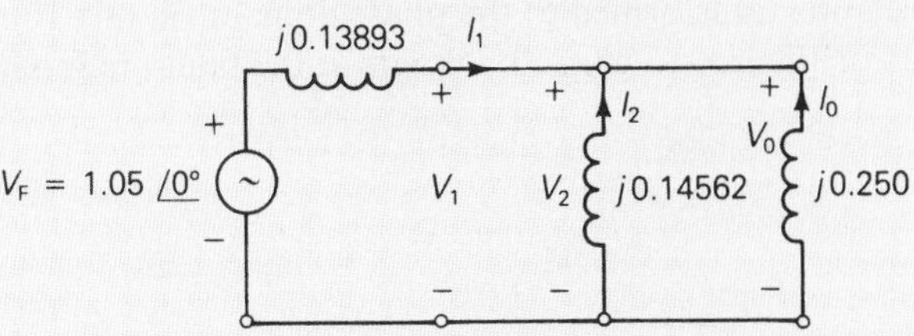

FIGURE 9.13 Example 9.5: Double line-to-ground fault at bus 2

b. The neutral fault current is

$$I_n = (I_b'' + I_c'') = 3I_0 = j5.0202 \quad \text{per unit}$$
$$= (j5.0202)(4.1837) = 21.00\underline{/90^\circ} \quad \text{kA}$$

c. Neglecting Δ–Y transformer phase shifts, the contributions to the fault current from the motor and transmission line can be obtained from Figure 9.4. From the zero-sequence network, Figure 9.4(a), the contribution to the zero-sequence fault current from the line is zero, due to the transformer connection. That is,

$$I_{\text{line }0} = 0$$
$$I_{\text{motor }0} = I_0 = j1.6734 \quad \text{per unit}$$

From the positive-sequence network, Figure 9.4(b), the positive terminals of the internal machine voltages can be connected, since $E_g'' = E_{m}''$. Then, by current division,

$$I_{\text{line1}} = \frac{X_m''}{X_m'' + (X_g'' + X_{T1} + X_{\text{line }1} + X_{T2})} I_1$$
$$= \frac{0.20}{0.20 + (0.455)}(-j4.5464) = -j1.3882 \quad \text{per unit}$$
$$I_{\text{motor }1} = \frac{0.455}{0.20 + 0.455}(-j4.5464) = -j3.1582 \quad \text{per unit}$$

From the negative-sequence network, Figure 9.4(c), using current division,

$$I_{\text{line }2} = \frac{0.21}{0.21 + 0.475}(j2.8730) = j0.8808 \quad \text{per unit}$$
$$I_{\text{motor }2} = \frac{0.475}{0.21 + 0.475}(j2.8730) = j1.9922 \quad \text{per unit}$$

Transforming to the phase domain with base currents of 0.41837 kA for the line and 4.1837 kA for the motor,

$$\begin{bmatrix} I''_{\text{line }a} \\ I''_{\text{line }b} \\ I''_{\text{line }c} \end{bmatrix} = \begin{bmatrix} 1 & 1 & 1 \\ 1 & a^2 & a \\ 1 & a & a^2 \end{bmatrix} \begin{bmatrix} 0 \\ -j1.3882 \\ j0.8808 \end{bmatrix}$$
$$= \begin{bmatrix} 0.5074\underline{/-90^\circ} \\ 1.9813\underline{/172.643^\circ} \\ 1.9813\underline{/7.357^\circ} \end{bmatrix} \quad \text{per unit}$$
$$= \begin{bmatrix} 0.2123\underline{/-90^\circ} \\ 0.8289\underline{/172.643^\circ} \\ 0.8289\underline{/7.357^\circ} \end{bmatrix} \quad \text{kA}$$

(*Continued*)

$$\begin{bmatrix} I''_{\text{motor } a} \\ I''_{\text{motor } b} \\ I''_{\text{motor } c} \end{bmatrix} = \begin{bmatrix} 1 & 1 & 1 \\ 1 & a^2 & a \\ 1 & a & a^2 \end{bmatrix} \begin{bmatrix} j1.6734 \\ -j3.1582 \\ j1.9922 \end{bmatrix}$$

$$= \begin{bmatrix} 0.5074\underline{/90^\circ} \\ 4.9986\underline{/153.17^\circ} \\ 4.9986\underline{/26.83^\circ} \end{bmatrix} \quad \text{per unit}$$

$$= \begin{bmatrix} 2.123\underline{/90^\circ} \\ 20.91\underline{/153.17^\circ} \\ 20.91\underline{/26.83^\circ} \end{bmatrix} \quad \text{kA}$$

The double line-to-line fault results for this example can be shown in PowerWorld Simulator by repeating the Example 9.3 procedure, with the exception that the Fault Type field value should be Double Line-to-Ground.

EXAMPLE 9.6

Effect of Δ–Y transformer phase shift on fault currents

Rework Example 9.5, with the Δ–Y transformer phase shifts included. Assume American standard phase shift.

SOLUTION

The sequence networks of Figure 9.4 are redrawn in Figure 9.14 with ideal phase-shifting transformers representing Δ–Y phase shifts. In accordance with the American standard, positive-sequence quantities on the high-voltage side of the transformers lead their corresponding quantities on the low-voltage side by 30°. Also, the negative-sequence phase shifts are the reverse of the positive-sequence phase shifts.

a. Recall from Section 3.1 and (3.1.26) that per-unit impedance is unchanged when it is referred from one side of an ideal phase-shifting transformer to the other. Accordingly, the Thévenin equivalents of the sequence networks in Figure 9.14, as viewed from fault bus 2, are the same as those given in Figure 9.5. Therefore, the sequence components as well as the phase components of the fault currents are the same as those given in Example 9.5(a).

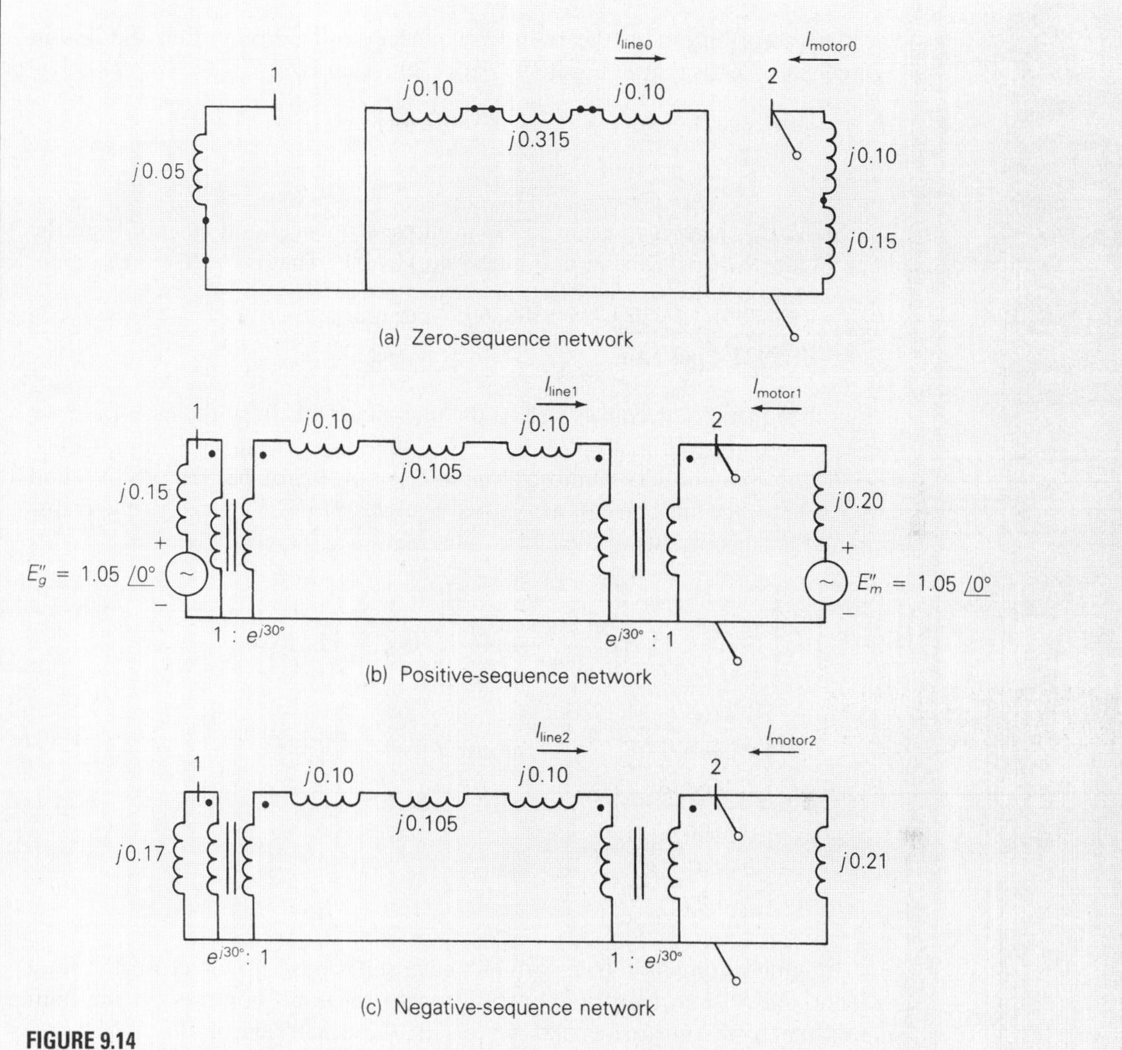

FIGURE 9.14

Sequence networks for Example 9.6

b. The neutral fault current is the same as that given in Example 9.5(b).

c. The zero-sequence network, Figure 9.14(a), is the same as that given in Figure 9.4(a). Therefore, the contributions to the zero-sequence fault current from the line and motor are the same as those given in Example 9.5(c).

$$I_{\text{line }0} = 0 \qquad I_{\text{motor }0} = I_0 = j1.6734 \quad \text{per unit}$$

(Continued)

The contribution to the positive-sequence fault current from the line in Figure 9.13(b) leads that in Figure 9.4(b) by 30°. That is,

$$I_{\text{line }1} = (-j1.3882)(1\underline{/30°}) = 1.3882\underline{/-60°} \quad \text{per unit}$$

$$I_{\text{motor }1} = -j3.1582 \quad \text{per unit}$$

Similarly, the contribution to the negative-sequence fault current from the line in Figure 9.14(c) lags that in Figure 9.4(c) by 30°. That is,

$$I_{\text{line }2} = (j0.8808)(1\underline{/-30°}) = 0.8808\underline{/60°} \quad \text{per unit}$$

$$I_{\text{motor }2} = j1.9922 \quad \text{per unit}$$

Thus, the sequence currents as well as the phase currents from the motor are the same as those given in Example 9.5(c). Also, the sequence currents from the line have the same magnitudes as those given in Example 9.5(c), but the positive- and negative-sequence line currents are shifted by +30° and −30°, respectively. Transforming the line currents to the phase domain,

$$\begin{bmatrix} I''_{\text{line }a} \\ I''_{\text{line }b} \\ I''_{\text{line }c} \end{bmatrix} = \begin{bmatrix} 1 & 1 & 1 \\ 1 & a^2 & a \\ 1 & a & a^2 \end{bmatrix} \begin{bmatrix} 0 \\ 1.3882\underline{/-60°} \\ 0.8808\underline{/+60°} \end{bmatrix}$$

$$= \begin{bmatrix} 1.2166\underline{/-21.17°} \\ 2.2690\underline{/180°} \\ 1.2166\underline{/21.17°} \end{bmatrix} \quad \text{per unit}$$

$$= \begin{bmatrix} 0.5090\underline{/-21.17°} \\ 0.9492\underline{/180°} \\ 0.5090\underline{/21.17°} \end{bmatrix} \quad \text{kA}$$

In conclusion, Δ–Y transformer phase shifts have no effect on the fault currents and no effect on the contribution to the fault currents on the fault side of the Δ–Y transformers. However, on the other side of the Δ–Y transformers, the positive- and negative-sequence components of the contributions to the fault currents are shifted by ±30°, which affects both the magnitude as well as the angle of the phase components of these fault contributions for unsymmetrical faults.

Figure 9.15 summarizes the sequence network connections for both the balanced three-phase fault and the unsymmetrical faults that have been considered. Sequence network connections for two additional faults, one-conductor-open and two-conductors-open, are also shown in Figure 9.15 and are left as an exercise for you to verify (see Problems 9.28 and 9.29).

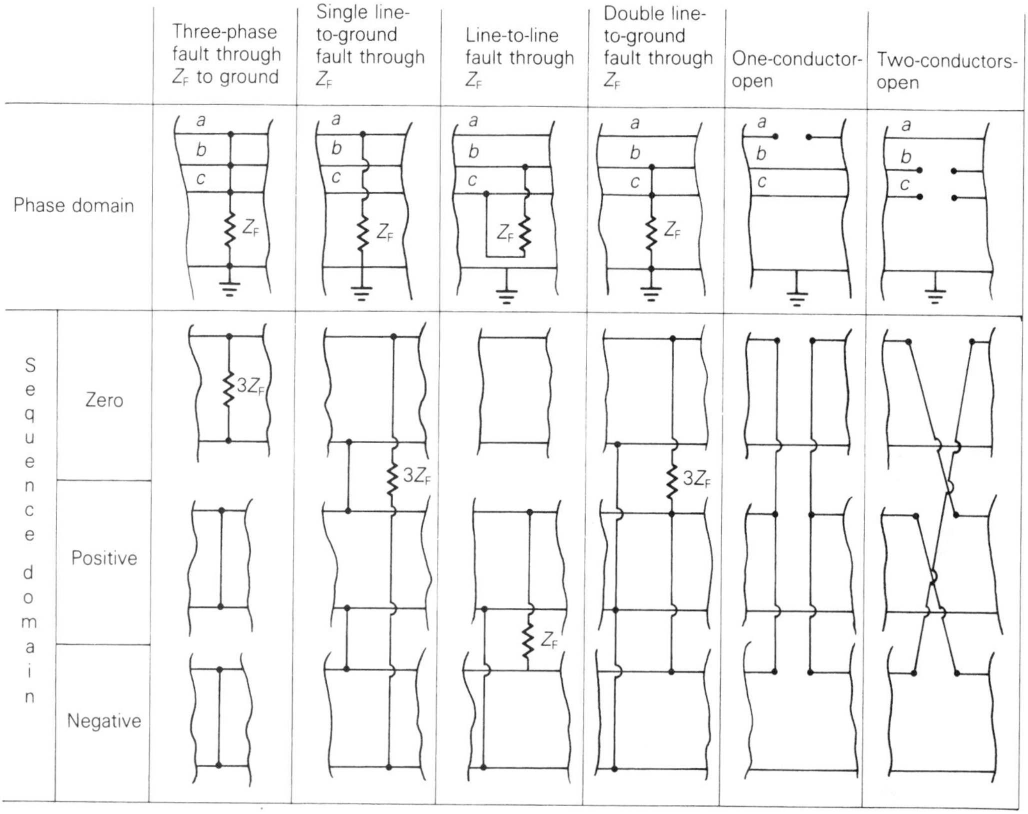

FIGURE 9.15

Summary of faults

9.5 SEQUENCE BUS IMPEDANCE MATRICES

The positive-sequence bus impedance matrix in Section 7.4 is used for calculating currents and voltages during balanced three-phase faults. This method is extended here to unsymmetrical faults by representing each sequence network as a bus impedance equivalent circuit (or as a rake equivalent). A bus impedance matrix can be computed for each sequence network by inverting the corresponding bus admittance network. For simplicity, resistances, shunt admittances, nonrotating impedance loads, and prefault load currents are neglected.

Figure 9.16 shows the connection of sequence rake equivalents for both symmetrical and unsymmetrical faults at bus n of an N-bus three-phase power system. Each bus impedance element has an additional subscript, 0, 1, or 2, that identifies the sequence rake equivalent in which it is located. Mutual impedances are not shown in the figure. The prefault voltage V_F is included in the positive-sequence rake equivalent. From the figure the sequence components of the fault current for each type of fault at bus n are as follows:

FIGURE 9.16 Connection of rake equivalent sequence networks for three-phase system faults (mutual impedances not shown)

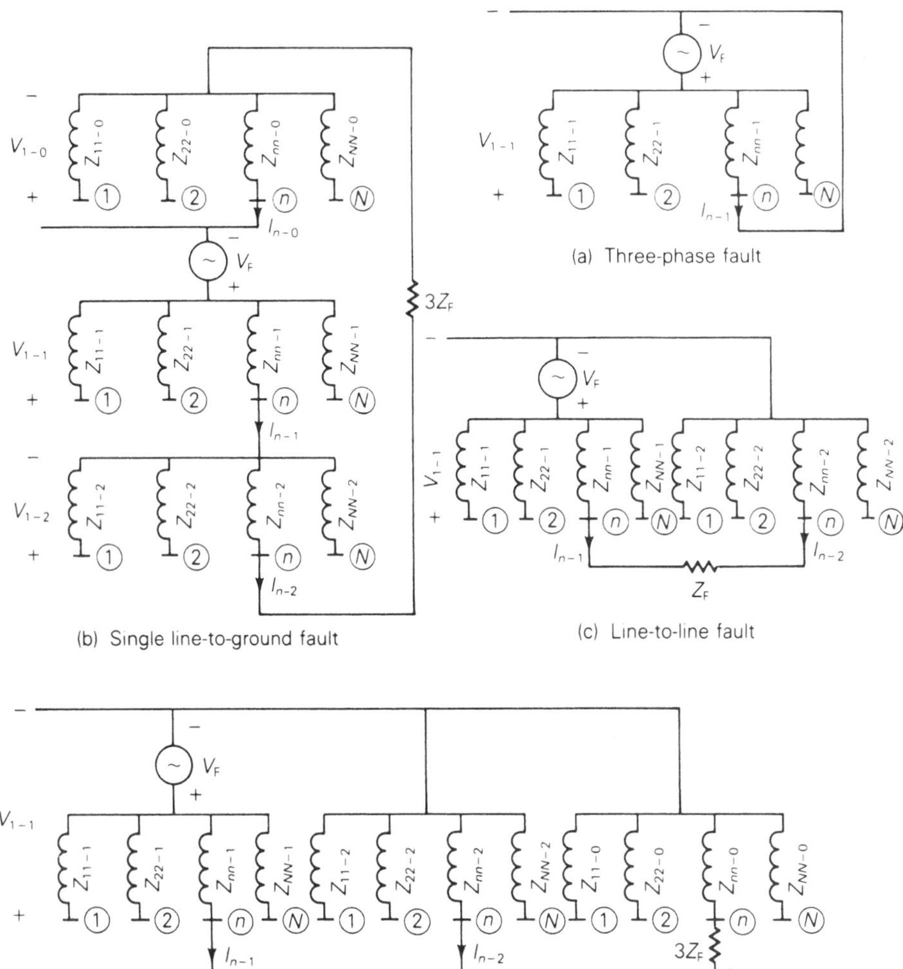

BALANCED THREE-PHASE FAULT:

$$I_{n-1} = \frac{V_F}{Z_{nn-1}} \tag{9.5.1}$$

$$I_{n-0} = I_{n-2} = 0 \tag{9.5.2}$$

SINGLE LINE-TO-GROUND FAULT (PHASE *a* TO GROUND):

$$I_{n-0} = I_{n-1} = I_{n-2} = \frac{V_F}{Z_{nn-0} + Z_{nn-1} + Z_{nn-2} + 3Z_F} \tag{9.5.3}$$

LINE-TO-LINE FAULT (PHASE *b* TO *c*):

$$I_{n-1} = -I_{n-2} = \frac{V_F}{Z_{nn-1} + Z_{nn-2} + Z_F} \tag{9.5.4}$$

$$I_{n-0} = 0 \tag{9.5.5}$$

DOUBLE LINE-TO-GROUND FAULT (PHASE *b* TO *c* TO GROUND):

$$I_{n-1} = \frac{V_F}{Z_{nn-1} + \left[\dfrac{Z_{nn-2}(Z_{nn-0} + 3Z_F)}{Z_{nn-2} + Z_{nn-0} + 3Z_F}\right]} \tag{9.5.6}$$

$$I_{n-2} = (-I_{n-1})\left(\frac{Z_{nn-0} + 3Z_F}{Z_{nn-0} + 3Z_F + Z_{nn-2}}\right) \tag{9.5.7}$$

$$I_{n-0} = (-I_{n-1})\left(\frac{Z_{nn-2}}{Z_{nn-0} + 3Z_F + Z_{nn-2}}\right) \tag{9.5.8}$$

Also from Figure 9.16, the sequence components of the line-to-ground voltages at any bus k during a fault at bus n are:

$$\begin{bmatrix} V_{k-0} \\ V_{k-1} \\ V_{k-2} \end{bmatrix} = \begin{bmatrix} 0 \\ V_F \\ 0 \end{bmatrix} - \begin{bmatrix} Z_{kn-0} & 0 & 0 \\ 0 & Z_{kn-1} & 0 \\ 0 & 0 & Z_{kn-2} \end{bmatrix} \begin{bmatrix} I_{n-0} \\ I_{n-1} \\ I_{n-2} \end{bmatrix} \tag{9.5.9}$$

If bus k is on the unfaulted side of a Δ–Y transformer, then the phase angles of V_{k-1} and V_{k-2} in (9.5.9) are modified to account for Δ–Y phase shifts. Also, the above sequence fault currents and sequence voltages can be transformed to the phase domain via (8.1.16) and (8.1.9).

EXAMPLE 9.7

Single line-to-ground short-circuit calculations using $Z_{bus\,0}$, $Z_{bus\,1}$, and $Z_{bus\,2}$

Faults at buses 1 and 2 for the three-phase power system given in Example 9.1 are of interest. The prefault voltage is 1.05 per unit. Prefault load current is neglected. (a) Determine the per-unit zero-, positive-, and negative-sequence bus impedance matrices. (b) Find the subtransient fault current in per-unit for a bolted single line-to-ground fault current from phase a to ground at bus 1, and (c) at bus 2. Find the per-unit line-to-ground voltages at (d) bus 1, and (e) bus 2 during the single line-to-ground fault at bus 1.

(*Continued*)

SOLUTION

a. Referring to Figure 9.4(a), the zero-sequence bus admittance matrix is

$$Y_{\text{bus }0} = -j\left[\begin{array}{c|c} 20 & 0 \\ \hline 0 & 4 \end{array}\right] \quad \text{per unit}$$

Inverting $\boldsymbol{Y}_{\text{bus }0}$,

$$Z_{\text{bus }0} = j\left[\begin{array}{c|c} 0.05 & 0 \\ \hline 0 & 0.25 \end{array}\right] \quad \text{per unit}$$

Note that the transformer leakage reactances and the zero-sequence transmission-line reactance in Figure 9.4(a) have no effect on $\boldsymbol{Z}_{\text{bus }0}$. The transformer Δ connections block the flow of zero-sequence current from the transformers to bus 1 and 2.

The positive-sequence bus admittance matrix, from Figure 9.4(b), is

$$Y_{\text{bus }1} = -j\left[\begin{array}{c|c} 9.9454 & -3.2787 \\ \hline -3.2787 & 8.2787 \end{array}\right] \quad \text{per unit}$$

Inverting $\boldsymbol{Y}_{\text{bus }1}$,

$$Z_{\text{bus }1} = j\left[\begin{array}{c|c} 0.11565 & 0.04580 \\ \hline 0.04580 & 0.13893 \end{array}\right] \quad \text{per unit}$$

Similarly, from Figure 9.4(c)

$$Y_{\text{bus }2} = -j\left[\begin{array}{c|c} 9.1611 & -3.2787 \\ \hline -3.2787 & 8.0406 \end{array}\right] \quad \text{per unit}$$

Inverting $\boldsymbol{Y}_{\text{bus }2}$,

$$Z_{\text{bus }2} = j\left[\begin{array}{c|c} 0.12781 & 0.05212 \\ \hline 0.05212 & 0.14562 \end{array}\right] \quad \text{per unit}$$

b. From (9.5.3), with $n=1$ and $Z_F = 0$, the sequence fault currents are

$$I_{1-0} = I_{1-1} = I_{1-2} = \frac{V_F}{Z_{11-0} + Z_{11-1} + Z_{11-2}}$$

$$= \frac{1.05\underline{/0°}}{j(0.05 + 0.11565 + 0.12781)} = \frac{1.05}{j0.29346} = -j3.578 \quad \text{per unit}$$

The subtransient fault currents at bus 1 are, from (8.1.16),

$$\begin{bmatrix} I''_{1a} \\ I''_{1b} \\ I''_{1c} \end{bmatrix} = \begin{bmatrix} 1 & 1 & 1 \\ 1 & a^2 & a \\ 1 & a & a^2 \end{bmatrix} \begin{bmatrix} -j3.578 \\ -j3.578 \\ -j3.578 \end{bmatrix} = \begin{bmatrix} -j10.73 \\ 0 \\ 0 \end{bmatrix} \quad \text{per unit}$$

c. Again from (9.5.3), with $n = 2$ and $Z_F = 0$,

$$I_{2-0} = I_{2-1} = I_{2-2} = \frac{V_F}{Z_{22-0} + Z_{22-1} + Z_{22-2}}$$

$$= \frac{1.05\underline{/0°}}{j(0.25 + 0.13893 + 0.14562)} = \frac{1.05}{j0.53455}$$

$$= -j1.96427 \quad \text{per unit}$$

and

$$\begin{bmatrix} I''_{2a} \\ I''_{2b} \\ I''_{2c} \end{bmatrix} = \begin{bmatrix} 1 & 1 & 1 \\ 1 & a^2 & a \\ 1 & a & a^2 \end{bmatrix} \begin{bmatrix} -j1.96427 \\ -j1.96427 \\ -j1.96427 \end{bmatrix} = \begin{bmatrix} -j5.8928 \\ 0 \\ 0 \end{bmatrix} \quad \text{per unit}$$

This is the same result as obtained in Example 9.3.

d. The sequence components of the line-to-ground voltages at bus 1 during the fault at bus 1 are, from (9.5.9), with $k=1$ and $n = 1$,

$$\begin{bmatrix} V_{1-0} \\ V_{1-1} \\ V_{1-2} \end{bmatrix} = \begin{bmatrix} 0 \\ 1.05\underline{/0°} \\ 0 \end{bmatrix} - \begin{bmatrix} j0.05 & 0 & 0 \\ 0 & j0.11565 & 0 \\ 0 & 0 & j0.12781 \end{bmatrix} \begin{bmatrix} -j3.578 \\ -j3.578 \\ -j3.578 \end{bmatrix}$$

$$= \begin{bmatrix} -0.1789 \\ 0.6362 \\ -0.4573 \end{bmatrix} \quad \text{per unit}$$

and the line-to-ground voltages at bus 1 during the fault at bus 1 are

$$\begin{bmatrix} V_{1-ag} \\ V_{1-bg} \\ V_{1-cg} \end{bmatrix} = \begin{bmatrix} 1 & 1 & 1 \\ 1 & a^2 & a \\ 1 & a & a^2 \end{bmatrix} \begin{bmatrix} -0.1789 \\ +0.6362 \\ -0.4573 \end{bmatrix}$$

$$= \begin{bmatrix} 0 \\ 0.9843\underline{/254.2°} \\ 0.9843\underline{/105.8°} \end{bmatrix} \quad \text{per unit}$$

(Continued)

e. The sequence components of the line-to-ground voltages at bus 2 during the fault at bus 1 are, from (9.5.9), with $k = 2$ and $n = 1$,

$$\begin{bmatrix} V_{2-0} \\ V_{2-1} \\ V_{2-2} \end{bmatrix} = \begin{bmatrix} 0 \\ 1.05\underline{/0^\circ} \\ 0 \end{bmatrix} - \begin{bmatrix} 0 & 0 & 0 \\ 0 & j0.04580 & 0 \\ 0 & 0 & j0.05212 \end{bmatrix} \begin{bmatrix} -j3.578 \\ -j3.578 \\ -j3.578 \end{bmatrix}$$

$$= \begin{bmatrix} 0 \\ 0.8861 \\ -0.18649 \end{bmatrix} \text{ per unit}$$

Note that since both bus 1 and 2 are on the low-voltage side of the Δ–Y transformers in Figure 9.3, there is no shift in the phase angles of these sequence voltages. From the above, the line-to-ground voltages at bus 2 during the fault at bus 1 are

$$\begin{bmatrix} V_{2-ag} \\ V_{2-bg} \\ V_{2-cg} \end{bmatrix} = \begin{bmatrix} 1 & 1 & 1 \\ 1 & a^2 & a \\ 1 & a & a^2 \end{bmatrix} \begin{bmatrix} 0 \\ 0.8861 \\ -0.18649 \end{bmatrix}$$

$$= \begin{bmatrix} 0.70 \\ 0.9926\underline{/249.4^\circ} \\ 0.9926\underline{/110.6^\circ} \end{bmatrix} \text{ per unit}$$

PowerWorld Simulator computes the symmetrical fault current for each of the following faults at any bus in an N-bus power system: balanced three-phase fault, single line-to-ground fault, line-to-line fault, or double line-to-ground fault. For each fault, the Simulator also computes bus voltages and contributions to the fault current from transmission lines and transformers connected to the fault bus.

Input data for the Simulator include machine, transmission-line, and transformer data, as illustrated in Tables 9.1, 9.2, and 9.3 as well as the prefault voltage V_F and fault impedance Z_F. When the machine positive-sequence reactance input data consist of direct axis subtransient reactances, the computed symmetrical fault currents are subtransient fault currents. Alternatively, transient or steady-state fault currents are computed when these input data consist of direct axis transient or synchronous reactances. Transmission-line positive- and zero-sequence series reactances are those of the equivalent π circuits for long lines or of the nominal π circuit for medium or short lines. Also, recall that the negative-sequence transmission-line reactance equals the positive-sequence transmission-line reactance. All machine, line, and transformer reactances are given in per-unit on a common MVA base. Prefault load currents are neglected.

Bus	X_0 per unit	$X_1 = X''_d$ per unit	X_2 per unit	Neutral Reactance X_n per unit
1	0.0125	0.045	0.045	0
3	0.005	0.0225	0.0225	0.0025

TABLE 9.1

Synchronous machine data for Example 9.8

Bus-to-Bus	X_0 per unit	X_1 per unit
2–4	0.3	0.1
2–5	0.15	0.05
4–5	0.075	0.025

TABLE 9.2

Line data for Example 9.8

Low-Voltage (connection) bus	High-Voltage (connection) bus	Leakage Reactance per unit	Neutral Reactance per unit
1(Δ)	5(Y)	0.02	0
3(Δ)	4(Y)	0.01	0

TABLE 9.3

Transformer data for Example 9.8

$S_{base} = 100$ MVA

$$V_{base} = \begin{cases} 15 \text{ kV at buses 1, 3} \\ 345 \text{ kV at buses 2, 4, 5} \end{cases}$$

The Simulator computes (but does not show) the zero-, positive-, and negative-sequence bus impedance matrices $\boldsymbol{Z}_{\text{bus } 0}$, $\boldsymbol{Z}_{\text{bus } 1}$, and $\boldsymbol{Z}_{\text{bus } 2}$, by inverting the corresponding bus admittance matrices.

After $\boldsymbol{Z}_{\text{bus } 0}$, $\boldsymbol{Z}_{\text{bus } 1}$, and $\boldsymbol{Z}_{\text{bus } 2}$ are computed, (9.5.1) through (9.5.9) are used to compute the sequence fault currents and the sequence voltages at each bus during a fault at bus 1 for the fault type selected by the program user (for example, three-phase fault, or single line-to-ground fault, and so on). Contributions to the sequence fault currents from each line or transformer branch connected to the fault bus are computed by dividing the sequence voltage across the branch by the branch sequence impedance. The phase angles of positive- and negative-sequence voltages are also modified to account for Δ–Y transformer phase shifts. The sequence currents and sequence voltages are then transformed to the phase domain via (8.1.16) and (8.1.9). All these computations are then repeated for a fault at bus 2, then bus 3, and so on to bus *N*.

Output data for the fault type and fault impedance selected by the user consist of the fault current in each phase, contributions to the fault current from each branch connected to the fault bus for each phase, and the line-to-ground voltages at each bus—for a fault at bus 1, then bus 2, and so on to bus *N*.

EXAMPLE 9.8

PowerWorld Simulator

Consider the five-bus power system whose single-line diagram is shown in Figure 6.2. Machine, line, and transformer data are given in Tables 9.1, 9.2, and 9.3. Note that the neutrals of both transformers and generator 1 are solidly grounded, as indicated by a neutral reactance of zero for this equipment. However, a neutral reactance equal to 0.0025 per unit is connected to the generator 2 neutral. The prefault voltage is 1.05 per unit. Using PowerWorld Simulator, determine the fault currents and voltages for a bolted single line-to-ground fault at bus 1, then bus 2, and so on to bus 5.

SOLUTION

Open PowerWorld Simulator case Example 9_8 to see this example. Tables 9.4 and 9.5 summarize the PowerWorld Simulator results for each of the faults. Note that these fault currents are subtransient currents, since the machine positive-sequence reactance input consists of direct axis subtransient reactances.

Fault Bus	Single Line-to-Ground Fault Current (Phase A) per unit/degrees	GEN LINE OR TRSF	Contributions to Fault Current Bus-to-Bus	Current Phase A (per unit/degrees)	Current Phase B (per unit/degrees)	Current Phase C (per unit/degrees)
1	46.02/−90.00	G1	GRND–1	34.41/−90.00	5.804/−90.00	5.804/−90.00
		T1	5–1	11.61/−90.00	5.804/90.00	5.804/90.00
2	14.14/−90.00	L1	4–2	5.151/−90.00	0.1124/90.00	0.1124/90.00
		L2	5–2	8.984/−90.00	0.1124/−90.00	0.1124/−90.00
3	64.30/−90.00	G2	GRND–3	56.19/−90.00	4.055/−90.00	4.055/−90.00
		T2	4–3	8.110/−90.00	4.055/90.00	4.055/90.00
4	56.07/−90.00	L1	2–4	1.742/−90.00	0.4464/90.00	0.4464/90.00
		L3	5–4	10.46/−90.00	2.679/90.00	2.679/90.00
		T2	3–4	43.88/−90.00	3.125/−90.00	3.125/−90.00
5	42.16/−90.00	L2	2–5	2.621/−90.00	0.6716/90.00	0.6716/90.00
		L3	4–5	15.72/−90.00	4.029/90.00	4.029/90.00
		T1	1–5	23.82/−90.00	4.700/−90.00	4.700/−90.00

TABLE 9.4

Fault currents for Example 9.8

$V_{prefault} = 1.05\angle 0$		Bus Voltages during Fault		
Fault Bus	Bus	Phase A	Phase B	Phase C
1	1	$0.0000\angle 0.00$	$0.9537\angle -107.55$	$0.9537\angle 107.55$
	2	$0.5069\angle 0.00$	$0.9440\angle -105.57$	$0.9440\angle 105.57$
	3	$0.7888\angle 0.00$	$0.9912\angle -113.45$	$0.9912\angle 113.45$
	4	$0.6727\angle 0.00$	$0.9695\angle -110.30$	$0.9695\angle 110.30$
	5	$0.4239\angle 0.00$	$0.9337\angle -103.12$	$0.9337\angle 103.12$
2	1	$0.8832\angle 0.00$	$1.0109\angle -115.90$	$1.0109\angle 115.90$
	2	$0.0000\angle 0.00$	$1.1915\angle -130.26$	$1.1915\angle 130.26$
	3	$0.9214\angle 0.00$	$1.0194\angle -116.87$	$1.0194\angle 116.87$
	4	$0.8435\angle 0.00$	$1.0158\angle -116.47$	$1.0158\angle 116.47$
	5	$0.7562\angle 0.00$	$1.0179\angle -116.70$	$1.0179\angle 116.70$
3	1	$0.6851\angle 0.00$	$0.9717\angle -110.64$	$0.9717\angle 110.64$
	2	$0.4649\angle 0.00$	$0.9386\angle -104.34$	$0.9386\angle 104.34$
	3	$0.0000\angle 0.00$	$0.9942\angle -113.84$	$0.9942\angle 113.84$
	4	$0.3490\angle 0.00$	$0.9259\angle -100.86$	$0.9259\angle 100.86$
	5	$0.5228\angle 0.00$	$0.9462\angle -106.04$	$0.9462\angle 106.04$
4	1	$0.5903\angle 0.00$	$0.9560\angle -107.98$	$0.9560\angle 107.98$
	2	$0.2309\angle 0.00$	$0.9401\angle -104.70$	$0.9401\angle 104.70$
	3	$0.4387\angle 0.00$	$0.9354\angle -103.56$	$0.9354\angle 103.56$
	4	$0.0000\angle 0.00$	$0.9432\angle -105.41$	$0.9432\angle 105.41$
	5	$0.3463\angle 0.00$	$0.9386\angle -104.35$	$0.9386\angle 104.35$
5	1	$0.4764\angle 0.00$	$0.9400\angle -104.68$	$0.9400\angle 104.68$
	2	$0.1736\angle 0.00$	$0.9651\angle -109.57$	$0.9651\angle 109.57$
	3	$0.7043\angle 0.00$	$0.9751\angle -111.17$	$0.9751\angle 111.17$
	4	$0.5209\angle 0.00$	$0.9592\angle -108.55$	$0.9592\angle 108.55$
	5	$0.0000\angle 0.00$	$0.9681\angle -110.07$	$0.9681\angle 110.07$

TABLE 9.5
Bus voltages for Example 9.8

MULTIPLE CHOICE QUESTIONS

SECTION 9.1

9.1 For power-system fault studies, it is assumed that the system is operating under balanced steady-state conditions prior to the fault, and sequence networks are uncoupled before the fault occurs.

(a) True (b) False

9.2 The first step in power-system fault calculations is to develop sequence networks based on the single-line diagram of the system, and then reduce them to their Thévenin equivalents, as viewed from the fault location.

(a) True (b) False

9.3 When calculating symmetrical three-phase fault currents, only ________ sequence network needs to be considered.

9.4 In order of frequency of occurrence of short-circuit faults in three-phase power systems, list those: ________, ________, ________, ________.

9.5 For a bolted three-phase-to-ground fault, sequence-fault currents ________ are zero, sequence fault voltages are ______, and line-to-ground voltages are ________.

SECTION 9.2

9.6 For a single-line-to-ground fault with a fault-impedance Z_F, the sequence networks are to be connected ________ at the fault terminals through the impedance ________; the sequence components of the fault currents are ________.

SECTION 9.3

9.7 For a line-to-line fault with a fault impedance Z_F, the positive-and negative-sequence networks are to be connected ________ at the fault terminals through the impedance of *1/2/3* times Z_F; the zero-sequence current is ________.

SECTION 9.4

9.8 For a double line-to-ground fault through a fault impedance Z_F, the sequence networks are to be connected ________, at the fault terminal; additionally, ________ is to be included in series with the zero-sequence network.

SECTION 9.5

9.9 The sequence bus-impedance matrices can also be used to calculate fault currents and voltages for symmetrical as well as unsymmetrical faults by representing each sequence network as a bus-impedance rake-equivalent circuit.

(a) True (b) False

PROBLEMS

SECTION 9.1

9.1 The single-line diagram of a three-phase power system is shown in Figure 9.17. Equipment ratings are given as follows:

Synchronous generators:

G1	1000 MVA	15 kV	$X''_d = X_2 = 0.18, X_0 = 0.07$ per unit
G2	1000 MVA	15 kV	$X''_d = X_2 = 0.20, X_0 = 0.10$ per unit
G3	500 MVA	13.8 kV	$X''_d = X_2 = 0.15, X_0 = 0.05$ per unit
G4	750 MVA	13.8 kV	$X''_d = 0.30, X_2 = 0.40, X_0 = 0.10$ per unit

Transformers:

T1	1000 MVA	15 kV Δ/765 kV Y	X = 0.10 per unit
T2	1000 MVA	15 kV Δ/765 kV Y	X = 0.10 per unit
T3	500 MVA	15 kV Y/765kV Y	X = 0.12 per unit
T4	750 MVA	15 kV Y/765 kV Y	X = 0.11 per unit

Transmission lines:

1–2	765 kV	$X_1 = 50\ \Omega,\ X_0 = 150\ \Omega$
1–3	765 kV	$X_1 = 40\ \Omega,\ X_0 = 100\ \Omega$
2–3	765 kV	$X_1 = 40\ \Omega,\ X_0 = 100\ \Omega$

The inductor connected to Generator 3 neutral has a reactance of 0.05 per unit using generator 3 ratings as a base. Draw the zero-, positive-, and negative-sequence reactance diagrams using a 1000-MVA, 765-kV base in the zone of line 1–2. Neglect the Δ–Y transformer phase shifts.

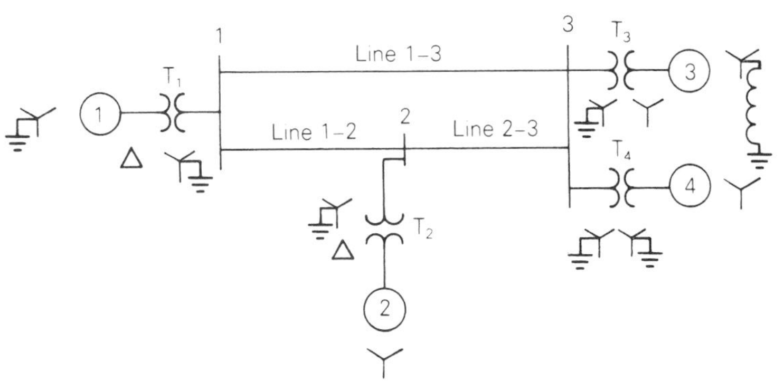

FIGURE 9.17 Problem 9.1

9.2 Faults at bus n in Problem 9.1 are of interest (the instructor selects $n = 1$, 2, or 3). Determine the Thévenin equivalent of each sequence network as viewed from the fault bus. Prefault voltage is 1.0 per unit. Prefault load currents and Δ–Y transformer phase shifts are neglected. (*Hint:* Use the Y–Δ conversion in Figure 2.33.)

9.3 Determine the subtransient fault current in per-unit and in kA during a bolted three-phase fault at the fault bus selected in Problem 9.2.

9.4 In Problem 9.1 and Figure 9.17, let 765 kV be replaced by 500 kV, keeping the rest of the data to be the same. Repeat (a) Problems 9.1, (b) 9.2, and (c) 9.3.

9.5 Equipment ratings for the four-bus power system shown in Figure 7.14 are given as follows:

Generator G1:	500 MVA, 13.8 kV, $X_d'' = X_2 = 0.20$, $X_0 = 0.10$ per unit
Generator G2:	750 MVA, 18 kV, $X_d'' = X_2 = 0.18$, $X_0 = 0.09$ per unit
Generator G3:	1000 MVA, 20 kV, $X_d'' = 0.17$, $X_2 = 0.20$, $X_0 = 0.09$ per unit
Transformer T1:	500 MVA, 13.8 kV Δ /500 kV Y, X = 0.12 per unit
Transformer T2:	750 MVA, 18 kV Δ/500 kV Y, X = 0.10 per unit
Transformer T3:	1000 MVA, 20 kV Δ/500 kV Y, X = 0.10 per unit
Each line:	$X_1 = 50$ ohms, $X_0 = 150$ ohms

The inductor connected to generator G3 neutral has a reactance of 0.028 Ω. Draw the zero-, positive-, and negative-sequence reactance diagrams using a 1000-MVA, 20-kV base in the zone of generator G3. Neglect Δ–Y transformer phase shifts.

9.6 Faults at bus n in Problem 9.5 are of interest (the instructor selects $n = 1$, 2, 3, or 4). Determine the Thévenin equivalent of each sequence network as viewed from the fault bus. Prefault voltage is 1.0 per unit. Prefault load currents and Δ–Y phase shifts are neglected.

9.7 Determine the subtransient fault current in per-unit and in kA during a bolted three-phase fault at the fault bus selected in Problem 9.6.

9.8 Equipment ratings for the five-bus power system shown in Figure 7.15 are given as follows:

Generator G1:	50 MVA, 12 kV, $X_d'' = X_2 = 0.20$, $X_0 = 0.10$ per unit
Generator G2:	100 MVA, 15 kV, $X_d'' = 0.2$, $X_2 = 0.23$, $X_0 = 0.1$ per unit
Transformer T1:	50 MVA, 10 kV Y/138 kV Y, X = 0.10 per unit
Transformer T2:	100 MVA, 15 kV Δ/138 kV Y, X = 0.10 per unit
Each 138-kV line:	$X_1 = 40$ ohms, $X_0 = 100$ ohms

Draw the zero-, positive-, and negative-sequence reactance diagrams using a 100-MVA, 15-kV base in the zone of generator G2. Neglect Δ–Y transformer phase shifts.

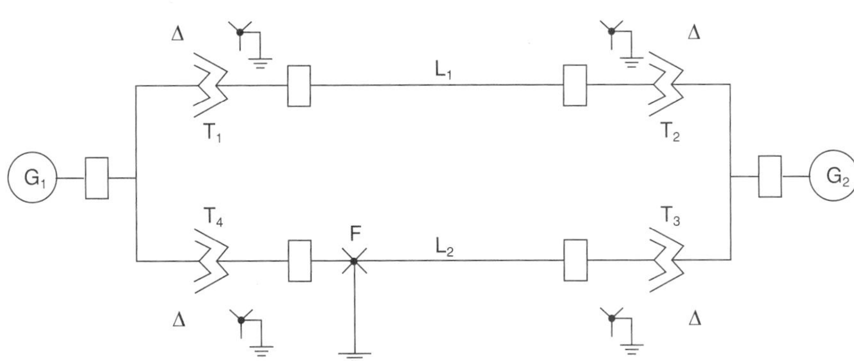

FIGURE 9.18

Problem 9.11

9.9 Faults at bus n in Problem 9.8 are of interest (the instructor selects n = 1, 2, 3, 4, or 5). Determine the Thévenin equivalent of each sequence network as viewed from the fault bus. Prefault voltage is 1.0 per unit. Prefault load currents and Δ–Y phase shifts are neglected.

9.10 Determine the subtransient fault current in per-unit and in kA during a bolted three-phase fault at the fault bus selected in Problem 9.9.

9.11 Consider the system shown in Figure 9.18. (a) As viewed from the fault at F, determine the Thévenin equivalent of each sequence network. Neglect Δ–Y phase shifts. (b) Compute the fault currents for a balanced three-phase fault at fault point F through three fault impedances $Z_{FA} = Z_{FB} = Z_{FC} = j0.5$ per unit. Equipment data in per-unit on the same base are given as follows:

Synchronous generators:

G1	$X_1 = 0.2$	$X_2 = 0.12$	$X_0 = 0.06$
G2	$X_1 = 0.33$	$X_2 = 0.22$	$X_0 = 0.066$

Transformers:

T1	$X_1 = X_2 = X_0 = 0.2$
T2	$X_1 = X_2 = X_0 = 0.225$
T3	$X_1 = X_2 = X_0 = 0.27$
T4	$X_1 = X_2 = X_0 = 0.16$

Transmission lines:

L1	$X_1 = X_2 = 0.14$	$X_0 = 0.3$
L1	$X_1 = X_2 = 0.35$	$X_0 = 0.6$

9.12 Equipment ratings and per-unit reactances for the system shown in Figure 9.19 are given as follows:

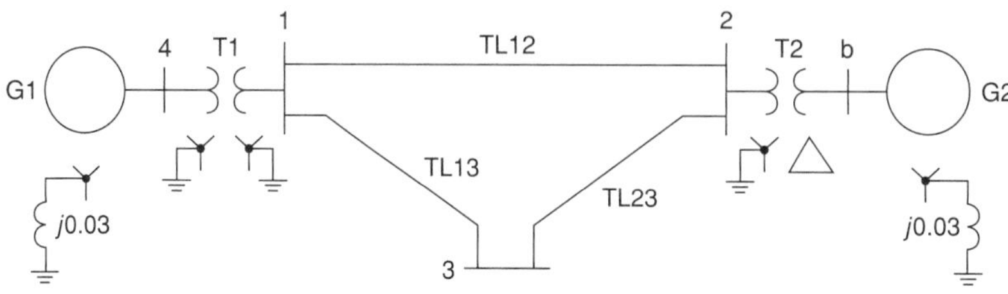

FIGURE 9.19

Problem 9.12

Synchronous generators:

G1	100 MVA	25 kV	$X_1 = X_2 = 0.2$	$X_0 = 0.05$
G2	100 MVA	13.8 kV	$X_1 = X_2 = 0.2$	$X_0 = 0.05$

Transformers:

T1	100 MVA	25/230 kV	$X_1 = X_2 = X_0 = 0.05$
T2	100 MVA	13.8/230 kV	$X_1 = X_2 = X_0 = 0.05$

Transmission lines:

TL12	100 MVA	230 kV	$X_1 = X_2 = 0.1$	$X_0 = 0.3$
TL13	100 MVA	230 kV	$X_1 = X_2 = 0.1$	$X_0 = 0.3$
TL23	100 MVA	230 kV	$X_1 = X_2 = 0.1$	$X_0 = 0.3$

Using a 100-MVA, 230-kV base for the transmission lines, draw the per-unit sequence networks and reduce them to their Thévenin equivalents, "looking in" at bus 3. Neglect Δ–Y phase shifts. Compute the fault currents for a bolted three-phase fault at bus 3.

9.13 Consider the oneline diagram of a simple power system shown in Figure 9.20. System data in per-unit on a 100-MVA base are given as follows:

Synchronous generators:

G1	100 MVA	20 kV	$X_1 = X_2 = 0.15$	$X_0 = 0.05$
G2	100 MVA	20 kV	$X_1 = X_2 = 0.15$	$X_0 = 0.05$

Transformers:

T1	100 MVA	20/220 kV	$X_1 = X_2 = X_0 = 0.1$
T2	100 MVA	20/220 kV	$X_1 = X_2 = X_0 = 0.1$

Transmission lines:

L12	100 MVA	220 kV	$X_1 = X_2 = 0.125$	$X_0 = 0.3$
L13	100 MVA	220 kV	$X_1 = X_2 = 0.15$	$X_0 = 0.35$
L23	100 MVA	220 kV	$X_1 = X_2 = 0.25$	$X_0 = 0.7125$

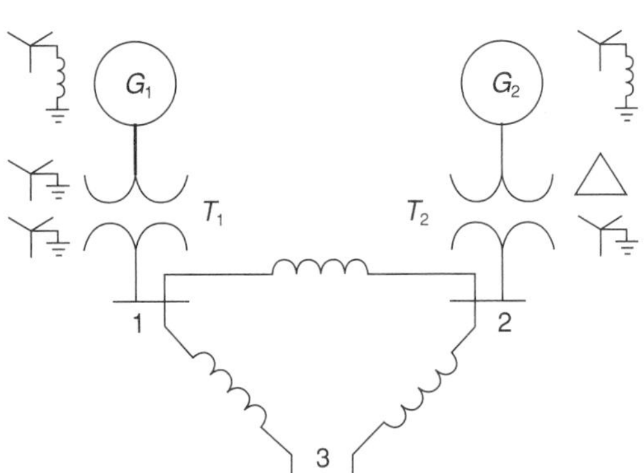

FIGURE 9.20

Problem 9.13

The neutral of each generator is grounded through a current-limiting reactor of 0.08333 per unit on a 100-MVA base. All transformer neutrals are solidly grounded. The generators are operating no-load at their rated voltages and rated frequency with their EMFs in phase. Determine the fault current for a balanced three-phase fault at bus 3 through a fault impedance $Z_F = 0.1$ per unit on a 100-MVA base. Neglect Δ–Y phase shifts.

SECTIONS 9.2–9.4

9.14 Determine the subtransient fault current in per-unit and in kA, as well as the per-unit line-to-ground voltages at the fault bus for a bolted single line-to-ground fault at the fault bus selected in Problem 9.2.

9.15 Repeat Problem 9.14 for a single line-to-ground arcing fault with arc impedance $Z_F = 15 + j0\ \Omega$.

9.16 Repeat Problem 9.14 for a bolted line-to-line fault.

9.17 Repeat Problem 9.14 for a bolted double line-to-ground fault.

9.18 Repeat Problems 9.1 and 9.14 including Δ–Y transformer phase shifts. Assume American standard phase shift. Also calculate the sequence components and phase components of the contribution to the fault current from generator n ($n = 1, 2,$ or 3 as specified by the instructor in Problem 9.2).

9.19 (a) Repeat Problem 9.14 for the case of Problem 9.4 (b).

(b) Repeat Problem 9.19(a) for a single line-to-ground arcing fault with arc impedance $Z_F = (15 + j0)\ \Omega$.

(c) Repeat Problem 9.19(a) for a bolted line-to-line fault.

(d) Repeat Problem 9.19(a) for a bolted double line-to-ground fault.

(e) Repeat Problems 9.4(a) and 9.19(a) including Δ–Y transformer phase shifts. Assume American standard phase shift. Also calculate the sequence components and phase components of the contribution to the fault current from generator n ($n = 1, 2,$ or 3) as specified by the instructor in Problem 9.4(b).

9.20 A 500-MVA, 13.8-kV synchronous generator with $X_d'' = X_2 = 0.20$ and $X_0 = 0.05$ per unit is connected to a 500-MVA, 13.8-kV Δ/500-kV Y transformer with 0.10 per-unit leakage reactance. The generator and transformer neutrals are solidly grounded. The generator is operated at no-load and rated voltage, and the high-voltage side of the transformer is disconnected from the power system. Compare the subtransient fault currents for the following bolted faults at the transformer high-voltage terminals: three-phase fault, single line-to-ground fault, line-to-line fault, and double line-to-ground fault.

9.21 Determine the subtransient fault current in per-unit and in kA, as well as contributions to the fault current from each line and transformer connected to the fault bus for a bolted single line-to-ground fault at the fault bus selected in Problem 9.6.

9.22 Repeat Problem 9.21 for a bolted line-to-line fault.

9.23 Repeat Problem 9.21 for a bolted double line-to-ground fault.

9.24 Determine the subtransient fault current in per-unit and in kA, as well as contributions to the fault current from each line, transformer, and generator connected to the fault bus for a bolted single line-to-ground fault at the fault bus selected in Problem 9.9.

9.25 Repeat Problem 9.24 for a single line-to-ground arcing fault with arc impedance $Z_F = 0 + j0.1$ per unit.

9.26 Repeat Problem 9.24 for a bolted line-to-line fault.

9.27 Repeat Problem 9.24 for a bolted double line-to-ground fault.

9.28 As shown in Figure 9.21 (a), two three-phase buses *abc* and *a′b′c′* are interconnected by short circuits between phases *b* and *b′* and between *c* and *c′*, with an open circuit between phases *a* and *a′*. The fault conditions in the phase domain are $I_a = I_{a'} = 0$ and $V_{bb'} = V_{cc'} = 0$. Determine the fault conditions in the sequence domain and verify the interconnection of the sequence networks as shown in Figure 9.15 for this one-conductor-open fault.

9.29 Repeat Problem 9.28 for the two-conductors-open fault shown in Figure 9.21(b). The fault conditions in the phase domain are

$$I_b = I_{b'} = I_c = I_{c'} = 0 \text{ and } V_{aa'} = 0$$

FIGURE 9.21 Problems 9.28 and 9.29: open conductor faults

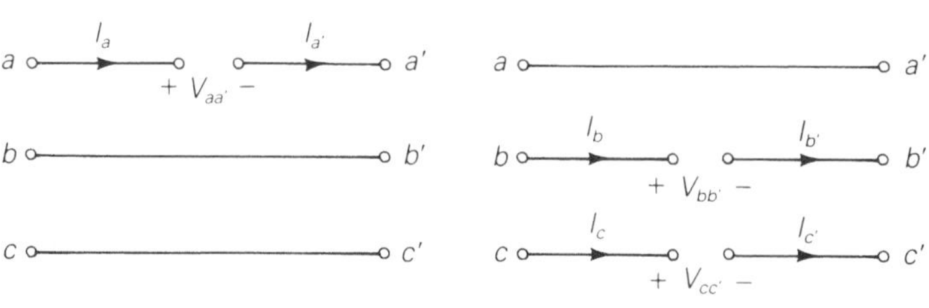

9.30 For the system of Problem 9.11, compute the fault current and voltages at the fault for the following faults at point F: (a) a bolted single line-to-ground fault; (b) a line-to-line fault through a fault impedance $Z_F = j0.05$ per unit; (c) a double line-to-ground fault from phase B to C to ground, where phase B has a fault impedance $Z_F = j0.05$ per unit, phase C also has a fault impedance $Z_F = j0.05$ per unit, and the common line-to-ground fault impedance is $Z_G = j0.033$ per unit.

9.31 For the system of Problem 9.12, compute the fault current and voltages at the fault for the following faults at bus 3: (a) a bolted single line-to-ground fault, (b) a bolted line-to-line fault, (c) a bolted double line-to-ground fault. Also, for the single line-to-ground fault at bus 3, determine the currents and voltages at the terminals of generators Gl and G2.

9.32 For the system of Problem 9.13, compute the fault current for the following faults at bus 3: (a) a single line-to-ground fault through a fault impedance $Z_F = j0.1$ per unit, (b) a line-to-line fault through a fault impedance $Z_F = j0.1$ per unit, (c) a double line-to-ground fault through a common fault impedance to ground $Z_F = j0.1$ per unit.

9.33 For the three-phase power system with single-line diagram shown in Figure 9.22, equipment ratings and per-unit reactances are given as follows:

Machines 1 and 2:	100 MVA	20 kV	$X_1 = X_2 = 0.2$
	$X_0 = 0.04$	$X_n = 0.04$	
Transformers 1 and 2:	100 MVA	20Δ/345Y kV	
	$X_1 = X_2 = X_0 = 0.08$		

Select a base of 100 MVA, 345 kV for the transmission line. On that base, the series reactances of the line are $X_1 = X_2 = 0.15$ and $X_0 = 0.5$ per unit. With a nominal system voltage of 345 kV at bus 3, machine 2 is operating as a motor drawing 50 MVA at 0.8 power factor lagging. Compute the change in voltage at bus 3 when the transmission line undergoes (a) a one-conductor-open fault, (b) a two-conductor-open fault along its span between buses 2 and 3.

9.34 At the general three-phase bus shown in Figure 9.7(a) of the text, consider a simultaneous single line-to-ground fault on phase a and line-to-line fault between phases b and c, with no fault impedances. Obtain the sequence-network interconnection satisfying the current and voltage constraints.

FIGURE 9.22

Problem 9.33

FIGURE 9.23 Problem 9.36

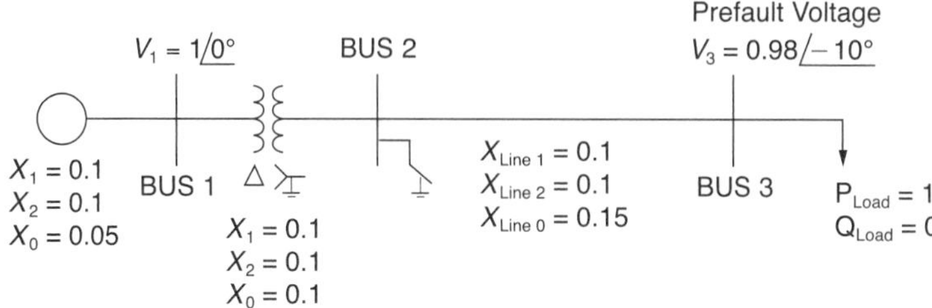

9.35 Thévenin equivalent sequence networks looking into the faulted bus of a power system are given with $Z_1 = j0.15$, $Z_2 = j0.15$, $Z_0 = j0.2$, and $E_1 = 1\underline{/0°}$ per unit. Compute the fault currents and voltages for the following faults occurring at the faulted bus:

(a) Balanced three-phase fault

(b) Single line-to-ground fault

(c) Line-line fault

(d) Double line-to-ground fault

Which is the worst fault from the viewpoint of the fault current?

9.36 The single-line diagram of a simple power system is shown in Figure 9.23 with per unit values. Determine the fault current at bus 2 for a three-phase fault. Ignore the effect of phase shift.

9.37 Consider a simple circuit configuration shown in Figure 9.24 to calculate the fault currents I_1, I_2, and I with the switch closed.

(a) Compute E_1 and E_2 prior to the fault based on the prefault voltage $V = 1\underline{/0°}$ and then, with the switch closed, determine I_1, I_2, and I.

(b) Start by ignoring prefault currents, with $E_1 = E_2 = 1\underline{/0°}$. Then superimpose the load currents, which are the prefault currents, $I_1 = -I_2 = 1\underline{/0°}$. Compare the results with those of part (a).

SECTION 9.5

9.38 The zero-, positive-, and negative-sequence bus impedance matrices for a three-bus three-phase power system are

FIGURE 9.24 Problem 9.37

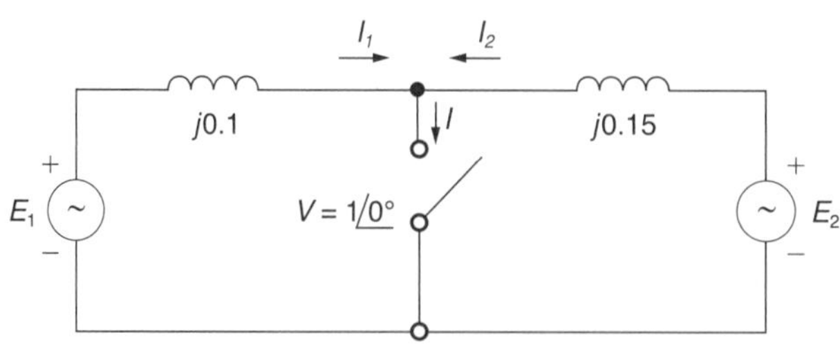

$$Z_{\text{bus }0} = j\begin{bmatrix} 0.10 & 0 & 0 \\ 0 & 0.20 & 0 \\ 0 & 0 & 0.10 \end{bmatrix} \quad \text{per unit}$$

$$Z_{\text{bus }1} = Z_{bus\ 2} = j\begin{bmatrix} 0.12 & 0.08 & 0.04 \\ 0.08 & 0.12 & 0.06 \\ 0.04 & 0.06 & 0.08 \end{bmatrix}$$

Determine the per-unit fault current and per-unit voltage at bus 2 for a bolted three-phase fault at bus 1. The prefault voltage is 1.0 per unit.

9.39 Repeat Problem 9.38 for a bolted single line-to-ground fault at bus 1.

9.40 Repeat Problem 9.38 for a bolted line-to-line fault at bus 1.

9.41 Repeat Problem 9.38 for a bolted double line-to-ground fault at bus 1.

9.42 (a) Compute the 3×3 per-unit zero-, positive-, and negative-sequence bus impedance matrices for the power system given in Problem 9.1. Use a base of 1000 MVA and 765 kV in the zone of line 1–2.

(b) Using the bus impedance matrices determined in Problem 9.42, verify the fault currents for the faults given in Problems 9.3, 9.14, 9.15, 9.16, and 9.17.

9.43 The zero-, positive-, and negative-sequence bus impedance matrices for a two-bus three-phase power system are

$$Z_{\text{bus }0} = j\left[\begin{array}{c|c} 0.10 & 0 \\ \hline 0 & 0.10 \end{array}\right] \quad \text{per unit}$$

$$Z_{\text{bus }1} = Z_{\text{bus }2} = j\left[\begin{array}{c|c} 0.20 & 0.10 \\ \hline 0.10 & 0.30 \end{array}\right] \quad \text{per unit}$$

Determine the per-unit fault current and per-unit voltage at bus 2 for a bolted three-phase fault at bus 1. The prefault voltage is 1.03 per unit.

9.44 Repeat Problem 9.43 for a bolted single line-to-ground fault at bus 1.

9.45 Repeat Problem 9.43 for a bolted line-to-line fault at bus 1.

9.46 Repeat Problem 9.43 for a bolted double line-to-ground fault at bus 1.

9.47 Compute the 3×3 per-unit zero-, positive-, and negative-sequence bus impedance matrices for the power system given in Problem 4(a). Use a base of 1000 MVA and 500 kV in the zone of line 1–2.

9.48 Using the bus impedance matrices determined in Problem 9.47, verify the fault currents for the faults given in Problems 9.4(b), 9.4(c), and 9.19 (a through d).

9.49 Compute the 4×4 per-unit zero-, positive-, and negative-sequence bus impedance matrices for the power system given in Problem 9.5. Use a base of 1000 MVA and 20 kV in the zone of generator G3.

9.50 Using the bus impedance matrices determined in Problem 9.42, verify the fault currents for the faults given in Problems 9.7, 9.21, 9.22, and 9.23.

9.51 Compute the 5×5 per-unit zero-, positive-, and negative-sequence bus impedance matrices for the power system given in Problem 9.8. Use a base of 100 MVA and 15 kV in the zone of generator G2.

9.52 Using the bus impedance matrices determined in Problem 9.51, verify the fault currents for the faults given in Problems 9.10, 9.24, 9.25, 9.26, and 9.27.

9.53 The positive-sequence impedance diagram of a five-bus network with all values in per-unit on a 100-MVA base is shown in Figure 9.25. The generators at buses 1 and 3 are rated 270 and 225 MVA, respectively. Generator reactances include subtransient values plus reactances of the transformers connecting them to the buses. The turns ratios of the transformers are such that the voltage base in each generator circuit is equal to the voltage rating of the generator. (a) Develop the positive-sequence bus admittance matrix $\boldsymbol{Y}_{\text{bus }1}$. (b) Using MATLAB or another computer program, invert $\boldsymbol{Y}_{\text{bus }1}$ to obtain $\boldsymbol{Z}_{\text{bus }1}$. (c) Determine the subtransient current for a three-phase fault at bus 4 and the contributions to the fault current from each line. Neglect prefault currents and assume a prefault voltage of 1.0 per unit.

9.54 For the five-bus network shown in Figure 9.25, a bolted single-line-to-ground fault occurs at the bus 2 end of the transmission line between buses 1 and 2. The fault causes the circuit breaker at the bus 2 end of the line to open, but all other breakers remain closed. The fault is shown in Figure 9.26. Compute the subtransient fault current with the circuit breaker at the bus-2 end of the faulted line open. Neglect prefault current and assume a prefault voltage of 1.0 per unit.

9.55 A single-line diagram of a four-bus system is shown in Figure 9.27. Equipment ratings and per-unit reactances are given as follows.

Machines 1 and 2:	100 MVA	20 kV	$X_1 = X_2 = 0.2$
	$X_0 = 0.04$	$X_n = 0.05$	
Transformers T_1 and T_2:	100 MVA	20Δ/345Y kV	
	$X_1 = X_2 = X_0 = 0.08$		

FIGURE 9.25

Problems 9.53 and 9.54

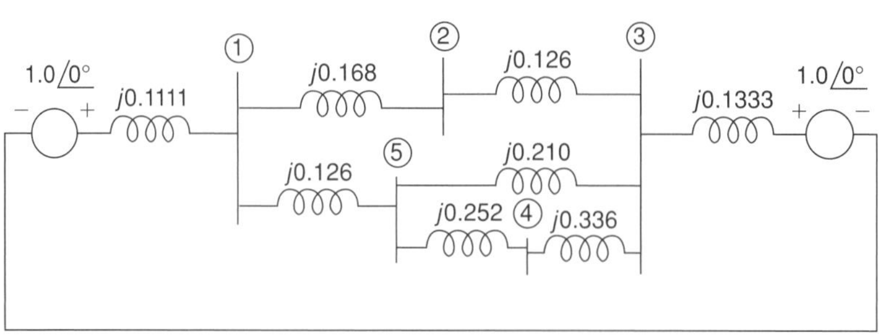

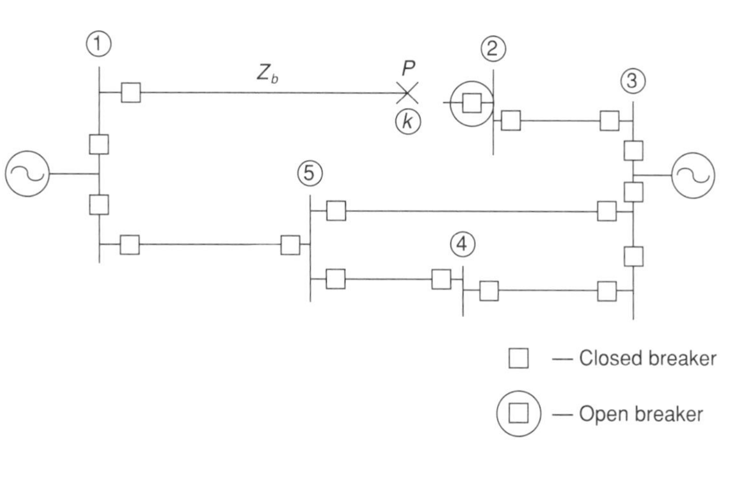

FIGURE 9.26

Problem 9.54

FIGURE 9.27

Problem 9.55

On a base of 100 MVA and 345 kV in the zone of the transmission line, the series reactances of the transmission line are $X_1 = X_2 = 0.15$ and $X_0 = 0.5$ per unit. (a) Draw each of the sequence networks and determine the bus impedance matrix for each of them. (b) Assume the system to be operating at nominal system voltage without prefault currents when a bolted line-to-line fault occurs at bus 3. Compute the fault current, the line-to-line voltages at the faulted bus, and the line-to-line voltages at the terminals of machine 2. (c) Assume the system to be operating at nominal system voltage without prefault currents, when a bolted double line-to-ground fault occurs at the terminals of machine 2. Compute the fault current and the line-to-line voltages at the faulted bus.

9.56 The system shown in Figure 9.28 is the same as in Problem 9.48 except that the transformers are now Y–Y connected and solidly grounded on both sides. (a) Determine the bus impedance matrix for each of the three sequence networks. (b) Assume the system to be operating at nominal system voltage without prefault currents when a bolted single-line-to-ground fault occurs on phase A at bus 3. Compute the fault current, the current out of phase C of machine 2 during the fault, and the line-to-ground voltages at the terminals of machine 2 during the fault.

FIGURE 9.28

Problem 9.56

9.57 The results in Table 9.5 show that during a phase *a* single line-to-ground fault the phase angle on phase *a* voltages is always zero. Explain why we would expect this result.

PW **9.58** The results in Table 9.5 show that during the single line-to-ground fault at bus 2 the *b* and *c* phase voltage magnitudes at bus 2 actually rise above the prefault voltage of 1.05 per unit. Use PowerWorld Simulator with case Example 9_8 to determine the type of in-line fault midway between bus 2 and 4 fault that gives the highest per-unit voltage magnitude.

PW **9.59** Using PowerWorld Simulator case Example 9_8, plot the variation in the bus 2 phase *a*, *b*, *c* voltage magnitudes during a single line-to-ground fault at bus 2 as the fault reactance is varied from 0 to 0.30 per unit in 0.05 per-unit steps (the fault impedance is specified on the Fault Options page of the Fault Analysis dialog).

PW **9.60** Using the Example 9_8 case determine the fault current in amps, except with a line-to-line fault at each of the buses. Compare the fault currents with the values given in Table 9.4.

PW **9.61** Using the Example 9_8 case determine the fault current in amps, except with a bolted double line-to-ground fault at each of the buses. Compare the fault currents with the values given in Table 9.4.

PW **9.62** Re-determine the Example 9_8 fault currents, except with a new line installed between buses 2 and 5. The parameters for this new line should be identical to those of the existing line between buses 2 and 5. The new line is not mutually coupled to any other line. Are the fault currents larger or smaller than the Example 9_8 values?

PW **9.63** Re-determine the Example 9_8 fault currents, except with a second generator added at bus 3. The parameters for the new generator should be identical to those of the existing generator at bus 3. Are the fault currents larger or smaller than the Example 9_8 values?

PW **9.64** Using PowerWorld Simulator case Chapter 9_Design, calculate the per-unit fault current and the current supplied by each of the generators for a single line-to-ground fault at the ORANGE69 bus. During the fault, what percentage of buses have voltage magnitude below 0.75 per unit?

PW **9.65** Repeat Problem 9.64, except place the fault at the POPLAR69 bus.

DESIGN PROJECT 3 (*CONTINUED*): POWER FLOW/ SHORT CIRCUITS

Additional time given: 3 weeks
Additional time required: 10 hours

This is a continuation of Design Project 3. Assignments 1 and 2 are given in Chapter 6. Assignment 3 is given in Chapter 7.

Assignment 4: Short Circuits—Breaker/Fuse Selection

For the single-line diagram that you have been assigned (Figure 6.22 or 6.23), convert the zero-, positive-, and negative-sequence reactance data to per-unit using the given system base quantities. Use subtransient machine reactances. Then using PowerWorld Simulator, create the generator, transmission line, and transformer input data files. Next run the Simulator to compute sub-transient fault currents for (1) single-line-to-ground, (2) line-to-line, and (3) double line-to-ground bolted faults at each bus. Also compute the zero-, positive-, and negative-sequence bus impedance matrices. Assume 1.0 per-unit prefault voltage. Also, neglect prefault load currents and all losses.

For students assigned to Figure 6.22: Select a suitable circuit breaker from Table 7.10 for *each* location shown on your single-line diagram. Each breaker that you select should: (1) have a rated voltage larger than the maximum system operating voltage, (2) have a rated continuous current at least 30% larger than normal load current (normal load currents are computed in Assignment 2), and (3) have a rated short-circuit current larger than the maximum fault current for any type of fault at the bus where the breaker is located (fault currents are computed in Assignments 3 and 4). This conservative practice of selecting a breaker to interrupt the entire fault current, not just the contribution to the fault through the breaker, allows for future increases in fault currents. *Note:* Assume that the (X/R) ratio at each bus is less than 15, such that the breakers are capable of interrupting the dc offset in addition to the subtransient fault current. Circuit breaker cost should also be a factor in your selection. Do *not* select a breaker that interrupts 63 kA if a 40-kA or a 31.5-kA breaker will do the job.

For students assigned to Figure 6.23: Enclosed [9, 10] are "melting time" and "total clearing time" curves for K rated fuses with continuous current ratings from 15 to 200 A. Select suitable branch and tap fuses from these curves for each of the following three locations on your single-line diagram: bus 2, bus 4, and bus 7. Each fuse you select should have a continuous current rating that is at least 15% higher but not more than 50% higher than the normal load current at that bus (normal load currents are computed in Assignment 2). Assume that cables to the load can withstand 50% continuous overload currents. Also, branch fuses should be coordinated with tap fuses; that is, for every fault current, the tap fuse should clear before the branch fuse melts. For each of the three buses, assume a reasonable X/R ratio and determine the asymmetrical fault current for a three-phase bolted fault (sub-transient current is computed in Assignment 3). Then for the fuses that you select from [9, 10], determine the clearing time CT of tap fuses and the melting time MT of branch fuses. The ratio MT/CT should be less than 0.75 for good coordination.

DESIGN PROJECT 4: CIRCUIT BREAKER SELECTION

Time given: 3 weeks
Approximate time required: 10 hours

As a protection engineer for Metropolis Light and Power (MLP) your job is to ensure that the transmission line and transformer circuit breaker ratings are sufficient to interrupt the fault current associated with any type of fault (balanced three phase, single line-to-ground, line-to-line, and double line-to-ground). The MLP power system is modeled in case Chapter9_DesignCase4. This case models the positive-, negative-, and zero-sequence values for each system device. Note that the 69/138 kV transformers are grounded Y on the low side and Δ on the high side; the 138 kV/345 kV transformers are grounded Y on both sides. In this design problem your job is to evaluate the circuit breaker ratings for the three 345 kV transmission lines and the six 345/138 kV transformers. You need not consider the 138 or 69 kV transmission lines, or the 138/69 kV transformers.

Design Procedure

1. Load Chapter9_DesignCase4 into PowerWorld Simulator. Perform an initial power flow solution to get the base case system operating point.
2. Apply each of the four fault types to each of the 345 kV buses and to the 138 kV buses attached to 345/138 kV transformers to determine the maximum fault current that each of the 345 kV lines and 345/138 kV transformers will experience.
3. For each device select a suitable circuit breaker from Table 7.10. Each breaker that you select should (a) have a rated voltage larger than the maximum system operating voltage, (b) have a rated continuous current at least 30% larger than the normal rated current for the line, and (c) have a rated short circuit current larger than the maximum fault current for any type of fault at the bus where the breaker is located. This conservative practice of selecting a breaker to interrupt the entire fault current, not just the contribution to the fault current through the breaker allows for future increases in fault currents. Since higher rated circuit breakers cost more, you should select the circuit breaker with the lowest rating that satisfies the design constraints.

Simplifying Assumptions

1. You need only consider the base case conditions given in the Chapter9_DesignCase4 case.
2. You may assume that the X/R ratios at each bus is sufficiently small (less than 15) so that the dc offset has decayed to a sufficiently low value (see Section 7.5 for details).
3. As is common with commercial software, including PowerWorld Simulator, the Δ–Y transformer phase shifts are neglected.

CASE STUDY QUESTIONS

a. What is the "three-position" scheme for MV switchgear? Describe the interlocking of the three positions.

b. Describe the shielded solid insulation system in MV switchgear.

REFERENCES

1. Westinghouse Electric Corporation, *Electrical Transmission and Distribution Reference Book,* 4th ed. (East Pittsburgh, PA, 1964).
2. Westinghouse Electric Corporation, *Applied Protective Relaying* (Newark, NJ, 1976).
3. P. M. Anderson, *Analysis of Faulted Power Systems* (Ames: Iowa State University Press, 1973).
4. J. R. Neuenswander, *Modern Power Systems* (New York: Intext Educational Publishers, 1971).
5. H. E. Brown, *Solution of Large Networks by Matrix Methods* (New York: Wiley, 1975).
6. W. D. Stevenson, Jr., *Elements of Power System Analysis,* 4th ed. (New York: McGraw-Hill, 1982).
7. C. A. Gross, *Power System Analysis* (New York: Wiley, 1979).
8. C. Ma et al., "Innovative MV Switchgear for Today's Applications," *2012 China International Conference on Electricity Distribution (2012 CICED),* Shanghai, September 5–6, 2012, Paper No. FP0112.
9. McGraw Edison Company, *Fuse Catalog,* R240-91-1 (Canonsburg, PA: McGraw Edison, April 1985).
10. Westinghouse Electric Corporation, *Electric Utility Engineering Reference Book: Distribution Systems* (Pittsburgh, PA: Westinghouse, 1959).

10 System Protection

Lightning slices through rainy skies above a city, in a time-exposure view (Jhaz Photography/ Shutterstock.com)

Short circuits occur in power systems when equipment insulation fails, due to system overvoltages caused by lightning or switching surges, to insulation contamination, or to other mechanical and natural causes. Careful design, operation, and maintenance can minimize the occurrence of short circuits but cannot eliminate them. Methods for calculating short-circuit currents for balanced and unbalanced faults were discussed in Chapters 7 and 9. Such currents can be several orders of magnitude larger than normal operating currents and, if allowed to persist, may cause insulation damage, conductor melting, fire, and explosion. Windings and busbars may also suffer mechanical damage due to high magnetic forces during faults. Clearly, faults must be quickly removed from a power system. Standard EHV protective equipment is designed to clear faults within 3 cycles, whereas lower-voltage protective equipment typically operates within 5–20 cycles.

This chapter provides an introduction to power system protection. Blackburn defines protection as "the science, skill, and art of applying and setting relays and/or fuses to provide maximum sensitivity to faults and undesirable conditions, but to avoid their operation on all permissible or tolerable conditions" [1]. The basic idea is to define the undesirable conditions and look for differences between the undesirable and permissible conditions that relays or fuses can sense. It is also important to remove only the faulted equipment from the system while maintaining as much of the unfaulted system as possible in service, in order to continue to supply as much of the load as possible.

Although fuses and reclosers (circuit breakers with built-in instrument transformers and relays) are widely used to protect primary distribution systems (with voltages in the 2.4–46 kV range), this chapter is primarily focused on circuit breakers and relays, which are used to protect HV (115–230 kV) and EHV (345–765 kV) power systems. The IEEE defines a relay as "a device whose function is to detect defective lines or apparatus or other power system conditions of an abnormal or dangerous nature and to initiate appropriate control action" [1]. In practice, a relay is a device that closes or opens a contact when energized. Relays are also used in low-voltage (600-V and below) power systems and almost anywhere that electricity is used. They are used in heating, air conditioning, stoves, clothes washers and dryers, refrigerators, dishwashers, telephone networks, traffic controls, airplane and other transportation systems, and robotics, as well as many other applications.

Problems with the protection equipment itself can occur. A second line of defense, called *backup* relays, may be used to protect the first line of defense, called *primary* relays. In HV and EHV systems, separate current- or voltage-measuring devices, separate trip coils on the circuit breakers, and separate batteries for the trip coils may be used. Also, the various protective devices must be properly coordinated such that primary relays assigned to protect equipment in a particular zone operate first. If the primary relays fail, then backup relays should operate after a specified time delay.

This chapter begins with a discussion of the basic system-protection components.

CASE STUDY

The following article traces the history of protective relays in power systems from electromechanical relays (since the early 1900s) to the introduction of solid-state relays (1960s and1970s), first-generation microprocessor-based digital (or numeric) relays (1980s), and now second-generation digital relays [14]. New-technology second-generation digital relays have more powerful microprocessors with faster analog-to-digital conversion, more reliable and robust surface-mount construction with longer-lasting components, improvements in the available number of protective components allowing for new protection schemes, increased memory for event recording, improved internal relay diagnostics, and enhanced relay security against cyber attacks.

Upgrading Relay Protection
Be Prepared for the Next Replacement or Upgrade Project

By Daniel L. Ransom

Digital Object Identifier 10.1109/MIAS.2013.2288404
Date of publication: 2 July 2014

There are many advantages to upgrading old electromechanical (EM), solid-state, and first-generation numeric relays with modern numeric relays. Reliability increases because there is less direct wiring and interconnection wiring, and the reliability and security of multifunction logic and settings are improved with the next-generation user interface software. Remote input–output modules, remote analog/digital inputs, and thermal measurement capabilities have expanded protection, control, and monitoring. New protection and monitoring features improve power system equipment life and increase personnel safety. Maintenance costs are reduced, while internal watchdogs alert the user if the relay has a problem. Settings groups can be changed instantaneously to adapt to varying power-system requirements. Modern second-generation numeric relays offer a variety of secure communications capabilities for interfacing with smart-grid controls, supervisory control, and data-acquisition systems, and business networks. Event memory is larger for more onboard, standardized oscillographs and event reporting. Relay security is in accord with the latest North American Electric Reliability Corporation (NERC) standards.

Initially, every relay upgrade seems simple and straightforward, but then come the details. Operating personnel have expectations for reading targets, resetting trips, ease of interface for settings and events, motor restarting, sync closing, and so on. Regulator requirements [NERC critical infrastructure protection (CIP), for example] must be implemented while maintaining smooth operations. Relay engineers must ensure that operational ease is retained with the new upgrade relay. Accurate oneline and connection drawings as well as good wiring documentation are essential. Escutcheon plates might be necessary, or perhaps switchgear will need modification (panel cutting, new doors, relocation, etc.). In addition, this is an opportunity to re-evaluate arc-flash hazards and possibly reduce the risks. These and other considerations are taken from actual relay-replacement projects.

This article provides guidance for your next replacement or upgrade project, including ways to reduce costs, save time, and minimize unexpected or unplanned complications.

History of Relays

Protective relaying in industrial and utility power systems has changed greatly since the beginning of system protection over a hundred years ago. At first, finely made "Swiss watch" precision EM relays were developed and installed by the hundreds of thousands.

In the 1970s, solid-state (or static) relays became available and replaced the EM relays with a slight change in scheme protection or relay function. These relays had the advantage of being cheaper and smaller, being less of a burden, being easier to maintain, and having less calibration drift and some programming. Typically, solid-state relays have no communication or event-recording functions. Some plug-and-play solid-state relays remain very popular because there is minimal installation labor, low outage time for replacements, and the scheme design remains simple. Some prefer to keep their systems in this configuration because no extensive computer software training is required to operate an effective power-system protective scheme.

Widespread growth of computer technology in the 1970s resulted in the introduction of microprocessor-based numeric (or digital) relays. This first generation of numeric relays brought innovations in developing new algorithms and the beginning of combining several protection functions in one multifunction relay package. Desktop and laptop computers configured the first-generation numeric relays via DE-9 and DB-25 serial ports (with all the complications of configuring these connections). Initially, first-generation numeric relays operated on ASCII protocol or proprietary serial commands; relay manufacturers later developed graphical user interface (GUI) software to generate the background serial commands to program the relay elements.

Now, these first-generation numeric relays are reaching an end-of-service-life state. A major reason for this situation is through-hole component-mounting design and poor solder joints. The heating and cooling of circuit boards also leads to problems, such as bad vias (electrical connections from one layer of the circuit board to another) and electrolytic capacitor failure, particularly in the power supplies. Because the first-generation numeric relays have built-in test routines (a "watchdog") that signal an alarm output, operators know when the numeric relay fails. EM designs and most solid-state designs did not have a watchdog alarm output.

The new-technology second-generation numeric relays are equipped with more powerful microprocessors, have a more reliable surface-mount construction, and have improved algorithms and elements. These developments have provided relay protection engineers with new protection schemes and have advanced the quest for more reliable, more secure, and more dependable operation. However, some manufacturers have provided so much programming capability in these second-generation products that extensive training is required to set and operate these relays. In some cases, the customer contracts with the manufacturer to develop setting files. In other cases, the manufacturer has simplified the setting process by using computing power to make setting a relay more straightforward and by providing setting intelligence in relay-setting software.

Operations personnel now have input into the second-generation numeric relay settings because the newer relay front panels have operating indicators and push buttons directly related to system operation.

The total costs of upgrading relays must be considered, including engineering, labor, testing, and commissioning costs. Also, you must consider the future values of longer-lasting relays and newer setting-checking features that help relay engineers to improve settings, avoid expensive outages, reduce paperwork, and avoid possible fines.

Replacing/Upgrading Relays

Knowing when to upgrade your relay protection should be a proactive function, and you should not operate protective relays to failure. Operating to failure might seem to be a cost-saving method, but this philosophy reduces the reliability, security, and dependability of the power system. Relevant standards for determining relay life are found in American National Standards Institute C37.90 [1] for the United States and in International Electrotechnical Commission (IEC) 60255 [2].

Replacement/Upgrade Timing

You must monitor the rise in relay failures so you can track and schedule the replacement/upgrade before an existing relay fails [3]. Monitoring is necessary, especially for older EM relays, solid-state relays, and first-generation numeric relays. Figure 1 shows the study results on relay longevity [3].

The longer a relay is in service, the more likely it is to fail, which is why primary/backup relaying is recommended. In general, the lifetime of a numeric relay is 15–20 years. This is based on the life of capacitors (loss of capacitance because of electrolyte drying and leakage) and by semiconductors (mainly integrated circuits) that degrade because of thermal vibration and humidity. Most of the first-generation numeric relays have already reached or passed normal lifetime. It is recommended that you schedule a replacement or upgrade to the second-generation relays (or replace the existing solid-state relays if you want to keep the solid-state relay scheme).

Protective Relay Stressors

Relays operated in severe conditions will need more maintenance and more record keeping to document the performance. Severe conditions include extremely hot and cold

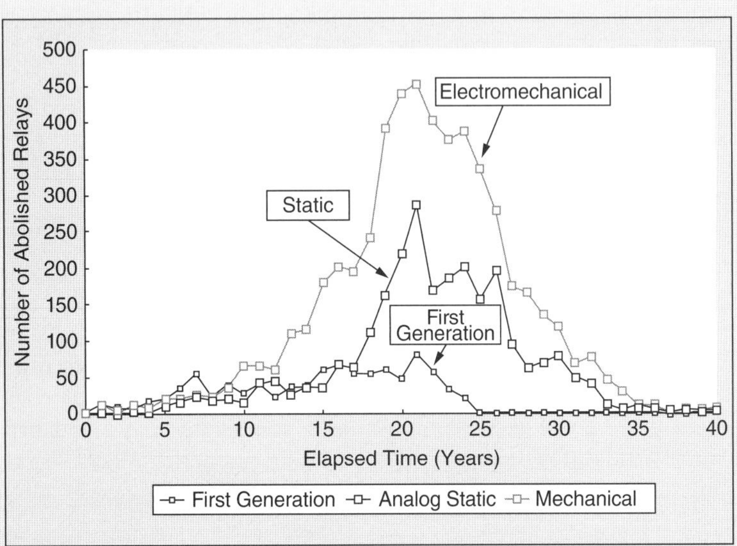

Figure 1 A study of relay longevity

environments, outdoor locations, humid areas, and environments that contain corrosive gases, like hydrogen sulfide.

Costs of Upgrading Relays

The costs associated with relay upgrades and replacements have many variables. Depending upon the importance you and your organization place on reliability, security, and dependability, optimizing relay upgrades and replacements might seem like a puzzle because of the many variables of relay operation. However, there are some interesting methods of quantifying upgrades and replacements with statistics and probability mathematics. These mathematical models use terms, such as mean time between failure (MTBF), mean time to failure (MTTF), mean time to repair (MTTR), reliability, security, and dependability.

Reliability is in the MTBF or MTTF parameters. Assuming that the modern second-generation numeric relays have surface-mounted components, which cannot be repaired to the component level, the model for MTTF is used. MTBF does not make sense in this case; relay users generally do not repair modern relays because the relay price is low, and replacing a failed unit brings more peace of mind.

One study on relay replacement is by Wang et al. [4], based on the least unit life-cycle cost with minimum maintenance model. Minimum maintenance is repairing a relay board subsystem only or replacing the relay, with no hardware refurbishment. These researchers found that relay longevity follows a lognormal distribution, with the following parameters for cost over the life cycle of the relay [$C(t)$]:

$$C(t) = \frac{C_L(t)}{t}, \quad (1)$$

where $C_L(t)$ is the unit life cycle cost, and

$$C_L(t) = C_d + C_x \int_0^t \rho(t)dt + C_t \frac{\text{MTTR}}{\text{MTTF}} \int_0^t \rho(t)dt, \quad (2)$$

where C_d is the price of a protection device (including installation), C_x is the average cost of maintenance, C_t is the cost effect when the power system is out of service because of a protective relay failure, and the function $\rho(t)$ is the probability distribution.

This equation shows the effects of outages, maintenance, and initial relay installation. The cost of downtime (the third term with C_t) can be greater than the cost of maintenance (the second term with C_x) if the MTTR is large and the MTTF is relatively small. C_t is the loss when the power system is out of service because of a protective relay failure. C_t/MTBF represents the probability that another device is also out of service. Note that the initial cost of the relay and installation (the first term with C_d) is not significant for a single relay when that relay is in service for a long period (large t). However, replacement/upgrade projects seldom have only one relay. By batching many relay replacements at one time, this term becomes significant. Shopping for a low relay price with a quick setup time (to reduce labor cost) is advised.

The lognormal probability distribution $\rho(t)$ indicates that once the relay has one failure, subsequent failures will occur with fewer and fewer time intervals. It is a good idea to replace/upgrade a relay after the first failure. High-reliability situations require that you make the decision to replace and/or upgrade before relay failure. Wang et al. found the optimal replacement period was every 18.4 years. In fact, power industry standards [5] specify that relay protection device life is generally not fewer than 12 years, and for devices operating in harsh conditions, the useful life should be shortened.

Another analysis of relay life by Montignies et al. [6] shows similar results. They describe reliability as the summation of the probability of relay failure over time t (the failure rate). They use the familiar "bathtub curve" to confirm the large infant mortality of brand-new relays, the useful life of in-service relays, and finally, the end-of-life obsolescence period, where relay failures accumulate rapidly (Figure 2). The IEC describes the useful life as "the time interval beginning at a given moment in time, and ending when the failure intensity becomes unacceptable or when the time is considered to be unrepairable as a result of a fault" (International Electrotechnical Vocabulary 191-19-06) [7].

During the middle useful-life, constant-aging period, the failure rate, $l(t)$, remains steady. The average operating time is

$$\text{MTTF} = \int_0^\infty R(t)dt = \int_0^\infty e^{-\lambda(t)} = \frac{1}{\lambda}, \quad (3)$$

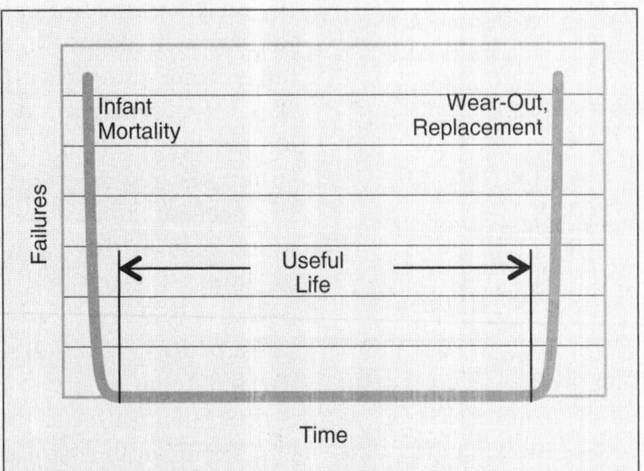

Figure 2 Relay failures over time

where $R(t)$ is the reliability at time t. The natural exponential function of the probability of failure $\lambda(t)$ also describes reliability $R(t)$.

Thus, the relationship of MTTF and the failure rate $\lambda(t)$ is inverse. When the failure rate goes up, the MTTF goes down, and vice versa. An MTTF of 300 years means that one relay out of 300 relays will fail during an operational year (0.33%/year).

Availability is the probability that a system operates properly at a given time. During the useful life and over a long period of use, availability (A) is

$$A = \frac{\text{MTTF}}{\text{MTTF} + \text{MTTR}} \ (\text{MTTF} \gg \text{MTTR}). \quad (4)$$

To increase the availability, the MTTR must be as brief as possible. Having on-the-shelf spares speeds repair and increases availability. In addition, be sure to apply power to spare relays for 30 minutes every year. Doing so keeps the internal electrolytic capacitors fresh, improving the usable life of shelf spares.

Primary/Backup Relay Systems

Protective relays should have a primary/backup arrangement with different manufacturers, especially when applied to critical protection areas, and in the severe conditions mentioned above. In this manner, mutual failure modes from hardware and from software algorithms used by one manufacturer are avoided. Relay manufacturers often use different algorithms, which have different accuracy and precision and different hardware. Using different manufacturers makes the primary/backup different-vendor solution more robust. Using two different relays can save valuable equipment if one relay does not detect a fault but the other relay does detect a fault. The added cost of training on another relay and added spares is offset by the savings realized in avoiding a fault.

Upgrading to the Second-Generation Numeric Relays

Upgrading to the latest or "second generation" of numeric protective relays gives a distinct advantage. There have been advances in hardware, software, element mix, and communication.

Hardware Advances

The second-generation protective relays have the following advantages over previous offerings.

Faster and Better Processors

Faster analog-to-digital (A/D) conversion benefits protection because more samples of the power system voltages and currents are available to the relay internal microprocessor. Thus, the relay has a more accurate representation of the actual conditions in the power system. The newer A/D converters also offer a wider range so small signals and large fault signals can be more accurately measured (if current transformers and voltage transformers are of good quality).

The processing power of the new microprocessors in protective relays mirrors the computing power improvements in ordinary office computing platforms. While the first generations could be compared with an old Intel 286 or 386, the latest protective relays have PowerPC (performance optimization with enhanced reduced instruction set computing—performance computing) or multicore-type processing chips. These new microprocessors are faster, smaller, and less expensive and have had an effect on the relay-design trend toward improved algorithms and simplicity.

Surface-Mount Reliability

Older relays employed "wire-lead" components that experienced vibration and heating issues, making the protective-relay circuit board unreliable over time. Power supplies were of the brute-force design, requiring large electrolytic capacitors, which are known to dry or leak, and fail. This drying effect is significant, especially with shelf spares that are not powered on a yearly basis.

Surface-mount technology has made protective-relay circuit boards more robust because the surface-mount components lay on the circuit board. The components have better thermal conduction to the circuit board so the components are less subject to heating and to vibration problems.

Power-supply design reliability has improved using switching designs, where dc and low-frequency ac are converted to a higher frequency and then smoothed to internal working power using smaller electrolytic capacitors with a longer lifetime. In addition, manufacturers have learned to optimize component selection for a more reliable working life.

Plug-Out Design More Reliable Than Drawout

There is a move to eliminate drawout protective relays in second-generation products. Many drawout designs damage the internal relay plug/jack interfaces with repeated use.

One popular solution to this problem is to use a "plug-out" design where, if the protective relay needs to be replaced, technicians detach the already-wired terminal blocks from the back of the relay. After the entire relay is removed from the panel, the technician replaces the already-wired terminal blocks (plugs) onto the back of the relay. This method has a lower force distributed over a number of connections, providing a more gentle connection scenario. Current-input circuits have shorting-type mechanisms for protecting personnel from open-current-transformer arcing. The reliability is increased with comparable downtimes to a drawout-case relay.

Ethernet-Based Connections

The second-generation hardware supports Ethernet-based communication. Usually, copper 100BASE-T and fiber-optic 100BASE-FX are provided. Properly managed for cybersecurity using the NERC CIP requirements [8], these interfaces achieve connections to a substation LAN and remote locations. New control and reporting protocols like IEC 61850 are available on the Ethernet interface.

Firmware and Software Advances

The second-generation numeric relay firmware offers improvements in the available number and type of protective elements, which allows new protection schemes, increases relay recording and reporting, provides improved relay internal diagnoses, and complies with the latest NERC security requirements. External setting software is GUI based and includes settings analysis and checking, along with template import and export, to improve and simplify the setting and testing processes. Some new relay software have innovations such as offline logic simulation, seen previously only in programmable logic controller software.

Number and Type of Elements

More processing power has allowed protective relay manufacturers to increase the number of traditional elements in a relay (for example, providing six or more 50 and six or more 51 overcurrent elements with any phase, sequence, or specialty input(s) for a feeder relay). In addition, newer elements are packaged in the relay for use in newer protection schemes; an example is 21 distance element(s) and 78 vector-jump element(s) in a generator relay. Hybrid-grounding generator schemes, fast-synchronization, and main-tie-main schemes are examples of newer protection schemes supported by the many elements in the secondgeneration protective relays.

Increased Relay Recording and Reporting

The second-generation numeric relays have more nonvolatile memory, allowing them to record longer and more event records and oscillography. Especially when synchronized by satellite Inter-Range Instrumentation Group or Ethernet time, a fault is much easier to analyze using time-aligned data across multiple relays. Ethernet connections allow faster downloading speeds from these large-data-volume relays.

Second-generation relays provide more elements that can be used for the early warning of impending failure and for through-fault and remaining-life estimates. Thus, you can realize long-term savings because of the ability to extend equipment maintenance intervals.

Improved Relay Internal Diagnoses

EM relays had no diagnostic features. You could test the EM relay, then put it back into service only for it to fail immediately. You would not discover this failure until the next testing interval, when it was time to test again. Some solid-state relays have power-supply alarms but no diagnostic testing. The first-generation of numeric relays have a microprocessor watchdog with a "deadman" alarm that would close alarm contacts upon suspension of algorithm execution and of power loss.

The second-generation protective relays employ more intricate self-tests that analyze not only when the microprocessor(s) is operating, but these relays also check memory status, measure accuracy of the A/D converter, and monitor power supply tolerances. If the relay exceeds any of these parameters, it sends an alarm via contacts or communications protocols to operators to check the relay. Again, long-term costs are lower and power-system protection is improved because of the relays alarm for failure.

Enhanced Relay Security

Worldwide, data security has become an important part of the protective relay arena. As shown in the Aurora effect [9], utilities and commercial/industrial power users must ensure that their processes are secure from ill-intentioned hackers. For example, in North America, the NERC security requirements, CIP-002 [8], require that protection relays connected to the bulk power system must log all entry attempts. The first-generation protective relays were designed before these requirements and do not provide adequate data security protection (without layering external security). The second-generation relays have extensive security provisions, including logging who accesses the relay and for what purpose.

Smart Software Increases Programming Simplicity and Provides Checking

Recent studies [10] have shown that a large amount of relay misoperations are caused by poor relay programming. Although it is good to have nearly endless computer-like programming in protective relays, it can be a pitfall to unknowing or inadequately trained protection engineers. The latest setting and monitoring software checks programmed logic and has templates for many protection schemes. These templates can be used "as is" or modified as a special settings or monitoring template for protection department use. The new smart software for the second-generation relays features step-by-step guided settings and summary pages when complete to check that all elements are programmed and set. Simplicity is improved, and misoperations are minimized.

In addition, the latest relay software provides monitoring capabilities for the power system, such as basic current and voltage measurements, sequence components, phasors, power quality, and harmonics. For motors and generators, there are start reports and calculated and actual temperature measurements. Trends are reported for crucial operating parameters, such as starting, transformer through-fault monitoring, and predicted end of life.

Automated testing is essential, and the newer relays coordinate with several available products to make testing and commissioning far easier than the first-generation numeric and previous relays.

Installation Data and Example

Experience from numerous relay installations over the years lends evidence to the need for relay upgrades and replacements. Historic data and installation examples make the benefits of upgrading apparent.

Historic Installation Data

Modern relays with active components (not EM construction) do not have the service life of EM relays. Figure 3 shows the expected, maximum and minimum service life of the four main relay types. The average of each type is also shown. The first-generation numeric relays, second-generation numeric relays, and solid-state relays offer more reliability, better calibration retention, and more repeatable protection than the older EM relays.

Note that advances in component technology and circuit board reliability have made a useful contribution to relay life. Components such as electrolytic capacitors continue to be a limiting factor. Improvements in component technology and proper manufacturer selection have made the

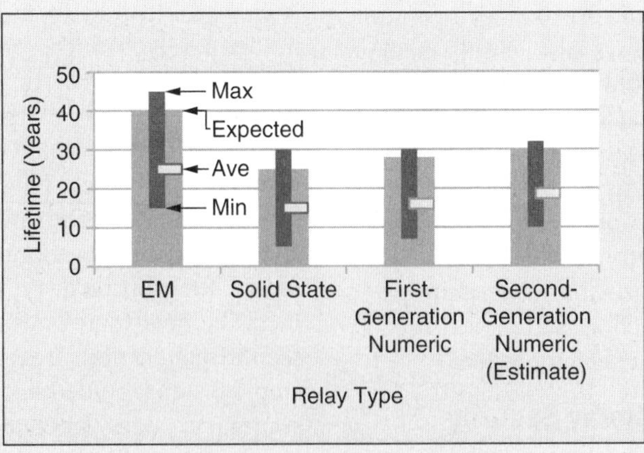

Figure 3 The relay longevity by type (years)

Figure 4 The old EM transformer differential relays (Photo courtesy of Basler Electric Company.)

second-generation numeric relays last longer (estimate based upon data to date in Figure 3).

EM Transformer Differential Scheme

Some practical examples of relay replacement show efficiency gains in having multiple functions in one relay, communications ease, and space savings. Figure 4 shows the "before" picture for a transformer differential scheme when EM relays were installed in the switchgear [11]. Not shown are the single-function overcurrent relays that backed up this differential scheme. The old installation had minimal protection without a modern dual-slope algorithm. There was no transformer life monitoring and no circuit-breaker and station-battery monitoring. In addition, the old scheme offered no communication, and therefore, operators and engineers could not monitor the transformer working in the power system. In addition, many EM differential relays consumed valuable front-panel space.

Upgrading the EM relays to numeric relays resulted in better protection, better relay communication, and a less-cluttered use of panel space (along with built-in metering, the company enjoyed increased ease of operation and less downtime). Figure 5 shows that this installation features new panels with a multifunction numeric relay replacement as well as test switches not available in the original installation. Backup protection could be added in the remaining panel blank at a later date.

Cost of Upgrading

Performing the actual relay upgrade and replacement begins with estimating the cost. A value must be placed on the operation of the power system with no changes and compared with the value of power system operation with protection upgrades.

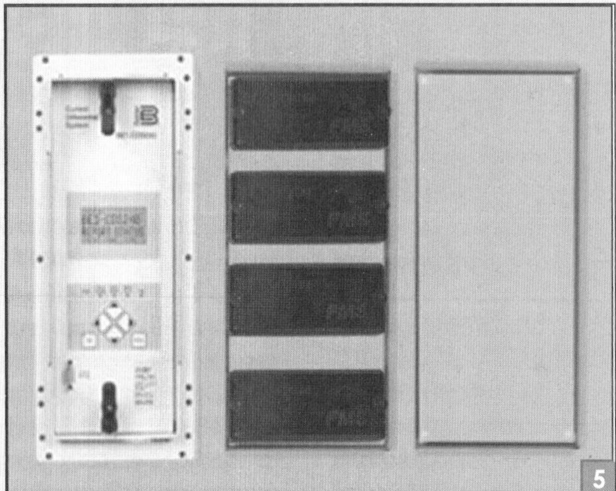

Figure 5 Smaller size and better protection are benefits of numeric relay upgrade replacement (Photo courtesy of Basler Electric Company.)

Making No Change

Continuing operation with older relays can make economic sense if the costs are acceptable for downtime and equipment repairs caused by nonworking protection. As was observed in the section "History of Relays," all of the EM, solid-state, and first-generation numeric relays are coming to the end of usable life. The loss of one power transformer for which the utility or commercial customer has no backup, or must reduce service/production to provide remediation, can be prohibitive. Management and the protection division must work together to identify the high-risk protection areas and get these relays upgraded as a high priority.

Upgrading Costs to Consider

Upgrading relay protection can save money in the long run. The future values of a relay upgrade far outweigh the initial cost of implementation (see the "History of Relays" section). Costs to consider include

- drawing changes
- physical mounting considerations (including building subpanels and rewiring to reach the new relays, if needed)
- upgrading communication infrastructure
- resizing battery requirements
- labor, including engineering, installation, settings, and commissioning
- system downtime and damaged equipment
- end-of-life maintenance.

Implementing a Relay Upgrade

When you have made the decision to upgrade protective relaying, these are the steps to follow.

- Obtain management approval and team acceptance.
- Update drawings for the new relays or panels.
- Order all required materials, tools, and labor.
- Develop an outage plan.
- Install, test, and commission.
- Complete drawings with as-is data.
- Responsibly dispose of old equipment.
- Report outcome to management.

Performing these steps helps to keep your project on time and within budget.

Obtain Management Approval and Team Acceptance

Organization management should be aware of your plans and grant approval for the upgrade project. Communicate the plans to the upgrade team of workers. Do not forget to account for the unexpected (incorrect relay or panel fit, door-swing problems with long-depth relays). A good value for unexpected costs is 10% (with thorough planning) to 25%.

Update Drawings for the New Relays or Panels

Planning begins with examining the existing drawings for the switchgear, substation, or line section. Be aware that the condition of existing system prints might not be up to date. Send a crew to examine the site, comparing the latest drawings to the present installed configuration; make any changes required.

Next, make demolition drawings. These drawings provide detail on the relays or panels to be removed.

Follow installation drawings, with detailed notes on relay or panel placement. These documents help to focus on the materials (metal stock, escutcheon plates, terminal blocks, panels, and relays, current transformer and voltage transformer replacements/connections, etc.) tools, and labor that will be required for a successful transition.

Consult the relay manufacturers' websites and contact the manufacturers' application engineers in your area for a final overview of your plans. This action helps to reveal any assumptions about the relay operation that might have been missed. In addition, you can find tools for settings, application notes, and white papers that pertain to your upgrade work. Developing a relationship with the manufacturer and manufacturers' field organizations now can save you money in the future by reducing rework.

Prepare a detailed list of step-by-step procedures for the work. This planning reduces cost, increases efficiency, reduces upgrade overall time, and makes the probability of excessive commissioning troubleshooting time much less.

Develop an Outage Plan

An important safety measure is performing the work when the associated areas of the power system are de-energized to avoid possible hazards to both your technical crew and power-system equipment. It is best to upgrade an out-of-service portion of the system. Doing "hot" upgrades and replacements is not recommended. Effective outage planning achieves the following goals:

- avoids outage cost increases and project delays
- avoids unforeseen circumstances during the outage because you have "walked through" the upcoming work
- simplifies coordinating your technical staff during inspections and commissioning
- increases personnel productivity.

Review work order historical data for the area of the power system getting the upgrade to determine whether there were existing problems and emerging trends. If you are using database software to schedule work orders, investigate whether it has an outage-planning module (many of these programs and options exist). Consult with electricians, the machine shop, and other trades on the nuances of the existing system. Document, order, and store these data so that all concerned parties have access (perhaps through a Web/Internet interface). Produce a draft of the work flow, estimate required hours, and identify resources. Add or subtract items in the plan addressed during the outage (time of year, provision for continuing customer or plant loads, backup protection during the outage, etc.).

Order All Required Materials, Tools, and Labor

Early ordering of all the required materials, tools, and arranging predicted labor needs make the upgrade go more smoothly. In general, protective relays have a three-to four-week lead time because of the many ordering options; manufacturers do not stock relays but build these to an exact option (or style) number. Although manufacturers can produce the relays faster, it might cost more.

Some upgrade solutions match the old terminal numbers to the new relay. In this manner, there are significantly fewer drawing changes and technicians simply wire from the old relay to the upgrade relay using the same terminal number. This method has been known not only to reduce cost, but it also reduces mistakes during the upgrade. Laser-cut plastic overlays on the back of the new relay or prewired terminal blocks marked with the old relay terminal numbers are a few examples. End-to-end wire testing would be reduced. If replacing an entire relay panel, be sure that some method of support is available for these heavy panels.

Getting key labor supervisors involved is important at this point. These men and women can make the job go well because of their special insights about the system hardware on which they and their subordinates work. Listen and learn. Not only is it important for a protection engineer to determine the protection settings, but also it is important to apply the right amount of effort on the user interface, screens, labels, and so on [12]. If you do not get labor buy-in, consulting labor later could cause additional work and an unsuitable working relationship that makes for late, over-budget upgrades.

Install, Test, and Commission

Refer to the installation drawings and procedures prepared earlier. Make sure that everyone on the installation team knows the order and scope of the work.

Labor can be your best help or a painful experience. As noted previously, the earlier that the labor pool is on board, the better the project implementation will be. These days, technicians are not as familiar with older relays, especially EM types. If installing these type of relays, make sure that the staff knows where to get the test-fixture extenders, where to attach the serial port, whether the port is "straight-through" or "crossed," the data rate, and all pertinent passwords. Many newer technicians are comfortable with USB ports, IP addresses, and Ethernet ports in the modern second-generation numeric relays but not so comfortable with older serial connections.

Put a dedicated landline telephone connection at the site. Not only is this telephone line a safety improvement, it is easier to use for contacting manufacturers' application engineers because there is no signal fading from nearby power-system noise as can be found with wireless or cellular telephones. Also, consider using a headset because noise levels at the site can make it difficult to hear someone on the telephone line.

Computer-aided testing is becoming standard throughout the industry to help existing technicians save time, providing more work at less cost (obviating the need to hire more personnel who are not available in the present workforce). Make sure that staff are trained for the relays in the upgrade [13].

Commissioning testing is essential to verify that all wiring is connected properly to all relay inputs and outputs (hardwired and virtual) and that the upgraded relays communicate as expected. Computer-aided testing can help save time at this point. Make sure to analyze the expected results against the scheme design to ensure trouble-free operation. When changes come (and changes are a certainty), be ready with a revision-tracking method. These days, there are many software library management programs that can assist in enforcing one copy of the changes as the working set.

Complete Drawings with As-Left Data

Documentation is essential for saving time and money. From good planning to smooth installation and final testing, complete, accurate documentation furthers efficiency. Updating project drawings should be an ongoing effort. Make final notes on protection design. File logic and software reports in a place where you can find these in the future. Capturing the as-left data now keeps those who follow you from guessing and wasting time on understanding the work.

Besides updating drawings, now is a good time to write a project summary and update project design documents. Those who come after you will thank you.

Dispose of Old Equipment Responsibly

Be responsible about disposing of older electronic equipment. Consult and follow the restriction of hazardous substances guidelines [14]. Currently, there is a large amount of lead, cadmium, mercury, and other dangerous chemicals in refuse landfills. The European Union spearheaded efforts in these areas, reducing contamination and making ground water safe.

Think about donating older equipment to a local vocational school or college. It is important that the new crop of protection engineers and technicians are familiar with older products so these students can learn valuable lessons about algorithms and test procedures. Mark the replaced old relays with painted or labeled warnings about reuse in the power system. Besides the enhanced goodwill that donations provide, perhaps your organization can earn a government tax deduction as well.

Report Outcome to Management

At the end of the relay upgrade, take the project summary report and ask to make a formal presentation to company management. Not only does this let management know what was accomplished, this effort pays back in increased favorable management buy-in when proposing the next project.

Reporting the project outcome also helps to complete upgrade documentation, provides an opportunity to point out helpful colleagues and staff, and gives you one last look at what was done. It is good to write a "lesson learned" document to pass along to the next project leader on the next upgrade.

Conclusions

Upgrading power-system protection relays allows you to see long-term cost reduction and increased operational efficiency. In these days of cost cutting and striving for doing more with less, relay upgrades, although initially requiring some cost and effort, make better use of funds. Whether it is a public utility district, an industry application, or a large utility, upgrading can alleviate many system problems and expensive relay failures.

Second-generation numeric relays (new multifunction microprocessor relays with multicore processing) are often less expensive per function than earlier relays. The new relays are rugged, with surface-mount technology that endures harsh operating environments better than older construction. The unreliable connections from older drawout designs are not a factor in newer relays, enhancing the already increased reliability.

More functions in one relay means lower cost and fewer replacements to improve protection schemes; the relay elements are already in the newer second-generation numeric relays. Even if these elements are not in use now, they will be available for later use if required by management, government, and protection mandates. Less panel wiring, longer maintenance intervals (from stable calibration and active watchdog reporting), and automated testing interfaces reduce overall costs. An unplanned outage can be more expensive than the cost of a relay-replacement project, not to mention the possible monetary fines (especially when connected to the bulk power system), loss of capital, loss of production, and loss of reputation.

Software for modern relays actually assists the protection engineer to make more consistent and better settings. In addition, second-generation numeric relay software provides summaries and internal logic checking to make certain that the protection elements are on, connected, and working.

Plan the upgrade project well. Keep management and labor informed on progress. Consult relay manufacturers for assistance. Make a final project summary, and report to management and all stakeholders (especially to all who worked on the project). This will promote future upgrade projects, which will improve power-system performance as well as the reputation of the relay protection engineer.

References

1. *Standard for Relays and Relay Systems Associated with Electric Power Apparatus*, ANSI/IEEE Standard C37.90, 2005.
2. *Measuring Relays and Protection Equipment—Part 1: Common Requirements*, IEC Standard 60255–1, 2009.
3. H. Kameda, S. Yoshiyama, G. Ushio, M. Usui, K. Sekiguchi, and C. Komatsu, "Estimation of replacement of numerical relay systems from reliability analysis," in *Proc. Int. Conf. Advanced Power System Automation Protection*, Oct. 16–20, 2011, vol. 1, pp. 114–119.
4. W. RuiChen, X. AnCheng, B. TianShu, and H. ShaoFeng, "Relay replacement strategy based on the least unit life cycle cost with minimum maintenance model," in *Proc. Int. Conf. Advanced Power System Automation Protection*, Oct. 16–20, 2011, vol. 1, pp. 609–614.
5. *Protection Devices Operation and Management Procedures, People's Republic of Electric Power Industry Standard [S]*, Standard DL/T587–2007.
6. P. Montignies, P. P. Basu, and F. Gruffaz, "Digital protective relays are designed for long life," in *Proc. 4th European Conf. Electrical Instrumentation Applications Petroleum Chemical Industry*, June 13–15, 2007, pp. 1–8.
7. *International Electrotechnical Vocabulary Chapter 191: Dependability and Quality of Service*, International Electrotechnical Committee Standard 60050–191, 2012.
8. National Electric Reliability Council. (2009). Critical Infrastructure Protection Standard CIP-002-3. [Online]. Available:

http://www.nerc.com/page.php?cid=2|20

9. M. Zeller, "Myth or reality—Does the Aurora vulnerability pose a risk to my generator?" in *Proc. 2011 64th Annu. Conf. Protective Relay Engineers*, Apr. 11–14, 2011, pp. 130–136.

10. National Electric Reliability Council. (2011). Protection system misoperation identification and correction. [Online]. Available: www.nerc.com/files/misoperations_webinar_master_deck_final.pdf

11. Basler Electric, "Multifunction relays fit in cutouts for transformer differential and overcurrent relays," Technical Resource Library, 2012.

12. J. Sperl, B. Carper, and W. Coe, "Upgrading your protective relays—When theory meets reality," in *Proc. 63rd Annu. Conf. Relay Engineers, Protective Relay Engineers*, Mar. 29–Apr. 1 2010, pp. 1–8.

13. A. Jezak and R. Garcia, "Aggressive electromechanical relay panel replacement project at TXU electric delivery," in *Proc. 58th Annu. Conf. Relay Engineers*, Apr. 5–7, 2005, pp. 189–193.

14. European Union, Directive on the restriction of the use of certain hazardous substances in electrical and electronic equipment 2002/95/EC (RoHS), 2003. ■

10.1 SYSTEM PROTECTION COMPONENTS

Protection systems have three basic components:

1. Instrument transformers
2. Relays
3. Circuit breakers

Figure 10.1 shows a simple overcurrent protection schematic with: (1) one type of instrument transformer—the current transformer (CT), (2) an overcurrent relay (OC), and (3) a circuit breaker (CB) for a single-phase line. The function of the CT is to reproduce in its secondary winding a current I' that is proportional to the primary current I. The CT converts primary currents in the kiloamp range to secondary currents in the 0–5 ampere range for convenience of measurement, with the following advantages.

Safety: Instrument transformers provide electrical isolation from the power system so that personnel working with relays will work in a safer environment.

Economy: Lower-level relay inputs enable relays to be smaller, simpler, and less expensive.

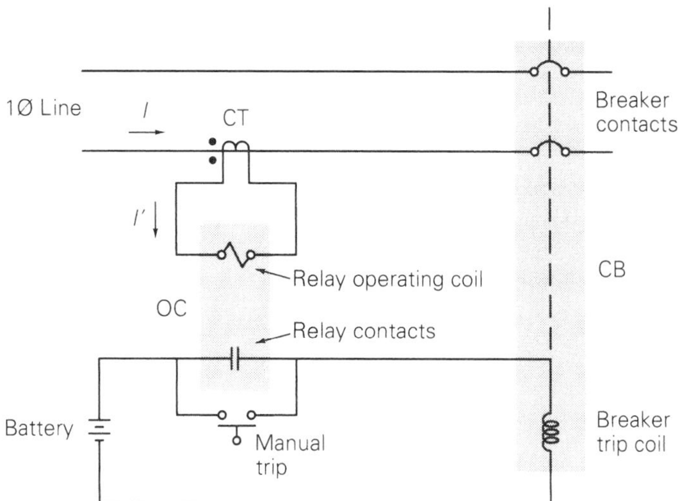

FIGURE 10.1

Overcurrent protection schematic

Accuracy: Instrument transformers accurately reproduce power system currents and voltages over wide operating ranges.

The function of the relay is to discriminate between normal operation and fault conditions. The OC relay in Figure 10.1 has an operating coil, which is connected to the CT secondary winding, and a set of contacts.

When $|I'|$ exceeds a specified "pickup" value, the operating coil causes the normally open contacts to close. When the relay contacts close, the trip coil of the circuit breaker is energized, which then causes the circuit breaker to open.

Note that the circuit breaker does not open until its operating coil is energized, either manually or by relay operation. Based on information from instrument transformers, a decision is made and "relayed" to the trip coil of the breaker, which actually opens the power circuit—hence the name *relay*.

System-protection components have the following design criteria [2]:

Reliability: Operate dependably when fault conditions occur, even after remaining idle for months or years. Failure to do so may result in costly damages.

Selectivity: Avoid unnecessary, false trips.

Speed: Operate rapidly to minimize fault duration and equipment damage. Any intentional time delays should be precise.

Economy: Provide maximum protection at minimum cost.

Simplicity: Minimize protection equipment and circuitry.

Since it is impossible to satisfy all these criteria simultaneously, compromises must be made in system protection.

10.2 INSTRUMENT TRANSFORMERS

There are two basic types of instrument transformers: voltage transformers (VTs), formerly called potential transformers (PTs), and current transformers (CTs). Figure 10.2 shows a schematic representation for the VT and CT.

The transformer primary is connected to or into the power system and is insulated for the power system voltage. The VT reduces the primary voltage, and the CT reduces the primary current to much lower, standardized levels suitable for operation of relays. Photos of VTs and CTs are shown in Figures 10.3–10.6.

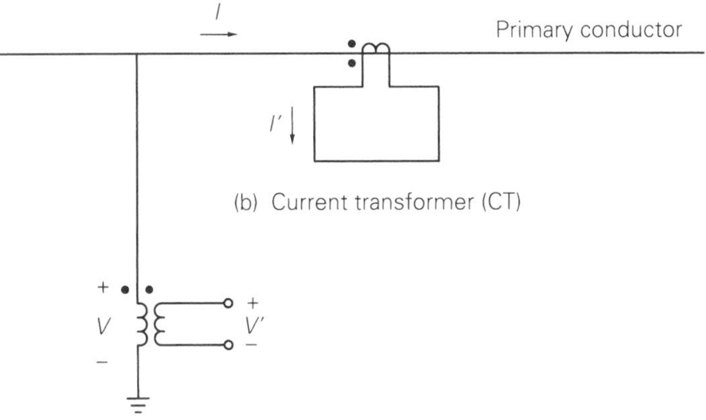

FIGURE 10.2

VT and CT schematic

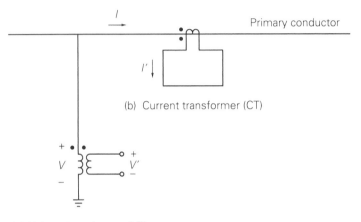

FIGURE 10.3

Three 34.5-kV voltage transformers with 34.5 kV: 115/67 volt VT ratios, at Lisle substation, Lisle, Illinois (Courtesy of Commonwealth Edison, an Exelon Company.)

For system-protection purposes, VTs are generally considered to be sufficiently accurate. Therefore, the VT is usually modeled as an ideal transformer, where

$$V' = (1/n)V \tag{10.2.1}$$

V' is a scaled-down representation of V and is in phase with V. A standard VT secondary voltage rating is 115 V (line-to-line). Standard VT ratios are given in Table 10.1.

Ideally, the VT secondary is connected to a voltage-sensing device with infinite impedance, such that the entire VT secondary voltage is across the sensing device. In practice, the secondary voltage divides across the high-impedance sensing device and the VT series leakage impedances. VT leakage impedances are kept low in order to minimize voltage drops and phase-angle differences from primary to secondary.

The primary winding of a current transformer usually consists of a single turn, obtained by running the power system's primary conductor through the CT core. The normal current rating of CT secondaries is standardized at 5 A in the United States,

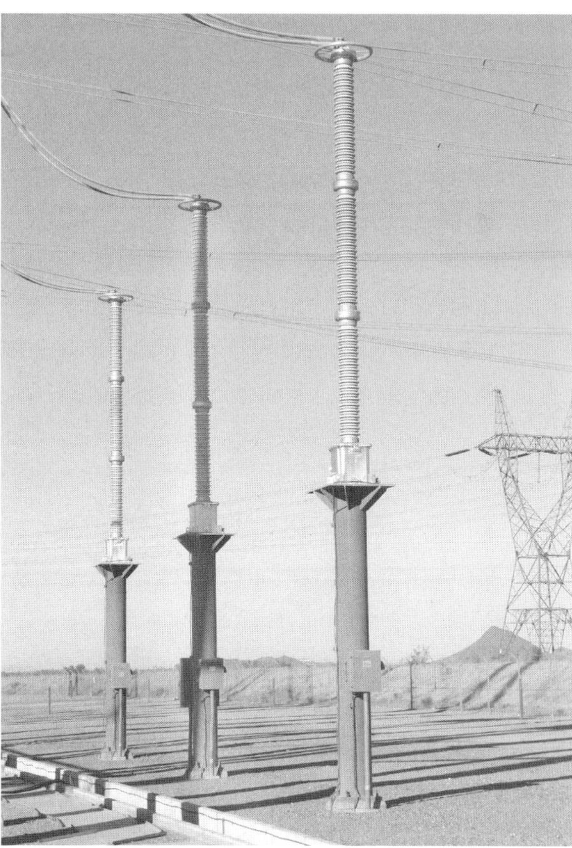

FIGURE 10.4
Three 500-kV coupling capacitor voltage transformers with 303.1 kV: 115/67 volt VT ratios, Westwing 500-kV Switching Substation (Courtesy of Arizona Public Service.)

whereas 1 A is standard in Europe and some other regions. Currents of 10 to 20 times (or greater) normal rating often occur in CT windings for a few cycles during short circuits. Standard CT ratios are given in Table 10.2.

Ideally, the CT secondary is connected to a current-sensing device with zero impedance, such that the entire CT secondary current flows through the sensing device. In practice, the secondary current divides, with most flowing through the low-impedance sensing device and some flowing through the CT shunt excitation impedance. CT excitation impedance is kept high in order to minimize excitation current.

An approximate equivalent circuit of a CT is shown in Figure 10.7, where

Z' = CT secondary leakage impedance
X_e = (Saturable) CT excitation reactance
Z_B = Impedance of terminating device (relay, including leads)

The total impedance Z_B of the terminating device is called the *burden* and is typically expressed in values of less than an ohm. The burden on a CT may also be expressed as volt-amperes at a specified current.

Associated with the CT equivalent circuit is an excitation curve that determines the relationship between the CT secondary voltage E' and excitation current I_e. Excitation curves for a multiratio bushing CT with ANSI classification C100 are shown in Figure 10.8.

FIGURE 10.5

Three 25 kV class current transformers-window design (Courtesy of Arizona Public Service.)

FIGURE 10.6

500-kV class current transformers with 2000:5 CT ratios in front of 500-kV SF6 circuit breakers. Westwing 500-kV Switching Substation (Courtesy of Arizona Public Service.)

Voltage Ratios						
1:1	2:1	2.5:1	4:1	5:1	20:1	40:1
60:1	100:1	200:1	300:1	400:1	600:1	800:1
1000:1	2000:1	3000:1	4500:1			

TABLE 10.1
Standard VT ratios

Current Ratios						
50:5	100:5	150:5	200:5	250:5	300:5	400:5
450:5	500:5	600:5	800:5	900:5	1000:5	1200:5
1500:5	1600:5	2000:5	2400:5	2500:5	3000:5	3200:5
4000:5	5000:5	6000:5				

TABLE 10.2
Standard CT ratios

Current transformer performance is based on the ability to deliver a secondary output current I' that accurately reproduces the primary current I. Performance is determined by the highest current that can be reproduced without saturation to cause large errors. Using the CT equivalent circuit and excitation curves, the following procedure can be used to determine CT performance.

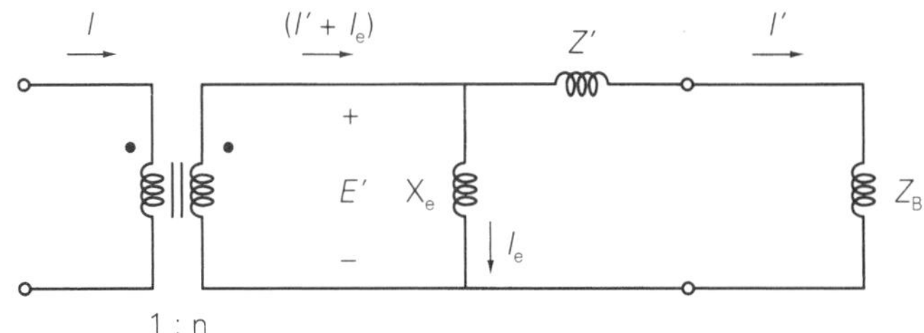

FIGURE 10.7 CT equivalent circuit

STEP 1 Assume a CT secondary output current I′.

STEP 2 Compute $E' = (Z' + Z_B)I'$.

STEP 3 Using E′, find I_e from the excitation curve.

STEP 4 Compute $I = n(I' + I_e)$.

STEP 5 Repeat Steps 1–4 for different values of I′, then plot I′ versus I.

For simplicity, approximate computations are made with magnitudes rather than with phasors. Also, the CT error is the percentage difference between $(I' + I_e)$ and I′, given by:

$$\text{CT error} = \frac{I_e}{I' + I_e} \times 100\% \tag{10.2.2}$$

The following examples illustrate the procedure.

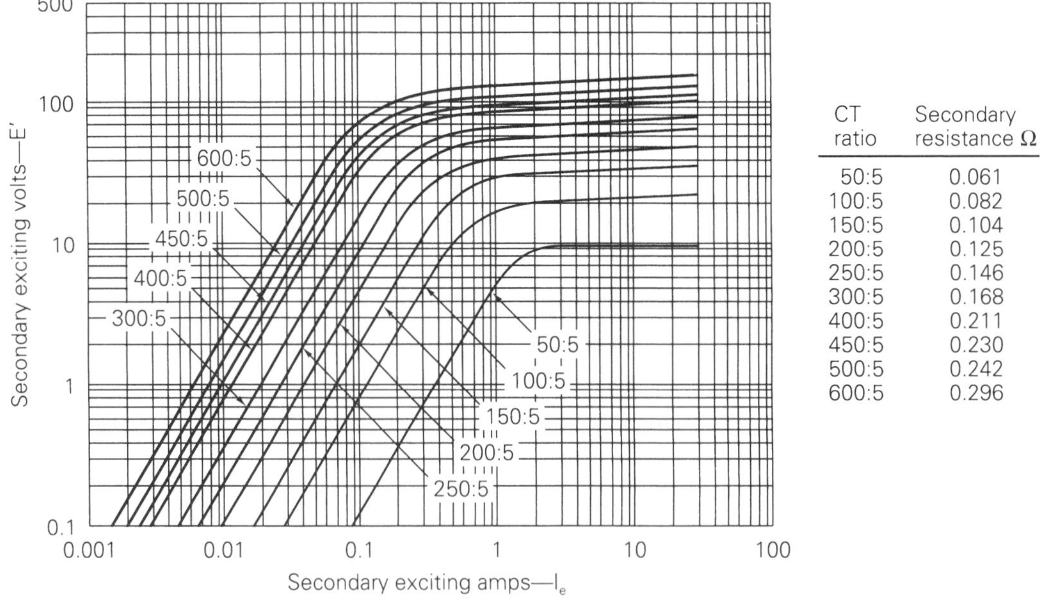

CT ratio	Secondary resistance Ω
50:5	0.061
100:5	0.082
150:5	0.104
200:5	0.125
250:5	0.146
300:5	0.168
400:5	0.211
450:5	0.230
500:5	0.242
600:5	0.296

FIGURE 10.8

Excitation curves for a bushing CT with a C100 ANSI accuracy classification [3] (Westinghouse Relay Manual, A New Silent Sentinels Publication (Newark, NJ: Westinghouse Electric Corporation, 1972).)

EXAMPLE 10.1

Current transformer (CT) performance

Evaluate the performance of the multiratio CT in Figure 10.8 with a 100:5 CT ratio, for the following secondary output currents and burdens: (a) $I' = 5$ A and $Z_B = 0.5\ \Omega$; (b) $I' = 8$ A and $Z_B = 0.8\ \Omega$; and (c) $I' = 15$ A and $Z_B = 1.5\ \Omega$. Also, compute the CT error for each output current.

SOLUTION

From Figure 10.8, the CT with a 100:5 CT ratio has a secondary resistance $Z' = 0.082\ \Omega$. Completing the above steps:

a. **STEP 1** $I' = 5$ A
STEP 2 From Figure 10.7,

$$E' = (Z' + Z_B)I' = (0.082 + 0.5)(5) = 2.91 \text{ V}$$

STEP 3 From Figure 10.8, $I_e = 0.25$ A
STEP 4 From Figure 10.7, $I = (100/5)(5 + 0.25) = 105$ A

$$\text{CT error} = \frac{0.25}{5.25} \times 100 = 4.8\%$$

b. **STEP1** $I' = 8$ A
STEP 2 From Figure 10.7,

$$E' = (Z' + Z_B)I' = (0.082 + 0.8)(8) = 7.06 \text{ V}$$

STEP 3 From Figure 10.8, $I_e = 0.4$ A
STEP 4 From Figure 10.7, $I = (100/5)(8 + 0.4) = 168$ A

$$\text{CT error} = \frac{0.4}{8.4} \times 100 = 4.8\%$$

c. **STEP 1** $I' = 15$ A
STEP 2 From Figure 10.7,

$$E' = (Z' + Z_B)I' = (0.082 + 1.5)(15) = 23.73 \text{ V}$$

STEP 3 From Figure 10.8, $I_e = 20$ A
STEP 4 From Figure 10.7, $I = (100/5)(15 + 20) = 700$ A

$$\text{CT error} = \frac{20}{35} \times 100 = 57.1\%$$

Note that for the 15-A secondary current in (c), high CT saturation causes a large CT error of 57.1%. Standard practice is to select a CT ratio to give a little less than 5-A secondary output current at maximum normal load. From (a), the 100:5 CT ratio and 0.5 Ω burden are suitable for a maximum primary load current of about 100 A. This example is extended in Problem 10.2 to obtain a plot of I' versus I.

EXAMPLE 10.2

Relay operation versus fault current and CT burden

An overcurrent relay set to operate at 8 A is connected to the multiratio CT in Figure 10.8 with a 100:5 CT ratio. Will the relay detect a 200-A primary fault current if the burden Z_B is (a) 0.8 Ω, (b) 3.0 Ω?

SOLUTION

Note that if an ideal CT is assumed, $(100/5) \times 8 = 160$-A primary current would cause the relay to operate.

a. From Example 10.1(b), a 168-A primary current with $Z_B = 0.8\ \Omega$ produces a secondary output current of 8 A, which would cause the relay to operate. Therefore, the higher 200-A fault current also causes the relay to operate.

b. **STEP 1** $I' = 8$ A

STEP 2 From Figure 10.7,

$$E' = (Z' + Z_B)I' = (0.082 + 3.0)(8) = 24.7 \text{ V}$$

STEP 3 From Figure 10.8, $I_e = 30$ A

STEP 4 From Figure 10.7, $I = (100/5)(8 + 30) = 760$ A

With a 3.0-Ω burden, 760 A is the lowest primary current that causes the relay to operate. Therefore, the relay will not operate for the 200-A fault current.

10.3 OVERCURRENT RELAYS

As shown in Figure 10.1, the CT secondary current I' is the input to the overcurrent relay operating coil. Instantaneous overcurrent relays respond to the magnitude of their input current, as shown by the trip and block regions in Figure 10.9. If the current magnitude $I' = |I'|$ exceeds a specified adjustable current magnitude I_p, called the *pickup* current, then the relay contacts close "instantaneously" to energize the circuit breaker trip coil. If I' is less than the pickup current I_p, then the relay contacts remain open, blocking the trip coil.

Time-delay overcurrent relays also respond to the magnitude of their input current, but with an intentional time delay. As shown in Figure 10.10, the time delay depends on the magnitude of the relay input current. If I' is a large multiple of the pickup current I_p, then the relay operates (or trips) after a small time delay. For smaller multiples of pickup, the relay trips after a longer time delay. And if $I' < I_p$, the relay remains in the blocking position.

Figure 10.11 shows two examples of a time-delay overcurrent relay: (a) Westinghouse electromechanical CO relay; and (b) Basler Electric digital relay. Characteristic

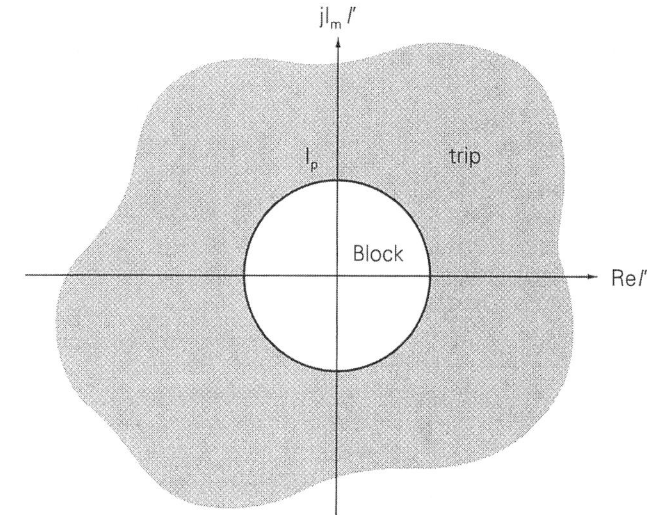

FIGURE 10.9 Instantaneous overcurrent relay block and trip regions

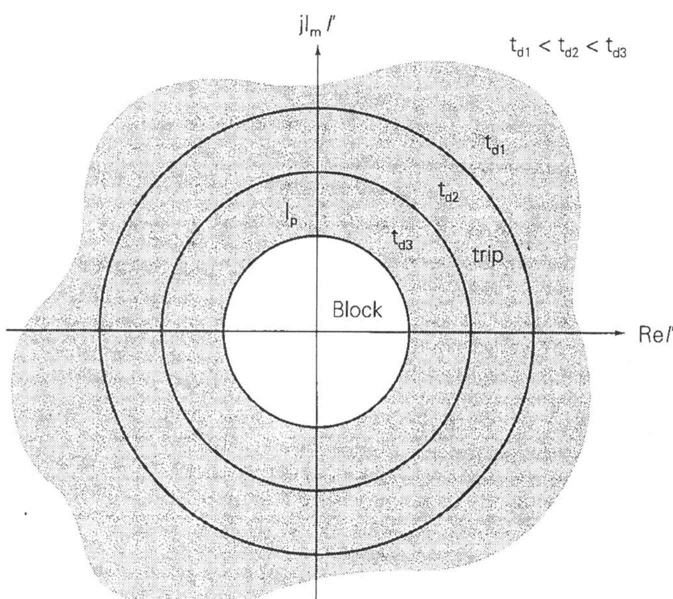

FIGURE 10.10 Time-delay overcurrent relay block and trip regions

curves of the Westinghouse CO-8 relay are shown in Figure 10.12. These relays have two settings:

Current tap setting: The pickup current in amperes.

Time-dial setting: The adjustable amount of time delay.

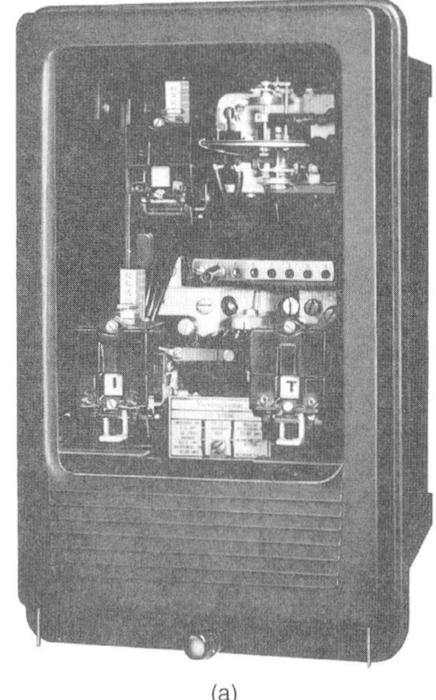

(a)

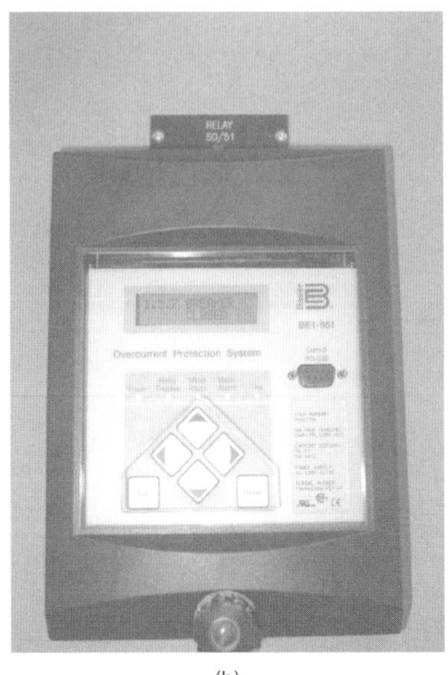

(b)

FIGURE 10.11

Time-delay overcurrent relays: (a) Westinghouse Electromechanical (Courtesy of ABB-Westinghouse.) (b) Basler Electric Digital (Courtesy Danvers Electric.)

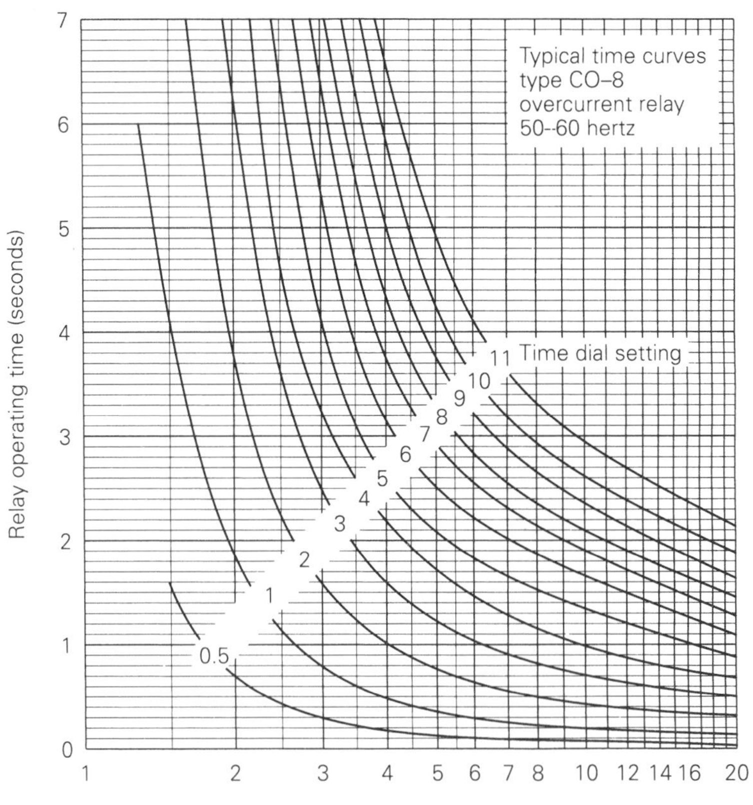

FIGURE 10.12

CO-8 time-delay overcurrent relay characteristics (Courtesy of Westinghouse Electric Corporation.)

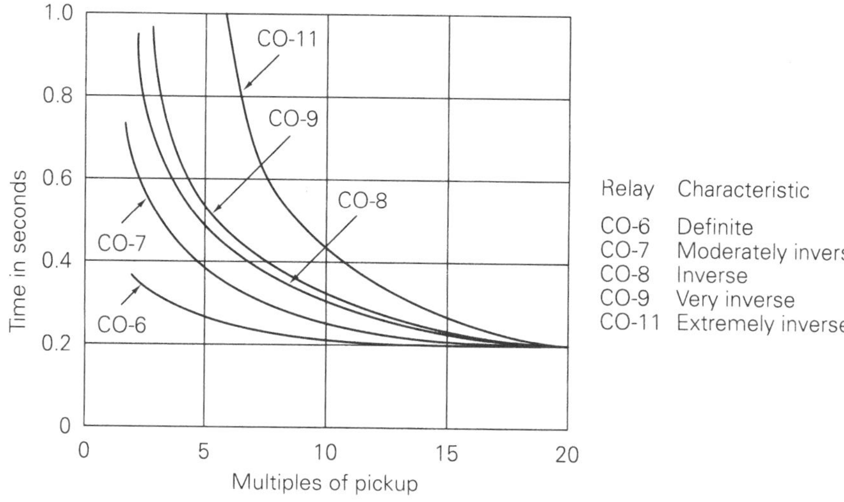

FIGURE 10.13

Comparison of CO relay characteristics (Courtesy of Westinghouse Electric Corporation.)

The characteristic curves are usually shown with operating time in seconds versus relay input current as a multiple of the pickup current. The curves are asymptotic to the vertical axis and decrease with some inverse power of current magnitude for values exceeding the pickup current. This inverse time characteristic can be shifted up or down by adjustment of the time-dial setting. Although discrete time-dial settings are shown in Figure 10.12, intermediate values can be obtained by interpolating between the discrete curves.

Figure 10.13 shows the time-current characteristics of five CO time-delay overcurrent relays used in transmission and distribution lines. The time-dial settings are

EXAMPLE 10.3

Operating time for a CO-8 time-delay overcurrent relay

The CO-8 relay with a current tap setting of 6 amperes and a time-dial setting of 1 is used with the 100:5 CT in Example 10.1. Determine the relay operating time for each case.

SOLUTION

a. From Example 10.1(a)

$$I' = 5\text{ A} \qquad \frac{I'}{I_p} = \frac{5}{6} = 0.83$$

The relay does not operate. It remains in the blocking position. *(Continued)*

b. $I' = 8\ A \qquad \frac{I'}{I_p} = \frac{8}{6} = 1.33$

Using curve 1 in Figure 10.12, $t_{operating} = 6$ seconds.

c. $I' = 15\ A \qquad \frac{I'}{I_p} = \frac{15}{6} = 2.5$

From curve 1, $t_{operating} = 1.2$ seconds.

selected in the figure so that all relays operate in 0.2 seconds at 20 times the pickup current. The choice of relay time-current characteristic depends on the sources, lines, and loads. The definite (CO-6) and moderately inverse (CO-7) relays maintain a relatively constant operating time above 10 times pickup. The inverse (CO-8), very inverse (CO-9), and extremely inverse (CO-11) relays operate respectively faster on higher fault currents.

Figure 10.14 illustrates the operating principle of an electromechanical time-delay overcurrent relay. The ac input current to the relay operating coil sets up a magnetic field that is perpendicular to a conducting aluminum disc. The disc can rotate and is restrained by a spiral spring. Current is induced in the disc, interacts with the magnetic field, and produces a torque. If the input current exceeds the pickup current, the disc rotates through an angle θ to close the relay contacts. The larger the input current, the larger is the torque and the faster the contact closing. After the current is removed or reduced below the pickup, the spring provides reset of the contacts.

A solid state relay panel between older-style electromechanical relays is shown in Figure 10.15.

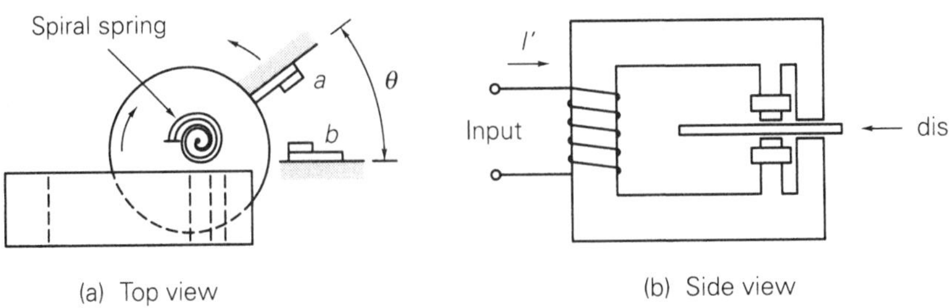

FIGURE 10.14 Electromechanical time-delay overcurrent relay—induction disc type

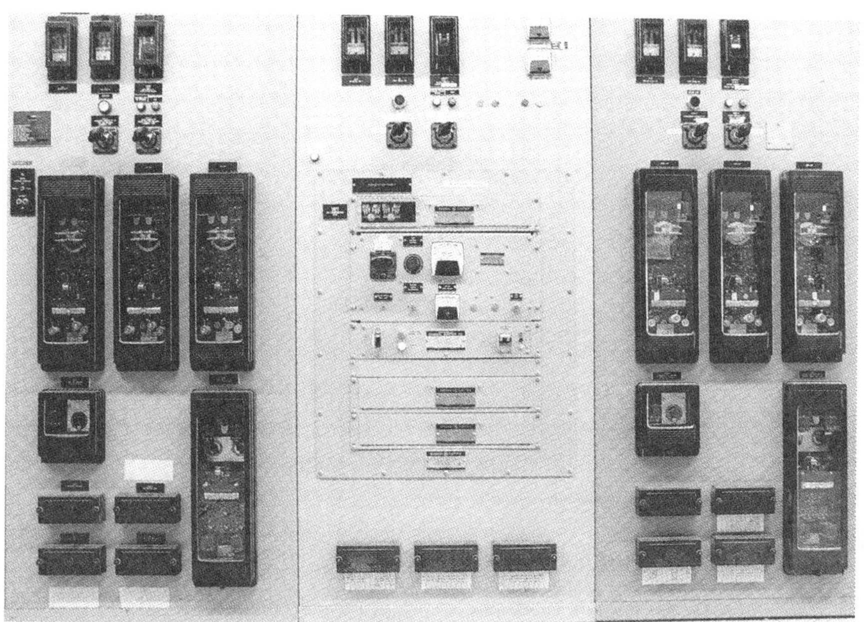

FIGURE 10.15

Solid-state relay panel (center) for a 345-kV transmission line, with electromechanical relays on each side, at Electric Junction Substation, Naperville, Illinois (Courtesy of Commonwealth Edison, an Exelon Company.)

10.4 RADIAL SYSTEM PROTECTION

Many radial systems are protected by time-delay overcurrent relays. Adjustable time delays can be selected such that the breaker closest to the fault opens, while other upstream breakers with larger time delays remain closed. That is, the relays can be coordinated to operate in sequence so as to interrupt minimum load during faults. Successful relay coordination is obtained when fault currents are much larger than normal load currents. Also, coordination of overcurrent relays usually limits the maximum number of breakers in a radial system to five or less; otherwise the relay closest to the source may have an excessive time delay.

Consider a fault at P_1 to the right of breaker B3 for the radial system of Figure 10.16. For this fault breaker B3 should open while B2 (and B1) remains closed. Under these conditions, only load L3 is interrupted. Select a longer time delay for the relay at B2, so that B3 operates first. Thus, for any fault to the right of B3, B3 provides primary protection. Only if B3 fails to open will B2 open, after a time delay, thus providing backup protection.

Similarly, consider a fault at P_2 between B2 and B3. B2 should open while B1 remains closed. Under these conditions, loads L2 and L3 are interrupted. Since the fault is closer to the source, the fault current is larger than for the previous fault considered. B2, set to open for the previous, smaller fault current after a time delay, will open more rapidly for this fault. Also selected is the B1 relay with a longer time delay than B2, so that B2 opens first. Thus, B2 provides primary protection for faults

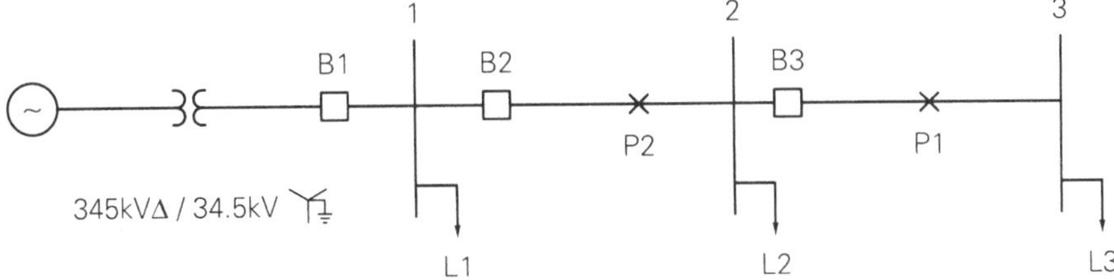

FIGURE 10.16 Single-line diagram of a 34.5-kV radial system

between B2 and B3, as well as backup protection for faults to the right of B3. Similarly, B1 provides primary protection for faults between B1 and B2, as well as backup protection for further downstream faults.

The *coordination time interval* is the time interval between the primary and remote backup protective devices. It is the difference between the time that the backup relaying operates and the time that circuit breakers clear the fault under primary relaying. Precise determination of relay operating times is complicated by several factors, including CT error, dc offset component of fault current, and relay overtravel. Therefore, typical coordination time intervals from 0.2 to 0.5 seconds are selected to account for these factors in most practical applications.

EXAMPLE 10.4

Coordinating time-delay overcurrent relays in a radial system

Data for the 60-Hz radial system of Figure 10.16 are given in Tables 10.3, 10.4, and 10.5. Select current tap settings (TSs) and time-dial settings (TDSs)

Bus	S MVA	Lagging p.f.
1	11.0	0.95
2	4.0	0.95
3	6.0	0.95

TABLE 10.3

Maximum loads—Example 10.4

Bus	Maximum Fault Current (Bolted Three-Phase) A	Minimum Fault Current (L–G or L–L) A
1	3000	2200
2	2000	1500
3	1000	700

TABLE 10.4

Symmetrical fault currents—Example 10.4

Breaker	Breaker Operating Time	CT Ratio	Relay
B1	5 cycles	400:5	CO-8
B2	5 cycles	200:5	CO-8
B3	5 cycles	200:5	CO-8

TABLE 10.5

Breaker, CT, and relay data—Example 10.4

to protect the system from faults. Assume three CO-8 relays for each breaker, one for each phase, with a 0.3-second coordination time interval. The relays for each breaker are connected as shown in Figure 10.17, so that all three phases of the breaker open when a fault is detected on any one phase. Assume a 34.5-kV (line-to-line) voltage at all buses during normal operation. Also, future load growth is included in Table 10.3, such that maximum loads over the operating life of the radial system are given in this table.

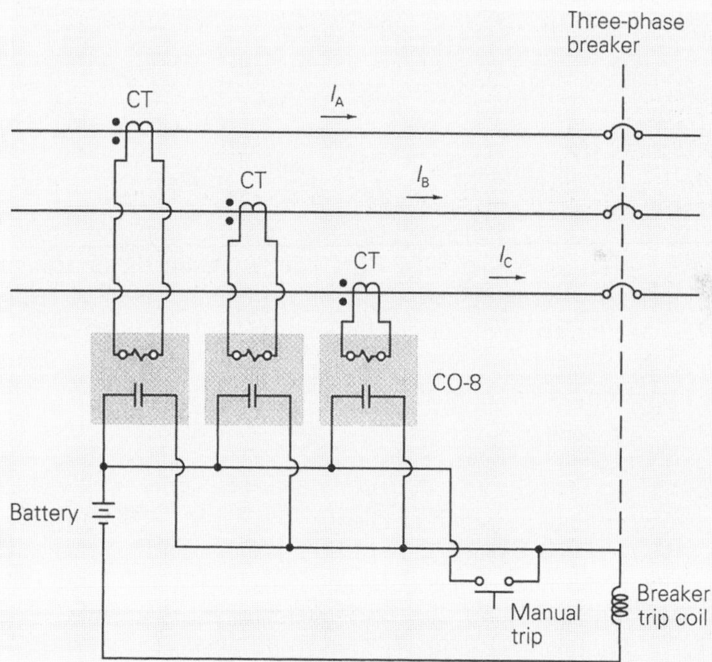

FIGURE 10.17

Relay connections to trip all three phases

SOLUTION

First, select TSs such that the relays do not operate for maximum load currents. Starting at B3, the primary and secondary CT currents for maximum load L3 are

(Continued)

$$I_{L3} = \frac{S_{L3}}{V_3\sqrt{3}} = \frac{6 \times 10^6}{(34.5 \times 10^3)\sqrt{3}} = 100.4 \text{ A}$$

$$I'_{L3} = \frac{100.4}{(200/5)} = 2.51\text{A}$$

From Figure 10.12, relay a 3-A TS is selected for the B3, which is the lowest TS above 2.51 A.

Note that $|S_{L2} + S_{L3}| = |S_{L2}| + |S_{L3}|$ because the load power factors are identical. Thus, at B2, the primary and secondary CT currents for maximum load are

$$I_{L2} = \frac{S_{L2} + S_{L3}}{V_2\sqrt{3}} = \frac{(4 + 6) \times 10^6}{(34.5 \times 10^3)\sqrt{3}} = 167.3 \text{ A}$$

$$I'_{L2} = \frac{167.3}{(200/5)} = 4.18 \text{ A}$$

From Figure 10.12, select for the B2 relay a 5-A TS, the lowest TS above 4.18 A. At B1,

$$I_{L1} = \frac{S_{L1} + S_{L2} + S_{L3}}{V_1\sqrt{3}} = \frac{(11 + 4 + 6) \times 10^6}{(34.5 \times 10^3)\sqrt{3}} = 351.4 \text{ A}$$

$$I'_{L1} = \frac{351.4}{(400/5)} = 4.39 \text{ A}$$

Select a 5-A TS for the B1 relay.

Next select the TDSs. First, coordinate for the maximum fault currents in Table 10.4, checking coordination for minimum fault currents later. Starting at B3, the largest fault current through B3 is 2000 A, which occurs for the three-phase fault at bus 2 (just to the right of B3). Neglecting CT saturation, the fault-to-pickup current ratio at B3 for this fault is

$$\frac{I'_{3\text{Fault}}}{TS3} = \frac{2000/(200/5)}{3} = 16.7$$

To clear faults as rapidly as possible, select a 1/2 TDS for the B3 relay. Then, from the 1/2 TDS curve in Figure 10.12, the relay operating time is T3 = 0.05 seconds. Adding the breaker operating time (5 cycles = 0.083 s), primary protection clears this fault in $T3 + T_{\text{breaker}} = 0.05 + 0.083 = 0.133$ seconds.

For this same fault, the fault-to-pickup current ratio at B2 is

$$\frac{I'_{2\text{Fault}}}{TS2} = \frac{2000/(200/5)}{5} = 10.0$$

Adding the B3 relay operating time (T3 = 0.05 s), breaker operating time (0.083 s), and 0.3 s coordination time interval, a B2 relay operating time is desirable:

$$T2 = T3 + T_{\text{breaker}} + T_{\text{coordination}} = 0.05 + 0.083 + 0.3 \approx 0.43 \text{ s}$$

Breaker	Relay	TS	TDS
B1	CO-8	5	3
B2	CO-8	5	2
B3	CO-8	3	1/2

TABLE 10.6
Solution—Example 10.4

From Figure 10.12, select TDS2 = 2.

Next select the TDS at Bl. The largest fault current through B2 is 3000 A, for a three-phase fault at bus 1 (just to the right of B2). The fault-to-pickup current ratio at B2 for this fault is

$$\frac{I'_{2\text{Fault}}}{\text{TS2}} = \frac{3000/(200/5)}{5} = 15.0$$

From the 2 TDS curve in Figure 10.12, T2 = 0.38 s. For this same fault,

$$\frac{I'_{1\text{Fault}}}{\text{TS1}} = \frac{3000/(400/5)}{5} = 7.5$$

$$\text{T1} = \text{T2} + \text{T}_{\text{breaker}} + \text{T}_{\text{coordination}} = 0.38 + 0.083 + 0.3 \approx 0.76\text{ s}$$

From Figure 10.12, select TDS1 = 3. The relay settings are shown in Table 10.6. Note that for reliable relay operation the fault-to-pickup current ratios with minimum fault currents should be greater than 2. Coordination for minimum fault currents listed in Table 10.4 is evaluated in Problem 10.11.

Note that separate relays are used for each phase in Example 10.4, and therefore these relays will operate for three-phase as well as line-to-line, single line-to-ground, and double line-to-ground faults. However, in many cases single line-to-ground fault currents are much lower than three-phase fault currents, especially for distribution feeders with high zero-sequence impedances. In these cases a separate ground relay with a lower current tap setting than the phase relays is used. The ground relay is connected to operate on zero-sequence current from three of the phase CTs connected in parallel or from a CT in the grounded neutral.

10.5 RECLOSERS AND FUSES

Automatic circuit reclosers are commonly used for distribution circuit protection. A *recloser* is a self-controlled device for automatically interrupting and reclosing an ac circuit with a preset sequence of openings and reclosures. Unlike circuit breakers, which have separate relays to control breaker opening and reclosing, reclosers have built-in controls. More than 80% of faults on overhead distribution circuits are temporary, caused by tree limb contact, by animal interference, by wind bringing bare conductors in contact, or by lightning. The automatic tripping-reclosing sequence of reclosers clears these temporary faults and restores service with only momentary outages, thereby significantly improving customer service. A disadvantage of reclosers is the increased hazard when a circuit is physically contacted by people—for example, in the case of a broken conductor at ground level that remains energized. Also, reclosing should be locked out during live-line maintenance by utility personnel.

Figure 10.18 shows a common protection scheme for radial distribution circuits utilizing fuses, reclosers, and time-delay overcurrent relays. Data for the 13.8-kV feeder in this figure is given in Table 10.7. There are three load taps protected by fuses. The recloser ahead of the fuses is set to open and reclose for faults up to and beyond the fuses. For temporary faults the recloser can be set for one or more instantaneous or time-delayed trips and reclosures in order to clear the faults and restore service. If faults persist, the fuses operate for faults to their right (downstream), or the recloser opens after a time delay and locks out for faults between the recloser and fuses. Separate time-delay overcurrent phase and ground relays open the substation breaker after multiple reclosures of the recloser.

Coordination of the fuses, recloser, and time-delay overcurrent relays is shown via the time-current curves in Figure 10.19. Type T (slow) fuses are selected because their time-current characteristics coordinate well with reclosers. The fuses are selected on the basis of maximum loads served from the taps. A 65 T fuse is selected for the bus 1 tap, which has a 60-A maximum load current, and 100 T is selected for the bus 2 and 3 taps, which have 95-A maximum load currents. The fuses should also have a rated voltage larger than the maximum bus voltage and an interrupting current rating larger than the maximum asymmetrical fault current at the fuse location. Type T fuses with voltage ratings of 15 kV and interrupting current ratings of 10 kA and higher are standard.

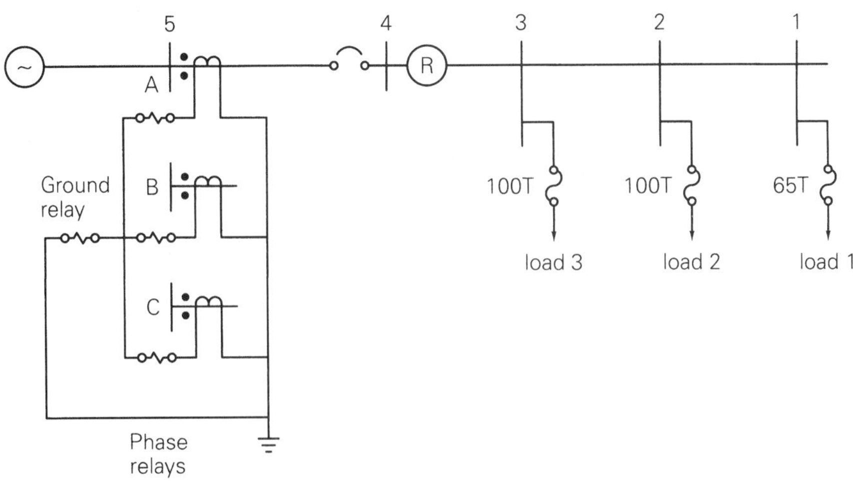

FIGURE 10.18 Single-line diagram of a 13.8-kV radial distribution feeder with fuse/recloser/relay protection

Bus	Maximum Load Current A	3ϕ Fault Current A	IL-G Fault Current A
1	60	1000	850
2	95	1500	1300
3	95	2000	1700
4	250	3000	2600
5	250	4000	4050

TABLE 10.7

Data for Figure 10.18

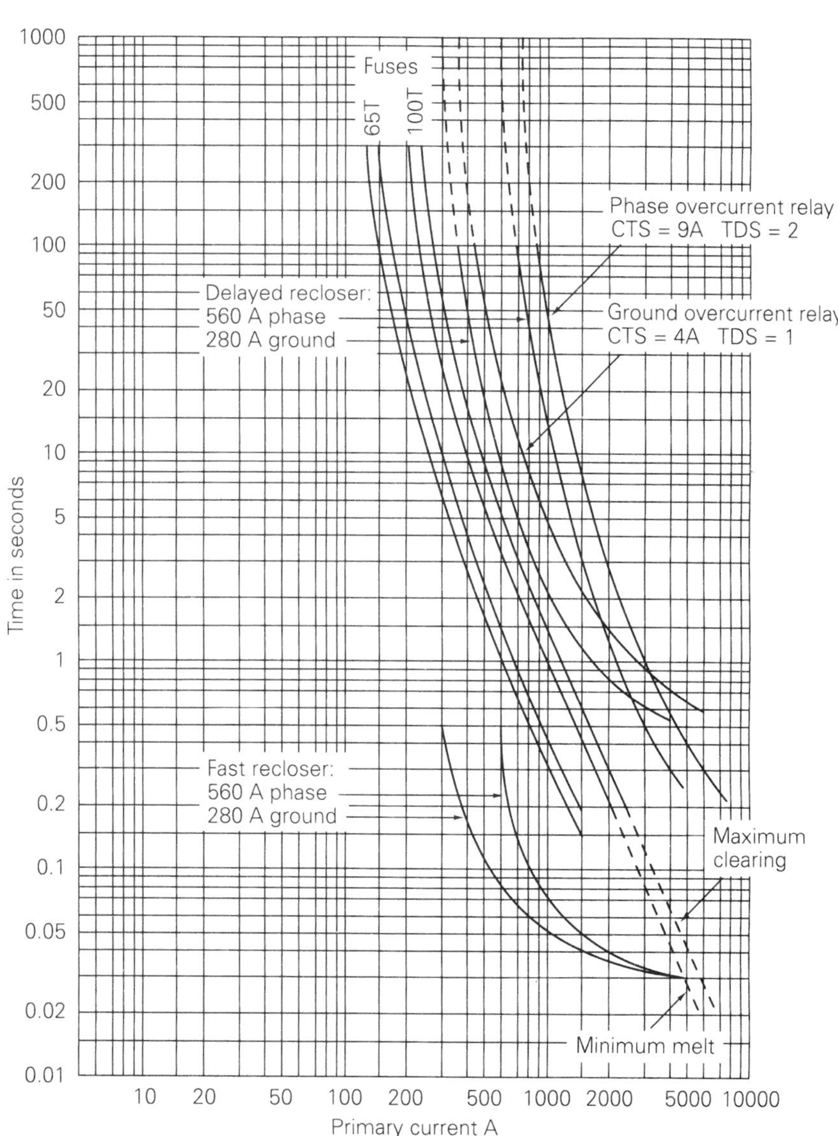

FIGURE 10.19

Time-current curves for the radial distribution circuit of Figure 10.18

Standard reclosers have minimum trip ratings of 50, 70, 100, 140, 200, 280, 400, 560, 800, 1120, and 1600 A, with voltage ratings up to 38 kV and maximum interrupting currents up to 16 kA. A minimum trip rating of 200–250% of maximum load current is typically selected for the phases, in order to override cold load pickup with a safety factor. The minimum trip rating of the ground unit is typically set at maximum load and should be higher than the maximum allowable load unbalance. For the recloser in Figure 10.18 that carries a 250-A maximum load, minimum trip ratings of 560 A for each phase and 280 A for the ground unit are selected.

A popular operation sequence for reclosers is two fast operations: without intentional time delay, followed by two delayed operations. The fast operations allow temporary faults to self-clear, whereas the delayed operations allow downstream fuses to clear permanent faults. Note that the time-current curves of the fast recloser lie below the fuse curves in Figure 10.19, such that the recloser opens before the fuses melt. The fuse curves lie below the delayed recloser curves, such that the fuses clear before the recloser opens. The recloser is typically programmed to reclose $\frac{1}{2}$ s after the first fast trip, 2 s after the second fast trip, and 5–10 s after a delayed trip.

Time-delay overcurrent relays with an extremely inverse characteristic coordinate with both reclosers and type T fuses. A 300:5 CT ratio is selected to give a secondary current of $250 \times (5/300) = 4.17$ A at maximum load. Relay settings are selected to allow the recloser to operate effectively to clear faults before relay operation. A current tap setting of 9 A is selected for the CO-11 phase relays so that minimum pickup exceeds twice the maximum load. A time-dial setting of 2 is selected so that the delayed recloser trips at least 0.2 s before the relay. The ground relay is set with a current tap setting of 4 A and a time-dial setting of 1.

EXAMPLE 10.5

Fuse/recloser coordination

For the system of Figure 10.18, describe the operating sequence of the protective devices for the following faults: (a) a self-clearing, temporary, three-phase fault on the load side of tap 2; and (b) a permanent three-phase fault on the load side of tap 2.

SOLUTION

a. From Table 10.7, the three-phase fault current at bus 2 is 1500 A. From Figure 10.19, the 560-A fast recloser opens 0.05 s after the 1500-A fault current occurs, and then recloses $\frac{1}{2}$ s later. Assuming the fault has self-cleared, normal service is restored. During the 0.05-s fault duration, the 100 T fuse does not melt.

b. For a permanent fault the fast recloser opens after 0.05 s, recloses $\frac{1}{2}$ s later into the permanent fault, opens again after $\frac{1}{2}$ s, and recloses into the fault a second time after a 2-s delay. Then the 560-A delayed recloser opens 3 seconds later. During this interval the 100 T fuse clears the fault. The delayed recloser then recloses 5 to 10 s later, restoring service to loads 1 and 3.

10.6 DIRECTIONAL RELAYS

Directional relays are designed to operate for fault currents in only one direction. Consider the directional relay D in Figure 10.20, which is required to operate only for faults to the right of the CT. Since the line impedance is mostly reactive, a fault at P_1 to the right of the CT will have a fault current I from bus 1 to bus 2 that lags the bus voltage V by an angle of almost 90°. This fault current is said to be in the forward direction. On the other hand, a fault at P_2, to the left of the CT, will have a fault current I that leads V by almost 90°. This fault current is said to be in the reverse direction.

The directional relay has two inputs: the reference voltage $V = \mathrm{V}\underline{/0^\circ}$, and current $I = \mathrm{I}\underline{/\phi}$. The relay trip and block regions, shown in Figure 10.21, can be described by

$$\begin{aligned} -180^\circ < (\phi - \phi_1) < 0^\circ \quad &\text{(Trip)} \\ \text{Otherwise} \quad &\text{(Block)} \end{aligned} \tag{10.6.1}$$

where ϕ is the angle of the current with respect to the voltage and ϕ_1, typically 2° to 8°, defines the boundary between the trip and block regions.

The contacts of the overcurrent relay OC and the directional relay D are connected in series in Figure 10.20, so that the breaker trip coil is energized only when the CT secondary current (1) exceeds the OC relay pickup value, and (2) is in the forward tripping direction.

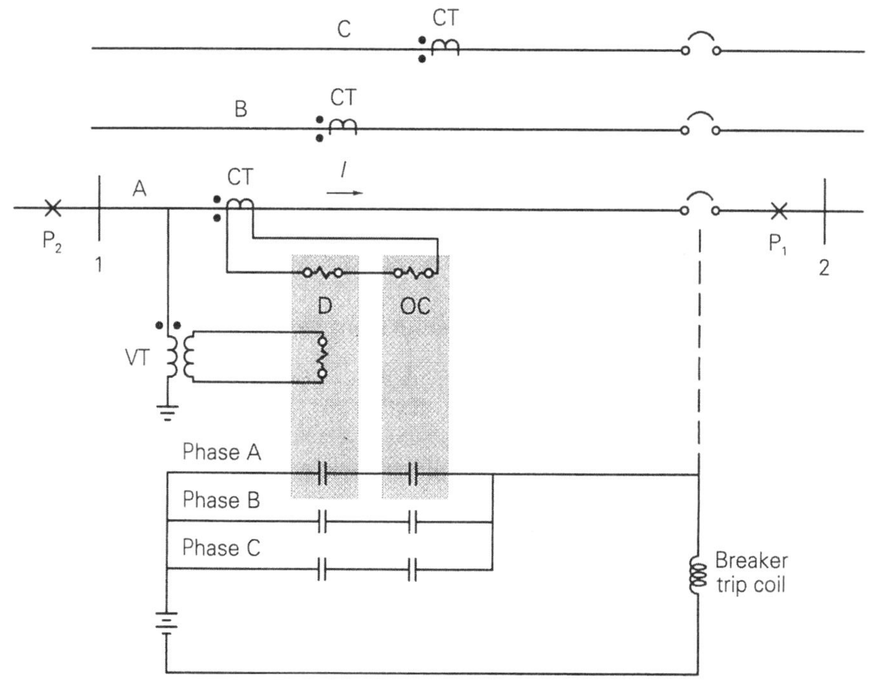

FIGURE 10.20 Directional relay in series with overcurrent relay (only phase A is shown)

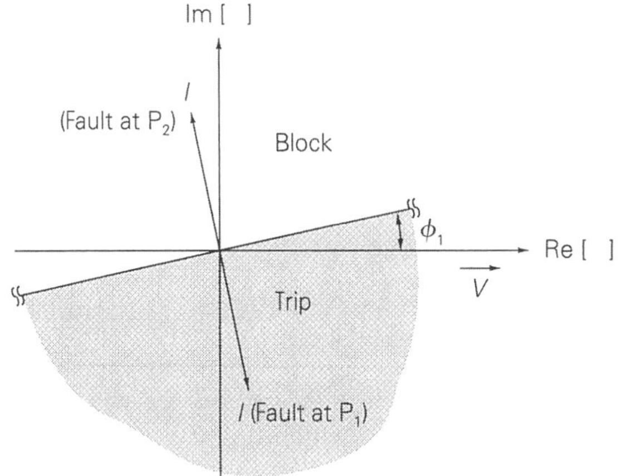

FIGURE 10.21 Directional relay block and trip regions in the complex plane

Although construction details differ, the operating principle of an electromechanical directional relay is similar to that of a watt-hour meter. There are two input coils, a voltage coil and a current coil, both located on a stator, and there is a rotating disc element. Suppose that the reference voltage is passed through a phase-shifting element to obtain $V_1 = \mathrm{V}\underline{/\phi_1 - 90°}$. If V_1 and $I = \mathrm{I}\underline{/\phi}$ are applied to a watt-hour meter, the torque on the rotating element is

$$\mathrm{T} = \mathrm{kVI}\cos(\phi_1 - \phi - 90°) = \mathrm{kVI}\sin(\phi_1 - \phi) \tag{10.6.2}$$

Note that for faults in the forward direction, the current lags the voltage, and the angle $(\phi_1 - \phi)$ in (10.6.2) is close to 90°. This results in maximum positive torque on the rotating disc, which would cause the relay contacts to close. On the other hand, for faults in the reverse direction the current leads the voltage, and $(\phi_1 - \phi)$ is close to −90°. This results in maximum negative torque tending to rotate the disc element in the backward direction. Backward motion can be restrained by mechanical stops.

10.7 PROTECTION OF A TWO-SOURCE SYSTEM WITH DIRECTIONAL RELAYS

It becomes difficult and in some cases impossible to coordinate overcurrent relays when there are two or more sources at different locations. Consider the system with two sources shown in Figure 10.22. Suppose there is a fault at P_1. B23 and B32 should clear the fault so that service to the three loads continues without interruption. Using time-delay overcurrent relays, set B23 faster than B21. Now consider a fault at P_2 instead. Breaker B23 will open faster than B21, and load L2 is disconnected. When a fault can be fed from both the left and right, overcurrent relays cannot be coordinated. However, directional relays can be used to overcome this problem.

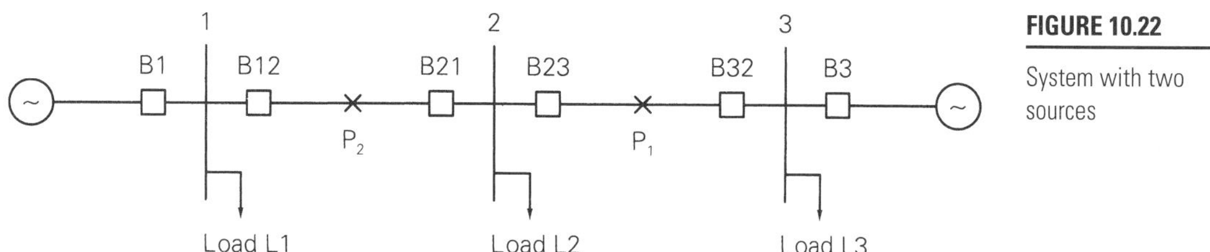

FIGURE 10.22

System with two sources

EXAMPLE 10.6

Two-source system protection with directional and time-delay overcurrent relays

Explain how directional and time-delay overcurrent relays can be used to protect the system in Figure 10.22. Which relays should be coordinated for a fault (a) at P_1, (b) at P_2? (c) Is the system also protected against bus faults?

SOLUTION

Breakers B12, B21, B23, and B32 should respond only to faults on their "forward" or "line" sides. Directional overcurrent relays connected as shown in Figure 10.20 can be used for these breakers. Overcurrent relays alone can be used for breakers B1 and B3, which do not need to be directional.

a. For a fault at P_1, the B21 relay would not operate; B12 should coordinate with B23 so that B23 trips before B12 (and B1). Also, B3 should coordinate with B32.

b. For a fault at P_2, B23 would not operate; B32 should coordinate with B21 so that B21 trips before B32 (and B3). Also, B1 should coordinate with B12.

c. Yes, the directional overcurrent relays also protect the system against bus faults. If the fault is at bus 2, relays at B21 and B23 will not operate, but B12 and B32 will operate to clear the fault. B1 and B21 will operate to clear a fault at bus 1. B3 and B23 will clear a fault at bus 3.

10.8 ZONES OF PROTECTION

Protection of simple systems has been discussed so far. For more general power system configurations, a fundamental concept is the division of a system into protective zones [1]. If a fault occurs anywhere within a zone, action is taken to isolate that zone from the rest of the system. Zones are defined for:

generators,

transformers,

buses,

transmission and distribution lines, and

motors.

Figure 10.23 illustrates the protective zone concept. Each zone is defined by a closed, dashed line. Zone 1, for example, contains a generator and connecting leads to a transformer. In some cases a zone may contain more than one component. For example, zone 3 contains a generator-transformer unit and connecting leads to a bus, and zone 10 contains a transformer and a line. Protective zones have the following characteristics:

Zones are overlapped.

Circuit breakers are located in the overlap regions.

For a fault anywhere in a zone, all circuit breakers in that zone open to isolate the fault.

Neighboring zones are overlapped to avoid the possibility of unprotected areas. Without overlap the small area between two neighboring zones would not be located in any zone and thus would not be protected.

Since isolation during faults is done by circuit breakers, they should be inserted between equipment in a zone and each connection to the system. That is, breakers should be inserted in each overlap region. As such, they identify the boundaries of protective zones. For example, zone 5 in Figure 10.23 is connected to

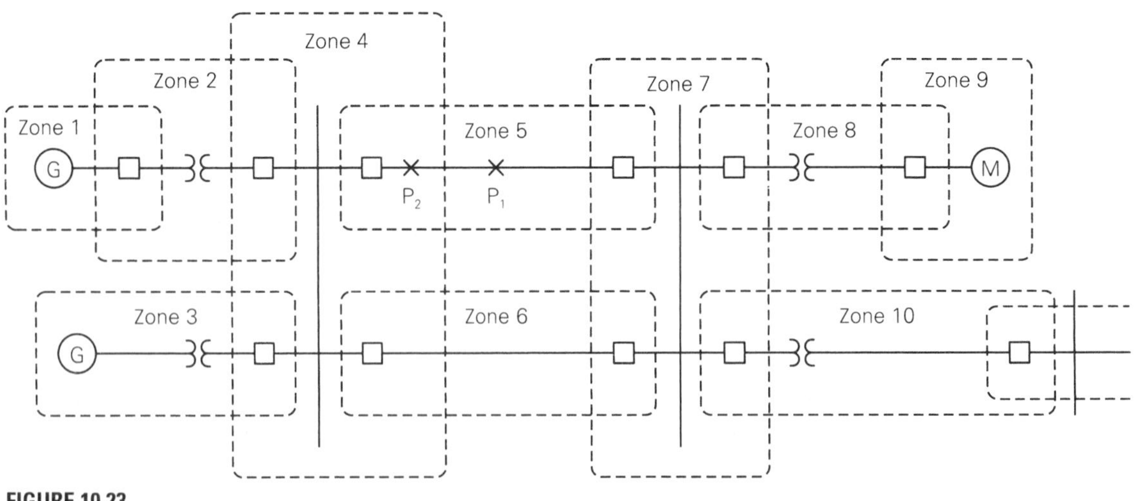

FIGURE 10.23

Power system protective zones

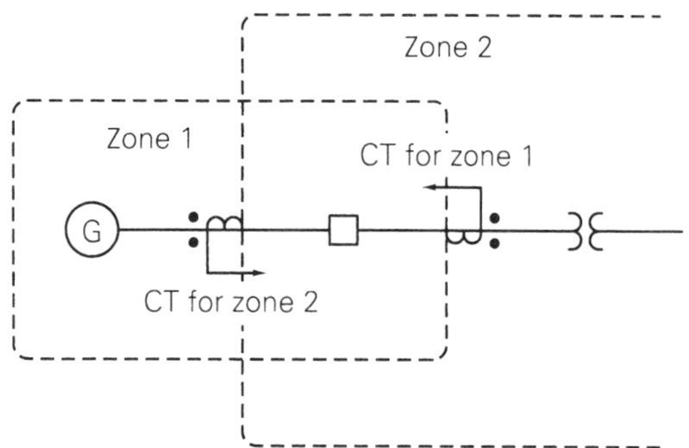

FIGURE 10.24 Overlapping protection around a circuit breaker

zones 4 and 7. Therefore, a circuit breaker is located in the overlap region between zones 5 and 4, as well as between zones 5 and 7.

If a fault occurs anywhere within a zone, action is taken to open all breakers in that zone. For example, if a fault occurs at P_1 on the line in zone 5, then the two breakers in zone 5 should open. If a fault occurs at P_2 within the overlap region of zones 4 and 5, then all five breakers in zones 4 and 5 should open. Clearly, if a fault occurs within an overlap region, two zones are isolated and a larger part of the system is lost from service. To minimize this possibility, overlap regions are kept as small as possible.

Overlap is accomplished by having two sets of instrument transformers and relays for each circuit breaker. For example, the breaker in Figure 10.24 shows two CTs, one for zone 1 and one for zone 2. Overlap is achieved by the order of the arrangement: first the equipment in the zone, second the breaker, and then the CT for that zone.

EXAMPLE 10.7

Zones of protection

Draw the protective zones for the power system shown in Figure 10.25. Which circuit breakers should open for a fault at P_1? at P_2?

SOLUTION

Noting that circuit breakers identify zone boundaries, protective zones are drawn with dashed lines as shown in Figure 10.26. For a fault at P_1, located in zone 5, breakers B24 and B42 should open. For a fault at P_2, located in the overlap region of zones 4 and 5, breakers B24, B42, B21, and B23 should open.

(Continued)

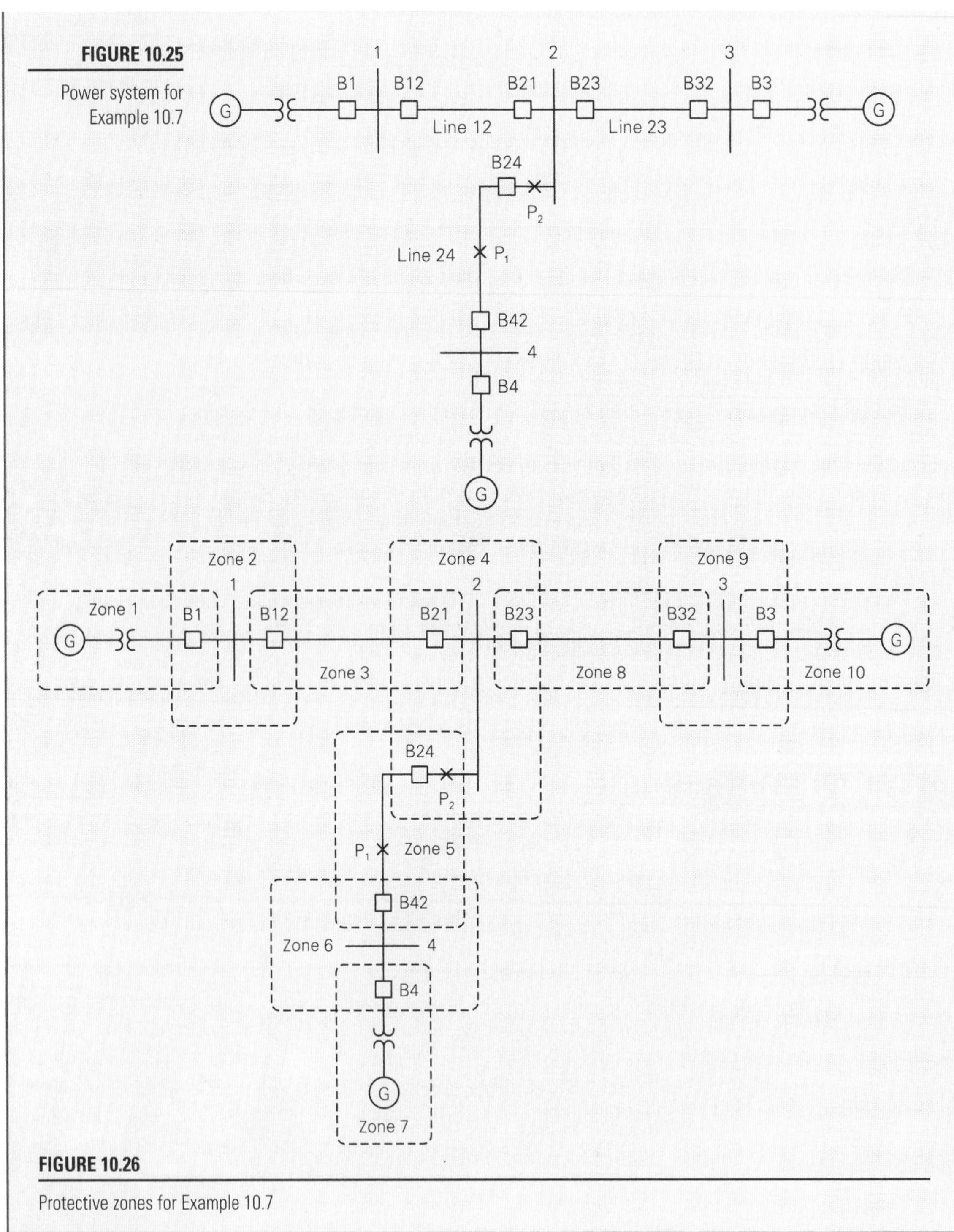

FIGURE 10.25

Power system for Example 10.7

FIGURE 10.26

Protective zones for Example 10.7

10.9 LINE PROTECTION WITH IMPEDANCE (DISTANCE) RELAYS

Coordinating time-delay overcurrent relays can also be difficult for some radial systems. If there are too many radial lines and buses, the time delay for the breaker closest to the source becomes excessive.

Also, directional overcurrent relays are difficult to coordinate in transmission loops with multiple sources. Consider the use of these relays for the transmission loop shown in Figure 10.27. For a fault at P_1, set the B21 relay to operate faster than the B32 relay. For a fault at P_2, B32 should be faster than B13. And for a fault at P_3, set B13 faster than B21. Proper coordination, which depends on the magnitudes of the fault currents, becomes a tedious process. Furthermore, when consideration is given to various lines or sources out of service, coordination becomes extremely difficult.

To overcome these problems, relays that respond to a voltage-to-current ratio can be used. Note that during a three-phase fault, current increases while bus voltages close to the fault decrease. If, for example, current increases by a factor of 5 while voltage decreases by a factor of 2, then the voltage-to-current ratio decreases by a factor of 10. That is, the voltage-to-current ratio is more sensitive to faults than current alone. A relay that operates on the basis of voltage-to-current ratio is called an *impedance* relay. It is also called a *distance* relay or a *ratio* relay.

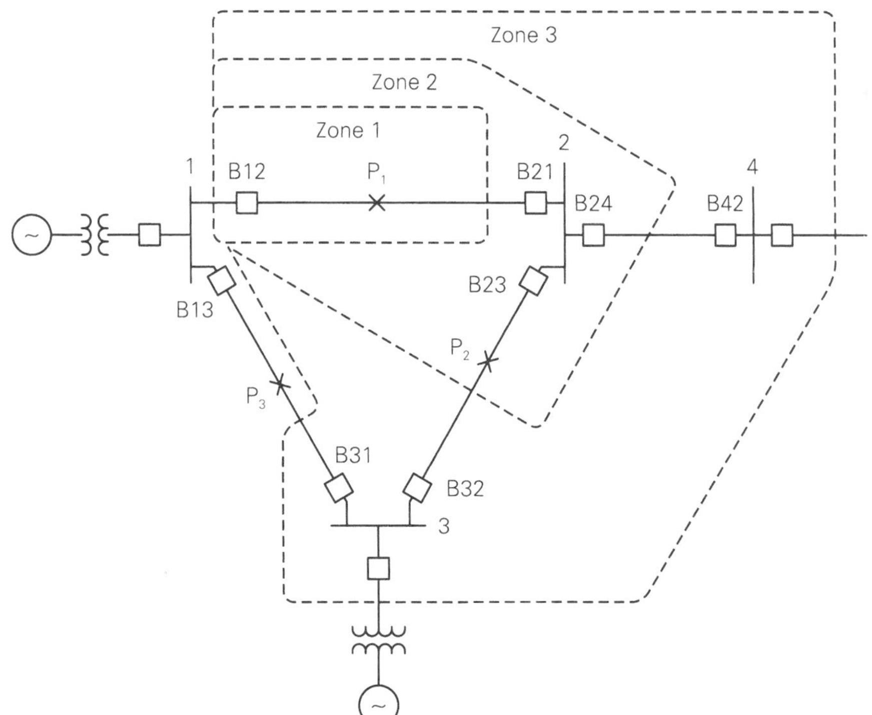

FIGURE 10.27 345-kV transmission loop

Impedance relay block and trip regions are shown in Figure 10.28, where the impedance Z is defined as the voltage-to-current ratio at the relay location. The relay trips for $|Z| < |Z_r|$, where Z_r is an adjustable relay setting. The impedance circle that defines the border between the block and trip regions passes through Z_r.

A straight line called the *line impedance locus* is shown for the impedance relay in Figure 10.28. This locus is a plot of positive sequence line impedances, predominantly reactive, as viewed between the relay location and various points along the line. The relay setting Z_r is a point in the R-X plane through which the impedance circle that defines the trip-block boundary must pass.

Consider an impedance relay for breaker B12 in Figure 10.27, for which $Z = V_1/I_{12}$. During normal operation, load currents are usually much smaller than fault currents, and the ratio Z has a large magnitude (and some arbitrary phase angle). Therefore, Z will lie outside the circle of Figure 10.28, and the relay will not trip during normal operation.

During a three-phase fault at P_1, however, Z appears to relay B12 to be the line impedance from the B12 relay to the fault. If $|Z_r|$ in Figure 10.28 is set to be larger than the magnitude of this impedance, then the B12 relay will trip. Also, during a three-phase fault at P_3, Z appears to relay B12 to be the negative of the line impedance from the relay to the fault. If $|Z_r|$ is larger than the magnitude of this impedance, the B12 relay will trip. Thus, the impedance relay of Figure 10.28 is not directional; a fault to the left or right of the relay can cause a trip.

Two ways to include directional capability with an impedance relay are shown in Figure 10.29. In Figure 10.29(a), an impedance relay with directional restraint is obtained by including a directional relay in series with an impedance relay, just as was done previously with an overcurrent relay. In Figure 10.29(b), a modified impedance relay is obtained by offsetting the center of the impedance circle from the origin.

FIGURE 10.28

Impedance relay block and trip regions

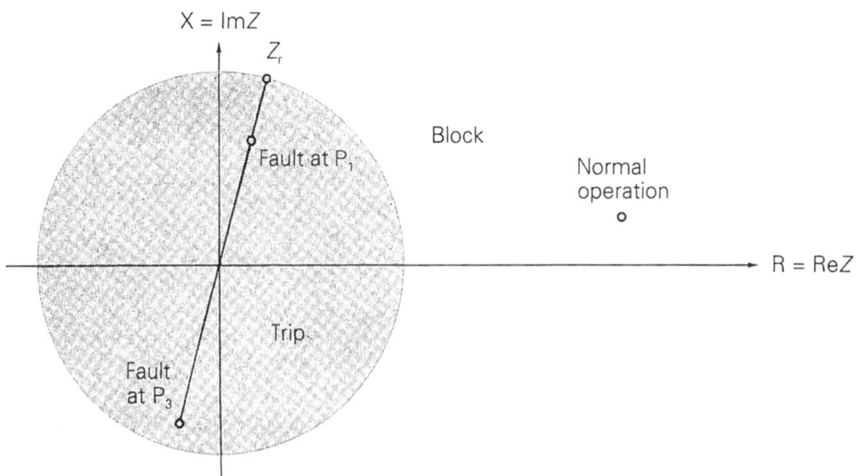

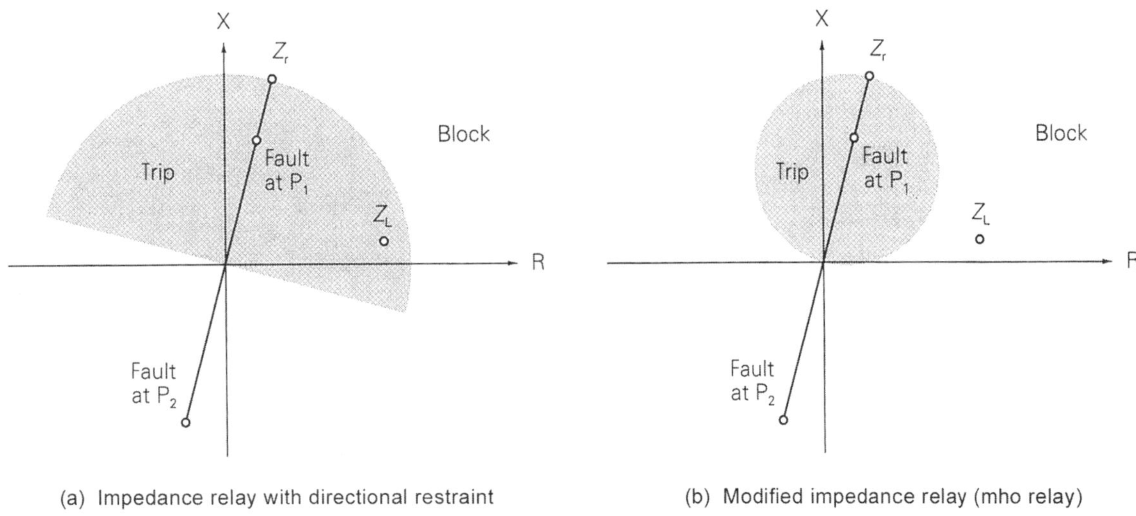

FIGURE 10.29

Impedance relays with directional capability

This modified impedance relay is sometimes called an *mho* relay. If either of these relays is used at B12 in Figure 10.27, a fault at P_1 will result in a trip decision, but a fault at P_3 will result in a block decision.

Note that the radius of the impedance circle for the modified impedance relay is half of the corresponding radius for the impedance relay with directional restraint. The modified impedance relay has the advantage of better selectivity for high power factor loads. For example, the high power factor load Z_L lies outside the trip region of Figure 10.29(b) but inside the trip region of Figure 10.29(a).

The *reach* of an impedance relay denotes how far down the line the relay detects faults. For example, an 80% reach means that the relay will detect any (solid three-phase) fault between the relay and 80% of the line length. This explains the term *distance* relay.

It is common practice to use three directional impedance relays per phase, with increasing reaches and longer time delays. For example, Figure 10.27 shows three protection zones for B12. The zone 1 relay is typically set for an 80% reach and instantaneous operation, in order to provide primary protection for line 1–2. The zone 2 relay is set for about 120% reach, extending beyond bus 2, with a typical time delay of 0.2 to 0.3 seconds. The zone 2 relay provides backup protection for faults on line 1–2 as well as remote backup for faults on line 2–3 or 2–4 in zone 2.

Note that in the case of a fault on line 2–3 it is desirable to set the B23 relay to trip, not the B12 relay. Since the impedance seen by B12 for faults near bus 2, either on line 1–2 or line 2–3, is essentially the same, do not set the B12 zone 1 relay for 100% reach. Instead, select an 80% reach to avoid instantaneous operation of B12 for a fault on line 2–3 near bus 2. For example, if there is a fault at P_2 on line 2–3, B23 should trip instantaneously; if it fails, B12 will trip after a time delay. Other faults at or near bus 2 also cause tripping of the B12 zone 2 relay after a time delay.

Reach for the zone 3 B12 relay is typically set to extend beyond buses 3 and 4 in Figure 10.27, in order to provide remote backup for neighboring lines. As such, the zone 3 reach is set for 100% of line 1–2 plus 120% of either line 2–3 or 2–4, whichever is longer, with an even larger time delay, typically one second.

Typical block and trip regions are shown in Figure 10.30 for both types of three-zone, directional impedance relays. Relay connections for a three-zone impedance relay with directional restraint are shown in Figure 10.31.

FIGURE 10.30 Three-zone, directional impedance relay

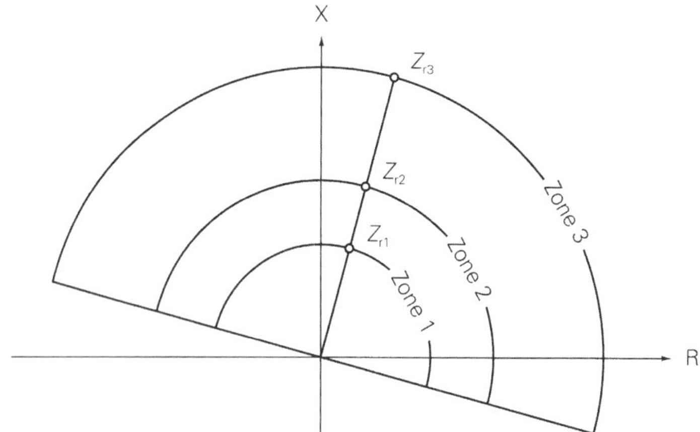

(a) Impedance relay with directional restraint

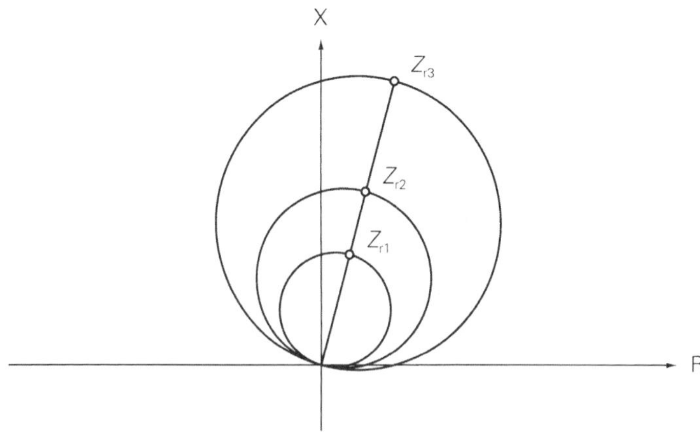

(b) Modified impedance relay (mho relay)

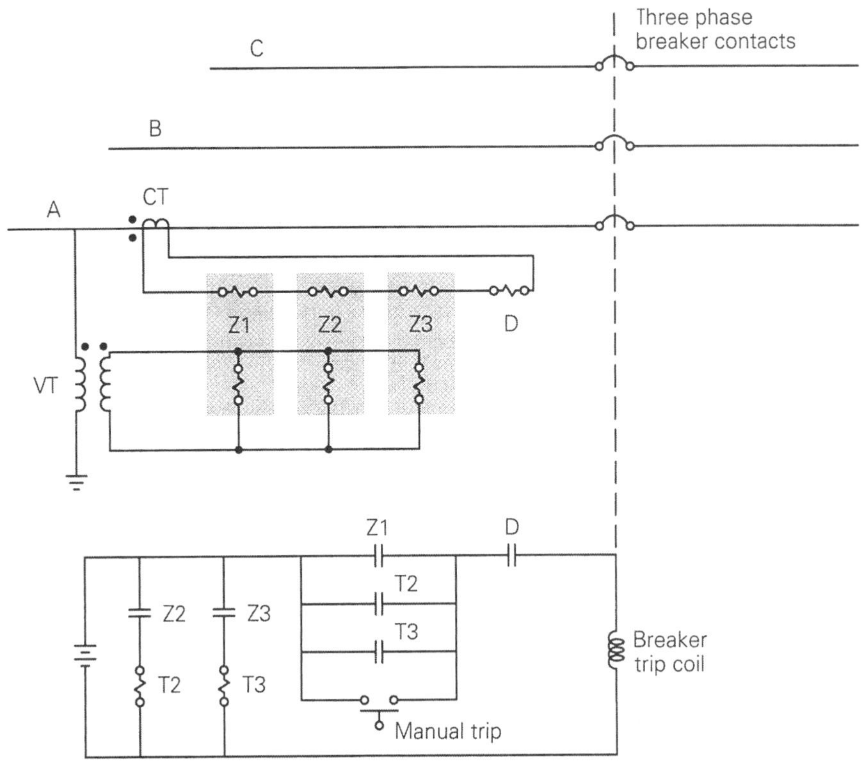

FIGURE 10.31

Relay connections for a three-zone directional impedance relay (only phase A is shown)

EXAMPLE 10.8

Three-zone impedance relay settings

Table 10.8 gives positive-sequence line impedances as well as CT and VT ratios at B12 for the 345-kV system shown in Figure 10.27. (a) Determine the settings Z_{r1}, Z_{r2}, and Z_{r3} for the B12 three-zone, directional impedance relays connected as shown in Figure 10.31. Consider only solid, three-phase faults.

(b) Maximum current for line 1–2 during emergency loading conditions is 1500 A at a power factor of 0.95 lagging. Verify that B12 does not trip during normal and emergency loadings.

(Continued)

Line	Positive-Sequence Impedance Ω	
1–2	8 +J50	
2–3	8 +J50	
2–4	5.3+J33	
1–3	4.3+J27	
Breaker	**CT Ratio**	**VT Ratio**
B12	1500:5	3000:1

TABLE 10.8
Data for Example 10.8

SOLUTION

a. Denoting V_{LN} as the line-to-neutral voltage at bus 1 and I_L as the line current through B12, the primary impedance Z viewed at B12 is

$$Z = \frac{V_{LN}}{I_L}\ \Omega$$

Using the CT and VT ratios given in Table 10.8, the secondary impedance viewed by the B12 impedance relays is

$$Z' = \frac{V_{LN}\Big/\left(\frac{3000}{1}\right)}{I_L\Big/\left(\frac{1500}{5}\right)} = \frac{Z}{10}$$

Setting the B12 zone 1 relay for 80% reach, that is, 80% of the line 1–2 (secondary) impedance:

$$Z_{r1} = 0.80(8 + j50)/10 = 0.64 + j4 = 4.05\underline{/80.9^\circ}\ \Omega \text{ secondary}$$

Setting the B12 zone 2 relay for 120% reach:

$$Z_{r2} = 1.2(8 + j50)/10 = 0.96 + j6 = 6.08\underline{/80.9^\circ}\ \Omega \text{ secondary}$$

From Table 10.8, line 2–4 has a larger impedance than line 2–3. Therefore, set the B12 zone 3 relay for 100% reach of line 1–2 plus 120% reach of line 2–4.

$$Z_{r3} = 1.0(8 + j50)/10 + 1.2(5.3 + j33)/10$$
$$= 1.44 + j8.96 = 9.07\underline{/80.9^\circ}\ \Omega \text{ secondary}$$

b. The secondary impedance viewed by B12 during emergency loading, using $V_{LN} = 345/\sqrt{3}\underline{/0^\circ} = 199.2\underline{/0^\circ}$ kV and $I_L = 1500\underline{/-\cos^{-1}(0.95)} = 1500\underline{/-18.19^\circ}$ A, is

$$Z' = Z/10 = \left(\frac{199.2 \times 10^3}{1500\underline{/-18.19^\circ}}\right)\Big/10 = 13.28\underline{/18.19^\circ}\ \Omega \quad \text{secondary}$$

Since this impedance exceeds the zone 3 setting of $9.07\underline{/80.9^\circ}\ \Omega$, the impedance during emergency loading lies outside the trip regions of the three-zone, directional impedance relay. Also, lower line loadings during normal operation will result in even larger impedances farther away from the trip regions. B12 will trip during faults but not during normal and emergency loadings.

Remote backup protection of adjacent lines using zone 3 of an impedance relay may be ineffective. In practice, buses have multiple lines of different lengths with sources at their remote ends. Contributions to fault currents from the multiple lines may cause the zone 3 relay to underreach. This "infeed effect" is illustrated in Problem 10.21.

The impedance relays considered so far use line-to-neutral voltages and line currents and are called *ground fault relays*. They respond to three-phase, single line-to-ground, and double line-to-ground faults very effectively. The impedance seen by the relay during unbalanced faults will generally not be the same as seen during three-phase faults and will not be truly proportional to the distance to the fault location. However, the relay can be accurately set for any fault location after computing impedance to the fault using fault currents and voltages. For other fault locations farther away (or closer), the impedance to the fault will increase (or decrease).

Ground fault relays are relatively insensitive to line-to-line faults. Impedance relays that use line-to-line voltages V_{ab}, V_{bc}, V_{ca} and line-current differences $I_a - I_b$, $I_b - I_c$, $I_c - I_a$ are called *phase relays*. Phase relays respond effectively to line-to-line faults and double line-to-ground faults but are relatively insensitive to single line-to-ground faults. Therefore, both phase and ground fault relays need to be used.

10.10 DIFFERENTIAL RELAYS

Differential relays are commonly used to protect generators, buses, and transformers. Figure 10.32 illustrates the basic method of differential relaying for generator protection. The protection of only one phase is shown. The method is repeated for the other two phases. When the relay in any one phase operates, all three phases of the main circuit breaker will open, as well as the generator neutral and field breakers (not shown).

For the case of no internal fault within the generator windings, $I_1 = I_2$, and, assuming identical CTs, $I_1' = I_2'$. For this case the current in the relay operating coil is zero, and the relay does not operate. On the other hand, for an internal fault such as a phase-to-ground or phase-to-phase short within the generator winding, $I_1 \neq I_2$,

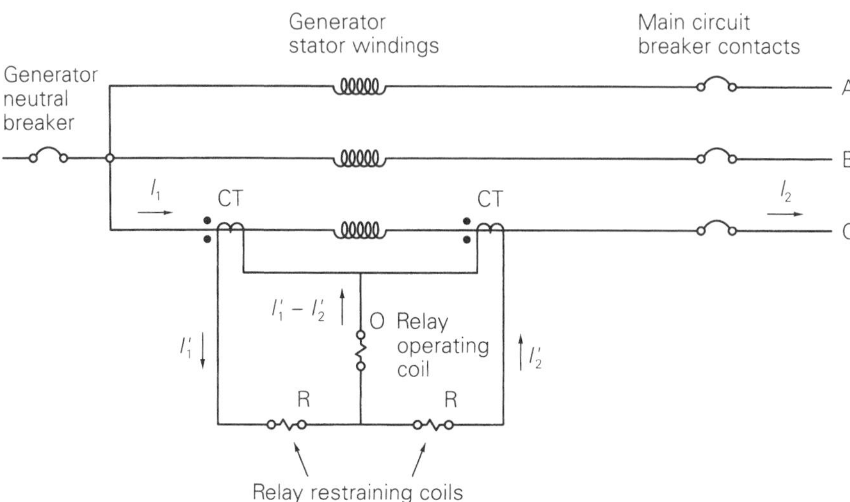

FIGURE 10.32 Differential relaying for generator protection (protection for one phase shown)

and $I'_1 \neq I'_2$. Therefore, a difference current $I'_1 - I'_2$ flows in the relay operating coil, which may cause the relay to operate. Since this relay operation depends on a *difference* current, it is called a *differential* relay.

An electromechanical differential relay called a *balance beam* relay is shown in Figure 10.33. The relay contacts close if the downward force on the right side exceeds the downward force on the left side. The electromagnetic force on the right, operating coil is proportional to the square of the operating coil mmf—that is, to $[N_0(I'_1 - I'_2)]^2$. Similarly, the electromagnetic force on the left, restraining coil is proportional to $[N_r(I'_1 + I'_2)/2]^2$. The condition for relay operation is then

$$[N_0(I'_1 - I'_2)]^2 > [N_r(I'_1 + I'_2)/2]^2 \tag{10.10.1}$$

Taking the square root:

$$|I'_1 - I'_2| > k|(I'_1 + I'_2)/2| \tag{10.10.2}$$

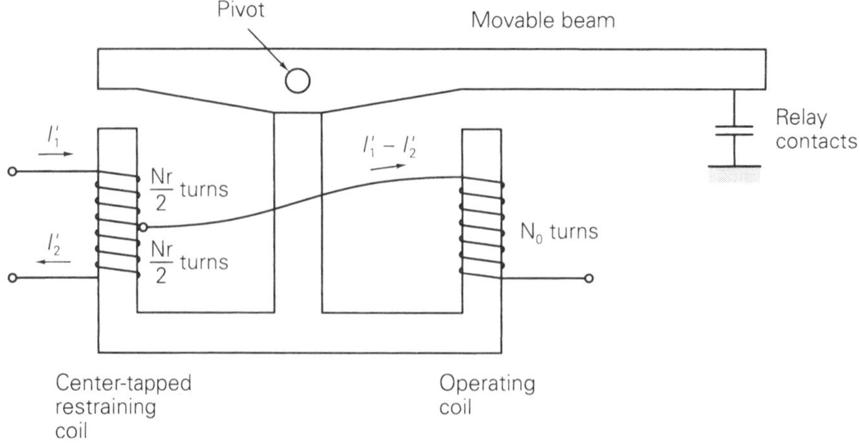

FIGURE 10.33 Balance beam differential relay

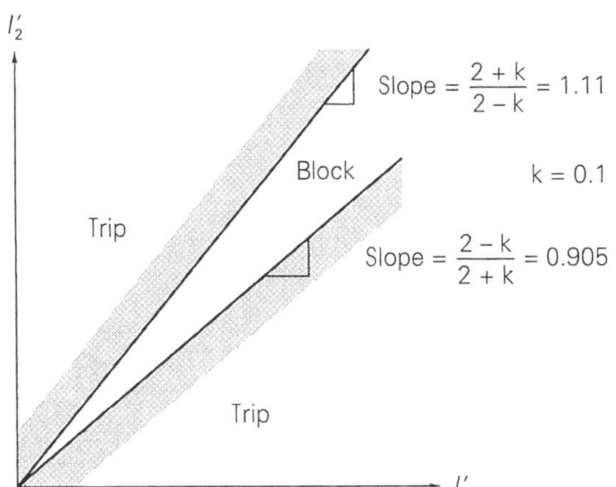

FIGURE 10.34

Differential relay block and trip regions

where

$$k = N_r/N_0 \tag{10.10.3}$$

Assuming I'_1 and I'_2 are in phase, (10.10.2) is solved to obtain

$$\begin{aligned} I'_2 &> \frac{2+k}{2-k} I'_1 \quad \text{for } I'_2 > I'_1 \\ I'_2 &< \frac{2-k}{2+k} I'_1 \quad \text{for } I'_2 < I'_1 \end{aligned} \tag{10.10.4}$$

Equation (10.10.4) is plotted in Figure 10.34 to obtain the block and trip regions of the differential relay for k = 0.1. Note that as k increases, the block region becomes larger; that is, the relay becomes less sensitive. In practice, no two CTs are identical, and the differential relay current $I'_1 - I'_2$ can become appreciable during external faults, even though $I_1 = I_2$. The balanced beam relay solves this problem without sacrificing sensitivity during normal currents, since the block region increases as the currents increase, as shown in Figure 10.34. Also, the relay can be easily modified to enlarge the block region for very small currents near the origin, in order to avoid false trips at low currents.

Note that differential relaying provides primary zone protection without backup. Coordination with protection in adjacent zones is eliminated, which permits high-speed tripping. Precise relay settings are unnecessary. Also, the need to calculate system fault currents and voltages is avoided.

10.11 BUS PROTECTION WITH DIFFERENTIAL RELAYS

Differential bus protection is illustrated by the single-line diagram of Figure 10.35. In practice, three differential relays are required, one for each phase. Operation of any one relay would cause all of the three-phase circuit breakers connected to the bus to open, thereby isolating the three-phase bus from service.

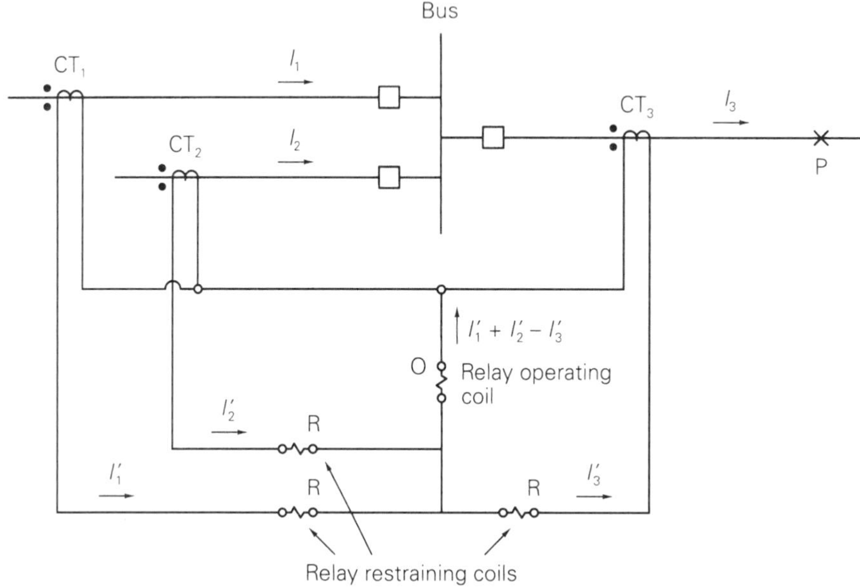

FIGURE 10.35 Single-line diagram of differential bus protection

For the case of no internal fault between the CTs—that is, no bus fault—$I_1 + I_2 = I_3$. Assuming identical CTs, the differential relay current $I_1' + I_2' - I_3'$ equals zero, and the relay does not operate. However, if there is a bus fault, the differential current $I_1' + I_2' - I_3'$, which is not zero, flows in the operating coil to operate the relay. Use of the restraining coils overcomes the problem of nonidentical CTs.

A problem with differential bus protection can result from different levels of fault currents and varying amounts of CT saturation. For example, consider an external fault at point P in Figure 10.35. Each of the CT_1 and CT_2 primaries carries part of the fault current, but the CT_3 primary carries the sum $I_3 = I_1 + I_2$. CT_3, energized at a higher level, will have more saturation, such that $I_3' \neq I_1' + I_2'$. If the saturation is too high, the differential current in the relay operating coil could result in a false trip. This problem becomes more difficult when there are large numbers of circuits connected to the bus. Various schemes have been developed to overcome this problem [1].

10.12 TRANSFORMER PROTECTION WITH DIFFERENTIAL RELAYS

The protection method used for power transformers depends on the transformer MVA rating. Fuses are often used to protect transformers with small MVA ratings, whereas differential relays are commonly used to protect transformers with ratings larger than 10 MVA.

The differential protection method is illustrated in Figure 10.36 for a single-phase, two-winding transformer. Denoting the turns ratio of the primary and secondary CTs by $1/n_1$ and $1/n_2$, respectively (a CT with 1 primary turn and n secondary turns has a turns ratio a = 1/n), the CT secondary currents are

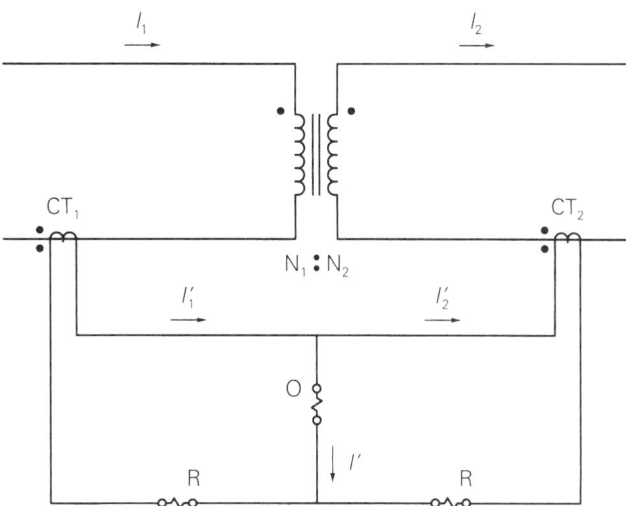

FIGURE 10.36

Differential protection of a single-phase, two-winding transformer

$$I'_1 = \frac{I_1}{\mathrm{n}_1} \qquad I'_2 = \frac{I_2}{\mathrm{n}_2} \tag{10.12.1}$$

and the current in the relay operating coil is

$$I' = I'_1 - I'_2 = \frac{I_1}{\mathrm{n}_1} - \frac{I_2}{\mathrm{n}_2} \tag{10.12.2}$$

For the case of no fault between the CTs—that is, no internal transformer fault—the primary and secondary currents for an ideal transformer are related by

$$I_2 = \frac{\mathrm{N}_1 I_1}{\mathrm{N}_2} \tag{10.12.3}$$

Using (10.12.3) in (10.12.2),

$$I' = \frac{I_1}{\mathrm{n}_1}\left(1 - \frac{\mathrm{N}_1/\mathrm{N}_2}{\mathrm{n}_2/\mathrm{n}_1}\right) \tag{10.12.4}$$

To prevent the relay from tripping for the case of no internal transformer fault, where (10.12.3) and (10.12.4) are satisfied, the differential relay current I′ must be zero. Therefore, from (10.12.4), select

$$\frac{\mathrm{n}_2}{\mathrm{n}_1} = \frac{\mathrm{N}_1}{\mathrm{N}_2} \tag{10.12.5}$$

If an internal transformer fault between the CTs does occur, (10.12.3) is not satisfied and the differential relay current $I' = I'_1 - I'_2$ is not zero. The relay will trip if the operating condition given by (10.10.4) is satisfied. Also, the value of k in (10.10.4) can be selected to control the size of the block region shown in Figure 10.34, thereby controlling relay sensitivity.

EXAMPLE 10.9

Differential relay protection for a single-phase transformer

A single-phase two-winding, 10-MVA, 80 kV/20 kV transformer has differential relay protection. Select suitable CT ratios. Also, select k such that the relay blocks for up to 25% mismatch between I'_1 and I'_2.

SOLUTION

The transformer-rated primary current is

$$I_{1\text{rated}} = \frac{10 \times 10^6}{80 \times 10^3} = 125 \quad \text{A}$$

From Table 10.2, select a 150:5 primary CT ratio to give $I'_1 = 125(5/150) = 4.17$ A at rated conditions. Similarly, $I_{2\text{rated}} = 500$ A. Select a 600:5 secondary CT ratio to give $I'_2 = 500(5/600) = 4.17$ A and a differential current $I = I'_1 - I'_2 = 0$ (neglecting magnetizing current) at rated conditions. Also, for a 25% mismatch between I'_1 and I'_2, select a 1.25 upper slope in Figure 10.34. That is,

$$\frac{2 + k}{2 - k} = 1.25 \qquad k = 0.2222$$

A common problem in differential transformer protection is the mismatch of relay currents that occurs when standard CT ratios are used. If the primary winding in Example 10.9 has a 138-kV instead of 80-kV rating, then $I_{1\text{rated}} = 10 \times 10^6/138 \times 10^3 = 72.46$ A, and a 100:5 primary CT would give $I'_1 = 72.46(5/100) = 3.62$ A at rated conditions. This current does not balance $I'_2 = 4.17$ A using a 5:600 secondary CT, nor $I'_2 = 3.13$ A using a 5:800 secondary CT. The mismatch is about 15%.

One solution to this problem is to use auxiliary CTs, which provide a wide range of turns ratios. A 5:5.76 auxiliary CT connected to the 5:600 secondary CT in the above example would reduce I'_2 to $4.17(5/5.76) = 3.62$ A, which does balance I'_1. Unfortunately, auxiliary CTs add their own burden to the main CTs and also increase transformation errors. A better solution is to use tap settings on the relays themselves, which have the same effect as auxiliary CTs. Most transformer differential relays have taps that provide for differences in restraining windings in the order of 2 or 3 to 1.

When a transformer is initially energized, it can draw a large "inrush" current, a transient current that flows in the shunt magnetizing branch and decays after a few cycles to a small steady-state value. Inrush current appears as a differential current since it flows only in the primary winding. If a large inrush current does occur upon transformer energization, a differential relay will see a large differential current and trip out the transformer unless the protection method is modified to detect inrush current.

One method to prevent tripping during transformer inrush is based on the fact that inrush current is nonsinusoidal with a large second-harmonic component. A filter can be used to pass fundamental and block harmonic components of the differential current I' to the relay operating coil. Another method is based on the fact that inrush current has a large dc component, which can be used to desensitize the relay. Time-delay relays may also be used to temporarily desensitize the differential relay until the inrush current has decayed to a low value.

Figure 10.37 illustrates differential protection of a three-phase Y–Δ two-winding transformer. Note that a Y–Δ transformer produces 30° phase shifts in the line currents. The CTs must be connected to compensate for the 30° phase shifts,

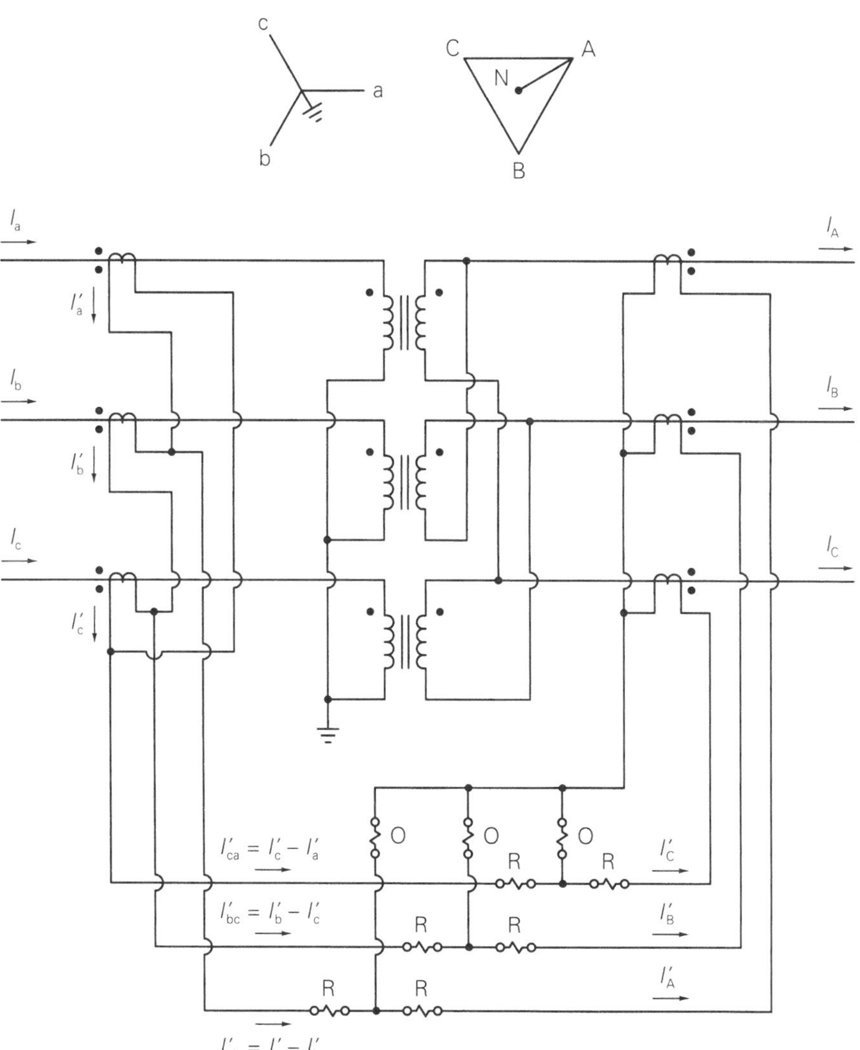

FIGURE 10.37

Differential protection of a three-phase, Y–Δ, two-winding transformer such that the CT secondary currents as seen by the relays are in phase. The correct phase-angle relationship is obtained by connecting CTs on the Y side of the transformer in Δ, and CTs on the Δ side in Y.

EXAMPLE 10.10

Differential relay protection for a three-phase transformer

A 30-MVA, 34.5 kV Y/138 kV Δ transformer is protected by differential relays with taps. Select CT ratios, CT connections, and relay tap settings. Also determine currents in the transformer and in the CTs at rated conditions. Assume that the available relay tap settings are 5:5, 5:5.5, 5:6.6, 5:7.3, 5:8, 5:9, and 5:10, giving relay tap ratios of 1.00, 1.10, 1.32, 1.46, 1.60, 1.80, and 2.00.

SOLUTION

As shown in Figure 10.37, CTs are connected in Δ on the (34.5-kV) Y side of the transformer, and CTs are connected in Y on the (138-kV) Δ side, in order to obtain the correct phasing of the relay currents.

Rated current on the 138-kV side of the transformer is

$$I_{A\ \text{rated}} = \frac{30 \times 10^6}{\sqrt{3}(138 \times 10^3)} = 125.51 \quad A$$

Select a 150:5 CT on the 138-kV side to give $I'_A = 125.51(5/150) = 4.184$ A in the 138-kV CT secondaries and in the righthand restraining windings of Figure 10.37.

Next, rated current on the 34.5-kV side of the transformer is

$$I_{a\ \text{rated}} = \frac{30 \times 10^6}{\sqrt{3}(34.5 \times 10^3)} = 502.04 \quad A$$

Select a 500:5 CT on the 34.5-kV side to give $I'_a = 502.0(5/500) = 5.02$ A in the 34.5-kV CT secondaries and $I'_{ab} = 5.02\sqrt{3} = 8.696$ A in the lefthand restraining windings of Figure 10.37.

Finally, select relay taps to balance the currents in the restraining windings. The ratio of the currents in the left- to righthand restraining windings is

$$\frac{I'_{ab}}{I'_A} = \frac{8.696}{4.184} = 2.078$$

The closest relay tap ratio is $T'_{AB}/T'_A = 2.0$, corresponding to a relay tap setting of $T'_A : T'_{ab} = 5 : 10$. The percentage mismatch for this tap setting is

$$\left| \frac{(I'_A/T'_A) - (I'_{ab}/T'_{ab})}{(I'_{ab}/T'_{ab})} \right| \times 100 = \left| \frac{(4.184/5) - (8.696/10)}{(8.696/10)} \right| \times 100 = 3.77\%$$

This is a good mismatch; since transformer differential relays typically have their block regions adjusted between 20% and 60% (by adjusting k in Figure 10.34), a 3.77% mismatch gives an ample safety margin in the event of CT and relay differences.

For three-phase transformers (Y–Y, Y–Δ, Δ–Y, Δ–Δ), the general rule is to connect CTs on the Y side in Δ and CTs on the Δ side in Y. This arrangement compensates for the 30° phase shifts in Y–Δ or Δ–Y banks. Note also that zero-sequence current cannot enter a Δ side of a transformer or the CTs on that side, and zero-sequence current on a grounded Y side cannot enter the Δ-connected CTs on that side. Therefore, this arrangement also blocks zero-sequence currents in the differential relays during external ground faults. For internal ground faults, however, the relays can operate from the positive- and negative-sequence currents involved in these faults.

Differential protection methods have been modified to handle multi-winding transformers, voltage-regulating transformers, phase-angle regulating transformers, power-rectifier transformers, transformers with special connections (such as zig-zag), and other, special-purpose transformers. Also, other types of relays such as gas-pressure detectors for liquid-filled transformers are used.

10.13 PILOT RELAYING

Pilot relaying refers to a type of differential protection that compares the quantities at the terminals via a communication channel rather than by a direct wire interconnection of the relays. Differential protection of generators, buses, and transformers considered in previous sections does not require pilot relaying because each of these devices is at one geographical location where CTs and relays can be directly interconnected. However, differential relaying of transmission lines requires pilot relaying because the terminals are widely separated (often by many kilometers). In actual practice, pilot relaying is typically applied to short transmission lines (up to 80 km) with 69 to 115 kV ratings.

Four types of communication channels are used for pilot relaying:

1. *Pilot wires*: Separate electrical circuits operating at dc, 50 to 60 Hz, or audio frequencies. These could be owned by the power company or leased from the telephone company.
2. *Power-line carrier*: The transmission line itself is used as the communication circuit, with frequencies between 30 and 300 kHz being transmitted. The communication signals are applied to all three phases using an L–C voltage divider and are confined to the line under protection by blocking filters called *line traps* at each end.

3. *Microwave*: A 2 to 12 GHz signal transmitted by line-of-sight paths between terminals using dish antennas.
4. *Fiber optic cable*: Signals transmitted by light modulation through electrically nonconducting cable. This cable eliminates problems due to electrical insulation, inductive coupling from other circuits, and atmospheric disturbances.

Two common fault detection methods are *directional comparison*, where the power flows at the line terminals are compared, and *phase comparison*, where the relative phase angles of the currents at the terminals are compared. Also, the communication channel can either be required for trip operations, which is known as a *transfer trip system*, or not be required for trip operations, known as a *blocking system*. A particular pilot-relaying method is usually identified by specifying the fault-detection method and the channel use. The four basic combinations are directional comparison blocking, directional comparison transfer trip, phase comparison blocking, and phase comparison transfer trip.

Like differential relays, pilot relays provide primary zone protection without backup. Thus, coordination with protection in adjacent zones is eliminated, resulting in high-speed tripping. Precise relay settings are unnecessary. Also, the need to calculate system fault currents and voltages is eliminated.

10.14 NUMERIC RELAYING

Previous sections covered the operating principle of relays built with electromechanical components, including the induction disc time-delay over-current relay, as in Figure 10.14; the directional relay, similar in operation to a watt-hour meter; and the balance-beam differential relay, as in Figure 10.33. These electromechanical relays, introduced in the early 1900s, have performed well over the years and continue in relatively maintenance-free operation today. Solid-state relays using analog circuits and logic gates, with block-trip regions similar to those of electromechanical relays and with newer types of block/trip regions, have been available since the late 1950s. Such relays, widely used in HV and EHV systems, offer the reliability and ruggedness of their electromechanical counterparts at a competitive price. Beyond solid-state analog relays, a new generation of numeric relays based on digital computer technology has been under development since the 1980s.

The numeric relay (or digital relay) is a protective relay that uses a microprocessor with software for purposes of fault detection. Benefits of numeric relays include the following [14, 16]:

- Compact size
- Programmable
- Low burden
- Multi-function capability
- Flexibility and Reliability
- Sensitivity and Speed
- Permits storage of fault data and disturbance records

- Self-checking diagnostics
- Adaptive relaying schemes
- Digital communications capabilities
- Costs

Numeric relays can be physically smaller and normally require less panel wiring than electromechanical or solid-state relays. Numeric relays also have the advantage that modifications to tripping characteristics, either changes in conventional settings or shaping of entirely new block/trip regions, can be made by updating the software from a remote computer terminal. For example, the relay engineer can reprogram tripping characteristics of field-installed, in-service relays without leaving the engineering office. Numeric relays have minimum burdens on instrument transformers. Numeric relays offer multiple functions and multiple characteristics with a wide range of operation, compared to electromechanical and solid-state relays with single functions and narrower ranges of operation. For numeric relays, software can be modified with the same hardware to achieve a flexible variety of protection functions, and significant improvement in reliability is obtained through use of fewer components and fewer interconnections.

Numeric relays offer greater sensitivity with high pickup ratio, and tripping times (not including circuit breaker opening times) of $\frac{1}{2}$ cycle or less. Numeric relays can assist in post-fault analysis by recording fault data including the type of fault, duration and magnitude of fault, circuit breaker problem, and CT saturation. Numeric relays have the capability to check whether return to a normal condition has been attained and then automatically reset. Numeric relay settings can be changed instantaneously to adapt to varying power-system requirements. Numeric relays offer secure communications capabilities for interface with smart-grid controls, supervisory control and data acquisition systems. By combining several functions that share hardware in one numeric relay, capital and maintenance costs are reduced compared to electromechanical and solid-state relays.

Numeric relays do have limitations. With digital communications, numeric relays are exposed to the potential risk of hacking, and therefore should be properly managed for cybersecurity. In North America, numeric relays should comply with NERC security requirements. Relay engineers should be aware of common failure modes that can affect multiple elements of protection. For example, the failure of a power supply or an input signal processor could disable a numeric relay that provides many different protection functions. Numeric relays may be exposed to external power-system transients that would not normally affect electromechanical or solid-state relays. And while numeric relays offer multiple functions as well as greater flexibility, sensitivity, and speed, that does not necessarily translate into better protection [16].

An important feature of power system protection is the decentralized, local nature of relays. Except for pilot relaying, each relay receives information from nearby local CTs and VTs and trips only local breakers. Interest in numeric relaying is not directed at replacing local relays by a central computer. Instead, local electromechanical or solid-state relays would be replaced by dedicated, local numeric relays with similar operating principles, such as time-delay overcurrent, impedance, or differential relaying. The central computer would interact with local digital relays in a supervisory role.

PROBLEMS

SECTION 10.2

10.1 The primary conductor in Figure 10.2 is one phase of a three-phase transmission line operating at 345 kV, 700 MVA, 0.95 power factor lagging. The CT ratio is 1200:5, and the VT ratio is 3000:1. Determine the CT secondary current I' and the VT secondary voltage V'. Assume zero CT error.

10.2 A CO-8 relay with a current tap setting of 5 amperes is used with the 100:5 CT in Example 10.1. The CT secondary current I′ is the input to the relay operating coil. The CO-8 relay burden is shown in the following table for various relay input currents.

CO-8 relay input current I′, A	5	8	10	13	15
CO-8 relay burden Z_B, Ω	0.5	0.8	1.0	1.3	1.5

Primary current and CT error are computed in Example 10.1 for the 5-, 8-, and 15-A relay input currents. Compute the primary current and CT error for (a) $I' = 10$ A and $Z_B = 1.0\ \Omega$, and for (b) $I' = 13$ A and $Z_B = 1.3\ \Omega$. (c) Plot I′ versus I for the above five values of I′. (d) For reliable relay operation, the fault-to-pickup current ratio with minimum fault current should be greater than two. Determine the minimum fault current for application of this CT and relay with 5-A tap setting.

10.3 An overcurrent relay set to operate at 10 A is connected to the CT in Figure 10.8 with a 500:5 CT ratio. Determine the minimum primary fault current that the relay will detect if the burden Z_B is (a) 1.0 Ω, (b) 4.0 Ω, and (c) 5.0 Ω.

10.4 Given the open-delta VT connection shown in Figure 10.38, both VTs having a voltage rating of 240 kV : 120 V, the voltages are specified as $V_{AB} = 230\underline{/0^\circ}$, $V_{BC} = 230\underline{/-120^\circ}$, and $V_{CA} = 230\underline{/120^\circ}$ kV. Determine V_{ab}, V_{bc}, and V_{ca} for the following cases: (a) The dots are shown in Figure 10.38. (b) The dot near c is moved to b in Figure 10.38.

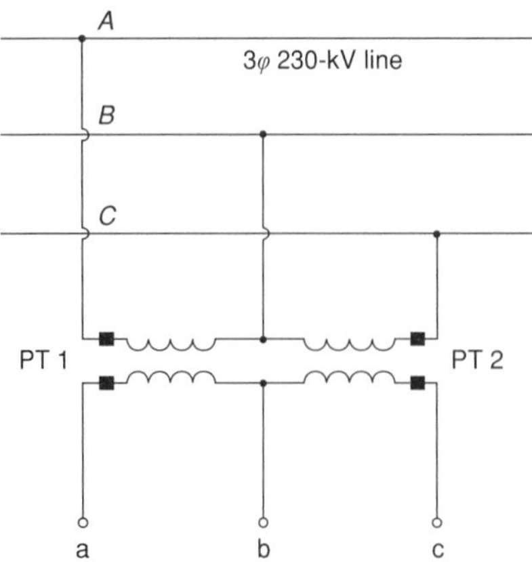

FIGURE 10.38 Problem 10.4

10.5 A CT with an excitation curve given in Figure 10.39 has a rated current ratio of 500:5 A and a secondary leakage impedance of $0.1 + j0.5\ \Omega$. Calculate the CT secondary output current and the CT error for the following cases: (a) The impedance of the terminating device is $4.9 + j0.5\ \Omega$ and the primary CT load current is 400 A. (b) The impedance of the terminating device is $4.9 + j0.5\ \Omega$ and the primary CT fault current is 1200 A. (c) The impedance of the terminating device is $14.9 + j1.5\ \Omega$ and the primary CT load current is 400 A. (d) The impedance of the terminating device is $14.9 + j1.5\ \Omega$ and the primary CT fault current is 1200 A.

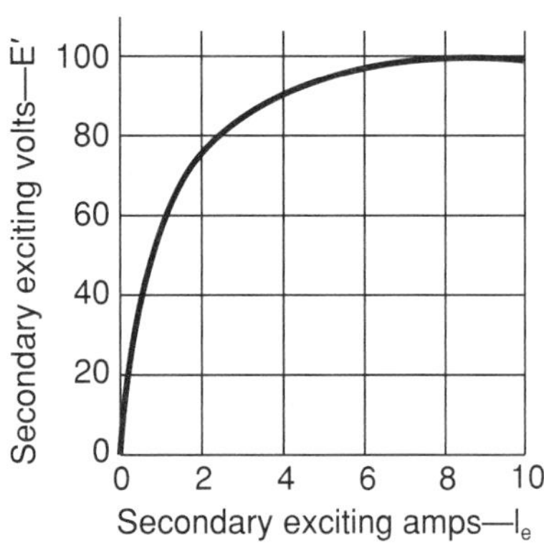

FIGURE 10.39

Problem 10.5

10.6 The CT of Problem 10.5 is utilized in conjunction with a current-sensitive device that will operate at current levels of 8 A or above. Check whether the device will detect the 1300-A fault current for cases (b) and (d) in Problem 10.5.

SECTION 10.3

10.7 The input current to a CO-8 relay is 10 A. Determine the relay operating time for the following current tap settings (TS) and time dial settings (TDS): (a) TS = 1.0, TDS = 1/2; (b) TS = 2.0, TDS = 1.5; (c) TS = 2.0, TDS = 7; (d) TS = 3.0, TDS = 7; and (e) TS = 12.0, TDS = 1.

10.8 The relay in Problem 10.2 has a time-dial setting of 4. Determine the relay operating time if the primary fault current is 400 A.

10.9 An RC circuit used to produce time delay is shown in Figure 10.40. For a step input voltage $v_i(t) = 2u(t)$ and $C = 10\ \mu F$, determine T_{delay} for the following cases: (a) $R = 100\ k\Omega$; and (b) $R = 1\ M\Omega$. Sketch the output $v_o(t)$ versus time for cases (a) and (b).

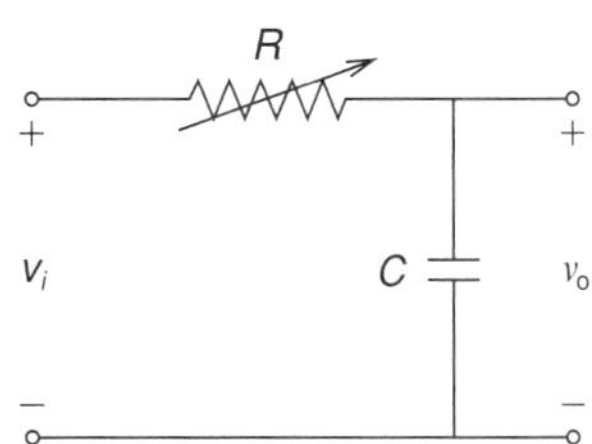

FIGURE 10.40

Problem 10.9

10.10 Reconsider case (b) of Problem 10.5. Let the load impedance $4.9 + j0.5\ \Omega$ be the input impedance to a CO-7 induction disc time-delay overcurrent relay. The CO-7 relay characteristic is shown in Figure 10.41. For a tap setting of 4 A and a time dial setting of 2, determine the relay operating time.

FIGURE 10.41

Problems 10.10 and 10.14

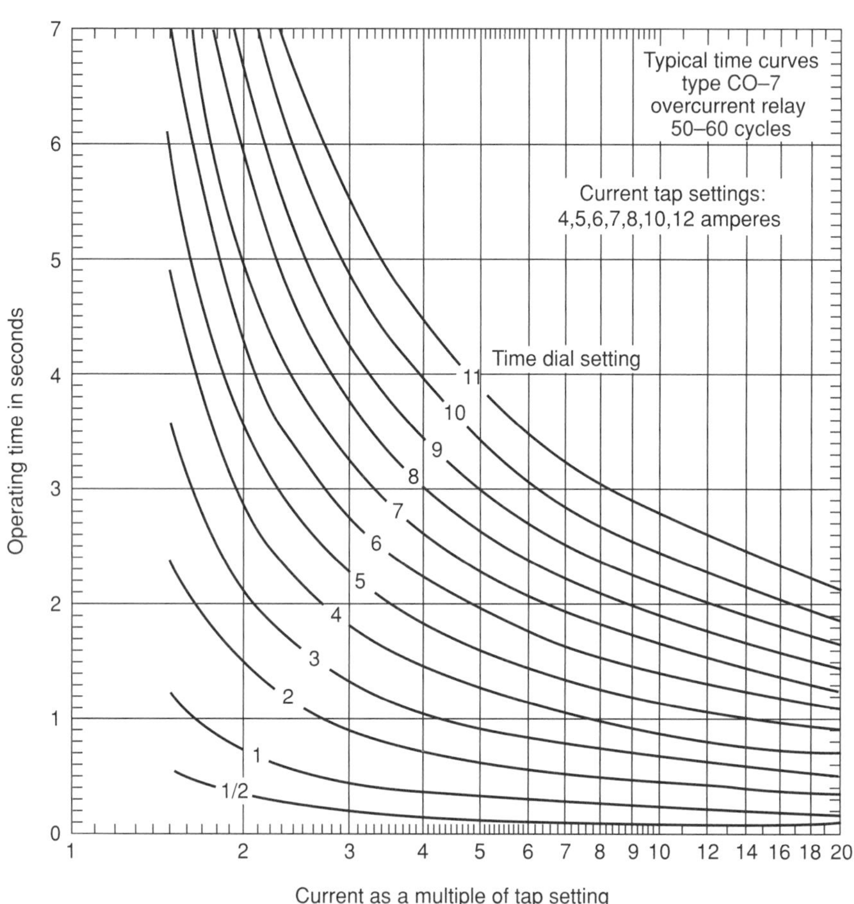

SECTION 10.4

10.11 Evaluate relay coordination for the minimum fault currents in Example 10.4. For the selected current tap settings and time dial settings, (a) determine the operating time of relays at B2 and B3 for the 700-A fault current. (b) Determine the operating time of relays at Bl and B2 for the 1500-A fault current. Are the fault-to-pickup current ratios $\geqslant 2.0$ (a requirement for reliable relay operation) in all cases? Are the coordination time intervals $\geqslant 0.3$ seconds in all cases?

10.12 Repeat Example 10.4 for the following system data. Coordinate the relays for the maximum fault currents.

	Maximum Load		Symmetrical Fault Current	
Bus	MVA	Lagging p.f.	Maximum A	Minimum A
1	9.0	0.95	5000	3750
2	9.0	0.95	3000	2250
3	9.0	0.95	2000	1500

Breaker	Breaker Operating Time	CT Ratio	Relay
B1	5 cycles	600:5	CO-8
B2	5 cycles	400:5	CO-8
B3	5 cycles	200:5	CO-8

10.13 Using the current tap settings and time dial settings that you have selected in Problem 10.12, evaluate relay coordination for the minimum fault currents. Are the fault-to-pickup current ratios $\geqslant 2.0$, and are the coordination time delays $\geqslant 0.3$ seconds in all cases?

10.14 An 11-kV radial system is shown in Figure 10.42. Assuming a CO-7 relay with relay characteristic given in Figure 10.41 and the same power factor for all loads, select relay settings to protect the system.

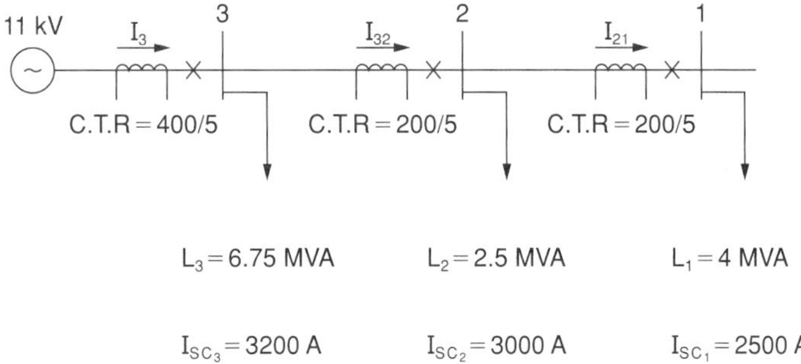

FIGURE 10.42

Problem 10.14

SECTION 10.5

10.15 Rework Example 10.5 for the following faults: (a) a three-phase, permanent fault on the load side of tap 3; (b) a single line-to-ground, permanent fault at bus 4 on the load side of the recloser; and (c) a three-phase, permanent fault at bus 4 on the source side of the recloser.

10.16 A three-phase 34.5-kV feeder supplying a 3.5-MVA load is protected by 80E power fuses in each phase, in series with a recloser. The time-current characteristic of the 80E fuse is shown in Figure 10.43. Analysis yields maximum and minimum fault currents of 1000 and 500 A, respectively, (a) To have the recloser clear the fault, find the maximum clearing time necessary for recloser operation. (b) To have the fuses clear the fault, find the minimum recloser clearing time. Assume that the recloser operating time is independent of fault current magnitude.

FIGURE 10.43

Problem 10.16

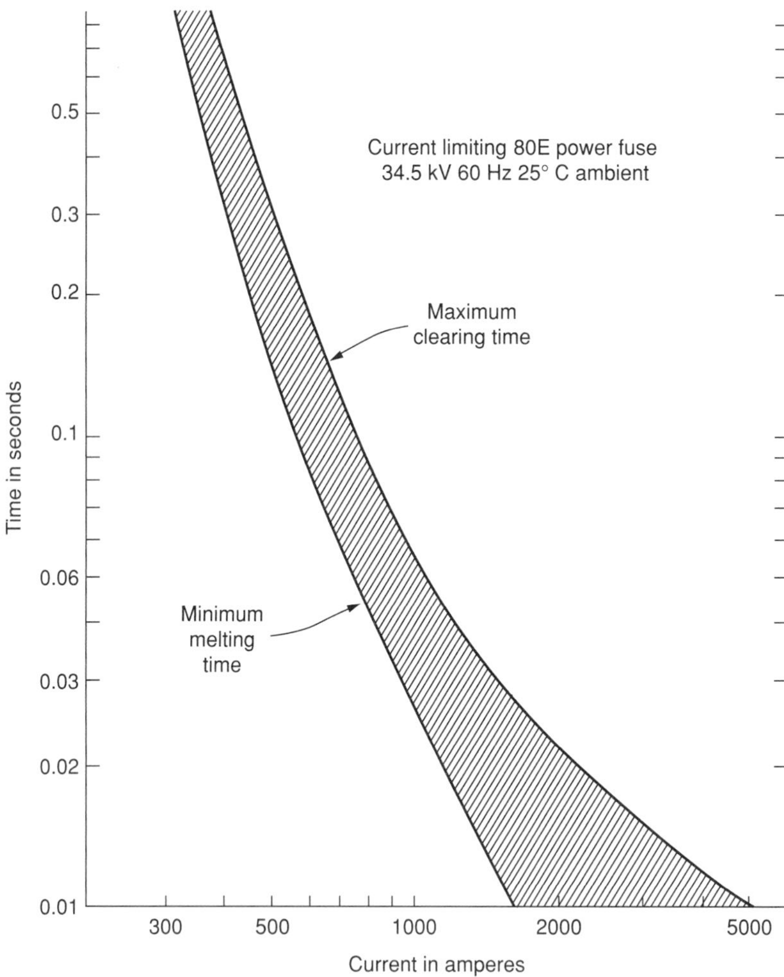

SECTION 10.7

10.17 For the system shown in Figure 10.44, directional overcurrent relays are used at breakers B12, B21, B23, B32, B34, and B43. Overcurrent relays alone are used at B1 and B4. (a) For a fault at P_1, which breakers do not operate? Which breakers should be coordinated? Repeat (a) for a fault at (b) P_2, (c) P_3. (d) Explain how the system is protected against bus faults.

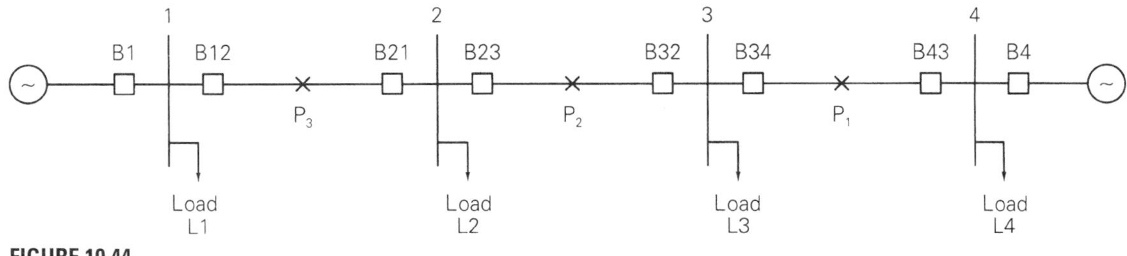

FIGURE 10.44

Problem 10.17

SECTION 10.8

10.18 Draw the protective zones for the power system shown in Figure 10.45. Which circuit breakers should open for a fault at (a) P_1, (b) P_2, and (c) P_3?

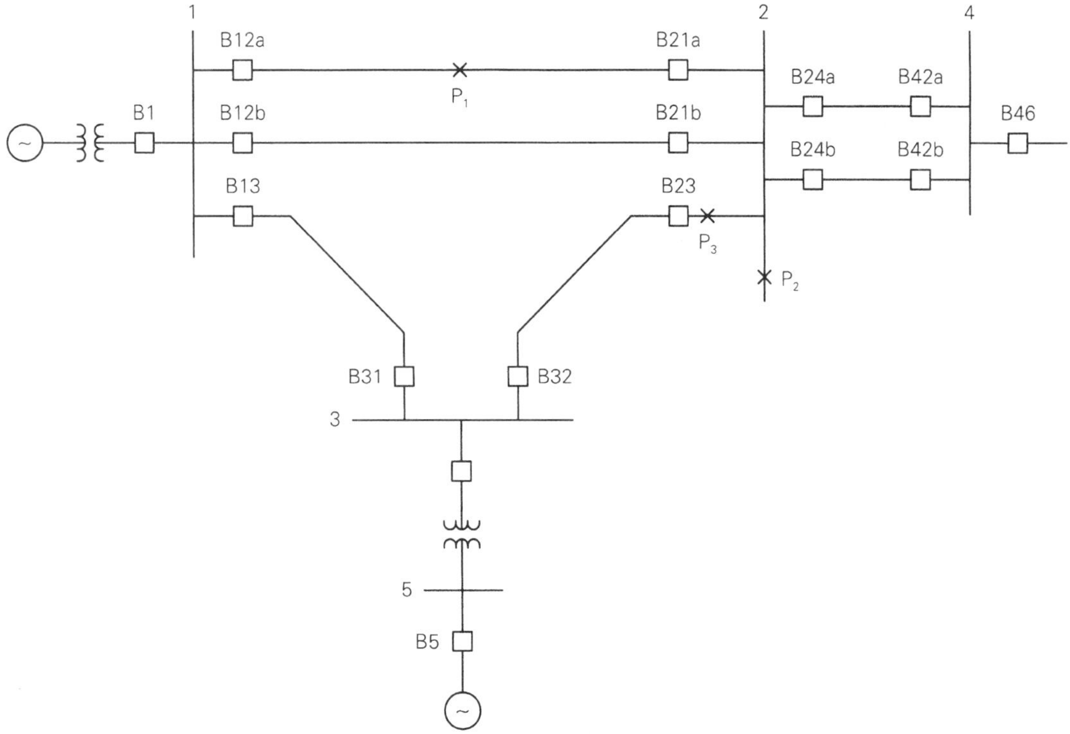

FIGURE 10.45

Problem 10.18

10.19 Figure 10.46 shows three typical bus arrangements. Although the number of lines connected to each arrangement varies widely in practice, four lines are shown for convenience and comparison. Note that the required number of circuit breakers per line is 1 for the ring bus, $1\frac{1}{2}$ for the breaker-and-a-half double-bus, and 2 for the double-breaker double-bus arrangement. For each arrangement: (a) Draw the protective zones. (b) Identify the breakers that open under primary protection for a fault on line 1. (c) Identify the lines that are removed from service under primary protection during a bus fault at P_1. (d) Identify the breakers that open under backup protection in the event a breaker fails to clear a fault on line 1 (that is, a stuck breaker during a fault on line 1).

FIGURE 10.46

Problem 10.19—typical bus arrangements

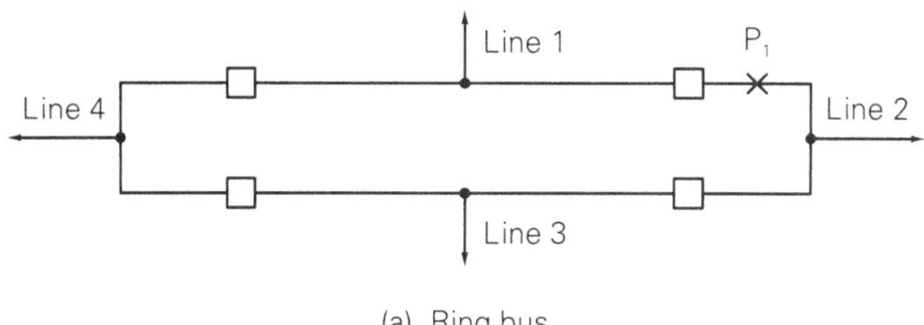

(a) Ring bus

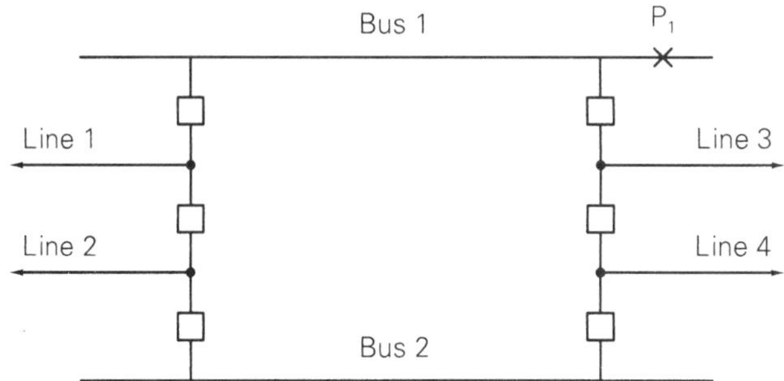

(b) Breaker-and-a-half double bus

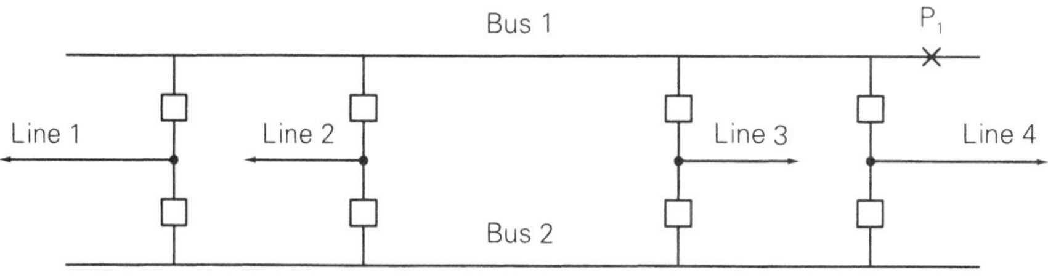

(c) Double-breaker double bus

SECTION 10.9

10.20 Three-zone mho relays are used for transmission line protection of the power system shown in Figure 10.25. Positive-sequence line impedances are given as follows.

Line	Positive-Sequence Impedance, Ω
1–2	$6 + j60$
2–3	$4 + j40$
2–4	$5 + j50$

Rated voltage for the high-voltage buses is 500 kV. Assume a 2500:5 CT ratio and a 4500:1 VT ratio at B12. (a) Determine the settings Z_{r1}, Z_{r2}, and Z_{r3} for the mho relay at B12. (b) Maximum current for line 1–2 under emergency loading conditions is 1400 A at 0.90 power factor lagging. Verify that B12 does not trip during emergency loading conditions.

10.21 Line impedances for the power system shown in Figure 10.47 are $Z_{12} = Z_{23} = 3.0 + j40.0\ \Omega$, and $Z_{24} = 6.0 + j80.0\ \Omega$. Reach for the zone 3 B12 impedance relays is set for 100% of line 1–2 plus 120% of line 2–4. (a) For a bolted three-phase fault at bus 4, show that the apparent primary impedance "seen" by the B12 relays is

$$Z_{\text{apparent}} = Z_{12} + Z_{24} + (I_{32}/I_{12})Z_{24}$$

where(I_{32}/I_{12}) is the line 2–3 to line 1–2 fault current ratio. (b) If $|I_{32}|I_{12}| > 0.20$, does the B12 relay see the fault at bus 4?

Note: This problem illustrates the "infeed effect." Fault currents from line 2–3 can cause the zone 3 B12 relay to underreach. As such, remote backup of line 2–4 at B12 is ineffective.

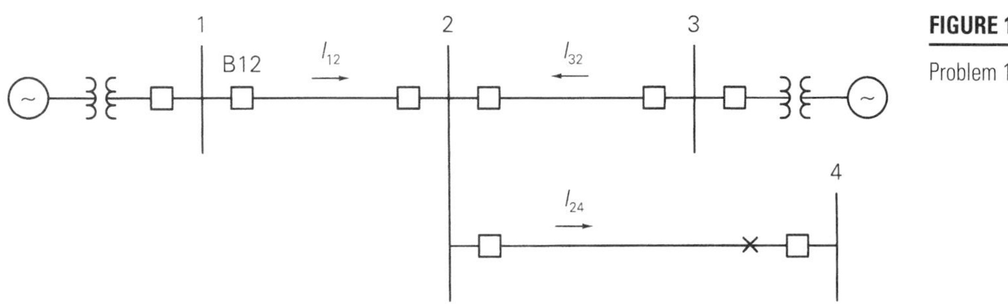

FIGURE 10.47

Problem 10.21

10.22 Consider the transmission line shown in Figure 10.48 with series impedance Z_L, negligible shunt admittance, and a load impedance Z_R at the receiving end. (a) Determine Z_R for the given conditions of $V_R = 1.0$ per unit and $S_R = 2 + j0.8$ per unit. (b) Construct the impedance diagram in the R-X plane for $Z_L = 0.1 + j0.3$ per unit. (c) Find Z_S for this condition and the angle δ between Z_S and Z_R.

FIGURE 10.48 Problem 10.22

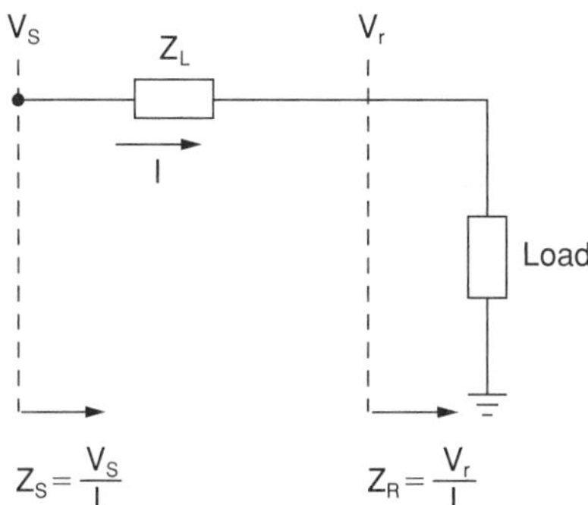

10.23 A simple system with circuit breaker-relay locations is shown in Figure 10.49. The six transmission-line circuit breakers are controlled by zone distance and directional relays, as shown in Figure 10.50. The three transmission lines have the same positive-sequence impedance of $j0.1$ per unit. The reaches for zones 1, 2, and 3 are 80, 120, and 250%, respectively. Consider only three-phase faults. (a) Find the settings Z_r in per unit for all distance relays. (b) Convert the settings in Ω if the VTs are rated 133 kV: 115 V and the CTs are rated 400:5 A. (c) For a fault at location X, which is 10% down line TL31 from bus 3, discuss relay operations.

FIGURE 10.49 Problem 10.23

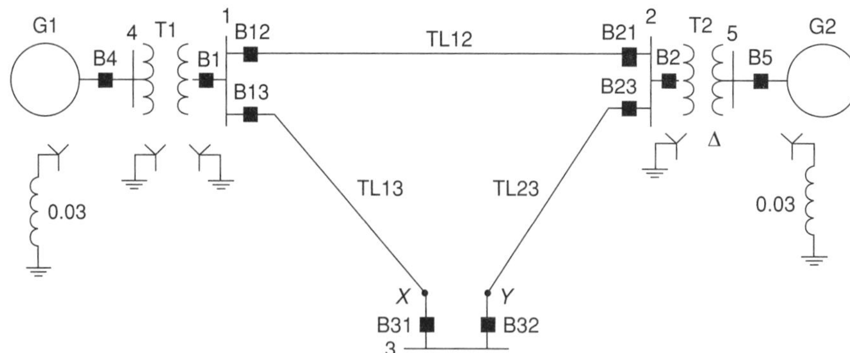

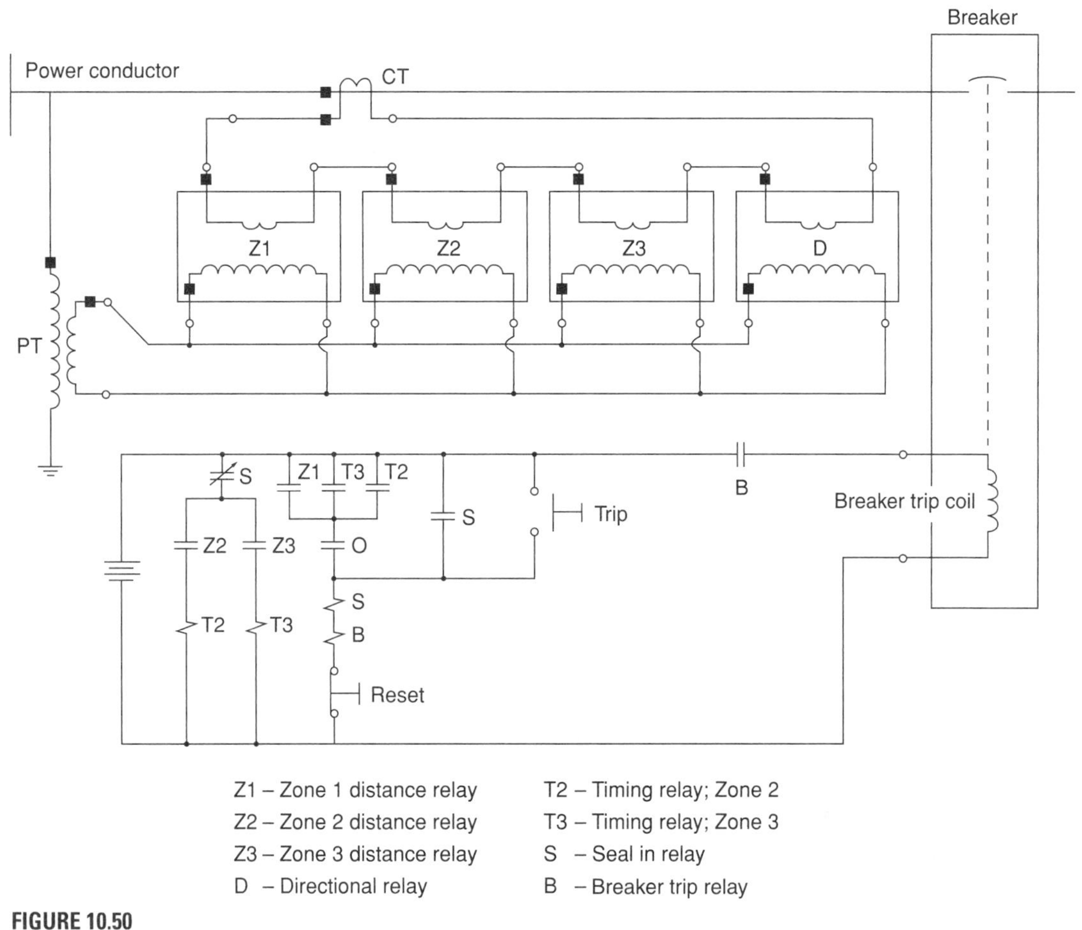

FIGURE 10.50

Three-zone distance-relay scheme (shown for one phase only) for Problem 10.23

SECTION 10.10

10.24 Select k such that the differential relay characteristic shown in Figure 10.34 blocks for up to 20% mismatch between I_1' and I_2'.

SECTION 10.11

10.25 Consider a protected bus that terminates four lines, as shown in Figure 10.51. Assume that the linear couplers have the standard $X_m = 5\ m\Omega$ and a three-phase fault externally located on line 3 causes the fault currents shown in Figure 10.51. Note that the infeed current on line 3 to the fault is $-j10$ kA. (a) Determine V_0. (b) Let the fault be moved to an internal location on the protected bus between lines 3 and 4. Find V_0 and discuss what happens. (c) By moving the external fault from line 3 to a corresponding point on (i) line 2 and (ii) line 4, determine V_0 in each case.

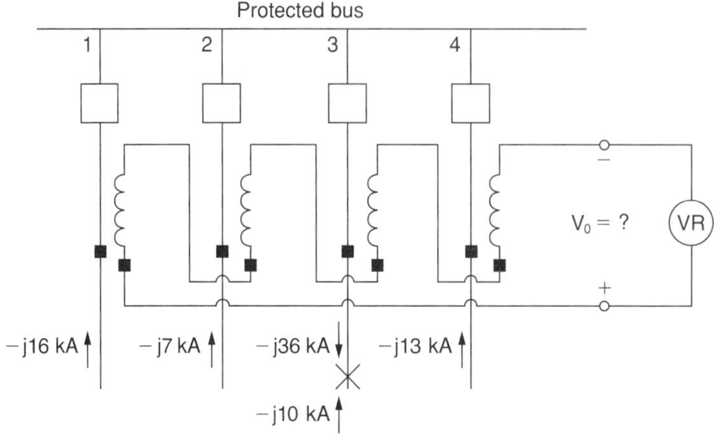

FIGURE 10.51

Problem 10.25—Bus differential protection using linear couplers

SECTION 10.12

10.26 A single-phase, 5-MVA, 20/8.66-kV transformer is protected by a differential relay with taps. Available relay tap settings are 5:5, 5:5.5, 5:6.6, 5:7.3, 5:8, 5:9, and 5:10, giving tap ratios of 1.00, 1.10, 1.32, 1.46, 1.60, 1.80, and 2.00. Select CT ratios and relay tap settings. Also, determine the percentage mismatch for the selected tap setting.

10.27 A three-phase, 500-MVA, 345 kV Δ /500 kV Y transformer is protected by differential relays with taps. Select CT ratios, CT connections, and relay tap settings. Determine the currents in the transformer and in the CTs at rated conditions. Also determine the percentage mismatch for the selected relay tap settings. Available relay tap settings are given in Problem 10.26.

10.28 For a Δ-Y connected, 15-MVA, 33 :11 kV transformer with differential relay protection and CT ratios shown in Figure 10.52, determine the relay currents at full load and calculate the minimum relay current setting to allow 125% overload.

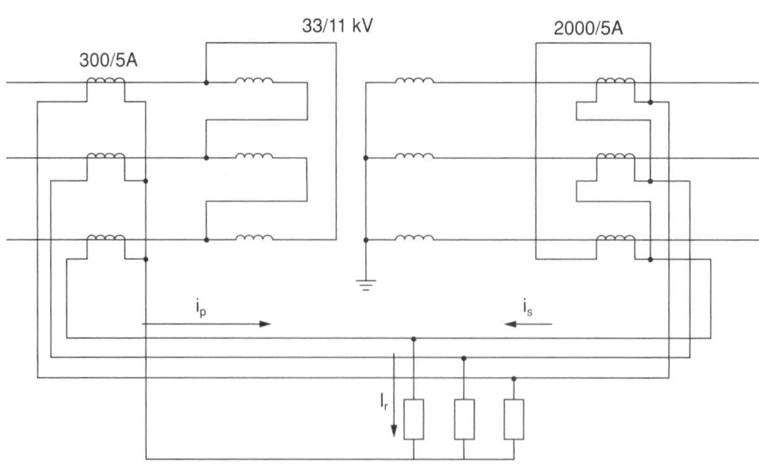

FIGURE 10.52

Problem 10.28

10.29 Consider a three-phase Δ–Y connected, 30-MVA, 33:11 kV transformer with differential relay protection. If the CT ratios are 500:5 A on the primary side and 2000:5 A on the secondary side, compute the relay current setting for faults drawing up to 200% of rated transformer current.

10.30 Determine the CT ratios for differential protection of a three-phase, Δ–Y connected, 10-MVA, 33:11 kV transformer, such that the circulating current in the transformer Δ does not exceed 5 A.

CASE STUDY QUESTIONS

a. What is the normal lifetime of a numeric relay? What is the basis of numeric relay lifetime?

b. What are the major reasons for failures of first-generation numeric relays?

c. In addition to the costs of the relays, what other costs should be considered when upgrading relays?

REFERENCES

1. J. L. Blackburn, *Protective Relaying* (New York: Dekker, 1997).
2. J. L. Blackburn et al., *Applied Protective Relaying* (Newark, NJ: Westinghouse Electric Corporation, 1976).
3. *Westinghouse Relay Manual, A New Silent Sentinels Publication* (Newark, NJ: Westinghouse Electric Corporation, 1972).
4. J. W. Ingleson et al., "Bibliography of Relay Literature. 1986–1987. IEEE Committee Report," *IEEE Transactions on Power Delivery*, 4, 3, pp. 1649–1658 (July 1989).
5. *IEEE Recommended Practice for Protection and Coordination of Industrial and Commercial Power Systems*—IEEE Buff Book, IEEE Standard 242-2001 (www.ieee.org, January 2001).
6. *Distribution Manual* (New York: Ebasco/Electrical World, 1990).
7. C. Russel Mason, *The Art and Science of Protective Relaying* (New York: Wiley, 1956).
8. C. A. Gross, *Power System Analysis* (New York: Wiley, 1979).
9. W. D. Stevenson, Jr., *Elements of Power System Analysis*, 4th ed. (New York: McGraw-Hill, 1982).
10. A. R. Bergen, *Power System Analysis* (Englewood Cliffs, NJ: Prentice-Hall, 1986).
11. S. H. Horowitz and A. G. Phadke, *Power System Relaying* (New York: Research Studies Press, 1992).

12. A. G. Phadke and J. S. Thorpe, *Computer Relaying for Power Systems* (New York: Wiley, 1988).

13. C. F. Henville, "Digital Relay Reports Verify Power System Models," *IEEE Computer Applications in Power*, 13, 1 (January 2000), pp. 26–32.

14. D. Ransom, "Upgrading Relay Protection," *IEEE Industry Applications Magazine*, 20, 5, ISSN 1077-2618, (September/October 2014), pp. 71–79.

15. D. Reimert, *Protective Relaying for Power Generation Systems* (Boca Raton, FL: CRC Press, 2005).

16. J. Parmar, "Types and Revolution of Electrical Relay," *Electrical Notes and Articles*, www.electricalnotes.com, December 1, 2012.

11 Transient Stability

1300-MW generating unit consisting of a cross-compound steam turbine and two 722-MVA synchronous generators (Courtesy of American Electric Power.)

Power system stability refers to the ability of a power system to move from one steady-state operating point following a disturbance to another steady-state operating point, without generators losing synchronism or having unacceptable voltage magnitude and frequency deviations [1]. There are three types of power system stability: steady-state, transient, and dynamic.

Steady-state stability, discussed in Chapter 5, involves slow or gradual changes in operating points. Steady-state stability studies, which are usually performed with a power-flow computer program (Chapter 6), ensure that phase angles across transmission lines are not too large, that bus voltages are close to nominal values, and that generators, transmission lines, transformers, and other equipment are not overloaded.

Transient stability, the main focus of this chapter, involves major disturbances such as loss of generation, line-switching operations, faults, and sudden load changes. Following a disturbance, synchronous machine frequencies undergo transient deviations from synchronous frequency (60 Hz in North America, 50 Hz in many other locations), and machine power angles change. The objective of a transient stability study is to determine whether or not the machines will return to synchronous frequency with new steady-state power angles. Changes in power flows and bus voltages are also of concern.

Elgerd [2] gives an interesting mechanical analogy to the power system transient stability program. As shown in Figure 11.1, a number of masses representing synchronous machines are interconnected by a network of elastic strings representing transmission lines. Assume that this network is initially at rest in steady-state, with the net force on each string below its break point, when one of the strings is cut, representing the loss of a transmission line. As a result, the masses undergo transient oscillations and the forces on the strings fluctuate. The system will then either settle down to a new steady-state operating point with a new set of string forces, or additional strings will break, resulting in an even weaker network and eventual system collapse. That is, for a given disturbance, the system is either transiently stable or unstable.

Transient stability studies are solved using a time-domain simulation for a specific disturbance that at each time point involves a solution of algebraic equations representing (primarily) the network power balance constraints, and differential equations representing the generator and sometimes the load dynamics. There are two main approaches for solving the system at each time point—either simultaneous in which the algebraic and differential equations are solved simultaneously, or partitioned in the program, alternately solves the algebraic and differential equations [15]. Both predisturbance, disturbance, and postdisturbance computations are performed. The program output includes power angles and frequencies of synchronous machines, bus voltages, and power flows versus time. Large-scale system studies can often involve many thousand algebraic equations and sometimes more than 100,000 differential equations.

Sometimes the transient stability is determined during the first swing of machine power angles following a disturbance. During the first swing, which typically lasts about 1 second, the mechanical output power and the internal voltage of a generating unit are often assumed constant. However, where multiswings lasting several seconds are of concern, models of turbine-governors and excitation systems (for example, see Figures 12.3 and 12.5) as well as more detailed machine models

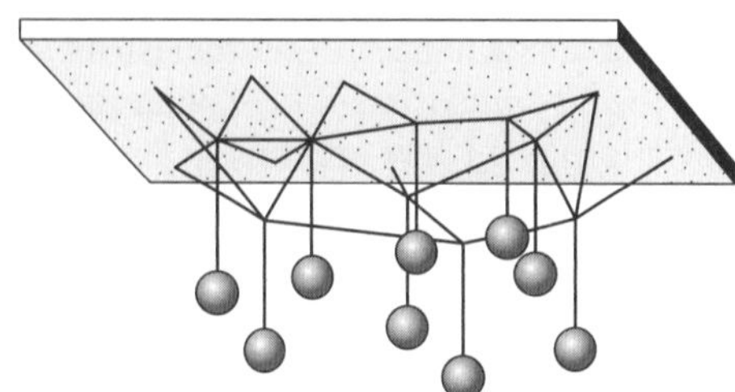

FIGURE 11.1 Mechanical analog of power system transient stability [2] (Based on Elgerd, Electric Energy Systems Theory: An Introduction, 1982. McGraw-Hill.)

and sometimes dynamic load models can be employed to obtain accurate transient stability results over the longer time period.

Dynamic stability involves an even longer time period, typically several minutes. It is possible for controls to affect dynamic stability even though transient stability is maintained. The action of turbine-governors, excitation systems, tap-changing transformers, and controls from a power system dispatch center can interact to stabilize or destabilize a power system several minutes after a disturbance has occurred.

To simplify transient stability studies, the following assumptions are commonly made:

1. Only balanced three-phase systems and balanced disturbances are considered. Therefore, only positive-sequence networks are employed.
2. Deviations of machine frequencies from synchronous frequency (60 Hz) are small, and dc offset currents and harmonics are neglected. Therefore, the network of transmission lines, transformers, and impedance loads is essentially in steady-state; and voltages, currents, and powers can be computed from algebraic power-flow equations.

In Section 11.1 introduces the swing equation, which determines synchronous machine rotor dynamics. In Section 11.2, a simplified model of a synchronous machine and a Thévenin equivalent of a system consisting of lines, transformers, loads, and other machines are given. Then, the equal-area criterion; that gives a direct method for determining the transient stability of one machine connected to a system equivalent is presented in Section 11.3. Numerical integration techniques for solving swing equations step by step are covered in Section 11.4, and they are used in Section 11.5 to determine multimachine stability. Section 11.6 introduces a more detailed synchronous generator model, while Section 11.7 discusses how wind turbines are modeled in transient stability studies. Finally, Section 11.8 discusses design methods for improving power system transient stability.

CASE STUDY

The following case study provides an overview of the various issues involved in power system restoration following a blackout [11]. Restoration involves regulatory, economic, and technical issues. The case study focuses mainly on the technical issues. Most outages involve only a portion of a power system that can be restored with assistance from neighboring grids. In this case, *top-down restoration* is typically applied, where tie-lines to neighboring grids can be used to first energize the high-voltage grid, followed by energizing the subtransmission system, which can begin to restore some system load and supply auxiliary power at power plants to bring generation back on line in the blacked-out portion. In the case of a widespread blackout, there may be no help from neighboring grids, and *bottom-up restoration* must be applied, where restoration begins at preselected generating units (called "black-start units") that do not require any external power sources to start up. In order to ensure that facilities and personnel are prepared, restoration plans and procedures must be developed and coordinated among the various parties that own and operate the generators, the transmission system, and the distribution systems. Restoration studies and analyses, including steady-state and dynamic analyses, are required. Technical issues include load-frequency control, voltage control, motor starting, self-excitation of black-start units, system stability, cold load pickup, and transient overvoltages. Development of thorough restoration plans and testing of those plans via simulation and drills will help to minimize disruption of service to loads and equipment damage following a blackout.

Down, but Not Out

By James Feltes and Carlos Grande-Moran

Modern power systems are highly reliable. They are operated to withstand the variability in system conditions that occur in the course of normal operations, including the daily changes in load levels, generation dispatch, and equipment availability. Modern power systems must also provide reliable service to load during unusual events, such as faults and the tripping of transmission lines or generating units. Extreme events can occur, however, that may lead to partial or total system blackouts. To recover from such catastrophic events, utilities thus have a responsibility to develop detailed plans and procedures for restoring the power system safely, efficiently, and as expeditiously as possible. This case study gives a brief overview of many issues involved in power system restoration.

Restoration challenges can be subdivided into three areas: regulatory, economic, and technical. Only a very brief overview of the first two will be given here, as our emphasis will be on the technical issues associated with the restoration of a power system following a partial or total blackout.

Regulatory issues are associated with directives from the authorities in charge of determining the required level of service reliability to be supplied by the utilities to end users. As part of this process, the regulatory authorities generally establish criteria and requirements for power system restoration. In the United States, standards and criteria are established by the North American Electric Reliability Corporation (NERC), the Nuclear Regulatory Commission, the NERC regional authorities, and the local independent system operator (ISO) or regional transmission operator (RTO). While the exact process is different in other parts of the world, the concepts are generally quite similar.

The NERC requirements are delineated in its emergency preparedness and operations (EOP) standards. The standards that cover the areas being discussed here are EOP-001 (*Emergency Operations Planning*) and EOP-005 (*System Restoration from Blackstart Resources*). EOP-001 covers the requirements to develop, maintain, and implement a set of plans to mitigate operating emergencies. The purpose of EOP-005 is to ensure plans, facilities, and personnel are prepared to enable system restoration from black-start resources so as to assure that reliability is maintained during restoration and a high priority is placed on restoring the interconnection.

Economic issues are handled differently in different regions of the United States and also throughout other parts of the world. In the traditional, vertically integrated utility structure, the utility was responsible for generation, transmission, and distribution and hence for all the system components required for restoration. In a market-based system, however, coordination among many parties is required, as different owners and operators for the generators, the transmission system, and the distribution systems are involved. Common to both structures is the need to set clear requirements and reimburse the various parties for fulfilling these requirements. This is often a difficult process. As with expenditures for life and medical insurance, we must spend money on something we hope never to have to actually use. It is also quite difficult to measure the quality of restoration services until it is too late and they are already being mobilized. We will not delve into these issues here other than to state that system restoration is a critical service, and the inducements for generation and transmission owners must be sufficient for them to not only offer these services but to diligently participate in the testing and training needed to implement them successfully. Below we discuss the main emphasis of this article: the technical issues that affect power system restoration.

General Philosophy

Most outages involve only a portion of the power system that can be restored with assistance from neighboring power grids. While such restoration can be a complex process, the restoration generally occurs in a relatively straightforward manner. First, the tie lines from the outside power system to the blacked-out area are energized. These ties are the starting points used to energize the transmission high-voltage grid. With significant circuits in the transmission network energized, it is relatively straightforward to pick up the subtransmission system, start to restore some system load, and supply auxiliary power to selected power plants to bring generation back online. This is an example of a *top-down restoration*, where the highvoltage grid is energized first and then used to energize the lower-voltage networks.

In the case of a widespread blackout, however, there may be no help from neighboring systems. In this case, system restoration must begin from preselected generating units that have the ability to start themselves, that is, that do not require any external power sources. These units, called black-start units (BSUs) or black-start resources (BSRs), are then used as the kernels with which to start the restoration process. This can be considered a *bottom-up restoration*, starting from an individual generating unit and emanating outward toward the critical system load. Of course, to speed the restoration process, this black-start sequence will probably occur using several generating units simultaneously and independently; these independent islands of generation and load will later be synchronized so as to restore the original power system.

Restoration can, of course, be accomplished by combining the approaches described above. In the case of a total blackout, it is important to decide how quickly to energize the high-voltage system. As will be discussed in more detail below, energizing the high-voltage system is more difficult and usually requires that large generating units be online. If a high-voltage overlay can be successfully established, however, it will generally lead to faster restoration of the loads.

Restoration Plans and Procedures

Restoration plans and procedures must be developed, clearly delineating the actions to be undertaken and the responsibilities of each of the parties. These procedures must meet established reliability criteria and be able to be implemented by system operators at a central control center, a local control center, or both.

Restoration plans define the sequence of steps and cranking paths needed to restore power to critical loads and generation facilities from BSRs. Ideally, the BSRs are able to start quickly and energize secure transmission paths. A backbone network is formed so that critical load can be restored—in particular, stabilizing loads for the BSRs and auxiliary loads at power plants without black-start capability.

The restoration plan focuses on restoring the key system elements (generating stations, transmission lines, substations, and loads) that facilitate restoration of additional facilities. In particular, flexibility is important because major equipment may be out of service and not be available to assist in the restoration process. For example, restoration of a generation plant or substation with multiple transmission outlets is preferable to restoration of a location with a single path. This also illustrates the need for simplicity: the more complex a plan is, the more chances there are that a key component or action cannot be performed when or in the manner required.

In order to be able to fully restore the system, the restoration plan must be able to reenergize the transmission system to major generating facilities in a timely manner and also reach interconnection points with neighboring systems to form a larger and usually more stable grid. The supply of off-site power to nuclear power plant auxiliary systems is also a priority for those systems with nuclear generation.

The restoration plans and procedures must also address a myriad of other topics, including:

- Staffing and communication requirements;
- The training of control center staff and field staff needed to perform switching actions;

- Communication protocols among control centers and transmission and generation facility operators;
- The requirements of the BSRs, including technical issues such as voltage and frequency control capabilities and managerial issues such as maintenance and adequate fuel supply capability; and
- The testing of the plans, including the starting of BSRs and drills testing the communication and coordination of the various power system operators.

Restoration Studies and Analysis

Each transmission operator is required to have plans, procedures, and resources available to restore the electric system following a partial or total shutdown of the system. The standards also require each transmission operator and balancing authority to verify their restoration procedure by actual testing or through simulation.

Restoration testing is a complex process. The testing of BSUs requires starting and running the units for a limited period of time, processes that involve straightforward procedures. The coordination of the test period, however, is complicated by considerations of staffing requirements, environmental emission restrictions, and the requirements of the overall system. The testing of line energizations is more complex, as it involves the deenergization of parts of the transmission system so they can be connected to the BSU. This must be accomplished without any adverse impact on loads, which may not always be possible. Restoration plans that require the energization of load at a particular step in the plan cannot be tested beyond that step because it is never acceptable to submit loads to outage and pickup as part of a test. In addition to any field testing performed, simulation is usually required to verify and validate the plan. A study including both steady-state and transient analysis is therefore performed.

The restoration plans must document the various cranking paths. For example, the plans show the number and switching sequence of transmission elements involved, including the initial switching requirements between each BSU and the unit to be started (i.e., its next-start unit).

In the case of a total system outage, system restoration must begin from the BSUs, with restoration of the power system proceeding outward toward critical system loads. As the BSUs themselves can only supply a small fraction of the system load, these units must be used to help start larger units that need their station service loads to be supplied by outside power sources. Full restoration of system load can only occur when these larger units can come online. The restoration plan following a system blackout should therefore include self-starting units that can be used to black-start large, steam turbine-driven plants located electrically close to these units. As mentioned above, another objective for many systems is the supply of auxiliary power to nuclear power stations in need of off-site power to supply critical station service loads. Other priority loads may include military facilities, law enforcement facilities, hospitals and other public health facilities, and communication facilities.

The typical black-start scenario includes the self-starting unit or units, the transmission lines that will transport the power supplied to the large motor loads in the power plant to be black-started, and at least three transformer units. These would include the generator step-up transformers of the black-start generating unit, the generator step-up transformers of the next-start unit involved (e.g., a large steam turbine unit), and one or more auxiliary transformers serving motor control centers at the next-start plant. The transmission lines used for the black start may be either overhead lines or high-voltage underground cables. The load to be black-started includes very large induction motors, ranging from a few hundred kilowatts to several megawatts as well as plant lighting and small-motor loads.

The key concerns are the control of voltage and frequency. Both voltage and frequency must be kept within a tight band around nominal values to guard against damage to equipment and to ensure restoration progress. Any equipment failure will severely hinder restoration and may require starting over with a revised plan. System protection operations can also occur if voltage or frequency strays outside acceptable ranges, again with the potential to set back or stop the restoration process. The following sections give an overview of several of the technical concerns that must be addressed.

Steady-State Concerns

The black-start plan describes the steps that the transmission operators need to take to restore the isolated power system from the BSU. This includes sequentially energizing transformers, transmission lines, and, potentially, shunt compensation and load pickup to supply power to the next-start unit auxiliary loads, allowing their associated units to begin operation. Once larger generating units are available, higher-voltage lines can be energized, again in a step-by-step sequence, to supply power to major substations where load can be picked up and further interconnections made until the grid is fully restored.

The steady-state analysis of this isolated power system includes:

- Voltage control and steady-state overvoltage (Ferranti effect) analysis;
- Capability of the BSUs to absorb reactive power (vars) produced by the charging capacitance of the transmission system;
- Step-by-step simulation of the black-start plan being tested to ensure its feasibility and compliance with required operational limits;
- Verification of the robustness of the tested black-start plan to ensure its ability to compensate

for the unavailability of key components to be used in the plan; and

- Demonstration of generation and load-matching capability.

Voltage control analysis determines the voltage reference set point of the black-starting generating unit and the off-nominal tap setting for all transformers that are part of the plan. This ensures proper control of voltage and provides the needed terminal voltage to start up the large induction motor loads at the black-started plant. Transformer tap settings that are appropriate for normal conditions that generally include significant current flows may result in high system voltages under the lightly loaded black-start condition. Since most taps on generator step-up transformers and station auxiliary transformers cannot be changed under load, the selection of transformer taps must be a balance between the needs of the black-start period and normal operation, when the power system is supplying a significant amount of load.

Load flow simulations can be used to calculate the receiving end bus voltage of the transmission lines when the black-starting unit energizes the unloaded generator, step-up transformer, and transmission lines. The charging current generated by an unloaded transmission line will result in a rise in voltage along the line. This is particularly true when underground cables are used as they have significantly more charging capacitance. The charging requirements can be large enough to result in the BSUs absorbing reactive power. There could be, under extreme conditions, the potential for self-excitation, which is discussed later.

The steady-state analysis of a black-start plan should include a step-by-step simulation of the plan to verify its compliance with required operational limits for voltage control and power flows. The robustness of the plan for the loss of a system component is also valuable knowledge because the events leading to the blackout could result in some equipment unavailability during the restoration period. Generally, thermal overloads are not a restoration issue because the system is lightly loaded. This may become a concern, however, as restoration progresses and load is picked up.

Dynamic Studies

Once the steady-state analysis has been completed, a dynamic analysis of the restoration plan is conducted. The dynamic analysis starts from an initial steady-state operating point representing a step in the plan. This initial system operating condition is usually obtained from the system steady-state analysis. One key simulation initially represents the isolated power system and then simulates the start-up of the largest induction motor load at the next-start generating unit. This verifies that the voltage supply is strong enough to start the motor and also that the voltage dip will not stall or cause the motor contactors of running motors to drop out.

The importance of accuracy in equipment modeling must be emphasized. The effect of the controls of an individual unit is generally not very significant under normal operation because a large number of units are sharing the control of system voltage and frequency. Both of these quantities are controlled solely by the BSU during the initial restoration period, however. The modeling of the generator, excitation system, and speed governor is therefore very important. The modeling of equipment that does not generally operate under normal conditions, such as over- and under-excitation limiters, can also be important. Governor modeling must take into account whether the machine is operated in an isochronous or droop control mode, as will be discussed later. The accuracy of the dynamic modeling parameters of any large motors to be started are also important to motor-starting simulations.

The dynamic analysis of a black-start plan includes some or all of the following functions:

- Load frequency control;
- Voltage control;
- Large induction motor starting;
- Motor-starting sequence assessment;
- Self-excitation assessment;
- System stability; and
- Transient overvoltages.

Because frequency may deviate significantly from its nominal value, the effect of frequency variation on system impedances must be modeled.

BSUs

As we have already explained, BSUs are units that do not require off-site power to start. Generally, these fall into four categories:

- **Hydroelectric units.** These units can be designed for black-start capability and have fast primary frequency response characteristics.
- **Diesel generator sets.** Diesel sets usually require only battery power and can be started very quickly. They are small in size and useful only for supplying the power needed to start larger units. They generally cannot be used to pick up any significant transmission system elements.
- **Aeroderivative gas turbine generator sets.** This type of gas turbine typically requires only local battery power to start. These units can usually be started using remote commands and can pick up load quickly.
- **Larger gas turbines operating in a simple cycle mode or steam turbine units.** These units are not in themselves black-start capable but are coupled with on-site diesel generator sets to make the plant a black-start source. The diesels are started and used to energize plant auxiliary buses and start either a gas turbine or a steam turbine. A gas turbine is generally quicker to bring online. The time to restart and available ramping capability are generally functions of how long the unit was off-line.

Load Frequency Control

When only a portion of the system is lost and is being restored using tie lines to a larger power system, load frequency control is not generally a large concern. The outside system generally has the capacity to absorb changes in load without significant frequency deviations.

When restoration without external resources is required, however, load frequency control is of critical importance. During the restoration process, the black-start generating unit will typically be used to pick up large induction motors associated with a larger power plant such as boiler feed pumps and forced-draft and induced-draft fan motors. The frequency of the black-start system will be controlled by the speed governor of the turbine driving the black-start synchronous generator.

Standard practice for units operating in parallel in multi-machine power systems under normal conditions is to operate all turbine speed governors in a droop-governing mode. This provides a stable sharing of the electric system load among all units. The proportional characteristic of the droop speed governor control, however, results in a steady-state frequency error remaining in the system. Automatic generation control (AGC) has the form of a pure integral controller and will follow the primary frequency control action of the speed governors to remove this undesirable steady-state frequency error. Typical steady-state regulating droop (R) for speed governors is 5% using a system frequency base of 50 Hz or 60 Hz and a power base equal to the turbine MW rating.

During a black-start event, AGC will not be operating, however, it is imperative that system frequency regain its scheduled value following the start-up of motors or pickup of other loads. This frequency control should be automatic, since the crew in charge of the BSU will be operating under extreme emergency conditions, which can lead to undesirable operating errors. The automation of the frequency control process can therefore be carried out by the prime-mover speed governor of the BSU, operated in a constant frequency or isochronous control mode. In this pure integral control mode, the steady-state frequency error is zero because of the resetting characteristic of the pure integral control. Most if not all modern diesel engines, gas turbines, and hydraulic turbines are furnished with digital speed governors with which a selection of either a droop or isochronous operating mode can be carried out by means of a simple change in command.

Once the system has more than one generating unit online, all speed governors should be operated in a droop control mode, unless it is decided in the restoration plan that one of the largest units should operate in isochronous control to maintain the control of the system frequency. As noted above, AGC would be disabled under this extreme condition.

In summary, the preferred control mode for speed governors associated with BSUs is isochronous or constant frequency control. When additional units are added, the preferred control mode for speed governors is droop control mode. In some cases, it may be preferable to keep one large unit in isochronous control mode. Units should not be operated in parallel with more than one unit in isochronous control mode.

Voltage Control

Control of voltage is obtained through the generator's excitation system. The excitation system must be operated in automatic control with the automatic voltage regulator (AVR) in service. The system voltage will be a function of the generator terminal voltage. The generator scheduled voltage may therefore need to be adjusted throughout the restoration process as load is picked up, and also coordinated with any changes in transformer tap positions. Such adjustments should be an integral part of the restoration plan. The changes in voltage that will occur with the starting of large motors or the pickup of large blocks of load require that the excitation system respond in a rapid, well-tuned manner.

Motor Starting

Motor starting is a concern during black-start restoration. A BSU's primary function is generally to start up the auxiliary load of a larger next-start unit. This auxiliary load is made up of lighting and motor load used, for the most part, in the start-up of steam generators and fuel systems. The motor load is made up of a large number of small- and medium-size motors and a few large motors ranging anywhere from several hundred kilowatts to several megawatts. Fuel and feedwater pump motors and forced- and induced-draft fan motors belong to this megawatt group. It is this latter group that presents the greatest challenge to the reactive power resources available in any well-designed black-start plan.

The method used for starting up these large motors is often a hard start, which applies full line-to-line voltage across the motor terminals. Occasionally, motors may also be soft-started by applying a reduced voltage during the starting period. It is therefore extremely important to properly identify the motor-starting method since this will greatly affect the depth of the dip in voltage resulting from the black-start process.

Accurate motor data are vital for conducting dynamic studies to verify the viability of a given black-start process. The information needed to establish the dynamic model for the large induction motors participating in the black-start process includes the inertia of the motor plus its mechanical load, the starting or locked rotor torque, the starting or locked rotor current and associated power factor, the pull-out torque, the full-load torque, and the full-load current and its associated power factor. All of these data should be at rated voltage and frequency. From these motor performance data, parameters for the stator and rotor circuits are estimated. The dynamic model for these large induction motors should include both inertial and rotor circuit flux dynamics. This dynamic model must closely match the speed-torque characteristic of the motors, particularly at starting, pull-out, and full-load operating points. In addition, it is important to include the mechanical load damping effect in the inertial model of the mechanical load, which for most centrifugal pumps and fans follows a quadratic speed-torque characteristic.

The motor-starting sequence is another variable that must be verified in any black-start process. The feasibility of plant start-up can be tested by dynamically simulating the various motor-starting sequences. The sequence must also accommodate the start-up requirements of the plant, which may require certain motors to be started before others.

The voltage dip caused by starting these large induction motors must be accurately quantified. This is because the motors already online have magnetic contactors that open at approximately 80% of the nominal bus voltage. IEEE Standard 399-1997 recommends a minimum terminal voltage of 80% of rated voltage. Occasionally there may be magnetic contactors that can hold their contacts with voltages as low as 70% of the rated value; the number of cycles for which this operating condition can be sustained is low, however. In addition, the life expectancy of the insulation of the stator and rotor windings is reduced as a result of the large currents circulating through these windings. In such situations, the motor manufacturer must be consulted to avoid motor damage, such as shorted turns or even catastrophic motor failure. Undervoltage protection settings should also be verified to avoid the opening of circuit protection caused by undervoltage relay action.

The accelerating time period required by an induction motor depends in great measure on the combined inertia of the motor and its mechanical load. The longer the accelerating period, the higher the heating experienced by the stator and rotor windings. When accelerating periods last a few tens of seconds, motor manufacturer data on allowed motor heating should be consulted to avoid a significant loss of useful operating life of the winding insulating material.

Self-Excitation

As noted above, energization of a transmission line or cable will result in a rise in voltage along the line or cable due to charging currents. The charging requirements can be large enough to result in the BSUs absorbing reactive power. There is the potential for self-excitation if the charging current is large relative to the size of the generating unit. The result can be an uncontrolled rise in voltage that could result in equipment failure. Such an undesirable operating condition may occur when the effective charging capacitive reactance of the transmission system used in the black-start operation, as seen by the BSU, is less than the q-axis generator reactance X_q. In generating units with no negative field current capability, self-excitation cannot be controlled by the excitation system, and thus the machine terminal voltage rises almost instantaneously for cases where the effective capacitive reactance is less than the d-axis reactance X_d. Generator excitation systems with negative field current capability delay but do not prevent the onset of self-excitation. It is worth noting that most generating units installed in the last 40 years do not have negative field current capability. It is therefore extremely important to verify the reactive power capability of the BSU when operated at a leading power factor.

Self-excitation can also occur from the load end for the inadvertent loss of supply resulting from the opening of a transmission line or cable at the sending end that leaves the line connected to a large motor or a group of motors.

System Stability

Power system stability was defined in 2004 by the IEEE/CIGRÉ Joint Task Force on Stability Terms and Definitions as "the ability of an electric power system, for a given initial operating condition, to regain a state of operating equilibrium after being subjected to a physical disturbance, with most system variables bounded so that practically the entire system remains intact." This simply means that the power system must be able to survive a disturbance and return to a sustainable operating point without a significant loss of equipment.

Power system stability can be further subdivided into:

- **Rotor angle stability:** the ability of synchronous machines of an interconnected power system to remain in synchronism after being subjected to a disturbance
- **Voltage stability:** the ability of a power system to maintain steady voltages at all buses in the system after being subjected to a disturbance
- **Frequency stability:** the ability of a power system to maintain steady frequency following a severe system disturbance resulting in a significant imbalance between generation and load

As discussed above, voltage and frequency control present the greatest concerns during restoration. Angular stability is generally not a major concern in the early stages of the restoration process. When the system is being restored from a total blackout, angular stability is assessed only when more than one generating unit is used in the black-start plan. Even when there are multiple units connected in the early stages of a restoration plan, the system is operating in a weakened state and stable performance is not expected for all system contingencies. Under normal conditions, there are generally multiple transmission paths between groups of generators such that a fault and trip of one of these paths does not result in instability. During restoration, however, there may only be one strong path, and hence a fault and outage of that path would cause instability. This period of exposure to possible but unlikely events cannot limit the restoration, however. It is simply a stage the system must pass through to reach a more robust operating condition. The exposure to events that could result in instability can be used as one of the criteria with which to rank restoration alternatives.

Cold Load Pickup

The purpose of the restoration process is, of course, to restore power supply to the loads and allow them to operate as they did prior to the outage. But the characteristics of the load immediately after reenergization may be quite different than the characteristics exhibited prior to the outage.

If the load has been deenergized for several hours or more, the inrush current upon reenergizing the load can be eight to ten times the normal load current. The magnitude and duration of the inrush current that flows when a feeder is reenergized after a prolonged outage is a function of the type of load served by the feeder. This could include: lighting; motors; and thermostatically controlled loads such as air conditioners, refrigerators, freezers, furnaces, and electric hot water heaters.

There are various components of the load that contribute to the total inrush current. One example is the component due to the filaments of incandescent lights. The resistance of the filament is very low until it warms to its operating temperature. This low resistance results in a very high inrush current—up to ten times the normal current. This high current flows for a short period, approximately one-tenth of a second.

Another component of the inrush current is the starting of motors when the load is picked up. When a motor starts, the current drawn will typically be five to six times normal, until the motor reaches its operating speed. This may take as long as several seconds for large, industrial-type motors.

A third component of inrush current is thermostatically controlled loads, which turn on and off automatically to hold temperature to a desired preset value. Under normal operating conditions, approximately one-third of these loads will likely be connected at any instant in time. But after a lengthy interruption of service, they will all have their thermostat contacts closed, waiting to run as soon as power is restored. As a result, these thermostatically controlled loads will be perhaps three times greater than they normally would be for the first half hour or so after being energized. Most thermostatically controlled loads also contain small, single-phase motors, which will draw five or six times running current until they are accelerated up to running speed in perhaps one-half of a second. This results in the initial current drawn by some thermostatically controlled loads being as high as 15 times normal current for the first one-half of a second following energization.

A summary of the magnitude and duration of the inrush for some of the various types of loads is shown in Figure 1.

Transient Overvoltages

Restoration of the power system is performed through a series of switching actions to sequentially reenergize system components. Energizing equipment during restoration conditions can result in higher overvoltages than during times of normal operation. These overvoltages can lead to equipment failure or damage that

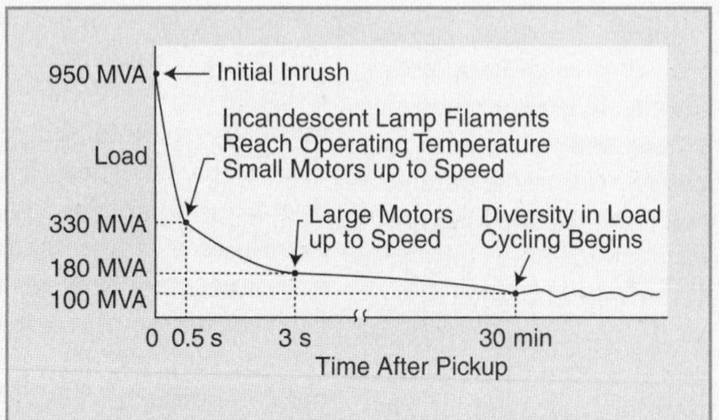

Figure 1 Load variation following cold load pickup

may hinder the successful implementation of the restoration plan.

Transient overvoltages include temporary overvoltages, switching surges, and lightning surges. Lightning surges, while an important design consideration, are usually not a concern that affects the restoration of a power system.

Switching surges are the transient overvoltages that immediately follow the opening or closing of a circuit breaker or other switching device. Switching surges have high-frequency components (from 100 Hz to 10 kHz) that decay quickly, typically within two to three cycles of the power frequency, and are followed by a normal steady-state voltage. Switching surges typically contain only one, or just a few, voltage peaks that are of interest, as shown in Figure 2. This sample waveform is from a simulation of the energizing of a typical overhead line.

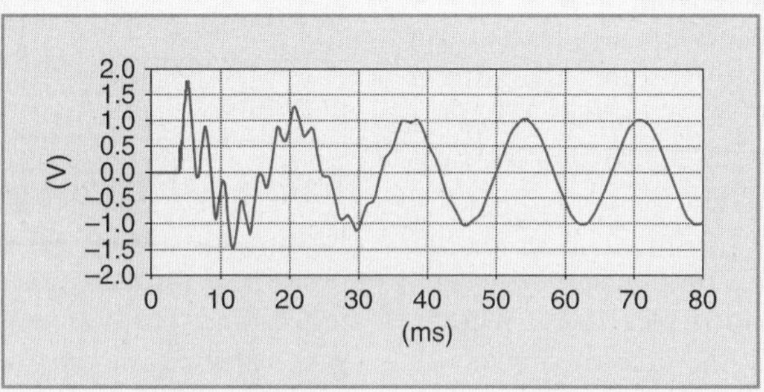

Figure 2 Switching surge from an electromechanical transients program simulation of the energizing of a typical overhead line. The voltage is shown per unit of the line's nominal voltage rating.

The magnitude and wave shape of the switching surge depends on the angle of the power frequency source voltage wave at the instant of circuit breaker closure. This requires that many simulations be performed with various closing times to obtain a statistical distribution of the overvoltage results. Surge arresters are effective in limiting the peak of the switching surges.

Temporary overvoltages (TOVs) include many types of events for which the voltage transient lasts longer than the surges discussed above, exceeding the rated value for three cycles or, potentially, significantly longer. TOVs encompass power frequency phenomena such as the Ferranti rise on an open-ended line or cable and the overvoltage on an unfaulted phase during a single line-to-ground fault.

TOVs can also follow switching surges. For example, a TOV can result from switching circuits that saturate the core of a power transformer, for example, when cables and transformers are energized together. The harmonic-rich transformer inrush currents can interact with the harmonic resonances of the power system. The resonant frequencies are a function of the series inductance associated with the system's short-circuit strength and the shunt capacitances of cables and lines. Higher inductances (a property of relatively weak systems, such as those often occurring during restoration) and higher capacitances (such as those due to long cables) yield lower resonant frequencies and a higher chance of TOVs.

Figure 3 shows an example of a TOV taken from a simulation of the energization of a large transformer. Like switching surges, this type of TOV can be dependent on the circuit breaker closing times. In contrast to

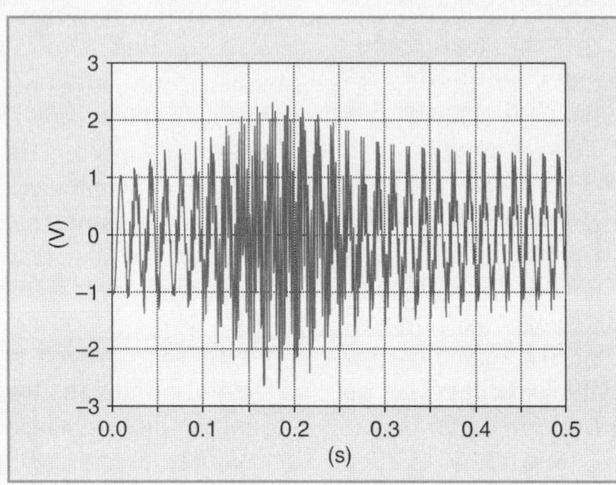

Figure 3 An example of a TOV; voltage is expressed per unit of nominal peak phase voltage

switching surges with one predominant peak, TOVs can have hundreds of peaks, all of about the same magnitude, if the TOV lasts for several seconds.

The expected TOV magnitude and duration is often a major concern for surge arresters. Metal-oxide varistor-type surge arresters have little effect on TOVs that are below about 1.6 per unit. Silicon carbide-type surge arresters are not affected by TOV levels below the 60-Hz spark-over level. If the TOV repeatedly exceeds the spark-over level, however, then the multiple discharges may result in excessive energy absorption and consequent arrester failure.

A Black-Start Example

In this section, we present a simulated black start. The scenario is one where a fast-starting, gas turbine-driven generating unit is used as a BSR to start up a combined-cycle power plant. The black-start system consists of the BSU's step-up transformer, underground high-voltage (HV) cables that connect the HV substation at the BSU to the HV substation at the combined-cycle plant, and the generator step-up and auxiliary transformers at the combined-cycle plant. Both the taps of transformers with tap-adjustment capability and the voltage reference set point of the BSU were chosen to ensure that terminal voltages at the large induction motor units used in the black start were close to their nominal values.

The black-start plan begins with the across-the-line starting of a 1.9-MW motor. The motor performance during the starting period is shown in Figure 4. Terminal voltage, motor reactive power, electrical torque, and motor slip are shown. Note the dip in motor terminal voltage and that the demand for reactive power rises during the period following the lowest voltage at the motor terminals. The motor's air gap torque increases significantly during the acceleration period, as expected, to overcome the mechanical load torque that opposes developed electromagnetic torque.

The dynamic response of the BSU during the starting of this large induction motor is shown in Figure 5. The performance of the excitation system is shown as it works to control the BSU's terminal voltage. Note the fast response and large field forcing applied to pull up the machine terminal voltage from the dip caused by the large reactive power demand imposed by the starting motor. The unit also sees a significant voltage rise caused by the rapid reduction in reactive power as the motor locks in to its operating speed. Electric power demand also increases during this period as the motor accelerates and moves toward its steady-state operating point.

Summary

Restoration actions involve very unusual conditions, especially for local generation used as a BSR. Important considerations for assuring that restoration plans are realizable include the ability to operate in islanded

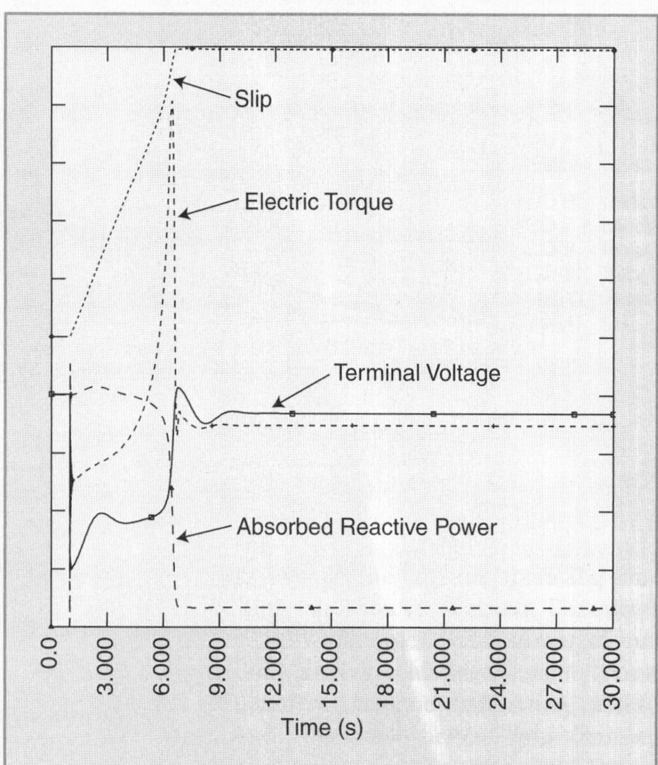

Figure 4 Motor performance during the starting period: motor terminal voltage, reactive power, electric torque, and slip

conditions with stable frequency and voltage control, the availability of synchronizing equipment at key substations to permit paralleling of separate sections, and the validity of assumptions used to assess the ability of synchronous generators to operate at the unusual points of their capability required during the restoration period.

Restoration actions performed to recover from a blackout may vary from those determined in the restoration studies. Restoration plans are based on a given set of assumptions, such as the available transmission, the amount of cold load to be picked up, and the many other conditions discussed in this case study. Although actual conditions could differ from these assumptions, restoration studies provide value by demonstrating the logic behind particular steps being taken, such as the reasoning behind the choice and sequence of operator actions and the

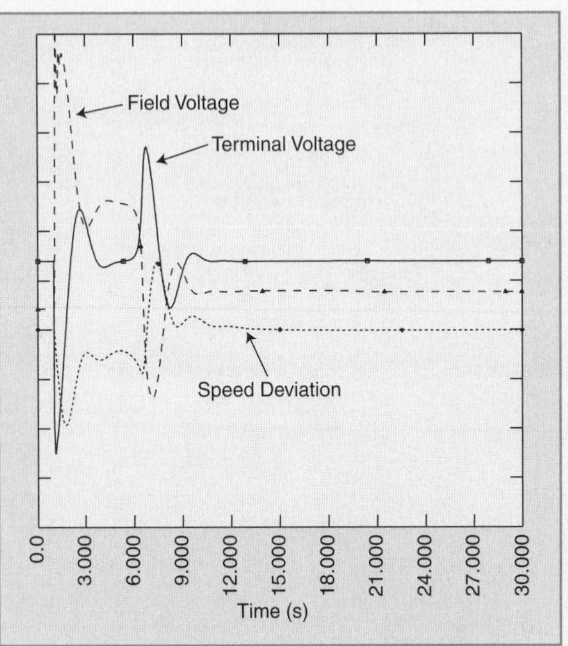

Figure 5 Dynamic response of the BSU as the large induction motor is started: field voltage, terminal voltage, and speed

expected results of those actions. With this understanding, the operating staff will be able to adapt to differences in the actual versus the assumed conditions.

This case study has described restoration operations and the studies that should be part of a restoration planning process. In particular, it attempted to describe technical issues such as system dynamics and control aspects of the black-start process. This overview should be helpful to utility staff involved in the development of restoration plans. Development of thorough restoration plans and the testing of those plans through simulation and drills will help to minimize disruption of service to loads and the risk of damage to equipment following partial or total power system blackouts.

For Further Reading

M. M. Adibi, "Power system restoration, methodologies, and implementation strategies," *IEEE Series on Power Engineering*, P. M. Anderson, Ed., 2000.

IEEE Committee Report, "System restoration-deploying the plan: current operational problems working group," *IEEE Trans.*, pp. 4623–4671, Nov. 1982.

J. W. Feltes and C. Grande-Moran, "Black start studies for

system restoration," presented at the 2008 IEEE Power and Energy Society General Meeting, Pittsburgh, PA, 20–24 July 2008.

J. W. Feltes, C. Grande-Moran, P. Duggan, S. Kalinowsky, M. Zamzam, V. C. Kotecha, and F. P. de Mello, "Some considerations in the development of restoration plans for electric utilities serving large metropolitan areas," *IEEE Trans. on Power Systems*, vol. 21, no. 2, pp. 909–915, May 2006.

IEEE Committee Report, "New approaches in power system restoration," *IEEE Trans. on Power Systems*, vol. 7, no. 4, pp. 1428–1434, Nov. 1992.

M. Henderson, E. Rappold, J. W. Feltes, C. Grande-Moran, D. Durbak, and O. Bileya, "Addressing restoration issues for the ISO New England system," 2012 IEEE Power and Energy Society General Meeting, San Diego, CA, July 2012.

Biographies

James Feltes is with Siemens Power Technologies International (Siemens PTI), Schenectady, New York.

Carlos Grande-Moran is with Siemens Power Technologies International (Siemens PTI), Schenectady, New York. ■

11.1 THE SWING EQUATION

Consider a generating unit consisting of a three-phase synchronous generator and its prime mover. The rotor motion is determined by Newton's second law, given by

$$\mathrm{J}\alpha_m(t) = \mathrm{T}_m(t) - \mathrm{T}_e(t) = \mathrm{T}_a(t) \tag{11.1.1}$$

where J = total moment of inertia of the rotating masses, kg-m^2

α_m = rotor angular acceleration, rad/s^2

T_m = mechanical torque supplied by the prime mover minus the retarding torque due to mechanical losses, N-m

T_e = electrical torque that accounts for the total three-phase electrical power output of the generator, plus electrical losses, N-m

T_a = net accelerating torque, N-m

Also, the rotor angular acceleration is given by

$$\alpha_m(t) = \frac{d\omega_m(t)}{dt} = \frac{d^2\theta_m(t)}{dt^2} \tag{11.1.2}$$

$$\omega_m(t) = \frac{d\theta_m(t)}{dt} \tag{11.1.3}$$

where ω_m = rotor angular velocity, rad/s

θ_m = rotor angular position with respect to a stationary axis, rad

T_m and T_e *are* positive for generator operation. In steady-state T_m equals T_e, the accelerating torque T_a is zero, and, from (11.1.1), the rotor acceleration α_m is zero, resulting in a constant rotor velocity called *synchronous speed.* When T_m is greater than T_e, T_a is positive and α_m is therefore positive, resulting in increasing rotor speed. Similarly, when T_m is less than T_e, the rotor speed is decreasing.

It is convenient to measure the rotor angular position with respect to a synchronously rotating reference axis instead of a stationary axis. Accordingly,

$$\theta_m(t) = \omega_{msyn}t + \delta_m(t) \tag{11.1.4}$$

where ω_{msyn} = synchronous angular velocity of the rotor, rad/s
δ_m = rotor angular position with respect to a synchronously rotating reference, rad

Using (11.1.2) and (11.1.4), (11.1.1) becomes

$$\mathrm{J}\frac{d^2\theta_m(t)}{dt^2} = \mathrm{J}\frac{d^2\delta_m(t)}{dt^2} = \mathrm{T}_m(t) - \mathrm{T}_e(t) = \mathrm{T}_a(t) \tag{11.1.5}$$

It is also convenient to work with power rather than torque, and to work in per-unit rather than in actual units. Accordingly, multiply (11.1.5) by $\omega_m(t)$ and divide by S_{rated}, the three-phase voltampere rating of the generator:

$$\frac{\mathrm{J}\omega_m(t)}{\mathrm{S}_{rated}}\frac{d^2\delta_m(t)}{dt^2} = \frac{\omega_m(t)\mathrm{T}_m(t) - \omega_m(t)\mathrm{T}_e(t)}{\mathrm{S}_{rated}}$$

$$= \frac{p_m(t) - p_e(t)}{\mathrm{S}_{rated}} = p_{mp.u.}(t) - p_{ep.u.}(t) = p_{ap.u.}(t) \tag{11.1.6}$$

where $p_{mp.u.}$ = mechanical power supplied by the prime mover minus mechanical losses, per unit
$p_{ep.u.}$ = electrical power output of the generator plus electrical losses, per unit

Finally, it is convenient to work with a normalized inertia constant, called the H constant, which is defined as

$$\mathrm{H} = \frac{\text{stored kinetic energy at synchronous speed}}{\text{generator voltampere rating}}$$

$$= \frac{\frac{1}{2}\mathrm{J}\omega_{msyn}^2}{\mathrm{S}_{rated}} \quad \text{joules/VA or per unit-seconds} \tag{11.1.7}$$

The H constant has the advantage that it falls within a fairly narrow range, normally between 1 and 10 p.u.-s, whereas J varies widely, depending on generating unit size and type. Solving (11.1.7) for J and using it in (11.1.6),

$$2\mathrm{H}\frac{\omega_m(t)}{\omega_{msyn}^2}\frac{d^2\delta_m(t)}{dt^2} = p_{mp.u.}(t) - p_{ep.u.}(t) = p_{ap.u.}(t) \tag{11.1.8}$$

Defining per-unit rotor angular velocity.

$$\omega_{\text{p.u.}}(t) = \frac{\omega_m(t)}{\omega_{m\text{syn}}} \tag{11.1.9}$$

Equation (11.1.8) becomes

$$\frac{2\text{H}}{\omega_{m\text{syn}}}\omega_{\text{p.u.}}(t)\frac{d^2\delta_m(t)}{dt^2} = p_{m\text{p.u.}}(t) - p_{e\text{p.u.}}(t) = p_{a\text{p.u.}}(t) \tag{11.1.10}$$

For a synchronous generator with P poles, the electrical angular acceleration α, electrical radian frequency ω, and power angle δ are

$$\alpha(t) = \frac{\text{P}}{2}\alpha_m(t) \tag{11.1.11}$$

$$\omega(t) = \frac{\text{P}}{2}\omega_m(t) \tag{11.1.12}$$

$$\delta(t) = \frac{\text{P}}{2}\delta_m(t) \tag{11.1.13}$$

Similarly, the synchronous electrical radian frequency is

$$\omega_{\text{syn}} = \frac{\text{P}}{2}\omega_{m\text{syn}} \tag{11.1.14}$$

The per-unit electrical frequency is

$$\omega_{\text{p.u.}}(t) = \frac{\omega(t)}{\omega_{\text{syn}}} = \frac{\frac{2}{\text{P}}\omega(t)}{\frac{2}{\text{P}}\omega_{\textbf{syn}}} = \frac{\omega_m(t)}{\omega_{m\text{syn}}} \tag{11.1.15}$$

Therefore, using (11.1.13–11.1.15), (11.1.10) can be written as

$$\frac{2\text{H}}{\omega_{\text{syn}}}\omega_{\text{p.u.}}(t)\frac{d^2\delta(t)}{dt^2} = p_{m\text{p.u.}}(t) - p_{e\text{p.u.}}(t) = p_{a\text{p.u.}}(t) \tag{11.1.16}$$

Frequently (11.1.16) is modified to also include a term that represents a damping torque anytime the generator deviates from its synchronous speed, with its value proportional to the speed deviation

$$\begin{aligned} &2\text{H}/\omega_{\text{syn}}\omega_{\text{p.u.}}(t)(d^2\delta(t)/(dt^2)) \\ &\quad = p_{m\text{p.u.}}(t) - p_{e\text{p.u.}}(t) - \text{D}/\omega_{\text{syn}}(d\delta(t)/(dt)) \\ &\quad = p_{a\text{p.u.}}(t) \end{aligned} \tag{11.1.17}$$

where D is either zero or a relatively small positive number with typical values between 0 and 2. The units of D are per-unit power divided by per-unit speed deviation.

Equation (11.1.17), called the per-unit *swing equation*, is the fundamental equation that determines rotor dynamics in transient stability studies. Note that it is nonlinear due to $p_{e\text{p.u.}}(t)$ which is shown in Section 11.2 to be a nonlinear function of δ. Equation (11.1.17) is also nonlinear due to the $\omega_{\text{p.u.}}(t)$ term. However, in practice the rotor speed does not vary significantly from synchronous speed during transients. That is, $\omega_{\text{p.u.}}(t) \simeq 1.0$, which is often assumed in (11.1.17) for hand calculations.

Equation (11.1.17) is a second-order differential equation that can be rewritten as two first-order differential equations. Differentiating (11.1.4), and then using (11.1.3) and (11.1.12) through (11.1.14), result in

$$\frac{d\delta(t)}{dt} = \omega(t) - \omega_{\text{syn}} \tag{11.1.18}$$

Using (11.1.18) in (11.1.17),

$$\frac{2\text{H}}{\omega_{\text{syn}}}\omega_{\text{p.u.}}(t)\frac{d\omega(t)}{dt} = p_{m\text{p.u.}}(t) - p_{e\text{p.u.}}(t) - \text{D}/\omega_{\text{syn}}\frac{d\delta(t)}{dt} = p_{a\text{p.u.}}(t) \tag{11.1.19}$$

Equations (11.1.18) and (11.1.19) are two first-order differential equations.

EXAMPLE 11.1

Generator per-unit swing equation and power angle during a short circuit

A three-phase, 60-Hz, 500-MVA, 15-kV, 32-pole hydroelectric generating unit has an H constant of 2.0 p.u.-s and D = 0. (a) Determine ω_{syn} and $\omega_{m\text{syn}}$. (b) Give the per-unit swing equation for this unit. (c) The unit is initially operating at $p_{m\text{p.u.}} = p_{e\text{p.u.}} = 1.0$, $\omega = \omega_{\text{syn}}$, and $\delta = 10°$ when a three-phase-to-ground bolted short circuit at the generator terminals causes $p_{e\text{p.u.}}$ to drop to zero for $t \geqslant 0$. Determine the power angle 3 cycles after the short circuit commences. Assume $p_{m\text{p.u.}}$ remains constant at 1.0 per unit. Also assume $\omega_{\text{p.u.}}(t) = 1.0$ in the swing equation.

SOLUTION

a. For a 60-Hz generator,

$$\omega_{\text{syn}} = 2\pi 60 = 377 \text{ rad/s}$$

and, from (11.1.14), with P = 32 poles,

$$\omega_{m\text{syn}} = \frac{2}{\text{P}}\omega_{\text{syn}} = \left(\frac{2}{32}\right)377 = 23.56 \quad \text{rad/s}$$

b. From (11.1.16), with H = 2.0 p.u.-s,

$$\frac{4}{2\pi 60}\omega_{\text{p.u.}}(t)\frac{d^2\delta(t)}{dt^2} = p_{m\text{p.u.}}(t) - p_{e\text{p.u.}}(t)$$

c. The initial power angle is

$$\delta(0) = 10° = 0.1745 \quad \text{radian}$$

Also, from (11.1.17), at $t = 0$,

$$\frac{d\delta(0)}{dt} = 0$$

Using $p_{mp.u.}(t) = 1.0$, $p_{ep.u.} = 0$, and $\omega_{p.u.}(t) = 1.0$, the swing equation from (b) is

$$\left(\frac{4}{2\pi 60}\right)\frac{d^2\delta(t)}{dt^2} = 1.0 \quad t \geqslant 0$$

Integrating twice and using the above initial conditions,

$$\frac{d\delta(t)}{dt} = \left(\frac{2\pi 60}{4}\right)t + 0$$

$$\delta(t) = \left(\frac{2\pi 60}{8}\right)t^2 + 0.1745$$

$$\text{At } t = 3 \text{ cycles} = \frac{3 \text{ cycles}}{60 \text{ cycles/second}} = 0.05 \text{ second,}$$

$$\delta(0.05) = \left(\frac{2\pi 60}{8}\right)(0.05)^2 + 0.1745$$

$$= 0.2923 \text{ radian} = 16.75°$$

EXAMPLE 11.2

Equivalent swing equation: two generating units

A power plant has two three-phase, 60-Hz generating units with the following ratings:

Unit 1: 500 MVA, 15 kV, 0.85 power factor, 32 poles, $H_1 = 2.0$ p.u.-s, D = 0

Unit 2: 300 MVA, 15 kV, 0.90 power factor, 16 poles, $H_2 = 2.5$ p.u.-s, D = 0

a. Give the per-unit swing equation of each unit on a 100-MVA system base.

b. If the units are assumed to "swing together," that is, $\delta_1(t) = \delta_2(t)$, combine the two swing equations into one equivalent swing equation.

SOLUTION

a. If the per-unit powers on the right-hand side of the swing equation are converted to the system base, then the H constant on the left-hand side must also be converted. That is,

(Continued)

$$H_{new} = H_{old} \frac{S_{old}}{S_{new}} \quad \text{per unit}$$

Converting H_1 from its 500-MVA rating to the 100-MVA system base,

$$H_{1new} = H_{1old} \frac{S_{old}}{S_{new}} = (2.0)\left(\frac{500}{100}\right) = 10 \quad \text{p.u.-s}$$

Similarly, converting H_2,

$$H_{2new} = (2.5)\left(\frac{300}{100}\right) = 7.5 \quad \text{p.u.-s}$$

The per-unit swing equations on the system base are then

$$\frac{2H_{1new}}{\omega_{syn}} \omega_{1p.u.}(t) \frac{d^2\delta_1(t)}{dt^2} = \frac{20.0}{2\pi 60} \omega_{1p.u.}(t) \frac{d^2\delta_1(t)}{dt^2}$$

$$= p_{m1p.u.}(t) - p_{e1p.u.}(t)$$

$$\frac{2H_{2new}}{\omega_{syn}} \omega_{2p.u.}(t) \frac{d^2\delta_2(t)}{dt^2} = \frac{15.0}{2\pi 60} \omega_{2p.u.}(t) \frac{d^2\delta_2(t)}{dt^2} = p_{m2p.u.}(t) - p_{e2p.u.}$$

b. Letting:

$$\delta(t) = \delta_1(t) = \delta_2(t)$$

$$\omega_{p.u.}(t) = \omega_{1p.u.}(t) = \omega_{2p.u.}(t)$$

$$p_{mp.u.}(t) = p_{m1p.u.}(t) + p_{m2p.u.}(t)$$

$$p_{ep.u.}(t) = p_{e1p.u.}(t) + p_{e2p.u.}(t)$$

and adding the above swing equations

$$\frac{2(H_{1new} + H_{2new})}{\omega_{syn}} \omega_{p.u.}(t) \frac{d^2\delta(t)}{dt^2}$$

$$= \frac{35.0}{2\pi 60} \omega_{p.u.}(t) \frac{d^2\delta(t)}{dt^2} = p_{mp.u.}(t) - p_{ep.u.}(t)$$

When transient stability studies involving large-scale power systems with many generating units are performed, computation time can be reduced by combining the swing equations of those units that swing together. Such units, which are called *coherent machines*, usually are connected to the same bus or are electrically close, and they are usually remote from the network disturbances under study.

11.2 SIMPLIFIED SYNCHRONOUS MACHINE MODEL AND SYSTEM EQUIVALENTS

Figure 11.2 shows a simplified model of a synchronous machine, called the classical model, that can be used in transient stability studies. As shown, the synchronous machine is represented by a constant internal voltage E' behind its direct axis transient reactance X'_d. This model is based on the following assumptions:

1. The machine is operating under balanced three-phase positive-sequence conditions.
2. Machine excitation is constant.
3. Machine losses, saturation, and saliency are neglected.

In transient stability programs, more detailed models can be used to represent exciters, losses, saturation, and saliency. However, the simplified model reduces model complexity while maintaining reasonable accuracy for some stability calculations.

Each generator in the model is connected to a system consisting of transmission lines, transformers, loads, and other machines. To a first approximation the system can be represented by an "infinite bus" behind a system reactance. An infinite bus is an ideal voltage source that maintains constant voltage magnitude, constant phase, and constant frequency.

Figure 11.3 shows a synchronous generator connected to a system equivalent. The voltage magnitude V_{bus} and 0° phase of the infinite bus are constant. The phase angle δ of the internal machine voltage is the machine power angle with respect to the infinite bus.

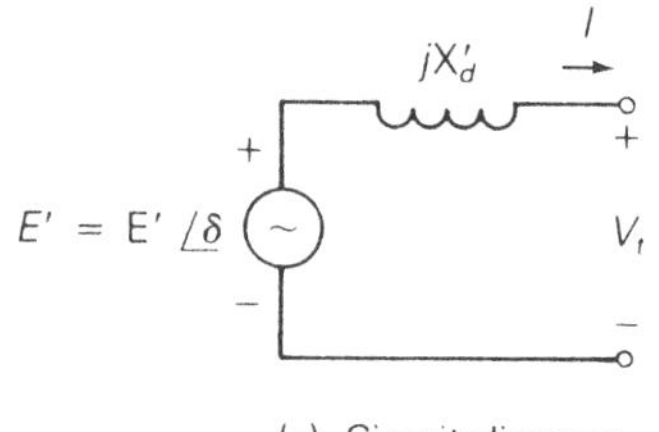

(a) Circuit diagram

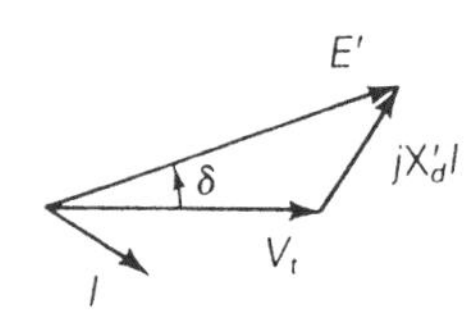

(b) Phasor diagram

FIGURE 11.2

Simplified synchronous machine model for transient stability studies

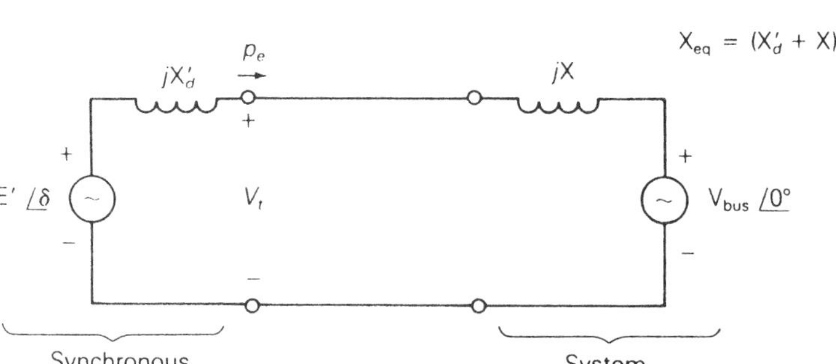

FIGURE 11.3

Synchronous generator connected to a system equivalent

The equivalent reactance between the machine internal voltage and the infinite bus is $X_{eq} = (X'_d + X)$. From (6.7.3), the real power delivered by the synchronous generator to the infinite bus is

$$p_e = \frac{E'V_{bus}}{X_{eq}} \sin \delta \tag{11.2.1}$$

During transient disturbances both E' and V_{bus} are considered constant in (11.2.1). Thus p_e is a sinusoidal function of the machine power angle δ.

EXAMPLE 11.3

Generator internal voltage and real power output versus power angle

Figure 11.4 shows a single-line diagram of a three-phase, 60-Hz synchronous generator, connected through a transformer and parallel transmission lines to an infinite bus. All reactances are given in per-unit on a common system base. If the infinite bus receives 1.0 per unit real power at 0.95 p.f. lagging, determine (a) the internal voltage of the generator and (b) the equation for the electrical power delivered by the generator versus its power angle δ.

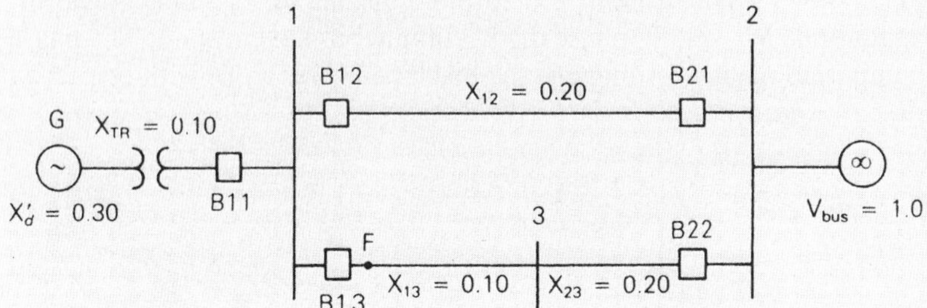

FIGURE 11.4 Single-line diagram for Example 11.3

SOLUTION

a. The equivalent circuit is shown in Figure 11.5, from which the equivalent reactance between the machine internal voltage and infinite bus is

$$X_{eq} = X'_d + X_{TR} + X_{12} \| (X_{13} + X_{23})$$

$$= 0.30 + 0.10 + 0.20 \| (0.10 + 0.20)$$

$$= 0.520 \quad \text{per unit}$$

The current into the infinite bus is

$$I = \frac{P}{V_{bus}(\text{p.f.})} \underline{/-\cos^{-1}(\text{p.f.})} = \frac{(1.0)}{(1.0)(0.95)} \underline{/-\cos^{-1} 0.95}$$

$$= 1.05263 \underline{/-18.195^\circ} \quad \text{per unit}$$

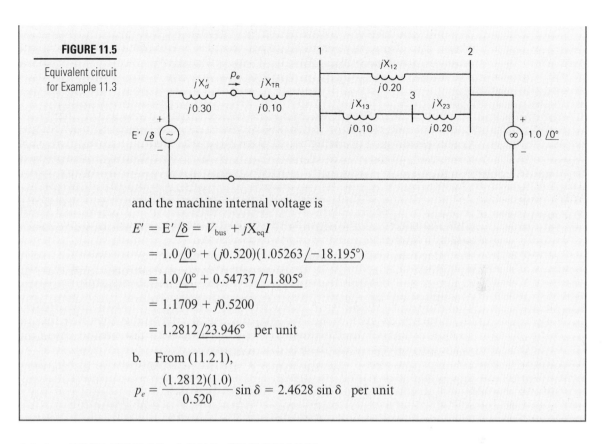

FIGURE 11.5

Equivalent circuit for Example 11.3

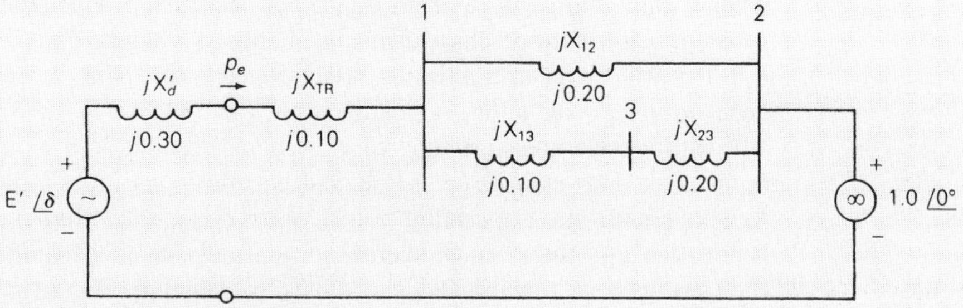

and the machine internal voltage is

$$E' = E'\underline{/\delta} = V_{\text{bus}} + jX_{\text{eq}}I$$

$$= 1.0\underline{/0^\circ} + (j0.520)(1.05263\underline{/-18.195^\circ})$$

$$= 1.0\underline{/0^\circ} + 0.54737\underline{/71.805^\circ}$$

$$= 1.1709 + j0.5200$$

$$= 1.2812\underline{/23.946^\circ} \quad \text{per unit}$$

b. From (11.2.1),

$$p_e = \frac{(1.2812)(1.0)}{0.520} \sin\delta = 2.4628 \sin\delta \quad \text{per unit}$$

11.3 THE EQUAL-AREA CRITERION

Consider a synchronous generating unit connected through a reactance to an infinite bus. Plots of electrical power p_e and mechanical power p_m versus power angle δ are shown in Figure 11.6. p_e is a sinusoidal function of δ, as given by (11.2.1).

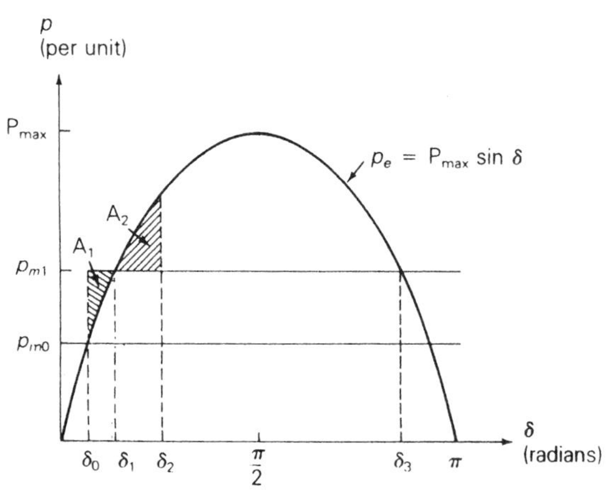

FIGURE 11.6

p_e and p_m versus δ

Suppose the unit is initially operating in steady-state at $p_e = p_m = p_{m0}$ and $\delta = \delta_0$, when a step change in p_m from p_{m0} to p_{m1} occurs at $t = 0$. Due to rotor inertia, the rotor position cannot change instantaneously. That is, $\delta_m(0^+) = \delta_m(0^-)$; therefore, $\delta(0^+) = \delta(0^-) = \delta_0$ and $p_e(0^+) = p_e(0^-)$. Since $p_m(0^+) = p_{m1}$ is greater than $p_e(0^+)$, the acceleration power $p_a(0^+)$ is positive and, from (11.1.16), $(d^2\delta)/(dt^2)(0^+)$ is positive. The rotor accelerates and δ increases. When δ reaches δ_1, $p_e = p_{m1}$ and $(d^2\delta)/(dt^2)$ becomes zero. However, $d\delta/dt$ is still positive and δ continues to increase, overshooting its final steady-state operating point. When δ is greater than δ_1, p_m is less than p_e, p_a is negative, and the rotor decelerates. Eventually, δ reaches a maximum value δ_2 and then swings back toward δ_1. Using (11.1.16), which has no damping, δ would continually oscillate around δ_1. However, damping due to mechanical and electrical losses causes δ to stabilize at its final steady-state operating point δ_1. Note that if the power angle exceeded δ_3, then p_m would exceed p_e and the rotor would accelerate again, causing a further increase in δ and loss of stability.

One method for determining stability and maximum power angle is to solve the nonlinear swing equation via numerical integration techniques using a digital computer. This method, which is applicable to multimachine systems, is described in Section 11.4. However, there is also a direct method for determining stability that does not involve solving the swing equation; this method is applicable for one machine connected to an infinite bus or for two machines. The method, called the *equal-area criterion*, is described in this section.

In Figure 11.6, p_m is greater than p_e during the interval $\delta_0 < \delta < \delta_1$, and the rotor is accelerating. The shaded area A_1 between the p_m and p_e curves is called the accelerating area. During the interval $\delta_1 < \delta < \delta_2$, p_m is less than p_e, the rotor is decelerating, and the shaded area A_2 is the decelerating area. At both the initial value $\delta = \delta_0$ and the maximum value $\delta = \delta_2$, $d\delta/dt = 0$. The equal-area criterion states that $A_1 = A_2$.

To derive the equal-area criterion for one machine connected to an infinite bus, assume $\omega_{\text{p.u.}}(t) = 1$ in (11.1.16), giving

$$\frac{2H}{\omega_{\text{syn}}}\frac{d^2\delta}{dt^2} = p_{m\text{p.u.}} - p_{e\text{p.u.}} \tag{11.3.1}$$

Multiplying by $d\delta/dt$ and using

$$\frac{d}{dt}\left[\frac{d\delta}{dt}\right]^2 = 2\left(\frac{d\delta}{dt}\right)\left(\frac{d^2\delta}{dt^2}\right)$$

(11.3.1) becomes

$$\frac{2H}{\omega_{\text{syn}}}\left(\frac{d^2\delta}{dt^2}\right)\left(\frac{d\delta}{dt}\right) = \frac{H}{\omega_{\text{syn}}}\frac{d}{dt}\left[\frac{d\delta}{dt}\right]^2 = (p_{m\text{p.u.}} - p_{e\text{p.u.}})\frac{d\delta}{dt} \tag{11.3.2}$$

Multiplying (11.3.2) by dt and integrating from δ_0 to δ,

$$\frac{H}{\omega_{\text{syn}}}\int_{\delta_0}^{\delta} d\left[\frac{d\delta}{dt}\right]^2 = \int_{\delta_0}^{\delta}(p_{m\text{p.u.}} - p_{e\text{p.u.}})d\delta$$

or

$$\frac{H}{\omega_{syn}}\left[\frac{d\delta}{dt}\right]^2\Bigg|_{\delta_0}^{\delta} = \int_{\delta_0}^{\delta}(p_{mp.u.} - p_{ep.u.})\,d\delta \qquad (11.3.3)$$

The above integration begins at δ_0 where $d\delta/dt = 0$, and continues to an arbitrary δ. When δ reaches its maximum value, denoted δ_2, $d\delta/dt$ again equals zero. Therefore, the left-hand side of (11.3.3) equals zero for $\delta = \delta_2$ and

$$\int_{\delta_0}^{\delta_2}(p_{mp.u.} - p_{ep.u.})d\delta = 0 \qquad (11.3.4)$$

Separating this integral into positive (accelerating) and negative (decelerating) areas, results in the equal-area criterion

$$\int_{\delta_0}^{\delta_1}(p_{mp.u.} - p_{ep.u.})d\delta + \int_{\delta_1}^{\delta_2}(p_{mp.u.} - p_{ep.u.})d\delta = 0$$

or

$$\int_{\delta_0}^{\delta_1}\underbrace{(p_{mp.u.} - p_{ep.u.})d\delta}_{A_1} = \int_{\delta_1}^{\delta_2}\underbrace{(p_{ep.u.} - p_{mp.u.})d\delta}_{A_2} \qquad (11.3.5)$$

In practice, sudden changes in mechanical power usually do not occur, since the time constants associated with prime mover dynamics are on the order of seconds. However, stability phenomena similar to that described above can also occur from sudden changes in electrical power, due to system faults and line switching. The following three examples are illustrative.

EXAMPLE 11.4

Equal-area criterion: transient stability during a three-phase fault

The synchronous generator shown in Figure 11.4 is initially operating in the steady-state condition given in Example 11.3, when a temporary three-phase-to-ground bolted short circuit occurs on line 1–3 at bus 1, shown as point F in Figure 11.4. Three cycles later the fault extinguishes by itself. Due to a relay misoperation, all circuit breakers remain closed. Determine whether stability is or is not maintained and determine the maximum power angle. The inertia constant of the generating unit is 3.0 per unit-seconds on the system base. Assume p_m remains constant throughout the disturbance. Also assume $\omega_{p.u.} = (t) = 1.0$ in the swing equation.

SOLUTION

Plots of p_e and p_m versus δ are shown in Figure 11.7. From Example 11.3 the initial operating point is $p_e(0^-) = p_m = 1.0$ per unit and $\delta(0^+) = \delta(0^-) = \delta_0 = 23.95° =$

(*Continued*)

FIGURE 11.7 p–δ plot for Example 11.4

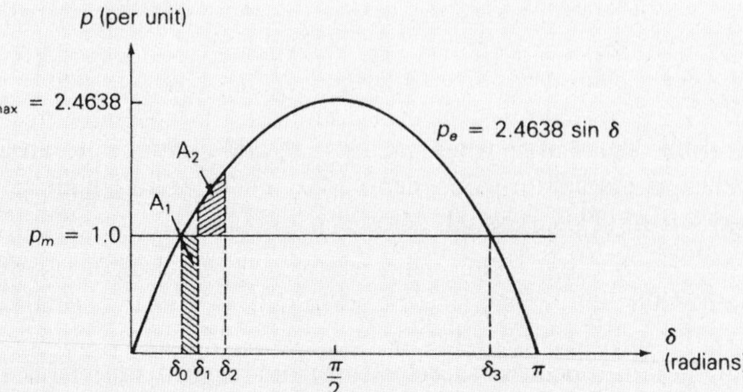

0.4179 radian. At $t = 0$, when the short circuit occurs, p_e instantaneously drops to zero and remains at zero during the fault since power cannot be transferred past faulted bus 1. From (11.1.16), with $\omega_{\text{p.u.}}(t) = 1.0$,

$$\frac{2H}{\omega_{\text{syn}}}\frac{d^2\delta(t)}{dt^2} = p_{m\text{p.u.}} \qquad 0 \leq t \leq 0.05 \text{ s}$$

Integrating twice with initial condition $\delta(0) = \delta_0$ and $\dfrac{d\delta(0)}{dt} = 0$,

$$\frac{d\delta(t)}{dt} = \frac{\omega_{\text{syn}}\,p_{m\text{p.u.}}}{2H}t + 0$$

$$\delta(t) = \frac{\omega_{\text{syn}}\,p_{m\text{p.u.}}}{4H}t^2 + \delta_0$$

At $t = 3$ cycles $= 0.05$ second,

$$\delta_1 = \delta(0.05 \text{ s}) = \frac{2\pi 60}{12}(0.05)^2 + 0.4179$$

$$= 0.4964 \text{ radian} = 28.44°$$

The accelerating area A_1, shaded in Figure 11.7, is

$$A_1 = \int_{\delta_0}^{\delta_1} p_m d\delta = \int_{\delta_0}^{\delta_1} 1.0 d\delta = (\delta_1 - \delta_0) = 0.4964 - 0.4179 = 0.0785$$

At $t = 0.05$ s the fault extinguishes and p_e instantaneously increases from zero to the sinusoidal curve in Figure 11.7. δ continues to increase until the decelerating area A_2 equals A_1. That is,

$$A_2 = \int_{\delta_1}^{\delta_2} (p_{\max} \sin\delta - p_m)d\delta$$

$$= \int_{0.4964}^{\delta_2} (2.4638 \sin\delta - 1.0)d\delta = A_1 = 0.0785$$

Integrating,

$$2.4638[\cos(0.4964) - \cos\delta_2] - (\delta_2 - 0.4964) = 0.0785$$

$$2.4638\cos\delta_2 + \delta_2 = 2.5843$$

The above nonlinear algebraic equation can be solved iteratively to obtain $\delta_2 = 0.7003$ radian $= 40.12°$

Since the maximum angle δ_2 does not exceed $\delta_3 = (180° - \delta_0) = 156.05°$, stability is maintained. In steady-state, the generator returns to its initial operating point $p_{ess} = p_m = 1.0$ per unit and $\delta_{ss} = \delta_0 = 23.95°$.

Note that as the fault duration increases, the risk of instability also increases. The *critical clearing time*, denoted t_{cr}, is the longest fault duration allowable for stability.

To see this case modeled in PowerWorld Simulator, open case Example 11_4 (see Figure 11.8). Then select **Add-Ons, Transient Stability**, which displays the Transient Stability Analysis Form. Notice that in the Transient Stability Contingency Elements list, a fault is applied to bus 1 at $t = 0$ s and cleared at $t = 0.05$ s (three cycles later). To see the time variation in the generator angle (modeled at bus 4 in PowerWorld Simulator), click the **Run Transient Stability** button. When the simulation is finished, a graph showing this angle automatically appears, as

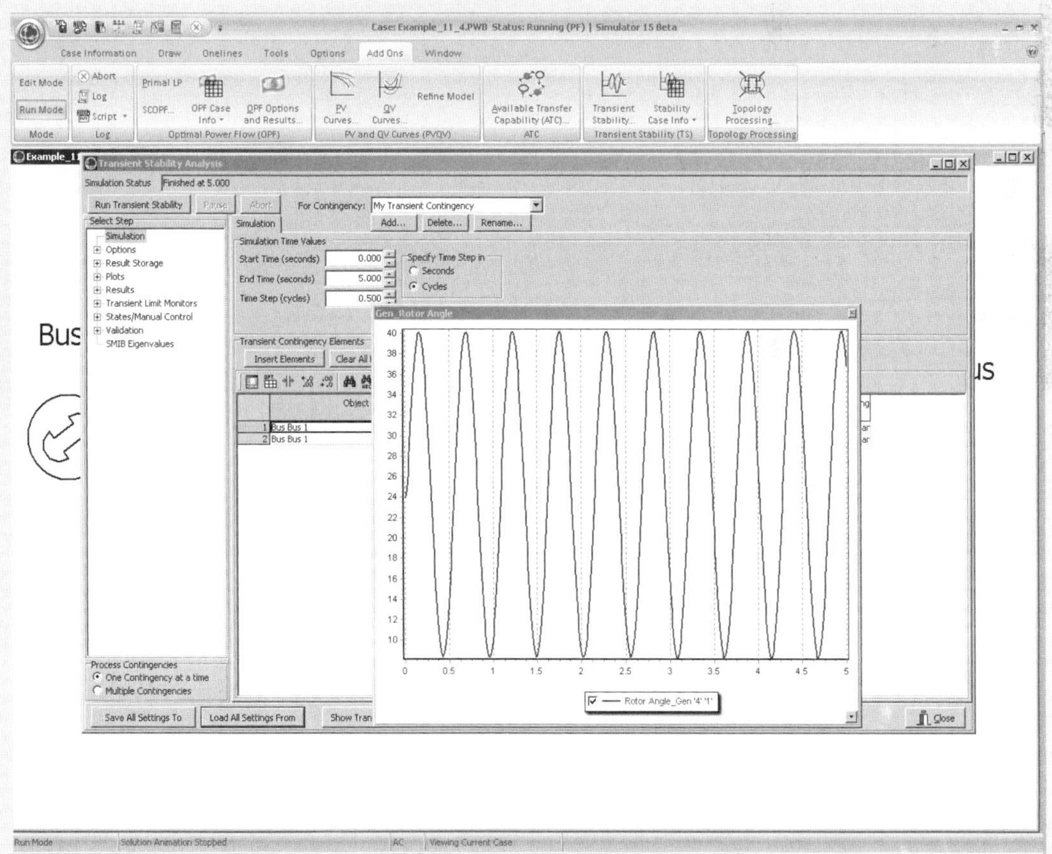

FIGURE 11.8

Variation in $\delta(t)$ without Damping

(Continued)

shown in the figure. More detailed results are also available by clicking on **Results** in the list on the left side of the form. To rerun the example with a different fault duration, modify the Time (second) field in the Transient Contingency Elements list, and then again click the **Run Transient Stability** button.

Notice that because this system is modeled without damping (i.e., D = 0), the angle oscillations do not damp out with time. To extend the example, right-click on the Bus 4 generator on the one-line diagram and select **Generator Information Dialog**. Then click on the **Stability, Machine Models** tab to see the parameters associated with the GENCLS model (i.e., a classical model—a more detailed machine model is introduced in Section 11.6). Change the D field to 1.0, select **OK** to close the dialog, and then rerun the transient stability case. The results are as shown in Figure 11.9. While the inclusion of damping did not significantly alter the maximum for δ(t), the magnitude of the angle oscillations is now decreasing with time. For convenience this modified example is contained in PowerWorld Simulator case Example 11_4b.

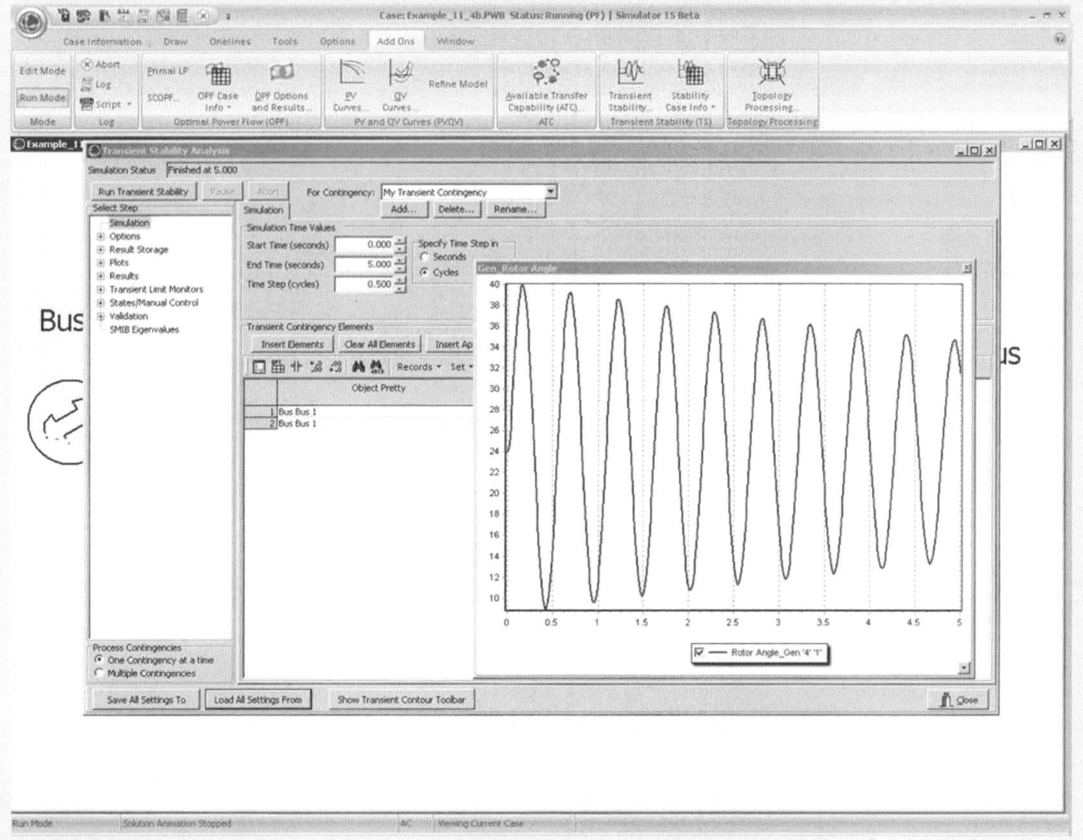

FIGURE 11.9

Variation in $\delta(t)$ with Damping

EXAMPLE 11.5

Equal-area criterion: critical clearing time for a temporary three-phase fault

Assuming the temporary short circuit in Example 11.4 lasts longer than 3 cycles, calculate the critical clearing time.

SOLUTION

The p–δ plot is shown in Figure 11.10. At the critical clearing angle, denoted δ_{cr}, the fault is extinguished. The power angle then increases to a maximum value $\delta_3 = 180° - \delta_0 = 156.05° = 2.7236$ radians, which gives the maximum decelerating area. Equating the accelerating and decelerating areas,

$$A_1 = \int_{\delta_0}^{\delta_{cr}} p_m \, d\delta = A_2 = \int_{\delta_{cr}}^{\delta_3} (P_{max} \sin\delta - p_m) d\delta$$

$$\int_{0.4179}^{\delta_{cr}} 1.0 \, d\delta = \int_{\delta_{cr}}^{2.7236} (2.4638 \sin\delta - 1.0) \, d\delta$$

Solving for δ_{cr},

$$(\delta_{cr} - 0.4179) = 2.4638[\cos\delta_{cr} - \cos(2.7236)] - (2.7236 - \delta_{cr})$$

$$2.4638 \cos\delta_{cr} = +0.05402$$

$$\delta_{cr} = 1.5489 \text{ radians} = 88.74°$$

From the solution to the swing equation given in Example 11.4,

$$\delta(t) = \frac{\omega_{syn} p_{mp.u.}}{4H} t^2 + \delta_0$$

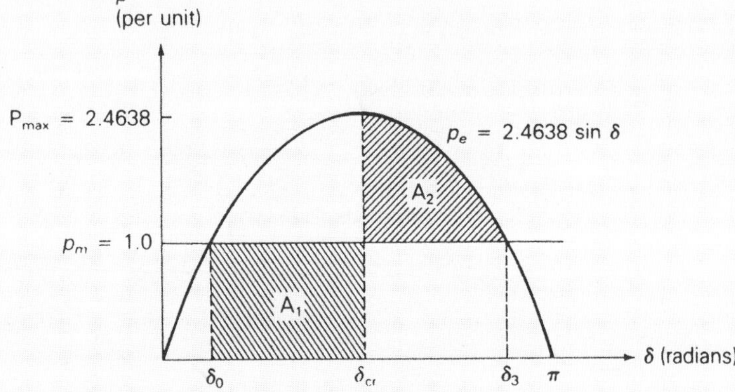

FIGURE 11.10 p–δ plot for Example 11.5

(Continued)

Solving

$$t = \sqrt{\frac{4H}{\omega_{\text{syn}} p_{m\text{p.u.}}}(\delta(t) - \delta_0)}$$

Using $\delta(t_{cr}) = \delta_{cr} = 1.5489$ and $\delta_0 = 0.4179$ radian,

$$t_{cr} = \sqrt{\frac{12}{(2\pi 60)(1.0)}(1.5489 - 0.4179)}$$
$$= 0.1897 \text{ s} = 11.38 \text{ cycles}$$

If the fault is cleared before $t = t_{cr} = 11.38$ cycles, stability is maintained. Otherwise, the generator goes out of synchronism with the infinite bus; that is, stability is lost.

To see a time-domain simulation of this case, open Example 11_5 in PowerWorld Simulator (see Figure 11.11). Again select **Add-Ons, Transient Stability**

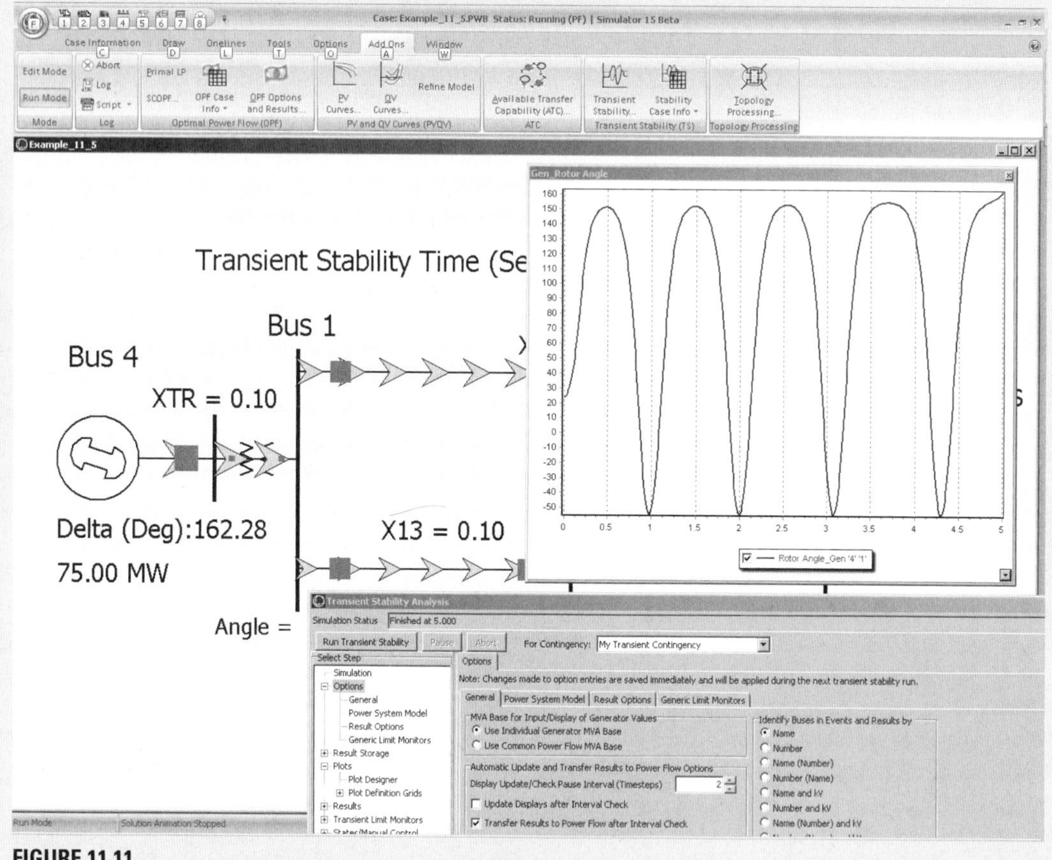

FIGURE 11.11

Variation in $\delta(t)$ for Example 11.5

to view the Transient Stability Analysis Form. In order to better visualize the results on a PowerWorld one-line diagram, there is an option to transfer the transient stability results to the one-line every n timesteps. To access this option, select the **Options** page from the list on the left side of the display, then the **General** tab, then check the **Transfer Results to Power Flow after Interval Check** field. With this option checked, click on the **Run Transient Stability**, which will run the case with a critical clearing time of 0.1895 seconds. The plot is set to dynamically update as well. The final results are shown in Figure 11.11. Because the oneline is reanimated every n time-steps (4 in this case), a potential downside to this option is it takes longer to run. Uncheck the option to restore full solution speed.

EXAMPLE 11.6

Equal-area criterion: critical clearing angle for a cleared three-phase fault

The synchronous generator in Figure 11.4 is initially operating in the steady-state condition given in Example 11.3 when a permanent three-phase-to-ground bolted short circuit occurs on line 1–3 at bus 3. The fault is cleared by opening the circuit breakers at the ends of line 1–3 and line 2–3. These circuit breakers then remain open. Calculate the critical clearing angle. As in previous examples, H = 3.0 p.u.-s, $p_m = 1.0$ per unit and $\omega_{\text{p.u.}} = 1.0$ in the swing equation.

SOLUTION

From Example 11.3, the equation for the prefault electrical power, denoted p_{e1} here, is $p_{e1} = 2.4638 \sin\delta$ per unit. The faulted network is shown in Figure 11.12 (a), and the Thévenin equivalent of the faulted network, as viewed from the generator internal voltage source, is shown in Figure 11.12(b). The Thévenin reactance is

$$X_{\text{Th}} = 0.40 + 0.20 \| 0.10 = 0.46666 \quad \text{per unit}$$

and the Thévenin voltage source is

$$\begin{aligned} V_{\text{Th}} &= 1.0\underline{/0^\circ}\left[\frac{X_{13}}{X_{13} + X_{12}}\right] = 1.0\underline{/0^\circ}\,\frac{0.10}{0.30} \\ &= 0.33333\underline{/0^\circ} \quad \text{per unit} \end{aligned}$$

From Figure 11.12(b), the equation for the electrical power delivered by the generator to the infinite bus during the fault, denoted p_{e2}, is

$$p_{e2} = \frac{E'V_{\text{Th}}}{X_{\text{Th}}}\sin\delta = \frac{(1.2812)(0.3333)}{0.46666}\sin\delta = 0.9152\sin\delta \quad \text{per unit}$$

(Continued)

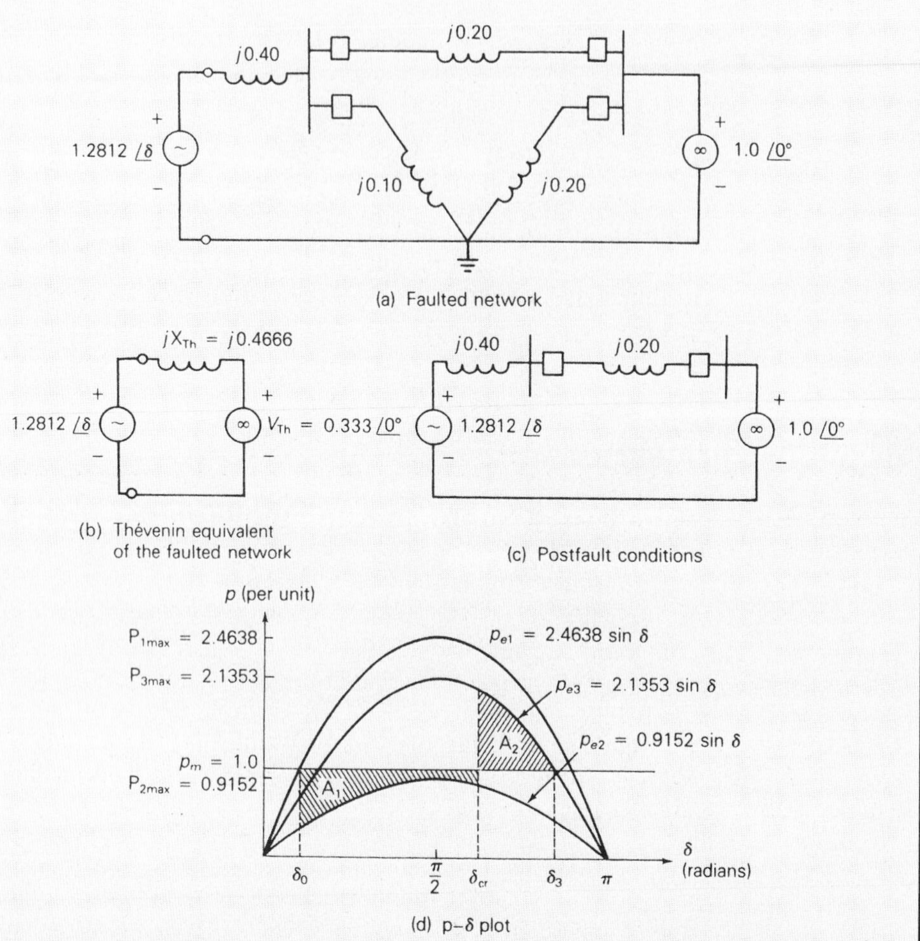

FIGURE 11.12 Example 11.6

The postfault network is shown in Figure 11.12(c), where circuit breakers have opened and removed lines 1–3 and 2–3. From this figure, the postfault electrical power delivered, denoted p_{e3}, is

$$p_{e3} = \frac{(1.2812)(1.0)}{0.60} \sin \delta = 2.1353 \sin \delta \quad \text{per unit}$$

The p–δ curves as well as the accelerating area A_1 and decelerating area A_2 corresponding to critical clearing are shown in Figure 11.12(d). Equating A_1 and A_2,

$$A_1 = \int_{\delta_0}^{\delta_{cr}} (p_m - P_{2max} \sin \delta) d\delta = A_2 = \int_{\delta_{cr}}^{\delta_3} (P_{3max} \sin \delta - p_m) d\delta$$

$$\int_{0.4179}^{\delta_{cr}} (1.0 - 0.9152 \sin \delta) d\delta = \int_{\delta_{cr}}^{2.6542} (2.1353 \sin \delta - 1.0) d\delta$$

Solving for δ_{cr},

$$(\delta_{cr} - 0.4179) + 0.9152(\cos\delta_{cr} - \cos 0.4179)$$
$$= 2.1353(\cos\delta_{cr} - \cos 2.6542) - (2.6542 - \delta_{cr})$$
$$-1.2201\cos\delta_{cr} = 0.4868$$
$$\delta_{cr} = 1.9812 \text{ radians} = 111.5°$$

If the fault is cleared before $\delta = \delta_{cr} = 111.5°$, stability is maintained. Otherwise, stability is lost. To see this case in PowerWorld Simulator open case Example 11_6.

11.4 NUMERICAL INTEGRATION OF THE SWING EQUATION

The equal-area criterion is applicable to one machine and an infinite bus or to two machines. For multimachine stability problems, however, numerical integration techniques can be employed to solve the swing equation for each machine.

Given a first-order differential equation

$$\frac{dx}{dt} = f(x) \tag{11.4.1}$$

one relatively simple integration technique is Euler's method [1], illustrated in Figure 11.13. The integration step size is denoted Δt. Calculating the slope at the beginning of the integration interval, from (11.4.1),

$$\frac{dx_t}{dt} = f(x_t) \tag{11.4.2}$$

The new value $x_{t+\Delta t}$ is calculated from the old value x_t by adding the increment Δx,

$$x_{t+\Delta t} = x_t + \Delta x = x_t + \left(\frac{dx_t}{dt}\right)\Delta t \tag{11.4.3}$$

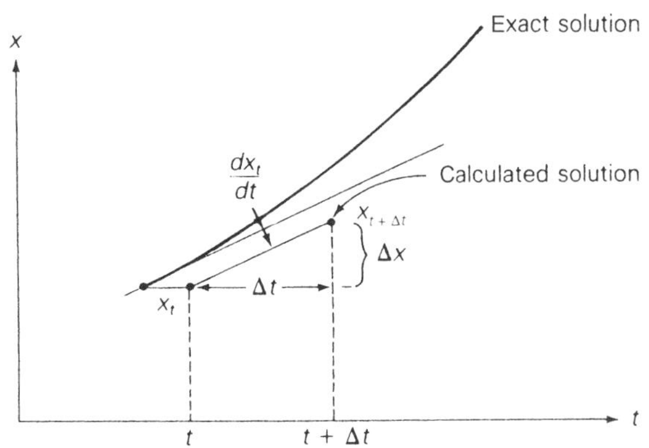

FIGURE 11.13

Euler's method

As shown in the figure, Euler's method assumes that the slope is constant over the entire interval Δt. An improvement can be obtained by calculating the slope at both the beginning and end of the interval, and then averaging these slopes. The modified Euler's method is illustrated in Figure 11.14. First, the slope at the beginning of the interval is calculated from (11.4.1) and used to calculate a preliminary value $\tilde{x}$ given by

$$\tilde{x} = x_t + \left(\frac{dx_t}{dt}\right)\Delta t \tag{11.4.4}$$

Next, the slope at $\tilde{x}$ is calculated:

$$\frac{d\tilde{x}}{dt} = f(\tilde{x}) \tag{11.4.5}$$

Then, the new value is calculated using the average slope:

$$x_{t+\Delta t} = x_t + \frac{\left(\dfrac{dx_t}{dt} + \dfrac{d\tilde{x}}{dt}\right)}{2}\Delta t \tag{11.4.6}$$

Now apply the modified Euler's method to calculate machine frequency ω and power angle δ. Letting x be either δ or ω, the old values at the beginning of the interval are denoted δ_t and ω_t. From (11.1.17) and (11.1.18), the slopes at the beginning of the interval are

$$\frac{d\delta_t}{dt} = \omega_t - \omega_{\text{syn}} \tag{11.4.7}$$

$$\frac{d\omega_t}{dt} = \frac{p_{a\text{p.u.}t}\,\omega_{\text{syn}}}{2\text{H}\omega_{\text{p.u.}t}} \tag{11.4.8}$$

where $p_{a\text{p.u.}t}$ is the per-unit accelerating power calculated at $\delta = \delta_t$ and $\omega_{\text{p.u.}t} = \omega_t/\omega_{\text{syn}}$. Applying (11.4.4), preliminary values are

$$\tilde{\delta} = \delta_t + \left(\frac{d\delta_t}{dt}\right)\Delta t \tag{11.4.9}$$

$$\tilde{\omega} = \omega_t + \left(\frac{d\omega_t}{dt}\right)\Delta t \tag{11.4.10}$$

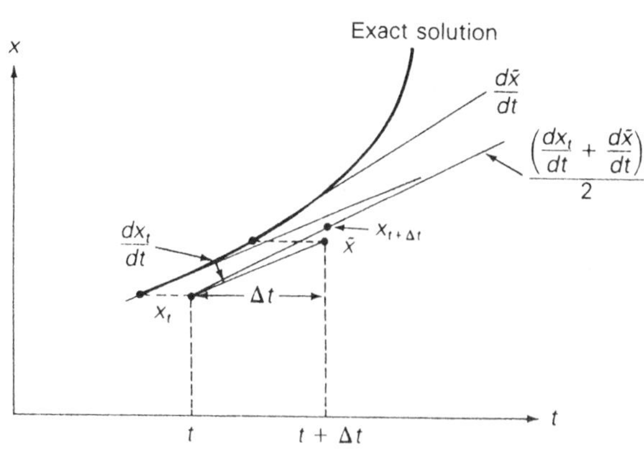

FIGURE 11.14 Modified Euler's method

Next, the slopes at $\tilde{\delta}$ and $\tilde{\omega}$ *are* calculated, again using (11.1.17) and (11.1.18):

$$\frac{d\tilde{\delta}}{dt} = \tilde{\omega} - \omega_{\text{syn}} \tag{11.4.11}$$

$$\frac{d\tilde{\omega}}{dt} = \frac{\tilde{p}_{a\text{p.u.}}\omega_{\text{syn}}}{2\text{H}\tilde{\omega}_{\text{p.u.}}} \tag{11.4.12}$$

where $\tilde{p}_{a\text{p.u.}}$ is the per-unit accelerating power calculated at $\delta = \tilde{\delta}$, and $\tilde{\omega}_{\text{p.u.}} = \tilde{\omega}/\omega_{\text{syn}}$. Applying (11.4.6), the new values at the end of the interval are

$$\delta_{t+\Delta t} = \delta_t + \frac{\left(\dfrac{d\delta_t}{dt} + \dfrac{d\tilde{\delta}}{dt}\right)}{2}\Delta t \tag{11.4.13}$$

$$\omega_{t+\Delta t} = \omega_t + \frac{\left(\dfrac{d\omega_t}{dt} + \dfrac{d\tilde{\omega}}{dt}\right)}{2}\Delta t \tag{11.4.14}$$

This procedure, given by (11.4.7) through (11.4.13), begins at $t = 0$ with specified initial values δ_0 and ω_0, and continues iteratively until $t = T$, a specified final time.

In addition to Euler's method, there are many other numerical integration techniques, such as Runge-Kutta, Picard's method, and Milne's predictor-corrector method [1]. Comparison of the methods shows a trade-off of accuracy versus computation complexity. The Euler method is a relatively simple method to compute, but requires a small step size Δt for accuracy. Some of the other methods can use a larger step size for comparable accuracy, but the computations are more complex.

To see this case in PowerWorld Simulator, open case Example 11_7, which plots both the generator angle and speed. However, rather than showing the speed in radians per second, Hertz is used. Also, in addition to plotting the angle and speed versus time, the case includes a "phase portrait" in which the speed is plotted as a function of the angle. Numeric results are available by clicking on the **Results** page. The results shown in PowerWorld differ slightly from those in the table because PowerWorld uses a more exact second order integration method.

EXAMPLE 11.7

Euler's method: computer solution to swing equation and critical clearing time

Verify the critical clearing angle determined in Example 11.6, and calculate the critical clearing time by applying the modified Euler's method to solve the swing equation for the following two cases:

Case 1 The fault is cleared at $\delta = 1.95$ radians $= 112°$ (which is less than δ_{cr})

Case 2 The fault is cleared at $\delta = 2.09$ radians $= 120°$ (which is greater than δ_{cr})

(*Continued*)

For calculations, use a step size $\Delta t = 0.01$ s, and solve the swing equation from $t = 0$ to $t = T = 0.85$ s.

SOLUTION

Equations (11.4.7) through (11.4.14) are solved by a computer program written in BASIC. From Example 11.6, the initial conditions at $t = 0$ are

$$\delta_0 = 0.4179 \text{ rad}$$
$$\omega_0 = \omega_{syn} = 2\pi 60 \text{ rad/s}$$

Also, the H constant is 3.0 p.u.-s, and the faulted accelerating power is

$$p_{ap.u.} = 1.0 - 0.9152 \sin\delta$$

The postfault accelerating power is

$$p_{ap.u.} = 1.0 - 2.1353 \sin\delta \quad \text{per unit}$$

The computer program and results at 0.02 s printout intervals are listed in Table 11.1. As shown, these results agree with Example 11.6, since the system is stable for Case 1 and unstable for Case 2. Also from Table 11.1, the critical clearing time is between 0.34 and 0.36 s.

Case 1 Stable			Case 2 Unstable		
Time s	Delta rad	Omega rad/s	Time s	Delta rad	Omega rad/s
0.000	0.418	376.991	0.000	0.418	376.991
0.020	0.426	377.778	0.020	0.426	377.778
0.040	0.449	378.547	0.040	0.449	378.547
0.060	0.488	379.283	0.060	0.488	379.283
0.080	0.541	379.970	0.080	0.541	379.970
0.100	0.607	380.599	0.100	0.607	380.599
0.120	0.685	381.159	0.120	0.685	381.159
0.140	0.773	381.646	0.140	0.773	381.646
0.160	0.870	382.056	0.160	0.870	382.056
0.180	0.975	382.392	0.180	0.975	382.392
0.200	1.086	382.660	0.200	1.086	382.660
0.220	1.202	382.868	0.220	1.202	382.868
0.240	1.321	383.027	0.240	1.321	383.027
0.260	1.443	383.153	0.260	1.443	383.153
0.280	1.567	383.262	0.280	1.567	383.262
0.300	1.694	383.370	0.300	1.694	383.370
0.320	1.823	383.495	0.320	1.823	383.495
0.340	1.954	383.658	0.340	1.954	383.658
	Fault Cleared		0.360	2.090	383.876
0.360	2.076	382.516		Fault Cleared	
0.380	2.176	381.510	0.380	2.217	382.915
0.400	2.257	380.638	0.400	2.327	382.138
0.420	2.322	379.886	0.420	2.424	381.546
0.440	2.373	379.237	0.440	2.511	381.135
0.460	2.413	378.674	0.460	2.591	380.902
0.480	2.441	378.176	0.480	2.668	380.844

Program Listing

```
10 REM EXAMPLE 13.7
20 REM SOLUTION TO SWING EQUATION
30 REM THE STEP SIZE IS DELTA
40 REM THE CLEARING ANGLE IS DLTCLR
50 DELTA+.01
60 DLTCLR = 1.95
70 J = 1
80 PMAX = .9152
90 PI = 3.1415927 #
100 T = 0
110 X1 =.4179
120 X2 = 2*PI*60
130 LPRINT "TIME DELTA OMEGA"
140 LPRINT"srad rad/s"
150 LPRINT USING "#####.###"; T;XI;X2
160 FORK = 1 TO 86
170 REM LINE 180 IS EQ (13.4.7)
180 X3 = X2 – (2*PI*60)
190 IF J = 2 THEN GOTO 240
200 IF X1 > DLTCLR OR XI = DLTCLR THEN
    PMAX = 2.1353
210 IF X1 > DLTCLR OR XI = DLTCLR THEN
    LPRINT "FAULT CLEARED"
220 IF XI > DLTCLR OR XI = DLTCLR THEN
    J = 2
```

0.500	2.460	377.726	0.500	2.746	380.969
0.520	2.471	377.307	0.520	2.828	381.288
0.540	2.473	376.900	0.540	2.919	381.824
0.560	2.467	376.488	0.560	3.022	382.609
0.580	2.453	376.056	0.580	3.145	383.686
0.600	2.429	375.583	0.600	3.292	385.111
0.620	2.396	375.053	0.620	3.472	386.949
0.640	2.351	374.446	0.640	3.693	389.265
0.660	2.294	373.740	0.660	3.965	392.099
0.680	2.221	372.917	0.680	4.300	395.426
0.700	2.130	371.960	0.700	4.704	399.079
0.720	2.019	370.855	0.720	5.183	402.689
0.740	1.884	369.604	0.740	5.729	405.683
0.760	1.723	368.226	0.760	6.325	407.477
0.780	1.533	366.773	0.780	6.941	407.812
0.800	1.314	365.341	0.800	7.551	406.981
0.820	1.068	364.070	0.820	8.139	405.711
0.840	0.799	363.143	0.840	8.702	404.819
0.860	0.516	362.750	0.860	9.257	404.934

```
230 REM LINES 240 AND 250 ARE EQ (13.4.8)
240 X4=I-PMAX*SIN(XI)
250 X5 = X4 * (2 * PI * 60) * (2 * PI * 60)/(6 * X2)
260 REM LINE 270 IS EQ (13.4.9)
270 X6 = X1 +X3* DELTA
280 REM LINE 290 IS EQ (I3.4.10)
290 X7 = X2 + X5* DELTA
300 REM LINE 3 10 IS EQ (13.4.1 1)
310 X8 = X7-2*PI*60
320 REM LINES 330 AND 340 ARE EQ (13.4.12)
330X9 = I-PMAX*SIN (X6)
340 X10 = X9 * (2 * PI * 60) * (2 * PI * 60)/(6 * X7)
350 REM LINE 360 IS EQ (13.4.13)
360 X1 = X1 + (X3 + X8) * (DELTA/2)
370 REM LINE 380 IS EQ (I3.4.14)
380 X2 = X2 + (X5 + X10) * (DELTA/2)
390 T = K* DELTA
400 Z = K/2
410 M = INT (Z)
420 IF M = Z THEN LPRINT USING
  "#####.###"; T;XI;X2
430 NEXT K
440 END
```

TABLE 11.1

Computer calculation of swing curves for Example 11.7

11.5 MULTIMACHINE STABILITY

The numerical integration methods discussed in Section 11.4 can be used to solve the swing equations for a multimachine stability problem. However, a method is required for computing machine output powers for a general network. Figure 11.15 shows a general N-bus power system with M synchronous machines. Each machine is the same as that represented by the simplified model of Figure 11.2, and the internal machine voltages are denoted $E'_1, E'_2, \ldots, E'_M$. The M machine terminals are connected

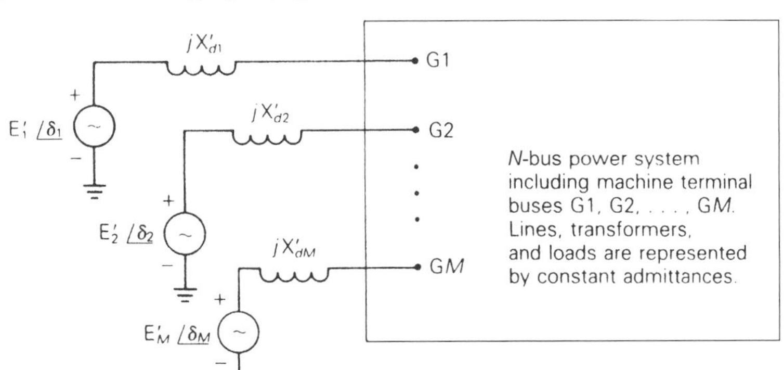

FIGURE 11.15

N-bus power-system representation for transient stability studies to system buses denoted Gl, G2,..., GM in Figure 11.15. All loads are modeled here as constant admittances. Writing nodal equations for this network,

$$\begin{bmatrix} \boldsymbol{Y}_{11} & \boldsymbol{Y}_{12} \\ \boldsymbol{Y}_{12}^{\mathrm{T}} & \boldsymbol{Y}_{22} \end{bmatrix} \begin{bmatrix} \boldsymbol{V} \\ \boldsymbol{E} \end{bmatrix} = \begin{bmatrix} \boldsymbol{0} \\ \boldsymbol{I} \end{bmatrix} \tag{11.5.1}$$

where

$$\boldsymbol{V} = \begin{bmatrix} V_1 \\ V_2 \\ \vdots \\ V_N \end{bmatrix} \text{ is the } N \text{ vector of bus voltages} \tag{11.5.2}$$

$$\boldsymbol{E} = \begin{bmatrix} E_1' \\ E_2' \\ \vdots \\ E_M' \end{bmatrix} \text{ is the } M \text{ vector of machine voltages} \tag{11.5.3}$$

$$\boldsymbol{I} = \begin{bmatrix} I_1 \\ I_2 \\ \vdots \\ I_M \end{bmatrix} \begin{array}{l} \text{is the } M \text{ vector of machine currents} \\ \text{(these are current sources)} \end{array} \tag{11.5.4}$$

$$\begin{bmatrix} \boldsymbol{Y}_{11} & \boldsymbol{Y}_{12} \\ \boldsymbol{Y}_{12}^{\mathrm{T}} & \boldsymbol{Y}_{22} \end{bmatrix} \text{ is an } (N + M) \times (N + M) \text{ admitance matrix} \tag{11.5.5}$$

The admittance matrix in (11.5.5) is partitioned in accordance with the N system buses and M internal machine buses, as follows:

$\boldsymbol{Y}_{11}$ is $N \times N$
$\boldsymbol{Y}_{12}$ is $N \times M$
$\boldsymbol{Y}_{22}$ is $M \times M$

$\boldsymbol{Y}_{11}$ is similar to the bus admittance matrix used for power flows in Chapter 6, except that load admittances and inverted generator impedances are included. That is, if a load is connected to bus n, then that load admittance is added to the diagonal element Y_{11nn}. Also, $(1/jX_{dn}')$ is added to the diagonal element Y_{11GnGn}.

$\boldsymbol{Y}_{22}$ is a diagonal matrix of inverted generator impedances; that is,

$$\boldsymbol{Y}_{22} = \begin{bmatrix} \frac{1}{jX_{d1}'} & & & 0 \\ & \frac{1}{jX_{d2}'} & & \\ & & \ddots & \\ 0 & & & \frac{1}{jX_{dM}'} \end{bmatrix} \tag{11.5.6}$$

Also, the kmth element of $\boldsymbol{Y}_{12}$ is

$$Y_{12km} = \begin{cases} \dfrac{-1}{jX'_{dn}} & \text{if } k = \text{G}n \text{ and } m = n \\ 0 & \text{otherwise} \end{cases} \tag{11.5.7}$$

Writing (11.5.1) as two separate equations,

$$\boldsymbol{Y}_{11}\boldsymbol{V} + \boldsymbol{Y}_{12}\boldsymbol{E} = \boldsymbol{0} \tag{11.5.8}$$

$$\boldsymbol{Y}_{12}^{\mathrm{T}}\boldsymbol{V} + \boldsymbol{Y}_{22}\boldsymbol{E} = \boldsymbol{I} \tag{11.5.9}$$

Assuming $\boldsymbol{E}$ is known, (11.5.8) is a linear equation in $\boldsymbol{V}$ that can be solved either iteratively or by Gauss elimination. Using the Gauss-Seidel iterative method given by (7.2.9), the kth component of $\boldsymbol{V}$ is

$$V_k(i+1) = \frac{1}{Y_{11kk}}\left[-\sum_{n=1}^{M} Y_{12kn}E_n - \sum_{n=1}^{k-1} Y_{11kn}V_n(i+1) - \sum_{n=k+1}^{N} Y_{11kn}V_n(i)\right] \tag{11.5.10}$$

After $\boldsymbol{V}$ is computed, the machine currents can be obtained from (11.5.9). That is,

$$\boldsymbol{I} = \begin{bmatrix} I_1 \\ I_2 \\ \vdots \\ I_M \end{bmatrix} = \boldsymbol{Y}_{12}^{T}\boldsymbol{V} + \boldsymbol{Y}_{22}\boldsymbol{E} \tag{11.5.11}$$

The (real) electrical power output of machine n is then

$$p_{en} = \mathrm{Re}[E_n I_n^*] \qquad n = 1,2,\ldots,M \tag{11.5.12}$$

The computation procedure for solving a transient stability problem alternately solves the swing equations representing the machines and the above algebraic power-flow equations representing the network. Use the modified Euler method of Section 11.4 to solve the swing equations and the Gauss-Seidel iterative method to solve the power-flow equations. The procedure is outlined in the following 11 steps.

TRANSIENT STABILITY COMPUTATION PROCEDURE

STEP 1 Run a prefault power-flow program to compute initial bus voltages V_k, $k = 1, 2,\ldots,N$, initial machine currents I_n, and initial machine electrical power outputs p_{en}, $n = 1, 2,\ldots,M$. Set machine mechanical power outputs, $p_{mn} = p_{en}$. Set initial machine frequencies, $\omega_n = \omega_{\text{syn}}$. Compute the load admittances.

STEP 2 Compute the internal machine voltages:

$$E_n = E_n \underline{/\delta_n} = V_{Gn} + (jX'_{dn})I_n \quad n = 1, 2, \ldots, M$$

where V_{Gn} and I_n are computed in Step 1. The magnitudes E_n will remain constant throughout the study. The angles δ_n *are* the initial power angles.

STEP 3 Compute Y_{11}. Modify the $(N \times N)$ power-flow bus admittance matrix by including the load admittances and inverted generator impedances.

STEP 4 Compute Y_{22} from (11.5.6) and Y_{12} from (11.5.7).

STEP 5 Set time $t = 0$.

STEP 6 Is there a switching operation, change in load, short circuit, or change in data? For a switching operation or change in load, modify the bus admittance matrix. For a short circuit, set the faulted bus voltage [in (11.5.10)] to zero.

STEP 7 Using the internal machine voltages $E_n = E_n \underline{/\delta_n}$, $n = 1, 2, \ldots, M$, with the values of δ_n at time t, compute the machine electrical powers p_{en} at time t from (11.5.10) to (11.5.12).

STEP 8 Using p_{en} computed in Step 7 and the values of δ_n and ω_n at time t, compute the preliminary estimates of power angles δ_n and machine speeds $\tilde{\omega}_n$ at time $(t + \Delta t)$ from (11.4.7) to (11.4.10).

STEP 9 Using $E_n = E_n \underline{/\tilde{\delta}_n}$, $n = 1, 2, \ldots, M$, compute the preliminary estimates of the machine electrical powers p_{en} at time $(t + \Delta t)$ from (11.5.10) to (11.5.12).

STEP 10 Using $\tilde{p}_{en}$ computed in Step 9, as well as $\tilde{\delta}_n$ and $\tilde{\omega}_n$ computed in Step 8, compute the final estimates of power angles δ_n and machine speeds ω_n at time $(t + \Delta t)$ from (11.4.11) to (11.4.14).

STEP 11 Set time $t = t + \Delta t$. Stop if $t > T$. Otherwise, return to Step 6.

An important transient stability parameter is the step size (time step), Δt, used in the numerical integration. Because the time required to solve a transient stability problem varies inversely with the time step, a larger value would be preferred. However, if too large a value is chosen, then the solution accuracy may suffer, and for some integration methods, such as Euler's, the solution can experience numeric instability. To see an example of numeric instability, re-do the PowerWorld Simulator Example 11_7, except change the time step to 0.02 seconds. A typical time step for commercial transient stability simulations is 1/2 cycle (0.00833 seconds for a 60 Hz system).

EXAMPLE 11.8

Modifying power-flow Y_{bus} for application to multimachine stability

Consider a transient stability study for the power system given in Example 6.9, with the 184-Mvar shunt capacitor of Example 6.14 installed at bus 2. Machine transient reactances are $X'_{d1} = 0.05$ and $X'_{d2} = 0.025$ per unit on the system base. Determine the admittance matrices Y_{11}, Y_{22}, and Y_{12}.

SOLUTION

From Example 6.9, the power system has $N = 5$ buses and $M = 2$ machines. The second row of the 5×5 bus admittance matrix used for power flows is calculated in Example 6.9. Calculating the other rows in the same manner, results in

$$Y_{bus} = \begin{bmatrix} (3.728 - j49.72) & 0 & 0 & 0 & (-3.728 + j49.72) \\ 0 & (2.68 - j26.46) & 0 & (-0.892 + j9.92) & (-1.784 + j19.84) \\ 0 & 0 & (7.46 - j99.44) & (-7.46 + j99.44) & 0 \\ 0 & (-0.892 + j9.92) & (-7.46 + j99.44) & (11.92 - j148.) & (-3.572 + j39.68) \\ (-3.728 + j49.72) & (-1.784 + j19.84) & 0 & (-3.572 + j39.68) & (9.084 - j108.6) \end{bmatrix} \text{per unit}$$

To obtain Y_{11}, Y_{bus} is modified by including load admittances and inverted generator impedances. From Table 6.1, the load at bus 3 is $P_{L3} - jQ_{L3} = 0.8 + j0.4$ per unit and the voltage at bus 3 is $V_3 = 1.05$ per unit. Representing this load as a constant admittance,

$$Y_{load\ 3} = \frac{P_{L3} - jQ_{L3}}{V_3^2} = \frac{0.8 - j0.4}{(1.05)^2} = 0.7256 - j0.3628 \quad \text{per unit}$$

Similarly, the load admittance at bus 2 is

$$Y_{load\ 2} = \frac{P_{L2} - jQ_{L2}}{V_2^2} = \frac{8 - j2.8 + j1.84}{(0.959)^2} = 8.699 - j1.044$$

where V_2 is obtained from Example 6.14 and the 184-Mvar (1.84 per unit) shunt capacitor bank is included in the bus 2 load.

The inverted generator impedances are: for machine 1 connected to bus 1,

$$\frac{1}{jX'_{d1}} = \frac{1}{j0.05} = -j20.0 \quad \text{per unit}$$

and for machine 2 connected to bus 3,

$$\frac{1}{jX'_{d2}} = \frac{1}{j0.025} = -j40.0 \quad \text{per unit}$$

(Continued)

To obtain Y_{11}, add $(1/jX'_{d1})$ to the first diagonal element of Y_{bus}, add Y_{load2} to the second diagonal element, and add $Y_{load\,3} + (1/jX'_{d2})$ to the third diagonal element. The 5×5 matrix Y_{11} is then

$$Y_{11} = \begin{bmatrix} (3.728-j69.72) & 0 & 0 & 0 & (-3.728-j49.72) \\ 0 & (11.38-j29.50) & 0 & (-0.892+j9.92) & (-1.784+j19.84) \\ 0 & 0 & (8.186-j139.80) & (-7.46+j99.44) & 0 \\ 0 & (-0.892+j9.92) & (-7.46+j99.44) & (11.92-j148.) & (-3.572-j39.68) \\ (-3.728+j49.72) & (-1.784+j19.84) & 0 & (-3.572+j39.68) & (9.084-j108.6) \end{bmatrix} \text{ per unit}$$

From (11.5.6), the 2×2 matrix Y_{22} is

$$Y_{22} = \begin{bmatrix} \frac{1}{jX'_{d1}} & 0 \\ 0 & \frac{1}{jX'_{d2}} \end{bmatrix} = \begin{bmatrix} -j20.0 & 0 \\ 0 & -j40.0 \end{bmatrix} \text{ per unit}$$

From Figure 6.2, generator 1 is connected to bus 1 (therefore, bus G1 = 1 and generator 2 is connected to bus 3 (therefore G2 = 3). From (11.5.7), the 5×2 matrix Y_{12} is

$$Y_{12} = \begin{bmatrix} j20.0 & 0 \\ 0 & 0 \\ 0 & j40.0 \\ 0 & 0 \\ 0 & 0 \end{bmatrix} \text{ per unit}$$

To see this case in PowerWorld Simulator, open case Example 11_8. To see the Y_{11} matrix entries, first display the Transient Stability Analysis Form, and then select **States/Manual Control, Transient Stability Ybus**. By default, this case is set to solve a self-clearing fault at bus 4 that extinguishes itself after three cycles (0.05 s). Both generators are modeled with H = 5.0 p.u.-s and D = 1.0 p.u.

For the bus 4 fault, Figure 11.16 shows the variation in the rotor angles for the two generators with respect to a 60 Hz synchronous reference frame. The angles are increasing with time because neither of the generators is modeled with a governor, and there is no infinite bus. While it is clear that the generator angles remain together, it is very difficult to tell from Figure 11.16 the exact variation in the angle differences. Therefore transient stability programs usually report angle differences, either with respect to the angle at a specified bus or with respect to the average of all the generator angles. The latter is shown in Figure 11.17 which displays the results from the PowerWorld Simulator Example 11_8A case.

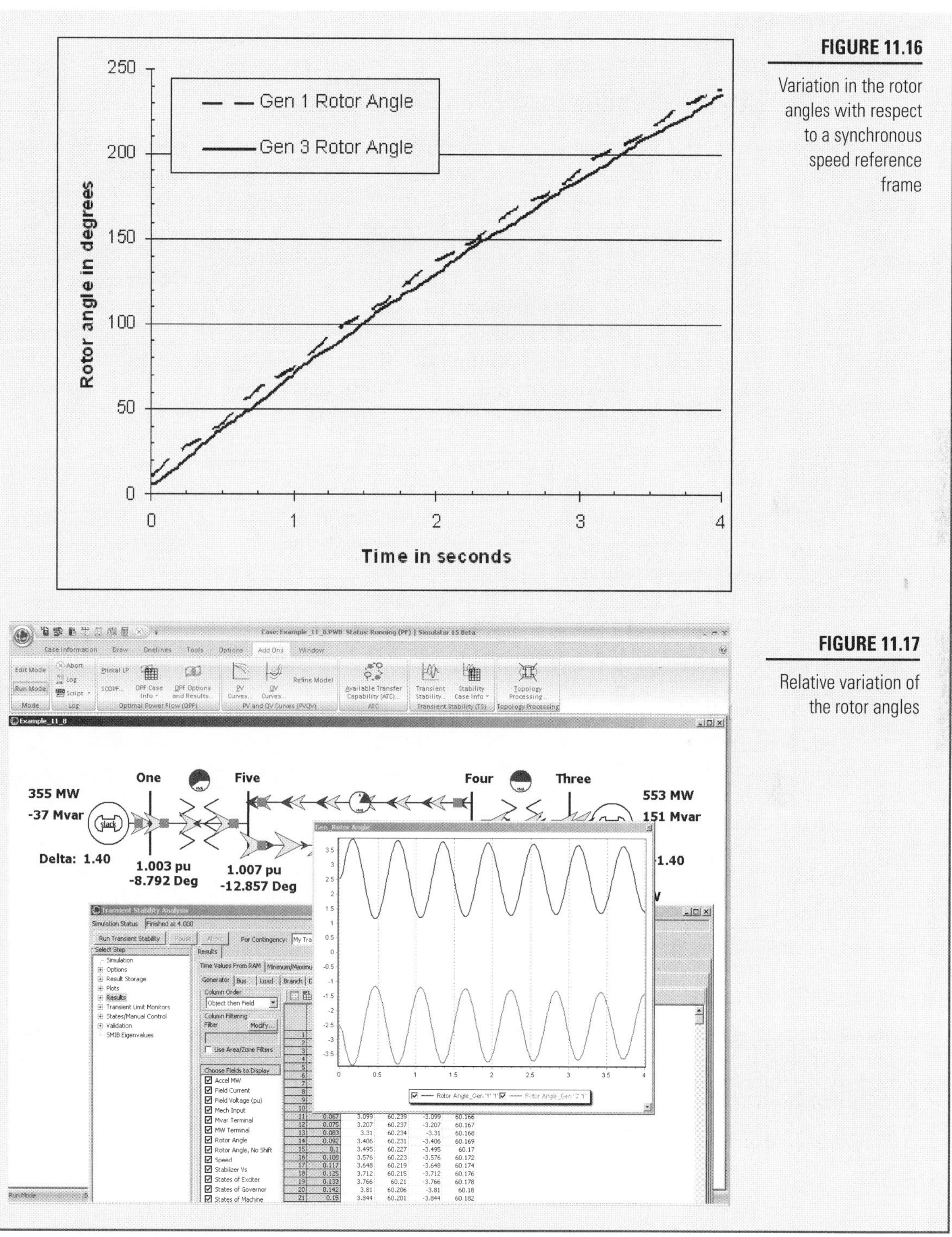

FIGURE 11.16 Variation in the rotor angles with respect to a synchronous speed reference frame

FIGURE 11.17 Relative variation of the rotor angles

EXAMPLE 11.9

Stability results for 37 bus, 9 generator system

PowerWorld Simulator case Example 11_9 demonstrates a transient stability solution using the 37 bus system introduced in Chapter 6 with the system augmented to include classical models for each of the generators. By default, the case models a transmission line fault on the 69 kV line between bus 44 (PEACH69) and bus 14 (REDBUD69) with the fault at the bus 44 end of the line. The fault is cleared after 0.1 seconds by opening this transmission line. The results from this simulation are shown in Figure 11.18, with the largest generator angle variation occurring (not surprisingly) at the bus 44 generator. Notice that during and initially after the fault, the bus 44 generator's angle increases relative to all the other angles in the system. The critical clearing time for this fault is about 0.262 seconds.

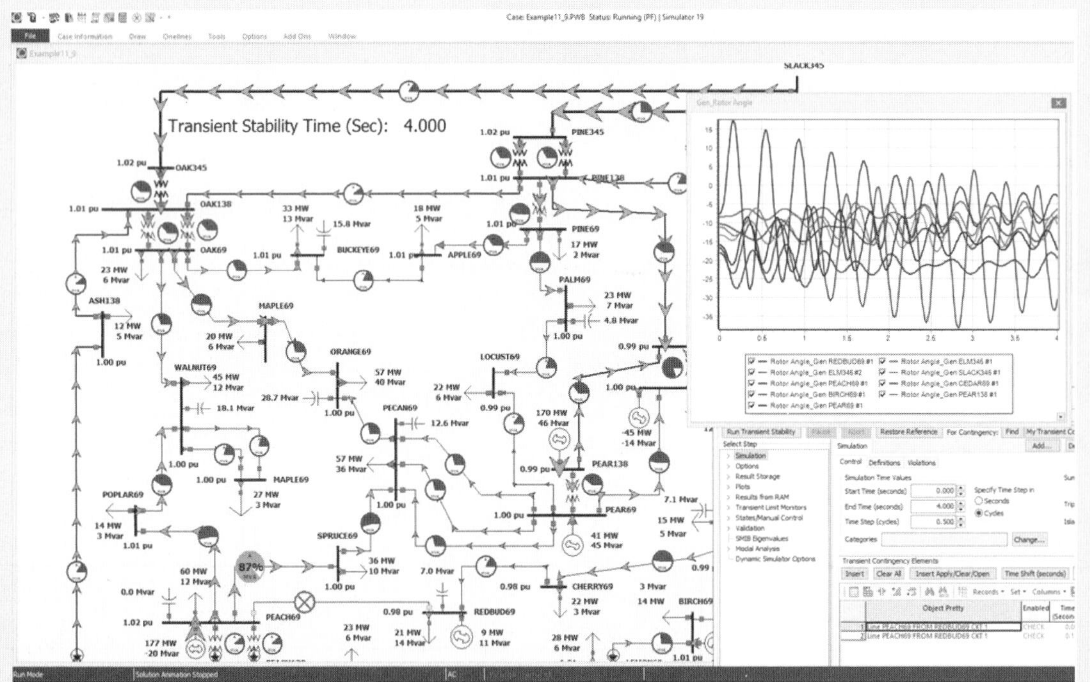

FIGURE 11.18

Rotor Angles for Example 11.9 case

11.6 A TWO-AXIS SYNCHRONOUS MACHINE MODEL

While the classical model for a synchronous machine provides a useful mechanism for introducing transient stability concepts, it is only appropriate for the most basic of system studies. Also, it is usually not coupled with the exciter and governor models that are introduced in the next chapter. In this section a more realistic synchronous machine model is introduced.

The analysis of more detailed synchronous machine models requires that each machine model be expressed in a frame of reference that rotates at the same speed as its rotor. The standard approach is to use a *d-q* reference frame in which the major "direct" (*d*) axis is aligned with the rotor poles, and the quadrature (*q*) axis leads the direct axis by 90°. The rotor angle δ is then defined as the angle by which the *q*-axis leads the network reference frame (see Figure 11.19). The equation for transforming the network quantities to the *d-q* reference frame is given by (11.6.1) and from the *d-q* reference frame by (11.6.2),

$$\begin{bmatrix} V_r \\ V_i \end{bmatrix} = \begin{bmatrix} \sin\delta & \cos\delta \\ -\cos\delta & \sin\delta \end{bmatrix} \begin{bmatrix} V_d \\ V_q \end{bmatrix} \tag{11.6.1}$$

$$\begin{bmatrix} V_d \\ V_q \end{bmatrix} = \begin{bmatrix} \sin\delta & -\cos\delta \\ \cos\delta & \sin\delta \end{bmatrix} \begin{bmatrix} V_r \\ V_i \end{bmatrix} \tag{11.6.2}$$

where the terminal voltage in the network reference frame is $V_T = V_r + jV_i$. A similar conversion is done for the currents.

Numerous different transient stability models exist for synchronous machines, most of which are beyond the scope of this text. The two-axis model, which

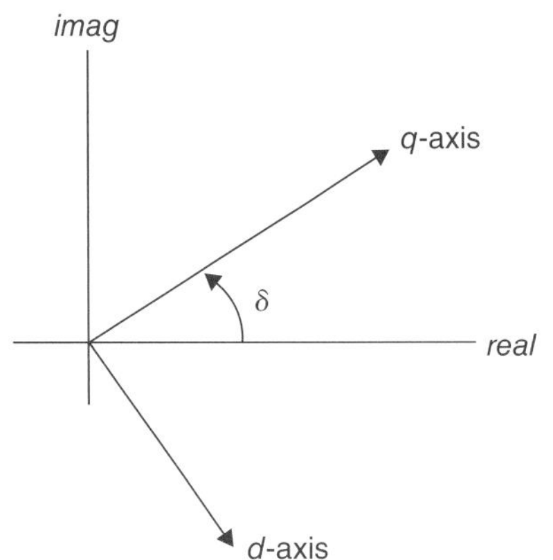

FIGURE 11.19

Reference frame transformations

models the dynamics associated with the synchronous generator field winding and one damper winding, while neglecting the faster subtransient damper dynamics and stator transients, provides a nice compromise. For accessibility, machine saturation is not considered. With the two-axis model, the electrical behavior of the generator is represented by two algebraic equations and two differential equations

$$E'_q = V_q + R_a I_q + X'_d I_d \tag{11.6.3}$$

$$E'_d = V_d + R_d I_d - X'_q I_q \tag{11.6.4}$$

$$\frac{dE'_q}{dt} = \frac{1}{T'_{do}}\left(-E'_q - (X_d - X'_d)I_d + E_{fd}\right) \tag{11.6.5}$$

$$\frac{dE'_d}{dt} = \frac{1}{T'_{qo}}\left(-E'_d + (X_q - X'_q)I_q\right) \tag{11.6.6}$$

where $V_d + jV_q$ and $I_d + jI_q$ are the generator's terminal voltage and current shifted into the generator's reference frame, and E_{fd} is proportional to the field voltage. The per unit electrical torque, T_{elec} is then

$$T_e = V_d I_d + V_q I_q + R_a(I_d^2 + I_q^2) \tag{11.6.7}$$

While $p_e = T_e\omega_{\text{p.u.}}$, it is often assumed that $\omega_{\text{p.u.}} = 1.0$ [10] with the result being an assumption that $p_e = T_e$. When (11.6.5) and (11.6.6) are combined with generator mechanical equations presented in (11.1.18) and (11.1.19), substituting (11.6.7) for $p_{e\text{p.u.}}$, the result is a synchronous generator model containing four first-order differential equations.

The initial value for δ can be determined by noting that in steady-state, δ is the same as the angle of the internal voltage [10],

$$E = V_T + jX_q I \tag{11.6.8}$$

Hence the initial value of δ is the angle on E. Once δ has been determined, (11.6.2) is used to transfer the generator terminal voltage and current into the generator's reference frame, and then (11.6.3), (11.6.4), and (11.6.5) (assuming the left-hand side is zero) are used to determine the initial values of E'_q, E'_d, and E_{fd}. In this chapter the field voltage, E_{fd}, will be assumed constant (the use of the generator exciters to control the field voltage is a topic covered in the next chapter).

EXAMPLE 11.10

Two-Axis Model Example

For the system from Example 11.3, with the synchronous generator modeled using a two-axis model, determine (a) the initial conditions, and then (b) use PowerWorld Simulator to determine the critical clearing time for the Example 11.6 fault (three phase fault at bus 3, cleared by opening lines 1–3 and 2–3). Assume H = 3.0 per unit-seconds, $R_a = 0$, $X_d = 2.1$, $X_q = 2.0$, $X'_d = 0.3$, $X'_q = 0.5$, all per unit using the 100 MVA system base.

SOLUTION

a. From Example 11.3, the current out of the generator is

$$I = 1.0526\underline{/-18.20^\circ} = 1 - j0.3288$$

which gives a generator terminal voltage of

$$V_T = 1.0\underline{/0^\circ} + (j0.22)(1.0526\underline{/-18.20^\circ}) = 1.0946\underline{/11.59^\circ} = 1.0723 + j0.220$$

From (11.6.8),

$$E = 1.0946\underline{/11.59^\circ} + (j2.0)(1.052\underline{/-18.2^\circ}) = 2.814\underline{/52.1^\circ}$$

$$\rightarrow \delta = 52.1^\circ$$

Using (11.6.2) gives

$$\begin{bmatrix} V_d \\ V_q \end{bmatrix} = \begin{bmatrix} 0.7889 & -0.6146 \\ 0.6146 & 0.7889 \end{bmatrix} \begin{bmatrix} 1.0723 \\ 0.220 \end{bmatrix} = \begin{bmatrix} 0.7107 \\ 0.8326 \end{bmatrix}$$

and

$$\begin{bmatrix} I_d \\ I_q \end{bmatrix} = \begin{bmatrix} 0.7889 & -0.6146 \\ 0.6146 & 0.7889 \end{bmatrix} \begin{bmatrix} 1.000 \\ -0.3287 \end{bmatrix} = \begin{bmatrix} 0.9909 \\ 0.3553 \end{bmatrix}$$

Then, solving (11.6.3), (11.6.4), and (11.6.5) gives

$$E'_q = 0.8326 + (0.3)(0.9909) = 1.1299$$
$$E'_d = 0.7107 - (0.5)(0.3553) = 0.5330$$
$$E_{fd} = 1.1299 + (2.1 - 0.3)(0.9909) = 2.9135$$

b. Open PowerWorld Simulator case Example 11_10 (see Figure 11.20). **Select Add-Ons, Transient Stability** to view the Transient Stability Analysis Form. Initially the bus 3 fault is set to clear at 0.05 seconds. Select **Run Transient Stability** to create the results shown in Figure 11.20. In comparing

(Continued)

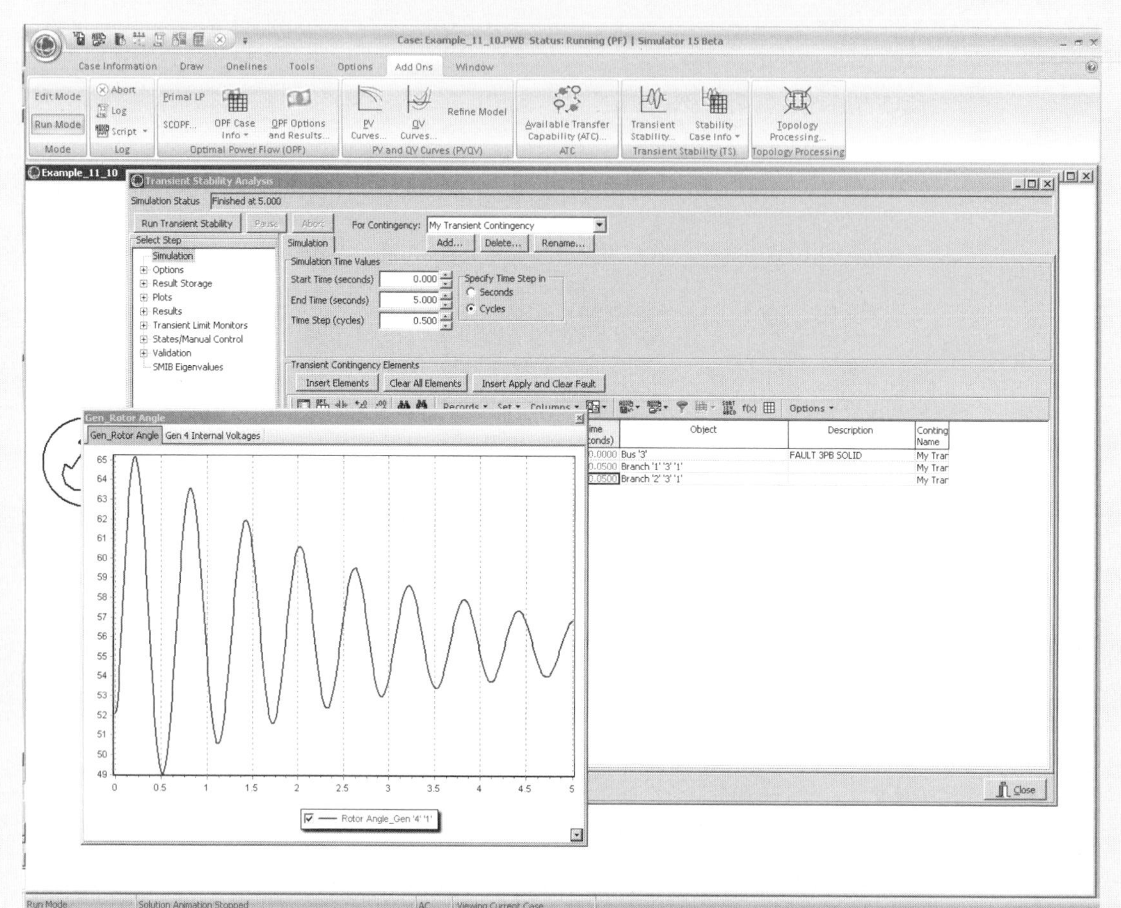

FIGURE 11.20

Variation in generator 4 rotor angle with a fault clearing time of 0.05 seconds

these results with those from Example 11.4, notice that while the initial value of δ is different, the initial angle increase of about 13° is similar to the increase of 16° in Example 11.4. A key difference between the two is the substantial amount of damping in the two-axis model case. This damping arises because of the explicit modeling of the field and damper windings with the two-axis model. The critical clearing time can be determined by gradually increasing the clearing time until the generator loses synchronism. This occurs at about 0.30 seconds, with the almost critically cleared angle shown in Figure 11.21. Since there are now two additional state variables for generator 4, E'_q, and E'_d, their values can also be shown. This is done in Figure 11.22, again for the 0.30 second clearing time.

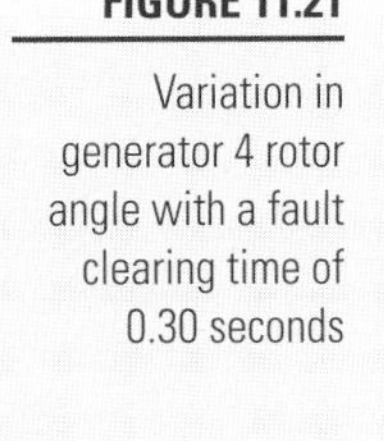

FIGURE 11.21 Variation in generator 4 rotor angle with a fault clearing time of 0.30 seconds

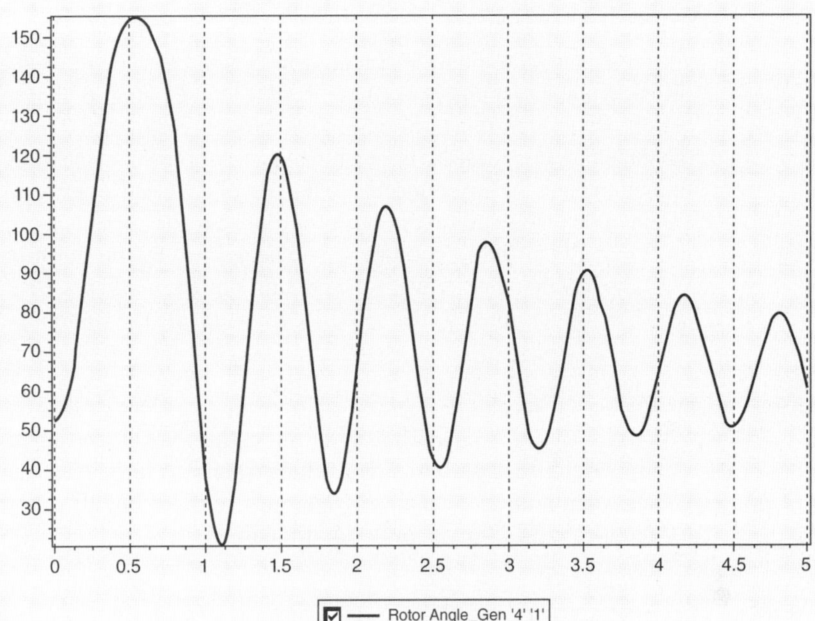

FIGURE 11.22 Variation in generator 4 E_q' and E_d' with a fault clearing time of 0.30 seconds

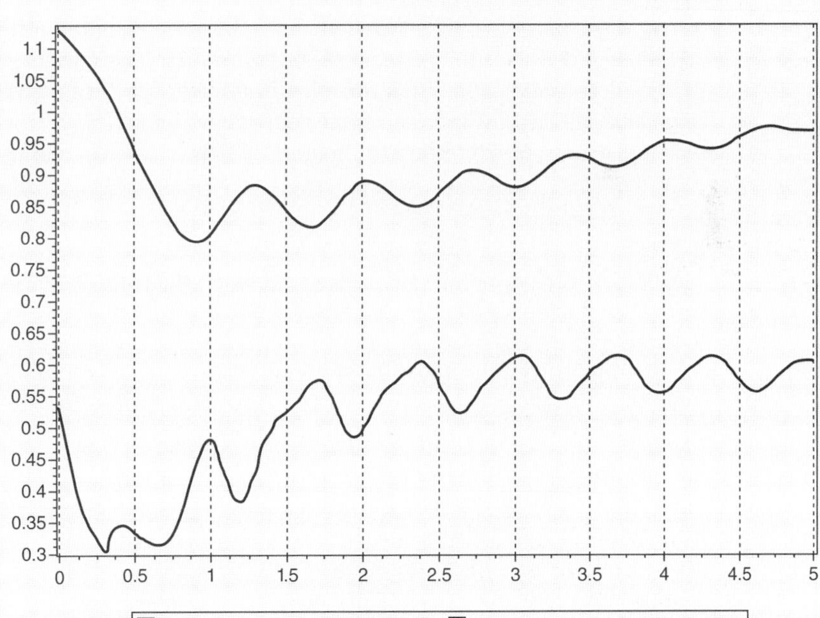

11.7 WIND TURBINE MACHINE MODELS

As wind energy continues its rapid growth, wind turbine models need to be included in transient stability analysis. As was introduced in Chapter 6, there are four main types of wind turbines that must be considered. Model types 1 and 2 are based on an induction machine models. As is the case with a synchronous machine, the stator windings of the induction machine are connected to the rest of the electric network. However, rather than having a dc field winding on the rotor, the ac rotor currents are induced by the relative motion between the rotating magnetic field setup by the stator currents, and the rotor. Usually the difference between the per unit synchronous speed, n_s, and the per unit rotor speed, n_r, is quantified by the slip (S), defined (using the standard motor convention), as

$$S = \frac{n_s - n_r}{n_s} \tag{11.7.1}$$

From (11.7.1), it is clear that if the machine were operating at synchronous speed, its slip would be 0, with positive values when it is operating as a motor and negative values when it is operating as a generator. Expressing all values in per unit, the mechanical equation for an induction machine is

$$\frac{dS}{dt} = \frac{1}{2H}(T_m - T_e) \tag{11.7.2}$$

where H is the inertia constant, T_m is the mechanical torque, and T_e the electrical torque, defined in (11.7.10).

The simplified electric circuit for a single-cage induction machine is shown in Figure 11.23, using the generator convention in which current out of the machine is assumed to be positive. Similar to what is done for synchronous machines, an induction machine can be modeled as an equivalent voltage behind the stator resistance and a transient reactance X'. Referring to Figure 11.23, the values used in this representation are

$$X' = X_a + \frac{X_1 X_m}{X_1 + X_m} \tag{11.7.3}$$

where X' is the apparent reactance seen when the rotor is locked (i.e., slip is 1),

$$X = X_a + X_m \tag{11.7.4}$$

X is the synchronous reactance, and

$$T'_0 = \frac{(X_1 + X_m)}{\omega_0 R_1} \tag{11.7.5}$$

is the open-circuit time constant for the rotor. Also, X_a is commonly called the leakage reactance.

FIGURE 11.23 Equivalent circuit for a single cage induction machine

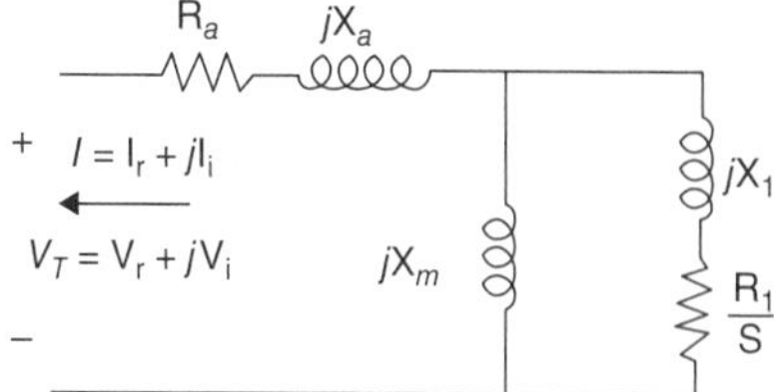

Electrically the induction machine is modeled using two algebraic and two differential equations. However, in contrast to synchronous machines, because the reactances of induction machines do not depend upon the rotor position values, they are specified in the network reference frame. The equations are

$$V_r = E'_r - R_a I_r + X' I_i \tag{11.7.6}$$

$$V_i = E'_i - R_a I_i - X' I_r \tag{11.7.7}$$

$$\frac{dE'_r}{dt} = \omega_o S E'_i - \frac{1}{T'_o}\left((E'_r - (X - X')I_i\right) \tag{11.7.8}$$

$$\frac{dE'_i}{dt} = -\omega_o S E'_r - \frac{1}{T'_o}\left((E'_i + (X - X')I_r\right) \tag{11.7.9}$$

The induction machine electric torque is then given by

$$T_e = (E'_r I_i + E'_i I_i)/\omega_o \tag{11.7.10}$$

and the terminal real power injection by

$$P_e = (V_r I_r + V_i I_i) \tag{11.7.11}$$

The transient stability initial conditions are determined by setting (11.7.8) and (11.7.9) to zero, and then using the power flow real power injection and terminal voltage as inputs to solve (11.7.6), (11.7.7), (11.7.8), (11.7.9), and (11.7.11) for the other variables. The Newton-Raphson approach (Section 6.3) is commonly used. Since the induction machine reactive power injection will not normally match the power flow value, the difference is modeled by including a shunt capacitor whose susceptance is determined to match the initial power flow conditions. The reactive power produced by the machine is given by

$$Q_e = (-V_r I_i + V_i I_r) \tag{11.7.12}$$

with the value negative since induction machines consume reactive power.

EXAMPLE 11.11

Induction Generator Example

For the system from Example 11.3, assume the synchronous generator is replaced with an induction generator and shunt capacitor in order to represent a wind farm with the same initial real and reactive power output as in Example 11.3. The induction generator parameters are H = 0.9 per unit-seconds, $R_a = 0.013$, $X_a = 0.067$, $X_m = 3.8$, $R_1 = 0.0124$, $X_1 = 0.17$ (all per unit using the 100 MVA system base). This system is modeled in PowerWorld Simulator case Example 11_11. (a) Use the previous equations to verify the initial conditions of $S = -0.0111$, $E'_r = 0.9314$, $E'_i = 0.4117$, $I_r = 0.7974$, $I_i = 0.6586$. (b) Plot the terminal voltage for the fault sequence from Example 11.6.

(Continued)

SOLUTION

a. Using (11.7.3) to (11.7.5) the values of X', X, and T_0' are determined to be 0.2297, 3.867, per unit and 0.85 seconds respectively. With $V_T = 1.0723 + j0.220$ and $P_e = 1.0$ per unit, (11.7.6) and (11.7.7) are verified as

$$V_r = 0.9314 - (0.013)\ (0.7974) + (0.2297(0.6586) = 1.0723$$
$$V_i = 0.4117 - (0.013)\ (0.6586) - (0.2297(0.7974) = 0.2200$$

And (11.7.8), (11.7.9) as

$$\frac{dE_r'}{dt} = 2\pi60(-0.0111)(0.4117) - \frac{1}{0.85}(0.9314 - (3.637)(0.6586)) = 0.0$$
$$\frac{dE_i'}{dt} = -2\pi60(-0.0111)(0.9314) - \frac{1}{0.85}(0.4117 + (3.637)(0.7974)) = 0.0$$
$$P_e = (1.0723)(0.7974) + (0.220)(0.6586) = 1.000$$
$$Q_e = -(1.0723)(0.6586) + (0.220)(0.7974) = -0.531$$

You can see the initial values in PowerWorld Simulator by first displaying the **States/Manual Control** page of the Transient Stability Analysis form, which initializes the transient stability. Then from the one-line diagram view the Generator Information Dialog for the generator at bus 4, and select **Stability, Terminal and State, Terminal Values**. Because the generator in the power flow is producing 57.2 Mvar, and the induction machine is consuming 53.1 Mvar, a shunt capacitor that produces 110.3 Mvar with a 1.0946 terminal voltage must be modeled.

b. Figure 11.24 plots the terminal voltage for the three cycle fault.

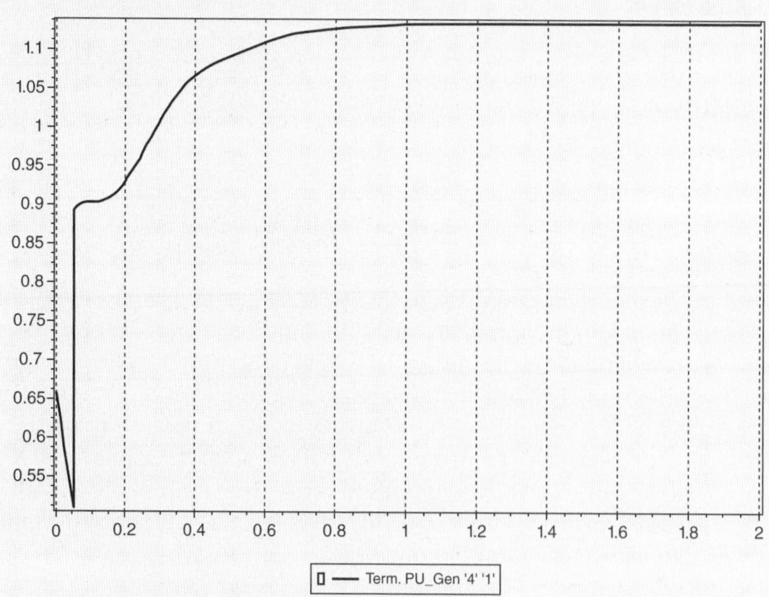

FIGURE 11.24 Example 11.11 Generator 4 voltage magnitude for a fault clearing time of 0.05 seconds

Both the Type 1 and 2 wind turbine models utilize induction generators, but whereas the Type 1 models have a conventional squirrel cage rotor with fixed rotor resistance, the Type 2 models are wound rotor induction machines that utilize a control system to vary the rotor resistance. The reason for this is to provide a more steady power output from the wind turbine during wind variation. From (11.7.5) it is clear that increasing this external resistance has the effect of decreasing the open circuit time constant. The inputs to the rotor resistance control system are turbine speed and electrical power output, while the output is the external resistance that is in series with R_1 from Figure 11.22. Figure 11.25 plots the variation in the real power output for the Example 11.11 generator as a function of speed for the original rotor resistance of 0.0124 and for a total rotor resistance of 0.05 per unit. With a total resistance of 0.05, the operating point slip changes to about –0.045, which corresponds to a per-unit speed of 1.045.

Most new wind turbines are either Type 3 or Type 4. Type 3 wind turbines are used to represent doubly-fed asynchronous generators (DFAGs), also sometimes referred to as doubly-fed induction generators (DFIGs). A DFAG consists of a traditional wound rotor induction machine, but with the rotor windings connected to the ac network through an ac-dc-ac converter—the machine is "doubly-fed" through both the stator and rotor windings (see Figure 11.26). The advantages of this arrangement are that it allows for separate control of both the real and reactive power (like a synchronous machine), and the ability to transfer power both ways through the rotor converter allows for a much wider speed range. Because the stator is directly connected to the ac grid, the rotor circuit converter need only be sized to about 30% of the machine's rated capacity. Another consequence of this design is the absence of an electrical coupling with the mechanical equation such as was seen with (11.1.10) for the synchronous machines and in (11.7.2) for the Type 1 and 2 induction machine models.

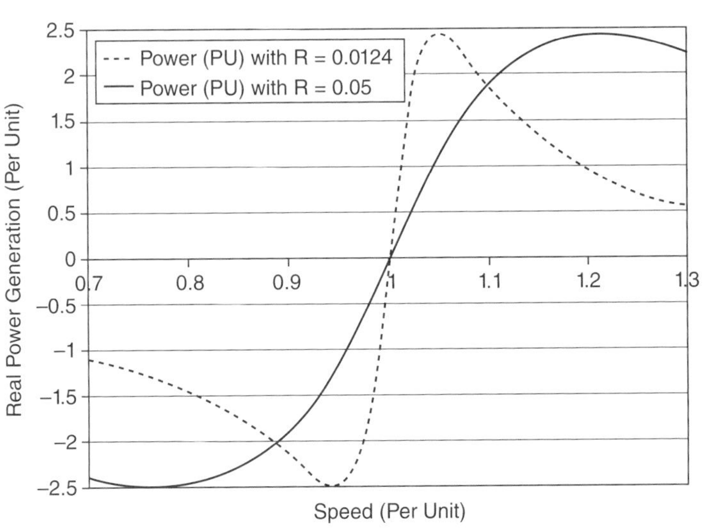

FIGURE 11.25

Effect of varying external resistance on an induction machine torque-speed curve

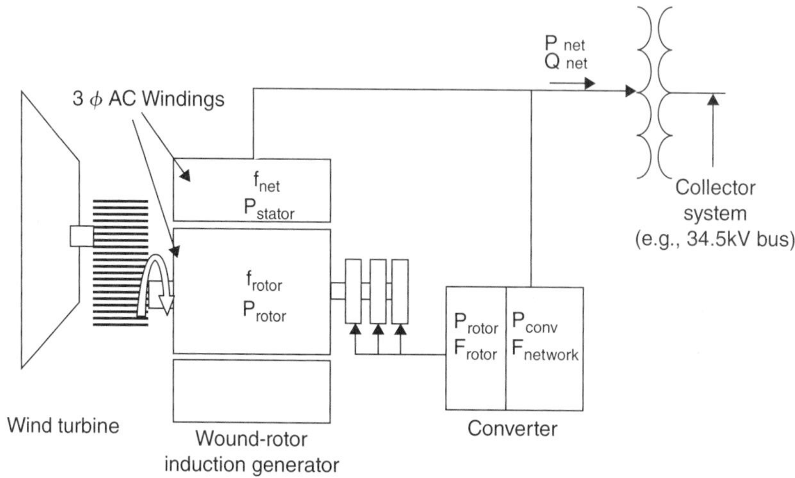

FIGURE 11.26 Doubly-fed asynchronous generator components

From a transient stability perspective, the DFAG dynamics are driven by the converter, with the result that the machine can be well approximated as a voltage-source converter (VSC). A VSC can be approximated as a synthesized current injection in parallel with an effective reactance, X_{eq}, (Figure 11.27) in which the current in phase with the terminal voltage, I_p, and the reactive power current, I_q, can be controlled independently. Low and high voltage current management is used to limit these values during system disturbances. With a terminal voltage angle of 0, the current injection on the network reference is

$$I_{sorc} = (\mathrm{I_p} + j\mathrm{I_q})\ (1\angle\theta) \tag{11.7.13}$$

And the reactive voltage is

$$E_q = -\mathrm{I_q} X_{eq} \tag{11.7.14}$$

Type 4 wind turbines utilize a completely asynchronous design in which the full output of the machine is connected to the ac network through an ac-dc-ac converter (see Figure 11.28). Because the converter completely decouples the electric generator from the rest of the network, there is considerable freedom in selecting the electric

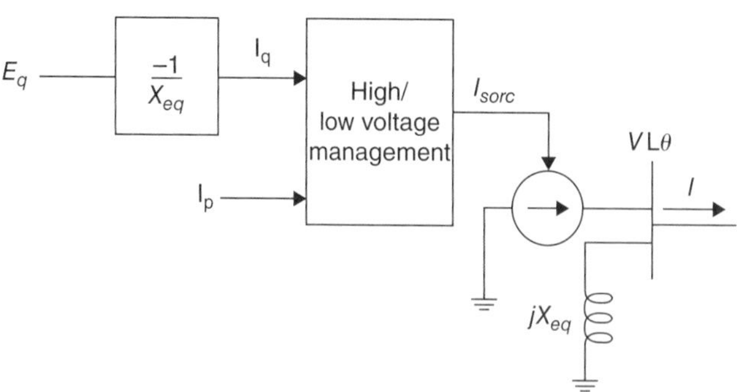

FIGURE 11.27 Type 3 DFAG model circuit diagram

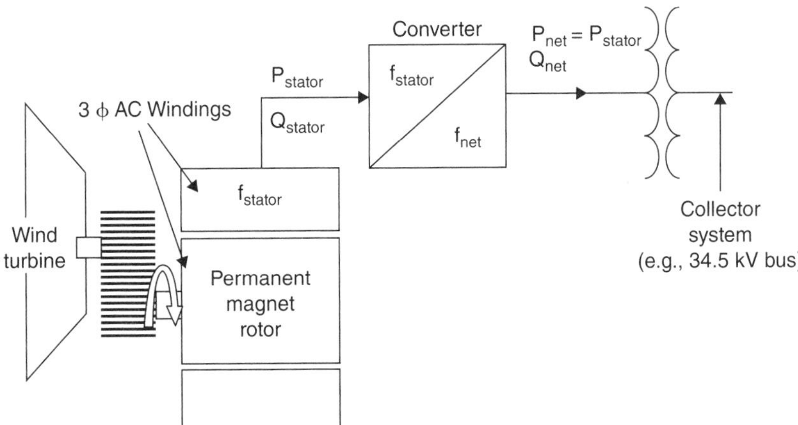

FIGURE 11.28

Type 4 full converter components

machine type: for example, a conventional synchronous generator, a permanent magnet synchronous generator, or even a squirrel cage induction machine.

From a transient stability perspective, electrically, the Type 4 model is similar to the Type 3 in that it can also be represented as a VSC. The key difference is lack of the effective reactance, with I_p and I_q being the direct control variables for the Type 4 model. As is the case with a DFAG, there is no electrical coupling with the turbine dynamics.

EXAMPLE 11.12

Doubly-Fed Asynchronous Generator Example

For the system from Example 11.3, assume the synchronous generator is replaced with a Type 3 DFAG generator in order to represent a wind farm with the initial current into the infinite bus set to 1.0 (unity power factor). The DFAG reactance $X_{eq} = 0.8$ per unit using a 100 MVA system base. Determine the initial values for I_p, I_q, and E_q.

SOLUTION

With $I = 1.0$ and an impedance of $j0.12$ between the machine's terminal and the infinite bus, the terminal voltage is

$$V_T = 1.0 + (1.0)(j0.22) = 1.0 + j0.22 = 1.0239\underline{/12.41^\circ}$$

The amount supplied by I_{sorc} is I plus the amount modeled as going into X_{eq}

$$I_{sorc} = I - \frac{V_T}{jX_{eq}} = 1.00 + \frac{1.0 + j0.220}{j0.8}$$

$$I_{sorc} = 1.275 - j1.25$$

(Continued)

The values of I_p and I_q are then calculated by shifting these values backward by the angle of the terminal voltage

$$I_p + jI_q = (1.275 - j1.25) * (1\underline{/12.41^\circ}) = 0.977 - j1.495$$

And then

$$E_q = -(-1.495)(0.8) = 1.196$$

You can see the initial values in PowerWorld Simulator by opening case Example 11_12 and then displaying the **States/Manual Control** page of the Transient Stability Analysis form and selecting **Reset to Start Time** which initializes the transient stability. Then from the one-line diagram view the Generator Information Dialog for the generator at bus 4, and select **Stability, Terminal and State, Terminal Values.**

11.8 DESIGN METHODS FOR IMPROVING TRANSIENT STABILITY

Design methods for improving power system transient stability include the following:

1. Improved steady-state stability
   a. Higher system voltage levels
   b. Additional transmission lines
   c. Smaller transmission-line series reactances
   d. Smaller transformer leakage reactances
   e. Series capacitive transmission-line compensation
   f. Static var compensators and flexible ac transmission systems (FACTS)
2. High-speed fault clearing
3. High-speed reclosure of circuit breakers
4. Single-pole switching
5. Larger machine inertia, lower transient reactance
6. Fast responding, high-gain exciters
7. Fast valving
8. Braking resistors

These design methods are discussed in the following paragraphs.

1. Increasing the maximum power transfer in steady-state can also improve transient stability, allowing for increased power transfer through the unfaulted portion of a network during disturbances. Upgrading voltage on existing transmission or opting for higher voltages on new

transmission increases line loadability (5.5.6). Additional parallel lines increase power-transfer capability. Reducing system reactances also increases power-transfer capability. Lines with bundled phase conductors have lower series reactances than lines that are not bundled. Oversized transformers with lower leakage reactances also help. Series capacitors reduce the total series reactances of a line by compensating for the series line inductance.

2. High-speed fault clearing is fundamental to transient stability. Standard practice for EHV systems is 1-cycle relaying and 2-cycle circuit breakers, allowing for fault clearing within 3 cycles (0.05 s). Ongoing research is presently aimed at reducing these to one-half cycle relaying and 1-cycle circuit breakers.
3. The majority of transmission-line short circuits are temporary, with the fault arc self-extinguishing within 5–40 cycles (depending on system voltage) after the line is deenergized. High-speed reclosure of circuit breakers can increase postfault transfer power, thereby improving transient stability. Conservative practice for EHV systems is to employ high-speed reclosure only if stability is maintained when reclosing into a permanent fault with subsequent reopening and lockout of breakers.
4. Since the majority of short circuits are single line-to-ground, relaying schemes and independent-pole circuit breakers can be used to clear a faulted phase while keeping the unfaulted phases of a line operating, thereby maintaining some power transfer across the faulted line. Studies have shown that single line-to-ground faults are self-clearing even when only the faulted phase is deenergized. Capacitive coupling between the energized unfaulted phases and the deenergized faulted phase is, in most cases, not strong enough to maintain an arcing short circuit [5].
5. Inspection of the swing equation, (11.1.16), shows that increasing the per-unit inertia constant H of a synchronous machine reduces angular acceleration, thereby slowing down angular swings and increasing critical clearing times. Stability is also improved by reducing machine transient reactances, which increases power-transfer capability during fault or postfault periods [see (11.2.1)]. Unfortunately, present-day generator manufacturing trends are toward lower H constants and higher machine reactances, which are a detriment to stability.
6. Modern machine excitation systems with fast thyristor controls and high amplifier gains (to overcome generator saturation) can rapidly increase generator field excitation after sensing low terminal voltage during faults. The effect is to rapidly increase internal machine voltages during faults, thereby increasing generator output power during fault and postfault periods. Critical clearing times are also increased [6].
7. Some steam turbines are equipped with fast valving to divert steam flows and rapidly reduce turbine mechanical power outputs. During faults near the generator, when electrical power output is reduced, fast valving action acts to balance mechanical and electrical power, providing

reduced acceleration and longer critical clearing times. The turbines are designed to withstand thermal stresses due to fast valving [7].

8. In power systems with generation areas that can be temporarily separated from load areas, braking resistors can improve stability. When separation occurs, the braking resistor is inserted into the generation area for a second or two, preventing or slowing acceleration in the generation area. Shelton et al. [8] describe a 3-GW-s braking resistor.

PROBLEMS

SECTION 11.1

11.1 A three-phase, 60-Hz, 500-MVA, 11.8-kV, 4-pole steam turbine-generating unit has an H constant of 5 p.u.-s. Determine: (a) ω_{syn} and ω_{msyn}; (b) the kinetic energy in joules stored in the rotating masses at synchronous speed; (c) the mechanical angular acceleration α_m and electrical angular acceleration α if the unit is operating at synchronous speed with an accelerating power of 500 MW.

11.2 Calculate J in kg-m^2 for the generating unit given in Problem 11.1.

11.3 Generator manufacturers often use the term WR^2, which is the weight in kilograms of all the rotating parts of a generating unit (including the prime mover) multiplied by the square of the radius of gyration in meters. WR^2 is then the total moment of inertia of the rotating parts in kg-m^2. (a) Determine a formula for the stored kinetic energy in J of a generating unit in terms of WR^2 and rotor angular velocity ω_m. (b) Show that

$$H = \frac{5.48 \times 10^{-3} \mathrm{WR}^2 (\mathrm{rpm})^2}{S_{\mathrm{rated}}} \quad \text{per unit-seconds}$$

where S_{rated} is the voltampere rating of the generator, and rpm is the synchronous speed in r/min. (c) Evaluate H for a three-phase generating unit rated 750 MVA, 3600 r/min, with WR^2 = 170 × 10^3 kg-m^2.

11.4 The generating unit in Problem 11.1 is initially operating at $p_{mp.u.} = p_{ep.u.} = 0.7$ per unit, $\omega = \omega_{syn}$, and $\delta = 12°$ when a fault reduces the generator electrical power output by 60%. Determine the power angle 3 five cycles after the fault commences. Assume that the accelerating power remains constant during the fault. Also assume that $\omega_{p.u.}(t) = 1.0$ in the swing equation.

11.5 How would the value of H change if a generator's assumed operating frequency is changed from 60 Hz to 55 Hz?

11.6 Repeat Example 11.1 except assume the number of poles is changed from 32 to 16, H is changed from 2.0 p.u.-s to 1.5 p.u.-s, and the unit is initially operating with an electrical and mechanical power of 0.5 p.u.

SECTION 11.2

11.7 Given that for a moving mass $W_{kinetic} = 1/2\,Mv^2$, how fast would a 80,000 kg diesel locomotive need to go to equal the energy stored in a 60-Hz, 100-MVA, 60-Hz, 2-pole generator spinning at synchronous speed with an H of 3.0 p.u.-s?

11.8 The synchronous generator in Figure 11.4 delivers 0.8 per-unit real power at 1.05 per-unit terminal voltage. Determine: (a) the reactive power output of the generator; (b) the generator internal voltage; and (c) an equation for the electrical power delivered by the generator versus power angle δ.

11.9 The generator in Figure 11.4 is initially operating in the steady-state condition given in Problem 11.8 when a three-phase-to-ground bolted short circuit occurs at bus 3. Determine an equation for the electrical power delivered by the generator versus power angle δ during the fault.

11.10 For the five bus system from Example 6.9, assume the transmission lines and transformers are modeled with just their per unit reactance (e.g., neglect their resistance and B shunt values). If bus one is assumed to be an infinite bus, what is the equivalent (Thévenin) reactance looking into the system from the bus three terminal? Neglect any impedances associated with the loads.

11.11 Repeat Problem 11.10, except assume there is a three-phase-to-ground bolted short circuit at bus five.

SECTION 11.3

11.12 The generator in Figure 11.4 is initially operating in the steady-state condition given in Example 11.3 when circuit breaker B12 inadvertently opens. Use the equal-area criterion to calculate the maximum value of the generator power angle δ. Assume $\omega_{p.u.}(t) = 1.0$ in the swing equation.

11.13 The generator in Figure 11.4 is initially operating in the steady-state condition given in Example 11.3 when a temporary three-phase-to-ground short circuit occurs at point F. Three cycles later, circuit breakers B13 and B22 permanently open to clear the fault. Use the equal-area criterion to determine the maximum value of the power angle δ.

11.14 If breakers B13 and B22 in Problem 11.13 open later than 3 cycles after the fault commences, determine the critical clearing time.

11.15 Building upon Problem 11.11, assume a 60 Hz nominal system frequency, that the bus fault actually occurs on the line between buses five and two but at the bus two end, and that the fault is cleared by opening breakers B21 and B52. Again, neglecting the loads, assume that the generator at bus three is modeled with the classical generator model having a per unit value (on its 800 MVA base) of $X'_d = 0.24$, and $H = 3$ p.u.-s. Before the fault occurs the generator is delivering 300 MW into the infinite bus at unity power factor (hence its terminal voltage is not 1.05 as was assumed in Example 6.9). Further, assume the fault is cleared after 3 cycles.

Determine: (a) the initial generator one power angle, (b) the power angle when the fault is cleared, and (c) the maximum value of the power angle using the equal area criteria.

11.16 Analytically determine whether there is a critical clearing time for Problem 11.15.

SECTION 11.4

11.17 Consider the first order differential equation, $\frac{dx_1}{dt} = -x_2$, with an initial value $x(0) = 10$. With an integration step size of 0.1 seconds, determine the value of $x(0.5)$ using (a) Euler's method, (b) the modified Euler's method.

11.18 The following set of differential equations can be used to represent that behavior of a simple spring-mass system, with $x_1(t)$ the mass's position and $x_2(t)$ its velocity:

$$\frac{dx_1}{dt} = x_2$$

$$\frac{dx_2}{dt} = -x_1$$

For the initial condition of $x_1(0) = 1.0$, $x_2(0) = 0$, and a step size 0.1 seconds, determine the values $x_1(0.3)$ and $x_2(0.3)$ using (a) Euler's method, (b) the modified Euler's method.

11.19 A 60 Hz generator is supplying 400 MW (and 0 Mvar) to an infinite bus (with 1.0 per unit voltage) through two parallel transmission lines. Each transmission line has a per unit impedance (100 MVA base) of $0.09j$. The per unit transient reactance for the generator is $0.0375j$, the per unit inertia constant for the generator (H) is 20 seconds, and damping is 0.1 per unit (all with a 100 MVA base). At time = 0, one of the transmission lines experiences a balanced three-phase short to ground one third (1/3) of the way down the line from the generator to the infinite bus. (a) Using the classical generator model, determine the prefault internal voltage magnitude and angle of the generator. (b) Express the system dynamics during the fault as a set of first order differential equations. (c) Using Euler's method, determine the generator internal angle at the end of the second timestep. Use an integration step size of one cycle.

PW **11.20** Open PowerWorld Simulator case Problem 11_20. This case models the Example 11.4 system with damping at the bus 1 generator, and with a line fault midway between buses 1 and 2. The fault is cleared by opening the line. Determine the critical clearing time for this fault.

PW **11.21** Open PowerWorld Simulator case Problem 11_21. This case models the Example 11.4 system with damping at the bus 1 generator, and with a line fault midway between buses 2 and 3. The fault is cleared by opening the line. Determine the critical clearing time (to the nearest 0.01 second) for this fault.

SECTION 11.5

11.22 Consider the six-bus power system shown in Figure 11.29, where all data are given in per-unit on a common system base. All resistances as well as transmission-line capacitances are neglected. (a) Determine the 6×6 per-unit bus admittance matrix $\boldsymbol{Y}_{\text{bus}}$ suitable for a power-flow computer program. (b) Determine the per-unit admittance matrices $\boldsymbol{Y}_{11}$, $\boldsymbol{Y}_{12}$, and $\boldsymbol{Y}_{22}$ given in (11.5.5), which are suitable for a transient stability study.

11.23 Modify the matrices $\boldsymbol{Y}_{11}$, $\boldsymbol{Y}_{12}$, and $\boldsymbol{Y}_{22}$ determined in Problem 11.22 for (a) the case when circuit breakers B32 and B51 open to remove line 3–5; and (b) the case when the load $P_{L3} + jQ_{L3}$ is removed.

PW **11.24** Open PowerWorld Simulator case Problem 11_24, which models the Example 6.9 with transient stability data added for the generators. Determine the critical clearing time (to the nearest 0.01 second) for a fault on the line between buses 4 and 5 at the bus 4 end which is cleared by opening the line.

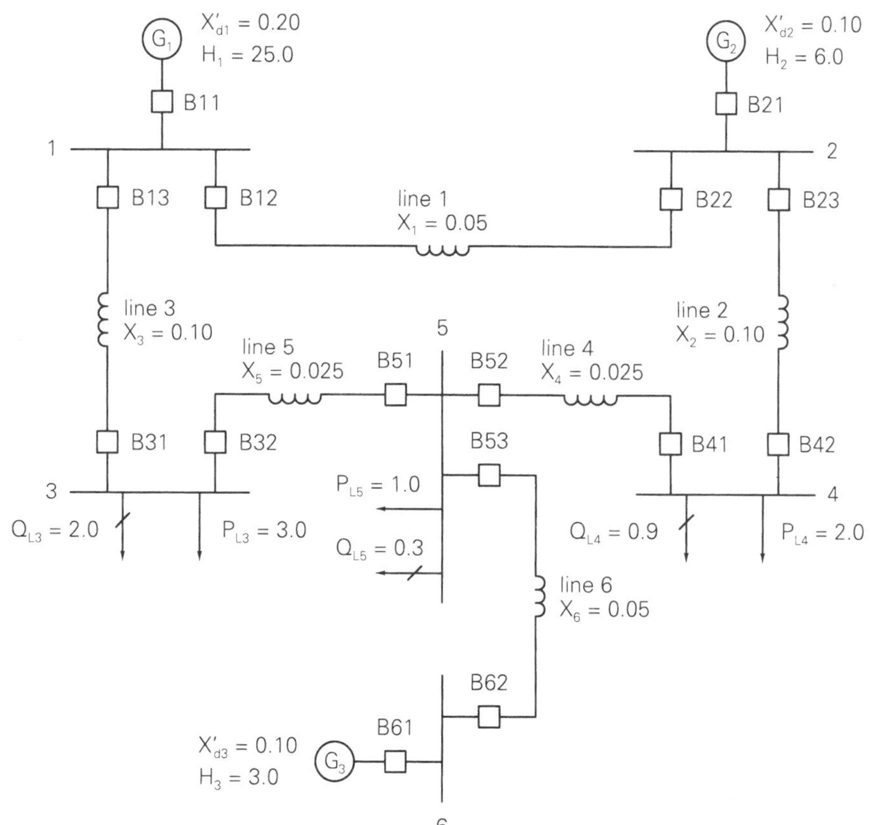

FIGURE 11.29 Single-line diagram of a six-bus power system (per-unit values are shown)

PW **11.25** With PowerWorld Simulator using the Example 11_9 case determine the critical clearing time (to the closest 0.01 second) for a transmission line fault on the transmission line between bus 44 (PEACH69) and bus 14 (REDBUD69), with the fault occurring near bus 44.

SECTION 11.6

PW **11.26** PowerWorld Simulator case Problem 11_26 duplicates Example 11.10, except with the synchronous generator initially supplying 75 MW at unity power factor to the infinite bus. (a) Derive the initial values for δ, E'_q, E'_d and E_{fd}. (b) Determine the critical clearing time for the Example 11.10 fault.

PW **11.27** PowerWorld Simulator case Problem 11_27 duplicates the system from Problem 11.24, except the generators are modeled using a two-axis model, with the same X'_d and H parameters as are in Problem 11.24. Compare the critical clearing time between this case and the Problem 11.24 case.

SECTION 11.7

PW **11.28** PowerWorld Simulator case Problem 11_28 duplicates Example 11.11 except the wind turbine generator is set so it is initially supplying 100 MW to the infinite bus at unity power factor. (a) Use the induction machine equations to verify the initial conditions of $S = -0.0129$, $E'_r = 0.8475$, $E'_i = 0.4230$, $I_r = 0.8433$, and $I_i = 0.7119$. (b) Plot the terminal voltage for the fault sequence from Example 11.6.

11.29 Redo Example 11.12 with the assumption the generator is supplying $100 + j10$ MVA to the infinite bus.

CASE STUDY QUESTIONS

a. What is a black-start generating unit? Name three types of black-start units. Are black-start units subject to self-excitation? If so, how?

b. In the case study, system stability is subdivided into what three categories? What is rotor angle stability?

c. Define cold-load pickup. What are the components of cold-load pickup?

REFERENCES

1. G. W. Stagg and A. H. El-Abiad, *Computer Methods in Power Systems* (New York: McGraw-Hill, 1968).
2. O. I. Elgerd, *Electric Energy Systems Theory*, 2d ed. (New York: McGraw-Hill, 1982).
3. C. A. Gross, *Power System Analysis* (New York: Wiley, 1979).

4. W. D. Stevenson, Jr., *Elements of Power System Analysis*, 4th ed. (New York: McGraw-Hill, 1982).

5. E. W. Kimbark, "Suppression of Ground-Fault Arcs on Single-Pole Switched EHV Lines by Shunt Reactors," *IEEE Trans PAS*, 83 (March 1964), pp. 285–290.

6. K. R. McClymont et al., "Experience with High-Speed Rectifier Excitation Systems," *IEEE Trans PAS*, vol. PAS-87 (June 1986), pp. 1464–1470.

7. E. W. Cushing et al., "Fast Valving as an Aid to Power System Transient Stability and Prompt Resynchronization and Rapid Reload after Full Load Rejection," *IEEE Trans PAS*, vol. PAS-90 (November/December 1971), pp. 2517–2527.

8. M. L. Shelton et al., "Bonneville Power Administration 1400 MW Braking Resistor," *IEEE Trans PAS*, vol. PAS-94 (March/April 1975), pp. 602–611.

9. P. W. Sauer and M. A. Pai, *Power System Dynamics and Stability* (Prentice Hall, 1997).

10. P. Kundar, *Power System Stability and Control* (McGraw-Hill, 1994).

11. J. Feltes and C. Grande-Moran, "Down but Not Out," *IEEE Power & Energy Magazine*, 12,1, (January/February 2014), pp. 34–43.

12. J. Arrillaga, C.P. Arnold, *Computer Analysis of Power Systems*, John Wiley & Sons Ltd., 1990.

13. K. Clark, N. W. Miller, J. J. Sanchez-Gasca, "Modeling of GE Wind Turbine-Generators for Grid Studies," Version 4.4, GE Energy, Schenectady, NY, September 2009.

14. E.H. Camm et al., "Characteristics of Wind Turbine Generators for Wind Power Plants," Proc. IEEE 2009 General Meeting, Calgary, AB, July 2009.

15. B. Stott, "Power System Dynamic Response Calculations," *Proc. IEEE*, vol. 67, No. 2 (February 1979), pp. 219–241.

12 Power System Controls

ISO New England's state-of-the-art control center helps to ensure the reliable operation of New England's bulk power generation and transmission system (Photograph © Adam Laipson)

Automatic control systems are used extensively in power systems. Local controls are employed at turbine-generator units and at selected voltage-controlled buses. Central controls are employed at area control centers.

Figure 12.1 shows two basic controls of a steam turbine-generator: the voltage regulator and turbine-governor. The voltage regulator adjusts the power output of the generator exciter in order to control the magnitude of generator terminal voltage V_t. When a reference voltage V_{ref} is raised (or lowered), the output voltage V_r of the

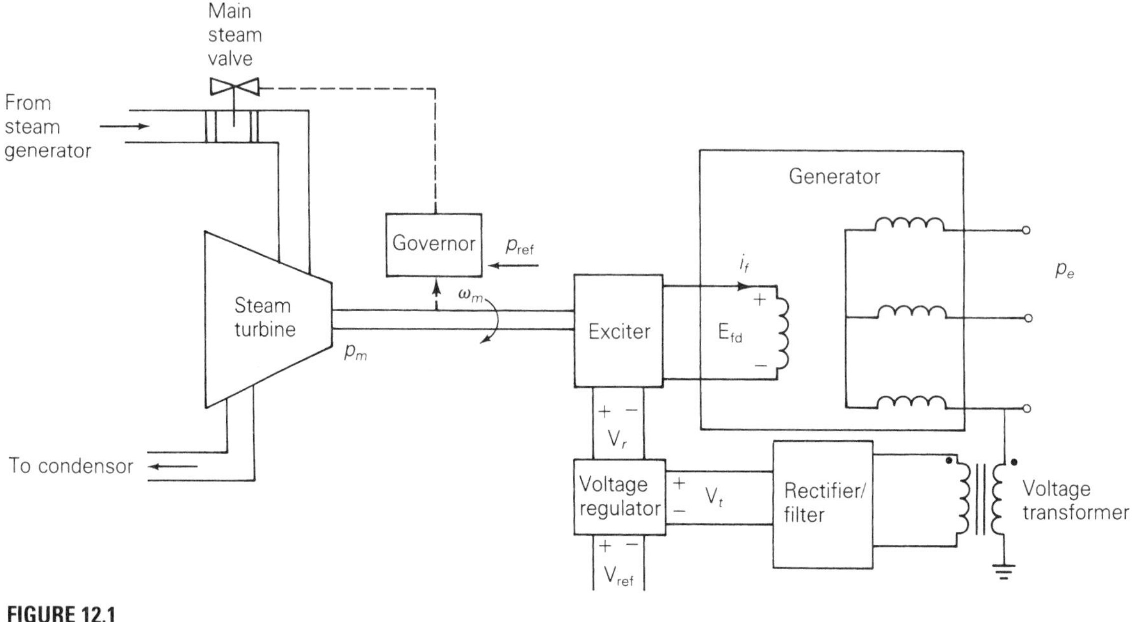

FIGURE 12.1

Voltage regulator and turbine-governor controls for a steam-turbine generator

regulator increases (or decreases) the exciter voltage E_{fd} applied to the generator field winding, which in turn acts to increase (or decrease) V_t. Also a voltage transformer and rectifier monitor V_t, which is used as a feedback signal in the voltage regulator. If V_t, decreases, the voltage regulator increases V_r to increase E_{fd}, which in turn acts to increase V_t.

The turbine-governor shown in Figure 12.1 adjusts the steam valve position to control the mechanical power output p_m of the turbine. When a reference power level p_{ref} is raised (or lowered), the governor moves the steam valve in the open (or close) direction to increase (or decrease) p_m. The governor also monitors rotor speed ω_m, which is used as a feedback signal to control the balance between p_m and the electrical power output p_e of the generator. Neglecting losses, if p_m is greater than p_e, ω_m increases, the governor moves the steam valve in the close direction to reduce p_m. Similarly, if p_m is less than p_e, ω_m decreases, the governor moves the valve in the open direction.

In addition to voltage regulators at generator buses, equipment is used to control voltage magnitudes at other selected buses. Tap-changing transformers, switched capacitor banks, and static var systems can be automatically regulated for rapid voltage control.

Central controls also play an important role in modern power systems. Today's systems are composed of interconnected areas, where each area has its own control

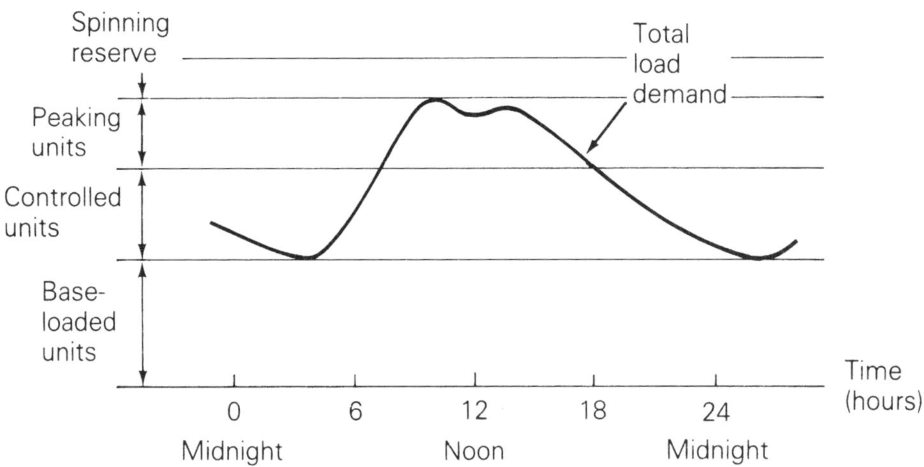

FIGURE 12.2

Daily load cycle

center. There are many advantages to interconnections. For example, interconnected areas can share their reserve power to handle anticipated load peaks and unanticipated generator outages. Interconnected areas can also tolerate larger load changes with smaller frequency deviations than an isolated area.

Figure 12.2 shows how a typical area meets its daily load cycle. The base load is carried by base-loaded generators running at 100% of their rating for 24 hours. Nuclear units and large fossil-fuel units are typically base-loaded. The variable part of the load is carried by units that are controlled from the central control center. Medium-sized fossil-fuel units and hydro units are used for control. During peak load hours, smaller, less efficient units such as gas-turbine or diesel-generating units are employed. Renewable generators, such as wind and solar, are usually operated to maximize their outputs for the given wind or solar conditions since their "fuel" is essentially free. In addition, generators operating at partial output (with *spinning reserve*) and standby generators provide a reserve margin.

The central control center monitors information including area frequency, generating unit outputs, and tie-line power flows to interconnected areas. This information is used by automatic *load-frequency control* (LFC) in order to maintain area frequency at its scheduled value (60 Hz) and net tie-line power flow out of the area at its scheduled value. Raise and lower reference power signals are dispatched to the turbine-governors of controlled units.

This chapter covers automatic controls employed in power systems under normal operation. Sections 12.1 and 12.2 describe the operation of the two generator controls: voltage regulator and turbine-governor, and load-frequency control is discussed in Section 12.3.

CASE STUDY

Beginning at 4:10 p.m Eastern Daylight Time on August 14, 2003, an enormous power disruption resulted in the loss of power to approximately 50 million people across the eastern Great Lakes region, the northeastern United States, and parts of eastern Canada. It took more than 24 hours to restore substantial load. The following article provides a restoration summary and reviews the challenges that had to be overcome to restore power [15]. A general pattern for sound restoration is inferred from the restoration process.

No Light in August: Power System Restoration Following the 2003 North American Blackout

E.H. Allen, R.B. Stuart, and T.E. Wiedman

On 14 August 2003, three 345-kV transmission circuits in northeastern Ohio contacted overgrown trees during a 40-minute time span, starting a chain of events that culminated in the collapse of the electrical grid across the eastern Great Lakes region, the northeastern United States, and parts of eastern Canada. In the aftermath of the disturbance, large portions of the Eastern Interconnection were blacked out, and several electrical islands were present. System operators faced a formidable task: to reassemble the grid and restore power to tens of millions of customers. The challenges that had to be overcome varied significantly from one state or province to another. New York, New England, Ontario, Michigan, and Ohio each had unique problems that operators had to address.

New York

Faced with a system restoration task, the priorities of the New York Independent System Operator (NYISO) restoration plan were to: (1) stabilize the remaining system, (2) extend the stabilized system into blacked-out areas for generation and load restoration, (3) connect energized islands to the stabilized system for restoration of frequency and voltage control, and (4) restore normal transmission operations. In keeping with the restoration plan, the highest priority operations were to: (1) energize the New York state power system, (2) synchronize the New York state power system with the Eastern Interconnection, and (3) restore off-site power to nuclear plants in New York. The operators first had to assess the state of the system and determine what was still energized. They found that New York had split electrically into two pieces, with the southeast part of the state blacked out and upstate New York remaining as an electrical island but with greatly reduced load and generation. Of the 28,700 MW of load in the New York control area just prior to the disturbance, only 5700 MW remained. Figure 1 is a representation of the New York power system in the following sequence of events.

An early step taken was the initiation of black-start procedures at the Gilboa pumped-storage plant in east-central New York (Figure 1). Two Gilboa units (250 MW each) were started at 4:27 p.m., but they could not be synchronized to the remaining power system until after 7:05 p.m. (when the Marcy–New Scotland 345-kV line was reclosed) due to excessive voltage disparity between the Gilboa and Fraser substations. Long Island Power Authority (LIPA) began restoring the Long Island system independently of NYISO, using gas turbines to energize the local system and pick up load.

After finding that the remaining system was an island, operators

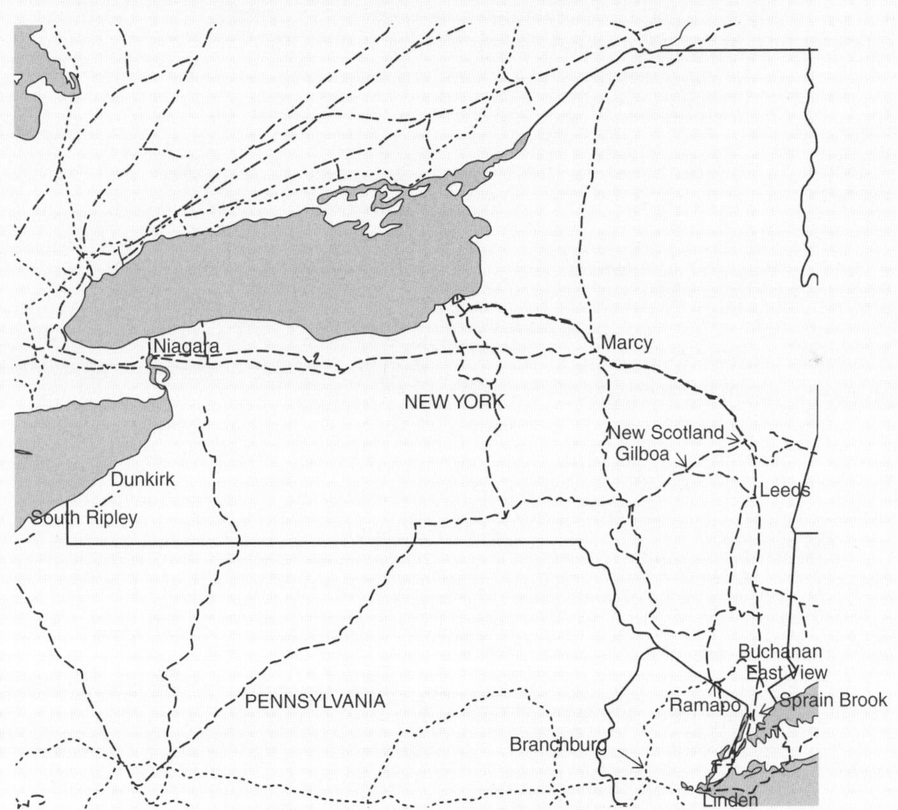

Figure 1 A representation of the New York Power System (dash-dot lines indicate 765 kV, dotted lines indicate 500 kV, and dashed lines indicate 345 kV)

concluded that in order to restore transmission circuits, they would need to restore load at the same time so as to prevent excessively high voltage. In order to restore load, however, they would need additional generation to serve the additional load and keep frequency under control. In order to bring generation online, the voltage and frequency in the New York island would need to be stable. Resynchronization of the island to the Eastern Interconnection therefore became the top priority for NYISO operators in order to achieve the restoration goals.

An initial attempt to synchronize the upstate New York island to the Eastern Interconnection at 6:02 p.m. by closing the Branchburg-Ramapo 500-kV circuit failed because the frequency difference between the two systems was too large. Interestingly, the synchronization was actually achieved without operators' knowledge at 6:52 p.m., as an automatic reclosing scheme on the South Ripley–Dunkirk 230-kV line at the far western end of the island detected that the two systems had come close enough in phase angle and voltage to permit a reclosing attempt, and the reclosing attempt was successful. When operators again attempted to reclose the Branchburg-Ramapo 500-kV circuit at 7:06 p.m., it was thus discovered that the two systems were already synchronized. The Branchburg-Ramapo 500-kV circuit was then successfully reclosed, providing a second, stronger transmission path for synchronizing the upstate New York island and stabilizing the frequency of the energized portion of the New York transmission system.

After resynchronization with the Eastern Interconnection was achieved, the energized system was extended into the New York City metropolitan area by a progressive reenergization of the 345-kV network from north to south. The Sprain Brook substation in Westchester County was energized at 7:56 p.m. from Ramapo via the Buchanan and Eastview stations. By 9:50 p.m., a second 345-kV transmission path from New Scotland down the east side of the Hudson River through Leeds to Sprain Brook had been restored, although the two transmission paths were not connected at Sprain Brook until 12:08 a.m. on August 15. A transmission path from New Jersey into Brooklyn and Staten Island was energized at 11:00 p.m.; this path also provided a connection for East Coast Power's Linden plant. A 345-kV path from Sprain Brook into the West 13th Street station in Manhattan was energized at 4:08 a.m. The LIPA system, which had been restored independently of NYISO, was synchronized with the rest of New York and the Eastern Interconnection at 5:12 a.m.

New England was resynchronized with New York and the Eastern Interconnection at 1:53 a.m. This resynchronization could not be accomplished until after the paralleling operation at Sprain Brook was completed to stabilize the voltage in eastern New York. New England provided substantial emergency power (up to 600 MW) throughout the remainder of the restoration. Restoration of generation, load, and additional transmission continued throughout the evening of August 15. By order of the U.S. Department of Energy, the Cross-Sound Cable was energized at 12:26 p.m., carrying 100 MW of emergency flow. This dc link between Connecticut and Long Island had been constructed but had not yet been authorized to begin operation.

By 12 a.m. on August 15, 40% of New York load had been restored, and 60% of the New York load was restored by 4 a.m. on that day. Because the morning load pickup exceeded the capability of the partially restored system, the Emergency Demand Response Program, which calls for voluntary load reduction, was invoked at 8:59 a.m. When this action proved insufficient, 300 MW of load west of Utica was shed at 9:33 a.m. Area control error at the time was –630 MW. Half of this load was restored at 10:02 a.m., and the rest was restored at 10:24 a.m. By 10:30 p.m. on August 15, 100% of New York load had been restored.

NYISO found that several lessons were learned from the experience:

- Staff duties in the event of such a calamity needed to be clearly defined.
- Communications needed to be improved.
- A recommendation was made to investigate a formal process for distributing information to transmission and generator owners.
- It was concluded that restoration training should be expanded.

New England

New England and the Maritimes remained intact, as a single island. The aggregate flow across the ac ties (i.e., excluding dc imports) was very small before the event, and the total generation loss and load loss during the event in these areas were nearly identical (around 3100 MW). The island was therefore able to remain in relatively stable operation. Southwestern Connecticut had separated along with southeast New York and was blacked out, while about 140 MW in northern Vermont was also deenergized, apparently in reaction to oscillations over a 4.5-s interval that drove voltage as low as 0.21 p.u. A small amount of load had been shed by underfrequency load-shedding (UFLS) relays that had acted above the standard threshold of 59.3 Hz for the first stage of UFLS in New England. Significant oscillations were reported, possibly due to a misalignment of frequency bias with actual generation frequency response. The status of the New England system after the event is shown in Figure 2.

Voltage was initially high after the separation but fell rapidly owing to the tripping of capacitors (both transmission and distribution) by overvoltage protection and a load increase over the next 7–10 minutes. At 4:16 p.m., all fast-start generation was ordered online. Due to flows on transmission lines over long-term

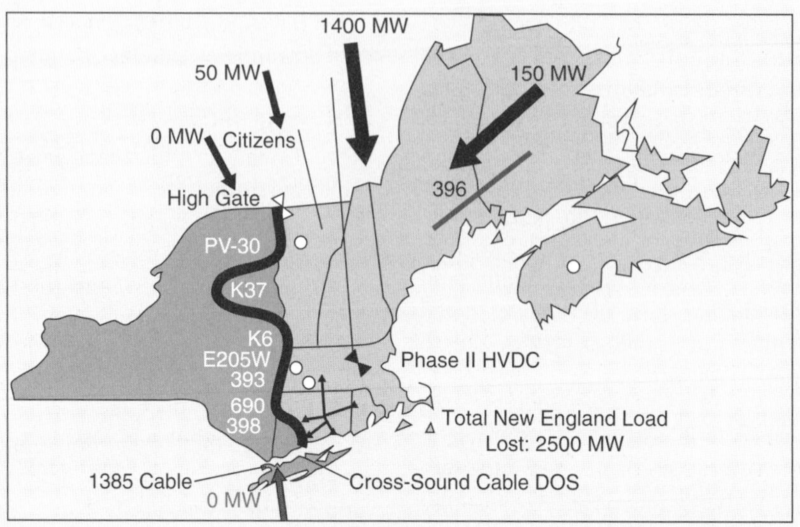

Figure 2 The New England power system island

emergency ratings and low voltages that prevented generation synchronization, CONVEX (a local transmission operator) ordered 500 MW of manual load shedding (400 MW in Connecticut and 100 MW in western Massachusetts), which was completed by 4:40 p.m. Over the next hour and 10 minutes, 400 MW of generation was synchronized to the system. The load that had been manually shed was restored between 5:42 p.m. and 7:28 p.m.

Restoration of power to southwestern Connecticut was an ongoing task. Stations were manually staffed in southwestern Connecticut. At 9:50 p.m., Danbury, Norwalk, and Stamford were reconnected. At 11:23 p.m., the Connecticut transmission system was restored except for the ties to New York and the connections to the Middle River substation, which were high-pressure, fluid-filled cables. By 1:35 a.m. on August 15, all distribution buses at bulk substations were energized except for Middle River. Restoration efforts in the area were complicated by a conductor splice failure on the Southington–Frost Bridge 345-kV line, which tripped at 5:44 a.m. Restoration of load was halted from 7 a.m. to noon, though no additional load shedding was needed. Load restoration was completed by the evening of August 15. The Norwalk-Northport 138-kV cable under Long Island Sound was not restored until August 24, however, due to the loss of cable insulation pressure.

Several lessons were learned from the New England experience:

- Some generators were not operating with automatic voltage regulators (AVRs) in voltage control mode, which hampered efforts to maintain stability of the islanded system and maintain voltages within emergency limits.

- It was concluded that operators should switch from tie line bias to flat-frequency automatic generation control (AGC) when the system becomes islanded.
- Generator governor response needed to be reviewed.
- Frequency and voltage match criteria for the resynchronization of islands needed to be reviewed.
- It was found that personnel are needed to transcribe critical decisions and actions.
- Communications improvements, such as the ability to keep a teleconference line between control centers open continuously, were also recommended.

Ontario

In restoring the system in Ontario, the first steps taken were to confirm the extent of the disturbance and to activate the Ontario Power System Restoration Plan, which includes establishing communications with other control areas, transmission owners, and market participants. Loads that have top priority in the restoration plan include class IV ac power to all nuclear sites, critical transmission and generation station service loads, and critical utility telecommunication facilities. The next priority is restoration of customer loads as necessary to control voltages and secure generators, and the priority that follows that one is synchronization of islands to each other and to neighboring power systems.

The Ontario system was completely blacked out, except for the portion of northern Ontario west of Wawa, two islands in the vicinity of Niagara and Cornwall (both of which were connected to upstate New York), and two other small islands (one north of Timmins and the other on the Ottawa River near Deep River). Transmission was restored from Niagara to the Bruce nuclear generating station starting at 4:42 p.m. The Ontario restoration paths are depicted in Figure 3. This path also provided potential for restarting the Nanticoke units. The three available Bruce units were restored from 7:13 p.m. to 9:13 p.m. Restoration continued toward the Toronto area, Pickering, and Lambton (on the border with Michigan). Controlled customer (load) restoration took place between London and Toronto to balance the transmission system. The Ontario restoration paths are depicted in Figure 3.

Meanwhile, an assessment of the Cornwall area was conducted between 4:11 p.m. and 5:15 p.m. Starting at 5:15 p.m., restoration began westward toward Pickering and Darlington. At 6:40 p.m., restoration of a path from Cornwall to Ottawa was begun in order to energize critical communication facilities. A Darlington unit was restored at 9:18 p.m. At 10:37 p.m., a transmission loop was completed around Lake Ontario.

Starting from a small remaining island, at 7:41 p.m. a path was energized south to Timmins. Load was restored, and the restoration continued south to Sudbury. This restoration

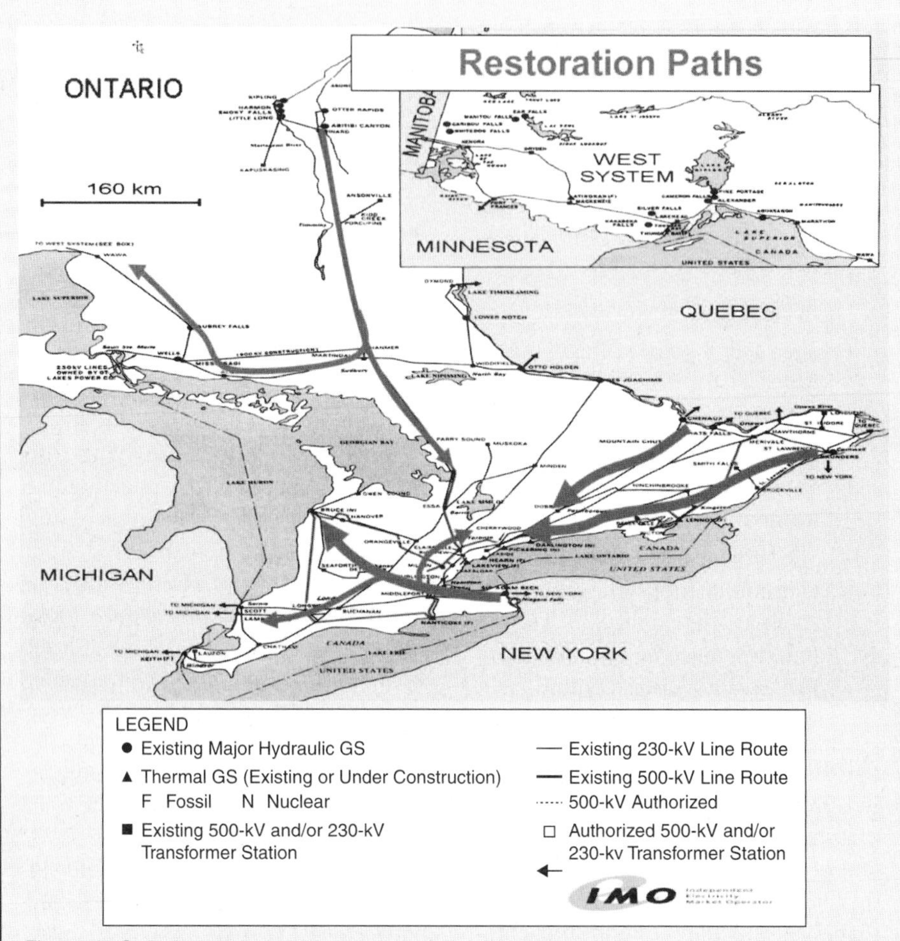

Figure 3 Ontario power system restoration paths

path reached southern Ontario at 3:43 a.m. on August 15. Northern Ontario was resynchronized to the rest of the province at Wawa at 5:20 a.m.

Immediately following the disturbance, 11 nuclear units were off-line. These units posed a number of special challenges. Fully powering down these reactors would have left them all off-line for several days, owing to the complications of restarting them. It was difficult, however, to maintain them at minimum power generation levels for the amount of time required to restore the transmission system and provide suitable outlets for their power. At Pickering B, the pumps in the emergency cooling system were unavailable for five and a half hours. In the end, three of four units at Bruce and one of four units at Darlington were able to come back online and support the system after several hours, but the remaining units were not available for several days. One unit at Bruce required repairs and was out for nine days, while one unit at Pickering did not return until two weeks after the initial disturbance. Ontario remained in a state of emergency operation for nine days because of capacity issues. Some parts of Ontario experienced rolling blackouts during this time.

The experience in Ontario provided a number of lessons learned:

- The devices in place to protect equipment operated as planned.
- The development and maintenance of a documented restoration plan, with training and rehearsals, were essential investments of time and money.
- Close cooperation among multiple entities, transmission owners and operators, generation owners and operators, the Independent Market Operator (now known as the Independent Electricity System Operator), market participants, and government, is needed.
- Communication protocols between different control areas and reliability coordinators are important.
- Emergency power at key facilities is crucial. A backup diesel generator at a Hydro One control center failed to start. The emergency operations center at the Canadian Nuclear Safety Commission in Ottawa had no backup power supply.

Detroit

The restoration effort in metropolitan Detroit was hampered by several instances of physical damage to equipment as well as a complete blackout of much of the area. Essentially, all of metropolitan Detroit and southeastern Michigan was blacked out, with only the northern part of the area known as the Thumb and a single connection from the west and south still energized.

The basic restoration plan started with the energization of the 120-kV network around Detroit from the still-energized grid to the west and north, which was synchronized with the Eastern Interconnection. Meeting this objective was followed by energization of the parallel 345-kV network. Starting from the north, energization was extended to the Harbor Beach unit from two 120-kV ties to Consumers Energy in central Michigan. Energization reached the 120-kV St. Clair station at 8:15 p.m. A stronger tie to Consumers Energy was established when the 345-kV Hampton-Pontiac line was closed in after midnight. The Remer station and the Dean power plant were reconnected at 3:18 a.m. on August 15. A 345-kV path from Belle River to St. Clair and Jewel was closed at 4:43 a.m. (see Figure 4).

On the south side, energization was extended from the Monroe 345-kV station to Brownstown and Fermi between 10:00 p.m. and 10:30 p.m. At 3 a.m. on August 15, a 120-kV connection to the Airport substation was energized, enabling restoration of critical load. But access to River Rouge and other southern generating plants was hampered by a fire at an oil refinery at the time of the blackout that forced an

Figure 4 Eastern Michigan power system restoration (dotted lines indicate 500 kV, dashed lines indicate 345 kV, and double-dashed dot lines indicate 230 kV)

evacuation of the area. Extension of energization to the Waterman station near downtown Detroit was therefore delayed until 8:00 a.m. With the closure of the Blackfoot-Madrid 345-kV line at 8:55 a.m. and the closure of the ring bus at Pontiac at 9:30 a.m., the 345-kV loop around the Detroit metro area was finally reestablished. A second 345-kV tie to Consumers Energy, the Thetford-Jewel circuit, was reclosed at 1:38 p.m. on August 15.

As restoration continued, Detroit Edison followed a practice of energizing whole distribution centers without isolating individual circuits. Up to this point, most of the restoration actions were preparatory to restoring load. Distribution crews were needed as transmission was restored in order to balance load against available generation. Substantial restoration of distribution started at about 2:30 p.m. on August 15. The restoration of the Essex 24-kV station at 5:30 p.m. picked up a significant portion of the load in the city of Detroit. The last of four critical pumps for the city's water and sanitation department was restored in the evening.

Detroit Edison conducted inspections of transmission lines and switchyards over the next several days to maintain system reliability. The inspections were conducted both by air and on the ground, and they were both visual and thermal. Storms delayed the start of the inspections. Approximately 2400 km of transmission were inspected between August 17 and 21.

Several instances of damage to generators were reported. Five units had failed rupture discs, which kept these units unavailable for more than 24 hours. Rupture discs are a pressure relief location for the steam condenser. They are prone to failure when the entire electrical network goes down, as a consequent loss of circulating water pumps and vacuum pumps results in a buildup of pressure in the condenser. The rupture discs, however, are designed to fail under stressed conditions and probably prevented serious physical damage to the units. The availability of spare discs so that units could be returned to service was an issue; the need for so many simultaneous rupture disc replacements had not been experienced previously. One unit suffered a loss of control air on the generator seal oil system valves, resulting in oil in the generator. The output of another unit was limited due to a ground in the no load/low load trip switch. Another unit had an overheated boiler circulating water pumps, which needed to be replaced. Finally, one unit (River Rouge 3) suffered bearing damage during the event, which left the unit out of service for weeks. Lessons learned from the restoration experience in and around Detroit include:

- The restoration effort was vulnerable to Detroit Edison's computerized "in-service application" function, used to prioritize and dispatch repair crews. This function was not operable due to the loss of power, forcing repair crews to make a number of extra trips back to service centers to pick up new work orders.
- Practices in maintaining spare rupture discs for generators were deficient for coping with system-wide disturbances and hence were changed.

Southern Michigan

The southeastern corner of the Consumers Energy system in southern Michigan had also blacked out. After the initial disturbance, operators received conflicting information about the amount of generation. Operators believed the Consumers Energy system was undergenerating, but the high frequency (60.2 Hz) suggested the opposite. Operators maintained generation levels until the status of the system could be ascertained. Restoration efforts commenced with an emergency page of transmission and distribution personnel and a conference call at 5:15 p.m. on August 14 to determine initial actions. After assessing the state of the transmission system and identifying the affected area, all of the breakers within that area were opened as preparation for restoration. The 138-kV network was restored by 7:25 p.m. Subsequently, the remaining 46-kV lines and customer load were restored; this work was completed by 10:05 p.m.

The area that was restored was fed by three 138-kV lines. One of the lines (Leoni-Beecher) tripped at 10:30 p.m. and did not reclose. The power increase on the other two lines caused them to trip as well, blacking out the area that had just been restored. A second restoration commenced quickly, bringing the 138-kV network back up by 12:55 a.m. on August 15 and the 46-kV network, with all customer load, by 1:35 a.m. A fault location was identified from relay data, but a cause for the fault was not found at the indicated location. The line was given a derated capability when it was reclosed at 11:00 p.m. Because of the limitations of the system caused by the derating, 40 MW of load, served by four 46-kV lines, was manually shed between 7 a.m. and 9:30 a.m. on August 15; the lines were restored when a Whiting generator came back online.

The initial attempt to return the Campbell 3 unit to service failed due to water hammer that developed after steam was turned back on. The water hammer damaged a number of piping hangers, delaying the unit's return to service.

Cleveland

The authors could not find any publicly available, comprehensive reports on the restoration process in the Cleveland area. A summary of the restoration in the region of the East Central Area Reliability Coordination Agreement (ECAR), which includes northern Ohio and southern Michigan, is provided from the ECAR final report dated July 7, 2004. In the region, 18,047 MW of load, 18,811 MW of generation, and 238 transmission circuits were lost. All load was restored by 9:00 a.m. on August 16. Restoration of 100% of the load took one day and 17 hours. The restoration strategy consisted of restoration of load from energized transmission lines and the use of black-start units not in black-start mode. These gas turbines were capable of black starting but were not used in this mode. These units can be started up very quickly to help restore load. Throughout the restoration, operators needed to be sure that transfers from American Electric Power (AEP) to Michigan did not overload the FirstEnergy transmission system in northern Ohio. Restoration of power to off-site nuclear power plant supply and to other generation resources off-line as a result of the blackout was a priority. Communications and fuel supply to critical infrastructure were major issues during the blackout restoration effort. The need to more effectively share restoration plans among reliability coordinators, transmission operators, and generation operators was identified. ECAR also identified the need for additional training and regular drills for blackout restoration.

State or Province	Time to Restore Substantial Load	Time to Restore 100% Load
Michigan	8/15/2003 at 5:30 p.m.	8/15/2003, evening
New England	8/15/2003 at 1:35 a.m.	8/15/2003, end of day
New York	8/15/2003 at 4:00 a.m.	8/15/2003 at 10:30 p.m.
Ohio	8/15/2003 at 7:00 a.m. (Cleveland water pumps)	8/16/2003 at 7:00 a.m. (as reported by ECAR)
Ontario	Data not publicly available	8/15/2003, end of day

TABLE 1

Time to restore substantial and 100% load for the various regions affected, as reported in publicly available documentation

Restoration Summary

The times to restore substantial and 100% load for the various regions affected by the blackout are given in Table 1.

Lessons Learned

A general pattern for sound restoration can be inferred from these examples. Restoration commences by first assessing the current state of the system. Then a skeletal high-voltage network is reassembled, providing a cranking path for generation, followed by the gradual connection of lower-voltage facilities and customer load. Of course, the process is not quite this simple. Some load is needed on the system throughout the restoration process in order to maintain voltage stability; otherwise, voltages become excessively high, to the point of potentially damaging equipment. Reactive devices alone are insufficient to regulate voltage on an unloaded transmission system. The system is vulnerable throughout the process: at least one instance was observed of a fully restored area collapsing, forcing utility personnel to start all over again.

The ability to tie to a large interconnection makes the restoration process easier and smoother. A large interconnection provides very stable frequency control and, to a lesser extent, voltage control. Maintaining voltage and frequency in a smaller island is considerably more difficult. In several such instances, operators found it necessary to manually shed load to prevent a complete collapse of the remaining system.

Operators must prioritize critical loads during restoration. Such loads, at the bulk power system level, include water and sanitation system pumps, off-site power for nuclear generating stations, and airports.

The old cliché "expect the unexpected" is certainly applicable to a system restoration effort. Despite the thoughtful development of restoration plans, a number of unforeseen difficulties arose throughout the process, including damaged generators, the inability to synchronize due to excessive voltage disparities, and an inability to physically reach substations due to industrial hazards spawned by the power outage. Communications was cited as an issue in nearly all areas; communication systems need to be able to function in the absence of the power grid. Similar flaws in emergency plans were seen following Hurricane Katrina: officials anticipated being able to use cellular phones following the storm, but many cellular towers were inoperable.

To summarize, the lessons learned included the following priorities and steps:

- Improve communications among responsible parties and communication systems.
- Develop and maintain a restoration plan, and always be prepared to "expect the unexpected."
- Restoration begins by first assessing the current state of the system.
- The next step is to reassemble a skeletal high-voltage network.
- Then a "cranking" path for generation must be provided so as to assure the system is stable and will be restored without sustaining excessive voltages.
- Gradually connect lower voltage facilities and customer load to maintain voltage stability. Reactive devices alone are insufficient to regulate voltage on an unloaded transmission system.
- The availability of connection to a large interconnection makes frequency control and voltage control more stable.
- Shed load if need be to maintain stability within an island being restored.
- Prioritize critical loads during restoration.

Regulatory Recommendations and Standards

Most of the recommendations that followed the postmortem analysis of the blackout focused on preventing a cascading outage as opposed to recovering from one. Recommendation 29 from the U.S.-Canada Power System Outage Task Force, however, was as follows: "Evaluate and disseminate lessons learned during system restoration." In the aftermath of the blackout, the North American Electric Reliability Council (NERC) developed a number of reliability standards to minimize the likelihood of major blackouts and mitigate their impact should they occur. Two particular standards—EOP-005-2 and EOP-006-2—address system restoration. EOP-005-2 provides standards regarding the use of black-start units to reliably restore the system after a blackout. Some of the lessons learned in the August 14 blackout, such as identifying key strategies for establishing cranking paths and adding load incrementally to stabilize voltage and frequency, were incorporated into EOP-005-2. NERC Standard EOP-006-2 addresses the need for coordination of restoration plans among reliability coordinators, the need for plans to achieve synchronization between islands formed by the blackout and the need for training and simulation drills at prescribed periods to ensure that the main ties among reliability coordinators are restored in a reliable manner. While the emphasis remains on preventing blackouts in the first place, it is important to have thoughtful and coordinated plans to restore the power system reliably, particularly in view of the ever present-threat of natural disasters such as fires, floods, and hurricanes.

Conclusion

On August 14, 2003, an enormous power disruption resulted in the loss of power to approximately 50 million people. A total of 61,800 MW of load was disconnected, along with 265 generating plants (see Figure 5). Three major metropolitan areas (Cleveland, Detroit, and New York), along with much of the province of Ontario, were blacked out. Two large areas (upstate New York and the combination of New England and the Maritime provinces) and several smaller areas became electrically islanded systems, disconnected from the Eastern Interconnection. Overall, given the difficulties confronted, the affected utilities and their dedicated personnel did an admirable job in overcoming those difficulties and restoring the power system following this historic power outage.

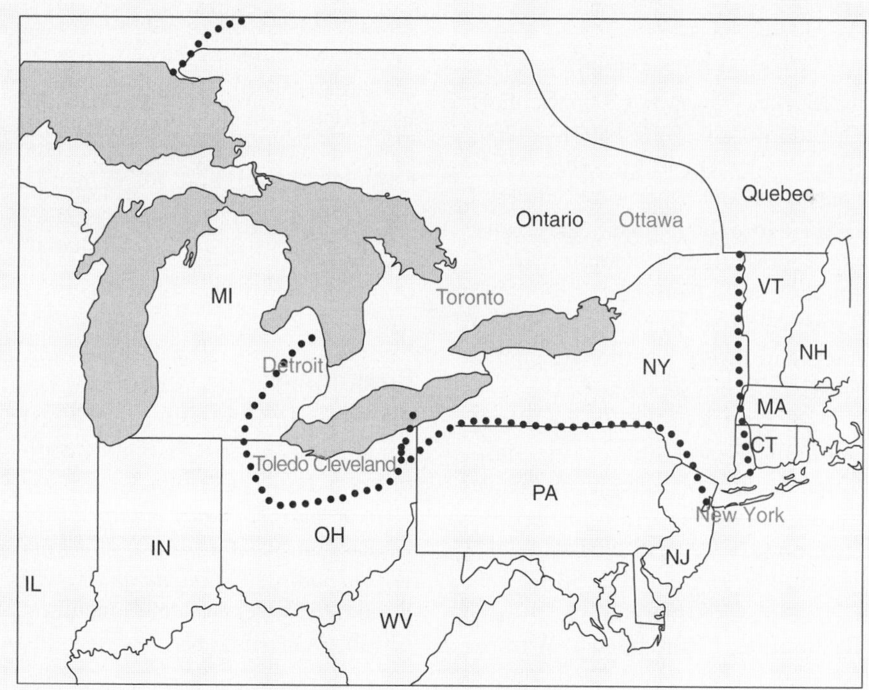

Figure 5 The area affected by August 14, 2003 blackout

For Further Reading

New York Independent System Operator. (2005, Feb.). Blackout August 14, 2003 final report. [Online]. Available: http://www.nyiso.com/public/webdocs/media_room/press_releases/2005/blackout_rpt_final.pdf

Independent System Operator New England. (2004, Feb.). Blackout 2003: Performance of the New England and Maritimes Power Systems During the August 14, 2003 Blackout. United States: ISO New England. [Online]. Available: http://www.iso-ne.com/pubs/spcl_rpts/2004/iso_august_blackout_report.pdf

Canada News Wire. (2003, Sept. 2). IMO releases details of power restoration effort. *T&D World Magazine* [Online]. Available: http://tdworld.com/energizing/imo-releases-details-power-restoration-effort

Ontario Independent Electricity Market Operator. (2003, Sept. 4). Testimony on August 14, 2003 Blackout by David Goulding President and Chief Operating Officer of IESO before the United States House of Representatives Committee on Energy and Commerce. [Online]. Available: http://www.ieso.ca/imoweb/pubs/corp/IMO-Testimony-Congress Hearings_ 2003Sep04.pdf

J. Spears (2013, Aug.). Blackout 2003: How Ontario went dark. *Toronto Star* [Online]. Available: http://www.thestar.com/business/economy/2013/08/13/blackout_2003_how_ontario_went_dark.html

NPCC Inter-Control Area Restoration Coordination Working Group (CO-11). Restoration of the NPCC areas following the power system collapse of August 14, 2003. Presented at the Joint Meeting of the NERC Market Committee, Operating Committee and Planning Committee, Vancouver, Canada, July 21, 2004. [Online]. Available: https://www.npcc.org/library/blackout%20recommendations/Restoration_of_the_NPCC_Areas.pdf

Michigan Public Service Commission, "Michigan public service commission report on August 14th blackout," Nov. 2003.

DTE Energy. (2004, Apr.). Detroit Edison's response to the August 14, 2003 blackout, Response to Michigan Public Service Commission Case No. U-13859. [Online]. Available: http://efile.mpsc.state.mi.us/efile/docs/13859/0006.pdf

U.S.-Canada Power System Outage Task Force. (2004, Apr.). Final report on the August 14, 2003 blackout in the United States and Canada: Causes and recommendations. [Online]. Available: http://www.ferc.gov/industries/electric/indus-act/reliability/blackout/ch1-3.pdf

Biographies

Eric H. Allen is with the North American Electric Reliability Corporation, Atlanta, Georgia.

Robert B. Stuart is with Stuart Consulting, Walnut Creek, California.

Thomas E. Wiedman is with Wiedman Power System Consulting, Elizabeth, Illinois. ■

12.1 GENERATOR-VOLTAGE CONTROL

The *exciter* delivers dc power to the field winding on the rotor of a synchronous generator. For older generators, the exciter consists of a dc generator driven by the rotor. The dc power is transferred to the rotor via slip rings and brushes. For newer generators, *static* or *brushless* exciters are often employed.

For static exciters, ac power is obtained directly from the generator terminals or a nearby station service bus. The ac power is then rectified via thyristors and transferred to the rotor of the synchronous generator via slip rings and brushes.

For brushless exciters, ac power is obtained from an "inverted" synchronous generator whose three-phase armature windings are located on the main generator rotor and whose field winding is located on the stator.

The ac power from the armature windings is rectified via diodes mounted on the rotor and is transferred directly to the field winding. For this design, slip rings and brushes are eliminated.

Block diagrams of several standard types of generator-voltage control systems have been developed by the IEEE Power and Energy Society, beginning in 1968 with [1] and most recently in 2005 with IEEE Std 421.5-2005. A block diagram for what is commonly known as the IEEE Type 1 exciter, which uses a shaft-driven dc generator to create the field current, is shown in Figure 12.3 (neglecting saturation).

In Figure 12.3, the leftmost block, $1/(1 + sT_r)$, represents the delay associated with measuring the terminal voltage V_t where s is the Laplace operator and T_r is the measurement time constant. Note that if a unit step is applied to a $1/(1 + sT_r)$ block, the output rises exponentially to unity with time constant T_r. The measured generator terminal voltage V_t is compared with a voltage reference V_{ref} to obtain a voltage error, ΔV, which in turn is applied to the voltage regulator. The voltage regulator is modeled as an amplifier with gain K_a and a time constant T_a, while the last forward block represents the dynamics of the exciter's dc generator. The output is the field voltage E_{fd}, which is applied to the generator field winding and acts to adjust the generator terminal voltage, as in (11.6.5). The feedback block in Figure 12.3 is used to improve the dynamic response of the exciter by reducing excessive overshoot. This feedback is represented by $(sK_f)/(1 + sT_f)$, which provides a filtered first derivative negative feedback.

For any transient stability study, the initial values for the state variables need to be determined. This is done by assuming that the system is initially operating in steady-state, and recognizing that in steady-state all the derivatives will be zero. Then, by knowing the initial field voltage (found as in Example 11.10) and terminal voltage, all the other variables can be determined.

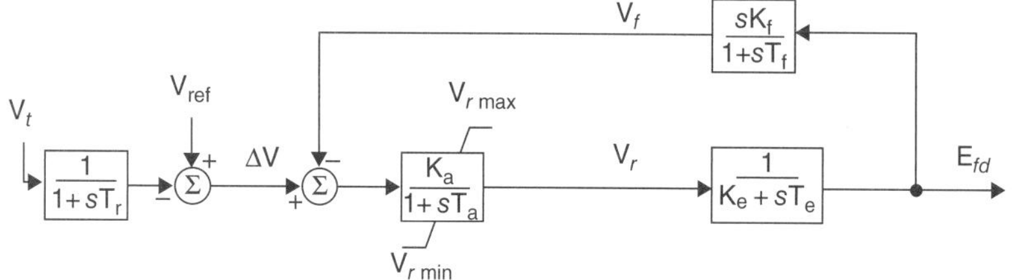

FIGURE 12.3
Block diagram for the IEEE Type 1 Exciter (neglecting saturation)

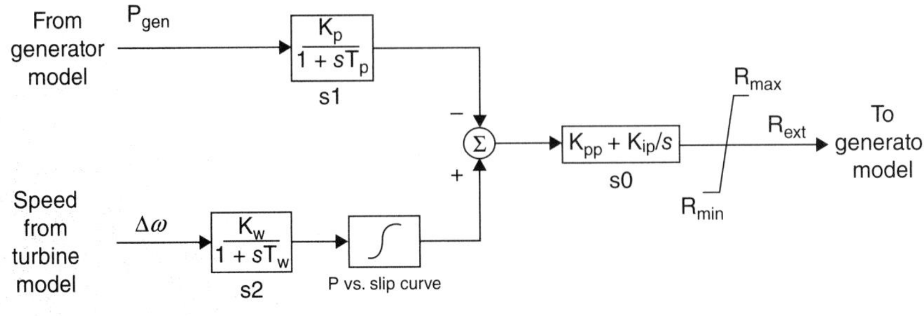

FIGURE 12.4 Simplified block diagram for a Type 2 wind turbine R_{ext} control system

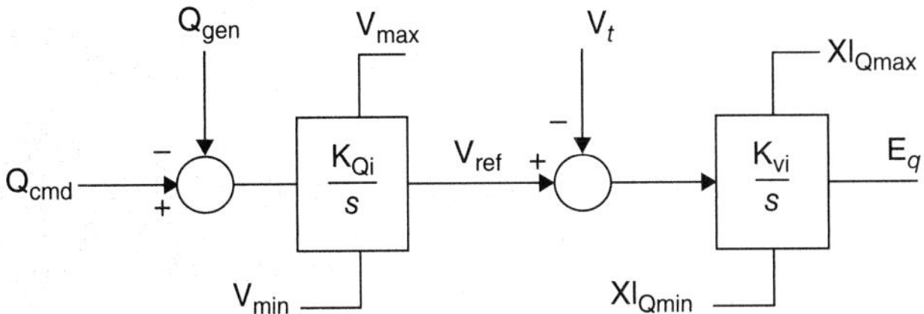

FIGURE 12.5 Simplified block diagram for a Type 3 wind turbine reactive power control system

For wind turbines, how their voltage is controlled depends upon the type of the wind turbines. Type 1 wind turbines, squirrel cage induction machines, have no direct voltage control. Type 2 wind turbines are wound rotor induction machines with variable external resistance. While they do not have direct voltage control, the external resistance control system is usually modeled as a type of exciter. The block diagram for such a model is shown in Figure 12.4. The purpose for this control is to allow for a more constant power output from the wind turbine. For example, if a wind gust were to cause the turbine blades to accelerate, this controller would quickly increase the external resistance, flatting the torque-speed curve as shown in Figure 11.25.

Similar to synchronous machines, the Type 3 and 4 wind turbines have the ability to perform voltage or reactive power control. Common control modes include constant power factor control, coordinated control across a wind farm to maintain a constant voltage at the interconnection point, and constant reactive power control. Figure 12.5 shows a simplified version of a Type 3 wind turbine exciter, in which Q_{cmd} is the commanded reactive power, V_t is the terminal voltage, and the output, E_q is the input to the DFAG model shown in Figure 11.27. For fixed reactive power Q_{cmd} is a constant, while for power factor control, Q_{cmd} varies linearly with the real power output.

EXAMPLE 12.1

Synchronous Generator Exciter Response

Using the system from Example 11.10, assume the two-axis generator is augmented with an IEEE Type 1 exciter with $T_r = 0$, $K_a = 100$, $T_a = 0.05$, $V_{rmax} = 5$, $V_{rmin} = -5$, $K_e = 1$, $T_e = 0.26$, $K_a = 0.01$ and $T_f = 1.0$. (a) Determine the initial values of V_r, V_f, and V_{ref}. (b) Using the fault sequence from Example 11.10, determine the bus 4 terminal voltage after 1 second and then after 5 seconds.

SOLUTION

a. The initial field voltage and terminal voltage, E_{fd} and V_t, do not depend on the exciter, so their values are equal to those found in Example 11.10, that is, 2.913 and 1.095 respectively. Since the system is initially in steady-state,

$$V_r = (K_e)(E_{fd}) = (1.0)(2.9135) = 2.9135$$

Because V_f is the output of the filtered derivative feedback, its initial value is zero. Finally, writing the equation for the second summation block in Figure 12.3

$$(V_{ref} - V_t - V_f)(K_a) = V_r$$

$$V_{ref} = \frac{V_r}{K_a} + V_t + V_f = \frac{2.9135}{100} + 1.0946 = 1.1237$$

b. Open PowerWorld Simulator case Example 12_1 and display the Transient Stability Analysis Form (see Figure 12.6). To see the initial conditions, select the

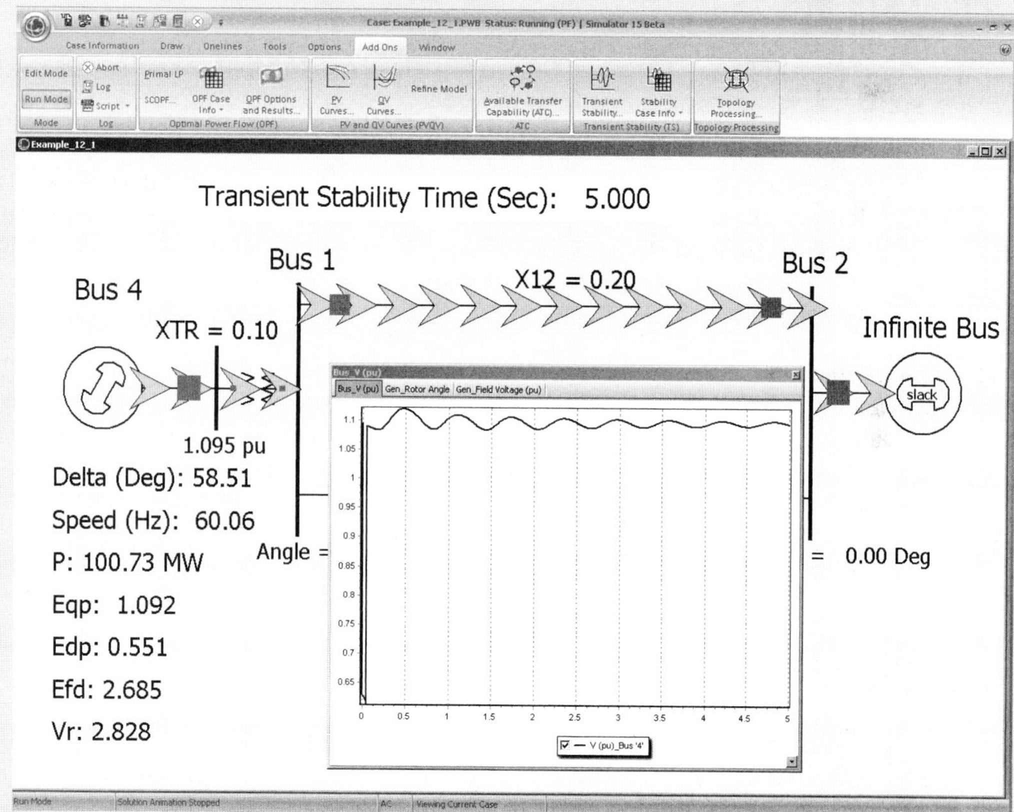

FIGURE 12.6

Example 12.1 results

(Continued)

States/Manual Control page, and then select the **Transfer Present State to Power Flow** button to update the oneline display. From this page, it is also possible to just do a specified number of timesteps by selecting the **Do Specified Number of Timesteps(s)** button or to run to a specified simulation time using the **Run Until Specified Time** button. To determine the terminal voltage after one second, select the **Run Until Specified Time** button. The value is 1.10 p.u. To finish the simulation, select the **Continue** button. The terminal voltage at five seconds is 1.095 p.u., which is close to the prefault voltage, indicating the exciter is restoring the voltage to its setpoint value. In contrast, the bus 4 terminal voltage after five seconds in the Example 11.10 case, which does not include an exciter, is 1.115 p.u.

EXAMPLE 12.2

Type 3 Wind Turbine Reactive Power Control

Assume the Type 3 wind turbine from Example 11.12 has a Figure 12.5 reactive power control system with $K_{Qi} = 0.4$, $K_{Vi} = 40$, $XI_{Qmax} = 1.45$, $XI_{Qmin} = 0.5$, $V_{max} = 1.1$, $V_{min} = 0.9$ (per unit using a 100 MVA base). For the Example 11.12 system conditions, determine the initial values for V_{ref}, Q_{cmd}, and estimate the maximum amount of reactive power this system could supply during a fault that depresses the terminal voltage to 0.5 p.u.

SOLUTION

Since in steady-state the inputs to each of the two integrator blocks in Figure 12.5 must be zero, V_{ref} is just equal to the initial terminal voltage magnitude from Example 11.12, that is, 1.0239 p.u., and Q_{cmd} is the initial reactive power output, which is 0.22 per unit (22 Mvar), found from the imaginary part of the product of the terminal voltage and the conjugate of the terminal current. During the fault with its low terminal voltage, the positive input into the K_{Vi}; integration block will cause E_q to rapidly rise to its limit $XI_{Qmax} = 1.45$. The reactive component of I_{sorc} will then be $-1.45/0.8 = -1.8125$ p.u. The total per unit reactive power injection with $V_t = 0.5$ during the fault is then

$$Q_{net} = (V_t)(1.8125) - \frac{V_t^2}{0.8} = 0.593 \text{ pu} = 59.3 \quad \text{Mvar}$$

This result can be confirmed by opening PowerWorld Simulator case Example 12_2 which models such a fault condition (see Figure 12.7). After the fault is cleared, the reactive power controller restores the machine's reactive power output to its prefault value.

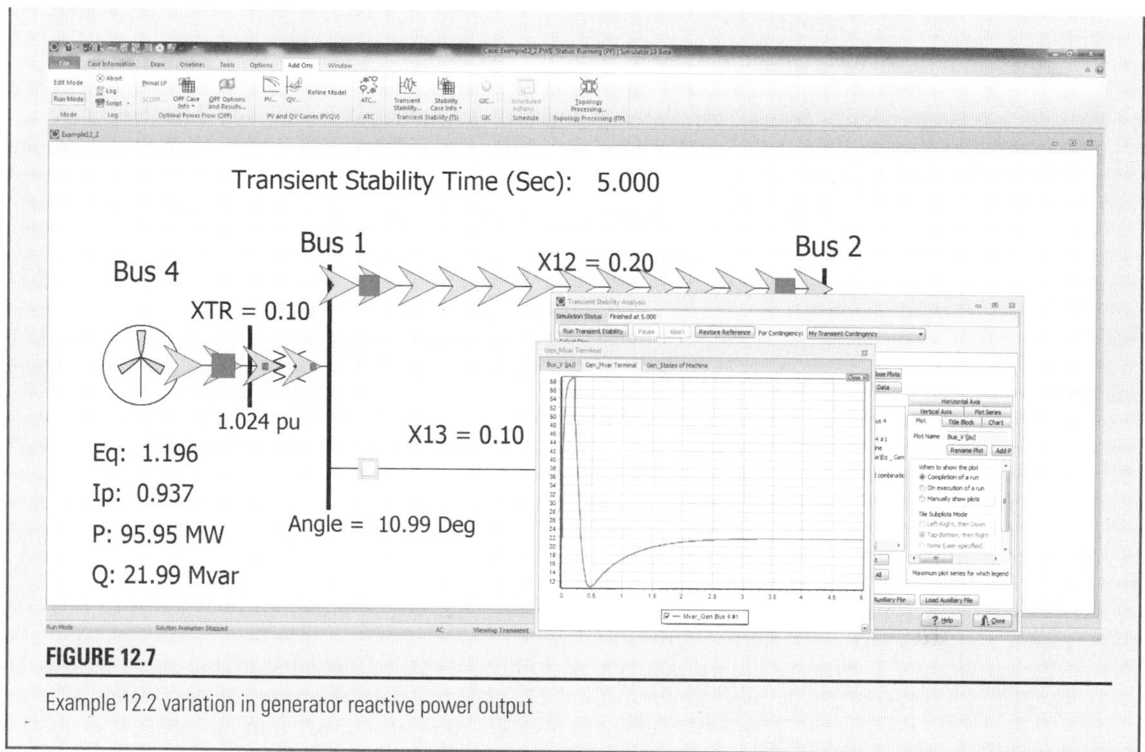

FIGURE 12.7

Example 12.2 variation in generator reactive power output

Block diagrams such as those shown in Figure 12.3 are used for computer representation of generator-voltage control in transient stability computer programs (see Chapter 11). In practice, high-gain, fast-responding exciters provide large, rapid increases in field voltage E_{fd} during short circuits at the generator terminals in order to improve transient stability after fault clearing. Equations represented in the block diagram can be used to compute the transient response of generator-voltage control.

12.2 TURBINE-GOVERNOR CONTROL

Turbine-generator units operating in a power system contain stored kinetic energy due to their rotating masses. If the system load suddenly increases, stored kinetic energy is released to initially supply the load increase. Also, the electrical torque T_e of each turbine-generating unit increases to supply the load increase, while the mechanical torque T_m of the turbine initially remains constant. From Newton's second law, $J\alpha = T_m - T_e$, the acceleration α is therefore negative. That is, each turbine-generator decelerates and the rotor speed drops as kinetic energy is released to supply the load increase. The electrical frequency of each generator, which is proportional to rotor speed for synchronous machines, also drops.

From this, we conclude that either rotor speed or generator frequency indicates a balance or imbalance of generator electrical torque T_e and turbine mechanical torque T_m. If speed or frequency is decreasing, then T_e is greater than T_m (neglecting generator losses). Similarly, if speed or frequency is increasing, T_e is less than T_m. Accordingly, generator frequency is an appropriate control signal for governing the mechanical output power of the turbine.

The steady-state frequency-power relation for turbine-governor control is

$$\Delta p_m = \Delta p_{\text{ref}} - \frac{1}{\text{R}}\Delta f \tag{12.2.1}$$

where Δf is the change in frequency, Δp_m is the change in turbine mechanical power output, and Δp_{ref} is the change in a reference power setting. R is called the *regulation constant.* The equation is plotted in Figure 12.8 as a family of curves, with Δp_{ref} as a parameter. Note that when Δp_{ref} is fixed, Δp_m is directly proportional to the drop in frequency.

Figure 12.8 illustrates a steady-state frequency-power relation. When an electrical load change occurs, the turbine-generator rotor accelerates or decelerates, and frequency undergoes a transient disturbance. Under normal operating conditions, the rotor acceleration eventually becomes zero, and the frequency reaches a new steady-state, shown in the figure.

The regulation constant R in (12.2.1) is the negative of the slope of the Δf versus Δp_m curves shown in Figure 12.8. The units of R are Hz/MW when Δf is in Hz and Δp_m is in MW. When Δf and Δp_m are given in per-unit, however, R is also in per-unit.

FIGURE 12.8 Steady-state frequency–power relation for a turbine-governor

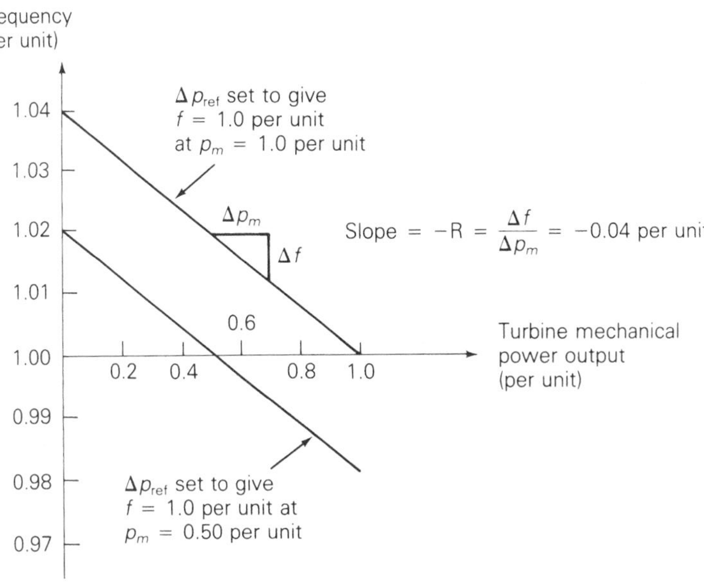

EXAMPLE 12.3

Turbine-governor response to a frequency change at a generating unit

A 500-MVA, 60-Hz turbine-generator has a regulation constant R = 0.05 per unit based on its own rating. If the generator frequency increases by 0.01 Hz in steady-state, what is the decrease in turbine mechanical power output? Assume a fixed reference power setting.

SOLUTION

The per-unit change in frequency is

$$\Delta f_{\text{p.u.}} = \frac{\Delta f}{f_{\text{base}}} = \frac{0.01}{60} = 1.6667 \times 10^{-4} \quad \text{per unit}$$

Then, from (12.2.1), with $\Delta p_{\text{ref}} = 0$,

$$\Delta p_{m\text{p.u.}} = \left(\frac{-1}{0.05}\right)(1.6667 \times 10^{-4}) = -3.3333 \times 10^{-4} \quad \text{per unit}$$

$$\Delta p_m = (\Delta p_{m\text{p.u.}})\text{S}_{\text{base}} = (-3.3333 \times 10^{-4})(500) = -1.6667 \quad \text{MW}$$

The turbine mechanical power output decreases by 1.67 MW.

The steady-state frequency-power relation for one area of an interconnected power system can be determined by summing (12.2.1) for each turbine-generating unit in the area. Noting that Δf is the same for each unit,

$$\begin{aligned}\Delta p_m &= \Delta p_{m1} + \Delta p_{m2} + \Delta p_{m3} + \cdots \\ &= (\Delta p_{\text{ref1}} + \Delta p_{\text{ref2}} + \cdots) - \left(\frac{1}{\text{R}_1} + \frac{1}{\text{R}_2} + \cdots\right)\Delta f \\ &= \Delta p_{\text{ref}} - \left(\frac{1}{\text{R}_1} + \frac{1}{\text{R}_2} + \cdots\right)\Delta f \end{aligned} \tag{12.2.2}$$

where Δp_m is the total change in turbine mechanical powers and Δp_{ref} is the total change in reference power settings within the area. We define the *area frequency response characteristic* β as

$$\beta = \left(\frac{1}{\text{R}_1} + \frac{1}{\text{R}_2} \cdots\right) \tag{12.2.3}$$

Using (12.2.3) in (12.2.2),

$$\Delta p_m = \Delta p_{\text{ref}} - \beta \Delta f \tag{12.2.4}$$

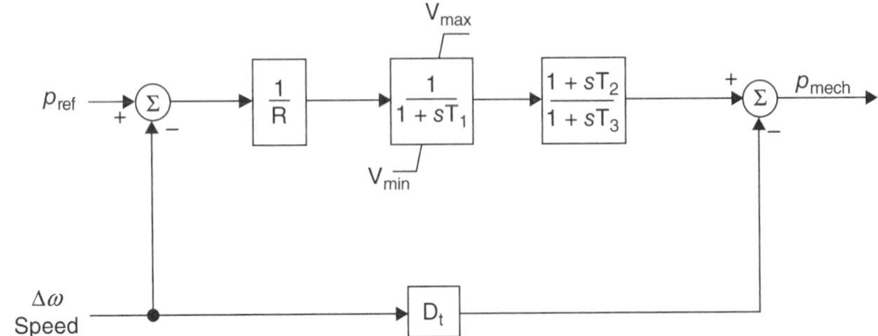

FIGURE 12.9 Turbine-governor block diagram

Equation (12.2.4) is the area steady-state frequency-power relation. The units of β are MW/Hz when Δf is in Hz and Δp_m is in MW. β can also be given in per-unit. In practice, β is somewhat higher than that given by (12.2.3) due to system losses and the frequency dependence of loads.

A standard value for the regulation constant is R = 0.05 per unit. When all turbine-generating units have the same per-unit value of R based on their own ratings, then each unit shares total power changes in proportion to its own rating. Figure 12.9 shows a block diagram for a simple steam turbine governor commonly known as the TGOV1 model. The $1/(1 + sT_1)$ models the time delays associated with the governor, where s is again the Laplace operator and T_1 is the time constant. The limits on the output of this block account for the fact that turbines have minimum and maximum outputs. The second block diagram models the delays associated with the turbine; for non-reheat turbines T_2 should be zero. Typical values are R = 0.05 p.u., T_1 =0.5 seconds, T_3 = 0.5 for a non-reheat turbine or T_2 = 2.5 and T_3 = 7.5 seconds otherwise. D_t, is a turbine damping coefficient that is usually 0.02 or less (often zero). Additional turbine block diagrams are available in [3].

EXAMPLE 12.4

Response of turbine-governors to a load change in an interconnected power system

An interconnected 60-Hz power system consists of one area with three turbine-generator units rated 1000, 750, and 500 MVA, respectively. The regulation constant of each unit is R = 0.05 per unit based on its own rating. Each unit is initially operating at one-half of its own rating, when the system load suddenly increases by 200 MW. Determine (a) the per-unit area frequency response characteristic β on a 1000 MVA system base, (b) the steady-state drop in area frequency, and (c) the increase in turbine mechanical power output of each unit. Assume that the reference power setting of each turbine-generator remains constant. Neglect losses and the dependence of load on frequency.

SOLUTION

a. The regulation constants are converted to per-unit on the system base using

$$R_{p.u.new} = R_{p.u.old} \frac{S_{base(new)}}{S_{base(old)}}$$

We obtain

$$R_{1p.u.new} = R_{1p.u.old} = 0.05$$

$$R_{2p.u.new} = (0.05)\left(\frac{1000}{750}\right) = 0.06667$$

$$R_{3p.u.new} = (0.05)\left(\frac{1000}{550}\right) = 0.10 \quad \text{per unit}$$

Using (12.2.3),

$$\beta = \frac{1}{R_1} + \frac{1}{R_2} + \frac{1}{R_3} = \frac{1}{0.05} + \frac{1}{0.06667} + \frac{1}{0.10} = 45.0 \quad \text{per unit}$$

b. Neglecting losses and dependence of load on frequency, the steady-state increase in total turbine mechanical power equals the load increase, 200 MW or 0.20 per unit. Using (12.2.4) with $\Delta p_{ref} = 0$,

$$\Delta f = \left(\frac{-1}{\beta}\right)\Delta p_m = \left(\frac{-1}{45}\right)(0.20) = -4.444 \times 10^{-3} \quad \text{per unit}$$

$$= (-4.444 \times 10^{-3})(60) = -0.2667 \quad \text{Hz}$$

The steady-state frequency drop is 0.2667 Hz.

c. From (12.2.1), using $\Delta f = -4.444 \times 10^{-3}$ per unit,

$$\Delta p_{m1} = \left(\frac{-1}{0.05}\right)(-4.444 \times 10^{-3}) = 0.08888 \quad \text{per unit}$$

$$= 88.88 \quad \text{MW}$$

$$\Delta p_{m2} = \left(\frac{-1}{0.06667}\right)(-4.444 \times 10^{-3}) = 0.06666 \quad \text{per unit}$$

$$= 66.66 \quad \text{MW}$$

$$\Delta p_{m3} = \left(\frac{-1}{0.10}\right)(-4.444 \times 10^{-3}) = 0.04444 \quad \text{per unit}$$

$$= 44.44 \quad \text{MW}$$

Note that unit 1, whose MVA rating is 33% larger than that of unit 2 and 100% larger than that of unit 3, picks up 33% more load than unit 2 and 100% more

(*Continued*)

load than unit 3. That is, each unit shares the total load change in proportion to its own rating.

PowerWorld Simulator case Example 12_4 contains a lossless nine bus, three generator system that duplicates the conditions from this example (see Figure 12.10). The generators at buses 1, 2 and 3 have ratings of 500, 1000, and 750 MVA respectively, with initial outputs of 300, 600, 500 MWs. Each is modeled with a two-axis synchronous machine model (see Section 11.6), an IEEET1 exciter and a TGOV1 governor model with the parameters equal to the defaults given earlier. At time t = 0.5 seconds, the load at bus 8 is increased from 200 to 400 MW. Figure 12.10 shows the results of a 10-second transient stability simulation. The final generator outputs are 344.5, 589.0, and 466.7 MWs, while the final frequency decline is 0.272 Hz, closely matching the results predicted in the example (the frequency decline exactly matches the 0.266 Hz prediction if the simulation is extended to 20 seconds).

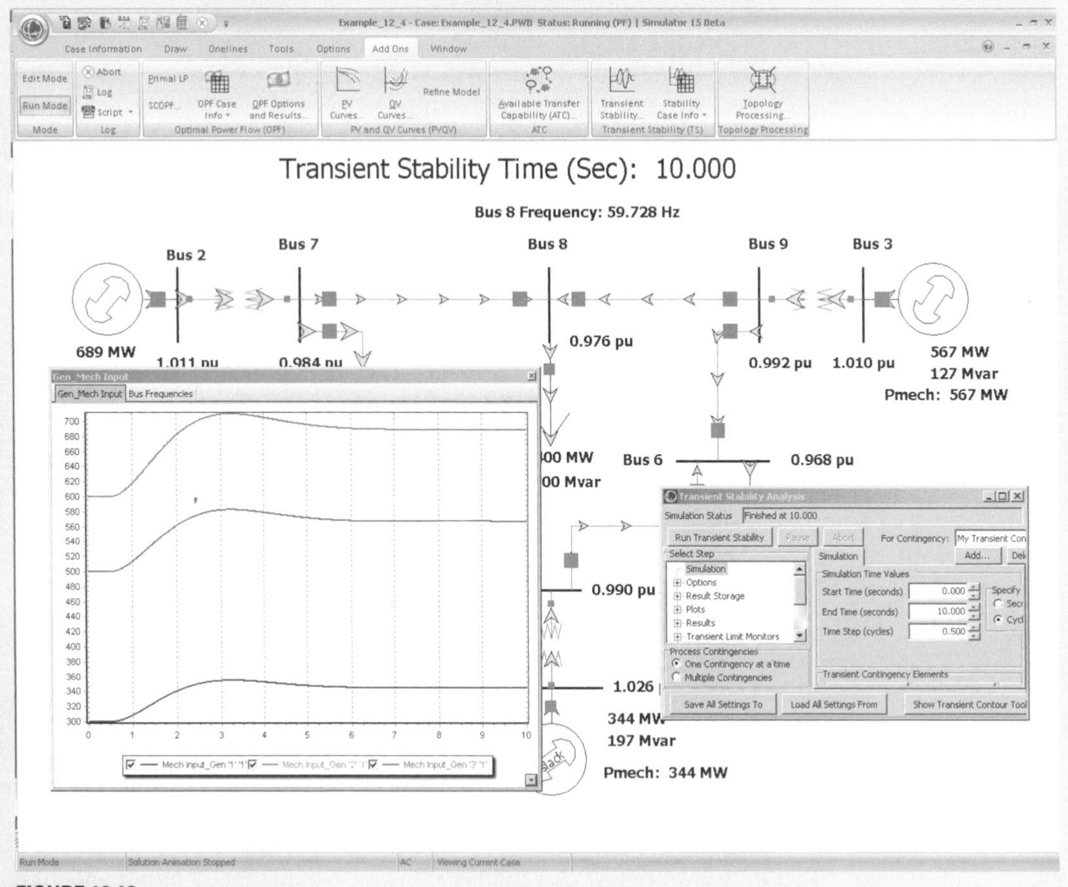

FIGURE 12.10

System oneline with generator mechanical power variation for Example 12.4

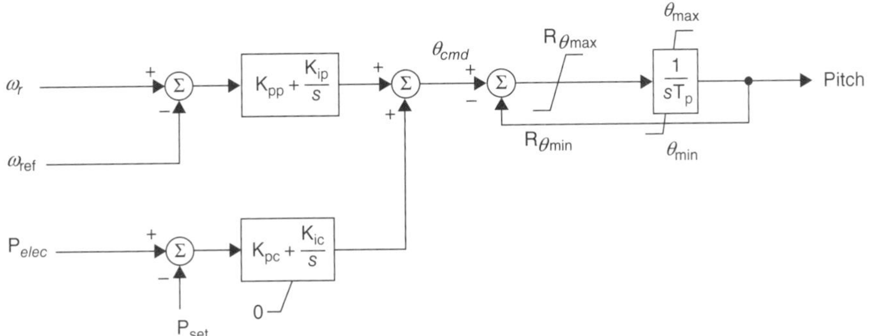

FIGURE 12.11

Pitch control for a Type 3 or 4 wind turbine model

The power output from wind turbines can be controlled by changing the pitch angle of the blades. When the available power in the wind is above the rating for the turbine, its blades are pitched to limit the mechanical power delivered to the electric machine. When the available power is less than the machine's rating, the pitch angle is set to its minimum. Figure 12.11 shows the generic pitch control model for Type 3 and 4 wind turbines, with the inputs being the per unit speed of the turbine, ω_r, the desired speed (normally 1.2 p.u.), the ordered per unit electrical output, and a setpoint power. These signals are combined as shown on the figure to produce a commanded angle for the blades, θ_{cmd}, expressed in degrees. The right side of the block diagram models the dynamics and limits associated with changing the pitch angle of the blades; R_θ is the rate at which the blades change their angle in degrees per second. Typical values are $T_p = 0.3$ seconds, $\theta_{min/max}$ between 0° and 27°, rate limits of ±10°/s, $K_{pp} = 150$, $K_{ip} = 25$, $K_{pc} = 3$, $K_{ic} = 30$.

In general, the larger the size of the interconnected system, the better the frequency response since there are more generators to share the task. However, "Owners/operators of generator units have strong economic reasons to operate generator units in many ways that prevent effective governing response." [16] For example, operating the unit at its full capacity, which maximizes the income that can be derived from the unit but prevents the unit from increasing its output. This is certainly an issue with wind turbines since their "fuel" is essentially free. Also, the Type 3 and 4 units do not contribute inertial response.

12.3 LOAD-FREQUENCY CONTROL

As shown in Section 12.2, turbine-governor control eliminates rotor accelerations and decelerations following load changes during normal operation. However, there is a steady-state frequency error Δf when the change in turbine-governor reference setting Δp_{ref} is zero. One of the objectives of load-frequency control (LFC), therefore, is to return Δf to zero.

In a power system consisting of interconnected areas, each area agrees to export or import a scheduled amount of power through transmission-line interconnections, or tie-lines, to its neighboring areas. Thus, a second LFC objective is to have each area absorb its own load changes during normal operation. This objective is achieved by maintaining the net tie-line power flow out of each area at its scheduled value.

The following summarizes the two basic LFC objectives for an interconnected power system:

1. Following a load change, each area should assist in returning the steady-state frequency error Δf to zero.
2. Each area should maintain the net tie-line power flow out of the area at its scheduled value, in order for the area to absorb its own load changes.

The following control strategy developed by N. Cohn [4] meets these LFC objectives. We first define the *area control error* (ACE) as follows:

$$\begin{aligned} \text{ACE} &= (p_{\text{tie}} - p_{\text{tie,sched}}) + \text{B}_f(f - 60) \\ &= \Delta p_{\text{tie}} + \text{B}_f \Delta_f \end{aligned} \tag{12.3.1}$$

where Δp_{tie} is the deviation in net tie-line power flow out of the area from its scheduled value $p_{\text{tie, sched}}$, and Δf is the deviation of area frequency from its scheduled value (60 Hz). Thus, the ACE for each area consists of a linear combination of tie-line error Δp_{tie} and frequency error Δf. The constant B_f is called a *frequency bias constant.*

The change in reference power setting $\Delta p_{\text{ref}i}$ of each turbine-governor operating under LFC is proportional to the integral of the area control error. That is,

$$\Delta p_{\text{ref}i} = -\text{K}_i \int \text{ACE}\, dt \tag{12.3.2}$$

Each area monitors its own tie-line power flows and frequency at the area control center. The ACE given by (12.3.1) is computed and a percentage of the ACE is allocated to each controlled turbine-generator unit. Raise or lower commands are dispatched to the turbine-governors at discrete time intervals of two or more seconds in order to adjust the reference power settings. As the commands accumulate, the integral action in (12.3.2) is achieved.

The constant K_i in (12.3.2) is an integrator gain. The minus sign in (12.3.2) indicates that if either the net tie-line power flow out of the area or the area frequency is low—that is, if the ACE is negative—then the area should increase its generation.

When a load change occurs in any area, a new steady-state operation can be obtained only after the power output of every turbine-generating unit in the interconnected system reaches a constant value. This occurs only when all reference power settings are zero, which in turn occurs only when the ACE of every area is zero. Furthermore, the ACE is zero in every area only when both Δp_{tie} and Δf are zero. Therefore, in steady-state, both LFC objectives are satisfied.

EXAMPLE 12.5

Response of LFC to a load change in an interconnected power system

As shown in Figure 12.12, a 60-Hz power system consists of two interconnected areas. Area 1 has 2000 MW of total generation and an area frequency response characteristic $\beta_1 = 700$ MW/Hz. Area 2 has 4000 MW of total generation and $\beta_2 = 1400$ MW/Hz. Each area is initially generating one-half of its total generation, at $\Delta p_{\text{tie1}} = \Delta p_{\text{tie2}} = 0$ and at 60 Hz when the load in area 1 suddenly increases by 100 MW. Determine the steady-state frequency error Δf and the steady-state tie-line error Δp_{tie} of each area for the following two cases: (a) without LFC, and (b) with LFC given by (12.3.1) and (12.3.2). Neglect losses and the dependence of load on frequency.

FIGURE 12.12 Example 12.5

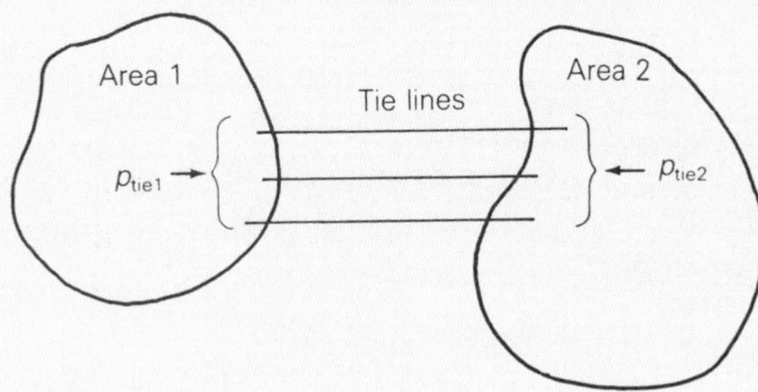

SOLUTION

a. Since the two areas are interconnected, the steady-state frequency error Δf is the same for both areas. Adding (12.2.4) for each area,

$$(\Delta p_{m1} + \Delta p_{m2}) = (\Delta p_{\text{ref1}} + \Delta p_{\text{ref2}}) - (\beta_1 + \beta_2)\Delta f$$

Neglecting losses and the dependence of load on frequency, the steady-state increase in total mechanical power of both areas equals the load increase, 100 MW. Also, without LFC, Δp_{ref1} and Δp_{ref2} *are* both zero. The above equation then becomes

$$100 = -(\beta_1 + \beta_2)\Delta f = -(700 + 1400)\Delta f$$
$$\Delta f = -100/2100 = -0.0476 \quad \text{Hz}$$

Next, using (12.2.4) for each area, with $\Delta p_{\text{ref}} = 0$,

$$\Delta p_{m1} = -\beta_1\Delta f = -(700)(-0.0476) = 33.33 \quad \text{MW}$$
$$\Delta p_{m2} = -\beta_2\Delta f = -(1400)(-0.0476) = 66.67 \quad \text{MW}$$

In response to the 100-MW load increase in area 1, area 1 picks up 33.33 MW and area 2 picks up 66.67 MW of generation. The 66.67-MW increase in area 2

(Continued)

generation is transferred to area 1 through the tie-lines. Therefore, the change in net tie-line power flow out of each area is

$$\Delta p_{\text{tie2}} = +66.67 \quad \text{MW}$$
$$\Delta p_{\text{tie1}} = -66.67 \quad \text{MW}$$

b. From (12.3.1), the area control error for each area is

$$\text{ACE}_1 = \Delta p_{\text{tie1}} + \text{B}_1\Delta f_1$$
$$\text{ACE}_2 = \Delta p_{\text{tie2}} + \text{B}_2\Delta f_2$$

Neglecting losses, the sum of the net tie-line flows must be zero; that is, $\Delta p_{\text{tie1}} + \Delta p_{\text{tie2}} = 0$ or $\Delta p_{\text{tie2}} = -\Delta p_{\text{tie1}}$. Also, in steady-state $\Delta f_1 = \Delta f_2 = \Delta f$.

Using these relations in the above equations,

$$\text{ACE}_1 = \Delta p_{\text{tie1}} + \text{B}_1\Delta f$$
$$\text{ACE}_2 = -\Delta p_{\text{tie1}} + \text{B}_2\Delta f$$

In steady-state, $\text{ACE}_1 = \text{ACE}_2 = 0$; otherwise, the LFC given by (12.3.2) would be changing the reference power settings of turbine-governors on LFC. Adding the above two equations,

$$\text{ACE}_1 + \text{ACE}_2 = 0 = (\text{B}_1 + \text{B}_2)\Delta f$$

Therefore, $\Delta f = 0$ and $\Delta p_{\text{tie1}} = \Delta p_{\text{tie2}} = 0$. That is, in steady-state the frequency error is returned to zero, area 1 picks up its own 100-MW load increase, and area 2 returns to its original operating condition—that is, the condition before the load increase occurred.

Note that the turbine-governor controls act almost instantaneously, subject only to the time delays shown in Figure 12.9. However, LFC acts more slowly. LFC raise and lower signals are dispatched from the area control center to turbine-governors at discrete time intervals of 2 or more seconds. Also, it takes time for the raise or lower signals to accumulate. Thus, case (a) represents the first action. Turbine-governors in both areas rapidly respond to the load increase in area 1 in order to stabilize the frequency drop. Case (b) represents the second action. As LFC signals are dispatched to turbine-governors, Δf and Δp_{tie} are slowly returned to zero.

The choice of the B_f and K_i constants in (12.3.1) and (12.3.2) affects the transient response to load changes—for example, the speed and stability of the response. The frequency bias B_f should be high enough such that each area adequately contributes to frequency control. Cohn [4] has shown that choosing B_f equal to the area frequency response characteristic, $\text{B}_f = \beta$, gives satisfactory performance of the interconnected system. The integrator gain K_i, should not be too high; otherwise, instability may result. Also, the time interval at which LFC signals are dispatched,

2 or more seconds, should be long enough so that LFC does not attempt to follow random or spurious load changes. A detailed investigation of the effect of B_f, K_i, and LFC time interval on the transient response of LFC and turbine-governor controls is beyond the scope of this text.

Two additional LFC objectives are to return the integral of frequency error and the integral of net tie-line error to zero in steady-state. By meeting these objectives, LFC controls both the time of clocks that are driven by 60-Hz motors and energy transfers out of each area. These two objectives are achieved by making temporary changes in the frequency schedule and tie-line schedule in (12.3.1).

Finally, note that LFC maintains control during normal changes in load and frequency—that is, changes that are not too large. During emergencies, when large imbalances between generation and load occur, LFC is bypassed and other, emergency controls are applied.

COORDINATION OF ECONOMIC DISPATCH WITH LFC

Both the load-frequency control (LFC) and economic dispatch objectives are achieved by adjusting the reference power settings of turbine-governors on control. Figure 12.13 shows an *automatic generation control* strategy for achieving both objectives in a coordinated manner. In this figure, P_{iD} is the desired output of each generator as computed from an economic dispatch program, which is discussed in

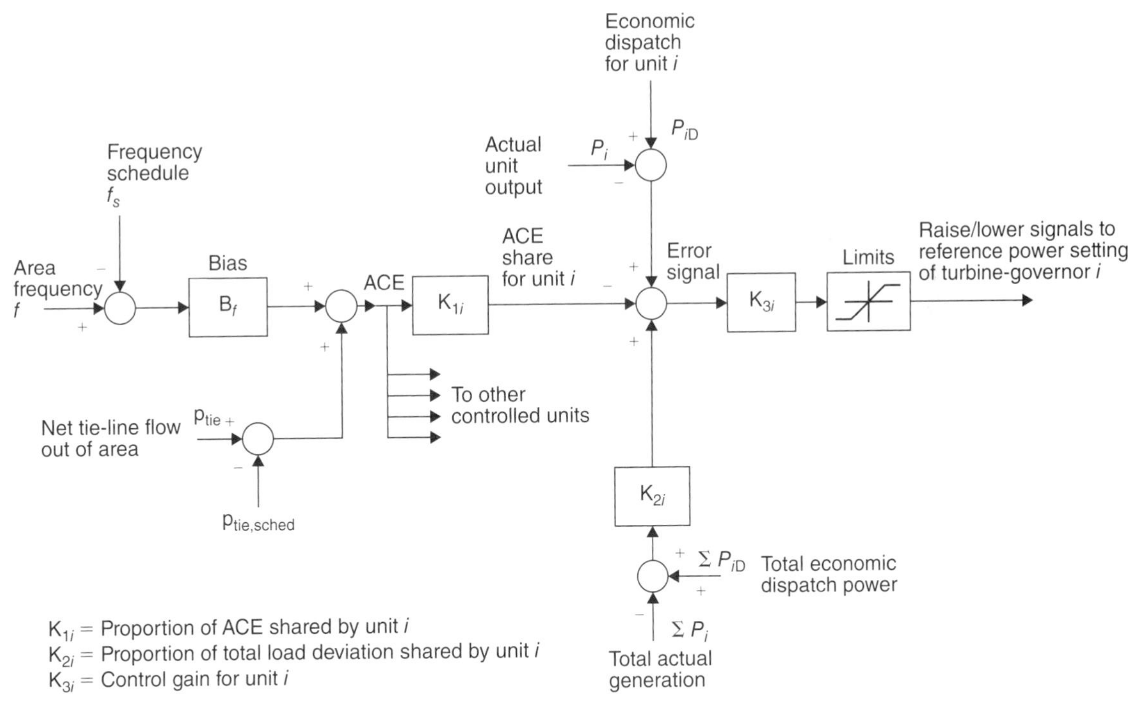

FIGURE 12.13

Automatic generation control [11] (Based on A.J. Wood and B.F. Wollenberg, Power Generation, Operation, and Control, 1989, John Wiley & Sons)

Section 6.12. As shown in Figure 12.13, the area control error (ACE) is first computed, and a share K_{1i} of the ACE is allocated to each unit. Second, the deviation of total actual generation from total desired generation is computed, and a share $K_{2i}\Sigma(P_{iD} - P_i)$ is allocated to unit *i*. Third, the deviation of actual generation from desired generation of unit *i* is computed, and $(P_{iD} - P_i)$ is allocated to unit *i*. An error signal formed from these three components and multiplied by a control gain K_{3i} determines the raise or lower signals that are sent to the turbine-governor of each unit *i* on control.

In practice, raise or lower signals are dispatched to the units at discrete time intervals of 2 to 10 seconds. The desired outputs P_{iD} of units on control, determined from the economic dispatch program, are updated at slower intervals, typically every 2 to 10 minutes.

PROBLEMS

SECTION 12.1

12.1 The block-diagram representation of a closed-loop automatic regulating system, in which generator voltage control is accomplished by controlling the exciter voltage, is shown in Figure 12.14. T_a, T_e, and T_f are the time constants associated with the amplifier, exciter, and generator field circuit, respectively. (a) Find the open-loop transfer function G(s). (b) Evaluate the minimum open-loop gain such that the steady-state error Δe_{ss} does not exceed 1%. (c) Discuss the nature of the dynamic response of the system to a step change in the reference input voltage.

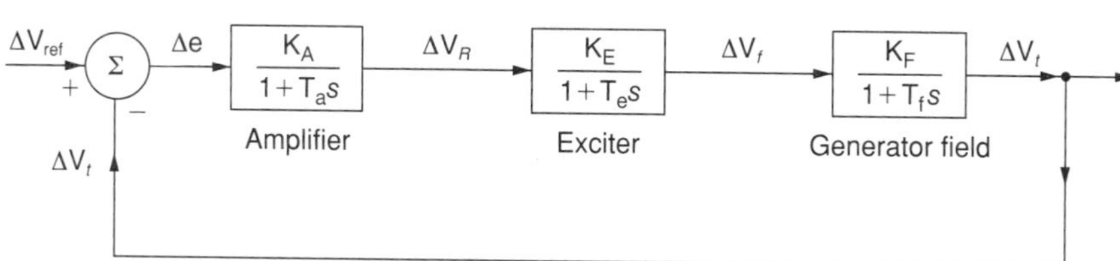

FIGURE 12.14 Problem 12.1

12.2 The Automatic Voltage Regulator (AVR) system of a generator is represented by the simplified block diagram shown in Figure 12.15, in which the sensor is modeled by a simple first-order transfer function. The voltage is sensed through a voltage transformer and then rectified through a bridge rectifier. Parameters of the AVR system are given as follows.

	Gain	Time Constant (seconds)
Amplifier	K_A	$T_A = 0.1$
Exciter	$K_E = 1$	$T_E = 0.4$
Generator	$K_G = 1$	$T_G = 1.0$
Sensor	$K_R = 1$	$T_R = 0.05$

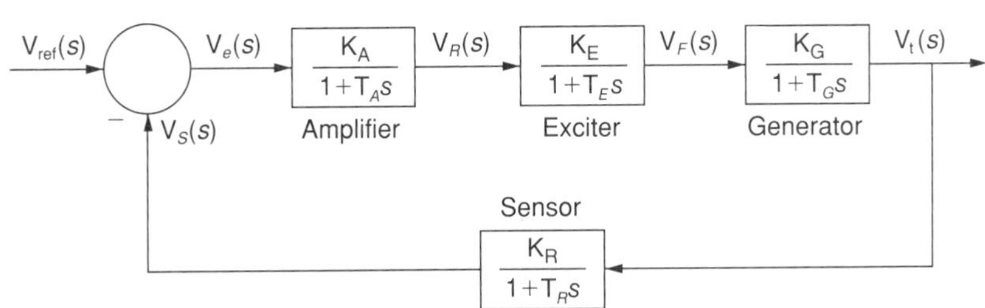

FIGURE 12.15 Problem 12.2

(a) Determine the open-loop transfer function of the block diagram and the closed-loop transfer function relating the generator terminal voltage $V_t(s)$ to the reference voltage $V_{ref}(s)$. (b) For the range of K_A from 0 to 12.16, comment on the stability of the system. (c) For $K_A = 10$, evaluate the steady-state step response and steady-state error.

PW **12.3** Open PowerWorld Simulator case Problem 12_3. This case models the system from Example 12.1 except with the rate feedback gain constant, K_f, has been set to zero and the simulation end time was increased to 30 seconds. Without rate feedback the system voltage response will become unstable if the amplifier gain, K_a, becomes too large. In the simulation this instability will be indicated by undamped oscillations in the terminal voltage (because of the limits on V_r the response does not grow to infinity but rather bounces between the limits). Using transient stability simulations, iteratively determine the approximate value of K_a at which the system becomes unstable. The value of K_a can be on the **Generator Information Dialog, Stability, Exciters** page.

PW **12.4** One of the disadvantages of the IEEET1 exciter is following a fault the terminal voltage does not necessarily return to its prefault value. Using PowerWorld Simulator case Problem 12_3 determine the prefault bus 4 terminal voltage and field voltage. Then use the simulation to determine the final, postfault values for these fields for $K_a = 100$, 200, 50, and 10. Referring to Figure 12.3, what is the relationship between the reference voltage, and the steady-state terminal voltage and the field voltage?

SECTION 12.2

12.5 An area of an interconnected 60-Hz power system has three turbine-generator units rated 200, 300, and 500 MVA. The regulation constants of the units are 0.03, 0.04, and 0.05 per unit, respectively, based on their ratings. Each unit is initially operating at one-half its own rating when the load suddenly decreases by 150 MW. Determine (a) the unit area frequency response characteristic β on a 100-MVA base, (b) the steady-state increase in area frequency, and (c) the MW decrease in mechanical power output of each turbine. Assume that the reference power setting of each turbine-governor remains constant. Neglect losses and the dependence of load on frequency.

12.6 Each unit in Problem 12.5 is initially operating at one-half its own rating when the load suddenly increases by 100 MW. Determine (a) the steady-state decrease in area frequency, and (b) the MW increase in mechanical power output of each turbine. Assume that the reference power setting of each turbine-generator remains constant. Neglect losses and the dependence of load on frequency.

12.7 Each unit in Problem 12.5 is initially operating at one-half its own rating when the frequency increases by 0.005 per unit. Determine the MW decrease of each unit. The reference power setting of each turbine-governor is fixed. Neglect losses and the dependence of load on frequency.

12.8 Repeat Problem 12.7 if the frequency decreases by 0.005 per unit. Determine the MW increase of each unit.

12.9 An interconnected 60-Hz power system consisting of one area has two turbine-generator units, rated 500 and 750 MVA, with regulation constants of 0.04 and 0.06 per unit, respectively, based on their respective ratings. When each unit carries a 300-MVA steady-state load, let the area load suddenly increase by 250 MVA. (a) Compute the area frequency response characteristic β on a 1000-MVA base. (b) Calculate Δf in per-unit on a 60-Hz base and in Hz.

PW **12.10** Open PowerWorld Simulator case Problem 12_10. The case models the system from Example 12.4 except 1) the load increases is a 50% rise at bus 6 for a total increase of 250 MW (from 500 MW to 750 MW), 2) the value of R for generator 1 is changed from 0.05 to 0.04 per unit. Repeat Example 12.4 using these modified values.

PW **12.11** Open PowerWorld Simulator case Problem 12_11, which includes a transient stability representation of the system from Example 6.13. Each generator is modeled using a two-axis machine model, an IEEE Type 1 exciter and a TGOV1 governor with R = 0.04 per unit (a summary of the generator models is available by selecting either **Stability Case Info, Transient Stability Generator Summary** which includes the generator MVA base, or **Stability Case Info, Transient Stability Case Summary**). The contingency is the loss of the generator at bus PEAR69, which initially has 65 MW of generation. Analytically determine the steady-state frequency error in Hz following this contingency. Use PowerWorld Simulator to confirm this result; also determine the magnitude and time of the largest bus frequency deviation.

PW **12.12** Repeat Problem 12.11 except first double the H value for each of the machines. This can be most easily accomplished by selecting **Stability Case Info, Transient Stability Case Summary** to view the summary form. Right click on the line corresponding to the Machine Model class, and then select Show Dialog to view an editable form of the model parameters. Compare the magnitude and time of the largest bus frequency deviations between Problem 12.12 and 12.11.

12.13 For a large, 60 Hz, interconnected electrical system assume that following the loss of two 1400 MW generators (for a total generation loss of 2800 MW) the change in frequency is −0.12 Hz. If all the on-line generators that are available to participate in frequency regulation have an R of 0.04 per unit (on their own MVA base), estimate the total MVA rating of these units.

SECTION 12.3

12.14 A 60-Hz power system consists of two interconnected areas. Area 1 has 1200 MW of generation and an area frequency response characteristic $\beta_1 = 400$ MW/Hz. Area 2 has 1800 MW of generation and $\beta_2 = 600$ MW/Hz. Each area is initially operating at one-half its total generation, at $\Delta p_{tie1} = \Delta p_{tie2} = 0$ and at 60 Hz, when the load in area 1 suddenly increases by 400 MW. Determine the steady-state frequency error and the steady-state tie-line error Δp_{tie} of each area. Assume that the reference power settings of all turbine-governors are fixed. That is, LFC is not employed in any area. Neglect losses and the dependence of load on frequency.

12.15 Repeat Problem 12.14 if LFC is employed in area 2 alone. The area 2 frequency bias coefficient is set at $B_{f2} = \beta_2 = 600$ MW/Hz. Assume that LFC in area 1 is inoperative due to a computer failure.

12.16 Repeat Problem 12.14 if LFC is employed in both areas. The frequency bias coefficients are $B_{f1} = \beta_1 = 400$ MW/Hz and $B_{f2} = \beta_2 = 600$ MW/Hz.

12.17 Rework Problems 12.15 through 12.16 when the load in area 2 suddenly decreases by 300 MW. The load in area 1 does not change.

12.18 On a 1000-MVA common base, a two-area system interconnected by a tie line has the following parameters:

Area	1	2
Area Frequency Response Characteristic	$\beta_1 = 0.05$ per unit	$\beta_2 = 0.0625$ per unit
Frequency-Dependent Load Coefficient	$D_1 = 0.6$ per unit	$D_2 = 0.9$ per unit
Base Power	1000 MVA	1000 MVA
Governor Time Constant	$\tau_{g1} = 0.25$ s	$\tau_{g2} = 0.3$ s
Turbine Time Constant	$\tau_{t1} = 0.5$ s	$\tau_{t2} = 0.6$ s

The two areas are operating in parallel at the nominal frequency of 60 Hz. The areas are initially operating in steady state with each area supplying 1000 MW when a sudden load change of 187.5 MW occurs in area 1. Compute the new steady-state frequency and change in tie-line power flow (a) without LFC, and (b) with LFC.

CASE STUDY QUESTIONS

a. As a result of the August 14, 2003 blackout in North America, what major electrical islands were formed?

b. What is the first step in restoration?

c. What lessons were learned from this blackout?

REFERENCES

1. IEEE Committee Report, "Computer Representation of Excitation Systems," *IEEE Transactions PAS,* vol. PAS-87 (June 1968), pp. 1460–1464.
2. M. S. Sarma, *Electric Machines* 2d ed. (Boston, PWS Publishing 1994).
3. IEEE Committee Report, "Dynamic Models for Steam and Hydro Turbines in Power System Studies," *IEEE Transactions PAS,* vol. PAS-92, no. 6 (November/December 1973), pp. 1904–1915.
4. N. Cohn, *Control of Generation and Power Flow on Interconnected Systems* (New York: Wiley, 1971).
5. L. K. Kirchmayer, *Economic Operation of Power Systems* (New York: Wiley, 1958).
6. L. K. Kirchmayer and G. W. Stagg, "Evaluation of Methods of Coordinating Incremental Fuel Costs and Incremental Transmission Losses," *Transactions AIEE*, vol. 71, part III (1952), pp. 513–520.
7. G. H. McDaniel and A. F. Gabrielle, "Dispatching Pumped Storage Hydro," *IEEE Transmission PAS*, vol. PAS-85 (May 1966), pp. 465–471.
8. E. B. Dahlin and E. Kindingstad, "Adaptive Digital River Flow Predictor for Power Dispatch," *IEEE Transactions PAS*, vol. PAS-83 (April 1964), pp. 320–327.
9. L. K. Kirchmayer, *Economic Control of Interconnected Systems* (New York: Wiley, 1959).
10. J. H. Drake et al., "Optimum Operation of a Hydrothermal System," *Transactions AIEE* (*Power Apparatus and Systems),* vol. 62 (August 1962), pp. 242–250.
11. A. J. Wood and B. F. Wollenberg, *Power Generation, Operation, and Control* (New York: Wiley, 1989).
12. B. Stott and J. L. Marinho, "Linear Programming for Power System Network Security Applications," *IEEE Trans, on Power Apparatus and Systems*, vol. PAS-98, (May/June 1979), pp. 837–848.

13. E. H. Allen, R. B. Stuart, and T. E. Wiedman, "No Light in August: Power System Restoration Following the 2003 North American Blackout," *IEEE Power and Energy Magazine,* 12,1 (January/February 2014) pp. 25–33.

14. IEEE PES Task Force Report, "Interconnected Power System Response to Generation Governing: Present Practice and Outstanding Concerns," IEEE 07TP180, May 2007.

15. K. Clark, N. W. Miller, and J. J. Sanchez-Gasca, "Modeling of GE Wind Turbine-Generators for Grid Studies," Version 4.4, GE Energy, Schenectady, NY, September 2009.

16. E. H. Camm et al., "Characteristics of Wind Turbine Generators for Wind Power Plants," Proc. IEEE 2009 General Meeting, Calgary, AB, July 2009.

13 Transmission Lines: Transient Operation

Power engineers with a Real Time Digital Simulator (RTDS) (Courtesy of Tim Yardley, University of Illinois.)

Transient overvoltages caused by lightning strikes to transmission lines and by switching operations are of fundamental importance in selecting equipment insulation levels and surge-protection devices. It is important, therefore, to understand the nature of transmission-line transients.

During our study of the steady-state performance of transmission lines in Chapter 5, the line constants R, L, G, and C were recognized as distributed rather than lumped constants. When a line with distributed constants is subjected to a disturbance such as a lightning strike or a switching operation, voltage and current waves arise and travel along the line at a velocity near the speed of light. When these waves arrive at the line terminals, reflected voltage and current waves arise and travel back down the line, superimposed on the initial waves.

Because of line losses, traveling waves are attenuated and essentially die out after a few reflections. Also, the series inductances of transformer windings effectively block the disturbances, thereby preventing them from entering generator windings. However, due to the reinforcing action of several reflected waves, it is possible for voltage to build up to a level that could cause transformer insulation or line insulation to arc over and suffer damage.

Circuit breakers, which can operate within 50 ms, are too slow to protect against lightning or switching surges. Lightning surges can rise to peak levels within a few microseconds and switching surges within a few hundred microseconds—fast enough to destroy insulation before a circuit breaker could open. However, protective devices are available. Called surge arresters, these can be used to protect equipment insulation against transient overvoltages. These devices limit voltage to a ceiling level and absorb the energy from lightning and switching surges.

This chapter begins with a discussion of traveling waves on single-phase lossless lines (Section 13.1). Section 13.2 presents boundary conditions, and the Bewley lattice diagram for organizing reflections is covered in Section 13.3. Discrete-time models of single-phase lines and of lumped RLC elements are derived in Section 13.4, and the effects of line losses and multiconductor lines are discussed in Sections 13.5 and 13.6. In Section 13.7 power system overvoltages including lightning surges, switching surges, and power-frequency overvoltages, followed by an introduction to insulation coordination are discussed in Section 13.8.

CASE STUDY 1

Two case-study reports are presented here. The first describes metal oxide varistor (MOV) arresters used by electric utilities to protect transmission and distribution equipment against transient overvoltages in power systems with rated voltages through 345 kV [22]. The second investigates the possible impact of large amounts of wind generation on frequency response in the United States Western and Eastern interconnections [23]. There is a fundamental difference between a wind turbine and the turbines (gas, steam, or hydro) supplying the majority of the world's traditional power plants. With traditional turbines, synchronous generators are employed and the rotational speed is nearly constant and locked to system frequency. With a wind turbine, rotational speed is not synchronized with the system frequency but is asynchronously controlled to maximize active power output. The investigation provides evidence that high levels of wind (and solar) generation can be tolerated with respect to frequency response in the U.S. Western and Eastern interconnections.

Surge Arresters

VariSTAR Type AZE Surge Arresters for Systems through 345 kV IEEE Std C62.11™ standard Certified Station Class Arresters*

General

The VariSTAR™ AZE Surge Arrester from Cooper Power Systems offers the latest in metal oxide varistor (MOV) technology for the economical protection of power and substation equipment. This arrester is gapless and constructed of a single series column of MOV disks. The arrester is designed and tested to the requirements of IEEE Std C62.11™ standard and is available in ratings suitable for the transient overvoltage protection of electrical equipment on systems through 345 kV.

Cooper Power Systems assures the design integrity of the AZE arrester through a rigorous testing program conducted at our Thomas A. Edison Technical Center and at the factory in Olean, New York. The availability of complete in-house testing facilities assures that as continuous process improvements are made, they are professionally validated to high technical standards.

VariSTAR Type AZE Surge Arresters for Systems through 345 kV IEEE Std C62.11TM standard Certified Station Class Arresters, © 2011 Cooper Industries, www.cooperpower.com

Table 1, shown below, contains information on some of the specific ratings and characteristics of AZE Series S (AZES) surge arresters.

Arrester Characteristic		Rating
System Application Voltages		3-345 kV
Arrester Voltage Ratings		3-360 kV
Rated Discharge Energy, (kJ/kV of MCOV)		
Arrester Ratings:	3-108 kV	3.4
	120-240 kV	5.6
	259-360 kV	8.9
System Frequency		50/60 Hz
Impulse Classifying Current		10 kA
High Current Withstand		100 kA
Pressure Relief Rating, kA rms sym		
Metal-Top Designs		65 kA
Cubicle-Mount Designs		40 kA
Cantilever Strength (in-lbs)*		
Metal-Top Designs:		
	3-240 kV	90000†
	258-360 kV	120,000†

TABLE 1

AZE Series S (AZES) Ratings and Characteristics

*Maximum working load should not exceed 40% of this value.

†90,000 in-lb = 10,000 N-m

†120,000 in-lb = 13,500 N-m

Construction

External

The Type AZE station class arrester is available in two design configurations—a metal-top design in ratings 3-360 kV and a cubicle-mount design

*(© 2011 Cooper Industries-www.cooperpower.com)

in ratings 3-48 kV. Cubicle-mount designs are ideally suited for confined spaces where clearances between live parts are limited.

The wet-process porcelain housing features an alternating shed design (ratings >48 kV) that provides excellent resistance to the effects of atmospheric housing contamination. AZE arresters are available with optional extra creepage porcelains for use in areas with extreme natural atmospheric and man-made pollution.

The dielectric properties of the porcelain are coordinated with the electrical protective characteristics of the arrester. The unit end castings are of a corrosion-resistant aluminum alloy configured for interchangeable mounting with other manufacturers' arresters for ease in upgrading to the VariSTAR arrester technology. This three-footed mounting is provided on a 21.6- to 25.4-cm diameter pattern for customer supplied 1.3-cm diameter hardware.

High Cantilever strength assures mechanical integrity (Table 1 lists the cantilever strength of metal-top AZES arresters). Cooper Power Systems recommends that a load limit of 440 newtons not be exceeded on the line terminal of cubicle mount designs. Loads exceeding this limit could cause a shortening of arrester life. Housings are available in standard grey or optional brown glaze color.

Standard line and ground terminal connectors accommodate up to a 1.9-cm diameter conductor. Insulating bases and discharge counters are optionally available for in-service monitoring of arrester discharge activity.

The end fittings and porcelain housing of each arrester unit are sealed and tested by means of a sensitive helium mass spectrometer; this assures that the quality and insulation protection provided by the arrester is never compromised over its

Figure 1 120 kV rated VariSTAR Type AZE surge arrester

lifetime by the entrance of moisture. A corrosion-resistant nameplate is provided and contains all information required by standards. In addition, stacking arrangement information is provided for multi-unit arresters. Voltage grading rings are included for arresters rated 172 kV and above.

Internal

VariSTAR AZE arresters are a totally gapless MOV design. Gapless

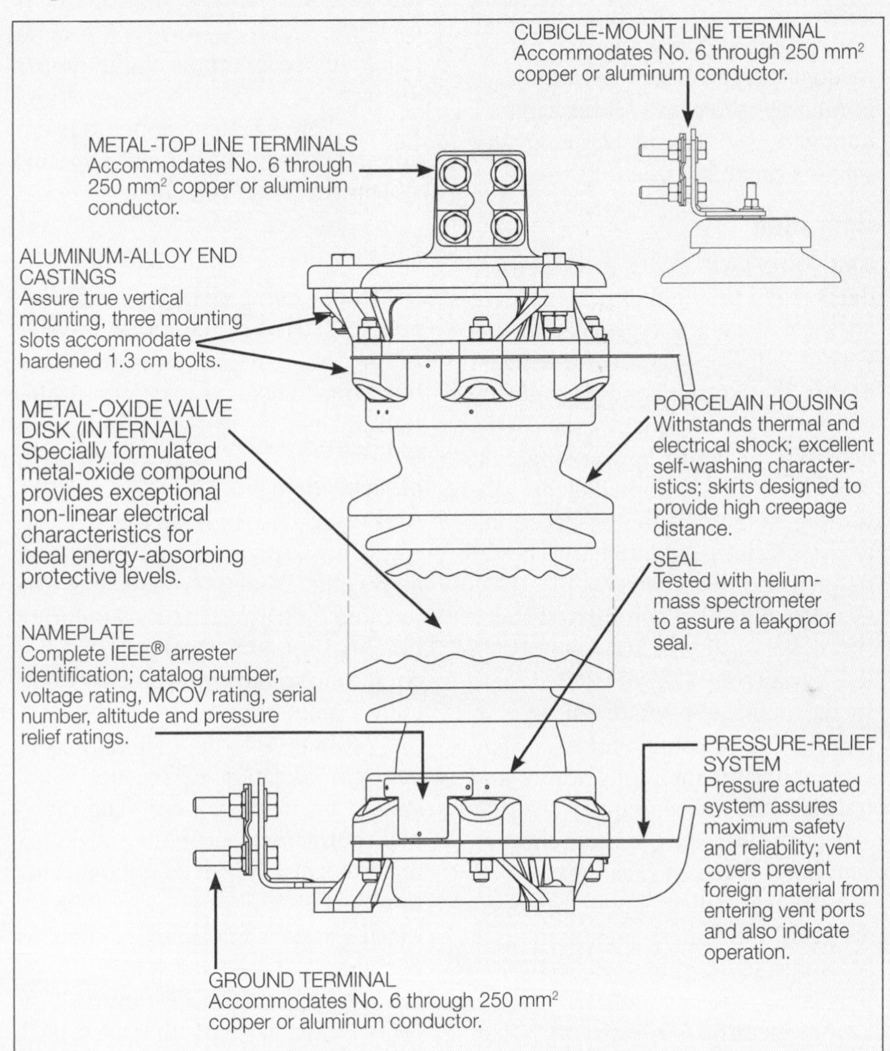
CUBICLE-MOUNT LINE TERMINAL
Accommodates No. 6 through 250 mm^2 copper or aluminum conductor.
METAL-TOP LINE TERMINALS
Accommodates No. 6 through 250 mm^2 copper or aluminum conductor.
ALUMINUM-ALLOY END CASTINGS
Assure true vertical mounting, three mounting slots accommodate hardened 1.3 cm bolts.
METAL-OXIDE VALVE DISK (INTERNAL)
Specially formulated metal-oxide compound provides exceptional non-linear electrical characteristics for ideal energy-absorbing protective levels.
PORCELAIN HOUSING
Withstands thermal and electrical shock; excellent self-washing characteristics; skirts designed to provide high creepage distance.
SEAL
Tested with helium-mass spectrometer to assure a leakproof seal.
NAMEPLATE
Complete IEEE® arrester identification; catalog number, voltage rating, MCOV rating, serial number, altitude and pressure relief ratings.
PRESSURE-RELIEF SYSTEM
Pressure actuated system assures maximum safety and reliability; vent covers prevent foreign material from entering vent ports and also indicate operation.
GROUND TERMINAL
Accommodates No. 6 through 250 mm^2 copper or aluminum conductor.

Figure 2 VariSTAR Type AZE arrester construction details. ***Note:*** *Multi-section arresters include an additional unit nameplate showing the order of assembly. For example, a 2-section arrester will have a bottom section with a unit nameplate labeled unit 1 of 2 and the top section with a unit nameplate labeled unit 2 of 2.*

construction makes a significant contribution to the performance of arresters through the elimination of gap reseal as a consideration associated with the discharge of switching surge currents. The specially formulated metal-oxide varistors, manufactured under Cooper Power System's exclusive quality control, provide exceptional non-linear protective characteristics, durability, and dependable energy dissipation capabilities.

Operation

The VariSTAR AZE arrester conducts only a few milliamperes of leakage current when energized at normal system voltage. When a surge event occurs, the arrester conducts only the current, and consequently the energy necessary to limit the overvoltage. It provides precise and predictable protection, minimizes the absorbed energy, and does not discharge power frequency systems currents.

A controlled and directed pressure relief system is incorporated in the VariSTAR AZE arrester design. In the unlikely event of an arrester failure, this pressure venting system rapidly relieves internal pressure and transfers the internal arc to the outside of the arrester porcelain through vent ports in the end castings.

When called upon to operate, this mechanism vents internal pressures in fractions of a cycle—preventing violent arrester failure. This mechanism is effective to system fault currents up to 65 kA in metal-top designs and 40 kA in cubicle-mount designs.

General Application Recommendations

The rating of an arrester is the power-frequency line-to-ground voltage at which the arrester is designed to pass the IEEE Std C62.11™ standard duty-cycle test. Table 2 provides a general guide for the selection of the proper arrester for a given system voltage. Cooper Power Systems application engineers are available to make specific system application recommendations.

Selection of Arrester Rating

In arrester rating selection it is preferable to determine the lowest arrester rating that will ensure satisfactory operation. This is the optimum solution because the arrester selected will not only provide the greatest margin of insulation protection but also be the most economical choice.

Increasing the arrester rating above the minimum increases the likelihood of arrester survival during potential system contingencies, but compromises the protection of equipment insulation.

Rating selection should begin with consideration of the maximum system operating voltage. The maximum power frequency voltage expected under normal system conditions (line-to-ground) should not exceed the selected arrester's maximum continuous operating voltage (MCOV).

The temporary overvoltage (TOV) capability of the VariSTAR AZE arrester is shown in Figure 3. The curves indicate the arrester's ability to withstand abnormal system power

System Voltage (kV rms)		Suggested Arrester Rating (kV rms)	
Nominal	Maximum	Solidly Grounded Neutral Circuits	High Impedance Grounded, Ungrounded, or Temporarily Ungrounded Circuits
2.4	2.52	—	3
4.16	4.37	3	6
4.8	5.04	—	6
6.9	7.24	—	9
12.47	13.2	9–10	—
13.2	14.0	10	15–18
13.8	14.5	10–12	15–18
20.7	21.8	15	—
23.0	24.2	—	24–27
24.9	26.4	18–21	—
27.6	29.0	21–24	27–30
34.5	36.5	27–30	36–39
46.0	48.3	—	48
69.0	72.5	54–60	66–72
115	121	90–96	108–120
138	145	108–120	132–144
161	169	120–144	144–168
230	242	172–192	228–240
345	362	258–312	288–360

TABLE 2

Commonly Applied Voltage Ratings of the VariSTAR Type AZE Arrester

frequency (sinusoidal) overvoltages for various durations. The values shown assume that the arrester has been energized at MCOV prior to an overvoltage event and that the arrester is in an ambient temperature of 60°C. After the overvoltage durations shown, the arrester will thermally recover when once again energized at MCOV.

Figure 3 also illustrates the arrester's TOV capabilities with and

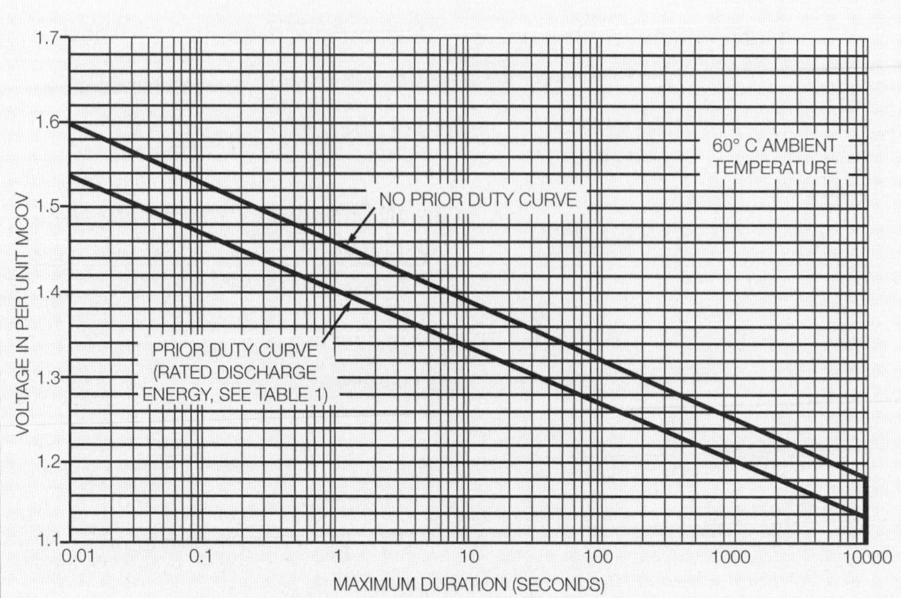

Figure 3 Temporary overvoltage capability of VariSTAR AZE surge arresters

without prior switching surge duties of up to the maximum capability of the arrester as listed in Table 1 (Rated Discharge Energy).

It is not recommended that the TOV curve be extended for periods in excess of 10,000 seconds (2.8 hrs.).

For ungrounded systems, systems utilizing high impedance or resonant grounding and other systems where the line-to-ground voltage may be elevated to line-to-line voltages for extended periods, arresters having an MCOV equal to line-to-line voltage may be required.

For non-sinusoidal transient voltages caused by system switching operations, a transient network analyzer (TNA) study is recommended; Cooper Power Systems engineers are available to make these studies.

To assure proper application, the following information is normally required:

1. Maximum system operating voltage.
2. System grounding conditions.
   - **A.** For four-wire circuits, grounding conditions depend upon whether the system is multi-grounded, whether it has neutral impedance, and whether common primary and secondary neutrals are used.
   - **B.** For three-wire circuits, grounding conditions depend upon whether the system is solidly grounded at the source, grounded through neutral impedance at the source, grounded through transformers, or ungrounded.

Where unusual conditions exist (high ground resistance, high capacitive load, unusual switching surge duty, etc.) the following supplementary information is required:

1. Type of unusual condition.
2. BIL of equipment and separation distance to protected equipment.
3. Type of construction (phase spacing, length of line, conductor size, etc.).
4. Grounding and phase-sequence components of source impedances.
5. Phase-sequence components of load impedances.
6. Available fault current.
7. Potential for loss of neutral grounding during system events.

Routine Tests

A complete production test program assures the quality of every VariSTAR AZE surge arrester. Each completed arrester is required to satisfactorily pass the following test regimen conducted in accordance with the procedures established in IEEE Std C62.11™ standard.

- Partial Discharge Test at 1.05 times MCOV;
- Power Frequency Test at 1.20 times MCOV;
- Discharge Voltage Test; and
- Sealing Effectiveness Test of Housing by helium mass spectrometer.

Standards

The VariSTAR AZE surge arrester has been tested and certified to IEEE Std C62.11™ standard. Guaranteed performance characteristics are specified in this catalog section and in the relevant "Design Certification Test Report" Cooper Power Systems bulletin 95028.

Dimensions and Mounting

Figure 4 illustrates an in-line mounting arrangement; the applicable minimum values of B and C may be found in Tables 4 and 5. Line and ground terminal details are shown in Figure 5; the supplied terminals accommodate aluminum and copper conductors to a maximum size of 1.9 cm.

For other conductors the terminal drilling pattern shown will

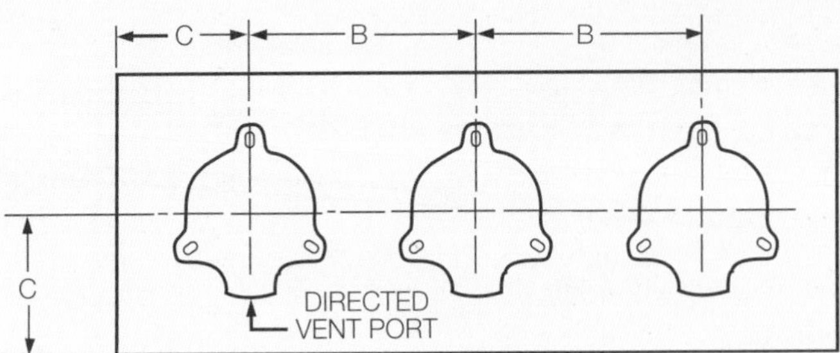

Figure 4 Three-phase in-line mounting, top view. ***Note:*** *Refer to Tables 4 and 5 for Dimensions B and C.*

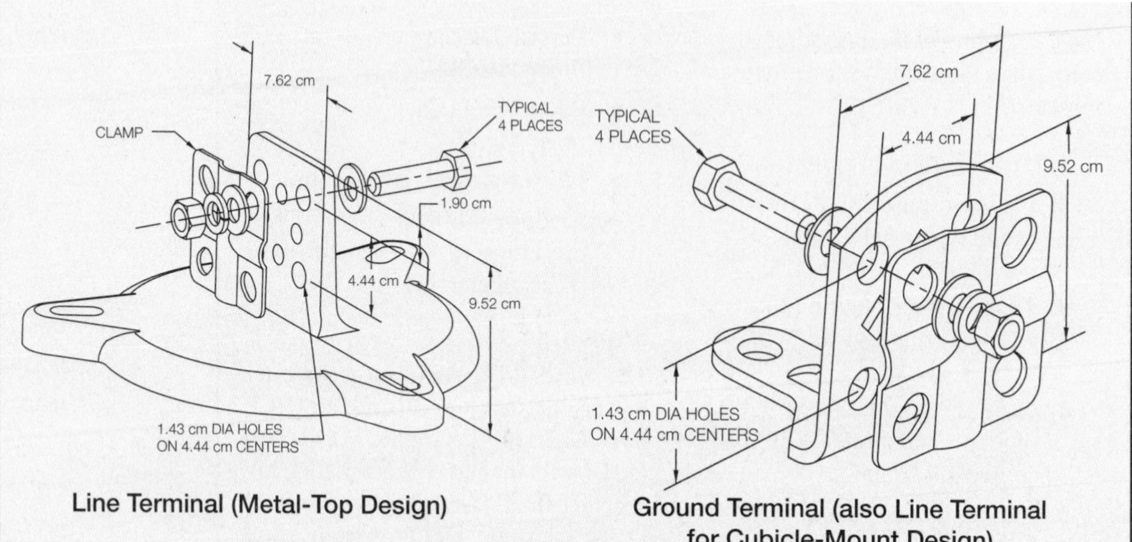

Figure 5 *Line and ground terminals (suitable for copper or aluminum conductors up to 1.9-cm diameter)*

Arrester Rating (kV rms)	Arrester MCOV (kV rms)	TOV* (kV rms)		Front-of-wave Protective Level** (kV Crest)	Maximum Discharge Voltage (kV crest) 8/20 μs Current Wave						Switching Surge Protective Level*** (kV crest)				
		1 sec	10 sec		1.5 kA	3 kA	5 kA	10 kA	20 kA	40 kA	125 A	250 A	500 A	1000 A	2000A
3	2.55	3.73	3.56	9.3	7.0	7.4	7.7	8.4	9.4	11.0	6.1	6.3	6.5	6.7	—
6	5.10	7.47	7.11	18.2	13.9	14.7	15.4	16.7	18.6	21.4	12.2	12.6	13.0	13.5	—
9	7.65	11.2	10.7	27.2	20.9	22.0	23.1	25.0	27.7	31.7	18.3	18.9	19.5	20.2	—
10	8.40	12.3	11.7	29.7	23.0	24.2	25.4	27.4	30.4	34.8	20.1	20.7	21.4	22.2	—
12	10.2	14.9	14.2	36.0	27.9	29.4	30.8	33.3	36.9	42.1	24.4	25.2	26.0	26.9	—
15	12.7	18.6	17.7	44.7	34.7	36.6	38.3	41.4	45.9	52.2	30.4	31.3	32.4	33.5	—
18	15.3	22.4	21.3	53.7	41.8	44.0	46.2	49.8	55.2	62.8	36.6	37.7	39.0	40.4	—
21	17.0	24.9	23.7	59.7	46.4	48.9	51.3	55.4	61.3	69.7	40.7	41.9	43.4	44.9	—
24	19.5	28.6	27.2	68.4	53.3	56.1	58.8	63.5	70.3	79.9	46.7	48.1	49.8	51.5	—
27	22.0	32.2	30.7	77.0	60.1	63.3	66.3	71.6	79.3	90.0	52.7	54.3	56.1	58.1	—
30	24.4	35.7	34.0	85.4	66.6	70.2	73.6	79.4	87.9	99.8	58.4	60.2	62.3	64.4	—
33	27.5	40.3	38.4	96.2	75.1	79.1	82.9	89.5	99.1	112	65.9	67.8	70.2	72.6	—
36	29.0	42.5	40.5	101	79.2	83.4	87.4	94.4	105	119	69.5	71.5	74.0	76.6	—

TABLE 3

Discharge Voltages—Maximum Guaranteed Protective Characteristics for AZES Surge Arresters

*Temporary overvoltage with prior duty energy surge.

**Based on a 10 kA current impulse that results in a discharge voltage cresting in a 0.5 μs.

***45–60 μs rise time current surge.

39	31.5	46.1	43.9	111	86.0	90.6	95.0	103	113	129	75.4	77.7	80.4	83.1	—
42	34.0	49.8	47.4	119	92.8	97.8	103	111	122	139	81.4	83.9	86.8	89.7	—
45	36.5	53.4	50.9	128	99.7	105	110	119	131	149	87.4	90.0	93.1	96.3	—
48	39.0	57.1	54.4	136	107	112	118	127	140	159	93.4	96.2	99.5	103	—
54	42.0	61.5	58.6	147	115	121	127	137	151	171	101	104	107	111	—
60	48.0	70.3	67.0	167	131	138	145	156	173	196	115	118	123	127	—
66	53.0	77.6	73.9	186	145	153	160	173	191	217	127	131	135	140	—
72	57.0	83.5	79.5	200	156	164	172	186	205	233	137	141	145	151	—
78	62.0	90.8	86.5	217	169	178	187	202	223	253	149	153	158	164	—
84	68.0	99.6	94.9	237	186	196	205	221	245	278	163	168	174	179	—
90	70.0	102.5	97.7	245	191	201	211	228	252	286	168	173	179	185	—
96	76.0	111.3	106.0	265	208	219	229	247	274	310	182	188	194	201	—
108	84.0	123.0	117.2	293	229	242	253	273	302	343	201	207	214	222	—
120	98.0	143.5	136.7	321	255	269	279	298	328	366	226	232	238	247	—
132	106	155.2	147.9	349	276	290	302	323	355	396	244	250	258	267	—
138	111	162.5	154.9	365	289	304	316	338	372	415	256	262	270	280	—
144	115	168.4	160.4	378	300	315	327	350	385	430	265	272	279	290	—
162	130	190.3	181.4	427	339	356	370	396	435	486	300	307	316	327	—
168	131	191.8	182.8	430	341	359	373	399	438	489	302	309	318	330	—
172	140	205.0	195.3	459	365	384	398	426	468	523	323	331	340	352	—
180	144	210.8	200.9	472	375	395	410	438	482	538	332	340	350	362	—
192	152	222.5	212.0	499	396	417	432	463	509	568	350	359	369	383	—
198	160	234.2	223.2	525	417	439	455	487	536	598	369	378	389	403	—
204	165	241.6	230.2	541	430	452	469	502	552	617	380	390	401	415	—
216	174	254.7	242.7	570	453	477	495	529	582	650	401	411	423	438	—
228	180	263.5	251.1	591	469	493	512	548	602	672	415	425	437	453	—
240	190	278.2	265.1	623	495	521	540	578	636	709	438	449	462	478	—
258	209	—	—	684	547	568	580	605	666	771	—	—	502	526	535
264	212	—	—	693	555	576	588	613	675	782	—	—	509	533	543
276	220	—	—	720	575	598	611	637	701	811	—	—	528	553	563
288	230	—	—	751	602	625	639	665	732	848	—	—	552	578	589
294	235	—	—	767	615	639	652	679	748	866	—	—	564	591	602
300	239	—	—	781	625	650	663	691	761	881	—	—	574	601	612
312	245	—	—	801	630	655	669	709	780	903	—	—	578	606	617
330	267	—	—	872	698	726	741	772	850	985	—	—	641	671	683
336	269	—	—	879	704	731	747	778	856	991	—	—	645	676	689
360	289	—	—	945	756	785	802	836	920	1064	—	—	693	727	740

TABLE 3

(Continued)

Arrester Rating (kV, rms)	Arrester MCOV (kV, rms)	Catalog Number	Figure 8 Dim. A (in.)	Figure 4 Dim. B Minimum Phase-to-Phase Clearance (in.)	Figure 4 Dim. C Minimum Phase-to-Ground Clearance (in.)	Housing Leakage Distance (in.)	Housing Strike Distance (in.)	Insulation Withstand Voltages: 1.2/50 μs Impulse (kV Crest)	Insulation Withstand Voltages: 60 Hz Dry, 60 sec (kV, rms)	Insulation Withstand Voltages: 60 Hz Wet, 10 sec (kV, rms)	Weight (lbs.)
3	2.55	AZES001G002003	18.6	12	6	9.2	5.2	130	75	40	42
6	5.10	AZES001G005006	18.6	12	7	9.2	5.2	130	75	40	42
9	7.65	AZES001G007009	18.6	13	7	9.2	5.2	130	75	40	43
10	8.40	AZES002G008010	21.1	13	7	16.0	7.7	170	95	65	48
12	10.2	AZES002G010012	21.1	14	8	16.0	7.7	170	95	65	49
15	12.7	AZES002G012015	21.1	14	9	16.0	7.7	170	95	65	49
18	15.3	AZES003G015018	24.8	15	10	26.2	11.4	230	125	95	56
21	17.0	AZES003G017021	24.8	16	11	26.2	11.4	230	125	95	57
24	19.5	AZES003G019024	24.8	16	11	26.2	11.4	230	125	95	58
27	22.0	AZES004G022027	28.6	17	12	36.3	15.2	265	150	125	65
30	24.4	AZES004G024030	28.6	18	13	36.3	15.2	265	150	125	65
33	27.5	AZES004G027033	28.6	20	14	36.3	15.2	265	150	125	66
36	29.0	AZES004G029036	28.6	20	14	36.3	15.2	265	150	125	66
39	31.5	AZES005G031039	33.6	21	15	49.9	20.2	320	190	165	77
42	34.0	AZES005G034042	33.6	22	16	49.9	20.2	320	190	165	78
45	36.5	AZES005G036045	33.6	23	17	49.9	20.2	320	190	165	78
48	39.0	AZES005G039048	33.6	24	18	49.9	20.2	320	190	165	78
54	42.0	AZES006G042054	36.6	25	20	64.8	23.6	365	200	170	86
60	48.0	AZES006G048060	36.6	28	22	64.8	23.6	365	200	170	87
66	53.0	AZES007G053066	39.4	30	24	73.7	26.5	385	235	200	96
72	57.0	AZES007G057072	39.4	31	26	73.7	26.5	385	235	200	97
78	62.0	AZES008G062078	48.0	33	28	100.0	35.0	505	305	260	116
84	68.0	AZES008G068084	48.0	36	30	100.0	35.0	505	305	260	118
90	70.0	AZES008G070090	48.0	36	31	100.0	35.0	505	305	260	118
96	76.0	AZES008G076096	48.0	39	33	100.0	35.0	505	305	260	119
108	84.0	AZES009G084108	56.5	42	36	127.0	43.6	650	370	300	161
120	98.0	AZES009G098120	56.5	43	37	127.0	43.6	650	370	300	175
132	106	AZES012G106132	71.5	46	40	138.5	50.1	735	425	360	191
138	111	AZES012G111138	71.5	47	42	138.5	50.1	735	425	360	193
144	115	AZES013G115144	74.4	49	43	147.4	52.9	770	475	400	202
162	130	AZES014G130162	80.1	54	48	164.8	58.6	865	490	415	217
168	131	AZES015G131168	82.9	54	48	173.7	61.5	880	545	455	225
172	140	AZES021G140172	83.2	72	59	173.7	59.4	920	535	450	237
180	144	AZES022G144180	88.9	73	60	191.8	60.9	920	515	440	270
192	152	AZES022G152192	88.9	76	63	191.8	60.9	920	515	440	273
198	160	AZES023G160198	91.8	79	66	200.7	63.8	930	535	480	283
204	165	AZES024G165204	100.3	81	68	227.0	72.3	1065	595	545	303

TABLE 4

Catalog Numbers, Dimensional Information, Insulation Withstand Voltages and Weights for AZES Metal-Top Surge Arresters

Note: All arresters are available in grey (standard) or brown porcelain glaze. For brown glaze, substitute B for G in the eighth position of the catalog number.

216	174	AZES024G174216	100.3	84	71	227.0	72.3	1065	595	545	306
228	182	AZES025G182228	108.9	86	73	254.0	80.9	1185	655	580	348
240	190	AZES025G190240	108.9	89	76	254.0	80.9	1185	655	580	351
258	209	AZES067G209258	123.9	105	84	265.5	83.9	1265	690	625	472
264	212	AZES067G212264	123.9	106	85	265.5	83.9	1265	690	625	473
276	220	AZES069G220276	132.5	108	87	291.8	92.4	1300	750	675	499
288	230	AZES069G230288	132.5	111	90	291.8	92.4	1300	750	675	503
294	235	AZES070G235294	135.3	113	92	300.7	95.3	1405	765	725	524
300	239	AZES070G239300	135.3	114	93	300.7	95.3	1405	765	725	526
312	245	AZES071G245312	141.0	116	95	318.8	100.9	1475	805	730	558
330	267	AZES074G267330	152.4	136	108	354.0	109.5	1440	810	790	605
336	269	AZES074G269336	152.4	136	109	354.0	109.5	1440	810	790	606
360	289	AZES075G289360	161.0	143	115	381.0	118.0	1535	860	840	655

TABLE 4

(Continued)

Arrester MCOV (kV, rms)	Arrester MCOV (kV, rms)	Catalog Number	Figure 7 Dim. "A" (in.)	Figure 4 Dim. "B" Minimum Phase-to-Phase Clearance (in.)	Figure 4 Dim. "C" Minimum Phase-to-Ground Clearance (in.)	Housing Leakage Distance (in.)	Housing Strike Distance (in.)	Insulation Withstand Voltages: 1.2/50 μs Impulse (kV Crest)	Insulation Withstand Voltages: 60 Hz Dry, 60 sec (kV, rms)	Insulation Withstand Voltages: 60 Hz Wet, 10 sec (kV, rms)	Weight (lbs.)
3	2.55	AZES091G002003	13.4	8	4	9.0	4.7	95	65	32	23
6	5.10	AZES091G005006	13.4	8	5	9.0	4.7	95	65	32	24
9	7.65	AZES091G007009	13.4	8	5	9.0	4.7	95	65	32	24
10	8.40	AZES092G008010	15.9	8	5	16.0	7.2	128	93	46	29
12	10.2	AZES092G010012	15.9	9	6	16.0	7.2	128	93	46	30
15	12.7	AZES092G012015	15.9	10	7	16.0	7.2	128	93	46	30
18	15.3	AZES093G015018	20.2	11	8	26.4	11.0	182	119	75	38
21	17.0	AZES093G017021	20.2	12	8	26.4	11.0	182	119	75	38
24	19.5	AZES093G019024	20.2	12	9	26.4	11.0	182	119	75	39
27	22.0	AZES094G022027	23.4	13	10	36.8	14.7	230	148	115	47
30	24.4	AZES094G024030	23.4	14	11	36.8	14.7	230	148	115	47
33	27.5	AZES094G027033	23.4	15	12	36.8	14.7	230	148	115	47
36	29.0	AZES094G029036	23.4	16	12	36.8	14.7	230	148	115	48
39	31.5	AZES095G031039	28.4	17	13	50.7	19.7	294	174	143	59
42	34.0	AZES095G034042	28.4	18	14	50.7	19.7	294	174	143	60
45	36.5	AZES095G036045	28.4	19	15	50.7	19.7	294	174	143	60
48	39.0	AZES095G039048	28.4	20	16	50.7	19.7	294	174	143	60

TABLE 5

Catalog Numbers, Dimensional Information, Insulation Withstand Voltages and Weights for AZES Cubicle-Mount Surge Arresters

Note: All arresters are available in grey (standard) or brown porcelain glaze. For brown glaze, substitute B for G in the eighth position of the catalog number.

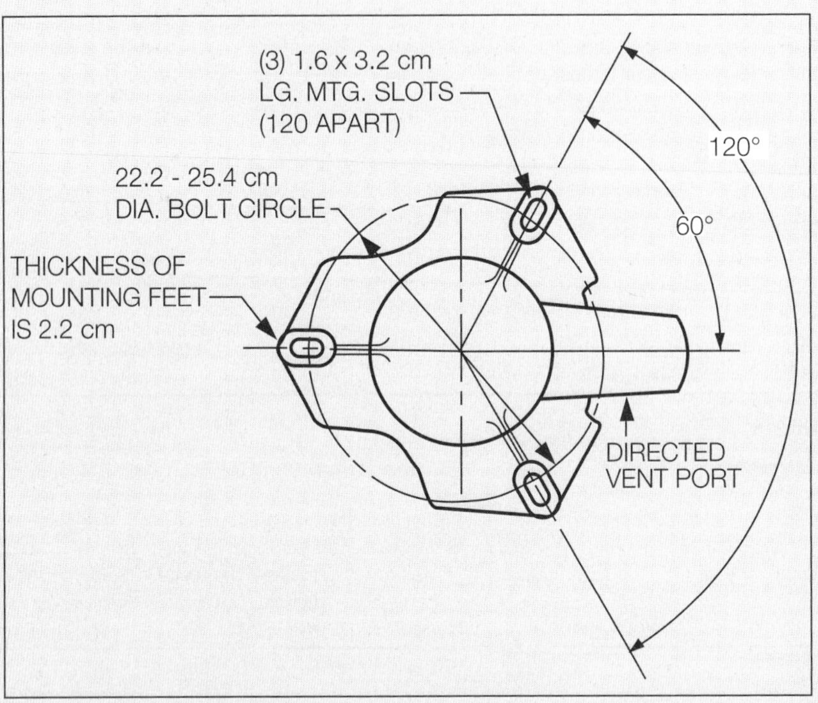

Figure 6 Base mounting. ***Note:*** *To develop rated cantilever strength use 25-cm bolt circle mounting diameter and 1.3-cm hardened bolts and flat washers.*

accommodate industry standard two (2) and four (4) hole flat pad connectors having a 4.4 cm spacing. Figure 6 provides the dimensional details for universal base mounting.

The vent port in the base must be directed away from any adjacent equipment to control and prevent ionized gases from damaging other equipment in the unlikely event of arrester failure.

Performance and Protective Characteristics

Table 3 displays the Arrester Rating, Maximum Continuous Operating Voltage (MCOV) and the guaranteed protective characteristics of the AZES surge arrester.

The Front-of-Wave protective level is the maximum discharge voltage for a 10 kA current impulse which results in a discharge voltage cresting in 0.5 microseconds. Lightning impulse Discharge Voltages represent the maximum voltage levels generated by the arrester when discharging lightning currents of the standard 8/20 microsecond wave-shape. The maximum Switching Impulse Discharge Voltages are based on a switching surge current having a time to crest of 45 to 60 microseconds. ■

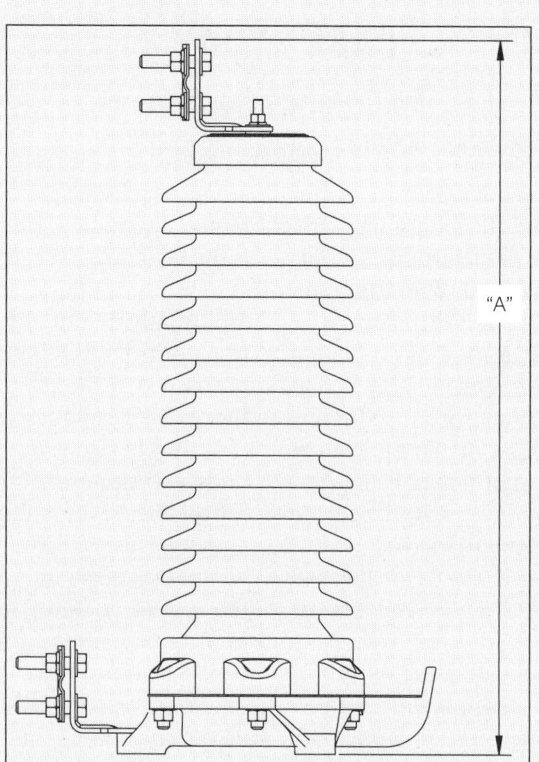

Figure 7 *Dimensions of VariSTAR Type AZE cubicle-mount surge arrester.* ***Note:*** *Refer to Table 5 for dimension A.*

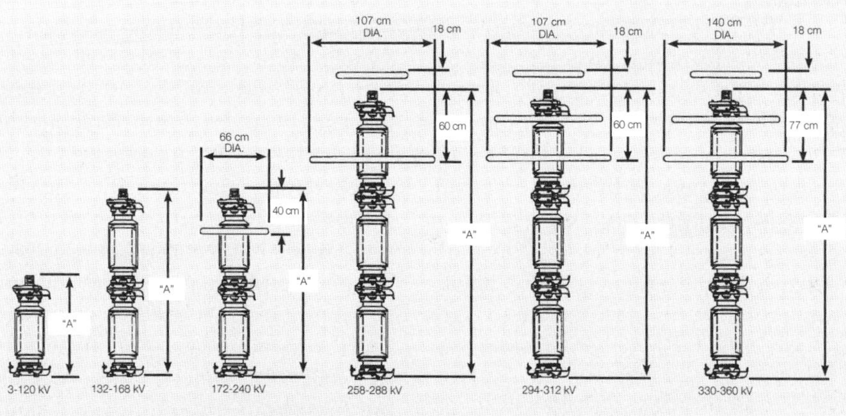

Figure 8 *Dimensions of VariSTAR Type AZE metal-top surge arresters.* ***Note:*** *Refer to Table 4 for dimension A.*

CASE STUDY 2

Emergency Response

By Nicholas Miller, Clyde Loutan, Miaolei Shao, and Kara Clark

The reliable operation of a power system depends on maintaining frequency within predetermined limits around the nominal operating frequency (60 Hertz in North America). A fundamental aspect of operating an electric power grid reliably is that the amount of power produced at any given instant must match almost exactly the amount of power being consumed. If extra power is produced, the frequency will tend to increase. If less power is produced, the frequency will tend to decrease. The frequency of the interconnected grid is mostly controlled by adjusting the output of generators in order to maintain a balance between generation and load. This balancing and frequency control occur over a continuum of time, using different resources that fall into the categories of primary, secondary, or tertiary controls.

Primary frequency control, or frequency response, depends on the rapid, autonomous action of resources, particularly generation, in response to significant changes in system frequency. Inherent inertial behavior and then primary frequency control actions are the first lines of defense for the system to avoid involuntary interruption of service to customers, which can occur within a few seconds following a system disturbance. Secondary frequency control is the fastest centralized control in the system. Secondary control actions are usually due to automatic generation control (AGC) instructions that are issued through a balancing authority's energy management system (EMS). They start within tens of seconds and dominate system response for the first several minutes following a disturbance. Tertiary control encompasses dispatch actions taken by the system operator to get resources in place to handle current and future contingencies. Reserve deployment and reserve restoration following a disturbance are common types of tertiary control.

In the first few seconds following the loss of a large generating plant, the frequency dynamics of the system are dominated by the inertial response of the online generation. Synchronous machines inherently contribute some of their stored inertial energy to the grid, reducing the initial rate of frequency decline. In synchronous machines, the inertial response is inherent to the machine physics and is not controllable. Slower initial rate of frequency decline can be helpful, allowing slower governor actions to nonetheless stabilize grid frequency. In some smaller systems, the initial rate of frequency decline is quite important. This does not appear to be as much of a concern in the major U.S. interconnections.

There is widespread and growing concern in North America about system response to underfrequency events. The North American Electric Reliability Corporation (NERC) has led a frequency response initiative. Its report, issued on October 30, 2012, examines the trends in and causes of the deteriorating frequency response that has been observed over the past two decades. It provides the rationale for establishing the responsibilities for maintaining adequate frequency performance laid out in a new standard on frequency response and bias that has recently (March 29, 2013) been submitted as a petition to the Federal Energy Regulatory Commission (FERC) for approval. In 2010, FERC sponsored work by the Lawrence Berkeley National Lab (LBNL) that identified metrics useful in planning and operations of a system with large amounts of variable generation. One focus is the concern that the addition of substantial amounts of wind power may exacerbate the observed decline in frequency response.

Wind Plant Frequency Response Today

There is a fundamental difference between a wind turbine and the turbines supplying motive power to the majority of the world's traditional power plants, be they gas, steam, or hydro turbines. With traditional turbines, the rotational speed is nearly constant and locked to system frequency. The speed of a wind turbine, however, is not synchronous with the grid and is controlled so as to maximize active power production. Wind turbine power production is therefore not inherently coupled to the system frequency and does not provide inertial or governor response unless specifically controlled to do so. Historically, wind plants have usually not been required to participate in frequency regulation.

With higher levels of wind and solar penetration, system operation changes. Thermal units may be decommitted as their output is displaced by lower-marginal-cost wind generation, or they may be dispatched to lower power levels. It can happen that the units most economical to displace also have the most desirable governor response. This leaves other resources running that do not provide the same governor characteristics. The remaining thermal units are also more frequently operated at a lower power output. This change in system operations is the primary reason for concerns that wind generation may exacerbate existing frequency performance challenges.

The Incumbent Generators

The frequency response of systems is dominated by the amount and type of generation committed and how it is dispatched. For performance investigations, it is useful to distinguish generators in accordance with how they respond to frequency excursions.

Governor-responsive units have speed governors with a droop characteristic and some available headroom (see the sidebar "Governor Basics"). These plants increase power output to support the grid when frequency declines. Other generation does not. This includes some generation that is

Governor Basics

Most synchronous generators have a speed governor that, when enabled, helps maintain system frequency. The speed of a synchronous machine is a good proxy for system frequency. In steady state, the speed is always an integer multiple of grid frequency—nominally 60 Hz in North America. The gray line in Figure S1 shows a representative governor characteristic. As frequency drops (moving left from nominal in the figure), once the deadband is exceeded, the generator output will increase along the droop line. The steepness of the line—the droop—is usually set so that a 1% drop in speed (i.e., 0.6 Hz) will result in a 20% increase in power output. The increase in power is limited by the physical limits of the generation: this is the headroom. Therefore, for a unit to usefully contribute to primary frequency response, the governor must be enabled and there must be some headroom.

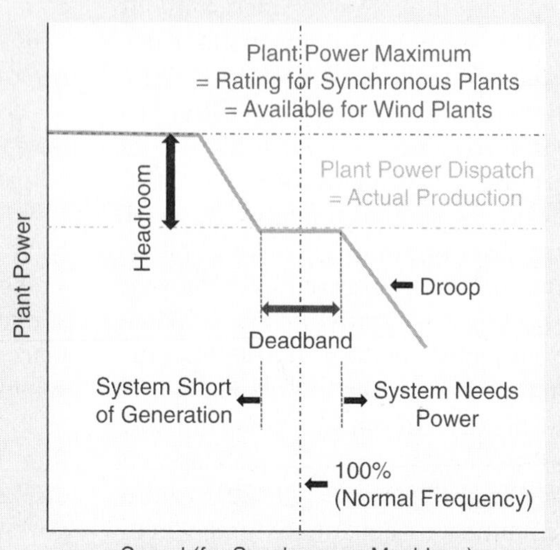

Figure S1 A representative governor characteristic

operated so that it cannot increase output for frequency decline but can decrease output for overfrequency conditions. In the Western Interconnection, these are referred to as base load units. Other generators may either lack governors or be operated with their governors disabled.

The ratio of generation that provides governor response to all generation running on the system is a useful metric used to quantify overall system readiness to provide frequency response. There are a number of physical and economic reasons why many power plants, particularly thermal plants, operate without underfrequency governor response. For example, steam plants are more efficient when operated with valves wide open, in so-called "boiler follow" mode. Nuclear plants in the United States are not permitted to have underfrequency response. The net result is that it is common in the major U.S. interconnections for only about one-third of generation operating at any given time to provide governor response. This means that careful attention must be given during operation to this important subset of synchronized generation.

Two New Investigations of Frequency Response

In this article, we present certain key findings of two new investigations of frequency response in the U.S. Western and Eastern interconnections. The work on the Western Interconnection (WI) system was done with California Independent System Operator (CAISO), with a focus on the performance of California. The Eastern Interconnection (EI) work was sponsored by the U.S. Department of Energy (DOE) Office of Electricity Delivery and Energy Reliability and managed by the National Renewable Energy Lab (NREL). The goals of both studies were to illustrate overall system frequency response, investigate the possible impact of large amounts of wind generation, and examine means to improve frequency response with the use of active power controls on wind plants.

California

This study was specifically designed to investigate the frequency response of California due to large loss-of-generation events of the type targeted by NERC Standard BAL-003-0.1b, Frequency Response and Bias, under system conditions with high levels of wind and solar generation, as envisioned in the near future. It addressed the overall frequency response of the Western grid, without considering any changes to the limits of stability-limited transmission paths that may be warranted at higher penetrations of variable energy resources.

For this work, CAISO created a number of credible load flow and stability base cases that represent the high penetrations of wind and solar generation expected in California in the near future. These cases were deliberately selected with the expectation that they would represent some of the most challenging conditions for CAISO with respect to frequency response. A weekend morning, high-wind and high-solar condition case represents an operating condition with a large number of synchronous generators displaced by variable renewable energy resources. In addition, some of the thermal power plants with synchronous generators were also assumed to have been retired due to once-through-cooling (OTC) regulation. At the snapshot of time represented here, the total WI load is 111 GW. The fraction of California generation coming from wind plants is 28% (8.6 GW total). With an additional 6.7 GW of solar generation included, California's total variable renewable generation reaches 50% (14.3 GW total). Most of the simulations focused on the trip of two units at the Palo Verde Nuclear Power Station. This 2690-MW event is recognized by NERC as the largest loss-of-generation event in the Western Interconnection for which involuntary load shedding and other stability consequences must be avoided.

Figure 1 shows the frequency response of the Western Interconnection to the Palo Verde generation tripping event imposed at 1 s for the base case. Measuring the frequency at a specific single node in the grid following a disturbance can be confusing and misleading. Here, an equivalent frequency based on the weighted speed of all the synchronous

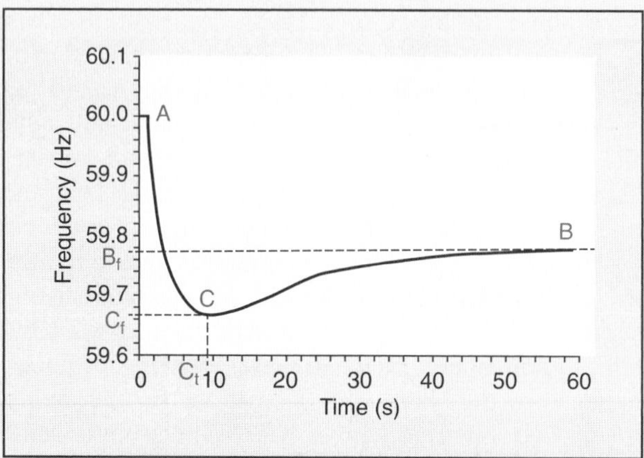

Figure 1 Frequency response to loss of generation (for the base case)

machines in the system is used. It filters out the local swings and can be regarded analytically as the common mode of the system. The *frequency nadir* (point C) occurs at about 10 s at a frequency of 59.7 Hz, and the *settling frequency* (point B) is 59.8 Hz. (For an explanation of these terms, see Table 1.) Figure 2 shows the electrical and mechanical power output of the synchronous machines with enabled governors. By the time of frequency nadir (point Cp), the units on governor control are delivering about an extra 2000 MW, and by 60 s, the interarea transients have settled out, frequency is largely uniform everywhere, and generator electrical and mechanical power are essentially equal. The change in generation by governor-responsive units represents the vast majority of system response and is nearly equal to the lost generation (2690 MW) after 60 s. The frequency nadir gives some margin above the first stage

Starting frequency	This is the system frequency before an upset; it is close to (but rarely exactly) 60 Hz.
Frequency nadir	This is the lowest frequency seen following an upset. It is usually expected to occur within several seconds of a big generation trip. It ought to stay comfortably above underfrequency load-shedding (UFLS) levels, i.e., 59.5 Hz in most of the United States.
Settling frequency	This is the frequency observed after the big swings are over and the primary frequency controls have acted but before secondary controls become significant. It ought to occur well before 1 min. after a big event.

TABLE 1

Frequency terminology

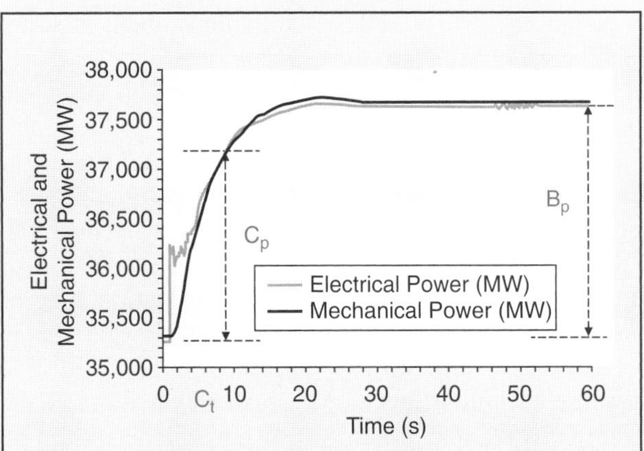

Figure 2 Base case electrical and mechanical power response

of underfrequency load shedding (UFLS) at 59.5 Hz.

A Higher-Wind Case

For the base case, the level of wind generation outside of California was relatively low, especially compared with the levels within California. In order to test conditions under which the rest of the WI is also host to significant amounts of wind generation, a higher-renewable-penetration case was developed. The change in commitment and dispatch of the synchronous generation as wind comes in is critical. In this case, changes were based on trends observed in the detailed economic evaluation of the Western Wind and Solar Integration Study (WWSIS). For every 3 MW of additional wind production, there is about 2 MW of reduction in thermal unit commitment and a 1 MW reduction in thermal unit dispatch. The committed thermal units that have the least annual operating time in the WWSIS study were selected to be displaced by wind turbine generators. Initially, the newly added wind turbine generators were assumed to have typical, nonfrequency-responsive controls and were also assumed to be operating at 50% of rated capacity, in order to capture the operational reality that all wind plants in a system are never operating simultaneously at rated power. This assumption for the incremental plants gives a reasonable, if somewhat simple, distribution of loadings on the wind plants in WECC. Thus in this case, 9508 MW of wind turbines were added to achieve an increased net wind dispatch of 4754 MW.

Coming up Short: Reduced Headroom

The system conditions in the higher-renewable-penetration case were further modified to reflect an operating condition with a reduced fraction of generation providing governor response and with those units having reduced headroom. In this case, the overall system headroom was reduced to about 8 GW. The system was deliberately stressed so as to consider the practical minimum headroom. It is worth noting that the condition from which this case was developed was already considered to be challenging and representative of a realistic commitment and dispatch for high wind and solar conditions. This case was designed to push headroom down to levels that we believe might occur relatively infrequently, that is, when the system is meeting just its minimum operating reserve obligation plus regulation needs.

In the most extreme case analyzed, the headroom was further reduced to 3 GW. This was essentially equal to the outage event size. The case was designed as a test and is not necessarily representative of a condition that is either practical or one that the system operator would regard as acceptable. Rather, we were looking to understand what might happen and establish a case for which remediation of extreme conditions might be tested. It is worth reemphasizing that this condition corresponds to about 20 GW of instantaneous production of wind and solar in the WI. Since one postulated cause of instantaneous operation at such acutely reduced headroom is from an unexpected drop in wind or solar production, one could reasonably suppose that this case is representative of a future condition in which much more than the 25 GW of nameplate rating of wind plus solar of this case is operating in the WI. The loss of 5 GW of headroom thus represents a drop of roughly 20% of the total generation. The statistical expectation of such an extreme drop is small. A final important point for this particular case is that the UFLS was disabled to allow the comparison of frequency response with other cases without the complexity of considering UFLS effects.

The frequency response for the three higher-renewable-penetration and reduced-headroom cases is shown in Figure 3. The highest trace shows the

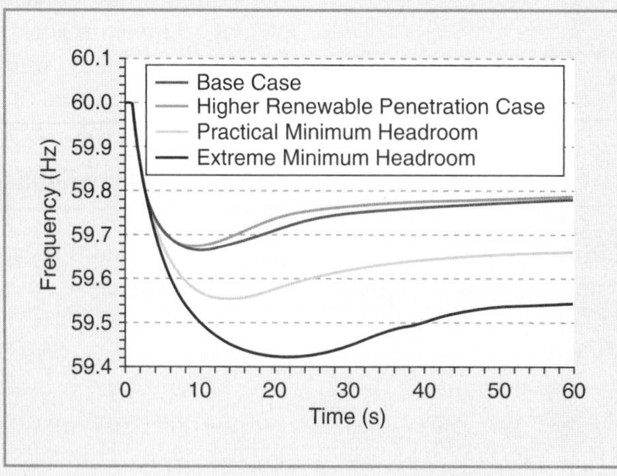

Figure 3 Frequency response to loss of generation

response for the higher-renewable-penetration case; the third highest trace shows the response for the higher-renewable-penetration and practical-minimum-headroom case; and the lowest trace shows the higher-renewable-penetration and extreme-minimum-headroom case. In comparison, the response of the base case is plotted as the second highest trace.

Compared with the base case, the higher-renewable-penetration case has better frequency response. This is because the dispatch increased headroom when wind generation was added. This effect is nowhere near in proportion to the amount of wind generation added, yet it helps the system response. This case illustrates the possibility that normal economic operation of the system may result in improved system frequency performance as wind power levels increase. The initial rate of change of frequency (ROCOF) is essentially unchanged between the cases, indicating that the reduction in system inertia that accompanies displacement of some synchronous generation has little impact.

For the higher renewable penetration and practical minimum headroom case, the frequency nadir is low and probably represents the lowest nadir that could be considered acceptable. For the extreme-minimum-headroom case, the frequency nadir would have impinged on the UFLS threshold—an unacceptable result. The settling frequency is barely above the minimum to avoid UFLS. This is evidence that the performance is limited by the generators' range of available response and static droop and not just the speed of response of the governors.

These cases indicate that extremely depleted headroom will result in unacceptable system performance. The fraction of generation providing governor response, while important, is insufficient to anticipate frequency performance. Headroom needs to be considered—at least when it is in short supply.

Turning to Wind for Help

Ultimately, wind plant controls can provide system operators with yet another tool to manage frequency (see sidebar "New System Tools: Active Power Control from Wind Plants"). Under conditions of high stress that result in a shortage of headroom and generation providing governor response, wind generation may well be the least costly means of achieving the necessary primary frequency response. In another test scenario, we start with the extreme-minimum-headroom case (the lowest trace in Figure 3). That case had unacceptable frequency performance, due to an insufficient supply of primary response. In the new case, about 41% of the wind generation is assumed to be operating slightly curtailed and with governorlike response enabled, using standard droop and deadbands. This adds 1812 MW of headroom.

A comparison of the cases is shown in Figure 4 for frequency response and in Figure 5 for the power output of those wind plants with the controls on. The performance of this case is dramatically better.

New System Tools: Active Power Control from Wind Plants

Since wind generation tends to displace thermal generation, under some conditions the system would benefit from the frequency-sensitive response of wind plants to (1) overfrequency events caused by load rejections, since thermal plants are less likely to have substantial down-regulation capability, and (2) underfrequency events caused by generation trips. Some wind plants with modern turbines and plant controls have the ability to control active power output in response to grid frequency. A number of grid operators around the world have started requiring that wind plants have these controls.

Inertial Controls

Most modern MW-class wind generation does not inherently exhibit inertial response. It is possible, however, to program the wind plant controls to provide a form of inertial response. For large underfrequency events, the inertial control feature temporarily increases the power output of the wind turbine by about 5–10% of rated power for several seconds. Below rated wind, stored kinetic energy from the turbine-generator rotor is temporarily donated to the grid but is recovered later. This inertial response is not quite energy-neutral, meaning that the period of increased power is followed by a longer period of decreased power.

Governorlike Controls

It is also possible to implement wind plant controls that provide something similar to governor response. Some active power controls that are commercially available are closely akin to governor controls for thermal and hydro generation. They respond to significant deviations in grid frequency, increasing or decreasing power output in response to low- or high-frequency events, respectively. To do this, the control alters the active power control reference targeted by the turbine controls. In order to allow for an increase of wind plant active power output in response to an underfrequency condition, some active power production must be kept in reserve—this is the "headroom" shown in the governor figure. The maximum power production of the wind plant is therefore constrained to a value less than that available from the wind. This results in unrecoverable loss of energy production (much like spilled water on hydro generation) and so has significant economic implications. (Alternatively, some energy could be stored in a different medium, such as a battery, for delivery at this critical time, but that is a discussion for a different article.) The potential for wind generation to respond quickly makes this resource effective in arresting and correcting frequency deviations, much as fast governor response does on thermal generation.

This case shows that, if necessary, primary frequency response from wind generation has the potential to greatly improve the system frequency performance of the entire grid. These cases do not include the inertial control that in general would be complementary to this control.

Eastern Frequency Response Study

A recently completed study of the EI investigated some of the same issues as the study of the WI, again focusing on the evaluation of frequency response and generation control with increasing wind penetration. While

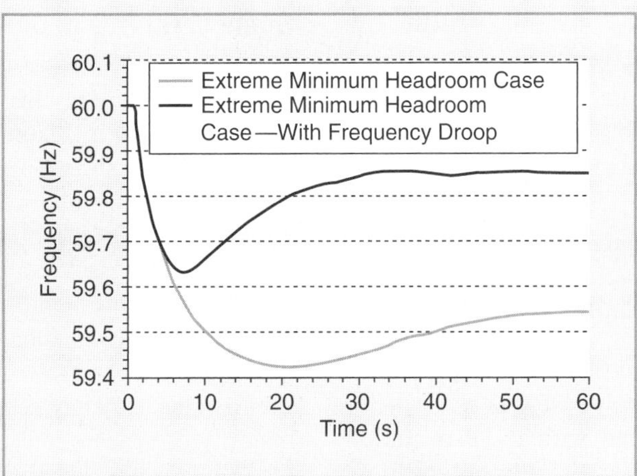

Figure 4 Frequency response, with and without wind plant frequency droop function

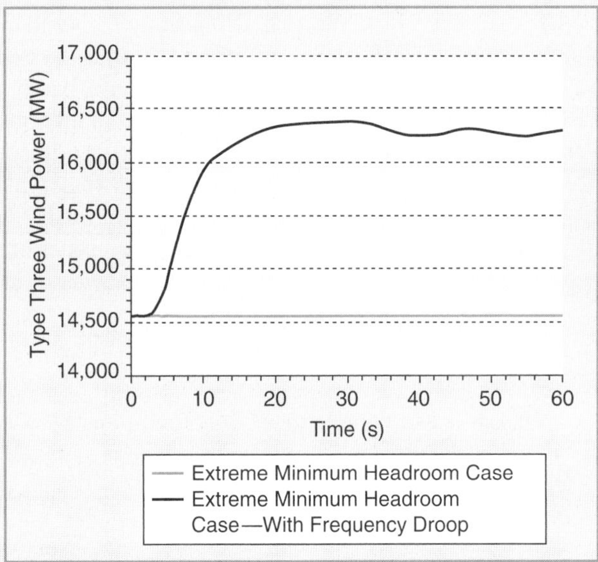

Figure 5 Wind power response, with and without frequency droop function

the studies were similar in focus, one particularly important difference required attention. Although there are efforts under way to improve matters, the dynamic representations of the EI that are in widespread use today are known to poorly reflect the interconnection's observed behavior. The main goals of this work were therefore to:

- Create a realistic baseline model of the EI for examining frequency response;

- Illustrate overall system frequency response;
- Investigate the possible impact of large amounts of wind generation; and
- Examine means of improving EI frequency response with the use of active power controls on wind plants.

For this investigation, a light-load power flow case was again deliberately selected with the expectation that it would represent one of the more challenging conditions for the EI with respect to frequency response. At the snapshot of time represented in this case, the total EI load was 272 GW. Most of the simulations focused on the trip of multiple thermal plants in the region of Rockport, Indiana. This 4455-MW event was patterned after the largest loss-of-generation event in the EI for which involuntary load shedding and other stability consequences must be avoided.

Some Wrinkles in the Model

A simulation of the loss of 4455 MW of generation using this widely used model is shown in Figure 6. The Frequency trace shows the frequency dropping slightly and rapidly recovering. Over 200 GW of generation is shown in the black trace to rapidly and effectively respond. This is nice performance. Unfortunately, it does not resemble the observed behavior of the EI. This is mainly because the vast majority of synchronous generation in the dynamic model of the EI has active governors modeled. Thus, the fraction of generation with governor response is 78.9% for the EI in this case. This value is significantly higher than the 30–40% of recent U.S. industry experience.

Based on a better understanding of thermal plant controls, the initial model was modified. Governors on plant types that typically do not have active governors were disabled. The fraction of generation with governor response was reduced to 32%.

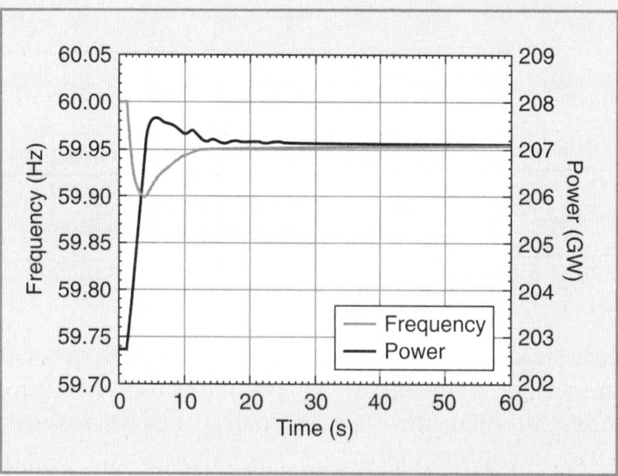

Figure 6 Frequency and governor response of the EI, using the old model

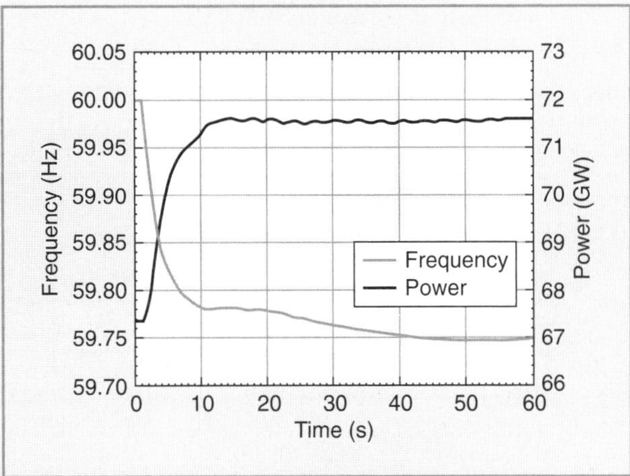

Figure 7 Frequency and governor response of the EI, using the new model

Further, plant controls representative of those that tend to override primary governor response creating "governor withdrawal" were added. These relatively common controls are intended to make certain that power plants adhere to their dispatch schedule. Governor action in response to frequency excursions drives machines away from their dispatch (see "Governor Basics"). These controls return the plants to schedule, regardless of whether system frequency has returned to normal.

The system response to the loss of 4455 MW of generation for the new base case is shown in Figure 7. The Frequency trace shows the frequency of the EI. The frequency drops initially and stays generally depressed. This is the so-called "lazy-L behavior" that has been observed for actual large-generation disturbances in the EI. This modified model was found to reasonably reproduce a large event that occurred on May 13, 2012. The power trace in Figure 7 shows the mechanical power output of the governor-responsive units. It presents quite a contrast to the power trace in Figure 6.

To examine the possible impact of high levels of wind penetration on the EI, a test case was devised with new wind generation of approximately 85 GW rating, operating at a total of 68 GW production, added across all of the NERC regions except those of the Southeast Electric Reliability Corporation (SERC) and the Florida Reliability Coordinating Council (FRCC). This represents an instantaneous penetration of about 40% for those regions and of 25% for the EI as a whole. As noted earlier, wind generation normally displaces thermal generation during operations. An initial high-wind case displaced governor-responsive thermal generation such that the fraction of generation providing frequency response was reduced to about 27%. That resulted in somewhat poorer

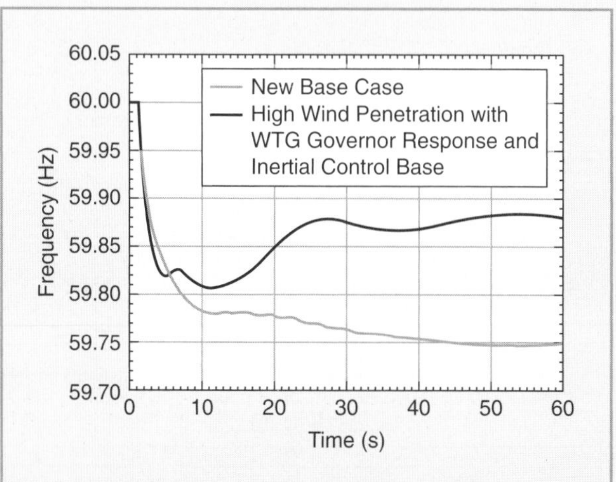

Figure 8 Frequency comparison: EI without wind versus high wind with governor and inertial response

frequency response. By modifying the case to bring the fraction of generation providing frequency response back to 32%, the frequency response was restored.

Wind Plant Controls: Governor and Inertial Controls Combined

To illustrate the potential benefits of frequency-sensitive wind plant control, most of the wind generation was again curtailed by 5% of the available wind power, and the governorlike controls were enabled. This condition added a total of 3.4 GW of headroom. Unlike the WI case shown in Figures 4 and 5, for this case inertial control on the wind turbines was enabled as well.

The system frequency, as shown in Figure 8, is substantially better in the high-wind case with these controls as compared with the no-wind base case. The electrical power of all the wind turbines is shown in Figure 9. The lighter trace shows the combined response of the governor and inertial controls. To help understand the impact of the inertial control, the darker trace shows only the contribution from the governor control. The energy associated with the inertial control is the area between the two curves. Notice that even though the inertial control is almost energy-neutral, it produces a performance benefit by delivering extra energy to the system earlier, as the frequency is declining. This can be a substantial benefit, if there is sufficient primary reserve on the system, as is the case here.

Conclusions

These investigations provide evidence that high levels of wind and solar generation can be well tolerated within the major U.S. interconnections with respect to system frequency response. The tendency for wind and solar

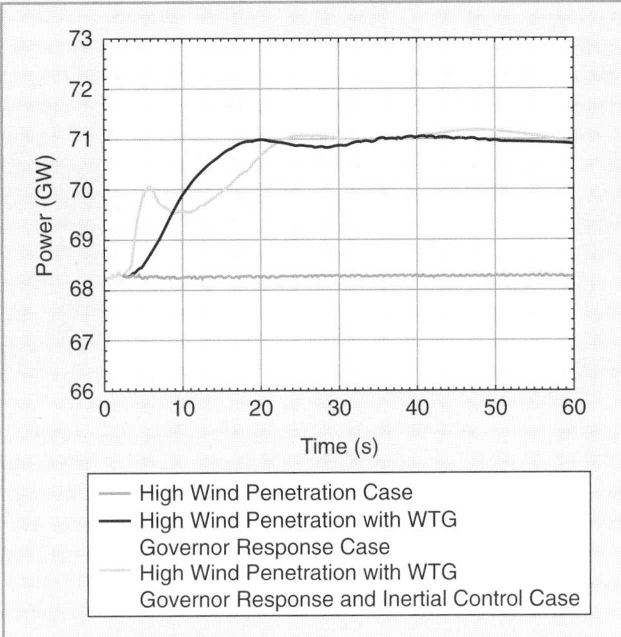

Figure 9 Wind turbine generator power comparison: three different controls

generation to displace other synchronous thermal generation that may contribute to healthy frequency response needs to be considered during system operation. These findings reinforce the work of others who have found that the fraction of generators providing governor response at any given time is a critical operational metric. Minimum levels on the order of 25–35% of total online generation are needed at all times. This is true independent of wind generation. There is evidence that these minimum levels are occasionally reached today. The fact that roughly two out of three operating generators may not contribute to primary frequency response under normal operating conditions is an economic reality of operation.

The headroom available for the fraction of generation that is contributing to primary frequency response can also be important under conditions of high stress. The system must maintain a minimum MW level of synchronized, responsive generation in order to provide adequate frequency performance. The speed of response of these reserves is important, since the frequency nadir following a major loss-of-generation event typically occurs within several seconds of the event. Contributions to primary frequency must be sustained until secondary frequency response, that is, automatic generation control, asserts control—typically, within the space of several minutes. The fact that many thermal plants have controls in place that quickly withdraw their primary response contributes to degraded frequency performance. The WI work showed a roughly 20% degradation in performance due to such controls.

The modeling of primary frequency response, including the capture of key elements such as which generators are actually contributing, which generators have controls that withdraw primary response, and the contribution of load behavior, requires careful attention. In the EI, today's commonly used dynamic models capture observed frequency behavior poorly. These studies showed that adjustments to the models can greatly improve model fidelity. But more work should be and is being done to improve the EI models.

As wind penetration increases, the potential participation of wind plants in frequency control becomes more important to the system. These investigations show that frequency-sensitive controls on wind plants can have a substantial beneficial impact on system performance. Inertial controls from wind generation provide fast transient support, via a controlled inertial response from wind turbines, that can significantly improve the system's frequency nadir. This is particularly the case if the frequency nadir is significantly lower than the settling frequency, which tends to be the case if the system has generators with adequate but slower control action. This means that these types of controls can reduce the need to worry about the speed of individual unit governor response. They do relatively little to correct a shortage in the amount of available response, however. Under normal conditions, these controls will add margin in avoiding UFLS. The participation of wind plants in providing primary frequency response, that is, by exploiting plants with governorlike controls, will have significant beneficial impacts on both the frequency nadir and the settling frequency. This should prove valuable under conditions when the system is short of other resources. To provide this function, wind plants must be dispatched below available wind power, causing an opportunity cost equal to the value of the lost production. Like spilling water over a hydro dam, this can be expensive unless there is a means to be paid for providing this service to the grid. Since the controls can be quite fast relative to conventional thermal and hydro generation, the benefit is greater. Other types of generation, active load controls, and energy storage devices may all be able to provide comparable benefits.

These investigations did not examine the operational practices or institutional mechanisms necessary to take advantage of the technical options identified, nor did they consider whether existing practice is sufficient to ensure adequate frequency response. The new NERC standards establish quantitative performance objectives. It is possible that new market mechanisms will be required to assure adequate frequency response. These should be structured to achieve economy of operations by allowing all qualifying resources, including variable renewable generation, to participate. These investigations, while extensive, are by no means exhaustive. Further experience and analysis will help steer practice as levels of wind generation increase.

For Further Reading

N. W. Miller, M. Shao, and S. Venkataraman, "California ISO (CAISO) frequency response study," GE Energy, 2011.

U.S. Department of Energy. (2013). Eastern frequency response study. [Online]. Available: www.osti.gov/servlets/purl/1083365/

"The reliability role of frequency response," NERC Frequency Response Initiative Report, NERC, Atlanta, GA, Oct. 30, 2012.

J. H. Eto, J. Undrill, P. Mackin, R. Daschmans, B. Williams, B. Haney, R. Hunt, J. Ellis, H. Illian, C. Martinez, M. O'Malley, K. Coughlin, and K. H. laCammare, "Use of frequency response metrics to assess the planning and operating requirements for reliable integration of variable renewable generation," Lawrence Berkeley National Laboratory, Report LBNL-4142E, Dec. 2010.

Biographies

Nicholas Miller is with GE Energy Consulting, Schenectady, New York.

Clyde Loutan is a California Independent System Operator, Folsom, California.

Miaolei Shao is with GE Energy Consulting, Schenectady, New York.

Kara Clark is with the National Renewable Energy Laboratory, Golden, Colorado. ■

13.1 TRAVELING WAVES ON SINGLE-PHASE LOSSLESS LINES

First consider a single-phase two-wire lossless transmission line. Figure 13.1 shows a line section of length Δx meters. If the line has a loop inductance L H/m and a line-to-line capacitance C F/m, then the line section has a series inductance L Δx H and shunt capacitance C Δx F, as shown. In Chapter 5, the direction of line position x was selected to be from the receiving end ($x = 0$) to the sending end ($x = l$); this selection was unimportant, since the variable x was subsequently eliminated when relating

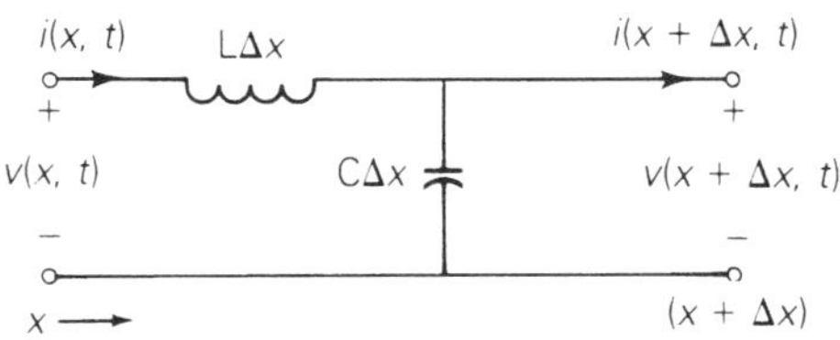

FIGURE 13.1 Single-phase two-wire lossless line section of length Δx

the steady-state sending-end quantities V_s and I_s to the receiving-end quantities V_R and I_R. Here, however, the interest is in voltages and current waveforms traveling along the line. Therefore, the direction of increasing x as being from the sending end ($x = 0$) toward the receiving end ($x = l$) is selected.

Writing a KVL and KCL equation for the circuit in Figure 13.1,

$$v(x + \Delta x, t) - v(x, t) = -L\Delta x \frac{\partial i(x, t)}{\partial t} \tag{13.1.1}$$

$$i(x + \Delta x, t) - i(x, t) = -C\Delta x \frac{\partial v(x, t)}{\partial t} \tag{13.1.2}$$

Dividing (13.1.1) and (13.1.2) by Δx and taking the limit as $\Delta x \rightarrow 0$, results in

$$\frac{\partial v(x, t)}{\partial x} = -L \frac{\partial i(x, t)}{\partial t} \tag{13.1.3}$$

$$\frac{\partial i(x, t)}{\partial x} = -C \frac{\partial v(x, t)}{\partial t} \tag{13.1.4}$$

Use partial derivatives here because $v(x, t)$ and $i(x, t)$ are differentiated with respect to both position x and time t. Also, the negative signs in (13.1.3) and (13.1.4) are due to the reference direction for x. For example, with a positive value of $\partial i/\partial t$ in Figure 13.1, $v(x, t)$ decreases as x increases. Taking the Laplace transform of (13.1.3) and (13.1.4),

$$\frac{dV(x, s)}{dx} = -sLI(x, s) \tag{13.1.5}$$

$$\frac{dI(x, s)}{dx} = -sCV(x, s) \tag{13.1.6}$$

where zero initial conditions are assumed. $V(x, s)$ and $I(x, s)$ are the Laplace transforms of $v(x, t)$ and $i(x, t)$. Also, ordinary rather than partial derivatives are used since the derivatives are now with respect to only one variable, x.

Next, differentiate (13.1.5) with respect to x and use (13.1.6), in order to eliminate $i(x, s)$:

$$\frac{d^2 V(x, s)}{dx^2} = -sL \frac{dI(x, s)}{dx} = s^2 LCV(x, s)$$

or

$$\frac{d^2 V(x, s)}{dx^2} - s^2 LCV(x, s) = 0 \tag{13.1.7}$$

Similarly, (13.1.6) can be differentiated in order to obtain

$$\frac{d^2 I(x, s)}{dx^2} - s^2 LCI(x, s) = 0 \tag{13.1.8}$$

Equation (13.1.7) is a linear, second-order homogeneous differential equation. By inspection, its solution is

$$V(x,s) = V^+(s)e^{-sx/v} + V^-(s)e^{+sx/v} \tag{13.1.9}$$

where

$$v = \frac{1}{\sqrt{LC}} \quad \text{m/s} \tag{13.1.10}$$

Similarly, the solution to (13.1.8) is

$$I(x,s) = I^+(s)e^{-sx/v} + I^-(s)e^{+sx/v} \tag{13.1.11}$$

You can quickly verify that these solutions satisfy (13.1.7) and (13.1.8). The "constants" $V^+(s)$, $V^-(s)$, $I^+(s)$, and $I^-(s)$, which in general are functions of s but are independent of x, can be evaluated from the boundary conditions at the sending and receiving ends of the line. The superscripts + and − refer to waves traveling in the positive x and negative x directions, soon to be explained.

Taking the inverse Laplace transform of (13.1.9) and (13.1.11), and recalling the time shift properly, $\mathcal{L}[f(t-\tau)] = F(s)e^{-s\tau}$, results in

$$v(x,t) = v^+\left(t - \frac{x}{v}\right) + v^-\left(t + \frac{x}{v}\right) \tag{13.1.12}$$

$$i(x,t) = i^+\left(t - \frac{x}{v}\right) + i^-\left(t + \frac{x}{v}\right) \tag{13.1.13}$$

where the functions $v^+(\ )$, $v^-(\ )$, $i^+(\ )$, and $i^-(\ )$, can be evaluated from the boundary conditions.

$v^+(t - x/v)$ represents a voltage wave traveling in the positive x direction with velocity $v = 1/\sqrt{LC}$ m/s. Consider any wave $f^+(u)$, where $u = t - x/v$. Suppose that this wave begins at $u = u_0$, as shown in Figure 13.2(a). At time $t = t_1$, the wavefront is at $u_0 = (t_1 - x_1/v)$, or at $x_1 = v(t_1 - u_0)$. At a later time, t_2, the wavefront is at $u_0 = (t_2 - x_2/v)$ or at $x_2 = v(t_2 - u_0)$. As shown in Figure 13.2(b), the wavefront has

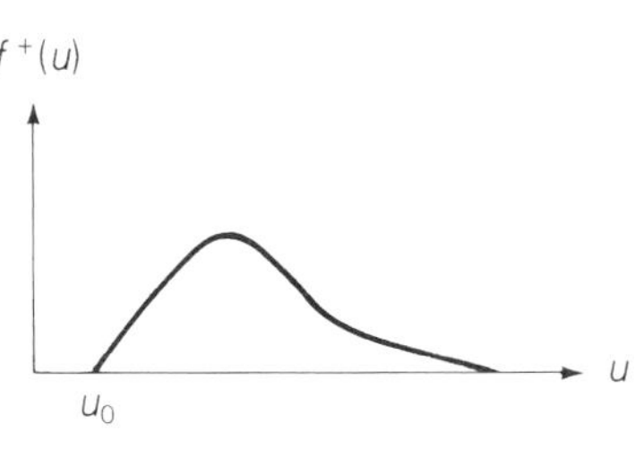

(a) $f^+(u)$ versus u

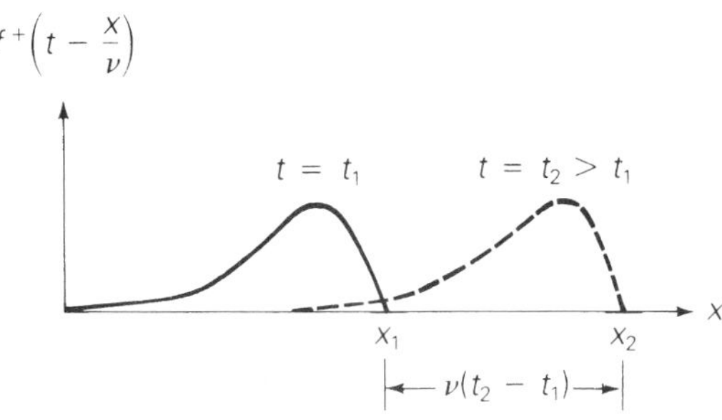

(b) $f^+\left(t - \frac{x}{v}\right)$ versus x

FIGURE 13.2 The function $f^+(u)$, where $u = \left(t - \frac{x}{v}\right)$

moved in the positive x direction a distance $(x_2 - x_1) = v(t_2 - t_1)$ during time $(t_2 - t_1)$. The velocity is therefore $(x_2 - x_1)/(t_2 - t_1) = v$.

Similarly, $i^+(t - x/v)$ represents a current wave traveling in the positive x direction with velocity v. The expressions $v^+(t - x/v)$ and $i^+(t - x/v)$ are called the *forward* traveling voltage and current waves. It can be shown analogously that $v^-(t + x/v)$ and $i^-(t + x/v)$ travel in the negative x direction with velocity v. Also, $v^-(t + x/v)$ and $i^-(t + x/v)$ are called the *backward* traveling voltage and current waves.

Recall from (5.4.16) that for a lossless line $f\lambda = 1/\sqrt{LC}$. It is now evident that the term $1/\sqrt{LC}$ in this equation is v, the velocity of propagation of voltage and current waves along the lossless line. Also, recall from Chapter 4 that L is proportional to μ and C is proportional to ε. For overhead lines, $v = 1/\sqrt{LC}$ is approximately equal to $1/\sqrt{\mu\varepsilon} = 1/\sqrt{\mu_0\varepsilon_0} = 3 \times 10^8$ m/s, the speed of light in free space. For cables, the relative permitivity $\varepsilon/\varepsilon_0$ may be 3 to 5 or even higher, resulting in a value of v lower than that for overhead lines.

Next, evaluate the terms $I^+(s)$ and $I^-(s)$. Using (13.1.9) and (13.1.10) in (13.1.6),

$$\frac{s}{v}[-I^+(s)e^{-sx/v} + I^-(s)e^{+sx/v}] = -sC[V^+(s)e^{-sx/v} + V^-(s)e^{+sx/v}]$$

Equating the coefficients of $e^{-sx/v}$ on both sides of this equation,

$$I^+(s) = (vC)V^+(s) = \frac{V^+(s)}{\sqrt{\frac{L}{C}}} = \frac{V^+(s)}{Z_c} \tag{13.1.14}$$

where

$$Z_c = \sqrt{\frac{L}{C}} \quad \Omega \tag{13.1.15}$$

Similarly, equating the coefficients of $e^{+sx/v}$,

$$I^-(s) = \frac{-V^-(s)}{Z_c} \tag{13.1.16}$$

Thus, (13.1.11) and (13.1.13) can be rewritten as

$$I(x, s) = \frac{1}{Z_c}[V^+(s)e^{-sx/v} - V^-(s)e^{+sx/v}] \tag{13.1.17}$$

$$i(x, t) = \frac{1}{Z_c}\left[v^+\left(t - \frac{x}{v}\right) - v^-\left(t + \frac{x}{v}\right)\right] \tag{13.1.18}$$

Recall from (5.4.3) that $Z_c = \sqrt{L/C}$ is the characteristic impedance (also called surge impedance) of a lossless line.

13.2 BOUNDARY CONDITIONS FOR SINGLE-PHASE LOSSLESS LINES

Figure 13.3 shows a single-phase two-wire lossless line terminated by an impedance $Z_R(s)$ at the receiving end and a source with Thévenin voltage $E_G(s)$ and with Thévenin impedance $Z_G(s)$ at the sending end. $V(x, s)$ and $I(x, s)$ are the Laplace transforms of the voltage and current at position x. The line has length l, surge impedance $Z_c = \sqrt{L/C}$, and velocity $v = 1/\sqrt{LC}$. Assume that the line is initially unenergized.

From Figure 13.3, the boundary condition at the receiving end is

$$V(l, s) = Z_R(s)I(l, s) \tag{13.2.1}$$

Using (13.1.9) and (13.1.17) in (13.2.1),

$$V^+(s)e^{-sl/v} + V^-(s)e^{+sl/v} = \frac{Z_R(s)}{Z_c}[V^+(s)e^{-sl/v} - V^-(s)e^{+sl/v}]$$

Solving for $V^-(l, s)$

$$V^-(l, s) = \Gamma_R(s)V^+(s)e^{-2s\tau} \tag{13.2.2}$$

where

$$\Gamma_R(s) = \frac{\dfrac{Z_R(s)}{Z_c} - 1}{\dfrac{Z_R(s)}{Z_c} + 1} \quad \text{per unit} \tag{13.2.3}$$

$$\tau = \frac{l}{v} \quad \text{seconds} \tag{13.2.4}$$

$\Gamma_R(s)$ is called the *receiving-end voltage reflection coefficient.* Also, τ, called the *transit time* of the line, is the time it takes a wave to travel the length of the line.

Using (13.2.2) in (13.1.9) and (13.1.17),

$$V(x, s) = V^+(s)[e^{-sx/v} + \Gamma_R(s)e^{s[(x/v)-2\tau]}] \tag{13.2.5}$$

$$I(x, s) = \frac{V^+(s)}{Z_c}[e^{-sx/v} - \Gamma_R(s)e^{s[(x/v)-2\tau]}] \tag{13.2.6}$$

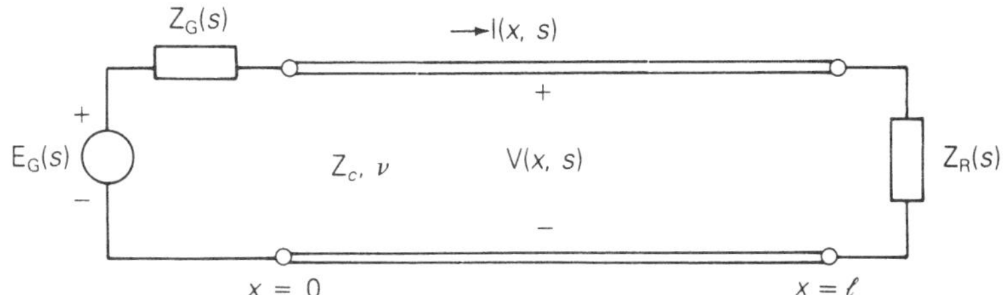

FIGURE 13.3 Single-phase two-wire lossless line with source and load terminations

From Figure 13.3 the boundary condition at the sending end is

$$V(0, s) = E_G(s) - Z_G(s)I(0, s) \tag{13.2.7}$$

Using (13.2.5) and (13.2.6) in (13.2.7),

$$V^+(s)[1 + \Gamma_R(s)e^{-2s\tau}] = E_G(s) - \left[\frac{Z_G(s)}{Z_c}\right]V^+(s)[1 - \Gamma_R(s)e^{-2s\tau}]$$

Solving for $V^+(s)$,

$$V^+(s)\left\{\left[\frac{Z_G(s)}{Z_c} + 1\right] - \Gamma_R(s)e^{-2s\tau}\left[\frac{Z_G(s)}{Z_c} - 1\right]\right\} = E_G(s)$$

$$V^+(s)\left[\frac{Z_G(s)}{Z_c} + 1\right]\{1 - \Gamma_R(s)\Gamma_S(s)e^{-2s\tau}\} = E_G(s)$$

or

$$V^+(s) = E_G(s)\left[\frac{Z_c}{Z_G(s) + Z_c}\right]\left[\frac{1}{1 - \Gamma_R(s)\Gamma_S(s)e^{-2s\tau}}\right] \tag{13.2.8}$$

where

$$\Gamma_S(s) = \frac{\dfrac{Z_G(s)}{Z_c} - 1}{\dfrac{Z_G(s)}{Z_c} + 1} \tag{13.2.9}$$

$\Gamma_S(s)$ is called the *sending-end voltage reflection coefficient.* Using (13.2.9) in (13.2.5) and (13.2.6), the complete solution is

$$V(x, s) = E_G(s)\left[\frac{Z_c}{Z_G(s) + Z_c}\right]\left[\frac{e^{-sx/v} + \Gamma_R(s)e^{s[(x/v)-2\tau]}}{1 - \Gamma_R(s)\Gamma_S(s)e^{-2s\tau}}\right] \tag{13.2.10}$$

$$I(x, s) = \left[\frac{E_G(s)}{Z_G(s) + Z_c}\right]\left[\frac{e^{-sx/v} - \Gamma_R(s)e^{s[(x/v)-2\tau]}}{1 - \Gamma_R(s)\Gamma_S(s)e^{-2s\tau}}\right] \tag{13.2.11}$$

where

$$\Gamma_R(s) = \frac{\dfrac{Z_R(s)}{Z_c} - 1}{\dfrac{Z_R(s)}{Z_c} + 1} \quad \text{per unit}$$

$$\Gamma_S(s) = \frac{\dfrac{Z_G(s)}{Z_c} - 1}{\dfrac{Z_G(s)}{Z_c} + 1} \quad \text{per unit} \tag{13.2.12}$$

$$Z_c = \sqrt{\frac{L}{C}} \ \Omega \qquad v = \frac{1}{\sqrt{LC}} \ \text{m/s} \qquad \tau = \frac{l}{v} \ \text{s} \tag{13.2.13}$$

The following four examples illustrate this general solution. All four examples refer to the line shown in Figure 13.3, which has length l, velocity v, characteristic impedance Z_c, and is initially unenergized.

EXAMPLE 13.1

Single-phase lossless-line transients: step-voltage source at sending end, matched load at receiving end

Let $Z_R = Z_c$ and $Z_G = 0$. The source voltage is a step, $e_G(t) = Eu_{-1}(t)$. (a) Determine $v(x, t)$ and $i(x, t)$. (b) Plot the voltage and current versus time t at the center of the line and at the receiving end.

SOLUTION

a. From (13.2.12) with $Z_R = Z_c$ and $Z_G = 0$,

$$\Gamma_R(s) = \frac{1-1}{1+1} = 0 \qquad \Gamma_S(s) = \frac{0-1}{0+1} = -1$$

The Laplace transform of the source voltage is $E_G(s) = E/s$. Then, from (13.2.10) and (13.2.11),

$$V(x, s) = \left(\frac{E}{s}\right)(1)(e^{-sx/v}) = \frac{Ee^{-sx/v}}{s}$$

$$I(x, s) = \frac{(E/Z_c)}{s} e^{-sx/v}$$

Taking the inverse Laplace transform,

$$v(x, t) = Eu_{-1}\left(t - \frac{x}{v}\right)$$

$$i(x, t) = \frac{E}{Z_c} u_{-1}\left(t - \frac{x}{v}\right)$$

b. At the center of the line, where $x = l/2$,

$$v\left(\frac{l}{2}, t\right) = Eu_{-1}\left(t - \frac{\tau}{2}\right) \qquad i\left(\frac{l}{2}, t\right) = \frac{E}{Z_c} u_{-1}\left(t - \frac{\tau}{2}\right)$$

(Continued)

At the receiving end, where $x = l$,

$$v(l, t) = Eu_{-1}(t - \tau) \qquad i(l, t) = \frac{E}{Z_c}u_{-1}(t - \tau)$$

These waves, plotted in Figure 13.4, can be explained as follows. At $t = 0$ the ideal step voltage of E volts, applied to the sending end, encounters Z_c, the characteristic impedance of the line. Therefore, a forward traveling step voltage wave of E volts is initiated at the sending end. Also, since the ratio of the forward traveling voltage to current is Z_c, a forward traveling step current wave of (E/Z_c) amperes is initiated. These waves travel in the positive x direction, arriving at the center of the line at $t = \tau/2$, and at the end of the line at $t = \tau$. The receiving-end load is *matched* to the line; that is, $Z_R = Z_c$, for a matched load, $\Gamma_R = 0$, and therefore no backward traveling waves are initiated. In steady-state, the line with matched load is energized at E volts with current E/Z_c amperes.

FIGURE 13.4 Voltage and current waveforms for Example 13.1

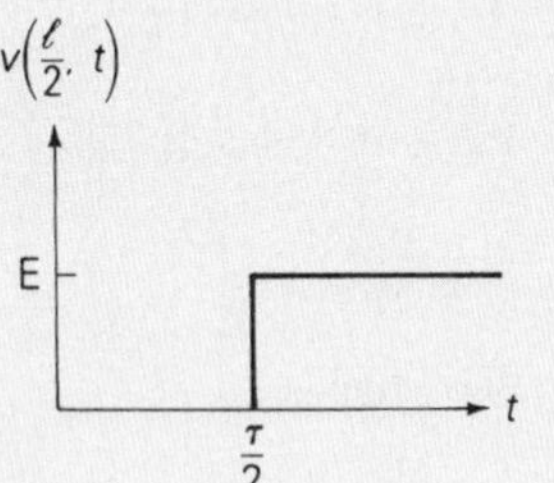

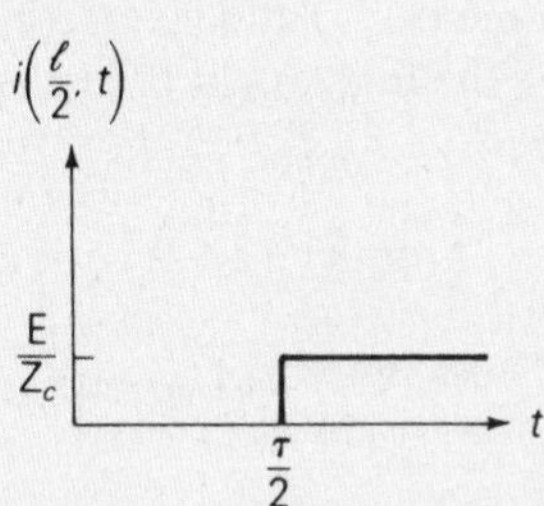

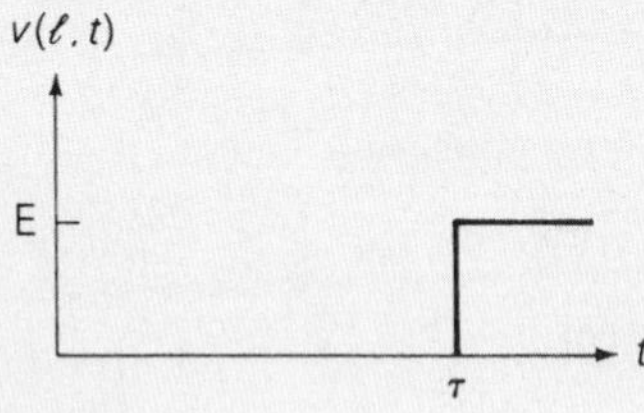

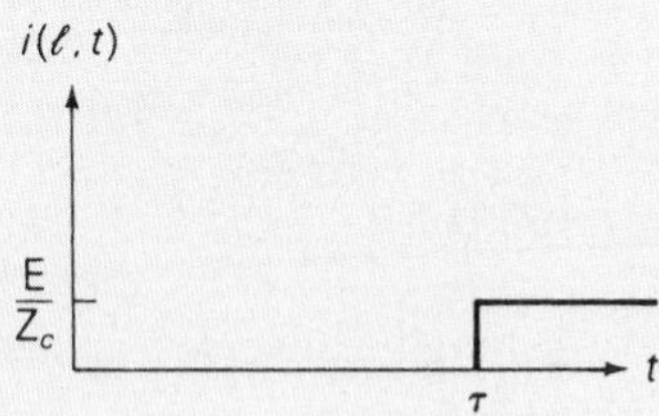

EXAMPLE 13.2

Single-phase lossless-line transients: step-voltage source matched at sending end, open receiving end

The receiving end is open. The source voltage at the sending end is a step $e_G(t) = Eu_{-1}(t)$, with $Z_G(s) = Z_c$. (a) Determine $v(x, t)$ and $i(x, t)$. (b) Plot the voltage and current versus time t at the center of the line.

SOLUTION

a. From (13.2.12),

$$\Gamma_R(s) = \lim_{Z_R\to\infty} \frac{\dfrac{Z_R}{Z_c} - 1}{\dfrac{Z_R}{Z_c} + 1} = 1 \qquad \Gamma_S(s) = \frac{1-1}{1+1} = 0$$

The Laplace transform of the source voltage is $E_G(s) = E/s$. Then, from (13.2.10) and (13.2.11),

$$V(x,s) = \frac{E}{s}\left(\frac{1}{2}\right)[e^{-sx/v} + e^{s[(x/v)-2\tau]}]$$

$$I(x,s) = \frac{E}{s}\left(\frac{1}{2Z_c}\right)[e^{-sx/v} - e^{s[(x/v)-2\tau]}]$$

Taking the inverse Laplace transform,

$$v(x,t) = \frac{E}{2}u_{-1}\left(t - \frac{x}{v}\right) + \frac{E}{2}u_{-1}\left(t + \frac{x}{v} - 2\tau\right)$$

$$i(x,t) = \frac{E}{2Z_c}u_{-1}\left(t - \frac{x}{v}\right) - \frac{E}{2Z_c}u_{-1}\left(t + \frac{x}{v} - 2\tau\right)$$

b. At the center of the line, where $x = l/2$,

$$v\left(\frac{l}{2}, t\right) = \frac{E}{2}u_{-1}\left(t - \frac{\tau}{2}\right) + \frac{E}{2}u_{-1}\left(t - \frac{3\tau}{2}\right)$$

$$i\left(\frac{l}{2}, t\right) = \frac{E}{2Z_c}u_{-1}\left(t - \frac{\tau}{2}\right) - \frac{E}{2Z_c}u_{-1}\left(t - \frac{3\tau}{2}\right)$$

These waves are plotted in Figure 13.5. At $t = 0$, the step voltage source of E volts encounters the source impedance $Z_G = Z_c$ in series with the characteristic impedance of the line, Z_c. Using voltage division, the sending-end voltage at $t = 0$ is $E/2$. Therefore, a forward traveling step voltage wave of $E/2$ volts and a forward traveling step current wave of $E/(2Z_c)$ amperes are initiated at the sending end. These waves arrive at the center of the line at $t = \tau/2$. Also, with $\Gamma_R = 1$, the backward traveling voltage wave equals the forward traveling voltage wave, and the backward traveling current wave is the negative of the forward traveling current wave. These backward traveling waves, which are initiated at the receiving end at $t = \tau$ when the forward traveling waves arrive there, arrive at the center of the line at $t = 3\tau/2$ and are superimposed on the forward traveling waves. No additional

(Continued)

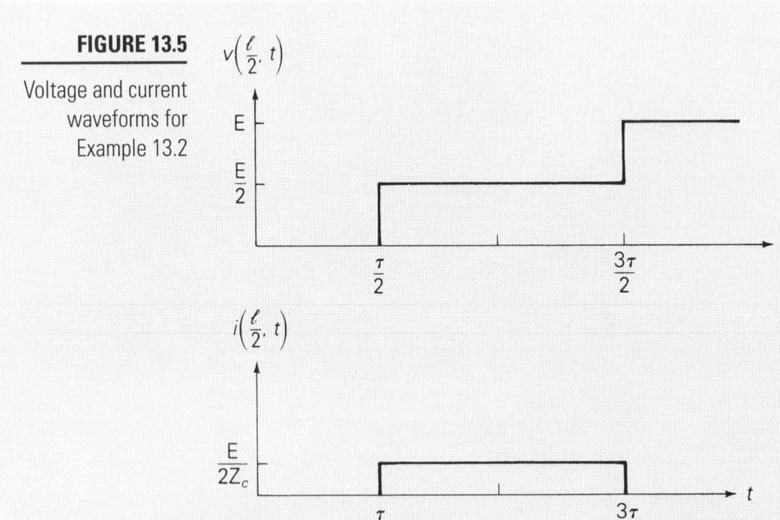

FIGURE 13.5 Voltage and current waveforms for Example 13.2

forward or backward traveling waves are initiated because the source impedance is matched to the line; that is, $\Gamma_S(s) = 0$. In steady-state, the line, which is open at the receiving end, is energized at E volts with zero current.

EXAMPLE 13.3

Single-phase lossless-line transients: step-voltage source matched at sending end, capacitive load at receiving end

The receiving end is terminated by a capacitor with C_R farads, which is initially unenergized. The source voltage at the sending end is a unit step $e_G(t) = Eu_{-1}(t)$, with $Z_G = Z_c$. Determine and plot $v(x, t)$ versus time t at the sending end of the line.

SOLUTION

From (13.2.12) with $Z_R = \dfrac{1}{sC_R}$ and $Z_G = Z_c$,

$$\Gamma_R(s) = \frac{\dfrac{1}{sC_R Z_c} - 1}{\dfrac{1}{sC_R Z_c} + 1} = \frac{-s + \dfrac{1}{Z_c C_R}}{s + \dfrac{1}{Z_c C_R}}$$

$$\Gamma_S(s) = \frac{1 - 1}{1 + 1} = 0$$

Then, from (13.2.10), with $E_G(s) = E/s$,

$$V(x,s) = \frac{E}{s}\left(\frac{1}{2}\right)\left[e^{-sx/v} + \left(\frac{-s + \frac{1}{Z_c C_R}}{s + \frac{1}{Z_c C_R}}\right)e^{s[(x/v)-2\tau]}\right]$$

$$= \frac{E}{2}\left[\frac{e^{-sx/v}}{s} + \frac{1}{s}\left(\frac{-s + \frac{1}{Z_c C_R}}{s + \frac{1}{Z_c C_R}}\right)e^{s[(x/v)-2\tau]}\right]$$

Using partial fraction expansion of the second term above,

$$V(x, s) = \frac{E}{2}\left[\frac{e^{-sx/v}}{s} + \left(\frac{1}{s} - \frac{2}{s + \frac{1}{Z_c C_R}}\right)e^{s[(x/v)-2\tau]}\right]$$

The inverse Laplace transform is

$$v(x, t) = \frac{E}{2}u_{-1}\left(t - \frac{x}{v}\right) + \frac{E}{2}[1 - 2e^{(-1/Z_c C_R)(t+x/v-2\tau)}]u_{-1}\left(t + \frac{x}{v} - 2\tau\right)$$

At the sending end, where $x = 0$,

$$v(0, t) = \frac{E}{2}u_{-1}(t) + \frac{E}{2}[1 - 2e^{(-1/Z_c C_R)(t-2\tau)}]u_{-1}(t - 2\tau)$$

$v(0, t)$ is plotted in Figure 13.6. As in Example 13.2, a forward traveling step voltage wave of $E/2$ volts is initiated at the sending end at $t = 0$. At $t = \tau$, when the forward traveling wave arrives at the receiving end, a backward traveling wave is initiated. The backward traveling voltage wave, an exponential with initial value $-E/2$, steady-state value $+E/2$, and time constant $Z_c C_R$, arrives at the sending end at $t = 2\tau$, where it is superimposed on the forward traveling wave. No additional waves are initiated, since the source impedance is matched to the line. In steady-state, the line and the capacitor at the receiving end are energized at E volts with zero current.

FIGURE 13.6

Voltage waveform for Example 13.3

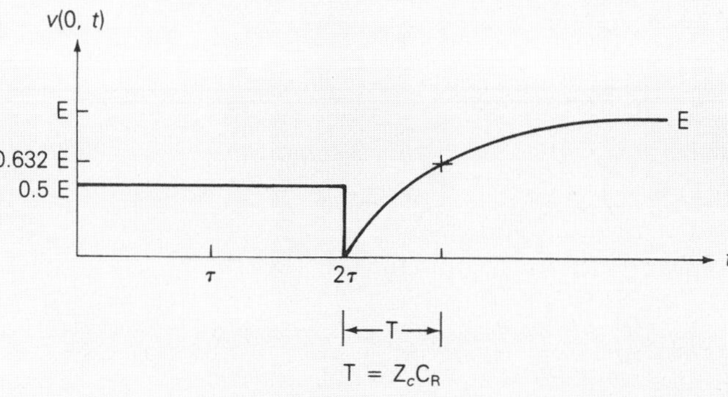

(*Continued*)

The capacitor at the receiving end can also be viewed as a short circuit at the instant $t = \tau$, when the forward traveling wave arrives at the receiving end. For a short circuit at the receiving end, $\Gamma_R = -1$, and therefore the backward traveling voltage wavefront is $-E/2$, the negative of the forward traveling wave. However, in steady-state the capacitor is an open circuit, for which $\Gamma_R = +1$, and the steady-state backward traveling voltage wave equals the forward traveling voltage wave.

EXAMPLE 13.4

Single-phase lossless-line transients: step-voltage source with unmatched source resistance at sending end, unmatched resistive load at receiving end

At the receiving end, $Z_R = Z_c/3$. At the sending end, $e_G(t) = Eu_{-1}(t)$ and $Z_G = 2Z_c$. Determine and plot the voltage versus time at the center of the line.

SOLUTION

From (13.2.12),

$$\Gamma_R = \frac{\frac{1}{3} - 1}{\frac{1}{3} + 1} = -\frac{1}{2} \qquad \Gamma_S = \frac{2 - 1}{2 + 1} = \frac{1}{3}$$

From (13.2.10), with $E_G(s) = E/s$,

$$V(x, s) = \frac{E}{s}\left(\frac{1}{3}\right)\frac{[e^{-sx/v} - \frac{1}{2}e^{s[(x/v)-2\tau]}]}{1 + \left(\frac{1}{6}e^{-2s\tau}\right)}$$

The preceding equation can be rewritten using the following geometric series:

$$\frac{1}{1 + y} = 1 - y + y^2 - y^3 + y^4 - \cdots$$

with $y = \frac{1}{6}e^{-2s\tau}$,

$$V(x, s) = \frac{E}{3s}\left[e^{-sx/v} - \frac{1}{2}e^{s[(x/v)-2\tau]}\right]$$

$$\times\left[1 - \frac{1}{6}e^{-2s\tau} + \frac{1}{36}e^{-4s\tau} - \frac{1}{216}e^{-6s\tau} + \cdots\right]$$

Multiplying the terms within the brackets,

$$V(x, s) = \frac{E}{3s}\left[e^{-sx/v} - \frac{1}{2}e^{s[(x/v)-2\tau]} - \frac{1}{6}e^{-s[(x/v)+2\tau]} + \frac{1}{12}e^{s[(x/v)-4\tau]}\right.$$

$$\left. + \frac{1}{36}e^{-s[(x/v)+4\tau]} - \frac{1}{72}e^{s[(x/v)-6\tau]} + \cdots\right]$$

Taking the inverse Laplace transform,

$$v(x, t) = \frac{E}{3}\left[u_{-1}\left(t - \frac{x}{v}\right) - \frac{1}{2}u_{-1}\left(t + \frac{x}{v} - 2\tau\right) - \frac{1}{6}u_{-1}\left(t - \frac{x}{v} - 2\tau\right) + \frac{1}{12}u_{-1}\left(t + \frac{x}{v} - 4\tau\right) + \frac{1}{36}u_{-1}\left(t - \frac{x}{v} - 4\tau\right) - \frac{1}{72}u_{-1}\left(t + \frac{x}{v} - 6\tau\right)\cdots\right]$$

At the center of the line, where $x = l/2$,

$$v\left(\frac{l}{2}, t\right) = \frac{E}{3}\left[u_{-1}\left(t - \frac{\tau}{2}\right) - \frac{1}{2}u_{-1}\left(t - \frac{3\tau}{2}\right) - \frac{1}{6}u_{-1}\left(t - \frac{5\tau}{2}\right) + \frac{1}{12}u_{-1}\left(t - \frac{7\tau}{2}\right) + \frac{1}{36}u_{-1}\left(t - \frac{9\tau}{2}\right) - \frac{1}{72}u_{-1}\left(t - \frac{11\tau}{2}\right)\cdots\right]$$

$v(l/2, t)$ is plotted in Figure 13.7(a). Since neither the source nor the load is matched to the line, the voltage at any point along the line consists of an infinite series of forward and backward traveling waves. At the center of the line, the first forward traveling wave arrives at $t = \tau/2$; then a backward traveling wave arrives at $3\tau/2$, another forward traveling wave arrives at $5\tau/2$, another backward traveling wave at $7\tau/2$, and so on.

The steady-state voltage can be evaluated from the final value theorem. That is,

$$v_{ss}(x) = \lim_{t\to\infty} v(x, t) = \lim_{s\to 0} sV(x, s)$$

$$= \lim_{s\to 0}\left\{s\left(\frac{E}{s}\right)\left(\frac{1}{3}\right)\frac{[e^{-sx/v} - \frac{1}{2}e^{s[(x/v)-2\tau]}]}{1 + \frac{1}{6}e^{-2s\tau}}\right\}$$

$$= E\left(\frac{1}{3}\right)\left(\frac{1 - \frac{1}{2}}{1 + \frac{1}{6}}\right) = \frac{E}{7}$$

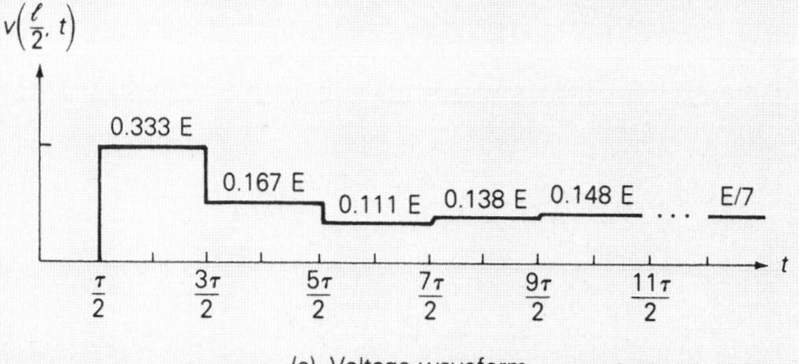

(a) Voltage waveform

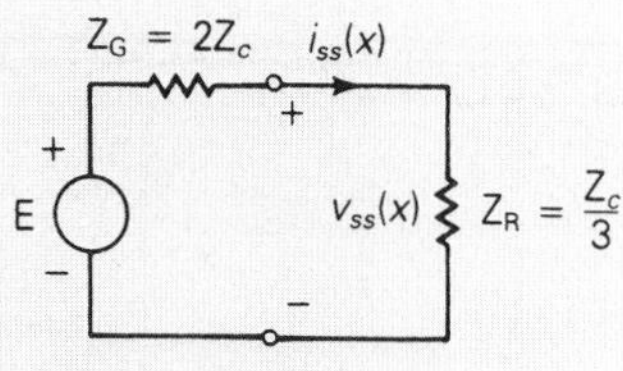

(b) Steady-state solution

FIGURE 13.7

Example 13.4

(Continued)

The steady-state solution can also be evaluated from the circuit in Figure 13.7(b). Since there is no steady-state voltage drop across the lossless line when a dc source is applied, the line can be eliminated, leaving only the source and load. The steady-state voltage is then, by voltage division,

$$v_{ss}(x) = E\left(\frac{Z_R}{Z_R + Z_G}\right) = E\left(\frac{\frac{1}{3}}{\frac{1}{3} + 2}\right) = \frac{E}{7}$$

13.3 BEWLEY LATTICE DIAGRAM

A lattice diagram developed by L. V. Bewley [2] conveniently organizes the reflections that occur during transmission-line transients. For the Bewley lattice diagram, shown in Figure 13.8, the vertical scale represents time and is scaled in units of τ, the transient time of the line. The horizontal scale represents line position x, and the diagonal lines represent traveling waves. Each reflection is determined by multiplying the incident wave arriving at an end by the reflection coefficient at that end.

FIGURE 13.8 Bewley lattice diagram for Example 13.5

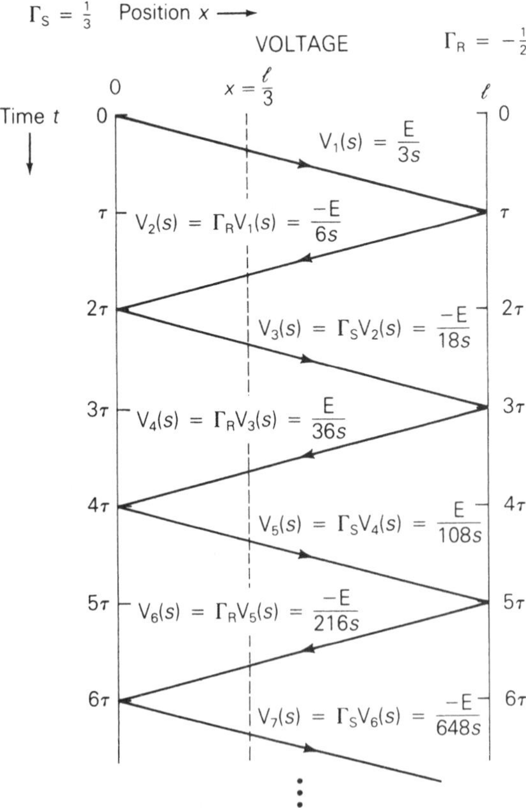

EXAMPLE 13.5

Lattice diagram: single-phase lossless line

For the line and terminations given in Example 13.4, draw the lattice diagram and plot $v(l/3, t)$ versus time t.

SOLUTION

The lattice diagram is shown in Figure 13.8. At $t = 0$, the source voltage encounters the source impedance and the line characteristic impedance, and the first forward traveling wave is determined by voltage division:

$$V_1(s) = E_G(s)\left[\frac{Z_c}{Z_c + Z_G}\right] = \frac{E}{s}\left[\frac{1}{1+2}\right] = \frac{E}{3s}$$

which is a step with magnitude (E/3) volts. The next traveling wave, a backward one, is $V_2(s) = \Gamma_R(s)V_1(s) = \left(-\frac{1}{2}\right)V_1(s) = -E/(6s)$, and the next wave, a forward one, is $V_3(s) = \Gamma_s(s)\,V_2(s) = \left(\frac{1}{3}\right)V_2(s) = -E/(18s)$. Subsequent waves are calculated in a similar manner.

The voltage at $x = l/3$ is determined by drawing a vertical line at $x = l/3$ on the lattice diagram, shown dashed in Figure 13.8. Starting at the top of the dashed line, where $t = 0$, and moving down, each voltage wave is added at the time it intersects the dashed line. The first wave V_1 arrives at $t = \tau/3$, the second v_2 arrives at $5\tau/3$, v_3 at $7\tau/3$, and so on. $v(l/3, t)$ is plotted in Figure 13.9.

FIGURE 13.9 Voltage waveform for Example 13.5

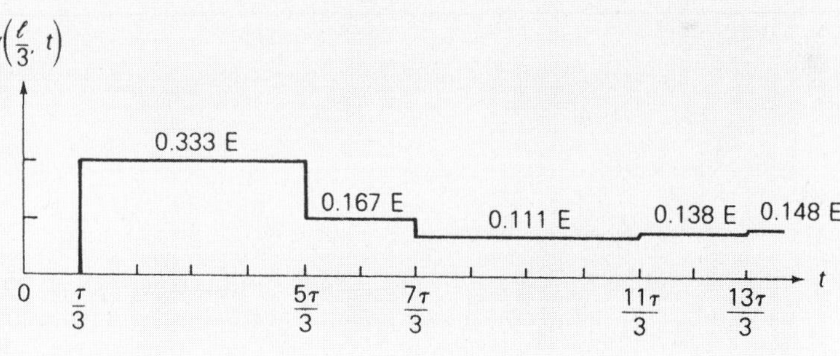

The voltage $v(x, t)$ at any point x and t on the diagram is determined by adding all the terms directly above that point.

Figure 13.10 shows a forward traveling voltage wave V_A^+ arriving at the junction of two lossless lines A and B with characteristic impedances Z_A and Z_B, respectively.

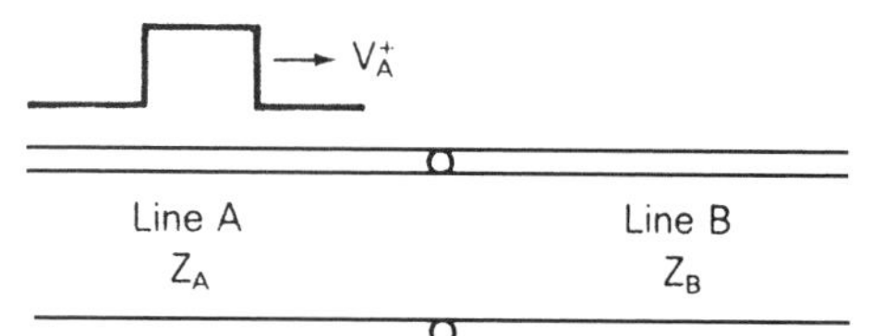

FIGURE 13.10 Junction of two single-phase lossless lines

This could be, for example, the junction of an overhead line and a cable. When $\mathbf{V}_A^+$ arrives at the junction, both a reflection $\mathbf{V}_A^-$ on line A and a refraction $\mathbf{V}_B^+$ on line B will occur. Writing a KVL and KCL equation at the junction,

$$V_A^+ + V_A^- = V_B^+ \tag{13.3.1}$$

$$I_A^+ + I_A^- = I_B^+ \tag{13.3.2}$$

Recall that $I_A^+ = V_A^+/Z_A$, $I_A^- = -V_A^-/Z_A$, and $I_B^+ = V_B^+/Z_B$. Using these relations in (13.3.2),

$$\frac{V_A^+}{Z_A} - \frac{V_A^-}{Z_A} = \frac{V_B^+}{Z_B} \tag{13.3.3}$$

Solving (13.3.1) and (13.3.3) for $\mathbf{V}_A^-$ and $\mathbf{V}_B^+$ in terms of $\mathbf{V}_A^+$ yields

$$V_A^- = \Gamma_{AA} V_A^+ \tag{13.3.4}$$

where

$$\Gamma_{AA} = \frac{\dfrac{Z_B}{Z_A} - 1}{\dfrac{Z_B}{Z_A} + 1} \tag{13.3.5}$$

and

$$V_B^+ = \Gamma_{BA} V_A \tag{13.3.6}$$

where

$$\Gamma_{BA} = \frac{2\left(\dfrac{Z_B}{Z_A}\right)}{\dfrac{Z_B}{Z_A} + 1} \tag{13.3.7}$$

Note that Γ_{AA}, given by (13.3.5), is similar to Γ_R, given by (13.2.12), except that Z_B replaces Z_R. Thus, for waves arriving at the junction from line A, the "load" at the receiving end of line A is the characteristic impedance of line B.

EXAMPLE 13.6

Lattice diagram: overhead line connected to a cable, single-phase lossless lines

As shown in Figure 13.10, a single-phase lossless overhead line with $Z_A = 400\ \Omega$, $v_A = 3 \times 10^8$ m/s, and $l_A = 30$ km is connected to a single-phase lossless cable with $Z_B = 100\ \Omega$, $v_B = 2 \times 10^8$ m/s, and $l_B = 20$ km. At the sending end of line A,

$e_g(t) = Eu_{-1}(t)$ and $Z_G = Z_A$. At the receiving end of line B, $Z_R = 2Z_B = 200\ \Omega$. Draw the lattice diagram for $0 \le t \le 0.6$ ms and plot the voltage at the junction versus time. The line and cable are initially unenergized.

SOLUTION

From (13.2.13),

$$\tau_A = \frac{30 \times 10^3}{3 \times 10^8} = 0.1 \times 10^{-3}\ \text{s} \qquad \tau_B = \frac{20 \times 10^3}{2 \times 10^8} = 0.1 \times 10^{-3}\ \text{s}$$

From (13.2.12), with $Z_G = Z_A$ and $Z_R = 2Z_B$,

$$\Gamma_S = \frac{1-1}{1+1} = 0 \qquad \Gamma_R = \frac{2-1}{2+1} = \frac{1}{3}$$

From (13.3.5) and (13.3.6), the reflection and refraction coefficients for waves arriving at the junction from line A are

$$\left.\Gamma_{AA} = \frac{\frac{100}{400} - 1}{\frac{100}{400} + 1} = \frac{-3}{5} \qquad \Gamma_{BA} = \frac{2\frac{100}{400}}{\frac{100}{400} + 1} = \frac{2}{5}\right\} \text{from line A}$$

Reversing A and B, the reflection and refraction coefficients for waves returning to the junction from line B are

$$\left.\Gamma_{BB} = \frac{\frac{400}{100} - 1}{\frac{400}{100} + 1} = \frac{3}{5} \qquad \Gamma_{AB} = \frac{2\frac{400}{100}}{\frac{400}{100} + 1} = \frac{8}{5}\right\} \text{from line B}$$

The lattice diagram is shown in Figure 13.11. Using voltage division, the first forward traveling voltage wave is

$$V_1(s) = E_G(s)\left(\frac{Z_A}{Z_A + Z_G}\right) = \frac{E}{s}\left(\frac{1}{2}\right) = \frac{E}{2s}$$

When v_1 arrives at the junction, a reflected wave v_2 and refracted wave v_3 are initiated. Using the reflection and refraction coefficients for line A,

$$V_2(s) = \Gamma_{AA}V_1(s) = \left(\frac{-3}{5}\right)\left(\frac{E}{2s}\right) = \frac{-3E}{10s}$$

$$V_3(s) = \Gamma_{BA}V_1(s) = \left(\frac{2}{5}\right)\left(\frac{E}{2s}\right) = \frac{E}{5s}$$

(Continued)

FIGURE 13.11

Lattice diagram for Example 13.6

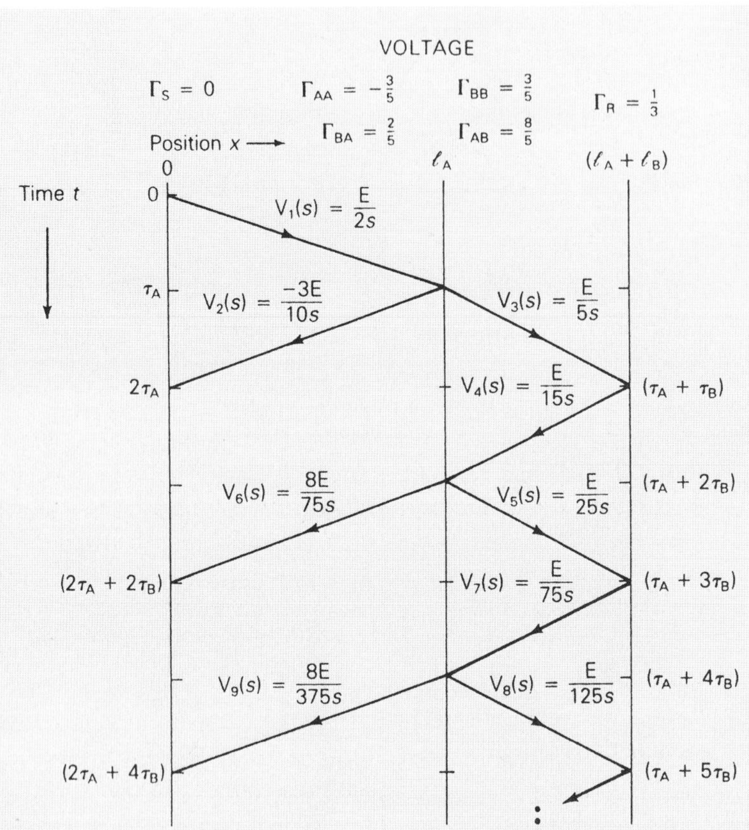

When v_2 arrives at the receiving end of line B, a reflected wave $V_4(s) = \Gamma_R V_3(s) = \frac{1}{3}(E/5s) = (E/15s)$ is initiated. When v_4 arrives at the junction, reflected wave v_5 and refracted wave v_6 are initiated. Using the reflection and refraction coefficients for line B,

$$V_5(s) = \Gamma_{BB} V_4(s) = \left(\frac{3}{5}\right)\left(\frac{E}{15s}\right) = \frac{E}{25s}$$

$$V_6(s) = \Gamma_{AB} V_4(s) = \left(\frac{8}{5}\right)\left(\frac{E}{15s}\right) = \frac{E}{75s}$$

Subsequent reflections and refractions are calculated in a similar manner.

The voltage at the junction is determined by starting at $x = l_A$ at the top of the lattice diagram, where $t = 0$. Then, moving down the lattice diagram, voltage waves either just to the left or just to the right of the junction are added when they occur. For example, looking just to the right of the junction at $x = l_A^+$, the voltage wave v_3, a step of magnitude E/5 volts occurs at $t = \tau_A$. Then at $t = (\tau_A + 2\tau_B)$,

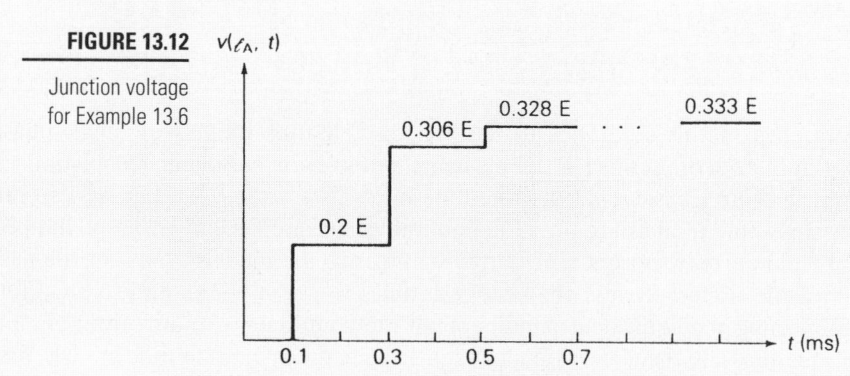

FIGURE 13.12 Junction voltage for Example 13.6

two waves v_4 and v_5, which are steps of magnitude E/15 and E/25, are added to v_3. $v(l_A, t)$ is plotted in Figure 13.12.

The steady-state voltage is determined by removing the lossless lines and calculating the steady-state voltage across the receiving-end load:

$$v_{ss}(x) = E\left(\frac{Z_R}{Z_R + Z_G}\right) = E\left(\frac{200}{200 + 400}\right) = \frac{E}{3}$$

The preceding analysis can be extended to the junction of more than two lossless lines, as shown in Figure 13.13. Writing a KVL and KCL equation at the junction for a voltage $\mathbf{V}_A^+$ arriving at the junction from line A,

$$V_A^+ + V_A^- = V_B^+ = V_C^+ = V_D^+ = \cdots \tag{13.3.8}$$

$$I_A^+ + I_A^- = I_B^+ + I_C^+ + I_D^+ + \cdots \tag{13.3.9}$$

Using $I_A^+ = V_A^+/Z_A$, $I_A^- = -V_A^-/Z_A$, $I_B^+ = V_B^+/Z_B$, and so on in (13.3.9),

$$\frac{V_A^+}{Z_A} - \frac{V_A^-}{Z_A} = \frac{V_B^+}{Z_B} + \frac{V_C^+}{Z_C} + \frac{V_D^+}{Z_D} + \cdots \tag{13.3.10}$$

Equations (13.3.8) and (13.3.10) can be solved for V_A^+, V_B^+, V_C^+, V_D^+, and so on in terms of $\mathbf{V}_A^+$. (See Problem 13.14.)

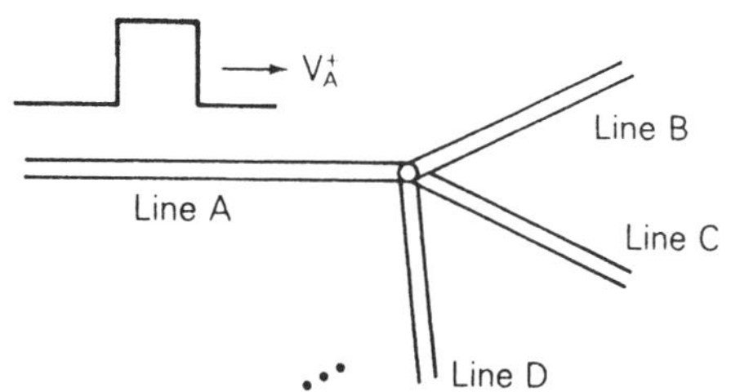

FIGURE 13.13 Junction of lossless lines A, B, C, D, and so on

13.4 DISCRETE-TIME MODELS OF SINGLE-PHASE LOSSLESS LINES AND LUMPED RLC ELEMENTS

Our objective in this section is to develop discrete-time models of single-phase lossless lines and of lumped RLC elements suitable for computer calculation of transmission-line transients at discrete-time intervals $t = \Delta t$, $2\Delta t$, $3\Delta t$, and so on. The discrete-time models are presented as equivalent circuits consisting of lumped resistors and current sources. The current sources in the models represent the past history of the circuit—that is, the history at times $t - \Delta t$, $t - 2\Delta t$, and so on. After interconnecting the equivalent circuits of all the components in any given circuit, nodal equations can then be written for each discrete time. Discrete-time models, first developed by L. Bergeron [3], are presented first.

SINGLE-PHASE LOSSLESS LINE

From the general solution of a single-phase lossless line, given by (13.1.12) and (13.1.18):

$$v(x, t) + Z_c i(x, t) = 2v^+\left(t - \frac{x}{v}\right) \tag{13.4.1}$$

$$v(x, t) - Z_c i(x, t) = 2v^-\left(t + \frac{x}{v}\right) \tag{13.4.2}$$

In (13.4.1), the left side $(v + Z_c i)$ remains constant when the argument $(t - x/v)$ is constant. Therefore, to a fictitious observer traveling at velocity v in the positive x direction along the line, $(v + Z_c i)$ remains constant. If τ is the transit time from terminal k to terminal m of the line, the value of $(v + Z_c i)$ observed at time $(t - \tau)$ at terminal k must equal the value at time t at terminal m. That is,

$$v_k(t - \tau) + Z_c i_k(t - \tau) = v_m(t) + Z_c i_m(t) \tag{13.4.3}$$

where k and m denote terminals k and m, as shown in Figure 13.14(a).

Similarly, $(v - Z_c i)$ in (13.4.2) remains constant when $(t + x/v)$ is constant. To a fictitious observer traveling at velocity v in the negative x direction, $(v - Z_c i)$ remains constant. Therefore, the value of $(v - Z_c i)$ at time $(t - \tau)$ at terminal m must equal the value at time t at terminal k. That is,

$$v_m(t - \tau) - Z_c i_m(t - \tau) = v_k(t) - Z_c i_k(t) \tag{13.4.4}$$

FIGURE 13.14 Single-phase two-wire lossless line

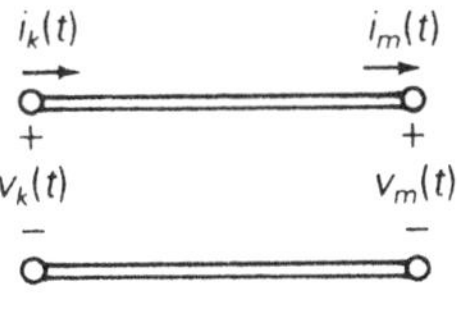

(a) Terminal variables

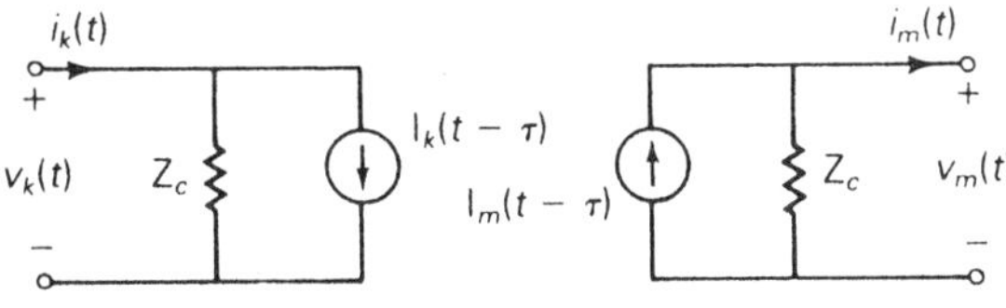

(b) Discrete-time equivalent circuit

Equation (13.4.3) is rewritten as

$$i_m(t) = \mathrm{I}_m(t - \tau) - \frac{1}{Z_c} v_m(t) \tag{13.4.5}$$

where

$$\mathrm{I}_m(t - \tau) = i_k(t - \tau) + \frac{1}{Z_c} v_k(t - \tau) \tag{13.4.6}$$

Similarly, (13.4.4) is rewritten as

$$i_k(t) = \mathrm{I}_k(t - \tau) + \frac{1}{Z_c} v_k(t) \tag{13.4.7}$$

where

$$\mathrm{I}_k(t - \tau) = i_m(t - \tau) - \frac{1}{Z_c} v_m(t - \tau) \tag{13.4.8}$$

Also, using (13.4.7) in (13.4.6),

$$\mathrm{I}_m(t - \tau) = \mathrm{I}_k(t - 2\tau) + \frac{2}{Z_c} v_k(t - \tau) \tag{13.4.9}$$

and using (13.4.5) in (13.4.8),

$$\mathrm{I}_k(t - \tau) = \mathrm{I}_m(t - 2\tau) - \frac{2}{Z_c} v_m(t - \tau) \tag{13.4.10}$$

Equations (13.4.5) and (13.4.7) are represented by the circuit shown in Figure 13.14(b). The current sources $\mathrm{I}_m(t - \tau)$ and $\mathrm{I}_k(t - \tau)$ shown in this figure, which are given by (13.4.9) and (13.4.10), represent the past history of the transmission line.

Note that in Figure 13.14(b) terminals k and m are not directly connected. The conditions at one terminal are "felt" indirectly at the other terminal after a delay of x seconds.

LUMPED INDUCTANCE

As shown in Figure 13.15 (a) for a constant lumped inductance L,

$$v(t) = \mathrm{L} \frac{di(t)}{dt} \tag{13.4.11}$$

Integrating this equation from time $(t - \Delta t)$ to t,

$$\int_{t-\Delta t}^{t} di(t) = \frac{1}{\mathrm{L}} \int_{t-\Delta t}^{t} v(t)dt \tag{13.4.12}$$

Using the trapezoidal rule of integration,

$$i(t) - i(t - \Delta t) = \left(\frac{1}{\mathrm{L}}\right)\left(\frac{\Delta t}{2}\right)[v(t) + v(t - \Delta t)]$$

Rearranging gives

$$i(t) = \frac{v(t)}{(2\mathrm{L}/\Delta t)} + \left[i(t - \Delta t) + \frac{v(t - \Delta t)}{(2\mathrm{L}/\Delta t)} \right]$$

FIGURE 13.15 Lumped inductance

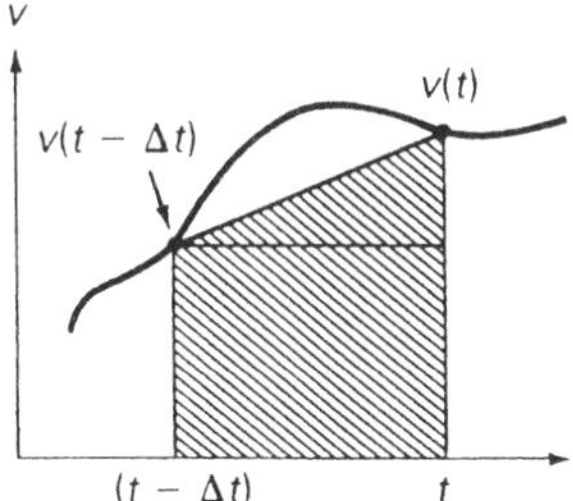

Trapezoidal Integration Rule

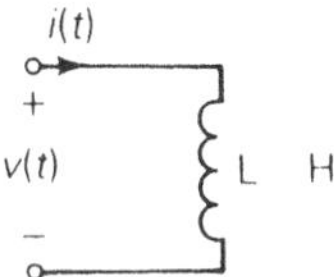

(a) Continuous time circuit

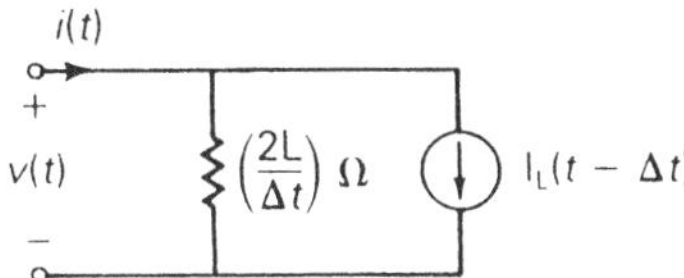

(b) Discrete-time circuit

or

$$i(t) = \frac{v(t)}{(2L/\Delta t)} + I_L(t - \Delta t) \tag{13.4.13}$$

where

$$I_L(t - \Delta t) = i(t - \Delta t) + \frac{v(t - \Delta t)}{(2L/\Delta t)} = I_L(t - 2\Delta t) + \frac{v(t - \Delta t)}{(L/\Delta t)} \tag{13.4.14}$$

Equations (13.4.13) and (13.4.14) are represented by the circuit shown in Figure 13.15(b). As shown, the inductor is replaced by a resistor with resistance $(2L/\Delta t)\ \Omega$. A current source $I_L(t - \Delta t)$ given by (13.4.14) is also included. $I_L(t - \Delta t)$ represents the past history of the inductor. Note that the trapezoidal rule introduces an error of the order $(\Delta t)^3$.

LUMPED CAPACITANCE

As shown in Figure 13.16 (a) for a constant lumped capacitance C,

$$i(t) = C\frac{dv(t)}{dt} \tag{13.4.15}$$

Integrating from time $(t - \Delta t)$ to t,

$$\int_{t-\Delta t}^{t} dv(t) = \frac{1}{C}\int_{t-\Delta t}^{t} i(t)\,dt \tag{13.4.16}$$

Using the trapezoidal rule of integration,

$$v(t) - v(t - \Delta t) = \frac{1}{C}\left(\frac{\Delta t}{2}\right)[i(t) + i(t - \Delta t)]$$

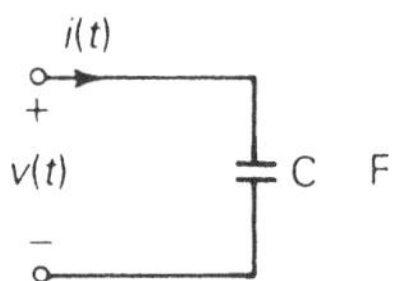

(a) Continuous time circuit

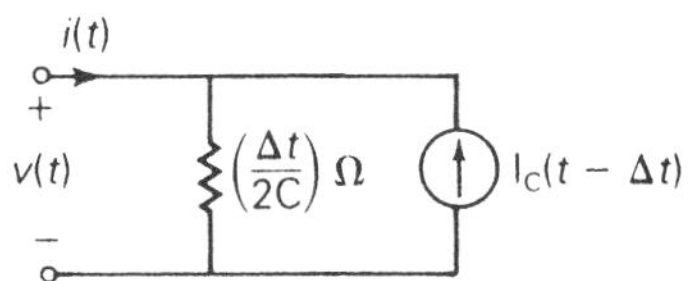

(b) Discrete-time circuit

FIGURE 13.16

Lumped capacitance

Rearranging gives

$$i(t) = \frac{v(t)}{(\Delta t/2C)} - I_C(t - \Delta t) \tag{13.4.17}$$

where

$$I_C(t - \Delta t) = i(t - \Delta t) + \frac{v(t - \Delta t)}{(\Delta t/2C)} = -I_C(t - 2\Delta t) + \frac{v(t - \Delta t)}{(\Delta t/4C)} \tag{13.4.18}$$

Equations (13.4.17) and (13.4.18) are represented by the circuit in Figure 13.16(b). The capacitor is replaced by a resistor with resistance $(\Delta t/2C)\ \Omega$. A current source $I_C(t - \Delta t)$, which represents the capacitor's past history, is also included.

i(t)
+
v(t)
−
R Ω

(a) Continuous time circuit

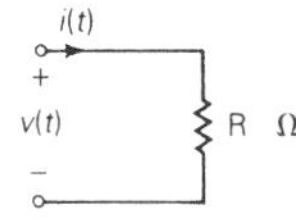

(b) Discrete-time circuit

FIGURE 13.17

Lumped resistance

LUMPED RESISTANCE

The discrete model of a constant lumped resistance R, as shown in Figure 13.17, is the same as the continuous model. That is,

$$v(t) = Ri(t) \tag{13.4.19}$$

NODAL EQUATIONS

A circuit consisting of single-phase lossless transmission lines and constant lumped RLC elements can be replaced by the equivalent circuits given in Figures 13.14(b), 13.15(b), 13.16(b), and 13.17(b). Then, writing nodal equations, the result is a set of linear algebraic equations that determine the bus voltages at each instant t.

EXAMPLE 13.7

Discrete-time equivalent circuit, single-phase lossless line transients, computer solution

For the circuit given in Example 13.3, replace the circuit elements by their discrete-time equivalent circuits and write the nodal equations that determine the sending-end and receiving-end voltages. Then, using a digital computer, compute the sending-end and receiving-end voltages for $0 \leqq t \leqq 9$ ms. For numerical calculations, assume $E = 100$ V, $Z_c = 400\ \Omega$, $C_R = 5\ \mu F$, $\tau = 1.0$ ms, and $\Delta t = 0.1$ ms.

(Continued)

SOLUTION

The discrete model is shown in Figure 13.18, where $v_k(t)$ represents the sending-end voltage $v(0, t)$ and $v_m(t)$ represents the receiving-end voltage $v(l, t)$. Also, the sending-end voltage source $e_G(t)$ in series with Z_G is converted to an equivalent current source in parallel with Z_G. Writing nodal equations for this circuit,

$$\left[\begin{array}{c|c} \left(\frac{1}{400}+\frac{1}{400}\right) & 0 \\ \hline 0 & \left(\frac{1}{400}+\frac{1}{10}\right) \end{array}\right]\begin{bmatrix} v_k(t) \\ v_m(t) \end{bmatrix} = \begin{bmatrix} \frac{1}{4} - I_k(t-1.0) \\ I_m(t-1.0) + I_C(t-0.1) \end{bmatrix}$$

Solving,

$$v_k(t) = 200[\tfrac{1}{4} - I_k(t-1.0)] \qquad \text{(a)}$$

$$v_m(t) = 9.75610[I_m(t-1.0) + I_C(t-0.1)] \qquad \text{(b)}$$

The current sources in these equations are, from (13.4.9), (13.4.10), and (13.4.18), with the argument $(t - \tau)$ replaced by t,

$$I_m(t) = I_k(t-1.0) + \frac{2}{400} v_k(t) \qquad \text{(c)}$$

$$I_k(t) = I_m(t-1.0) - \frac{2}{400} v_m(t) \qquad \text{(d)}$$

$$I_C(t) = -I_C(t-0.1) + \frac{1}{5} v_m(t) \qquad \text{(e)}$$

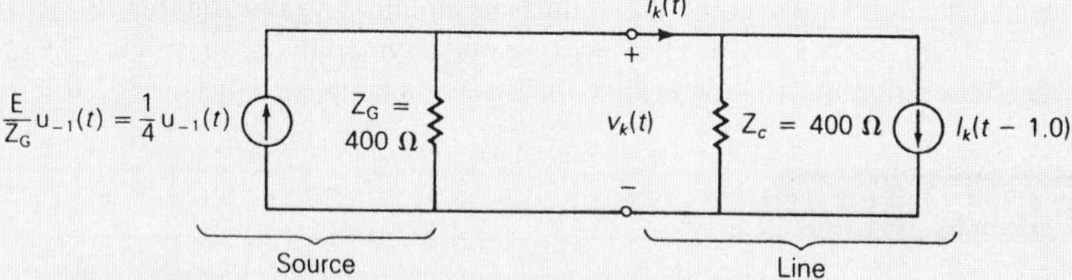

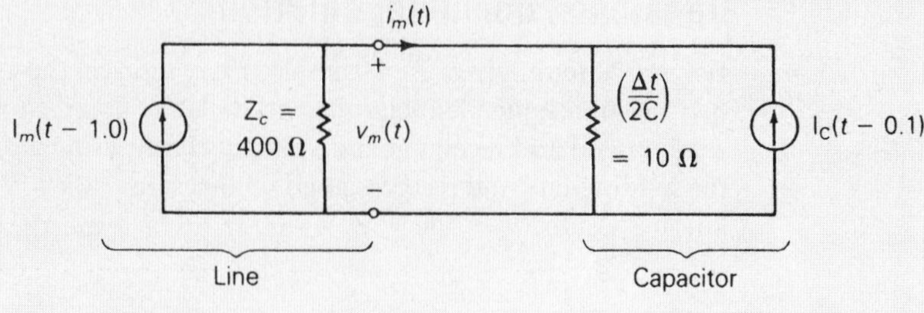

FIGURE 13.18 Discrete-time equivalent circuit for Example 13.7

FIGURE 13.19

Example 13.7

Output

Time ms	VK Volts	VM Volts
0.00	50.00	0.00
0.20	50.00	0.00
0.40	50.00	0.00
0.60	50.00	0.00
0.80	50.00	0.00
1.00	50.00	2.44
1.20	50.00	11.73
1.40	50.00	20.13
1.60	50.00	27.73
1.80	50.00	34.61
2.00	2.44	40.83
2.20	11.73	46.46
2.40	20.13	51.56
2.60	27.73	56.17
2.80	34.61	60.34
3.00	40.83	64.12
3.20	46.46	67.53
3.40	51.56	70.62
3.60	56.17	73.42
3.80	60.34	75.95
4.00	64.12	78.24
4.20	67.53	80.31
4.40	70.62	82.18
4.60	73.42	83.88
4.80	75.95	85.41
5.00	78.24	86.80
5.20	80.31	88.06
5.40	82.18	89.20
5.60	83.88	90.22
5.80	85.41	91.15
6.00	86.80	92.00
6.20	88.06	92.76
6.40	89.20	93.45
6.60	90.22	94.07
6.80	91.15	94.64
7.00	92.00	95.15
7.20	92.76	95.61
7.40	93.45	96.03
7.60	94.07	96.40
7.80	94.64	96.75
8.00	95.15	97.06
8.20	95.61	97.34
8.40	96.03	97.59
8.60	96.40	97.82
8.80	96.75	98.03
9.00	97.06	98.22

Computer Program Listing

```
10 REM EXAMPLE 12.7
20 LPRINT "TIME VK VM"
30 LPRINT "ms Volts Volts"
40 IC = 0
50 T = 0
60 KPRINT = 2
65 REM T IS TIME. KPRINT DETERMINES THE PRINTOUT INTERVAL.
70 REM LINES 1 10 to 210 COMPUTE EQS(a)−(e) FOR THE FIRST
80 REM TEN TIME STEPS (A TOTAL OF ONE ms) DURING WHICH
90 REM THE CURRENT SOURCES ON THE RIGHT HAND SIDE
100 REM OF THE EQUATIONS ARE ZERO. TEN VALUES OF
105 REM CURRENT SOURCES IK(J) AND IM(J) ARE STORED.
110 FOR J = 1 TO 10
120 VK = 200/4
130 VM = 9.7561*IC
140 IM(J) = (2/400)*VK
150 IK(J) = (−2/400)*VM
160 IC=−IC + (I/5)*VM
170 Z = (J−1)/KPRINT
180 M = INT (Z)
190 IF M = Z THEN LPRINT USING "*****"; T,VK,VM
200 T = T + .I
210 NEXT J
220 REM LINES 250 to 420 COMPUTE EQS(a)−(e) FOR TIME T
230 REM EQUAL TO AND GREATER THAN 1.0 ms. THE PAST TEN
240 REM VALUES OF IK(J) AND IM(J) ARE STORED
250 FOR J = 1 TO 10
260 REM LINE 270 IS EQ(a).
270 VK = 200*((1 /4)−IK (J))
280 REM LINE 290 IS EQ (b).
290 VM = 9.756 1*(IM (J) + IC)
300 REM LINE 310 IS EQ (e).
310 IC=−IC + (1/5)*VM
320 REM LINES 330−360 ARE EQS (c) and (d).
330 C1 = IK(J) + (2/400)*VK
340 C2 = IM(J)−(2/400)*VM
350 IM(J) = CI
360 IK(J) = C2
370 Z = (J−1)/KPRINT
380 M = INT(Z)
390 IF M = Z THEN LPRINT USING "*****"; T,VK,VM
400 T = T +.I
410 NEXT J
420 IF T < 9.0 THEN GOTO 250
430 STOP
```

(*Continued*)

FIGURE 13.19

(Continued)

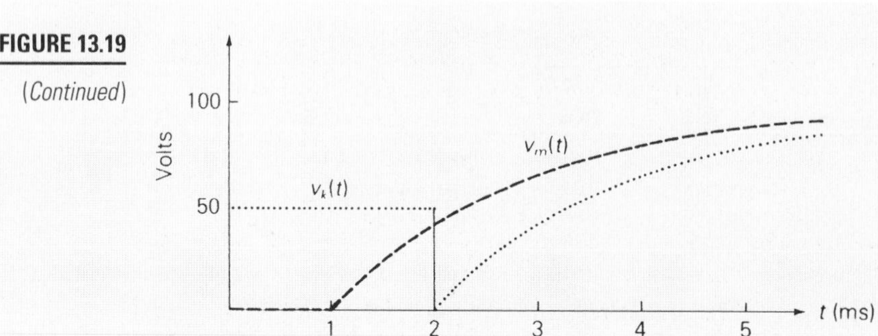

Equations (a) through (e) above are in a form suitable for digital computer solution. A scheme for iteratively computing v_k and v_m is as follows, starting at $t = 0$:

1. Compute $v_k(t)$ and $v_m(t)$ from equations (a) and (b).
2. Compute $I_m(t)$, $I_k(t)$, and $I_C(t)$ from equations (c), (d), and (e). Store $I_m(t)$ and $I_k(t)$.
3. Change t to $(t + \Delta t) = (t + 0.1)$ and return to (1) above.

Note that since the transmission line and capacitor are unenergized for time t less than zero, the current sources $I_m(\ )$, $I_k(\ )$, and $I_C(\ )$ are zero whenever their arguments ( ) are negative. Note also from equations (a) through (e) that it is necessary to store the past ten values of $I_m(\ )$ and $I_k(\ )$.

A program written in BASIC that executes the above scheme and the computational results are shown in Figure 13.19. The plotted sending-end voltage $v_k(t)$ can be compared with the results of Example 13.3.

Example 13.7 can be generalized to compute bus voltages at discrete-time intervals for an arbitrary number of buses, single-phase lossless lines, and lumped RLC elements. When current sources instead of voltage sources are employed, the unknowns are all bus voltages, for which nodal equations $\mathbf{YV} = \mathbf{I}$ can be written at each discrete-time instant. Also, the dependent current sources in $\mathbf{I}$ are written in terms of bus voltages and current sources at prior times. For computational convenience, the time interval Δt can be chosen constant so that the bus admittance matrix $\mathbf{Y}$ is a constant real symmetric matrix as long as the RLC elements are constant.

13.5 LOSSY LINES

Transmission-line series resistance or shunt conductance causes the following:

1. Attenuation
2. Distortion
3. Power losses

A brief discussion of these effects follows.

ATTENUATION

When constant series resistance R Ω/m and shunt conductance G S/m are included in the circuit of Figure 13.1 for a single-phase two-wire line, (13.1.3) and (13.1.4) become

$$\frac{\partial v(x,t)}{\partial x} = -\mathrm{R}\, i(x,t) - \mathrm{L}\frac{\partial i(x,t)}{\partial t} \tag{13.5.1}$$

$$\frac{\partial i(x,t)}{\partial x} = -\mathrm{G}v(x,t) - \mathrm{C}\frac{\partial v(x,t)}{\partial t} \tag{13.5.2}$$

Taking the Laplace transform of these, equations analogous to (13.1.7) and (13.1.8) are

$$\frac{d^2\mathrm{V}(x,s)}{dx^2} - \gamma^2(s)\,\mathrm{V}(x,s) = 0 \tag{13.5.3}$$

$$\frac{d^2\mathrm{I}(x,s)}{dx^2} - \gamma^2(s)\mathrm{I}(x,s) = 0 \tag{13.5.4}$$

where

$$\gamma(s) = \sqrt{(\mathrm{R} + s\mathrm{L})(\mathrm{G} + s\mathrm{C})} \tag{13.5.5}$$

The solution to these equations is

$$\mathrm{V}(x,s) = \mathrm{V}^+(s)e^{-\gamma(s)x} + \mathrm{V}^-(s)e^{+\gamma(s)x} \tag{13.5.6}$$

$$\mathrm{I}(x,s) = \mathrm{I}^+(s)e^{-\gamma(s)x} + \mathrm{I}^-(s)e^{+\gamma(s)x} \tag{13.5.7}$$

In general, it is impossible to obtain a closed form expression for $v(x,t)$ and $i(x,t)$, which are the inverse Laplace transforms of these equations. However, for the special case of a *distortionless* line, which has the property R/L = G/C, the inverse Laplace transform can be obtained as follows. Rewrite (13.5.5) as

$$\gamma(s) = \sqrt{\mathrm{LC}[(s+\delta)^2 - \sigma^2]} \tag{13.5.8}$$

where

$$\delta = \frac{1}{2}\left(\frac{\mathrm{R}}{\mathrm{L}} + \frac{\mathrm{G}}{\mathrm{C}}\right) \tag{13.5.9}$$

$$\sigma = \frac{1}{2}\left(\frac{\mathrm{R}}{\mathrm{L}} - \frac{\mathrm{G}}{\mathrm{C}}\right) \tag{13.5.10}$$

For a distortionless line, $\sigma = 0$, $\delta = \mathrm{R}/\mathrm{L}$, and (13.5.6) and (13.5.7) become

$$\mathrm{V}(x,s) = \mathrm{V}^+(s)e^{-\sqrt{\mathrm{LC}}[s+(\mathrm{R}/\mathrm{L})]x} + \mathrm{V}^-(s)e^{+\sqrt{\mathrm{LC}}[s+(\mathrm{R}/\mathrm{L})]x} \tag{13.5.11}$$

$$\mathrm{I}(x,s) = \mathrm{I}^+(s)e^{-\sqrt{\mathrm{LC}}[s+(\mathrm{R}/\mathrm{L})]x} + \mathrm{I}^-(s)e^{+\sqrt{\mathrm{LC}}[s+(\mathrm{R}/\mathrm{L})]x} \tag{13.5.12}$$

Using $\mathrm{v}=1/\sqrt{\mathrm{LC}}$ and $\sqrt{\mathrm{LC}}(\mathrm{R}/\mathrm{L}) = \sqrt{\mathrm{RG}} = \alpha$ for the distortionless line, the inverse transform of these equations is

$$v(x,t) = e^{-\alpha x}v^+\left(t - \frac{x}{\mathrm{v}}\right) + e^{+\alpha x}v^-\left(t + \frac{x}{\mathrm{v}}\right) \tag{13.5.13}$$

$$i(x, t) = e^{-\alpha x} i^{+}\left(t - \frac{x}{v}\right) + e^{\alpha x} i^{-}\left(t + \frac{x}{v}\right) \tag{13.5.14}$$

These voltage and current waves consist of forward and backward traveling waves similar to (13.1.12) and (13.1.13) for a lossless line. However, for the lossy distortionless line, the waves are attenuated versus x due to the $e^{\pm\alpha x}$ terms. Note that the attenuation term $\alpha = \sqrt{\text{RG}}$ is constant. Also, the attenuated waves travel at constant velocity $v = 1/\sqrt{\text{LC}}$ Therefore, waves traveling along the distortionless line do not change their shape; only their magnitudes are attenuated.

DISTORTION

For sinusoidal steady-state waves, the propagation constant $\gamma(j\omega)$ is, from (13.5.5), with $s = j\omega$

$$\gamma(j\omega) = \sqrt{(\text{R} + j\omega\text{L})(\text{G} + j\omega\text{C})} = \alpha + j\beta \tag{13.5.15}$$

For a lossless line, R = G = 0; therefore, $\alpha = 0$, $\beta = \omega\sqrt{\text{LC}}$, and the phase velocity $v = \omega/\beta = 1/\sqrt{\text{LC}}$ is constant. Thus, sinusoidal waves of all frequencies travel at constant velocity v without attenuation along a lossless line.

For a distortionless line (R/L) = (G/C), and $\gamma(j\omega)$ can be rewritten, using (13.5.8) through (13.5.10), as

$$\begin{aligned}\gamma(j\omega) &= \sqrt{\text{LC}\left(j\omega + \frac{\text{R}}{\text{L}}\right)^2} = \sqrt{\text{LC}}\left(j\omega + \frac{\text{R}}{\text{L}}\right)\\ &= \sqrt{\text{RG}} + j\frac{\omega}{v} = \alpha + j\beta\end{aligned} \tag{13.5.16}$$

Since $\alpha = \sqrt{\text{RG}}$ and $v = 1/\sqrt{\text{LC}}$ are constant, sinusoidal waves of all frequencies travel along the distortionless line at constant velocity with constant attenuation—that is, without distortion.

It can also be shown that for frequencies above 1 MHz, practical transmission lines with typical constants R, L, G, and C tend to be distortionless. Above 1 MHz, α and β can be approximated by

$$\alpha \simeq \frac{\text{R}}{2}\sqrt{\frac{\text{C}}{\text{L}}} + \frac{\text{G}}{2}\sqrt{\frac{\text{L}}{\text{C}}} \tag{13.5.17}$$

$$\beta \simeq \omega\sqrt{\text{LC}} = \frac{\omega}{v} \tag{13.5.18}$$

Therefore, sinusoidal waves with frequencies above 1 MHz travel along a practical line undistorted at constant velocity $v = 1/\sqrt{\text{LC}}$ with attenuation α given by (13.5.17).

At frequencies below 1 MHz, these approximations do not hold, and lines are generally not distortionless. For typical transmission and distribution lines, (R/L) is much greater than (G/C) by a factor of 1000 or so. Therefore, the condition (R/L) = (G/C) for a distortionless line does not hold.

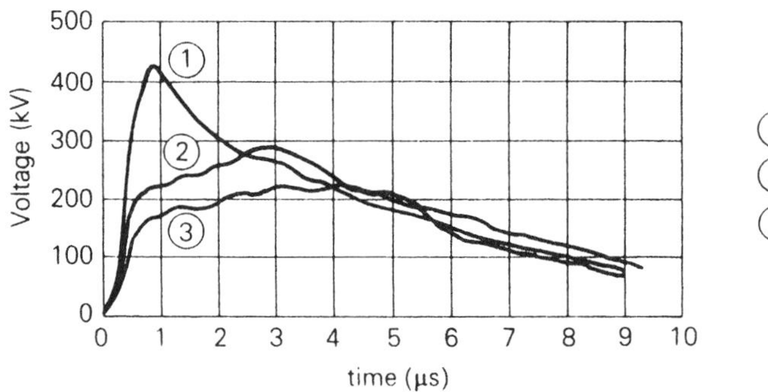

FIGURE 13.20

Distortion and attenuation of surges on a 132-kV overhead line [4] (Lacey, H.M., "The lightning protection of high-voltage overhead transmission and distribution systems," *Proceedings of the IEE - Part II: Power Engineering*, vol. 96, no. 50, pp. 287–299, April 1949. Reprinted by permission of Institution of Engineering and Technology.)

Figure 13.20 shows the effect of distortion and attenuation of voltage surges based on experiments on a 132-kV overhead transmission line [4]. The shapes of the surges at three points along the line are shown. Note how distortion reduces the front of the wave and builds up the tail as it travels along the line.

POWER LOSSES

Power losses are associated with series resistance R and shunt conductance G. When a current I flows along a line, I^2R losses occur, and when a voltage V appears across the conductors, V^2G losses occur. V^2G losses are primarily due to insulator leakage and corona for overhead lines, and to dielectric losses for cables. For practical lines operating at rated voltage and rated current, I^2R losses are much greater than V^2G losses.

As discussed above, the analysis of transients on single-phase two-wire lossy lines with constant parameters R, L, G, and C is complicated. The analysis becomes more complicated when skin effect is included, which means that R is not constant but frequency-dependent. Additional complications arise for a single-phase line consisting of one conductor with earth return, where Carson [5] has shown that both series resistance and inductance are frequency-dependent.

In view of these complications, the solution of transients on lossy lines is best handled via digital computation techniques. A single-phase line of length l can be approximated by a lossless line with half the total resistance $(Rl/2)$ Ω lumped in series with the line at both ends. For improved accuracy, the line can be divided into various line sections, and each section can be approximated by a lossless line section, with a series resistance lumped at both ends. Simulations have shown that accuracy does not significantly improve with more than two line sections.

Discrete-time equivalent circuits of a single-phase lossless line, shown in Figure 13.14, together with a constant lumped resistance, shown in Figure 13.17, can be used to approximate a lossy line section, as shown in Figure 13.21. Also, digital techniques for modeling frequency-dependent line parameters [6, 7] are available but are beyond the scope of this book.

FIGURE 13.21 Approximate model of a lossy line segment

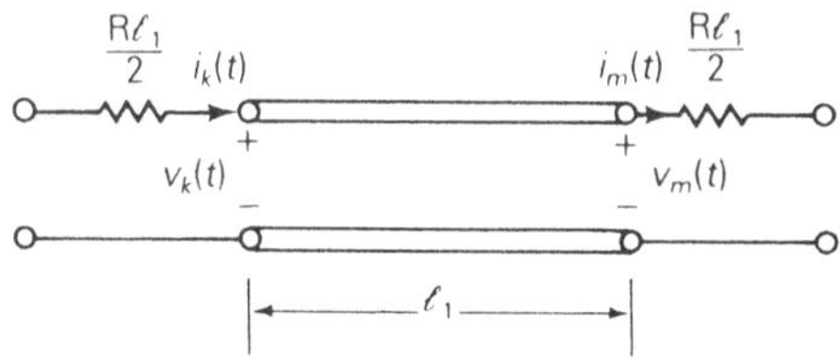

(a) Lossless line segment of length ℓ_1 with lumped line resistance

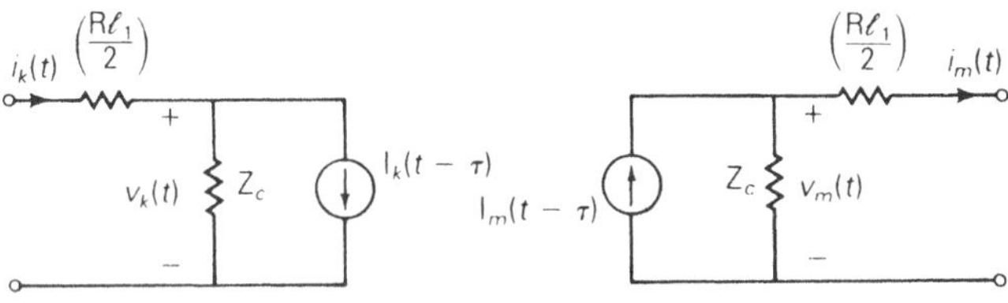

(b) Discrete-time model

13.6 MULTICONDUCTOR LINES

Transients on single-phase two-wire lines have been considered until now. For a transmission line with n conductors above a ground plane, waves travel in n "modes," where each mode has its own wave velocity and its own surge impedance. In this section "model analysis" for a relatively simple three-phase line [8] is illustrated.

Given a three-phase, lossless, completely transposed line consisting of three conductors above a perfectly conducting ground plane, the transmission-line equations are

$$\frac{d\mathbf{V}(x, s)}{dx} = -s\mathbf{L}\mathbf{I}(x, s) \tag{13.6.1}$$

$$\frac{d\mathbf{I}(x, s)}{dx} = -s\mathbf{C}\mathbf{V}(x, s) \tag{13.6.2}$$

where

$$\mathbf{V}(x, s) = \begin{bmatrix} \mathrm{V}_{ag}(x, s) \\ \mathrm{V}_{bg}(x, s) \\ \mathrm{V}_{cg}(x, s) \end{bmatrix} \qquad \mathbf{I}(x, s) = \begin{bmatrix} \mathrm{I}_{a}(x, s) \\ \mathrm{I}_{b}(x, s) \\ \mathrm{I}_{c}(x, s) \end{bmatrix} \tag{13.6.3}$$

Equations (13.6.1) and (13.6.2) are identical to (13.1.5) and (13.1.6) except that scalar quantities are replaced by vector quantities. $\mathbf{V}(x, s)$ is the vector of line-to-ground voltages and $\mathbf{I}(x, s)$ is the vector of line currents. For a completely transposed line, the 3×3 inductance matrix $\mathbf{L}$ and capacitance matrix $\mathbf{C}$ are given by

$$\mathbf{L} = \begin{bmatrix} \mathrm{L}_s & \mathrm{L}_m & \mathrm{L}_m \\ \mathrm{L}_m & \mathrm{L}_s & \mathrm{L}_m \\ \mathrm{L}_m & \mathrm{L}_m & \mathrm{L}_s \end{bmatrix} \quad \mathrm{H/m} \tag{13.6.4}$$

$$\mathbf{C} = \begin{bmatrix} C_s & C_m & C_m \\ C_m & C_s & C_m \\ C_m & C_m & C_s \end{bmatrix} \quad \text{F/m} \tag{13.6.5}$$

For any given line configuration, **L** and **C** can be computed from the equations given in Sections 4.7 and 4.11. Note that L_s, L_m, and C_s are positive, whereas C_m is negative.

Now transform the phase quantities to modal quantities. First, define

$$\begin{bmatrix} V_{ag}(x, s) \\ V_{bg}(x, s) \\ V_{cg}(x, s) \end{bmatrix} = \mathbf{T}_V \begin{bmatrix} V^0(x, s) \\ V^+(x, s) \\ V^-(x, s) \end{bmatrix} \tag{13.6.6}$$

$$\begin{bmatrix} I_a(x, s) \\ I_b(x, s) \\ I_c(x, s) \end{bmatrix} = \mathbf{T}_I \begin{bmatrix} I^0(x, s) \\ I^+(x, s) \\ I^-(x, s) \end{bmatrix} \tag{13.6.7}$$

$V^0(x, s)$, $V^+(x, s)$, and $V^-(x, s)$ are denoted *zero-mode*, *positive-mode*, and *negative-mode* voltages, respectively. Similarly, $I^0(x, s)$, $I^+(x, s)$, and $I^-(x, s)$ are *zero-*, *positive-*, and *negative-mode* currents. $\mathbf{T}_V$ and $\mathbf{T}_I$ are 3×3 constant transformation matrices, soon to be specified. Denoting $\mathbf{V}_m(x, s)$ and $\mathbf{I}_m(x, s)$ as the modal voltage and modal current vectors,

$$\mathbf{V}(x, s) = \mathbf{T}_V \mathbf{V}_m(x, s) \tag{13.6.8}$$

$$\mathbf{I}(x, s) = \mathbf{T}_I \mathbf{I}_m(x, s) \tag{13.6.9}$$

Using (13.6.8) and (13.6.9) in (13.6.1),

$$\mathbf{T}_V \frac{d\mathbf{V}_m(x, s)}{dx} = -s\mathbf{L}\mathbf{T}_I\mathbf{I}_m(x, s)$$

or

$$\frac{d\mathbf{V}_m(x, s)}{dx} = -s(\mathbf{T}_V^{-1}\mathbf{L}\mathbf{T}_I)\mathbf{I}_m(x, s) \tag{13.6.10}$$

Similarly, using (13.6.8) and (13.6.9) in (13.6.2),

$$\frac{d\mathbf{I}_m(x, s)}{dx} = -s(\mathbf{T}_I^{-1}\mathbf{C}\mathbf{T}_V)\mathbf{V}_m(x, s) \tag{13.6.11}$$

The objective of the modal transformation is to diagonalize the matrix products within the parentheses of (13.6.10) and (13.6.11), thereby decoupling these vector equations. For a three-phase completely transposed line, T_V and T_I are given by

$$\mathbf{T}_V = \mathbf{T}_I = \begin{bmatrix} 1 & 1 & 1 \\ 1 & -2 & 1 \\ 1 & 1 & -2 \end{bmatrix} \tag{13.6.12}$$

Also, the inverse transformation matrices are

$$\mathbf{T}_V^{-1} = \mathbf{T}_I^{-1} = \frac{1}{3}\begin{bmatrix} 1 & 1 & 1 \\ 1 & -1 & 0 \\ 1 & 0 & -1 \end{bmatrix} \tag{13.6.13}$$

Substituting (13.6.12), (13.6.13), (13.6.4), and (13.6.5) into (13.6.10) and (13.6.11) yields

$$\frac{d}{dx}\begin{bmatrix} V^0(x,s) \\ V^+(x,s) \\ V^-(x,s) \end{bmatrix} = \begin{bmatrix} -s(L_s + 2L_m) & 0 & 0 \\ 0 & -s(L_s - L_m) & 0 \\ 0 & 0 & -s(L_s - L_m) \end{bmatrix} \times \begin{bmatrix} I^0(x,s) \\ I^+(x,s) \\ I^-(x,s) \end{bmatrix} \tag{13.6.14}$$

$$\frac{d}{dx}\begin{bmatrix} I^0(x,s) \\ I^+(x,s) \\ I^-(x,s) \end{bmatrix} = \begin{bmatrix} -s(C_s + 2C_m) & 0 & 0 \\ 0 & -s(C_s - C_m) & 0 \\ 0 & 0 & -s(C_s - C_m) \end{bmatrix} \times \begin{bmatrix} V^0(x,s) \\ V^+(x,s) \\ V^-(x,s) \end{bmatrix} \tag{13.6.15}$$

From (13.6.14) and (13.6.15), the zero-mode equations are

$$\frac{dV^0(x,s)}{dx} = -s(L_s + 2L_m)I^0(x,s) \tag{13.6.16}$$

$$\frac{dI^0(x,s)}{dx} = -s(C_s + 2C_m)V^0(x,s) \tag{13.6.17}$$

These equations are identical in form to those of a two-wire lossless line, (13.1.5) and (13.1.6). By analogy, the zero-mode waves travel at velocity

$$v^0 = \frac{1}{\sqrt{(L_s + 2L_m)(C_s + 2C_m)}} \quad \text{m/s} \tag{13.6.18}$$

and the zero-mode surge impedance is

$$Z_c^0 = \sqrt{\frac{L_s + 2L_m}{C_s + 2C_m}} \quad \Omega \tag{13.6.19}$$

Similarly, the positive- and negative-mode velocities and surge impedances are

$$v^+ = v^- = \frac{1}{\sqrt{(L_s - L_m)(C_s - C_m)}} \quad \text{m/s} \tag{13.6.20}$$

$$Z_c^+ = Z_c^- = \sqrt{\frac{L_s - L_m}{C_s - C_m}} \quad \Omega \tag{13.6.21}$$

These equations can be extended to more than three conductors—for example, to a three-phase line with shield wires or to a double-circuit three-phase line. Although the details are more complicated, the modal transformation is straightforward. There is one mode for each conductor, and each mode has its own wave velocity and its own surge impedance.

The solution of transients on multiconductor lines is best handled via digital computer methods, and such programs are available [9, 10]. Digital techniques are also available to model the following effects:

1. Nonlinear and time-varying RLC elements [8]
2. Lossy lines with frequency-dependent line parameters [6, 7, 12]

13.7 POWER SYSTEM OVERVOLTAGES

Overvoltages encountered by power system equipment are of three types:

1. Lightning surges
2. Switching surges
3. Power frequency (50 or 60 Hz) overvoltages

LIGHTNING

Cloud-to-ground (CG) lightning is the greatest single cause of overhead transmission and distribution line outages. Data obtained over a 14-year period from electric utility companies in the United States and Canada and covering 40,000 kilometers of transmission show that CG lightning accounted for about 26% of outages on 230-kV circuits and about 65% of outages on 345-kV circuits [13]. A similar study in Britain, also over a 14-year period, covering 50,000 faults on distribution lines, shows that CG lightning accounted for 47% of outages on circuits up to and including 33 kV [14].

The electrical phenomena that occur within clouds leading to a lightning strike are complex and not totally understood. Several theories [15, 16, 17] generally agree, however, that charge separation occurs within clouds. Wilson [15] postulates that falling raindrops attract negative charges and therefore leave behind masses of positively charged air. The falling raindrops bring the negative charge to the bottom of the cloud, and upward air drafts carry the positively charged air and ice crystals to the top of the cloud, as shown in Figure 13.22. Negative charges at the bottom of the cloud induce a positively charged region, or "shadow," on the earth directly below the cloud. The electric field lines shown in Figure 13.22 originate from the positive charges and terminate at the negative charges.

When voltage gradients reach the breakdown strength of the humid air within the cloud, typically 5 to 15 kV/cm, an ionized path or downward *leader* moves from the cloud toward the earth. The leader progresses somewhat randomly along an

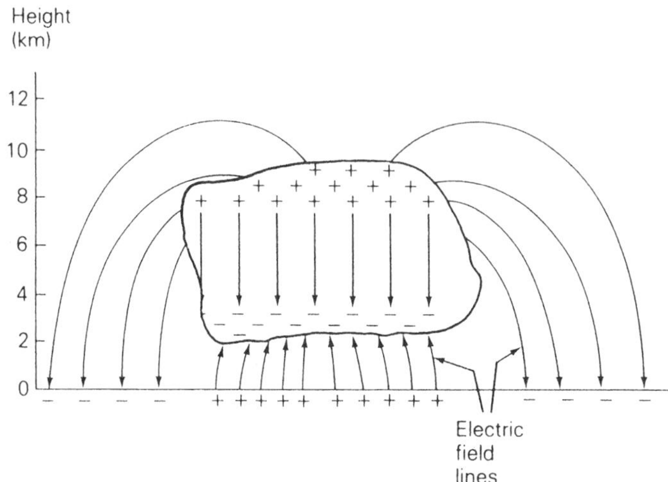

FIGURE 13.22 Postulation of charge separation within clouds [16] (Based on G. B. Simpson and F. J. Scrase, "The Distribution of Electricity in Thunderclouds", *Proceedings of the Royal Society A: Mathematical, Physical and Engineering Sciences*, 161 (1937), p. 309)

irregular path, in steps. These leader steps, about 50 m long, move at a velocity of about 10^5 m/s. As a result of the opposite charge distribution under the cloud, another upward leader may rise to meet the downward leader. When the two leaders meet, a lightning discharge occurs, which neutralizes the charges.

The current involved in a CG lightning stroke typically rises to a peak value within 1 to 10 μs, and then diminishes to one-half the peak within 20 to 100 μs. The distribution of peak currents is shown in Figure 13.23 [20]. This curve represents the percentage of strokes that exceed a given peak current. For example, 50% of all strokes have a peak current greater than 45 kA. In extreme cases, the peak current can exceed 200 kA. Also, test results indicate that approximately 90% of all strokes are negative.

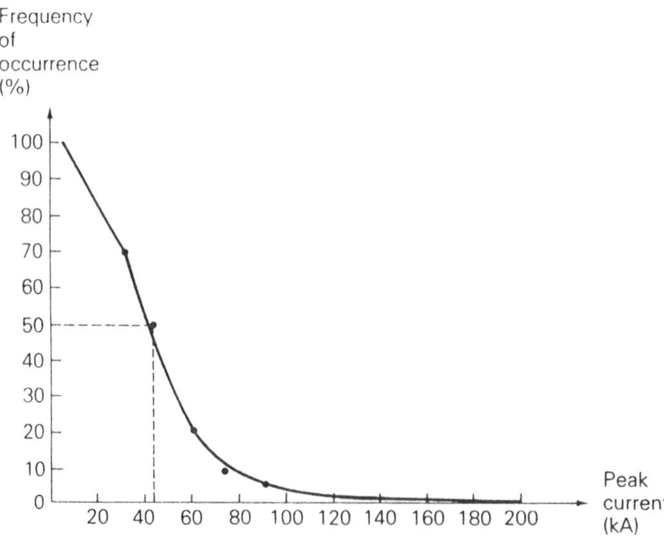

FIGURE 13.23 Frequency of occurrence of lightning currents that exceed a given peak value [20] (IEEE Std C62.22-1997: *IEEE Guide for the Application of Metal-Oxide Surge Arresters for Alternating-Current Systems*. Reprinted with permission from IEEE.)

It has also been shown that what appears to the eye as a single flash of lightning is often the cumulative effect of many strokes. A typical flash consists of typically 3 to 5, and occasionally as many as 40, strokes, at intervals of 50 ms.

The U.S. National Lightning Detection Network® (NLDN), owned and operated by Vaisala-GAI, Inc., is a system that senses the electromagnetic fields radiated by individual return strokes in CG flashes. As of 2001, the NLDN employed more than 100 ground-based sensors geographically distributed throughout the 48 contiguous United States. The sensors transmit lightning data to a network control center in Tucson, Arizona via a satellite communication system. Data from the remote sensors are recorded and processed in real time at the network control center to provide the time, location, polarity, and an estimate of the peak current in each return stroke. The real-time data are then sent back through the communications network for satellite broadcast dissemination to real-time users, all within 30–40 seconds of each CG lightning flash. Recorded data are also reprocessed off-line within a few days of acquisition and stored in a permanent database for access by users who do not require real-time information. NLDN's archive data library contains over 160 million flashes dating from 1989 [25, www.LightningStorm.com].

Figure 13.24 shows a lightning flash density contour map providing a representation of measured annual CG flash density detected by the NLDN from 1989 to 1998. As shown, average annual CG lightning flash densities range from about 0.1 flashes/km^2/year near the West Coast to more than 14 flashes/km^2/year in portions

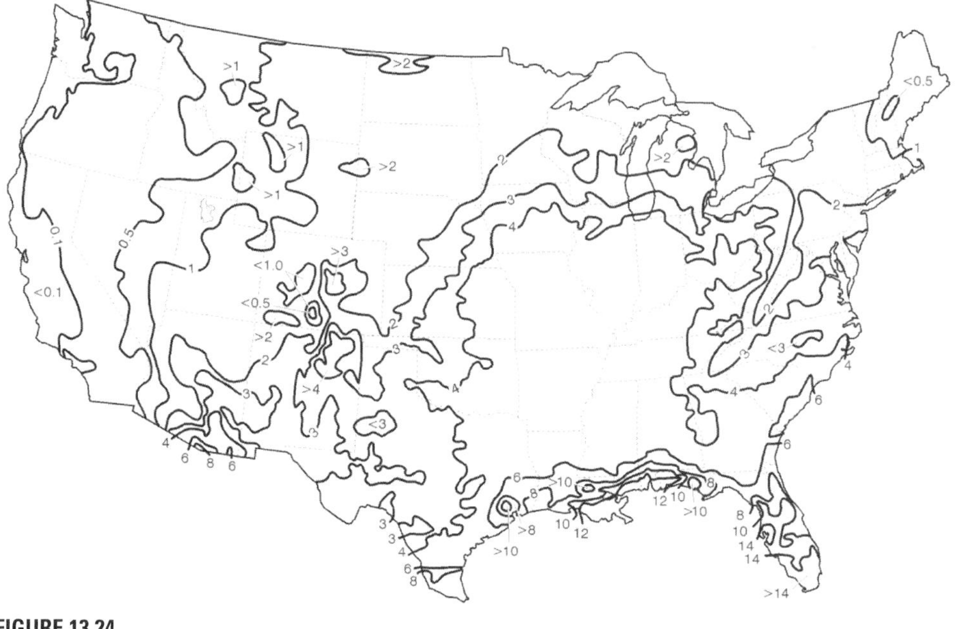

FIGURE 13.24

Lightning flash density contour map showing annual CG flash densities (#flashes/km^2/year) in the contiguous United States, as detected by the NLDN from 1989 to 1998 (Courtesy of Vaisala, Inc.)

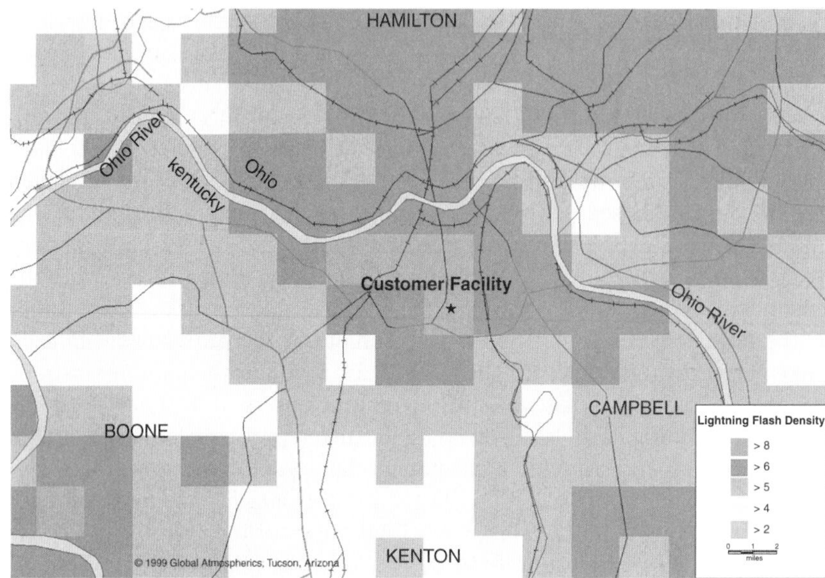

FIGURE 13.25 High-resolution CG flash density map in grid format (Courtesy of Vaisala, Inc.)

of the Florida peninsula. Figure 13.25 shows a high-resolution, 2 mi × 2 mi, CG flash density map in grid format.

Figure 13.26 shows an "asset exposure map" of all CG strikes in 1995 in a region that contains a 15-mile 69-kV transmission line. This map provides an indication of the level of exposure to lightning within an exposure area that surrounds the transmission line. By combining this data with estimates of peak stroke currents and transmission line fault records, the lightning performance of the line and individual line segments can be quantified. Improvements in line design and line protection can also be evaluated.

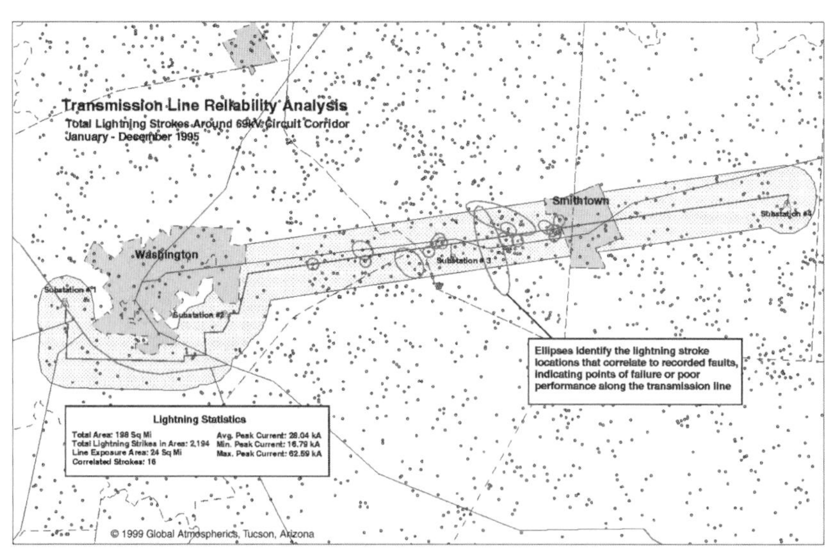

FIGURE 13.26 Asset exposure map showing CG strikes in 1995 in a region that contains a 15-mile (24-km) 69-kV transmission line (Courtesy of Vaisala, Inc.)

Electric utilities use real-time lightning maps to monitor the approach of lightning storms, estimate their severity, and then either position repair crews in advance of storms, call them out, or hold them over as required. Utilities also use real-time lightning data together with on-line monitoring of circuit breakers, relays, and/or substation alarms to improve operations, minimize or avert damage, and speed up the restoration of their systems.

A typical transmission-line design goal is to have an average of less than 0.50 lightning outages per year per 100 miles of transmission. For a given overhead line with a specified voltage rating, the following factors affect this design goal:

1. Tower height
2. Number and location of shield wires
3. Number of standard insulator discs per phase wire
4. Tower impedance and tower-to-ground impedance

It is well known that lightning strikes tall objects. Thus, shorter, H-frame structures are less susceptible to lightning strokes than taller, lattice towers. Also, shorter span lengths with more towers per kilometer can reduce the number of strikes.

Shield wires installed above phase conductors can effectively shield the phase conductors from direct lightning strokes. Figure 13.27 illustrates the effect of shield wires. Experience has shown that the chance of a direct hit to phase conductors located within $\pm 30^\circ$ arcs beneath the shield wires is reduced by a factor of 1000 [18]. Some lightning strokes are, therefore, expected to hit these overhead shield wires. When this occurs, traveling voltage and current waves propagate in both directions along the shield wire that is hit. When a wave arrives at a tower, a reflected wave returns toward the point where the lightning hit, and two refracted waves occur. One refracted wave moves along the shield wire into the next span. Since the shield wire is electrically connected to the tower, the other refracted wave moves down the tower, its energy being harmlessly diverted to ground.

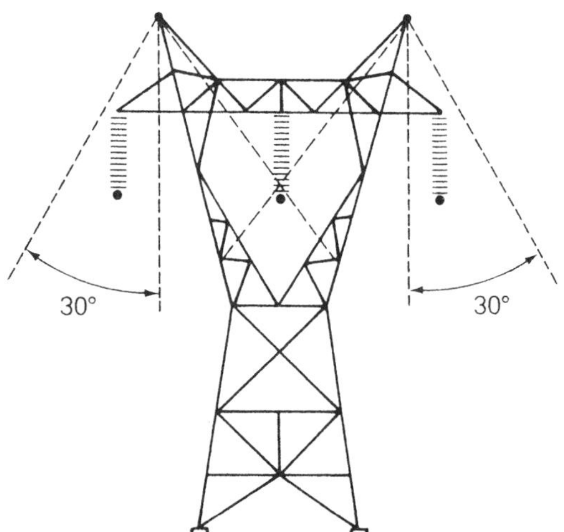

FIGURE 13.27

Effect of shield wires

However, if the tower impedance or tower-to-ground impedance is too high, IZ voltages that are produced could exceed the breakdown strength of the insulator discs that hold the phase wires. The number of insulator discs per string (see Table 4.1) is selected to avoid insulator flashover. Also, tower impedances and tower footing resistances are designed to be as low as possible. If the inherent tower construction does not give a naturally low resistance to ground, driven ground rods can be employed. Sometimes buried conductors running under the line (called *counterpoise*) are employed.

SWITCHING SURGES

The magnitudes of overvoltages due to lightning surges are not significantly affected by the power system voltage. On the other hand, overvoltages due to switching surges are directly proportional to system voltage. Consequently, lightning surges are less important for EHV transmission above 345 kV and for UHV transmission, which has improved insulation. Switching surges become the limiting factor in insulation coordination for system voltages above 345 kV.

One of the simplest and largest overvoltages can occur when an open-circuited line is energized, as shown in Figure 13.28. Assume that the circuit breaker closes at the instant the sinusoidal source voltage has a peak value $\sqrt{2}$ V. Assuming zero source impedance, a forward traveling voltage wave of magnitude $\sqrt{2}$ V occurs. When this wave arrives at the open-circuited receiving end, where $\Gamma_R = +1$, the reflected voltage wave superimposed on the forward wave results in a maximum voltage of $2\sqrt{2}$ V = 2.83 V. Even higher voltages can occur when a line is reclosed after momentary interruption.

In order to reduce overvoltages due to line energizing or reclosing, resistors are almost always preinserted in circuit breakers at 345 kV and above. Resistors ranging from 200 to 800 Ω are preinserted when EHV circuit breakers are closed, and subsequently bypassed. When a circuit breaker closes, the source voltage divides across the preinserted resistors and the line, thereby reducing the initial line voltage. When the resistors are shorted out, a new transient is initiated, but the maximum line voltage can be substantially reduced by careful design.

Dangerous overvoltages can also occur during a single line-to-ground fault on one phase of a transmission line. When such a fault occurs, a voltage equal and opposite to that on the faulted phase occurs at the instant of fault inception. Traveling waves are initiated on both the faulted phase and, due to capacitive coupling, the unfaulted phases. At the line ends, reflections are produced and are superimposed on the normal operating voltages of the unfaulted phases. Kimbark and Legate [19] show that a line-to-ground fault can create an overvoltage on an unfaulted phase as high as 2.1 times the peak line-to-neutral voltage of the three-phase line.

FIGURE 13.28 Energizing an open-circuited line

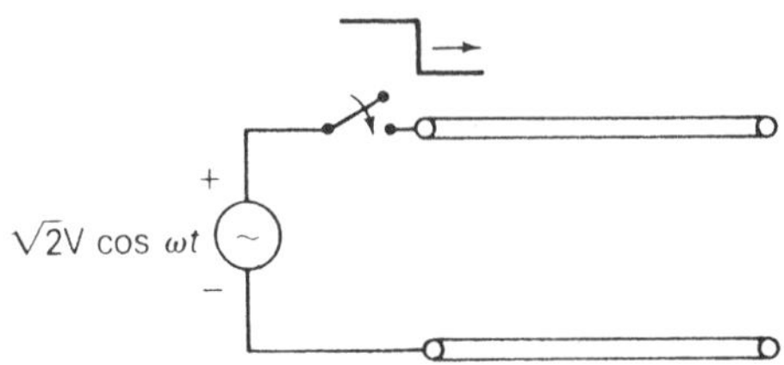

POWER FREQUENCY OVERVOLTAGES

Sustained overvoltages at the fundamental power frequency (60 Hz in the United States) or at higher harmonic frequencies (such as 120 Hz, 180 Hz, and so on) occur due to load rejection, to ferroresonance, or to permanent faults. These overvoltages are normally of long duration, seconds to minutes, and are weakly damped.

13.8 INSULATION COORDINATION

Insulation coordination is the process of correlating electric equipment insulation strength with protective device characteristics so that the equipment is protected against expected overvoltages. The selection of equipment insulation strength and the protected voltage level provided by protective devices depends on engineering judgment and cost.

As shown by the top curve in Figure 13.29, equipment insulation strength is a function of time. Equipment insulation can generally withstand high transient overvoltages only if they are of sufficiently short duration. However, determination of insulation strength is somewhat complicated. During repeated tests with identical voltage waveforms under identical conditions, equipment insulation may fail one test and withstand another.

For purposes of insulation testing, a standard impulse voltage wave, as shown in Figure 13.30, is defined. The impulse wave shape is specified by giving the time T_1 in microseconds for the voltage to reach its peak value and the time T_2 for the

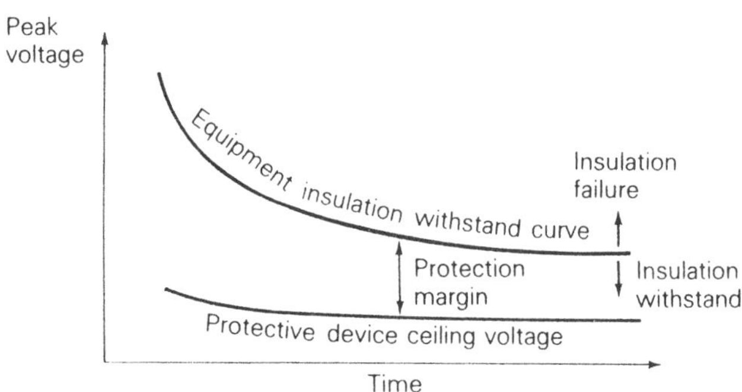

FIGURE 13.29

Equipment insulation strength

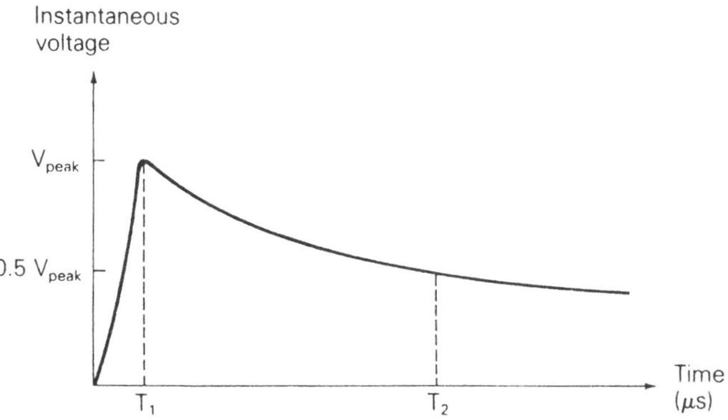

FIGURE 13.30

Standard impulse voltage waveform

voltage to decay to one-half its peak. One standard wave is a 1.2 × 50 wave, which rises to a peak value at $T_1 = 1.2\ \mu s$ and decays to one-half its peak at $T_2 = 50\ \mu s$.

Basic insulation level or BIL is defined as the peak value of the standard impulse voltage wave in Figure 13.30. Standard BILs adopted by the IEEE are shown in Table 13.1. Equipment conforming to these BILs must be capable of withstanding repeated applications of the standard waveform of positive or negative polarity without insulation failure. Also, these standard BILs apply to equipment regardless of how it is grounded. For nominal system voltages 115 kV and above, solidly grounded equipment with the reduced BILs shown in the table have been used.

BILs are often expressed in per-unit, where the base voltage is the maximum value of nominal line-to-ground system voltage. Consider for example a 345-kV system, for which the maximum value of nominal line-to-ground voltage is $\sqrt{2}(345/\sqrt{3}) = 281.7$ kV. The 1550-kV standard BIL shown in Table 13.1 is then $(1550/281.7) = 5.5$ per unit.

Nominal System Voltage kV rms	Standard BIL kV	Reduced BIL* kV
1.2	45	
2.5	60	
5.0	75	
8.7	95	
15	110	
23	150	
34.5	200	
46	250	
69	350	
92	450	
115	550	450
138	650	550
161	750	650
196	900	750
230	1050	825–900
287	1300	1000–1100
345	1550	1175–1300
500		1300–1800
765		1675–2300

TABLE 13.1

Standard and reduced basic insulation levels [18]

*For solidly grounded systems

These BILs are based on 1.2 × 50 μs voltage waveforms. They apply to internal (or non-self-restoring) insulation such as transformer insulation, as well as to external (or self-restoring) insulation, such as transmission-line insulation, on a statistical basis.

(Based on Westinghouse Electric Corporation, Electrical Transmission and Distribution Reference Book, 4th ed. (East Pittsburgh, PA: 1964).)

Note that overhead-transmission-line insulation, which is external insulation, is usually self-restoring. When a transmission-line insulator string flashes over, a short circuit occurs. After circuit breakers open to deenergize the line, the insulation of the string usually recovers, and the line can be rapidly reenergized. However, transformer insulation, which is internal, is not self-restoring. When transformer insulation fails, the transformer must be removed for repair or replaced.

To protect equipment such as a transformer against overvoltages higher than its BIL, a protective device, such as that shown in Figure 13.31, is employed. Such protective devices are generally connected in parallel with the equipment from each phase to ground. As shown in Figure 13.29, the function of the protective device is to maintain its voltage at a ceiling voltage below the BIL of the equipment it protects. The difference between the equipment breakdown voltage and the protective device ceiling voltage is the *protection margin.*

Protective devices should satisfy the following four criteria:

1. Provide a high or infinite impedance during normal system voltages, to minimize steady-state losses;
2. Provide a low impedance during surges, to limit voltage;
3. Dissipate or store the energy in the surge without damage to itself; and
4. Return to open-circuit conditions after the passage of a surge.

One of the simplest protective devices is the rod gap, two metal rods with a fixed air gap, which is designed to spark over at specified overvoltages. Although it satisfies the first two protective device criteria, it dissipates very little energy and it cannot clear itself after arcing over.

A surge arrester, consisting of an air gap in series with a nonlinear silicon carbide resistor, satisfies all four criteria. The gap eliminates losses at normal voltages and arcs during overvoltages. The resistor has the property that its resistance decreases sharply as the current through it increases, thereby limiting the voltage across the resistor to a specified ceiling. The resistor also dissipates the energy in the surge. Finally, following the passage of a surge, various forms of arc control quench the arc within the gap and restore the surge arrester to normal open-circuit conditions.

The "gapless" surge arrester, consisting of a nonlinear metal oxide resistor with no air gap, also satisfies all four criteria. At normal voltages the resistance is extremely high, limiting steady-state currents to microamperes and steady-state losses to a few watts. During surges, the resistance sharply decreases, thereby limiting

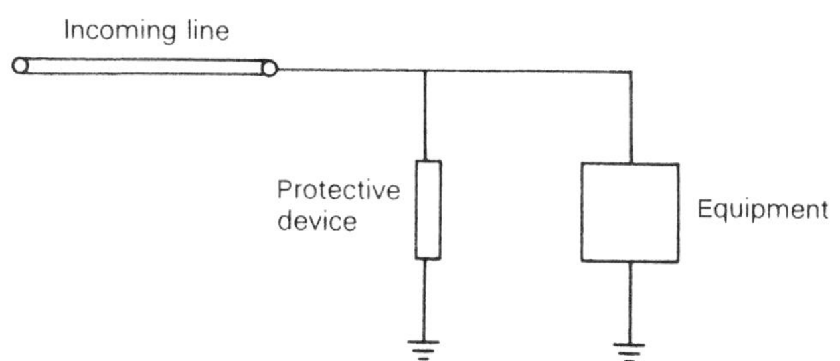

FIGURE 13.31

Single-line diagram of equipment and protective device

overvoltage while dissipating surge energy. After the surge passes, the resistance naturally returns to its original high value. One advantage of the gapless arrester is that its ceiling voltage is closer to its normal operating voltage than is the conventional arrester, thus permitting reduced BILs and potential savings in the capital cost of equipment insulation.

There are four classes of surge arresters: station, intermediate, distribution, and secondary. Station arresters, which have the heaviest construction, are designed for the greatest range of ratings and have the best protective characteristics. Intermediate arresters, which have moderate construction, are designed for systems with nominal voltages of 138 kV and below. Distribution arresters are employed with lower-voltage transformers and lines, where there is a need for economy. Secondary arresters are used for nominal system voltages below 1000 V. A summary of the protective characteristics of station- and intermediate-class metal-oxide surge arresters is given in Table 13.2 [20]. This summary is based on manufacturers' catalog information.

Note that arrester currents due to lightning surges are generally less than the lightning currents shown in Figure 13.23. In the case of direct strokes to transmission-line phase conductors, traveling waves are set up in two directions from the point of the stroke. Flashover of line insulation diverts part of the lightning current from the arrester. Only in the case of a direct stroke to a phase conductor very near an arrester, where no line flash-over occurs, does the arrester discharge the full lightning current. The probability of this occurrence can be significantly

EXAMPLE 13.8

Metal-oxide surge arrester selection

Consider the selection of a station-class metal-oxide surge arrester for a 345-kV system in which the maximum 60-Hz voltage under normal system conditions is 1.04 per unit. (a) Select a station-class arrester from Table 13.2 with a maximum continuous operating voltage (MCOV) rating that exceeds the 1.04 per-unit maximum 60-Hz voltage of the system under normal system conditions. (b) For the selected arrester, determine the protective margin for equipment in the system with a 1300-kV BIL, based on a 10-kA impulse current wave cresting in 0.5 μs.

SOLUTION

a. The maximum 60 Hz line-to-neutral voltage under normal system conditions is $1.04(345/\sqrt{3}) = 207$ kV. From Table 13.2, select a station-class surge arrester with a 209-kV MCOV. This is the lowest MCOV rating that exceeds the 207 kV providing the greatest protective margin as well as economy.

b. From Table 13.2 for the selected surge arrester, the maximum discharge voltage for a 10-kA impulse current wave cresting in 0.5 μs ranges from 2.19 to 2.36 in per-unit of MCOV, or 457–493 kV, depending on arrester manufacturer. Therefore, the protective margin ranges from $(1300 - 493) = 807$ kV to $(1300 - 457) = 843$ kV.

Station Class									
Steady-state operation system voltage and arrester ratings effectively grounded systems (NOTE 1)				Protective levels range of industry maxima per unit (crest of 60hz) of MCOV			Durability characteristics IEEE Std C62.11-2005		
Max system rms L-L voltage kV*	Max system rms L-G voltage kV*	Min rms MCOV rating kV	Duty-cycle rms voltage rating kV	0.5 μs FOW protective level (NOTE 2)	8/20 μs protective level (NOTE 3)	Switching surge protective level (NOTE 4)	High crest current withstand A	Trans. line discharge Miles†	Pressure relief rms (symmetrical) current kA (NOTE 5)
4.37	2.52	2.55	3	2.32–2.48	2.10–2.20	1.70–1.85	65 000	150	40–80
8.73	5.04	5.1	6–9	2.33–2.48	1.97–2.23	1.70–1.85	65 000	150	40–80
13.1	7.56	7.65	9–12	2.33–2.48	1.97–2.23	1.70–1.85	65 000	150	40–80
13.9	8.00	8.4	10–15	2.33–2.48	1.97–2.23	1.70–1.85	65 000	150	40–80
14.5	8.37	8.4	10–15	2.33–2.48	1.97–2.23	1.70–1.85	65 000	150	40–80
26.2	15.1	15.3	18–27	2.33–2.48	1.97–2.23	1.70–1.85	65 000	150	40–80
36.2	20.9	22	27–36	2.43–2.48	1.97–2.23	1.70–1.85	65 000	150	40–80
48.3	27.8	29	36–48	2.43–2.48	1.97–2.23	1.70–1.85	65 000	150	40–80
72	41.8	42	54–72	2.19–2.40	1.97–2.18	1.64–1.84	65 000	150	40–80
121	69.8	70	90–120	2.19–2.40	1.97–2.18	1.64–1.84	65 000	150	40–80
145	83.7	84	108–144	2.19–2.39	1.97–2.17	1.64–1.84	65 000	150	40–80
169	97.5	98	120–172	2.19–2.39	1.97–2.17	1.64–1.84	65 000	175	40–80
242	139	140	172–240	2.19–2.36	1.97–2.15	1.64–1.84	65 000	175	40–80
362	209	209	258–312	2.19–2.36	1.97–2.15	1.71–1.85	65 000	200	40–80
550	317	318	396–564	2.01–2.47	2.01–2.25	1.71–1.85	65 000	200	40–80
800	461	462	576–612	2.01–2.47	2.01–2.25	1.71–1.85	65 000	200	40–80
Intermediate class									
4.37–145	2.52–83.7	2.8–84	3–144	2.38–2.85	2.28–2.55	1.71–1.85	65 000	100	16.1

TABLE 13.2

Typical characteristics of station- and intermediate-class metal-oxide surge arresters [20]

*Voltage Range A, ANSI C84.1-2006

NOTE 1—See Table 8 and 6.4 for typical arrester ratings for systems with noneffectively grounded neutral.

NOTE 2—Equivalent FOW producing a voltage wave cresting in 0.5 μs.

NOTE 3—The protective level is the maximum discharge voltage for a 10 kA impulse current wave on arrester duty-cycle rating through 312 kV, 15 kA for duty-cycle ratings 396 kV to 564 kV, and 20 kA for duty-cycle ratings 576 kV to 612 kV, per IEEE Std C62.11-2005.

NOTE 4—Switching surge characteristics based on maximum switching surge classifying current (based on an impulse current wave a time to actual crest of 45 μs to 60μs) of 500 A on arrester duty-cycle ratings 3 kV to 108 kV, 1000 A on duty-cycle ratings 120 kV to 240 kV, and 2000 A on duty-cycle ratings above 240 kV, per IEEE Std C62.11-2005.

NOTE 5—Test values for arresters with procelain tops have not been standardized, Pressure relief classification is in 5 kA. steps.

(IEEE std. C62.22- 2009. IEEE Guide for the Application of Metal-Oxide Surge Arresters for Alternating-Current Systems. Reprinted with permission from IEEE.)

† 1 mile = 1.6 km

reduced by using overhead shield wires to shield transmission lines and substations. Recommended practice for substations with unshielded lines is to select an arrester discharge current of at least 20 kA (even higher if the isokeraunic level is above 40 thunderstorm days per year). For substations with shielded lines, lower arrester discharge currents, from 5 to 20 kA, have been found satisfactory in most situations [20].

When selecting a metal-oxide surge arrester, it is important that the arrester MCOV exceeds the maximum 60-Hz system voltage (line-to-neutral) under normal conditions. In addition to considerations affecting the selection of arrester MCOV, metal-oxide surge arresters should also be selected to withstand temporary overvoltages in the system at the arrester location—for example, the voltage rise on unfaulted phases during line-to-ground faults. That is, the temporary overvoltage (TOV) capability of metal-oxide surge arresters should not be exceeded. Additional considerations in the selection of metal-oxide surge arresters are discussed in reference [22] (see www.cooperpower.com).

PROBLEMS

SECTION 13.2

13.1 From the results of Example 13.2, plot the voltage and current profiles along the line at times $\tau/2$, τ, and 2τ. That is, plot $v(x, \tau/2)$ and $i(x, \tau/2)$ versus x for $0 \leq x \leq 1$; then plot $v(x, \tau)$, $i(x,x)$, $v(x,2\tau)$, and $i(x,2x)$ versus x.

13.2 Rework Example 13.2 if the source voltage at the sending end is a ramp, $e_G(t) = Eu_{-2}M = Etu_{-1}(t)$, with $Z_G = 2Z_c$.

13.3 Referring to the single-phase two-wire lossless line shown in Figure 13.3, the receiving end is terminated by an inductor with $2L_R$ henries. The source voltage at the sending end is a step, $e_G(t) = Eu_{-1}(t)$ with $Z_G = Z_c$. Both the line and inductor are initially unenergized. Determine and plot the voltage at the center of the line $v(l/2, t)$ versus time t.

13.4 Rework Problem 13.3 if $Z_R = Z_c$ at the receiving end and the source voltage at the sending end is $e_G(t) = Eu_{-1}(t)$, with an inductive source impedance $Z_G(s) = s2L_G$. Both the line and source inductor are initially unenergized.

13.5 Rework Example 13.4 with $Z_R = 5Z_c$ and $Z_G = Z_c/3$.

13.6 The single-phase, two-wire lossless line in Figure 13.3 has a series inductance $L = (1/3) \times 10^{-6}$ H/m, a shunt capacitance $C = (1/3) \times 10^{-10}$ F/m, and a 50-km line length. The source voltage at the sending end is a step $e_G(t) = 100u_{-1}(t)$ kV with $Z_G(s) = 100\ \Omega$. The receiving-end load consists of a 100-Ω resistor in parallel with a 2-mH inductor. The line and load are initially unenergized. Determine (a) the characteristic impedance in ohms, the wave velocity in m/s, and the transit time in ms for this line; (b) the sending-and receiving-end voltage reflection coefficients in per-unit; (c) the Laplace transform of the receiving-end current, $I_R(s)$; and (d) the receiving-end current $i_R(t)$ as a function of time.

13.7 The single-phase, two-wire lossless line in Figure 13.3 has a series inductance $L = 2 \times 10^{-6}$ H/m, a shunt capacitance $C = 1.25 \times 10^{-11}$ F/m, and a 100-km line length. The source voltage at the sending end is a step $e_G(t) = 100u_{-1}(t)$ kV with a source impedance equal to the characteristic impedance of the line. The receiving-end load consists of a 100-mH inductor in series with a 1-μF capacitor. The line and load are initially unenergized. Determine (a) the characteristic impedance in Ω, the wave velocity in m/s, and the transit time in ms for this line; (b) the sending- and receiving-end voltage reflection coefficients in per-unit; (c) the receiving-end voltage $v_R(t)$ as a function of time; and (d) the steady-state receiving-end voltage.

13.8 The single-phase, two-wire lossless line in Figure 13.3 has a series inductance $L = 0.999 \times 10^{-6}$ H/m, a shunt capacitance $C = 1.112 \times 10^{-11}$ F/m, and a 60-km line length. The source voltage at the sending end is a ramp $e_G(t) = Etu_{-1}(t) = Eu_{-2}(t)$ kV with a source impedance equal to the characteristic impedance of the line. The receiving-end load consists of a 150-Ω resistor in parallel with a 1-μF capacitor. The line and load are initially unenergized. Determine (a) the characteristic impedance in Ω, the wave velocity in m/s, and the transit time in ms for this line; (b) the sending- and receiving-end voltage reflection coefficients in per-unit; (c) the Laplace transform of the sending-end voltage, $V_S(s)$; and (d) the sending-end voltage $v_S(t)$ as a function of time.

SECTION 13.3

13.9 Draw the Bewley lattice diagram for Problem 13.5.

13.10 Rework Problem 13.9 if the source voltage is a pulse of magnitude E and duration $\tau/10$; that is, $e_G(t) = E[u_{-1}(t) - u_{-1}(t - \tau/10)]$. $Z_R = 5Z_c$ and $Z_G = Z_c/3$ are the same as in Problem 13.9. Also plot $v(l/3, t)$ versus time t for $0 \le t < 6\tau$.

13.11 Rework Example 13.6 if the source impedance at the sending end of line A is $Z_G = Z_A/4 = 100\ \Omega$, and the receiving end of line B is short-circuited, $Z_R = 0$.

13.12 Rework Example 13.6 if the overhead line and cable are interchanged. That is, $Z_A = 100\ \Omega$, $v_A = 2 \times 10^8$ m/s, $l_A = 20$ km, $Z_B = 400\ \Omega$, $v_B = 3 \times 10^8$ m/s, and $l_B = 30$ km. The step voltage source $e_G(t) = Eu_{-1}(t)$ is applied to the sending end of line A with $Z_G = Z_A = 100\ \Omega$, and $Z_R = 2Z_B = 800\ \Omega$, at the receiving end. Draw the lattice diagram for $0 \le t \le 0.6$ ms and plot the junction voltage versus time t.

13.13 As shown in Figure 13.32, a single-phase two-wire lossless line with $Z_c = 400\ \Omega$, $v = 3 \times 10^8$ m/s, and $l = 100$ km has a 400-Ω resistor, denoted R_J, installed across the center of the line, thereby dividing the line into two sections, A and B. The source voltage at the sending end is a pulse of magnitude 100 V and duration 0.1 ms. The source impedance is $Z_G = Z_c = 400\ \Omega$, and the receiving end of the line is short-circuited, (a) Show that

FIGURE 13.32

Circuit for Problem 13.13

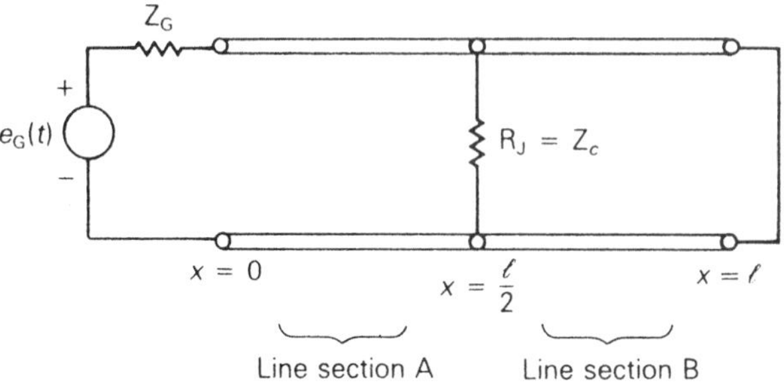

for an incident voltage wave arriving at the center of the line from either line section, the voltage reflection and refraction coefficients are given by

$$\Gamma_{BB} = \Gamma_{AA} = \frac{\left(\frac{Z_{eq}}{Z_c}\right) - 1}{\left(\frac{Z_{eq}}{Z_c}\right) + 1} \qquad \Gamma_{AB} = \Gamma_{BA} = \frac{2\left(\frac{Z_{eq}}{Z_c}\right)}{\left(\frac{Z_{eq}}{Z_c}\right) + 1}$$

where

$$Z_{eq} = \frac{R_J Z_c}{R_J + Z_c}$$

(b) Draw the Bewley lattice diagram for $0 \le t \le 6\tau$. (c) Plot $v(l/2, t)$ versus time t for $0 \le t \le 6\,\tau$ and plot $v(x, 6\tau)$ versus x for $0 \le x \le l$.

13.14 The junction of four single-phase two-wire lossless lines, denoted A, B, C, and D, is shown in Figure 13.13. Consider a voltage wave v_A^+ arriving at the junction from line A. Using (13.3.8) and (13.3.9), determine the voltage reflection coefficient Γ_{AA} and the voltage refraction coefficients Γ_{BA}, Γ_{CA}, and Γ_{DA}.

13.15 Referring to Figure 13.3, the source voltage at the sending end is a step $e_G(t) = Eu_{-1}(t)$ with an inductive source impedance $Z_G(s) = sL_G$, where $L_G/Z_c = \tau/3$. At the receiving end, $Z_R = Z_c/4$. The line and source inductance are initially unenergized. (a) Draw the Bewley lattice diagram for $0 \le t \le 5\tau$. (b) Plot $v(l, t)$ versus time t for $0 \le t \le 5\tau$.

13.16 As shown in Figure 13.33, two identical, single-phase, two-wire, lossless lines are connected in parallel at both the sending and receiving ends. Each line has a 400-Ω characteristic impedance, 3×10^8 m/s velocity of propagation, and 100-km line length. The source voltage at the sending end is a 100-kV step with source impedance $Z_G = 100\ \Omega$. The receiving end is shorted ($Z_R = 0$). Both lines are initially unenergized. (a) Determine the first forward traveling voltage waves that start at time $t = 0$ and travel on each line toward the receiving end. (b) Determine the sending- and receiving-end voltage reflection coefficients in per-unit,

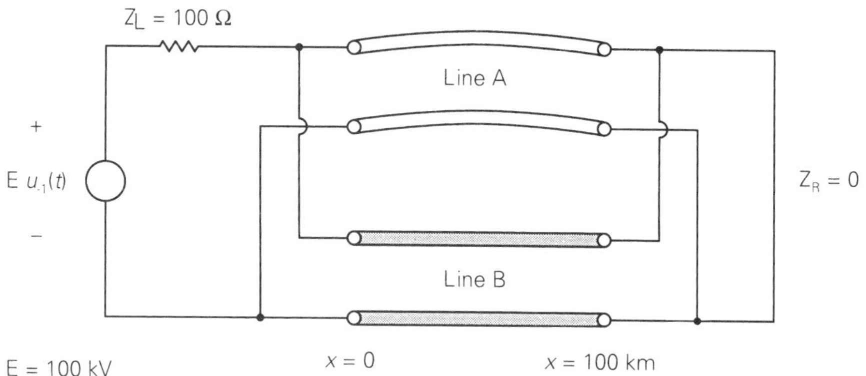

FIGURE 13.33 Circuit for Problem 13.16

(c) Draw the Bewley lattice diagram for $0 \leq t \leq 2.0$ ms. (d) Plot the voltage at the center of one line versus time t for $0 \leq t \leq 2.0$ ms.

13.17 As shown in Figure 13.34, an ideal current source consisting of a 10-kA pulse with 50-μs duration is applied to the junction of a single-phase, lossless cable and a single-phase, lossless overhead line. The cable has a 200-Ω characteristic impedance, 2×10^8 m/s velocity of propagation, and 20-km length. The overhead line has a 300-Ω characteristic impedance, 3×10^8 m/s velocity of propagation, and 60-km length. The sending end of the cable is terminated by a 400-Ω resistor, and the receiving end of the overhead line is terminated by a 100-Ω resistor. Both the line and cable are initially unenergized. (a) Determine the voltage reflection coefficients Γ_S, Γ_R, Γ_{AA}, Γ_{AB}, Γ_{BA}, and Γ_{BB} (b) Draw the Bewley lattice diagram for $0 \leq t \leq 0.8$ ms. (c) Determine and plot the voltage $v(0, t)$ at $x = 0$ versus time t for $0 \leq t \leq 0.8$ ms.

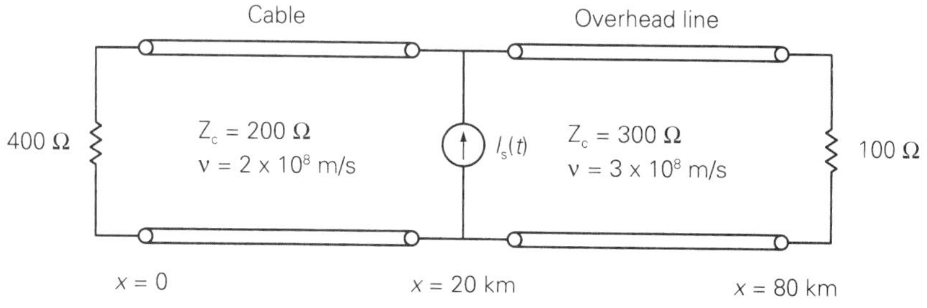

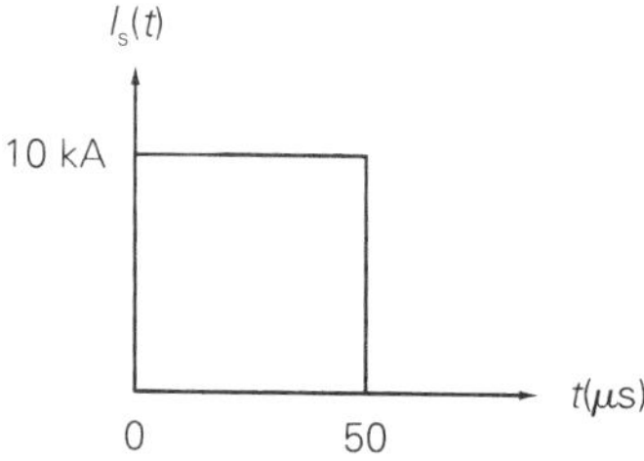

FIGURE 13.34 Circuit for Problem 13.17

SECTION 13.4

13.18 For the circuit given in Problem 13.3, replace the circuit elements by their discrete-time equivalent circuits and write nodal equations in a form suitable for computer solution of the sending-end and receiving-end voltages. Give equations for all dependent sources. Assume $E = 1000$ V, $L_R = 20$ mH, $Z_c = 100\ \Omega$, $v = 2 \times 10^8$ m/s, $l = 40$ km, and $\Delta t = 0.02$ ms.

13.19 Repeat Problem 13.18 for the circuit given in Problem 13.13. Assume $\Delta t = 0.03333$ ms.

13.20 For the circuit given in Problem 13.7, replace the circuit elements by their discrete-time equivalent circuits. Use $\Delta t = 100\ \mu s = 1 \times 10^{-4}$ s. Determine and show all resistance values on the discrete-time circuit. Write nodal equations for the discrete-time circuit, giving equations for all dependent sources. Then solve the nodal equations and determine the sending- and receiving-end voltages at the following times: $t = 100, 200, 300, 400, 500$, and $600\ \mu s$.

13.21 For the circuit given in Problem 13.8, replace the circuit elements by their discrete-time equivalent circuits. Use $\Delta t = 50\ \mu s = 5 \times 10^{-5}$ s and $E = 100$ kV. Determine and show all resistance values on the discrete-time circuit. Write nodal equations for the discrete-time circuit, giving equations for all dependent sources. Then solve the nodal equations and determine the sending-and receiving-end voltages at the following times: $t = 50, 100, 150, 200, 250$, and $300\ \mu s$.

SECTION 13.5

13.22 Rework Problem 13.18 for a lossy line with a constant series resistance $R = 0.3\ \Omega$/km. Lump half of the total resistance at each end of the line.

SECTION 13.8

13.23 Repeat Example 13.8 for a 500-kV system with a 1.08 per-unit maximum 60-Hz voltage under normal operating conditions and with a 2000-kV BIL.

13.24 Select a station-class metal-oxide surge arrester from Table 13.2 for the high-voltage side of a three-phase 400 MVA, 345-kV Y/13.8-kV Δ transformer. The maximum 60-Hz operating voltage of the transformer under normal operating conditions is 1.03 per unit. The high-voltage windings of the transformer have a BIL of 1300 kV and a solidly grounded neutral. A minimum protective margin of 1.4 per unit based on a 10-kA impulse current wave cresting in 0.5 μs is required. (*Note:* Additional considerations for the selection of metal-oxide surge arresters are given in reference [22] (www.cooperpower.com).

CASE STUDIES QUESTIONS

a. Why are circuit breakers and fuses ineffective in protecting against transient overvoltages due to lightning and switching surges?

b. Where are surge arresters located in power systems?

c. How does one select a surge arrester to protect specific equipment?

d. What is the largest loss-of-generation event in the U.S. Western Interconnection as recognized by the North America Electric Reliability Corporation?

e. What is "frequency nadir"?

REFERENCES

1. A. Greenwood, *Electrical Transients in Power Systems*, 2d ed. (New York: Wiley Interscience, 1991).
2. L. V. Bewley, *Travelling Waves on Transmission Systems*, 2d ed. (New York: Wiley, 1951).
3. L. Bergeron, *Water Hammer in Hydraulics and Wave Surges in Electricity* (New York: Wiley, 1961).
4. H. M. Lacey, "The Lightning Protection of High-Voltage Overhead Transmission and Distribution Systems," *Proc. IEE*, 96 (1949), p. 287.
5. J. R. Carson, "Wave Propagation in Overhead Wires with Ground Return," *Bell System Technical Journal* 5 (1926), pp. 539–554.
6. W. S. Meyer and H. W. Dommel, "Numerical Modelling of Frequency-Dependent Transmission Line Parameters in an Electromagnetic Transients Program," *IEEE Transactions PAS*, vol. PAS-99 (September/October 1974), pp. 1401–1409.
7. A. Budner, "Introduction of Frequency-Dependent Line Parameters into an Electromagnetics Transients Program," *IEEE Transactions PAS*, vol. PAS-89 (January 1970), pp. 88–97.
8. D. E. Hedman, "Propagation on Overhead Transmission Lines: I—Theory of Modal Analysis and II—Earth Conduction Effects and Practical Results," *IEEE Transactions PAS* (March 1965), pp. 200–211.
9. H. W. Dommel, "A Method for Solving Transient Phenomena in Multiphase Systems," *Proceedings 2nd Power Systems Computation Conference*, Stockholm, 1966.
10. H. W. Dommel, "Digital Computer Solution of Electromagnetic Transients in Single- and Multiphase Networks," *IEEE Transactions PAS*, vol. PAS-88 (1969), pp. 388–399.

11. H. W. Dommel, "Nonlinear and Time-Varying Elements in Digital Simulation of Electromagnetic Transients," *IEEE Transactions PAS*, vol. PAS-90 (November/December 1971), pp. 2561–2567.

12. S. R. Naidu, *Transitorios Electromagneticos em Sistemas de Potencia*, Eletrobras/UFPb, Brazil, 1985.

13. "Report of Joint IEEE-EEI Committee on EHV Line Outages," *IEEE Transactions PAS*, 86 (1967), p. 547.

14. R. A. W. Connor and R. A. Parkins, "Operations Statistics in the Management of Large Distribution System," *Proc. IEEE*, 113 (1966), p. 1823.

15. C. T. R. Wilson, "Investigations on Lightning Discharges and on the Electrical Field of Thunderstorms," *Phil. Trans. Royal Soc*, Series A, 221 (1920), p. 73.

16. G. B. Simpson and F. J. Scrase, "The Distribution of Electricity in Thunderclouds," *Proc. Royal Soc*, Series A, 161 (1937), p. 309.

17. B. F. J. Schonland and H. Collens, "Progressive Lightning," *Proc. Royal Soc*, Series A, 143 (1934), p. 654.

18. Westinghouse Electric Corporation, *Electrical Transmission and Distribution Reference Book*, 4th ed. (East Pittsburgh, PA: 1964).

19. E. W. Kimbark and A. C. Legate, "Fault Surge Versus Switching Surge, A Study of Transient Voltages Caused by Line to Ground Faults," *IEEE Transactions PAS*, 87 (1968), p. 1762.

20. *IEEE Guide for the Application of Metal-Oxide Surge Arresters for Alternating-Current Systems*, IEEE std. C62.22-2009 (New York: The Institute of Electrical and Electronics Engineers, http://standards.ieee.org)

21. C. Concordia, "The Transient Network Analyzer for Electric Power System Problems," Supplement to *CIGRE Committee No.* 13 *Report*, 1956.

22. *Varistar Type AZE Surge Arresters for Systems through* 345 *kV*, Cooper Power Systems Catalog 235–87 (Waukesha, WI: Cooper Power Systems. http://www.cooperpower.com, July 2011).

23. N. Miller et al., "Emergency Response," *IEEE Power & Energy* Magazine, 11, 6 (November/December 2013), pp. 63–71.

24. W. R. Newcott, "Lightning, Nature's High-Voltage Spectacle," *National Geographic.* 184, 1 (July 1993), pp. 83–103.

25. K. L. Cummins, E. P. Krider, and M. D. Malone, "The U.S. National Lightning Detection Network TM and Applications of Cloud-to-Ground Lightning Data by Electric Power Utilities," *IEEE Transactions on Electromagnetic Compatibility*, 40, 4 (November 1998), pp. 465–480.

14 Power Distribution

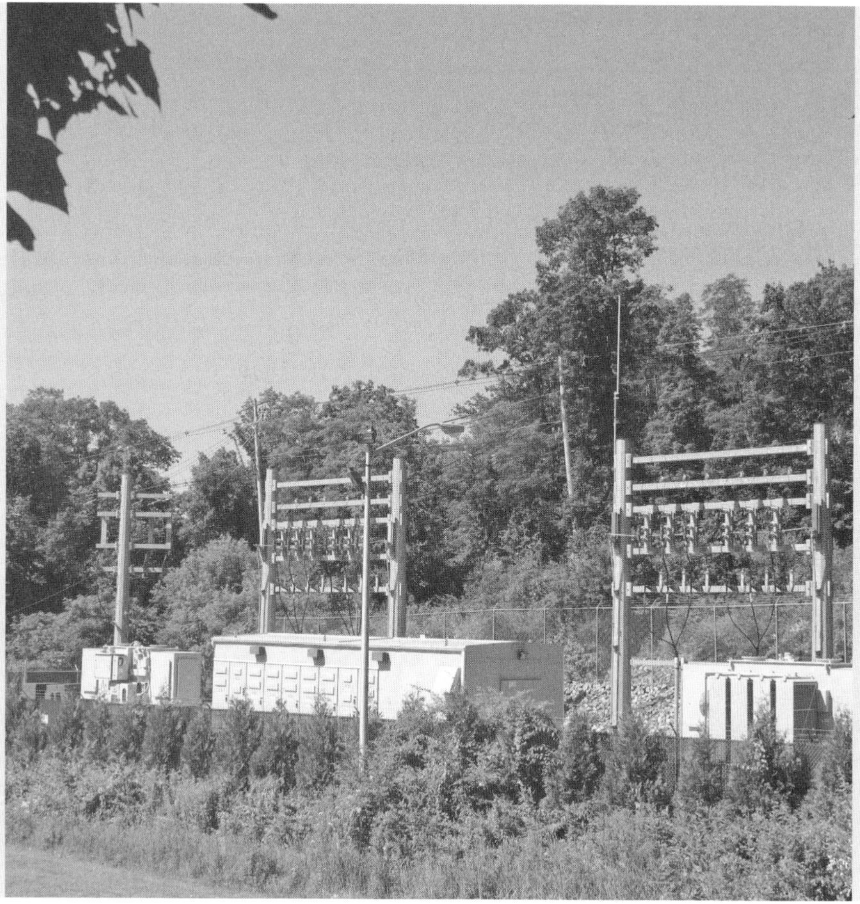

Distribution substation fed by two 22.9-kV (one overhead and one underground) lines through two 15-MVA, 22.9Δ/4.16Y-kV distribution substation transformers. The transformers are located on either side of the switchgear building shown in the center. This substation feeds six 4.16-kV radial primary feeders through 5-kV, 1200-A vacuum circuit breakers (Courtesy of Danvers Electric.)

Major components of an electric power system are generation, transmission, and distribution. Distribution, including primary and secondary distribution, is that portion of a power system that runs from distribution substations to a customer's service entrance equipment. In 2013 in the United States, distribution systems served approximately 146 million customers that consumed 3.73 trillion kWh (www.eia.doe.gov).

This chapter provides an overview of distribution. Sections 14.1–14.3 introduce the basic configurations and characteristics of distribution including primary and secondary distribution. Sections 14.4 and 14.5 discuss the application of transformers and capacitors in distribution systems. Then in Sections 14.6–14.9 distribution software, distribution reliability, distribution automation, and smart grids are introduced.

CASE STUDY

During the last five years distribution grids in the United States and in many other countries have been changing. There has been a notable installation of distributed energy resources (DERs) on distribution grids, either at medium- or low-voltage levels. For example, in 2013 alone in the United States, 155,000 grid-connected solar photovoltaic (PV) systems were installed, 94% of those on distribution grids. This is expected to nearly triple by 2016. The following article describes an integrated, proactive approach to utility planning of distribution grids with DERs [20]. The article discusses the impacts of DERs on the following technical issues: voltage, protection, energy, capacity, and reliability.

It's All in the Plans

By Jeff Smith, Matthew Rylander, Lindsey Rogers & Roger Dugan

Distribution planners are facing a new reality: the vast majority of new generation currently being connected to the grid is through the distribution system. This "edge" of the grid, where utilities have the least amount of visibility and controllability, is also where most of the change is occurring. The result is a new set of challenges associated with further integrating these ever-increasing levels of distributed energy resources (DERs). To meet these challenges, an integrated approach for planning is needed.

The U.S. electricity landscape is undergoing a gradual but far-reaching makeover. The existing power system was designed to connect a relatively small number of large generation plants to a large number of consumers through an extensive high-voltage transmission system and a medium- to low-voltage distribution system. In recent years, there has been a notable influx of DERs onto the grid—particularly on the distribution system, either at the medium- or low-voltage level. Specifically, installations of solar photovoltaic (PV) systems are growing rapidly in the United States, with 12.1 GW installed by the end of 2013, according to the Interstate Renewable Energy Council, accounting for 471,000 total installations. In 2013 alone in the United States, 155,000 grid-connected PV systems were installed, with distribution-connected systems accounting for 94% of the additions. This is expected to nearly triple by 2016, translating to increased end-user self-generation and greater grid management expense. This shift alters the manner in which electricity is being generated, transmitted, and managed, thus necessitating a change in how utilities plan and integrate this resource.

The Immense Scale of Distribution

An entire distribution service territory often consists of multiple large planning areas in which substations and feeders have widely varying design and control parameters. Due to unique design and operating criteria developed over the years, two adjoining planning areas may have unique planning and operational requirements. Within each planning area, utilities may have tens or even hundreds of substations that connect and deliver energy from the transmission system to serve hundreds or thousands of different distribution feeders. Each of these feeders is outfitted with equipment for providing both voltage control and system protection; this equipment is operated using custom settings to enable the utility to serve all customers in an efficient and reliable manner.

Within each feeder there are tens or hundreds of service transformers that convert power from the medium voltage down to a more usable, low-voltage service level. These transformers distribute this service through multiple secondary systems that connect each service transformer with individual residences, commercial buildings, and industrial complexes.

Therefore, customers located at the very "edge" of the grid—and distribution utilities often have hundreds of thousands or even millions of customers—are served by a vast and diverse network of feeders, substations, planning areas, and, ultimately, an entire distribution service territory (see Figure 1). In evaluating the grid from the transmission system all the way down to each individual customer, the amount of visibility (measurement data, models, and so on), controllability (voltage and protection), and flexibility (automation) lessens. Therefore, the vast majority of the change is occurring where utilities have the least amount of information.

Regardless of whether the DER installations are connected to the grid at the customer side of the meter or directly to the medium-voltage grid, to properly evaluate the impact of DERs the complexity of the distribution system and the uniqueness of each location and individual feeder must be considered.

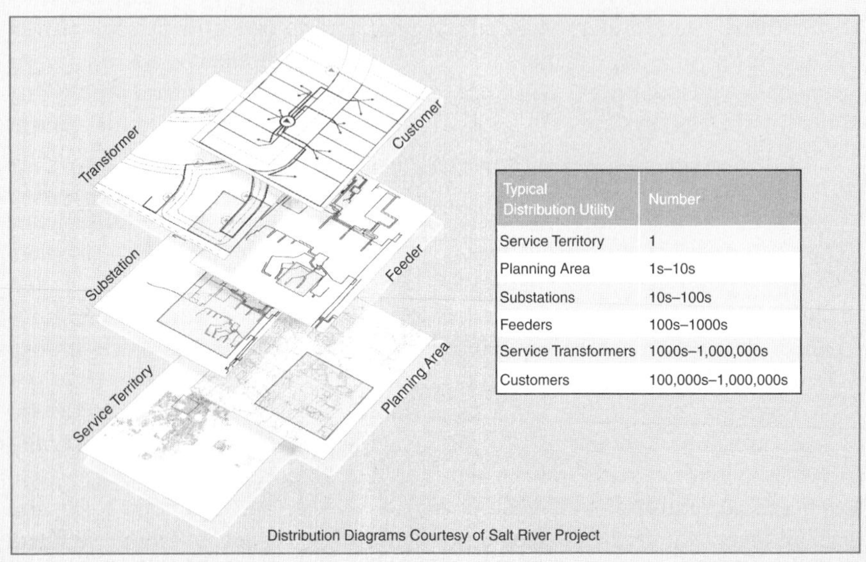

Typical Distribution Utility	Number
Service Territory	1
Planning Area	1s–10s
Substations	10s–100s
Feeders	100s–1000s
Service Transformers	1000s–1,000,000s
Customers	100,000s–1,000,000s

Figure 1 The large "scale" of distribution systems

The Grid Is Changing

Given the breadth of current DER additions, conflicts have already arisen between distribution systems designed for one-way power flow and DERs that want to force power flow in the opposite direction. The structural diversity of the distribution system poses an even greater challenge, as each distribution feeder can have a unique response to this new resource. Screening methods exist to avoid adverse impacts due to DERs, but this is addressing the abundance of DER interconnect requests and can result in higher overall costs if the resource isn't fully integrated and located appropriately.

One example of this shift in planning philosophy is in designing a voltage regulation scheme for a feeder. Without DERs, a planner designs for a voltage *drop* from the substation to the feeder extremities. If the voltage is projected to drop too low during peak demand periods, a capacitor bank or voltage regulator is added to boost the voltage. Feeder voltage control is designed to yield voltages within ANSI C84.1. With DERs, the planner must also design for the voltage *rise* resulting from DER power output and must consider *time* in the analysis. This requires new or enhanced planning tools. An example of the time and location evaluation of PVs on a distribution feeder is captured in Figure 2. The interaction of control devices and variable resources is evident in the embedded video (to view the video, scan the QR code with a smartphone).

The daylight hours are simulated at a 1-s resolution to capture the solar variability impacts on voltage magnitude and on voltage regulator and capacitor operations. Three snapshots captured during the day are

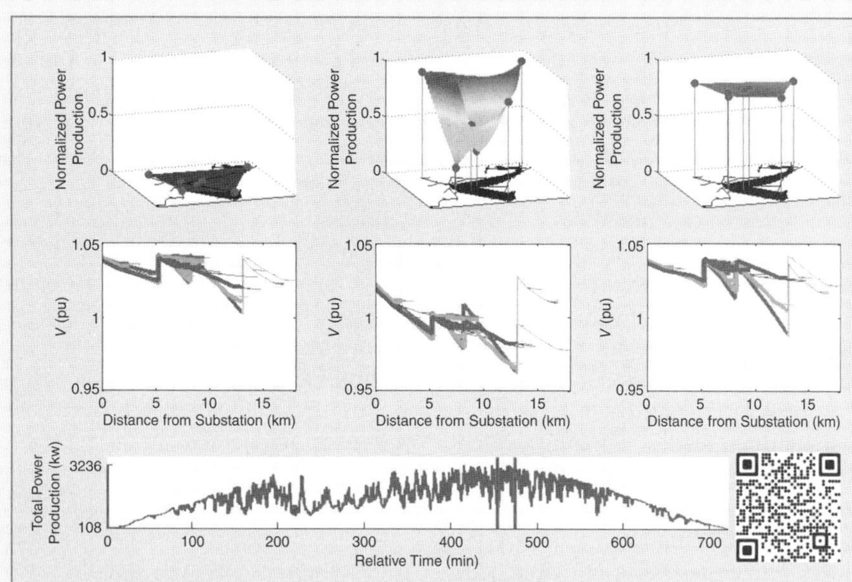

Figure 2 Time and location impacts of variable solar PV on voltage control

shown in the figure to represent the impact of solar generation on voltage profiles. At the beginning of the day, the solar output is low, as shown by the surface plot of location-based solar resource potential for power production. The surface represents the solar resource as measured from several ground-based measurement units, indicated by vertical lines. The actual power production is dependent on the location of the actual PV systems on the feeder. During the day, the solar resource and total power production fluctuate as clouds pass overhead. The impact this variability has on the feeder affects the voltage profile, which at times exceeds 105% (126 V on 120-V base). Increased voltage regulator and capacitor operations are observed as well.

Another example is the design of feeder protection. One reason most distribution feeders in the United States are built in a radial configuration is that such a design yields the lowest-cost protection system while also providing safe and reliable service. A meshed system, like that found in transmission, may better support DERs but requires an entirely different planning paradigm for distribution. New types of line equipment would be required to protect a new system configuration.

Given that the grid is becoming more complex, utility planning needs to change. Distribution utilities must transform existing planning methods into new planning methods that can accommodate a more integrated system.

A New Approach Is Needed

To effectively plan and operate the distribution grid under this changing landscape, a more proactive approach must be taken that accounts for the actual characteristics of the DERs as well as where the resources are connecting to the grid. While on the surface these aspects of the problem may seem obvious, they are quite challenging to capture.

What does a proactive planning approach look like? A proactive approach considers the true impact, both positive and negative, on distribution. This approach considers how the location of DERs affects each distribution feeder differently and identifies where DERs can be located such that minimum system upgrades are necessary. Similarly, it also considers where the DERs can most benefit a distribution system, considering the unique assets and performance of each distribution feeder. A proactive approach can be applied across an entire distribution system to account for all feeders and their unique limitations and capabilities.

Why is this more proactive approach important? If a more proactive approach is not taken, there could potentially be increased cost or decreased reliability and power quality for all customers. If a proactive planning approach is instituted, costs can be minimized, reliability and power quality can be maintained, and system performance can be improved by taking advantage of the DERs and their capabilities.

What Matters Most

While planning functions do not change, certain additional items should be considered to better integrate DERs and quantify the overall impact of such resources more precisely. The main factors that need to be considered in a proactive planning approach with DERs are (1) DER size and location, (2) the distribution system's response characteristics, and (3) DER technologies. All of these factors are critically important when planning the integrated grid.

Size and Location

Centralized, utility-scale DERs have a specific impact on the grid that is based on where they are located along a distribution feeder. Dispersed, customer-based generation has a quite different impact, even though the same amount of electricity may be produced.

Distribution Feeder Response Characteristics

Even though feeders may appear to have similar characteristics (such as voltage class, topology, and load), analyses performed by the Electric Power Research Institute (EPRI) have shown that easily observed feeder characteristics alone cannot definitively and accurately determine how many DERs a feeder can accommodate. A distribution feeder's response characteristics (like its voltage profile and short-circuit strength) can be used, however, to help determine what level of DERs can be accommodated and where.

DER Technologies

Variable generation such as wind and solar can have widely varying impacts on voltage and capacity value compared with forms of generation that are dispatchable. The differences primarily emanate from the timing in which the electricity is generated and the character of the energy output. Rotating machines and static inverter interfaces have different impacts on system protection, for example.

Considering all of these factors can be rather challenging, particularly due to:

- the sheer number of feeders associated with analyzing entire systems;
- limited manpower and data with which to perform detailed analysis of an entire distribution system;
- the reduced visibility in terms of what is happening at the edge of the grid (important for models and measurements); and
- the wide range of results that can be seen based on the particular scenarios investigated (location, PV size, control, and so on).

To work around these challenges, planners have tried to estimate DERs impacts based on load levels alone or have simplified the problem by limiting the analysis to only a few "typical" distribution feeders. For the latter solution, feeders are chosen using clustering, or the grouping of feeders by topological characteristics. The downside of this approach is that it does not take into account the unique distribution system response as a whole, since each individual feeder dictates how the distribution system responds and changes as the system evolves over time.

These complex associations are the reason why one feeder can host more DERs than another. The hosting capacity of a feeder is defined as the amount of DERs a feeder can support under its existing topology, configuration, and physical response characteristics. When this hosting capacity is reached, any further additions will result in a deterioration of service if remedial actions are not taken. If done properly, a hosting capacity assessment provides a range of information, including:

- how many DERs can be accommodated without system upgrades
- what issues arise at the hosting capacity limits
- where DERs can be sited so that problems can be avoided
- the locations where additional DERs are likely to cause issues on the grid.

Figure 3 illustrates results for these four topics, in this case specifically related to distributed PVs. The amount of PVs that can be supported for 28 different feeders is illustrated in Figure 3(a), where the colored regions represent no issues (impact below threshold), issues dependent on location, and issues regardless of location (impact above threshold). In many cases, a range of impacts is observed even where PVs can be more or less optimally placed, based on its location. The issues that arise are because of the increase in impact that comes along with the increased penetration of DERs, as illustrated in Figure 3(b).

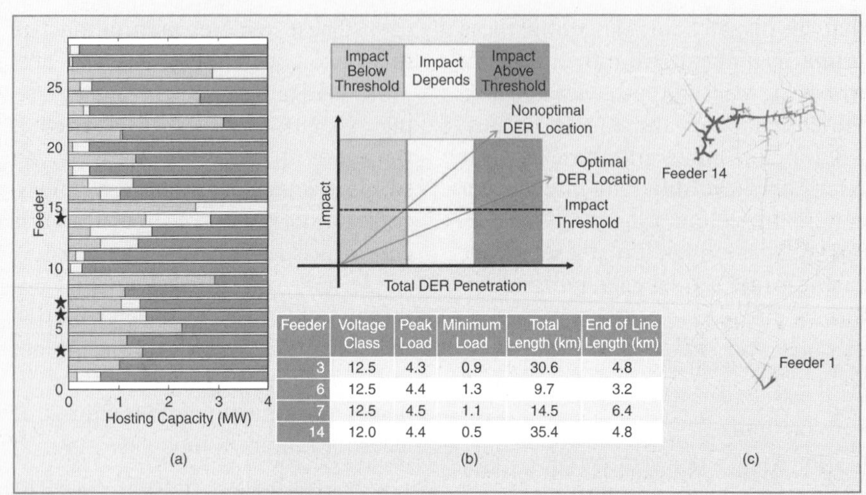

Feeder	Voltage Class	Peak Load	Minimum Load	Total Length (km)	End of Line Length (km)
3	12.5	4.3	0.9	30.6	4.8
6	12.5	4.4	1.3	9.7	3.2
7	12.5	4.5	1.1	14.5	6.4
14	12.0	4.4	0.5	35.4	4.8

Figure 3 (a) Feeder hosting capacity, (b) impact of DER penetration and table showing the characteristics of four apparently similar feeders, and (c) determining the optimal location for DERs

The two images in Figure 3(c) offer a visual interpretation of the hosting capacity plotted against the schematic of the feeder. The darkness and thickness of the schematic indicate where the DERs can be sited such that adverse issues are avoided and, conversely, where the DERs are more likely to be problematic.

One observation that can readily be made is that there is a considerable range in the amount of DERs that can be easily accommodated across all the feeders without taking preventative measures. Feeders 3, 6, 7, and 14, in particular, all have similar peak loads of 4.3–4.5 MW, as shown in Figure 3(b). The minimum hosting capacity—as indicated by the left area in Figure 3(a)—for these four feeders, however, ranges from 0.6–1.5 MW. Therefore, indicators such as peak load fail to provide adequate estimates of DER hosting capacity.

A slightly more detailed approach would consider several feeder characteristics to better understand their impact on a feeder's ability to accommodate PVs. The four feeders in the previous example are also in the same voltage class, but it is clear that basing the analysis on those two characteristics alone will not improve a determination of impact. Additional characteristics such as end-of-feeder length could be used in the clustering, but there is not much variation in those characteristics for the four feeders: they are 3.2–6.4 km long. Out of the 28-feeder data set, the end-of-line length ranges from 1.6 to 43.5 km, so including line length in a clustering methodology may still group these four feeders together as being "similar."

In short, such techniques for screening feeders and/or reducing the number of feeders analyzed reduces the time and effort that go into analysis. Both these techniques result in less-than-optimal visibility in terms of how DERs will potentially impact the grid, however, particularly when it comes to location. Feeder clustering can be used, however, when the clustering is based on *the actual feeder response* rather than topological data. Taking into consideration such things as where the DERs are located as well as how the feeder responds provides a better metric for clustering.

There is a complex interrelationship among the forces that determine how electricity flows in a conventional distribution system, and that complexity deepens when DERs produce two-way flows. The location, size, and DER technology are important considerations and are interdependent with system strength and customer loads. A holistic plan must take into account the unique performance of the entire distribution system as well as the characteristics of DERs and how they affect the grid. Such an integrated approach is the foundation of a new methodology developed by EPRI for incorporation into distribution planning.

The Plan: An Integrated Approach

Distribution system assessment for DERs must focus on incorporating DERs while maintaining established standards of reliability and power quality. In other words, the first step in the plan is to establish the distribution feeder's ability to host DERs (determine its hosting capacity). As discussed previously, hosting capacity is the amount of DERs that can be accommodated under current grid conditions without affecting power quality or reliability.

Understanding the hosting capacity under the system's current arrangement is, however, only the first step. To provide the proper analytical perspective, the analysis must be taken further to:

- analyze the ability to expand hosting capacity through a series of mitigation options (if necessary); and
- investigate the benefits that DERs can provide to the particular distribution system.

EPRI has developed and demonstrated methods for conducting feeder-specific analyses that take all of these factors into account. Based on experience from detailed individual feeder analyses, EPRI has developed a new methodology that makes possible the practical characterization of the impacts of various levels and arrangements of DERs on all of a utility's feeders. The methodology uses commercially available analysis tools for most of its steps, in particular, for determining the hosting capacity based on voltage and protection evaluations.

EPRI's methods for characterizing the benefits and impacts on distribution of DERs are contained in the following five root-cause categories, as shown in Figure 4:

- voltage
- protection
- energy
- capacity
- reliability

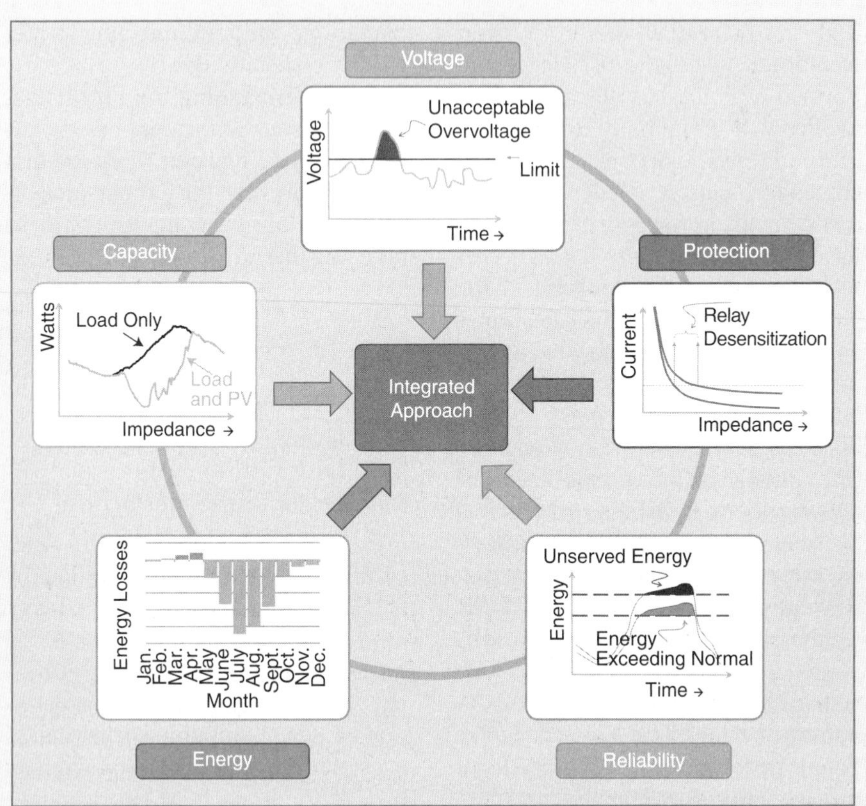

Figure 4 The five categories used in EPRI's performance assessment methodology for characterizing distribution feeders

Voltage

Utilities design the distribution system so as to maintain primary voltages within standard ranges; typically, primary voltages are kept within ±5% of the nominal rating. DER installations have the potential to change voltage along a distribution feeder because of the power they inject into the grid. For variable generation, such as PVs and wind, unacceptable voltages can occur on either the primary or secondary systems, causing overvoltages and/or excessive voltage deviations that can affect regulation equipment and result in flicker. Alternatively, when coordinated with utility regulation, additional voltage support can be achieved by means of dispatchable DERs and DERs with reactive power control capability.

Standard load flow analyses coupled with appropriate approximations for the distribution of DERs can effectively analyze the potential voltage benefits and impacts. Voltage magnitude, voltage deviation, and voltage imbalance are examined at all feeder primary and secondary nodes as well as at voltage regulation points. Various load levels, which also dictate baseline response, are considered as well to better capture the range of impacts that can be expected.

Protection

Utilities must retain the ability to detect and isolate faults as well as provide service restoration to all customers in a timely fashion. Additional DERs can affect the utility's ability to perform these functions. Common impacts from integration of DERs include: nuisance fuse blowing, misoperation of equipment, increased short-circuit current, unintentional islanding, and sympathetic tripping of the feeder.

Standard fault current analysis can be used to compare the fault response with and without the DERs to evaluate the potential impact on system protection. Impact assessment includes the examination of feeder fault current magnitudes that can cause inadvertent or faulty operation of protection devices.

Energy

DER installations have the potential to reduce distribution losses because the generation is provided closer to where the energy is consumed. The extent to which DERs can reduce distribution losses depends on the location of the resource and the length of time for which the energy is provided to the grid. Alternatively, a change in customer voltage may counteract some of the change in losses.

Time-series analysis captures the time-coincident nature, or lack thereof, of the DER and load profiles so as to quantify the energy benefits and impacts. Annual 8760-hour simulation analyses are conducted with sequential load flow analyses that incorporate the varying load and generation profiles along with the automatic control devices such as load tap changers.

Capacity

Distribution systems are designed to provide service to all customers, especially at peak load periods when assets are most constrained. A potential benefit of integrating DERs into the distribution system is their ability to reduce net feeder demand and relieve capacity on existing distribution infrastructure, potentially deferring distribution-capacity upgrades. For any resource to potentially provide distribution-capacity relief, it must be analyzed in terms of its ability to provide capacity during peak load periods when assets are most constrained and when feeder capacity is a limiting factor.

Capacity analysis requires specific information regarding thermal characteristics, the associated load profile characteristics, and projected load growth. Hence, capacity is analyzed separately for each feeder and substation to identify the potential benefits (in terms of asset investment deferral) arising from power being generated locally, as well as any adverse consequences of two-way power flows on feeder carrying capacity. Utilities have developed different approaches to evaluating capacity, and therefore the specific metrics and methods used

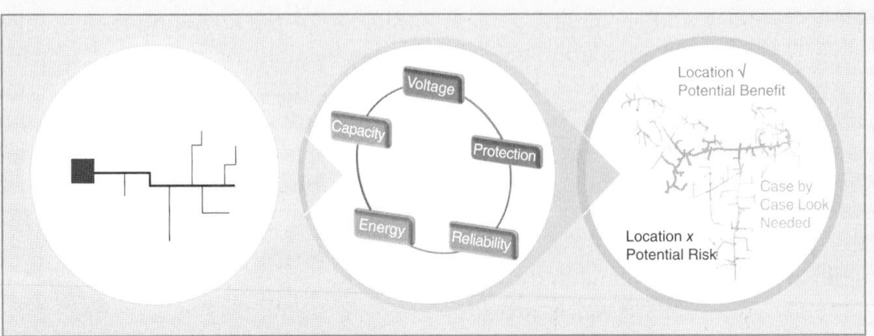

Figure 5 Integrated approach to determining the optimal placement of DERs

to evaluate asset loading and loss of life are taken into account.

Reliability

Reliability is a measure of the number and duration of interruptions of electrical service experienced by consumers. DER installations have the potential to improve reliability, but they must use dependable technologies and be sited in locations on the distribution system where they can effectively deliver power during system failure events. As with capacity, DER output must be available at the time of need to improve distribution reliability.

Quantifying the impact of DERs on reliability is a difficult task even when the DERs are dispatchable. The analysis is by nature probabilistic because failures have a high degree of uncertainty. Researchers have come up with various approaches to probabilistic planning that, for the most part, have not been embraced by utility planners. Utility distribution planners prefer more deterministic methods that are based on average failure rates and assume predictable interactions among circuit components. EPRI's plan has yet to embrace a preferred method.

An overall assessment must consider the incremental impacts—both positive and negative—that result from DER additions. The overall assessment must also consider the interrelationship of the benefits and impacts from each of the five categories. Once the full set of assessments have been performed, the result is a clearer picture of the full range of response that can be expected on a distribution feeder, one that considers unique feeder characteristics, constraints, and capabilities as well as the location-specific impacts of DERs (see Figure 5).

Performing such a thorough analysis can be challenging. The greatest challenge lies in analyzing all feeders within a service territory, which is necessary to better quantify the impacts of DERs across a planning area and/or system and thus determine the overall costs and benefits that should be captured in an integration study.

Systemwide Application

In 2010, EPRI began a detailed investigation into the impact that increasing levels of DERs, specifically PVs, could have on distribution system performance. Approximately 6 million uniquely different solar deployment cases and the resulting feeder response outcomes have been analyzed across 34 feeders throughout North America using the open-source distribution system simulator OpenDSS. A glimpse of the overall hosting capacities was provided in Figure 3.

From the outset, the goal of performing such a thorough, detailed analysis was that lessons should be learned and trends observed to develop better methods and tools. These improved methods would then let utilities efficiently and confidently answer such questions as: "What level of PVs is going to affect my system, and where will these issues occur?"

Throughout 2013 and 2014, EPRI analyzed a vast quantity of feeder hosting capacity results to identify trends. These trends form the basis for a streamlined method of determining, on a feeder-by-feeder basis, four important items:

- the hosting capacities for individual feeders;
- the locations where PVs can be placed without requiring system upgrades (optimal versus nonoptimal locations);
- the issues that can arise due to increased levels of PVs; and
- mitigation solutions for integrated DERs beyond the base hosting capacity level.

This streamlined method for evaluating the hosting capacity of individual feeders can also be performed across an entire distribution system in an automated fashion, as shown in Figure 6. Efficient implementation also enables qualified users to evaluate mitigation solutions on a feeder-by-feeder basis. As the distribution system changes over time, the assessment method can be repeated as needed to account for changes to the

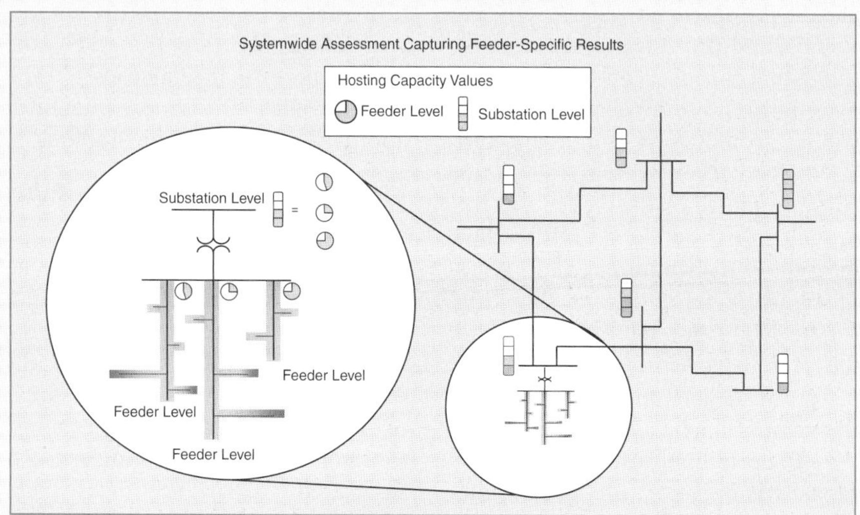

Figure 6 An illustration of system-level hosting capacity results

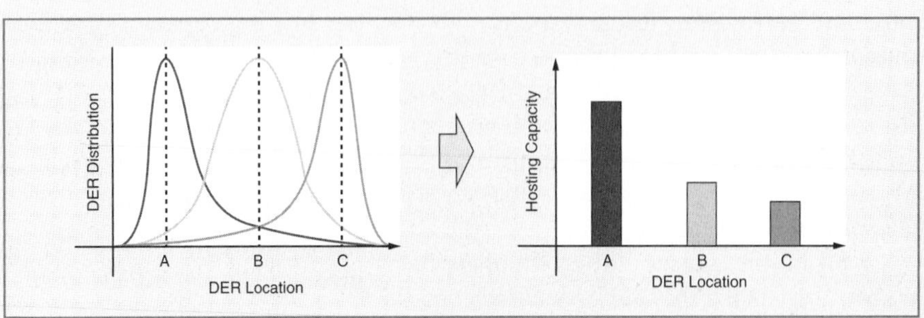

Figure 7 A simplistic illustration of the location and hosting capacity for a streamlined approach. The distribution describes what was evaluated, and the hosting capacity describes the outcome.

distribution system, such as reconfigurations implemented to accommodate load growth or for contingency reasons. This aspect is critical when there is a need to understand the impacts (in terms of cost and value) of DERs spread across an entire distribution service territory.

This streamlined hosting capacity method utilizes the feeder model to determine voltage profiles, short-circuit impedances, feeder topologies, and customer characteristics. Voltage and protection issues related to hosting capacity can be effectively determined by correlating several voltage profiles corresponding to different load levels with the shortcircuit impedance of the feeder and the probable distribution of DERs. The locational distributions and resultant hosting capacities are shown in Figure 7. The number of laterals and the overall feeder topology are also factors instrumental in determining the appropriate hosting capacity.

Determining a feeder's hosting capacity is only part of the solution, however. Additional systemwide benefits and impacts can be quantified and garnered through proper analysis. In a manner similar to the streamlined method for performing hosting capacity assessments with respect to voltage and protection impacts, energy and capacity assessments across the fleet can be performed as well.

Various applications can be realized after evaluating the distribution system in a more holistic manner by applying such a method across an entire distribution system and taking into consideration: (1) DER characteristics, (2) locational impacts, and (3) grid performance metrics. Several key outcomes of this more integrated approach are described in Table 1.

Programmatic Implementation of a Proactive Plan

This new knowledge lets utilities develop strategies that can influence the locational deployment of DERs. Distribution utilities can tactically direct DER development to areas of the grid that are better suited to accommodate the resource. Costs associated with the deployment of DERs in suboptimal areas can be avoided. At the same time, the

Feeder-specific hosting capacities	Determine the existing hosting capacity for each feeder across the entire distribution system under current grid configurations
Optimal DER location	Identify locations that can minimize the upgrades necessary to accommodate DERs
Improved fast-screening techniques	Improve screening techniques that efficiently account for the proposed DERs and associated grid capacity at that location
Substation-level hosting capacities	Improve visibility into substation-level capacity for accommodating PVs connected at the individual feeder level
Range of issues and costs for accommodating DERs across the distribution system	Provide better insight into the specific issues that can arise and where and how often they may occur throughout the system
Range of values that can be garnered through widespread adoption of DERs	Provide better insight about the true value of widespread DER adoption throughout the system
Aggregate amount of DERs across a system for bulk system analysis	Identify locations and aggregate DERs for improved bulk system studies
Impact of DERs on system reconfiguration	Streamline processes that can be repeated in an automated fashion for additional distribution configurations

TABLE 1

The Potential Outcomes of an Integrated Approach

benefits of DERs can be used to support utility operations.

The targeted siting of future installations could potentially be influenced through a variety of mechanisms that leverage rates, grid interconnection processes, enhanced information systems, and/or distribution system upgrades. Examples include interconnection process improvements, locational incentives or interconnection costs, and holistic distribution upgrades. Strategies vary in effectiveness and ease of implementation; many of these solutions require that challenges due to regulatory constraints, concerns about customer equity, and trade-offs between DER and non-DER customers be overcome.

There are a number of industry activities under way that are attempting to develop and/or implement a broader, more proactive planning approach at both the regulatory and utility levels. They include:

- **California Public Utilities Commission's distribution resource plans:** In August 2014, the California Public Utilities Commission opened a rule-making process that seeks to establish policies and procedures for developing investor-owned utility distribution resource plans in the state that integrate DERs into long-range grid plans by considering a range of locational values, including energy, capacity, power quality, reliability, and resiliency.

- **Tennessee Valley Authority's (TVA's) Distributed Generation Integrated Value (DG-IV) initiative:** In 2014, TVA launched a process to develop a methodology to adequately gauge the value DG provides to the electric grid. The goal of this effort is to develop a method that takes into account locational impact and cost.
- **EPRI's Integrated Grid framework:** The Integrated Grid framework recently developed by EPRI is intended to provide a more comprehensive understanding of DERs impacts to present a more nuanced range of cost-effective integration strategies to utilities and grid operators.

Moving Forward

The electric power system is evolving from large, central generating stations interconnected with customers through grids of transmission and distribution lines into a system that includes substantial DERs. Consumers of all sizes are installing DERs with technical and economic attributes that differ radically from the central energy resources that have traditionally dominated the electric power system. In many settings, DER installations have already had a sizable impact on the operation of the electricity grid.

If utilities had the luxury of investing in all the power delivery capacity they needed to support whatever might be connected to the distribution system, planning would be simply a matter of constructing new wire-delivery facilities as needed. But they do not, and utility engineers increasingly find their systems stretched to the limits. Control and planning must be done with ever greater accuracy for utilities to operate successfully.

New engineering graduates will require training to analyze the integrated distribution grid, in which power flows in multiple directions and can change quickly. For example, most power engineering curricula today teach how to perform a static power flow analysis with simple load and generator models, but such deterministic approaches are no longer sufficient. Engineers must also learn how to capture the dynamic nature of a system that contains rapidly varying DERs and identify the time-dependent benefits and costs of DERs as well as their location-specific impacts as derived from traditional static methods. This will require a collaborative effort on the part of utility, research, and academic organizations to determine the new curricula.

Planners have to take into account how DER devices behave as well as the complexity of the distribution system. This will require new capabilities in planning tools and new approaches to distribution planning. It will not be easy to make the transition in one leap: with a massive investment in models and data, existing tools will slowly evolve from their present state toward a new paradigm.

A planning approach for the electric grid that is integrated overall will fully realize the value of DERs while cost-effectively serving all customers at established standards of quality and reliability. Such a holistic approach to distribution planning is one of the core components in what EPRI refers to as an integrated grid.

For Further Reading

"The integrated grid, phase II: A framework," EPRI, Palo Alto, CA , 2015.

"Distribution feeder hosting capacity: What matters most in planning for DER ?" EPRI, Palo Alto, CA, Tech. Rep. 3002004777, 2015.

"A new method for characterizing distribution system hosting capacity for DER: A streamlined approach for PV," EPRI, Palo Alto, CA, Tech. Rep. 3002003278, 2014.

Streamlined Methods for Determining Feeder Hosting Capacity for PV. Palo Alto, CA: EPRI, 2014.

"Distributed photovoltaic feeder analysis: Preliminary findings from hosting capacity analysis of 18 distribution feeders," EPRI, Palo Alto, CA, Tech. Rep. 3002001245, 2013.

"Analysis of high-penetration solar PV impacts for distribution planning: Stochastic and time-series methods for determining feeder hosting capacity," EPRI, Palo Alto, CA, Tech. Rep. 1026640, 2012.

S. K. Price and R. C. Dugan, "Including distributed resources in distribution planning," in *Proc. IEEE Power Systems Conf. Expo., PES* 2005, New York, NY, Oct. 2004.

Biographies

Jeff Smith is with the Electric Power Research Institute, Knoxville, Tennessee.

Matthew Rylander is with the Electric Power Research Institute, Knoxville, Tennessee.

Lindsey Rogers is with the Electric Power Research Institute, Knoxville, Tennessee.

Roger Dugan is with the Electric Power Research Institute, Knoxville, Tennessee. ■

14.1 INTRODUCTION TO DISTRIBUTION

Figure 14.1 shows the basic components of an electric power system [1–9]. Power plants convert energy from fuel (coal, gas, nuclear, oil, etc.) and from water, wind, or other forms into electric energy. Power plant generators, with typical ratings varying from 50 to 1300 MVA, are of three-phase construction, with three-phase armature windings embedded in the slots of stationary armatures. Generator terminal voltages, which are limited by material and insulation capabilities, range from a few kV for older and smaller units up to 20 kV for newer and larger units.

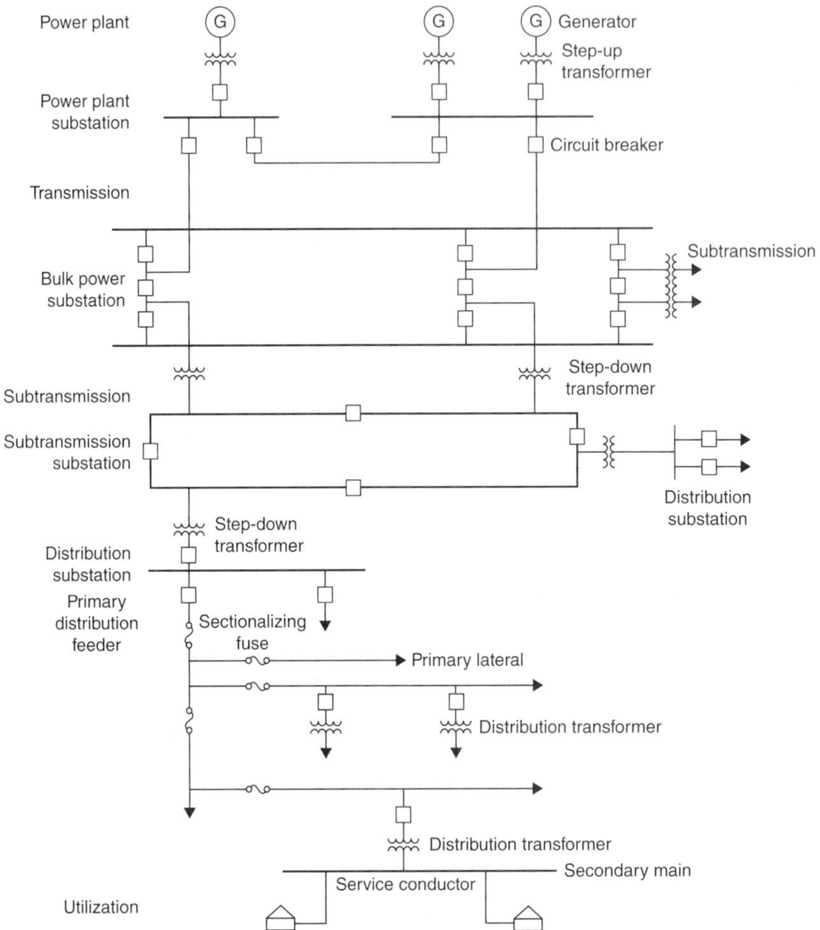

FIGURE 14.1

Basic components of an electric power system (J. D. Glover, "Electric Power Distribution," *Encyclopedia of Energy Technology and The Environment*, John Wiley & Sons, New York, NY, 1995. Copyright © 1995, John Wiley and Sons.)

To reduce transmission energy losses, generator step-up (GSU) transformers at power plant substations increase voltage and decrease current. Both the GSU transformers and the buses in these substations are protected by circuit breakers, surge arresters, and other protection equipment.

The transmission system serves three basic functions:

1. It delivers energy from generators to the system.
2. It provides for energy interchange among utilities.
3. It supplies energy to the subtransmission and distribution system.

The transmission system consists of a network of three-phase transmission lines and transmission substations, also called bulk power substations. Typical transmission voltages range from 230 up to 765 kV. Single-circuit three-phase ratings vary from 400 MVA at 230 kV up to 4000 MVA at 765 kV. In some cases, HVDC lines with solid-state converters are embedded in the transmission system as well as back-to-back ac-dc links.

The subtransmission system consists of step-down transformers, substations, and subtransmission lines that connect bulk power substations to distribution substations. In some cases, a subtransmission line may be tapped, usually through a circuit breaker, to supply a single-customer distribution load such as a large industrial plant. Typical subtransmission voltages range from 69 to 138 kV.

Distribution substations include step-down transformers (distribution substation transformers) that decrease subtransmission voltages to primary distribution voltages in the 2.2- to 46-kV range for local distribution. These transformers connect through associated circuit breaker and surge arrester protection to substation buses, which in turn connect through circuit breakers to three-phase primary distribution lines called distribution circuits or feeders. Each substation bus usually supplies several feeders. Typical distribution substation ratings vary from 15 MVA for older substations to 200 MVA or higher for newer installations. Distribution substations may also include equipment for regulating the primary voltage, such as load tap changers (LTCs) on the distribution substation transformers or separate voltage regulators.

Typical primary distribution feeder ratings include 4 MVA for 4.16 kV, 12 MVA for 13.8 kV, 20 MVA for 22.9 kV, and 30 MVA for 34.5-kV feeders. Feeders are usually segregated into several three-phase sections connected through sectionalizing fuses or switches. Each feeder section may have several single-phase laterals connected to it through fuses. Three-phase laterals may also be connected to the feeders through fuses or reclosers. Separate, dedicated primary feeders supply industrial or large commercial loads.

Feeders and laterals run along streets, as either overhead lines or underground cables, and supply distribution transformers that step the voltage down to the secondary distribution level (120 to 480 V). Distribution transformers, typically rated 5 to 5000 kVA, are installed on utility poles for overhead lines, on pads at ground level, or in vaults for underground cables. Distribution transformers are protected from overloads and faults by fuses or circuit breakers on the primary and/or the secondary side. From these transformers, energy flows through secondary mains and service conductors to supply single- or three-phase power directly to customer loads (residential, commercial, and light industrial).

Service conductors connect through meters, which determine kilowatt-hour consumption for customer billing purposes as well as other data for planning and operating purposes, to service panels located on customers' premises. Customers' service panels contain circuit breakers or fuses that connect to wiring that in turn supplies energy for utilization devices (lighting, appliances, motors, heating-ventilation-air conditioning, etc.).

Distribution of electric energy from distribution substations to meters at customers' premises has two parts:

1. Primary distribution, which distributes energy in the 2.2- to 46-kV range from distribution substations to distribution transformers, where the voltage is stepped down to customer utilization levels.
2. Secondary distribution, which distributes energy at customer utilization voltages of 120 to 480 V to meters at customers' premises.

14.2 PRIMARY DISTRIBUTION

Table 14.1 shows typical primary distribution voltages in the United States [1–9]. Primary voltages in the "15-kV class" predominate among U.S. utilities. The 2.5- and 5-kV classes are older primary voltages that are gradually being replaced by 15-kV class primaries. In some cases, higher 25- to 34.5-kV classes are used in new high-density load areas as well as in rural areas that have long feeders.

The three-phase, four-wire multigrounded primary system is the most widely used. Under balanced operating conditions, the voltage of each phase is equal in magnitude and 120° out of phase with each of the other two phases. The fourth wire in these Y-connected systems is used as a neutral for the primaries, or as a common neutral when both primaries and secondaries are present. Usually the windings of distribution substation transformers are Y-connected on the primary distribution side, with the neutral point grounded and connected to the common neutral wire. The neutral is also grounded at frequent intervals along the primary, at distribution transformers, and at customers' service entrances. Sometimes distribution substation transformers are grounded through an impedance (approximately one ohm) to limit short-circuit currents and improve coordination of protective devices.

The three-wire delta primary system is also popular, although not as widely used as the four-wire multigrounded primary system. Three-wire delta primary systems are not being actively expanded. They are generally older and lower in voltage than the four-wire multigrounded type. They are also popular in industrial systems.

Rural areas with low-density loads are usually served by overhead primary lines with distribution transformers, fuses, switches, and other equipment mounted on poles. Urban areas with high-density loads are served by underground cable systems with

Class, kV	Voltage, kV
2.5	2.4
5	4.16
8.66	7.2
15	12.47
	13.2
	13.8
25	22.9
	24.94
34.5	34.5

TABLE 14.1

Typical Primary Distribution Voltages in the United States

Based on (J. D. Glover, "Electric Power Distribution," *Encyclopedia of Energy Technology and The Environment,* John Wiley & Sons, New York, NY, 1995.)

distribution transformers and switchgear installed in underground vaults or in ground-level cabinets. There is also an increasing trend toward underground residential distribution (URD), particularly single-phase primaries serving residential areas. Underground cable systems are highly reliable and usually unaffected by weather. But the installation costs of underground distribution are significantly higher than overhead.

Primary distribution includes three basic systems:

1. Radial
2. Loop
3. Primary network systems

PRIMARY RADIAL SYSTEMS

The primary radial system, as illustrated in Figure 14.2, is a widely used, economical system often found in low-load-density areas [1, 3, 4, and 9]. It consists of separate three-phase feeder mains (or feeders) emanating from a distribution substation in a radial fashion, with each feeder serving a given geographical area. The photograph at the beginning of this chapter shows a distribution substation that supplies six radial feeders for a suburban residential area. A three-phase feeder main can be as short

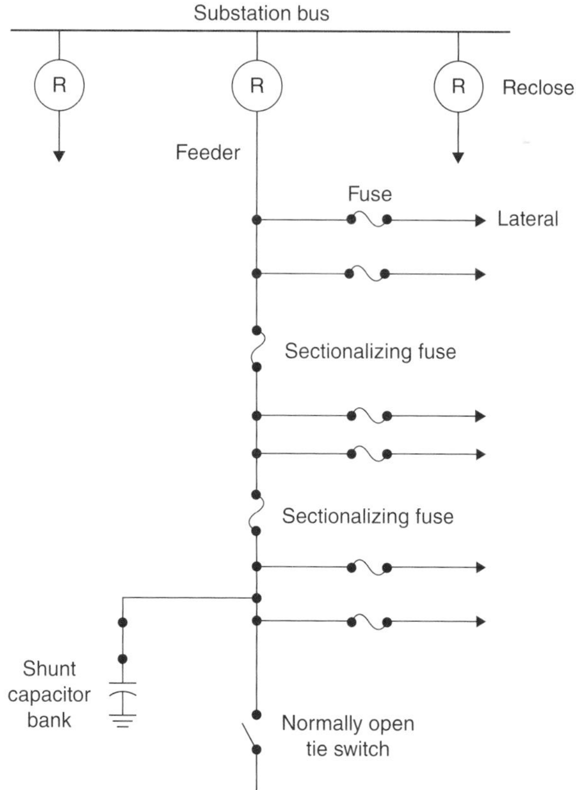

FIGURE 14.2 Primary radial system (J. D. Glover, "Electric Power Distribution," *Encyclopedia of Energy Technology and The Environment*, John Wiley & Sons, New York, NY, 1995. Copyright © 1995, John Wiley and Sons.)

as a mile or two or as long as 48 km. Single-phase laterals (or branches) are usually connected to feeders through fuses, so that a fault on a branch can be cleared without interrupting the feeder. Single-phase laterals are connected to different phases of the feeder, so as to balance the loading on the three phases.

To reduce the duration of interruptions, overhead feeders can be protected by automatic reclosing devices located at the distribution substation, at the first overhead pole, or at other locations along the feeder [11]. As an example, Figure 14.3 shows a pole-mount recloser for a 22.9-kV circuit. Studies have shown that the large majority of faults on overhead primaries are temporary, caused by lightning flashover of line insulators, momentary contact of two conductors, momentary bird or animal contact, or momentary tree limb contact. The recloser or circuit breaker with reclosing relays opens the circuit either "instantaneously" or with intentional time delay when a fault occurs, and then recloses after a short period of time. The recloser can repeat this open-and-reclose operation if the fault is still on the feeder. A popular reclosing sequence is two instantaneous openings (to clear temporary faults), followed by two delayed openings (allowing time for fuses to clear persistent downstream faults), followed by opening and lockout for persistent faults between the recloser and fuses.

FIGURE 14.3

Pole-mount recloser for a three-phase 22.9-kV circuit. This recloser has an 800-A continuous current rating and a 16-kA interrupting rating. The 22.9-kV feeder is located near the top of the pole. There are two three-phase 4.16-kV circuits below the recloser. An antenna located below the 4.16-kV circuits is for remote control of the recloser from the dispatch center. A normally open bypass switch located on the top crossarm can be manually operated if the recloser fails to reclose. (Courtesy of Danvers Electric.)

For safety purposes, the reclosing feature is bypassed during live line maintenance. Reclosing is not used on circuits that are primarily underground.

To further reduce the duration and extent of customer interruptions, sectionalizing fuses are installed at selected intervals along radial feeders. In the case of a fault, one or more fuses blow to isolate the fault, and the unfaulted section upstream remains energized. In addition, normally open tie switches to adjacent feeders are incorporated, so that during emergencies, unfaulted sections of a feeder can be tied to the adjacent feeder. Spare capacity is often allocated to feeders to prevent overloads during such emergencies, or there may be enough diversity between loads on adjacent feeders to eliminate the need for spare capacity. Many utilities have also installed automatic fault-locating equipment and remote controlled sectionalizers (controlled switches) at intervals along radial lines, so that faulted sections of a feeder can be isolated and unfaulted sections reenergized rapidly from a dispatch center, before the repair crew is sent out. Figure 14.4 shows a radio-controlled sectionalizing switch on a 22.9-kV circuit.

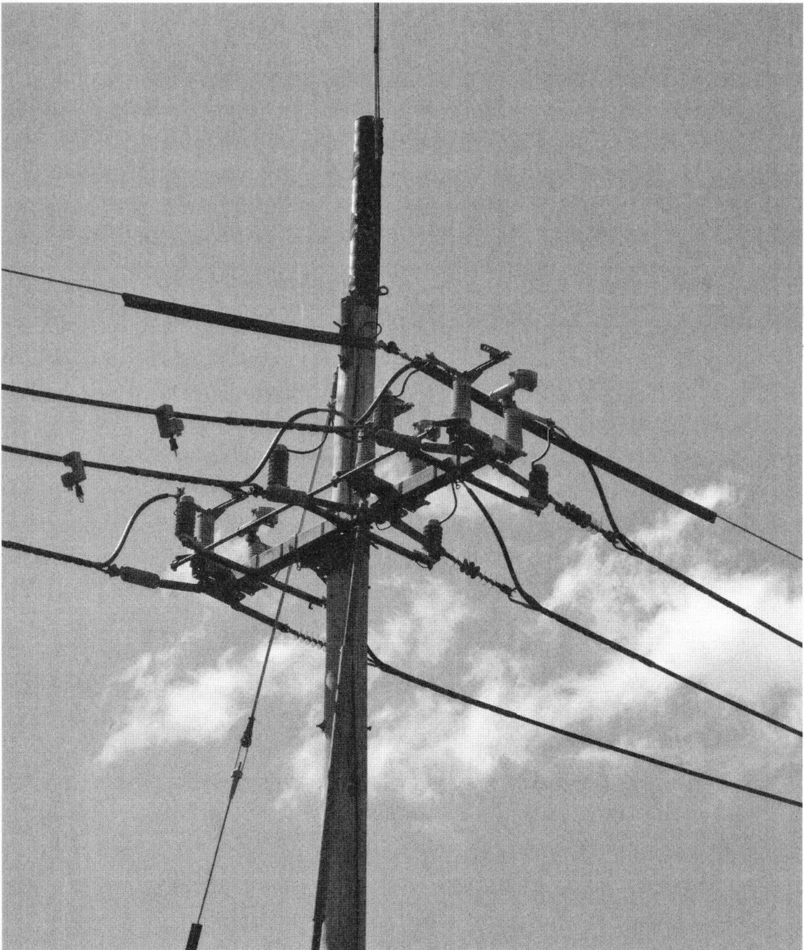

FIGURE 14.4

S&C normally open radio-controlled sectionalizing switch on a 22.9-kV circuit. This switch has a 1200-A load-break capability. (Courtesy of Danvers Electric.)

FIGURE 14.5 Pole-mount three-phase 1800 kvar shunt capacitor bank for a 22.9-kV circuit. The capacitor bank, which is protected by 50-A type K fuses, is a switched bank. (Courtesy of Danvers Electric.)

Shunt capacitor banks including fixed and switched banks are used on primary feeders to reduce voltage drop, reduce power losses, and improve power factor. Capacitors are typically switched off during the night for light loads and switched on during the day for heavy loads. Figure 14.5 shows a pole-mount switched capacitor bank. Computer programs are available to determine the number, size, and location of capacitor banks that optimize voltage profile, power factor, and installation and operating costs. In some cases, voltage regulators are used on primary feeders.

One or more additional, independent feeders along separate routes may be provided for critical loads, such as hospitals that cannot tolerate long interruptions. Switching from the normal feeder to an alternate feeder can be done manually or automatically with circuit breakers and electrical interlocks to prevent the connection of a good feeder to a faulted feeder. Figure 14.6 shows a primary selective

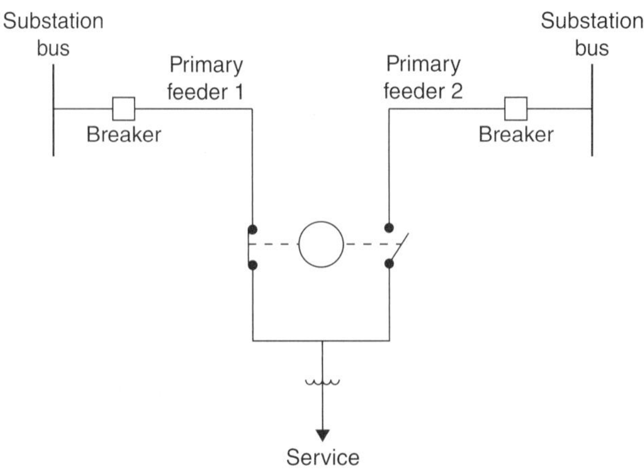

FIGURE 14.6 Primary selective system (J. D. Glover, "Electric Power Distribution," *Encyclopedia of Energy Technology and The Environment*, John Wiley & Sons, New York, NY, 1995. Copyright © 1995, John Wiley and Sons.)

system, often used to supply concentrated loads over 300 kVA [3, 4, and 9]. There are two primary feeders with automatic switching in front (upstream) of the distribution transformer. In case of feeder loss, automatic transfer to the other feeder is rapid and does not require fault locating before transfer.

PRIMARY LOOP SYSTEMS

The primary loop system, as illustrated in Figure 14.7 for overhead, is used where high service reliability is important [1, 3, 4 and 9]. The feeder loops around a load area and returns to the distribution substation, especially providing two-way feed from the substation. The size of the feeder conductors, which are kept the same throughout the loop, is usually selected to carry the entire load connected to the loop, including future load growth. Reclosers and tie switches (sectionalizers) are used to reduce customer interruptions and isolate faulted sections of the loop. The loop is normally operated with the tie switch (or tie recloser) open. Power to a customer at any one time is supplied through a single path from the distribution substation, depending on the open/close status of the reclosers/sectionalizers. Each of the circuit breakers at the distribution substation can be connected to separate bus sections and fed from separate distribution substation transformers.

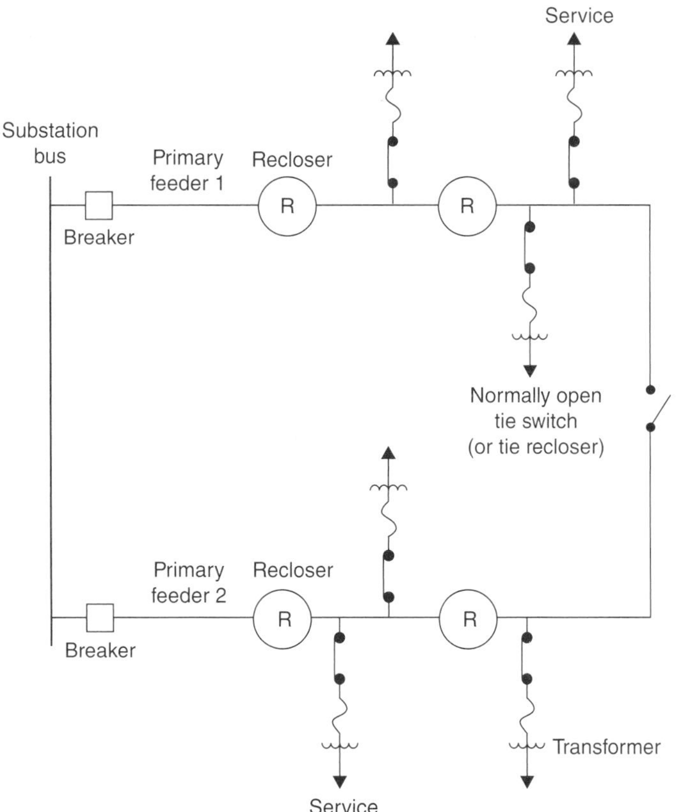

FIGURE 14.7

Overhead primary loop
(J. D. Glover, "Electric Power Distribution," *Encyclopedia of Energy Technology and The Environment*, John Wiley & Sons, New York, NY, 1995. Copyright © 1995, John Wiley and Sons.)

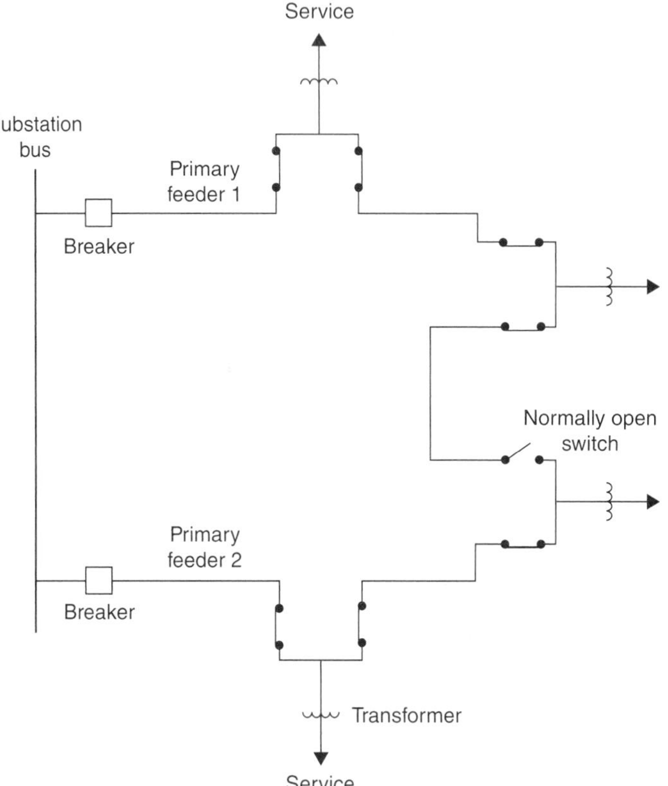

FIGURE 14.8

Underground primary loop (J. D. Glover, "Electric Power Distribution," *Encyclopedia of Energy Technology and The Environment*, John Wiley & Sons, New York, NY, 1995. Copyright © 1995, John Wiley and Sons.)

Figure 14.8 shows a typical primary loop for underground residential distribution (URD). The size of the cable, which is kept the same throughout the loop, is selected to carry the entire load, including future load growth. Underground primary feeder faults occur far less frequently than in overhead primaries, but are generally permanent. The duration of outages caused by primary feeder faults is the time to locate the fault and perform switching to isolate the fault and restore service. Fault locators at each distribution substation transformer help to reduce fault locating times.

PRIMARY NETWORK SYSTEMS

Although the primary network system, as illustrated in Figure 14.9, provides higher service reliability and quality than a radial or loop system, only a few primary networks remain in operation in the United States today [1, 3, 4, and 9]. They are typically found in downtown areas of large cities with high load densities. The primary network consists of a grid of interconnected feeders supplied from a number of substations. Conventional distribution substations can be replaced by smaller, self-contained unit substations at selected network locations. Adequate voltage is maintained at utilization points by voltage regulators at distribution substations and by locating distribution transformers close to major load centers on the grid. However, it is difficult to maintain adequate voltage everywhere on the primary grid

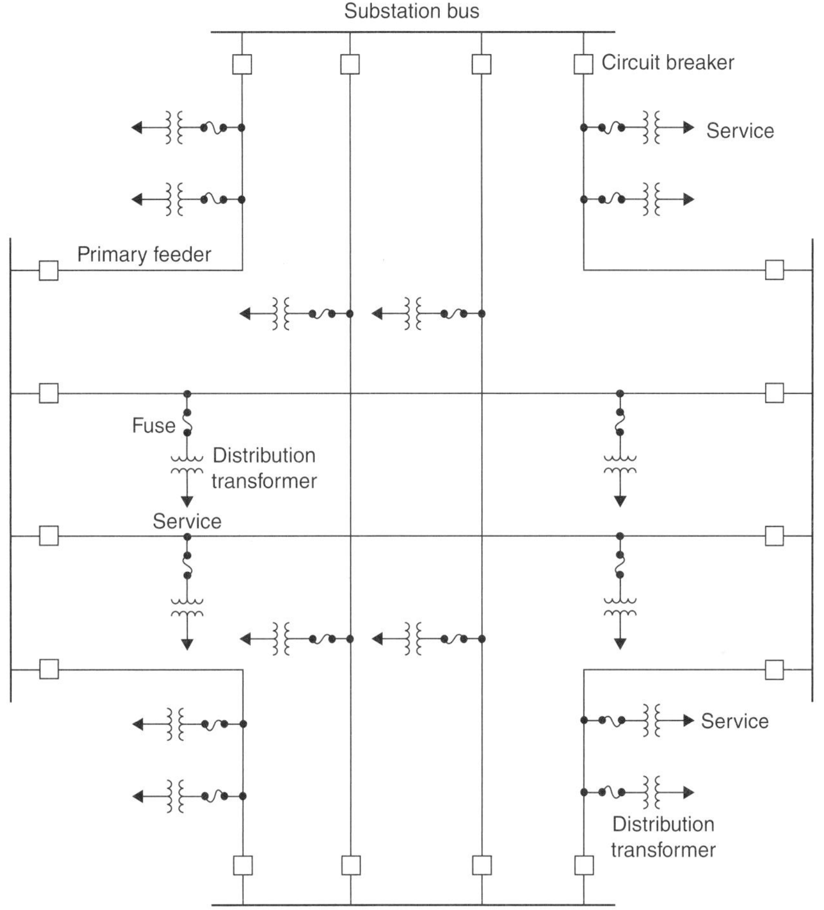

FIGURE 14.9

Primary network (J. D. Glover, "Electric Power Distribution," *Encyclopedia of Energy Technology and The Environment,* John Wiley & Sons, New York, NY, 1995. Copyright © 1995, John Wiley and Sons.)

under various operating conditions. Faults on interconnected grid feeders are cleared by circuit breakers at distribution substations, and in some cases, by fuses on the primary grid. Radial primary feeders protected by circuit breakers or fuses can be tapped off the primary grid or connected directly at distribution substations.

14.3 SECONDARY DISTRIBUTION

Secondary distribution distributes energy at customer utilization voltages from distribution transformers up to meters at customers' premises. Table 14.2 shows typical secondary voltages and applications in the United States [1–9]. In residential areas, 120/240-V, single-phase, three-wire service is the most common, where lighting loads and outlets are supplied by 120-V, single-phase connections, and large household appliances such as electric ranges, clothes dryers, water heaters, and electric space heating are supplied by 240-V, single-phase connections. In urban areas serving high-density residential and commercial loads, 108Y/120-V, three-phase, four-wire service is common,

Voltage	# Phases	# Wires	Application
120/240 V	Single-phase	Three	Residential
208Y/120 V	Three-phase	Four	Residential/Commercial
480Y/277 V	Three-phase	Four	Commercial/Industrial/High Rise

TABLE 14.2

Typical Secondary Distribution Voltages in the United States

(Based on J. D. Glover, "Electric Power Distribution," *Encyclopedia of Energy Technology and The Environment,* John Wiley & Sons, New York, NY, 1995.)

where lighting, outlets, and small motor loads are supplied by 120-V, single-phase connections, and larger motor loads are supplied by 208-V, three-phase connections. In areas with very high-density commercial and industrial loads as well as high-rise buildings, 480Y/277-V, three-phase, four-wire service is common, with fluorescent lighting supplied by 277-V, single-phase connections and motor loads supplied by 480-V, three-phase connections. Separate 120-V radial systems fed by small transformers from the 480-V system are used to supply outlets in various offices, retail stores, or rooms.

Figure 14.10 shows a typical residential customer voltage profile along a radial feeder. In accordance with ANSI standards, during normal conditions utilities in the United States are required to maintain customer voltage at the customer's service panel between 114 and 126 volts (±5%) based on a 120-V nominal secondary voltage. As shown in Figure 14.10, the first customer, closest to the substation, has the highest voltage, and the last customer, furthest from the substation, has the lowest voltage.

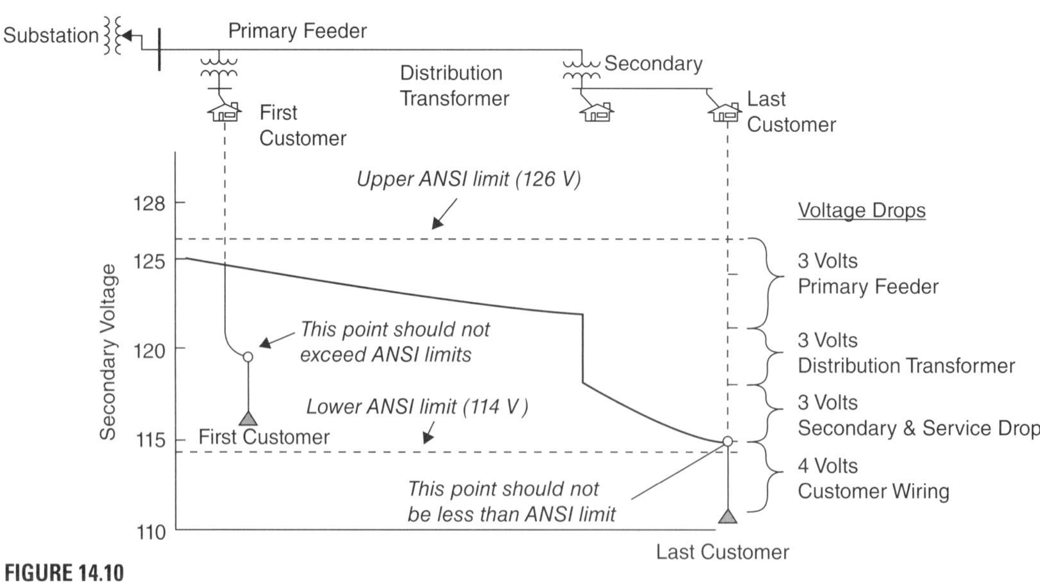

FIGURE 14.10

Typical residential customer voltage profile along a radial feeder, assuming no shunt capacitors or voltage regulators along the feeder

Proper distribution design dictates that the first customer's voltage is less than 126 V during light loads and the last customer's voltage is greater than 114 V during peak loads, so that all customers remain within 120 V ±5% under all normal loading conditions. Load-tap-changing distribution substation transformers and voltage regulators (see Section 14.4) as well as shunt capacitors (see Section 14.5) are used to maintain customer voltages within ANSI limits.

There are four general types of secondary systems:

1. Individual distribution transformer per customer
2. Common secondary main
3. Secondary network
4. Spot network

INDIVIDUAL DISTRIBUTION TRANSFORMER PER CUSTOMER

Figure 14.11 shows an individual distribution transformer with a single service supplying one customer, which is common in rural areas where distances between customers are large and long secondary mains are impractical [3 and 4]. This type of system may also be used for a customer that has an unusually large load or for a customer that would otherwise have a low-voltage problem with a common secondary main. Although transformer installation costs and operating costs due to no-load losses are higher than those of other types of secondary systems, the installation costs of secondary mains are avoided.

COMMON SECONDARY MAIN

Figure 14.12 shows a primary feeder connected through one or more distribution transformers to a common secondary main with multiple services to a group of customers [3 and 4]. This type of secondary system takes advantage of diversity among customer demands that allows a smaller capacity of the transformer supplying a group compared to the sum of the capacities of individual transformers for each customer in the group. Also, the large transformer supplying a group can handle motor starting currents and other abrupt, load changes without severe voltage drops.

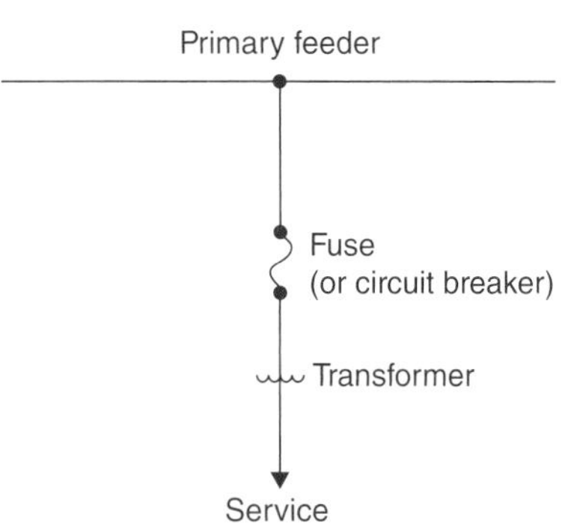

FIGURE 14.11

Individual distribution transformer supplying single-service secondary (J. D. Glover, "Electric Power Distribution," *Encyclopedia of Energy Technology and The Environment*, John Wiley & Sons, New York, NY, 1995. Copyright © 1995, John Wiley and Sons.)

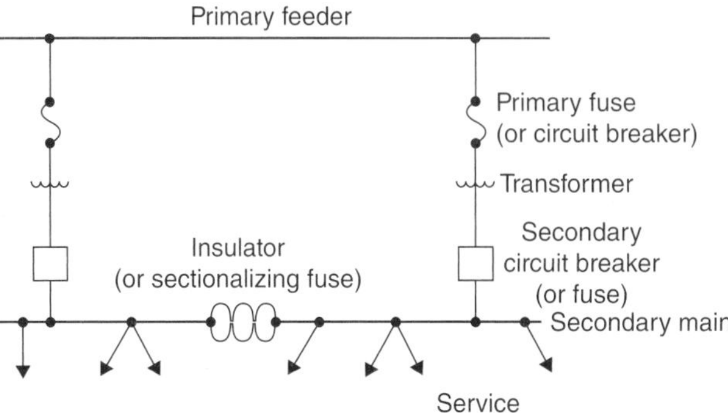

FIGURE 14.12 Common secondary main (J. D. Glover, "Electric Power Distribution," *Encyclopedia of Energy Technology and The Environment*, John Wiley & Sons, New York, NY, 1995. Copyright © 1995, John Wiley and Sons.)

In most cases, the common secondary main is divided into sections, where each section is fed by one distribution transformer and is also isolated from adjacent sections by insulators. In some cases, fuses are installed along a continuous secondary main, which results in banking of distribution transformers, also called banked secondaries.

SECONDARY NETWORK

Figure 14.13 shows a secondary network or secondary grid, which may be used to supply high-density load areas in downtown sections of cities, where the highest degree of reliability is required and revenues justify grid costs [1, 3, 4, and 9].

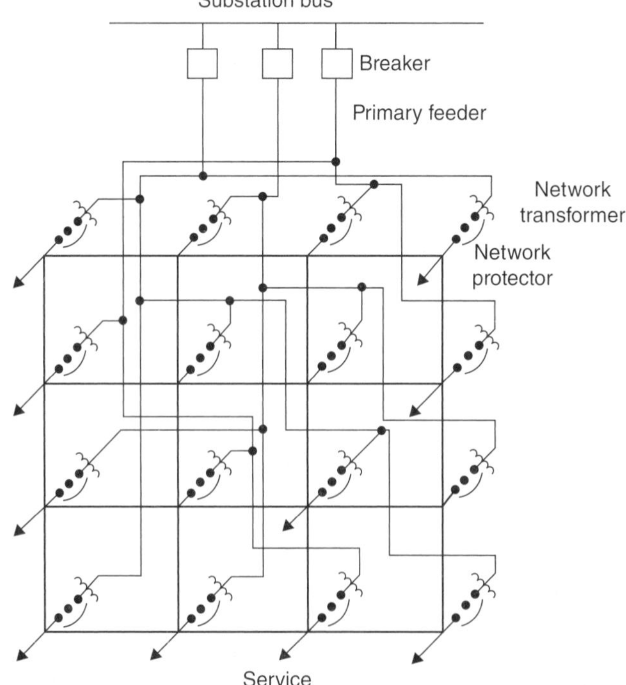

FIGURE 14.13 Secondary network (J. D. Glover, "Electric Power Distribution," *Encyclopedia of Energy Technology and The Environment*, John Wiley & Sons, New York, NY, 1995. Copyright © 1995, John Wiley and Sons.)

The underground secondary network is supplied simultaneously by two or more primary feeders through network transformers. Most networks are supplied by three or more primary feeders with transformers that have spare capacity, so that the network can operate with two feeders out of service.

Secondary grids operate at either 208Y/120-V or 480Y/277-V in the United States. Commonly used secondary cable sizes range from 4/0 to 500 kcmil (250 mm^2) AWG [5].

More than 260 cities in the United States have secondary networks [5]. New York City has the largest secondary network system in the United States with approximately 23,000 network transformers feeding various secondary networks and an online monitoring system that continuously monitors transformer loadings. Some of the secondary networks in New York City are fed by as many as 24 primary feeders operating in parallel [9].

Network transformers are protected by network protectors between the transformers and secondary mains. A network protector is an electrically operated low-voltage air circuit breaker with relays and auxiliary devices that automatically opens to disconnect the transformer from the network when the transformer or the primary feeder is faulted, or when there is a power flow reversal. The network protector also has the ability to close automatically when a feeder is energized [5]. Fuses may also be used for backup of network protectors.

In many cases especially on 208Y/120-V secondary networks, main protection of secondary cables has come from the ability of the cable system to "burn clear" with no fuse or other protective device. However, in many instances for 480Y/277-V secondary networks, this practice was not able to successfully burn clear, resulting in fires and considerable damage. As a solution, special fuses called cable limiters are commonly used at tie points in the secondary network to isolate faulted secondary cables. Cable limiters, which are designed with restricted sections of copper which act like a fuse, do not limit the magnitude of fault current like current limiting fuses. In high short-circuit locations on the secondary network, current limiting fuses may be used instead of cable limiters.

In secondary network systems, a forced or scheduled outage of a primary feeder does not result in customer outages. Because the secondary mains provide parallel paths to customer loads, secondary cable failures usually do not result in customer outages, either. Also, each network is designed to share the load equally among transformers and to handle large motor starting and other abrupt load changes without severe voltage drops.

SPOT NETWORK

Figure 14.14 shows a spot network consisting of a secondary network supplying a single, concentrated load such as a high-rise building or shopping center, where a high degree of reliability is required [1, 3, 4, and 9]. The secondary spot network bus is supplied simultaneously by two or more primary feeders through network transformers. In some cases, a spot network load as large as 25 MVA may be fed by up to six primary feeders. Most all spot networks in the United States operate at a 480Y/277-V secondary voltage [5]. Separate 120-V radial systems fed by small transformers from the 480-V system are used to supply outlets in various offices, retail stores, or rooms.

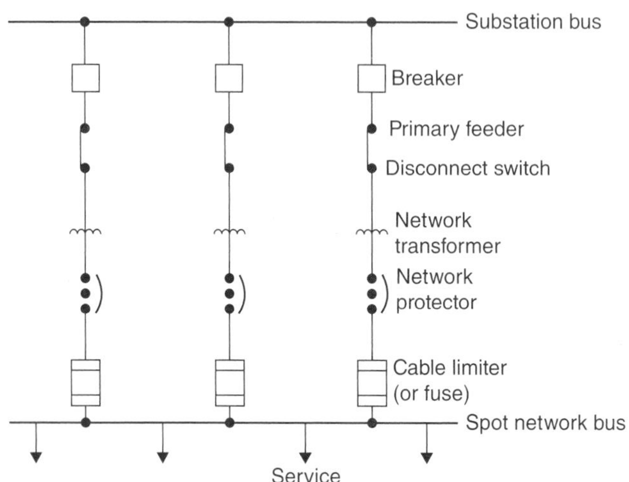

FIGURE 14.14 Secondary spot network (J. D. Glover, "Electric Power Distribution," *Encyclopedia of Energy Technology and The Environment*, John Wiley & Sons, New York, NY, 1995. Copyright © 1995, John Wiley and Sons.)

High service reliability and operating flexibility are achieved with a spot network fed by two or more primary feeders through network transformers. The secondary bus is continuously energized by all network transformers. Network protectors are used to automatically disconnect transformers from the spot network bus for transformer/feeder faults or for power-flow reversal, and cable limiters or fuses are used to protect against overloads and faults on secondary cables. Scheduled or forced outages of primary feeders occur without customer interruption or involvement. Spot networks also provide a very compact and reliable arrangement of components [5].

14.4 TRANSFORMERS IN DISTRIBUTION SYSTEMS

Transformers in distribution systems include distribution substation transformers and distribution transformers.

DISTRIBUTION SUBSTATION TRANSFORMERS

Distribution substation transformers come in a wide variety of ratings. Some of the typical characteristics of distribution substation transformers are given in Table 14.3.

Distribution substation transformers usually contain mineral oil for insulating and cooling purposes (older transformers manufactured prior to 1978 originally contained askarels with high PCB content, but many of these have either been retired or reclassified as non-PCB transformers using per-chloroethylene). In some units, an inert gas such as nitrogen fills the space above the oil, in order to keep moisture and air out of the oil, and the transformer tank is sealed. Some sealed transformers have a pressure-relief diaphragm that is designed to rupture when the internal pressure exceeds a specified value, indicating possible deterioration of the insulation. Sealed transformers may also have a sudden pressure relay to either provide an alarm or de-energize the transformer when the internal pressure suddenly increases above a specified threshold [4].

Class, kV	Voltage, kV
Rating of High Voltage Winding	34.5 to 230 kV
Rating of Low Voltage Winding	2.4 to 46 kV
MVA Rating (OA)	2.5 to 75 MVA
Transformer Impedance	5 to 12 %
Number of Transformers in Substation	1 to 4
Loading	OA, OA/FA, OA/FA/FOA, OA/FA/FA
High Side Protection	Circuit Switches, Circuit Breakers, Fuses
Relay Protection	Overcurrent, Differential, Under-Frequency
Feeder Protection	Circuit Breakers, Reclosers

TABLE 14.3

Typical Characteristics of Distribution Substation Transformers [5]

Many distribution substation transformers have load tap changers (LTCs) that automatically regulate voltage levels based on loading conditions. Figure 14.15 shows a distribution substation transformer that has an internal LTC on the low-voltage side. Some distribution substations have distribution substation transformers with fixed taps and separate voltage regulators. A voltage regulator is basically an autotransformer with taps that automatically raise or lower voltage, operating in a similar way as LTCs on distribution substation transformers. Figure 14.16 shows a voltage regulator at a distribution substation. In addition to voltage regulators for distribution substations, there are also pole-mount voltage regulators that can be placed on feeders.

FIGURE 14.15

Three-phase 22.9 kVΔ/4.16 kVY distribution substation transformer rated 12 MVA OA/16 MVA FA1/20 MVA FA2. The transformer has fixed taps on the high-voltage side and an LTC on the low-voltage side (Courtesy of Danvers Electric.)

FIGURE 14.16

Voltage regulators at the 69/13.8 kV Lunenburg distribution substation, Lunenburg MA. The regulators are General Electric Type VR1 with ratings of 7.96 kV (line-to neutral), 437 A (Courtesy of Unitil Corporation.)

Some outdoor distribution substation transformers are equipped with a tank on the top of the transformer called a "conservator," in which expansion and contraction of the oil takes place. Condensation of moisture and formation of sludge occur within the conservator, which is also provided with a sump pump to draw off the moisture and sludge [4].

Distribution substation transformers have MVA ratings that indicate the continuous load that the transformers carry without exceeding a specified temperature rise of either 55°C (for older transformers) or 65°C (for newer transformers) above a specified ambient temperature (typically 40°C). Also, distribution substation transformers are typically equipped with external radiators with fans and/or oil circulating pumps, in order to dissipate heat generated by copper and core losses. These transformers have multiple MVA ratings that include the following:

1. OA rating (passive convection with oil circulating pumps and fans off).
2. FA rating (with fans on but oil circulating pumps off).
3. FOA rating (with both fans and oil circulating pumps on). Some units, such as the one shown in Figure 14.15, may have two FA ratings, a lower FA rating with one of two sets of fans on, and a higher FA rating with both sets of fans on. Also, some units have water-cooled heat exchangers. The nameplate transformer impedance is usually given in percent using the OA rating as the base MVA [5].

Most utilities have a planning and operating policy of loading distribution substation transformers within their nameplate OA/FA/FOA ratings during normal conditions, but possibly above their nameplate ratings during short-term emergency conditions. If one transformer has a scheduled or forced outage, the remaining transformer or transformers can continuously carry the entire substation load.

EXAMPLE 14.1

Distribution Substation Transformer Rated Current and Short-Circuit Current

A three-phase 230 kVΔ/34.5 kV Y distribution substation transformer rated 75 MVA OA/100 MVA FA/133 MVA FOA has a 7% impedance. (a) Determine the rated current on the low-voltage side of the transformer at its OA, FA, and FOA ratings. (b) Determine the per unit transformer impedance using a system base of 100 MVA and 34.5 kV on the low-voltage side of the transformer. (c) Calculate the short-circuit current on the low-voltage side of the transformer for a three-phase bolted fault on the low-voltage side. Assume that the prefault voltage is 34.5 kV.

SOLUTION

a. At the OA rating of 75 MVA,

$$I_{OA,\,L} = 75/(\sqrt{3} \times 34.5) = 7.372 \text{ kA per phase}$$

Similarly,

$$I_{FA,\,L} = 100/(\sqrt{3} \times 34.5) = 9.829 \text{ kA per phase}$$

$$I_{FOA,\,L} = 133/(\sqrt{3} \times 34.5) = 13.07 \text{ kA per phase}$$

b. The transformer impedance is 7% or 0.07 per unit based on the OA rating of 75 MVA. Using (3.3.11), the transformer per unit impedance on a 100 MVA system base is:

$$Z_{\text{puSystem Base}} = 0.07(100/75) = 0.09333 \text{ per unit}$$

c. For a three-phase bolted fault, using the transformer ratings as the base quantities,

$$\begin{aligned} Isc3\phi &= 1.0/(0.07) = 14.286 \text{ per unit} \\ &= (14.286)(7.372) \\ &= 105.31 \text{ kA/phase} \end{aligned}$$

Note that in (c) above, the OA rating is used to calculate the short-circuit current, because the transformer manufacturer gives the per unit transformer impedance using the OA rating as the base quantity.

Typically there are two emergency loading criteria for distribution substation transformers:

1. A two-hour emergency rating, which gives time to perform switching operations and reduce loadings.
2. A longer-duration emergency rating (10 to 30 days), which gives time to replace a failed transformer with a spare that is in stock.

As one example, the distribution substation shown in the photograph at the beginning of this chapter has two transformers rated 9 MVA OA/12 MVA FA1/ 15 MVA FA2. The practice of the utility that owns the substation is to normally operate the substation at or below 15 MVA. As such, if there is a forced or scheduled outage of one transformer, the other transformer can supply all six 4.16-kV feeders without being loaded above its FA2 nameplate of 15 MVA. For this conservative operating practice, emergency transformer ratings above nameplate are not used.

Some utilities operate their distribution substation transformers above nameplate ratings during normal operating conditions, as well as during emergency conditions. ANSI/IEEE C-57.91-1995 entitled, *IEEE Guide for Loading Mineral-Oil-Immersed Transformers* identifies the risks of transformer loads in excess of nameplate rating and establishes limitations and guidelines, the application of which are intended minimize the risks to an acceptable limit [21, 22].

EXAMPLE 14.2

Distribution Substation Normal, Emergency, and Allowable Ratings

As shown in Figure 14.17, a distribution substation is served by two 138-kV subtransmission lines, each connected to a 40 MVA (FOA nameplate rating) 138 kVΔ/ 12.5 kV Y distribution substation transformer, denoted TR1 and TR2. Both TR1

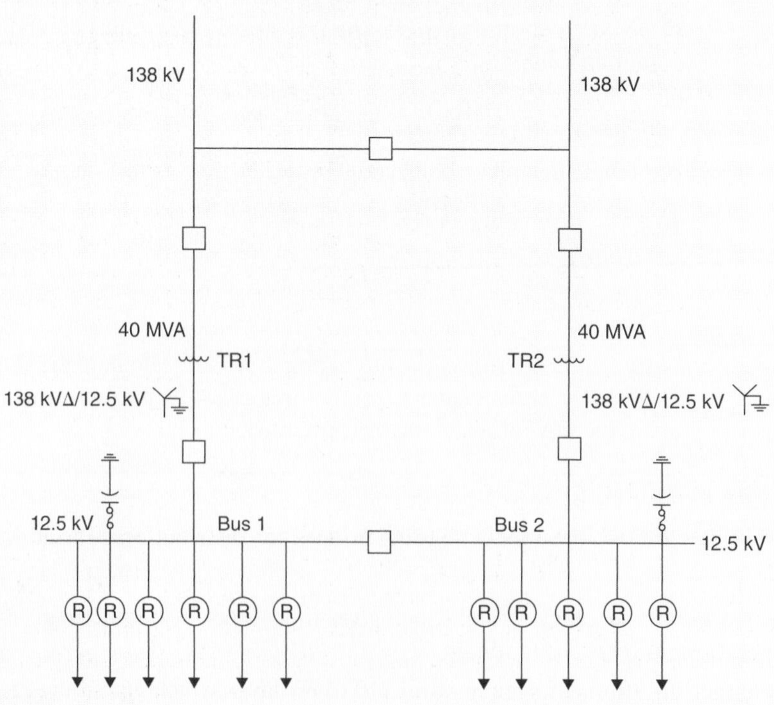

FIGURE 14.17 Distribution substation for Example 14.2

and TR2 are relatively new transformers with insulation systems designed for 65°C temperature rises under continuous loading conditions. Shunt capacitor banks are also installed at 12.5-kV bus 1 and bus 2. The utility that owns this substation has the following transformer loading criteria based on a percentage of nameplate rating:

1. 128% for normal summer loading;
2. 170% during a two-hour summer emergency; AND
3. 155% during a 30-day summer emergency.

(a) Assuming a 5% reduction for unequal transformer loadings, determine the summer "normal" rating of the substation. (b) Determine the "allowable" summer rating of the substation under the single-contingency loss of one transformer. (c) Determine the 30-day summer emergency rating of the substation under the single-contingency loss of one transformer.

SOLUTION

a. During normal operations, both transformers are in service. Using a 5% reduction to account for unequal transformer loadings, the summer normal substation rating is $1.28 \times (40 + 40) \times 0.95 = 97$ MVA. With both transformers in service, the substation can operate as high as 97 MVA without exceeding the summer normal rating of 128% or 51.2 MVA for each transformer.

b. The summer allowable substation rating, based on the single-contingency loss of one transformer, is $1.7 \times 40 = 68$ MVA. The transformer that remains in service is allowed to operate at 170% of its nameplate rating (68 MVA) for two hours, which gives time to perform switching operations to reduce the transformer loading to its 30-day summer emergency rating. Note that, even though the normal summer substation rating is 97 MVA, it is only allowed to operate up to 68 MVA so that a transformer will not exceed its two-hour emergency rating in case the other transformer has an outage.

c. The 30-day summer emergency rating of the substation is $1.55 \times 40 = 62$ MVA. When one transformer has a permanent failure, the other can operate at 62 MVA for 30 days, which gives time to replace the failed transformer with a spare that is in stock.

DISTRIBUTION TRANSFORMERS

Distribution transformers connect the primary system (2.4 to 34.5 kV) to the secondary system (480 V and lower). Distribution transformers may be installed outdoors on overhead poles (pole-mount), outdoors at ground level on pads (padmount transformers), indoors within buildings, or underground in manholes and vaults.

Pole-mount transformers for overhead distribution are liquid-filled transformers that can be either single-phase or three-phase, depending on the load requirements and the primary supply configuration. Pole-mount distribution transformers may be manufactured as conventional transformers with no integral surge protection, overload protection, or short-circuit protection, or alternatively as completely self-protected (CSP) transformers.

For conventional pole-mount transformers, the protective devices are mounted external to the transformer. Typically a fuse cutout, which is a combination of a fuse and a switch, is installed adjacent to the conventional distribution transformer to disconnect it from the primary under overload conditions or an internal transformer failure. Similarly, a surge arrester is installed adjacent to the conventional transformer primary to protect it against transient overvoltages due to switching and lightning surges. Figure 14.18 shows three conventional single-phase pole-mount distribution transformers that are wired to form a three-phase bank rated 75 kVA supplying 120/208 V overhead secondary service for commercial customers.

For CSP transformers, a primary fuse is located within the transformer tank. The surge arrester is mounted outside the tank, but connected to the primary bushing. Circuit breakers on the secondary side of CSP transformers provide protection from overloads and are coordinated with primary fuses.

Padmount transformers for underground distribution are liquid-filled or dry-type transformers that can be either single-phase or three-phase, outdoors or indoors. Single-phase padmount distribution transformers are typically designed for underground residential and commercial distribution systems where safety, reliability and aesthetics are especially important. Three-phase padmount distribution transformers are compact power centers usually for large commercial or industrial applications. Figure 14.19 shows a three-phase liquid-filled padmount transformer

FIGURE 14.18

Three conventional single-phase pole-mount 25-kVA transformers. The transformers are wired to form a three-phase bank rated 75 kVA, 4160VΔ-208/120 V grounded Y, which supplies secondary service for commercial customers. The transformers are supplied from a 4160- V primary through fused cutouts, with surge arresters mounted vertically on the sides of the transformer tanks. (Courtesy of Danvers Electric.)

(a)

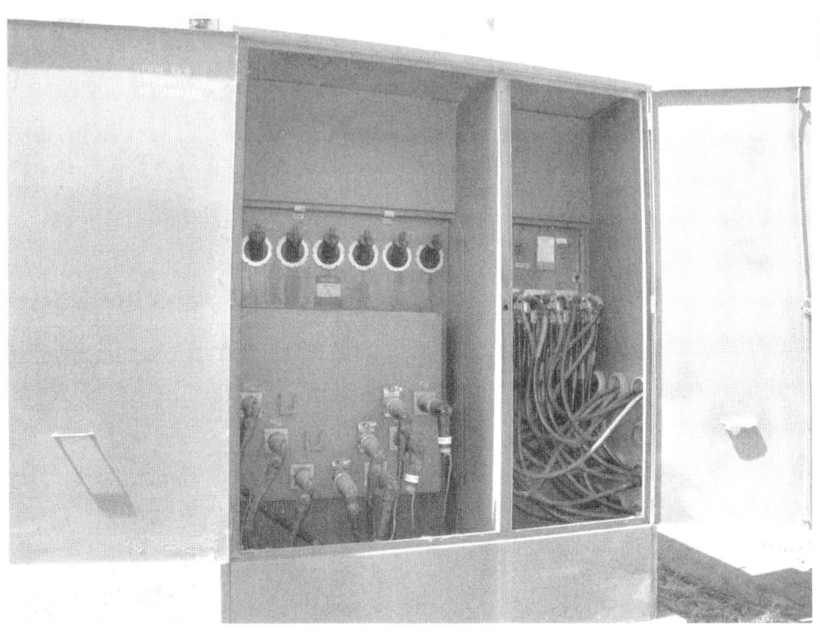
(b)

FIGURE 14.19

Three-phase oil-filled padmount transformer shown with doors closed (a) and open (b). This padmount, rated 1,500 kVA OA, kVΔ-480/277 V grounded Y with internal fuses on the high-voltage side, supplies secondary service to an industrial plant. Courtesy of Danvers Electric.)

that supplies 480/277 V underground secondary service to an industrial plant. Dry-type padmount distribution transformers, whose insulation is solid (for example glass, silica, epoxy, or polyester resins) are primarily used where safety is a major concern, in close proximity to people such as at schools, hospitals, commercial buildings, and industrial plants, both indoors and outdoors.

Network transformers are large (300 to 2500 kVA) liquid-filled, three-phase distribution transformers that are designed for use in underground vaults or in specially designed rooms within buildings to supply power to either secondary networks or spot networks. Their voltage ratings vary from 4.16 to 34.5 kVΔ or grounded Y for the high-voltage windings, and either 216 grounded Y/125 V or 480 grounded Y/277 V for the low-voltage windings. Network transformers are designed to be connected through network protectors that are integrally mounted on the transformer. Figure 14.20 shows a network transformer from utility stock. Network transformers are built as either "vault type" (suitable for occasional submerged operation) or "subway type" (suitable for continuous submerged operation).

Table 14.4 shows typical kVA ratings of distribution transformers. The kVA ratings of distribution transformers are based on the continuous load the transformers can carry without exceeding a specified temperature rise of either 55°C (for older transformers) or 65°C (for newer transformers above a specified ambient temperature (usually 40°C). When in service, distribution transformers are rarely loaded continuously at their rated kVA as they go through a daily load cycle. Oil-filled distribution transformers have a relatively long thermal time constant; that is, the load temperature rises slowly during load increases. As such, it is possible to load these transformers above their kVA ratings without compromising the life expectancy of

FIGURE 14.20

General Electric 500 kVA, 13.8 kVΔ-120/208 V grounded Y network transformer from utility stock (Courtesy of Unitil Corporation.)

Single-Phase	Three-Phase
kVA	kVA
5	30
10	45
15	75
25	112.5
38	150
50	225
75	300
100	500
167	750
250	1000
333	1500
500	2500
	3000
	3750
	5000

TABLE 14.4

Standard Distribution Transformer kVA Ratings

J. D. Glover, "Electric Power Distribution," *Encyclopedia of Energy Technology and The Environment*, John Wiley & Sons, New York, NY, 1995. Reprinted with permission of John Wiley & Sons, Inc. (Based on ANSI/IEEE C.57 Standard)

the transformer. ANSI/EEE Std. C57.92-1981 is entitled *IEEE Guide for Loading Mineral-Oil-Immersed Overhead and Pad-Mounted Distribution Transformers Rated 500 kVA and Less with 65°C or 55°C Average Winding Rise* [23]. Table 14.5 shows a typical loading guide, based on this standard.

Period of Increased Loading, Hours	Average Initial Load in Per Unit of Transformer Rating		
	0.9	0.7	0.5
	Maximum Load in Per Unit of Transformer Rating		
0.5	1.59	1.77	1.89
1.0	1.40	1.54	1.60
2.0	1.24	1.33	1.37
4.0	1.12	1.17	1.19
8.0	1.06	1.08	1.08

TABLE 14.5

Permissible Daily Short-Time Loading of Liquid-Filled Distribution Transformers Based on Normal Life Expectancy [5]

Power distribution engineering: fundamentals and applications by Burke. Copyright 1994 by TAYLOR & FRANCIS GROUP LLC - BOOKS. Reproduced with permission of TAYLOR & FRANCIS GROUP LLC - BOOKS in the format Textbook via Copyright Clearance Center (Based on Burke, Power distribution engineering: fundamentals and applications , 1994 Taylor & Francis.)

Note that in accordance with Table 14.5, short-time loadings can be as high as 89% above the nameplate kVA rating for short durations. Also note that dry-type distribution transformers, which are not considered as rugged as liquid-filled units of the same rating, are not normally loaded above their kVA ratings.

14.5 SHUNT CAPACITORS IN DISTRIBUTION SYSTEMS

Loads in electric power systems consume real power (MW) and reactive power (Mvar). At power plants, many of which are located at long distances from load centers, real power is generated and reactive power may either be generated, such as during heavy load periods, or absorbed as during light load periods. Unlike real power (MW), the generation of reactive power (Mvar) at power plants and transmission of the reactive power over long distances to loads is not economically feasible. Shunt capacitors, however, are widely used in primary distribution to supply reactive power to loads. They draw leading currents that offset the lagging component of currents in inductive loads. Shunt capacitors provide an economical supply of reactive power to meet reactive power requirements of loads as well as transmission and distribution lines operating at lagging power factor. They can also reduce line losses and improve voltage regulation.

EXAMPLE 14.3

Shunt Capacitor Bank at End of Primary Feeder

Figure 14.21 shows a single-line diagram of a 13.8-kV primary feeder supplying power to a load at the end of the feeder. A shunt capacitor bank is located at the load bus. Assume that the voltage at the sending end of the feeder is 5% above rated and that the load is Y-connected with $R_{Load} = 20\ \Omega$/phase in parallel with load $jX_{Load} = j\ 40\ \Omega$/phase. (a) With the shunt capacitor bank out of service, calculate the following: (1) line current; (2) voltage drop across the line; (3) load voltage; (4) real and reactive power delivered to the load; (5) load power factor; (6) real and reactive line losses; and (7) real power, reactive power, and apparent power delivered by the distribution substation. (b) The capacitor bank is Y-connected with a reactance $X_C = 40\ \Omega$/phase. With the shunt capacitor bank in service, redo the calculations. Also calculate the reactive power supplied by the capacitor bank. (c) Compare the results of (a) and (b).

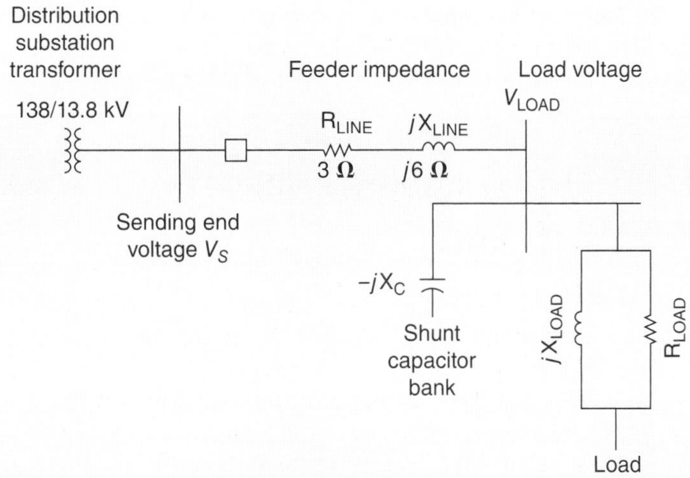

FIGURE 14.21 Single-line diagram of a primary feeder for Example 14.3

SOLUTION

a. Without the capacitor bank, the total impedance seen by the source is

$$Z_{\text{TOTAL}} = \text{R}_{\text{LINE}} + j\text{X}_{\text{LINE}} + \frac{1}{\frac{1}{\text{R}_{\text{LOAD}}} + \frac{1}{j\text{X}_{\text{LOAD}}}}$$

$$Z_{\text{TOTAL}} = 3 + j6 + \frac{1}{\frac{1}{20} + \frac{1}{j40}}$$

$$Z_{\text{TOTAL}} = 3 + j6 + \frac{1}{0.0559\underline{/-26.57^\circ}}$$

$$= 3 + j6 + 17.89\underline{/26.56^\circ}$$

$$Z_{\text{TOTAL}} = 3 + j6 + 16 + j8 = 19 + j14$$

$$= 23.60\underline{/36.38^\circ}\ \Omega/\text{phase}$$

1. The line current is

$$I_{\text{LINE}} = V_{\text{SLN}}/Z_{\text{TOTAL}} = \frac{1.05(13.8/\sqrt{3})\underline{/0^\circ}}{23.60\underline{/36.38^\circ}}$$

$$= 0.3545\underline{/-36.38^\circ}\ \text{kA/phase}$$

(Continued)

2. The voltage drop across the line is

$$V_{DROP} = Z_{LINE}\, I_{LINE} = (3 + j6)(0.3545\angle{-36.38°})$$
$$= (6.708\angle 63.43°)(0.3545\angle{-36.38°})$$
$$= 2.378\angle 27.05° \text{ kV}$$
$$|V_{DROP}| = 2.378 \text{ kV}$$

3. The load voltage is

$$V_{LOAD} = V_{SLN} - Z_{LINE}I_{LINE} = 1.05(13.8/\sqrt{3})\angle 0° - 2.378\angle 27.05°$$
$$= 8.366 - (2.117 + j1.081) = 6.249 - j1.081$$
$$= 6.342\angle{-9.814°} \text{ kV}_{LN}$$
$$|V_{LOAD}| = 6.342\sqrt{3} = 10.98 \text{ kV}_{LL}$$

4. The real and reactive power delivered to the three-phase load is

$$P_{LOAD3\phi} = 3(V_{LOADLN})^2/R_{LOAD} = 3(6.342)^2/20 = 6.033 \text{ MW}$$
$$Q_{LOAD3\phi} = 3(V_{LOADLN})^2/X_{LOAD} = 3(6.342)^2/40 = 3.017 \text{ Mvar}$$

5. The load power factor is

$$\text{p.f.} = \cos[\tan^{-1}(Q/P)]$$
$$= \cos[\tan^{-1}(3.017/6.033)]$$
$$= 0.89 \text{ lagging}$$

6. The real and reactive line losses are

$$P_{LINELOSS3\phi} = 3\, I_{LINE}^2 R_{LINE} = 3(0.3545)^2(3) = 1.131 \text{ MW}$$
$$Q_{LINELOSS3\phi} = 3\, I_{LINE}^2 X_{LINE} = 3(0.3545)^2(6) = 2.262 \text{ Mvar}$$

7. The real power, reactive power, and apparent power delivered by the distribution substation are

$$P_{SOURCE3\phi} = P_{LOAD3\phi} + P_{LINELOSS3\phi} = 6.033 + 1.131 = 7.164 \text{ MW}$$
$$Q_{SOURCE3\phi} = Q_{LOAD3\phi} + Q_{LINELOSS3\phi} = 3.017 + 2.262 = 5.279 \text{ Mvar}$$
$$S_{SOURCE3\phi} = \sqrt{(7.164^2 + 5.279^2)} = 8.899 \text{ MVA}$$

b. With the capacitor bank in service, the total impedance seen by the source is

$$Z_{TOTAL} = R_{LINE} + jX_{LINE} + \frac{1}{\frac{1}{R_{LOAD}} + \frac{1}{jX_{LOAD}} - \frac{1}{jX_C}}$$

$$Z_{TOTAL} = 3 + j6 + \frac{1}{\frac{1}{20} + \frac{1}{j40} - \frac{1}{j40}}$$

$$Z_{TOTAL} = 3 + j6 + \frac{1}{0.05} = 23 + j6 = 23.77\underline{/14.62^\circ}\ \Omega\text{/phase}$$

1. The line current is

$$I_{LINE} = V_{SLN}/Z_{TOTAL} = \frac{1.05(13.8/\sqrt{3})/0^\circ}{23.77\underline{/14.62^\circ}}$$

$$= 0.3520\underline{/-14.62^\circ}\text{ kA/phase}$$

2. The voltage drop across the line is

$$V_{DROP} = Z_{LINE}\,I_{LINE} = (6.708\underline{/63.43^\circ})(0.3520\underline{/-14.62^\circ})$$

$$= 2.361\underline{/48.81^\circ}\text{ kV}$$

$$|V_{DROP}| = 2.361\text{ kV}$$

3. The load voltage is

$$V_{LOAD} = V_{SLN} - Z_{LINE}\,I_{LINE}$$

$$= 1.05(13.8/\sqrt{3})\underline{/0^\circ} - 2.361\underline{/48.81^\circ}$$

$$= 8.366 - (1.555 + j1.778)$$

$$= 6.81 - j1.778$$

$$= 7.038\underline{/-14.62^\circ}\text{ kV}_{LN}$$

$$|V_{LOAD}| = 7.038\sqrt{3} = 12.19\text{ kV}_{LL}$$

4. The real and reactive power delivered to the three-phase load is

$$P_{LOAD3\phi} = 3(V_{LOADLN})^2/R_{LOAD} = 3(7.038)^2/20 = 7.430\text{ MW}$$

$$Q_{LOAD3\phi} = 3(V_{LOADLN})^2/X_{LOAD} = 3(7.038)^2/40 = 3.715\text{ Mvar}$$

5. The load power factor is

$$\begin{aligned} \text{p.f.} &= \cos[\tan^{-1}(Q/P)] \\ &= \cos[\tan^{-1}(3.715/7.430)] \\ &= 0.89 \text{ lagging} \end{aligned}$$

6. The real and reactive line losses are

$$P_{LINELOSS3\phi} = 3\, I_{LINE}^2 R_{LINE} = 3(0.3520)^2(3) = 1.115 \text{ MW}$$

$$Q_{LINELOSS3\phi} = 3\, I_{LINE}^2 X_{LINE} = 3(0.3520)^2(6) = 2.230 \text{ Mvar}$$

7. The reactive power delivered by the shunt capacitor bank is

$$Q_C = 3(V_{LOADLN})^2/X_C = 3(7.038)^2/40 = 3.715 \text{ Mvar}$$

8. The real power, reactive power, and apparent power delivered by the distribution substation are

$$P_{SOURCE3\phi} = P_{LOAD3\phi} + P_{LINELOSS3\phi} = 7.430 + 1.115 = 8.545 \text{ MW}$$

$$\begin{aligned} Q_{SOURCE3\phi} &= Q_{LOAD3\phi} + Q_{LINELOSS3\phi} - Q_C \\ &= 3.715 + 2.230 - 3.715 \\ &= 2.230 \text{ Mvar} \end{aligned}$$

$$S_{SOURCE3\phi} = \sqrt{(8.545^2 + 2.230^2)} = 8.675 \text{ MVA}$$

c. Comparing the results of (a) and (b), with the shunt capacitor bank in service, the real power delivered to the load increases by 23% (from 6.033 to 7.430 MW) while at the same time

- The line current decreases
- The real and reactive line losses decrease
- The voltage drop across the line decreases
- The reactive power delivered by the source decreases
- The load voltage increases

The above benefits are achieved by having the shunt capacitor bank (instead of the distribution substation) deliver reactive power to the load.

The location of a shunt capacitor bank along a primary feeder is important. If there were only one load on the feeder, the best location for the capacitor bank would be directly at the load, so as to minimize I^2R losses and voltage drops on the feeder. Note that if shunt capacitors were placed at the distribution substation, I^2R feeder losses and feeder voltage drops would not be reduced because the total power including MW and Mvar would still have to be sent from the substation all the way to the load. Shunt capacitors at distribution substations, however, can be effective in reducing I^2R losses and voltage drops on the transmission or subtransmission lines that feed the distribution substations.

For a primary feeder that has a uniformly distributed load along the feeder, a common application is the "two-thirds" rule; that is, place 2/3 of the required reactive power 2/3 of the way down the feeder. Locating shunt capacitors 2/3 of the way down the feeder allows for good coordination between LTC distribution substation transformers or voltage regulators at the distribution substation. For other load distributions, computer software is available for optimal placement of shunt capacitor banks. Note that capacitors are rarely applied to secondary distribution systems due to their small economic advantage [3, 5].

During the daily load cycle, reactive power requirements change as a function of time. To meet the changing reactive power requirements, many utilities use a combination of fixed and switched capacitor banks. Fixed capacitor banks can be used to compensate for reactive power requirements at light loads, and switched capacitor banks can be added during heavy load conditions. The goal is to obtain a close-to-unity power factor throughout the day by switching capacitor banks on when needed and off when not needed. Methods of controlling switched capacitor banks include the following:

1. Voltage control
2. Current control
3. Var control
4. Time control
5. Temperature control
6. Radio dispatch/SCADA control

14.6 DISTRIBUTION SOFTWARE

Computer programs are available for the planning, design, and operation of electric power distribution systems. Program functions include:

1. Arc flash hazard and fault analysis
2. Capacitor placement optimization
3. Circuit breaker duty
4. Conductor and conduit sizing—ampacity and temperature calculations
5. Database management
6. Demand management
7. Distribution reliability evaluation
8. Distribution short circuit calculations
9. Fault detection and location

10. Graphics for single-line diagrams and mapping systems
11. Harmonic analysis
12. Motor starting
13. Outage management
14. Power factor correction
15. Power flow/voltage drop computations
16. Power loss computations and costs of losses
17. Power quality and reliability
18. Relay and protective device coordination
19. Switching optimization
20. Tie capacity optimization
21. T & D modeling and analysis
22. Transformer sizing—load profile and life expectancy
23. Voltage/var optimization

Some of the vendors that offer distribution software packages are given as follows:

- ABB Network Control, Ltd., Switzerland
- ASPEN, San Mateo, CA
- Cooper Power Systems, Pittsburgh, PA
- Cyme International, Burlington, MA
- EDSA Corporation, Bloomfield, MI
- Electrocon International Inc., Ann Arbor, MI
- Operation Technology, Irvine, CA
- Milsoft Utility Solutions Inc., Abilene, TX
- RTDS Technologies, Winnipeg, Manitoba, Canada

14.7 DISTRIBUTION RELIABILITY

Reliability in engineering applications, as defined in the *The Authoritative Dictionary of IEEE Standard Terms* (IEEE 100), is the probability that a device will function without failure over a specified time period or amount of usage. In the case of electric power distribution, reliability concerns have come from customers who want uninterrupted continuous power supplied to their facilities at minimum cost [11–17].

A typical goal for an electric utility is to have an overall average of one interruption of no more than two hours' duration per customer year. Given 8760 hours in a non-leap year, this goal corresponds to an Average Service Availability Index (ASAI) greater than or equal to 8758 service hours/8760 hours = 0.999772 = 99.9772%.

IEEE Standard 1366-2012 entitled, *IEEE Guide for Electric Power Distribution Reliability Indices,* defines the following distribution reliability indices [24]:

System Average Interruption Frequency Index (SAIFI):

$$\text{SAIFI} = \frac{\Sigma \text{ Total Number of Customers Interrupted}}{\text{Total Number of Customers Served}} \tag{14.7.1}$$

System Average Interruption Duration Index (SAIDI):

$$\text{SAIDI} = \frac{\Sigma \text{ Customer Minutes of Interruption}}{\text{Total Number of Customers Served}} \tag{14.7.2}$$

Customer Average Interruption Duration Index (CAIDI):

$$\begin{aligned}\text{CAIDI} &= \frac{\Sigma \text{ Customer Minutes of Interruption}}{\text{Total Number of Customers Interrupted}} \\ &= \frac{\text{SAIDI}}{\text{SAIFI}}\end{aligned} \tag{14.7.3}$$

Average Service Availability Index (ASAI):

$$\text{ASAI} = \frac{\text{Customer Hours Service Availability}}{\text{Customer Hours Service Demands}} \tag{14.7.4}$$

In accordance with IEEE Std. 1366-2012, when calculating the above reliability indices, momentary interruption events are not included. A momentary interruption event has an interruption duration that is limited to the time required to restore service by an interrupting device (including multiple re-closures of reclosers or circuit breakers). Switching operations must be completed within five minutes for a momentary interruption event. As such, customer interruption durations less than five minutes are excluded when calculating the reliability indices. IEEE Std. 1366-2012 also includes a method, when calculating reliability indices, for excluding major events, such as severe storms, for which the daily SAIDI exceeds a specified threshold.

The above formulas for reliability indices use customers' out-of-service and customer-minutes out-of-service data. Electric utilities with outage management systems including geographical information systems (GIS) and customer information systems (CIS) are able to very accurately keep track of this data. Some utilities in the United States are required to report distribution reliability indices to state public service commissions, while other utilities may voluntarily report these indices to regional power associations. Typical values for these indices are given in Table 14.6 [24].

SAIDI	SAIFI	CAIDI	ASAI
90 minutes/year	1.1 Interruptions/year	76 minutes/year	99.982%

TABLE 14.6

Typical Values of Reliability Indices [24] (Based on IEEE Std. 1366-2003, IEEE Guide for Electric Power Distribution Reliability Indices, 2004.)

The following example uses outage data given in IEEE Std. 1366 [24].

EXAMPLE 14.4

Distribution Reliability Indices

Table 14.7 gives 2010 annual outage data (sustained interruptions) from a utility's CIS database for one feeder. This feeder (denoted circuit 7075) serves 2000 customers with a total load of 4 MW. Excluding momentary interruption events (less than five minutes duration) and major events, which are omitted from Table 14.7, calculate the SAIFI, SAIDI, CAIDI, and ASAI for this feeder.

Outage Date	Time at Beginning of Outage	Outage Duration (minutes)	Circuit	Number of Customers Interrupted
3/17/2010	12:12:20	8.17	7075	200
5/5/2010	00:23:10	71.3	7075	600
6/12/2010	09:30:10	30.3	7075	25
8/20/2010	15:45:39	267.2	7075	90
8/31/2010	08:20:00	120	7075	700
9/03/2010	17:10:00	10	7075	1500
10/07/2010	10:15:00	40	7075	100

TABLE 14.7

Customer Outage Data for Example 14.4 [24] (Based on IEEE Std. 1366-2012, IEEE Guide for Electric Power Distribution Reliability Indices.)

SOLUTION

Using the outage data from Table 14.7 in (14.7.1) through (14.7.4):

$$\text{SAIFI} = \frac{200 + 600 + 25 + 90 + 700 + 1500 + 100}{2000}$$

$$= 1.6075 \text{ interruptions/year}$$

$$\text{SAIDI} = \frac{(8.17 \times 200) + (71.3 \times 600) + (30.3 \times 25) + (267.2 \times 90) + (120 \times 700) + (10 \times 1500) + (40 \times 100)}{2000}$$

$$\text{SAIDI} = 86.110 \text{ minutes/year}$$

$$\text{CAIDI} = \frac{\text{SAIDI}}{\text{SAIFI}} = \frac{86.110}{1.6075} = 53.57 \text{ minutes/year}$$

ASAI

$$= \frac{8760 \times 2000 - [(8.17 \times 200) + (71.3 \times 600) + (30.3 \times 25) + (267.2 \times 90) + (120 \times 700) + (10 \times 1{,}500) + (40 \times 100)]/60}{8760 \times 2000}$$

$$\text{ASAI} = 0.999836 = 99.9836\%$$

Table 14.8 lists basic outage reporting information recommended by an IEEE committee [15]. Table 14.9 lists generic and specific causes of outages, based on a U.S. Department of Energy study [16]. Many electric utilities routinely prepare distribution outage reports monthly, quarterly, and annually by town (municipality) or by district. The purposes of the reports are to monitor and evaluate distribution reliability, uncover weaknesses and potential problems, and make recommendations for improving reliability. These reports may include:

1. Frequency and duration reports, which provide data for the number of interruptions on distribution circuits together with power interrupted, average interruption duration, and causes.
2. Annual reports that sort outages according to cause of failure and according to circuit classification (for example, sort for each primary voltage; or sort for each conductor type including overhead open wire, overhead spacer cable, underground direct-burial, and cable in conduit).
3. Five- or ten-year trends for reliability indices, and outage trends for specific causes such as tree-contact outages for overhead distribution or dig-in outages for underground distribution.
4. Lists of problem circuits such as the 20 "worst" (lowest ASAI) circuits in a district, or all circuits with repeated outages during the reporting period.

Methods for improving distribution reliability include replacement of older distribution equipment, upgrades of problem circuits, crew staffing and training for fast responses to outages and rapid restoration of service, formal maintenance programs, and public awareness programs to reduce hazards in the vicinity of distribution equipment such as contractor dig-ins. Reliability evaluation has also become an important component of bid selections to procure new distribution equipment. Also, great strides in distribution reliability have come through distribution automation.

1. Type, design, manufacturer, and other descriptions for classification purposes
2. Date of installation, location on system, length in case of a line
3. Mode of failure (short circuit, false operation, etc.)
4. Cause of failure (lightning, tree, etc.)
5. Times (both out of service and back in service, rather than outage duration alone), date, meteorological conditions when the failure occurred
6. Type of outage, forced or scheduled, transient or permanent (momentary or sustained)

TABLE 14.8

Basic Outage Reporting Information [15]

Weather	Miscellaneous	System Components	System Operation
Blizzard/snow	Airplane/helicopter	Electrical & mechanical:	System conditions:
Cold	Animal/bird/snake	Fuel supply	Stability
Flood	Vehicle:	Generating unit failure	High/low voltage
Heat	Automobile/truck	Transformer failure	High/low frequency
Hurricane	Crane	Switchgear failure	Line overload
Ice	Dig-in	Conductor failure	Transformer overload
Lightning	Fire/explosion	Tower, pole attachment	Unbalanced load
Rain	Sabotage/vandalism	Insulation failure:	Neighboring power
		Transmission line	system
Tornado	Tree	Substation	Public appeal:
Wind	Unknown	Surge arrester	Commercial & industrial
Other	Other	Cable failure	All customers
		Voltage control equipment:	Voltage Reduction:
		Voltage regulator	0–2% voltage reduction
		Automatic tap changer	Greater than 2–8%
		Capacitor	voltage reduction
		Reactor	Rotating Blackout
		Protection and control:	Utility personnel:
		Relay failure	System operator error
		Communication signal error	Power plant operator error
		Supervisory control error	Field operator error
			Maintenance error
			Other

TABLE 14.9

Generic and Specific Causes of Outages [16]

© 1976 IEEE. Reprinted, with permission, from IEEE Committee Report, "List of Transmission and Distribution Components for Use in Outage Reporting and Reliability Calculations," IEEE Transactions on Power Apparatus and Systems, PAS 95(4) (July/August 1976), pp. 1210–1215.

14.8 DISTRIBUTION AUTOMATION

Throughout its existence, the electric power industry has been a leader in the application of electric, electronic, and later computer technology for monitoring, control, and automation. Initially, this technology consisted of simple meters to show voltages and flows, and telephones to call the manned substations to do control operations. Yet as early as the 1950s, supervisory control and the associated telemetering equipment was in widespread use, with a 1955 AIEE (American Institute of Electrical Engineers) report [25] indicating 31% of transmission level stations (switching stations) had such control, and that the U.S. electric industry had more than 48,000 channel kilometers for communication, and that continuous monitoring of watts, vars, and voltage was widespread. By the late 1960s increasingly sophisticated *energy management systems* (EMSs) were beginning to be deployed in electric utility control centers with applications that included automatic generation control, alarming, state estimation, on-line power flows, and contingency analysis. Initially all of this monitoring, control, and automation was confined to the generators and the transmission level substations. Because of its larger number of devices and more diffuse nature, the costs of monitoring and automating the distribution system could not be justified. However, this is now rapidly changing.

The monitoring, control, and automation of the distribution system is known under the general rubric of *distribution automation* (DA). While prototype DA systems date back to the 1970s, it has only been in the last 15 years or so, as communication and computer costs have continued to decrease, that they have started to become widespread.

A primary reason for DA is to reduce the duration of customer outages. As was mentioned in Section 14.2, distribution systems are almost always radial, with sectionalizing fuses and circuit breakers used to clear faults, avoiding having prolonged outages for most customers upstream from the fault location. Then sectionalizing switches can be used to further isolate the faulted area, and by closing normally open switches the unfaulted downstream sections of the feeder can be fed from adjacent feeders. Prior to DA these switches were manually controlled, requiring delays of potentially more than an hour for a line crew to travel to the switch location. DA can greatly reduce the time necessary to complete this process by either

1. providing the distribution system operators with the ability to remotely control the various sectionalizing switches, or
2. automating the entire process.

Real-time monitoring of the voltages and power flows is used to ensure there is sufficient capacity to pickup the unfaulted load on the adjacent feeders.

An example of this situation is illustrated in Figure 14.22, which can also be seen in PowerWorld Simulator by loading case Figure 14_22. This case represents a feeder system modeled using the primary loop approach from Section 14.2 with a

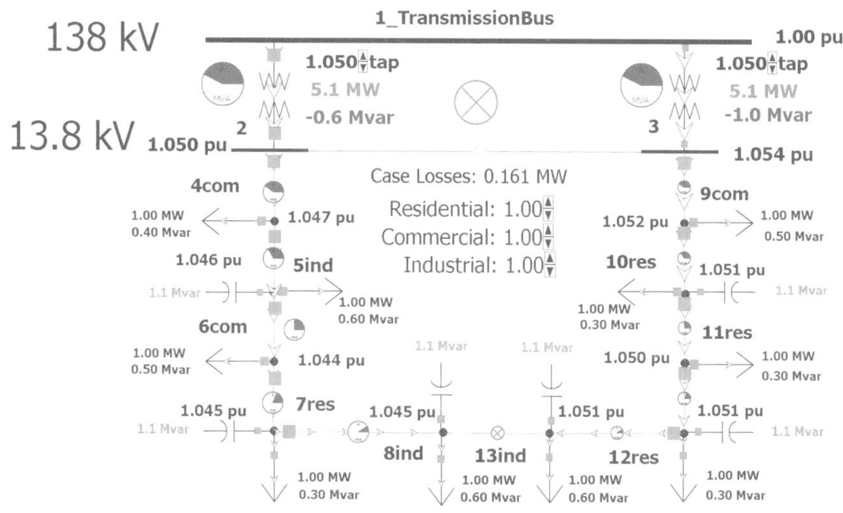

FIGURE 14.22 Primary loop feeder example, before fault

nominal 13.8 kV feeder voltage. A total of 10 loads are represented, with each load classified as either primarily residential ('res'), primarily commercial ('com'), or industrial ('ind'); the bus name suffix indicates the type. The left side of the loop goes from the substation distribution bus 2 to bus 8ind; the right side of the loop goes from the substation distribution bus 3 to bus 13ind. Substation buses 2 and 3 are connected by a normally open bus-tie breaker, while buses 8ind and 13ind are joined together by a normally open feeder segment to complete the loop. Each feeder line segment is assumed to use 336,400 26/7 ACSR line conductors and to be 0.6 miles in length giving a per unit impedance (on a 100 MVA base) of $0.0964 + j0.1995$. The two 12 MVA 138/13.8 kV transformers have an impedance of $0.1 + j0.8$ per unit.

Assume a persistent fault occurs immediately downstream from the bus 3 breaker. This fault would be cleared by the bus 3 breaker, outaging all of the customers of the right branch feeder. Without DA, a line crew would need to be dispatched to locate the fault and then manually change the status of the appropriate sectionalizers to restore service to most customers on this feeder.

In contrast, even with a simple DA, which just consisted of having the ability to remotely control the sectionalizers and monitor flow values/voltages, service could be more quickly restored to most customers on the feeder. This could be accomplished by first opening the sectionalizer downstream from bus 9com, then closing the sectionalizer between buses 8ind and 13ind, then closing the distribution substation bus-tie breaker between buses 2 and 3 (after first balancing their taps to prevent circulating reactive power). This new configuration is shown in Figure 14.23.

Another important application of DA is the use of voltage control devices, such as switched capacitor banks, load tap changers (LTCs) on the distribution substation transformers, or separate voltage regulators, to minimize distribution losses and to better manage the customer voltage. This is known as volt-var optimization. Since the feeder load is continually varying, these devices often need to be changed in order to maintain the desired feeder voltages and minimize feeder

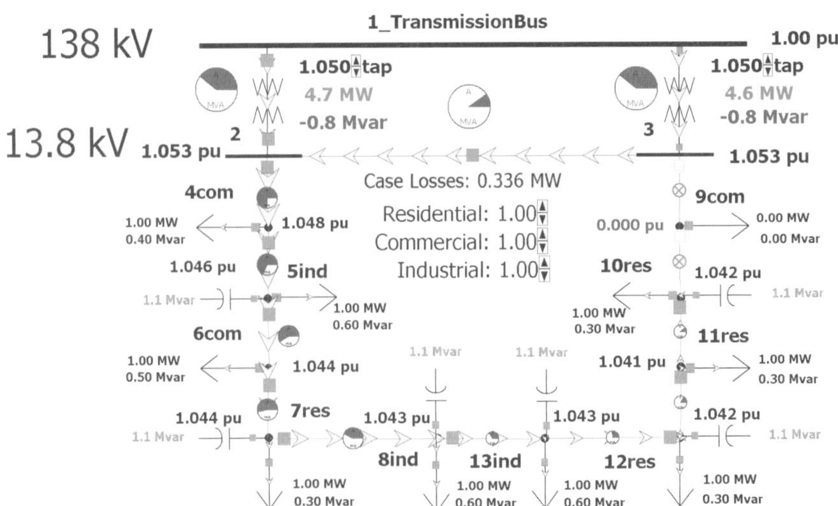

FIGURE 14.23 Primary loop feeder example, after fault and switching

losses. Without DA various techniques using only local information for local control have been employed. These included changing LTC transformer taps using the voltage sensed at the transformer, switching capacitors using local voltage, current, temperature, reactive power sensors, or just simple timers. While these approaches are better than nothing, they all have limitations. By providing a more global view of the entire feeder, DA can greatly improve the feeder voltage profile while minimizing losses.

To illustrate, again consider the Figure 14.22 case, which includes six 1.0 Mvar (nominal voltage) switched capacitors. The one-line display also shows the total system losses, and allows variation in the load multiplier for each of the three customer classes (residential, commercial, industrial) with a typical value for each ranging between 0.5 and 1.25. Initially all of the capacitors are in-service with total losses of 0.161 MW. However by manually opening the capacitors at buses 7res and 12res the losses can be decreased modestly to 0.153 MW. Under lighter loading situations the loss reduction from capacitor optimization can be even more significant.

DA can also help to reduce peak electric demands, potentially saving energy, by what is known as conservation voltage reduction (CVR). Because the power used by many loads is voltage dependent, the idea of CVR is to reduce the feeder voltage profile, while maintaining all customer voltages within the ANSI limits. A drop in the feeder voltage profile can result in an immediate decrease in power consumption. However, because some loads have their own closed-loop control systems, such as thermostats for electric heating, over the long-term these loads tend to behave more as constant energy resources. Overall energy savings from CVR are feeder specific and are estimated to be between 0.3% and 1% per percent voltage reduction.

While the benefits of DA can be substantial, what had been holding back more widespread adoption of DA was the costs, both for the initial updated equipment installations and the ongoing costs associated with maintaining the monitoring and control infrastructure, such as communication costs. However many of these costs have continued to decrease resulting in much more widespread adoption of DA technology.

14.9 SMART GRIDS

Over the last several years, the term "smart grid" has taken the electric power industry by storm, with its use being further cemented in the power industry lexicon with the launch of the *IEEE Transactions on Smart Grid Journal* in 2010. While a new word, the smart grid actually represents an evolutionary advancement on the technological innovation that has been present in the power industry since its inception in the 1880s. Such pervasive innovation over more than a century resulted in electrification being named the top engineering technology of the 20th century by the U.S. National Academy of Engineering in 2000. The smart grid represents a continuation of this application of advanced technology into the 21st century to take advantage of near ubiquitous computing and communication.

As defined in [29], "A smart grid uses digital technology to improve reliability, security and efficiency (both economic and energy) of the electric system from large generation, through the delivery system to electricity consumers and a growing number of distributed-generation and storage resources." Probably the best definition of the attributes of the smart grid is also given in [29] with its listing of six key characteristics:

1. Enables informed participation by customers;
2. Accommodates all generation and storage options;
3. Enables new products, services, and markets;
4. Provides the power quality for the range of needs;
5. Optimizes asset utilization and operating efficiency; and
6. Operates resiliency to disturbances, attacks, and natural disasters.

While the smart grid covers large generation and high-voltage transmission, it is most germane to the distribution system and ultimately the end-use customer. The distribution system that, quoting from [28], "has traditionally been characterized as the most unglamorous component" of the power grid is suddenly front and center. Rather than just being a passive, radial conduit for power to flow from the networked transmission system, it is the means for supporting a bidirectional flow of information and energy to customers who are no longer content to just receive a monthly electric bill. The large, new load potential of electric vehicles requires that the home electric meter and the distribution system become smarter, since the grid cannot reasonably accommodate charging a large number of car batteries as people return to their garages in the early evening when the remainder of the electric load is at peak demand. Providing real-time prices to some loads can provide additional control flexibility. In many locations distributed energy resources such as solar photovoltaics are resulting in situations in which distribution feeders have reversible power flows, looking like net generation when the sun is shining and load when it is not.

Underlying the smart grid is the need for a trustworthy and resilient cyber infrastructure. As more smart grid technologies are deployed, the result will be a power grid increasingly dependent on communication and computing. Disruptions in this cyber infrastructure, either due to accidents, bugs, or deliberate attacks could result in wide scale blackouts.

PROBLEMS

SECTION 14.2

14.1 Are laterals on primary radial systems typically protected from short circuits? If so, how (by fuses, circuit breakers, or reclosers)?

14.2 What is the most common type of grounding on primary distribution systems?

14.3 What is the most common primary distribution voltage class in the United States?

14.4 Why are reclosers used on overhead primary radial systems and overhead primary loop systems? Why are they not typically used on underground primary radial systems and underground primary loop systems?

SECTION 14.3

14.5 What are the typical secondary distribution voltages in the United States?

14.6 What are the advantages of secondary networks? Name two disadvantages.

14.7 Using the Internet, name three cities in the Western Interconnection of the United States that have secondary network systems.

SECTION 14.4

14.8 A three-phase 138 kVΔ/13.8 kV Y distribution substation transformer rated 40 MVA OA/50 MVA FA/65MVA FOA has an 9% impedance. (a) Determine the rated current on the primary distribution side of the transformer at its OA, FA, and FOA ratings. (b) Determine the per-unit transformer impedance using a system base of 100 MVA and 13.8 kV on the primary distribution side of the transformer. (c) Calculate the short-circuit current on the primary distribution side of the transformer for a three-phase bolted fault on the primary distribution side. Assume that the prefault voltage is 13.8 kV.

14.9 As shown in Figure 14.24, an urban distribution substation has one 30-MVA (FOA) and three 33.3 MVA (FOA), 138 kVΔ/12.5 kV Y transformers denoted TR1-TR4, which feed through circuit breakers to a ring

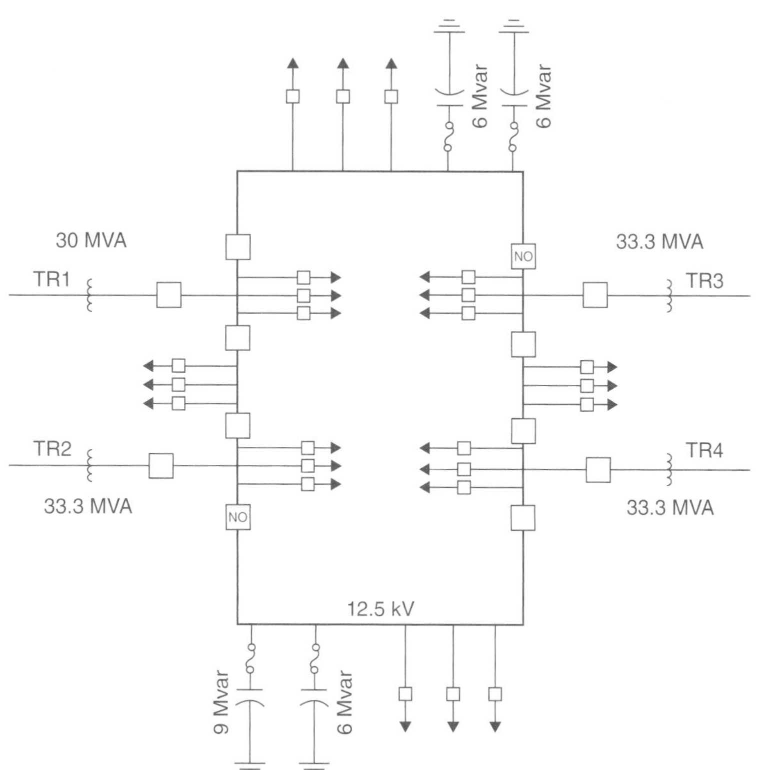

FIGURE 14.24 Distribution substation for Problems 14.9 and 14.10

bus. The transformers are older transformers designed for 55°C temperature rise. The ring bus contains eight bus-tie circuit breakers, two of which are normally open (NO), so as to separate the ring bus into two sections. TR1 and TR2 feed one section, and TR3 and TR4 feed the other section. Also, four capacitor banks, three banks rated at 6 Mvar and one at 9 Mvar, are connected to the ring bus. Twenty-four 12.5-kV underground primary feeders are served from the substation, 12 from each section. The utility that owns this substation has the following transformer summer loading criteria based on a percentage of nameplate rating:

a. 120% for normal summer loading.

b. 150% during a two-hour emergency.

c. 130% 30-day emergency loading.

Determine the following summer ratings of this substation: (a) the normal summer rating with all four transformers in service; (b) the allowable substation rating assuming the single-contingency loss of one transformer; and (c) the 30-day emergency rating under the single-contingency loss of one transformer. Assume that during a two-hour emergency, switching can be performed to reduce the total substation load by 10% and to approximately balance the loadings of the three transformers remaining in service. Assume a 5% reduction for unequal transformer loadings.

14.10 For the distribution substation given in Problem 14.9, assume that each of the four circuit breakers on the 12.5-kV side of the distribution substation transformers has a maximum continuous current of 2000A/phase during both normal and emergency conditions. Determine the summer allowable substation rating under the single-contingency loss of one transformer, based on not exceeding the maximum continuous current of these circuit breakers at 12.5-kV operating voltage. Assume a 5% reduction for unequal transformer loadings. Comparing the results of this problem with Problem 14.9, what limits the substation allowable rating, the circuit breakers or the transformers?

SECTION 14.5

14.11 (a) How many Mvars of shunt capacitors are required to increase the power factor on a 10 MVA load from 0.85 to 0.9 lagging? (b) How many Mvars of shunt capacitors are required to increase the power factor on a 10 MVA load from 0.90 to 0.95 lagging? (c) Which requires more reactive power, improving a low power-factor load or a high power-factor load?

14.12 Rework Example 14.3 with $R_{Load} = 40\ \Omega$/phase, $X_{Load} = 60\ \Omega$/phase, and $X_C = 60\ \Omega$/phase.

SECTION 14.7

14.13 Table 14.10 gives 2010 annual outage data (sustained interruptions) from a utility's CIS database for feeder 8050. This feeder serves 4500 customers

Outage Date	Time at Beginning of Outage	Outage Duration (minutes)	Circuit	Number of Customers Interrupted
1/15/2010	01:24:20	14.4	8050	342
4/4/2010	08:14:20	151.2	8050	950
7/08/2010	07:15:46	89.8	8050	125
9/10/2010	15:45:39	654.6	8050	15
10/11/2010	07:40:59	32.7	8050	2200
11/04/2010	22:30:00	10,053*	8050	4000*
12/01/2010	14:18:07	40	8050	370

TABLE 14.10

Customer Outage Data for Problem 14.13

* Major event due to ice storm

with a total load of 9 MW. Table 14.10 includes a major event that began on 04/11/2010 with 4000 customers out of service for approximately six days (10,053 minutes) due to an ice storm. Momentary interruption events (less than five minutes' duration) are excluded from Table 14.10. Calculate the SAIFI, SAIDI, CAIDI, and ASAI for this feeder: (a) including the major event; and (b) excluding the major event.

14.14 Assume that a utility's system consists of two feeders: feeder 7075 serving 2000 customers and feeder 8050 serving 4000 customers. Annual outage data during 2010 is given in Table 14.6 and 14.10 for these feeders. Calculate the SAIFI, SAIDI, CAIDI, and ASAI for the system, excluding the major event.

SECTION 14.8

PW **14.15** Open PowerWorld Simulator case Problem 14_15, which represents a lower load scenario for the Figure 14.22 case. Determine the optimal status of the six switched shunts to minimize the system losses.

PW **14.16** Open PowerWorld Simulator case Problem 14_16, which represents a lower load scenario for the Figure 14.22 case and has the LTC transformer taps each changed to 1.025. Determine the optimal status of the six switched shunts to minimize the system losses for this case.

PW **14.17** Open PowerWorld Simulator case Problem 14_17 and note the case losses. Then close the bus tie breaker between buses 2 and 3. How do the losses change? How can the case be modified to reduce the system losses?

PW **14.18** Usually in power flow studies the load is treated as being independent of the bus voltage. That is, a constant power model is used. However, in reality the load usually has some voltage dependence, so that decreasing the feeder voltage (such as for conversation voltage reduction) will result in less power consumption, at least temporarily. Open PowerWorld Simulator case Problem 14_18, which contains the Figure 14.22 system except the load model is set so 50% of the load is modeled as constant power, and 50% of the load is modeled as constant impedance (i.e., the load varies with the square of the bus voltage magnitude). By adjusting the tap positions for the two substation transformers and the capacitor banks, determine the operating configuration that minimizes the total load plus losses (shown on the display), with the constraint that all bus voltage must be at least 0.97 per unit.

PW **14.19** Repeat Problem 14.18, except using PowerWorld Simulator case Problem 14_19 (which has a different load level from the Problem 14.18 case).

SECTION 14.9

14.20 Select one of the smart grid characteristics from the list given in this section. Write a one page (or other instructor-selected length) summary and analysis paper on a current news story that relates to this characteristic.

CASE STUDY QUESTIONS

a. Do DER installations reduce distribution losses? If so, why?

b. Why are most distribution feeders in the United States built in a radial configuration?

c. What are the challenges associated with large penetrations of DERs in distribution systems?

REFERENCES

1. J. D. Glover, "Electric Power Distribution," *Encyclopedia of Energy Technology and The Environment* (John Wiley & Sons, New York, NY, 1995).
2. D. G. Fink and H. W. Beaty, *Standard Handbook for Electrical Engineers*, 11th ed. (McGraw-Hill, New York, 1978), Sec 18.
3. T. Gonen, *Power Distribution Engineering* (Wiley, New York, 1986).
4. A. J. Pansini, *Electrical Distribution Engineering*, 2nd ed., The Fairmont Press (Liburn, GA, 1992).
5. J. J. Burke, *Power Distribution Engineering*, Marcel Dekker (New York, 1994).
6. Various co-workers, *Electric Distribution Systems Engineering Handbook, Ebasco Services Inc.*, 2nd ed., McGraw-Hill (New York, 1987).
7. Various co-workers, *Electrical Transmission & Distribution Reference Book*, Westinghouse Electric Corporation (Pittsburgh, 1964).

8. Various co-workers, *Distribution Systems Electric Utility Engineering Reference Book*, Vol. 3, Westinghouse Electric Corporation (Pittsburgh, 1965).

9. Various co-workers, *Underground Systems Reference Book*, Edison Electric Institute (New York, 1957).

10. R. Settembrini, R. Fisher, and N. Hudak, "Seven Distribution Systems: How Reliabilities Compare," *Electrical World*, 206, 5 (May 1992), pp. 41–45.

11. J. L. Blackburn, *Protective Relaying*, Marcel Dekker (New York, 1987).

12. R. Billinton, *Power System Reliability Evaluation*, Gordon and Breach (1988).

13. R. Billinton, R. N. Allan, and L. Salvaderi, *Applied Reliability Assessment in Electric Power Systems*, Institute of Electrical and Electronic Engineers (New York, 1991).

14. R. Billinton and J. E. Billinton, "Distribution Reliability Indices," *IEEE Transactions on Power Delivery*, 4, 1 (January 1989), pp. 561–568.

15. D. O. Koval and R. Billinton, "Evaluation of Distribution Circuit Reliability," Paper F77 067-2, *IEEE Winter Power Meeting* (New York, NY, January/February 1977).

16. IEEE Committee Report, "List of Transmission and Distribution Components for Use in Outage Reporting and Reliability Calculations," *IEEE Transactions on Power Apparatus and Systems*, PAS 95, 54 (July/August 1976), pp. 1210–1215.

17. U. S. Department of Energy, *The National Electric Reliability Study: Technical Reports*, DOE/EP-0003, (April 1981).

18. J. B. Bunch and co-workers, *Guidelines for Evaluating Distribution Automation*, EPRI-EL-3728, Project 2021-1, Electric Power Research Institute (Palo Alto, CA, November 1984).

19. T. Desmond, "Distribution Automation: What is it? What does it do?" *Electrical World*, 206, 2 (February 1992), pp. 56 & 57.

20. J. Smith et al., "It's All in the Plans," *IEEE Power & Energy Magazine*, 13, 2, (March/April 2015), pp. 20–29.

21. ANSI/IEEE Std. C57.91-1995, *IEEE Guide for Loading Mineral-Oil-Immersed Transformers*, Approved June 14, 1995.

22. ANSI/EEE Std. C57.92-1981, *IEEE Guide for Loading Mineral-Oil-Immersed Power Transformers Up to and Including 100 MVA with 55°C or 65°C Average Winding Rise*, Approved January 12, 1981.

23. ANSI/EEE Std. C57.92-1981, *IEEE Guide for Loading Mineral-Oil-Immersed Overhead and Pad-Mounted Distribution Transformers Rated 500 kVA and Less with 65°C or 55°C Average Winding Rise*, Approved November 19, 1980.

24. IEEE Std. 1366, *IEEE Guide for Electric Power Distribution Reliability Indices*, 2003 and 2012 editions.

25. AIEE Committee Report, "Supervisory Control and Associated Telemetering Equipment—A Survey of Current Practice," *AIEE Transactions Power Apparatus and Systems, Part III*, (January 1955), pp. 36–68.

26. Technical and System Requirements for Advance Distribution Automation, EPRI Report 1010915, June 2004.

27. H. L. Willis, *Power Distribution Planning Reference Book*, 2nd Edition, CRC Press, Boca Raton, FL, 2004.

28. W. H. Kersting, Distribution System Modeling and Analysis, 3rd Edition, CRC Press, Boca Raton, FL, 2012.

29. *Smart Grid System Report*, U.S. DOE, July 2009.

30. Z. Wang and J. Wang, "Review on Implementation and Assessment of Conservation Voltage Reduction," *IEEE Trans. on Power Systems*, vol. 29, May 2014, pp. 1306–1315.

APPENDIX

TABLE A.1

Typical average values of synchronous-machine constants

Constant (units)	Type	Symbol	Turbo-Generator (solid rotor)	Water-Wheel Generator (with dampers)	Synchro-nous Condenser	Synchro-nous Motor
Reactances (per unit)	Synchronous	X_d	1.1	1.15	1.80	1.20
		X_q	1.08	0.75	1.15	0.90
	Transient	X'_d	0.23	0.37	0.40	0.35
		X'_q	0.23	0.75	1.15	0.90
	Subtransient	X''_d	0.12	0.24	0.25	0.30
		X''_q	0.15	0.34	0.30	0.40
	Negative-sequence	X_2	0.13	0.29	0.27	0.35
	Zero-sequence	X_0	0.05	0.11	0.09	0.16
Resistances (per unit)	Positive-sequence	R (dc)	0.003	0.012	0.008	0.01
		R (ac)	0.005	0.012	0.008	0.01
	Negative-sequence	R_2	0.035	0.10	0.05	0.06
Time constants (seconds)	Transient	T'_{d0}	5.6	5.6	9.0	6.0
		T'_d	1.1	1.8	2.0	1.4
	Subtransient	$T''_d = T''_q$	0.035	0.035	0.035	0.036
	Armature	T_a	0.16	0.15	0.17	0.15

(Adapted from E. W. Kimbark, *Power System Stability: Synchronous Machines* (New York: Dover Publications, 1956/1968), Chap. 12)

TABLE A.2

Typical transformer leakage reactances

Rating of Highest Voltage Winding kV	BIL of Highest Voltage Winding kV	Leakage Reactance per unit*	
Distribution Transformers			
2.4	30	0.023–0.049	
4.8	60	0.023–0.049	
7.2	75	0.026–0.051	
12	95	0.026–0.051	
23	150	0.052–0.055	
34.5	200	0.052–0.055	
46	250	0.057–0.063	
69	350	0.065–0.067	
Power Transformers 10 MVA and Below			
8.7	110	0.050–0.058	
25	150	0.055–0.058	
34.5	200	0.060–0.065	
46	250	0.060–0.070	
69	350	0.070–0.075	
92	450	0.070–0.085	
115	550	0.075–0.100	
138	650	0.080–0.105	
161	750	0.085–0.011	
Power Transformers Above 10 MVA			
		Self-Cooled or Forced-Air-Cooled	Forced-Oil-Cooled
8.7	110	0.050–0.063	0.082–0.105
34.5	200	0.055–0.075	0.090–0.128
46	250	0.057–0.085	0.095–0.143
69	350	0.063–0.095	0.103–0.158
92	450	0.060–0.118	0.105–0.180
115	550	0.065–0.135	0.107–0.195
138	650	0.070–0.140	0.117–0.245
161	750	0.075–0.150	0.125–0.250
230	900	0.070–0.160	0.120–0.270
345	1300	0.080–0.170	0.130–0.280
500	1550	0.100–0.200	0.160–0.340
765		0.110–0.210	0.190–0.350

*Per-unit reactances are based on the transformer rating

TABLE A.3 Characteristics of copper conductors, hard drawn, 97.3% conductivity

Size of Conductor: Circular Mils	Size of Conductor: A.W.G. or B. & S.	Number of Strands	Diameter of Individual Strands (inches)	Outside Diameter (inches)	Breaking Strength (pounds)	Weight (pounds per mile)	Approx. Current Carrying Capacity* (amps)	Geometric Mean Radius at 60 Hz (feet)	r_a Resistance (Ohms per Conductor per Mile) 25°C (77°F) dc	25°C 25 Hz	25°C 50 Hz	25°C 60 Hz	50°C (122°F) dc	50°C 25 Hz	50°C 50 Hz	50°C 60 Hz	x_a Inductive Reactance (ohms per conductor per mile at 1 ft spacing) 25 Hz	x_a 50 Hz	x_a 60 Hz	x'_a Shunt Capacitive Reactance (megohms per conductor per mile at 1 ft spacing) 25 Hz	x'_a 50 Hz	x'_a 60 Hz
1 000 000	...	37	0.1644	1.151	43 830	16 300	1 300	0.0368	0.0585	0.0594	0.0620	0.0634	0.0640	0.0648	0.0672	0.0685	0.1666	0.333	0.400	0.216	0.1081	0.0901
900 000	...	37	0.1560	1.092	39 510	14 670	1 220	0.0349	0.0650	0.0658	0.0682	0.0695	0.0711	0.0718	0.0740	0.0752	0.1693	0.339	0.406	0.220	0.1100	0.0916
800 000	...	37	0.1470	1.029	35 120	13 040	1 130	0.0329	0.0731	0.0739	0.0760	0.0772	0.0800	0.0806	0.0826	0.0837	0.1722	0.344	0.413	0.224	0.1121	0.0934
750 000	...	37	0.1424	0.997	33 400	12 230	1 090	0.0319	0.0780	0.0787	0.0807	0.0818	0.0853	0.0859	0.0878	0.0888	0.1739	0.348	0.417	0.226	0.1132	0.0943
700 000	...	37	0.1375	0.963	31 170	11 410	1 040	0.0308	0.0836	0.0842	0.0861	0.0871	0.0914	0.0920	0.0937	0.0947	0.1759	0.352	0.422	0.229	0.1145	0.0954
600 000	...	37	0.1273	0.891	27 020	9 781	940	0.0285	0.0975	0.0981	0.0997	0.1006	0.1066	0.1071	0.1086	0.1095	0.1799	0.360	0.432	0.235	0.1173	0.0977
500 000	...	37	0.1162	0.814	22 510	8 151	840	0.0260	0.1170	0.1175	0.1188	0.1196	0.1280	0.1283	0.1296	0.1303	0.1845	0.369	0.443	0.241	0.1205	0.1004
500 000	...	19	0.1622	0.811	21 590	8 151	840	0.0256	0.1170	0.1175	0.1188	0.1196	0.1280	0.1283	0.1296	0.1303	0.1853	0.371	0.445	0.241	0.1206	0.1005
450 000	...	19	0.1539	0.770	19 750	7 336	780	0.0243	0.1300	0.1304	0.1316	0.1323	0.1422	0.1426	0.1437	0.1443	0.1879	0.376	0.451	0.245	0.1224	0.1020
400 000	...	19	0.1451	0.726	17 560	6 521	730	0.0229	0.1462	0.1466	0.1477	0.1484	0.1600	0.1603	0.1613	0.1619	0.1909	0.382	0.458	0.249	0.1245	0.1038
350 000	...	19	0.1357	0.679	15 590	5 706	670	0.0214	0.1671	0.1675	0.1684	0.1690	0.1828	0.1831	0.1840	0.1845	0.1943	0.389	0.466	0.254	0.1269	0.1058
350 000	...	12	0.1708	0.710	15 140	5 706	670	0.0225	0.1671	0.1675	0.1684	0.1690	0.1828	0.1831	0.1840	0.1845	0.1918	0.384	0.460	0.251	0.1253	0.1044
300 000	...	19	0.1257	0.629	13 510	4 891	610	0.01987	0.1950	0.1953	0.1961	0.1966	0.213	0.214	0.214	0.215	0.1982	0.396	0.476	0.259	0.1296	0.1080
300 000	...	12	0.1581	0.657	13 170	4 891	610	0.0208	0.1950	0.1953	0.1961	0.1966	0.213	0.214	0.214	0.215	0.1957	0.392	0.470	0.256	0.1281	0.1068
250 000	...	19	0.1147	0.574	11 360	4 076	540	0.01813	0.234	0.234	0.235	0.235	0.256	0.256	0.257	0.257	0.203	0.406	0.487	0.266	0.1329	0.1108
250 000	...	12	0.1443	0.600	11 130	4 076	540	0.01902	0.234	0.234	0.235	0.235	0.256	0.256	0.257	0.257	0.200	0.401	0.481	0.263	0.1313	0.1094
211 600	4/0	19	0.1055	0.528	9 617	3 450	480	0.01668	0.276	0.277	0.277	0.278	0.302	0.303	0.303	0.303	0.207	0.414	0.497	0.272	0.1359	0.1132
211 600	4/0	12	0.1328	0.552	9 483	3 450	490	0.01750	0.276	0.277	0.277	0.278	0.302	0.303	0.303	0.303	0.205	0.409	0.491	0.269	0.1343	0.1119
211 600	4/0	7	0.1739	0.522	9 154	3 450	480	0.01579	0.276	0.277	0.277	0.278	0.302	0.303	0.303	0.303	0.210	0.420	0.503	0.273	0.1363	0.1136
167 800	3/0	12	0.1183	0.492	7 556	2 736	420	0.01559	0.349	0.349	0.349	0.350	0.381	0.381	0.382	0.382	0.210	0.421	0.505	0.277	0.1384	0.1153
167 800	3/0	7	0.1548	0.464	7 366	2 736	420	0.01404	0.349	0.349	0.349	0.350	0.381	0.381	0.382	0.382	0.216	0.431	0.518	0.281	0.1405	0.1171
133 100	2/0	7	0.1379	0.414	5 926	2 170	360	0.01252	0.440	0.440	0.440	0.440	0.481	0.481	0.481	0.481	0.222	0.443	0.532	0.289	0.1445	0.1205
105 500	1/0	7	0.1228	0.368	4 752	1 720	310	0.01113	0.555	0.555	0.555	0.555	0.606	0.607	0.607	0.607	0.227	0.455	0.546	0.298	0.1488	0.1240
83 690	1	7	0.1093	0.328	3 804	1 364	270	0.00992	0.699	0.699	0.699	0.699	0.765				0.233	0.467	0.560	0.306	0.1528	0.1274
83 690	1	3	0.1670	0.360	3 620	1 351	270	0.01016	0.692	0.692	0.692	0.692	0.757				0.232	0.464	0.557	0.299	0.1495	0.1246
66 370	2	7	0.0974	0.292	3 045	1 082	230	0.00883	0.881	0.882	0.882	0.882	0.964				0.239	0.478	0.574	0.314	0.1570	0.1308
66 370	2	3	0.1487	0.320	2 913	1 071	240	0.00903	0.873				0.955				0.238	0.476	0.571	0.307	0.1537	0.1281
66 370	2	1	...	0.258	3 003	1 061	220	0.00836	0.864				0.945				0.242	0.484	0.581	0.323	0.1614	0.1345
52 630	3	7	0.0867	0.260	2 433	858	200	0.00787	1.112				1.216				0.245	0.490	0.588	0.322	0.1611	0.1343
52 630	3	3	0.1325	0.285	2 359	850	200	0.00805	1.101				1.204				0.244	0.488	0.585	0.316	0.1578	0.1315
52 630	3	1	...	0.229	2 439	841	190	0.00745	1.090		Same as dc		1.192		Same as dc		0.248	0.496	0.595	0.331	0.1656	0.1380
41 740	4	3	0.1180	0.254	1 879	674	180	0.00717	1.388				1.518				0.250	0.499	0.599	0.324	0.1619	0.1349
41 740	4	1	...	0.204	1 970	667	170	0.00663	1.374				1.503				0.254	0.507	0.609	0.339	0.1697	0.1415
33 100	5	3	0.1050	0.226	1 505	534	150	0.00638	1.750				1.914				0.256	0.511	0.613	0.332	0.1661	0.1384
33 100	5	1	...	0.1819	1 591	529	140	0.00590	1.733				1.895				0.260	0.519	0.623	0.348	0.1738	0.1449
26 250	6	3	0.0935	0.201	1 205	424	130	0.00568	2.21				2.41				0.262	0.523	0.628	0.341	0.1703	0.1419
26 250	6	1	...	0.1620	1 280	420	120	0.00526	2.18				2.39				0.265	0.531	0.637	0.356	0.1779	0.1483
20 820	7	1	...	0.1443	1 030	333	110	0.00468	2.75				3.01				0.271	0.542	0.651	0.364	0.1821	0.1517
16 510	8	1	...	0.1285	826	264	90	0.00417	3.47				3.80				0.277	0.554	0.665	0.372	0.1862	0.1552

*For conductor at 75°C, air at 25°C, wind 1.4 miles per hour (2 ft/sec), frequency = 60 Hz.

TABLE A.4 Characteristics of aluminum cable, steel, reinforced (Aluminum Company of America)—ACSR

Code Word	Circular Mils Aluminum	Aluminum			Steel		Outside Diameter (inches)	Copper Equivalent* Circular Mils or A.W.G.	Ultimate Strength (pounds)	Weight (pounds per mile)	Geometric Mean Radius at 60 Hz (feet)	Approx. Current Carrying Capacity† (amps)	r_a Resistance (Ohms per Conductor per Mile)								x_a Inductive Reactance (ohms per conductor per mile at 1 ft spacing all currents)	x'_a Shunt Capacitive Reactance (megohms per conductor per mile at 1 ft spacing)
				Strand Diameter (inches)		Strand Diameter (inches)							25°C (77°F) Small Currents				50°C (122°F) Current Approx. 75% Capacity‡					
													dc	25 Hz	50 Hz	60 Hz	dc	25 Hz	50 Hz	60 Hz	60 Hz	60 Hz
Joree	2 515 000	76	..	0.1819	19	0.0849	1.880		61 700		0.0621									0.0450	0.337	0.0755
Thrasher	2 312 000	76	..	0.1744	19	0.0814	1.802		57 300		0.0595									0.0482	0.342	0.0767
Kiwi	2 167 000	72	4	0.1735	7	0.1157	1.735		49 800		0.0570									0.0511	0.348	0.0778
Bluebird	2 156 000	84	4	0.1602	19	0.0961	1.762		60 300		0.0588									0.0505	0.344	0.0774
Chukar	1 781 000	84	4	0.1456	19	0.0874	1.602		51 000		0.0534									0.0598	0.355	0.0802
Falcon	1 590 000	54	3	0.1716	19	0.1030	1.545	1 000 000	56 000	10 777	0.0520	1 380	0.0587	0.0588	0.0590	0.0591	0.0646	0.0656	0.0675	0.0684	0.359	0.0814
Parrot	1 510 500	54	3	0.1673	19	0.1004	1.506	950 000	53 200	10 237	0.0507	1 340	0.0618	0.0619	0.0621	0.0622	0.0680	0.0690	0.0710	0.0720	0.362	0.0821
Plover	1 431 000	54	3	0.1628	19	0.0977	1.465	900 000	50 400	9 699	0.0493	1 300	0.0652	0.0653	0.0655	0.0656	0.0718	0.0729	0.0749	0.0760	0.365	0.0830
Martin	1 351 000	54	3	0.1582	19	0.0949	1.424	850 000	47 600	9 160	0.0479	1 250	0.0691	0.0692	0.0694	0.0695	0.0761	0.0771	0.0792	0.0803	0.369	0.0838
Pheasant	1 272 000	54	3	0.1535	19	0.0921	1.382	800 000	44 800	8 621	0.0465	1 200	0.0734	0.0735	0.0737	0.0738	0.0808	0.0819	0.0840	0.0851	0.372	0.0847
Grackle	1 192 500	54	3	0.1486	19	0.0892	1.338	750 000	43 100	8 082	0.0450	1 160	0.0783	0.0784	0.0786	0.0788	0.0862	0.0872	0.0894	0.0906	0.376	0.0857
Finch	1 113 000	54	3	0.1436	19	0.0862	1.293	700 000	40 200	7 544	0.0435	1 110	0.0839	0.0840	0.0842	0.0844	0.0924	0.0935	0.0957	0.0969	0.380	0.0867
Curlew	1 033 500	54	3	0.1384	7	0.1384	1.246	650 000	37 100	7 019	0.0420	1 060	0.0903	0.0905	0.0907	0.0909	0.0994	0.1005	0.1025	0.1035	0.385	0.0878
Cardinal	954 000	54	3	0.1329	7	0.1329	1.196	600 000	34 200	6 479	0.0403	1 010	0.0979	0.0980	0.0981	0.0982	0.1078	0.1088	0.1118	0.1128	0.390	0.0890
Canary	900 000	54	3	0.1291	7	0.1291	1.162	566 000	32 300	6 112	0.0391	970	0.104	0.104	0.104	0.104	0.1145	0.1155	0.1175	0.1185	0.393	0.0898
Crane	874 500	54	3	0.1273	7	0.1273	1.146	550 000	31 400	5 940	0.0386	950	0.107	0.107	0.107	0.108	0.1178	0.1188	0.1218	0.1228	0.395	0.0903
Condor	795 000	54	3	0.1214	7	0.1214	1.093	500 000	28 500	5 399	0.0368	900	0.117	0.118	0.118	0.119	0.1288	0.1308	0.1358	0.1378	0.401	0.0917
Drake	795 000	26	2	0.1749	7	0.1360	1.108	500 000	31 200	5 770	0.0375	900	0.117	0.117	0.117	0.117	0.1288	0.1288	0.1288	0.1288	0.399	0.0912
Mallard	795 000	30	2	0.1628	19	0.0977	1.140	500 000	38 400	6 517	0.0393	910	0.117	0.117	0.117	0.117	0.1288	0.1288	0.1288	0.1288	0.393	0.0904
Crow	715 500	54	3	0.1151	7	0.1151	1.036	450 000	26 300	4 859	0.0349	830	0.131	0.131	0.131	0.132	0.1442	0.1452	0.1472	0.1482	0.407	0.0932
Starling	715 500	26	2	0.1659	7	0.1290	1.051	450 000	28 100	5 193	0.0355	840	0.131	0.131	0.131	0.131	0.1442	0.1442	0.1442	0.1442	0.405	0.0928
Redwing	715 500	30	2	0.1544	19	0.0926	1.081	450 000	34 600	5 865	0.0372	840	0.131	0.131	0.131	0.131	0.1442	0.1442	0.1442	0.1442	0.399	0.0920
Flamingo	666 600	54	3	0.1111	7	0.1111	1.000	419 000	24 500	4 527	0.0337	800	0.140	0.140	0.141	0.141	0.1541	0.1571	0.1591	0.1601	0.412	0.0943
Rook	636 000	54	3	0.1085	7	0.1085	0.977	400 000	23 600	4 319	0.0329	770	0.147	0.147	0.148	0.148	0.1618	0.1638	0.1678	0.1688	0.414	0.0950
Grosbeak	636 000	26	2	0.1564	7	0.1216	0.990	400 000	25 000	4 616	0.0335	780	0.147	0.147	0.147	0.147	0.1618	0.1618	0.1618	0.1618	0.412	0.0946
Egret	636 000	30	2	0.1456	19	0.0874	1.019	400 000	31 500	5 213	0.0351	780	0.147	0.147	0.147	0.147	0.1618	0.1618	0.1618	0.1618	0.406	0.0937
Peacock	605 000	54	3	0.1059	7	0.1059	0.953	380 500	22 500	4 109	0.0321	750	0.154	0.155	0.155	0.155	0.1695	0.1715	0.1755	0.1775	0.417	0.0957
Squab	605 000	26	2	0.1525	7	0.1186	0.966	380 500	24 100	4 391	0.0327	760	0.154	0.154	0.154	0.154	0.1700	0.1720	0.1720	0.1720	0.415	0.0953
Dove	556 500	26	2	0.1463	7	0.1138	0.927	350 000	22 400	4 039	0.0313	730	0.168	0.168	0.168	0.168	0.1849	0.1859	0.1859	0.1859	0.420	0.0965
Eagle	556 500	30	2	0.1362	7	0.1362	0.953	350 000	27 200	4 588	0.0328	730	0.168	0.168	0.168	0.168	0.1849	0.1859	0.1859	0.1859	0.415	0.0957
Hawk	477 000	26	2	0.1355	7	0.1054	0.858	300 000	19 430	3 462	0.0290	670	0.196	0.196	0.196	0.196	0.216				0.430	0.0988
Hen	477 000	30	2	0.1261	7	0.1261	0.883	300 000	23 300	3 933	0.0304	670	0.196	0.196	0.196	0.196	0.216				0.424	0.0980
Ibis	397 500	26	2	0.1236	7	0.0961	0.783	250 000	16 190	2 885	0.0265	590	0.235				0.259				0.441	0.1015
Lark	397 500	30	2	0.1151	7	0.1151	0.806	250 000	19 980	3 277	0.0278	600	0.235		Same as dc		0.259		Same as dc		0.435	0.1006
Linnet	336 400	26	2	0.1138	7	0.0855	0.721	4/0	14 050	2 442	0.0244	530	0.278				0.306				0.451	0.1039
Oriole	336 400	30	2	0.1059	7	0.1059	0.741	4/0	17 040	2 774	0.0255	530	0.278				0.306				0.445	0.1032
Ostrich	300 000	26	2	0.1074	7	0.0835	0.680	188 700	12 650	2 178	0.0230	490	0.311				0.342				0.458	0.1057
Piper	300 000	30	2	0.1000	7	0.1000	0.700	188 700	15 430	2 473	0.0241	500	0.311				0.342				0.462	0.1049
Partridge	266 800	26	2	0.1013	7	0.0788	0.642	3/0	11 250	1 936	0.0217	460	0.350				0.385				0.465	0.1074

*Based on copper 97%, aluminum 61% conductivity.
†For conductor at 75°C, air at 25°C, wind 1.4 miles per hour (2 ft/sec), frequency = 60 Hz.
‡"Current Approx. 75% Capacity" is 75% of the "Approx. Current Carrying Capacity in Amps" and is approximately the current which will produce 50°C conductor temp. (25°C rise) with 25°C air temp., wind 1.4 miles per hour.

Index

AAAC. *See* All-aluminum alloy conductor (AAAC)
AAC. *See* All-aluminum conductor (AAC)
ABCD parameters, 274–275
ABCD parameters and nominal π circuit, 263–265
ABCD parameters of common networks, 261
ACCR. *See* Aluminum conductor composite reinforced (ACCR)
ac fault current, 439
ACSR. *See* Aluminum conductor, steel-reinforced (ACSR)
Active power, 44
Actual quantity, 108
Advantages, balanced three-phase versus single-phase systems, 68–69
Aeroderivative gas turbine sets, 678
All-aluminum alloy conductor (AAAC), 175
All-aluminum conductor (AAC), 175
Aluminum conductor
 composite reinforced (ACCR), 175
 steel-reinforced (ACSR), 174
Ampacity, 68
Ampere's law, 96
Area control error, 768
Area frequency response characteristic, 763
Areas, 376
Armature time constant, 440
Asset exposure map, 844
Asymmetrical fault current, 436
Asynchronous back-to-back tie, 2 HVDC transmission technologies, 42
Asynchronous ties, HVDC transmission technologies, 248–249
Attenuation, 835–836
Automatic generation control strategy, 771
Autotransformers, 129–131
 ideal single-phase transformers, 129
 single-phase autotransformers, 130–131
Average power, 44

Back substitution, 331
Backup relays, 594
Balance beam relay, 646
Balanced *abc* currents, 497–498
Balanced-Δ and-Y loads, 61–62
Balanced-Δ loads, 58–60
Balanced line currents, 57–58
Balanced line-to-line voltages, 56–57
Balanced line-to-neutral voltages, 56, 496–497
Balanced three-phase circuits, 55–62
 balanced-Δ and-Y loads, 61–62
 balanced-Δ loads, 58–60
 balanced line currents, 57–58
 balanced line-to-line voltages, 56–57
 balanced line-to-neutral voltages, 56
 balanced-Y connections, 55–56
 Δ–Y conversion for balanced loads, 60–61
 equivalent line-to-neutral diagrams, 62
Balanced three-phase fault, 568
Balanced three-phase transformer, voltage drop and fault current, 124–125
Balanced-Y and balanced-Δ loads, 503
Balanced-Y connections, 55–56
Balanced-Y impedance load, 499
Base quantities, 108
Base quantities, rules for selecting, 121
Basic single-phase two-winding transformer, 95
B coefficient, 387
Bewley lattice diagram, 822–827
 junction, lossless lines, 827
 junction, two single-phase lossless lines, 823
 overhead line connected to cable, single-phase lossless lines lattice diagram, 824–827
 single-phase lossless line lattice diagram, 823
Black-start, 686–687
Block and trip regions, 646
Bottom-up restoration, 671–672
Boundary conditions for single-phase lossless lines, 813–822
 receiving-end voltage reflection coefficient, 813
Boundary conditions for single-phase lossless lines
 step-voltage source at sending end, matched load at receiving end, 815–816
 step-voltage source matched at sending end, capacitive load at receiving end, 818–820
 step-voltage source matched at sending end, open receiving end, 816–818

Boundary conditions for single-phase lossless lines (*Continued*)
  step-voltage source with unmatched source resistance at sending end, unmatched resistive load at receiving end, 820–822
  transit time, 813
Brushless exciters, 757
Bundle, 175
Bundle conductor configurations, 196
Bus impedance equivalent circuit, 449
Bus impedance matrix, 445–454
  bus impedance equivalent circuit, 449
  mutual impedances, 448
  parallel connection, unloaded synchronous machine internal-voltage sources, 449
  self-impedances, 448
  Z_{bus} use to compute three-phase short-circuit currents, 447–448
Bus protection with differential relays, 647–648
Bus types, 345–346

California frequency response investigation, 797–802
Capacitance, 205–211
  single-phase line, capacitance, admittance, reactive power supplied, 210
  single-phase two-wire line, 206
  three-phase line, capacitance, shunt admittance; charging current, reactive power supplied, 210–211
  three-phase line with two conductors per bundle, 209
  three-phase three-wire line with equal phase spacing, 207
Capacitive load, 43
Characteristic impedance, 267
Charge separation within clouds. 842
Circuit breaker, 455
Circuit breaker selection, 455–459
  circuit breaker, 455
  current ratings, 458
  low-voltage circuit breakers, 455
  outdoor circuit breakers ratings, 456
  power circuit breakers, 455
  related required capabilities, 457
  voltage ratings, 458
Class current transformers, 617
Coal units, capacity committed and dispatched, 320
CO-8 time-delay overcurrent relay
  characteristics, 622
  operating time, 623–624
Coherent machines, 694
Cold load pickup, 683–684
Collector system, 418
Common identities involving $a = 1\underline{/120°}$, 494
Common secondary main, 887–888
Compensated transmission-line section, 290
Completely transposed three-phase line
  capacitances, 510
  sequence impedances, 509
Complex power, 47–52
  balanced three-phase generators, 64–65
  balanced three-phase motors, 65
  balanced-Y and balanced-Δ impedance loads, 65–66
  generator convention, 48
  load convention, 48, 49
  negative reactive power, 49
  positive reactive power, 49
  power triangle and power factor correction, 50–52
  real and reactive power, delivered or absorbed, 48–49
Conductance, 182
Conductors, design considerations, transmission lines, 174–175
  all-aluminum alloy conductor (AAAC), 175
  all-aluminum conductor (AAC), 175
  aluminum conductor composite reinforced (ACCR), 175
  aluminum conductor, steel-reinforced (ACSR), 174
  bundle, 175
  corona loss, 182
  gap-type ZT-aluminum conductor (GTZACSR), 175
Consumption growth, 13
Continuous current, 458
Continuous current rating, 461
Control and operating principles, HVDC transmission technologies, 255–257
Conventional HVDC, 252–253
Conventional power plants versus wind power plants, 417–418
Converter station, 239
Coordination, fuse/recloser, 632
Coordination of economic dispatch, 771
CO relay characteristics comparison, 623
Core loss current, 102
Corona loss, 182
Costs, comparative of HVDC and EHV AC transmission alternatives, 247
CO_2
  average emissions by plant type, 322
  emissions by scenario, 321
Critical clearing angle for cleared three-phase fault, 705–707
Critical clearing time for temporary three-phase fault, 703–705
CT. *See* Current transformers
Current in series R–L circuit with ac voltage source, 436
Current ratings, 458–459
  continuous current, 458
  interrupting MVA, 458–459
  interrupting time, 458
  momentary current, 458
  short-circuit current, 458
Currents in sequence networks, 513–514
Current transformers
  equivalent circuit, 618
  performance, 619
  performance determination, 617–618
  window design, 616
Customer outage data, 908

dc and ac resistance, 181–182
dc offset current, 436
dc power flow, 373–374
Δ–Δ transformer features, 519
Δ–Y conversion for balanced loads, 60–61
Δ–Y transformer phase shift effect on fault currents, 564–566
Density contour map, 843
Deregulation, 2–9
  degrees of regulation, 6
  demand, 9
  free markets, 2–3
  infrastructure, 8
  investment environment, 7
  Public Utility Regulatory Policies Act of 1978, 4
  regulation attempts, 3–4
  variables in price, 5
DERs. *See* Distributed energy sources (DERs)
Design considerations, transmission lines, 174–179
  conductors, 174–175
  economic factors, 179
  electrical factors, 178
  environmental factors, 178
  insulators, 175–176
  mechanical factors, 178
  shield wires, 176
  support structures, 176
  transmission-line characteristics, 177
Diagonal sequence impedances, 505
Diesel generator sets, 678
Differential current, 646
Differential relays, 645–647
  balance beam relay, 646
  block and trip regions, 646
  differential current, 646
Direct axis short-circuit subtransient time constant, 439
Direct axis short-circuit transient time constant, 440
Directional relays, 633–634
  block and trip regions in complex plane, 634
  series with overcurrent relay, 633
Direct solutions to linear algebraic equations: Gauss elimination, 330–334
  back substitution, 331
  Gauss elimination and back substitution, 332–333
  Gauss elimination: triangularizing a matrix, 333–334
Discrete-time equivalent circuit, single-phase lossless line transients, 831–834
Discrete-time models, 828–834
  discrete-time equivalent circuit, single-phase lossless line transients, 831–834
  lumped capacitance, 830–831
  lumped inductance, 829–830
  lumped resistance, 831
  nodal equations, 831
  single-phase lossless line, 828–829
Distance relay, 641
Distortion, 836–837
Distributed energy sources (DERs), 860–875
  distribution feeder response characteristics, 864
  integrated approach, 867–870
  proactive approach, 863–864
  programmatic implementation of proactive plan, 872–874
  scale of distribution, 861–862
  size and location, 864
  structural diversity of distribution system, 862–863
  systemwide application, 870–872
  technologies, 864–867
Distribution automation, 910–913
  primary loop feeder after fault, 912
  primary loop feeder before fault, 911
Distribution reliability, 906–910
  customer outage data, 908
  indices, 908
  outage causes, 910
  outage reporting information, 909
  values of reliability indices, 907
Distribution substation transformers, 890–895
Distribution transformer kVA ratings, 899
Distribution transformers, 895–900
Double-circuit line, single-line diagram, 219
Double line-to-ground fault, 560–567
  Δ–Y transformer phase shift effect on fault currents, 564–566
  faults summary, 567
  short-circuit calculations using sequence networks, 562–564
Double line-to-ground fault (phase *b* to *c* to ground), 569
Doubly-fed asynchronous generator, 729–730
Doubly-fed asynchronous generator components, 728
Driving-point admittance, 54

Eastern frequency response study, 802–806
Economic dispatch, 376–389, 741
  areas, 376
  B coefficient, 387
  economic dispatch solution including generator limits, 381–382
  economic dispatch solution including generator limits and line losses, 385–386
  economic dispatch solution neglecting generator limits and line losses, 379–381
  fossil-fuel generating unit incremental operating cost versus real power output, 378
  fossil-fuel generating unit operating cost versus real power output, 377
  fossil-fuel units, no inequality constraints or transmission losses, 377–381
  heat rate, 377
  inequality constraints effect, 381–382
  loss coefficient, 387
  transmission losses effect, 382–388
Economic factors, design considerations, transmission lines, 179
Effective resistance, 180

Electrical factors, design considerations, transmission lines, 178
Electric energy generation by types, 17
Electric field and voltage, solid cylindrical conductor, 202–205
 Gauss's law, 203
 M solid cylindrical conductors array, 204
 perfectly conducting solid cylindrical conductor with uniform charge distribution, 203
Electric field strength, conductor surfaces and ground level, 216–219
 ground-level electric field strength, overhead conductor and image, 218
 maximum ground-level electric field strength versus transmission-line voltage, 218
 single-phase line, conductor surface, ground-level field strengths, 218–219
 vector addition, electric fields at surface of one conductor in bundle, 217
Electric utility industry, 1–29
 computers in, 21–22
 deregulation, 2–9
 history, 10–16
 PowerWorld simulator, 22–28
 structure, 20–21
 trends, 17–20
Electromechanical time-delay overcurrent relay, 624
Emergency response, 794–809
 base case electrical and mechanical power response, 799
 California frequency response investigation, 797–802
 Eastern frequency response study, 802–806
 frequency and governor response, 804, 805
 frequency comparison, 806
 frequency response to loss of generation, 798, 800
 frequency terminology, 798
 governor basics, 796
 incumbent generators, 795–796
 power control from wind plants, 802
 wind assistance, 801–802
 wind plant controls, 806
 wind plant frequency response, 795
 wind power response, 803
 wind turbine generator power comparison, 807
Emissions
 CO_2, average emissions by plant type, 322
 CO_2 emissions by scenario, 321
 NO_x emissions by scenario, 323
 penalty for part-load operation, 313
 ramping, 314
 reductions, 321–323
 SO_2 emissions by scenario, 324
 solar and wind, emissions
 start-up, 314
Environmental factors, design considerations, transmission lines, 178
Equal-area criterion, 697–707
 critical clearing angle for cleared three-phase fault, 705–707
 critical clearing time for temporary three-phase fault, 703–705
 three-phase fault, transient stability during, 699–702
Equipment insulation strength, 847
Equivalent circuit for single cage induction machine, 724
Equivalent circuits for practical transformers, 101–107
 core loss current, 102
 versus ideal transformer, 101–102
 inrush current, 106
 magnetizing current, 102
 nonsinusoidal exciting current, 107
 practical single-phase two-winding transformer, 102, 103
 saturation, 105–106
 surge phenomena, 107
 transformer short-circuit and open-circuit tests, 104–105
Equivalent line-to-neutral diagrams, 62
Equivalent π circuit, 271–273, 275
 long line equivalent π circuit, 273
 transmission line equivalent π circuit, 271
Equivalent Y-representation of balanced-Δ load, 502
Euler's method, 707
Euler's method and computer solution to swing equation and critical clearing time, 709–710
Euler's method, modified, 708
Excitation curves for bushing CT, 618
Exciter, 757
External magnetic field, 184

Faraday's law, 98
Fast decoupled power flow, 372
Faults
 currents, R–L circuit with ac source, 438
 summary, 567
 terminals, 548
Flash density map, 844
Flexibility and cycling the conventional fleet, 310–330
 CO_2, average emissions by plant type, 322
 CO_2 emissions by scenario, 321
 coal units, capacity committed and dispatched, 320
 contingency, regulating, and flexibility reserves, 317
 data sets, 312–315
 emission reductions, 321–323
 emissions penalty for part-load operation, 313
 emissions per MW of capacity, ramping, 314
 emissions per MW of capacity, start-up, 314
 five-minute dispatch stacks, 318
 forecast errors, 328
 lower-bound cycling cost, 326
 median costs, cycling for various generation types, 315
 NO_x emissions by scenario, 323
 production costs, 324

production costs, change for each scenario, 327
production costs, cycling, 325
renewable energy penetration levels, 316
solar and wind, emissions avoided per MWh, 322
SO_2 emissions by scenario, 324
western grid operations, simulating, 317
western wind and solar integration study, 312
wind and solar, generation displaced, 319
wind- and solar-induced cycling operational costs, 323–327
wind and solar, system operation with high penetrations, 317–320
wind uncertainty and solar variability, 327–328
wind versus solar, 315–316

Fossil-fuel generating unit
incremental operating cost versus real power output, 378
operating cost versus real power output, 377

Fossil-fuel units, no inequality constraints or transmission losses, 377–381

Frequency
California frequency response investigation, 797–802
comparison, 806
Eastern frequency response study, 802–806
and governor response, 804, 805
overvoltages, 847
response to loss of generation, 798, 800
terminology, 798

Frequency of occurrence of lightning currents, 842

Fuse/recloser coordination, 632

Fuse selection, 460–462
continuous current rating, 461
interrupting current rating, 461
time response, 461, 462
voltage rating, 461

Gap-type ZT-aluminum conductor (GTZACSR), 175

Gas-insulated substations (GIS), 476–492
advantages, 478–479
causes of failure, 487
components, 477
components involved in failures, 485
construction, 477–478
defects in insulation system, 487
development trends, 482–484
dielectric failure rates, 486
environmental impact, 484
failures, 485–486
four-interrupter and closing resistor, 484
history and technological progress, 479–482
partial discharge mitigation methods, 488–490
partial discharge monitoring systems, 490–491
partial discharge origins, 486–488
percentage evolution and development of HV substations in market, 481
short-circuit breaking current rating, 480

Gauss elimination, 330–334

Gauss elimination and back substitution, 332–333

Gauss elimination: triangularizing a matrix, 333–334

Gauss-Seidel method, divergence, 339

Gauss-Seidel method, iterative solutions to linear algebraic equations, 337–338

Gauss-Seidel power flow solution, 351–353

General three-phase impedance load, 504

Generator convention, 48

Generator internal voltage and real power output versus power angle, 696–697

Generator per-unit swing equation and power angles during short circuit, 692–693

Generator variation, rotator angle and clearing time, 722–723

Generator-voltage control, 757–761
brushless exciters, 757
exciter, 757
static exciters, 757
synchronous generator exciter response, 758–760
wind turbine (type 3) reactive power control, 760–761

GIS. *See* Gas-insulated substations (GIS)

Grid congestion, 168–174
back story, 168–169
costs, 169–170
definition, 168
efficiency, 173
forecast, 171–172
impact, 170–171
leveraging technology, 173
modern conductors, 173
progress, 171
strategy flaws, 172

Grounded Y-network transformer, 898

Ground fault relays, 645

Ground-level electric field strength, overhead conductor and image, 218

GTZACSR. *See* Gap-type ZT-aluminum conductor (GTZACSR)

Heat rate, 377

High voltage dc (HVDC) transmission technologies, 238–258
asynchronous back-to-back tie, 242
asynchronous ties, 248–249
comparative costs of HVDC and EHV AC transmission alternatives, 247
control and operating principles, 255–257
conventional HVDC, 252–253
converter station, 239
IGBT valve converter arrangement, 244
line-commutated current source converter, 239–242
long-distance bulk power transmission, 245–246
multiterminal systems, 250

High voltage dc (HVDC) (*Continued*)
offshore transmission, 249–250
power delivery to large urban areas, 250–251
reactive power compensation, 242
self-commuted voltage source, 243–244
solid-state converter development, 243
system configurations and operating modes, 251–252
thyristor valve arrangement, 241
transmission line, 239
underground and submarine cable transmission, 246, 248
voltage source converter operating range, 244
voltage source converters, 240
VSC-based HVDC, 253–255
History, electric utility industry, 10–16
consumption growth, 13
generator size growth, 13
interconnected operations, 15
natural gas generation, 11–12
nuclear plants, 11
renewable energy sources, 12
steam plants, 11
three-phase transmission voltage increases, 14
transmission in North America, 16
HVDC transmission technologies. *See* High voltage dc (HVDC) transmission technologies
Hydroelectric units, 678

Ideal single-phase transformers, 129
Ideal transformer, 95–101
Ampere's law, 96
assumptions, 96
basic single-phase two-winding transformer, 95
Faraday's law, 98
Ohm's law, 97
versus practical single-phase two-winding transformer, 101–102
single-phase, phase-shifting transformer, 101
single-phase, two-winding transformer, 99–100
Ideal Y–Y transformer, 121, 517
IGBT valve converter arrangement, 244
Impedance loads sequence networks, 499–506
balanced-Y and balanced-Δ loads, 503
balanced-Y impedance load, 499
diagonal sequence impedances, 505
equivalent Y-representation of balanced-Δ load, 502
general three-phase impedance load, 504
negative-sequence impedance, 501
negative-sequence networks, 501
off-diagonal sequence impedances, 505–506
positive-sequence impedance, 501
positive-sequence networks, 501
three-phase symmetrical impedance load, 506
uncoupled equations, 500
zero-sequence impedance, 501
zero-sequence networks, 501
Impedance relay block and trip regions, 640
Impedance relays with directional capability, 641
Impulse voltage waveform, standard, 847
Impulse withstand voltage, 458
Individual distribution transformer per customer, 887
Inductance, 187–199
bundle conductor configurations, 196
inductive reactance, single-phase line, 193
inductive reactance, three-phase line, 195
inductive reactance, three-phase line with bundled conductors, 196–197
single-phase two-conductor line, 191–193
single-phase, two-conductor line with composite transistors, 190
single-phase, two-wire line inductance, 187–188
three-phase three-wire line with equal phase spacing inductance, 189
three-phase transmission line with earth replaced by earth return conductors, 198
transposed three-phase line, 194
transposition, 194
Inductance, solid cylindrical conductor, 182–187
external magnetic field, 184
internal magnetic field, 183
M solid cylindrical conductors, 186
permeability determination, 182
Induction generator, 725–726
Induction machine, external resistance variation effect on torque-speed curve, 727
Inductive load, 43
Inductive reactance, single-phase line, 193
Inductive reactance, three-phase line, 195
Inductive reactance, three-phase line with bundled conductors, 196–197
Inequality constraints effect, 381–382
Inrush current, 106
Installed generating capability by principal fuel types, 18
Instantaneous overcurrent relay, 621
Instantaneous power, balanced three-phase generators, 63–64
Instantaneous power, balanced three-phase motors and impedance loads, 64
Instantaneous power in single-phase ac circuits, 42–47
active power, 44
average power, 44
capacitive load, 43
inductive load, 43
instantaneous, real, and reactive power, 45–47
lagging power factor, 44

leading power factor, 44
power factor, 44–45
reactive power, 45
real and reactive power, physical significance of, 47
real power, 44
resistive load, 42–43
RLC load, 43–44
Instantaneous, real, and reactive power, 45–47
Instrument transformers, 614–620
class current transformers, 617
coupling capacitor voltage transformers, 615
CT equivalent circuit, 618
CT performance, 619
CT performance determination, 617–618
current transformer window design, 616
excitation curves for bushing CT, 618
relay operation versus fault current and CT burden, 620
standard CT ratios, 617
standard VT ratios, 617
voltage transformer, 614
VT and CT schematic, 614
Insulation coordination, 847–852
characteristics of station- and intermediate-class metal-oxide surge arresters, 851
equipment insulation strength, 847
metal-oxide surge arrester selection, 850
protective devices and equipment, 848
standard and reduced basic insulation levels, 848
standard impulse voltage waveform, 847
Insulators, design considerations, transmission lines, 175–176
Interconnected operations, 15
Intermediate substations effect on number of lines required for power transfer, 287–289
Internal magnetic field, 183
Iterative solutions to linear algebraic equations: Jacobi and Gauss-Seidel, 334–340
divergence, Gauss-Seidel method, 339
Gauss-Seidel method, iterative solutions to linear algebraic equations, 337–338
Jacobi method, iterative solutions to linear algebraic equations, 336–337
Iterative solutions to nonlinear algebraic equations: Newton-Raphson, 340–344
Newton Raphson method in four steps, 344
Newton-Raphson method: solution to nonlinear algebraic equations, 342–343
Newton-Raphson method: solution to polynomial equations, 340–341
Interrupting current rating, 461
Interrupting MVA, 458–459
Interrupting time, 458

Jacobian matrix and power flow solution by Newton-Raphson, 356–357
Jacobian matrix elements, 355
Jacobi method, iterative solutions to linear algebraic equations, 336–337
Junction, two single-phase lossless lines, 823

KCV. *See* Kirchoff's current law
Kirchoff's current law, 52
Kirchoff's voltage law, 52
KVL. *See* Kirchoff's voltage law

Lagging power factor, 44
Larger gas turbines units, 678
Leading power factor, 44
LFC. *See* Load-frequency control (LFC)
Lightning, 841–846
asset exposure map, 844
charge separation within clouds, 842
density contour map, 843
flash density map, 844
frequency of occurrence of lightning currents, 842
shield wires effect, 845
Line-commutated current source converter, 239–242
Line impedance locus, 640
Line loadability, 284–289
effect, intermediate substations on number of lines required for power transfer, 287–289
practical line loadability and percent voltage regulation, long line, 284–286
selection, transmission line voltage and number of lines for power transfer, 286–287
Line protection with impedance (distance) relays, 639–645
distance relay, 641
ground fault relays, 645
impedance relay block and trip regions, 640
impedance relays with directional capability, 641
line impedance locus, 640
modified impedance relays, 641
ratio relay, 639
reach of impedance relay, 641
345-kV transmission loop, 639
three-zone directional impedance relay, 642
three-zone impedance relay settings, 643–645
Line-to-line fault, 557–559
phase *b* to *c*, 569
short-circuit calculations using sequence networks, 559
Line-to-line (LL) faults, 431–432
Line-to-line-to-ground(LLG) faults, 431–432
LL faults. *See* Line-to-line (LL) faults
LLG faults. *See* Line-to-line-to-ground (LLG) faults
Load change in interconnected system, response to, 769–770
Load convention, 48, 49
Load flow, 310
Load-frequency control (LFC), 679–680, 741, 767–772
area control error, 768
automatic generation control strategy, 771
coordination of economic dispatch, 771

Load-frequency control (LFC) (*Continued*)
  load change in interconnected system, response to, 769–770
Long-distance bulk power transmission, 245–246
Long line
  equivalent π circuit, 273
  practical line loadability and percent voltage regulation, 284–286
  theoretical maximum power delivered, 283–284
  theoretical steady-state stability limit, 281–282
Loss coefficient, 387
Lossless lines, 274–282
  ABCD parameters, 274–275
  equivalent π circuit, 275
  steady-state stability limit, 279–282
  surge impedance, 274
  surge impedance loading (SIL), 276–278
  theoretical steady-state stability limit, long line, 281–282
  transmission line loadability curve, 280
  voltage profiles, 278–279
  wavelength, 276
Lossy lines, 834–838
  attenuation, 835–836
  distortion, 836–837
  power losses, 837–838
Low frequency withstand voltage, 458
Low-voltage circuit breakers, 455
Lumped capacitance, 830–831
Lumped inductance, 829–830
Lumped resistance, 831

Magnetizing current, 102
Maximum ground-level electric field strength versus transmission-line voltage, 218
Maximum voltage, 458
Mechanical factors, design considerations, transmission lines, 178
Median costs, cycling for various generation types, 315
Medium and short line approximations, 258–265
  ABCD parameters and nominal π circuit, 263–265
  ABCD parameters of common networks, 261
  medium-length transmission line, 260
  short transmission line, 259
  short transmission line phasor diagram, 262
  two-port network representation, 258
  voltage regulation, 262
Medium-length transmission line, 260
Medium voltage switchgear, 540–547
  air-insulated masonry cubicles, 542
  AIS panel with drawout technology, 542
  challenges, 544
  circuit breaker technological evolution, 541–542
  dedicated cable test device, 546
  drawout oil circuit breaker with arc control, 541
  drawout vacuum and SF_6 circuit breaker, 542
  evolution of technology, 541–542
  innovations, 544–546
  primary GIS panel, 543
  protection relays and sensors evolution, 544
  SF_6 Ring Main Unit with circuit breaker, 543
  shielded solid insulation system, 545–546
  single line diagram evolution, 542–543
  smart grid challenges, 544
  smart grid readiness, 546
  three-position diagram, 545
Metal-oxide surge arrester selection, 850
Method of images, 211, 212
Microgrids, 32–40
  benefits, 32–33
  Chevron Energy Solutions project, 35
  Smart Grid R&D Program, 33–40
Modified impedance relays, 641
Momentary current, 458
Motor starting, 680–681
M solid cylindrical conductors array, 186, 204
Multiconductor lines, 838–841
Multimachine stability, 711–718
  computational procedure, 713–714
  N-bus power system, 711
  rotor angle variation, 716–717
  stability results, 37 bus 9 generator system, 718
  Y_{bus} power-flow modification, 714–716
Multiterminal systems, HVDC transmission technologies, 250
Mutual admittance, 54
Mutual impedances, 448

Natural gas generation, 11–12
N-bus power system, 711
Negative reactive power, 49
Negative-sequence
  components, 493
  impedance, 501
  networks, 501
Negative-sequence phasor diagram for Y–Δ transformer bank construction, 119
Network equations, 52–55
  driving-point admittance, 54
  Kirchoff's current law, 52
  Kirchoff's Voltage law, 52
  mutual admittance, 54
  nodal equations steps, 53–54
  self-admittance, 54
  transfer admittance, 54
Networks of practical transformers, 518
Newton Raphson method
  in four steps, 344
  solution to nonlinear algebraic equations, 342–343
  solution to polynomial equations, 340–341
Newton-Raphson power flow solution, 353–362
  Jacobian matrix and power flow solution by Newton-Raphson, 356–357

Jacobian matrix elements, 355
power flow program, change in generation, 359–360
power flow program: 37-bus system, 361–362
Nodal equations, 831
Nodal equations steps, 53–54
Nonsinusoidal exciting current, 107
North America's power grid, 162–167
electricity exports and imports between Canada and the U.S., 165
electricity imports and exports, 167
grid integration benefits, 166–167
major transmission interconnections, 164
US and Canada working relationship maximization, 162–163
US-Canada electricity relationship, 163–166
NO_x emissions by scenario, 323
Nuclear plants, 11
Numerical integration, swing equation, 707–711
Euler's method, 707
Euler's method and computer solution to swing equation and critical clearing time, 709–710
Euler's method, modified, 708
Numeric relaying, 654–655

Off-diagonal sequence impedances, 505–506
Offshore transmission, HVDC transmission technologies, 249–250
Ohm's law, 97
Oil-filed padmount transformer, 897
Optimal power flow, 389–391, 741
Outage causes, 910
Outage reporting information, 909
Outdoor circuit breakers ratings, 456
Overcurrent relays, 620–625
CO-8 time-delay overcurrent relay characteristics, 622
CO-8 time-delay overcurrent relay operating time, 623–624
CO relay characteristics comparison, 623
electromechanical time-delay overcurrent relay, 624
instantaneous overcurrent relay, 621
pickup current, 620
solid-state relay panel, 625
time-delay overcurrent relay block and trip regions, 621
time-delay overcurrent relays, 622
Overhead line connected to cable, single-phase lossless lines lattice diagram, 824–827
Overhead primary loop, 883
Overlapping protection around circuit breaker, 637

Parallel circuit three-phase lines, 219–221
double-circuit line, single-line diagram, 219
Parallel connection, unloaded synchronous machine internal-voltage sources, 449
% conductivity, resistivity, and temperature constant of conductor metals, 180
Permeability determination, 182
Permissible daily short-time loading of liquid-filled distribution transformers, 899
Per-unit and actual currents, balanced three-phase networks, 114–115
Per-unit circuit, three-zone single-phase network, 111–113
Per-unit equivalent circuits of balanced three-phase two-winding transformers, 120–125
balanced three-phase transformer, voltage drop and fault current, 124–125
ideal Y–Y transformer, 121
rules for selecting base quantities, 121
voltage calculations, balanced Y–Y and Δ–Y transformers, 122–123
Per-unit equivalent circuits, single-phase, two-winding transformer, 110
Per-unit impedance, single-phase transformer, 108–109
Per-unit sequence models of three-phase three-winding transformers, 522–524
Per-unit sequence models of three-phase two-winding transformers, 516–521
Δ–Δ transformer features, 519
ideal Y–Y transformer, 517
networks of practical transformers, 518
symmetrical series impedances, 519
symmetrical Y-load, 519
three-phase networks with transformers using per-unit sequence components solution, 520–521
Y–Δ transformer features, 519
Per-unit system, 107–115
actual quantity, 108
base quantities, 108
per-unit and actual currents, balanced three-phase networks, 114–115
per-unit circuit, three-zone single-phase network, 111–113
per-unit equivalent circuits, single-phase, two-winding transformer, 110
per-unit impedance, single-phase transformer, 108–109
quantity calculation, 107–108
Phase-angle-regulating transformer, 136
Phase shift in Δ–Y transformers, 118
Phasors, 40–42
conversion from polar to rectangular form, 41
effective value, 40
relationships summary, 42
Pickup current, 620
Pilot relaying, 653–654
Pitch control, Type 3 or 4 wind turbine model, 767
Pole-mount recloser, 880
Pole-mount shunt capacitor bank, 882

Pole-mount transformer, 896
Positive reactive power, 49
Positive-sequence
  components, 493
  impedance, 501
  networks, 501
  phasor diagram construction, 116–118
Power circuit breakers, 455
Power delivery to large urban areas, HVDC transmission technologies, 250–251
Power distribution, 859–920
  automation, 910–913
  components of electric power system, 876
  distributed energy sources (DERs), 860–875
  primary distribution, 878–885
  distribution reliability, 906–910
  secondary distribution, 885–890
  shunt capacitors in distribution systems, 900–905
  smart grids, 913–914
  software, 905–906
  transformers in distribution systems, 890–900
Power factor, 44–45
Power flow control, 363–369
  generator Thévenin equivalent, 363
  means of control, 363
  shunt capacitor bank, effect of adding to power system bus, 364
  shunt capacitor banks, effect of, 365–369
Power flow problem, 344–351
  bus types, 345–346
  power flow input data and Y_{bus}, 346–350
Power flows, 309–413
  control of power flow, 363–369
  dc power flow, 373–374
  direct solutions to linear algebraic equations: Gauss elimination, 330–334
  economic dispatch, 376–389
  fast decoupled power flow, 372
  flexibility and cycling the conventional fleet, 310–330
  Gauss-Seidel power flow solution, 351–353
  input data and Y_{bus}, 346–350
  iterative solutions to linear algebraic equations: Jacobi and Gauss-Seidel, 334–340
  iterative solutions to nonlinear algebraic equations: Newton-Raphson, 340–344
  Newton-Raphson power flow solution, 353–362
  optimal power flow, 389–391
  power flow problem, 344–351
  program, change in generation, 359–360
  program: 37-bus system, 361–362
  sparsity techniques, 369–372
  wind generation modeling, 374–376
Power frequency overvoltages, 847
Power in balanced three-phase circuits, 63–68
  complex power, balanced three-phase generators, 64–65
  complex power, balanced three-phase motors, 65
  complex power, balanced-Y and balanced-Δ impedance loads, 65–66
  instantaneous power, balanced three-phase generators, 63–64
  instantaneous power, balanced three-phase motors and impedance loads, 64
  power in balanced three-phase system, 66–68
Power in sequence networks, 524–526
Power losses, 837–838
Power system controls, 739–777
  generator-voltage control, 757–761
  load-frequency control, 767–772
  restoration of system after blackout, 742–756
  turbine-governor control, 761–767
Power system overvoltages, 841–847
  lightning, 841–846
  power frequency overvoltages, 847
  switching surges, 846
Power system protective zones, 636
Power system restoration, 671–689
  aeroderivative gas turbine sets, 678
  black-start, 686–687
  bottom-up restoration, 671–672
  cold load pickup, 683–684
  diesel generator sets, 678
  dynamic studies, 677–678
  general philosophy, 673–674
  hydroelectric units, 678
  larger gas turbines units, 678
  load frequency control, 679–680
  motor starting, 680–681
  plans and procedures, 674–675
  self-excitation, 681–682
  steady-stage concerns, 676–677
  studies and analysis, 675–676
  system stability, 682–683
  top-down restoration, 671
  transient overvoltages, 684–686
  voltage control, 680
Power-system sequence networks and Thévenin equivalents, 549–551
Power system three-phase short circuits, 442–445
  calculation assumptions, 442
  synchronous generator feeding synchronous motor, 443
  three-phase short-circuit currents, power system, 444–445
Power transfer
  effect, intermediate substations on number of lines required, 287–289
  selection, transmission line voltage and number of lines, 286–287
Power transformers, 87–159
  autotransformers, 129–131
  coils, 91
  condition monitoring and maintenance program, 94
  equivalent circuits for practical transformers, 101–107
  ideal transformer, 95–101
  installation and commissioning, 90, 92
  life cycle, 91
  life extension, 89
  life management and extension, 88–95

life of, 88
manufacturing, 90
operation and maintenance, 92–94
per-unit equivalent circuits of balanced three-phase two-winding transformers, 120–125
per-unit system, 107–115
specifying and purchasing, 89
three-phase transformer connections and phase shift, 115–120
three-winding transformers, 125–129
transformers with off-nominal turns ratios, 131–139
Power triangle and power factor correction, 50–52
PowerWorld simulator, 22–28
economic dispatch, including generator limits, 383–384
economic dispatch, including generator limits and line losses, 386–387
edit mode, 25–27
fault current, three-phase faults, 450–453, 453–454
introduction, 23–25
optimal power flow, 390–391
run mode, 27–28
symmetrical faults, 572–575
Practical line loadability and percent voltage regulation, long line, 284–286
Practical single-phase two-winding transformer, 102, 103
Primary distribution, 878–885
overhead primary loop, 883
pole-mount recloser, 880
pole-mount shunt capacitor bank, 882
primary loop systems, 883–884
primary network system, 884–885
primary radial systems, 879–883
primary selective system, 882
radio-controlled sectionalizing switch, 881
underground primary loop, 884
US voltage distribution, 878
Primary loop feeder
after fault, 912
before fault, 911
Primary loop systems, 883–884
Primary network system, 884–885
Primary radial systems, 879–883
Primary relays, 594
Primary selective system, 882
Propagation constant, 266
Protective devices and equipment, 848
Public Utility Regulatory Policies Act of 1978, 4

Quantity calculation, 107–108

Radial system protection, 625–629
34.5-kV radial system, 626
time-delay overcurrent relay coordination, 626–629
Radio-controlled sectionalizing switch, 881
Rake-equivalent sequence networks, 568
Rated current and short circuit current, 893
Ratio relay, 639
Reach of impedance relay, 641
Reactive compensation techniques, 289–294
compensated transmission-line section, 290
series capacitive compensation to increase transmission-line loadability, 292–294
shunt reactive compensation to improve transmission-line voltage regulation, 290–292
static var compensators, 289
subsynchronous resonance, 290
Reactive power, 45
Reactive power compensation, HVDC transmission technologies, 242
Reactive power flow, 139
Real and reactive power, delivered or absorbed, 48–49
Real and reactive power, physical significance of, 47
Real power, 44
Real power flow, 139
Receiving-end voltage reflection coefficient, 813
Reclosers and fuses, 629–632
coordination, fuse/recloser, 632
13.8-kV radial distributor feeder, 630
time-current curves for radial distribution circuit, 631
Reference frame transformations, 719
Regulation constant, 762
Regulatory recommendations and standards, blackouts, 754–755
Related required capabilities, 457
Relay operation versus fault current and CT burden, 620
Relay protection upgrading, 595–612
costs, 598–600, 605–606
drawings with as-left data completion, 609–610
enhanced relay security, 603
firmware and software advances, 602–603
hardware advances, 601–602
history of relays, 596–597
implementation, 606–607
installation data, 604–605
longevity, 598
primary/backup relay systems, 600–601
replacing versus upgrading, 597–598
second-generation numeric relays, 601–604
smart software, 603–604
update drawings, 607–609
Reliability indices values, 907
Renewable energy
penetration levels, 316
sources, 12
Residential customer voltage profile, 886
Resistance, 179–182
dc and ac resistance, 181–182
effective resistance, 180
% conductivity, resistivity, and temperature constant of conductor metals, 180
SI and English units comparison for calculating conductor resistance, 179
skin effect, 181
Resistive load, 42–43

Resolving phase voltages, 493
Response to load change in interconnected power system, 764–766
Restoration of system after blackout, 742–756
  Cleveland, 752–753
  Detroit, 749–751
  lessons learned, 752–754
  New England, 745–747
  New York, 742–745
  Ontario, 747–749
  regulatory recommendations and standards, 754–755
  Southern Michigan, 751–752
R–L circuits, 418–420
RLC load, 43–44
Rotating machines sequence networks, 510–516
  currents in sequence networks, 513–514
  subtransient impedance, 512
  synchronous impedance, 511, 512
  three-phase motors sequence networks, 512
  transient impedance, 512
  unbalanced three-phase networks solutions using sequence components, 514–516
  Y-connected synchronous generator, 511
Rotor angle variation, 716–717

Saturation, 105–106
SCC comparison for symmetrical faults, 428–429
SCC from Type 1 WTG, 420–424
SCC from Type 2 WTG, 424–425
SCC from Type 3 WTG, 425–427
SCC for Type 4 WTG under different faults, 433
Secondary distribution, 885–890
  common secondary main, 887–888
  individual distribution transformer per customer, 887
  residential customer voltage profile, 886
  secondary network, 888
  spot network, 889–990
Secondary network, 888
Self-admittance, 54
Self-commuted voltage source, HVDC transmission technologies, 243–244
Self-excitation, 681–682
Self-impedances, 448
Sequence bus impedance matrices, 567–575
  balanced three-phase fault, 568
  double line-to-ground fault (phase *b* to *c* to ground), 569
  line-to-line fault (phase *b* to *c*), 569
  rake-equivalent sequence networks, 568
  single line-to-ground fault (phase *a* to ground), 568
  single line-to-ground short-circuit calculations using Z_{bus}, Z_{bus1}, and Z_{bus2}, 569–572
Sequence networks, 475
  impedance loads sequence networks, 499–506
  power in sequence networks, 524–526
  rotating machines sequence networks, 510–516
  series impedances sequence networks, 506–508
  three-phase lines sequence networks, 508–510
Series capacitive compensation to increase transmission-line loadability, 292–294
Series impedances sequence networks, 506–508
  three-phase series impedances, 507
  three-phase symmetrical series impedances, 508
Series impedances, three-phase line, neutral conductors, earth return, 197–202
  circuit representation of series-phase impedances, 200
  earth resistives and 60-Hz equivalent conductor distances, 199
  neutral wire versus shield wire, 197
Series R–L circuit transients, 435–438
  asymmetrical fault current, 436
  current in series R–L circuit with ac voltage source, 436
  dc offset current, 436
  fault currents, R–L circuit with ac source, 438
  short-circuit current—series R–L circuit, 437
  steady-state fault current, 436
  symmetrical fault current, 436
Shield wires
  design considerations, transmission lines, 176
  effect, 845
Short circuit behaviors under symmetrical faults, 418
Short circuit current, 458. *See also SCC entries*
Short-circuit current—series R–L circuit, 437
Short circuit modeling of wind power plant, 416–435
  collector system, 418
  conventional power plants versus wind power plants, 417–418
  line-to-line (LL) faults, 431–432
  R–L circuits, 418–420
  SCC comparison for symmetrical faults, 428–429
  SCC from Type 1 WTG, 420–424
  SCC from Type 2 WTG, 424–425
  SCC from Type 3 WTG, 425–427
  SCC for Type 4 WTG under different faults, 433
  short circuit behaviors under symmetrical faults, 418
  single line-to-ground (SLG) faults, 430–431
  Type 4 WTGs, 427–428
  unsymmetrical faults, 429–430
  wind turbine generators operation, 418
Short transmission line, 259
Short transmission line phasor diagram, 262
Shunt admittances, lines with neutral conductors and earth return, 211–216
  method of images, 211, 212
  single-phase line, effect of earth on capacitance, 212–213
  three-phase line, neutral

conductors and earth plane replaced by conductors, 214
Shunt capacitors in distribution systems, 900–905
control methods, 905
at end of primary feeder, 900–904
Shunt reactive compensation to improve transmission-line voltage regulation, 290–292
SI and English units comparison for calculating conductor resistance, 179
SIL. *See* Surge impedance loading (SIL)
Simplified synchronous machine model and system equivalents, 695–697
generator internal voltage and real power output versus power angle, 696–697
synchronous generator connected to system equivalent, 695
Single line-to-ground fault (phase a to ground), 568
Single line-to-ground short-circuit calculations using Z_{bus}, Z_{bus1}, and Z_{bus2}, 569–572
Single line-to-ground (SLG) faults, 430–431, 553–557
short-circuit calculations using sequence networks, 554–557
Single-phase autotransformers, 130–131
Single-phase line, conductor surface, ground-level field strengths, 218–219
Single-phase line, effect of earth on capacitance, 212–213
Single-phase lossless line, 828–829
Single-phase lossless line lattice diagram, 823
Single-phase, phase-shifting transformer, 101
Single-phase three-winding transformer, 126
Single-phase three-winding transformer, per-unit impedances, 127–128
Single-phase transformer differential relay protection, 650
Single-phase two-conductor line, 191–193
Single-phase, two-conductor line with composite transistors, 190
Single-phase, two-winding transformer, 99–100
Single-phase, two-winding transformer differential relay protection, 649
Single-phase, two-wire line inductance, 187–188
Skin effect, 181
SLG faults. *See* Single line-to-ground (SLG) faults
Smart Grid R&D Program, 33–40
key R&D areas, 39
ongoing activities, 33–37
performance targets, 37
workshops, 37–38
Smart grids, 913–914
Smart grid technology, 19
Software for distribution, 905–906
Solid-state converter development, HVDC transmission technologies, 243
Solid-state relay panel, 625
SO_2 emissions by scenario, 324
Sparsity techniques, 369–372
in 37-bus system, 371–372
Spinning reserve, 741
Spot network, 889–990
Stability results, 37 bus 9 generator system, 718
Standard CT ratios, 617
Standard VT ratios, 617
Static exciters, 757
Static var compensators, 289
Station- and intermediate-class metal-oxide surge arresters characteristics, 851
Steady-state
fault current, 436
frequency–power regulation, 762
stability limit, 279–282
Steam plants, 11
Step-voltage source
matched at sending end, capacitive load at receiving end, 818–820
matched at sending end, open receiving end, 816–818
at sending end, matched load at receiving end, 815–816
with unmatched source resistance at sending end, unmatched resistive load at receiving end, 820–822
Substation normal, emergency, and allowable ratings, 894–895
Subtransient fault current, 439
Subtransient impedance, 512
Support structures, design considerations, transmission lines, 176
Surge arresters, 781–793
application recommendations, 784–792
dimensions and mounting, 787–792
discharge voltages, 788–789
external construction, 781–783
internal construction, 783–784
operation, 784
performance and protective characteristics, 792
routine tests, 787
selection of arrester rating, 784–787
specifications, 790–791
standards, 787
voltage ratings, 785
Surge impedance, 274
Surge impedance loading (SIL), 276–278
Surge phenomena, 107
Swing equation, 689–694
coherent machines, 694
generator per-unit swing equation and power angles during short circuit, 692–693
two generating units equivalent swing equation, 693–694
Switching surges, 846
Symmetrical components, 475–537
balanced *abc* currents, 497–498
balanced line-to-neutral voltages, 496–497
common identities involving $a = 1\angle 120°$, 494
gas-insulated substations (GIS), 476–492
impedance loads sequence networks, 499–506

Symmetrical components (*Continued*)
  negative-sequence components, 493
  per-unit sequence models of three-phase three-winding transformers, 522–524
  per-unit sequence models of three-phase two-winding transformers, 516–521
  positive-sequence components, 493
  power in sequence networks, 524–526
  resolving phase voltages, 493
  rotating machines sequence networks, 510–516
  series impedances sequence networks, 506–508
  three-phase lines sequence networks, 508–510
  unbalanced currents, 498–499
  zero-sequence components, 493
Symmetrical fault current, 436
Symmetrical faults, 415–473
  bus impedance matrix, 445–454
  circuit breaker selection, 455–459
  fuse selection, 460–462
  power system three-phase short circuits, 442–445
  series R–L circuit transients, 435–438
  short circuit modeling of wind power plant, 416–435
  three-phase short circuit, unloaded synchronous machine, 438–442
Symmetrical series impedances, 519
Symmetrical Y-load, 519
Synchronous generator connected to system equivalent, 695
Synchronous generator exciter response, 758–760
Synchronous generator feeding synchronous motor, 443
Synchronous impedance, 511, 512
System configurations and operating modes, HVDC transmission technologies, 251–252
System protection, 593–668
  backup relays, 594
  bus protection with differential relays, 647–648
  design components, 612–613
  differential relays, 645–647
  directional relays, 633–634
  instrument transformers, 614–620
  line protection with impedance (distance) relays, 639–645
  numeric relaying, 654–655
  overcurrent relays, 620–625
  pilot relaying, 653–654
  primary relays, 594
  radial system protection, 625–629
  reclosers and fuses, 629–632
  relay protection upgrading, 595–612
  transformer protection with differential relays, 648–653
  two-source system with directional relays protection, 634–635
  zones of protection, 635–638
System representation, 547–552
  fault terminals, 548
  power-system sequence networks and Thévenin equivalents, 549–551
  three-phase bus, general, 548
  three-phase bus, general, balanced system, 549
  three-phase short-circuit calculations using sequence networks, 551–552
System stability, 682–683

Tap-changing three-phase transformer, per-unit positive-sequence network, 133–135
Theoretical maximum power delivered, long line, 283–284
Theoretical steady-state stability limit, long line, 281–282
345-kV transmission loop, 639
13.8-kV radial distributor feeder, 630
34.5-kV radial system, 626
Three-phase bus, general, 548
Three-phase bus, general, balanced system, 549
Three-phase fault, transient stability during, 699–702
Three-phase line, neutral conductors and earth plane replaced by conductors, 214
Three-phase lines sequence networks, 508–510
  completely transposed three-phase line capacitances, 510
  completely transposed three-phase line sequence impedances, 509
Three-phase motors sequence networks, 512
Three-phase networks with transformers using per-unit sequence components solution, 520–521
Three-phase series impedances, 507
Three-phase short-circuit
  calculations using sequence networks, 551–552
  currents, power system, 444–445
  currents, unloaded synchronous generator, 441–442
Three-phase short circuit, unloaded synchronous machine, 438–442
  ac fault current, 439
  armature time constant, 440
  direct axis short-circuit subtransient time constant, 439
  direct axis short-circuit transient time constant, 440
  subtransient fault current, 439
  three-phase short-circuit currents, unloaded synchronous generator, 441–442
  transient fault current, 440
Three-phase symmetrical impedance load, 506
Three-phase symmetrical series impedances, 508
Three-phase three-wire line with equal phase spacing inductance, 189
Three-phase transformer connections and phase shift, 115–120
  negative-sequence phasor diagram for Y–Δ transformer bank construction, 119
  phase shift in Δ–Y transformers, 118

positive-sequence phasor diagram construction, 116–118
three-phase two-winding Y–Δ transformer bank, 117
three-phase two-winding Y–Y transformer bank, 116
transformer core configurations, 120
Three-phase transformer differential relay protection, 652–653
Three-phase transmission line with earth replaced by earth return conductors, 198
Three-phase transmission voltage increases, 14
Three-phase two-winding Y–Δ transformer bank, 117
Three-phase two-winding Y–Y transformer bank, 116
Three-phase, Y–Δ, two-winding transformer differential relay protection, 651
Three-winding, three-phase transformer, balanced operation, 128–129
Three-winding transformers, 125–129
single-phase three-winding transformer, 126
single-phase three-winding transformer, per-unit impedances, 127–128
three-winding, three-phase transformer, balanced operation, 128–129
Three-zone directional impedance relay, 642
Three-zone impedance relay settings, 643–645
Thyristor valve arrangement, HVDC transmission technologies, 241
Time-current curves for radial distribution circuit, 631
Time-delay overcurrent relays, 622, 635
block and trip regions, 621
coordination, 626–629
Time response fuse, 461, 462
Top-down restoration, 671
Transfer admittance, 54
Transformer core configurations, 120
Transformer protection with differential relays, 648–653
single-phase transformer differential relay protection, 650
single-phase, two-winding transformer differential relay protection, 649
three-phase transformer differential relay protection, 652–653
three-phase, Y–Δ, two-winding transformer differential relay protection, 651
Transformer short-circuit and open-circuit tests, 104–105
Transformers in distribution systems, 890–900
distribution substation transformers, 890–895
distribution transformer kVA ratings, 899
distribution transformers, 895–900
grounded Y-network transformer, 898
oil-filed padmount transformer, 897
permissible daily short-time loading of liquid-filled distribution transformers, 899
pole-mount transformer, 896
rated current and short circuit current, 893
substation normal, emergency, and allowable ratings, 894–895
voltage regulators, 892
Transformers with off-nominal turns ratios, 131–139
phase-angle-regulating transformer, 136
reactive power flow, 139
real power flow, 139
tap-changing three-phase transformer, per-unit positive-sequence network, 133–135
two transformers connected in parallel, 131
voltage-magnitude-regulating transformer, 136
voltage-regulating and phase-shifting three-phase transformers, 137–139
Transient fault current, 440
Transient impedance, 512
Transient operation in transmission lines, 779–858
Bewley lattice diagram, 822–827
boundary conditions for single-phase lossless lines, 813–822
discrete-time models, 828–834
emergency response, 794–809
insulation coordination, 847–852
lossy lines, 834–838
multiconductor lines, 838–841
power system overvoltages, 841–847
surge arresters, 781–793
traveling waves on single-phase lossless lines, 809–812
Transient overvoltages, 684–686
Transient stability, 669–737
design methods for improvement of, 730–732
equal-area criterion, 697–707
multimachine stability, 711–718
numerical integration, swing equation, 707–711
power system restoration, 671–689
simplified synchronous machine model and system equivalents, 695–697
swing equation, 689–694
two-axis synchronous machine model, 719–723
wind turbine machine models, 724–730
Transit time, 813
Transmission in North America, 16
Transmission-line *ABCD* parameters
long line, 269–271
summary, 269
Transmission-line characteristics, design considerations, transmission lines, 177
Transmission-line differential equations, 265–271
characteristic impedance, 267
propagation constant, 266

Transmission-line differential
  transmission line $ABCD$ parameters, long line, 269–271
  transmission line $ABCD$ parameters, summary, 269
  transmission-line section of length Δx, 265
Transmission line equivalent π circuit, 271
Transmission line, HVDC transmission technologies, 239
Transmission line loadability curve, 280
Transmission line parameters, 161–236
  capacitance, 205–211
  conductance, 182
  design considerations, 174–179
  electric field and voltage, solid cylindrical conductor, 202–205
  electric field strength, conductor surfaces and ground level, 216–219
  grid congestion, 168–174
  inductance, 187–199
  inductance, solid cylindrical conductor, 182–187
  North America's power grid, 162–167
  parallel circuit three-phase lines, 219–221
  resistance, 179–182
  series impedances, three-phase line, neutral conductors, earth return, 197–202
  shunt admittances, lines with neutral conductors and earth return, 211–216
Transmission-line section of length Δx, 265
Transmission lines: steady state operation, 237–307
  equivalent π circuit, 271–273
  HVDC transmission technologies, 238–258
  line loadability, 284–289
  lossless lines, 274–282
  maximum power flow, 282–284
  medium and short line approximations, 258–265
  reactive compensation techniques, 289–294
  transmission-line differential equations, 265–271
Transmission line voltage and number of lines for power transfer selection, 286–287
Transmission losses effect, 382–388
Transposed three-phase line, 194
Transposition, 194
Traveling waves on single-phase lossless lines, 809–812
Trends, electric utility industry, 17–20
  electric energy generation by types, 17
  installed generating capability by principal fuel types, 18
  smart grid technology, 19
Turbine-generator response to frequency change at generating unit, 763
Turbine-governor control, 761–767
  area frequency response characteristic, 763
  pitch control, Type 3 or 4 wind turbine model, 767
  regulation constant, 762
  response to load change in interconnected power system, 764–766
  steady-state frequency–power regulation, 762
  turbine-generator response to frequency change at generating unit, 763
Two-axis synchronous machine model, 719–723
  generator variation, 722–723
  reference frame transformations, 719
Two generating units equivalent swing equation, 693–694
Two-port network representation, 258
Two-source system with directional relays protection, 634–635
  and time-delay overcurrent relays, 635
  two sources, 635
Two transformers connected in parallel, 131
Type 4 WTGs, 427–428
Type 3 DFAG model circuit diagram, 728

Unbalanced currents, 498–499
Unbalanced three-phase networks solutions using sequence components, 514–516
Uncoupled equations, 500
Underground and submarine cable transmission, HVDC transmission technologies, 246, 248
Underground primary loop, 884
Unsymmetrical faults, 429–430, 539–586
  double line-to-ground fault, 560–567
  line-to-line fault, 557–559
  medium voltage switchgear, 540–547
  sequence bus impedance matrices, 567–575
  single line-to-ground fault, 553–557
  system representation, 547–552
U.S. voltage distribution, 878

Vector addition, electric fields at surface of one conductor in bundle, 217
Voltage calculations, balanced Y–Y and Δ–Y transformers, 122–123
Voltage control, 680
Voltage-magnitude-regulating transformer, 136
Voltage profiles, 278–279
Voltage range factor K, 458
Voltage rating, 458, 461
  impulse withstand voltage, 458
  low frequency withstand voltage, 458
  maximum voltage, 458
  voltage range factor K, 458
Voltage-regulating and phase-shifting three-phase transformers, 137–139
Voltage regulation, 262

Voltage regulator and turbine-governor controls for steam-turbine generator, 740
Voltage regulators, 892
Voltage source converter operating range, HVDC transmission technologies, 244
Voltage source converters, HVDC transmission technologies, 240
Voltage transformer, 614
VSC-based HVDC, 253–255
VT. *See* Voltage transformer
VT and CT schematic, 614

Wavelength, 276
Wind
  assistance, 801–802
  plant controls, 806
  plant frequency response, 795
  power response, 803
  turbine generator power comparison, 807
Wind and solar
  emissions avoided per MWh, 322
  generation displaced, 319
  induced cycling operational costs, 323–327
  system operation with high penetrations, 317–320
  uncertainty of wind and solar variability, 327–328
  wind versus solar, 315–316
Wind generation modeling, 374–376
  wind plant collector system topology, 375
  wind speed versus power curve, 376
Wind turbine generator. *See WTG entries*
Wind turbine generators operation, 418
Wind turbine machine models, 724–730
  doubly-fed asynchronous generator, 729–730
  doubly-fed asynchronous generator components, 728
  equivalent circuit for single cage induction machine, 724
  induction generator, 725–726
  induction machine, external resistance variation effect on torque-speed curve, 727
  type 3 DFAG model circuit diagram, 728
Wind turbine (type 3) reactive power control, 760–761

Y_{bus} power-flow modification, 714–716
Y-connected synchronous generator, 511
Y–Δ transformer features, 519

Z_{bus} use to compute three-phase short-circuit currents, 447–448
Zero-sequence components, 493
Zero-sequence impedance, 501
Zero-sequence networks, 501
Zones of protection, 635–638
  overlapping protection around circuit breaker, 637
  power system protective zones, 636